KU-196-255

MOLECULAR BIOLOGY AND BIOTECHNOLOGY OF PLANT ORGANELLES

Molecular Biology and Biotechnology of Plant Organelles

Chloroplasts and Mitochondria

Edited by

Henry Daniell
Department of Microbiology and Molecular Biology, University of Central Florida, Orlando, Florida, U.S.A.

and

Christine Chase
Horticultural Sciences Department, University of Florida, Gainesville, Florida, U.S.A.

Springer

A C.I.P. Catalogue record for this book is available from the Library of Congress.

ISBN 1-4020-2713-3 (HB)
ISBN 1-4020-3166-1 (e-book)

Published by Springer,
P.O. Box 17, 3300 AA Dordrecht, The Netherlands.

Sold and distributed in North, Central and South America
by Springer,
101 Philip Drive, Norwell, MA 02061, U.S.A.

In all other countries, sold and distributed
by Springer,
P.O. Box 322, 3300 AH Dordrecht, The Netherlands.

Printed on acid-free paper

Printed in the Netherlands.

CONTENTS

Preface

We have taught plant molecular biology and biotechnology at the undergraduate and graduate level for over 20 years. In the past few decades, the field of plant organelle molecular biology and biotechnology has made immense strides. From the green revolution to golden rice, plant organelles have revolutionized agriculture. Given the exponential growth in research, the problem of finding appropriate textbooks for courses in plant biotechnology and molecular biology has become a major challenge. After years of handing out photocopies of various journal articles and reviews scattered through out the print and electronic media, a serendipitous meeting occurred at the 2002 IATPC World Congress held in Orlando, Florida.

After my talk and evaluating several posters presented by investigators from my laboratory, Dr. Jacco Flipsen, Publishing Manager of Kluwer Publishers asked me whether I would consider editing a book on Plant Organelles. I accepted this challenge, after months of deliberations, primarily because I was unsuccessful in finding a text book in this area for many years. I signed the contract with Kluwer in March 2003 with a promise to deliver a camera-ready textbook on July 1, 2004. Given the short deadline and the complexity of the task, I quickly realized this task would need a co-editor. Dr. Christine Chase was the first scientist who came to my mind because of her expertise in plant mitochondria, and she readily agreed to work with me on this book. Both of us made a list of the topics and the eminent researchers in chloroplasts and mitochondria.

Fortunately almost all invited scientists agreed to contribute a chapter on their topic of expertise. We are extremely thankful to the authors who contributed their knowledge to this book, and their commitment to educate students of future generations. Through meticulous work they ensured that the topics were thoroughly researched, up to date and well presented. We are thankful that they took time from their very busy schedules and submitted chapters in a timely manner. Unfortunately, we lost two of our valued colleagues during the preparation of this book. Professor Rainer Maier died in April 2004 after completing Chapter 5. Professor Lawrence Bogorad died in December 2003 before he could write an overview of this field. He has been Henry Daniell's mentor and friend for two decades, and will be missed very much.

The transformation of the manuscript into a book, in a camera-ready format, involved a myriad of tasks that required the help of many. Jacco Flipsen and Noeline Gibson of Springer Press were patient with our questions regarding style, graphics, and format. Henry Daniell would like to acknowledge the assistance of his lab colleagues Amit Dhingra, Paul Cohill, and Shashi Kumar; Vijay Pattisapu and Suneel Modani were unstinting in their efforts. He is deeply indebted to his wife Shobana for help in editing this book and constant support. Christine Chase would like to thank her first mentor in science, Dr. Rolf Benzinger, for introducing her to the field of genetic transformation. She also thanks her colleagues Drs. Daryl Pring, Susan Gabay-Laughnan, and Ken Cline for countless discussions of organelle biology and genetics over the years and for their insights regarding the content of this volume.

We dedicate this book to the memory of Professors Lawrence Bogorad and Rainer Maier. Both of them made major contributions and continue to inspire many scientists to conduct research in chloroplast molecular biology.

Henry Daniell
Christine Chase

About the Editors

Henry Daniell was born and educated in India. He moved to the United States as a post-doctoral fellow at the University of Illinois, Urbana-Champaign in 1980. Since then he has served on the faculty of Washington State University, University of Idaho Auburn University and University of Central Florida (as Pegasus Professor & Trustee Chair). He has published over 100 research articles on several areas, including DNA replication in chloroplast and mitochondria, identification of new genes in the mitochondrial genome, promiscuous DNA and their evolutionary significance, maternal inheritance, transgene containment, photosynthesis (Rubisco, electron transport), chlorophyll biosynthesis, chloroplast development, in organello protein synthesis, transcription, RNA processing, RNA stability, translation, protein import, proteolysis and regulation of these processes. He pioneered the chloroplast genetic engineering approach in the 1980s and advanced this concept to confer useful agronomic traits (for biotic & abiotic stress, photoremediation, cytoplasmic male sterility, etc.) and to express vaccine antigens, biopharmaceuticals and biomaterials in transgenic plants. He has extended this technology to major crops, including cotton. Numerous undergraduates, graduates and post-doctoral fellows were trained to use this experimental system. Based on his oldest patents in this field, he is the technical founder of Chlorogen Inc., which is now evaluating chloroplast transgenic crops in the field. He is an editor of the Plant Biotechnology Journal and has received several Distinguished Researcher and Teaching Excellence Awards. Dr. Daniell has served as a consultant to the United Nations in biotechnology and was elected to the National Academy of Sciences, Italy in 2004, a rare honor bestowed on 14 Americans in the past 222 years.

Christine Chase was born and raised in rural, upstate New York. She earned a B.S. degree in Science Education from Cornell University in 1973 and a Ph.D. in Biology from the University of Virginia in 1981. During her graduate studies, she was supported by a NIH Traineeship in Genetics awarded through the Biology Department. Her introduction to organelle research began with her postdoctoral training (1981-1984) in the laboratory of Daryl Pring, USDA-ARS, at the University of Florida. She joined the faculty of the University of Florida Institute of Food and Agricultural Sciences in 1985, and she currently holds the rank of Professor in the Horticultural Sciences Department. She was a founding member of the University of Florida Interdepartmental Graduate Program in Plant Molecular and Cellular Biology, and she served as Co-Director (1993-1995) and Director (1995-1997) of that program. Dr. Chase teaches in graduate-level Advanced Genetics and Plant Molecular Biology courses. Her research exploits the unique biology and genetics of S cytoplasmic male sterility in maize to study plant mitochondrial biogenesis and function. This research platform presents an array of interesting biological questions that can be addressed through a variety of experimental approaches, and numerous undergraduate and graduate students have been introduced to research via this experimental system.

CONTRIBUTORS

ADAM, Z. - Institute of Plant Sciences, The Hebrew University of Jerusalem, Israel

ADAMSON, S. W. - Department of Chemistry and Biochemistry, University of Southern Mississippi, Mississippi, Hattiesburg, USA

ALLEN, J. - Biological Sciences, University of Missouri, Columbia, Missouri, USA

BAISAKH, N. - Plant Breeding, Genetics, and Biochemistry Division, International Rice Research Institute, Manila, Philippines

BARKAN, A. - Institute of Molecular Biology, University of Oregon, Eugene, Oregon, USA

BINDER, S. - Allgemeine Botanik, Universitäät Ulm, Germany

BOHMERT, K. - Metabolix, Inc., Cambridge, Massachusetts, USA

BONEN, L. - Biology Department, University of Ottawa, Canada

CAI, X. D. - National Key Laboratory of Crop Genetic Improvement, Huazhong Agricultural University, Wuhan, China

CANNON, G. C. - Department of Chemistry and Biochemistry, University of Southern Mississippi, Hattiesburg, Mississippi, USA

CHASE, C. - Horticultural Sciences Department, University of Florida, Gainesville, Florida, USA

CHERRY, J. H. - Auburn University, Auburn, Alabama, USA

CHI-HAM, C. L. - Department of Chemistry and Biochemistry, University of Southern Mississippi, Hattiesburg, USA

CLIFTON, S. - Genome Sequencing Center, Washington University, St. Louis, Missouri, USA

COHILL, P. R. - Department of Microbiology and Molecular Biology, University of Central Florida, Orlando, Florida, USA

COLAS DES FRANCS-SMALL, C. - Institut de Biotechnologie des Plants, Université Paris-Sud, Orsay, France

DANIELL, H. - Department of Microbiology and Molecular Biology, University of Central Florida, Orlando, Florida, USA

DATTA, S. - Plant Breeding, Genetics, and Biochemistry Division, International Rice Research Institute, Manila, Philippines

DHINGRA, A. - Department of Microbiology and Molecular Biology, University of Central Florida, Orlando, Florida, USA

DUFOURMANTEL, N. - CEA Cadarache, DSV, DEVM, CNRS-CEA, Aix-Marseille II, Saint-Paul-lez-Durance, France

FAURON, C. - Eccles Institute of Genetics, University of Utah, Salt Lake City, Utah, USA

GABAY-LAUGHNAN, S. - Department of Plant Biology, University of Illinois, Urbana-Champaign, Illinois, USA

GLASER, E. - Department of Biochemistry and Biophysics, Stockholm University, Sweden

GRAY, M. W. - Department of Biochemistry and Molecular Biology, Dalhousie University, Halifax, Canada

GROSSER, J. W. - Citrus Research and Education Center, University of Florida, Lake Alfred, Florida, USA

GUO, W. W. - National Key Laboratory of Crop Genetic Improvement, Huazhong Agricultural University, Wuhan, China

HAGEMANN, R. - Institut für Genetik, Martin-Luther-Universität, Halle/S, Germany

HAVEY, M. J. - USDA, ARS Vegetable Crops Unit, Department of Horticulture, University of Wisconsin, Madison, Wisconsin, USA

HEINHORST, S. - Department of Chemistry and Biochemistry, University of Southern Mississippi, Hattiesburg, Mississippi, USA

KUMAR, S. - Department of Microbiology and Molecular Biology, University of Central Florida, Orlando, Florida, USA

MAIER, R. M. - Cell Biology, Philipps University, Marburg, Germany

MARCHFELDER, A. - Molekulare Botanik, Universitat Ulm, Germany

MULLIGAN, R. M. - Department of Developmental and Cell Biology, University of California, Irvine, California, USA

NEWTON, K. - Biological Sciences, University of Missouri, Columbia, Missouri, USA

NIELSEN, B. L. - Department of Microbiology and Molecular Biology, Brigham Young University, Provo, Utah, USA

NISHIDA, I. - Graduate School of Sciences, University of Tokyo, Japan

PEOPLES, O. P. - Metabolix, Inc., Cambridge, Massachusetts, USA

SCHMITZ-LINNEWEBER, C. - Institute of Molecular Biology, University of Oregon, Eugene, Oregon, USA

SOLL, J. - Department für Biologie I, Botanik, Ludwig-Maximilians-Universität, München, Germany

SMALL, I. - Unite de Recherche en Genomique Vegetale, Cremieux, France

SNELL, K. D. - Metabolix, Inc., Cambridge, Massachusetts, USA

SZUREK, B. - Unité de Recherche en Génomique Végétale, Cremieux, France

WEIHE, A. - Institut füür Biologie, Humboldt-Universitäät, Berlin, Germany

ZERGES, W. - Biology Department, Concordia University, Montreal, Canada

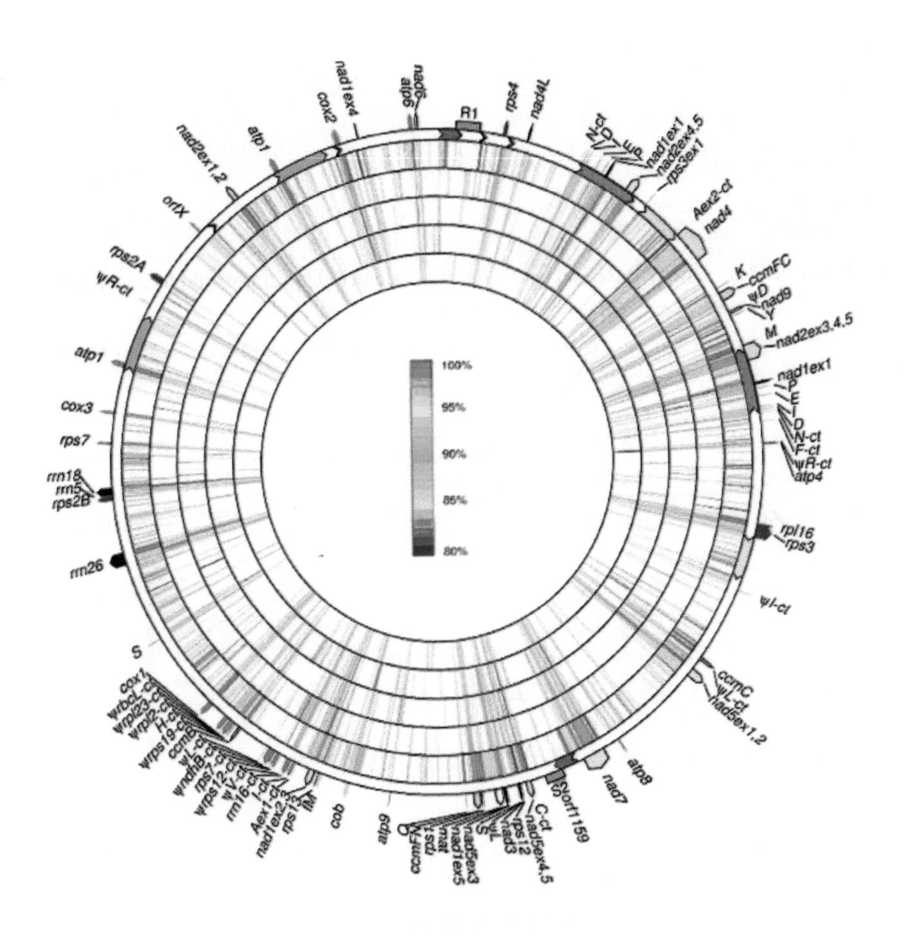

Plate 1: **Comparison of various plant mitochondrial genomes.**
See Chapter 6, Figure 3.

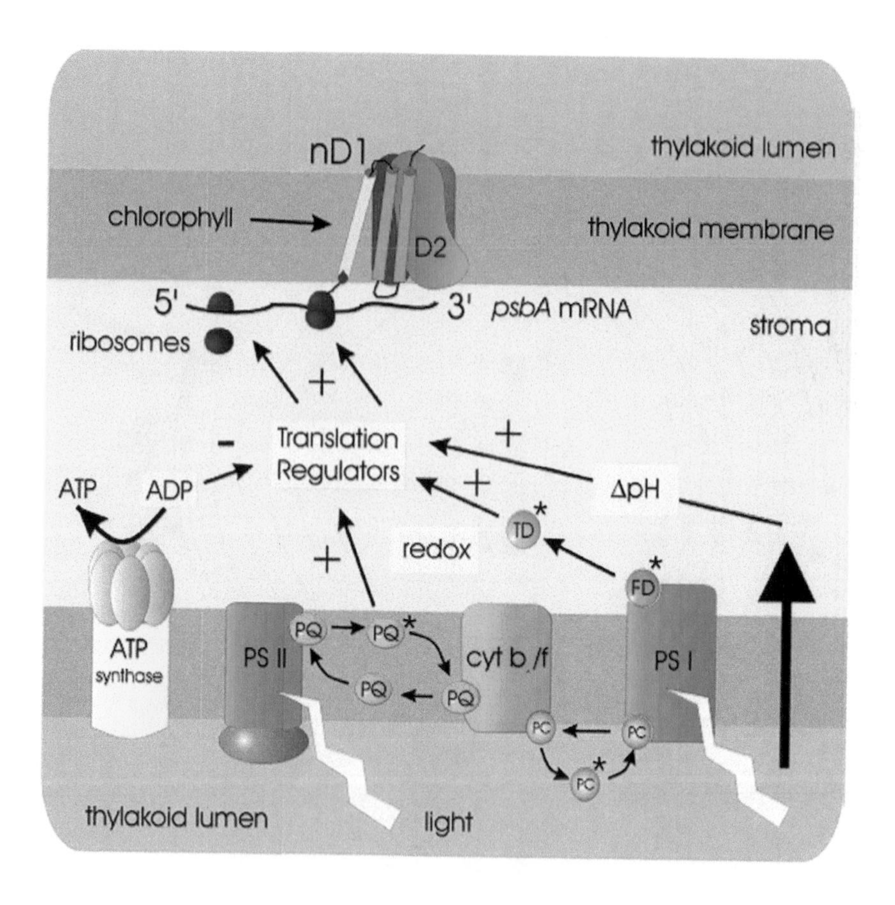

See Chapter 13, Figure 1.

Plate 2: Light regulation of translation through photosynthesis.

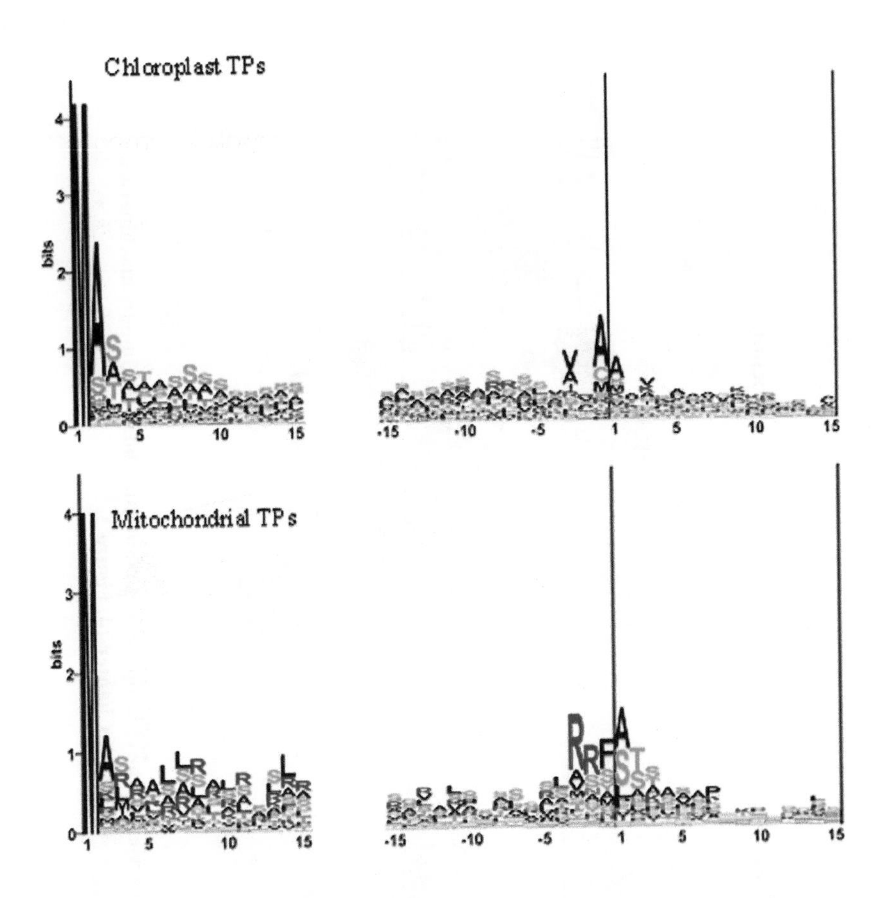

Plate 3: Sequence logos of chloroplast and plant mitochondrial signal peptides. See Chapter 14, Figure 1.

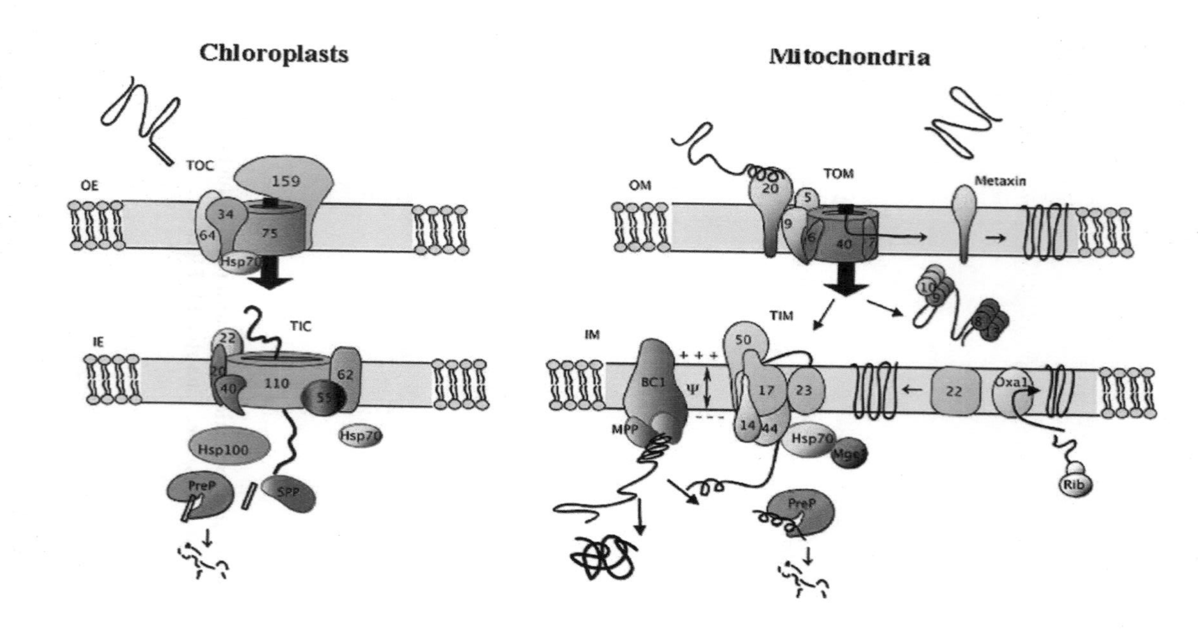

Plate 4: Graphical overview of the chloroplast and plant mitochondrial import machineries.
See Chapter 14, Figure 2.

Plate 5: **Maternal inheritance of transgenes engineered via the chloroplast genome. See Chapter 16, Figure 2.**

Plate 6: Engineering the carrot chloroplast genome to confer salt tolerance. See Chapter 16, Figure 3.

Plate 7: Engineering the cotton chloroplast genome.
See Chapter 16, Figure 4.

Plate 8: **Engineering the soybean chloroplast genome.**
See Chapter 16, Figure 5.

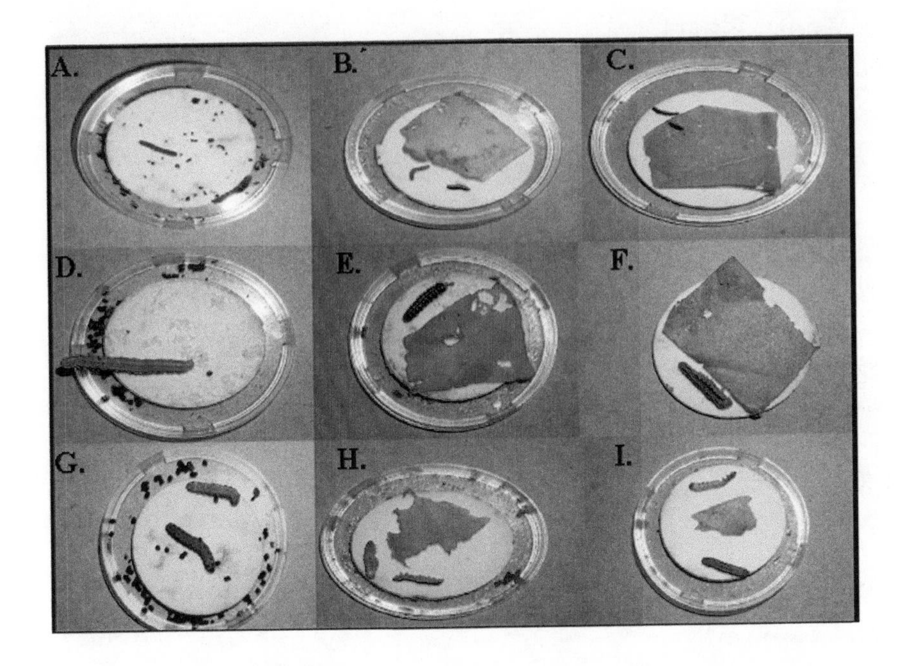

Plate 9: **Insect bioassays of control (A, D, G) and chloroplast transgenic plants with tobacco bud worm (A-C), cotton boll worm (D-F), and beet army worm (G-I). See Chapter 16, Figure 7.**

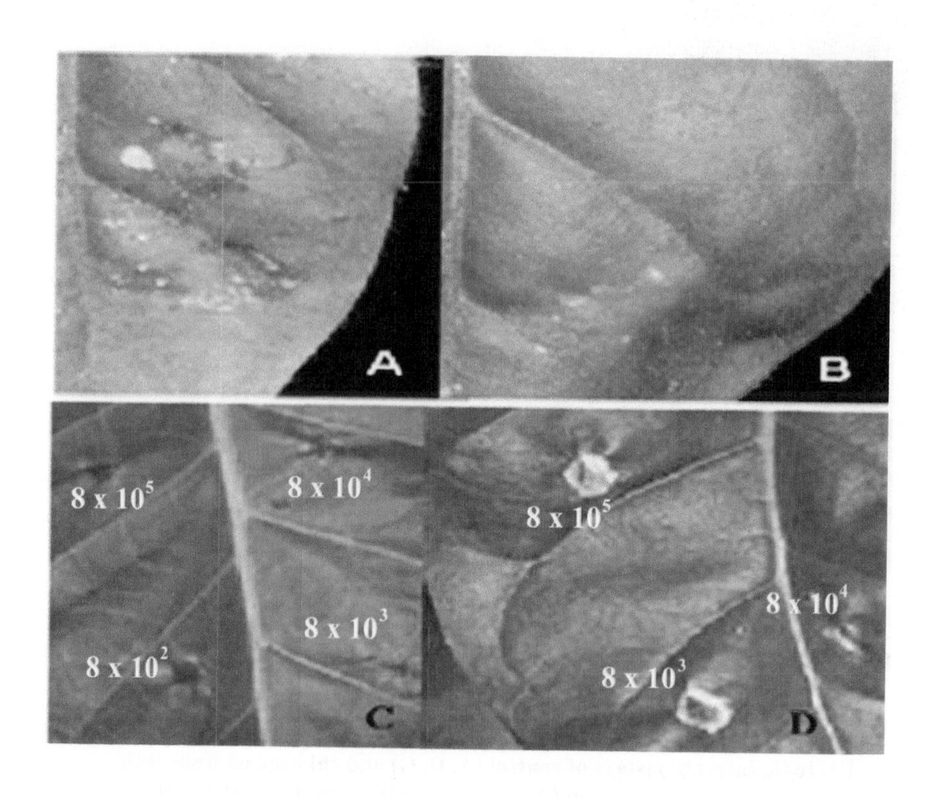

Plate 10: **Bioassays for fungal (A, B) and bacterial (C, D) disease resistance in control (A, D) and chloroplast transgenic lines (B, C)**
See Chapter 16, Figure 8.

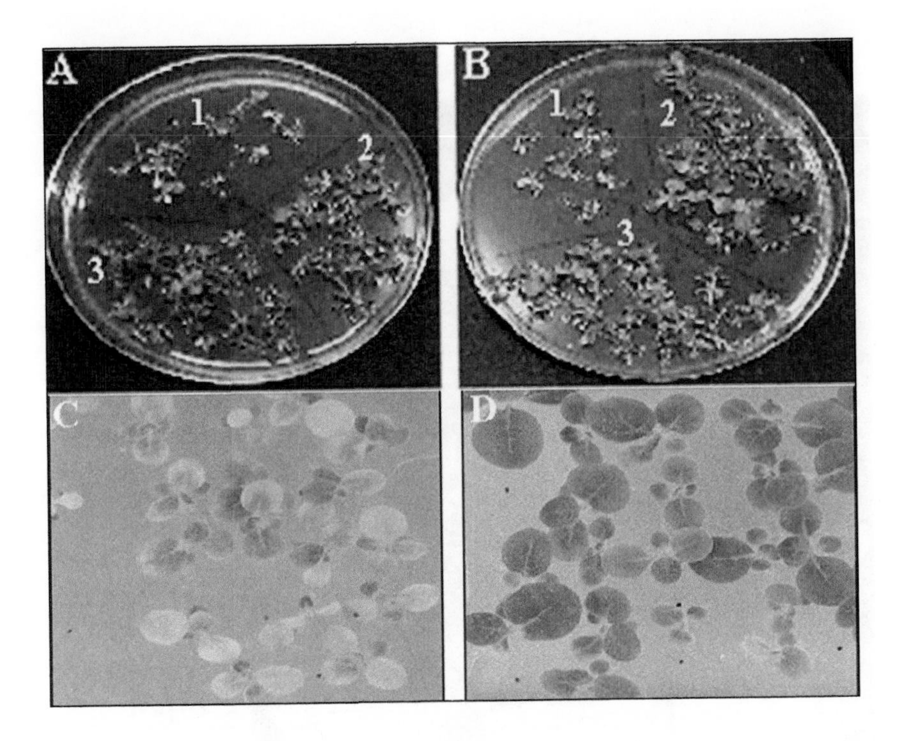

Plate 11: Drought tolerance assays in control (A1, B1, C) and chloroplast transgenic lines (A2, A3, B2, B3, D). See Chapter 16, Figure 9.

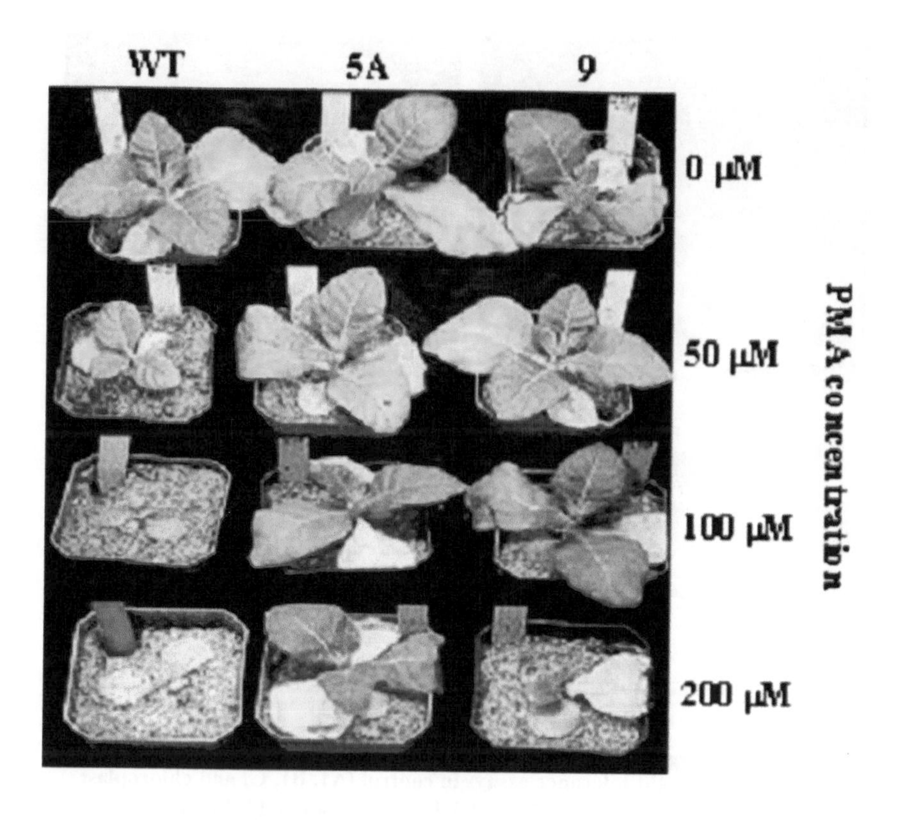

Plate 12: **Effect of phenyl mercuric acetate on control (WT) and chloroplast transgenic lines (5A, 9). See Chapter 16, Figure 10.**

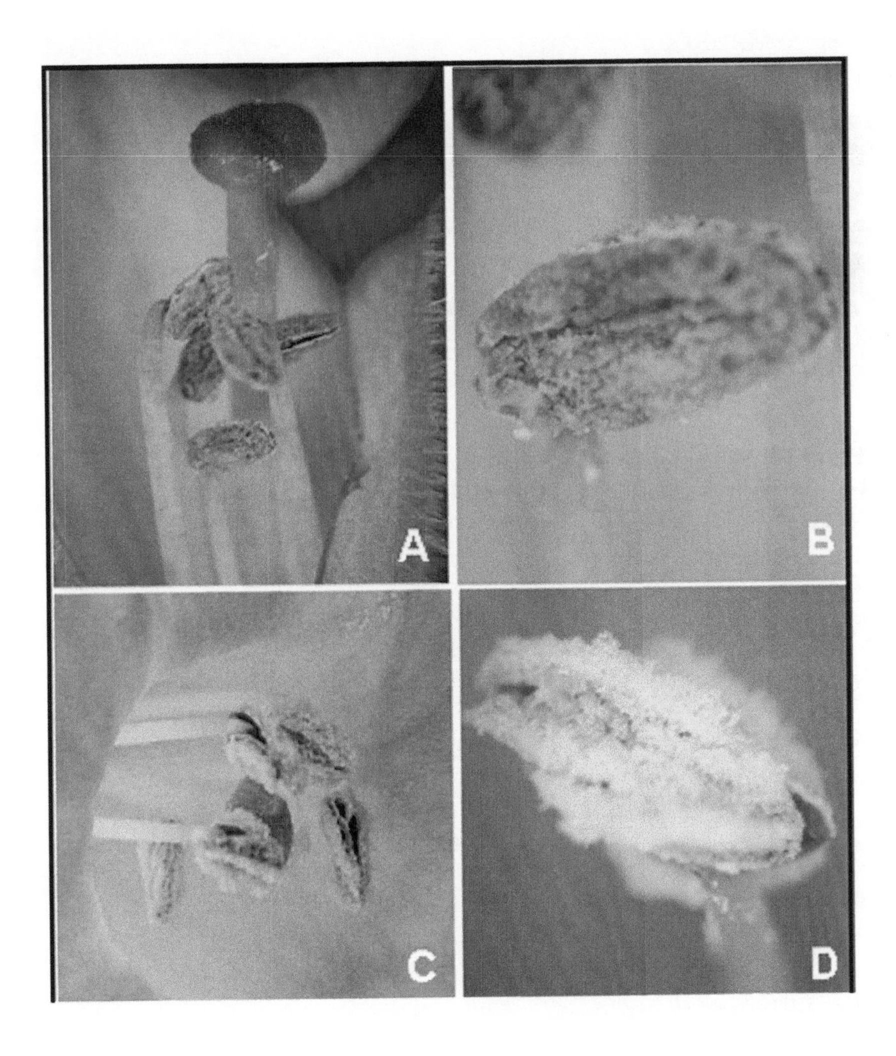

Plate 13: **Cytoplasmic male sterility engineered via the chloroplast genome. Control (C, D) and chloroplast transgenic lines (A, B).**
See Chapter 16, Figure 11.

Plate 14: **Pleiotropic effect on the accumulation of trehalose. 1: control, 2-5: nuclear transgenic lines, 6: control, 7: chloroplast transgenic line. See Chapter 16, Section 6.3.**

Plate 15: **Herbicide resistance in control (B) and chloroplast transgenic line (A). See Chapter 17, Figure 6.**

Plate 16: **Golden rice - Mendelian segregation of genes for ß-carotene biosynthetic pathway.** **See Chapter 19, Figure 3.**

CHAPTER 1

INTRODUCTION TO THE MOLECULAR BIOLOGY AND BIOTECHNOLOGY OF PLANT ORGANELLES

HENRY DANIELL[1] AND CHRISTINE D. CHASE[2]

[1]*Dept. Molecular Biology & Microbiology, University of Central Florida, Orlando, FL, USA. E-mail:* daniell@mail.ucf.edu
[2]*Horticultural Sciences Department, University of Florida Institute of Food and Agricultural Sciences, Gainesville, FL, USA. E-mail:* ctdc@mail.ifas.ufl.edu

Abstract. The fields of organelle biology and plant biotechnology have now been inter-related for a century. Plant organelles have intrigued biologists since their defiance of Mendelian inheritance and their endosymbiont origins became apparent. The first application of organelle biotechnology was the role played by cytoplasmic male sterility in hybrid seed production, a contribution towards the "Green Revolution". In modern times, plant organelles are again leading the way for the creation of genetically modified crops. On a global scale, 75% of GM crops are engineered for herbicide resistance and most of these herbicides target pathways that reside within plastids. Several thousand proteins are imported into chloroplasts that participate in biosynthesis of fatty acids, amino acids, pigments, nucleotides and numerous metabolic pathways including photosynthesis. Thus, from green revolution to golden rice, plant organelles have played a critical role in revolutionizing agriculture.

This book details not only the basic concepts and current understanding of plant organelle genetics and molecular biology but also focuses on the synergy between basic biology and biotechnology. Forty-four authors from nine countries have contributed 24 chapters, containing 52 figures and 28 tables. Section one on organelle genomes & proteomes discusses molecular features of plastid and mitochondrial genomes, evolutionary origins, somatic and sexual inheritance, proteomics, bioinformatics and functional genomics. Section two on organelle gene expression and signalling discusses transcription, translation, RNA processing, RNA editing, introns and splicing, protein synthesis, proteolysis, import of proteins into chloroplast and mitochondria and regulation of all these processes. Section three on organelle biotechnology discusses the genetic manipulation of organelles by somatic cell genetics, the use of cytoplasmic male sterility for hybrid seed production and the exciting applications of chloroplast and nuclear genetic engineering for biotic/abiotic stress tolerance, improved fatty acid/amino acid biosynthesis, and for the production of biopharmaceuticals, biopolymers and biomaterials.

1. HISTORICAL PERSPECTIVE

Plant organelles have fascinated geneticists since the rediscovery of Mendel's laws. In 1904, Correns reported maternally inherited male sterility in savory, and similar observations were soon reported for other plant species (reviewed by Edwardson, 1956; Duvick, 1959; and Havey, chapter 23 of this volume). In 1907, Correns

H. Daniell and C.D. Chase (eds.), Molecular Biology and Biotechnology of Plant Organelles, 1—12. © 2004 *Springer. Printed in the Netherlands.*

reported maternal inheritance of mutant plastids in Mirabilis, and Bauer reported non-Mendelian, bi-parental inheritance of mutant plastids in Pelargonium (reviewed by Hagemann, 2000; Hagemann, chapter 4 of this volume; Maier and Schmitz-Linneweber, chapter 5 of this volume). At the same time, the stage was being set for the first application of organelle genetics to biotechnology. In 1908, Shull and East rediscovered hybrid vigour, or heterosis, in maize, and this ultimately led to the commercial development of F1 hybrid maize (reviewed by Duvick, 2001). The production of F1 hybrid seed on a commercial scale requires large, uniform populations of male-sterile plants. For many crops, this has been achieved primarily through the use of cytoplasmically inherited male sterility (CMS) (reviewed by Duvick, 1959; Havey, chapter 23). Long before molecular approaches demonstrated that CMS resulted from gain-of-function mutations in the mitochondrial genome (reviewed by Chase and Gabay-Laughnan, chapter 22 of this volume), this organelle-encoded trait was extensively exploited in commercial agriculture. Since their discovery, mitochondrial CMS genes have served as useful reporter genes in basic studies of plant mitochondrial gene expression. (See chapter 22 for examples). Daniell et al. (chapter 16) review the first chloroplast-encoded CMS trait, engineered using the *phaA* gene coding for β-ketothiolase. Thus the basic science and practical applications of organelle biology have now been synergistically linked for a century.

2. EVOLUTION AND MOLECULAR BIOLOGY

Questions concerning the evolution of organelles have been a key force driving studies of organelle molecular biology. The endosymbiont origin of cellular organelles was proposed as early as 1882 and this hypothesis was further developed early in the twentieth century (reviewed by Margulis, 1970). Molecular features of organelle genes and genomes (reviewed by Gray, chapter 2 of this volume) provided the confirming data for the endosymbiont hypothesis. Molecular research over the subsequent decades revealed many prokaryotic features in the modern-day plant organelles, including some aspects of organelle division, genome organization and coding content, transcription, translation, RNA processing, and protein turn-over. These features are discussed in chapters 3, 5 and 6, 8, 10, 12 and 15 of this volume, respectively. For genome maps, models or graphical representations of pathways, readers are referred to color plates inserted at the beginning of this book.

While confirming the basic endosymbiont hypothesis, molecular investigations also revealed many surprises and raised new questions as to how evolution has shaped the modern-day organelles. Present-day mitochondrial genomes are extremely diverse (Gray, chapter 2), and the mitochondrial genomes of the land plants (reviewed by Fauron et al., chapter 6) are uniquely expanded over the ancestral type. The mitochondrial genomes, and even the relatively conserved plastid genomes, vary in coding content among land plant species (Maier and Schmitz-Linneweber, chapter 5; Fauron et al., chapter 6). The functional transfer of genes from organelle to nuclear genomes has occurred frequently and recently (on an evolutionary scale) in the plant lineages. Group I and group II introns are abundant in plant organelle genomes, but are found only rarely in the modern bacterial relatives of the organelle progenitors. The gain of introns by plant organelles was accompanied by the recruitment of an interesting spectrum of proteins that facilitate intron splicing (reviewed by Barkan, chapter 11). One of the greatest surprises of plant organelle biology was the C-to-U and U-to-C RNA editing process that occurs in both plastids and mitochondria. In many cases RNA editing is required to restore the coding of highly conserved amino acid sequences, but specific RNA editing sites are not always conserved, even among closely related plant species. The molecular mechanisms that underlie the selection and editing of these sites are only beginning to be understood (reviewed by Mulligan, chapter 9).

3. ORGANELLES AND THE NUCLEUS

The endosymbiont origin of the organelles was accompanied by the transfer of significant genetic information to the nuclear genome. In the case of the mitochondria, this transfer may have been simultaneous with the origin of the organelle (Gray, chapter 2). The majority of organelle proteins are translated on cytosolic ribosomes, and sophisticated targeting and import machinery (reviewed by Glaser and Soll, chapter 14) is required to ensure these proteins find their ultimate destination in the cell. The analysis of complete nuclear genome sequences with organelle targeting prediction programs (reviewed by Colas des Francs-Small et al., chapter 7; Glasser and Soll, chapter 14) estimates that 10% and 15% of plant nuclear genes predict proteins targeted to mitochondria and chloroplasts, respectively. Organelle function is therefore dependent upon interactions between nuclear and organelle genetic systems.

The identification of nuclear genes having organelle targeting signals does not always, however, provide information regarding the role of the gene or its product in the organelle. Systematic functional genomics approaches (reviewed by Colas des Francs-Small et al., chapter 7) will doubtless provide further insights. Forward genetic analysis has also identified nuclear genes key to organelle processes such as RNA processing, intron splicing, protein synthesis and protein complex assembly. (See chapters 10, 11, 12, 13 and 22 for examples.)

Much remains to be learned regarding the regulation of the many nuclear genes relating to organelle function. The "retrograde" regulation of nuclear genes by signals originating in the organelle was first discovered in yeast (Parikh et al., 1987), and many examples of retrograde signalling have now been uncovered in plants. The reader is referred to recent reviews (McIntosh et al., 1998; Rodermel, 2001; Surpin et al., 2002; Pfannschmidt et al., 2003; Gray et al., 2003; Juszczuk and Rychter, 2003; Pfannschmidt, 2003; and Rodermel and Park, 2003) for details on these systems. The recent finding that some cases of CMS result from retrograde regulation of floral homeotic genes by the mitochondria (reviewed by Zubko, 2004; Chase and Gabay-Laughnan, chapter 22) indicates that the nuclear gene targets of organelle regulatory signals are likely to extend beyond those genes directly associated with organelle functions.

Organelle proteomics (reviewed by Colas des Francs-Small et al., chapter 7) provide an excellent complement to genetics and genome analysis for the study of organelle biogenesis and function. Organelle targeting signals are not yet predicted with high confidence. Some may be overlooked or misidentified, and there are several interesting observations of proteins targeted to both plastids and mitochondria. (See chapters 7 and 14 for examples.) Organelles are well suited to proteomic studies, because these membrane-bound components are relatively easy to separate from other cellular constituents. Recent studies of plant organelle proteomes have provided new insights into many aspects of organelle gene expression and metabolism. This, in turn, reveals new targets for the genetic improvement of plants via modification of organelle functions.

4. GENETIC ENGINEERING OF PLASTIDS

In contrast to modifying plastids via the nuclear genome, direct plastid transformation has several advantages. Chloroplasts serve as ideal hosts for the expression of transgenes. Once integrated via homologous recombination, these transgenes express large amounts of protein (up to 46% of total leaf protein) due to the high copy number of the chloroplast genome in each plant cell, and chloroplast transgenic plants display their normal phenotype despite protein accumulation in such large quantities (DeCosa et al., 2001). Foreign proteins that are toxic when present in the cytosol, such as vaccine antigens, trehalose and xylanase, are non-toxic when sequestered within transgenic plastids (Daniell et al., 2001; Lee et al., 2003; Leelavathi et al., 2003). Because transgenes are maternally inherited, there is little danger of cross-pollination with wild-type plants (as discussed by Hagemann in chapter 4). The chloroplast genome has been successfully engineered to confer herbicide, insect and disease resistance, drought and salt tolerance, and phytoremediation (as discussed by Daniell et al., in chapter 16). More recently, highly efficient soybean and cotton plastid transformation have been accomplished via somatic embryogenesis using species-specific vectors (Kumar et al., 2004b; Dufourmantel et al., 2004). Chloroplast transgenic carrot plants withstand salt concentrations that only halophytes could tolerate (Kumar et al., 2004a). In these studies, for the first time, transgenes were successfully expressed in green as well as non-green plastids using endogenous (green) and heterologous (non-green) 5' and 3' untranslated regions. The chloroplast 16S rRNA promoter with binding sites for both the nuclear encoded (non-green) RNA polymerase and chloroplast encoded (green) RNA polymerases were utilized in these studies. Extension of chloroplast genetic engineering technology to other useful crops will depend on the availability of the plastid genome sequences and the ability to regenerate transgenic events and advance them towards homoplasmy. Successful plastid transformation in two among three major GM crops is undoubtedly an amazing advancement in this field. For actual pictures of chloroplast transgenic plants with valuable agronomic traits, readers are referred to color plates inserted at the beginning of this book.

Previously an exclusively mitochondrial-encoded trait, cytoplasmic male sterility is now possible through □-ketothiolase expression via the chloroplast genome (chapters 16, 23). This is a valuable tool for nuclear transgene containment, in addition to the maternal inheritance of transgenes integrated into the chloroplast genome, in most crops (Daniell, 2002). Crops such as tobacco have expressed transgenes for a variety of biopharmaceuticals, vaccines and biomaterials. Due to the high biomass of tobacco plants (~40 mtons/acre), large amounts of vaccines preventing anthrax, plague, tetanus and cholera can be manufactured cheaply (Daniell et al., 2004 a,b; Chebolu and Daniell, 2004). One acre of chloroplast transgenic plants can produce up to 400 million doses of clean and fully functional anthrax vaccine (Daniell et al., 2004b). Production of pharmaceuticals like human somatotropin, serum albumin, interferons and insulin-like growth factor may not be prohibitively expensive when using transgenic chloroplasts as bioreactors (Daniell et al., 2004a). Such high levels of expression were achieved primarily through light regulation of the chloroplast *psbA* 5' and 3' untranslated regions. The chloroplast also contains machinery that allows for correct folding and disulfide bond formation, resulting in fully functional human blood proteins or vaccine antigens (Daniell et al., 2004 a,b; Chebolu and Daniell, 2004).

Additionally, expression of the Rubisco small subunit gene (R*bc*S) within the chloroplast restored normal photosynthetic activity in a nuclear *rbc*S antisense line, a goal that has been elusive for decades (Dhingra et al., 2004). Multigene operons engineered via the chloroplast genome (DeCosa et al., 2001; Ruiz et al., 2003) do not require processing of polycistrons to monocistrons for efficient translation. Secondary structures formed by intergenic spacer regions in bacterial operons are efficiently recognized by the chloroplast processing machinery; when such processing occurs, 3' UTRs are not required for transcript stability. Thus, the ground rules governing transcript processing and stability appear to be different for native and foreign polycistrons (chapter 16).

In addition to biotechnology applications, plastid transformation has become a powerful tool for the study of plastid biogenesis and function. This approach has been used to investigate plastid DNA replication origins (Chapter 3), RNA editing elements (Chapter 9), promoter elements (Chapter 8), RNA stability determinants (Chapter 10), intron maturases (Chapter 11), translation elements (Chapters 12 and 13), protein import machinery (chapter 14), proteolysis (Chapter 15), transgene movement and evolution (Chapters 4, 16) and transcription & translation of polycistrons (chapter 16).

5. BIOTECHNOLOGY AND MOLECULAR BIOLOGY

Molecular investigations of plastid function, in turn, have significant implications for biotechnology. Maier and Schmitz-Linneweber (chapter 5) point out that, currently, more than 30 plastid genomes from various species of embryophytes and eukaryotic algae have been completely sequenced. However, only five of these sequenced genomes are from crop plant species. Authors point out that this is in particular lamentable, as an important prerequisite for plastid transformation is a detailed knowledge of intergenic spacer regions in plastid chromosomes. These spacers are targets for the integration of transgenes by homologous recombination, a technique already proven fruitful for various biotechnological applications (see chapter 16). Identity between vector sequences and target sequence is necessary (Daniell et al., 2004c; Kumar and Daniell, 2004), as heterologous transformation vectors have not yielded high frequency transformations so far even in tobacco, in which plastid transformation is highly efficient (e.g. Degray et al., 2001, also see chapter 16). Therefore further genome sequencing projects of crop plant plastid chromosomes is one of the more pressing needs in this field.

Plastid transformation was recently used to investigate the rate of transgene movement from chloroplast to nuclear genomes, and a very high rate of transgene movement from plastid to nuclear genomes has been reported (Huang et al., 2003). It should be mentioned, however, that these transfer frequencies are not in congruence with the number and size of organelle DNA insertions characterized by recent genome projects (Daniell and Parkinson, 2003, chapters 5,16). No study has yet determined how frequently plastid transgenes move to the nucleus, in a manner that supports their expression, and therefore, such transfer to the nucleus without expression is inconsequential in biotechnology applications (Daniell and Parkinson, 2003). Both sides of these issues are discussed in detail in chapters 5 and 16.

A detailed understanding of plastid gene expression is critical to the use of plastid transformation in biotechnology. Selectable marker genes, simultaneously driven by regulatory signals in the light and in the dark, in both proplastids and chloroplasts, provided continuous protection for transformed developing plastids/chloroplasts from the selectable agent; this helped to achieve a high frequency plastid transformation of cotton, a highly recalcitrant crop (Chapter 16, Kumar et al., 2004b). In contrast to non-green plastids in which the rate of transcription and transcript stability play a critical role in translation, transcript abundance is not correlated with transgene expression in mature chloroplasts (Dhingra et al., 2004, chapter 16). Enhanced translation of the *badh* gene in transgenic chloroplasts, chromoplasts and proplastids resulted in very high levels of salt tolerance (up to 400 mM) and accumulated as much betaine as in halophytes (Kumar et al., 2004a; also see chapter 12). Light regulated translation mediated by the chloroplast *psbA* 5' UTR (see chapter 13) was instrumental in achieving 500-fold increase in translation of human serum albumin in transgenic plastids (Fernandez-San Millan et al., 2003). Chaperones and proteases (see chapter 15) also have tremendous influence on yields of transgene-encoded proteins in the plastid. The accumulation of the insecticidal Cry2Aa2 protein was dramatically enhanced 128-fold, when this protein was co-expressed with a chaperone that folded Cry2Aa2 such that it can be packaged into cuboidal crystals (DeCosa et al., 2001). Further enhancements of biotechnology achieved through applications of plastid molecular biology are described in chapter 16.

6. MODIFICATION OF ORGANELLE FUNCTIONS VIA THE NUCLEAR GENOME

Two major molecular genetic approaches have been used to modify functions of plant organelles. The first approach is to express transgenes integrated into the nuclear genome but the synthesized foreign proteins are targeted to plastids. For example, about 75% of the transgenic crops planted around the world are now engineered for herbicide resistance. Most of the herbicide targets reside within the plastids and directly or indirectly affect plastid functions. Therefore, it is not a coincidence that the first identified herbicide (atrazine) inhibited the D-1 polypeptide of photosystem II. In order to provide crop species an advantage over weeds, crops have been genetically engineered with mutant versions of target enzymes functional within plastids, which are insensitive to herbicide or enzymes that detoxify the herbicide. In chapter 17, Dhingra and Daniell provide several examples of chloroplast-specific targets and engineering crops for herbicide tolerance via the nuclear genome using herbicide insensitive enzymes, or via metabolic detoxification, including the first successful demonstration (Shah et al., 1986). Authors also points out the concerns of transgene containment for nuclear transgenic crops. They describe the first example of engineering herbicide resistance via the chloroplast genome (Daniell et al., 1998) and give several additional examples of using this approach for enhanced tolerance and containment of transgenes via maternal inheritance.

Chapter 18 discusses metabolic engineering of chloroplasts for abiotic stress tolerance. The authors point out that there are chloroplast localized heat shock proteins that confer thermotolerance in plants (Osteryoung and Vierling, 1994). For example, following an initial heat episode and recovery, HSP 21 was observed in chloroplast stroma. Furthermore, heat stress resulted in temperature dependant redistribution of HSP 21 from a soluble form to an insoluble chloroplast protein fraction. Cherry and Nielsen also provide several examples of osmoprotectants like glycine betaine or trehalose that accumulate within plastids and confer salt or drought tolerance. Authors also point out successful engineering of the chloroplast genome to confer high levels of salt tolerance and accumulation of betaine as much as in halophytes (Kumar et al., 2004a). The formation of oxygen radicals by partial reduction of molecular oxygen is an unfortunate consequence of aerobic life. Active oxygen species (AOS), such as the superoxide anion and hydrogen peroxide are natural byproducts of metabolism and increase to toxic levels (oxidative stress) during a wide range of environmental stress (such as ozone, drought and salt stress). When transgenic maize plants were created by targeting iron superoxide dismutase (FeSOD) to chloroplasts, enhanced oxidative stress tolerance was observed (Van Breusegem et al, 1999). Similarly, MnSOD, localized in chloroplasts of bundle sheath cells conferred stress tolerance. Authors also discuss early light induced low molecular weight proteins (elips) localized in thylakoid membranes that are ideal candidates to confer stability against photoinhibition. Transgenes that confer tolerance to abiotic stress should not outcross with weeds and permanently transfer these valuable traits to the weed nuclear genome. Therefore, authors recommend that maternal inheritance of transgenes that confer abiotic stress tolerance via the chloroplast genome may be ideal for this purpose.

Chapter 19 deals with metabolic engineering for nutrition enrichment. Baisakh and Datta give several examples of nutritional enhancement including provitamin A, vitamin E and C, lysine, polyhydroxybutyrate, inulin, flavinoids, etc. Most of these metabolic pathways are compartmentalized within the plastids; though the enzymes involved in the pathway are nuclear encoded, precursor proteins are targeted to plastids. Authors explain the creation of golden rice by manipulating the carotenoid biosynthetic pathway within plastids (shown in color plate). Authors elaborate on the pathway of tocopherol biosynthesis, which is physiologically the most important for humans, with maximal vitamin A activity. Both cyclization and methylation of dimethylphytyl quinol to tocopherols are localized in plastid membrane. Similarly, the function of vitamin C (L-ascorbic acid –AsA) in human nutrition is well understood, including its antioxidation activity. In plants, AsA is involved in regeneration of tocopherol and zeaxanthin and PH-mediated modulation of PSII activity. The critical importance of AsA in photosynthetic metabolism is emphasized by its abundance in chloroplasts. Because AsA cannot be stored in the body, a regular supply through plant dietary source is required but not many foods are rich in AsA. Authors discuss several successful genetic engineering approaches to enhance AsA biosynthesis in plastids.

Authors of chapter 19 also discuss approaches to increase lysine, which is limiting in cereals that form the diet of a sizeable world population. The entire aspartate family pathway except the last step of methionine production occurs within plastids and is regulated by several feedback inhibition steps. The lysine feed back insensitive *dapA* gene and a mutant *lysC* gene were linked to the chloroplast transit peptide and expressed from a seed-specific promoter in transgenic canola and soybean seeds. Expression of DHPS and lysine insensitive AK in soybean caused several hundred fold increase in free lysine and five fold increase in total seed lysine content. Thus, chapter 19 shows the critical role played by plastids in nutritional enhancement.

The fatty acid biosynthesis in plants takes place almost exclusively within plastids. The plastid fatty acid synthesis requires acetyl-CoA, ATP and NAD(P)H as substrates and multiple routes exist to provide these substrates. Chain extension of fatty acids take place by C2 unit derived from malonyl-CoA, which is synthesized from acetyl CoA by acetyl CoA-carboxylase (ACCase). The fact that ACCase is a regulatory enzyme in fatty acid synthesis has stimulated a number of molecular and transgenic studies to elucidate their regulatory mechanisms and improve oil production. In chapter 20, Nishida provides an in-depth overview of plant fatty acids, fatty acid synthesis in plastids, requirements of fatty acid synthesis, key enzymes involved in generation of substrates, key genes for fatty acid synthesis in plastids, genes required for biosynthesis of polyunsaturated fatty acids in plastids, genetic analysis of oil accumulation and perspective of plastid fatty acid metabolism in plant lipid biotechnology. After compiling a detailed basic knowledge of plastid fatty acid metabolism, Nishida recaps current accomplishment in genetic engineering and prospects for the future. ACCase over-expression in tobacco by plastome transformation doubled the seed yield and improved leaf longevity in tobacco. Although it should be tested in other plants such as *B. napus*, author points out that the plastome transformation has advantage over the nuclear transformation, alleviating the risk of gene dispersal.

Bohmert et al. in chapter 21 contemplate metabolic engineering with a focus of using plastids as the site of production for polyhydroxyalkonates (PHAs), bacterially derived biopolymers composed of repeating units of ®-hydroxy acids and other biopolymers to illustrate the utility of plastids as bioreactors to produce a range of biomaterials. Authors give examples of biomaterials produced in plastids, PHAs produced in transgenic plastids and pathways for PHA synthesis and describe various crops used to produce biomaterials including tobacco, potato, Brassica, alfalfa, sugar beet, corn, sugarcane, oil palm and flax. Authors also discuss direct plastid transformation, copolymer production and plastid biochemistry relevant for PHA formaion. Bohmert et al. also conclude that production of PHAs in plastids illustrates the utility of this organelle for novel biopolymer production. The highest levels of PHA production in transgenic plants have been achieved in plastids and attempts to produce PHAs in other cellular compartments, including the cytosol were unsuccessful.

All of the examples illustrated in chapters 17-21 are ideal targets for direct plastid transformation because of hyper-expression capabilities and containment of transgenes via maternal inheritance. This should provide ample opportunities for research advancement in this field for years ahead.

7. GENETIC MODIFICATION OF PLANT MITOCHONDRIA

Although mitochondrial transformation has been achieved in yeast (Fox et al., 1988; Johnston et al., 1988) and Chlamydomonas (Randolph-Anderson, 1993), this milestone has not yet been accomplished for any vascular plant. This may stem, in part, from the challenges of selecting transformed cells. Plants are obligate aerobes, whereas yeast and Chlamydomonas cells can survive in the absence of mitochondrial respiration.

While the stable genetic modification of plant mitochondria has not been achieved, DNA can be introduced into isolated plant mitochondria (Farré and Araya, 2001; Koulintchenko et al., 2003). This new tool has already been exploited for the study of plant mitochondrial RNA editing (Farré et al., 2001; Staudinger and Kempken, 2003) and intron splicing (Farré and Araya, 2001). Plant mitochondria import tRNAs (reviewed by Dietrich et al., 1996), and modified tRNAs expressed from the nuclear genome currently represent one strategy for introducing nucleic acids into plant mitochondria *in vivo* (Small et al., 1992).

Somatic cell fusions (reviewed by Guo et al. in chapter 24 of this volume) present an alternative to transformation strategies for the genetic modification of organelle functions. Although we do not fully understand the factors that govern the organelle genotypes recovered following plant cell fusion events, this technology has been used to transfer organelle-encoded traits such as CMS and herbicide resistance between sexually incompatible plants, and to generate novel combinations of organelle genomes that could not be recovered from genetic crosses of sexually compatible plants.

8. ACKNOWLEDGEMENTS

Oganelle research in the Chase lab has been funded by the United States Department of Agriculture National Research Initiative Awards 00-35300-9409 and 2001-0534-10888, and by the Florida Agricultural Experiment Station. This work was approved for publication as Journal Series No. N-02500. Research in the Daniell laboratory was supported by USDA, NSF, DOE and DOD grants, and is currently supported by the United States Department of Agriculture (grant 3611-21000-017-00D) and the National Institutes of Health (grant R01 GM 63879).

9. REFERENCES

Chebolu, S & Daniell, H. (2004). Chloroplast derived vaccine antigens and biopharmaceuticals: expression, folding, assembly and functionality. Current Trends in Microbiology and Immunology, in press.

Daniell, H. (2002). Molecular strategies for gene containment in transgenic crops. *Nat. Biotechnol., 20,* 581-586.

Daniell, D., Carmona-Sanchez, O., & Burns, B. B. (2004a). Chloroplast derived antibodies, biopharmaceuticals and edible vaccines. In R. Fischer & S. Schillberg (Eds.) *Molecular Farming* (pp.113-133). Weinheim: WILEY-VCH Verlag.

Daniell, H., Chebolu, S., Kumar, S., Singleton, M., & Falconer, R. (2004b). Chloroplast-derived vaccine antigens and other therapeutic proteins. *Vaccine*, in press

Daniell, H., Datta, R., Varma, S., Gray, S., & Lee, S. B. (1998). Containment of herbicide resistance through genetic engineering of the chloroplast genome. *Nat. Biotechnol., 16,* 345-348.

Daniell, H. & Parkinson, C.L. (2003). Jumping genes and containment. *Nat. Biotechnol.* 21: 374 - 375.

Daniell, H., Lee, S. B., Panchal, T., & Wiebe, P. O. (2001). Expression of cholera toxin B subunit gene and assembly as functional oligomers in transgenic tobacco chloroplasts. *J. Mol. Biol., 311,* 1001-1009.

Daniell, H., Ruiz, O.N. and Dhingra, A. (2004c). Chloroplast genetic engineering to improve agronomic traits in transgenic Plants. Methods Mol. Biol. 286: 111-137.

DeCosa, B., Moar, W., Lee, S. B., Miller, M., & Daniell, H. (2001). Overexpression of the *Bt* cry2Aa2 operon in chloroplasts leads to formation of insecticidal crystals. *Nat. Biotechnol., 19,* 71-74.

DeGray, G., Rajasekaran, K., Smith, F., Sanford, J., & Daniell, H. (2001). Expression of an antimicrobial peptide via the chloroplast genome to control phytopathogenic bacteria and fungi. *Plant Physiology, 127,* 852-862.

Dufourmantel, N., Pelissier, B., Garçon, F.,Peltier, J. M., & Tissot, G. (2004). Generation of fertile transplastomic soybean. *Plant Mol. Biol.*, in press.

Dhingra, A., Portis, A.R. & Daniell, H. (2004). Enhanced translation of a chloroplast expressed RbcS gene restores SSU levels and photosynthesis in nuclear antisense *RbcS* plants. *Proc. Natl. Acad. Sci., U.S.A., 101,* 6315-6320.

Dietrich, A., Small, I., Cosset, A., Weil, J. H., & Marechal-Drouard, L. (1996) Editing and import: strategies for providing plant mitochondria with a complete set of functional transfer RNAs. *Biochimie, 78,* 518-529.

Duvick, D. N. (1959). The use of cytoplasmic male-sterility in hybrid seed production. *Econ. Bot., 13,* 167-195.

Duvick, D. N. (2001). Biotechnology in the 1930s: the development of hybrid maize. *Nature Rev. Genet., 2,* 69-74.

Edwardson, J. N. (1956). Cytoplasmic male sterility. *Bot. Rev., 22,* 696-738.

Farré, J. –C., & Araya, A. (2001). Gene expression in isolated plant mitochondria: high fidelity of transcription, splicing and editing of a transgene product in electroporated organelles. *Nucl. Acids. Res., 29*, 2484-2491.

Farré, J. –C., & Araya, A. (2002). RNA splicing in higher plant mitochondria: determination of functional elements in group II intron from a chimeric *coxII* gene in electroporated wheat mitochondria. *Plant J., 29*, 203-213.

Farré, J. –C., Leon, G., Jordana, X., and Araya, A. (2001) cis recognition elements in plant mitochondrion RNA editing. *Mol. Cell. Biol., 21*, 6731-6737.

Fernandez-San Millan, A., Mingeo-Castel, A. M., Miller, M., & Daniell, H. (2003). A chloroplast transgenic approach to hyper-express and purify human serum albumin, a protein highly susceptible to proteolytic degradation. *Plant Biotechnol. J., 1*, 71-79.

Fox, T. D., Sanford, J. C., & McMulling, T. W. (1988). Plasmids can stably transform yeast mitochondria lacking endogenous mtDNA. *Proc. Natl. Acad. Sci. USA, 85*, 7288-7292.

Gray, J. C., Sullivan, J. A., Wang, J. H., Jerome, C. A., & MacLean, D. (2003). *Philos. Trans. R. Soc. Lond. B Biol. Sci., 358*, 135-144.

Huang, C. Y., Ayliffe, M. A., & Timmis, J. N. (2003). Direct measurement of the transfer rate of chloroplast DNA into the nucleus. *Nature, 422*, 72-76.

Hagemann, R. (2000). Erwin Baur or Carl Correns: who really created the theory of plastid inheritance? *J. Hered., 91*, 435-440.

Johnston, S. A., Anziano, P. Q., Shark, K., Sanford, J. C., & Butow, R. A. (1988). Mitochondrial transformation in yeast by bombardment with microprojectiles. *Science, 140*, 1538-1541.

Juszczuk, I. M. & Rychter, A. M. (2003). Alternative oxidase in higher plants. *Acta Biochim. Pol., 50*, 1257-1271.

Koulintchenko, M., Konstantinov, Y., & Dietrich, A. (2003). Plant mitochondria actively import DNA via the permeability transition pore complex. *EMBO J., 22*, 1245-1254.

Kumar S, Daniell H (2004) Engineering the chloroplast genome for hyper-expression of human therapeutic proteins and vaccine antigens. Methods Mol Biol **267:** 365-383

Kumar, S., Dhingra, A. & Daniell, H. (2004a). Plastid expressed *betaine aldehyde dehydrogenase* gene in carrot cultured cells, roots and leaves confers enhanced salt tolerance. *Plant Physiol.,* in press.

Kumar, S., Dhingra, A., & Daniell, H. (2004b). Manipulation of gene expression facilitates cotton plastid transformation of cotton by somatic embryogenesis & maternal inheritance of transgenes. *Plant Mol. Biol.*, in press.

Lee, S. B., Kwon, H. B., Kwon, S. J., Park, S. C., Jeong, M. J., Han, S. E., Byun, M.O., Daniell, H. (2003). Accumulation of trehalose within transgenic chloroplasts confers drought tolerance. *Mol. Breed., 11*, 1-13.

Leelavathi, S., Gupta, N., Maiti, S., Ghosh, A., & Reddy, V. S. (2003). Overproduction of an alkali- and thermo-stable xylanase in tobacco chloroplasts and efficient recovery of the enzyme. *Mol.Breed., 11*, 59-67.

Madoka, Y., Tomizawa, K.-I., Mizoi, J., Nishida, I., Nagano, Y., & Sasaki, Y. (2002) Chloroplast transformation with modified *accD* operon increases acetyl-CoA carboxylase and causes extension of leaf longevity and increase in seed yield in tobacco. *Plant Cell Physiol. 43*, 1518-1525.

Margulis, L. (1970). *Origin of Eukaryotic Cells.* New Haven: Yale University Press.

McIntosh, L., Eichler, T., Gray, G., Mazewll, D., Nickels, R., & Wang, Y. (1998). Biochemical and genetic controls exerted by plant mitochondria. *Biochim. Biophys. Acta, 1365*, 278-284.

Osteryoung, K.W. & Vierling, E. (1994). Dynamics of small heat shock protein distribution within the chloroplasts of higher plants. *J. Biol. Chem.*, 269, 28676-28682.

Parikh, V. S., Morgan, M. M., Scott, R., Clements, L. S., & Butow, R. A. (1987). The mitochondrial genotype can influence nuclear gene expression in yeast. *Science, 235*, 576-580.

Pfannschmidt, T. (2003). Chloroplast redox signals: how photosynthesis controls its own genes. *Trends Plant Sci., 8*, 33-41.

Pfannschmidt, T., Schutxe, K., Fey, V., Sherameti, I., & Oelmuller, R. (2003). Chloroplast redox control of nuclear gene expression – a new class of plastid signals in interorganellar communication. *Antioxid. Redox Signal. 5*, 95-101.

Randolph-Anderson, B. L., Boynton, J. E., Gillham, N. W., Harris, E. H., Johnson, A. M., Dorthu, M. – P., & Matagne, R. F. (1993) Further characterization of a respiratory deficient *dum-1* mutation of *Chlamydomonas reinhardtii* and its use as a recipient for mitochondrial transformation. *Mol. Gen. Genet., 236*, 235-244.

Rodermel, S. (2001). Pathways of plastid-to-nucleus signaling. *Trends Plant Sci. 10*, 471-478.

Rodermel, S., & Park, S. (2003). Pathways of intracellular communication: tetrapyrroles and plastid-to-nucleus signaling. *Bioessays, 25*, 631-636.

Ruiz, O. N., Hussein, H., Terry, N., & Daniell, H. (2003). Phytoremediation of organomercurial compounds via chloroplast genetic engineering. *Plant Physiol.*, *132*, 1344-1352.

Shah, D. M., Horsch, R. B., Klee, H. J., Kishore, G. M., Winter, J. A., Tumer, N. E., et al. (1986). Engineering herbicide tolerance in transgenic plants. *Science,* 233, 478-481.

Small, I., Marechal-Drouard, L., Masson, J., Pelletier, G., Cosset, A., Weil, J. H., & Dietrich, A. (1992). *In vivo* import of a normal or mutagenized heterologous transfer RNA into the mitochondria of transgenic plants: towards novel way of influencing mitochondrial gene expression? *EMBO J., 11*, 1291-1296.

Staudinger, M., & Kempken, F. (2003). Electroporation of isolated higher-plant mitchondria: transcripts of an introduced *cox2* gene, but not an *atp6* gene, are edited *in organello*. *Mol. Genet. Genom., 269*, 553-561.

Surpin, M., Larkin, R. M., & Chory, J. (2002). Signal transduction between chloroplast and the nucleus. *Plant Cell, 14S*, 237-238.

Van Breusegem, F., Slooten, L., Stassart, J.-M., Moens, T., Van Montagu, M., & Inze, D. (1999). Overproduction of *Arabidopsis thaliana* FeSOD confers oxidative stress tolerance to transgenic maize. *Plant Cell Physiol.*, 40, 98-129.

Ye, X., Al-Babili, S., Kloti, A., Zhang, J., Lucca, P., Beyer, P., & Potrykus, I. (2000) Engineering the provitamin A (□-carotene) biosynthetic pathway into (carotenoid-free) rice endosperm. *Science 287,* 303-330.

Zubko, M. K. (2004). Mitochondrial tuning fork in nuclear homeotic functions. *Trends Plant Sci., 9,* 61-64.

Section 1

Organelle Genomes and Proteomics

CHAPTER 2

THE EVOLUTIONARY ORIGINS OF PLANT ORGANELLES

MICHAEL W. GRAY

Department of Biochemistry and Molecular Biology, Dalhousie University, Halifax, Nova Scotia B3H 1X5, Canada

Abstract. Alone among plant organelles, mitochondria and plastids contain genetic information, in the form of organellar genomes. These genomes specify a limited set of protein and RNA components that are essential for the normal biogenesis and functioning of these two organelles. Importantly, organellar genomes are now known to be remnant eubacterial genomes, reflecting an origin of mitochondria and plastids as eubacterial symbionts in eukaryotic cells. Through analysis of organellar gene content, gene sequence and gene organization, we have been able to trace the origin of mitochondrial and plastid genomes to particular groups of extant eubacteria, with which these organelles share a specific evolutionary ancestor. Over the past several decades, our burgeoning knowledge of organellar genomes, culminating in the complete sequences of a diverse collection of mitochondrial and plastid DNAs, has brought solid evidence to bear on over a century of speculation and debate about the nature and origin of mitochondria and plastids, a subject that is of fundamental importance to our understanding of the evolutionary emergence of the eukaryotic cell as a whole. Our current view of organelle evolution is that mitochondria (almost certainly) and plastids (very likely) each originated only once, the former from within the α-Proteobacteria and the latter from within the Cyanobacteria. In the case of plastids, a single *primary endosymbiosis* event, involving a eukaryotic host cell and a cyanobacterium, was followed by a number of *secondary endosymbioses*, each between a (presumably) non-photosynthetic, heterotrophic eukaryotic host and a plastid-containing eukaryotic symbiont. Three lineages (green algae + land plants, red algae, glaucocystophytes) appear to contain primary plastids, whereas all other groups of algae contain secondary plastids. Early on, data derived from analysis of plant organellar DNAs was particularly persuasive in affirming a eubacterial origin of mitochondrial DNA (mtDNA) specifically from within the α-Proteobacteria. At the same time, our emerging appreciation of the extraordinary structural variability of mtDNA, especially within the lineage of green algae and land plants, prompted questions about whether these markedly different mitochondrial genomes had separate evolutionary origins. With the availability of additional mtDNA sequences, this issue now seems settled in favour of a single (monophyletic) origin of mitochondria, with mitochondrial genomes in particular lineages (e.g., chlorophycean green algae) having highly derived genomes, unrepresentative of the ancestral state.

H. Daniell and C.D. Chase (eds.), Molecular Biology and Biotechnology of Plant Organelles, 15—36.

1. INTRODUCTION

Mitochondria and plastids are now known to be the descendants of eubacterial symbionts. More specifically, the genomes present in these two organelles are clearly the remnants of eubacterial genomes. The most compelling data leading to this conclusion has come from comprehensive analyses of organellar DNAs, culminating in the complete sequencing of a diverse collection of mitochondrial and plastid genomes. Comparison of organellar molecular data relating to gene sequence, gene organization and gene expression with similar data from a diverse collection of eubacteria has pinpointed the evolutionary origin of mitochondria and plastids and identified those extant eubacterial lineages to which they are most closely related (see Gray, 1999). We can say with confidence that mitochondria are derived from within the α-Proteobacteria, having as their closest living relatives a sub-group comprising genera such as Rickettsia, Ehrlichia, Anaplasma and Wolbachia, all obligate intracellular parasites (Gray et al., 1999; see also Wu et al., 2004). Plastids originate from within the Cyanobacteria, although it has so far not been possible to identify a particular cyanobacterial lineage as being specifically related to plastids (Archibald and Keeling, 2002).

With the evolutionary source of mitochondria and plastids established, many other questions continue to challenge workers in this field. How many times did each organelle originate in evolution (i.e., are mitochondria and plastids *monophyletic* or *polyphyletic*)? How were the eubacterial ancestors of mitochondria and plastids incorporated into the eukaryotic cell? In particular in the case of the mitochondrion, do eukaryotic lineages exist that never had this organelle, i.e., that are *primitively amitochondriate*? What was the ancestral form of the mitochondrial and plastid genomes, and how have these DNAs evolved in different eukaryotic lineages? How much of the mitochondrial and plastid proteome (the collection of proteins that makes up each organelle) can be traced to their eubacterial ancestors? If there are non-eubacterial proteins in these organelles, where did they come from, and when and how?

This chapter summarizes current views on the evolutionary origin of plastids and of mitochondria, with an emphasis on plant mitochondria. More detailed compilations and critical analysis of the body of data that underpins these views can be found in Archibald and Keeling (2002, 2004), Burger and Lang (2003), Burger et al. (2003), Emelyanov (2001, 2003), Gray (1989, 1992, 1993, 1999, 2004), Gray and Doolittle (1982), Gray and Spencer (1996), Gray et al. (1998a, 1999, 2001), Lang et al. (1999), Andersson et al. (2003) and McFadden (2001, 2003).

2. EUKARYOTES WITHOUT MITOCHONDRIA: AMITOCHONDRIATE EUKARYOTES

According to classical endosymbiosis theory, the α-proteobacterial ancestor of mitochondria was taken up by a host cell lacking this organelle: i.e., by an amitochondriate eukaryote (Margulis, 1970; de Duve, 1996). Rather than being digested, the endocytosed eubacterium established a symbiotic relationship with the host cell, a process that culminated in massive reduction of the endosymbiont genome, with wholesale loss of genes and/or their transfer to the nuclear genome of the host (Timmis et al., 2004). Classical theory further posits that, at a later point in evolutionary history, the plastid was acquired in essentially the same way, via a second endosymbiotic event involving a cyanobacterium.

Initial phylogenetic trees of the eukaryotic lineage (domain Eucarya) seemed to provide evidence in support of the foregoing scenario, in that several unicellular eukaryotes (protists) that do not have recognizable mitochondria were seen to emerge as the first branches on such trees (Sogin, 1991). In terms of intracellular structure, such amitochondriate protists tend to be primitive, lacking not only mitochondria but also other characteristic eukaryotic organelles such as an endomembrane system with associated Golgi dictyosomes. A reasonable interpretation at the time was that these structurally simple, amitochondriate groups (termed 'Archezoa' by Cavalier-Smith, 1983) had diverged away from the main line of eukaryotic evolution prior to the acquisition of mitochondria: i.e., that these lineages are primitively without mitochondria (Cavalier-Smith, 1987).

Over the past decade, the concept of primitively amitochondriate lineages has been challenged by several important findings. First, more sophisticated phylogenetic analyses, employing expanded data sets, more comprehensive taxon sampling, and a variety of refined and more rigorous reconstruction algorithms, have indicated that the basal branching position of many amitochondriate lineages is a 'long-branch attraction' artefact, attributable to rapid sequence evolution in the macromolecules (notably, small subunit rRNA) typically used to test phylogenetic relationships (Embley and Hirt, 1998; Philippe and Adoutte, 1998; Philippe et al., 2000). In particular Microsporidia, a clade of intracellular parasites that infect other protist groups as well as animals, is now generally acknowledged to be a derived fungal lineage (Keeling and Fast, 2002). Other amitochondriate protists, such as *Entamoeba histolytica* and *Mastigamoeba balamuthi*, are seen to nest robustly within mitochondria-containing lineages (Bapteste et al., 2002).

Second, mitochondrial gene relicts have now been identified in many supposed amitochondriate lineages. In mitochondria-containing organisms, the homologous genes encode proteins that are targeted to and function in mitochondria. Chief among the identified amitochondriate 'mitochondrial genes' are ones specifying heat shock proteins (e.g., Hsp70) and chaperonins (e.g., Cpn60) that are typically localized and function in mitochondria, although genes encoding other proteins typical of mitochondria have also been identified in amitochondriate protists (Embley and Hirt, 1998; Roger, 1999). More tellingly, when such proteins are included in phylogenetic trees, they generally cluster robustly with their counterparts from mitochondriate eukaryotes, as a sister group to the specific α-proteobacterial relatives of mitochondria (e.g., Clark and Roger, 1995). These findings have been interpreted as evidence of secondary loss of mitochondria in amitochondriate lineages that contain mitochondrial gene relicts.

Although the presumption is that mitochondrial gene relicts originally resided in a pre-mitochondrial endosymbiont genome and were subsequently transferred to the nuclear genome as the evolving endosymbiont genome was reduced, it is worth pointing out that in no case have any of these genes been found in any characterized mitochondrial genome. Hence, evolutionary transfer to the nucleus, if it indeed occurred, would have had to happen very early in the transition from endosymbiont to organelle. Alternatively, transient associations, short of full-blown endosymbioses, might have provided an opportunity for acquisition of a 'mitochondrial' gene from a source other than the endosymbiont that was destined to become the mitochondrion. Or, a 'mitochondrial' gene might have been acquired via ingestion and digestion of foreign genetic material, in a 'you-are-what-you-eat' scenario (Doolittle, 1998). Admittedly, phylogenetic trees consistently point to the mitochondrial ancestor of α-Proteobacteria as the source of mitochondrial gene relicts in amitochondriate protists, strengthening (but ultimately not proving) the thesis that these genes arrived in the nuclear genome by way of mitochondrial genomes that no longer exist.

A third source of evidence weighing against a primitively amitochondriate condition is the finding of what are regarded as remnant mitochondria (as opposed to mitochondrial genetic relicts) in most of the eukaryotic lineages that were once supposed never to have had mitochondria: i.e., these 'amitochondriate' protists have been found to harbor cryptic organelles, evidently derived from mitochondria (Williams and Keeling, 2003; Dyall et al., 2004). Examples include hydrogenosomes in various protists (Dyall and Johnson, 2000; Embley et al., 2003a,b; Rotte et al., 2000) and tiny organelles now collectively dubbed 'mitosomes' in *Entamoeba histolytica* (Mai et al., 1999; Tovar et al., 1999), the microsporidian *Trachipleistophora hominis* (Williams et al., 2002), the apicomplexan *Cryptosporidium parvum* (Riordan et al., 2003) and, most recently, the diplomonad *Giardia lamblia* (Tovar et al., 2003). The fact that proteins encoded by mitochondrial genetic relicts have been shown to be targeted to these newly discovered organelles (Williams and Keeling, 2003) strengthens the view that hydrogenosomes and mitosomes are derived mitochondria, and that none of the organisms containing them is primitively amitochondriate. These observations have substantially reduced the number of candidate amitochondriate protist lineages (Roger and Silberman, 2002) to two poorly studied groups: retortamonads and oxymonads (see Roger, 1999; Simpson and Roger, 2004). However, because oxymonads and retortamonads tend to associate within lineages of hydrogenosome- or mitosome-containing 'amitochondriate' protists in phylogenetic trees (Silberman et al., 2002; Simpson and Roger, 2004), even this possibility seems increasingly unlikely.

If we are left without candidate protists that are primitively amitochondriate, what does this mean? One possibility is that such organisms do exist, but that they are exceedingly rare and occupy such specialized biological niches that we simply have not discovered them yet. Alternatively, primitively amitochondriate lineages, predating the advent of mitochondria, may once have existed but have all become extinct. A third possibility is that the origin of mitochondria was not subsequent to the origin of the eukaryotic host cell (as required by classical endosymbiont theory) but was, in fact, coincident with it: as, e.g., posited in the 'hydrogen hypothesis' of Martin and Müller (1998). At the moment, whereas the ultimate source of the mitochondrial genome (an α-

Proteobacteria-like symbiont) is clear, the details of how this symbiont was inducted into the eukaryotic cell remain obscure (Martin et al., 2001; Dyall et al., 2004).

3. EVOLUTIONARY ORIGIN OF MITOCHONDRIA: A MITOCHONDRIAL GENOMICS PERSPECTIVE

3.1. Mitochondrial genomes: extraordinary structural diversity, conservative genetic function

Attempting to formulate a concept of mtDNA structure is akin to the exercise described by the American poet John Godrey Saxe (1816-1887) in his piece "The Blind Men and the Elephant", based on an old fable of India. Depending on which part of the elephant each of six learned (but unsighted) men examined, each came away with a very different impression of the beast: wall, spear, snake, tree, fan or rope. The upshot was that although each disputant "was partly in the right," nevertheless "all were in the wrong."

In an analogous fashion, our impression of the size, coding capacity and organization of mtDNA depends very much on the organismal source of our 'mitochondrial elephant'. Plant mitochondrial genomes present us with a radically different view of mtDNA structure than do animal mitochondrial genomes. In fact, no consensus mitochondrial genome exists but rather an extraordinary and most intriguing diversity of size, form, coding capacity, organization and mode of expression (Gray et al., 1998a; Gray et al., 1999; Lang et al., 1999; Burger et al., 2003), all of which bespeaks radically different evolutionary pathways and mechanisms in different eukaryotic lineages.

The most eubacteria-like and gene-rich mitochondrial genome identified to date is found in the jakobid flagellate, *Reclinomonas americana* (Lang et al., 1997). This 69,034-bp mtDNA is circular-mapping, mostly (>90%) coding, and contains an array of genes that at the time of its characterization had not been found in any other completely sequenced mtDNA. Among the most significant of these novel mtDNA-encoded genetic functions are components of a multi-subunit eubacterial-type (□$_2$□□'□) RNA polymerase. In non-jakobid eukaryotes, the mtDNA does not encode RNA polymerase genes; in these cases, the mitochondrial RNA polymerase is a nucleus-encoded, single-subunit, T3 phage-like enzyme (Gray and Lang, 1998) of uncertain evolutionary origin (Cermakian et al., 1997). At the other end of the spectrum is the 6-kb mitochondrial genome of apicomplexan parasites such as *Plasmodium falciparum* (the causative agent of malaria), which encodes only three protein genes and fragmented large subunit (LSU) and small subunit (SSU) rRNA genes.

Despite this structural and coding disparity, the genetic role of mtDNA is fundamentally the same in all eukaryotes. Mitochondrial genomes have a primary role in specifying a limited number (3-26) of inner membrane components involved in the generation of ATP through coupled electron transport-oxidative phosphorylation, plus a few proteins (up to 6) functioning in protein transport. As well, mtDNA contains a variable number (2-63) of informational genes encoding mostly components of the mitochondrial protein-synthesizing system (rRNAs, tRNAs and ribosomal proteins), which serves to translate mtDNA-encoded mRNAs.

3.2. Origin of the mitochondrial genome: monophyletic or polyphyletic?

Several observations support the inference that the mitochondrial genome (and therefore the mitochondrion itself) originated only once, i.e., that the mitochondrial genome is monophyletic. In the first place, the genes contained in completely sequenced mitochondrial genomes (although variable in number) are seen to be a subset of the collection of genes present in the most gene-rich mitochondrial genome, that of *R. americana* (Gray et al., 1999). Because even the smallest genomes of extant, free-living α-Proteobacteria encode at least an order of magnitude more genes than the most gene-rich mtDNAs sequenced to date, the proto-mitochondrial genome must initially have contained a much greater number of genes than do contemporary mtDNAs. Hence, mitochondrial genome evolution must have involved an initial phase of massive gene loss, culminating in the retention of a relatively small number of the genes originally contributed by the eubacteria-like proto-mitochondrial genome. It is difficult to imagine how and why a convergent process of massive genome reduction, occurring independently in separately acquired symbionts, would have resulted in basically the same small collection of mtDNA-encoded genes in polyphyletic lineages.

Nevertheless, Stiller et al. (2003) have recently argued in the case of plastids that similarities in gene content are better explained by convergent evolution due to constraints on gene loss than by a shared evolutionary history. More compelling than overall gene content are aspects of gene organization that may be considered to constitute shared derived characters. Mitochondrial gene order is remarkably fluid, but certain evolutionarily conserved clusters have been identified. In particular, physically linked ribosomal protein genes are encoded in plant and some protist mtDNAs, and in several cases these gene clusters are co-linear with the homologous operons in eubacterial genomes. Due to evolutionary gene loss, the mitochondrial clusters show gene deletions relative to the corresponding eubacterial clusters. For example, in the contiguous S10-*spc* operons of *Rickettsia prowazekii* (a specific α-proteobacterial relative of mitochondria), one finds the gene order -S19-L22-S3-L16-L29-S17-L14-L24-L5-. We now know of more than 12 examples where all or part of this cluster is represented in different mtDNAs, except that in all of the mitochondrial cases, L22, L29, S17 and L24 are absent. These mitochondrion-specific gene deletions are best interpreted as shared derived (rather than convergent) characters already present in the common ancestor of these particular mitochondrial genomes (Gray et al., 1999; Lang et al., 1999).

Although shared derived characters can be used to argue for the monophyly of the mitochondrial genomes containing them, their presence (and therefore their phylogenetic utility) is generally limited. For example, ribosomal protein genes are totally absent from animal mtDNAs and almost entirely absent from fungal mtDNAs (Lang et al., 1999). On the other hand, mitochondrial genomes encode a generally conserved set of proteins whose concatenated sequences can be used in comprehensive phylogenetic reconstructions. In these types of analysis, mitochondrial genomes (including those of animals and fungi) are seen to cluster together as a monophyletic group (clade), to the exclusion of the eubacterial outgroup (Gray et al., 1999; Lang et al., 1999).

In summary, several types of data (gene content, gene organization and phylogenetic reconstruction) support the view that the mitochondrial genome is monophyletic throughout the eukaryotic lineage. Conversely, there is no compelling evidence supporting the alternative view of a polyphyletic origin.

3.3. Mitochondrial genome structure and organization in the green algal relatives of land plants

On the basis of phylogenetic reconstructions, supported by biochemical, physiological and ultrastructural data, green plants (Viridiplantae *sensu* Cavalier-Smith, 1981) can be apportioned between two sister phyla: the Chlorophyta (Sluiman, 1985), containing most of the green algae; and the Streptophyta (Bremer, 1985), containing land plants as well as their specific green algal relatives, the Charophyceae (*sensu* Mattox and Stewart, 1984). Phylum Chlorophyta comprises the monophyletic classes Chlorophyceae, Trebuxiophyceae and Ulvophyceae, as well as a paraphyletic assemblage of primitive green algae (Prasinophyceae) (for discussion, see Gray et al., 1998b). The branching position of a key primitive green alga, *Mesostigma viride*, remains uncertain, with some analyses placing it as the most basal lineage in the Streptophyta (Karol et al., 2001) or Chlorophyta (Nozaki et al., 2003b), whereas other reports suggest that it emerged prior to the split between Chlorophyta and Streptophyta (i.e., as a sister group to all green algae; Lemieux et al., 2000).

Green plants exemplify the three main types of mitochondrial genome arrangement (*ancestral*, *reduced* and *expanded*) that have been defined through comparison of completely sequenced mtDNAs. *Ancestral* mitochondrial genomes, which retain many of the properties of a minimally derived mitochondrial genome (and therefore most closely resemble the archetype eubacteria-like mtDNA of *Reclinomonas americana*) include those of the trebuxiophyte alga *Prototheca wickerhamii* (Wolff et al., 1994), the prasinophyte alga *Nephroselmis olivacea* (Turmel et al., 1999b), and the charophyte algae *Chaetosphaeridium globosum* (Turmel et al., 2002b) and *Chara vulgaris* (Turmel et al., 2003). An ancestral type of mtDNA is also found in *M. viride* (Turmel et al., 2002a). These particular mitochondrial genomes map as circular molecules and range in size from 42.4 kb (*M. viride*) to 67.7 kb (*C. vulgaris*).

Ancestral green algal mtDNAs are relatively gene-rich, encoding 18-20 proteins that function in coupled oxidative phosphorylation, 1-4 proteins involved in import of proteins into mitochondria, and 13-15 mitoribosomal proteins, as well as eubacteria-like LSU, SSU and 5S rRNAs. These genomes also specify a set of 26-28 different eubacteria-like tRNAs capable of decoding all or almost all codons in the mtDNA-encoded protein genes. Like *R. americana* mtDNA, the *N. olivacea* mitochondrial genome carries a gene (*rnpB*) for the RNA component of an RNase P.

In marked contrast, chlorophycean green algae are characterized by having *reduced* mitochondrial genomes, as in *Chlamydomonas reinhardtii* (Boer and Gray, 1991; Vahrenholz et al. 1993), *Chlamydomonas eugametos* (Denovan-Wright et al., 1998), *Chlorogonium elongatum* (Kroymann and Zetsche, 1998) and *Polytomella parvum* (Fan and Lee, 2002), as well *Pedinomonas minor* (Turmel et al., 1999b), a chlorophyte alga of uncertain phylogenetic placement. These relatively small (16-25 kb) and highly derived mtDNAs feature both circular-mapping forms (*C. eugametos, C. elongatum, P. minor*) and linear forms (*C. reinhardtii*, with *P. parvum* mtDNA comprising two linear molecules of 13.5 and 3.5 kb). They have a substantially reduced complement of respiratory chain genes (only 7-10) and completely lack any protein translocation or ribosomal protein genes, as well as a 5S rRNA gene. In all of these chlorophycean reduced mitochondrial genomes, LSU and (with the exception of *P. minor*) SSU rRNA genes are fragmented and rearranged to different extents (Nedelcu et al., 2000), a striking pattern originally described for *C. reinhardtii* mtDNA (Boer and Gray, 1988). Only a few (1-8) tRNA genes are found in these reduced chlorophycean mtDNAs, necessitating import of nucleus-encoded tRNAs to support mitochondrial protein synthesis. Paralleling the radical divergence in gene structure and overall genome organization is a very rapid evolution of mtDNA-encoded protein and rRNA sequences in these organisms, leading to difficulties in correctly positioning chlorophycean algae in mitochondrial phylogenetic trees (e.g., Turmel et al., 1999b).

A somewhat larger (43 kb) chlorophycean mtDNA, present in the green alga *Scenedesmus obliquus* (Nedelcu et al., 2000; Kück et al., 2000), reflects an intermediate stage in the evolution of the green algal mitochondrial genome. Although *S. obliquus* mtDNA displays features characteristic of smaller reduced derived chlorophycean mtDNAs (including complete absence of ribosomal protein genes, lack of a 5S rRNA gene, fragmented and rearranged LSU and SSU rRNA genes), it does encode a few additional protein genes (notably *cox2* and *cox3*) as well as an almost complete set of 27 tRNA genes. Notably, the *S. obliquus* mitochondrial translation system uses UAG to code for leucine while also employing UCA (which normally specifies serine) as a termination codon.

A surprisingly large mtDNA (95.9 kb) has been newly described in the ulvophycean alga, *Pseudendoclonium akinetum*: the first ulvophyte mtDNA sequence to be determined (Pombert et al., 2004). Despite being larger than typical ancestral chlorophyte mtDNAs, *Pseudendoclonium* mtDNA actually encodes a slightly smaller number of proteins, and no 5S rRNA gene is detectable. On the other hand, the 27 mtDNA-encoded tRNAs are sufficient to decode all of the codons in the protein-coding genes of *Pseudendoclonium* mtDNA. Phylogenetic analyses of seven mtDNA-encoded proteins revealed a sister-group relationship between the ulvophyte and chlorophytes displaying reduced mtDNAs.

One may suggest that *P. akinetum* and *S. obliquus* mtDNAs represent a continuum (characterized by gene loss) in a gradual evolutionary transition from the ancestral type of mtDNA found in prasinophytes and trebuxiophytes to the reduced mtDNA type found in most chlorophycean algae examined to date. On the other hand, *P. akinetum* mtDNA also exemplifies aspects of mitochondrial genome expansion that characterize the very large mtDNAs of land plants.

3.4. Mitochondrial genome structure and organization in land plants

The mitochondrial genome of land plants (reviewed by Fauron et al. in Chapter 6 of this volume) represents an *expanded* type of organization that is characteristic of and largely restricted to this lineage. Complete mitochondrial genome sequences are now available for six land plants: the liverwort (bryophyte) *Marchantia polymorpha* (Oda et al. 1992); the dicot angiosperms *Arabidopsis thaliana* (Unseld et al., 1997), *Beta vulgaris* (Kubo et al., 2000) and *Brassica napus* (rapeseed; Handa, 2003); and the monocot angiosperms *Oryza sativa* (rice; Notsu et al. 2002) and *Zea mays* (maize; NCBI acc. no. AY506529). In these six examples, mitochondrial genome size ranges from 187 kb in *M. polymorpha* to 570 kb in *Z. mays*, some 3.5- to 10-fold higher than the size of a typical ancestral-type mtDNA (~55 kb). These are not even the smallest plant mtDNAs, as sizes ranging up to 2400 kb□ the size of a small bacterial genome□ have been reported for some species of cucurbit (Ward et al., 1981).

Despite its large size, the land plant mitochondrial genome does not encode a proportionately greater number of genes than the ancestral-type mtDNAs of its green algal relatives: in fact, somewhat fewer genes are present. Moreover, progressive evolutionary loss of ribosomal protein and tRNA genes is particularly evident as one moves from primitive land plants like bryophytes to advanced flowering plants. For example, whereas *Marchantia* mtDNA encodes 16 ribosomal proteins and 27 tRNAs (Oda et al., 1992), the *Arabidopsis* mitochondrial genome specifies only 7 and 17, respectively. When the four angiosperm mtDNAs are compared, these two gene classes show the greatest degree of variation in terms of whether they are present or absent and whether they are pseudogenes or active genes (Notsu et al., 2002; Handa, 2003). Gene loss and transfer to the nucleus (mostly ribosomal protein genes but occasionally respiratory chain genes, as well) has occurred repeatedly and frequently during evolution of the angiosperm mitochondrial genome (Adams et al., 2000, 2001, 2002). Widespread transfer of mitochondrial genes between distantly related flowering plant species has even been documented (Bergthorsson et al., 2003).

Expansion of the plant mitochondrial genome is accounted for by larger intergenic regions, the appearance of repeated segments of substantial size, a marked increase in the number of introns and intron ORFs, and incorporation of foreign DNA (plastid, nuclear and plasmid). Size expansion is accompanied by the appearance of additional open reading frames, but because these are not conserved among plant mitochondrial genomes, they are likely to be fortuitous. Plastid and nuclear DNA sequences are not evident in the *Marchantia* mitochondrial genome, so evolutionary incorporation of foreign DNA was presumably initiated at a later stage in land plant mtDNA evolution. Notably, 'promiscuous' chloroplast DNA in plant mtDNA is actually functional in some cases, encoding so-called 'chloroplast-like' tRNAs that substitute for 'missing' native mitochondrial tRNAs. Even so, some nucleus-encoded tRNAs must be imported into angiosperm mitochondria to complement an otherwise incomplete set of native and chloroplast-like tRNAs (Maréchal-Drouard et al., 1995).

3.5. A common evolutionary origin of the mitochondrial genome within green algae and land plants

By virtue of their highly eubacteria-like LSU and SSU rRNAs (in both primary and secondary structure) and other features such as the presence of a mtDNA-encoded 5S rRNA, plant mitochondria have provided particularly compelling molecular evidence pointing to a eubacterial origin of the mitochondrial genome (Bonen et al., 1977; Schnare and Gray, 1982; Spencer et al., 1984). Because their sequences evolve relatively slowly, plant mitochondrial rRNAs could be used in phylogenetic reconstructions to pinpoint the evolutionary source of the mitochondrial genome: namely, the α-proteobacterial lineage of eubacteria (Yang et al., 1985).

Although a monophyletic origin of mitochondria is now strongly supported, this was not the case when a much less comprehensive and representative collection of mtDNA sequences was available. Comparison of available information about plant mitochondrial genomes with data for the first green algal mtDNA to be investigated in detail, that of the *Chlamydomonas reinhardtii* (Chlorophyceae), provided little or no indication of a common evolutionary origin; moreover, plant mitochondria did not cluster together with those of other eukaryotes in phylogenetic trees based on SSU and LSU rRNA sequences (Gray et al., 1984; Cedergren et al., 1988). In considering this dichotomy, we (Gray et al., 1989) proposed that the rRNA genes in plant mitochondria might be of more recent evolutionary origin than the rRNA genes in other mitochondria, raising the possibility that the entire plant mitochondrion might have been derived in a second endosymbiosis. In making this suggestion, we noted (Gray et al., 1989): "Validation or rejection of the secondary acquisition hypothesis ... will obviously require further comparative data: in particular, additional mitochondrial genomes within the chlorophyte-metaphyte grouping must be examined. If the thesis developed here is correct, then we should expect to see an abrupt transition from a prototypical chlorophyte mitochondrial genome (if such can be defined) to a prototypical metaphyte mitochondrial genome somewhere within the green algal lineage leading directly to higher plants."

In fact, the very next chlorophyte mtDNA to be sequenced (that of *Prototheca wickerhamii*; Wolff et al., 1994) effectively quashed this secondary symbiosis proposal, because *P. wickerhamii* mtDNA displays substantially greater similarity to plant mtDNA than to that of its green algal relative, *C. reinhardtii*. These results suggested that it was the mtDNA of *C. reinhardtii*, not that of land plants, that was likely atypical. Further additions of streptophyte and chlorophyte mtDNA sequences have reinforced the inference of a monophyletic green algal/land plant mitochondrion, with the radically different pattern of mtDNA organization and sequence in chlorophycean green algae being attributable to extensive evolutionary divergence away from an ancestral pattern to a reduced derived state, as outlined earlier. Rapid sequence evolution in chlorophycean mitochondrial gene sequences accounts for the fact that they do not typically form a clade with land plants in phylogenetic reconstructions based on mtDNA-encoded sequences (e.g., Gray and Spencer, 1996; Turmel et al., 1999b).

Comparison of gene content and organization in ancestral-type green algal and land plant mtDNAs has revealed that the mitochondrial genome of the charophyte *Chara vulgaris*, a member of the Charales, strikingly resembles *Marchantia polymorpha* (liverwort) mtDNA in gene content, A+T content, gene sequences, codon usage and gene order. For example, all except 9 of 68 conserved genes are colinear in the two mtDNAs (Turmel et al., 2003). In all of these features, *Chara* mtDNA is clearly more similar to its *Marchantia* counterpart than is the mtDNA of a second charophyte alga, *Chaetosphaeridium globosum.* In agreement with trees based on a four-gene set representing nuclear, plastid and mitochondrial lineages (Karol et al., 2001), phylogenetic analysis of a concatenated data set of 23 mitochondrial protein sequences provides unequivocal support for a sister-group relationship between the Charales and land plants (Turmel et al., 2003), to the exclusion of the Coleochaetales, in which *C. globosum* is typically placed (e.g., Karol et al., 2001). Importantly, *Chara* and *Marchantia* mtDNAs uniquely share a number of derived features of gene organization, including common pairs of overlapping genes, that serve to reinforce their relative phylogenetic placement.

Both *C. vulgaris* and *C. globosum* mtDNAs are typical compact, ancestral-type mtDNAs that stand in marked contrast to the expanded genome pattern characteristic of land plant mtDNAs, including that of *M. polymorpha*. This observation indicates that mitochondrial genome expansion must have been a late event in streptophyte evolution, perhaps coinciding with the emergence of land plants *per se*. On the other hand, incomplete sequencing of the mtDNA of a basally diverging charophyte, *Klebsormidium flaccidum*, indicates that this mitochondrial genome is an expanded one, having an estimated genome size approximating that of *M. polymorpha*: without, however, evidence of the marked conservation of gene order seen between *C. vulgaris* and *M. polymorpha* mtDNAs (Organelle Genome Megaequencing Program, unpublished data). These results are suggestive of independent events of mitochondrial genome expansion within Streptophyta.

Although the foregoing conclusions are persuasive, it should be cautioned that they are based on knowledge of relatively few charophyte and primitive green algal mtDNA sequences. Phylogeny within the charophytes and deep relationships within the Viridiplantae (e.g., the precise branching position of the primitive green alga, *Mesotisgma viride*) remain unsettled; in fact, one recent analysis proposes a new class of streptophyte green algae comprising the genera Mesostigma and Chaetosphaeridium (Marin and Melkonian, 1999). We should therefore be prepared for a few surprises as additional green algal and land plant mtDNA sequences are determined. In particular, in order to understand fully the evolution of the land plant mitochondrial genome, both with respect to its origin from a common ancestor shared with the charophyte mitochondrial genome and its subsequent divergence, it will be essential to pursue a more comprehensive program of complete mitochondrial genome sequencing in non-angiosperm plants.

4. EVOLUTIONARY ORIGIN OF PLASTIDS

4.1. Certainties, uncertainties and controversies in plastid evolution

One certainty in plastid evolution is that these organelles originated as endosymbionts from within the cyanobacterial lineage of eubacteria. As in the case of mitochondria, but even more persuasively, organellar gene and genome data argue that plastid DNA (plDNA) is of direct eubacterial, and specifically cyanobacterial, ancestry (see Gray, 1992; Morden et al., 1992; Bhattacharya and Medlin, 1995). Plastid genomes (reviewed by Maier et al. in Chapter 5 of this volume) are generally larger (~100-200 kb) and contain more genes (~100-250) than mitochondrial genomes; they are much less diverse in structure; and they tend to be more conservative in their retention of ancestral features of their eubacterial ancestor. In particular, gene clusters reminiscent of eubacterial operons are considerably more prominent in plastid than in mitochondrial DNAs (Stoebe and Kowallik, 1999). Notable exceptions to the typical picture of the plastid genome include the reduced (~70-kb) plDNAs in certain non-photosynthetic plants (Wolfe et al., 1992) and algae (Gockel and Hachtel, 2000); the 35-kb genome in the plastid remnant (apicoplast) of *Plasmodium falciparum* and other apicomplexans (Wilson et al., 1996); and the small (2- to 3-kb) single-gene circles that carry plastid gene sequences in dinoflagellates (Zhang et al., 1999).

A comprehensive treatment of plastid origins and evolution is beyond the scope of this chapter, and in any event this subject has been thoroughly reviewed in the recent literature (Archibald and Keeling, 2002, 2004; Bhattacharya et al., 2003; Cavalier-Smith, 2002; Funes et al., 2004; Palmer, 2003; Stiller et al., 2003; Stiller, 2003). However, a few comments are appropriate here concerning more general aspects of plastid evolution.

4.2. The origin of plastids: monophyletic or polyphyletic?

Although the weight of evidence continues to support a single (monophyletic) plastid origin (Palmer, 2003), this question continues to be debated. In particular, Stiller et al. (2003) suggest that a scenario of multiple plastid origins is at least as plausible as the single-origin hypothesis in explaining evidence generally regarded as favouring a monophyletic plastid origin. Although the issue should not be considered absolutely settled (Palmer, 2003), the fact remains that there is no strong positive support for the alternative, multiple origins hypothesis. At present, a single-origin scenario remains the favoured hypothesis (Archibald and Keeling, 2004).

4.3. Primary endosymbioses in plastid evolution

The initial acquisition of plastids by the eukaryotic lineage is considered to have been via a *primary endosymbiosis* in which a cyanobacterium was taken up by a phagotrophic eukaryotic host. Plastids that trace their origin directly to this event are called primary plastids, and are found in three eukaryotic lineages: red algae, green algae and glaucophytes (glaucocystophytes). The latter group comprises a small assemblage of algae with only a few known genera, but it has the distinction of possessing plastids that retain vestiges of the peptidoglycan wall of their cyanobacterial ancestor. The relationships among these three plastid groups remain ambiguous; the relative branching order of these clades must be rigorously defined before we are able draw solid inferences about genome evolution in primary plastids. Analysis of a concatenated set of plastid-encoded protein sequences has suggested that glaucophytes (represented by *Cyanophora paradoxa*) diverged basally, prior to the separation of red and green algae (Martin et al., 2002). Phylogenetic reconstructions using nucleus-encoded genes have been notoriously unsuccessful in generating robust phylogenies that unite red, green and glaucophyte algae, let alone revealing the relative branching order among the three. This failure has largely been ascribed to lack of phylogenetic resolution provided by the available data sets (Archibald and Keeling, 2004).

Of course, failure to unite the host (i.e., nuclear) component of the three types of algae in phylogenetic analyses could imply a very early (rather than relatively later) acquisition of a primary plastid followed by diversification and subsequent plastid loss in many eukaryotic lineages than are now aplastidic (Nozaki et al., 2003a). Thus, many protist groups currently considered to be primitively aplastidic might actually be secondarily so, depending on how early in eukaryotic cell evolution the primary plastid endosymbiosis occurred. On the other hand, phylogenies based on a concatenated set of mtDNA-encoded proteins robustly unite red and green algae, to the exclusion of other eukaryotes (Burger et al., 1999), an observation supporting a relatively late origin of primary plastids.

4.4. Secondary endosymbioses in plastid evolution

Many eukaryotic algal lineages obtained their plastids via a *secondary endosymbiosis* (Archibald and Keeling, 2002, 2004), a process in which a eukaryotic host cell took up not a cyanobacterium but another eukaryote: a photosynthetic alga already containing a plastid. In this way, plastids have been spread horizontally between eukaryotic lineages, rather than through strictly vertical descent. One of the defining features of secondary plastids is that they have three or four bounding membranes; the innermost two are considered to represent the inner and outer membranes of the plastid present in the algal endosymbiont, the next (third) the plasma membrane of the symbiont, and the outermost (fourth, when present) the phagosomal membrane of the engulfing host cell. In most cases of plastid acquisition by secondary endosymbiosis, nothing remains of the algal endosymbiont but its plastid and cell membrane. However, in at least two cases (cryptophytes□ also called cryptomonads□ and chlorarachniophytes), a residual symbiont nucleus (termed a nucleomorph) remains. The nucleomorph contains the remnant and highly reduced nuclear genome of the algal symbiont (Gilson et al., 1997; Douglas et al., 2001).

How many times have secondary plastids arisen? The minimum number is two, because some secondary plastids are obviously derived from the green lineage of primary plastids, whereas others are evidently of red algal origin. Green secondary plastids are found in euglenid protists such as *Euglena gracilis* and in chlorarachniophytes such as *Chlorarachnion*. The existing evidence suggests that these two groups represent independent endosymbiotic events involving different hosts and different green algae (Archibald and Keeling, 2002, 2004; but see Cavalier-Smith, 2000).

Three major groups contain secondary plastids of red algal origin and having four bounding membranes: heterokonts (or stramenopiles), a diverse and abundant group of unicellular and multicellular algae; haptophytes, equally abundant and ecologically important; and cryptomonads. Although biochemical and ultrastructural arguments favor a single secondary endosymbiosis as the source of the red algal-type plastid in these three groups, only recently has molecular evidence supported the idea of a single rather than three separate secondary acquisitions (Fast et al., 2001; Yoon et al., 2002; Harper and Keeling, 2003).

Two other eukaryotic groups, dinoflagellates and apicomplexans (members of the alveolates and sister clades to one another) possess red algal plastids, although in the case of the parasitic apicomplexans, this inference is still quite controversial. The evolutionary history of the residual apicoplast genome has been extremely difficult to decipher (Funes et al., 2002, 2003, 2004; Waller et al., 2003; Palmer, 2003). Nevertheless, molecular data uniting apicomplexa with heterokonts, haptophytes and cryptomonads have recently been published (Van der Peer and De Wachter, 1997; Ben Ali et al., 2001; Baldauf et al., 2000; Yoon et al., 2002), with these groups constituting a supergroup referred to as the 'chromalveolates' (Cavalier-Smith, 1999). Hence, the accumulating data seem to be pointing to a single, ancient origin for alveolate, haptophyte, heterokont and cryptomonad plastids.

4.5. Evolution of the plastid genome in land plants

As of this writing, complete plastid genome sequences for 35 eukaryotic species are listed in the Organelle Genome Resources section on the website of the National Center for Biotechnology Information (NCBI). Of these 35 sequences, 25 are from Viridiplantae, with 20 representing land plants. Excluding the reduced (70 kb) plastid genome of the non-photosynthetic angiosperm, *Epifagus virginiana*, sequenced plDNAs in land plants range in size from 117 kb in *Pinus koraiensis* (a gymnosperm) to 164 kb in *Oenothera elata* subsp. *hokkeri* (a dicot angiosperm), with a mean of 143 kb. Gene content varies between ~100 and ~120, with some ORFs, including a number of conserved ones, remaining unidentified.

Of particular relevance to the issue of plastid genome evolution in land plants, complete plDNA sequences are also available for three chlorophyte algae: *Chlorella vulgaris* (151 kb; Trebuxiophyceae; Wakasugi et al., 1997), *Nephroselmis olivacea* (201 kb; Prasinophyceae; Turmel et al., 1999a), and *Chlamydomonas reinhardtii* (203 kb; Chlorophyceae; Maul et al., 2002). As well, complete plDNA sequences have been determined for *Mesostigma viride* (118 kb; Lemieux et al., 2000), a green alga whose phylogenetic placement (as noted earlier) is uncertain, and *Chaetosphaeridium globosum*, a charophyte alga (131 kb; Turmel et al., 2002b).

Comparison of these chlorophyte and streptophyte plDNA sequences has provided considerable insight into plastid genome evolution within the Viridiplantae. Most surprising, perhaps, is the finding that despite its small size, the plDNA of *M. viride*, which according to Lemieux et al. (2000) belongs to the earliest diverging Viridiplantae lineage discovered to date, has the largest gene repertoire (135 genes) so far reported among green algal and land plant plDNAs. Moreover, the organization of *Mesostigma* plDNA is strikingly similar to that of land plant plDNAs, with 81% of its genes being found in clusters that are shared with land plant plDNAs. A retained ancestral feature that is particularly notable is the presence of an inverted repeat containing the rRNA genes, which partitions the plastid genomes into two repeat and two single-copy regions (a quadripartite structure). Remarkably, with few exceptions, the genes shared between *Mesostigma* and land plant plDNAs are localized in corresponding regions of the genome, and most are found in conserved clusters. These observations indicate that the structure and organization of the plastid genome has been extremely well conserved in the lineage leading to land plants.

The plDNA sequence of the charophyte, *C. globosum*, shares a number of features specifically with land plant plDNAs, consistent with the established sister group relationship between embryophytes and charophytes (Karol et al., 2001). At the levels of gene content (124 genes), intron composition (18 introns) and gene order, *Chaetosphaeridium* plDNA is remarkably similar to the plDNAs of land plants (Turmel et al., 2002b). Virtually all genes in *Chaetosphaeridium* plDNA that are shared with land plant and *Mesostigma* plDNAs are partitioned within the same genomic regions. Further, the 17 group II introns and single group I intron found in *Chaetosphaeridium* plDNA are all similar positionally and structurally to land plant plastid introns. In contrast, the plDNAs of *M. viride* and *N. olivacea* lack introns altogether. These observations imply that the ancestral plDNA in Viridiplantae had few if any of the introns that characterize the plDNA of land plants; instead, these introns must have invaded the streptophyte lineage at some point during the evolution of charophytes, after the divergence of the latter from other green algae.

Although the plDNA of the prasinophyte *N. olivacea* exhibits much of the ancestral character evident in *Mesostigma* and *Chaetosphaeridium* plDNAs (Turmel et al., 1999a), plDNA evolution and divergence has evidently accelerated in later-diverging chlorophytes. The plDNA of *Chorella vulgaris*, for example, lacks the characteristic large inverted repeat, having only one copy of the rRNA gene cluster (Wakasugi et al., 1997). *C. reinhardtii* plDNA, despite its large size, has the smallest plDNA gene content found so far (99 genes). More than 20% of the *C. reinhardtii* plastid genome consists of repetitive DNA, with the majority of intergenic regions comprising numerous classes of short dispersed repeats (Maul et al., 2002).

Considering that *Porphyra purpurea* (red algal) plDNA contains 251 genes (Reith and Munholland, 1995) but *M. viride* plDNA□ the most gene-rich among members of the Viridiplantae□ contains only 135 genes, massive loss of genes from plDNA must have occurred in the green lineage following the divergence of the red and green algae from a common ancestor. Ongoing determination of complete plDNA sequences from a variety of green and non-green eukaryotes will allow a rigorous assessment of evolutionary gene loss, providing such losses can be mapped to robust phylogenetic trees that accurately describe the evolutionary relationships among the compared plastid genomes. This approach not only reveals the nature and timing of gene losses but also provides evidence of multiple independent losses of the same gene in different lineages (Turmel et al., 1999b, 2002; Martin et al., 2002; Maul et al., 2002).

5. CONCLUDING REMARKS

From the perspective of the genomes they contain, the evolutionary origin of mitochondria and plastids is clear. However, because they encode only a small fraction of the proteins that make up the organelle proteome, the evolutionary information that organelle genomes provide is necessarily incomplete. In fact, the evolutionary perspective that will be gleaned from the organelle proteome is likely to be considerably more complex. In the case of yeast, estimates of the number of proteins in the mitochondrial proteome range up to about 850 in four recent analyses (Karlberg et al., 2000; Marcotte et al., 2000; Kumar et al., 2002; Sickmann et al., 2003). About 50-60% of the yeast mitochondrial proteome is classified as 'prokaryote-specific', having homologs in eubacteria and archaebacteria; however, only a small fraction of this category can be traced with confidence to the α-proteobacterial ancestor of mitochondria. Another portion (~20-30%) of yeast mitochondrial proteins is 'eukaryote-specific', with homologs identifiable only in other eukaryotes. About 20% of yeast mitochondrial proteins are unique, with no identifiable homologs. These results indicate that the mitochondrial proteome has multiple evolutionary origins and a complex evolutionary history (Kurland and Andersson, 2000; Gray et al., 2001).

Mass-spectrometry based proteomics approaches (Aebersold and Mann, 2003), targeted to organelles (Dreger, 2003; Taylor et al., 2003), are beginning to be applied to plant mitochondria (Kruft et al., 2001; Millar et al., 2001; Giegé et al., 2003; Heazlewood et al., 2003, 2004) and plastids (Zabrouskov et al., 2003; Baginsky and Gruissem, 2004; Kleffmann et al., 2004). While the swelling flood of proteomics data (reviewed by Francs-Small et al. in Chapter 7 of this volume) will undoubtedly provide much insight into the function of plant mitochondria and plastids, they will have the happy consequence of giving those of us interested in organelle origins and evolution invaluable molecular data with which to continue to assess what has happened to mitochondria and plastids since their eubacterial endosymbionts joined the eukaryotic lineage.

6. REFERENCES

Adams, K.L., Daley, D.O., Qiu, Y.-L., Whelan, J., & Palmer, J.D. (2000). Repeated, recent and diverse transfers of a mitochondrial gene to the nucleus in flowering plants. *Nature, 408*, 354-357.

Adams, K.L., Rosenblueth, M., Qiu, Y.-L., & Palmer, J.D. (2001). Multiple losses and transfers to the nucleus of two mitochondrial succinate dehydrogenase genes during angiosperm evolution. *Genetics, 158*, 1289-1300.

Adams, K.L., Qiu, Y.-L., Stoutmeyer, M., & Palmer, J.D. (2002). Punctuated evolution of mitochondrial gene content: high and variable rates of mitochondrial gene loss and transfer to the nucleus during angiosperm evolution. *Proc. Natl. Acad. Sci. USA, 99*, 9905-9912.

Aebersold, R., & Mann, M. (2003). Mass spectrometry-based proteomics. *Nature, 422*, 198-207.

Andersson, S.G.E., Karlberg, O., Canbäck, B., & Kurland, C.G. (2003). On the origin of mitochondria: a genomics perspective. *Phil. Trans. R. Soc. Lond. B, 358*, 165-179.

Archibald, J.M., & Keeling, P.J. (2002). Recycled plastids: a 'green movement' in eukaryotic evolution. *Trends Genet., 18*, 577-584.

Archibald, J.M., & Keeling, P.J. (2004). On the origin and evolution of plastids. In J. Sapp & L. Margulis (Eds.), *Microbial Evolution: Concepts and Controversies* (in press). Oxford: Oxford University Press.

Baginsky, S., & Gruissem, W. (2004). Chloroplast proteomics: potentials and pitfalls. *J. Exp. Bot., 55*, 1213-1220.

Baldauf, S.L., Roger, A.J., Wenk-Siefert, I., & Doolittle, W.F. (2000). A kingdom-level phylogeny of eukaryotes based on combined protein data. *Science, 290*, 972-977.

Bapteste, E., Brinkmann, H., Lee, J.A., Moore, D.V., Sensen, C.W., Gordon, P., et al. (2002). The analysis of 100 genes supports the grouping of three highly divergent amoebae: *Dictyostelium*, *Entamoeba*, and *Mastigamoeba*. *Proc. Natl. Acad. Sci. USA, 99*, 1414-1419.

Ben Ali, A., De Baere, R., Van der Auwera, G., De Wachter, R., & Van de Peer, Y. (2001). Phylogenetic relationships among algae based on complete large-subunit rRNA sequences. *Int. J. Syst. Evol. Microbiol., 51*, 737-749.

Bergthorsson, U., Adams, K.L., Thomason, B., & Palmer, J.D. (2003). Widespread horizontal transfer of mitochondrial genes in flowering plants. *Nature, 424*, 197-201.

Bhattacharya, D., & Medlin, L. (1995). The phylogeny of plastids: a review based on comparisons of small subunit ribosomal RNA coding regions. *J. Phycol., 31*, 489-498.

Bhattacharya, D., Yoon, H.S., & Hackett, J.D. (2003). Photosynthetic eukaryotes unite: endosymbiosis connects the dots. *BioEssays, 26*, 50-60.

Boer, P.H., & Gray, M.W. (1988). Scrambled ribosomal RNA gene pieces in Chlamydomonas reinhardtii mitochondrial DNA. *Cell, 55*, 399-411.

Boer, P.H., & Gray, M.W. (1991). Short dispersed repeats localized in spacer regions of *Chlamydomonas reinhardtii* mitochondrial DNA. *Curr. Genet., 19*, 309-312.

Bonen, L., Cunningham, R.S., Gray, M.W., & Doolittle, W.F. (1977). Wheat embryo mitochondrial 18S ribosomal RNA: evidence for its prokaryotic nature. *Nucleic Acids Res., 4*, 663-671.

Bremer, K. (1985). Summary of green plant phylogeny and classification. *Cladistics, 1*, 369-385.

Burger, G., & Lang, B.F. (2003). Parallels in genome evolution in mitochondria and bacterial symbionts. *IUBMB Life, 55*, 205-212.

Burger, G., Gray, M.W., & Lang, B.F. (2003). Mitochondrial genomes: anything goes. *Trends Genet., 19*, 709-716.

Burger, G., Saint-Louis, D., Gray, M.W., & Lang, B.F. (1999). Complete sequence of the mitochondrial DNA of the red alga *Porphyra purpurea*: cyanobacterial introns and shared ancestry of red and green algae. *Plant Cell, 11*, 1675-1694.

Cavalier-Smith, T. (1981). Eukaryote kingdoms: seven or nine? *Biosystems, 14*, 461-481.

Cavalier-Smith, T. (1983). A 6-kingdom classification and a unified phylogeny. In W. Schwemmler & H.E.A. Schenk (Eds.), *Endocytobiology II* (pp. 1027-1034). Berlin: De Gruyter.

Cavalier-Smith, T. (1987). Eukaryotes with no mitochondria. *Nature, 326*, 322-323.

Cavalier-Smith, T. (1999). Principles of protein and lipid targeting in secondary symbiogenesis: euglenoid, dinoflagellate, and sporozoan plastid origins and the eukaryote family tree. *J. Eukaryot. Microbiol., 46*, 347-366.

Cavalier-Smith, T. (2000). Membrane heredity and early chloroplast evolution. *Trends Plant Sci., 5*, 174-182.

Cavalier-Smith, T. (2002). Chloroplast evolution: secondary symbiogenesis and multiple losses. *Curr. Biol., 12*, R62-R64.

Cedergren, R., Gray, M.W., Abel, Y., & Sankoff, D. (1988). The evolutionary relationships among known life forms. *J. Mol. Evol., 28*, 98-112.

Cermakian, N., Ikeda, T.M., Miramontes, P., Lang, B.F., Gray, M.W., & Cedergren, R. (1997). On the evolution of the single-subunit RNA polymerases. *J. Mol. Evol., 45*, 671-681.

Clark, C.G., & Roger, A.J. (1995). Direct evidence for secondary loss of mitochondria in *Entamoeba histolytica*. *Proc. Natl. Acad. Sci. USA, 92*, 6518-6521.
de Duve, C. (1996). The birth of complex cells. *Sci. Am., 274*, 38-45.
Denovan-Wright, E.M., Nedelcu, A.M., & Lee, R.W. (1998). Complete sequence of the mitochondrial DNA of *Chlamydomonas eugametos*. *Plant Mol. Biol., 36*, 285-295.
Doolittle, W.F. (1998). You are what you eat: a gene transfer ratchet could account for bacterial genes in eukaryotic nuclear genomes. *Trends Genet., 14*, 307-311.
Douglas, S., Zauner, S., Fraunholz, M., Beaton, M., Penny, S., Deng, L.-T., et al. (2001). The highly reduced genome of an enslaved algal nucleus. *Nature, 410*, 1091-1096.
Dreger, M. (2003). Sub-cellular proteomics. *Mass. Spectrom. Rev., 22*, 27-56.
Dyall, S.D., & Johnson, P.J. (2000). Origins of hydrogenosomes and mitochondria: evolution and organelle biogenesis. *Curr. Opin. Microbiol., 3*, 404-411.
Dyall, S.D., Brown, M.T. & Johnson, P.J. (2004). Ancient invasions: from endosymbionts to organelles. *Science, 304*, 253-257.
Embley, T.M., & Hirt, R.P. (1998). Early branching eukaryotes? *Curr. Opin. Genet. Dev., 8*, 624-629.
Embley, T.M., van der Giezen, M., Horner, D.S., Dyal, P.L., & Foster, P. (2003a). Mitochondria and hydrogenosomes are two forms of the same fundamental organelle. *Phil. Trans. R. Soc. Lond. B, 358*, 191-203.
Embley, T.M., van der Giezen, M., Horner, D.S., Dyal, P.L., Bell, S., & Foster, P.G. (2003b). Hydrogenosomes, mitochondria and early eukaryotic evolution. *IUBMB Life, 55*, 387-395.
Emelyanov, V.V. (2001). Evolutionary relationship of Rickettsiae and mitochondria. *FEBS Lett., 501*, 11-18.
Emelyanov, V.V. (2003). Mitochondrial connection to the origin of the eukaryotic cell. *Eur. J. Biochem., 270*, 1599-1618.
Fan, J., & Lee, R.W. (2002). Mitochondrial genome of the colorless green alga *Polytomella parva*: two linear DNA molecules with homologous inverted repeat termini. *Mol. Biol. Evol., 19*, 999-1007.
Fast, N.M., Kissinger, J.C., Roos, D.S., & Keeling, P.J. (2001). Nuclear-encoded, plastid-targeted genes suggest a single common origin for apicomplexan and dinoflagellate plastids. *Mol. Biol. Evol., 18*, 418-426.
Funes, S., Davidson, E., Reyes-Prieto, A., Magallón, S., Herion, P., King, M.P., & González-Halphen, D. (2002). A green algal apicoplast ancestor. *Science, 298*, 2155.
Funes, S., Davidson, E., Reyes-Prieto, A., Magallón, S., Herion, P., King, M.P., & González-Halphen, D. (2003). Response to comment on "A green algal apicoplast ancestor". *Science, 301*, 49b.
Funes, S., Reyes-Prieto, A., Pérez-Martínez, X., & González-Halphen, D. (2004). On the evolutionary origins of apicoplasts: revisiting the rhodophyte vs. chlorophyte controversy. *Microbes Infect., 6*, 305-311.
Giegé, P., Heazlewood, J.L., Roessner-Tunali, U., Millar, A.H., Fernie, A.R., Leaver, C.J., & Sweetlove, L.J. (2003). Enzymes of glycolysis are functionally associated with the mitochondrion in Arabidopsis cells. *Plant Cell, 15*, 2140-2151.
Gilson, P.R., Maier, U.-G., & McFadden, G.I. (1997). Size isn't everything: lessons in genetic miniaturization from nucleomorphs. *Curr. Opin. Genet. Dev., 7*, 800-806.
Gockel, G., & Hachtel, W. (2000). Complete gene map of the plastid genome of the nonphotosynthetic euglenoid flagellate *Astasia longa*. *Protist, 151*, 347-351.
Gray, M.W. (1989). The evolutionary origins of organelles. *Trends Genet., 5*, 294-299.
Gray, M.W. (1992). The endosymbiont hypothesis revisited. *Int. Rev. Cytol., 141*, 233-357.
Gray, M.W. (1993). Origin and evolution of organelle genomes. *Curr. Opin. Genet. Dev., 3*, 884-890.
Gray, M.W. (1998). Rickettsia, typhus and the mitochondrial connection. *Nature, 396*, 109-110.
Gray, M.W. (1999). Evolution of organellar genomes. *Curr. Opin. Genet. Dev., 9*, 678-687.
Gray, M.W. (2004). Contemporary issues in mitochondrial origins and evolution. In J. Sapp & L. Margulis (Eds.), *Microbial Evolution: Concepts and Controversies* (In press). Oxford: Oxford University Press.
Gray, M.W., & Doolittle, W.F. (1982). Has the endosymbiont hypothesis been proven? *Microbiol. Rev., 46*, 1-42.
Gray, M.W., & Lang, B.F. (1998). Transcription in chloroplasts and mitochondria: a tale of two polymerases. *Trends Microbiol., 6*, 1-3.

Gray, M.W., & Spencer, D.F. (1996). Organellar evolution. In D. M. Roberts, P. Sharp, G. Alderson & M. Collins (Eds.), *Evolution of Microbial Life* (pp. 109-126). Cambridge: Cambridge University Press.

Gray, M.W., Burger, G., & Lang, B.F. (1999). Mitochondrial evolution. *Science, 283*, 1476-1481.

Gray, M.W., Burger, G., & Lang, B.F. (2001). The origin and early evolution of mitochondria. *Genome Biol., 2*, reviews 1018.1-1018.5.

Gray, M.W., Sankoff, D., & Cedergren, R.J. (1984). On the evolutionary descent of organisms and organelles: a global phylogeny based on a highly conserved core in small subunit ribosomal RNA. *Nucleic Acids Res., 12*, 5837-5852.

Gray, M.W., Cedergren, R., Abel, Y., & Sankoff, D. (1989). On the evolutionary origin of the plant mitochondrion and its genome. *Proc. Natl. Acad. Sci. USA, 86*, 2267-2271.

Gray, M.W., Lang, B.F., Cedergren, R., Golding, G.B., Lemieux, C., Sankoff, D., et al. (1998a). Genome structure and gene content in protist mitochondrial DNAs. *Nucleic Acids Res., 26*, 865-878.

Gray, M.W., Lemieux, C., Burger, G., Lang, B.F., Otis, C., Plante, I., & Turmel, M. (1998b). Mitochondrial genome organization and evolution within the green algae and land plants. In I. M. Møller , P. Gardeström, K. Glimelius & E. Glaser (Eds.), *Plant Mitochondria: From Gene to Function* (pp. 1-8). Leiden: Backhuys Publishers.

Handa, H. (2003). The complete nucleotide sequence and RNA editing content of the mitochondrial genome of rapeseed (*Brassica napus* L.): comparative analysis of the mitochondrial genomes of rapeseed and *Arabidopsis thaliana*. *Nucleic Acids Res., 31*, 5907-5916.

Harper, J.T., & Keeling, P.J. (2003). Nucleus-encpded, plastid-targeted glyceraldehyde-3-phosphate dehydrogenase (GAPDH) indicates a single origin for chromalveolate plastids. *Mol. Biol. Evol., 20*, 1730-1735.

Heazlewood, J.L., Howell, K.A., Whelan, J., & Millar, A.H. (2003). Towards an analysis of the rice mitochondrial proteome. *Plant Physiol., 132*, 230-242.

Heazlewood, J.L., Tonti-Filippini, J.S., Gout, A.M., Day, D.A., Whelan, J., & Millar, A.H. (2004). Experimental analysis of the Arabidopsis mitochondrial proteome highlights signaling and regulatory components, provides assessment of targeting prediction programs, and indicates plant-specific mitochondrial proteins. *Plant Cell, 16*, 241-256.

Karlberg, O., Canbäck, B., Kurland, C.G., & Andersson, S.G.E. (2000). The dual origin of the yeast mitochondrial proteome. *Yeast, 17*, 170-187.

Karol, K.G., McCourt, R.M., Cimino, M.T., & Delwiche, C.F. (2001). The closest living relatives of land plants. *Science, 294*, 2351-2353.

Keeling, P.J., & Fast, N. (2002). Microsporidia: biology and evolution of highly reduced intracellular parasites. *Annu. Rev. Microbiol., 56*, 93-116.

Kleffmann, T., Russenberger, D., von Zychlinski, A., Christopher, W., Sjölander, K., Gruissem, W., & Baginsky, S. (2004). The *Arabidopsis thaliana* chloroplast proteome reveals pathway abundance and novel protein functions. *Curr. Biol., 14*, 354-362.

Kroymann, J., & Zetsche, K. (1998). The mitochondrial genome of *Chlorogonium elongatum* inferred from the complete sequence. *J. Mol. Evol., 47*, 431-440.

Kruft, V., Eubel, H., Jänsch, L., Werhahn, W., & Braun, H.-P. (2001). Proteomic approach to identify novel mitochondrial proteins in Arabidopsis. *Plant Physiol., 127*, 1694-1710.

Kubo, T., Nishizawa, S., Sugawara, A., Itchoda, N., Estiati, A., & Mikami, T. (2000). The complete nucleotide sequence of the mitochondrial genome of sugar beet (*Beta vulgaris* L.) reveals a novel gene for $tRNA^{Cys}$(GCA). *Nucleic Acids Res., 28*, 2571-2576.

Kück, U., Jekosch, K., & Holzamer, P. (2000). DNA sequence analysis of the complete mitochondrial genome of the green alga *Scenedesmus obliquus*: evidence for UAG being a leucine and UCA being a non-sense codon. *Gene, 253*, 13-18.

Kumar, A., Agarwal, S., Heyman, J.A., Matson, S., Heidtman, M., Piccirillo, S., et al. (2002). Subcellular localization of the yeast proteome. *Genes Dev., 16*, 707-719.

Kurland, C.G., & Andersson, S.G.E. (2000). Origin and evolution of the mitochondrial proteome. *Microbiol. Mol. Biol. Rev., 64*, 786-820.

Lang, B.F., Burger, G., O'Kelly, C.J., Cedergren, R., Golding, G.B., Lemieux, C., et al. (1997), An ancestral mitochondrial DNA resembling a eubacterial genome in miniature. *Nature, 387*, 493-497.

Lang, B.F., Gray, M.W., & Burger, G. (1999). Mitochondrial genome evolution and the origin of eukaryotes. *Annu. Rev. Genet., 33*, 351-397.

Lemieux, C., Otis, C., & Turmel, M. (2000). Ancestral chloroplast genome in *Mesostigma viride* reveals an early branch of green plant evolution. *Nature, 403*, 649-652.

Mai, Z., Ghosh, S., Frisardi, M., Rosenthal, B., Rogers, R., & Samuelson, J. (1999). Hsp60 is targeted to a cryptic mitochondrion-derived organelle ("crypton") in the microaerophilic protozoan parasite *Entamoeba histolytica. Mol. Cell. Biol., 19*, 2198-2205.

Marcotte, E.M., Xenarios, I., van der Bliek, A.M., & Eisenberg, D. (2000). Localizing proteins in the cell from their phylogenetic profiles. *Proc. Natl. Acad. Sci. USA, 97*, 12115-12120.

Maréchal-Drouard, L., Small, I., Weil, J.-H., & Dietrich, A. (1995). Transfer RNA import into plant mitochondria. *Methods. Enzymol., 260*, 310-327.

Margulis, L. (1970). *Origin of Eukaryotic Cells.* New Haven: Yale University Press.

Marin, B., & Melkonian, M. (1999). Mesostigmatophyceae, a new class of streptophyte green algae revealed by rRNA sequence comparisons. *Protist, 150*, 399-417.

Martin, W., & Müller, M. (1998). The hydrogen hypothesis for the first eukaryote. *Nature, 392*, 37-41.

Martin, W., Hoffmeister, M., Rotte, C., & Henze, K. (2001). An overview of endosymbiotic models for the origins of eukaryotes, their ATP-producing organelles (mitochondria and hydrogenosomes), and their heterotrophic lifestyle. *Biol. Chem., 382*, 1521-1539.

Martin, W., Rujan, T., Richly, E., Hansen, A., Cornelsen, S., Lins, T., et al. (2002). Evolutionary analysis of *Arabidopsis*, cyanobacterial, and chloroplast genomes reveals plastid phylogeny and thousands of cyanobacterial genes in the nucleus. *Proc. Natl. Acad. Sci. USA, 99*, 12246-12251.

Mattox, K.R., & Stewart, K.D. (1984). Classification of the green algae: a concept based on comparative cytology. In D.E.G. Irvine & D.M. John (Eds.), *The Systematics of the Green Algae* (pp. 29-72). London: Academic Press.

Maul, J.E., Lilly, J.W., Cui, L., dePamphilis, C.W., Miller, W., Harris, E.H., & Stern, D.B. (2002). The *Chlamydomonas reinhardtii* plastid chromosome: islands of genes in a sea of repeats. *Plant Cell, 14*, 2659-2679.

McFadden, G.I. (2001). Primary and secondary endosymbiosis and the origin of plastids. *J. Phycol., 37*, 951-959.

McFadden, G.I. (2003). Plastids, mitochondria, and hydrogenosomes. In J. Marr, T.W. Nilsen & R.W. Komuniecki (Eds.), *Molecular Medical Parasitology* (pp. 277-294). London/San Diego: Academic Press.

Millar, A.H., Sweetlove, L.J., Giegé, P., & Leaver, C.J. (2001). Analysis of the Arabidopsis mitochondrial proteome. *Plant Physiol., 127*, 1711-1727.

Morden, C.W., Delwiche, C.F., Kuhsel, M., & Palmer, J.D. (1992). Gene phylogenies and the endosymbiotic origin of plastids. *Biosystems, 28*, 75-90.

Nedelcu, A.M., Lee, R.W., Lemieux, C., Gray, M.W., & Burger, G. (2000). The complete mitochondrial DNA sequence of *Scenedesmus obliquus* reflects an intermediate stage in the evolution of the green algal mitochondrial genome. *Genome Res., 10*, 819-831.

Notsu, Y., Masood, S., Nishikawa, T., Kubo, N., Akiduki, G., Nakazono, M., et al. (2002). The complete sequence of the rice (*Oryza sativa* L.) mitochondrial genome: frequent DNA sequence acquisition and loss during the evolution of flowering plants. *Mol. Gen. Genom., 268*, 434-445.

Nozaki, H., Matsuzaki, M., Takahara, M., Misumi, O., Kuroiwa, H., Hasegawa, M., et al. (2003a). The phylogenetic position of red algae revealed by multiple nuclear genes from mitochondria-containing eukaryotes and an alternative hypothesis on the origin of plastids. *J. Mol. Evol., 56*, 485-497.

Nozaki, H., Ohta, N., Matsuzaki, M., Misumi, O., & Kuroiwa, T. (2003b). Phylogeny of plastids based on cladistic analysis of gene loss inferred from complete plastid genome sequences. *J. Mol. Evol., 57*, 377-382.

Oda, K., Yamato, K., Ohta, E., Nakamura, Y., Takemura, M., Nozato, N., et al. (1992). Gene organization deduced from the complete sequence of liverwort *Marchantia polymorpha* mitochondrial DNA. *J. Mol. Biol., 223*, 1-7.

Palmer, J.D. (2003). The symbiotic birth and spread of plastids: how many times and whodunit? *J. Phycol., 39*, 4-11.

Philippe, H., & Adoutte, A. (1998). The molecular phylogeny of Eukaryota: solid facts and uncertainties. In G.H. Coombs, K. Vickerman, M.A. Sleigh & A. Warren (Eds.), *Evolutionary Relationships Among Protozoa* (pp. 25-56). London: Chapman & Hall.

Philippe, H., Germot, A., & Moreira, D. (2000). The new phylogeny of eukaryotes. *Curr. Opin. Genet. Dev., 10*, 596-601.

Pombert, J.-F., Otis, C., Lemieux, C., & Turmel, M. (2004). The complete mitochondrial DNA sequence of the green alga *Pseudendoclonium akinetum* (Ulvophyceae) highlights distinctive evolutionary trends in the Chlorophyta and suggests a sister-group relationship between the Ulvophyceae and Chlorophyceae. *Mol. Biol. Evol., 21*, 922-935.

Reith, M.E., & Munholland, J. (1995). Complete nucleotide sequence of the *Porphyra purpurea* chloroplast genome. *Plant Mol. Biol. Rep., 13*, 333-335.

Riordan, C.E., Ault, J.G., Langreth, S.G., & Keithly, J.S. (2003). *Cryptosporidium parvum* Cpn60 targets a relict organelle. *Curr. Genet., 44*, 138-147.

Roger, A.J. (1999). Reconstructing early events in eukaryotic evolution. *Am. Nat., 154*, S146-S163.

Roger, A.J., & Silberman, J.D. (2002). Mitochondria in hiding. *Nature, 418*, 827-829.

Rotte, C., Henze, K., Müller, M., & Martin, W. (2000). Origins of hydrogenosomes and mitochondria. *Curr. Opin. Microbiol., 3*, 481-486.

Schnare, M.N., & Gray, M.W. (1982). 3′-Terminal sequence of wheat mitochondrial 18S ribosomal RNA: further evidence of a eubacterial evolutionary origin. *Nucleic Acids Res., 10*, 3921-3932.

Sickmann, A., Reinders, J., Wagner, Y., Joppich, C., Zahedi. R., Meyer, H.E., et al. (2003). The proteome of *Saccharomyces cerevisiae* mitochondria. *Proc. Natl. Acad. Sci. USA, 100*, 13207-13212.

Silberman, J.D., Simpson, A.G.B., Kulda, J., Cepicka, I., Hampl, V., Johnson, P.J., & Roger, A.J. (2002). Retortamonad flagellates are closely related to diplomonads□ implications for the history of mitochondrial function in eukaryote evolution. *Mol. Biol. Evol., 19*, 777-786.

Simpson, A.G.B., & Roger, A.J. (2004). Excavata and the origin of amitochondriate eukaryotes. In R.P. Hirt & D.S. Horner (Eds.), *Organelles, Genomes and Eukaryote Phylogeny: An Evolutionary Synthesis in the Age of Genomics* (pp. 27-53). Boca Raton: Taylor & Francis Books, Inc., CRC Press.

Sluiman, H.J. (1985). A cladistic evaluation of the lower and higher green plants (*Viridiplantae*). *Plant Syst. Evol., 149*, 217-232.

Sogin, M.L. (1991). Early evolution and the origin of eukaryotes. *Curr. Opin. Genet. Dev., 1*, 457-463.

Spencer, D.F., Schnare, M.N., & Gray, M.W. (1984). Pronounced structural similarities between the small subunit ribosomal RNA genes of wheat mitochondria and *Escherichia coli*. *Proc. Natl. Acad. Sci. USA, 81*, 493-497.

Stiller, J.W. (2003). Weighing the evidence for a single origin of plastids. *J. Phycol., 39*, 1283-1285.

Stiller, J.W., Reel, D.C., & Johnson, J.C. (2003). A single origin of plastids revisited: convergent evolution in organellar genome content. *J. Phycol., 39*, 95-105.

Stoebe, B., & Kowallik, K.V. (1999). Gene-cluster analysis in chloroplast genomics. *Trends Genet., 15*, 344-347.

Taylor, S.W., Fahy, E., & Ghosh, S.S. (2003). Global organellar proteomics. *Trends Biotechnol., 21*, 82-88.

Timmis, J.N., Ayliffe, M.A., Huang, C.Y., & Martin, W. (2004). Endosymbiotic gene transfer: organelle genomes forge eukaryotic chromosomes. *Nat. Rev. Genet., 5*, 123-135.

Tovar, J., Fischer, A., & Clark, C.G. (1999). The mitosome, a novel organelle related to mitochondria in the amitochondrial parasite *Entamoeba histolytica*. *Mol. Microbiol., 32*, 1013-1021.

Tovar, J., León-Avila, G., Sánchez, L.B., Sutak, R., Tachezy, J., van der Giezen, M., et al. (2003). Mitochondrial remnant organelles of *Giardia* function in iron-sulphur protein maturation. *Nature, 426*, 172-176.

Turmel, M., Otis, C., & Lemieux, C. (1999a). The complete chloroplast DNA sequence of the green alga *Nephroselmis olivacea*: insights into the architecture of ancestral chloroplast genomes. *Proc. Natl. Acad. Sci. USA, 96*, 10248-10253.

Turmel, M., Lemieux, C., Burger, G., Lang, B.F., Otis, C., Plante, I., & Gray, M.W. (1999b). The complete mitochondrial DNA sequences of *Nephroselmis olivacea* and *Pedinomonas minor*: two radically different evolutionary patterns within green algae. *Plant Cell, 11*, 1717-1729.

Turmel, M., Otis, C., & Lemieux, C. (2002a). The complete mitochondrial DNA sequence of *Mesostigma viride* identifies this green alga as the earliest green plant divergence and predicts a highly compact mitochondrial genome in the ancestor of all green plants. *Mol. Biol. Evol., 19*, 24-38.

Turmel, M., Otis, C., & Lemieux, C. (2002b). The chloroplast and mitochondrial genome sequences of the charophyte *Chaetosphaeridium globosum*: insights into the timing of the events that restructured organelle DNAs within the green algal lineage that led to land plants. *Proc. Natl. Acad. Sci. USA, 99*, 11275-11280.

Turmel, M., Otis, C., & Lemieux, C. (2003). The mitochondrial genome of *Chara vulgaris*: insights into the mitochondrial DNA architecture of the last common ancestor of green algae and land plants. *Plant Cell, 15*, 1888-1903.

Unseld, M., Marienfeld, J.R., Brandt, P., & Brennicke, A. (1997). The mitochondrial genome of *Arabidopsis thaliana* contains 57 genes in 366,924 nucleotides. *Nat. Genet., 15*, 57-61.

Vahrenholz, C., Riemen, G., Pratje, E., Dujon, B., & Michaelis, G. (1993). Mitochondrial DNA of *Chlamydomonas reinhardtii*: the structure of the ends of the linear 15.8-kb genome suggests mechanisms for DNA replication. *Curr. Genet., 24*, 241-247.

Van de Peer, Y., & De Wachter, R. (1997). Evolutionary relationships among eukaryotic crown taxa taking into account site-to-site variation in 18S rRNA. *J. Mol. Evol., 45*, 619-630.

Wakasugi, T., Nagai, T., Kapoor, M., Sugita, M., Ito, M., Ito, S., et al. (1997). Complete nucleotide sequence of the chloroplast genome from the green alga *Chlorella vulgaris*: the existence of genes possibly involved in chloroplast division. *Proc. Natl. Acad. Sci. USA, 94*, 5967-5972.

Waller, R.F., Keeling, P.J., van Dooren, G.G., & McFadden, G.I. (2003). Comment on "A green algal apicoplast ancestor". *Science, 301*, 49a.

Ward, B.L., Anderson, R.S., & Bendich, A.J. (1981). The mitochondrial genome is large and variable in a family of plants (Cucurbitaceae). *Cell, 25*, 793-803.

Williams, B.A.P., & Keeling, P.J. (2003). Cryptic organelles in parasitic protists and fungi. *Adv. Parasitol., 54*, 9-68.

Williams, B.A.P., Hirt, R.P., Lucocq, J.M., & Embley, T.M. (2002). A mitochondrial remnant in the microsporidian *Trachipleistophora hominis*. *Nature, 418*, 865-869.

Wilson, R.J.M., Denny, P.W., Preiser, P.R., Rangachari, K., Roberts, K., Roy, A., et al. (1996). Complete gene map of the plastid-like DNA of the malaria parasite *Plasmodium falciparum*. *J. Mol. Biol., 261*, 155-172.

Wolfe, K.H., Morden, C.W., & Palmer, J.D. (1992). Function and evolution of a minimal plastid genome from a nonphotosynthetic parasitic plant. *Proc. Natl. Acad. Sci. USA, 89*, 10648-10652.

Wolff, G., Plante, I., Lang, B.F., Kück, U., & Burger, G. (1994). Complete sequence of the mitochondrial DNA of the chlorophyte alga *Prototheca wickerhamii*. Gene content and genome organization. *J. Mol. Biol., 237*, 75-86.

Wu, M., Sun, L.V., Vamathevan, J., Riegler, M., Deboy, R., Browlie, J.C., et al. (2004). Phylogenomics of the reproductive parasite *Wolbachia pipientis* wMel: a streamlined genome overrun by mobile genetic elements. *PLoS Biol., 2*, 0327-0341.

Yang, D., Oyaizu, Y., Oyaizu, H., Olsen, G.J., & Woese, C.R. (1985). Mitochondrial origins. *Proc. Natl. Acad. Sci. USA, 82*, 4443-4447.

Yoon, H.S., Hackett, J.D., Pinto, G., & Bhattacharya, D. (2002). The single, ancient origin of chromist plastids. *Proc. Natl. Acad. Sci. USA, 99*, 15507-15512.

Zabrouskov, V., Giacomelli, L., van Wijk, K.J., & McLafferty, F.W. (2003). A new approach for plant proteomics. Characterization of chloroplast proteins of *Arabidopsis thaliana* by top-down mass spectrometry. *Mol. Cell. Proteomics, 2*, 1253-1260.

Zhang, Z., Green, B.R., & Cavalier-Smith, T. (1999). Single gene circles in dinoflagellate chloroplast genomes. *Nature, 400*, 155-159.

CHAPTER 3

THE SOMATIC INHERITANCE OF PLANT ORGANELLES

S. HEINHORST, C.L. CHI-HAM, S.W. ADAMSON AND G.C. CANNON

Department of Chemistry and Biochemistry, The University of Southern Mississippi, Hattiesburg, MS 39406, USA

Abstract. Plastids and mitochondria fulfill important metabolic functions that greatly affect plant growth and productivity. One can therefore easily envision that division of the organelles themselves, as well as replication, maintenance and partitioning of their genomes must be carefully controlled processes that ensure even organelle distribution during cell division and coordinate the organellar metabolic processes with the needs of the cell, tissues and the entire plant. This chapter reviews the combined cytological, biochemical, genetic and genomics approaches that have led to novel insights into key players that mediate or regulate these processes.

1. INTRODUCTION

It is widely accepted that plastids and mitochondria arose from the symbiotic capture of free-living prokaryotes by ancestors of modern eukaryotic cells. During the evolution of the symbiotic relationship, the engulfed organelle predecessors lost some genes that were necessary for free living and relinquished others to the host nucleus, thus giving rise to modern organelles with genomes that are considerably smaller than those of their prokaryotic ancestors. Plastids and mitochondria divide by binary fission, a sequence of exquisitely choreographed events that must be closely coordinated with cytokinesis to insure adequate distribution of these essential energy producing structures among the daughter cells. The nuclear/cytoplasmic compartment exerts control over most aspects of organelle division and DNA metabolism by supplying structural proteins and enzymatic components necessary to replicate, maintain and partition the organellar genome copies and bring about organellokinesis. This chapter focuses on three aspects that

H. Daniell and C.D. Chase (eds.), Molecular Biology and Biotechnology of Plant Organelles, 37—92.

are crucial to the somatic inheritance of plastids and mitochondria: a) the actual process of organelle division, which for plastids seems to have maintained many prokaryotic features but for mitochondria appears to be mediated largely by eukaryotic components; b) the structure and function of nucleoids, the protein/DNA complexes that are thought to play an essential role in the distribution of organellar DNA and in regulating the expression of its genes; c) organelle DNA replication and repair processes that are essential to inheritance of the genetic information on the organelle genome but are not well understood at all on the molecular level. We hope to have provided a comprehensive review of the field by summarizing the established concepts that were derived from conventional genetic, biochemical and molecular biological approaches. This review also highlights some very recent insights from novel experimental approaches such as genomics and proteomics, which complement, extend and challenge some of the current paradigms and provide exciting new avenues for future research directions.

2. ORGANELLE DIVISION

2.1. Control of plastid division

Chloroplasts, like their prokaryotic evolutionary predecessors, the cyanobacteria, divide by binary fission, as evidenced by light microscopic images of dumbbell-shaped, centrally constricted organelles in the process of dividing (Kuroiwa et al., 1998; Pyke, 1997; 1999). Comparatively little is known about the division of the precursors of green chloroplasts, since the small, undifferentiated proplastids of meristematic cells are difficult to study by conventional microscopy. Based on ultrastructural analyses, 10-14 and 20-40 proplastids have been estimated to exist in shoot and root meristems, respectively, and for very young chloroplasts in postmitotic mesophyll cells at the onset of the leaf cell expansion phase (Pyke, 1997). Plastid numbers in mature leaf mesophyll cells are much higher and presumably reflect the need of the growing plant for increased photosynthate. Two- to six-fold increases in plastid numbers between developing mesophyll cells at the leaf base and older cells at the tip of young leaves, and between tips of young and old leaves have been reported for mesophyll cells of several mono-and dicotyledonous plant species (Baumgartner et al., 1989; Boffey et al., 1979; Ellis et al., 1983; Lawrence and Possingham, 1986; Miyamura et al., 1986; Possingham and Smith, 1972; Scott and Possingham, 1980; Sodmergen Kawano et al., 1989; Tymms et al., 1983), indicating that plastids undergo several rounds of division as the leaf expands and matures. Although the cellular and environmental cues that trigger and control plastid division are not known, chloroplasts do not seem to divide before reaching a certain threshold size. Furthermore, a positive correlation between chloroplast number, cell size and nuclear ploidy has been observed, and other factors that have been implicated in participating in the control of plastid division are light quality and nuclear genotype (Pyke, 1997). In higher plants and *Chlamydomonas reinhardtii*, plastid division is independent of the cell division cycle (Boffey, 1985; Harper and John, 1986; Pyke, 1997) and probably responds to fundamentally different regulatory strategies than those that ensure that division of the single *Cyanidioschyzon merolae* chloroplast proceeds in synchrony with cytokinesis (Kuroiwa et al., 1998). However, even in this organism the chloroplast continues to divide when nuclear DNA synthesis is inhibited by treatment of the alga with aphidicolin, while mitochondrial and nuclear division cease (Itoh et al., 1996).

2.2. The arc mutants

The *arc* (accumulation and replication of chloroplasts) chloroplast division mutants of Arabidopsis were identified in T-DNA insertion mutant populations or isolated after treatment of plants with ethylmethane sulfonate. The mutant phenotypes are stably inherited in a Mendelian fashion and segregate as independent single recessive nuclear traits (Marrison et al., 1999; Pyke, 1999). Isolated leaf mesophyll cells of candidate mutant plants were screened microscopically for chloroplast numbers that differ significantly from the 80 to 120 organelles per mesophyll cell of wild type Arabidopsis (Marrison et al., 1999; Pyke and Leech, 1992; 1994). Chloroplast numbers range from an average of only two very large organelles per cell in the most severe *arc* mutants to as many as 140 smaller chloroplasts in others (Pyke, 1997). Interestingly, growth, pigmentation and flowering characteristics of even those *arc* mutants that feature the largest changes in organelle copy numbers are only mildly affected, and no difference in fertility was observed between wild type Arabidopsis and any of the *arc* mutants. Apparently, the *arc* mutants compensate for the change in chloroplast copy number by an equivalent and opposite change in organelle dimensions that maintains the overall chloroplast compartment size, thereby presumably preventing a compromise of their photosynthetic competency. A similar correlation exists between total chloroplast area and cell size, implying tight cellular controls that prevent extensive chloroplast growth during leaf development and ensure initiation of division once the chloroplast has reached a certain size (Pyke, 1997; 1999). Although the necessity for such rigorous control of chloroplast size might not be immediately obvious, gravitropism of *arc6* mutants that contain between one and three giant chloroplasts per mesophyll cell was shown to be compromised (Yamamoto et al., 2002). Furthermore, multiple smaller organelles are more easily able to orient themselves in response to the incident light intensity for protection against photodamage and for optimal light use (Jeong et al., 2002; Pyke, 1999).

Phenotypic analysis of single and double *arc* mutants has established epistatic relationships between the affected proteins and, prior to the very recent cloning of the *arc5, arc6* and *arc3* genes, has allowed the first tentative assignment of functions to individual ARC proteins. Mesophyll cells of *arc1* Arabidopsis plants contain an elevated number of chloroplasts that are approximately half as large as those of the wild type (Pyke and Leech, 1994). Because all double mutants involving *arc1* contain more chloroplasts than would be expected from the phenotypes of even those crossing partners that feature greatly reduced chloroplast numbers, *arc1* is believed to act by accelerating proplastid division prior to chloroplast differentiation (Marrison et al., 1999), possibly to compensate for a reduction in chloroplast growth (Pyke, 1999).

The *arc3* mutant contains an average of 16 irregularly shaped, 4-6 fold larger than wild type chloroplasts per leaf cell. This number of chloroplasts is very similar to the estimated number of proplastids in mitotic leaf cells and suggests that the chloroplasts of *arc3* mutants do not divide in expanding mesophyll cells. Although the *arc5* mutant features similar mesophyll chloroplast numbers and sizes and, like *arc3*, is not affected in proplastid division (Marrison et al., 1999; Robertson et al., 1996), most *arc5* chloroplasts are dumbbell shaped and contain a central constriction reminiscent of dividing wild type organelles. Since such division intermediates are absent from *arc3arc5* double mutants, the ARC5 protein appears to act downstream from ARC3 late in the chloroplast division process and seems to play a vital role in the completion of organelle fission (Marston and Errington, 1999; Pyke, 1997; Pyke and Leech, 1992,Pyke, 1994 #337; Robertson et al., 1996).

The most extreme plastid division mutants are *arc6* and *arc12*, which contain an average of two giant, heterogeneously shaped chloroplasts per mesophyll cell (Pyke, 1999; Pyke and Leech, 1994). The curled leaves of mature *arc6* plants are likely to be a reflection of the ensuing changes in mesophyll cell size and morphology and might be indicative of aberrant mesophyll cell expansion. The *arc6* defect affects division of practically all plastid types, including the few enlarged and irregularly shaped proplastids of meristematic tissue (Robertson et al., 1995). Interestingly, *arc6* proplastids contain thylakoids that are more developed than those of wild type proplastids at the same stage, suggesting that the ARC6 protein plays a role in early proplastid development. All tested *arc6* double mutants segregate with the *arc6* phenotype and suggest that ARC6 acts upstream from ARC3, ARC5 and ARC11 in plastid division.

The on average 29 large mesophyll chloroplasts of *arc11* plants are indicative of only a single round of division in postmitotic leaf cells. The *arc11* chloroplasts fall into two distinct size ranges, while in postmitotic young mesophyll cells asymmetrically constricted organelles are prevalent, suggesting a role for ARC11 in a very early stage of chloroplast division or in the regulation of organelle expansion (Marrison et al., 1999).

2.3. *Plastid division proteins*

Several proteins associated with plastid division have now been characterized. These are discussed below and summarized in Table 1.

2.3.1. *FtsZ*

Without a doubt the most thoroughly characterized plastid division proteins are the FtsZ (*f*ilamentous *t*emperature *s*ensitive) proteins, which are derived from the presumed prokaryotic ancestors of eukaryotic tubulins. Bacterial FtsZ polymerizes into filaments and assembles into a contractile ring below the cytoplasmic membrane at the central constriction site during bacterial cell division (reviewed in Jacobs and Shapiro, 1999; Ma et al., 1996; Rothfield and Justice, 1997). The GTPase activity of the protein is thought to generate the force needed for constriction of the FtsZ ring and for the eventual severing of the bacterial membranes (Erickson, 1995; Gilson and Beech, 2001). Whereas division of *Escherichia coli* and most other eubacteria requires only a single FtsZ gene, multiple FtsZ genes have been identified in higher plants, in the nonvascular plant *Physcomitrella patens* and in the green alga *C. reinhardtii* (Fujiwara and Yoshida, 2001; Gilson and Beech, 2001; Kiessling et al., 2000; Osteryoung et al., 1998; Wang, D. et al., 2003). Phylogenetic analysis of FtsZ at the protein and cDNA level has established that most plant and green algal FtsZ proteins belong to two distinct clades (FtsZ1 and FtsZ2) that appear to have diverged early in the evolution of photosynthetic eukaryotes (Gilson and Beech, 2001; Stokes and Osteryoung, 2003; Wang, D. et al., 2003). Both FtsZ genes of *P. patens* encode FtsZ2-ype proteins (Kiessling et al., 2000). A third family of eukaryotic FtsZ proteins is represented by red and brown algal FtsZ (Beech et al., 2000; Stokes and Osteryoung, 2003).

In vitro chloroplast import studies have revealed that both FtsZ1 and FtsZ2 of Arabidopsis are targeted to the plastid through N-terminal transit sequences (Fujiwara and Yoshida, 2001; Gaikwad et al., 2000; Kiessling et al., 2000; McAndrew et al., 2001; Osteryoung and Vierling, 1995). The fact that all known eukaryotic FtsZ1 and FtsZ2 proteins contain the conserved tubulin signature motif associated with the GTP binding site and GTPase activity of eukaryotic tubulins and bacterial FtsZ (Erickson, 1995; Stokes and Osteryoung, 2003) strongly suggests that the function of FtsZ in bacterial cytokinesis has at least in part been preserved in the organellar division process. This premise is supported by the observation that pea FtsZ is able to rescue the mutant filamentous phenotype of *ftsZ*^{ts} *E. coli* (Gaikwad et al., 2000). Like bacterial FtsZ (Erickson et al., 1996; Mukherjee, A. and Lutkenhaus, 1998), recombinant pea FtsZ protein rapidly polymerizes *in vitro* in a GTP-dependent fashion, whereas a mutant version of the protein that contains an altered GTP binding site is unable to form multimers (Gaikwad et al., 2000).

The phenotypes of FtsZ knockout mutants in *P. patens* (Strepp et al., 1998) and of Arabidopsis FtsZ1 antisense lines (Osteryoung et al., 1998) established that the eukaryotic FtsZ proteins play a crucial role in plastid division. The most severely affected *Physcomitrella* knockout mutants contain a single, very long chloroplast per cell, as opposed to the 50 or so lens-shaped organelles of the wild type (Strepp et al., 1998) and suggest that organelle constriction is inhibited in a manner similar to that seen in bacterial *ftsZ* mutants (Erickson, 1995). Arabidopsis FtsZ1 and FtsZ2 antisense lines fall into two distinct phenotypic classes, of which the moderate one is characterized by 10-30 chloroplasts of intermediate size per mesophyll cell that are indicative of an inhibition of chloroplast amplification during leaf expansion. A second, more extreme phenotype is represented by plants whose mesophyll cells contain only 1-3 greatly enlarged organelles. Since mitotic cells of wild type Arabidopsis leaves are thought to contain 10-14 proplastids, the greatly reduced chloroplast number in these antisense lines points to an effect of reduced FtsZ protein levels on proplastid division (Osteryoung et al., 1998). Overexpression of Arabidopsis FtsZ1 (Stokes et al., 2000) and FtsZ2 (McAndrew et al., 2001) genes under the control of the strong cauliflower mosaic virus (CaMV) 35S promoter yields phenotypes similar to those of the antisense lines and strongly suggest that FtsZ1 and FtsZ2 have nonredundant functions in plastokinesis. This idea is further supported by the apparent independent regulation of their expression, as manifested in the lack of an effect on the level of one FtsZ protein by ectopic expression or antisense repression of the other (Stokes et al., 2000), and by the formation of FtsZ filaments in these lines (Vitha et al., 2001).

The localization of the two FtsZ proteins in chloroplasts was shown by immunofluorescence microscopy with antibodies that specifically recognize either FtsZ1 or FtsZ2 (Vitha et al., 2001). Both antibodies detected rings at the organelle midpoint in chloroplasts at all division stages from Arabidopsis, tobacco and pea, and at the constriction sites of the division-arrested *arc5* chloroplasts. The FtsZ ring in *Lilium longiflorum* was shown by immunoelectron microscopy to reside on the stromal side of the chloroplast constriction site (Mori et al., 2001). The FtsZ1 and FtsZ2 rings appear to be permanent fixtures in mesophyll chloroplasts, contrary to the bacterial FtsZ ring, which is the earliest recognizable sign of cell division at the cell midpoint, remains a stable structure throughout the division process but disappears after cytokinesis (reviewed in (Jacobs and Shapiro, 1999; Rothfield and Justice, 1997). Overexpressed fusion constructs of both *P. patens* FtsZ proteins with green fluorescent protein (GFP) form a network of filaments in the plastids, for which the term "plastoskeleton" has been suggested to indicate that the structure might play a role in determining plastid shape in the moss. A similar filamentous FtsZ structure does not seem to exist in chloroplasts of higher plants (Fujiwara and Yoshida, 2001; Mori et al., 2001; Vitha et al., 2001) and may be an artifact of FtsZ overexpression.

Image overlays of dually fluorescently labeled mesophyll chloroplasts precisely colocalize the rings formed by FtsZ1 and FtsZ2 in wild type (Vitha et al., 2001) and in the abnormal filaments of *arc* mutant Arabidopsis (McAndrew et al., 2001). The need for two FtsZ proteins of different function yet very similar structure in plastid division of higher plants is not immediately obvious in light of the single FtsZ protein that suffices to bring about bacterial cytokinesis. A subtle C-terminal difference in primary structure between the two plant FtsZ proteins might explain their different roles in plastokinesis. A short, conserved amino acid stretch at the C-terminus of FtsZ2, but not FtsZ1, is related to a motif in the amino acid sequence of *E. coli* FtsZ that is important for the protein's interaction with the two essential division proteins FtsA and ZipA (Ma and Margolin, 1999). Although there is no evidence for FtsA and ZipA homologues in plants, the C-terminal amino acid motif of FtsZ2 might be important for the recruitment of as yet unidentified protein partners to the plastid division ring.

The differential expression of FtsZ in various plant tissues suggests that the steady state levels of FtsZ transcripts is correlated with chloroplast numbers and organellar division activity (Gaikwad et al., 2000). The abundance of FtsZ mRNA in pea is higher in leaves than in stems and roots, and young leaves with actively dividing chloroplasts express considerably more FtsZ RNA than old leaves in which the organelles no longer divide. The observed accumulation of FtsZ transcripts upon illumination of dark-grown pea seedlings suggests a role for light in induction of FtsZ transcription or in stabilization of its mRNA (Gaikwad et al., 2000). The levels of FtsZ1 and FtsZ2 transcripts and protein in non-synchronously growing *Nicotiana tabacum* BY2 cells remain constant for eight days of culture encompassing logarithmic growth and stationary phase. In synchronized cultures, the expression of both FtsZ proteins follows a cyclic pattern with transcript and protein levels peaking during mitosis (El-Shami et al., 2002). Since plastid division in the context of the cell cycle was not addressed in this study, it is unclear whether the observed oscillations in FtsZ expression are indicative of parallel waves of plastid division or of additional functions these proteins may have. The accumulation of mRNA levels for the chloroplast-targeted FtsZ gene (*CmFtsZ2*) immediately before the onset of cell and organelle division in synchronized cultures of the primitive red alga *C. merolae* suggests coupling between FtsZ expression and plastokinesis in this organism (Takahara et al., 2000).

2.3.2. MinD and MinE

The existence of *MinD* and *MinE* homologs in plants was first established for the algae *Chlorella vulgaris* and *Guillardia theta* (Douglas and Penny, 1999; Wakasugi et al., 1997). While in these two organisms the genes are plastome-encoded and, like their counterparts in *E. coli*, arranged adjacent to each other, *MinD* and *MinE* are nuclear genes in Arabidopsis and other higher plants (Colletti et al., 2000; Itoh et al., 2001a; Maple et al., 2003; Kanamaru et al., 2000). MinD, but not MinE is plastome-encoded in several other unicellular algae (Colletti et al., 2000; Lemieux et al., 2000; Turmel et al., 1999).

A GFP-tagged fusion protein derived from the single Arabidopsis *MinE* gene (*AtMinE1*) produces a plastid-targeted protein in transgenic Arabidopsis plants that is localized at the organellar poles in a pattern suggestive of the dynamic oscillatory intracellular pattern of prokaryotic MinE and of an AtMinE1/GFP fusion protein in *E. coli* (Maple et al., 2003). Overexpression of the plant protein in its bacterial host leads to aberrant cell division and minicell formation, whereas MinE overexpression in Arabidopsis yields plants with fewer, very large and often elongated mesophyll chloroplasts of variable size that feature multiple misplaced constriction sites. These results suggest that MinE functions as a topological specificity factor similar to its prokaryotic counterpart, and that it is likely to play a cardinal role in directing the placement of the plastid division apparatus to the midpoint of the organelle.

In vitro translated MinD precursor is imported into the stromal compartment of isolated chloroplasts and processed to the expected 35.6 kDa product (Colletti et al., 2000), and a GFP/MinD fusion protein is targeted to plastids *in vivo* (Kanamaru et al., 2000). Mesophyll cells of Arabidopsis MinD (*AtMinD1*) antisense lines contain fewer, much larger chloroplasts of variable size, and their leaf epidermis and petals harbor plastids with asymmetric constriction sites, suggesting that MinD plays an important role in the correct placement of the chloroplast division ring. Ectopic expression of *AtMinD1* in Arabidopsis (Colletti et al., 2000; Kanamaru et al., 2000) and tobacco (Dinkins et al., 2001) yields plants with greatly enlarged and irregularly shaped mesophyll chloroplasts whose average number per cell is lower than that estimated for proplastids in leaf meristem cells, and points to an involvement of MinD in chloroplast as well as proplastid division (Colletti et al., 2000).

MinD appears to inhibit plastid FtsZ ring placement and/or stability, since MinD overexpression and antisense repression in Arabidopsis lead to fragmented FtsZ filaments and multiple FtsZ rings at several sites in the organelle, respectively (Vitha et al., 2003). Although at present no direct evidence exists for a physical association between chloroplast MinD and MinE, their almost identical organellar localization patterns and their structural and functional similarity to prokaryotic MinD and MinE suggest that the two proteins interact *in vivo* to regulate FtsZ ring placement in plastids (Maple et al., 2003).

Although the short, evolutionarily conserved C-terminal amino acid sequence motif of MinD is predicted to form an amphipathic helix that allows the prokaryotic protein to interact with the bacterial inner membrane (Szeto et al., 2002), biochemical proof for an association of plastid MinD with the inner envelope, and for a role of the putative membrane targeting motif in mediating this interaction, is lacking. It is also unknown whether the conserved nucleotide binding motif GKGGVGKT of eukaryotic and prokaryotic MinD proteins enables plant MinD protein to hydrolyze ATP. The ATPase activity of *E. coli* MinD is stimulated by MinE in the presence of phospholipids, and nucleotide hydrolysis is believed to drive the oscillatory movement of the MinC/D division inhibition complex between the two poles of the bacterial cell, thereby preventing FtsZ ring formation at sites other than the cell midpoint. Since the Arabidopsis genome does not encode an obvious *MinC* homolog, either a different mechanism for constriction site selection or a novel, eukaryotic MinD protein partner must be evoked for plastids. The position of the bacterial nucleoid is known to contribute to the placement of the FtsZ ring (Sun and Margolin, 2001); it will be interesting to determine whether the intraorganellar distribution of the plastid nucleoids plays an analogous role in the organelle.

2.3.3. ARTEMIS

The ARTEMIS (*Ar*abidopsis *t*haliana *e*nvelope *m*embrane *i*ntegra*s*e) protein is an integral chloroplast inner envelope protein. The mature protein is characterized by an N-terminal domain that is reminiscent of receptor protein kinases, a glycine-rich GTP binding domain, and a C-terminal domain that is related to prokaryotic YidC and organellar Alb3/Oxa1p protein translocases (Fulgosi et al., 2002). Mutants carrying a maize transposon in the *ARTEMIS* gene and antisense repression lines contain elongated, sometimes tubular chloroplasts that feature apparently doubled thylakoid membrane systems with central constrictions but lack the envelope invaginations that appear late in the plastid division cycle. This phenotype points to the independence of thylakoid duplication from the plastid envelope constriction process and suggests a possible role for ARTEMIS in recruitment of plastid division components to the envelope constriction site via the protein's translocase domain, or in the proper placement of the constriction rings in the dividing organelle. The cyanobacterium *Synechocystis* PCC6803 contains a protein that is phylogenetically related to ARTEMIS. The cell division phenotype of the cyanobacterial *slr1471* mutant resembles that of the Arabidopsis ARTEMIS mutant and can be rescued by the plant ARTEMIS translocase domain, suggesting that the functions of both proteins in their respective organisms are similar or, at the very least, show considerable overlap.

2.3.4. ARC5, ARC6 and ARC3

Although the genes responsible for the aberrant chloroplast replication and accumulation phenotypes of the *arc* mutants of Arabidopsis are largely unknown, the genes affected in the *arc 5, arc 6* and *arc3* mutants have been identified recently (Gao et al., 2003; Shimada et al., 2003; Vitha et al., 2003). ARC6 contains an N-terminal J-domain characteristic for DnaJ cochaperones, a transmembrane domain that anchors the protein in the inner plastid envelope, and a conserved C-terminal domain that protrudes into the intermembrane space (Vitha et al., 2003). In transgenic Arabidopsis, an ARC6/GFP fusion protein localizes to the middle of the plastid, where it is associated with a ring that is also evident in highly constricted dividing plastids, suggesting a role for ARC6 throughout the organelle division process and a possible interaction of ARC6 with the FtsZ ring. Since the large chloroplasts of *arc6* plants contain multiple short FtsZ filaments, whereas long FtsZ filaments are evident in ARC6 overexpressing lines, a role for ARC6 in fostering assembly and/or increasing stability of the FtsZ ring has been proposed (Vitha et al., 2003). The suggestion that ARC6 may interact with Hsp70 through its J-domain is intriguing, since *E. coli* Hsp70 is known to interact with FtsZ (Uehara et al., 2001). Like the ARTEMIS protein, ARC6 has a homolog *(ftn2)* in cyanobacteria (Koksharova and Wolk, 2002). *Synechococcus* sp. PCC7942 *ftn2* mutants are elongated and appear to divide less frequently than wild type cells.

To date the only known purely eukaryotic “acquisition” of the plastid division apparatus is represented by the ARC5 protein (Gao et al., 2003), which is related to the dynamin family of GTPases that participate in fusion and fission of intracellular membrane structures and are important in endocytosis and Golgi vesicle trafficking (Danino and Hinshaw, 2001). The primary structure of dynamins is conserved and consists of an N-terminal GTPase domain, a central pleckstrin homology (PH) domain that mediates membrane association (Lemmon et al., 2002), and a C-terminal domain that can interact with the GTPase domain of other dynamin molecules and is thought to mediate oligomerization (Danino and Hinshaw, 2001). Consistent with its presumed eukaryotic origin, Arabidopsis ARC5 is not taken up by isolated chloroplasts *in vitro* and is so far the only known plastid division protein that is not targeted to the interior of the organelle. The protein appears to have an affinity for the outer envelope, and a GFP/ARC5 fusion protein forms a discontinuous ring-like structure on the cytosolic side of the chloroplast constriction site. Although the exact role ARC5 plays in plastokinesis is not yet known, Gao *et al.* (Gao et al., 2003) suggest that the protein might have a mechanochemical function, generating the force necessary for organelle constriction. The primitive red alga *C. merolae* also contains a dynamin-related protein (CmDnm2) that forms a ring on the cytosolic side of the chloroplast constriction site (Miyagishima et al., 2003b). Synchronization of chloroplast division revealed that CmDnm2 is maintained in cytoplasmic patches, from which it is recruited to the organellar division site. CmDnm2 is maximally expressed during the chloroplast division phase, but the timing of CmDnm2 ring formation late in the process after constriction has already been initiated precludes the possibility that this dynamin is a component of the PD ring, which forms at a much earlier stage of plastokinesis in the alga (Miyagishima et al 2003). The filaments comprising the outer PD ring of *C. merolae* are apparently formed by an as yet unidentified 56 kDa protein (Miyagishima et al., 2001).

The *Arc3* gene of Arabidopsis encodes a protein of mixed evolutionary ancestry. The N-terminal portion of ARC3 is related to prokaryotic FtsZ, whereas its C-terminal domain is homologous to eukaryotic phosphatidylinositol-4-phosphate 5-kinases (PIP5K) of the phosphatidyl inositol signaling pathway. ARC3 does not seem to possess PIP5K activity. The fact that ARC3 can be phosphorylated by a chloroplast extract *in vitro* and that the phosphorylation efficiency appears to be controlled by light prompted the suggestion that the protein might be part of a signal transduction pathway that would allow the nucleus to regulate organelle division (Shimada et al., 2003). Since the majority of dynamin-like proteins are able to bind phosphoinositides (Lemmon et al., 2002), it will be interesting to determine whether ARC3 and ARC5 interact in this postulated signaling pathway, or in mediating the formation of the ARC5 ring.

Table 1. *Plastid division proteins*

Protein	*Protein Type / Domains*	*Localization*	*Mutant Phenotype*	*Function*	*Organism*
FtsZ1	Tubulin-like GTPase	Ring on stromal side at chloroplast midpoint	Few large chloroplasts	Contractile ring formation	Several higher plants, eukaryotic algae, *P. patens*
FtsZ2	Tubulin-like GTPase	Ring on stromal side at chloroplast midpoint	Few large chloroplasts	Contractile ring formation	Several higher plants, eukaryotic algae, *P. patens*
MinD	ATPase?	Chloroplast poles	Few large, asymmetrically constricted chloroplasts	FtsZ ring placement and/or stability	Several higher plants, eukaryotic algae
MinE	ATPase?	Chloroplast poles	Few large, asymmetrically constricted chloroplasts	FtsZ ring placement; topological specificity factor	Several higher plants, eukaryotic algae

Table 1. (cont.)

Table 1. (cont.)

Protein	*Protein Type / Domains*	*Localization*	*Mutant Phenotype*	*Function*	*Organism*
ARTEMIS	Receptor protein kinase, GTP binding, And protein trans-locase domains	Inner envelope, integral membrane protein	Elongated chloroplast; doubled, centrally constricted thylakoid system; unconstricted envelope	Division ring placement? Recruitment of division apparatus to envelope?	*A. thaliana*
ARC3	FtsZ and Phosphatidyl inositol-4-phosphate 5-kinase domain		Average 16 irregularly shaped chloroplasts	Signal transduction for regulation of division?	*A. thaliana*
ARC5	Dynamin-like	Discontin-uous ring on cytosolic side of constriction site	Few, constricted chloroplasts	Force-generating for final constriction?	*A. thaliana*
CmDmn2	Dynamin-like	Patches on cytosolic side of chloroplast constriction site		Constriction late in division?	*C. merolae*
ARC6	J, Trans - membrane and C-terminal domains	Inner envelope at central division site	1-3 very large chloroplasts	Stabilization, membrane anchoring and/or assembly of FtsZ ring?	*A. thaliana* Cyanobac-teria? (Ftn2)

2.4. Control of mitochondrial division

The scarcity of cytological studies that address plant mitochondrial division is, in part, a reflection of the small size of the organelle in higher plants and the difficulty of obtaining microscopic images of sufficient resolution. Perhaps not surprisingly, the number of mitochondria per cell appears to be correlated with the cell's demand for energy. Electron micrographs of developing *Nicotiana sylvestris* pollen show a high number of mitochondria per cell during microsporogenesis (De Paepe et al., 1993). Likewise, the few, rod-shaped mitochondria in cells of the shoot meristem and leaf primordium apparently become fragmented into more numerous, smaller organelles during leaf development (Fujie et al., 1994). Increases in the number of mitochondria per cell were also inferred from the expression patterns of respiratory chain polypeptides that accompany floral development in tobacco (Huang et al., 1994). An investigation of young wheat plants (Robertson et al., 1995) and a more recent large-scale study of nine plant species (Griffin et al., 2001) showed two-to three-fold increases in mitochondrial number per cell in plants that were grown in elevated CO_2. Despite the circumstantial evidence for a connection between energy demand and mitochondrial numbers, the cellular mechanisms that control mitochondrial division are unknown.

Recent progress in visualizing mitochondria in transgenic plants that express organelle-targeted GFP has revealed dynamic changes in shape and intracellular distribution of plant mitochondria (Hanson and Koehler, 2001; Logan and Leaver, 2000; Sheahan et al., 2004; Stickens and Verbelen, 1996; Van Gestel et al., 2002; Van Gestel and Verbelen, 2002). Whereas the organelles are large and adopt a tubular shape in expanding cultured *N. tabacum* protoplasts, they become more numerous, smaller and round in cultured protoplasts that have been induced to divide (Sheahan et al., 2004; Stickens and Verbelen, 1996). Logan et al. (Logan et al., 2003) have applied this organelle labeling approach to screen ethyl methane sulfonate-treated Arabidopsis plants for mutants with altered mitochondrial morphology and intracellular distribution. They have identified the first gene, termed FMT (*f*riendly *mit*ochondria phenotype) that controls distribution of mitochondria in the cell. Mutants in this gene contain large, unevenly distributed mitochondrial aggregates.

2.5. Mitochondrial division proteins

Proteins that have been associated with mitochondrial division are discussed below and summarized in Table 2.

2.5.1. CmFtsZ1

The first documented mitochondrial division proteins of photosynthetic eukaryotes are the nuclear-encoded FtsZ homologs of the red alga *C. merolae* (Takahara et al., 2000) and of the chromophyte *Mallomonas splendens* (Beech et al., 2000). Of the two *ftsZ* genes present in *C. merolae*, *CmftsZ2* encodes a protein that co-purifies with a fraction enriched in dividing algal chloroplasts (Takahara et al., 2000). The *CmftsZ1* gene, on the other hand, encodes a protein that localizes to the constriction site of dividing mitochondria (Takahara et al., 2001). Both proteins are highly expressed during the organellar division phase in synchronized algal cultures (Takahara et al., 2000). Two *FtsZ* genes also exist in the genome of *M. splendens*, of which *MsFtsZ-cp* is believed to encode the plastidal FtsZ version (Beech et al.,

2000), whereas MsFtsZ-mt protein was shown to localize to the middle and the poles of constricted and unconstricted algal mitochondria. The protein's N-terminal presequence targets translationally fused GFP to the mitochondria of transgenic yeast. These results strongly suggest a function for mitochondrial FtsZ in organellar division in these algae, probably analogous to the role played by FtsZ in plastokinesis.

2.5.2. CmDnm1

In addition to the FtsZ homolog that is presumed to participate in mitochondrial division, *C. merolae* also contains the dynamin homolog CmDnm1 (Nishida et al., 2003) that is closely related to the Dnm1p and Drp1 proteins that participate in mitochondrial fission in yeast and animals, respectively (Osteryoung and Nunnari, 2003). Like its relative CmDnm2, CmDnm1 is located in cytoplasmic patches between organelle divisions and appears to be recruited from its cytoplasmic location to already constricted dividing mitochondria late in the organellar division process. The protein assembles into a ring that is located between the cytosolic mitochondrial dividing (MD) ring and the mitochondrial outer membrane and suggests a role for CmDnm1 in the final stage of mitochondrial fission after constriction of the FtsZ and MD rings (Nishida et al., 2003).

2.5.3. ADL2B

The ADL2B protein from Arabidopsis is the first protein from higher plants that was shown to play a role in mitochondrial fission (Arimura and Tsutsumi, 2002). The protein is most closely related to the dynamins Drp1 and Dnm1p that regulate mitochondrial fission in animals and yeast, respectively (Bleazard et al., 1999; Labrousse et al., 1999). ADL2B lacks the central pleckstrin homology domain of most dynamins but does contain their signature GTPase and GTPase effector domains. In transiently transformed tobacco protoplasts, GFP/ADL2B fusion protein localizes to the mitochondrial poles and to the organellar constriction sites and is assumed to accumulate on the cytosolic side of the outer mitochondrial membrane, since this plant dynamin does not contain an N-terminal extension that could serve as an organellar targeting signal. Dominant negative mutants of ADL2B that carry a nonfunctional GTPase domain and cause morphological changes of mitochondria in transformed Arabidopsis and tobacco cells support the premise that this protein plays a role in organellar division.

2.5.4. ADL1C and ADL1E

Two novel isoforms, ADL1C and ADL1E, of the dynamin-like protein ADL1 from Arabidopsis colocalize in a punctuate pattern in transgenic plants that express green and red fluorescent protein fusion constructs of these proteins (Jin et al., 2003). A dominant negative mutant version of ADLC1 causes an apparent organellar division defect in transiently transformed protoplasts that manifests itself in elongated mitochondria and can at least be partially rescued by coexpressed wild type ADL1C. Likewise, an Arabidopsis mutant carrying a T-DNA insertion in the ADL1E coding sequence contains normal chloroplasts but elongated mitochondria. Wild type ADL1E and ADL1C transiently expressed in adl1e mutant protoplasts are both able to rescue the mutant phenotype, suggesting redundant or overlapping roles for both proteins. Although this notion is difficult to reconcile with the presence of abnormal mitochondria in the single mutants of ADL1C and ADL1E, the elongated mitochondrial phenotype might be caused by an imbalance in the expression of both proteins that is reminiscent of the plastid division defects seen in plants that overexpress or repress expression of individual plastid division proteins. Since only a portion of ADL1C and ADL1E appears to be associated with mitochondria, these proteins may be targeted to multiple organelles or cytoplasmic structures, such as the cytoplasmic CmDnm reservoirs seen in synchronized C. merolae (Nishida et al., 2003), or may fulfill multiple functions in the plant cell, since ADL1E is known to participate in cytokinesis and polar cell growth (Kang, B.-H. et al., 2003).

Table 2. *Mitochondrial division proteins*

Protein	*Type or Domain*	*Localization*	*Mutant Phenotype*	*Function*	*Organism*
MsFtsZ-mt	FtsZ	Medial, peripheral		Mitochondrial (mt) division	*M. splendens*; *C. merolae*
CmDnm1	Dynamin-like	Middle of mt dividing ring, on cytoplasmic side		Final mt severance; present in cytosolic patches during early stages of mt division	*C. merolae*
ADL2b	Dynamin-like	Tips and constriction sites	Fewer, tubular mitochondria	mt division	*A. thaliana*
ADL1C ADL1E	Dynamin-like	Mitochondria and other intracellular location	Tubular mitochondria	mt division?	*A. thaliana*
MKRP1 MKRP2	Kinesin-like	Mitochondria		mt nucleoid segregation?	*A. thaliana*

2.6. Evolution of the organellar division machineries

The chloroplast and mitochondrial division proteins of higher plants that have been identified and characterized during the past decade have revealed that the evolution of the molecular machineries that divide the two organelles has taken different paths. Clearly, the plastid division apparatus consists of many proteins with predominantly prokaryotic characteristics. FtsZ, MinD and MinE proteins play prominent roles in prokaryotic cell division (Erickson, 1995; Miyagishima et al., 2003a), and homologs of these proteins that function in plastid division exist in all autotrophic eukaryotes so far examined. Likewise, the ARTEMIS and ARC6 proteins have prokaryotic counterparts, although their distribution apart from higher plants appears to be limited to the cyanobacteria (Osteryoung and Nunnari, 2003). The fact that the genes for all of these proteins are nuclear-encoded in higher plants indicates that the original autotrophic endosymbiont relinquished and transferred control of its own division to the eukaryotic nucleus during the evolution of higher plants. Furthermore, since only *MinD* and/or *MinE* genes still reside on the plastid genome of some primitive eukaryotic algae, the transfer of plastid division genes to the nucleus must have been an early event in the evolution of eukaryotic autotrophs (Douglas and Penny, 1999; Wakasugi et al., 1997). The recent discovery that chloroplast division proteins with dynamin-like characteristics exist in eukaryotic algae and in higher plants (Gao et al., 2003; Miyagishima et al., 2003b), however, has challenged the earlier-held view of an exclusively prokaryotic origin of the plastid division apparatus. The picture of plastid binary fission has become more complex and, in addition to the prokaryote-derived proteins that operate on the inside of the organelle, now includes components derived from the genetic complement of the former eukaryotic host that act on the cytoplasmic side of the outer envelope (Osteryoung and Nunnari, 2003). Interestingly, the close phylogenetic relatedness of CmDnm2 and ARC5 (Miyagishima et al., 2003a) indicates that dynamins must have been recruited for the division of photosynthetic organelles before the divergence of red and green algae (McFadden and Ralph, 2003).

While some primitive eukaryotes have retained the FtsZ proteins of the prokaryotic cell division apparatus for severance of their mitochondria (Beech et al., 2000; Gilson et al., 2003; Takahara et al., 2000), the fact that FtsZ homologs do not exist in the completely sequenced genomes of *Saccharomyces cerevisiae* and *Caenorhabditis elegans* implies that in these organisms mitochondrial fission must proceed by a different mechanism (Osteryoung, 2000). Indeed, division of mitochondria in most eukaryotes, including higher plants, is brought about by a set of outer membrane-associated dynamin-like proteins. Unless a more thorough search of eukaryotic genomes reveals the widespread existence of distantly related homologs of prokaryotic cell division proteins, one has to conclude that the prokaryotic characteristics of organellar division have been lost prior to the divergence of most eukaryotic branches (Osteryoung and Nunnari, 2003).

2.7. Summary and outlook

Almost all known plastid division mutants and transgenic lines that under- or overexpress individual division proteins are normal in appearance, growth rate and fertility. This remarkable tolerance towards large changes in plastid numbers is most likely due to the compensatory adjustment of organelle size that maintains the overall compartment size and the plant's photosynthetic competency (reviewed in

Pyke, 1997). Although microscopy studies (Fujie et al., 1994) suggest that an inverse relationship between mitochondrial size and number may exist in Arabidopsis leaf cells during development, more careful measurements are needed to determine whether the mitochondrial compartment size is regulated in a similar manner. This is particularly important since division of mitochondria and plastids appears to proceed by different mechanisms. Related to that question is the severance of the thylakoid membrane system during chloroplast division, a plant-specific process that likely requires proteins without homologs in heterotrophic prokaryotes but might have features used by cyanobacteria, the phylogenetic ancestors of present-day plastids.

The composition of the prominent PD and MD rings is still unknown, and studies with synchronized *C. merolae* cultures have established that timing of formation and existence of these rings during organellar division rule out any of the organellar division proteins characterized so far as structural components. The recent isolation of the outer PD ring from *C. merolae* (Miyagishima et al., 2001) and the identification of a putative protein component underscore the need for more biochemical studies that can help to resolve the molecular constituents of this and related poorly defined organelle division structures.

To gain a more complete understanding of organelle division in higher plants, *in vivo* physical interactions between individual proteins will have to be established and protein partners identified that might take over the roles of known bacterial division proteins (e.g. MinC) not present in plants. The structural, enzymatic, signal transduction or regulatory roles of currently known plant organelle division proteins have to be more closely investigated and the structural basis for their activity determined. Organelle division is a prerequisite for the maintenance of the subcellular compartments and is a particularly crucial process in rapidly dividing cells, yet the mechanisms that control initiation and termination of organelle division and lead to amplification of plastid numbers during leaf development, as well those that ensure the maintenance of overall compartment size seen in the *arc* mutants, still remain elusive. The fact that some *arc* mutants are only affected in the division of plastids at a particular developmental stage indicates that differences in the regulation or mechanism of proplastid and chloroplast division exist, for which the molecular basis needs to be determined. Finally, the role of cytoskeleton components (Kwok and Hanson, 2003; Logan, 2003; Nebenfuehr et al., 2000; Sheahan et al., 2004; Van Gestel et al., 2002), particularly motor proteins, in organelle movement and partitioning during cell division needs to be defined to gain a more complete picture of molecular mechanisms and regulation of organelle division in plants.

3. ORGANELLAR NUCLEOIDS

The genetic material in semi-autonomous organelles is associated with nucleic acid binding proteins to yield characteristic nucleoprotein complexes termed nucleoids, which are the presumed units of inheritance of the organellar genome. Nucleoids are thought to be active in all aspects of organellar genome metabolism, to accommodate changes in genome function as the organelles mature, and function as vehicles to properly partition the multiple organellar genome copies to ensure continuity of plastids and mitochondria during organelle division. Documented dynamic changes in nucleoid architecture and composition likely reflect such changes in function and represent a form of epigenetic regulation that controls organellar genome replication and gene expression.

3.1. Nucleoid changes during organelle development

Numerous ultrastructural and cytological studies have established that plastid nucleoids undergo dynamic changes during plastid development (reviewed in Kuroiwa, 1991). In higher plants, the nucleoids in proplastids of meristems and in etioplasts of non-photosynthetic tissues are large and diffuse, associate with the inner envelope at the organellar periphery. Depending on the plant species and meristem type, proplasitds generally contain 1-10 nucleoids (Kuroiwa, 1991; Kuroiwa et al., 1981; Lindbeck et al., 1987; Miyamura et al., 1990; Miyamura et al., 1986; Nemoto et al., 1988; Sato et al., 1993; Sodmergen Kawano et al., 1989). Likewise, chromoplasts of *Narcissus pseudonarcissus* harbor only a few, peripherally located nucleoids (Hansmann et al., 1985). Before the thylakoid system forms in developing chloroplasts, nucleoids become more numerous, smaller and tighter, migrate to the chloroplast interior and are eventually distributed throughout the chloroplast. This nucleoid distribution pattern in the mature organelle appears to be dictated by the space available between the grana stacks, to which the nucleoids are attached (Fujie et al., 1994; Lindbeck et al., 1987; Miyamura et al., 1986; Sato et al., 1993; Sodmergen Kawano et al., 1989). Estimates for the number of nucleoids per chloroplast range from six to 40 for most land plants, and up to 300 for some algae (Hansmann et al., 1985; Kuroiwa et al., 1981; Nemoto et al., 1990; Yurina et al., 1995b).

The DNA content per nucleoid varies between one and 20 ctDNA copies, depends on plastid developmental stage, plant species and tissue type (Hansmann et al., 1985; Kuroiwa, 1991; Lindbeck et al., 1989; Miyamura et al., 1986; Oldenburg and Bendich, 2004; Scott and Possingham, 1980; Sodmergen Kawano et al., 1989), and seems to be correlated with nucleoid size (Kuroiwa, 1991; Lindbeck et al., 1989). Diameters of 0.3-4 μm have been reported for plastid nucleoids stained with 4',6'diamidino-2-phenylindole (DAPI) (Hansmann et al., 1985; Kuroiwa, 1991). Fluorescence and electron microscopy of isolated nucleoids from various plastid types has revealed fibrous, sometimes beaded structures reminiscent of, albeit considerably smaller than, bacterial nucleoids (Hansmann et al., 1985; Nakano et al., 1993; Nemoto et al., 1988; Yurina et al., 1995a; Yurina et al., 1995b). *In vitro* proteolysis leads to an unraveling of these structures (Hansmann et al., 1985; Nakano et al., 1993; Nemoto et al., 1990; Nemoto et al., 1989; Yurina et al., 1995a), whereas treatment with DNase eliminates the fibers protruding from the nucleoid core (Hansmann et al., 1985; Nemoto et al., 1989; Yurina et al., 1995a), consistent with the presence of proteins in plastid nucleoids that compact the organellar DNA. The observed changes in nucleoid morphology during plastid development are reflected by differences in their protein to DNA ratios (Hansmann et al., 1985; Nemoto et al., 1989; Yurina et al., 1995a; Yurina et al., 1995b) and in their protein complement (Baumgartner and Mullet, 1991; Hansmann et al., 1985; Nakano et al., 1993; Nemoto et al., 1990; Nemoto et al., 1991; Nemoto et al., 1988; Oleskina et al., 2001; Oleskina et al., 1999; Sato et al., 1997; Yurina et al., 1995a; Yurina et al., 1995b) but have not been correlated with changes in template accessibility and nucleoid function.

Apart from the establishment of restriction fragment patterns for nucleoid DNA, which are identical to those obtained from ctDNA that was isolated by other means (Nakano et al., 1993; Nemoto et al., 1988), almost nothing is known about the physical state (linear, circular, concatenated) of ctDNA in nucleoids. Recent results obtained with chloroplasts embedded in agarose gels (Oldenburg and Bendich, 2004), however, suggest that plastid nucleoids contain multi-genome length linear, branched and catenated forms of the plastome.

Not many studies have addressed structure and composition of mitochondrial nucleoids in plants beyond the level of microscopic observations. The mitochondrial nucleoids of most land plants are considerably smaller than the smallest chloroplast nucleoids and were originally reported not to increase in number during development beyond the 1-2 nucleoids that are present in actively dividing cells (Kuroiwa et al., 1981). A more recent study, however, has revealed that mitochondrial nucleoids also undergo dynamic changes during leaf and root development. Mitochondria in cells of the shoot and root meristems of Arabidopsis and in cells close to the quiescent center of the tobacco root apex contain a few large, elongated mitochondrial nucleoids. In more mature root and shoot cells, mitochondrial nucleoids are more numerous and considerably smaller (Fujie et al., 1993a; Fujie et al., 1994; Suzuki et al., 1992). How these morphological changes relate to the function(s) of mitochondrial nucleoids is unknown.

3.2. Nucleoid functions in organellar DNA metabolism

3.2.1. DNA replication

The nucleoid has been implicated as the site of organellar DNA replication based on the high incidence of replication forks in thylakoid-associated nucleoids and colocalization with sites of [^{3}H]-thymidine incorporation (Lindbeck and Rose, 1990; Liu and Rose, 1992; Rose and Possingham, 1976). Furthermore, nucleoids isolated from developing barley chloroplasts and from proplastids of cultured tobacco cells synthesize DNA *in vitro* (Baumgartner and Mullet, 1991; Sakai et al., 1999), and in the unicellular alga *Ochromonas danica*, sites of bromo-deoxyuridine incorporation into newly synthesized DNA coincide with the single ring-shaped plastid nucleoid (Nerozzi and Coleman, 1997). By contrast, replicating DNA is not associated with the multiple smaller nucleoids of permeabilized chloroplasts from cultured soybean cells (Cannon et al., 1999), possibly because of differences in nucleoid architecture and function between cultured higher plant cells and algae, or because of an uncoupling of DNA replication from packaging of newly replicated DNA into nucleoids in soybean chloroplasts.

Plant mitochondrial nucleoids also support DNA replication, as demonstrated by the *in vitro* run-on DNA synthesis activity of purified mitochondrial nucleoids from cultured tobacco cells (Sakai et al., 1999) and of the rather ill-defined active chromosomes from potato and tobacco mitochondria (Fey et al., 1999).

3.2.2. Transcription

A significant RNA component has been reported for plastid nucleoids from some sources (Hansmann et al., 1985; Yurina et al., 1995a; Yurina et al., 1995b), although RNA was not detectable in highly purified nucleoids from proplastids and chloroplasts of tobacco (Nemoto et al., 1990; Nemoto et al., 1988). Furthermore, highly purified chloroplast and proplastid nucleoids from tobacco (Nakano et al., 1993; Sakai et al., 1998a), developing chloroplasts from barley (Baumgartner and Mullet, 1991), mitochondrial nucleoids from tobacco BY2 cells (Sakai et al., 1998b) and mitochondrial active chromosomes (Fey et al., 1999) show considerable activity in run-on transcription assays. Plastid-derived transcriptionally active chromosomes (TACs) (Baumgartner and Mullet, 1991; Briat et al., 1982; Briat and Mache, 1980; Krause and Krupinska, 2000; Reiss and Link, 1985; Rushlow et al., 1980; Sakai, 2001; Suck et al., 1996) probably represent partially unwound nucleoids or portions of the protein/DNA complexes (Sakai, 2001), since TACs tend to have a more diffuse appearance when stained with DAPI, lower sedimentation coefficients, a higher protein and RNA content and lower transcription and DNA synthesis activity than nucleoids.

3.2.3. Organellar genome partitioning

Nucleoids have to be partitioned between the daughter organelles during plastid and mitochondrial division. In bacteria, the partitioning of the replicated chromosome is an active process that involves an interplay of nucleoid-organizing proteins with enzymes that decatenate and resolve the replicated genome copies (Lewis, 2001). Although the molecular players that separate the organellar nucleoids in plants are unknown, Rose and coworkers have reported that the plastid nucleoid is attached to plastidal membranes, presumably through specific interactions with DNA binding proteins that might aid in the segregation of the protein/DNA complexes (Lindbeck and Rose, 1990; Liu and Rose, 1992). More recently, the inner envelope protein PEND has been proposed to play a role in nucleoid segregation in very young chloroplasts (Sato et al., 1993), and the integral thylakoid protein MFP1 might aid in nucleoid partitioning in more mature organelles (Jeong et al., 2003), although at present there is no direct evidence for a role of either protein in the process.

The *Plasmodium falciparum* apicoplast genome, the plastomes of several red algae and the cyanelle genome of *Cyanophora paradoxa* contain an ortholog of the bacterial gene *ycf24* (Law et al., 2000). Disruption of this gene in *Synechocystis* PCC6803 yields cells that contain less DNA than the wild type and often have aberrantly partitioned nucleoids. Ycf24 overexpressed in *E. coli* localizes to both side of the nucleoid, suggesting that the protein plays a role in nucleoid partitioning. The Arabidopsis nuclear genome encodes a YCF24 ortholog that is predicted to be targeted to plastids, but it is not known whether this protein plays a role in plastid nucleoid segregation.

Treatment of *C. merolae* with nalidixic acid inhibits plastid and mitochondrial DNA synthesis and leads to asymmetric division of the plastid, but not the mitochondrial nucleoid (Itoh et al., 1997). Division of neither mitochondrion nor chloroplast is affected by this treatment. These results suggest that a DNA gyrase-like protein might play a role in plastid nucleoid partitioning in the red alga, possibly by decatenating the replicated plastome copies and thereby permitting them to be separated into daughter nucleoids. Since treatment with nalidixic acid has no effect on division of the mitochondrial nucleoid, this process must proceed by a different mechanism. It remains to be determined whether a topoisomerase II functions in a similar capacity in chloroplasts of higher plants.

Itoh and colleagues (Itoh et al., 2001b) have identified Arabidopsis nuclear genes for the two kinesin-related proteins MKRP1 and MKRP2. Both proteins feature N-terminal mitochondrial targeting signals that direct reporter proteins to the organelles in transiently transformed tobacco cells, a conserved N-terminal kinesin motor domain and a C-terminal coiled coil domain. Their structural similarity to the *E. coli* chromosome partitioning protein MukB suggests that they might participate in plant mitochondrial nucleoid segregation.

3.3. Plastid nucleoid proteins

3.3.1. HU protein

The existence of a gene for an HU-like protein on the plastomes of two primitive eukaryotic algae is well established (Kobayashi et al., 2002; Wang, S. and Liu, 1991). The plastidal proteins HlpA of the cryptomonad *Gilliardia theta* (formerly *Cryptomonas* Φ) and HC of the rhodophyte *C. merolae* share significant amino acid homology with bacterial HU proteins, particularly in the region that encompasses the DNA binding “arm” of HU-like proteins. Not surprisingly, homology is greatest to HU-like proteins from cyanobacteria, the phylogenetic ancestors of plastids. Both algal proteins have been shown to complement bacterial mutants lacking a functional HU-like protein (Grasser et al., 1997; Kobayashi et al., 2002). Recombinant HlpA protein, like its eubacterial counterpart, binds double stranded DNA in a non-sequence specific manner, and is able to bend DNA and introduce supercoils into covalently closed circular DNA species in the presence of topoisomerase I (Grasser et al., 1997; Wu, H. and Liu, 1997). The protein prefers four-way junctions and DNA minicircles to linear DNA fragments, suggesting that its *in vivo* function might extend beyond that of a simple DNA compacting nucleoid protein. The HU protein of *E. coli* affects DNA replication, recombination and repair, as well as chromosome partitioning and gene expression (Aki and Adhya, 1997; Bramhill and Kornberg, 1988; Dri et al., 1992; Hashimoto et al., 2003; Huisman et al., 1989; Jaffe et al., 1997; Li, S. and Waters, 1998), and HlpA might play similar roles in the algal plastid. The related HC protein of *C. merolae* is able to condense DNA *in vitro* and is likely to be an important structural determinant of the plastid nucleoid (Kobayashi et al., 2002). Whether the protein has additional functions related to organellar DNA metabolism in the alga awaits further study.

Briat and colleagues (Briat et al., 1984) detected among the nucleoid proteins of spinach plastids a 17 kDa protein that crossreacts with antisera against HU protein from *E. coli* and *Synechocystis*, whereas a protein that is immunologically related to HU in spinach chloroplasts and mitochondria has an estimated molecular weight of 26,000 (Henschke and Nuecken, 1989). Yurina *et al.* (Yurina et al., 1995b) predicted the presence of histone-like proteins in plastid nucleoids from pea and *Azolla pinnata* based on similarities in amino acid composition. More recent attempts to detect HU-like proteins in nucleoids from higher plants, however, have been unsuccessful. Although proteins of the molecular weights expected for the HU monomer (17 kDa) and its potential dimer (35 kDa) are present in nucleoids from soybean chloroplasts, neither protein crossreacts with antisera against *E. coli* and cyanobacterial HU (Cannon et al., 1999). Sato *et al.* (Sato, N. et al., 1999) were also unable to detect an HU homolog in nucleoids of pea chloroplasts, and the completely sequenced higher plant plastomes and nuclear genomes do not carry a gene for an obvious HU ortholog (Sato, S. et al., 1999).

3.3.2. CND41

The CND41 (chloroplast nucleoid DNA binding) protein is one of the best characterized plastid nucleoid components. The protein was discovered on southwestern blots of highly purified nucleoids from the *N. tabacum* NII cell line (Nakano et al., 1993) and has since emerged as a much more important regulatory player in plant development than would have been predicted from its non-sequence specific DNA binding ability. CND41 contains a conserved motif found in the active site of aspartic proteases. *In vitro* at low pH, purified CND41 possesses proteolytic activity that is largely resistant to conventional protease inhibitors but is sensitive to the presence of ribo- and deoxyribonucleoside triphosphates, NADPH and Fe^{3+} (Murakami et al., 2000; Nakano et al., 1997). Although present in chloroplasts, CND41 protein seems to be most abundant in plastids of non-photosynthesizing tissues, and a clear correlation exists between its abundance and the levels of plastome-encoded transcripts. Antisense tobacco plants that produce significantly reduced levels of CND41 show a dwarf phenotype, accelerated plastid development and lower levels of active gibberellins than their wild type counterparts (Nakano et al., 1997; Nakano et al., 2003). Despite an attempt to identify CND41's mode of action at the molecular level, the role this nucleoid protein plays in regulating gibberellin biosynthesis and/or plastid development has yet to be elucidated and awaits the availability of suitable mutants. Interestingly, the transcript level of a CND41 homolog from tobacco is moderately (2-fold) upregulated in response to cytokinin (Schaefer et al., 2000), but the significance of this finding for the regulation of nucleoid structure and function is not clear.

3.3.3. PEND

Southwestern blotting was also used to search for pea plastid envelope proteins with affinity for ctDNA (Sato et al., 1993). A 130 kDa polypeptide of the inner envelope was discovered and later named PEND (plastid envelope DNA binding). A more detailed analysis of PEND's DNA binding specificity using cloned ctDNA fragments revealed several preferred A/T-rich regions on the pea plastome that were subsequently narrowed down to the canonical TAAGAAGT octamer. Cloning of the cDNA for the PEND protein revealed an open reading frame for a protein of only 70 kDa, suggesting that the protein exists as a homodimer *in vivo*. Dimerization and DNA binding ability are mediated by the protein's N-terminal basic region/leucine zipper (bZIP) domain (Sato and Ohta, 2001). Consistent with the observed location of nucleoids on the periphery of developing plastids, PEND is abundant in developing leaves but is not present in mature chloroplasts, however, proof for its suggested roles as the nucleoid anchor and as a participant in plastome replication and segregation requires a thorough functional analysis.

3.3.4. DCP68/SiR

Our own lab has described a DNA binding protein of 68 kDa (DCP68) that is an abundant DNA binding protein in chloroplast nucleoids of the *Glycine max* SB-M cell line (Cannon et al., 1999) and of pea plants (Chi-Ham et al., 2000). DCP68 binds DNA non-sequence specifically, compacts DNA *in vitro* to yield particulate, nucleoid-like structures, and inhibits DNA synthesis by a partially purified plastid protein extract. The protein was identified as ferredoxin:sulfite reductase (SiR), an enzyme of the reductive sulfur assimilation pathway in plants (Leustek et al., 2000; Leustek and Saito, 1999), and shown to co-localize with ctDNA in nucleoids of soybean and pea chloroplasts (Chi-Ham et al., 2000; 2002). Like CND41, DCP68/SiR constitutes a bifunctional organelle nucleoid protein. It is likely that DCP68/SiR affects the template function of ctDNA *in vivo* through its DNA compacting activity. Sekine et al. (Sekine et al., 2002) have presented evidence to suggest that RNA synthesis in purified pea chloroplast nucleoids is negatively correlated with DNA compaction by recombinant SiR from maize, but whether a similar mode of transcriptional regulation by template accessibility is also operative *in vivo* is unknown. Interesting in this context is the increased affinity for DNA of purified DCP68/SiR after dephosphorylation *in vitro* (Chi-Ham et al., 2002), which suggests a potential strategy for regulating template accessibility of the plastome replication and transcription enzymes. Although the biological significance of utilizing a sulfur assimilation enzyme as a major plastid nucleoid component is not immediately obvious, Ilv5p of yeast mitochondria (see below) is another example of a bifunctional organellar nucleoid protein that links mitochondrial DNA (mtDNA) maintenance and organellar nucleoid structure to another metabolic pathway.

3.3.5. MFP1

The 82 kDa MFP1 protein from Arabidopsis, originally believed to be a matrix attachment region DNA binding protein of the nuclear envelope, was recently shown to reside in plastids (Jeong et al., 2003). This integral thylakoid protein preferentially binds double stranded DNA regardless of sequence via its long C-terminal coiled coil domain and co-purifies with plastid nucleoids. Unlike the PEND, MFP1 accumulates as the organelle matures. Consistent with the documented nucleoid repositioning during plastid development from the periphery to the photosynthetic membranes (reviewed in Kuroiwa, 1991), MFP1 has been suggested to be the predicted thylakoid nucleoid anchor. The normal appearance and distribution of chloroplast nucleoids in homozygous MFP1 knockouts are not easily reconciled with this hypothesis but may reflect functional redundancy between MFP1 and six additional genes for related coiled coil proteins in Arabidopsis, or may be related to the multiple plastome copies in mature leaf cells, which are no longer replicated and might therefore tolerate the lack of a thylakoid anchor during organelle division. Analysis of single and double mutants defective in the related coiled coil proteins will help to differentiate between these possible scenarios.

3.3.6. CpPTP

A coiled coil protein that belongs to a small gene family was detected in the resurrection plant, *Craterostigma plantagineum* (Phillips et al., 2002). The plastid targeted protein CpPTP is inducible by treatment with abscissic acid or by dehydration and accumulates predominantly in the palisade layer of the leaf. Recombinant GFP/CpPTP fusion protein localizes to the plastids in transgenic tobacco plants and accumulates in punctuate structures that are consistent with proplastid nucleoids in transiently transformed cultured tobacco cells. Although rigorous proof that CpPTP co-localizes with plastid nucleoids and binds ctDNA is still outstanding, since recombinant glutathione-S-transferase (GST)/CpPTP partially protects DNA from digestion by DNaseI *in vitro*, a biological role for the protein in protecting ctDNA from damage during desiccation was suggested.

3.4. Potential mitochondrial nucleoid proteins

No *bona fide* protein components of plant mitochondrial nucleoids are known, although some DNA binding proteins might reasonably be assumed to be associated with mtDNA. The mitochondrial active chromosomes from potato and maize described by Fey et al (Fey et al., 1999) copurify with RNA and DNA polymerase activities, and a handful of mitochondrial proteins that bind specifically to certain promoter elements have been described (Chang and Stern, 1999; Hatzack et al., 1998; Ikeda and Gray, 1999), but none of these proteins are likely to represent structural elements that might compact DNA and/or shape the nucleoid. *In silico* data mining of the Arabidopsis and rice genomes for predicted organellar proteins that are likely involved in nucleic acid metabolism (Elo et al., 2003) has identified homologs of mitochondrial proteins from yeast and vertebrates that might be structural and functional constituents of the mitochondrial nucleoid in plants.

3.4.1. Histone H3

Vermel et al. (Vermel et al., 2002) identified a histone H3 among a set of single-stranded DNA binding proteins from potato tuber mitochondria, but were unable to show *in vitro* import of the protein into purified mitochondria. A *Brassica napus* histone H3 homolog features a potential C-terminal nuclear localization signal as well as an N-terminal amino acid extension that is a predicted mitochondrial targeting signal (Kourtz et al., 1996). Histone H3 was also detected in the proteome of soybean chloroplast nucleoids (S. Heinhorst and G. Cannon, unpublished observations) but was dismissed as a nuclear chromatin contaminant, like a histone H4 that was identified in the *Sesbania rostrata* thylakoid proteome (Peltier et al., 2000). The Arabidopsis genome contains several histone H3 genes that encode proteins predicted to be organelle-targeted (Elo et al., 2003; Chi-Ham, unpublished observation). However, the known unreliability of organelle targeting predictions and the fact that a histone H3 homolog is obviously not a component of mitochondrial nucleoids from yeast (Kaufman et al., 2000), *Xenopus laevis* (Bogenhagen et al., 2003) and humans (Garrido et al., 2003) underscore the need for rigorous biochemical and cytological proofs to ascertain co-localization of histones with plant mitochondrial nucleoids and define the proteins' potential architectural and regulatory functions.

3.4.2. mtSSB

A prominent constituent of the mitochondrial nucleoid in yeast and vertebrates is a single-stranded DNA binding protein of 13,000-18,000 kDa (mtSSB, or Rim1p in yeast) that is structurally and functionally related to *E. coli* SSB (Bogenhagen et al., 2003; Garrido et al., 2003; Kaufman et al., 2000; Pavco and Van Tuyle, 1985; Thoemmes et al., 1995; Van Dyck et al., 1992; Van Tuyle and Pavco, 1985). The protein stimulates the activity and increases the processivity of mtDNA polymerase γ (Farr et al., 1999; Mikhailov and Bogenhagen, 1996; Thoemmes et al., 1995) and its homotetramers preferentially bind to the single-stranded mitochondrial displacement loop (D-loop) (Li, K. and Williams, 1997; Van Tuyle and Pavco, 1985). The protein is essential for the maintenance of mtDNA in yeast (Van Dyck et al., 1992) and *Drosophila melanogaster* (Maier et al., 2001). The potential plant equivalent of this protein is conspicuously absent from the set of mitochondrial single stranded nucleic acid binding proteins isolated by Vermel *et al.* (Vermel et al., 2002) but is predicted to be organelle-targeted (Elo et al., 2003). It is possible that mtSSB is not a constituent of the plant mitochondrial nucleoid and that a histone or HMG2-like protein fulfills that function.

3.4.3. HMG2-like protein

The abundant mitochondrial protein Abf2p, a relative of the nuclear high mobility group (HMG) DNA binding proteins, is a non-specific high-affinity DNA binding protein that is able to bend and wrap DNA (Diffley and Stillman, 1991) and is a predominant structural determinant of the yeast mitochondrial nucleoid (Kaufman et al., 2000; Newman et al., 1996). In addition, the protein plays important roles in mtDNA recombination, segregation and copy number control (MacAlpine et al., 1998; Zelenaya-Troitskaya et al., 1998). Although structurally unrelated, the *E. coli* nucleoid protein HU is a functional homolog of Abf2p, as shown by its ability to complement the phenotype of a yeast *abf2* null mutant (Megraw and Chae, 1993). The vertebrate equivalent, TFAM, is a mitochondrial transcription factor that was recently shown to co-purify with nucleoids (Alam et al., 2003; Bogenhagen et al., 2003; Garrido et al., 2003). A homolog of Abf2p/TFAM was identified in the genomes of Arabidopsis and rice (Elo et al., 2003). Although the plant HMG2-like protein is predicted to be targeted to mitochondria, its localization in the organelle, association with mtDNA and the mitochondrial nucleoid, as well as a role(s) in mtDNA metabolism remain to be ascertained. Of particular interest is the question whether the HMG2-like protein of plants, like its vertebrate counterparts, is an important transcriptional regulator in the organelle, or whether its main function, like that of yeast Abf2p, is to ensure mtDNA stability and proper inheritance.

3.4.4. *Hsp60*

Mitochondrial HSP60p of yeast, a homolog of *E. coli* GroEL, is encoded by an essential gene and is needed for the proper folding of imported mitochondrial proteins (Cheng et al., 1989). The protein can be crosslinked to mtDNA by formaldehyde treatment of purified nucleoids. Furthermore, recombinant histidine-tagged Hsp60p specifically binds ssDNA and has a particular preference for the template strand of *ori5*, the active origin of replication of yeast mtDNA (Kaufman et al., 2000). Mitochondrial DNA is unstable and organellar nucleoids display an altered morphology in *hsp60* temperature sensitive mutants (Kaufman et al., 2003). The collective evidence suggests a role for HSP60p in the transmission of nucleoids and mtDNA during yeast cell division. It is not clear if homologs of HSP60p are also constituents of the vertebrate mitochondrial nucleoid, since this protein was not reported to be associated with purified *Xenopus laevis* and human nucleoids (Bogenhagen et al., 2003; Garrido et al., 2003). The genomes or Arabidopsis and rice contain a gene encoding the plant equivalent of HSP60p, which is predicted to be located in mitochondria (Elo et al., 2003) and appears to be abundant in the organelle, since it was identified in three mitochondrial proteome studies (Bardel et al., 2002; Heazlewood et al., 2003; Kruft et al., 2001) and as a ssDNA-binding protein of potato tuber mitochondria (Vermel et al., 2002). It is unclear whether plant HSP60 plays multiple roles in the organelle as a chaperonin, a nucleoid component and/or a determinant of organellar genome maintenance and transmission.

3.4.5. *Ilv5p*

The abundant yeast mitochondrial matrix protein Ilv5p is an acetohydroxy acid reductoisomerase in the branched chain amino acid biosynthesis pathway. The protein is also a multicopy suppressor of the mtDNA instability phenotype of Δ*abf2* mutants during growth on a fermentable carbon source (Zelenaya-Troitskaya et al., 1995) and is a mitochondrial nucleoid protein (Kaufman et al., 2000). The number of yeast mitochondrial nucleoids and the parsing of mtDNA into nucleoids parallel the level of Ilv5p expression, which in turn is controlled by the general amino acid control pathway (MacAlpine et al., 2000; Zelenaya-Troitskaya et al., 1995). Mutations that affect either the enzymatic activity or the mtDNA maintenance function of the yeast protein map to two different domains of the Ilv5p ortholog from spinach, whose crystal structure is known (Bateman et al., 2002). The protein products of the Arabidopsis and rice orthologs are also predicted to reside in mitochondria (Elo et al., 2003), and the Ilv5 protein was recently identified in a study of the Arabidopsis mitochondrial proteome (Heazlewood et al., 2004). Proof, however, that this protein is associated with the plant organellar nucleoid is still outstanding. It is noteworthy that highly purified mitochondrial nucleoids from *X. laevis* oocytes contain the lipoyl-containing subunit of the branched chain α-ketoacid dehydrogenase complex (Bogenhagen et al., 2003), suggesting a possible connection between amino acid metabolism and nucleoid structure and function in vertebrate mitochondria as well.

3.4.6. Other protein candidates

Among the proteins that are associated with mitochondrial nucleoids from animals and/or yeast are the helicase Twinkle (Garrido et al., 2003), aconitase and aldolase (Kaufman et al., 2000), the adenine nucleotide translocator (Bogenhagen et al., 2003; Garrido et al., 2003), prohibitin 2 (Bogenhagen et al., 2003), DNA polymerase γ (Garrido et al., 2003), and subunits of pyruvate dehydrogenase (Bogenhagen et al., 2003; Kaufman et al., 2000), α-ketoglutarate dehydrogenase (Kaufman et al., 2000), ATPase (Bogenhagen et al., 2003; Kaufman et al., 2000) and cytochrome oxidase (Bogenhagen et al., 2003; Garrido et al., 2003). Since nucleoids are the units of inheritance and sites of replication of the mitochondrial genome, the presence of a DNA polymerase and a helicase is not surprising. The preponderance of seemingly unrelated proteins in nucleoid preparations, however, is unexpected and emphasizes the need for consensus purification protocols and, most importantly, an operational definition of the mitochondrial nucleoid (Bogenhagen et al., 2003) to distinguish between artefactual and biologically significant associations. The enzymes of respiratory, biosynthetic and biodegradative pathways that are found in mitochondrial nucleoids probably reflect the proteins' multiple roles in the organelle and point to novel regulatory circuits that determine structure, function and maintenance of the mitochondrial genome.

3.5. Summary and outlook

Several protein components of the plastid nucleoid have been identified and characterized to varying extents, but colocalization with nucleoid DNA *in vivo* or *in situ* has only been shown for DCP68/SiR (Chi-Ham et al., 2002). Likewise, *in planta* overexpression or antisense repression studies are lacking for all but two plastid nucleoid proteins (CND41 and MFP1), and an effect of the transgene on nucleoid structure, function and distribution within the organelle was either not investigated or not observed. Compared to our current knowledge of the proteins that constitute plastid nucleoids, the composition of plant mitochondrial nucleoids is still an enigma. A reliable method of purifying mitochondrial protein/DNA complexes that are likely to constitute the organellar nucleoid has yet to be developed, and consequently no *bona fide* mitochondrial nucleoid proteins have been identified and characterized to date.

The association of multifunctional "moonlighting" proteins (Copley, 2003; Jeffery, 1999; 2003) with the plastid nucleoid is intriguing (Chi-Ham et al., 2002; Murakami et al., 2000; Nakano et al., 1997), particularly since such proteins have also been identified as components of mitochondrial nucleoids from animals and yeast. The comprehensive list of cellular components that influence mtDNA stability, integrity and inheritance in yeast (Contamine and Picard, 2000) suggests that a few surprises might yet be in store for the composition of plant organellar nucleoids.

Although the power of proteomics approaches to catalog the protein complements of plant sub-cellular structures is well documented (Calikowski et al., 2003; Gomez et al., 2002; Schroeder and Kieselbach, 2003; Schubert et al., 2002; van Wijk, 2001; Werhahn and Braun, 2002; Zabrouskov et al., 2003; reviewed by Francs-Small in Chapter 7 of this volume), several recently published organellar proteomes are devoid of any proteins (with the possible exception of mitochondrial HSP60 and a plastid nucleic acid binding protein) that might be structural components of the organellar nucleoids, and a disappointingly small number was identified as belonging to the organellar transcription, translation and DNA replication apparatus (Bardel et al., 2002; Ferro et al., 2003; Ferro et al., 2002; Gomez et al., 2002; Heazlewood et al., 2003; Heazlewood et al., 2004; Kruft et al., 2001; Lonosky et al., 2004; Millar et al., 2001; Peltier et al., 2000). Technical difficulties related to the physical properties of these proteins and their generally very low abundance are likely reasons for their absence from the organellar proteome catalogs. Determination of the nucleoid protein complement of purified protein/DNA complexes might be a more promising strategy to elucidate which structural and enzymatic entities contribute to nucleoid morphology and function(s).

The remodeling of nucleoid structure and changes in intraorganellar distribution during organelle and plant development are well documented, however, the accompanying changes in the role the DNA/protein complexes play in organellar genome replication and transcription have not been addressed at the molecular level. Likewise, the cellular signals that initiate these changes are unknown, as is the role of membranes and their protein components in nucleoid and genome partitioning during organelle division. Although the limited mutant studies that have been performed so far have not proven particularly helpful in elucidating the role individual nucleoid proteins play in these processes or in their regulation, approaches such as the induction of post-transcriptional gene silencing at specific developmental stages or during imposed experimental conditions might yield some insight into the molecular mechanisms that shape organellar nucleoids and determine their functions.

4. ORGANELLE DNA REPLICATION AND REPAIR

4.1. Control of Chloroplast DNA Replication

The number of plastome copies per organelle and cell is specific for a particular tissue (reviewed in Heinhorst and Cannon, 1993). In general, chloroplasts of photosynthetically active cells contain more ctDNA than the undeveloped plastids of non-green tissue (Cannon et al., 1985; Cannon et al., 1986b; Dubell and Mullet, 1995; Lamppa and Bendich, 1979; Lau et al., 2000; Lawrence and Possingham,

1986; Miyamura et al., 1990; Scott and Possingham, 1983), which is in keeping with the presumed reliance of the plant's photosynthetic capacity on a high dosage of plastome-encoded genes that ensures sufficient organellar transcription and translation rates (Bendich, 1987). This premise is supported by the well documented amplification of plastome copies that takes place in the developing leaf of higher plants. Through this process, organellar ploidy increases from the 10-30 plastome copies per proplastid of meristematic cells to as many as 200-300 plastome molecules in young chloroplasts, which translates to many thousand copies per developing mesophyll cell (Baumgartner et al., 1989; Lamppa and Bendich, 1979; Lawrence and Possingham, 1986; Miyamura et al., 1986; Scott and Possingham, 1980; 1983; Sodmergen Kawano et al., 1989; Tymms et al., 1982; 1983). The regulatory circuits that trigger plastome amplification during leaf development are unknown but might involve components of the phytochrome signaling system (Dubell and Mullet, 1995).

Organelle division and the replication of its DNA are not tightly coupled processes. Plastome amplification and organelle division tend to keep pace with each other in early postmitotic leaf cells. In older leaf sections the division rate of the young chloroplasts overtakes the plastome replication rate, which leads to a re-distribution of existing plastome copies among the dividing chloroplasts and to a reduction in organelle ploidy. No ctDNA replication occurs in fully developed leaves, and the overall amount of ctDNA per cell appears to remain constant (Baumgartner and Mullet, 1991; Boffey et al., 1979; Boffey and Leech, 1982; Dubell and Mullet, 1995; Lamppa and Bendich, 1979; Lawrence and Possingham, 1986; Miyamura et al., 1986; Scott and Possingham, 1980; 1983; Sodmergen Kawano et al., 1989; Tymms et al., 1982) or may be reduced in some species (Baumgartner et al., 1989). In contrast, amyloplast DNA synthesis in the developing wheat endosperm occurs even after the organelle has ceased to divide (Catley et al., 1987), suggesting different regulatory mechanisms for the two plastid types.

Plastome replication in higher plants and many eukaryotic algae is independent of the nuclear DNA synthesis (S) phase and proceeds throughout the cell division cycle (Cannon et al., 1986a; Heinhorst et al., 1985a; Heinhorst and Cannon, 1993; Heinhorst et al., 1985b; Lawrence and Possingham, 1986; Nerozzi and Coleman, 1997; Rose et al., 1975; Rose and Possingham, 1976; Suzuki et al., 1992; Takeda et al., 1992; Turmel et al., 1981; Tymms et al., 1982; Yasuda et al., 1988). In *C. merolae*, on the other hand, plastid and mtDNA S phases are well defined (reviewed by Kuroiwa et al., 1998). Blocking the nuclear S phase in the red alga with aphidicolin also inhibits organellar DNA synthesis (Itoh et al., 1996) and points to differences between the regulation of ctDNA replication in the single chloroplast of *C. merolae* and in the multiple organelles of higher plants.

4.2. Control of mitochondrial DNA replication

In contrast to the well documented changes in plastome numbers during organelle and plant development, not much is known about changes in mtDNA copies. Suzuki et al. (Suzuki et al., 1992) followed BrdU incorporation into nuclear, plastid and mtDNA and determined that organelle DNA is preferentially synthesized in the absence of concomitant nuclear DNA synthesis in cells close to the quiescent center of the tobacco root apex, whereas only nuclear DNA synthesis activity could be detected in cells that are located at a greater distance above the quiescent center, suggesting independence of mtDNA replication from that in the nucleus and from cell division (Fujie et al., 1993b). The observed high levels of mtDNA synthesis activity precede the fragmentation of the large, elongated nucleoids into multiple smaller entities with low DNA synthesis activity and suggest that mtDNA is selectively amplified and re-distributed among multiple nucleoids in preparation for subsequent cell divisions that are accompanied by little or no organelle DNA synthesis. Similar observations were made for mitochondria of the shoot apical meristem and the leaf primordium of Arabidopsis (Fujie et al., 1994), which contain considerably more DNA than the organelles in the mature leaf and corroborate earlier reports of decreases in mtDNA copy numbers during leaf development in pea (Heinhorst et al., 1990b; Lamppa and Bendich, 1984); rev in (Heinhorst et al., 1990b)). In contrast, no changes in mtDNA content per embryo and in the ratio of mitochondrial to nuclear DNA were observed during germination of maize embryos over a 48 h imbibation period (Logan et al., 2001), suggesting that mitochondrial, like ctDNA replication is regulated in a manner that is a specific for a particular tissue and/or developmental stage.

4.3. Mechanism of chloroplast DNA replication

The currently accepted model of plastome replication is based on electron microscopy of replicating pea and maize ctDNA (Kolodner and Tewari, 1975a; b). The observed theta- and sigma-shaped molecules were interpreted as Cairns-type and rolling circle replication intermediates. Two D-loops located on opposite DNA strands several thousand base pairs apart are the presumed initiation sites (*ori*s) of plastome replication and, through unidirectional extension toward each other followed by bidirectional fork movement, are thought to give rise to the observed theta intermediates. Sigma structures are proposed to arise after a completed round of Cairns-type replication by a switch in replication mode to rolling circle amplification 180° from the *ori* region.

Electron microscopy has been used to map D-loops on the plastomes of several higher plants and algae. Although the predicted two D-loops (Kolodner and Tewari, 1975a; b) have rarely been observed on a single ctDNA fragment, between one and four D-loops have been reported for plastomes from various plant species and organelles at different developmental stages (Chiu and Sears, 1992; Koller and Delius, 1982; Meeker et al., 1988; Ravel-Chapuis et al., 1982; Takeda et al., 1992; Waddell et al., 1984). Considering the highly conserved structure and gene content of ctDNA, the observed differences in number and location of plastome D-loops are somewhat surprising. Two-dimensional gel electrophoresis of *in vivo* and *in vitro* ctDNA replication intermediates and other experimental approaches to further define plastome *ori* regions (Hedrick et al., 1993; Kunnimalaiyaan and Nielsen, 1997b; Kunnimalaiyaan et al., 1997; Lu et al., 1996; Nielsen et al., 1993; Wang, Y.

et al., 2002) for the most part support the presence of two D-loops but also failed to establish a consensus *ori* location on the plastome of higher plants. In keeping with these results is the general lack of primary structure homology between those *ori* regions that have been sequenced (Kunnimalaiyaan and Nielsen, 1997b; Lu et al., 1996; Nielsen et al., 1993; Takeda et al., 1992; Wu, M. et al., 1986); rev in (Heinhorst and Cannon, 1993; Kunnimalaiyaan and Nielsen, 1997a).

Plastome origins do, however, share certain structural features, such as an abundance of (A+T)-rich regions flanked by (G+C)-rich motifs, and several short inverted repeats that are predicted to form stable stem/loop structures (Kunnimalaiyaan and Nielsen, 1997b; Nielsen et al., 1993; Wu, M. et al., 1986). Such secondary structure elements may be important for initiation site selection, like the *in vitro* D-loop extension site that was mapped to the base of a (G+C)-rich potential stem/loop in the cloned *oriA* region of the tobacco plastome (Lu et al., 1996). Hairpin structures might also mediate specific interactions with the organellar DNA replication machinery, since gel mobility shift analysis revealed that the cloned *oriB* from tobacco (Kunnimalaiyaan et al., 1997) and *oriA* from *Chlamydomonas* (Nie et al., 1987) specifically interact with a small subset of chloroplast proteins. One of the algal *ori*-binding proteins was identified as a plastome-encoded 18 kDa Fe-S protein that is related to ferredoxins and to a subunit of NADH dehydrogenase (Wu, M. et al., 1989). Such interaction between *ori* and a redox-active protein might explain the observed sensitivity of *Chlamydomonas* ctDNA copy numbers to the redox state of the cell (Lau et al., 2000).

The different DNA replication initiation sites used in proplastids versus leaf chloroplasts of rice and tobacco (Kunnimalaiyaan and Nielsen, 1997b; Kunnimalaiyaan et al., 1997; Lu et al., 1996; Takeda et al., 1992; Wang, Y. et al., 2003) raise the interesting possibility that plastome *ori* regions and maybe replication mode might change during organelle differentiation. One can envision that the rapid plastome amplification in very young, developing chloroplasts proceeds by a different mechanism and initiates from a different origin than the slower maintenance replication in proplastids and in more mature chloroplasts. A similar switch in replication mode is well documented for the yeast 2-micron circle (Kornberg and Baker, 1992). Likewise, hours after addition of novobiocin to a *C. reinhardtii* culture the mode of plastome replication switches from a novobiocin-sensitive mechanism that is thought to involve a topoisomerase II to a novobiocin-insensitive mode that initiates a cryptic origin in the inverted repeat region and proceeds unidirectionally, thereby replicating only a portion of the plastome (Woelfle et al., 1993). Since the initiation region of this alternative DNA replication mode was mapped to a recombination hotspot on the *C. reinhardtii* plastome, the possibility of a recombination-dependent replication mechanism was invoked.

Of the collective experimental evidence that addresses the plastome replication mechanism only a fraction supports the expected unidirectional D-loop extension (Kunnimalaiyaan and Nielsen, 1997b; Kunnimalaiyaan et al., 1997; Lu et al., 1996; Nielsen et al., 1993), whereas results from several other studies are consistent with bi-directional fork movement (Chiu and Sears, 1992; Ravel-Chapuis et al., 1982; Takeda et al., 1992; Waddell et al., 1984; Wang, Y. et al., 2002). Some of these apparent discrepancies can probably be explained by the presence of a single *versus* two cloned *ori* regions in *in vitro* DNA replication templates (Kunnimalaiyaan and Nielsen, 1997b; Kunnimalaiyaan et al., 1997; Lu et al., 1996; Nielsen et al., 1993; Reddy et al., 1994) and *per se* may not contradict the accepted plastome replication model. Recent results from a series of studies challenge the premise that most plastome molecules exist in circular, covalently closed form and call for a re-thinking of the almost three decades-old plastome replication model.

Using the powerful chloroplast transformation approach, Muehlbauer et al. (Muehlbauer et al., 2002) generated tobacco deletion mutants for *oriA* and *oriB*. Deletion of both *oriA* copies yielded phenotypically normal, homoplastomic transformants that strongly suggest *oriA* is dispensable. Although *oriB* can successfully be deleted from one inverted repeat (B), no homoplastomic mutants lacking the *oriB* copy in inverted repeat A were obtained, probably because this *oriB* copy colocalizes with an essential gene or because one functional copy of *oriB* appears to be necessary for plastome replication. One can conclude from these results that either the double D-loop model needs to be rethought, or that the tobacco plastome employs a different mode of replication when a functional *oriA* region is not present.

Evidence that the *in vivo* structure of the higher plant plastome is considerably more complex than that represented by covalently closed, circular monomers is supported by results from a variety of experimental approaches. The predominant plastome forms in several higher plants appear to be multimers (Backert et al., 1995; Bendich and Smith, 1990; Deng et al., 1989; Lilly et al., 2001; Oldenburg and Bendich, 2004), consisting of head-to-tail linear concatemers, tangled fibers and complex branched structures (Bendich and Smith, 1990; Lilly et al., 2001; Oldenburg and Bendich, 2004). The ends of the linear concatemers of maize ctDNA map to the inverted repeats (Oldenburg and Bendich, 2004), rather than to an area in the large single copy region 180° from the *ori* regions, as would be expected if rounds of rolling circle replication were initiated following a complete round of Cairns-type replication (Kolodner and Tewari, 1975a; b). Based on their mapping results, Oldenburg and Bendich (Oldenburg and Bendich, 2004) proposed that plastome replication initiates at one of five *ori* regions on linear plastome concatemers, rather than on monomeric circles. The preponderance of larger-than-genome-size, branched DNA structures that resemble replicative forms of herpes simplex virus DNA (reviewed in Sandri-Goldin, 2003)) suggests that homologous recombination likely plays an important role in the replication of ctDNA. The observed complex, connected plastome multimers can also explain the disparity between the high chloroplast ploidy and the much lower estimates for the number of units of plastome inheritance.

4.4. *Mechanism of mitochondrial DNA replication*

Following early *in organello* and *in vitro* nucleotide incorporation studies that were unable to reveal a mechanism for plant mtDNA replication (Bedinger and Walbot, 1986; Carlson et al., 1986), two mtDNA fragments from *Petunia* were selected based on their ability to autonomously replicate in yeast. Based on their structural similarity to the two yeast and mammalian mtDNA *ori* regions (de Haas et al., 1991) and on replication intermediates generated from these fragments *in vitro*, a mechanism was inferred for *Petunia* mtDNA replication similar to that reported for mammalian mtDNA. More recently, however, results from a combination of experimental approaches have necessitated a reassessment of this model.

Like the plastome, the mitochondrial genome is considerably more complex than had been assumed based on the original restriction mapping results. The predicted predominance of unit genome size master circles and, in most species, multiple subgenomic circles that are the products of intra- and intermolecular homologous recombination events (reviewed in Backert et al., 1997), has not been corroborated by experimental evidence. Instead, plant mtDNA is comprised of mainly larger than genome length linear and complex branched and concatenated molecular species (Backert and Boerner, 2000; Backert et al., 1995; Backert et al., 1996a; Backert et al., 1996b; Bendich and Smith, 1990; Oldenburg and Bendich, 1996; 1998; 2001; Scissum-Gunn et al., 1998). In the liverwort *Marchantia polymorpha*, mtDNA consists of circularly permutated linear concatemers in exclusively head-to-tail arrangement and with 5' ends that are blocked by a bound protein, possibly a denatured topoisomerase II (Oldenburg and Bendich, 2001). The well-bound fraction of mtDNA from cultured tobacco and liverwort cells was shown to incorporate the bulk of [^{3}H]-thymidine *in vitro* (Oldenburg and Bendich, 1996; 1998) and, for tobacco, to give rise to linear products of 50–150 kbp (Oldenburg and Bendich, 1996). The few circular DNA species present were not labeled. Taken with the failure to detect structures that are consistent with theta-type replication bubbles by electron microscopy, the labeling pattern of nascent mtDNA does not support a Cairns-type replication mechanism. Instead, the predominant molecular species observed by microscopy or derived from biochemical experiments suggest that the plant mitochondrial genome is replicated by a mechanism that involves multiple homologous recombination events and is reminiscent of that employed by bacteriophage T4 (Backert and Boerner, 2000; Oldenburg and Bendich, 1998; 2001). Although the exact functions of the organellar RecA proteins in chloroplast and mtDNA metabolism are not well defined, these proteins are likely candidates to mediate the multiple DNA strand exchange reaction associated with such a mechanism. The role of structures that could be rolling circle intermediates or their derivatives (Backert and Boerner, 2000; Backert et al., 1995; Backert et al., 1996a) in a T4-like replication scheme or in an alternative mode of replication remains to be resolved. In cultured cells of tobacco (Oldenburg and Bendich, 1996) and *M. polymorpha* (Oldenburg and Bendich, 1998), mtDNA synthesis begins within hours after transfer into fresh medium, reaches a maximum rate and subsequently declines to a low basal level that is maintained through stationary phase. It is possible that this change from high to low mtDNA replication activity is indicative of a switch from recombination-based to rolling circle mode. Alternatively, linear and circular DNA species might be replicated by different replication mechanisms.

4.5. Potential organellar DNA replication enzymes

Several organellar proteins that have been proposed to play a role in organellar DNA metabolism have been isolated and purified to varying extents. Although the physical properties and enzymatic activities of these putative DNA replication proteins are fairly well characterized, for most of them direct biochemical and genetic evidence for their purported *in vivo* roles is scarce or lacking altogether. The earlier prediction, based on biochemical and cytological evidence (reviewed in Heinhorst and Cannon, 1993; Heinhorst et al., 1990b), that all ctDNA replication proteins are nuclear-encoded, has been corroborated by the completely sequenced plastomes of several higher plants but does not completely hold true for some algae (Simpson and Stern, 2002).

4.5.1. DNA helicases

DNA helicases are molecular motors that, powered by nucleotide hydrolysis, translocate along DNA and unwind the double helix ahead of the advancing replication forks, and during recombination and repair (Lohman et al., 1998). Plastids of higher plants contain several DNA helicases that display DNA-dependent ATPase activity in the presence of a divalent cofactor (Tuteja, 2000). The partially purified enzyme from cultured soybean cells (Cannon and Heinhorst, 1990) and the highly purified helicase I from pea (Tuteja et al., 1996) are able to unwind fully annealed, blunt ended oligonucleotides, whereas helicase II requires fork-like structures (i.e. a single-stranded extension on an otherwise double stranded DNA). Both enzymes from pea unwind DNA duplex structures in the 3' to 5' direction (Tuteja and Phan, 1998a; b; Tuteja et al., 1996) and are inhibited by daunorubicin and nogalamycin. Helicase I is a homodimer of two 68 kDa subunits. It is not known whether helicase II, which consists of 78 kDa monomers, also exists in multimeric form. The structural and functional relationships between the three organellar DNA helicases are not known.

The plastid genomes of the red algae *Cyanidium caldarium*, *C. merolae* and *Porphyra purpurea*, the diatom *Odontella sinensis* and the cryptomonad *G. theta* encode a homolog of *E. coli* DNA replication and nucleotide excision repair helicase DnaB, which is thought to play similar roles in the algal chloroplast (Leipe et al., 2000; Odintsova and Yurina, 2003). Homologs of DnaB and of the yeast mtDNA helicase PIF1 are encoded by the nuclear genome of Arabidopsis (Elo et al., 2003; Leipe et al., 2000). Contrary to the algal DnaB, Arabidopsis DnaB protein contains an N-terminal DNA primase domain and is most closely related to phage-type helicases (Elo et al., 2003; Leipe et al., 2000). Although both Arabidopsis homologs are predicted to be organelle targeted, their intracellular localization and role(s) have yet to be determined.

4.5.2. DNA topoisomerases

DNA topoisomerases change the topological state of DNA by generating a nick or double strand break and passing one or both strands of duplex DNA, respectively, through the break(s) in the DNA backbone (Wang, J.C., 2002). The enzymes, which can be separated into two classes based on their mode of action and effect on substrate topology, play important roles in practically all aspects of DNA metabolism, where they resolve the topological strain that builds up during replication, transcription, repair and recombination, and decatenate newly replicated molecules.

A 115 kDa topoisomerase I activity was partially purified from spinach chloroplasts (Siedlecki et al., 1983). The enzyme relaxes negatively, but not positively, supercoiled DNA in the presence of Mg^{++}, is unaffected by ATP and brings about a unit change in topological linking number. An enzyme of nearly identical characteristics was purified to apparent homogeneity from pea chloroplasts and shown to be resistant to the topoisomerase II inhibitors nalidixic acid and novobiocin (Nielsen and Tewari, 1988). The enzyme, an apparent monomer of 112 kDa, is able to stimulate DNA synthesis by a chloroplast protein extract that is devoid of endogenous topoisomerase I activity. A considerably smaller (54 and 69 kDa) eukaryotic topoisomerase I activity that is able to relax negatively and positively supercoiled DNA and is sensitive to berenil and camptothecin has been isolated from chloroplasts of cauliflower and pea, respectively (Fukata et al., 1991; Mukherjee, S.K. et al., 1994), and both eukaryotic and prokaryotic topoisomerase I activities copurify with chloroplast nucleoids from soybean (S. Lewis, S. Heinhorst and G.C. Cannon, unpublished observations).

A DNA topoisomerase II activity that is resistant to novobiocin was identified in extracts from pea chloroplasts (Lam and Chua, 1987) and *C. reinhardtii* cells (Thompson and Mosig, 1985). The algal enzyme introduces supercoils into DNA in the presence of ATP. Although this topoisomerase II was purified from a cellular extract, its observed sensitivity *in vitro* to the *E. coli* DNA gyrase inhibitors novobiocin and nalidixic acid and the effect of these inhibitors on chloroplast transcript abundance (Thompson and Mosig, 1985; Thompson and Mosig, 1987) and DNA replication (Woelfle et al., 1993) *in vivo* suggest that the enzyme is derived from the chloroplast. Results from psoralen crosslinking assays that assess the extent of DNA supercoiling further support the existence of a DNA topoisomerase II activity in the algal chloroplast and establish a role for light in modulating enzyme activity (Thompson and Mosig, 1990) and organellar transcription rates (Salvador et al., 1998). Two polypeptides of 96 and 101 kDa, respectively, that crossreact with antiserum against yeast topoisomerase II and colocalize with DAPI-stained ctDNA were detected in wheat chloroplasts (Pyke et al., 1989). The abundance of the protein increases during the ctDNA replication phase just prior to organelle division (Marrison and Leech, 1992).

Topoisomerase activities in plant mitochondria have been documented, but the enzymes have not been highly purified. Wheat germ and maize mitochondria contain an activity that is able to relax negatively and positively supercoiled DNA in the presence of Mg^{++} (Daniell et al., 1995; Echeverria et al., 1986). The activity of the enzyme from wheat germ is not affected by ATP and is insensitive to novobiocin, ethidium bromide, berenil and nalidixic acid, suggesting that this enzyme is a eukaryotic type I topoisomerase, were it not for its insensitivity to camptothecin, a known inhibitor of class II topoisomerases. Meissner et al. (Meissner et al., 1992) described a topoisomerase I activity in a mitochondrial extract from *Chenopodium album* suspension cells that can relax negatively supercoiled DNA in the presence of Mg^{++}, is resistant to nalidixic acid and camptothecin but inhibited by ethidium bromide and by the DNA gyrase inhibitor novobiocin. These inhibition patterns are inconsistent with the current classification scheme for topoisomerases and either suggest that plant mitochondrial enzymes are fundamentally different than their counterparts from other sources or, a probably more likely scenario, reflect the combined properties of several different topoisomerases.

The observed inhibition of plastid and mtDNA synthesis in *C. merolae* by nalidixic acid (Itoh et al., 1997) suggests that a DNA gyrase activity participates in DNA replication in both organelles of the red alga. The Arabidopsis and rice nuclear genomes contain *gyrA* and *gyrB* genes for presumed organelle proteins that are related to the two subunits of prokaryotic DNA gyrase (Elo et al., 2003). A very recent analysis of the Arabidopsis mitochondrial proteome by a combination of liquid chromatography and tandem mass spectrometry confirmed the presence of both DNA gyrase subunits in mitochondria (Heazlewood et al., 2004). It is unknown at present whether this DNA gyrase is exclusively targeted to mitochondria or dually targeted to both organelles.

4.5.3. DNA primase

The synthesis of new DNA strands by DNA polymerases requires an existing 3' hydroxyl, which is generally provided by an RNA polymerase or DNA primase through the synthesis of a transcript or a short oligonucleotide primer, respectively.

A DNA primase activity copurifies with the presumed ctDNA polymerase from *C. reinhardtii* through several chromatography steps (Nie, Z. and Wu, 1999). The enzyme synthesizes short RNA primers on single stranded DNA that carries the cloned plastome *oriA*, and mapping of the initiation sites indicates a preference for the oligo(dA) tract that coincides with an area of DNA bending. Nielsen et al. (Nielsen et al., 1991) partially purified a 112 kDa protein from pea chloroplasts that preferentially synthesizes primers at the D-loop initiation sites of the cloned pea plastome origins of replication and, like the enzyme from *C. reinhardtii*, is able to stimulate *in vitro* DNA synthesis by primase-depleted chloroplast extracts. The primase can be distinguished from the plastome-encoded RNA polymerase by its resistance to tagetitoxin. However, since a plastid RNA polymerase activity that is not inhibited by tagetitoxin exists (Sakai et al., 1998b), it is not clear that the *in vitro* synthesized primers are the products of a *bona fide* DNA primase activity.

4.5.4. DNA polymerases

DNA polymerases synthesize a new DNA strand that is complementary to an existing template strand. Several enzymes with different functions in DNA replication and repair exist in prokaryotes and the eukaryotic nucleus. Only a single DNA polymerase of the γ class is believed to exist in chloroplasts and mitochondria, although an additional, β-type enzyme has been reported to reside in mitochondria (Kaguni, 2004; Kang, D. and Hamasaki, 2002).

Chloroplast DNA polymerases have been purified to various extents from different plant sources (Gaikwad et al., 2002; Sakai et al., 1999; Wu, H. and Liu, 1997; reviewed in Heinhorst and Cannon, 1993) and, with one exception, are considered to be γ-type DNA polymerases despite some unusual biochemical properties exhibited by some representatives. Consensus features are a moderate molecular mass of 70–116 kDa, a dependence of the polymerase activity on 100-125 mM KCl, sensitivity to the sulfhydryl agent N-ethylmaleimide (NEM), resistance to the nuclear DNA polymerase α inhibitor aphidicolin, and only moderate sensitivity to high concentrations of chain terminating dideoxynucleotides. The most highly purified enzymes are unable to utilize covalently closed circular and unprimed single stranded templates.

Since the nucleotide substitution rate of the plastome is considerably lower than that of plant nuclear DNA (Wolfe et al., 1987), a strong DNA polymerase editing function might be evoked that would ensure a high fidelity during DNA replication. Although exonuclease activity was detected in various ctDNA polymerase fractions, it is unclear at present whether the nuclease associates with the polymerase *in vivo*. The partially purified enzyme from the cyanelles of *C. paradoxa* is devoid of a 3' to 5' exonuclease but copurifies with a 5' to 3' exonuclease activity (White et al., 1997). Such an activity is apparently not associated with polymerases from higher plant sources. The enzyme from spinach and one form of DNA polymerase from soybean chloroplasts copurify with a 3' to 5' exonuclease activity (Bailey et al., 1995; Keim and Mosbaugh, 1991), although the gradual loss of nuclease activity during successive purification steps suggests that both enzymatic activities are located on different polypeptides. The spinach exonuclease prefers mismatched 3' termini, suggesting that it has a proofreading function (Keim and Mosbaugh, 1991). The fidelity of the soybean ctDNA polymerase, however, is unaffected by the associated 3' to 5' exonuclease activity (Bailey et al., 1995). Although the ctDNA polymerase from pea has been purified to apparent homogeneity (Gaikwad et al., 2002; McKown and Tewari, 1984), it is not clear whether it possesses 3' to 5' exonuclease activity. McKown and Tewari (McKown and Tewari, 1984) did not detect any nuclease activity in the homogenous enzyme fraction but a more recent study (Gaikwad et al., 2002) presents strong evidence for the presence of polymerase and 3' to 5' exonuclease activities on the same 70 kDa polymerase polypeptide. Nevertheless, the ctDNA polymerase exhibits only moderate fidelity with a five fold higher error rate (2.5×10^{-6} errors per base) than that of the Klenow fragment of *E. coli* DNA polymerase I. The processivity of the purified pea ctDNA polymerase on a singly primed M13 template is 1500-3000 nucleotides per catalytic cycle, a value that is considerably higher than the 80 and 260 nucleotides, respectively that are incorporated by two highly purified forms of the enzyme from cultured soybean cells (Bailey, 1997). The reason for this discrepancy is not clear, since the possibility that the processivity enhancing protein p43 is present in the rice enzyme preparation was excluded (Gaikwad et al., 2002). The p43 protein is a 43 kDa non-sequence specific DNA binding protein in pea leaf chloroplasts that specifically interacts with and stimulates the activity of the organellar DNA polymerase (Chen, W. et al., 1996) through its extensively Oarabinosylated N-terminal domain. At present, the role p43 plays in regulating DNA synthesis in the pea chloroplast is unknown.

Very few plant mtDNA polymerases have been described, and none of them has been purified to apparent homogeneity (Christophe et al., 1981; Daniell et al., 1995; Fukasawa and Chou, 1980; Heinhorst et al., 1990a; Meissner et al., 1993). Four of the five enzymes have characteristics that are similar to those of their plastid counterparts. The enzyme from *C. album* was reported to co-purify with a 3' to 5' exonuclease activity, but the strength of the potential protein interaction was not examined. The enzyme from wheat embryos seems to be different from the other three mitochondrial enzymes based on its significantly higher molecular weight (180,000 as opposed to 80,000-90,000) and its resistance to NEM (Christophe et al., 1981), but further work is needed to establish that the observed differences are not caused by contaminating activities in the enzyme fractions.

Although the first plant organellar DNA polymerase activity was detected in 1969 (reviewed in Heinhorst et al., 1990b), it has taken until 2003 to identify the nuclear genes in Arabidopsis that encode two highly homologous γ-type DNA polymerases (Elo et al., 2003). The protein product of a fusion between the *AtPolγ2* organelle targeting region and the GFP coding region is directed to the plastids of transgenic Arabidopsis, while the product of a similar reporter construct with *AtPolγ1* is targeted to plastids as well as mitochondria. This finding is in keeping with the identical physical and biochemical characteristics of chloroplast and mtDNA polymerases that have been demonstrated for several higher plants (Heinhorst et al., 1990a; Kimura et al., 2002; Sakai et al., 1999).

An interesting DNA polymerase gene was discovered in the rice genome, which encodes an enzyme with high homology in primary and predicted tertiary structure to *E. coli* DNA polymerase I (Kimura et al., 2002). The predicted 100 kDa enzyme carries a 60 kDa polymerase domain that, when expressed in *E. coli*, is active in the bacterium. The recombinant polymerase fragment differs from all other known ctDNA polymerases in its sensitivity to even moderate KCl concentrations and its resistance to NEM. The localization of this enzyme in plastids was not rigorously proven, but the protein was shown to copurify with isolated chloroplasts from young rice seedlings. The preferential expression of this enzyme in meristematic tissue suggests an involvement in proplastid DNA replication.

A recent mitochondrial proteome study detected a protein that is related to the B subunit of the nuclear replicative DNA polymerase α (Heazlewood et al., 2004). Since DNA polymerases are low abundance proteins and the mitochondria used in the study were shown to be essentially free of nuclear contamination, this finding raises the intriguing possibility that another DNA polymerase in addition to the γ-type enzyme might play a role in plant mtDNA metabolism, although only a single polymerizing activity was detected in biochemical studies (Christophe et al., 1981; Fukasawa and Chou, 1980; Heinhorst et al., 1990a; Meissner et al., 1993; Sakai et al., 1999). The plastome and mitochondrial genome of the algae *P. purpurea* and *O. danica*, and the linear plasmid of cytoplasmic male sterile maize mitochondria encode putative proteins that are related to DNA polymerases (Burger et al., 1999; Paillard et al., 1985), but the *P. purpurea* gene is almost certainly a pseudogene, and it is unclear at present if the other two homologs produce functional protein products that participate in replication and/or repair of organellar DNA.

4.6. Organelle DNA repair

Because of their aerobic metabolism, plastids and mitochondria are likely to suffer DNA damage induced by reactive oxygen species. In addition, plants cannot avoid exposure to ultraviolet (UV) radiation and the resulting cyclobutane pyrimidine dimers (CPDs) and pyrimidine (6,4) pyrimidone photoproducts (6,4 photoproducts). Despite the existence of several protective mechanisms (Britt, 1996), there is evidence that plants rely on DNA damage repair systems to cope with DNA lesions that, if not removed, would have detrimental effects on their viability (Stapleton et al., 1997). Although there is a general dearth of information concerning repair pathways in plant organelles, biochemical and genetic evidence has steadily been accumulating for the existence of repair pathways in organelles. Like organellar DNA replication proteins the putative chloroplast and mtDNA repair proteins are encoded by nuclear DNA in higher plants.

4.6.1. Photoreactivation

Irradiation of a *Euglena gracilis* culture with UV light leads to bleaching of the cells, degradation of ctDNA, and severely compromised cell viability (Nicolas et al., 1980). Post-irradiation exposure of the culture to visible light is able to reverse these detrimental effects, suggesting the existence of a photoreactivation and the absence of a dark repair pathway in the chloroplast of the alga. *Chlamydomonas* chloroplasts, on the other hand, appear to possess both light-dependent and -independent repair pathways (Small and Greimann, 1977). The *PHR2* gene from *Chlamydomonas* (Petersen, J.L. et al., 1999) encodes a class II DNA photolyase that is targeted to and active in both nucleus and chloroplast when overexpressed in transgenic *Chlamydomonas* (Petersen, J.G. and Small, 2001). For full activity, the photolyase requires the product of the *PHR1* gene, which may aid in FAD incorporation into PHR2 or in the proper folding of the photolyase (Petersen, J.G. and Small, 2001).

Whether higher plant chloroplasts and mitochondria are able to remove UV photoproducts from their respective genomes remains an open question. Cannon *et al.* (Cannon et al., 1995) irradiated cultured green soybean cells with UV light and, using a quantitative PCR assay, showed that the resulting plastome lesions are efficiently removed in the light, but not in the dark. By contrast, no evidence for repair of CPDs was found in plastids and mitochondria of light-or dark grown young Arabidopsis seedlings employing an assay that tests susceptibility of the damaged DNA to cleavage by T4 endonuclease V (Chen, J.-J. et al., 1996). Likewise, photorepair could not be detected in UVB-irradiated isolated leaves and chloroplasts of spinach, using monoclonal antibodies specific for CPDs and pyrimidine (6-4) pyrimidone photoproducts (Hada et al., 1998). Draper and Hays (Draper and Hays, 2000) followed replication and PCR amplification efficiency of nuclear and organellar DNA after UVB irradiation of cultured detached Arabidopsis leaves and saw evidence for efficient blue light-dependent removal of photoproducts from all three genomes over a period of several days. The apparent slow pace of light-dependent photoproduct repair in organellar DNA (Cannon et al., 1995; Draper and Hays, 2000) may have prevented detection of this process in some studies (Chen, J.-J. et al., 1996; Hada et al., 1998). Stapleton et al. (Stapleton et al., 1997), however, observed essentially identical kinetics of light-dependent photodamage removal from nuclear, chloroplast and mtDNA of irradiated *Z. mays* seedlings. The observed light-dependence of UV-induced lesion repair in plastids and mitochondria may simply reflect a requirement for energy (Cannon et al., 1995) and, instead of involving a photolyase activity, may proceed via a RecA-mediated homologous recombination mechanism (see below). This possibility is consistent with the lack of a gene in the Arabidopsis genome for an obviously organelle-targeted photolyase (Draper and Hays, 2000). Apparently no or very inefficient dark repair of UV-induced lesions occurs in organellar DNA of higher plants (Cannon et al., 1995; Chen, J.-J. et al., 1996; Draper and Hays, 2000; Stapleton et al., 1997).

4.6.2. RecA and other recombination proteins

Several lines of evidence support the existence of an active homologous recombination mechanism in both chloroplasts and mitochondria of higher plants (Cao et al., 1997; Khazi et al., 2003). A crucial enzyme in this process is RecA, whose strand transfer activity mediates exchange of genetic information and the repair of damaged DNA. A *recA* homolog was identified in Arabidopsis by hybridization with a cyanobacterial *recA* probe (Cerutti et al., 1992). The plant RecA preprotein is imported into isolated pea chloroplasts and processed to a 42 kDa product (Cao et al., 1997). Recombinant Arabidopsis RecA protein overexpressed in a *recA⁻ E. coli* displays strand transfer activity (Cao et al., 1997), and an analogous activity was detected with several different assays in stromal extracts from pea chloroplasts (Cerutti and Jagendorf, 1993). The RecA protein from pea and Arabidopsis is induced by exposure of cultured leaves or protoplasts to DNA damaging agents (Cao et al., 1997; Cerutti et al., 1993), suggesting a role of the protein in ctDNA damage repair.

A later search of the Arabidopsis genome database revealed the presence of several putative *recA* homologs, among them a gene whose product is predicted to be targeted to mitochondria (Khazi et al., 2003). The protein's presequence directs the uptake of a fused GFP reporter protein into isolated mitochondria, rather than chloroplasts, and recombinant Arabidopsis mitochondrial RecA protein can functionally complement the defect of a *recA⁻ E. coli* mutant (Khazi et al., 2003). It is not known how the steady state level of the mitochondrial RecA protein is regulated.

Based on these functional complementation studies, a role for the plant organellar RecA proteins in DNA damage repair can be inferred, which was experimentally shown for RecA from *Chlamydomonas* (Cerutti et al., 1995) Expression of dominant negative mutant RecA protein from *E. coli* in the algal chloroplast yields cells that are more sensitive to DNA damaging agents, repair alkali-labile lesions in ctDNA more slowly, and are not as active in recombination of ctDNA than wild type algae. Employing an elegant test for the repair of double strand breaks (DSB), Duerrenberger et al (Duerrenberger et al., 1996) presented evidence for DSB-induced recombinational repair in the algal chloroplast.

In light of the evidence supporting recombination-based replication mechanisms for plastid and mtDNA, the organellar RecA proteins are also likely to play crucial roles in organellar DNA replication (Backert and Boerner, 2000; Oldenburg and Bendich, 1996; 1998; 2001; 2004). However, although the Arabidopsis genome contains several *recA* homologs (Cerutti et al., 1992; Khazi et al., 2003), candidate genes related to *E. coli recBCD*, whose products participate in homologous recombination in prokaryotes, have not been identified in plants (Khazi et al., 2003).

In plant mitochondria, ectopic recombination mediates substoichiometric shifting, which gives rise to small, subgenomic mtDNA species of very low copy number (reviewed by Mackenzie and McIntosh, 1999). Apart from having profound effects on mitochondrial gene expression, this process also affects transmission of the mitochondrial genome. Although the nuclear genes *CHM* (*chloroplast mutator*) in Arabidopsis and *Fr* (*fertility restorer*) in common bean had been pinpointed as key players in this process (Janska et al., 1998; Martinez-Zapater et al., 1992; Sakamoto et al., 1996), *CHM* was only recently cloned, identified as a homolog of *E. coli MutS* and yeast *MSH1*, and shown to encode an apparent mitochondrial protein (Abdelnoor et al., 2003). Although CHM contains the expected ATPase and mismatch recognition and binding motifs, the protein lacks other features that are characteristic of MutS-like proteins. Considering that the Arabidopsis nuclear genome does not encode obvious organelle-targeted *recQ* candidates (Hartung et al., 2000) and that *chm* T-DNA insertion mutants do not accumulate mitochondrial point mutations, CHM's main function in plant mitochondria might be the maintenance of substoichiometric intermediates through an as yet ill-defined role in their replication or in ectopic recombination (Abdelnoor et al., 2003).

4.6.3. Uracil-DNA glycosylases

A uracil-DNA glycosylase was partially purified from chloroplasts and mitochondria of Z. mays, indicating that a base excision repair pathway exists in plant organelles (Bensen and Warner, 1987a; b). Both enzymes show a slight preference for single stranded over double stranded DNA and are sensitive to NEM. With estimated molecular weights of 18,000, the enzymes from maize are considerably smaller than the 42 kDa B subunit of glyceraldehyde-3-phosphate dehydrogenase from pea chloroplasts that possesses uracil-DNA glycosylase activity. This enzyme is different from the maize organellar enzymes in its response to changes in pH (Wang, X. et al., 1999). Elo et al. (Elo et al., 2003) reported an Arabidopsis gene encoding a putative uracil-DNA glycosylase that is predicted to be targeted to mitochondria, although its intracellular localization in plants has not been confirmed.

4.6.4. Other potential DNA repair proteins

Hays and coworkers tested an Arabidopsis expression library in *E. coli* DNA repair and recombination mutants for functional complementation of the mutant phenotypes and identified several *DRT* (DNA damage repair toleration) genes (Pang et al., 1992; 1993a; Pang et al., 1993b). The protein products of *DRT100*, *DRT 101*, *DRT111* and *DRT112* carry N-terminal extensions that predict targeting to chloroplasts, but intraorganellar localization of these proteins has not been verified. Although not closely related to RecA-like proteins in primary structure, DRT100 expressed in *recA$^-$ E. coli* mutants is able to significantly increase the cells' tolerance to DNA damage and to augment their homologous recombination activity, but is unable to induce the SOS response (Pang et al., 1992).

The DRT101 protein functionally complements the phenotypes of *E. coli uvrB* and *uvrC* and of *S. cerevisiae rad2* mutants, suggesting that it might function in nucleotide excision repair (Pang et al., 1993b). Transcription of *DRT100* and *DRT101* is stimulated by exposure of Arabidopsis plants to UV light and to DNA damaging agents (Pang et al., 1993b).

Recombinant DRT111 and DRT112 proteins increase the resistance of *E. coli ruvC$^-$* and *ruvC$^-$recG$^-$*, but not of *recA$^-$* mutants to DNA damage (Pang et al., 1993a). In addition, DRT111, but not DRT112 can complement the UV and mitomycin C sensitive phenotype of a *recG$^-$* mutant, and both proteins are able to restore recombination in *ruvC$^-$* single and *ruvC$^-$ recG$^-$* double mutants. Interestingly, the primary structure of DRT112 is highly homologous to that of plastocyanin, whereas DRT111 contains an RNA recognition motif, a glycine-rich G-patch and an SF45 motif that is similar to a motif in human splicing factor 45 and might indicate multiple intracellular roles of the protein. DRT111 has apparent orthologs in many other organisms, among them several apicomplexan parasites (Dendouga et al., 2002).

The product of the Arabidopsis *thi1* gene is able to complement several *E. coli* DNA repair mutants (Machado et al., 1996). Although the biochemical role the Thi1 protein plays in repair of or tolerance to plant organellar DNA damage is unknown, the protein appears to fulfill dual functions in plants and in yeast. The Arabidopsis protein is able to complement a yeast mutant that is deficient in the synthesis of the thiamine precursor thiazol and in mtDNA damage tolerance (Machado et al., 1997). Interestingly, in Arabidopsis Thi1 is targeted to both mitochondria and chloroplasts through the differential use of two in-frame translational start codons (Chabregas et al., 2003), suggesting that the protein might contribute to DNA damage tolerance in both organelles.

4.7. Summary and outlook

Although physiological studies of organellar DNA replication have quite a long history and have revealed basic trends for changes in DNA copy number in mitochondria and plastid during plant growth and development, the molecular players that regulate organellar DNA amplification are completely unknown. Likewise, although several putative DNA replication proteins have been characterized, proof for their *in vivo* role has been difficult to establish, since these enzymes tend to be of very low abundance, difficult to purify, and their activity is often unstable. Although the use of rapidly dividing plant cell lines has alleviated these problems to some extent, cultured cells do not replace *in planta* studies, particularly in questions of tissue- and developmental stage-specific regulatory mechanisms that govern organellar DNA replication. Answers to those questions will likely be provided by genetic and biochemical studies with loss- and gain-of-function mutants of the relevant genes. The evidence for dual targeting of some nucleic acid metabolism enzymes to both organelles is particularly intriguing in this context, since it raises interesting questions about the regulatory strategies the cell employs to coordinate nucleic acid metabolism in plastids and mitochondria. The finding that a DNA helicase (*E. coli* RecG ortholog) and FtsZ were recovered from a thioredoxin affinity column in a recent plastid proteome study that set out to identify previously unknown potential thioredoxin targets (Balmer et al., 2003) raises the interesting possibility that the plastid redox state might play a role in the regulation of organelle division, DNA replication and/or repair.

Another open question is that of the *in vivo* structure of the organellar genomes. Recent studies have challenged our existing view that mitochondrial and ctDNA exists in predominantly circular form and have indicated that plant organellar DNA is replicated by a more complex mechanism than had been assumed. These findings require a re-thinking of the existing paradigms, and the application of novel experimental approaches to elucidate key players and mechanisms that govern organellar genome replication in plants. Knowledge of the functional structure of organellar genomes, their *in vivo* replication origin(s) and mechanism(s) has important implications for ongoing efforts to devise more efficient organelle transformation strategies (see chapter 16 of this volume).

5. ACKNOWLEDGEMENTS

S.H. and G.C.C. gratefully acknowledge support for portions of their work from the National Science Foundation (MCB-0131269 and Mississippi EPSCoR program) and from the U.S. Department of Agriculture (2002-35301-12065).

6. REFERENCES

Abdelnoor, R. V., Yule, R., Elo, A., Christensen, A. C., Meyer-Gauen, G., & Mackenzie, S. (2003). Substoichiometric shifting in the plant mitochondrial genome is influenced by a gene homologous to *MutS*. *Proc. Natl. Acad. Sci. USA, 100*, 5968-5973.

Aki, T., & Adhya, S. (1997). Repressor induced site-specific binding of HU for transcriptional regulation. *Embo J., 16*, 3666-3674.

Alam, T. I., Kanki, T., Muta, T., Ukaji, K., Abe, Y., Nakayama, H., et al. (2003). Human mitochondrial DNA is packaged with TFAM. *Nucleic Acids Res., 31*, 1640-1645.

Arimura, S., & Tsutsumi, N. (2002). A dynamin-like protein (ADL2b), rather than FtsZ, is involved in Arabidopsis mitochondrial division. *Proc. Natl. Acad. Sci. USA, 99*, 5727-5731.

Backert, S., & Boerner, T. (2000). Phage T4-like intermediates of DNA replication and recognition in the mitochondria of the higher plant *Chenopodium album* (L.). *Curr. Genet., 37*, 304-314.

Backert, S., Doerfel, P., & Boerner, T. (1995). Investigation of plant organellar DNAs by pulsed-field gel electrophoresis. *Curr. Genet., 28*, 390-399.

Backert, S., Doerfel, P., Lurz, R., & Boerner, T. (1996a). Rolling-circle replication of mitochondrial DNA in the higher plant *Chenopodium album* (L.). *Mol. Cell Biol., 16*, 6285-6294.

Backert, S., Lurz, R., & Boerner, T. (1996b). Electron microscopic investigation of mitochondrial DNA from *Chenopodium album* (L.). *Curr. Genet., 29*, 427-436.

Backert, S., Nielsen, B. L., & Boerner, T. (1997). The mystery of the rings: structure and replication of mitochondrial genomes from higher plants. *Trends Plant Sci., 2*, 477-483.

Bailey, J. C. (1997). Studies of the chloroplast DNA polymerase: Association of exonucleases, processivity, fidelity and identification of the catalytically active polypeptide. The University of Southern Mississippi, Hattiesburg, MS, USA.

Bailey, J. C., Heinhorst, S., & Cannon, G. C. (1995). Accuracy of deoxynucleotide incorporation by soybean chloroplast DNA polymerases is independent of the presence of a 3' to 5' exonuclease. *Plant Physiol., 107*, 1277-1284.

Balmer, Y., Koller, A., del Val, G., Manieri, W., Schuermann, P., & Buchanan, B. (2003). Proteomics gives insight into the regulatory function of chloroplast thioredoxins. *Proc. Natl. Acad. Sci. USA, 100*, 370-375.

Bardel, J., Louwagie, M., Jaquinod, M., Jourdain, A., Luche, S., Rabilloud, T., et al. (2002). A survey of the plant mitochondrial proteome in relation to development. *Proteomics, 2*, 880-898.

Bateman, J. M., Perlman, P. S., & Butow, R. A. (2002). Mutational dissection of the mitochondrial DNA stability and amino acid biosynthetic functions. *Genetics, 161*, 1043-1052.

Baumgartner, B. J., & Mullet, J. E. (1991). Plastid DNA synthesis and nucleic acid binding proteins in developing barley chloroplasts. *J. Photochem. Photobiol. B, 11*, 203-218.

Baumgartner, B. J., Rapp, J. C., & Mullet, J. E. (1989). Plastid transcription activity and DNA copy number increases early in barley chloroplast development. *Plant Physiol., 89*, 1011-1018.

Bedinger, P., & Walbot, V. (1986). DNA synthesis in purified maize mitochondria. *Curr. Genet., 10*, 631-637.

Beech, P. L., Nheu, T., Schultz, T., Herbert, S., Lithgow, T., Gilson, P. R., & McFadden, G. I. (2000). Mitochondrial FtsZ in a chromophyte alga. *Science, 287*, 1276-1279.

Bendich, A. J. (1987). Why do chloroplasts and mitochondria contain so many copies of their genome? *Bioessays, 6*, 279-282.

Bendich, A. J., & Smith, S. B. (1990). Moving pictures and pulsed-field gel electrophoresis show linear DNA molecules from chloroplasts and mitochondria. *Curr. Genet., 17*, 421-425.

Bensen, R. J., & Warner, H. R. (1987a). The partial purification and characterization of nuclear and mitochondrial uracil-DNA glycosylase activities from *Zea mays* seedlings. *Plant Physiol., 83*, 149-154.

Bensen, R. J., & Warner, H. R. (1987b). Partial purification and characterization of uracil-DNA glycosylase activity from chloroplasts of *Zea mays* seedlings. *Plant Physiol., 84*, 1102-1106.

Bleazard, W., McCaffery, J. M., King, E. J., Bale, S., Mozdy, A., Tieu, Q., et al. (1999). The dynamin-related GTPase Dnm1 regulates mitochondrial fission in yeast. *Nat. Cell Biol., 1*, 298-304.

Boffey, S. A. (1985). The chloroplast division cycle and its relationship to the cell division cycle. *Soc. Exp. Bot. Sem. Ser., 26*, 233-246.

Boffey, S. A., Ellis, J. R., Sellden, G., & Leech, R. M. (1979). Chloroplast division and DNA synthesis in light-grown wheat leaves. *Plant Physiol., 64*, 502-505.

Boffey, S. A., & Leech, R. M. (1982). Chloroplast DNA levels and the control of chloroplast division in light-grown wheat leaves. *Plant Physiol., 69*, 1387-1391.

Bogenhagen, D. F., Wang, Y., Shen, E. L., & Kobayashi, R. (2003). Protein composition of mitochondrial DNA nucleoids in higher eukaryotes. *Mol. Cell. Proteomics, 2*, 1205-1216.

Bramhill, D., & Kornberg, A. (1988). A model for initiation at origins of DNA replication. *Cell, 54*, 915-918.

Briat, J.-F., Gigot, C., Laulhere, J. P., & Mache, R. (1982). Visualization of a spinach plastid transcriptionally active DNA-protein complex in a highly condensed structure. *Plant Physiol., 69*, 1205-1211.

Briat, J.-F., Letoffe, S., Mache, R., & Rouviere-Yaniv, J. (1984). Similarity between the bacterial histone-like protein HU and a protein from spinach chloroplasts. *FEBS Lett., 172*, 75-79.

Briat, J.-F., & Mache, R. (1980). Properties and characterization of a spinach chloroplast RNA polymerase isolated from a transcriptionally active DNA-protein complex. *Eur. J. Biochem., 111*, 503-509.

Britt, A. B. (1996). DNA damage and repair in plants. *Annu. Rev. Plant. Physiol. Plant. Mol. Biol., 47*, 75-100.

Burger, G., Saint-Louis, D., Gray, M. W., & Lang, B. F. (1999). Complete sequence of the mitochondrial DNA of the red alga *Porphyra purpurea*: Cyanobacterial introns and shared ancestry of red and green algae. *Plant Cell, 11*, 1675-1694.

Calikowski, T. T., Meulia, T., & Meier, I. (2003). A proteomic study of the Arabidopsis nuclear matrix. J *Cell Biochem., 90*, 361-378.

Cannon, G. C., Hedrick, L. A., and Heinhorst, S. (1995). Repair mechanisms of UV induced DNA-damage in soybean chloroplasts. *Plant Mol. Biol., 29*, 1167-1177.

Cannon, G. C., & Heinhorst, S. (1990). Partial purification and characterization of a DNA helicase from chloroplasts of *Glycine max*. *Plant Mol. Biol.*, 15, 457-464.

Cannon, G. C., Heinhorst, S., Siedlecki, J., & Weissbach, A. (1985). Chloroplast DNA synthesis in light and dark grown cultured *Nicotiana tabacum* cells as determined by molecular hybridization. *Plant Cell Rep., 4*, 41-45.

Cannon, G. C., Heinhorst, S., & Weissbach, A. (1986a). Organellar DNA synthesis in permeabilized soybean cells. *Plant Mol. Biol., 7*, 331-341.

Cannon, G. C., Heinhorst, S., & Weissbach, A. (1986b). Plastid DNA content in a cultured soybean line capable of photoautotrophic growth. *Plant Physiol., 80*, 601-603.

Cannon, G. C., Ward, L. N., Case, C. I., & Heinhorst, S. (1999). The 68 kDa DNA compacting nucleoid protein from soybean chloroplasts inhibits DNA synthesis *in vitro*. *Plant Mol. Biol., 39*, 835-845.

Cao, J., Combs, C., & Jagendorf, A. T. (1997). The chloroplast-located homolog of bacterial DNA recombinase. *Plant Cell Physiol., 38*, 1319-1325.

Carlson, J. E., Brown, G. L., & Kemble, R. J. (1986). *In organello* mitochondrial DNA and RNA synthesis in fertile and cytoplasmic male sterile *Zea mays* L. *Curr. Genet., 11*, 151-160.

Catley, M. A., Bowman, C. M., Bayliss, M. W., & Gale, M. D. (1987). The pattern of amyloplast DNA accumulation during wheat endosperm development. *Planta, 171*, 416-421.

Cerutti, H., Ibrahim, H. Z., & Jagendorf, A. T. (1993). treatment of pea (*Pisum sativum* L.) protoplasts with DNA-damaging agents induces a 39-kilodalton chloroplast protein immunologically related to *Escherichia coli* RecA. *Plant Physiol., 102*, 155-163.

Cerutti, H., & Jagendorf, A. T. (1993). DNA strand-transfer activity in pea (*Pisum sativum* L.) chloroplasts. *Plant Physiol., 102*, 145-153.

Cerutti, H., Johnson, A. M., Boynton, J. E., & Gillham, N. W. (1995). Inhibition of chloroplast DNA recombination and repair by dominant negative mutants of *Escherichia coli* RecA. *Mol. Cell Biol., 15*, 3003-3011.

Cerutti, H., Osman, M., Grandoni, P., & Jagendorf, A. T. (1992). A homolog of *Escherichia coli* RecA protein in plastids of higher plants. *Proc. Natl. Acad. Sci. USA, 89*, 8068-8072.

Chabregas, S. M., Luche, D. D., Van Sluys, M.-A., Menck, C. F. M., & Silva-Filho, M. C. (2003). Differential usage of two in-frame translational start codons regulates subcellular localization of *Arabidopsis thaliana* THI1. *J. Cell Sci., 116*, 285-291.

Chang, C.-C., & Stern, D. B. (1999). DNA-binding factors assemble in a sequence-specific manner on the maize mitochondrial *atpA* promoter. *Curr. Genet., 35*, 506-511.

Chen, J.-J., Jiang, C.-Z., & Britt, A. B. (1996). Little or no repair of cyclobutyl pyrimidine dimers is observed in the organellar genomes of the young Arabidopsis seedling. *Plant Physiol., 111*, 19-25.

Chen, W., Gaikwad, A., Mukherjee, S. K., Choudhary, N. R., Kumar, D., & Tewari, K. K. (1996). A 43 kDa DNA binding protein from the pea chloroplast interacts with and stimulates the cognate DNA polymerase. *Nucleic Acids Res., 24*, 3953-3961.

Cheng, M. Y., Hartl, H.-U., Marton, J., Pollock, R. A., Kalousek, F., Neupert, W., et al. (1989). Mitochondrial heat shock protein hsp60 is essential for assembly of proteins imported into yeast mitochondria. *Nature, 337*, 620-625.

Chi-Ham, C. L., Keaton, M. A., Cannon, G. C., & Heinhorst, S. (2000). Soybean chloroplast nucleoid proteins and their interactions with DNA. Paper presented at the 6th International Congress of Plant Molecular Biology, Quebec, Canada.

Chi-Ham, C. L., Keaton, M. A., Cannon, G. C., & Heinhorst, S. (2002). The DNA-compacting protein DCP68 from soybean chloroplasts is ferredoxin:sulfite reductase and co-localizes with the organellar nucleoid. *Plant Mol. Biol., 49*, 621-631.

Chiu, W.-L., & Sears, B. B. (1992). Electron microscopic localization of replication origins in *Oenothera* chloroplast DNA. *Mol. Gen. Genet., 232*, 33-39.

Christophe, L., Tarrago-Litvak, L., Castroviejo, M., & Litvak, S. (1981). Mitochondrial DNA polymerase from wheat embryos. *Plant Sci. Lett., 21*, 181-192.

Colletti, K. S., Tattersall, E. A., Pyke, K. A., Froelich, J. E., Stokes, K. D., & Osteryoung, K. W. (2000). A homologue of the bacterial cell division site-determining factor MinD mediates placement of the chloroplast division apparatus. *Curr. Biol., 10*, 507-516.

Contamine, V., & Picard, M. (2000). Maintenance and integrity of the mitochondrial genome: a plethora of nuclear genes in the budding yeast. *Microbiol. Mol. Biol. Rev., 64*, 281-315.

Copley, S. D. (2003). Enzymes with extra talents: moonlighting functions and catalytic promiscuity. *Curr. Opin. Chem. Biol., 7*, 265-272.

Daniell, H., Zheng, D., & Nielsen, B. L. (1995). Isolation and characterization of an *in vitro* DNA replication system from maize mitochondria. *Biochem. Biophys. Res. Commun., 208*, 287-294.

Danino, D., & Hinshaw, J. E. (2001). Dynamin family of mechanoenzymes. *Curr. Opin. Cell Biol., 13*, 454-460.

de Haas, J. M., Hille, J., Kors, F., van der Meer, B., Kool, A. J., Folkerts, O., & Nijkamp, H. J. (1991). Two potential *Petunia hybrida* mitochondrial DNA replication origins show structural and *in vitro* functional homology with the animal mitochondrial DNA heavy and light strand replication origins. *Curr. Genet., 20*, 503-513.

De Paepe, R., Forchioni, A., Chetrit, P., & Vedel, F. (1993). Specific mitochondrial proteins in pollen: presence of an additional ATP synthase □ subunit. *Proc. Natl. Acad. Sci. USA, 90*, 5934-5938.

Dendouga, N., Callebaut, I., & Tomavo, S. (2002). A novel DNA repair enzyme containing RNA recognition, G-patch and specific splicing factor 45-like motifs in the protozoan parasite *Toxoplasma gondii. Eur. J. Biochem., 269*, 3393-3401.

Deng, X.-W., Wing, R. A., & Gruissem, W. (1989). The chloroplast genome exists in multimeric forms. *Proc. Natl. Acad. Sci. USA, 86*, 4156-4160.

Diffley, J. F., & Stillman, B. (1991). A close relative of the nuclear, chromosomal high-mobility group protein HMG1 in yeast mitochondria. *Proc. Natl. Acad. Sci. USA, 88*, 7864-7868.

Dinkins, R., Reddy, M. S. S., Leng, M., & Collins, G. B. (2001). Overexpression of the *Arabidopsis thaliana MinD1* gene alters chloroplast size and number in transgenic tobacco plants. *Planta, 214*, 180-188.

Douglas, S., & Penny, S. L. (1999). The plastid genome of the cryptophyte *Guillardia theta*: Complete sequence and conserved synteny groups confirm its common ancestry with red algae. *J. Mol. Evol., 48*, 236-244.

Draper, C. K., & Hays, J. B. (2000). Replication of chloroplast, mitochondrial and nuclear DNA during growth of unirradiated and UVB-irradiated Arabidopsis leaves. *Plant J., 23*, 255-265.

Dri, A.-M., Moreau, P., & Rouviere-Yaniv, J. (1992). Involvement of the histone-like proteins OsmZ and HU in homologous recombination. *Gene, 120*, 11-16.

Dubell, A. N., & Mullet, J. E. (1995). Continuous far-red light activates plastid DNA synthesis in pea leaves but not full cell enlargement or an increase in plastid number per cell. *Plant Physiol., 109*, 95-103.

Duerrenberger, F., Thompson, A. J., Herrin, D. L., & Rochaix, J.-D. (1996). Double strand break-induced recombination in *Chlamydomonas reinhardtii* chloroplasts. *Nucleic Acids Res., 24*, 3323-3331.

Echeverria, M., Martin, M. T., Ricard, B., & Litvak, S. (1986). A DNA topoisomerase type I from wheat embryo mitochondria. *Plant Mol. Biol., 6*, 417-427.

Ellis, J. R., Jellings, A. J., & Leech, R. M. (1983). Nuclear DNA content and the control of chloroplast replication in wheat leaves. *Planta, 157*, 376-380.

Elo, A., Lyznik, A., Gonzalez, D. O., Kachman, S. D., & Mackenzie, S. A. (2003). Nuclear genes that encode mitochondrial proteins for DNA and RNA metabolism are clustered in the Arabidopsis genome. *Plant Cell, 15*, 1619-1631.

El-Shami, M., El-Kafadi, S., Falconet, D., & Lerbs-Mache, S. (2002). Cell cycle-dependent modulation of *FtsZ* expression in synchronized tobacco BY2 cells. *Mol. Genet. Genom., 267*, 254-261.

Erickson, H. P. (1995). FtsZ, a prokaryotic homolog of tubulin? *Cell, 80*, 367-370.

Erickson, H. P., Taylor, D. W., Taylor, K. A., & Bramhill, D. (1996). Bacterial cell division protein FtsZ assembles into protofilament sheets and minirings, structural homologs of tubulin polymers. *Proc. Natl. Acad. Sci. USA, 93*, 519-523.

Farr, C. L., Wang, Y., & Kaguni, L. S. (1999). Functional interaction of mitochondrial DNA polymerase and single-stranded DNA-binding protein. *J. Biol. Chem., 274*, 14779-14785.

Ferro, M., Salvi, D., Brugiere, S., Miras, S., Kowalski, S., Louwagie, M., Garin, J., Joyard, J., and Rolland, N. (2003). Proteomics of the chloroplast envelope membranes from *Arabidopsis thaliana. Mol. Cell. Proteomics, 2*, 325-345.

Ferro, M., Salvi, D., Riviere-Rolland, H., Vermat, T., Seigneurin-Berny, D., Grunwald, D., et al. (2002). Integral membrane proteins of the chloroplast envelope: Identification and subcellular localization of new transporters. *Proc. Natl. Acad. Sci. USA, 99*, 11487-11492.

Fey, J., Vermel, M., Grienenberger, J., Marechal-Drouard, L., & Gualberto, J. M. (1999). Characterization of a plant mitochondrial active chromosome. *FEBS Lett., 458*, 124-128.

Fujie, M., Kuroiwa, H., Kawano, S., & Kuroiwa, T. (1993a). Behavior of organelles and their nucleoids in the root apical meristem of *Arabidopsis thaliana* (L.). *Planta, 189*, 443-452.

Fujie, M., Kuroiwa, H., Kawano, S., Mutoh, S., & Kuroiwa, T. (1994). Behavior of organelles and their nucleoids in the shoot apical meristem during leaf development in *Arabidopsis thaliana* L. *Planta, 194*, 395-405.

Fujie, M., Kuroiwa, H., Suzuki, T., Kawano, S., & Kuroiwa, T. (1993b). Organelle DNA synthesis in the quiescent centre of *Arabidopsis thaliana* (Col.). *J. Exp. Bot., 44*, 689-693.

Fujiwara, M., & Yoshida, S. (2001). Chloroplast targeting of chloroplast division FtsZ2 proteins in *Arabidopsis*. *Biochem. Biophys. Res. Commun., 287*, 462-467.

Fukasawa, H., & Chou, M.-Y. (1980). Mitochondrial DNA polymerase from cauliflower inflorescence. *Japan J. Genet., 55*, 441-445.

Fukata, H., Mochida, A., Maruyama, N., & Fukasawa, H. (1991). Chloroplast DNA topoisomerase I from cauliflower. *J. Biochem., 109*, 127-131.

Fulgosi, H., Gerdes, L., Westphal, S., Glockmann, C., & Soll, J. (2002). Cell and chloroplast division requires ARTEMIS. *Proc. Natl. Acad. Sci. USA, 99*, 11501-11506.

Gaikwad, A., Babbarwal, V., Pant, V., & Mukherjee, S. K. (2000). Pea chloroplast FtsZ can from multimers and correct the thermosensitive defect of an *Escherichia coli ftsZ* mutant. *Mol. Gen. Genet., 263*, 213-221.

Gaikwad, A., Hop, D. V., & Mukherjee, S. K. (2002). A 70-kDa chloroplast DNA polymerase from pea (*Pisum sativum*) that shows high processivity and displays moderate fidelity. *Mol. Gen. Genom., 267*, 45-56.

Gao, H., Kadirjan-Kalbach, D., Froehlich, J. E., & Osteryoung, K. W. (2003). ARC5, a cytosolic dynamin-like protein from plants, is part of the chloroplast division machinery. *Proc. Natl. Acad. Sci. USA, 100*, 4328-4333.

Garrido, N., Griparic, L., Jokitalo, E., Wartiovaara, J., van der Bliek, A., & Spelbrink, J. N. (2003). Composition and dynamics of human mitochondrial nucleoids. *Mol. Biol. Cell, 14*, 1583-1596.

Gilson, P. R., & Beech, P. L. (2001). Cell division protein FtsZ: Running rings around bacteria, chloroplasts and mitochondria. *Res. Microbiol., 152*, 3-10.

Gilson, P. R., Yu, X.-C., Hereld, D., Barth, C., Savage, A., Kiefel, B. R., et al. (2003). Two Dictyostelium orthologs of the prokaryotic cell division protein FtsZ localize to mitochondria and are required for the maintenance of normal mitochondrial morphology. *Eukaryotic Cell, 2*, 1315-1326.

Gomez, S. M., Nishio, J. N., Faull, K. F., & Whitelegge, J. P. (2002). The chloroplast grana proteome defined by intact mass measurements from liquid chromatography mass spectrometry. *Mol Cell Proteomics, 1*, 46-59.

Grasser, K. D., Ritt, C., Krieg, M., Fernandez, S., Alonso, J. C., & Grimm, R. (1997). The recombinant product of the *Cryptomonas* □ plastid gene *hlpA* is an architectural HU-like protein that promotes the assembly of complex nucleoprotein structures. *Eur. J. Biochem., 249*, 70-76.

Griffin, K. L., Anderson, O. R., Gastrich, M. D., Lewis, J. D., Lin, G., Schuster, W., et al. (2001). Plant growth in elevated CO_2 alters mitochondrial number and chloroplast fine structure. *Proc. Natl. Acad. Sci. USA, 98*, 2473-2478.

Hada, M., Hashimoto, T., Nikaido, O., & Shin, M. (1998). UVB-induced DNA damage and its photorepair in nuclei and chloroplasts of *Spinacea oleracea* L. *Photochem. Photobiol., 68*, 319-322.

Hansmann, P., Falk, H., Ronai, K., and Sitte, P. (1985). Structure, composition, and distribution of plastid nucleoids in *Narcissus pseudonarcissus*. *Planta, 164*, 459-472.

Hanson, M. R., & Koehler, R. H. (2001). GFP imaging: methodology and application to investigate cellular compartmentation in plants. *J. Exp. Bot., 52*, 529-539.

Harper, J. D. I., & John, P. C. L. (1986). Coordination of division events in the *Chlamydomonas* cell cycle. *Protoplasma, 131*, 118-130.

Hartung, F., Plchova, H., & Puchta, H. (2000). Molecular characterization of RecQ homologues in *Arabidopsis thaliana*. *Nucl. Acids Res., 28*, 4275-4282.

Hashimoto, M., Imhoff, B., Ali, M. M., & Kow, Y. W. (2003). HU protein of *Escherichia coli* has a role in the repair of closely opposed lesions in DNA. *J. Biol. Chem., 278*, 28501-28507.

Hatzack, F., Dombrowski, S., Brennicke, A., & Binder, S. (1998). Characterization of DNA-binding proteins from pea mitochondria. *Plant Physiol., 116*, 519-528.

Heazlewood, J. L., Howell, K. A., Whelan, J., & Millar, A. H. (2003). Towards and analysis of the rice mitochondrial proteome. *Plant Physiol., 132*, 230-242.
Heazlewood, J. L., Tonti-Filippini, J. S., Gout, A. M., Day, D. A., Whelan, J., & Millar, A. H. (2004). Experimental analysis of the Arabidopsis mitochondrial proteome highlights signaling and regulatory components, provides assessment of targeting prediction programs, and indicates plant-specific mitochondrial proteins. *Plant Cell, 16*, 241-256.
Hedrick, L. A., Heinhorst, S., White, M. A., &Cannon, G. C. (1993). Analysis of soybean chloroplast DNA replication by two-dimensional gel electrophoresis. *Plant Mol. Biol.*, 23, 779-792.
Heinhorst, S., Cannon, G., & Weissbach, A. (1985a). Plastid and nuclear DNA synthesis are not coupled in suspension cells of *Nicotiana tabacum*. *Plant Mol. Biol., 4*, 3-12.
Heinhorst, S., & Cannon, G. C. (1993). DNA replication in chloroplasts. *J. Cell Sci.*, 104, 1-9.
Heinhorst, S., Cannon, G. C., and Weissbach, A. (1985b). Chloroplast DNA synthesis during the cell cycle in cultured cells of *Nicotiana tabacum*: inhibition by nalidixic acid and hydroxyurea. *Arch. Biochem. Biophys., 239*, 475-479.
Heinhorst, S., Cannon, G. C., and Weissbach, A. (1990a). Chloroplast and mitochondrial DNA polymerases from cultured soybean cells. *Plant Physiol., 92*, 939-945.
Heinhorst, S., Cannon, G. C., and Weissbach, A. (1990b). DNA replication in higher plants. In K. W. Adolph (Ed.), *Chromosomes: Eukaryotic, Prokaryotic, and Viral* (Vol. II, pp. 129-152). Boca Raton, FL: CRC Press.
Henschke, R. B., and Nuecken, E. J. (1989). Proteins from organelles of higher plants with homology to the bacterial DNA-binding protein HU. *J. Plant Physiol., 134*, 110-112.
Huang, J., Struck, F., Matzinger, D. F., and Levings, C. S. I. (1994). Flower-enhanced expression of a nuclear-encoded mitochondrial respiratory protein is associated with changes in mitochondrion number. *Plant Cell, 6*, 439-448.
Huisman, O., Faelen, M., Girard, D., Jaffe, A., Toussaint, A., and Rouviere-Yaniv, J. (1989). Multiple defects in *Escherichia coli* mutants lacking HU protein. *J. Bacteriol., 171*, 3704-3712.
Ikeda, T. M., and Gray, M. W. (1999). Characterization of a DNA-binding protein implicated in transcription in wheat mitochondria. *Mol. Cell Biol., 19*, 8113-8122.
Itoh, R., Fujiwara, M., Nagata, N., & Yoshida, S. (2001a). A chloroplast protein homologous to the eubacterial topological specificity factor MinE plays a role in chloroplast division. *Plant Physiol., 127*, 1644-1655.
Itoh, R., Fujiwara, M., & Yoshida, S. (2001b). Kinesin-related proteins with a mitochondrial targeting signal. *Plant Physiol., 127*, 724-726.
Itoh, R., Takahashi, H., Toda, K., Kuroiwa, H., & Kuroiwa, T. (1996). Aphidicolin uncouples the chloroplast division cycle from the mitotic cycle of the unicellular alga *Cyanidioschyzon merolae*. *Eur. J. Cell Biol., 71*, 303-310.
Itoh, R., Takahashi, H., Toda, K., Kuroiwa, H., & Kuroiwa, T. (1997). DNA gyrase involvement in chloroplast-nucleoid division in *Cyanidioschyzon merolae*. *Eur. J. Cell Biol., 73*, 252-258.
Jacobs, C., & Shapiro, L. (1999). Bacterial cell division: A moveable feast. *Proc. Natl. Acad. Sci. USA, 96*, 5891-5893.
Jaffe, A., Vinella, D., & D'Ari, R. (1997). The *Escherichia coli* histone-like protein HU affects DNS initiation, chromosome partitioning via MukB, and cell division via MinCDE. *J. Bacteriol., 179*, 3493-3499.
Janska, H., Sarria, R., Woloszynska, M., & Arrieta-Montiel, M. (1998). Stoichiometric shifts in the common bean mitochondrial genome leading to male sterility and spontaneous reversion to fertility. *Plant Cell, 10*, 1163-1180.
Jeffery, C. J. (1999). Moonlighting proteins. *Trends Biochem. Sci., 24*, 8-11.
Jeffery, C. J. (2003). Multifunctional proteins: examples of gene sharing. *Ann. Med., 35*, 28-35.
Jeong, S. Y., Park, Y.-I., Sugh, K. H., Raven, J. A., Yoo, O. J., & Liu, J. R. (2002). A large population of small chloroplasts in tobacco leaf cells allows more effective chloroplast movement than a few enlarged chloroplasts. *Plant Physiol., 129*, 112-121.
Jeong, S. Y., Rose, A., & Meier, I. (2003). MFP1 is a thylakoid-associated, nucleoid-binding protein with a coiled coil structure. *Nucl. Acids Res., 31*, 5175-5185.
Jin, J. B., Bae, H., Kim, S. J., Jin, Y. H., Goh, C.-H., Kim, D. H., et al. (2003). The Arabidopsis dynamin-like proteins ADL1C and ADL1E play a critical role in mitochondrial morphogenesis. *Plant Cell, 15*, 2357-2369.
Kaguni, L. S. (2004). DNA polymerase, the mitochondrial replicase. Annual Review of Biochemistry, 73.

Kanamaru, K., Fujiwara, M., Kim, M., Nagashima, A., Nakazato, E., Tanaka, K., & Takahashi, H. (2000). Chloroplast targeting, distribution and transcriptional fluctuation of AtMinD1, a eubacteria-type factor critical for chloroplast division. *Plant Cell Physiol., 41*, 1119-1128.

Kang, B.-H., Busse, J. S., & Bednarek, S. Y. (2003). Members of the Arabidopsis dynamin-like gene family, ADL1, are essential for plant cytokinesis and polarized cell growth. *Plant Cell, 15*, 899-913.

Kang, D., & Hamasaki, N. (2002). Maintenance of mitochondrial DNA integrity: Repair and degradation. *Curr. Genet., 41*, 311-322.

Kaufman, B. A., Kolesar, J. E., Perlman, P. S., & Butow, R. A. (2003). A function for the mitochondrial chaperonin Hsp60 in the structure and transmission of mitochondrial DNA nucleoids in *Saccharomyces cerevisiae*. *J. Cell Biol., 163*, 457-461.

Kaufman, B. A., Newman, S. M., Hallberg, R. L., Slaughter, C. A., Perlman, P. S., & Butow, R. A. (2000). *In organello* formaldehyde crosslinking of proteins to mtDNA: Identification of bifunctional proteins. *Proc. Natl. Acad. Sci. USA, 97*, 7772-7777.

Keim, C. A., & Mosbaugh, D. W. (1991). Identification and characterization of a 3' to 5' exonuclease associated with spinach chloroplast DNA polymerase. *Biochemistry, 30*, 11109-11118.

Khazi, F. R., Edmondson, A. C., & Nielsen, B. L. (2003). An Arabidopsis homologue of bacterial *recA* that complements an *E. coli recA* deletion is targeted to plant mitochondria. *Mol. Gen. Genom., 269*, 454-463.

Kiessling, J., Kruse, S., Rensing, S. A., Harter, K., Decker, E. L., & Reski, R. (2000). Visualization of a cytoskeleton-like FtsZ network in chloroplasts. *J. Cell. Biol., 151*, 945-950.

Kimura, S., Uchiyama, Y., Kasai, N., Namekawa, S., Saotome, A., Ueda, T., Ando, T., Ishibashi, T., Oshige, M., Furukawa, T., Yamamoto, T., Hashimoto, J., & Sakaguchi, K. (2002). A novel DNA polymerase homologous to *Escherichia coli* DNA polymerase I from a higher plant, rice (*Oryza sativa* L.). *Nucl. Acids Res., 30*, 1585-1592.

Kobayashi, T., Takahara, M., Miyagishima, S.-Y., Kuroiwa, H., Sasaki, N., Ohta, N., Matsuzaki, M., & Kuroiwa, T. (2002). Detection and localization of a chloroplast-encoded HU-like protein that organizes chloroplast nucleoids. *Plant Cell, 14*, 1579-1589.

Koksharova, O. A., & Wolk, C. P. (2002). A novel gene that bears a DnaJ motif influences cyanobacterial cell division. *J. Bacteriol., 184*, 5524-5528.

Koller, B., & Delius, H. (1982). Origin of replication in chloroplast DNA of *Euglena gracilis* located close to the region of variable size. *EMBO J., 1*, 995-998.

Kolodner, R. D., & Tewari, K. K. (1975a). Chloroplast DNA from higher plants replicates by both the Cairns and the rolling circle mechanism. *Nature, 256*, 708-711.

Kolodner, R. D., & Tewari, K. K. (1975b). Presence of displacement loops in the covalently closed circular chloroplast deoxyribonucleic acid from higher plants. *J. Biol. Chem., 250*, 8840-8847.

Kornberg, A., & Baker, T. A. (1992). DNA replication (2nd ed.). New York: W.H. Freeman and Co.

Kourtz, L., Pollett, J., & Ko, K. (1996). Nucleotide sequence of *Brassica napus* histone H3 homolog. *Plant Physiol., 111*, 652.

Krause, K., & Krupinska, K. (2000). Molecular and functional properties of highly purified transcriptionally active chromosomes from spinach chloroplasts. *Physiol. Plant., 109*, 188-195.

Kruft, V., Eubel, H., Jaensch, L., Werhahn, W., & Braun, H.-P. (2001). Proteomic approach to identify novel mitochondrial proteins in Arabidopsis. *Plant Physiol, 127*, 1694-1710.

Kunnimalaiyaan, M., & Nielsen, B. L. (1997a). Chloroplast DNA replication: Mechanism, enzymes and replication origins. J. *Plant Biochem. Biotech., 6*, 1-7.

Kunnimalaiyaan, M., & Nielsen, B. L. (1997b). Fine mapping of replication origins (*ori*A and *ori*B) in *Nicotiana tabacum* chloroplast DNA. *Nucl. Acids Res., 25*, 3681-3686.

Kunnimalaiyaan, M., Shi, F., & Nielsen, B. L. (1997). Analysis of the tobacco chloroplast DNA replication origin (*ori*B) downstream of the 23S rRNA gene. *J. Mol. Biol., 268*, 273-283.

Kuroiwa, T. (1991). The replication, differentiation, and inheritance of plastids with emphasis on the concept of organelle nuclei. *Int. Rev. Cytol., 128*, 1-62.

Kuroiwa, T., Kuroiwa, H., Takahashi, H., Toda, K., & Itoh, R. (1998). The division apparatus of plastids and mitochondria. *Int. Rev. Cytol., 181*, 1-41.

Kuroiwa, T., Suzuki, T., Ogawa, K., & Kawano, S. (1981). The chloroplast nucleus: Distribution, number, size, and shape, and a model for the multiplication of the chloroplast genome during chloroplast development. *Plant Cell Physiol., 22*, 381-396.

Kwok, E. V., & Hanson, M. R. (2003). Microfilaments and microtubules control the morphology and movement of non-green plastids and stromules in *Nicotiana tabacum*. *Plant J., 35*, 16-26.

Labrousse, A. M., Zappaterra, M. D., Rube, D. A., & van der Bliek, A. M. (1999). *C. elegans* dynamin-related protein DRP-1 controls severing the mitochondrial outer membrane. *Mol. Cell, 4*, 815-826.

Lam, E., & Chua, N. H. (1987). Chloroplast DNA gyrase and *in vitro* regulation of transcription by template topology and novobiocin. *Plant Mol. Biol., 8*, 415-424.

Lamppa, G. K., & Bendich, A. J. (1979). Changes in chloroplast DNA levels during development of pea (*Pisum sativum*). *Plant Physiol., 64*, 126-130.

Lamppa, G. K., & Bendich, A. J. (1984). Changes in mitochondrial DNA levels during development of pea (*Pisum sativum* L.). *Planta, 162*, 463-468.

Lau, K. W., Ren, J., & Wu, M. (2000). Redox modulation of chloroplast DNA replication in *Chlamydomonas reinhardtii*. *Antioxid. Redox Signal., 2*, 529-535.

Law, A. E., Mullineaux, C. W., Hirst, E. M. A., Saldanha, J., & Wilson, R. J. M. (2000). bacterial orthologues indicate the malarial plastid gene *ycf24* is essential. *Protist, 151*, 317-327.

Lawrence, M. E., & Possingham, J. V. (1986). Microspectrofluorometric measurement of chloroplast DNA in dividing and expanding leaf cells of *Spinacea oleracea*. *Plant Physiol., 81*, 708-710.

Leipe, D. A., Aravind, L., Grishin, N. V., & Koonin, E. V. (2000). The bacterial replicative helicase DnaB evolved from a RecA duplication. *Genome Res., 10*, 5-16.

Lemieux, C., Otis, C., & Turmel, M. (2000). Ancestral chloroplast genome in *Mesostigma viride* reveals an early branch of green plant evolution. *Nature, 403*, 649-652.

Lemmon, M. A., Ferguson, K. M., & Abrams, C. S. (2002). Pleckstrin homology domains and the cytoskeleton. *FEBS Lett., 513*, 71-76.

Leustek, T., Martin, M., Bick, J., & Davies, J. P. (2000). Pathways and regulation of sulfur metabolism revealed through molecular and genetics studies. *Annu. Rev. Plant Physiol. Plant Mol. Biol., 51*, 141-165.

Leustek, T., & Saito, K. (1999). Sulfate transport and assimilation in plants. *Plant Physiol., 120*, 637-644.

Lewis, P. J. (2001). Bacterial chromosome segregation. *Microbiol., 147*, 519-526.

Li, K., & Williams, R. S. (1997). Tetramerization and single-stranded DNA binding properties of native and mutated forms of murine mitochondrial single-stranded DNA-binding proteins. *J. Biol. Chem., 272*, 8686-8694.

Li, S., & Waters, R. (1998). *Escherichia coli* strains lacking HU are UV sensitive due to a role for HU in homologous recombination. *J. Bacteriol., 180*, 3750-3756.

Lilly, J. W., Harvey, M. J., Jackson, S. A., & Jiang, J. (2001). Cytogenomic analyses reveal the structural plasticity of the chloroplast genome in higher plants. *Plant Cell, 13*, 245-254.

Lindbeck, A. G., & Rose, R. J. (1990). Thylakoid-bound chloroplast DNA from spinach is enriched for replication forks. *Biochem. Biophys. Res. Commun., 172*, 204-210.

Lindbeck, A. G., Rose, R. J., Lawrence, M. E., & Possingham, J. V. (1987). The role of chloroplast membranes in the location of chloroplast DNA during the greening of *Phaseolus vulgaris* etioplasts. *Protoplasma, 139*, 92-99.

Lindbeck, A. G., Rose, R. J., Lawrence, M. E., & Possingham, J. V. (1989). The chloroplast nucleoids of the bundle sheath and mesophyll cells of *Zea mays*. *Physiol. Plant., 75*, 7-12.

Liu, J. W., & Rose, R. J. (1992). The spinach chloroplast chromosome is bound to the thylakoid membrane in the region of the inverted repeat. *Biochem. Biophys. Res. Commun., 184*, 993-1000.

Logan, D. C. (2003). Mitochondrial dynamics. *New Phytol., 160*, 463-478.

Logan, D. C., & Leaver, C. J. (2000). Mitochondria-targeted GFP highlights the heterogeneity of mitochondrial shape, size and movement within living plant cells. *J. Exp. Bot., 51*, 865-871.

Logan, D. C., Millar, A. H., Sweetlove, L. J., Hill, S. A., & Leaver, C. J. (2001). Mitochondrial biogenesis during germination in maize embryos. *Plant Physiol., 125*, 662-672.

Logan, D. C., Scott, I., & Tobin, A. K. (2003). The genetic control of plant mitochondrial morphology and dynamics. *Plant J., 36*, 500-509.

Lohman, T. M., Thorn, K., & Vale, R. D. (1998). Staying on track: Common features of DNA helicases and microtubule motors. *Cell, 93*, 9-12.

Lonosky, P. M., Zhang, X., Honovar, V. G., Dobbs, D. L., Fu, A., & Rodermel, S. R. (2004). A proteomic analysis of maize chloroplast biogenesis. *Plant Physiol., 134*, 560-574.

Lu, Z., Kunnimalaiyaan, M., & Nielsen, B. L. (1996). Characterization of replication origins flanking the 23S rRNA gene in tobacco chloroplast DNA. *Plant Mol. Biol., 32*, 693-706.

Ma, X., Ehrhardt, D. W., & Margolin, W. (1996). Colocalization of cell division proteins FtsZ and FtsA to cytoskeletal structures in living *Escherichia coli* cells by using green fluorescent protein. *Proc. Natl. Acad. Sci. USA, 93*, 12998-13003.

Ma, X., & Margolin, W. (1999). Genetic and functional analyses of the conserved C-terminal core domain of *Escherichia coli* FtsZ. *J. Bacteriol., 181*, 7531-7544.

MacAlpine, D. M., Perlman, P. S., & Butow, R. A. (1998). The high mobility group protein Abf2p influences the level of yeast mitochondrial DNA recombination intermediates *in vivo*. *Proc. Natl. Acad. Sci. USA, 95*, 6739-6743.

MacAlpine, D. M., Perlman, P. S., & Butow, R. A. (2000). The numbers of individual mitochondrial DNA molecules and mitochondrial DNA nucleoids in yeast are co-regulated by the general amino acid control pathway. *EMBO J., 19*, 767-775.

Machado, C. R., Costa de Oliveira, R., Boiteux, S., Praekelt, U. M., Meacock, O. A., & Menck, C. F. M. (1996). Thi1, a thiamine biosynthetic gene in *Arabidopsis thaliana*, complements bacterial defects in DNA repair. *Plant Mol. Biol., 31*, 585-593.

Machado, C. R., Praekelt, U. M., Costa de Oliveira, R., Barbosa, A. C. C., Byrne, K. L., Meacock, O. A., & Menck, C. F. M. (1997). Dual role for the yeast *THI4* gene in thiamine biosynthesis and DNA damage tolerance. *J. Mol. Biol., 273*, 114-121.

Mackenzie, S., & McIntosh, L. (1999). Higher plant mitochondria. *Plant Cell, 11*, 571-585.

Maier, D., Farr, C. L., Poeck, B., Alahari, A., Vogel, M., Fischer, S., Kaguni, L. S., and Schneuwly, S. (2001). Mitochondrial single-stranded DNA-binding protein is required for mitochondrial DNA replication and development in *Drosophila melanogaster*. *Mol. Biol. Cell, 12*, 821-830.

Maple, J., Chua, N. H., & Moeller, S. (2003). The topological specificity factor AtMinE1 is essential for correct plastid division site placement in *Arabidopsis*. *Plant J., 31*, 269-277.

Marrison, J. L., & Leech, R. M. (1992). Co-immunolocalization of topoisomerase II and chloroplast DNA in developing, dividing and mature wheat chloroplasts. *Plant J., 2*, 783-790.

Marrison, J. L., Rutherford, S. M., Robertson, E. J., Lister, C., Dean, C., & Leech, R. M. (1999). The distinctive roles of five different *ARC* genes in the chloroplast division process in Arabidopsis. *Plant J., 18*, 651-662.

Marston, A. L., & Errington, J. (1999). Dynamic movement of the ParA-like Soj protein of *B. subtilis* and its dual role in nucleoid organization and developmental regulation. *Mol. Cell, 4*, 673-682.

Martinez-Zapater, J. M., Gil, P., Capel, J., & Somerville, C. E. (1992). Mutations at the Arabidopsis *CHM* locus promote rearrangements of the mitochondrial genome. *Plant Cell, 4*, 889-899.

McAndrew, R. S., Froehlich, J. E., Vitha, S., Stokes, K. D., & Osteryoung, K. W. (2001). Colocalization of plastid division proteins in the chloroplast stromal compartment establishes a new functional relationship between FtsZ1 and FtsZ2 in higher plants. *Plant Physiol., 127*, 1656-1666.

McFadden, G. I., & Ralph, S. A. (2003). Dynamin: The endosymbiosis ring of power? *Proc. Natl. Acad. Sci. USA, 100*, 3557-3559.

McKown, R. L., & Tewari, K. K. (1984). Purification and properties of a pea chloroplast DNA polymerase. *Proc. Natl. Acad. Sci. USA, 81*, 2354-2358.

Meeker, R., Nielsen, B., &Tewari, K. K. (1988). Localization of replication origins in pea chloroplast DNA. *Mol. Cell. Biol., 8*, 1216-1223.

Megraw, T., & Chae, C.-B. (1993). Functional complementarity between the HMG1-like yeast mitochondrial histone HM and the bacterial histone-like protein HU. *J. Biol. Chem., 268*, 12758-12763.

Meissner, K., Doerfel, P., & Boerner, T. (1992). Topoisomerase activity in mitochondrial lysates of a higher plant (*Chenopodium album* L.). *Biochem. Internatl., 27*, 1119-1125.

Meissner, K., Heinhorst, S., Cannon, G. C., & Boerner, T. (1993). Purification and characterization of a □-like DNA polymerase from *Chenopodium album* L. *Nucl. Acids Res., 21*, 4893-4899.

Mikhailov, V. S., & Bogenhagen, D. F. (1996). Effects of Xenopus laevis mitochondrial single-stranded DNA-binding protein on primer-template binding and 3'-->5' exonuclease activity of DNA polymerase □ *J. Biol. Chem., 271*, 18939-18946.

Millar, A. H., Sweetlove, L. J., Giege, P., & Leaver, C. J. (2001). Analysis of the Arabidopsis mitochondrial proteome. *Plant Physiol., 127*, 1711-1727.

Miyagishima, S.-Y., Nishida, K., & Kuroiwa, T. (2003a). An evolutionary puzzle: chloroplast and mitochondrial division rings. *Trends Plant Sci., 8*, 432-438.

Miyagishima, S.-Y., Nishida, K., Mori, T., Matzusaki, M., Higashiyama, T., Kuroiwa, H., & Kuroiwa, T. (2003b). A plant-specific dynamin-related protein forms a ring at the chloroplast division site. *Plant Cell, 15*, 655-665.

Miyagishima, S.-Y., Takahara, M., & Kuroiwa, T. (2001). Novel filaments 5 nm in diameter constitute the cytosolic ring of the plastid division apparatus. *Plant Cell, 13*, 707-721.

Miyamura, S., Kuroiwa, T., & Nagata, T. (1990). Multiplication and differentiation of plastid nucleoids during development of chloroplasts and etioplasts from proplastids in *Triticum aestivum*. *Plant Cell Physiol., 31*, 597-602.

Miyamura, S., Nagata, T., & Kuroiwa, T. (1986). Quantitative fluorescence microscopy on dynamic changes of plastid nucleoids during wheat development. *Protoplasma, 133*, 66-72.

Mori, T., Kuroiwa, H., Takahara, M., Miyagishima, S.-Y., & Kuroiwa, T. (2001). Visualization of an FtsZ ring in chloroplasts of *Lilium longiflorum* leaves. *Plant Cell Physiol., 42*, 555-559.

Muehlbauer, S. K., Loessl, A., Tzekova, L., Zou, Z., & Koop, H.-U. (2002). Functional analysis of plastid DNA replication origins in tobacco by targeted inactivation. *Plant J., 32*, 175-184.

Mukherjee, A., & Lutkenhaus, J. (1998). Dynamic assembly of FtsZ regulated by GTP hydrolysis. *EMBO J., 17*, 462-469.

Mukherjee, S. K., Reddy, M. K., Kumar, D., & Tewari, K. K. (1994). Purification and characterization of a eukaryotic type I topoisomerase from pea chloroplast. *J. Biol. Chem., 269*, 3793-3801.

Murakami, S., Kondo, Y., Nakano, T., & Sato, F. (2000). Protease activity of CND41, a chloroplast nucleoid DNA-binding protein, isolated from cultured tobacco cells. *FEBS Lett., 468*, 15-18.

Nakano, T., Murakami, S., Shoji, T., Yoshida, S., Yamada, Y., & Sato, F. (1997). A novel protein with DNA binding activity from tobacco chloroplast nucleoids. *Plant Cell, 9*, 1673-1682.

Nakano, T., Nagata, N., Kimura, T., Sekimoto, M., Kawaide, H., Murakami, S., Kaneko, Y., Matsushima, H., Kamiya, Y., Sato, F., & Yoshida, S. (2003). CND41, a chloroplast nucleoid protein that regulates plastid development, causes reduced gibberellin content and dwarfism in tobacco. *Physiol. Plant., 117*, 130-136.

Nakano, T., Sato, F., & Yamada, Y. (1993). Analysis of nucleoid-proteins in tobacco chloroplasts. *Plant Cell Physiol., 34*, 873-880.

Nebenfuehr, A., Frohlick, J. A., & Staehelin, L. A. (2000). Redistribution of Golgi stacks and other organelles during mitosis and cytokinesis in plant cells. *Plant Physiol., 124*, 135-151.

Nemoto, Y., Kawano, S., Kondoh, K., Nagata, T., & Kuroiwa, T. (1990). Studies on plastid-nuclei (nucleoids) in *Nicotiana tabacum* L. III. Isolation of chloroplast-nuclei from mesophyll protoplasts and identification of chloroplast DNA-binding proteins. *Plant Cell Physiol., 31*, 767-776.

Nemoto, Y., Kawano, S., Nagata, T., & Kuroiwa, T. (1991). Studies on plastid-nuclei (nucleoids) in *Nicotiana tabacum* L. IV. Association of chloroplast-DNA with proteins at several specific sites in isolated chloroplast-nuclei. *Plant Cell Physiol., 32*, 131-141.

Nemoto, Y., Kawano, S., Nakamura, S., Mita, T., Nagata, T., & Kuroiwa, T. (1988). Studies on plastid-nuclei (nucleoids) in *Nicotiana tabacum* L. I. Isolation of proplastid-nuclei from cultured cells and identification of proplastid-nuclear proteins. *Plant Cell Physiol., 29*, 167-177.

Nemoto, Y., Nagata, T., & Kuroiwa, T. (1989). Studies on plastid-nuclei (nucleoids) in *Nicotiana tabacum* L. II. Disassembly and reassembly of proplastid-nuclei isolated from cultured cells. *Plant Cell Physiol., 30*, 445-454.

Nerozzi, A. M., & Coleman, A. W. (1997). Localization of plastid DNA replication on a nucleoid structure. *Am. J. Bot., 84*, 1028-1041.

Newman, S. M., Zelenaya-Troitskaya, O., Perlman, P. S., & Butow, R. A. (1996). Analysis of mitochondrial DNA nucleoids in wild-type and a mutant strain of *Saccharomyces cerevisiae* that lacks the mitochondrial HMG box protein Abf2p. *Nucl. Acids Res., 24*, 386-393.

Nicolas, P., Hussein, Y., Heizmann, P., & Nigon, V. (1980). Comparative studies of chloroplastic and nuclear DNA repair abilities after ultraviolet irradiation of *Euglena gracilis*. *Mol. Gen. Genet., 178*, 567-572.

Nie, Z., & Wu, M. (1999). The functional role of a DNA primase in chloroplast DNA replication in *Chlamydomonas reinhardtii*. *Arch. Biochem. Biophys., 369*, 174-180.

Nie, Z. Q., Chang, D. Y., & Wu, M. (1987). Protein-DNA interaction within one cloned chloroplast DNA replication origin of *Chlamydomonas*. *Mol. Gen. Genet., 209*, 265-269.

Nielsen, B. L., Lu, Z., & Tewari, K. K. (1993). Characterization of the pea chloroplast DNA *OriA* region. *Plasmid, 30*, 197-211.

Nielsen, B. L., Rajasekhar, V. K., & Tewari, K. K. (1991). Pea chloroplast DNA primase: characterization and role in initiation of replication. *Plant Mol. Biol., 15*, 1019-1034.

Nielsen, B. L., & Tewari, K. K. (1988). pea chloroplast topoisomerase I: Purification, characterization, and role in replication. *Plant Mol. Biol., 11*, 3-14.

Nishida, K., Takahara, M., Miyagishima, S.-Y., Kuroiwa, H., Matsuzaki, M., & Kuroiwa, T. (2003). Dynamic Recruitment of dynamin for final mitochondrial severance in a primitive red alga. *Proc. Natl. Acad. Sci. USA, 100*, 2146-2151.

Odintsova, M. S., & Yurina, N. P. (2003). Plastid genomes of higher plants and algae: Structure and functions. *Mol. Biol., 37*, 649-662.

Oldenburg, D. J., & Bendich, A. J. (1996). Size and structure of replicating mitochondrial DNA in cultured tobacco cells. *Plant Cell, 8*, 447-461.

Oldenburg, D. J., & Bendich, A. J. (1998). The structure of mitochondrial DNA from the liverwort, *Marchantia polymorpha*. *J. Mol. Biol., 276*, 745-758.

Oldenburg, D. J., & Bendich, A. J. (2001). Mitochondrial DNA from the liverwort *Marchantia polymorpha*: Circularly permutated linear molecules, head-to-tail concatemers, and a 5' protein. *J. Mol. Biol., 310*, 549-562.

Oldenburg, D. J., & Bendich, A. J. (2004). Most chloroplast DNA of maize seedlings in linear molecules with defined ends and branched forms. *J. Mol. Biol., 335*, 953-970.

Oleskina, Y. P., Yurina, N. P., Mel'nik, S. M., Belkina, G. G., &Odintsova, M. S. (2001). DNA-binding proteins and structural characteristics of chloroplast nucleoids. Russ. *J. Plant Physiol., 48*, 487-492.

Oleskina, Y. P., Yurina, N. P., Odintsova, T. I., Egorov, T. A., Otto, A., Wittmann-Liebold, B., & Odintsova, M. S. (1999). Nucleoid proteins of pea chloroplasts: detection of a protein homologous to ribosomal protein. Biochem. *Mol. Biol. Int., 47*, 757-763.

Osteryoung, K. W. (2000). Organelle fission: crossing the evolutionary divide. *Plant Physiol., 123*, 1213-1216.

Osteryoung, K. W., & Nunnari, J. (2003). The division of endosymbiotic organelles. *Science, 302*, 1698-1704.

Osteryoung, K. W., Stokes, K. D., Rutherford, S. M., Percival, A. L., & Lee, W. Y. (1998). Chloroplast division in higher plants requires members of two functionally divergent gene families with homology to bacterial *ftsZ*. *Plant Cell, 10*, 1991-2004.

Osteryoung, K. W., & Vierling, E. (1995). Conserved cell and organelle division. *Nature, 376*, 473-474.

Paillard, M., Sederoff, R. R., & Levings, C. S. I. (1985). Nucleic acid sequence of the S-1 mitochondrial DNA from the S cytoplasm of maize. *EMBO J., 4*, 1125-1128.

Pang, Q., Hays, J. B., & Rajagopal, I. (1992). A plant cDNA that partially complements *Escherichia coli recA* mutations predicts a polypeptide not strongly homologous to RecA proteins. *Proc. Natl. Acad. Sci. USA, 89*, 8073-8077.

Pang, Q., Hays, J. B., & Rajagopal, I. (1993a). Two cDNA clones from the plant *Arabidopsis thaliana* that partially restore recombination proficiency and DNA-damage resistance to *E. coli* mutants lacking recombination-intermediate-resolution activities. *Nucleic Acids Res., 21*, 1647-1653.

Pang, Q., Hays, J. B., Rajagopal, I., & Schaefer, T. (1993b). Selection of Arabidopsis cDNAs that partially correct phenotypes of *Escherichia coli* DNA-damage-sensitive mutants and analysis of two plant cDNAs that appear to express UV-specific dark repair activities. *Plant Mol. Biol., 22*, 411-426.

Pavco, P. A., & Van Tuyle, G. C. (1985). Purification and General Properties of the DNA-binding protein (P16) from rat liver mitochondria. *J. Cell Biol., 100*, 258-264.

Peltier, J. B., Friso, G., Kalume, D. E., Roepstorff, P., Nilsson, F., Adamska, I., & van Wijk, K. J. (2000). Proteomics of the chloroplast: systematic identification and targeting analysis of lumenal and peripheral thylakoid proteins. *Plant Cell, 12*, 319-341.

Petersen, J. G., & Small, G. D. (2001). A gene required for the novel activation of a class II DNA photolyase in *Chlamydomonas*. Nu*cl. Acids Res., 29*, 4472-4481.

Petersen, J. L., Lang, D. W., & Small, G. D. (1999). Cloning and characterization of a class II DNA photolyase from *Chlamydomonas*. *Plant Mol. Biol., 40*, 1063-1070.

Phillips, J. R., Hilbricht, T., Salamini, F., & Bartels, D. (2002). A novel abscissic acid- and dehydration-responsive gene family from the resurrection plant *Craterostigma plantagineum* encodes a plastid-targeted protein with DNA-binding activity. *Planta, 215*, 258-266.

Possingham, J. V., & Smith, J. W. (1972). Factors affecting chloroplast replication. *J. Exp. Bot., 23*, 1050-1059.

Pyke, K. A. (1997). The genetic control of plastid division in higher plants. Am. J. Bot., 84, 1017-1027.

Pyke, K. A. (1999). Plastid division and development. *Plant Cell, 11*, 549-556.

Pyke, K. A., & Leech, R. M. (1992). Nuclear mutations radically alter chloroplast division and expansion in *Arabidopsis thaliana*. *Plant Physiol., 99*, 1005-1008.

Pyke, K. A., and Leech, R. M. (1994). A genetic analysis of chloroplast division and expansion in *Arabidopsis thaliana*. *Plant Physiol., 104*, 201-207.

Pyke, K. A., Marrison, J. L., & Leech, R. M. (1989). Evidence for a type II topoisomerase in wheat chloroplasts. *FEBS Lett., 242*, 305-308.

Ravel-Chapuis, P., Heizmann, P., & Nigon, V. (1982). Electron microscopic localization of the replication origin of *Euglena gracilis* chloroplast DNA. *Nature, 300*, 78-81.

Reddy, M. K., Choudhury, N. R., Kumar, D., Mukherjee, S. K., & Tewari, K. K. (1994). Characterization and mode of in vitro replication of pea chloroplast *OriA* sequences. *Eur. J. Biochem., 220*, 933-941.

Reiss, T., & Link, G. (1985). Characterization of transcriptionally active DNA-protein complexes from chloroplasts and etioplasts of mustard (*Sinapis alba* L.). *Eur. J. Biochem., 148*, 207-212.

Robertson, E. J., Rutherford, S. M., & Leech, R. M. (1996). Characterization of chloroplast division using the Arabidopsis mutant *arc5*. *Plant Physiol., 112*, 149-159.

Robertson, E. J., Williams, M., Harwood, J. L., Lindsay, J. G., Leaver, C. J., & Leech, R. M. (1995). Mitochondria increase three-fold and mitochondrial proteins and lipid change dramatically in postmeristematic cells in young wheat leaves grown in elevated CO_2. *Plant Physiol, 108*, 469-474.

Rose, R. J., Cran, D. G., &Possingham, J. V. (1975). Changes in DNA synthesis during cell growth and chloroplast replication in greening spinach leaf disks. *J. Cell Sci., 17*, 27-41.

Rose, R. J., & Possingham, J. V. (1976). The localization of [^{3}H]thymidine incorporation in the DNA of replicating spinach chloroplasts by electron-microscope autoradiography. *J. Cell Sci., 20*, 341-355.

Rothfield, L. I., & Justice, S. S. (1997). Bacterial cell division: The cycle of the ring. *Cell, 88*, 581-584.

Rushlow, K. E., Orozco, E. M. J., Lipper, C., & Hallick, R. B. (1980). Selective *in vitro* transcription of *Euglena* chloroplast ribosomal RNA genes by transcriptionally active chromosomes. *J. Biol. Chem., 255*, 3786-3799.

Sakai, A. (2001). *In vitro* transcription/DNA synthesis using isolated organelle-nuclei: Application to the analysis of the mechanisms that regulate organelle genome function. *J. Plant Res., 114*, 199-211.

Sakai, A., Suzuki, T., & Kuroiwa, T. (1998a). Comparative analysis of plastid gene expression in tobacco chloroplasts and proplastids: relationship between transcription and transcript accumulation. *Plant Cell Physiol., 39*, 581.

Sakai, A., Suzuki, T., Miyazawa, Y., & Kuroiwa, T. (1998b). Simultaneous isolation of cell-nuclei, plastid-nuclei and mitochondrial-nuclei from cultured tobacco cells; comparative analysis of their transcriptional activities *in vitro*. *Plant Sci., 133*, 17-31.

Sakai, A., Suzuki, T., Nagata, N., Sasaki, N., Miyazawa, Y., Saito, C., et al. (1999). Comparative analysis of DNA synthesis activity in plastid-nuclei and mitochondrial-nuclei simultaneously isolated from cultured tobacco cells. *Plant Sci., 140*, 11-24.

Sakamoto, W., Kondo, H., Murata, M., & Motoyoshi, F. (1996). Altered mitochondrial gene expression in a maternal distorted leaf mutant of Arabidopsis induced by *chloroplast mutator*. *Plant Cell, 8*, 1377-1390.

Salvador, M. L., Klein, U., &Bogorad, L. (1998). Endogenous fluctuations of DNA topology in the chloroplast of *Chlamydomonas reinhardtii*. *Mol.Cell Biol., 18*, 7235-7242.

Sandri-Goldin, R. M. (2003). Replication of the herpes simplex virus genome: Does it really go around in circles? *Proc. Natl. Acad. Sci. USA, 100*, 7428-7429.

Sato, N., Albrieux, C., Joyard, J., Douce, R., & Kuroiwa, T. (1993). Detection and characterization of a plastid envelope DNA-binding protein which may anchor plastid nucleoids. *EMBO J., 12*, 555-561.

Sato, N., Misumi, O., Shinada, Y., Sasaki, M., & Yoine, M. (1997). Dynamics of localization and protein composition of plastid nucleoids in light-grown pea seedlings. *Protoplasma, 200*, 163-173.

Sato, N., &Ohta, N. (2001). DNA-binding specificity and dimerization of the DNA-binding domain of the PEND protein in the chloroplast envelope membrane. *Nucleic Acids Res., 29*, 2244-2250.

Sato, N., Rolland, N., Block, M. A., & Joyard, J. (1999). Do plastid envelope membranes play a role in the expression of the plastid genome? *Biochimie, 81*, 619-629.

Sato, S., Nakamura, Y., Kaneko, T., Azamizu, E., & Tabata, S. (1999). Complete structure of the chloroplast genome of *Arabidopsis thaliana*. *DNA Res., 6*, 283-290.

Schaefer, S., Krolzik, S., Romanov, G. A., & Schmuelling, T. (2000). Cytokinin-regulated transcripts in tobacco cell culture. *Plant Growth Reg., 32*, 307-313.

Schroeder, W. P., & Kieselbach, T. (2003). Update on chloroplast proteomics. *Photosynth Res., 78*, 181-193.

Schubert, M., Petersson, U. A., Haas, B. J., Funk, C., Schroeder, W. P., & Kieselbach, T. (2002). Proteome map of the chloroplast lumen of *Arabidopsis thaliana*. *J. Biol. Chem., 277*, 8354-8365.

Scissum-Gunn, K. D., Gandhi, M., Backert, S., & Nielsen, B. L. (1998). Separation of different conformations of plant mitochondrial DNA molecules by filed inversion gel electrophoresis. Plant *Mol. Biol. Rep., 16*, 219-229.

Scott, N. S., & Possingham, J. V. (1980). Chloroplast DNA in expanding spinach leaves. *J. Exp. Bot., 31*, 1081-1092.

Scott, N. S & Possingham, J. V. (1983). Changes in chloroplast DNA levels during growth of spinach leaves. *J. Exp. Bot., 34*, 1756-1767.

Sekine, K., Hase, T., & Sato, N. (2002). Reversible DNA compaction by sulfite reductase regulates transcriptional activity of chloroplast nucleoids. *J. Biol. Chem., 277*, 24399-24404.

Sheahan, M. B., Rose, R. J., & McCurdy, D. W. (2004). Organelle inheritance in plant cell division: The actin cytoskeleton is required for unbiased inheritance of chloroplasts, mitochondria and endoplasmic reticulum in dividing protoplasts. *Plant J., 37*, 379-390.

Shimada, H., Koizumi, M., Kuroki, K., Mariko, M., Fujimoto, H., Masuda, T., et al. (2003). ARC3, a regulation factor of chloroplast division, is a junction point of prokaryotic and eukaryotic regulatory system. Paper presented at the Annual Meeting of the American Society of Plant Biologists, Honolulu, Hawaii.

Siedlecki, J., Zimmerman, W., & Weissbach, A. (1983). Characterization of a prokaryotic topoisomerase I activity in chloroplast extracts from spinach. *Nucl. Acids Res., 11*, 1523-1536.

Simpson, C. L., & Stern, D. B. (2002). The treasure trove of algal chloroplast genomes. Surprises in architecture and gene content, and their functional implications. *Plant Physiol., 129*, 957-966.

Small, G. D., & Greimann, C. S. (1977). Photoreactivation and dark repair of ultraviolet light-induced pyrimidine dimers in chloroplast DNA. *Nucl. Acids Res., 4*, 2893-2902.

Sodmergen Kawano, S., Tano, S., & Kuroiwa, T. (1989). Preferential digestion of chloroplast nuclei (nucleoids) during senescence of the coleoptile of *Oryza sativa*. *Protoplasma, 152*, 65-68.

Stapleton, A. E., Thornber, C. S., & Walbot, V. (1997). UV-B component of sunlight causes measurable damage in flied-grown maize (*Zea mays* L.): Developmental and cellular heterogeneity of damage and repair. *Plant Cell Environ., 20*, 279-290.

Stickens, D., & Verbelen, J.-P. (1996). Spatial structure of mitochondria and ER denotes changes in cell physiology of cultured tobacco protoplasts. *Plant J., 9*, 85-92.

Stokes, K. D., McAndrew, R. S., Figueroa, R., Vitha, S., & Osteryoung, K. W. (2000). Chloroplast division and morphology are differentially affected by overexpression of *FtsZ1* and *FtsZ2* genes in Arabidopsis. *Plant Physiol., 124*, 1668-1677.

Stokes, K. D., & Osteryoung, K. W. (2003). Early divergence of the *FtsZ1* and *FtsZ2* plastid division gene families in photosynthetic eukaryotes. *Gene, 320*, 97-108.

Strepp, R., Scholz, S., Kruse, S., Speth, V., & Reski, R. (1998). Plant nuclear gene knockout reveals a role in plastid division for the homolog of the bacterial cell division protein FtsZ, and ancestral tubulin. *Proc. Natl. Acad. Sci. USA, 95*, 4368-4373.

Suck, R., Zeltz, P., Falk, J., Acker, A., Koessel, H., & Krupinska, K. (1996). Transcriptionally active chromosomes (TACs) of barley chloroplasts contain the □-subunit of plastome-encoded RNA polymerase. *Curr. Genet., 30*, 515-521.

Sun, Q., & Margolin, W. (2001). Influence of the nucleoid on placement of FtsZ and MinE rings in *Escherichia coli*. *J. Bacteriol., 183*, 1413-1422.

Suzuki, T., Kawano, S., Sakai, A., Fujie, M., Kuroiwa, H., Nakamura, H., & Kuroiwa, T. (1992). Preferential mitochondrial and plastid DNA synthesis before multiple cell divisions in *Nicotiana tabacum*. *J. Cell Sci., 103*, 831-837.

Szeto, T. H., Rowland, S. L., Rothfield, L. I., & King, G. F. (2002). Membrane localization of MinD is mediated by a C-terminal motif that is conserved across eubacteria, archaea and chloroplasts. *Proc. Natl. Acad. Sci. USA, 99*, 15693-15698.

Takahara, M., Kuroiwa, H., Miyagishima, S.-Y., Mori, T., & Kuroiwa, T. (2001). Localization of the mitochondrial FtsZ protein in a dividing mitochondrion. *Cytologia, 66*, 421-425.

Takahara, M., Takahashi, H., Matsunaga, S., Miyagishima, S.-Y., Takano, H., Sakai, A., Kawano, S., and Kuroiwa, T. (2000). A putative mitochondrial ftsZ gene is present in the unicellular primitive red alga *Cyanidioschyzon merolae*. *Mol. Gen. Genet., 264*, 452-460.

Takeda, Y., Hirokawa, H., & Nagata, T. (1992). The replication origin of proplastid DNA in cultured cells of tobacco. *Mol. Gen. Genet., 232*, 191-198.

Thoemmes, P., Farr, C. L., Marton, R. F., Kaguni, L. S., & Cotterill, S. (1995). Mitochondrial single stranded DNA-binding protein from *Drosophila* embryos. *J. Biol. Chem., 270*, 21137-21143.

Thompson, R. J., & Mosig, G. (1985). An ATP-dependent supercoiling topoisomerase of *Chlamydomonas reinhardtii* affects accumulation of specific chloroplast transcripts. *Nucleic Acids Res.,* 13, 873-891.

Thompson, R. J., & Mosig, G. (1987). Stimulation of *Chlamydomonas* chloroplast promoter by novobiocin *in situ* in *E. coli* implies regulation by torsional stress in the chloroplast DNA. *Cell, 48*, 281-287.

Thompson, R. J., & Mosig, G. (1990). Light affects the structure of *Chlamydomonas* chloroplast chromosomes. *Nucleic Acids Res., 18*, 2625-2631.

Turmel, M., Lemieux, C., & Lee, R. W. (1981). Net synthesis of chloroplast DNA throughout the synchronized vegetative cell cycle of *Chlamydomonas*. *Curr. Genet., 2*, 229-232.

Turmel, M., Otis, C., & Lemieux, C. (1999). The complete chloroplast DNA sequence of the green alga *Nephroselmis olivacea*: Insights into the architecture of ancestral chloroplast genomes. *Proc. Natl. Acad. Sci. USA, 96*, 10248-10253.

Tuteja, N. (2000). Plant cell and viral helicases: essential enzymes for nucleic acid transactions. *Crit. Rev. Plant Sci., 19*, 449-478.

Tuteja, N., & Phan, T.-N. (1998a). A chloroplast DNA helicase II from pea that prefers fork-like replication structures. *Plant Physiol., 118*, 1029-1039.

Tuteja, N., & Phan, T.-N. (1998b). Inhibition of pea chloroplast DNA helicase unwinding and ATPase activities by DNA-interacting ligands. *Biochem. Biophys. Res. Commun., 244*, 861-867.

Tuteja, N., Phan, T.-N., & Tewari, K. K. (1996). Purification and characterization of a DNA helicase from pea chloroplast that translocates in the 3'-to-5' direction. *Eur. J. Biochem., 238*, 54-63.

Tymms, M. J., Scott, N. S., & Possingham, J. V. (1982). Chloroplast and nuclear DNA content of cultured spinach leaf discs. *J. Exp. Bot., 33*, 831-837.

Tymms, M. J., Scott, N. S., & Possingham, J. V. (1983). DNA content of *Beta vulgaris* chloroplasts during leaf cell expansion. *Plant Physiol., 71*, 785-788.

Uehara, T., Matsuzawa, H., & Nishimura, A. (2001). HscA is involved in the dynamics of FtsZ-ring formation in *Escherichia coli* K12. *Genes Cells, 6*, 803-814.

Van Dyck, E., Foury, F., Stillman, B., & Brill, S. J. (1992). A single-stranded DNA binding protein required for mitochondrial DNA replication in *S. cerevisiae* is homologous to *E. coli* SSB. *EMBO J., 11*, 3421-3430.

Van Gestel, K., Koehler, R. H., & Verbelen, J.-P. (2002). Plant mitochondria move on F-actin, but their positioning in the cortical cytoplasm depends on both F-actin and microtubules. *J. Exp. Bot., 53*, 659-667.

Van Gestel, K., & Verbelen, J.-P. (2002). Giant mitochondria are a response to low oxygen pressure in cells of tobacco (*Nicotiana tabacum* L.). *J. Exp. Bo.t, 53*, 1215-1218.

Van Tuyle, G. C., & Pavco, P. A. (1985). The rat liver mitochondrial DNA-protein complex: displaced single strands of replicative intermediates are protein coated. *J. Cell Biol., 100*, 251-257.

van Wijk, K. J. (2001). Challenges and prospects of plant proteomics. *Plant Physiol., 126*, 501-508.

Vermel, M., Guermann, B., Delage, L., Grienenberger, J.-M., Marechal-Drouard, L., and Gualberto, J. M. (2002). A family of RRM-type RNA-binding proteins specific to plant mitochondria. *Proc. Natl. Acad. Sci. USA, 99*, 5866-5871.

Vitha, S., Froehlich, J. E., Koksharova, O., Pyke, K. A., van Erp, H., & Osteryoung, K. W. (2003). ARC6 is a J-domain plastid division protein and an evolutionary descendant of the cyanobacterial cell division protein Ftn2. *Plant Cell, 15*, 1918-1933.

Vitha, S., McAndrew, R. S., & Osteryoung, K. W. (2001). FtsZ ring formation at the chloroplast division site in plants. J. Cell Biol., 153, 111-119.

Waddell, J., Wang, X.-M., & Wu, M. (1984). Electron microscopic localization of the chloroplast DNA replicative origins in *Chlamydomonas reinhardtii*. *Nucleic Acids Res., 12*, 3843-3856.

Wakasugi, T., Nagai, T., Kapoor, M., Sugita, M., Ito, M., Ito, S., et al. (1997). Complete nucleotide sequence of the chloroplast genome from the green alga *Chlorella vulgaris*: The existence of genes possibly involved in chloroplast division. *Proc. Natl. Acad. Sci. USA, 94*, 5967-5972.

Wang, D., Kong, D., Wang, Y., Hu, Y., He, Y., & Sun, J. (2003). Isolation of two plastid *ftsZ* genes from *Chlamydomonas reinhardtii* and its evolutionary implication for the role of FtsZ in plastid division. *J. Exp. Bot., 54*, 1115-1116.

Wang, J. C. (2002). Cellular roles of DNA topoisomerases: A molecular perspective. Nature Rev. Mol. Cell Biol. 3, 430-440.

Wang, S., & Liu, X.-Q. (1991). The plastid genome of *Cryptomonas* □ encodes an hsp70-like protein, a histone-like protein, and an acyl carrier protein. *Proc. Natl. Acad. Sci. USA, 88*, 10783-10787.

Wang, X., Sirover, M. A., & Anderson, L. E. (1999). Pea chloroplast glyceraldehyde-3-phosphatedehyrdogenase has uracil glycosylase activity. *Arch. Biochem. Biophys., 367*, 348-353.

Wang, Y., Saitoh, Y., Sato, T., Hidaka, S., & Tsutsumi, K. (2003). Comparison of plastid DNA replication in different cells and tissues of the rice plant. *Plant Mol. Biol., 52*, 905-913.

Wang, Y., Tamura, K., Saitoh, Y., Sato, T., Hidaka, S., & Tsutsumi, K. (2002). Mapping major replication origins on the rice plastid DNA. *Plant Biotechnol., 19*, 27-35.

Werhahn, W., & Braun, H.-P. (2002). Biochemical dissection of the mitochondrial proteome from *Arabidopsis thaliana* by three-dimensional gel electrophoresis. *Electrophoresis, 23*, 640-646.

White, M. A., Bailey, J. C., Cannon, G. C., & Heinhorst, S. (1997). Partial purification and characterization of the DNA polymerase from the cyanelles of *Cyanophora paradoxa*. *FEBS Lett., 410*, 509-514.

Woelfle, M. A., Thompson, R. J., & Mosig, G. (1993). Roles of novobiocin-sensitive topoisomerases in chloroplast DNA replication in *Chlamydomonas reinhardtii*. *Nucl. Acids Res., 21*, 4231-4238.

Wolfe, K. H., Li, W. H., & Sharp, P. M. (1987). Rates of nucleotide substitution vary greatly among plant mitochondrial, chloroplast and nuclear DNA. *Proc. Natl. Acad. Sci. USA, 84*, 9054-9058.

Wu, H., & Liu, X. Q. (1997). DNA binding and bending by a chloroplast-encoded HU-like protein overexpressed in *Escherichia coli*. *Plant Mol. Biol., 34*, 339-343.

Wu, M., Lou, J. K., Chang, D. Y., Chang, C. H., & Nie, Z. Q. (1986). Structure and function of a chloroplast DNA replication origin of *Chlamydomonas reinhardtii*. *Proc. Natl. Acad. Sci. USA, 83*, 6761-6765.

Wu, M., Nie, Z. Q., & Yang, J. (1989). The 18-kD protein that binds to the chloroplast DNA replicative origin is an iron-sulfur protein related to a subunit of NADH dehydrogenase. *Plant Cell, 1*, 551-557.

Yamamoto, K., Pyke, K. A., & Kiss, J. Z. (2002). Reduced gravitropism in inflorescence stems and hypocotyls, but not roots, of *Arabidopsis* mutants with large plastids. *Physiol. Plant., 114*, 627-636.

Yasuda, T., Kuroiwa, T., & Nagata, T. (1988). Preferential synthesis of plastid DNA and increased replication of plastids in cultured tobacco cells following medium renewal. *Planta, 174*, 235-241.

Yurina, N. P., Belkina, G. G., Karapetyan, N. V., & Odintsova, M. S. (1995a). Nucleoids of pea chloroplasts: microscopic and chemical characterization, occurrence of histone-like proteins. *Biochem. Mol. Biol. Int., 36*, 145-154.

Yurina, N. P., Belkina, G. G., Oleskina, Y. P., Karapetyan, N. V., & Odintsova, M. S. (1995b). Composition and organization of chloroplast nucleoids. Appl Biochem Microbiol, 31, 32-38.

Zabrouskov, V., Giacomelli, L., van Wijk, K. J., & McLafferty, F. W. (2003). A new approach for plant proteomics. *Mol. Cell. Proteomics,* 2, 1253-1260.

Zelenaya-Troitskaya, O., Newman, S. M., Okamoto, K., Perlman, P. S., & Butow, R. A. (1998). Functions of the high mobility group protein, Abf2p, in mitochondrial DNA segregation, recombination and copy number in *Saccharomyces cerevisiae*. *Genetics, 148*, 1763-1776.

Zelenaya-Troitskaya, O., Perlman, P. S., & Butow, R. A. (1995). An enzyme in yeast mitochondria that catalyzes a step in branched-chain amino acid biosynthesis also functions in mitochondrial DNA stability. *EMBO J., 14*, 3268-3276.

CHAPTER 4

THE SEXUAL INHERITANCE OF PLANT ORGANELLES

RUDOLPH HAGEMANN

Genetik, Martin-Luther-Universität, Halle/S., Germany

Abstract. In plants three modes of inheritance (transmission) of plasmatic organelles - plastids and mitochondria -- are found: (1) uniparental maternal, (2) biparental and (3) uniparental paternal organelle inheritance. The numerical proportion of these three modes of inheritance is very different in various taxa of angiosperms, gymnosperms, yeasts and algae. Regarding the plastids, the majority of angiosperms exhibit the maternal mode, whereas in gymnosperms the paternal or the biparental modes are prevailing. In *Chlamydomonas* a maternal inheritance (via the mt^+ parent) is the rule, but the biparental mode is an important exception, the frequency of which can intensely be increased experimentally. All three modes are found within all groups of higher taxa. Regarding the mitochondria the three modes of organelle inheritance are found in varying frequencies among different taxa. The different kinds of experimental procedures are characterized. The determination what mode of organelle inheritance is found in a particular species needs careful observations and experimental studies.

1. INTRODUCTION

The eukaryotic cell consists of the cell nucleus and many plasmatic organelles. Two of these organelles are, from the genetic point of view, of special interest, because they carry primary genetic information. These are the plastids, containing specific plastid DNA (ptDNA), and the mitochondria with their specific mitochondrial DNA (mtDNA). It is of great interest to find out, in which ways these two types of plasmatic organelles and their specific DNAs are transmitted into the next generation. Many research activities during the past decades have elucidated a great variety of cytological and molecular mechanisms that determine different ways and modes of organelle transmission from one generation to the next. Unfortunately, even nowadays several textbooks and reviews state in an unconcerned mode that plastids and mitochondria of plants are transmitted to the next generation only by the

H. Daniell and C.D. Chase (eds.), Molecular Biology and Biotechnology of Plant Organelles, 93—113.

mother, by "maternal inheritance". Such a statement is simply wrong and inadmissible.

The aim of this contribution is to describe in some detail, the ways in which transmission of plastids and mitochondria in different taxa of plants have been found and proved by careful cytological and molecular methods. First a short overview of the different methods used in these investigations will be given. The following presentation will be classified under two aspects: plastids (or chloroplasts) versus mitochondria, and seed plants (angio- and gymnosperms) versus algae and yeast.

2. METHODS USED IN ELUCIDATION THE MODES OF ORGANELLE TRANSMISSION

2.1. Genetic analyses

The basic method for elucidating the transmission of organelles is genetic analysis, the crossing of organisms that differ in the genetic information of their plasmatic organelles and the careful analysis of the progenies. Several traits with clear phenotypic effects can be used.

Differences in plastid genes are expressed in at least three types of phenotypic effects – deficiencies in the light and dark reactions of photosynthesis, or larger deletions or rearrangements of ptDNA, connected with changes in the leaf color (light green, yellow, creme or white leaves or leaf sectors); herbicide resistances (e.g. atrazine resistance) or resistance against phytotoxins; and antibiotic resistances. Moreover, there exist molecular differences in ptDNA between different taxa (lines, varieties, species) and with no deleterious effects – single nucleotide polymorphisms (SNPs) or larger ptDNA differences in non-coding sequences. These can be identified and analyzed by restriction fragment length polymorphism (RFLP), by polymerase chain reaction (PCR) amplification of ptDNA followed by RFLP analyses, or by similar methods. These differences can be used to identify the origin of specific ptDNAs after suitable crosses.

Differences in mitochondrial genes can also be expressed in phenotypic effects – several different respiration deficiencies, many antibiotic resistances and several types of cytoplasmic male sterility. In mtDNAs there are, as in ptDNAs, small sequence differences with no deleterious effects, which can be used to distinguish the mtDNAs of different taxa by RFLP analyses and to trace back the origin of a specific mtDNA after (reciprocal) crosses.

2.2. Light and electron microscopy

While genetic studies allow the clear identification of organelle DNA (maternal or paternal or both) found in the progeny of crosses (F_1, F_2, etc.) these studies have to be complemented by microscopic studies to determine the cytological or molecular mechanisms that underlie this transmission. What are the cytological processes during micro- and mega-sporogenesis and micro- and mega-gametogenesis and during fertilization, which determine the organelle transmission into the next generation ?

A very suitable method has been and is the transmission electron microscopic investigation of these processes. The high optical resolution of electron microscopy allows the identification of developmental processes and the mechanisms during sporogenesis, gametogenesis and the fertilization process leading to equal or unequal organelle distribution or to a pathway of plastid exclusion or degeneration. It will be shown in the following paragraphs how successful these studies have been.

Electron microscopy requires a lot of effort regarding specimen preparation and microscopic analysis. For the screening of many samples, the use of light microscopic techniques is recommended. For this purpose plastids were stained with the DNA specific fluorochrome DAPI (4'-6-diamidino-2-phenylindole). The organelles can be identified by the use of a fluorescence microscope. Such investigations give a first hint at the possible mode of plastid transmission, but they have to be combined with other methods (e.g. genetic analyses or electron microscopy) in order to provide reliable results. Nevertheless, DAPI staining has been used by several research groups with interesting results.

2.3. In situ hybridization with DNA probes for plastid and mitochondrial genes

The presence of plastid or mitochondrial genes in generative and sperm cells of pollen or in the egg cell can be detected by *in situ* hybridization with suitable probes. Tests have been made using the probes for the plastid genes *rbcL* (large subunit of rubisco), *psaA*, *psaB* and *psbA*, *psbC* and *psbD*. Similar tests can be performed with probes for the mitochondrial genes *cox1*, *cox2*, *cox3* and *cob*. When the tests for these genes are positive, then this result can be considered as a proof for the presence of plastids or mitochondria, respectively, in these cells. Although this procedure has so far not been used in many investigations, its potential should be kept in mind. The combination of the methods outlined in Section 2 led to the results that will be outlined in the following paragraphs.

3. DIFFERENT MODES OF PLASTID INHERITANCE (TRANSMISSION) IN ANGIOSPERMS

In different species and genera of seed plants (angiosperms and gymnosperms) several modes of plastid transmission into the next generation have been found during the past decades. Three modes of plastid inheritance, uniparental maternal, biparental, and uniparental paternal, are realized during the processes of sexual reproduction.

3.1. Uniparental maternal plastid inheritance

In the majority of angiosperms there is a uniparental, purely maternal inheritance of plastids. This mode of plastid inheritance was first described for *Mirabilis jalapa* by

Correns (1909), for *Antirrhinum majus* by Baur (1910) and since that time for many other species and genera including *Beta, Hordeum, Zea, Lycopersicon, Gossypium, Triticum* (compare the lists in Correns and v. Wettstein, 1937; Tilney-Bassett, 1975; Hagemann and Schroeder, 1989; Hagemann, 1992 and Mogensen, 1996). This mode of transmission, which may give the impression of a simple and uniform event, is in reality caused by several different cytological mechanisms.

3.1.1. Plastid exclusion from the generative cell during the microspore mitosis, often referred to as the first haploid pollen mitosis

During male pollen development (microsporogenesis and formation of male gametes) the microspore mitosis is extremely unequal regarding the distribution of the plastids. All plastids are distributed into the vegetative cell; therefore the generative cell is free of plastids. Thus the two sperm cells, formed from the generative cell, are free of plastids, too. We proposed the term "Lycopersicon type" for this mode of distribution (Hagemann and Schröder, 1989; Hagemann, 1992). It is found in *Lycopersicon, Antirrhinum, Gasteria, Chlorophytum, Tulbaghia* etc. (Figure 1).

3.1.2. Plastid degeneration during maturation of the generative cell

In some species the young generative cell gets a few plastids, but during maturation of the generative cell these plastids degenerate; therefore the mature generative cell and the sperm cell are free of plastids. This "Solanum type" of plastid distribution was found in *Solanum chacoense, S. tuberosum, Hosta japonica* and *Convallaria majalis* (Clauhs and Grun, 1977; Schroeder, 1986; Hagemann 1992).

3.1.3. Plastid exclusion during sperm cell maturation and during the fertilization process

In several cereals, as wheat, triticale, barley and maize, both generative and sperm cells regularly contain plastids, but they are not transmitted into the egg cell. Obviously they are stripped off the sperm nucleus just before or during the process of fertilization. In *Triticum* genetic differences in plastids (green-white striping) and in mitochondria (cytoplasmic male sterility) are inherited in a uniparental maternal mode. This shows that the paternal plastids and the mitochondria are not transmitted into the egg cells ("Triticum type").

The exclusion of organelles from sperm cells may occur even earlier than during the fertilization process. Mogensen and Rusche (1985) reported a decrease in the number of mitochondria within maturing sperm cells. The mitochondria are enclosed in evaginations of the sperm cell plasmalemma and are afterwards transferred into the vegetative cytoplasm.

The reduction of the cytoplasm and the cytoplasmic organelles surrounding the sperm nucleus may be a rather widespread characteristic of sperm cell maturation (compare Mogensen, 1996). In the described case it is acting on mitochondria; however, in other cases it may also act on plastids.

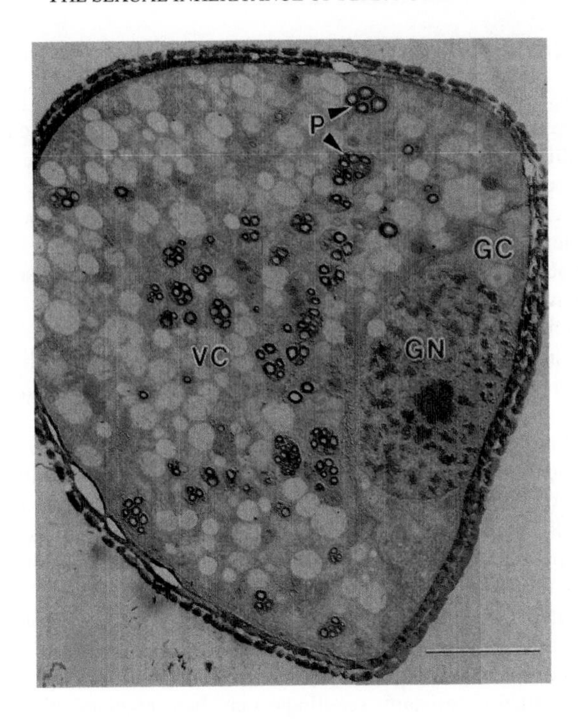

Figure 1. *Transmission electron micrograph of a pollen grain from Tulbaghia violacea. The newly formed generative cell (GC) with the generative nucleus (GN) do not contain plastids. However, the vegetative cell (VC) contains numerous plastids (P) with several small starch granules. Bar = 5 μm. Reproduced with permission of M.-B. Schroeder, Geisenheim.*

3.2. Importance for biotechnology and containment of ecological risks

The cytological mechanisms, characterized under the headings 3.1.1, 3.1.2 and 3.1.3 lead to a uniparental maternal inheritance of plastids. This situation is found in many angiospermous species, among them are many crop plants. This feature is of great importance for the public discussions about possible ecological risks of the use of transgenic crop plants. When the transgenes, e.g. for herbicide resistance (Daniell et al., 1998), for resistance against pathogens (DeGray et al., 2001), insects (McBride et al., 1995; DeCosa et al., 2001) or other traits of economic value (drought tolerance, Lee et al., 2003; salt tolerance, Kumar et al., 2004; phytoremediation, Ruiz et al., 2003) are integrated into the plastid genome of such species, then there is no longer any danger of transgene transmission via the pollen to neighboring fields of cultivated plants or weedy relatives of crop plants - because the pollen does not transmit plastids! For more information on biotechnology applications, the reader is referred to chloroplast genetic engineering, chapter 16 of this book.

3.3. Caution: Single exceptions are always possible!

Biological processes may always exhibit "seldom exceptions". Furthermore, cytological processes do not act with 100% efficiency. This means in this connection, the different possibilities for plastid exclusion from the sperm cells and during the fertilization process (characterized in section 3.1) are clearly defined and established in careful investigations. Thus, the classification of different species into the "types" (as defined in section 3.1) is fully justified. Nevertheless "exceptions" are possible and occur with very low frequency. Van Went and Willemse (1984) and Russell (1992) have described in great detail, how multifarious the numerous steps in the fertilization processes in different angiosperm species are. It is easily conceivable that small disturbances are able to cause rare deviations that may have consequences with regard to the transmission of paternal plastids or mitochondria into the zygote.

Diers (1967, 1971) found in snapdragon, *Antirrhinum majus*, after very extensive reciprocal crosses of wild type plants ("Sippe 50") with particular plastome mutants ("prasinizans" and "yellow-green prasinizans") a rare transmission of paternal plastids into the next generation. A total of 97,613 F_1 plants were tested, and 1 plant with paternal plastids was found in a progeny of 4,066 plants. Recently Wang et al. (2004) reported comparable results in foxtail, *Setaria italica*. They used plants with a plastid-inherited atrazine resistance (and a nuclear dominant red leaf base marker) as the male parent and five male-sterile, yellow- or green-leafed herbicide- susceptible lines as the female parents. Among 787,329 hybrid seedlings they found 241 atrazine-resistant seedlings that had received their herbicide resistance from the father; thus the frequency of resistant seedlings was 3.1×10^{-4}.

There are still more reports about "leakage of male plastids" (Mogensen 1996). Medgyesy et al. (1986) and Horlow et al. (1990) reported the occasional transmission of paternal plastids in tobacco, *Nicotiana tabacum*. Cornu and Dulieu (1988) observed rare male plastid transmission in *Petunia hybrida*, and Schmitz and Kowallik (1986) found rare paternal plastid transmission in *Epilobium*. Wang et al. (2003) express the view that one should not carelessly accept reports about a uniparental maternal inheritance of plastids when these reports are based on only a small number of tested plants. When the question is asked if plant traits of economical value (e.g. herbicide resistances or transgenes in "model plants" or in cultivated plants) are really inherited in an exclusively maternal way, then such a conclusion should be based on large-scale genetic experiments. (We will later discuss - in paragraph 3.5 - the cases of a uniparentally *paternal* plastid inheritance, where we are confronted with the same problem - only viewed from the other side.)

3.4. Biparental plastid inheritance

In a minority of angiospermous genera and species a clear biparental plastid inheritance is found, i.e. both the mother and the father plant regularly contribute plastids into the zygote. The first report about this mode of plastid inheritance is the classic genetic paper of Erwin Baur (1909) on the inheritance patterns of white-margined plants of Pelargonium zonale (compare Hagemann 1964, 2000, 2002).

Other intensely studied genera of this type are Oenothera, Hypericum and Medicago (Hagemann, 1964; Smith, 1989).The cytological basis for the mode of plastid transmission in these plants is obvious. During pollen development there is an equal distribution of plastids during the first pollen mitosis into the generative and the vegetative cells. Thus the generative cell contains many plastids, and the sperm cells (at least those which fuse with the egg cell) contain plastids, too. The sperm cells transmit the plastids with great regularity into the egg cell. We proposed the term "Pelargonium type" for this type of plastid transmission. This fact has been proven both by genetic experiments and by cytological investigations (Guo and Hu, 1995) (Figures 2 and 3).

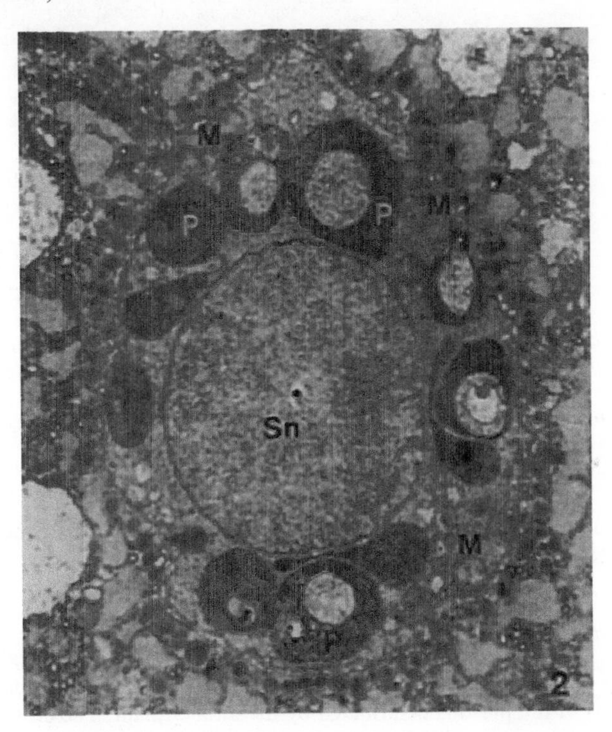

Figure 2. *Transmission electron micrograph of a sperm cell in the mature pollen of Pelargonium hortorum (= P. zonale) with many plastids (P), which are electron-dense, often cup-shaped and devoid of starch granules. The sperm mitochondria (M) are much smaller and usually circular in section. Sn = sperm nucleus. Reproduced by permission of Springer Verlag from Guo & Hu (1995).*

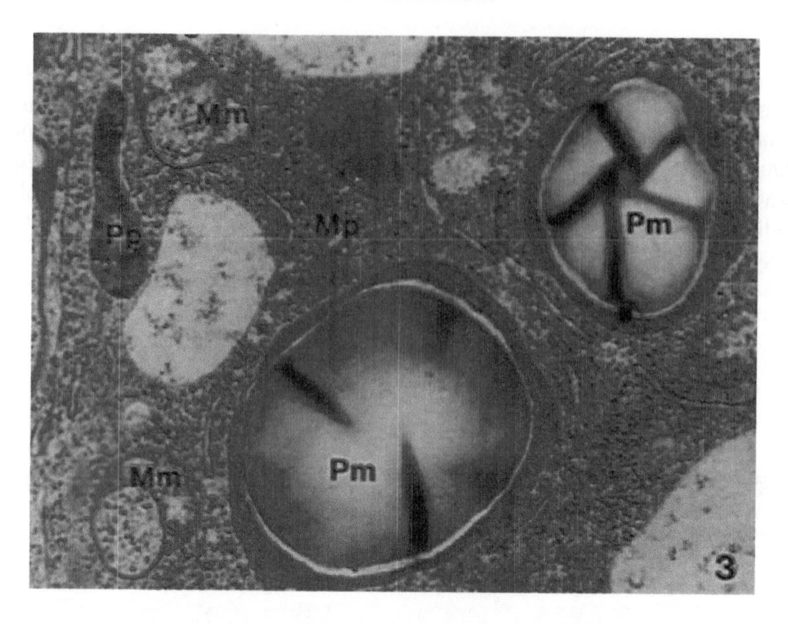

Figure 3. Plastids and mitochondria in a zygote of Pelargonium hortorum 24 h after pollination. The paternal plastids Pp are electron-dense and without starch granules. The maternal plastids Pm are much bigger and contain starch granules. The paternal mitochondria Mp are small and electron-dense. The maternal mitochondria Mm are bigger, less electron-dense and similar to the mitochondria in the egg cell. Reproduced by permission of Springer Verlag from Guo & Hu (1995).

Within the group of plants with biparental plastid inheritance it seems to be possible to distinguish between three subtypes.

3.4.1. Biparental plastid inheritance subtype a

In the genera *Oenothera* (evening primrose) and *Hypericum* reciprocal crosses between green and white plants or branches show distinct reciprocal differences with a clear bias of the plastids contributed by the mother plant. The cross 'green x white' yields many green seedlings, a number of variegated and (almost) no white seedlings. In contrast, the cross 'white x green' leads to variegated, many white and (almost) no green seedlings. (The detailed numbers from many crosses are given by Hagemann, 1964; Kirk and Tilney-Bassett, 1967; and Renner 1936).

3.4.2. Biparental plastid inheritance subtype b

In *Pelargonium* (Figure 3) there is often a rather equal contribution of plastids from the mother and the father; but the results of different reciprocal crosses (using parents of varying nuclear constitutions) vary widely in that respect (Baur, 1909;

Hagemann, 1964; Herrmann and Hagemann, 1971; Börner et al., 1972; Tilney-Bassett, 1994; Kuroiwa et al. 1993).

3.4.3. Biparental plastid inheritance subtype c

In *Medicago sativa* (alfalfa) several investigators described a biparental plastid inheritance with a strong predominance of the paternal plastids (Smith et al., 1986; Lee et al., 1988; Smith, 1989; Schumann and Hancock, 1989; Masoud et al., 1990; Zhu et al. 1992, 1993). This subtype seems to be rather rare among angiosperms. It has also been found in gymnosperms, for *Cryptomeria japonica*, Taxodiaceae (Ohba et al. 1971; see section 4).

3.4.4 . Mechanisms leading to biparental plastid inheritance

Corriveau and Coleman (1988) developed a rapid screening method (using the fluorescent dye DAPI) for the detection of plastid DNA in generative and sperm cells of the pollen. They could find male plastids in these cells in many (43) species and concluded that these species have the potential of biparental inheritance. This conclusion has to be dealt with distinct reserve. All plants of the “Triticum type”, as defined in Section 3.1.3, contain plastids in the sperm cells, but nevertheless do not transmit their plastids into the egg cell. Only plants of the “Pelargonium type” transmit their plastids into the egg cell. Furthermore, DAPI is a DNA binding fluorochrome staining plastid and mtDNA. Because of the optical resolution of fluorescence microscopes it is impossible to make a clear distinction between mitochondria and DAPI-stained proplastids. As mentioned in paragraph 2.2., these studies have to be combined with genetic or electron microscopic investigations in order to get exact results.

Polarized plastid distribution during sperm cell formation is a special situation, which leads to a biparental plastid inheritance. In *Plumbago zeylanica* a very interesting sperm cell heteromorphism (Figure 4) has been found (reviewed by Russell, 1984, 1992). It was shown by serial sectioning and three dimensional reconstruction that the two sperm cells, formed by pollen mitosis (the second haploid pollen mitosis), are extremely dimorphic (Figure 4). They differ in size, morphology and organelle content. The larger sperm cell is closely associated with the vegetative nucleus (VN); it contains many mitochondria and no (or only very few) plastids. The smaller sperm cell is not associated with the vegetative nucleus; it contains numerous (up to 46) plastids and relatively few mitochondria. In more than 94% of the fertilization events analyzed, the smaller sperm cell with many plastids fuses with the egg cell and thus transmits the plastids to the next generation. This results in a biparental plastid inheritance. In contrast, the larger mitochondria-rich sperm cell (without plastids) fuses with the central cell and so gives rise to the nutritive endosperm. In terms of plastid distribution, the pollen mitosis (second haploid pollen mitosis) in *Plumbago* is extremely unequal, it leads to a polarized plastid distribution. It is easy to imagine the opposite situation – the exclusion of plastids from the sperm cell which fuses with the egg cell. It may well be that such events may regularly occur in other species.

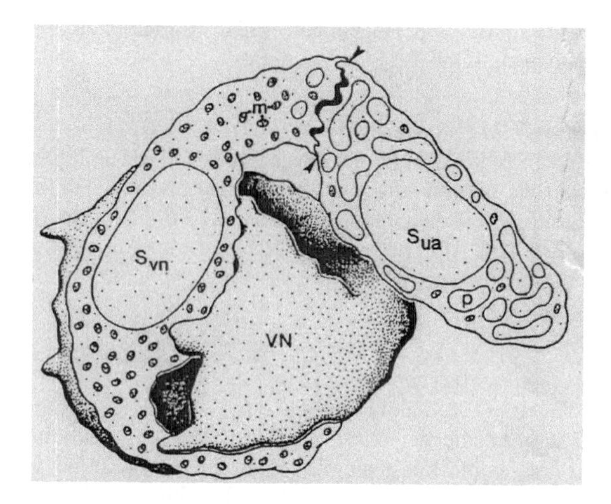

Figure 4. Reconstruction of the dimorphic sperm cells within a mature pollen grain of Plumbago zeylanica with superimposed profiles of plastids (p) and mitochondria (m). The larger sperm cell Svn contains the majority of the mitochondria and no (or very seldom a few) plastids. In contrast, the smaller sperm cell Sua contains numerous plastids and relatively few mitochondria. VN = vegetative nucleus. Reproduced with permission of Springer Verlag from Russell (1984).

3.5. Uniparental paternal inheritance

The first report on uniparental plastid inheritance in angiosperms was only published in 1995 for the kiwi plant *Actinidia deliciosa* (Cipriani et al. 1995). This interesting result was confirmed by several investigations (Testolin and Cipriani, 1997; Cipriani et al., 1998; Chat et al. 1999).

3.6. Genetic studies versus RFLP analyses

A great many of the results, reported so far, are based on genetic studies, e.g. the outcome of reciprocal crosses between plants that differ in the genetic constitution of the plastids (green versus white or yellow or yellow-green). Other results are based on RFLP analyses or comparable marker analyses. The validity of both procedures is not equal. The validity of genetic studies is much higher. Diers (1967, 1971) reported, as already mentioned in section 3.3, the results of numerous reciprocal crosses in *Antirrhinum majus* using wild type green plants and two special yellow and yellow-green plastome mutants (prasinizans). Among 97,613 F_1-plants he observed 24 yellow-green variegated plants. These very rare events could be observed because the genetic method of analysis, e.g. looking at many individual plants, allowed the observation of such rare variegated plants with yellow-colored leaf spots as a phenotypic marker. Nobody would be willing or able to perform

RFLP analyses on 50,000 or 100,000 plants. Moreover, most of the variegated plants analyzed by Diers contain more green plastids (with a possible "wild type" RFLP pattern) than yellow plastids (with a differing RFLP pattern). Certainly part of the variegated plants would escape their discovery by RFLP analyses. Thus reports based on RFLP analyses of just one dozen or a few dozen plants cannot be accepted as an unequivocal proof of strict paternal plastid inheritance (compare paragraph 3.5).

3.7. Plastid transmission and distribution on the female side

In angiosperms, all electron microscopic investigations published so far report the presence of plastids in the egg cells (see reviews by Schnarf, 1941; Heslop-Harrison, 1972; Jensen, 1972l; Knox, 1984; van Went and Willemse, 1984; Russell, 1992). But obviously the number of plastids in the egg cells of different angiosperms may be different. Mogensen (1972) reported that in *Quercus gambelii* plastids in the egg cells are very scarce. Electron microscopic studies of the female gametophyte and the egg apparatus and its development in *Actinidia* - where the genetic results point to the entire loss of plastids on the female side - have so far not been published.

Unequal distribution of plastids during the development of the female gametophyte was described in several species of angiosperms. In *Epilobium palustre* (Bednara and Rodkiewicz, 1974) and in orchids (Corti and Cecchi, 1970) it was found that after meiosis the haploid cell, which later develops into the embryo sac, received the majority of plastids, whereas the other 3 haploid cells, which degenerate later, received almost no plastids. This is obviously due to an unequal plastid distribution before or during meiosis. It is conceivable that unequal plastid distribution in the opposite direction may, in other species, lead to a distinct diminution or even the absence of plastids in egg cells.

In egg cells of *Daucus muricatus* Hause (1991) could find only a few intact plastids, containing small starch grains. However, at the periphery of the cell he observed myelin-like structures, which possibly represent degenerating plastids because they contain starch-like particles.

3.8. General conclusions on angiosperms

In angiosperms there are distinct differences regarding the distribution of plastids during the development of the male gametophyte, the male gamete formation and the fertilization process. In fact all possibilities of plastid distribution and inheritance (uniparental maternal, biparental, and uniparental paternal) have been found in different angiosperm species. The mechanisms that lead to these differences are primarily cytological mechanisms. Hints at enzymatic, restriction-modification type mechanisms, which lead to the destruction of male plastids as was described in *Chlamydomonas,* have not been found in angiosperms. The only indication for the action of a physiological mechanism is the degeneration and decay of plastids during the maturation of generative cells in plants of the "Solanum type", as described in paragraph 3.1.2. for *Solanum tuberosum* and *Convallaria majalis* (Clauhs and Grun, 1977; Schroeder, 1986).

4. DIFFERENT MODES OF PLASTID INHERITANCE (TRANSMISSION) IN GYMNOSPERMS

In angiosperms all possible types of plastid inheritance have been found, as described in Section 3; the most frequent type is purely maternal plastid inheritance. The different classes of gymnosperms (Coniferophytina) have not been studied as intensely as the angiosperms. Nevertheless some general statements can be made. The three possible types of plastid transmission, maternal, biparental and paternal, are found in gymnosperms, but their relative frequency is distinctly different from that found in angiosperms.

For Cycadopsida there are only older light microscopic studies on the genus *Zamia*; they point to a maternal transmission of plastids. *Ginkgo biloba*, the only species of Ginkgoopsida, was studied by light and electron microscopy; these studies also report a maternal transmission of plastids. The Gnetopsida *Ephedra distachya* was investigated by electron microscopy; these studies point to a maternal transmission of plastids, as well (Mogensen 1996).

Intense and thorough investigations have been made in Coniferopsida. In the gymnospermous family of Taxodiaceae, the species *Cryptomeria japonica* (D. Dorn) was investigated by Ohba et al. (1971); they reported a very interesting type of biparental plastid inheritance. Reciprocal crosses between green and yellowish-white shoots showed a very strong bias of paternal plastids. The cross 'green x white' gave 111 green : 284 variegated : 3,386 white seedlings, and the cross 'white x green' gave 8,618 green : 49 variegated: 22 white plants. These results are a convincing example of our statement in Section 3.5, that thorough genetic crossing experiments have a much higher power of resolution than RFLP analyses (compare Yamada et al., 1993). There are good arguments that an RFLP analysis with even 100 F_1 plants would not have found the few plants with maternal plastids in *Cryptomeria*.

There are many reports on uniparental paternal plastid inheritance in many families, genera and species of Coniferopsida. These studies are almost exclusively based on analyses of RFLPs in ptDNA. This type of plastid transmission was reported for species of the genera *Calocedrus* (Cupressaceae), *Sequoia* (Taxodiaceae), *Abies*, *Pseudotsuga*, *Pinus*, *Picea* and *Larix* (Pinaceae) (Szmidt et al., 1987; Neale et al., 1986, 1989a,b, 1991; Wagner, 1992; Wagner et al., 1989; Sutton et al., 1991; Dong et al., 1992; Salaj et al., 1998; additional references in Mogensen, 1996). Szmidt et al. (1987) found, in some F_1 progenies of crosses between *Larix decidua* and *L. leptolepsis*, a mixture of maternal and paternal plastid DNA patterns, and in one progeny only a maternal plastid DNA pattern.

These results indicate the frequent and prevailing transmission of paternal plastids into the next generation. There is no doubt that in many cases we are dealing with a uniparental paternal plastid inheritance. But there are indications that, side by side, there may in some species also be a biparental plastid inheritance with a strong bias of the paternal plastids. The results obtained for *Cryptomeria japonica*, reported above, may give an indication for the amount of the paternal bias.

The prevailing or entirely paternal inheritance of plastids in many gymnosperms is only possible if on the female side there is a massive reduction of plastids in the egg cells or their exclusion in the course of megasporogenesis and megagametogenesis or during the fertilization process. Several species, e.g. *Abies alba, Larix decidua, Biota orientalis, Chamaecyparis lawsonia* and *Cephalotaxus drupacea*, have been carefully studied in electron microscopic investigations. They revealed the entire (or almost entire) absence of plastids in the egg cells (Camefort, 1969; Chesnoy, 1969, 1971, 1987; Chesnoy and Thomas, 1971; Singh, 1978; Gianordoli, 1974).

Detailed and careful investigations on the cytological basis of the paternal plastid inheritance (and the maternal mitochondrial inheritance) in the Douglas fir, *Pseudotsuga menziesii*, were performed by Owens and Morris (1990, 1991). They found that at the beginning of the fertilization process there is a modification (the start of the disintegration) of the female plastids within the egg cell. These modified plastids are not closely associated with the egg nucleus, whereas the female mitochondria tightly surround the egg nucleus. Later on, the pollen tube discharges the sperm nuclei and the male cytoplasm, containing the male plastids, into the egg cell. A sperm nucleus and the male cytoplasm come in close contact with the female nucleus, which fuse afterwards. Finally, the embryonic plasm is derived from a combination of both parents, but includes almost exclusively male plastids and female mitochondria.

Another careful investigation of the fertilization process in the oriental arborvitae, *Biota orientalis* (Cupressaceae), performed by Chesnoy (1977), revealed in detail the processes that lead to the paternal inheritance of both plastids and mitochondria. Two modes seem to be acting – unequal plastid distribution during the course of megasporogenesis, megagametogenesis and the fertilization process, combined with the degeneration and lysis of plastids during these processes. The multifarious processes and events, taking place in several species, have been carefully described by Mogensen (1996).

5. PLASTID INHERITANCE IN CHLAMYDOMONAS

5.1. Uniparental maternal plastid inheritance

The green alga *Chlamydomonas reinhardtii* is the most thoroughly investigated object of plastid genetics among all classes of algae. Important steps have been the discovery of the uniparental mode of plastid inheritance in 1954 by Sager, and the proof of recombination of plastid genes in 1965 by Sager, Ramanis and Gillham.

The haploid *C. reinhardtii* forms two morphologically indistinguishable isogametes – mating type plus (mt^+) or mating type minus (mt^-). The cell of *C. reinhardtii* contains one large plastid per cell. A characteristic feature of the fertilization process is the fact that, after the fusion of a mt^+ and a mt^- cell, there is also a regular fusion of the plastids contributed by both parents.

In the course of time, several types of antibiotic resistance and other traits were found to be inherited in a non-mendelian mode. These traits resulted from plastid mutations. Plastid-encoded traits in *C. reinhardtii* include resistance against streptomycin, erythromycin, neomycin, carbomycin, spiramycin, and spectinomycin; photosynthetic deficiency and acetate requirement; and resistance against growth inhibitors. All these plastid mutations exhibit, as a rule, a very characteristic mode of non-mendelian, uniparental maternal inheritance (if one considers mt^+ as the "female" and mt^- as the "male" isogamete).

In a cross between a streptomycin-resistant mt^+ strain and a streptomycin-sensitive mt^- strain, the F_1 cells only exhibit the streptomycin resistance. The paternal trait, streptomycin sensitivity, is not found in the progeny. Also in the following generations, the only plastid trait that is transmitted is the trait present in the mt^+ parent (be it the "mutant" or the "wild type" trait). The paternal plastid genes are disappearing, although there is a fusion of the two isogametes and a fusion of the plastids contributed by both parents. This mode of inheritance is found for all plastid mutations, mentioned above, and the respective wild type alleles.

Detailed genetic and molecular analyses in several laboratories led to the conclusion that in zygotes *of Chlamydomonas* enzymatic actions take place which lead to the destruction of the paternal plastid DNA. In Sections 3 and 4 we described that in seed plants (angiosperms and gymnosperms) the exclusion of plastids from male or female gametes is taking place for the most part by cytological mechanisms. In contrast, in *Chlamydomonas* enzymes are acting to destroy the genetic plastid information of the paternal, the mt^- parent.

Initially Sager and Ramanis (1973) put forward the idea that in *Chlamydomonas* a restriction-modification mechanism is acting as is found in bacteria, and that the plastid DNA of mt^+ and mt^- cells differ in the degree of methylation (Sager and Grabowy, 1983). But typical restriction enzymes do not exist in eukaryotes. Nevertheless comparable eukaryotic systems (nucleases) act in this alga.

Not all details of this interesting phenomenon have been elucidated so far. But already a number of interesting observations have been worked out. Using the DNA-specific fluorochrome DAPI (4'-6-diamidino-2-phenylindole) Kuroiwa et al. (1982) and Kuroiwa (1985) reported that soon (40 minutes) after the beginning of fusion of the mt^+ and mt^- isogametes and zygote formation, the DNA nucleoids in the mt^- plastid become smaller and disappear completely within 10 minutes. They could no longer be detected by DAPI staining. There are indications that the ptDNA of the mt^- cell is being degraded by specific nucleases within a time period of 10 minutes; this is obviously taking place before the fusion of both plastids (Nishimura et al., 1999).

C. reinhardtii contains a cluster of nuclear genes, in the center of which is the mating-type locus, mt; it codes for specific functions that are activated only in cells undergoing meiosis. These mt-controlled enzymatic functions (including nuclease C) are responsible for the destruction and decay of the ptDNA of the mt^- cell, but simultaneously cause the protection of the ptDNA of the mt^+ cell (Boynton et al., 1992; Nishimura et al., 1999). It is highly remarkable that this mt gene cluster has an inverse action on the mitochondrial DNA. It causes the protection of the mt^-

mitochondrial DNA and simultaneously causes the destruction of the mt^+ mitochondrial DNA.

5.2. Biparental and paternal plastid inheritance

It has been emphasized at the beginning of Section 5 that in *Chlamydomonas,* as a rule, a uniparental maternal mode of inheritance of plastid genes is observed. But there are interesting and far reaching exceptions to this rule – biparental or even paternal plastid inheritance. The mechanism for the destruction of male ptDNA does not work with 100% efficiency. Spontaneously, in a few per cent of meiotic zygotes (1 - 9%), the male plastid DNA is not destroyed. Thus a biparental plastid inheritance is taking place. Later on it was observed that a UV irradiation of the mt^+ gametes increases the frequency of zygotes with biparental plastid inheritance. Such "biparental zygotes" contain both maternal and paternal plastid genes and transmit them, after meiosis, into the haploid gonospores.

In 1965 Ruth Sager proved that in such "biparental zygotes" genetic recombination of plastid genes is regularly taking place. During the following years the labs of Ruth Sager as well as of Gillham, Boynton and Harris worked out the characteristics of the recombination processes of plastid genes (Sager and Ramanis, 1970; Boynton et al., 1992; Gillham, 1994).

After the application of very high doses of UV on mt^+ gametes even a complete inactivation of maternal plastid genes may occur which results in a uniparental paternal plastid inheritance.

6. INHERITANCE OF MITOCHONDRIA IN SEED PLANTS AND LOWER PLANTS

6.1. Seed plants (angiosperms and gymnosperms)

The inheritance of mitochondria exhibits a variety of different modes in different taxa. Angiospermous plants with a biparental plastid inheritance, e.g. *Pelargonium* and *Oenothera,* also show a biparental inheritance of mitochondria. In *Pelargonium zonale* Kuroiwa et al. (1993) could, after DAPI staining combined with fluorescence imaging, follow the transmission of plastids and mitochondria separately. They demonstrated the inheritance of both mitochondria and plastids from the generative and sperm cells into the egg cells. Guo and Hu (1995) published convincing electron microscopic investigations which proved the biparental inheritance of mitochondria in *Pelargonium*, too (Figure 3).

In contrast, in plants with a uniparental maternal plastid inheritance, a uniparental maternal inheritance of the mitochondria was observed. A good marker, e.g. in *Zea mays* and *Triticum aestivum*, is the maternal inheritance pattern of "cytoplasmic male sterility" which is caused by mutations in the mtDNA (reviewed in Chapters 22 and 23 of this volume). In these cases the uniparental maternal inheritance of the (mutant) mitochondria is obvious.

In other species a transmission of paternal mitochondria into the egg cell is possible. After crossing two races of *Brassica napus*, one of which contained a mitochondrial plasmid and the other no plasmid, the transmission of the mitochondrial plasmid from the paternal plant into 10% of the F_1 plants was shown by Erickson and Kemble (1990). In *Musa acuminata* a mainly paternal transmission of the mitochondria was reported by Faure et al. (1994), and simultaneously a predominantly maternal inheritance of the plastids.

In gymnosperms, a variety of mitochondrial inheritance patterns were observed. In conifers, mainly paternal inheritance of plastids was found (as described in Section 4), but several modes of mitochondrial transmission were reported. In *Sequoia sempervirens* and in *Calocedrus decurrens* there is a paternal inheritance of mitochondria, as also for the plastids (Neale et al., 1989; Neale et al., 1991). However, in *Pinus rigida* and *Pinus taeda,* a maternal inheritance of the mitochondria was found, accompanied by a paternal plastid inheritance (Neale & Sederoff, 1989). In *Pinus banksiana* and *Pinus contorta*, a biparental inheritance of mitochondria with a bias for the maternal side was described (Wagner et al., 1991). Mogensen (1996) discussed several further aspects of the transmission of mitochondria.

An interesting finding, made in vertebrates, may perhaps be informative also for plant geneticists. In cattle and in rhesus apes, a sperm is transmitting about 100 paternal mitochondria into the egg cell. But these mitochondria have already been marked, in the male seminal tract, with ubiquitin, a recycling marker. Consequently these mitochondria are degraded during the first embryonic divisions and are soon lost (Sutovsky et al., 1999). Whether similar or comparable mechanisms may also be acting in plants, is an absolutely open question.

6.2. Yeast

Haploid cells of yeast, *Saccharomyces cerevisae*, belong to one of the two mating types, a and α. In crosses of cells of different mating types, mitochondria are transmitted from both cells into the zygote. This creates the possibility for the fusion of mitochondria and allows the recombination of mitochondrial genes from different parents. The recombination of mitochondrial genes, coming from different parents, is a normal phenomenon in yeast, and of course a clear proof that the mitochondria and the genetic mitochondrial information from both parents are transmitted into the yeast zygote. (For further details, see Beale and Knowles, 1978; Gillham, 1994.)

6.3. Chlamydomonas

The physical structure of the mitochondrial DNA of *C. reinhardtii* has been elucidated, and its DNA sequence was determined (Boer and Gray, 1989; Riemen et al., 1993; Lang et al., 1999). Several mitochondrial mutations have been found and used in experiments (Boynton et al., 1992). As already described in paragraph 5.1 there is, as a rule, a uniparental maternal inheritance of the ptDNA (via the mt^+ parent) in *C. reinhardtii*. This mode of inheritance is controlled by a cluster of nuclear genes, in the center of which is the mating-type locus, mt. The mt^+-controlled enzymatic activity causes the destruction of the mt^- ptDNA, but simultaneously protects the mitochondrial DNA contributed by the mt^- cell and destructs the mitochondrial DNA contributed by the mt^+ parent. It was shown that UV irradiation of the mt^+ gamete before cell fusion with a mt^- gamete causes the intense weakening of the enzymatic destruction system and leads to a biparental inheritance of the ptDNAs of both parents (Sager and Ramanis, 1967; Boynton et al., 1992; Gillham, 1994). There are indications that the weakening of the enzymatic destruction system for the ptDNA has also has an influence on the transmission of the mitochondrial DNA (Gillham, 1994), which under these conditions may also be transmitted in a biparental manner.

7. ACKNOWLEDGEMENTS

Acknowledgement: I thank Dr. Gerd Hause, Biocenter, Martin-Luther-University, Halle, for valuable discussions.

8. REFERENCES

Baur, E. (1909). Das Wesen und die Erblichkeitsverhältnisse der „Varietates albomarginatae hort." von *Pelargonium zonale*. *Zeitschrift für Induktive Abstammungs- und Vererbungslehre, 1*, 330-351.

Baur, E. (1910a).Vererbungs- und Bastardierungsversuche mit *Antirrhinum*. *Zeitschrift für Induktive Abstammungs- und Vererbungslehre, 3*, 34-98.

Baur, E. (1910b).Untersuchungen über die Vererbung von Chromatophorenmerkmalen bei *Melandrium, Antirrhinum und Aquilegia*. *Zeitschrift für Induktive Abstammungs- und Vererbungslehre, 4*, 81-102.

Bednara, J., & Rodkiewicz, B. (1974). Megasporocyte and megaspore ultrastructure in *Epilobium*. *Bulletin Academiae Poloniae Scientiarum, II. 2*.847-850.

Birky, C.W. (2001). The inheritance of genes in mitochondria and chloroplasts: Laws, mechanisms, and models. *Annu. Rev. Genet., 35*, 125-148.

Boer, P.H., & Gray, M.W. (1989). Nucleotide sequence of a region encoding subunit 6 of NADH dehydrogenase (ND6) and $tRNA^{Trp}$ in *Chlamydomonas reinhardtii* mitochondrial DNA. *Nucleic Acids Res., 17*, 3993.

Boynton, J., Gillham, N.W., Newman, S.M., & Harris, E.H. (1992). Organelle genetics and transformation of *Chlamydomonas*. In R.G. Herrmann (Ed.), *Cell Organelles* (pp.3-64). Wien, New York: Springer Verlag.

Camefort, H. (1967). Observations sur les mitochondries et les plastes de la cellule centrale et de l'oosphere du *Larix decidua* Mill. (*Larix europea* D.C.). *Comptes Rendus del Academie des Sciences, Paris, 265*,1293-1296.

Chat, J., Chalak, L., & Petit, R.J. (1999). Strict paternal inheritance of chloroplast DNA and maternal inheritance of mitochondrial DNA in intraspecific crosses of kiwifruit. *Theor. Appl. Genet., 99*, 314-322.

Chesnoy, L. (1987). La reproduction sexuee des Gymnospermes. *Bulletin de la Societe Botanique de France, Actualites Botaniques, 134*, 63-85.

Chesnoy, L., & Thomas, M..J. (1971). Electron microscopy studies on gametogenesis and fertilization in gymnosperms. *Phytomorphol., 21*, 50-63.

Cipriani, G., Testolin, R., & Morgante, M. (1995). Paternal inheritance of plastids in interspecific hybrids of the genus *Actinidia* revealed by PCR-amplification of chloroplast DNA fragments. *Mol. Gen. Genet., 247*, 693-697.

Cipriani, G., Testolin, R., & Gardner, R. (1998). Restriction-site variation of PCR-amplified chloroplast DNA regions and its implication for the evolution and taxonomy of *Actinidia*. *Theor. Appl. Genet., 94*, 389-396.

Clauhs, R.P., & Grun, P. (1977). Changes in plastid and mitochondrion content during maturation of generative cells of *Solanum* (Solanaceae). *Amer. J. Bot., 64,* 377-383.

Cornu, A. & Dulieu, H. (1988). Pollen transmission of plastid-DNA under genotypic control in *Petunia hybrida* hort. *J. Hered., 79*, 40-44.

Correns C. (1909). Vererbungsversuche mit blass(gelb)gruenen und buntblaettrigen Sippen bei *Mirabilis jalapa, Urtica pilulifera* und *Lunaria annua. Zeitschrift für Induktive Abstammungs- u Vererbungslehre, 1,* 291-329.

Correns C., Hrsg. von Wettstein, F. (1937). *Nicht mendelnde Vererbung. (Handbuch der Vererbungswissenschaft,. Band II H)*, Berlin: Gebrueder Borntraeger.

Corriveau, J.S., & Coleman, A.W. (1988). Rapid screening method to detect potential biparental inheritance of plastid DNA and results for over 200 angiosperm species. *Amer. J. Bot., 75*, 1443-1458.

Corti, E.F.,& Cecchi, A.F. (1970). The behaviour of the cytoplasm during the megasporogenesis in *Paphiodition spicerianum* (RcWb.f.) Pfitzer. *Caryologia, 23,* 715.

Daniell, H., Datta, R., Varma, S., Gray, S., & Lee, S. B. (1998). Containment of herbicide resistance through genetic engineering of the chloroplast genome. *Nat. Biotechnol., 16*, 345-348.

DeCosa, B., Moar, W., Lee, S. B., Miller, M., & Daniell, H. (2001). Overexpression of the *Bt* cry2Aa2 operon in chloroplasts leads to formation of insecticidal crystals. *Nat. Biotechnol., 19*, 71-74.

DeGray, G., Rajasekaran, K., Smith, F., Sanford, J., & Daniell, H. (2001). Expression of an antimicrobial peptide via the chloroplast genome to control phytopathogenic bacteria and fungi. *Plant Physiology, 127*, 852-862.

Diers. L. (1967). Uebertragung von Plastiden durch den Pollen bei *Antirrhinum majus. Mol. Gen. Genet., 100,* 56-62.

Diers L (1971). Uebertragung von Plastiden durch den Pollen bei *Antirrhinum majus*. II. Der Einfluss verschiedener Temperaturen auf die Zahl der Schecken. *Mol. Gen. Genet., 113*,150-153.

Dong, J., Wagner, D.B., Yanchuk, A.D., Carlson, M.R., Magnussen, S., Wang XR. et al. (1992). Paternal chloroplast DNA inheritance in *Pinus contorta* and *Pinus banksiana*: Independence of parental species or cross direction. *J. Hered. 83*, 419-422.

Erickson, L.,& Kemble, R. (1990). Paternal inheritance of mitochondria in rapeseed (*Brassica napus*). *Mol. Gen. Genet., 222*, 135-139.

Faure, S., Noyer, J-L., Carreal, F., Horry, J-P., Bakry, F.,& Lanaud, C. (1994). Maternal inheritance of chloroplast genome and paternal inheritance of mitochondrial genome in bananas (*Musa acuminata*). *Curr. Genet., 25*, 265-269.

Gillham, N.W. (1963). The nature of exceptions to the pattern of uniparental inheritance for high level streptomycin resistance in *Chlamydomonas reinhardi. Genetics, 48*, 431-439.

Gillham, N.W. (1965). Linkage and recombination between nonchromosomal mutations in *Chlamydomonas reinhardi. Proc. Natl. Acad. Sci. USA, 54*, 1560-1567.

Gillham N.W., Boynton, J.E.& Lee, R.W. (1974). Segregation and recombination of non-Mendelian genes in *Chlamydomonas. Genetics, 78,* 439-457.

Gillham, N.W. (1994). *Organelle Genes and Genomes.* New York, Oxford: Oxford University Press.

Guo, F.L., & Hu, S.Y. (1995). Cytological evidence of biparental inheritance of plastids and mitochondria in *Pelargonium. Protoplasma*, 186, 201-207.

Hagemann R. (1964). *Plasmatische Vererbung*. Jena: Gustav Fischer Verlag.

Hagemann R. (1965). Advances in the field of plastid inheritance in higher plants. In S.J. Geerts (Ed), *Genetics Today. Proceedings of the XI International Congress of Genetics* (vol 3, pp. 613-625). The Hague, London: Pergamon Press.

Hagemann, R. (1979). Genetics and molecular biology of plastids of higher plants. *Stadler Genetics Symposia (University of Missouri, Columbia), 11*, 91-116,

Hagemann R (1992). Plastid genetics in higher plants. In R.G.Herrmann (Ed), *Cell Organelles* (pp. 65-96). Wien, New York: Springer Verlag

Hagemann, R. (2000a). Erwin Baur (1875-1933). *Pionier der Genetik und Züchtungsforschung. Seine wissenschaftlichen Leistungen und ihre Ausstrahlung auf Genetik, Biologie und Züchtungsforschung von heute.* Eichenau: Verlag Roman Kovar.

Hagemann, R. (2000b). Erwin Baur or Carl Correns: Who really created the theory of plastid inheritance? *J. Hered., 91*, 435-440.

Hagemann, R. (2002). Milestones in plastid genetics of higher plants. *Prog. Bot. , 63*, 5-51.

Hagemann, R., & Schroeder, M.-B. (1985). New results about the presence of plastids in generative and sperm cells of Gramineae. In *Sexual reproduction in seed plants, ferns and mosses* (pp. 53-55). Wageningen: PUDOC.

Hagemann, R., & Schroeder, M.-B. (1989). The cytological basis of the plastid inheritance in angiosperms. *Protoplasma, 153*, 57-64

Herrmann, R.G. (Ed.) (1992). *Cell Organelles. (Plant Gene Research).* Wien, New York: Springer Verlag.

Heslop-Harrison, J.S. (1972). Sexuality of angiosperms. In F. C. Steward (Ed.), *Plant Physiology* (pp. 133-289). New York: Academic Press.

Hess, W., & Boerner, T. (1999). Organellar RNA polymerases of higher plants. *Internatl. Rev. Cytol., 190*, 1-59.

Hipkins, V.D., Krutovskii, S.H., & Strauss, S.H. (1994). Organelle genomes in conifers: structure, evolution, and diversity. *For. Genet. 1*, 179-189.

Horlow, C., Goujand, J., Lepingle, A., Missonier, C., & Bourigin, J-P. (1990). Transmission of paternal chloroplasts in tobacco (*Nicotiana tabacum*.). *Plant Cell Rep., 9*, 249-252.

Jensen, W.A. (1972). *The embryo sac and fertilization in angiosperms.* Honolulu, Hawaii: Harold L. Lyon Arboretum.

Kirk, J.T.O.,& Tilney-Bassett, R.A.E. (1967). *The Plastids. Their Chemistry, Structure, Growth and Inheritance.* San Francisco: W.H.Freeman and Company.

Knox, R.B. (1984). The pollen grain. In B.M. Johri (Ed.), *Embryology of angiosperms* (pp. 197-272). Berlin, Heidelberg, New York, Tokyo: Springer Verlag.

Kumar, S., Dhingra, A., & Daniell, H. (2004). Highly efficient stable genetic transformation of carrot plastid via somatic embryogenesis to confer salt tolerance. *Plant Phys. in press*

Kuroiwa, T. (1985). Mechanism of maternal inheritance of chloroplast DNA: an active digestion hypothesis. *Microbiolog. Sci., 2*, 267-272.

Kuroiwa, T., Kawazu, T., Uchida, H., Ohta, T., & Kuroiwa, H. (1993). Direct evidence of plastid DNA and mitochondrial DNA in sperm cells in relation to biparental inheritance of organelle DNA in *Pelargonium zonale* by fluorescence/electron microscopy. *Eur. J. Cell Biol., 62*, 307-313.

Lang, B.F., Gray, M.W., & Burger, G. (1999). Mitochondrial genome evolution and the origin of eukaryotes. *Annu. Rev. Genet., 33*, 351-397.

Lee, S. B., Kwon, H. B., Kwon, S. J., Park, S. C., Jeong, M. J., Han, S. E., Byun, M.O. & Daniell, H. (2003). Accumulation of trehalose within transgenic chloroplasts confers drought tolerance. *Mol. Breeding, 11*, 1-13.

Masoud, S.A., Johnson, L.B., & Sorensen, E.L. (1990). High transmission of paternal plastid DNA in alfalfa plants demonstrated by restriction fragment polymorphic analysis. *Theor. Appl. Genet., 79*, 49-55.

McBride, K. E., Svab, Z., Schaaf, D. J., Hogan, P. S., Stalker, D. M., & Maliga, P. (1995). Amplification of a chimeric *Bacillus* gene in chloroplasts leads to an extraordinary level of an insecticidal protein in tobacco. *Bio/Technology 13*, 362-365.

Medgyesy, P., Pay, A., & Marton, L. (1986). Transmission of paternal chloroplasts in *Nicotiana. Mol. Gen. Genet., 204*, 195-198.

Mogensen, H.L. (1988). Exclusion of male mitochondria and plastids during syngamy in barley as a basis for maternal inheritance. *Proc. Natl. Acad. Sci. USA, 85*, 2594-2597.

Mogensen, H.L. (1992). The male germ unit: concept, composition, and significance. *Intl. Rev. Cytol., 140*, 129-147.

Mogensen, H.L. (1996). The Hows and the Whys of cytoplasmic inheritance in seed plants. *Amer. J. Bot., 83*, 383-404.

Mogensen, H.L., & Rusche (1985). Quantitative ultrastructural analysis of barley sperm. I. Occurrence and mechanism of cytoplasm and organelle reduction and the question of sperm dimorphism. *Protoplasma, 128*, 1-13.

Mogensen, H.L., Wagner, V.T., & Dumas, C. (1990). Quantitative, three-dimensional ultrastructure of isolated corn (*Zea mays*) sperm cells. *Protoplasma, 153*, 136-140

Neale, D.B., Marshall, K.A., & Harry, D.E. (1991). Inheritance of chloroplast and mitochondrial DNA in incense-cedar *(Calocedrus decurrens.) Can. J. For., 21*, 717-720

Neale, D.B., Marshall, K.A., & Sederoff, R.R. (1989). Chloroplast and mitochondrial DNA are paternally inherited in *Sequoia sempervirens* don Endl. *Proc. Natl. Acad. Sci. USA, 86*, 9347-9349.

Neale, D.B., & Sederoff, R.R. (1989). Paternal inheritance of chloroplast DNA and maternal inheritance of mitochondrial DNA in loblolly-pine. *Theor. Appl. Genet., 77*,212-216.

Neale, D.B., Wheeler, N.C. & Allard, R.W. (1986). Paternal inheritance of chloroplast DNA in Douglas-fir. *Can. J. For. Res., 16*, 1152-1154.

Nishimura, Y., Miszumi, O., Matsunaga, S., Higashiyama, T., Yokota, & Kuroiwa, T. (1999). The active digestion of uniparental chloroplast DNA in a single zygote of *Chlamydomonas reinhardtii* is revealed by using the optical tweezer. *Proc. Natl. Acad. Sci. USA, 96*, 12577-12582.

Ohba, K., Iwakawa, M., Ohada, Y., & Murai, M. (1971). Paternal transmission of a plastid anomaly in some reciprocal crosses of Suzi, *Cryptomeria japonica D.* Don. *Silvae Genet., 210*, 101-107.

Owens, J.N., & Morris, S.J. (1990). Cytological basis for cytoplasmic inheritance in *Pseudotsuga menziesii*. I. Pollen tube and archegonial development. *Amer. J. Bot., 77*, 433-445.

Owens, J.N., & Morris, S.J. (1991). Cytological basis for cytoplasmic inheritance in *Pseudotsuga menziesii*. II Fertilization and proembryo. *Amer. J. Bot., 78*, 1515-1527.

Reboud, X., & Zeyl, C. (1994). Organelle inheritance in plants. *Heredity, 72*, 132-140.

Renner, O. (1936). Zur Kenntnis der nichtmendelnden Buntheit der Laubblaetter. *Flora, 130*, 218-290.

Riemen, G., Lisowsky, T., Maggouta, F., Michaelis, G., & Pratje, E. (1993). Extranuclear inheritance: Mitochondrial genetics. *Prog. Bot., 54*, 318-333.

Rochaix J.-D. (1994). *Chlamydomonas reinhardtii* as the photosynthetic yeast. *Annu. Rev. Genet., 29*, 209-230.

Ruiz, O. N., Hussein, H., Terry, N., & Daniell, H. (2003). Phytoremediation of organomercurial compounds via chloroplast genetic engineering. *Plant Physiol., 132*, 1-9.

Russell, S.D. (1984). Ultrastructure of the sperm cell of *Plumbago zeylanica*. II. Quantitative cytology and three-dimensional organization. *Planta, 162*, 385-391.

Russell, S.D. (1987). Quantitative cytology of the egg and central cell of *Plumbago zeylanica* and its impact an cytoplasmic inheritance patterns. *Theor. Appl. Genet., 74*, 693-699.

Russell, S.D. (1992). Double fertilization. *Intl. Rev. Cytol., 140*, 357-388.

Sager, R. (1954). Mendelian and non-Mendelian inheritance of streptomycin resistance in *Chlamydomonas reinhardi. Proc. Natl. Acad. Sci. USA, 40*, 356-362.

Sager, R. (1955). Inheritance in the green alga *Chlamydomomas reinhardi. Genetics, 40*, 476-489.

Sager, R. (1960). Genetic systems in *Chlamydomonas. Science, 132*, 1459-1465.

Sager, R. (1972). *Cytoplasmic Genes and Organelles*. New York, London: Academic Press.

Sager, R., & Grabowy, C. (1983). Differential methylation of chloroplast DNA regulates maternal inheritance in a methylated mutant of *Chlamydomonas. Proc. Natl. Acad. Sci. USA, 80*, 3025-3029.

Sager, R., & Ishida, M.R. (1963). Chloroplast DNA in *Chlamydomonas. Proc. Natl. Acad. Sci. USA, 50*, 725-730.

Sager, R., & Lane, D. (1972). Molecular basis of maternal inheritance. *Proc. Natl. Acad. Sci. USA, 69*, 2410-2413.

Sager, R., & Ramanis, Z. (1965). Recombination of non-chromosomal genes in *Chlamydomonas. Proc. Natl. Acad. Sci. USA, 53*, 1053- 1061.

Sager, R, & Ramanis, Z. (1967). Biparental inheritance of non-chromosomal genes induced by ultraviolet irradiation. *Proc. Natl. Acad. Sci. USA, 58*, 931-937.

Salaj, J., Kosova, A., Kormutak, A., & Walles B. (1998). Ultrastructural and molecular study of plastid inheritance in *Abies alba* and some *Abies* hybrids. *Sex. Plant Reprod., 11*, 284-291.

Schmitz, U.K., & Kowallik, K-V. (1986). Plastid inheritance in *Epilobium. Curr. Genet., 11*, 1-5.

Schnarf, K. (1941). *Vergleichende Cytologie des Geschlechtsapparates der Kormophyten*. Berlin: Gebrueder Borntraeger.

Schroeder, M-B. (1986). Ultrastructural studies on plastids of generative and vegetative cells in Liliaceae. 4. Plastid distribution during generative cell maturation in *Convallaria majalis* L. *Biologisches Zentralblatt, 105*, 427-433.

Schroeder, M-B., & Oldenburg, H: (1990). Ultrastructural studies on plastids of generative and vegetative cells in Liliaceae. 7. Plastid distribution during generative cell development in *Tulbaghia violacea* Harv. *Flora, 184*, 131-136.

Schumann, C.M., & Hancock, J.F. (1989). Paternal inheritance of plastids in *Medicago sativa*. *Theor. Appl. Genet., 78*, 863-866.

Sears, B.B. (1980). Elimination of plastids during spermatogenesis and fertilization in the plant kingdom. *Plasmid, 4*, 233-255.

Smith, S.E., Bingham E.T., & Fulton, R.W. (1986). Transmission of chlorophyll deficiencies in *Medicago sativa*: evidence for biparental inheritance of plastids. *J. Hered., 77*, 35-38.

Stine, M., Sears, B.B., & Keathley, D.E. (1989). Inheritance of plastids in interspecific hybrids of blue spruce and white spruce. *Theor. Appl. Genet., 78*, 768-774.

Sutovsky, P., Moreno, R.D., Ramalho-Santos, J., Dominko, T. Simerly, C., & Schattan, G. (1999). Ubiquitin tag for sperm mitochondria. *Nature, 402*, 371-372.

Sutton, B.C.S., Flanagan, D.J., Gawley, J.R., Newton, C.H., Lester, D.T., & Elkassaby, Y.A. (1991). Inheritance of chloroplast and mitochondria DNA in *Picea* and composition of hybrids from introgression zones. *Theor. Appl. Genet., 82*, 242-248.

Szmidt, A.E., Alden. T., & Hallgren, J.E. (1987). Paternal inheritance of chloroplast DNA in *Larix*. *Plant Mol. Biol., 9*, 59-64.

Testolin, R., & Cipriani, G. (1997). Paternal inheritance of chloroplast DNA and maternal inheritance of mitochondrial DNA in the genus *Actinidia*. *Theor. Appl. Genet., 94*, 897-903.

Thomas, M.J., & Chesnoy, L. (1969). Observations relatives aux mitochondries Feulgen positive de la zone perinucleaire de l'oosphere du *Pseudotsuga menziesii* (Mirb.) Franco. *Revue de Cytologie et de Biologie Vegetales, 32*, 165-182.

Tilney-Bassett, R.A.E. (1994). Nuclear control of chloroplast inheritance in higher plants. *J. Hered., 85*, 347-354.

Van Went, J.L., & Willemse, M.T.M. (1984). Fertilization. In B.N. Johri (Ed.), *Embryology of Angiosperms* (pp. 273-317). Berlin, Heidelberg, New York, Tokyo: Springer-Verlag.

Wagner, D.B. (1992). Nucleolar, chloroplast, and mitochondrial DNA polymorphisms as biochemical markers in population genetic analysis in forest trees. *New For., 6*, 373-390.

Wagner, D.B., Dong, J., Carlson, M.R. & Yanchuk, A.D. (1991). Paternal leakage of mitochondrial DNA in *Pinus*. *Theor. Appl. Genet., 82*, 510-514.

Wagner, D.B., Govindaraju, D.R., Yeatman, C.W. & Pitel, J.A. (1989). Paternal chloroplast DNA inheritance in a diallel cross of jack pine *(Pinus banksiana* Lamb.). *J. Hered., 80*,483-485

Wang, T., Li, Y., Shi, Y., Reboud, X., Darmency, H. & Gressel, J. (2003). Low frequency transmission of a plastid encoded trait in *Setaria italica*. *Theor. Appl. Genet., 108,* 315-320.

Yamada, A.S., Miyamura, S.& Hori, T. (1993). Cytological study on plastid inheritance of *Cryptomeria japonica* D. Don. *Plant Morphol., 5,* 19-29.

Zhu, T., Mogensen, H.L. & Smith, S.E. (1992). Heritable paternal cytoplasmic organelles in alfalfa sperm cells: ultrastructural reconstruction and quantitative cytology. *Eur. J.Cell Biol., 59,* 211-218.

Zhu, T., Mogensen, H.L & Smith, S.E. (1993). Quantitative, three-dimensional analysis of alfalfa sperm cells in two genotypes: implications for biparental plastid inheritance. *Planta, 190*, 143-150.

CHAPTER 5

PLASTID GENOMES

R.M. MAIER[1] AND C. SCHMITZ-LINNEWEBER[2]

[1]*Cell Biology, Philipps University, Marburg, Germany*
[2]*Institute of Molecular Biology, University of Oregon, Eugene, Oregon, USA*

Abstract. Plastids possess their own genome, the plastome, and a specific machinery to decode its genetic information. The first evidence for the presence of heritable material in plastids was reported at the beginning of the last century and was based on observations of non-Mendelian inheritance of variegated leaf phenotypes. More than 50 years later, specific plastid-localized DNA was identified. The first complete plastid genome sequences were published in 1986. Since then, more than 30 genomes from embryophytes and eukaryotic algae have been deciphered. The typical plastid genome consists of multiple copies of a basic unit of double-stranded DNA of species-specific length. Circular and linear molecules in monomeric and multimeric forms have been observed. Generally, several plastid DNA molecules are organized into nucleo/protein complexes, so-called nucleoids. Whereas the physical organization of plastid genomes of embryophytes is highly conserved, algal plastid genomes appear highly divergent. The plastid genome of embryophytes typically consists of units of 120 to 160 kbp in length, with each unit generally subdivided in four sections with two of the sections made up of two identical copies of a large inverted repeat region. In contrast, plastid genome units of algae show large size variations from less than 100 kbp to more than 1.5 Mbp and, instead of a single genome unit, certain dinoflagellates contain several 2 to 3 kbp minicircles mostly encoding only a single gene. Moreover, in some protozoan parasites, relic non-photosynthetic plastid-like organelles of algal origin, so-called apicoplasts, contain small 35 kbp genomes. The evolutionary origin of plastid genomes traces back to the engulfment of a free-living cyanobacterium by a eukaryotic host more than 1.2 billion years ago. The functional and genetic integration of the former autarkic cyanobacterium into the newly emerging photosynthetic eukaryotic cell was accompanied by an intermixing and restructuring of genomes. During the course of this process the original cyanobacterial genome, encoding several thousand genes, was intensely reduced such that the present-day plastid genome contains only 100 to 250 genes. A large proportion of the original genes have been translocated to the nucleus and most gene products for plastid functions have to be reimported into the organelle. Gene transfer from the plastid genome to the mitochondrial genome also took place. However, no evidence for retrograde gene transfer from the nuclear or mitochondrial genome to the plastid genome has been found. Since DNA transfer from plastids to the nuclear genome is still an ongoing process, the question arises why it was not driven to completion. Most of the genes remaining in the plastid genome encode either components of the photosynthetic apparatus or of the transcription/translation apparatus of the organelle. Probably a core set of plastid genes must be maintained because of their regulatory properties. In this respect, it has been hypothesized that to prevent the production of harmful oxygen radicals, it is favourable to maintain certain redox-regulated genes inside the plastid compartment.

H. Daniell and C.D. Chase (eds.), Molecular Biology and Biotechnology of Plant Organelles, 115—150.

1. INTRODUCTION

The concept that plastids contain genetic information dates back to the beginning of the last century, when Erwin Baur and Carl Correns independently reported on observations of non-Mendelian inheritance of variegated leaf phenotypes in *Pelargonium zonale* (Baur, 1909) and *Mirabilis jalapa* (Correns, 1909). Baur correctly interpreted this phenomenon as a consequence of plastid-localized hereditary material which is able to mutate (Baur, 1909). According to Baur's concept, leaf variegation is based on somatic segregation and random sorting-out of nonmutated and mutated plastids. Species-specific variations of non-Mendelian inheritance of leaf colour traits as observed by Baur and Correns were concluded to be based on different modes of plastid transmission to the zygote (biparental, i.e. by both the egg cell and the sperm cell, in *P. zonale*, and maternal, i.e. only by the egg cell, in *M. jalapa* (Winge, 1919; reviewed by Hagemann in Chapter 4 of this volume).

The diversification of the genetic constitution of plastids during evolution has been inferred from the finding that plastids of certain *Oenothera* species are impaired in development when combined with the nuclear genome of certain other species, a phenomenon described as hybrid deficiency ("Bastardbleichheit"; (Renner, 1924; 1929). Otto Renner introduced the term "plastome" to define the plastid hereditary material of a cell (Renner, 1934). Genetic analysis of plastome mutants and genome/plastome (in)compatibilities in *Oenothera* revealed that the plastome is composed of several functional subunits (Schötz, 1958; Stubbe, 1959). A detailed historical account of these early stages of plastid genetics has been given by Hagemann (2000; 2002).

Soon after the discovery that genetic information in living organisms is stored in the form of nucleic acids, the presence of DNA in plastids was suggested by the light microscopic detection of faint Feulgen-positiv material (Chiba, 1951). In 1962, Hans Ris and Walter Plaut reported the electron microscopic detection of uranyl acetate labelled 25 - 30 Å fibrils within plastids of *Chlamydomonas moewusii* that were DNAse-sensitive (Ris & Plaut, 1962). Such fibrils, reminiscent of DNA fibrils in the nucleoplasm of eubacteria, were subsequently detected in plastids of various embryophytes and eukaryotic algae. Results of DNAse-sensitive incorporation of radioactivity into plastids of cells incubated with [^{3}H]thymidine also strongly suggested the presence of plastid localized DNA (Brachet, 1959). In 1963, it was shown that CsCl density gradient centrifugation of total cell DNA allows the separation of distinct minor satellite DNA fractions from the main fraction of nuclear DNA (Chun et al., 1963). The identification of a specific satellite DNA fraction with a buoyant density different from nuclear DNA, which was enriched severalfold in purified chloroplast fractions of *Chlamydomonas reinhardtii* (Sager & Ishida, 1963), together with the finding that this DNA and corresponding DNA fractions of various other species differ significantly in their GC-content from nuclear DNA (Kirk, 1963), led to the general acceptance of the presence of unique plastid-specific DNA. This was also supported by the absence of a specific satellite DNA in UV-radiated chloroplast-less cells of *Euglena gracilis* (Leff et al., 1963). Further investigations revealed that, in contrast to nuclear DNA, plastid DNA lacks 5-methyl-cytosine and exhibits specific reassociation kinetics different from nuclear DNA. For plastid DNA from lettuce, a kinetic complexity of only 1.25 x 10^8 daltons (or ~120 kbp) was determined indicating that the plastome either consists of repeated DNA sequences or multiple copies of relatively small DNA molecules (Wells & Birnstiel, 1969). Such small DNA molecules were subsequently identified by electron microscopy of lysed chloroplasts from *Euglena gracilis* (Manning et al., 1971). A large fraction of the observed DNAs were circular with a length of about 40 μm (or ~140 kbp), whereas the remainder were linear molecules of different lengths. It was suggested that each circular molecule contains the total genetic information of the plastome (Manning et al., 1971).

Besides monomeric circles, plastid DNA molecules from various higher plant species were observed as circular dimers, in head-to-tail and head-to-head conformation of two monomers (Kolodner & Tewari, 1972; Herrmann et al., 1975; Kolodner & Tewari, 1975). Further analyses revealed that the lengths of monomeric plastid DNA molecules show only slight variations (40 – 46 μm) between higher plant species but can be highly divergent (36 – 62 μm) amongst algae. Up to 100 copies of a plastid DNA molecule have been found in mature leaf chloroplasts, in some algal chloroplasts even more than 100 have been noted (summarized in Kirk & Tilney-Bassett, 1987). Over the intervening years, the findings (i) that plastid DNA is replicated in the plastids themselves and not derived from the nucleus, (ii) that plastids contain specific DNA-dependent RNA polymerases, which transcribe plastid DNA and (iii) that plastids contain characteristic ribosomes, aminoacyl tRNA synthetases and tRNAs involved in synthesis of plastid proteins (summarized in Tewari, 1971; Kirk & Tilney-Bassett, 1987) dispelled any doubt that the plastid-localized DNA molecules represent an active genome. The persistent uncertainty of the molecular basis of the phenomenon of cytoplasmic inheritance that Boris Ephrussi expressed as "there are two kinds of genetics – *nuclear* and *unclear*" was finally over.

The development of new techniques, in particular the application of restriction endonucleases and improved DNA sequencing methods, opened up new possibilities to explore the plastid genome. Restriction fragment length polymorphisms (RFLPs) between plastid genomes of various species, even between those of the same genus (*Nicotiana*), were demonstrated (Vedel et al., 1976). The first restriction maps of plastid chromosomes were soon determined for *Zea mays* (Bedbrook & Bogorad, 1976), *Spinacia oleracea* (Herrmann et al., 1976) and *Euglena gracilis* (Gray & Hallick, 1977). Shortly after, the first plastid-encoded genes, 16S and 23S rDNA from maize, were cloned and mapped (Bedbrook et al., 1977). The first sequence information on plastid nucleic acid molecules were derived by sequencing of small oligonucleotide fragments of T1 RNAse digested 16S rRNAs from *Porphyridium cruentum* (Bonen & Doolittle, 1975) and *Euglena gracilis* (Zablen et al., 1975). In 1979, the group of Hans Kössel determined the very first plant DNA sequence in deciphering a part of the sequence of the maize chloroplast 16S rRNA gene (Schwarz & Kössel, 1979). The complete sequence of 16S rDNA was published

soon after (Schwarz & Kössel, 1980). In the same year, the first complete gene sequence for a plastid-encoded protein, the large subunit of RuBisCO from maize, was reported (McIntosh et al., 1980). During the following years, many plastid genes were identified and sequenced in various species of embryophytes and eukaryotic algae (reviewed in Bogorad, 2003). Finally, in 1986, a milestone in plastid genetics was reached when Masahiro Sugiura and coworkers reported the complete sequence of the plastid genome of *Nicotiana tabacum* (Shinozaki et al., 1986) and Kanji Ohyama and coworkers reported the sequence of the liverwort *Marchantia polymorpha* (Ohyama et al., 1986). Since then, an ever-growing number (currently more than 30) of plastid genomes from various species of embryophytes and eukaryotic algae have been completely sequenced (for the actual list see http://www.ncbi.nlm.nih.gov/genomes/ORGANELLES/plastids_tax.html).
However, only five of these belong to crop plant species. This is in particular lamentable, as an important prerequisite for plastid transformation is a detailed knowledge of intergenic spacers in plastid chromosomes. These spacers are targets for the integration of transgenes by homologous recombination, a technique which has already proven extraordinarily fruitful for various biotechnological applications (see chaper 16). Identity between vector sequences and target sequence is necessary Daniell et al., 2004; Kumar and Daniell, 2004), as heterologous transformation vectors have not yielded high frequency transformations so far (e.g. DeGray et al., 2001, see also chapter 16). Therefore, further genome sequencing projects of crop plant plastid chromosomes is one of the more pressing tasks of this field.

In this chapter we review the current knowledge of the physical organization, coding content, and variation in coding content of plastid genomes. Intracellular and interspecific horizontal transfer of plastid genetic information will also be discussed. Finally, we will consider the question of why plastids have retained genomes during evolution.

2. PHYSICAL ORGANIZATION OF PLASTID GENOMES

Physical maps of more than 1000 plastid genomes have been determined (Palmer, 1991) and more than 30 complete sequences are available (see the website mentioned above). This comprehensive set of indirect data on the structural organization of plastid chromosomes is complemented by direct data obtained by means of electron microscopy (e.g. Kolodner & Tewari, 1972; Kowallik & Herrmann, 1972), fluorescence microscopy (Oldenburg and Bendich, 2004), fluorescence in situ hybridization (FISH) (Lilly et al., 2001; Maul et al., 2002) and pulsed-field gel electrophoresis (Deng et al., 1989; Backert et al., 1995; Oldenburg & Bendich, 2004; see Figure 1). These analyses revealed that the typical plastome consists of multiple copies of a basic unit of double-stranded DNA with species-specific lengths of typically 100 to 250 kbp. Closed circles as well as linear molecules, both in monomeric and multimeric form, were observed, and clusters of plastid DNA molecules have been found to be organized into nucleo/protein complexes, so-called nucleoids.

2.1. Nucleoids

Plastid DNA molecules are not evenly dispersed in the stroma, but are organized in particulate structures, nucleoids, consisting of plastid DNA, RNA and various proteins. Nucleoids of mature leaf chloroplasts of embryophytes have a diameter of approximately 0.2 μm and contain about 10 plastid chromosomes (Sato et al., 2003).

The morphology of nucleoids is phylogeneticaly diverse with respect to shape, size and distribution. In unicellular red algae, nucleoids are located in the center of the chloroplast and resemble bacterial nucleoids. In multicellular red algae, nucleoids are scattered along the periphery of chloroplasts. In brown algae, nucleoids form a large ring-shaped structure and in green algae nucleoids are distributed in the stroma as small particles (Sato et al., 2003). In embryophytes, the number and location of nucleoids changes during plastid development. In proplastids, only a single small nucleoid is located in the center of the organelle. In developing plastids, nucleoids are attached to the envelope membrane, whereas in chloroplasts of mature leaf cells, nucleoids are located in the stroma associated to the thylakoid membrane (Kowallik & Herrmann, 1972; Kuroiwa & Suzuki, 1981). These developmental changes in the distribution of nucleoids are accompanied by qualitative and quantitative changes in protein composition (Sato et al., 2003). For instance, the anchoring of nucleoids to the envelope membrane or thylakoid membrane is thought to be accomplished by different membrane-specific nucleoid proteins (Sato et al., 1998; Jeong et al., 2003). It further appears that development-specific changes, in particular in plastid transcription, replication, and transmission of genetic information are linked to structural remodeling of nucleoids. Besides various DNA-binding proteins involved in organization and maintenance of nucleoid structure (Kobayashi et al., 2002), nucleoids have been shown to contain enzymatic activities for replication and transcription of the plastid genome (Sakai, 2001). Moreover, nucleoid proteins have been identified, which enhance or inhibit transcription or replication (Chi-Ham et al., 2002; Sato et al., 2003). For detailed reviews on the role of nucleoids in plastid inheritance, their organization, developmental dynamics and evolution see Kuroiwa (1991), Sato et al. (2003) and Chapter 3 in this volume.

2.2. Plastid chromosomes

Commonly, plastid chromosomes are represented as closed circles (Figure 2). This depiction is not only deduced indirectly from restriction mapping and sequencing but matches with what is actually visible by microscopy (e.g. Figure 1). The overall picture of plastid genomes, however, is more complex. In embryophytes and most eukaryotic algae, in addition to monomeric molecules, multimers in both circular and linear form have been observed. Plastid genomes also exhibit some phylogenic diversity. For instance, the plastid genome of certain dinoflagellates has been observed to be composed of many small circular molecules generally encoding only a single gene. Moreover, the plastid DNA of some apicomplexan parasites primarily exists as tandem arrays of linear units.

2.2.1. Embryophytes

The plastid genome of embryophytes (vascular plants and mosses, hornworts, liverworts) typically consists of basic units of double stranded DNA of 120 to 160 kbp length, arranged in monomeric and multimeric circles (e.g. Kolodner & Tewari, 1972; Kowallik & Herrmann, 1972; Lilly et al., 2001) as well as in linear molecules (Oldenburg & Bendich, 2004).

Generally, a basic unit is subdivided in four sections with two identical copies of a (20 – 30 kbp) inverted repeat region (IR_A and IR_B) separating a large (LSC) and small (SSC) single copy region (Figure 2). They are all identical with respect to gene content but exist in two equimolar conformations differing only in the relative orientation of the single copy regions. These structural isomers result from intramolecular (flip-flop) recombination between the IR-regions of circular molecules (Palmer 1983) or, alternatively, from recombination events during replication (Oldenburg & Bendich, 2004). Multimers of these basic units exist in head-to-head and head-to-tail conformation (Kolodner & Tewari, 1979).

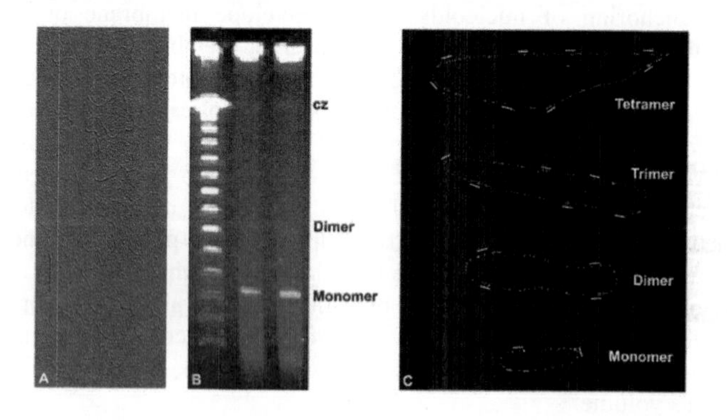

Figure 1. *Plastid DNA molecules from land plants visualized by (A) electron microscopy, (B) pulsed-field electrophoresis and (C) fluorescence in situ hybridization. Size marker in (B) are concatemers of 48.5 kbp λ-DNA molecules. cz marks the compression zone of the gel. Locations of IR-regions in (C) are marked by bars. [(A) from Kleinig & Maier, 1999, ©Spektrum Academic Publishers, Heidelberg, Berlin, (C) from Lilly et al., 2001, ©American Society of Plant Biologists; reprinted with permission.]*

Different results have been obtained on the relative amount of circular and linear plastid DNA molecules. Electron microscopic analysis of CsCl gradient purified plastid DNA revealed up to 80% circular molecules with only a low percentage in dimeric form (Herrmann et al., 1975). Pulsed-field gel electrophoresis (PFGE) revealed a significantly higher amount of multimeric plastid DNA with relative amounts of monomer to dimer, trimer and tetramer of 1:3, 1:9 and 1:27, respectively (Deng et al., 1989). Since high molecular circular DNA usually remained immobile even in PFGE (Levene & Zimm, 1987), the DNA molecules migrating through the gel were considered as linear, which has been directly verified via UV light microscopy of gel slices (Bendich & Smith, 1990). Investigation of the immobile fraction revealed up to 80% circular molecules (Bendich, 1991). Cytogenomic analysis by fluorescence in situ hybridization (FISH) revealed the presence of approximately 45% circular molecules in addition to a fraction of linear molecules with both fractions existing in up to tetrameric form in leaf chloroplasts of 4-week-old tobacco and *Arabidopsis* plants (Lilly et al., 2001). Quite in contrast, a recent work based on PFGE, restriction fragment mapping and fluorescence microscopy of in-gel prepared plastid DNA showed that most of the molecules in leaf chloroplasts of 14-day-old maize seedlings were linear or in complexed branched forms and only 3 - 4% exist as a circle (Oldenburg & Bendich, 2004). Obviously, the composition and intactness of plastid DNA molecules seems to be highly sensitive to the isolation method applied. Since different DNA conformations may be linked to the developmental stage of plastids, e.g. linear molecules may represent replication intermediates that are overrepresented in rapidly dividing cells, the conformations of plastid chromosomes found may be development/tissue-specific as well.

Compared to the nuclear and mitochondrial genomes, the plastid genome is quite conserved among embryophytes. Only few exceptions from the general architecture and lengths have been observed. Species specific differences in plastid chromosome size are mainly attributed to evolutionary expansions and contractions of the IR-regions. In some gymnosperms like *Pinus thunbergii* the IR is shrunk to only 495 bp, in some legumes, like *Pisum sativum*, it is even completely lost. On the other hand, in *Pelargonium hortorum* the IR is bloated to some 76 kbp. Since the IR-organization is also realized in several algae, it seems likely that it is an ancient feature which has been later lost in individual branches during evolution (Palmer, 1991). Characteristically, the IR-region contains a complete rRNA operon. Duplicated rRNA operons are also observed in cyanobacterial genomes which argues for a selective pressure to increase rRNA gene number (Palmer, 1991). The loss of IRs in certain lineages, however, suggest that they are not, or at least not always, obligatory for plastid function. Possibly, the IR structure itself is resistant to recombinational loss and therefore perpetuates itself without any benefit to the plastid. On the other hand, it has been observed that plastid chromosomes of such legumes which have lost the IR have undergone more rearrangements than those that have not, although alternatively, the loss of one IR copy may be simply the result of an overall increase in the frequency of DNA rearrangements in legumes (Palmer & Thompson, 1982). Speculatively, the IR-organization may play a direct role in maintaining the conserved structure of the chloroplast chromosome and also in directly conserving genes encoded by the IR, as these genes characteristically have lower rates of nucleotide substitutions than those encoded by the single copy regions (Curtis & Clegg, 1984; Wolfe et al., 1987). A gene conversion system operating with a slight bias in favour of wild-type sequences would lead to differences in the nucleotide substitution rates between genes of different copy number and might explain the divergent evolutionary rates observed (Birky & Walsh, 1992; Perry & Wolfe, 2002).

Only little divergence is observed in gene order of embryophytic plastid chromosomes. Genes are generally organized in operons and of the few rearrangements found most have their boundaries between operons (Palmer, 1991). Destruction of operon structures would require the subsequent installation of new regulatory elements, and thus should be counterselected. Most of the few large scale differences found can be explained by inversions. These inversions have occurred by intramolecular homologous recombination events between short inversely orientated repeat sequences. In several instances such repeats have been identified close to the endpoints of inversions (Palmer, 1991). Sometimes the endpoints of inversions are associated with tRNA genes. For instance, in grasses, a tRNA-fM/G pseudogene is found at one endpoint of a large 28 kbp inversion and it has been proposed that intermolecular recombination between tRNA genes can be responsible for inversion events (Hiratsuka et al., 1989).

Most of the length mutations found between closely related plastid genomes have an extension of only 1 to 10 bp. Many of them are flanked by or close to short direct repeats or occur within homopolymer tracts and probably result from slippage and mispairing during DNA replication or repair (Palmer, 1991). Less frequent larger deletions, which can comprise up to several hundred base pairs, are most probably generated by unequal crossing-over between misaligned tandem repeats or by intramolecular recombination between short direct repeats. Plastid genomes of embryophytes are a target of a very active recombination system, and without strong selection DNA sequences are rapidly lost. In contrast to the nuclear and mitochondrial genomes, plastid genomes seem to be immune to a durable integration of futile foreign DNA segments. This constraint to eliminate unnecessary sequences is further exemplified by parasitic non-photosynthetic plants, like *Epifagus virginiana*, whose plastid genomes have lost most or all genes for photosynthetic functions, whereas its mitochondria retain previously transferred non-functional copies of photosynthetic genes (Palmer, 1990). Thus, with respect to size and gene content, plastid genomes of embryophytes appear to evolve according to the rule "use it or lose it", resulting in the maintenance of a compact, largely genic, evolutionary highly conservative genome (Palmer, 1990; 1991).

2.2.2. Eukaryotic algae

In contrast to those of embryophytes, algal plastid genomes appear highly divergent, putting into perspective the view of the plastid chromosome as a structurally conservative molecule. The completely sequenced plastid genomes of the charophyte *Chaetospaeridium globosum* (Turmel et al., 2002a) and of the prasinophytes *Mesostigma viridis* (Lemieux et al., 2000) and *Nephroselmis olivacea* (Turmel et al., 1999) structurally resemble plastid genomes of embryophytes, reflecting their close phylogenetic relationships. Similarly, they posses a quadripartite structure with two rDNA containing IR-regions. This genome architecture, however, is not realized in all members of the green lineage (like *Chlorella vulgaris*), but can be found in individual members of the red lineage (like the diatom *Odontella sinensis* and the cryptomonade *Guillardia theta*), which suggests that the IR has been gained and lost on multiple occasions (Simpson & Stern, 2002).

Conspicuous length variations are observed within the group of green algae. Beside many genera with genome sizes of 100 to 200 kbp, plastid genomes of members of e.g. *Acetabularia*, *Nitella* and *Pandorina* can comprise some 400 kbp, and individual members of *Acetabularia* have been described to contain plastid chromosomes of even up to 1.5 Mbp (Palmer, 1991; Simpson & Stern, 2002). Variations from 187 to 292 kbp have been observed between distinct *Chlamydomonas* species, and even intraspecific differences from 150 to more than 400 kbp have been noted for sexually incompatible, but morphologically indistinguishable, members of the same 'species', *Pandorina morum* (Palmer, 1991). Other green algae species, like *Codium fragile* (89 kbp) have relatively small plastid chromosomes (Simpson & Stern, 2002). In *Chlamydomonas* interspecies length differences, in part, trace back to large intergenic insertions (Simpson & Stern, 2002). Another remarkable feature driving interspecific length divergences between plastid genomes is the accumulation of repetitive DNA. Complete sequence determination of the *Chlamydomonas reinhardtii* plastome revealed more than 20,000 short dispersed repeats (SDRs) located in intergenic regions. These repeated sequences which average 30 bp in length make up more than 20% of the complete genome (Maul et al., 2002; Jiao et al., 2004). The evolutionary dynamics of SDRs is conspicuously high. They were not found in every *Chlamydomonas* species, but were present in *Chlorella vulgaris*. Interestingly, SDRs were even polymorphic among different laboratory strains of *Chlamydomonas reinhardtii* that evolved separately for only 30 to 40 years (Maul et al., 2002). Detailed structural analyses revealed that the plastid genome of *Chlamydomonas reinhardtii* consists of a population of monomeric and multimeric circular and linear molecules, with the latter predominating. Unlike plastid genomes of embryophytes, only few molecules larger than trimer size could be detected, and all of these were linear. A large amount of linear molecules of less than monomeric size was hypothesized to originate from intramolecular recombination events via SDR sequences (Maul et al., 2002).

The organization of the plastid genome of the green flagellate *Euglena gracilis* is quite different from that of green algae. Instead of the standard IR-architecture, it is equipped with three tandem direct repeats, each containing a rDNA gene cluster (Hallick et al., 1993). More conspicuously, however, is the large number of introns dispersed throughout the genome. Altogether 155 introns have been found, some of them even located inside other introns, resulting in so-called twintrons. It has been suggested that the large number of introns are descendants of mobile genetic elements that have invaded this plastid genome (Hallick et al., 1993).

Astasia longa, a colourless heterotrophic flagellate closely related to photoautotrophic *Euglena* species, contains a circular 73 kbp plastid genome that is only about half the size of the genome of *Euglena gracilis* (Gockel & Hachtel, 2000). Similar to the holoparasitic plant *Epifagus virginiana*, the reduced genome size results from the preferential loss of no longer required photosynthesis genes.

In the red lineage, sizes of plastid chromosomes seem to be more consistent mostly varying in the range of 120 to 190 kbp. Among the red algae, plastid genomes of some species, like *Porphyra purpurea*, contain large direct repeats coding for a rDNA cluster whereas others, like *Cyanidioschyzon merolae* and *Cyanidium caldarum*, do not. Strikingly, the plastid genome of *Cyanidium caldarum*, a unicellular red alga living in acidic high temperature environments and considered very ancient, contains a 1.2 kbp gene-free region with a small G-rich stem-loop structure that is flanked by short direct repeats. This structural element was proposed to have acted as seed for the development of larger direct repeats as observed in *Porphyra purpurea* (Glockner et al., 2000). In contrast to this, the plastids of *Odontella sinensis* and *Guillardia theta*, which were obtained by secondary endosymbiosis of red algae, contain rDNA clusters arranged in an IR manner, which may have evolved secondarily by rearrangements of preexisting direct repeats.

A most peculiar structure of plastid genomes is observed in several species of peridinin-containing dinoflagellates. Instead of a single genome unit, they contain several 2 to 3 kbp minicircles (Zhang et al., 1999). These molecules mostly encode a single gene, sometimes two, but circles with only fragmentary genes or even 'empty' circles have been observed, too. Minicircles contain a conserved core region, which is similar between the molecules of a given species, but very different even between closely related species. Less than twenty minicircle-coded genes have so far been identified, suggesting that the dinoflagellate plastid genome is in the final stages of gene transfer to the nucleus (Howe et al., 2002; Hackett et al., 2004). Based on structural similarities to bacterial integrons, it was speculated that dinoflagellate minicircles are part of a system for gene transfer, moving genes from plastids to the nucleus (Howe et al., 2002). Quite interestingly, in the dinoflagellate *Ceratium horridum*, minicircles are primarily, if not exclusively, localized in the nucleus whereas plastids contain a mysterious, possibly non-coding high molecular weight DNA as shown by the analysis of fractionated cells (Laatsch et al., 2004).

Protozoan parasites of the phylum Apicomplexa, including the causative organisms of malaria, *Plasmodium* ssp., possess relic plastid-like organelles termed apicoplasts. These non-photosynthetic organelles most probably originated from a secondary endosymbiosis of a plastid-containing alga. Whether this alga was a member of the green or red lineage is still controversial (for review see Foth & McFadden, 2003; Wilson et al., 2003). The apicoplast genomes exhibit the IR-architecture and have been reduced in size to approximately 35 kbp. Complete apicoplast genome sequences from *Plasmodium falciparum* (Wilson et al., 1996), *Toxoplasma gondii* (Kohler et al., 1997) and *Eimeria tenella* (Cai et al., 2003) have been determined. All genes for photosynthetic functions have been lost. The exact role of apicoplasts in parasites is unknown, but probably, the parasite is dependent on retained plastid functions, such as fatty acid, isoprenoid and heme synthesis. Structural analyses revealed the apicoplast genome to consist of a population of circular molecules and linear tandem arrays. The relative proportion of both conformations, however, seems to differ between apicomplexean species. Whereas the apicoplast genome of *Plasmodium falciparum* comprises over 90% circular molecules, in *Toxoplasma gondii* primarily linear tandem arrays of the basic 35 kbp units have been observed that could extend to more than 400 kbp (Williamson et al., 2001). It was suggested that the linear arrayed molecules are not copied *per se*, but originate from rolling circle replication and are therefore not stable genetic units (Williamson et al., 2001). It remains to be answered to what extent this concept also holds for other systems.

3. CODING CONTENT OF PLASTID GENOMES

Plastid chromosomes are packed with genes, with open reading frames often spaced by only a few nucleotides or even overlapping. This, together with the fact that plastid genes are organized in operons, very much resembles the situation in cyanobacteria, which have early been shown to be the plastids' closest relatives by quantitative DNA hybridization (Pigott & Carr, 1972), as well as T1 RNAse oligomer analysis (Bonen & Doolittle, 1975; Zablen et al., 1975; Bonen & Doolittle, 1976). Of the more than 30 plastid genomes completely sequenced, about half fall into the embryophytes. A brief overview of genes found in this group set in contrast to those found in eukaryotic algae will highlight evolutionary trends in plastid genome development.

3.1. Embryophytes

The more than 120 plastid genes described in embryophytes fall with few exceptions into three major classes: (i) genes for components of the photosynthetic apparatus, (ii) genes for factors with functions in biosyntheses and (iii) genes for components of the genetic system of the organelle (Table 1). The majority of these genes are shared among all sequenced members of this group, with the only exception being *Epifagus virginiana*, a non-photosynthetic parasitic angiosperm, which has lost all genes for components of the photosynthetic apparatus (Wolfe et al., 1992a). In the following discussion of gene content in plastid genomes of embryophytes, neither conserved reading frames of unknown function (*ycfs*) nor species-specific reading frames of unknown function (ORFs) will be considered.

3.1.1. Genes for subunits of the photosynthetic apparatus

Of the approximately 60 subunits constituting the four major protein complexes in the thylakoid membrane, the photosystems I and II, the cytochrome b_6/f complex and the ATP synthase, about half are encoded in the plastid genome (Herrmann et al., 1992). Most of the genes have been initially identified by means of purification and characterization of their corresponding proteins (for details the reader is referred to a historical review by Bogorad, 2003). A handful of genes has been characterized based on loss-of-function mutagenesis. For example, using this method, *ycf3* and *ycf4* gene products were found to be essential for the correct assembly of photosystem I (Boudreau et al., 1997) and *petN* and *psbZ* have been found to be subunits of the cytochrome b_6/f complex and photosystem II, respectively (Hager et al., 1999; Ruf et al., 2000; Swiatek et al., 2001). Today, five plastid genes for photosystem I, sixteen for photosystem II, six for the cytochrome b_6/f complex and six for the ATP synthase are known in embryophytes, excluding assembly factors. These genes are present in all sequenced plastid chromsomes of photosynthetic members in this group. In addition, plastid genomes of all photosynthetic embryophytes contain the gene for the large subunit of RuBisCO (*rbcL*), the only component of the photosynthetic dark-reaction still encoded in the plastid. Finally, a set of 11 *ndh* genes for components of a thylakoid-located NADH dehydrogenase is present in most embryophytes with the exception of several gymnosperms and non-photosynthetic parasites (Wolfe et al., 1992a; Wakasugi et al., 2001).

3.1.2. Genes for components of the genetic apparatus

Unlike the photosynthesis-related genes, genes of the genetic apparatus were initially identified not by biochemical means, but by sequence homology to eubacterial genes. A detailed functional biochemical and/or genetic analysis has only recently been achieved for ribosomal proteins (Yamaguchi & Subramanian, 2000; Yamaguchi et al., 2000), the plastid-encoded RNA polymerase (Allison et al., 1996; De Santis-Maciossek et al., 1999) and some tRNAs (e.g. Willows et al., 1995; Cheng et al., 1997; Vogel & Hess, 2001). Functional aspects of plastid transcription and translation are reviewed in Chapters 8 and 12, respectively, of this volume.

Plastid genomes of embryophytes contain a complete set of genes for ribosomal RNAs. The plastid homologue to the eubacterial 23S rRNA is encoded by two genes, 23S rDNA and 4.5S rDNA with the 4.5S rRNA being a structural equivalent of the 3' terminal region of eubacterial 23S rRNA (Edwards & Kössel, 1981; Strittmatter & Kössel, 1984).

In plastid genes, all 61 possible codons can be found (Shinozaki et al., 1986). The typically 30 tRNAs encoded by the plastid genome can only serve all of these codons if a two-out-of-three and a U:N wobble mechanism are valid in addition to normal wobble base-pairing (Sugiura et al., 1998). Import of tRNAs into plastids as shown for mitochondria (Delage et al., 2003) still awaits experimental proof, although the rudimentary set of only 17 tRNA genes in the plastid genome of *Epifagus virginiana* suggests that such a mechanism might well exist (Wolfe et al., 1992b). It is noteworthy that tRNA-Glu (UUC) is required as a cofactor for the formation of the initial precursor of tetrapyrrol rings (Jahn, 1992).

Only one non-tRNA and non-rRNA gene coding for a non-mRNA (*sprA*) has been found in plastids of embryophytes, although its function remains elusive (Sugita et al., 1997; Wakasugi et al., 2001). Other RNA genes have been proposed in intergenic spacers based on transcription profiling of the complete tobacco plastid chromosome, but genetic or biochemical confirmation of their existence is lacking (Nakamura et al., 2003). The 20 to 21 ribosomal protein subunits encoded in the plastid chromosome represent roughly one-third of all ribosomal subunits. Several genes for ribosomal proteins are only plastid-encoded in a subset of embryophytes like *rpl21*, which is restricted to the plastids of ferns and mosses. Similarly, another gene involved in plastid translation, the gene for the initiation factor InfA, has a patchy distribution in angiosperm phylogeny (Millen et al., 2001). As well as translation, transcription in plastids depends on both the plastid and the nuclear compartment. First, in addition to the four plastid-encoded subunits of the eubacteria-type RNA polymerase, nuclear-encoded sigma factors are presumably needed for polymerase function (Allison, 2000). The four plastid genes *rpoA*, *rpoB*, *rpoC1* and *rpoC2* are highly conserved in the plastids of embryophytes, and only in the moss *Physcomitrella patens* is the gene for subunit □ (*rpoA*) missing from the plastid chromosome. Secondly, a whole different transcription machinery, represented by nuclear-encoded phage-type RNA polymerases has been found in angiosperms and mosses (Allison et al., 1996; Hedtke et al., 1997; Richter et al., 2002). No interactions of this phage-type enzymes with plastid-encoded proteins has been reported.

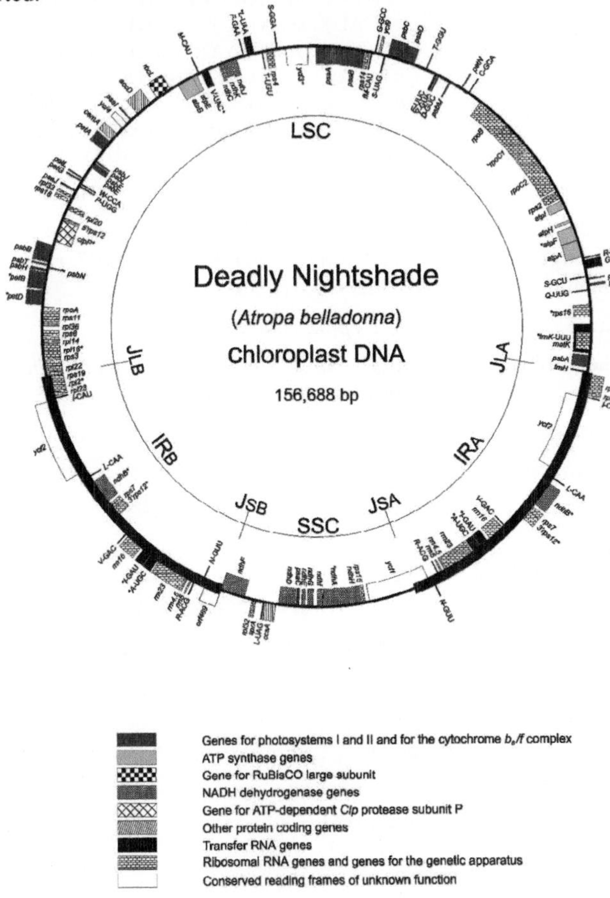

Figure 2. *Gene organization of a typical plastid chromosome from embryophytic plants, here from Atropa belladonna, in its circular monomeric form (Schmitz-Linneweber et al., 2002). Large and small single copy regions (LSC, SSC) are separated by the inverted repeats IR_A and IR_B (bold), respectively. Intron-containing genes are marked by asterisks. Genes drawn inside the circle are transcribed clockwise, those outside anti-clockwise. Genes belonging to a functional group are marked by the same gray-scale/pattern. For abbreviations and nomenclature of genes, see Stoebe et al. (1998).*

Table 1. *Genes encoded by the Atropa belladonna plastid genome*

Product(s)	*Gene(s)*
23 S, 16 S, 5 S, 4.5 S ribosomal RNAs	*rrn23*[#], *rrn16*[#], *rrn5*[#], *rrn4.5*[#]
tRNAs	*trnA*(UGC)[#]*, *trnC*(GCA), *trnD*(GUC), *trnE*(UUC), *trnF*(GAA), *trnG*(GCC), *trnG*(UCC)*, *trnH*(GUG), *trnI*(CAU)[#], *trnI*(GAU)[#]*, *trnK*(UUU)*, *trnL*(CAA)[#], *trnL*(UAA)*, *trnL*(UAG), *trnfM*(CAU), *trnM*(CAU), *trnN*(GUU), *trnP*(UGG), *trnQ*(UUG), *trnR*(ACG)[#], *trnR*(UCU), *trnS*(GCU), *trnS*(GGA), *trnS*(UGA), *trnT*(GGU), *trnT*(UGU), *trnV*(GAC)[#], *trnV*(UAC)*, *trnW*(CCA), *trnY*(GUA)
Proteins of the small ribosomal subunit	*rps2*, *rps3*, *rps4*, *rps7*[#], *rps8*, *rps11*, *rps12*[§+], *rps14*, *rps15*, *rps16**, *rps18*, *rps19*
Proteins of the large ribosomal subunit	*rpl2*[#]*, *rpl14*, *rpl16*[#]*, *rpl20*, *rpl22*, *rpl23*, *rpl32*, *rpl33*, *rpl36*
RNA polymerase subunits	*rpoA*, *rpoB*, *rpoC1**, *rpoC2*
Subunits of the NADH dehydrogenase	*ndhA**, *ndhB*[#]*, *ndhC*, *ndhD*, *ndhE*, *ndhF*, *ndhG*, *ndhH*, *ndhI*, *ndhJ*, *ndhK*
Subunits of photosystem I	*psaA*, *psaB*, *psaC*, *psaI*, *psaJ*
Subunits of photosystem II	*psbA*, *psbB*, *psbC*, *psbD*, *psbE*, *psbF*, *psbH*, *psbI*, *psbJ*, *psbK*, *psbL*, *psbM*, *psbN*, *psbT*, *psbZ*
Large subunit of RuBisCO	*rbcL*
Cytochrome b_6f subunits	*petA*, *petB**, *petD**, *petG*, *petL*, *petN*
ATP synthase subunits	*atpA*, *atpB*, *atpE*, *atpF**, *atpH*, *atpI*
Clp protease subunit	*clpP*[§]
Acetyl CoA carboxylase subunit	*accD*
Involved in assembly/stability of photosystem I	*ycf3*[§], *ycf4*
in cytochrome *c* biogenesis	*ccsA*
in inorganic carbon uptake	*cemA*
Unknown function	*sprA*, *ycf1*, *ycf2*[#], *ycf14 (matK*; partial homologous to intron maturases)

*Intron-containing gene; [§]gene containing two introns, [+]gene containing transspliced exons, [#]two gene copies due to the inverted repeat; for abbreviations and nomenclature of genes, see Stoebe et al. (1998).

3.1.3. Genes for metabolic functions and miscellaneous genes

Plastid genomes of embryophytes have lost almost all genes for the different biosyntheses, although many metabolic pathways still take place in the plastid compartment, entirely run by nuclear-encoded enzymes. At least ten genes remain in plastid DNA, and of these only *rbcL*, *clpP* for a protease subunit and *cemA* for a transporter of inorganic carbon are ubiquitous. Others are restricted to certain groups, like the genes *chlB*, *chlL* and *chlN* for chlorophyll biosynthesis that are only found in the liverwort *Marchantia polymorpha* and gymnosperms. Less sporadic is *accD*, the product of which is involved in fatty acid biosynthesis, and that is only absent in some monocot families (Konishi et al., 1996). Unlike many other genes lost from the plastid genome, *accD* has not been transferred to the nucleus. Instead, in certain monocots its function has been taken over by a different nuclear-encoded enzyme (Konishi et al., 1996). *ccsA*, specifying a protein involved in cytochrome c biogenesis is absent from *Physcomitrella patens*, but otherwise ubiquitous in embryophytes. Finally, two genes putatively involved in sulfate transport are restricted to the most basal land plant *Marchantia polymorpha* and are otherwise found only outside of embryophytes in green algae (Wakasugi et al., 2001).

3.2. Eukaryotic algae

Eukaryotic algae are a very diverse group of organisms that contain simply structured unicellular life forms like *Chlorella vulgaris* and gigantic thallophytes like the sea kelps. The phylogenetic distance of different taxons unified by the term algae is far greater than that between the most divergent embryophytes. Hence, in contrast to the homogeneity of embryophyte plastid genomes, it is not surprising that algae plastid chromosomes vary greatly not only in size and organization but also in gene content.

The closest algal relative to the embryophytes are the charophytes (Graham et al., 2000), and both together form the streptophytes. The sister group to the streptophytes are the chlorophytes, whereas other algal groups like the red alga, the glaucocystophytes, the heterokontophytes and the dinophytes are more distantly related to embryophytes. The plastid genome of a member of the charophytes, *Chaetosphaeridium globosum*, has been sequenced recently (Turmel et al., 2002a) as well as the plastid genome of *Mesostigma viridis*, a green alga believed to be at the base of the split between streptophytes and chlorophytes (Lemieux et al., 2000). Interestingly, both genomes are highly similar in gene organization to plastid genomes of embryophytes and more than 80% of the genes found in both groups reside in conserved clusters. However, the extent of gene and intron conservation differs between *C. globosum* and *M. viridis* compared to embryophytes. Whereas only 4 genes have been lost in plastids of embryophytes relative to *C. globosum*, *M. viridis* has 20 genes neither found in the charophyte nor in embryophytes and with 135 genes altogether boasts the largest plastid gene repertoire in the viridiplantae (chlorophytes and streptophytes) found to date. Also, *M. viridis* has no introns, whereas *C. globosum* has - with only 4 exceptions - the entire set of introns found in embryophytes.

Table 2. *Comparison of the number of plastid genes in selected functional groups among photosynthetic organisms*

		Glauco-cysto-phyte	Red-algae		Green-algae		Embryo-phytes	
		C. paradoxa[0]	*P. purpurea*[1]	*C. merolae*[2]	*C. rheinhardtii*[3]	*M. viridis*[4]	*P. patens*[5]	*A. belladonna*[6]
Genetic system	DNA maintenance and division	1	2	3	0	3	0	0
	Transcription, transcription regulators	7	7	8	5	4	3	4
	Translation	1	6	4	1	2	1	0
	Ribosomal proteins	37	47	46	17	24	21	21
	Protein quality-control and assembly	8	6	7	4	4	4	4
Photo-synthesis	Phycobilisomes	8	10	9	0	0	0	0
	Photosystem I	8	11	11	4	6	6	5
	Photosystem II	18	17	18	14	15	15	15
	Cyt. b_6/f complex	7	6	8	5	6	6	6
	NADH dehydrogenase	0	0	0	0	11	11	11
	ATP synthase	7	8	8	6	6	6	6
Meta-bolism		12	28	33	4	5	5	2
Trans-port		1	3	6	1	3	1	1
RNAs	rRNAs	3	3	3	5	3	4	4
	tRNAs	33	37	31	30	32	31	30
	Other	1	1	2	1	0	0	0

[0](Stirewalt et al., 1995), [1](Reith & Munholland, 1995), [2](Ohta et al., 2003), [3](Maul et al., 2002), [4](Lemieux et al., 2000), [5](Sugiura et al., 2003), [6](Schmitz-Linneweber et al., 2002). Genes of unknown function are not considered.

Introns are absent from the plastid genomes from various other algal lineages like the red algae *Porphyra pupurea*, the diatom *Odontella sinensis* and the cryptomonade *Guillardia theta*. From this it has been concluded that introns spread in streptophytes after the divergence of the lineage leading to *Mesostigma* and are also a derived feature in certain chlorophytes like *Chlamydomonas*, where they reside in entirely different locations than those in embryophytes (Palmer & Delwiche, 1997; Wakasugi et al., 2001).

The high order of gene cluster conservation is restricted to streptophytes and *Mesostigma viridis*, whereas other green algae like *Nephroselmis olivacea* or *Chlorella vulgaris* do have far less gene clusters in common with embryophytes (Lemieux et al., 2000; Turmel et al., 2002b). These results suggest that gene rearrangements are rather rare in streptophytes, in particular when considering that *Mesostigma* and embryophytes evolved independently for more than 800 million years (Lemieux et al., 2000).

In contrast to this, the plastid genomes of *Chlorella vulgaris* (Wakasugi et al., 1997), *Chlamydomonas reinhardtii* (Maul et al., 2002) and the fragmentary plastid DNA sequence data available from other chlorophytes indicate that green algal plastid genomes evolve in a much less conservative fashion compared with their counterparts from embryophytes (Palmer, 1991). In accordance with this, genome size is highly variable (see Section 2.2.2.).

Unlike green algae, red algae have a far more narrow range of genome size and have the highest gene density of all plastid chromosomes, with for example 40% of all genes overlapping in *Cyanidioschyzon merolae* (Ohta et al., 2003). Also, red algae contain considerably more genes, like the 243 annotated genes in *Cyanidioschyzon merolae* and the 251 in *Porphyra purpurea*, which starkly contrasts the 135 genes in *Mesostigma viridis* (the maximum number found for a member of the viridiplantae) and the 182 genes in the glaucocystophyte *Cyanophora paradoxa*.

The difference in gene content between the major plant groups is not due to a general increase in gene number distributed evenly over all functional classes of plastid genes, like one would suspect in case the red algae had simply lost genetic information to the nucleus more slowly than the viridiplantae and glaucocystophytes. In fact, gene numbers for most functional classes are not that different between red algae, glaucocystophytes, green algae and embryophytes (Table 2). For example, the number of genes for photosystem II, the cytochrome b_6/f complex, the ATP synthase, tRNAs and rRNAs are not strikingly higher in red algae than in the other groups (Table 2). The overall difference in gene number can mainly be accounted for by genes for ribosomal proteins that are about twice as abundant in red algae than in the green lineage and in particular by genes with functions in different metabolic pathways. Metabolism-related genes in the green lineage are only represented by *rbcL* and *accD* and genes for proteins involved in chlorophyll metabolism (*chl*). In contrast, red algae plastid chromosomes boast 6 to 10 times more genes, covering parts of anabolic pathways like nucleotide, lipid, carbohydrate, amino acid and cofactor biosynthesis that are not represented at all in the viridiplantae. Interestingly, all these syntheses still take place inside plastids of embryophytes (Halliwell, 1978; Kirk & Tilney-Bassett, 1987) and presumably in their related chlorophytes, but in contrast to red algae all the necessary enzymes have to be imported. Hence, it is not as if the viridiplantae have lost these pathways from the plastid compartment and therefore could dispense with keeping the genes in the plastid chromosome. It is also not admissible to argue that having a lot of genes is a 'primitive' trait and the plastids of embryophytes represent a more advanced state, and that if we waited a few million years, the red algae would lose their 'metabolic genes' as well. The fact that both plastid types had ample time to lose genes since the split of these two lineages, and the finding that the earliest branching, 'most primitive' lineage, the glaucocystophytes, have far less metabolic genes than the red algae precludes this assumption. So, at the moment it seems reasonable to conclude that there must be some kind of selective pressure on the maintenance of 'metabolic genes' and also on the larger array of ribosomal proteins in red algae. It has been argued that an important selective force keeping plastid genes from moving to the nucleus is that the expression of these genes has to be tightly controlled by a plastid-derived signal, the thylakoid's redox state (Allen, 2003). Possibly, the number of genes that are under some sort of immediate control from a plastid-borne signal, be it a redox or some other type of signal, is larger in red algae than in viridiplantae.

As evidence for this, the red algae harbour several open reading frames with considerable homology to the well-described transcriptional regulators *ompR*, *ntcA* and *lysR* in eubacteria. Plastid-internal regulatory circuits based on these regulators could well prevent downstream metabolic target genes from being transferred to the nucleus. A plastid transformation system for a red algae would greatly facilitate the elucidation of the role of these potential transcriptional regulators in order to determine downstream targets and extend the redox-control hypothesis for the maintenance of plastid genes (Allen, 2003; see also Section 6.).

4. INTRACELLULAR TRANSFER OF PLASTID GENES

When plastids arose more than 1.2 billion years ago by the endosymbiotic acquisition of a cyanobacterium, the genome of both host and endosymbiont faced the challenge of a drastically changed intragenomic competition. Each cyanobacterial gene not only had to assert its usefulness inside a singular genome, but had to prove its value in comparison to genes of the host. Vice versa, host proteins were tested by selection against their cyanobacterial counterparts. Although stochastic processes may have contributed to this, intragenomic competition led to the replacement of many host genes by those from the endosymbiont and concomitantly to an invasion of host proteins into the new organellar compartment. This situation is further complicated by the fact that the host already underwent a first wave of prokaryotic genetic information after the acquisition of the mitochondrion. Little is known about whether this first endosymbiotic event paved the road for the uptake of the cyanobacterium or whether these events are completely independent.

4.1. *Plastid to nucleus gene transfer*

Obviously, cyanobacterial genes for processes no longer needed inside the host are not found in present-day plant cells (e.g. motility-related genes). In other cases, the initial cell-conglomerate had two sets of genes with identical functions such as those involved in primary synthetic pathways, an obvious redundancy that could be solved by discarding one of the two sets. As the plastid genome with its roughly 100 to 200 genes is rather small in comparison to the about 3,000 to 4,000 genes found in a typical cyanobacterial genome, the answer to this question at first sight seemed to be that it had been the cyanobacterial genes that were discarded. However, phylogenetic analyses of all genes in the major primary synthetic pathways have shown that most of the enzymes are of eubacterial, endosymbiotic origin, either having their closest relative in the ancestry of the plastid, that is cyanobacteria, or the mitochondrion, that is □-proteobacteria. In addition, it became clear that the plastid's proteome despite its tiny genome was with 1,000 to 5,000 proteins of comparable size as a cyanobacterial proteome (Martin & Herrmann, 1998; Rujan & Martin, 2001).

Detailed analysis of homologies between modern plastid and nuclear genomes revealed substantial amounts of plastid-derived DNA in the nucleus. This has been observed for spinach (Timmis & Steele Scott, 1983; Cheung & Steele Scott, 1989), various chenopod species (Ayliffe et al., 1988), potato (du Jardin, 1990), tomato (Pichersky et al., 1991), tobacco (Ayliffe & Timmis, 1992), rice, and *Arabidopsis thaliana* (Shahmuradov et al., 2003). For the latter species, not only comparisons of nuclear and plastid sequences have been carried out, but also homology analyses of nuclear genes with representatives of the putative ancestor of plastids, the cyanobacteria *Synechocystis* and *Nostoc*. These phylogenetic comparisons of many nuclear proteins targeted to the plastid revealed that they were of cyanobacterial origin and it was concluded that they were transferred from the endosymbiont to the nucleus (Martin et al., 1998; Rujan & Martin, 2001; Martin et al., 2002). Recently, Martin et al. (2002) have carried out a genome-wide phylogenetic survey of known *A. thaliana* proteins to estimate the number of genes that stem from cyanobacteria. Their results indicated that about 4,500 genes, that is 18% of all *A. thaliana* genes, have a cyanobacterial origin. This number matches the estimated 5,000 proteins in a plastid proteome and would fit neatly the assumption that the gene products of cyanobacterial genes transferred to the nucleus are redirected to the plastid, the so-called product-specificity corollary of the endosymbiotic theory (Weeden, 1981). However, upon analysing the predicted location of the 'cyanobacterial' genes in *A. thaliana*, Martin and colleagues found that less than half of them end up in the plastid, and many remain in the cytosol or are secreted from the cell.

The above findings have two important implications. (i) A major part of the transferred genetic cyanobacterial information is obviously used for a variety of host purposes. Sometimes host functions are simply replaced by plastid-derived genes as in the case of enzymes involved in glycolysis or gluconeogenesis (Martin & Schnarrenberger, 1997) and sometimes the host may use them as raw material for new functions, such as in disease resistance (Martin et al., 2002). (ii) More than half of the proteins received by the plastid are not of cyanobacterial origin and hence make the organelle more eukaryote-like. For instance, the curious complexity of plastid gene expression, with its abundance in RNA polymerases, RNA splicing and RNA editing is rather atypical for prokaryotic organisms and could be explained by the 'hostile take over' of the organelle by host proteins. In fact, among the big contributions of non-cyanobacterial like proteins in the plastid proteome are various RNA binding proteins with RRM or PPR domains (Small & Peeters, 2000). Among these proteins are some with functions completely new to the 'prokaryotic' organelle like the endonucleolytic processing of polycistronic transcripts mediated by CRP1 or HCF152 (Fisk et al., 1999; Meierhoff et al., 2003). Others take over existing plastid functions like the nuclear-encoded ribosomal protein that replaced plastid-encoded Rpl23 in spinach and relatives (Bubunenko et al., 1994; Yamaguchi & Subramanian, 2000) or the nuclear-encoded enzyme that replaced the eubacterial-type acetyl-CoA-carboxylase (composed by plastid-encoded AccD and several nuclear-encoded subunits) in certain monocots (Konishi et al., 1996). Even mitochondria-derived nuclear genes can replace plastid homologues as exemplified by *rpl21*, which has replaced the plastid counterpart in all embryophytes (Martin et al., 1990).

Most of the transfer of plastid genes to the nucleus probably occurred in the twilight interval between the initial endosymbiotic event and the last common ancestors of all extant plastids, which is believed to have harbored around 210 genes (Martin et al., 1998). After this initial burst, the flow of genetic information dwindled considerably, although comparison of complete plastid chromosome sequences from different plant lineages has revealed that gene loss is an ongoing process with certain genes lost many times independently in different lineages (Martin & Herrmann, 1998). For a few of these lost genes, a transfer to the nucleus has been verified. *rpl22* for example has been transferred in legumes and acquired proper targeting signals (Gantt et al., 1991). An unusually unstable plastid gene is *infA*, which has been lost 24 times in 300 angiosperm species examined (Millen et al., 2001). Four independent transfers of this gene to the nucleus have been proposed and their gene products are targeted back to the plastid. One of these transfers occurred recently, in the caryophyllid lineage, which shows that plastid-to-nucleus transfer of genes is an ongoing process (Millen et al., 2001).

In contrast to gene transfer from organelles to the nucleus, a retrograde transfer is a very rare event. Among all the mitochondrial sequences available, only a partial cytosolic 18S rDNA sequence and a *mutS* gene have been found in the mitochondrial genome of the evening primrose (Schuster & Brennicke, 1987) and the coral *Sarcophyton glaucum* (Pont-Kingdon et al., 1998), respectively, but so far no sequences of nuclear origin have been detected in plastids.

For the successful transfer and replacement of genetic information originally housed in the plastid, a number of criteria must be met. First, the escape of genetic material has to occur at a reasonable frequency. Much progress has been made of late regarding this issue, uncovering an astounding propensity of plastid DNA to travel to the nucleus. In two independent studies, Huang et al. (2003) and Stegemann et al. (2003) used plastid transformation to introduce a modified *npt* gene into the tobacco plastid chromosome that could only establish resistance to kanamycin when transferred to the nucleus as it was equipped with nuclear expression signals and/or nuclear introns. They found that 1 in 16,000 pollen grains or 1 in 50,000 somatic cells contained a *npt* insert, thus establishing that plastid to nucleus gene transfer is a very frequent occurrence (Huang et al., 2003; Stegemann et al., 2003). This is similar to frequencies observed for mitochondria to nucleus transfer (Thorsness & Fox, 1990). It should however not be failed to mention that the transfer frequencies found are not in congruence with the total amount of insertions of organelle DNA described in recent genome projects (Project, 2000, RiceSequencingConsortium, 2003, Daniell and Parkinson 2003). Also, it should be stressed that the mere transfer of genetic information does not necessarily entail its proper expression (see below).

Most of these transfers have no resemblance of an mRNA intermediate, that is, operon structures are not maintained while introns are, it is assumed that most transfers occur directly as DNA, probably originating from lysed organelles (Martin, 2003). That disintegrating organelles are the prime source for transferred DNA is indirectly evidenced by the absence of any transferred plastid DNA in the nucleus of the unicellular algae *Chlamydomonas rheinhardtii* and also the absence of any transfer in a '*Chlamydomonas* experiment' with a marker gene as described above for tobacco (Lister et al., 2003). *Chlamydomonas* has only one plastid and hence cannot afford to lose it by lysis.

Although there is now overwhelming evidence for direct transfer of DNA from organelles to the nucleus (Shahmuradov et al., 2003), in mitochondria, in a few cases a transfer via an RNA intermediate seems possible, as transferred sequences resembled processed RNAs rather than DNA templates (Schuster & Brennicke, 1987; Nugent & Palmer, 1988; 1991; Grohmann et al., 1992; Wischmann & Schuster, 1995). In contrast to mitochondria, no conclusive evidence for RNA-mediated transfer of plastid genetic information has been presented so far.

Even though these data make clear that random transfer of plastid DNA is a common occurrence in higher plants, almost nothing is known about the mechanism of integration of these organellar DNA pieces into nuclear chromosomes. One can at the moment only speculate that chromosome breaks and the nuclear DNA repair machinery play a role here (Pichersky et al., 1991). Plastid DNA fragments that are incorporated into the nuclear genome are then faced with the toughest challenge of all: in order to replace the organellar copy, (i) the transferred plastid genes have to be expressed by the nuclear transcription apparatus, (ii) the mRNA has to be translated by cytosolic ribosomes, and (iii) the gene product has to be targeted back to the plastid. Only upon retargeting of the gene product to the plastid can the nuclear copy compete and eventually replace the organellar copy. To achieve targeting, the protein has to obtain a signal peptide recognized by the plastid import machinery. Analysis of transferred genes has shown that transit peptides can be added to a coding region by exon shuffling as in the case of *rpl22* in legumes and *rpoA* in *Physcomitrella patens* (Gantt et al., 1991; Sugiura et al., 2003). In both cases, the coding region for the transit peptide is totally unrelated and separated by an intron from the rest of the plastid-like reading frame. Interestingly, the transfer of *rpl22* occurred at least 100 million years before its paralogue was lost from the plastid chromosome. This great time gap between transfer and replacement might reflect the great difficulties of correct expression and retargeting of the nuclear copy. On the other hand, a gene inert over such a long time interval would be expected to become a pseudogene, and hence, nuclear *rpl22* likely became active much earlier than the loss of its plastid counterpart. In other words, over an extended time period, two *rpl22* gene copies were likely active in the ancestors of legumes. A similar scenario with both organellar and nuclear gene copy active has been proposed for *tufA* (Baldauf & Palmer, 1990). Indeed, many plastid genes transferred to the nucleus seem to have remained intact over long time periods and exhibit a strong bias of synonymous to non-synonymous nucleotide substitutions, an indicator of purifying selection and hence functionality, despite the presence of a plastid counterpart. Whether these nuclear genes are active side-by-side with the plastid copy over a transitional period until loss of the plastid gene remains to be determined.

The acquisition of targeting signals by exon shuffling as in the case of *rpl22* and *rpoA* seems to be a general rule, as it has also been observed for mitochondria-to-nucleus gene transfers (Wischmann & Schuster, 1995; Kadowaki et al., 1996; Sánchez et al., 1996). In addition, mitochondrial genes have been reported to piggyback on already existing mitochondrial-targeted nuclear genes, utilizing the preexisting promoter as well as the signal peptide by means of alternative splicing (Figueroa et al., 1999; Kubo et al., 1999).

In summary, current data on mechanistic aspects of plastid-to-nucleus transfer suggest that there is a constant rain of organellar DNA onto the host nucleus, resulting in many transfer events per generation. However, the bottleneck before an organellar gene can be replaced by such a transfer is the gain of the proper expression signals. These seem to be acquired in very different ways, varying for each individual transfer event.

4.2. Plastid to mitochondrion gene transfer

The transfer of plastid genes to the mitochondrion seems to be restricted to higher plants, where it is found frequently, but is absent from mosses (Oda et al., 1992), green algae (Wolff et al., 1994; Turmel et al., 2003) and red algae (Leblanc et al., 1995). Plastid DNA fragments in the mitochondrion exhibit signs of accelerated evolution, degeneration and thus, transferred genes usually become pseudogenes (e.g. Zheng et al., 1997; Cummings et al., 2003). Therefore, although large chunks of plastid DNA of more than 10 kbp have been reported in mitochondria (Stern & Lonsdale, 1982), predominantly much smaller pieces are found, supposedly remnants of larger transferred pieces of plastid DNA (e.g. Stern & Palmer, 1986; Nakazono & Hirai, 1993; Watanabe et al., 1994). In rare cases, a plastid DNA fragment becomes stably integrated into the mitochondrial chromosome, which is taken as a clear sign of functionality. For example in maize, a plastid *trnM* gene has been inserted into mitochondria and is strongly expressed (Sangare et al., 1989). Other transferred plastid-like tRNAs in maize are expressed as well, and in several cases it has been shown that they are essential for mitochondrial translation (Marechal-Drouard et al., 1995; Miyata et al., 1998). Transferred chloroplast tRNA genes have also been utilized for other mitochondrial functions besides translation. For instance, in a transfer of several plastid genes early in grass evolution the sequences of *rps19* and *trnH* were probably recycled as a promoter for a mitochondrial gene (Kanno et al., 1997a; b). Similarly, in rice, a plastid sequence of more then 4 kbp serves as a target for the mitochondrial RNA polymerase to transcribe the downstream *nad9* gene (Nakazono et al., 1996). Even older, but without any assigned function, is the integration of a plastid tRNA gene cluster into the mitochondrial genome of the common ancestor of monocots, which is given here as a representative of many more plastid-like tRNA genes found in mitochondria of plants (Kubo et al., 1995).

The dynamics and frequency of plastid-to-mitochondria transfers have been most thoroughly studied for the gene *rbcL* (Nugent & Palmer, 1988; Cummings et al., 2003). This particular gene has been transferred at least five times in angiosperm evolution. Among these transfers is a rather ancient one in an early crucifer, whereas a very recent one took place in the morning glory species *Ipomoea coccinea* (Cummings et al., 2003). Many other plastid-to-mitochondrial transfers have been inferred to be recent as well (Marienfeld et al., 1999).

Both data on plastid-to-nucleus transfer and plastid-to-mitochondrion transfer show that the intercompartmental exchange of genetic information is an ongoing process and decisively contributed to the structure of the overall plant genome. To make things even more complicated, recent studies convincingly demonstrate that some genes that ultimately have a plastid (cyanobacterial) origin are transferred not intracellularly, but between species.

5. HORIZONTAL TRANSFER OF PLASTID GENES

If horizontal gene transfer (HGT) is considered as the transfer of genetic information between species, then plastid to mitochondria or plastid to nucleus transfer ceased to be HGT as soon as the initial endosymbiosis was established and crossed the threshold to becoming a good species (a matter of definition of course). HGT however did not stop there. In the dinoflagellate *Gonyaulax polyedra*, the gene for plastid targeted glyceraldehyde-3-phosphate dehydrogenase has likely been acquired laterally from a cryptomonade (Fagan et al., 1998). Furthermore, a recent analysis of plastid targeted nuclear genes of the amoeboflagellate alga *Bigelowiella natans* revealed that at least 21% of these genes were acquired by horizontal gene transfer (Archibald et al., 2003). Contrasting this, in the nuclear genome of the green alga *Chlamydomonas rheinhardtii*, no evidence for HGT has been found at all. Interestingly, *B. natans* is mixotrophic and it has been suggested that the foreign genes originate from food-algae and food-bacteria, a source of genetic information not accessible to autotrophic *C. rheinhardtii*. These data support an earlier hypothesis that eukaryotic cells capable of phagocytosis will incorporate their food source's genetic information (Doolittle, 1998). The origin of the non-chlorophyte plastid genes in *B. natans* is diverse, with the majority stemming from other algae. As proteins targeted to a plastid acquired by secondary endosymbiosis (which is the case for *B. natans*) not only need a transit peptide to cross the plastid envelope but also a signal peptide to pass the ER-like membrane wrapped around the plastid, it is presumably much easier to make use of genes with these sequences than invent them anew. Nevertheless, transfer of genes initially devoid of any targeting signals has occurred as well, as both prokaryotic genes as well as a plastid-coded gene have managed to successfully integrate into the *B. natans* nuclear genome (Archibald et al., 2003). The steady increase in genome sequences of plastid-harbouring eukaryotes will show, whether HGT is a common theme for nuclear-encoded, plastid-targeted proteins.

Though plastid genes have been found in mitochondria and the nucleus and have also been laterally transferred, the plastid itself seems a poor recipient for foreign genetic information. In fact, there seems to be no exchange of genetic information between plastids of distinguishable genetic constitution when present in the same cell. For instance, no recombination of plastid genetic information has been observed in *Oenothera* and other species with biparental plastid inheritance, although heteroplastidic cells have been found (Hagemann, 2002). Also, in artificial combinations of plastids from different species in one cell, so-called cybrids, transfer of genetic information between plastids could only be obtained by heavy selective pressure and only at very low frequency (Medgyesy et al., 1985). These observation suggest that there is some kind of genetic or mechanical barrier to uptake of DNA into plastids, which is surprising as plastid transformation both by mechanical and chemical means is a standard technique (Boynton et al., 1988; Daniell et al., 2004; Kumar and Daniell, 2004; Kindle et al., 1991; Golds et al., 1993) and has become an important tool not only for basic but also for applied research (reviewed in Bogorad, 2000; Bock, 2001; Daniell et al., 2002; Daniell et al., 2004 a-c; Chebolu and Daniell, 2004; Devine and Daniell, 2004, Chapters 16 in this volume). Still, a number of examples for lateral gene transfer into plastids have been described, like the integration of two large pieces of DNA of uncertain origin into the *Nephroselmis olivacea* plastid chromosome (Turmel et al., 1999) or the 21 kbp insertion found in two *Chlamydomonas* species (Lemieux et al., 1985). Also in the genus *Chlamydomonas*, the so-called Wendy-element has been found, which shows similarities to transposase and integrase genes and hence represents the only case for the invasion of a plastid chromosome by a transposon (Fan et al., 1995). Finally, several gene phylogenies incongruent with the widely accepted monophylie of all plastid chromosomes have been found that could be explained by ancient horizontal gene transfers (Morden et al., 1992). The best studied example among these is *rbcL*, of which two types are found with a very patchy distribution among plastids and eubacteria. It has been postulated that red algae plastids and related secondary plastid lineages received their *rbcL* from a proteobacterium-like donor genome, most likely the mitochondrion present in the common progenitor of brown and red algae (Delwiche & Palmer, 1996). Lateral transfers like this that occurred at deep branches in plastid phylogeny are difficult to resolve and are likely underappreciated.

6. WHY HAVE PLASTIDS RETAINED GENOMES?

The chloroplast offers a particularly unfriendly environment for DNA. The chemistry of photosynthesis generates high concentrations of various oxygen species that are highly mutagenic (Allen & Raven, 1996). As plastids are asexual, mutations should accumulate rapidly, because they cannot be eliminated by sexual recombination. Therefore, plastid genes may be expected to 'escape' these disfavourable conditions and end up in the nucleus, a predominantly sexual genome (Moran, 1996). This scenario, called Muller's ratchet, may have been one of the major driving forces for gene transfer from the endosymbiont to the nucleus in the past, although in present-day plants the plastid mutation rate is lower than the nuclear one (Wolfe et al., 1992b; Lynch, 1997). Whatever the selective pressures have been that reduced the plastid genome to its modern size, the question why this was not driven to completion is still open. Several hypothesis have been put forward to answer this question. Quite early, it has been argued that several of the organelle-encoded proteins are highly hydrophobic and hence would not easily cross the plastid envelop when translated in the cytoplasm (von Heijne, 1986; Palmer, 1997). In opposition to this, the highly hydrophobic light-harvesting chlorophyll (Lhc) proteins are universally nuclear-encoded and the hydrophilic large subunit (RbcL) of RuBisCO, with few exceptions, is plastid-encoded. Furthermore, allotopic expression of *rbcL* in the nucleus can functionally complement the deletion of the plastid-encoded *rbcL* (Kanevski & Maliga, 1994). Other explanations for the maintenance of the plastid chromosome are that plastid proteins could be toxic in the cytosol (Martin & Herrmann, 1998) and it has also been proposed that – as gene transfer is an ongoing process – we just need to wait a little more and the last remnants of the plastid chromosome will have disappeared (Herrmann, 1997).

The most elaborate hypothesis on the persistence of organellar genetic information in general has been suggested by Allen (1993; 2003). The hypothesis states that a core set of plastid (and also mitochondrial) genes is maintained because of their regulatory properties. Specifically, they are expressed according to the redox state of their gene products or – more indirectly – according to the redox state of electron carriers like plastoquinone, that interact with their gene products. The redox state of thylakoid membrane proteins directly reflects environmental signals in particular light/shading conditions. Under certain conditions, it is desirable for the chloroplast to have more photosystem II (PS II) than photosystem I (PS I), whereas other conditions favour the production of PSI but not PSII. If the two photosystems are not balanced well, electron jams occur that produce harmful oxygen radicals (Allen & Raven, 1996). Therefore, in order to rapidly adjust PSI/PSII ratios, it seems more effective to keep redox-regulated genes in the chloroplast than to have to forward signals to a nuclear-localized gene. This hypothesis is named CORR for CO-localization of genes and gene products for Redox Regulation of gene expression (Allen, 2003). CORR is basically supported by two observations: first, when comparing the more than 30 plastid genomes sequenced to date, there indeed emerges a common subset of genes both for subunits of PSI and PS II and for components of the plastid's genetic apparatus (Martin et al., 1998; Allen, 2003). Secondly, redox control of plastid transcription seems to apply for several of these 'permanent resident' plastid genes involved in photosynthesis (Pfannschmidt et al., 1999; Tullberg et al., 2000). For example, the gene *psbA*, which codes for a light-labile core subunit of PSII, is more strongly transcribed when a plant is irradiated with light specific for PSI, probably because the intermediate plastoquinone pool is oxidized and the system needs more electrons to be pumped into this pool by PSII.

Complementary to this, transcription of PSI-genes are activated when the plant is irradiated with PSII-light and the plastoquinone pool is reduced (Pfannschmidt, 2003).

The other class of genes retained in plastid genomes, those coding for the transcriptional and translational machinery of the cell, does not necessarily have to be redox-controlled to have a reason for staying in the organelle. According to CORR, these genes are retained only secondarily, to ensure that the redox-controlled photosynthesis genes have the means to be expressed. Interestingly, the subset of plastid genes with a function in gene expression is far more diverse between taxa than those that are retained for photosynthesis are (Allen, 2003; see Table 1 and Table 2). Does this mean that it is not very important *which* genes of the genetic apparatus are retained as long as at least *some* are? Also, plastid ribosomes seem to be expressed rather constitutively and in fact, ribosomal rRNAs are used to normalize RNA analysis experiments. So why should it matter according to CORR, whether genes for the transcription/translation apparatus are in the nucleus or in the plastid? As long as there is no data supporting that these genes of the genetic apparatus are themselves redox-controlled, their retention can not easily be explained by CORR. The same problem applies to a situation, where photosynthetic genes are discarded from the plastid chromosome like in the heterotrophic unicellular alga *Astasia longa* or the parasites *Epifagus virginiana*, *Cuscuta reflexa* and also in *Plasmodium falciparum* and *Toxoplasma gondii*. In all these species, genes for the genetic apparatus are maintained, whereas genes with products important for photosynthesis are not. This clearly does not match CORR's proposition, that the genetic apparatus is still partly encoded in the plastid as a consequence of the retention of photosynthesis-related genes. Speculatively, the large ribosomal RNAs may not easily be imported into plastids, in particular, as no RNA import into this organelle has been directly demonstrated. This contrasts to mitochondria, where tRNA import from the cytosol occurs (Delage et al., 2003). If this idea is true, the retention of RNA polymerase genes would be a consequence of the retention of the rDNAs that need to be transcribed. Maintenance of ribosomal proteins may be explained by the complex assembly map of the ribosome, which requires sequential addition of ribosomal proteins (Culver & Noller, 2000), a task best achieved if the primary ribosomal proteins are produced according to their demand. The question remains, why a genetic apparatus has remained at all. Obviously, it is there to express the remaining genes even after the organelle has been stripped off of genes for photosynthesis. In *Epifagus virginiana*, four genes aside of the genetic apparatus are maintained, namely *clpP* encoding a protease subunit, *accD*, the product of which is involved in fatty acid synthesis, and the enigmatic genes *ycf1* and *ycf2* (Wolfe et al., 1992a). *ycf1*, *ycf2* and *clpP* have been shown to be essential for plant survival (Drescher et al., 2000; Kuroda & Maliga, 2003). In *Plasmodium falciparum* and *Toxoplasma gondii*, the *suf* gene is believed to account for the survival of plastid genetic information. The *suf* gene is involved in FE/S cluster formation and protection against oxidative stress (Wilson et al., 2003). *Astasia longa* retains RuBisCO that might have been recruited for metabolic pathways other than the Calvin cycle (Gockel & Hachtel, 2000). Why these few genes are maintained can not be answered at the moment, but in these spare samples of non-photosynthetic plastids, they appear to be the raison d'être for the genetic apparatus and plastid DNA as a whole. This shows that the question why organelles have retained genomes has to be studied case by case, although CORR provides a general model for the presence of at least photosynthesis-related genes.

7. ACKNOWLEDGEMENTS

We thank Kenneth Watkins for comments on the manuscript, Jiming Jiang for providing the fiber-FISH figure and Peter Poltnigg for technical assistance. Work on this paper was supported by grants from the Deutsche Forschungsgemeinschaft (SFB-TR1) to RMM. CSL is a recipient of a fellowship from the Deutsche Forschungsgemeinschaft.

8. REFERENCES

Allen, J. F. (1993). Control of gene expression by redox potential and the requirement for chloroplast and mitochondrial genomes. *J. Theor. Biol.*, *165*, 609-631.

Allen, J. F. (2003). The function of genomes in bioenergetic organelles. *Philos. Trans. R. Soc. Lond. B Biol. Sci.*, *358*, 19-37; discussion 37-18.

Allen, J. F., & Raven, J. A. (1996). Free-radical-induced mutation vs redox regulation: costs and benefits of genes in organelles. *J. Mol. Evol.*, *42*, 482-492.

Allison, L. A. (2000). The role of sigma factors in plastid transcription. *Biochimie, 82*, 537-548.

Allison, L. A., Simon, L. D., & Maliga, P. (1996). Deletion of *rpoB* reveals a second distinct transcription system in plastids of higher plants. *EMBO J., 15*, 2802-2809.

Archibald, J. M., Rogers, M. B., Toop, M., Ishida, K., & Keeling, P. J. (2003). Lateral gene transfer and the evolution of plastid-targeted proteins in the secondary plastid-containing alga *Bigelowiella natans*. *Proc. Natl. Acad. Sci. USA, 100*, 7678-7683.

Ayliffe, M. A., & Timmis, J. N. (1992). Plastid DNA sequence homologies in the tobacco nuclear genome. *Mol. Gen. Genet.*, *236*, 105-112.

Ayliffe, M. A., Timmis, J. N., & Steele Scott, N. (1988). Homologies to chloroplast DNA in the nuclear DNA of a number of Chenopod species. *Theor. Appl. Genet.*, *75*, 282-285.

Backert, S., Dörfel, P., & Börner, T. (1995). Investigation of plant organellar DNAs by pulsed-field gel electrophoresis. *Curr. Genet.*, *28*, 390-399.

Baldauf, S. L., & Palmer, J. D. (1990). Evolutionary transfer of the chloroplast *tufA* gene to the nucleus. *Nature, 344*, 262-265.

Baur, E. (1909). Das Wesen und die Erblichkeitsverhältnisse der "*Varietates albomarginatae hort.*" von *Pelargonium zonale*. *Z. Indukt. Abstammungs- Vererbungsl.*, *1*, 330-351.

Bedbrook, J. R., & Bogorad, L. (1976). Endonuclease recognition sites mapped on *Zea mays* chloroplast DNA. *Proc. Natl. Acad. Sci. USA, 73*, 4309-4313.

Bedbrook, J. R., Kolodner, R., & Bogorad, L. (1977). *Zea mays* chloroplast ribosomal RNA genes are part of a 22,000 base pair inverted repeat. *Cell, 11*, 739-749.

Bendich, A. J. (1991). Moving pictures of DNA released upon lysis from bacteria, chloroplasts, and mitochondria. *Protoplasma, 160*, 121-130.

Bendich, A. J., & Smith, S. B. (1990). Moving pictures and pulsed-field electrophoresis show linear DNA molecules from chloroplasts and mitochondria. *Curr. Genet.*, *17*, 421-425.

Birky, C. W., Jr. & Walsh, J. B. (1992). Biased gene conversion, copy number, and apparent mutation rate differences within chloroplast and bacterial genomes. *Genetics, 130*, 677-683.

Bock, R. (2001). Transgenic plastids in basic research and plant biotechnology. *J. Mol. Biol.*, *312*, 425-438.

Bogorad, L. (2000). Engeneering chloroplasts: an alternative site for foreign genes, proteins, reactions and products. *Trends Biotechnol.*, *18*, 257-263.

Bogorad, L. (2003). Photosynthesis research: advances through molecular biology - the beginnings, 1975-1980s and on..... *Photosynth. Res.*, *76*, 13-33.

Bonen, L., & Doolittle, W. F. (1975). On the prokaryotic nature of red algal chloroplasts. *Proc. Natl. Acad. Sci. USA, 72*, 2310-2314.

Bonen, L., & Doolittle, W. F. (1976). Partial sequences of 16S rRNA and the phylogeny of blue-green algae and chloroplasts. *Nature, 261*, 669-673.

Boudreau, E., Takahashi, Y., Lemieux, C., Turmel, M., & Rochaix, J. D. (1997). The chloroplast *ycf3* and *ycf4* open reading frames of *Chlamydomonas reinhardtii* are required for the accumulation of the photosystem I complex. *EMBO J., 16*, 6095-6104.

Boynton, J. E., Gillham, N. W., Harris, E. H., Hosler, J. P., Johnson, A. M., Jones, A. R., et al. (1988). Chloroplast transformation in *Chlamydomonas* with high velocity microprojectiles. *Science, 240*, 1534-1538.

Brachet, J. (1959). New observations on biochemical interactions between nucleus and cytoplasm in *Amoeba* and *Acetabularia*. *Exp. Cell Res. Suppl., 6*, 78-96.

Bubunenko, M. G., Schmidt, J., & Subramanian, A. R. (1994). Protein substitution in chloroplast ribosome evolution. A eukaryotic cytosolic protein has replaced its organelle homologue (L23) in spinach. *J. Mol. Biol., 240*, 28-41.

Cai, X., Fuller, A. L., McDougald, L. R., & Zhu, G. (2003). Apicoplast genome of the coccidian *Eimeria tenella*. *Gene, 321*, 39-46.

Chebolu, S & Daniell, H. (2004). Chloroplast derived vaccine antigens and biopharmaceuticals: expression, folding, assembly and functionality. Current Trends in Microbiology and Immunology, in press.

Cheng, Y. S., Lin, C. H., & Chen, L. J. (1997). Transcription and processing of the gene for spinach chloroplast threonine tRNA in a homologous *in vitro* system. *Biochem. Biophys. Res. Commun., 233*, 380-385.

Cheung, W. Y., & Steele Scott, N. (1989). A contigous sequence in spinach nuclear DNA is homologous to three separated sequences in chloroplast DNA. *Theor. Appl. Genet., 77*, 625-633.

Chiba, Y. (1951). Cytochemical studies on chloroplasts. I. Cytologic demonstration of nucleic acids in chloroplasts. *Cytologia, 16*, 259-264.

Chi-Ham, C. L., Keaton, M. A., Cannon, G. C., & Heinhorst, S. (2002). The DNA-compacting protein DCP68 from soybean chloroplasts is ferredoxin: sulfite reductase and co-localizes with the organellar nucleoid. *Plant Mol. Biol., 49*, 621-631.

Chun, E. H., Vaughan, M. H., Jr., & Rich, A. (1963). The isolation and characterization of DNA associated with chloroplast preparations. *J. Mol. Biol., 41*, 130-141.

Correns, C. (1909). Vererbungsversuche mit blaß(gelb)grünen und buntblättrigen Sippen bei *Mirabilis jalapa*, *Urtica pililifera* und *Lunaria annua*. *Z. Indukt. Abstammungs- Vererbungsl., 1*, 291-329.

Culver, G. M., & Noller, H. F. (2000). *In vitro* reconstitution of 30S ribosomal subunits using complete set of recombinant proteins. *Methods Enzymol., 318*, 446-460.

Cummings, M. P., Nugent, J. M., Olmstead, R. G., & Palmer, J. D. (2003). Phylogenetic analysis reveals five independent transfers of the chloroplast gene *rbcL* to the mitochondrial genome in angiosperms. *Curr. Genet., 43*, 131-138.

Curtis, S. E., & Clegg, M. T. (1984). Molecular evolution of chloroplast DNA sequences. *Mol. Biol. Evol., 1*, 291-301.

Daniell, H., Khan, M. S., & Allison, L. (2002). Milestones in chloroplast genetic engineering: an environmentally friendly era in biotechnology. *Trends Plant Sci., 7*, 84-91.

Daniell, D., Carmona-Sanchez, O., & Burns, B. B. (2004a). Chloroplast derived antibodies, biopharmaceuticals and edible vaccines. In R. Fischer & S. Schillberg (Eds.) *Molecular Farming* (pp.113-133). Weinheim: WILEY-VCH Verlag.

Daniell, H., Chebolu, S., Kumar, S., Singleton, M., Falconer, R. (2004b). Chloroplast-derived vaccine antigens and other therapeutic proteins. *Vaccine*.

Daniell, H. & Parkinson, C. L. (2003). Jumping genes and containment. *Nature Biotechnol., 21*, 374-375.

Daniell, H., Ruiz, O.N. and Dhingra, A. 2004. Chloroplast genetic engineering to improve agronomic traits in transgenic Plants. Methods Mol. Biol. 286: 111-137.

DeGray, G., Rajasekaran, K., Smith, F., Sanford, J., & Daniell, H. (2001). Expression of an antimicrobial peptide via the chloroplast genome to control phytopathogenic bacteria and fungi. *Plant Physiology, 127*, 852-862.

Delage, L., Dietrich, A., Cosset, A., & Marechal-Drouard, L. (2003). *In vitro* import of a nuclearly encoded tRNA into mitochondria of *Solanum tuberosum*. *Mol. Cell. Biol., 23*, 4000-4012.

Delwiche, C. F., & Palmer, J. D. (1996). Rampant horizontal transfer and duplication of rubisco genes in eubacteria and plastids. *Mol. Biol. Evol., 13*, 873-882.
Deng, X. W., Wing, R. A., & Gruissem, W. (1989). The chloroplast genome exists in multimeric forms. *Proc. Natl. Acad. Sci. USA, 86*, 4556-4160.
De Santis-Maciossek, G., Kofer, W., Bock, A., Schoch, S., Maier, R. M., Wanner, G., et al. (1999). Targeted disruption of the plastid RNA polymerase genes *rpoA*, *B* and *C1*: molecular biology, biochemistry and ultrastructure. *Plant J., 18*, 477-489.
Devine, A. L., & Daniell, H. (2004). Chloroplast genetic engineering. In S. Moller (Ed.), *Plastids* (pp. 283-320). United Kingdom: Blackwell Publishers.
Doolittle, W. F. (1998). You are what you eat: a gene transfer ratchet could account for bacterial genes in eukaryotic nuclear genomes. *Trends Genet., 14*, 307-311.
Drescher, A., Ruf, S., Calsa, T., Jr., Carrer, H., & Bock, R. (2000). The two largest chloroplast genome-encoded open reading frames of higher plants are essential genes. *Plant J., 22*, 97-104.
du Jardin, P. (1990). Homologies to plastid DNA in the nuclear and mitochondrial genome of potato. *Theor. Appl. Genet., 79*, 807-812.
Edwards, K., & Kössel, H. (1981). The rRNA operon from *Zea mays* chloroplasts: nucleotide sequence of 23S rDNA and its homology with *E.coli* 23S rDNA. *Nucleic Acids Res., 9*, 2853-2869.
Fagan, T., Hastings, J. W., & Morse, D. (1998). The phylogeny of glyceraldehyde-3-phosphate dehydrogenase indicates lateral gene transfer from cryptomonades to dinoflagellates. *J. Mol. Biol., 47*, 633-639.
Fan, W. H., Woelfle, M. A., & Mosig, G. (1995). Two copies of a DNA element, 'Wendy', in the chloroplast chromosome of *Chlamydomonas reinhardtii* between rearranged gene clusters. *Plant Mol. Biol., 29*, 63-80.
Figueroa, P., Gomez, I., Holuigue, L., Araya, A., & Jordana, X. (1999). Transfer of *rps14* from the mitochondrion to the nucleus in maize implied integration within a gene encoding the iron-sulphur subunit of succinate dehydrogenase and expression by alternative splicing. *Plant J., 18*, 601-609.
Fisk, D. G., Walker, M. B., & Barkan, A. (1999). Molecular cloning of the maize gene *crp1* reveals similarity between regulators of mitochondrial and chloroplast gene expression. *EMBO J., 18*, 2621-2630.
Foth, B. J., & McFadden, G. I. (2003). The apicoplast: a plastid in *Plasmodium falciparum* and other Apicomplexan parasites. *Int. Rev. Cytol., 224*, 57-110.
Gantt, J. S., Baldauf, S. L., Calie, P. J., Weeden, N. F., & Palmer, J. D. (1991). Transfer of *rpl22* to the nucleus greatly preceded its loss from the chloroplast and involved the gain of an intron. *EMBO J., 10*, 3073-3078.
Glockner, G., Rosenthal, A., & Valentin, K. (2000). The structure and gene repertoire of an ancient red algal plastid genome. *J. Mol. Evol., 51*, 382-390.
Gockel, G., & Hachtel, W. (2000). Complete gene map of the plastid genome of the nonphotosynthetic euglenoid flagellate *Astasia longa*. *Protist, 151*, 347-351.
Golds, T., Maliga, P., & Koop, H. U. (1993). Stable plastid transformation in PEG-treated protoplasts of *Nicotiana tabacum*. *Biotechnology, 11*, 95-97.
Graham, L. E., Cook, M. E., & Busse, J. S. (2000). The origin of plants: body plan changes contributing to a major evolutionary radiation. *Proc. Natl. Acad. Sci. USA, 97*, 4535-4540.
Gray, P. W., & Hallick, R. B. (1977). Restriction endonuclease map of *Euglena gracilis* chloroplast DNA. *Biochemistry, 16*, 1665-1671.
Grohmann, L., Brennicke, A., & Schuster, W. (1992). The gene for mitochondrial ribosomal protein *S12* has been transferred to the nuclear genome of *Oenothera*. *Nucleic Acids Res., 20*, 5641-5646.
Hackett, J. D., Yoon, H. S., Soares, M. B., Bonaldo, M. F., Casavant, T. L., Scheetz, T. E., et al. (2004). Migration of the plastid genome to the nucleus in a peridinin dinoflagellate. *Curr. Biol. 14*, 213-218.
Hagemann, R. (2000). Erwin Baur or Carl Correns: Who really created the theory of plastid inheritance? *J. Hered., 91*, 435-440.
Hagemann, R. (2002). Milestones in plastid genetics in higher plants. *Prog. Bot., 63*, 1-51.
Hager, M., Biehler, K., Illerhaus, J., Ruf, S., & Bock, R. (1999). Targeted inactivation of the smallest plastid genome-encoded open reading frame reveals a novel and essential subunit of the cytochrome b(6)f complex. *EMBO J., 18*, 5834-5842.
Hallick, R. B., Hong, L., Drager, R. G., Favreau, M. R., Monfort, A., Orsat, B., et al. (1993). Complete sequence of *Euglena gracilis* chloroplast DNA. *Nucleic Acids Res., 21*, 3537-3544.

Halliwell, B. (1978). The chloroplast at work. A review of modern developments in our understanding of chloroplast metabolism. *Prog. Biophys. Mol. Biol., 33*, 1-54.

Hedtke, B., Börner, T., & Weihe, A. (1997). Mitochondrial and chloroplast phage-type RNA polymerases in *Arabidopsis*. *Science, 277*, 809-811.

Herrmann, R. G. (1997). Eukaryotism, towards a new interpretation. In H. E. Schenk, K. W. Jeon, N. E. Muller, W. Schwemmler, K. W. Jeon (Eds.), *Eukaryotism and Symbiosis* (pp. 73-118). Springer: Berlin.

Herrmann, R. G., Westhoff, P., & Link, G. (1992). Biogenesis of plastids in higher plants. In R.G. Herrmann (Ed.), *Plant Gene Research* (pp. 275-349). Wein, New York: Springer.

Herrmann, R. G., Bohnert, H. J., Kowallik, K. V., & Schmitt, J. M. (1975). Size, conformation and purity of chloroplast of some higher plants. *Biochim. Biophys. Acta, 378*, 305-307.

Herrmann, R. G., Bohnert, H. J., Diesel, A., & Hobom, G. (1976). The location of rRNA genes on the restriction endonuclease map of the *Spinacea oleraceae* chloroplast DNA. In T. Bucher, S. Werner, W. Sebald (Eds.), *Genetics and Biogenesis of Chloroplasts and Mitochondria* (pp. 351-359). Amsterdam: North-Holland Publishers.

Hiratsuka, J., Shimada, H., Whittier, R., Ishibashi, T., Sakamoto, M., Mori, M., et al. (1989). The complete sequence of the rice (*Oryza sativa*) chloroplast genome: intermolecular recombination between distinct tRNA genes accounts for a major plastid DNA inversion during the evolution of the cereals. *Mol. Gen. Genet., 217*, 185-194.

Howe, C. J., Barbrook, A. C., Koumandou, V. L., Nisbet, R. E., Symington, H. A., & Wightman, T. F. (2002). Evolution of the chloroplast genome. *Philos. Trans. R. Soc. Lond. B Biol. Sci., 358*, 99-107.

Huang, C. Y., Ayliffe, M. A., & Timmis, J. N. (2003). Direct measurement of the transfer rate of chloroplast DNA into the nucleus. *Nature, 422*, 72-76.

Jahn, D. (1992). Complex formation between glutamyl-tRNA synthetase and glutamyl-tRNA reductase during the tRNA-dependent synthesis of 5-aminolevulinic acid in *Chlamydomonas reinhardtii*. *FEBS Lett., 314*, 77-80.

Jeong, S. Y., Rose, A., & Meier, I. (2003). MFP1 is a thylakoid-associated, nucleoid-binding protein with a coiled-coil structure. *Nucleic Acids Res., 31*, 5175-5185.

Jiao, H. S., Hicks, A., Simpson, C., & Stern, D. B. (2004). Short dispersed repeats in the *Chlamydomonas chloroplast* genome are collocated with sites for mRNA 3' end formation. *Curr. Genet., 45*, 311-322.

Kadowaki, K., Kubo, N., Ozawa, K., & Hirai, A. (1996). Targeting presequence acquisition after mitochondrial gene transfer to the nucleus occurs by duplication of existing targeting signals. *EMBO J., 15*, 6652-6661.

Kanevski, I., & Maliga, P. (1994). Relocation of the plastid *rbcL* gene to the nucleus yields functional ribulose-1,5-bisphosphate carboxylase in tobacco chloroplasts. *Proc. Natl. Acad. Sci. USA, 91*, 1969-1973.

Kanno, A., Nakazono, M., Hirai, A., & Kameya, T. (1997a). A chloroplast derived *trnH* gene is expressed in the mitochondrial genome of gramineous plants. *Plant Mol. Biol., 34*, 353-356.

Kanno, A., Nakazono, M., Hirai, A., & Kameya, T. (1997b). Maintenance of chloroplast-derived sequences in the mitochondrial DNA of Gramineae. *Curr. Genet., 32*, 413-419.

Kindle, K. L., Richards, K. L., & Stern, D. B. (1991). Engineering the chloroplast genome: techniques and capabilities for chloroplast transformation in *Chlamydomonas reinhardtii*. *Proc. Natl. Acad. Sci. USA, 88*, 1721-1725.

Kirk, J. T. O. (1963). The deoxyribonucleic acid of broad bean chloroplasts. *Biochem. Biophys. Acta., 76*, 417-424.

Kirk, J. T. O., & Tilney-Bassett, R. A. E. (1987). *The plastids: chemistry, structure, growth and inheritance*. New York, Amsterdam: Elsevier-North Holland.

Kleinig, H. & Maier, U. G. (1999). *Kleinig/Sitte, Zellbiologie*. Heidelberg, Berlin: Spektrum Academic Publishers.

Kobayashi, T., Takahara, M., Miyagishima, S. Y., Kuroiwa, H., Sasaki, N., Ohta, N., et al. (2002). Detection and localization of a chloroplast-encoded HU-like protein that organizes chloroplast nucleoids. *Plant. Cell, 14*, 1579-1589.

Kohler, S., Delwiche, C. F., Denny, P. W., Tilney, L. G., Webster, P., Wilson, R. J., et al. (1997). A plastid of probable green algal origin in Apicomplexan parasites. *Science, 275*, 1485-1489.

Kolodner, R., & Tewari, K. K. (1972). Molecular size and conformation of chloroplast deoxyribonucleic acid from pea leaves. *J. Biol. Chem., 247*, 6355-6364.

Kolodner, R., & Tewari, K. K. (1975). The molecular size and conformation of the chloroplast DNA from higher plants. *Biochim. Biophys. Acta, 402*, 372-390.

Kolodner, R., & Tewari, K. K. (1979). Inverted repeats in chloroplasts from higher plants. *Proc. Natl. Acad. Sci. USA, 76*, 41-45.

Konishi, T., Shinohara, K., Yamada, K., & Sasaki, Y. (1996). Acetyl-CoA carboxylase in higher plants: most plants other than gramineae have both the prokaryotic and the eukaryotic forms of this enzyme. *Plant Cell Physiol., 37*, 117-122.

Kowallik, K. V., & Herrmann, R. G. (1972). Variable amounts of DNA related to the size of chloroplasts. IV. Three-dimensional arrangement of DNA in fully differentiated chloroplasts of *Beta vulgaris* L. *J. Cell Sci., 11*, 357-377.

Kubo, N., Harada, K., Hirai, A., & Kadowaki, K. (1999). A single nuclear transcript encoding mitochondrial RPS14 and SDHB of rice is processed by alternative splicing: common use of the same mitochondrial targeting signal for different proteins. *Proc. Natl. Acad. Sci. USA, 96*, 9207-9211.

Kubo, T., Yanai, Y., Kinoshita, T., & Mikami, T. (1995). The chloroplast *trnP-trnW-petG* gene cluster in the mitochondrial genomes of *Beta vulgaris, B. trigyna* and *B. webbiana*: evolutionary aspects. *Curr. Genet., 27*, 285-289.

Kumar S, Daniell H (2004) Engineering the chloroplast genome for hyper-expression of human therapeutic proteins and vaccine antigens. Methods Mol Biol **267:** 365-383

Kuroda, H., & Maliga, P. (2003). The plastid clpP1 protease gene is essential for plant development. *Nature, 425*, 86-89.

Kuroiwa, T. (1991). The replication, differentiation, and inheritance of plastids with emphasis on the concept of organelle nuclei. *Int. Rev. Cytol., 128*, 1-62.

Kuroiwa, T., & Suzuki, T. (1981). Circular nucleoids isolated from chloroplasts in a brown alga *Ectocarpus siliculosus*. *Exp. Cell Res., 134*, 457-461.

Laatsch, L., Zauner, S., Stoebe-Maier, B., Kowallik, K. V., & Maier, U. G. (2004). Plastid-derived single gene minicircles of the dinoflagellate *Ceratium horridum* are localized in the nucleus. *Mol. Biol. Evol., 21*, 1318-1322.

Leblanc, C., Boyen, C., Richard, O., Bonnard, G., Grienenberger, J. M., & Kloareg, B. (1995). Complete sequence of the mitochondrial DNA of the rhodophyte *Chondrus crispus* (Gigartinales). Gene content and genome organization. *J. Mol. Biol., 250*, 484-495.

Leff, J., Mandel, M., Epstein, H. P., & Schiff, J. A. (1963). DNA satellites from cells of green and aplastidic algae. *Biochem. Biophys. Res. Commun., 13*, 126-130.

Lemieux, C., Otis, C., & Turmel, M. (2000). Ancestral chloroplast genome in *Mesostigma viride* reveals an early branch of green plant evolution. *Nature, 403*, 649-652.

Lemieux, C., Turmel, M., Lee, R. W., & Bellemare, G. (1985). A 21 kilobase-pair deletion/addition difference in the inverted repeat sequence of *Chlamydomonas eugametos* and *C. moewusii* chloroplast DNA. *Plant Mol. Biol., 5*, 77-84.

Levene, S. D., & Zimm, B. H. (1987). Separations of open-circular DNA using pulsed-field electrophoresis. *Proc. Natl. Acad. Sci. USA, 84*, 4054-4057.

Lilly, J. W., Havey, M. J., Jackson, S. A., & Jiang, J. (2001). Cytogenomic analyses reveal the structural plasticity of the chloroplast genome in higher plants. *Plant Cell, 13*, 245-254.

Lister, D. L., Bateman, J. M., Purton, S., & Howe, C. J. (2003). DNA transfer from chloroplast to nucleus is much rarer in *Chlamydomonas* than in tobacco. *Gene, 316*, 33-38.

Lynch, M. (1997). Mutation accumulation in nuclear, organelle, and prokaryotic transfer RNA genes. *Mol. Biol. Evo.l, 14*, 914-925.

Manning, J. E., Wolstenholme, D. R., Ryan, R. S., Hunter, J. A., & Richards, O. C. (1971). Circular chloroplast DNA from *Euglena gracilis*. *Proc. Natl. Acad. Sci. USA, 68*, 1169-1173.

Marechal-Drouard, L., Small, I., Weil, J. H., & Dietrich, A. (1995). Transfer RNA import into plant mitochondria. *Methods Enzymol., 260*, 310-327.

Marienfeld, J., Unseld, M., & Brennicke, A. (1999). The mitochondrial genome of *Arabidopsis* is composed of both native and immigrant information. *Trends Plant Sci., 4*, 495-502.

Martin, W. (2003). Gene transfer from organelles to the nucleus: frequent and in big chunks. *Proc. Natl. Acad. Sci. USA, 100*, 8612-8614.

Martin, W., & Schnarrenberger, C. (1997). The evolution of the Calvin cycle from prokaryotic to eukaryotic chromosomes: a case study of functional redundancy in ancient pathways through endosymbiosis. *Curr. Genet., 32*, 1-18.

Martin, W., & Herrmann, R. G. (1998). Gene transfer from organelles to the nucleus: how much, what happens, and Why? *Plant Physiol., 118*, 9-17.

Martin, W., Lagrange, T., Li, Y. F., Bisanz-Seyer, C., & Mache, R. (1990). Hypothesis for the evolutionary origin of the chloroplast ribosomal protein L21 of spinach. *Curr. Genet., 18*, 553-556.

Martin, W., Stoebe, B., Goremykin, V., Hapsmann, S., Hasegawa, M., & Kowallik, K. V. (1998). Gene transfer to the nucleus and the evolution of chloroplasts. *Nature, 393*, 162-165.

Martin, W., Rujan, T., Richly, E., Hansen, A., Cornelsen, S., Lins, T., et al. (2002). Evolutionary analysis of *Arabidopsis*, cyanobacterial, and chloroplast genomes reveals plastid phylogeny and thousands of cyanobacterial genes in the nucleus. *Proc. Natl. Acad. Sci. USA, 99*, 12246-12251.

Maul, J. E., Lilly, J. W., Cui, L., dePamphilis, C. W., Miller, W., Harris, E. H., & Stern, D. B. (2002). The *Chlamydomonas reinhardtii* plastid chromosome: islands of genes in a sea of repeats. *Plant Cell, 14*, 2659-2679.

McIntosh, L., Poulsen, C., & Bogorad, L. (1980). Chloroplast gene sequence for the large subunit of ribulose bisphosphate carboxylase of maize. *Nature, 288*, 556-560.

Medgyesy, P., Fejes, E., & Maliga, P. (1985). Interspecific chloroplast recombination in a *Nicotiana* somatic hybrid. *Proc. Natl. Acad. Sci. USA, 82*, 6960-6964.

Meierhoff, K., Felder, S., Nakamura, T., Bechtold, N., & Schuster, G. (2003). HCF152, an *Arabidopsis* RNA binding pentatricopeptide repeat protein involved in the processing of chloroplast *psbB-psbT-psbH-petB-petD* RNAs. *Plant Cell, 15*, 1480-1495.

Millen, R. S., Olmstead, R. G., Adams, K. L., Palmer, J. D., Lao, N. T., Heggie, L., et al. (2001). Many parallel losses of *infA* from chloroplast DNA during angiosperm evolution with multiple independent transfers to the nucleus. *Plant Cell, 13*, 645-658.

Miyata, S., Nakazono, M., & Hirai, A. (1998). Transcription of plastid-derived tRNA genes in rice mitochondria. *Curr. Genet., 34*, 216-220.

Moran, N. A. (1996). Accelerated evolution and Muller's rachet in endosymbiotic bacteria. *Proc. Natl. Acad. Sci. USA, 93*, 2873-2878.

Morden, C. W., Delwiche, C. F., Kuhsel, M., & Palmer, J. D. (1992). Gene phylogenies and the endosymbiotic origin of plastids. *Biosystems, 28*, 75-90.

Nakamura, T., Furuhashi, Y., Hasegawa, K., Hashimoto, H., Watanabe, K., Obokata, J., et al. (2003). Array-based analysis on tobacco plastid transcripts: preparation of a genomic microarray containing all genes and all intergenic regions. *Plant Cell Physiol., 44*, 861-867.

Nakazono, M., & Hirai, A. (1993). Identification of the entire set of transferred chloroplast DNA sequences in the mitochondrial genome of rice. *Mol. Gen. Genet., 236*, 341-346.

Nakazono, M., Nishiwaki, S., Tsutsumi, N., & Hirai, A. (1996). A chloroplast-derived sequence is utilized as a source of promoter sequences for the gene for subunit 9 of NADH dehydrogenase (*nad9*) in rice mitochondria. *Mol. Gen. Genet., 252*, 371-378.

Nugent, J. M., & Palmer, J. D. (1988). Location, identity, amount and serial entry of chloroplast DNA sequences in crucifer mitochondrial DNAs. *Curr. Genet., 14*, 501-509.

Nugent, J. M., & Palmer, J. D. (1991). RNA-mediated transfer of the gene *coxII* from the mitochondrion to the nucleus during flowering plant evolution. *Cell, 66*, 473-481.

Oda, K., Yamato, K., Ohta, E., Nakamura, Y., Takemura, M., Nozato, N., et al. (1992). Transfer RNA genes in the mitochondrial genome from a liverwort, *Marchantia polymorpha*: the absence of chloroplast-like tRNAs. *Nucleic Acids Res., 20*, 3773-3777.

Ohta, N., Matsuzaki, M., Misumi, O., Miyagishima, S. Y., Nozaki, H., Tanaka, K., et al. (2003). Complete sequence and analysis of the plastid genome of the unicellular red alga *Cyanidioschyzon merolae*. *DNA Res., 10*, 67-77.

Ohyama, K., Fukuzawa, H., Kohchi, T., Shirai, H., Sano, T., Sano, S., et al. (1986). Chloroplast gene organization deduced from complete sequence of liverwort *Marchantia polymorpha* chloroplast DNA. *Nature, 322*, 572-574.

Oldenburg, D. J., & Bendich, A. J. (2004). Most chloroplast DNA of maize seedlings in linear molecules with defined ends and branched forms. *J. Mol. Biol., 335*, 953-970.

Palmer, J. D. (1990). Contrasting modes and tempos of genome evolution in land plant organelles. *Trends Genet., 6*, 115-120.

Palmer, J. D. (1991). Plastid chromosomes: structure and evolution. In L. Bogorad and I. K. Vasil (Eds.), *The molecular biology of plastids* (pp. 5-53). San Diego: Academic Press.

Palmer, J. D. (1997). Organelle genomes: going, going, gone! *Science, 275*, 790-791.

Palmer, J. D., & Thompson, W. F. (1982). Chloroplast DNA rearrangements are more frequent when a large inverted repeat sequence is lost. *Cell, 29*, 537-550.

Palmer, J. D., & Delwiche, C. F. (1997). The origin of plastids and their genomes. In *Molecular Systematics of Plants* (pp. 375-409). Norwell: Chapman and Hall/Kluwer.

Perry, A. S., & Wolfe, K. H. (2002). Nucleotide substitution rates in legume chloroplast DNA depend on the presence of the inverted repeat. *J. Mol. Evol., 55*, 501-508.

Pfannschmidt, T. (2003). Chloroplast redox signals: how photosynthesis controls its own genes. *Trends Plant Sci. 8*, 33-41.

Pfannschmidt, T., Nilsson, A., Tullberg, A., Link, G., & Allen, J. F. (1999). Direct transcriptional control of the chloroplast genes *psbA* and *psaAB* adjusts photosynthesis to light energy distribution in plants. *IUBMB Life, 48*, 271-276.

Pichersky, E., Logsdon, J. M., Jr., McGrath, J. M., & Stasys, R. A. (1991). Fragments of plastid DNA in the nuclear genome of tomato: prevalence, chromosomal location, and possible mechanism of integration. *Mol. Gen. Genet., 225*, 453-458.

Pigott, G. & Carr, N. (1972). Homology between nucleic acids of blue-green algae and chloroplasts of *Euglena gracilis*. *Science, 175*, 1259-1261.

Pont-Kingdon, G., Okada, N. A., Macfarlane, J. L., Beagley, C. T., Watkins-Sims, C. D., Cavalier-Smith, T., et al. (1998). Mitochondrial DNA of the coral *Sarcophyton glaucum* contains a gene for a homologue of bacterial *MutS*: a possible case of gene transfer from the nucleus to the mitochondrion. *J. Mol. Evol., 46*, 419-431.

Project, T. A. G. (2000). Analysis of the genome sequence of the flowering plant *Arabidopsis thaliana*. *Nature, 408*, 796-815.

Reith, M. E., & Munholland, J. (1995). Complete nucleotide sequence of the *Porphyra purpurea* chloroplast genome. *Plant Mol. Biol. Rep., 13*, 333-335.

Renner, O. (1924). Die Scheckung der Oenotherenbastarde. *Biol. Zentralbl., 27*, 309-336.

Renner, O. (1929). *Artbastarde bei Pflanzen*. Bornträger, Berlin.

Renner, O. (1934). Die pflanzlichen Plastiden als selbstständige Elemente der genetischen Konstitution. *Ber. Sächs Akad. Wiss. Math. Phys. Kl., 86*, 214-266.

Rice Sequencing Consortium. (2003). In-depth view of structure, activity, and evolution of rice chromosome 10. *Science, 300*, 1566-1569.

Richter, U., Kiessling, J., Hedtke, B., Decker, E., Reski, R., Börner, T. & Weihe, A. (2002). Two RpoT genes of *Physcomitrella patens* encode phage-type RNA polymerases with dual targeting to mitochondria and plastids. *Gene, 290*, 95-105.

Ris, H., & Plaut, W. (1962). Ultrastructure of DNA-containing areas in the chloroplast of *Chlamydomonas*. *J. Cell Biol., 13*, 383-391.

Ruf, S., Biehler, K., & Bock, R. (2000). A small chloroplast-encoded protein as a novel architectural component of the light-harvesting antenna. *J. Cell Bio.l, 149*, 369-378.

Rujan, T., & Martin, W. (2001). How many genes in Arabidopsis come from cyanobacteria? An estimate from 386 protein phylogenies. *Trends Genet., 17*, 113-120.

Sager, R., & Ishida, M. R. (1963). Chloroplast DNA in *Chlamydomonas*. *Proc. Natl. Acad. Sci. USA, 50*, 725-730.

Sakai, A. (2001). In vitro transcription/DNA synthesis system using isolated organelle-nuclei: Application to the analysis on the mechanisms regulating the function of organelle genomes. *J. Plant Res., 114*, 199-211.

Sánchez, H., Fester, T., Kloska, S., Schröder, W., & Schuster, W. (1996). Transfer of *rps19* to the nucleus involves the gain of an RNP-binding motif which may functionally replace RPS13 in *Arabidopsis* mitochondria. *EMBO J., 15*, 2138-2149.

Sangare, A., Lonsdale, D., Weil, J. H., & Grienenberger, J. M. (1989). Sequence analysis of the tRNA(Tyr) and tRNA(Lys) genes and evidence for the transcription of a chloroplast-like tRNA(Met) in maize mitochondria. *Curr. Genet., 16*, 195-201.

Sato, N., Terasawa, K., Miyajima, K., & Kabeya, Y. (2003). Organization, developmental dynamics and evolution of plastid nucleoids. *Int. Rev. Cytol., 232*, 217-262.

Sato, N., Ohshima, K., Watanabe, A., Ohta, N., Nishiyama, Y., Joyard, J., & Douce, R. (1998). Molecular characterization of the PEND protein, a novel bZIP protein present in the envelope membrane that is the site of nucleoid replication in developing plastids. *Plant Cell, 10*, 859-872.

Schmitz-Linneweber, C., Regel, R., Du, T. G., Hupfer, H., Herrmann, R. G., & Maier, R. M. (2002). The plastid chromosome of *Atropa belladonna* and its comparison with that of *Nicotiana tabacum*: the role of RNA editing in generating divergence in the process of plant speciation. *Mol. Biol. Evol., 19*, 1602-1612.

Schötz, F. (1958). Periodische Ausbleichungserscheinungen des Laubes bei *Oenothera. Planta, 43*, 182-240.

Schuster, W., & Brennicke, A. (1987). Plastid, nuclear and reverse transcriptase sequences in the mitochondrial genome of *Oenothera*: is genetic information tranferred between organelles via RNA? *EMBO J., 6*, 2857-2863.

Schwarz, Z., & Kössel, H. (1979). Sequencing of the 3'-terminal region of a 16 S rRNA gene from *Zea mays* chloroplast reveals homology with *E. coli* 16 S rRNA. *Nature, 279*, 520-522.

Schwarz, Z., & Kössel, H. (1980). The primary structure of the 16S rDNA from *Zea mays* chloroplast is homologous top E. coli 16S rRNA. *Nature, 283*, 739-742.

Shahmuradov, I. A., Akbarova, Y. Y., Solovyev, V. V., & Aliyev, J. A. (2003). Abundance of plastid DNA insertions in nuclear genomes of rice and Arabidopsis. *Plant. Mol. Biol., 52*, 923-934.

Shinozaki, K., Ohme, M., Tanaka, M., Wakasugi, T., Hayashida, N., Matsubayashi, T., et al. (1986). The complete nucleotide sequence of tobacco chloroplast genome:its gene organization and expression. *EMBO J., 5*, 2043-2049.

Simpson, C. L., & Stern, D. B. (2002). The treasure trove of algal chloroplast genomes. Surprises in architecture and gene content, and their functional implications. *Plant Physiol., 129*, 957-966.

Small, I., & Peeters, N. M. (2000). The PPR motif - a TPR-related motif prevalent in plant organellar proteins. *Trends Biochem. Sci., 25*, 46-47.

Stegemann, S., Hartmann, S., Ruf, S., & Bock, R. (2003). High-frequency gene transfer from the chloroplast genome to the nucleus. *Proc. Natl. Acad. Sci. USA, 100*, 8828-8833.

Stern, D. B., & Lonsdale, D. M. (1982). Mitochondrial and chloroplast genomes of maize have a 12-kilobase DNA sequence in common. *Nature, 299*, 698-702.

Stern, D. B., & Palmer, J. D. (1986). Tripartite mitochondrial genome of spinach: physical structure, mitochondrial gene mapping, and locations of transposed chloroplast DNA sequences. *Nucleic Acids Res., 14*, 5651-5666.

Stirewalt, V. L., Michalowski, C. B., Löffelhardt, W., & Bohnert, H. J. (1995). Nucleotide sequence of the cyanelle genome from *Cyanophora paradoxa. Plant Mol. Biol. Rep., 13*, 327-332.

Stoebe, B., Martin, W., & Kowallik K. V. (1998). Distribution and nomenclature of protein-coding genes in 12 sequenced chloroplast genomes. *Plant Mol. Biol. Rep., 16*, 243-255.

Strittmatter, G., & Kössel, H. (1984). Cotranscription and processing of 23S, 4.5S and 5S rRNA in chloroplasts from *Zea mays. Nucleic Acids Res., 12*, 7633-7647.

Stubbe, W. (1959). Genetische Analyse des Zusammenwirkens von Genom und Plastom bei *Oenothera. Z. Vererbungsl., 90*, 288-298.

Sugita, M., Svab, Z., Maliga, P., & Sugiura, M. (1997). Targeted deletion of *sprA* from the tobacco plastid genome indicates that the encoded small RNA is not essential for pre-16S rRNA maturation in plastids. *Mol. Gen. Genet., 257*, 23-27.

Sugiura, C., Kobayashi, Y., Aoki, S., Sugita, C., & Sugita, M. (2003). Complete chloroplast DNA sequence of the moss *Physcomitrella patens*: evidence for the loss and relocation of *rpoA* from the chloroplast to the nucleus. *Nucleic Acids Res., 31*, 5324-5331.

Sugiura, M., Hirose, T., & Sugita, M. (1998). Evolution and mechanism of translation in chloroplasts. *Annu. Rev. Genet., 32*, 437-459.

Swiatek, M., Kuras, R., Sokolenko, A., Higgs, D., Olive, J., Cinque, G., et al. (2001). The chloroplast gene *ycf9* encodes a photosystem II (PSII) core subunit, PsbZ, that participates in PSII supramolecular architecture. *Plant Cell, 13*, 1347-1367.

Tewari, K. K. (1971). Genetic autonomy of extranuclear organelles. *Ann Rev Plant Physiol, 22*, 141-168.

Thorsness, P. E., & Fox, T. D. (1990). Escape of DNA from mitochondria to the nucleus in *Saccharomyces cerevisiae. Nature, 346*, 376-379.

Timmis, J. N. & Steele Scott, N. (1983). Sequence homology between spinach nuclear and chloroplast genomes. *Nature, 305*, 65-67.

Tullberg, A., Alexciev, K., Pfannschmidt, T., & Allen, J. F. (2000). Photosynthetic electron flow regulates transcription of the *psaB* gene in pea (*Pisum sativum* L.) chloroplasts through the redox state of the plastoquinone pool. *Plant Cell Physiol., 41*, 1045-1054.

Turmel, M., Otis, C., & Lemieux, C. (1999). The complete chloroplast DNA sequence of the green alga *Nephroselmis olivacea*: insights into the architecture of ancestral chloroplast genomes. *Proc. Natl. Acad. Sci. USA, 96*, 10248-10253.

Turmel, M., Otis, C., & Lemieux, C. (2002a). The chloroplast and mitochondrial genome sequences of the charophyte *Chaetosphaeridium globosum*: insights into the timing of the events that restructured organelle DNAs within the green algal lineage that led to land plants. *Proc. Natl. Acad. Sci. USA, 99*, 11275-11280.

Turmel, M., Otis, C., & Lemieux, C. (2002b). The complete mitochondrial DNA sequence of *Mesostigma viride* identifies this green alga as the earliest green plant divergence and predicts a highly compact mitochondrial genome in the ancestor of all green plants. *Mol. Biol. Evol., 19*, 24-38.

Turmel, M., Otis, C., & Lemieux, C. (2003). The mitochondrial genome of *Chara vulgaris*: insights into the mitochondrial DNA architecture of the last common ancestor of green algae and land plants. *Plant Cell, 15*, 1888-1903.

Vedel, F., Quetier, F., & Bayen, M. (1976). Specific cleavage of chloroplast DNA from higher plants by *Eco*RI restriction nuclease. *Nature, 263*, 440-442.

Vogel, J., & Hess, W. R. (2001). Complete 5' and 3' end maturation of group II intron-containing tRNA precursors. *RNA, 7*, 285-292.

von Heijne, G. (1986). Why mitochondria need a genome. *FEBS Lett., 198*, 1-4.

Wakasugi, T., Tsudzuki, J., & Sugiura, M. (2001). The genomics of land plant chloroplasts: gene content and alteration of genomic information by RNA editing. *Photosynthesis Res., 70*, 107-118.

Wakasugi, T., Nagai, T., Kapoor, M., Sugita, M., Ito, M., Ito, S., et al. (1997). Complete nucleotide sequence of the chloroplast genome from the green alga *Chlorella vulgaris*: the existence of genes possibly involved in chloroplast division. *Proc. Natl. Acad. Sci. USA, 94*, 5967-5972.

Watanabe, N., Nakazono, M., Kanno, A., Tsutsumi, N., & Hirai, A. (1994). Evolutionary variations in DNA sequences transferred from chloroplast genomes to mitochondrial genomes in the Gramineae. *Curr. Genet., 26*, 512-518.

Weeden, N. F. (1981). Genetic and biochemical implications of the endosymbiotic origin of the chloroplast. *J. Mol. Evol., 17*, 133-139.

Wells, R. & Birnstiel, M. (1969). Kinetic complexity of chloroplastal deoxyribonucleic acid and mitochondrial deoxyribonucleic acid from higher plants. *Biochem. J., 112*, 777-786.

Williamson, D. H., Denny, P. W., Moore, P. W., Sato, S., McCready, S., & Wilson, R. J. (2001). The *in vivo* conformation of the plastid DNA of *Toxoplasma gondii*: implications for replication. *J. Mol. Biol., 306*, 159-168.

Willows, R. D., Kannangara, C. G., & Pontoppidan, B. (1995). Nucleotides of tRNA (Glu) involved in recognition by barley chloroplast glutamyl-tRNA synthetase and glutamyl-tRNA reductase. *Biochim. Biophys. Acta., 1263*, 228-234.

Wilson, R. J., Rangachari, K., Saldanha, J. W., Rickman, L., Buxton, R. S., & Eccleston, J. F. (2003). Parasite plastids: maintenance and functions. *Philos. Trans. R. Soc. Lond. B Biol. Sci., 358*, 155-162; discussion 162-154.

Wilson, R. J., Denny, P. W., Preiser, P. R., Rangachari, K., Roberts, K., Roy, A., et al. (1996). Complete gene map of the plastid-like DNA of the malaria parasite *Plasmodium falciparum*. *J. Mol. Biol., 261*, 155-172.

Winge, Ö. (1919). On the non-mendelian inheritance in variegated plants. *C. R. Trav. Labor Carlsberg, 14*, 1-20.

Wischmann, C., & Schuster, W. (1995). Transfer of *rps10* from the mitochondrion to the nucleus in *Arabidopsis thaliana*: evidence for RNA-mediated transfer and exon-shuffling at the integration site. *FEBS Lett., 374*, 152-156.

Wolfe, K. H., Li, W. H., & Sharp, P. M. (1987). Rates of nucleotide substitution vary greatly among plant mitochondrial, chloroplast, and nuclear DNAs. *Proc. Natl. Acad. Sci. USA, 84*, 9054-9058.

Wolfe, K. H., Morden, C. W., & Palmer, J. D. (1992a). Function and evolution of a minimal plastid genome from a nonphotosynthetic parasitic plant. *Proc. Natl. Acad. Sc. USA, 89*, 10648-10652.

Wolfe, K. H., Morden, C. W., Ems, S. C., & Palmer, J. D. (1992b). Rapid evolution of the plastid translational apparatus in a nonphotosynthetic plant: loss or accelerated sequence evolution of tRNA and ribosomal protein genes. *J. Mol. Evol., 35*, 304-317.

Wolff, G., Plante, I., Lang, B. F., Kuck, U., & Burger, G. (1994). Complete sequence of the mitochondrial DNA of the chlorophyte alga *Prototheca wickerhamii*. Gene content and genome organization. *J. Mol. Biol., 237*, 75-86.

Yamaguchi, K., & Subramanian, A. R. (2000). The plastid ribosomal proteins. Identification of all the proteins in the 50 S subunit of an organelle ribosome (chloroplast). *J. Biol. Chem., 275*, 28466-28482.

Yamaguchi, K., von Knoblauch, K., & Subramanian, A. R. (2000). The plastid ribosomal proteins. Identification of all the proteins in the 30 S subunit of an organelle ribosome (chloroplast). *J. Biol. Chem., 275,* 28455-28465.

Zablen, L. B., Kissil, M. S., Woese, C. R., & Buetow, D. E. (1975). Phylogenetic origin of the chloroplast and prokaryotic nature of its ribosomal RNA. *Proc. Natl. Acad. Sci. USA, 72,* 2418-2422.

Zhang, Z., Green, B. R., & Cavalier-Smith, T. (1999). Single gene circles in dinoflagellate chloroplast genomes. *Nature, 400*, 155-159.

Zheng, D., Nielsen, B. L., & Daniell, H. (1997). A 7.5-kbp region of the maize (T cytoplasm) mitochondrial genome contains a chloroplast-like *trnI* (CAT) pseudo gene and many short segments homologous to chloroplast and other known genes. *Curr. Genet., 32*, 125-131.

CHAPTER 6

PLANT MITOCHONDRIAL GENOMES

C. FAURON[1], J. ALLEN[2], S. CLIFTON[3] AND K. NEWTON[2]

[1]*University of Utah, Eccles Institute of Genetics, Salt Lake City, UT 84112*
[2]*University of Missouri, Division of Biological Sciences, Columbia, MO 65211*
[3]*Washington University, Genome Sequencing Center, 4444 Forest Park Boulevard, St. Louis, MO 63108*

Abstract. Plant mitochondria share a functional role with their fungal and animal counterparts, and their genomes have a number of genes in common. In almost every other aspect, however, plant mitochondrial genomes differ. Whereas animal genomes are very small (14-18 kb) and fungal genomes somewhat larger (70-100 kb), known plant mitochondrial genomes range in size from 187 kb to over 2400 kb. Plant mitochondrial genomes have a variety of other distinctive or unique features. They often have multiple large repeated regions, some of which recombine frequently to yield a multipartite genome structure. Some subgenomes are present at extremely low (substoichiometric) levels, but can become amplified to become major constituents. Plant mitochondrial genes contain mainly group II introns, some of which are *trans*-spliced. The transcripts of many mitochondrial protein-coding genes undergo C-to-U RNA editing. Several mitochondrial tRNAs are known to be transcribed from nuclear genes and imported into the mitochondrion. While only a few mitochondrial genomes from plants have been completely sequenced to date, at least half of the DNA in each of these genomes is still of unknown origin. Comparisons between monocot and dicot mitochondrial genomes show that, in general, only genic exons are conserved. Even between rice and maize, or between Arabidopsis and rapeseed, most of the intergenic space is not conserved. On relatively short evolutionary time scales, plant mitochondrial genomes are in flux, taking up exogenous DNA, losing portions of their DNA and rearranging the order of their sequences.

1. INTRODUCTION

The functions of mitochondria are well conserved across species, but this homogeneity of function contrasts with the diversity of mitochondrial genomes in size, gene content, gene organization and gene expression. Mitochondria generally encode proteins for oxidative phosphorylation and parts of their translational apparatus and, in some species, proteins involved in transcription, RNA maturation and protein import.

As of April 2004, complete mitochondrial genome sequences were available for 536 organisms (http://megasun.bch.umontreal.ca/ogmp/projects/

H. Daniell and C. D. Chase (eds.), Molecular Biology and Biotechnology of Plant Organelles, 151—177.

other/mt_list.html). The sequenced genomes range in size from 5967 bp in the parasitic apicomplexans protist *Plasmodium falciparum* (Conway et al., 2000) to 569,630 bp in the angiosperm *Zea mays* (Clifton et al., 2004). Most animal mitochondrial genomes are small, circular DNA molecules of less than 20 kb. The human mitochondrial genome, the first to be sequenced, is a 16.5 kb circle (Anderson et al., 1981). Fungal mitochondrial genomes also commonly map as single, circular DNA molecules, although both linear and segmented mitochondrial genomes have been identified (Forget et al., 2002; Fungal Mitochondrial Genome Project http://megasun.bch.umontreal.ca/People/lang/FMGP/seqprojects.html). Protist mitochondrial genome sizes are more diverse, and can exhibit unusual genome architectures (reviewed by Burger et al., 2003; Nosek and Tomaska, 2003).

Mitochondrial genomes in plants vary considerably in size, ranging from less than 200 kb to over 2000 kb (Oda et al., 1992; Ward et al., 1981), although there is no correlation between the sizes of mitochondrial and nuclear genomes within species. Intriguingly, variation in genomes size has been seen even within the same species (Fauron et al., 1995a). Land-plant mitochondrial DNAs (mtDNAs) appear to have expanded considerably in size. Some of the size variation can be accounted for by insertions of foreign DNA within the genome and by the presence of large duplicated segments of the genome sequences (Notsu et al., 2002; Clifton et al., 2004) but most of the variation appears to be through the acquisition of large amounts of variable, non-coding spacer DNA of unknown origin.

As of April 2004, complete mtDNA sequences had been reported from one lower plant–the liverwort *Marchantia polymorpha* (Oda et al., 1992), three dicots–*Arabidopsis thaliana* (Unseld et al., 1997), *Beta vulgaris* (Kubo et al., 2000) and *Brassica napus* (Handa, 2003), and two monocots–*Oryza sativa* (Notsu et al., 2002) and *Zea mays* NB (GenBank accession AY506529; Clifton et al., 2004).

2. GENE CONTENT IN MITOCHONDRIAL GENOMES

Common genes in both plant and animal mitochondrial genomes include those for components of oxidative phosphorylation—nine subunits of the respiratory-chain complex-I NADH dehydrogenase (*nad*1, *nad*2, *nad*3, *nad*4, *nad*4L, *nad*5, *nad*6, *nad*7); one subunit of complex III—apocytochrome b (*cob)*; three subunits of complex IV—cytochrome oxidase (*cox*1, *cox*2, *cox*3); and two subunits of complex V—ATP-synthase subunits (*atp*6, *atp*8). They also include the ribosomal RNA (rRNA) genes and a set of transfer RNA (tRNA) genes. Plant mitochondrial genomes usually specify additional proteins as seen in *Table* 1. Unlike animal and most fungal mtDNAs, plant mitochondrial genomes encode several ribosomal proteins, as well as proteins for cytochrome c biogenesis (*ccm*) and a *mtt*-B-like transporter.

Only four genes appear to be coded for by all known mitochondrial genomes: two components of the electron transport chain—apocytochrome b (*cob*) and cytochrome oxidase subunit I (*cox*1)—plus the large (26S) and small (18S) ribosomal RNAs (Feagin, 2000). The mitochondrial genome of the green alga *Chlamydomonas reinhardtii* is only 16 kb and contains six protein-coding and three tRNA genes, in addition to *cob*, *cox*1 and the two rRNA genes (GenBank accession NC_001638, *Table* 1).

The most gene-rich mitochondrial genome known is that of the protozoan *Reclinomonas americana* (Lang et al., 1997; *Table* 1). Its 69-kb genome is thought to contain a subset of the 1.1 Mb genome of the □-proteobacterium, *Rickettsia prowazekii*, an obligate intracellular pathogen that has a reduced bacterial genome size (Gray et al., 1999), and that is the closest-known relative of the likely progenitor of mitochondria. The *R. americana* genome contains all of the genes already identified in all of the other mitochondrial genomes, and encodes at least 18 more proteins, including additional subunits of ribosomes and of respiratory complexes II, IV and V (Lang et al., 1997). In contrast to other sequenced mitochondrial genomes, it encodes a translation elongation factor, the RNA component of RNaseP and a eubacterial-type RNA polymerase. In other mitochondria, including those of plants, the genes are transcribed by nuclear-encoded, single-subunit, phage-like RNA polymerases (Ikeda and Gray, 1999; Chang et al., 1999).

In contrast to mammalian mitochondrial genomes, linear gene order is not conserved in the mitochondrial genomes of plants. Considerable variation in genome organization can occur within a single species. Despite the larger coding capacities of higher plant mitochondrial DNAs, their known gene sets are not much larger than those of animals. Indeed, the largest number of mitochondrial genes occurs in the most "primitive" plant, *M. polymorpha*, which has the smallest mitochondrial genome among those sequenced to date.

Table 1. *Gene content of plant mitochondrial genomes*

	***Zm* 570**	***Os* 491**	***Bv* 369**	***At* 367**	***Bn* 222**	***Mp* 187**	***Ra* 69**	***Cr* 16**	**gene product**
Energy production									
Complex I									
*nad*1	*	*	*	*	*	*	*	*	NADH dehydrogenase subunit 1
*nad*2	*	*	*	*	*	*	*	*	NADH dehydrogenase subunit 2
*nad*3	*	*	*	*	*	*	*		NADH dehydrogenase subunit 3
*nad*4	*	*	*	*	*	*	*	*	NADH dehydrogenase subunit 4
*nad*4L	*	*	*	*	*	*	*		NADH dehydrogenase subunit 4L
*nad*5	*	*	*	*	*	*	*	*	NADH dehydrogenase subunit 5
*nad*6	*	*	*	*	*	*	*	*	NADH dehydrogenase subunit 6
*nad*7	*	*	*	*	*	□	*		NADH dehydrogenase subunit 7
*nad*8							*		NADH dehydrogenase subunit 8
*nad*9	*	*	*	*	*	*	*		NADH dehydrogenase subunit 9
*nad*10							*		NADH dehydrogenase subunit 10
*nad*11							*		NADH dehydrogenase subunit 11
Complex II									
*sdh*2							*		succinate oxido-reductase subunit 2
*sdh*3						*	*		succinate oxido-reductase subunit 3
*sdh*4				□	□	*	*		succinate oxido-reductase subunit 4

Table 1 *(cont.)*

Table 1 (cont.)

	Zm	*Os*	*Bv*	*At*	*Bn*	*Mp*	*Ra*	*Cr*	gene product
Complex III									
cob	*	*	*	*	*	*	*	*	apocytochrome b
Complex IV									
*cox*1	*	*	*	*	*	*	*	*	cytochrome c oxidase 1
*cox*2	*	*	*	*	*	*	*		cytochrome c oxidase 2
*cox*3	*	*	*	*	*	*	*		cytochrome c oxidase 3
*cox*11							*		cytochrome c oxidase 11
Complex V									
*atp*1	*	*	*	*	*	*	*		ATP synthase F1 subunit 1
*atp*3							*		ATP synthase F1 subunit 3
*atp*4 (*orf*25)	*	*	*	*	*				ATP synthase F1 subunit 4
*atp*6	*	*	*	*	*	*	*		ATP synthase F1 subunit 6
*atp*8 (*orf*B)	*	*	*	*	*	*	*		ATP synthase F1 subunit 8
*atp*9	*	*	*	*	*	*	*		ATP synthase F1 subunit 9
Cytochrome *c* biogenesis									
*ccm*A - *yej*W	*	*	*	*	*	*	*		ABC transporter for cytochrome c subunit A
*ccm*B - *yej*V	*	*	*	*	*	*	*		ABC transporter for cytochrome c subunit B
*ccm*C - *yej*U	*	*	□	*	*	*	*		ABC transporter for cytochrome c subunit C
*ccm*F - *yej*R	*	*	*	*	*	*	*		ABC transporter for cytochrome c subunit F
Transcription									
*rpo*A							*		RNA polymerase subunit alpha
*rpo*B							*		RNA polymerase subunit beta
*rpo*C							*		RNA polymerase subunit beta'
*rpo*D							*		transcription initiation factor
mat-r	*	*	*	*	*		*		maturase
rtl								*	Reverse transcriptase-like
Translation									
*rps*1	*	*				*	*		ribosomal protein S1
*rps*2	*	*			*	*	*		ribosomal protein S2
*rps*3	*	*	*	*	*	*	*		ribosomal protein S3
*rps*4	*	*	*	*	*	*	*		ribosomal protein S4
*rps*7	*	*	*	*	*	*	*		ribosomal protein S7
*rps*8						*	*		ribosomal protein S8
*rps*10						*	*		ribosomal protein S10
*rps*11		□				*	*		ribosomal protein S11
*rps*12	*	*	*	*	*	*	*		ribosomal protein S12
*rps*13	*	*	*			*	*		ribosomal protein S13
*rps*14		□		□	*	*	*		ribosomal protein S14
*rps*19		*		□		*	*		ribosomal protein S19
*rpl*1							*		ribosomal protein L1
*rpl*2		*		*	*	*	*		ribosomal protein L2

Table 1 (cont.)

Table 1 (cont.)

	Zm	***Os***	***Bv***	***At***	***Bn***	***Mp***	***Ra***	***Cr***	**gene product**
*rpl*5		*	*	*	*	*	*		ribosomal protein L5
*rpl*6						*	*		ribosomal protein L6
*rpl*10							*		ribosomal protein L10
*rpl*11							*		ribosomal protein L11
*rpl*14							*		ribosomal protein L14
*rpl*16	*	*		*	*	*	*		ribosomal protein L16
*rpl*18							*		ribosomal protein L18
*rpl*19							*		ribosomal protein L19
*rpl*20							*		ribosomal protein L20
*rpl*27							*		ribosomal protein L27
*rpl*31							*		ribosomal protein L31
*rpl*32							*		ribosomal protein L32
*rpl*34							*		ribosomal protein L34
*tuf*A							*		Elongation factor
Transporter protein									
*sec*Y - *orf*X - *mtt*-B	*	*	*	*	*	*	*		transporter protein
tRNA									
*trn*A						*	*		tRNA-Ala
*trn*C	*	*	*	*	*	*			tRNA-Cys
*trn*D	*	*	*	*	*	*	*		tRNA-Asp
*trn*E	*	*	*	*	*	*	*		tRNA-Glu
*trn*F	*	*	*			*	*		tRNA-Phe
*trn*G			*	*	*	*	*	*	tRNA-Gly
*trn*H	*	*	*	*	*	*	*		tRNA-His
*trn*I	*	*	*	*	*	*	*		tRNA-Ile
*trn*K	*	*	*	*	*	*	*		tRNA-Lys
*trn*L						*	*		tRNA-Leu
*trn*M	*	*	*	*	*	*	*	*	tRNA-Met
*trn*fM	*	*	*	*	*	*	*		tRNA-fMet
*trn*N	*	*	*	*	*	*	*		tRNA-Asn
*trn*P	*	*	*	*	*	*	*		tRNA-Pro
*trn*Q	*	*	*	*	*	*	*		tRNA-Gln
*trn*R						*	*		tRNA-Arg
*trn*S	*	*	*	*	*	*	*		tRNA-Ser
*trn*T						*			tRNA-Thr
*trn*V						*	*		tRNA-Val
*trn*W	*	*	*	*	*	*	*	*	tRNA-Trp
*trn*Y	*	*	*	*	*	*	*		tRNA-Tyr
rRNA									
*rrn*5	*	*	*	*	*	*	*		5S ribosomal RNA
*rn*18	*	*	*	*	*	*	*	*	18S ribosomal RNA
*rrn*26	*	*	*	*	*	*	*	*	26S ribosomal RNA

*Zm, Zea mays NB; Os, Oryza sativa; Bv, Beta vulgaris; Bn, Brassica napus; Mp Marchantia polymorpha; Ra, Reclinomonas americana; Cr, Chlamydomonas reinhardtii. Numbers indicate genome sizes in kb. *, gene present; □, pseudogene.*

3. CODING CONTENT OF PLANT MITOCHONDRIAL GENOMES

3.1. GC content

The base composition of mtDNA is nearly constant among all sequenced plant mitochondrial genomes. The average GC content is: *Z. mays NB,* 43.9%; *O. sativa,* 43.9%; *B. vulgaris,* 43.9%; *A. thaliana,* 44.8%; *B. napus,* 45.2%; *M. polymorpha,* 42.4%. Surprisingly, the variable intergenic regions do not exhibit lower or higher average GC content than does the rest of the genome. Furthermore, no discrete regions of higher or lower GC content are present in the genomes.

3.2. Protein-coding genes, transfer RNA genes, ribosomal RNA genes

The mitochondrial genome of *M. polymorpha* (liverwort) is a single chromosome of 187 kb encoding 66 identified genes, including rRNAs and tRNAs (Oda et al., 1992; Figure 1A). The smallest of the sequenced angiosperm mitochondrial genomes is the 222 kb genome of *B. napus* (rapeseed), which has 54 known, apparently functional genes, including 34 protein-coding, 3 rRNA and 17 tRNA genes. The sequence data generated a circle (Handa, 2003; Figure 1B). The 367 kb *A. thaliana* mitochondrial genome was reported to include 59 identified genes (Unseld et al., 1997; Figure 1C). The only difference in protein-gene content between rapeseed and Arabidopsis, which are members of the same family, is the presence of an intact *rps*14 gene in rapeseed, which is a pseudogene in Arabidopsis. Like Arabidopsis, the rapeseed mtDNA has tRNAs specifying 15 species of amino acids; however, there are five fewer tRNA genes in the rapeseed mtDNA (Handa, 2003).

A more distantly related dicot, *B. vulgaris* (sugar beet), has a 369-kb mitochondrial genome, similar in size to that of Arabidopsis (Kubo et al. 2000; Figure 1D). However, the sugar beet mtDNA contains *rps*13 and two tRNA genes (*trn*F-GAA and *trn*C2-GCA) that are not present in the Arabidopsis mitochondrial genome (Kubo et al., 2000). Conversely, the *rpl*2, *rpl*16 and *trn*Y2-GUA genes in Arabidopsis appear to be absent or truncated in sugar beet. In Arabidopsis and *B. napus*, the *ccm*FN gene is present as two separate genes, *ccm*FN1 and *ccm*FN2.

Two monocot mtDNAs have been sequenced, both of which are larger than any other sequenced mitochondrial genomes. Despite the increased sizes, both rice (491 kb; Notsu et al., 2002; Figure 1E) and maize (NB cytotype, 570 kb; Clifton et al., 2004; Figure 1F) contain a number of functional genes similar to those of the dicot mitochondrial genomes.

All six of the sequenced plant mitochondrial genomes include *mtt*-B (formerly *orf*X), a presumed *tat*C homolog (Bonnard and Grienenberger, 1995; Bogsch et al., 1998); *mat-r*, a gene encoding a protein with significant similarity to retroviral reverse transcriptases that is located in a *nad*1 intron (Wahleithner et al., 1990) is found in all angiosperms mtDNAs but absent in liverwort mtDNA. The angiosperm mitochondrial genomes do not contain a complete set of tRNA genes (see review by Maréchal-Drouard et al., 1993).

Most of the variation among plant mitochondrial genomes is in their ribosomal protein gene content. The *rpl*2 and *rpl*16 genes found in rice, Arabidopsis and rapeseed are not found in sugar beet mtDNA. The *rps*13 gene, which is present in rice and sugar beet, is not found in Arabidopsis or rapeseed. The *rps*1, *rps*2 and *rps*19 genes have been found only in rice. The liverwort mitochondrial genome contains all of the above-mentioned ribosomal protein genes. A functional *rps*11 is also present in *Marchantia* but is not found in the other plant mtDNAs. Although an *rps*11 pseudogene is present in rice mtDNA, the functional gene is located in the nucleus (Notsu et al., 2002). The *rps*19 gene, present in *Marchantia* and rice, is only a pseudogene in Arabidopsis, and has been lost entirely from the maize NB mitochondrial genome. Overall, these observations agree with the proposition that transfer of the ribosomal protein genes to the nucleus is an on-going process in higher plants (Adams et al., 2002).

3.3. Unidentified open reading frames (ORFS)

The definition of unidentified open reading frames (ORFs) is somewhat arbitrary. A commonly used definition for an ORF is a reading frame of at least 100 codons with in-frame start and stop codons. The published numbers of ORFs vary depending on the size limits imposed. Based on the published reports and sequences, there are 121 such defined ORFs in maize NB, 113 in rice, 93 in sugar beet, 85 in Arabidopsis, 45 in rapeseed and 32 in liverwort, which represent 10.4%, 9.4%, 10.4%, 10.2%, 9.0% and 7.5% of their respective genomes (Notsu et al., 2002; Handa, 2003; Clifton et al., 2004; Kubo et al., 2000; Marienfeld et al., 1999; Oda, 1992).

The significance of these ORFs is unclear, but it is unlikely that very many of them are expressed as proteins. Indeed, analyses of cDNAs in Arabidopsis mitochondria suggest that few, if any of these ORFs are expressed as stable mRNAs (Giege and Brennicke, 2001). Skovgaard et al., (2001) examined 34 bacterial genomes and concluded that 30-90% of the annotated ORFs, particularly those that were not present in other species, do not actually encode proteins. It is not unreasonable to expect that most of the ORFs in the endosymbiotic mitochondrial genomes, particularly those not conserved across taxa, are not expressed.

Figure 1*. (see pages 158-160) Circular maps of the sequenced plant mitochondrial genomes. Circular maps generated from sequence data of the following mitochondrial genomes: A, Marchantia polymorpha; B, Beta vulgaris; C, Brassica napus; D, Arabidopsis thaliana; E, Oryza sativa; F, Zea mays NB. Known protein-coding genes, tRNA genes, rRNA genes and repeats are shown on the outside circle. Gene functions: Complex I, nad; Complex III, cob; Complex IV, cox; Complex V, atp; cytochrome assembly, ccm; ribosomal proteins, rpl, rps; maturase, mat-r; rRNAs, rrn; and genes transferred from chloroplast, -ct. Single letter designations indicate tRNAs. Large repeats are shown within the outer ring. The R1- and S2/R2-homologous regions in maize are indicated by grey blocks. The inner circle shows positions of ORFs (plus strand outward and minus strand inward) with predicted sizes of ≥99 amino acids.*

1A

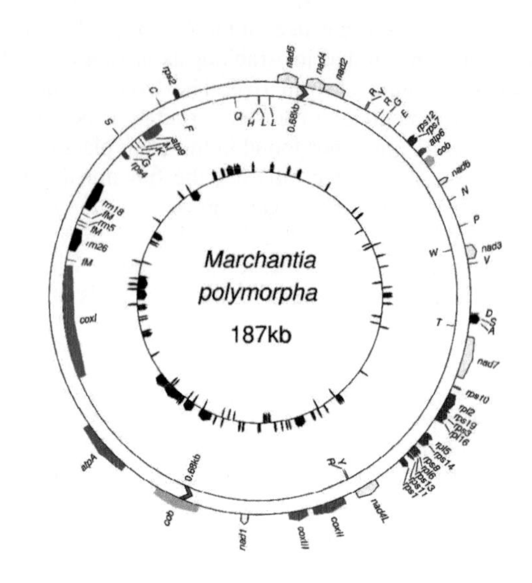
Marchantia
polymorpha
187kb
nad5
nad4
nad2
coxI
atpA
cob
nad1
coxIII
coxII
nad4L
nad7
nad3
0.68kb

1B

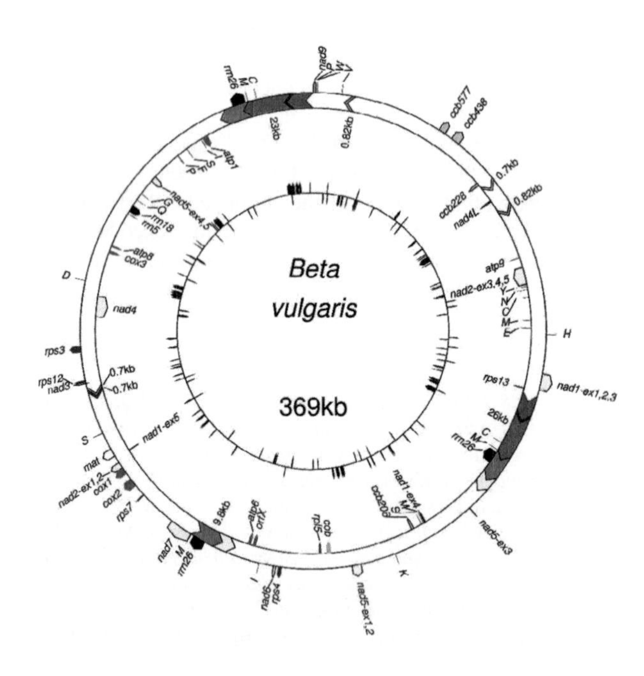
Beta
vulgaris
369kb
23kb
0.82kb
0.7kb
26kb
9.8kb
nad4
rps3
rps12
nad3
nad1-ex5
nad1-ex1,2,3
nad5-ex3
nad5-ex1,2
nad2-ex3,4,5
atp9
nad4L
ccb577
ccb438
ccb228
rps13
nad7
rrn26

1C

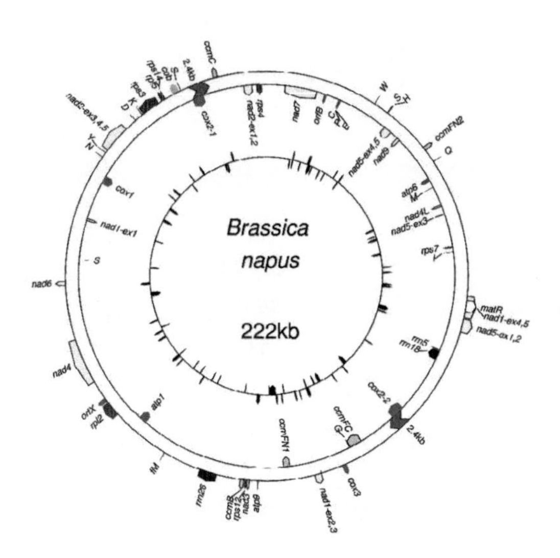
Brassica
napus
222kb
ccmC
2.4kb
cox2-1
rps4
nad2-ex1,2
nad7
orfB
W
ccmFN2
Q
nad5-ex4,5
nad9
atp6
M
nad4L
nad5-ex3
rps7
matR
nad1-ex4,5
nad5-ex1,2
rrn5
rrn18
cox2-2
2.4kb
ccmFC
G
ccmFN1
cox3
nad1-ex2,3
atp9
rrn26
fM
orfX
rpl2
atp1
nad4
nad6
S
nad1-ex1
cox1
nad2-ex3,4,5

1D

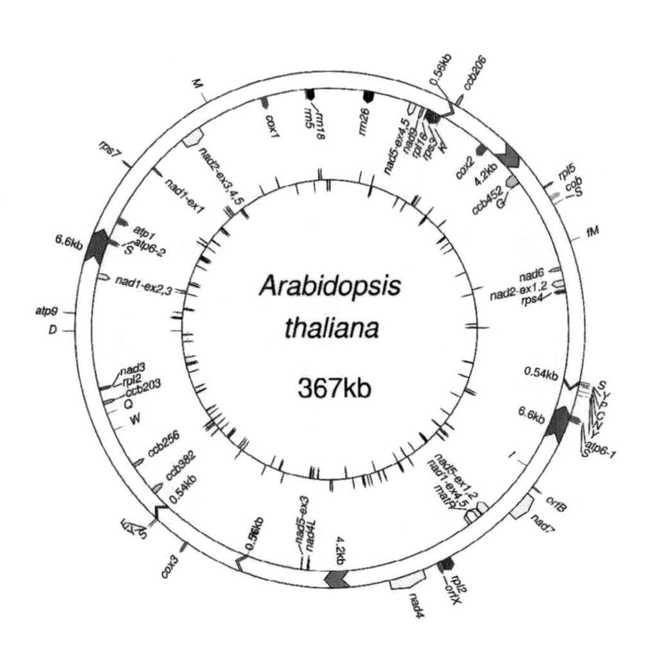
Arabidopsis
thaliana
367kb
M
cox1
rrn18
rrn5
rrn26
0.55kb
ccb206
nad5-ex4,5
nad9
rpl16
rps3
K
cox2
4.2kb
ccb452
G
rpl5
cob
S
fM
nad6
nad2-ex1,2
rps4
0.54kb
6.6kb
atp6-1
S
orfB
nad7
rpl2
orfX
nad4
4.2kb
nad5-ex3
nad4L
cox3
0.54kb
ccb382
ccb256
W
Q
ccb203
rpl2
nad3
atp9
D
nad1-ex2,3
6.6kb
atp1
atp6-2
S
rps7
nad1-ex1
nad2-ex3,4,5
nad5-ex1,2
nad1-ex4,5
matR

1E

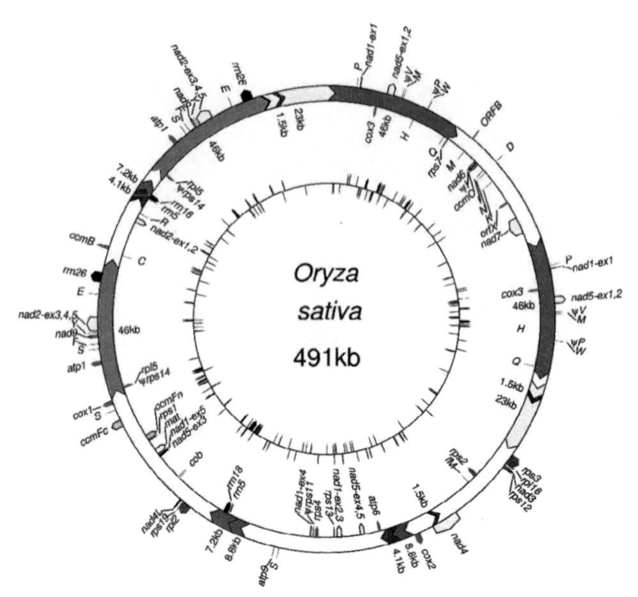
Oryza
sativa
491kb
nad1-ex1
nad5-ex1,2
cox3
46kb
1.5kb
23kb
ccmB
rrn26
nad2-ex3,4,5
atp1
cox1
ccmFc
nad4
cox2

1F

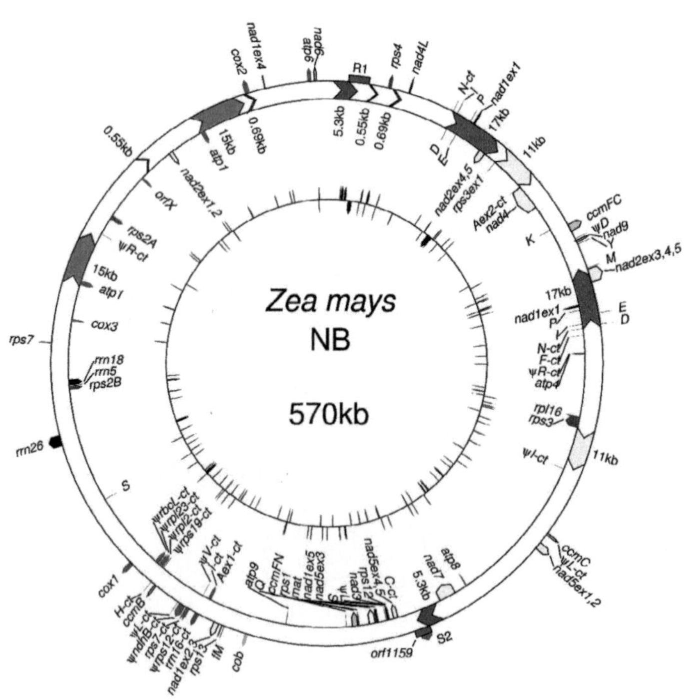
Zea mays
NB
570kb
R1
rps7
rrn26
S
17kb
11kb
nad1ex1
E
D
ccmFC
nad2ex3,4,5
rpl16
rps3
ccmC
nad5ex1,2
S2
orf1159
cox1
cob

3.4. Chimeric genes, pseudogenes and gene fragments

The presence of novel chimeric genes has been reported in all plant mitochondrial genomes. A chimeric or mosaic gene is composed of linked stretches of DNA having similarity to other regions of the genome or to DNA sequences of unknown origin. They are presumably formed as the result of recombination. The production of proteins resulting from such chimeric genes is directly associated with cytoplasmic male sterility (CMS), a maternally inherited trait (reviewed in Chapters 22 and 23 of this volume). The first and most unusual chimeric gene identified was *urf*13 in the CMS-T mitochondrial genome of maize (Dewey et al., 1986); *urf*13 is the cause of the sterility. Other chimeric genes of maize were identified by Levings and coworkers (Dewey et al., 1991; reviewed by Fauron et al., 1995b).

Pseudogenes are defective versions of genes that no longer produce a functional product. Plant mitochondrial genomes usually contain at least some pseudogenes. In addition to chimeric and pseudogenes, fragments of genes have been found in all mitochondrial genomes, and are thought to be the result of mitochondrial recombination processes.

3.5. Introns

The first identification of introns in mitochondrial genes was in *Saccharomyces cerevisiae* (Lazowska et al., 1980). Two major classes of introns have been characterized: group I and group II. Each group has characteristic secondary structures and is spliced by a different molecular mechanism. Group II introns are further divided into two classes based on differences in secondary structure. Group I introns are released as linear molecules with covalently added G residues, whereas group II introns are released as lariat structures. (see http://www.fp.ucalgary.ca/group2introns/index.html; Qin and Pyle, 1998; Bonen and Vogel, 2001; Zimmerly et al., 2001; Chapter 11 of this volume). Although the liverwort mitochondrial genome contains both group I and group II introns (Ohta et al., 1993), angiosperm mitochondrial genes contain mostly group II introns. Interestingly a *cox*1 intron is the only reported group I intron in the mitochondrial genomes of vascular plants (Cho et al., 1998). The intron is thought to have been acquired by lateral transfer from a fungal source. From a survey of more than 300 diverse land plants, Cho and Palmer (1999) concluded that the most parsimonious explanation for the presence of this intron in the 48 angiosperm genera in which it was found required 32 separate horizontal transfer events.

Some group I and group II introns will self-splice *in vitro*. However, most higher plant organelle introns require trans-acting protein and/or RNA factors for splicing *in vivo*. Some group II introns are split introns, *i.e.* parts of the intron (and adjacent exons) are located at different positions in the genome. Each of the independently transcribed segments come together to form the classic group II secondary structure, hence the process is termed "*trans*-splicing".

Intron content is almost identical among the five sequenced angiosperm mitochondrial genomes, and the introns are all group II. *Trans*-splicing introns are found in *nad*1, *nad*2 and *nad*5, and *cis*-splicing introns are found in *cox*2, *nad*1, *nad*2, *nad*4, *nad*5, *nad*7, *ccm*FC and *rps*3. Among the sequenced angiosperm mtDNAs, the only difference in protein-gene intron content is that sugar beet lacks the second intron of *nad*4 and the *rps*3 intron found in the other mtDNAs. The higher-plant rRNA and tRNA mitochondrial genes lack introns, with the exception of a maize mitochondrial tRNA-Ile gene, which contains a 951-bp group II intron.

Plant organelle introns do not fold into good secondary structures (Michel et al., 1989; Bonen and Vogel, 2001), and they have not been shown to self-splice *in vitro*. The presence of marginal secondary structures and poor catalytic function suggests that plant mitochondria might have lost intron structural motifs and intron-encoded ORFs that aid in splicing, and that plant mitochondrial introns must rely on nuclear splicing factors. Several nuclear-encoded splicing factors have been identified in the chloroplast (Ostheimer et al., 2003).

3.6. Transfer of mitochondrial genes to the nucleus

Mitochondria are thought to be descendents of free-living organisms that were engulfed by a proto-eukaryotic cell, and whose genomes have subsequently been drastically reduced (reviewed by Gray in Chapter 2 of this volume). The genes missing from endosymbiotic organelles were lost either because 1) they were not needed in this intracellular environment, 2) nuclear genes took over their functions, or 3) the functional mitochondrial gene was transferred to the nucleus (Selosse et al., 2001). For an organelle gene to be functionally transferred to the nuclear genome, it must have acquired a promoter and a transit sequence that allows the gene product to be targeted to the mitochondrion. Current mitochondrial genomes have far fewer genes than their ancestral genomes: 13 protein-coding genes in human mtDNA, 35 in maize mtDNA and 62 in *R. americana* mtDNA, versus 832 in their closest proteobacterial relative, *Rickettsia prowazekii*. A massive transfer of organelle genes into the nuclear genome occurred very early in the evolution of mitochondrial genomes (Adams et al., 2002; Adams and Palmer, 2003). This process continues only in the plant kingdom, although the extent of the transfer differs among plant species and does not seem to have reached equilibrium yet. Some genes have been transferred very frequently, apparently hundreds of times during plant evolution, whereas some genes never seem to have been transferred (Adams et al., 2000).

4. RNA EDITING

RNA editing (reviewed by Mulligan in Chapter 9 of this volume) was first described for trypanosome mitochondrial *cox*2 transcripts (Benne et al., 1986). Several types of RNA editing are known, but in plant mitochondrial genomes, only one type is present, in which genomically encoded C residues are converted post-transcriptionally to U residues (or occasionally U is converted to C). These edited nucleotides are identified by comparing genomic and cDNA sequences. Plant mitochondrial genes are translated according to the universal code, and RNA editing is required to allow many of their transcripts to code for functional proteins. Editing occurs primarily within coding regions (Giege and Brennicke, 1999). Editing events within the coding sequences improve the overall conservation of the predicted proteins when compared to the same proteins in other organisms. Thus, RNA editing functions as a copy-correction mechanism. RNA editing can create new initiation codons (ACG to AUG) and new termination codons (CGA to UGA). For example, the *nad*1 initiation codon is created by editing of ACG to AUG in all known angiosperm mitochondrial genomes. In general, however, editing sites vary among the mitochondrial genomes of different plant species. Editing sites are also found in tRNAs, where editing is required for efficient tRNA processing and correction of mismatches to allow the formation of functional secondary structures (Fey et al., 2002).

RNA editing has been shown to occur in introns in both animals and plants. Lippok et al. (1994) have shown that the introns of *nad*1 (a/b) and *nad*2 (c/d) in *Oenothera* (evening primrose) mitochondria are closely related. There is a 48 bp element that is identical in both group II introns. However, the C-to-U editing occurs only in the *nad*2 intron to stabilize its secondary structure, *i.e.* there is differential RNA editing. In another instance, Zanlungo and colleagues (1995) showed that there was a C-to-U change in the 774-base-pair class-II intron of unspliced cDNAs of the *rps*10 transcript of potato mitochondria and suggested that it was involved in the control of the splicing reaction.

The mechanism of editing in plant organelles appears to be the enzymatic deamination of thymidine to produce uridine (Covello and Gray, 1989; Gualberto et al., 1989; Hiesel et al., 1989). The particular molecular mechanism that specifies the bases to be edited is not known. In general, however, at least the 25 nucleotides 5' to the editing site are critical to specify editing (Kubo and Kadowaki, 1997;Williams et al., 1998), and some sites appear to have additional requirements for a limited number of 3' nucleotides (Farré et al., 2001). Partially edited transcripts have been identified, and the question remains as to whether these transcripts are translated into functional proteins (Lu and Hanson, 1996; Phreaner et al., 1996).

5. CHLOROPLAST SEQUENCE INSERTIONS

Plant mitochondrial genomes have been described as promiscuous, donating DNA to the nuclear genome and accepting DNA from the nuclear and the plastid genomes. Sequences from the plastid genome are more unambiguously identifiable, and have been noted in all angiosperm mitochondrial genomes in which they have been searched for (Notsu et al., 2002; Handa, 2003; Clifton et al., 2004; Kubo et al., 2000; Marienfeld et al., 1999). The proportion of plastid DNA (ctDNA) that has translocated to the mitochondrial genome appears to follow a taxonomic gradient. Oda reported finding essentially no recognizable plastid sequences in the mitochondrial genome of liverwort (Oda et al., 1992b). The three sequenced dicot species contain from 4,400 to 13,000 bp of plastid DNA in the non-repetitive portion of their mitochondrial genomes, while the two sequenced monocot species each contain about 23,000 bp in their non-repetitive portions. The transferred plastid sequences compose 7.0% of the rice mitochondrial genome, 5.9% of rapeseed, 4.4%, of maize, 2.1% of sugar beet, 2.1% of Arabidopsis and 0% of liverwort (Notsu et al., 2002; Handa, 2003; Clifton et al., 2004; Kubo et al., 2000; Marienfeld et al., 1999; Oda, 1992).

Figure 2 shows that there is little similarity among the taxa in the specific plastid sequences that have found their way to the mitochondrial genome, which is consistent with either ctDNA transfer being a rare event, or frequent DNA loss after transfer. Evidence from an unusual insertion in maize indicates that a third mechanism is also at work (Clifton et al., 2004). Maize, rice, wheat and barley mitochondrial genomes contain a 4.1-kb plastid insertion composed of two distinct plastid sequences, an unusual configuration that argues for a common origin. However, phylogenetic analyses indicate that for each species, the mitochondrial version of the plastid sequence is most closely related to its own species' plastid sequence, implying that it was recently transferred. One way to explain these contradictory results is that plastid sequences often make their way into mitochondria, but integrate rarely for lack of recombination sites. However, if homologous sequences are present, recombination occurs, leading to copy correction of the plastid insertion (Clifton et al., 2004).

6. COMPARATIVE ANALYSES

6.1. Genic and intergenic regions

Comparative analyses of mitochondrial genes have shown that, with rare exceptions, the sequences of protein-coding genes are highly conserved. Indeed, the rates of nucleotide substitution in plant mitochondrial protein-coding genes are lower than those of chloroplast genes or of plant or animal nuclear genes (reviewed by Palmer, 1990). In contrast, the sequences between genes can be highly variable.

Although these intergenic regions can include retrotransposons of nuclear origin and integrated chloroplast sequences, approximately 50% of each of the sequenced angiosperm mtDNAs could not be found in the extant databases (Unseld et al., 1997; Marienfeld et al., 1999; Kubo et al., 2000; Handa, 2003; Clifton et al., 2004). Within cucurbits, where mitochondrial genomes show an eight-fold size variation, an expansion of short, degenerate repeats has been proposed to account for much of the size differences (Lilly and Havey, 2001).

Only 39.5% of the rice mitochondrial sequence is at least 90% identical to maize NB mtDNA (Figure 3; minimum match of 20 contiguous bp). Conversely, 27.9% of maize NB mitochondrial genome is equally similar to rice mtDNA. Only 9.5% of the maize NB genome is shared with Arabidopsis mtDNA and 13.8% of the Arabidopsis mitochondrial genome is shared with maize. Therefore, most of the mitochondrial genome does not seem to have a common ancestry, even between two closely related species as maize and rice. The only common sequences between a monocot and a dicot seem to be the genes.

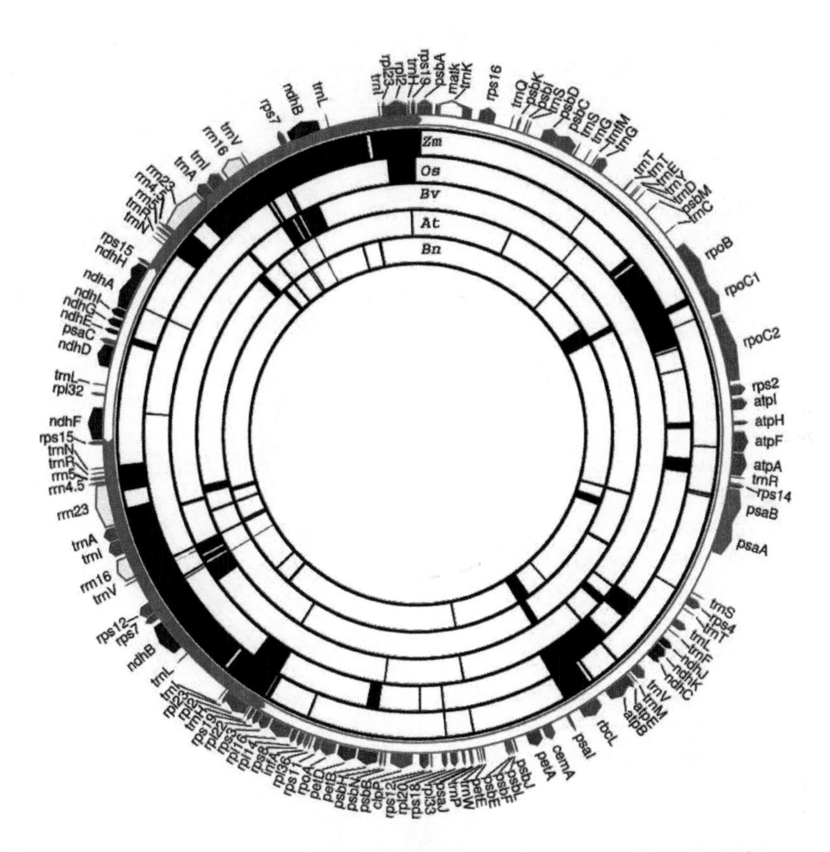

Figure 2. *Circular map showing chloroplast DNA sequences present in various plant mitochondrial genomes. The outer ring represents the maize chloroplast genome (Maier et al., 1995), with all identified genes labelled on the outside and the large inverted repeat represented within the ring. The next 5 rings from outside to inside represent the mitochondrial genomes of Zea mays NB, Zm; Oryza sativa, Os; Beta vulgaris, Bv; Arabidopsis thaliana, At; and Brassica napus, Bn.*

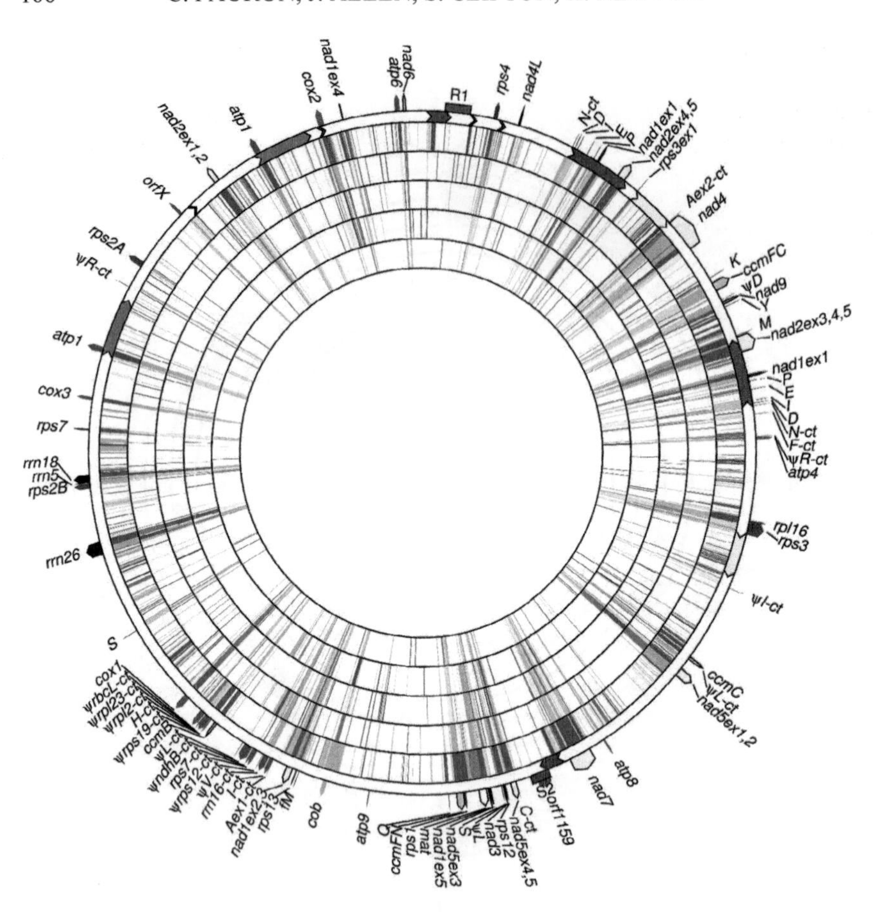

Figure 3. *Circular map showing regions of various plant mitochondrial genomes homologous to maize NB mtDNA. (For a color version of this figure, see the color figure section of the book.) The outer ring represents the maize mitochondrial genome, with all identified genes labelled on the outside. Repeats larger than 0.5 kb are shown within the thin ring. The 5 thick rings represent, from outside to inside, mtDNAs of Oryza sativa, Beta vulgaris, Arabidopsis, thaliana, Brassica napus, and Marchantia polymorpha. Sequences with ≥60% similarity to maize are shown. It should not be inferred from this comparison that the genomes are the same size or that the gene order is identical in all genomes. The figure retains the gene order of the outer ring reference genome maize NB, and aligns the other genomes with it, irrespective of the order that the sequences occur in those genomes (see Color Plate 1).*

6.2. MultiPipMaker—Graphical comparisons of the sequenced plant mtDNAs

Similarity between a reference plant genome and other plant mtDNA sequences serves as a useful guide for identifying homologous sequences in the genomes being compared. The software *MultiPipMaker* computes alignments of similar regions in two or more DNA sequences (Schwartz et al., 2000; Schwartz et al., 2003). The resulting alignments are summarized as a "percent identity plot" (pip). *MultiPipmaker* retains the linear order of the reference genome, displaying it on the horizontal axis, and aligns the other genomes with it, irrespective of the order that the sequences occur in those genomes. The percent identity of each of the other genomes to the reference genome is represented graphically on the vertical axis. Figure 3 uses the rice mitochondrial genome as the reference genome (with genes and ORFs indicated by the black boxes) and illustrates the high level of conservation of genes relative to the non-genic regions. Indeed, coding-region conservation can be very high even in comparisons with mtDNA of *M. polymorpha* or *R. americana*, a protist with the most bacterial-like mtDNA. It also shows that most of the ORF-containing regions are not conserved, even between rice and maize.

Although genes are generally not clustered in plant mtDNA, there are regions that are relatively richer or poorer in known genes. Intron sequences tend to be well conserved between maize and rice (*e.g.* the *nad5* intron), although these similarities disappear with increasing phylogenetic distance. Some regions identified as containing ORFs in rice mtDNA are not conserved in the other taxa. Still, some of the mtDNA lacking known genes, but encompassing maize ORFs or intergenic regions, was found to be shared between maize and rice mtDNA. Nevertheless, most of the intergenic regions in rice are not conserved even with mtDNA from another grass, and many showed no sequence similarity to any other known sequences.

7. REPEATED SEQUENCES, RECOMBINATION AND GENOMIC ORGANIZATION

7.1. Repeated sequences in plant mitochondrial genomes

The role of repeated sequences is an interesting and controversial topic for evolutionary studies. Accumulation of repeated sequences appears to be a factor in the expansion of plant mitochondrial genomes. A summary of the repeats present within each plant mitochondrial genome is shown in Table 2.

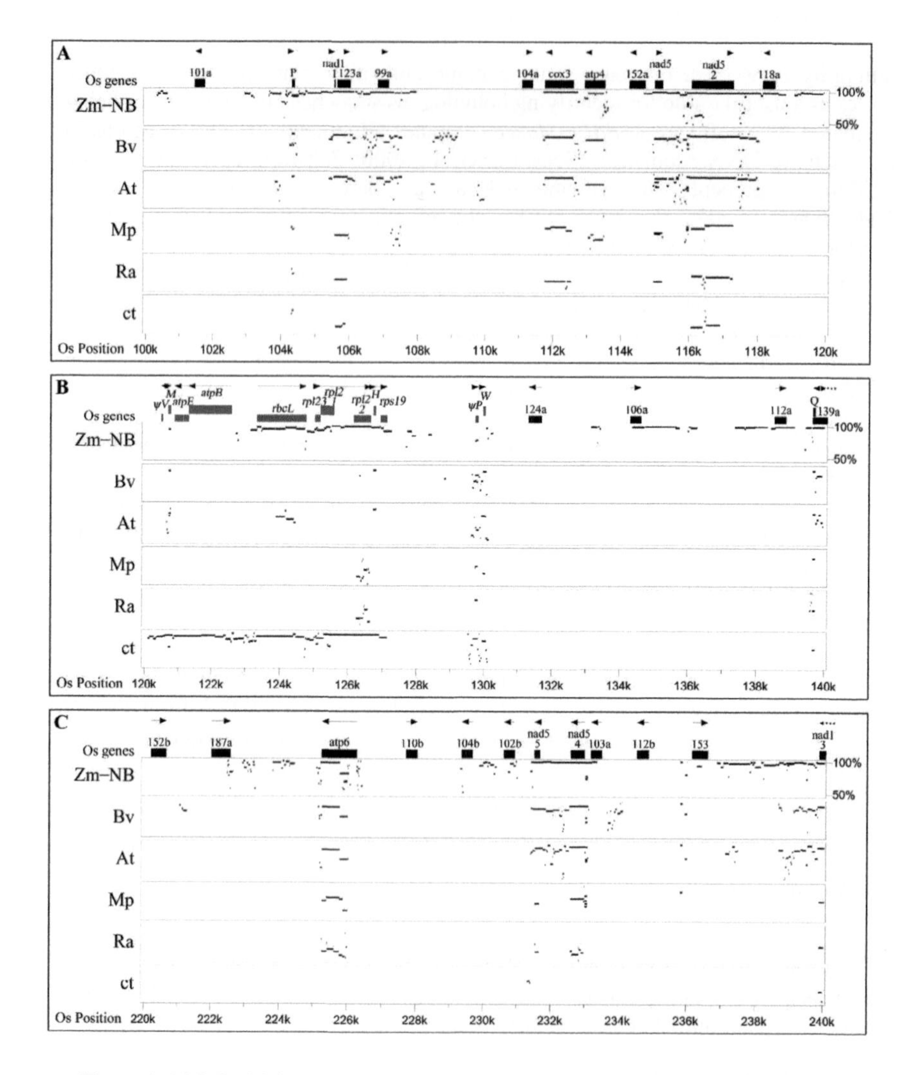

Figure 4. *MultiPipMaker analysis of several sequenced mtDNAs. Comparisons of the reference Oryza sativa, Os, mitochondrial genome with other mitochondrial genomes: Zea mays NB, Zm–NB; Beta vulgaris, Bv; Arabidopsis thaliana, At; Marchantia polymorpha, Mp; Reclinomonas americana, Ra; O. sativa, Os; zea mays chloroplast DNA, ct. At the top of the figure, positions of rice mitochondrial genes are indicated by black boxes and their orientations are indicated by arrows.*

***Table* 2.** *Numbers of repeats present in plant mitochondrial genomes*

Repeat size	***Zm NB*** 570 kb		***Os*** 491 kb		***Bv*** 369 kb		***At*** 367 kb		***Bn*** 222 kb		***Mp*** 187 kb	
	#	%	#	%	#	%	#	%	#	%	#	%
10kb to100kb	6	15.1%	6	46.8%	2	11.0%	--	--	--	--	--	--
1kb to 10kb	2	1.9%	10	6.6%	10	7.7%	4	5.9%	2	2.2%	--	--
100bp to 1kb	57	1.9%	66	2.2%	131	4.8%	90	4.5%	37	2.2%	90	8.3%
50bp to 100bp	213	2.0%	274	1.9%	164	2.5%	237	3.4%	122	3.0%	355	7.6%

These repeats were identified using Miropeats 1.1 (Parsons, 1995) employing the smallest possible match score and eliminating all with similarities under 80%. Zea mays NB, Zm NB; Oryza sativa, Os;Beta vulgaris, Bv,; Brassica napus, Bn; Arabidopsis thaliana, At; Marchantia polymorpha, Mp. Genome sizes are below the species designations. # indicates the number of repeats; % indicates the percent of the genome they represent.

Even among close relatives, the regions of the mitochondrial genomes that have duplicated are different (Fauron et al., 1995a). In general, large repeats are more frequent in larger mitochondrial genomes, while smaller repeats are more frequent in smaller mitochondrial genomes. More than 50% of the 369-kb rice mitochondrial genome is composed of 16 repeats at least 1 kb long (five pairs and two triplets). The 570 kb maize NB mitochondrial genome contains only 8 repeats 1 kb or longer, which represent only 17% of the genome. In addition to the larger repeats, smaller repeats have also been identified in all six sequenced plant mitochondrial genomes. Smaller repeats between 100 bp and 1000 bp long are more frequent within the smaller mitochondrial genomes. Repeats between 100 bp and 1 kb comprise 8.3% of the liverwort mtDNA (usually two copies but occasionally three).

A BLAST similarity search (Asltschul et al., 1990) using all repeated sequences identified in the rice mitochondrial genome (which represents more than 50% of the genome) against the entire set of repeated sequences from the other five sequenced plant mitochondrial genomes revealed that only a very small fraction (less than 5%) of the rice repeated sequences is also found as repeats in the other plants.

7.2. Recombination and the multipartite structure

Although a "master circle", as shown on Figure 1, can be constructed for all sequenced angiosperm plant mitochondrial genomes, the genomes have been predicted to exist as a multipartite structure arising by a process of repeat-mediated homologous recombination (Palmer and Shields, 1984; Fauron et al., 1995a, b). The repeated sequences present in the genomes can be in direct or inverted orientation with respect to each other. Some repeats are involved in high-frequency inter- and intra-molecular recombination and others are not. Recombination between a pair of direct repeats on a circular chromosome results in fragmentation of the genome into two subgenomic molecules, while recombination across a pair of inverted repeats generates an isomeric form of the original molecule (reviewed by Fauron et al.,1995a, b). This complexity of genome structure is somewhat dependent of the recombination activity present within the mitochondria. Overall, an equilibrium among the genomic subforms must be maintained because the mtDNA restriction enzyme profiles are constant from generation to generation (Oro et al., 1985).

The mechanism of recombination in plant mitochondria is not yet known. One of the most common enzymes involved in homologous DNA strand transfer is *RecA*. It is highly conserved in bacteria and has homologs in eukaryotes. Recently an *E. coli RecA* homolog with a putative mitochondrial targeting sequence *(mtRec*A) has been identified in Arabidopsis (Khazi et al., 2003).

7.3. Other implications of recombination: mutations

A combination of aberrant, rare and normal, high-frequency recombination within mitochondrial genomes and between mitochondrial subgenomes has been invoked to explain the origin of deletion mutations such as those responsible for the non-chromosomal-stripe (NCS) abnormal growth mutants and for reversions of cytoplasmic-male-sterile (CMS) plants to fertility (Hunt and Newton, 1991; Fauron et al., 1990). Although most of the deletion mutations are thought to result from a combination of aberrant, rare recombination at short repeats and normal, high-frequency recombination between the large repeats, single recombination events resulting in deletions have been reported in *Nicotiana sylvestris* (Chétrit et al., 1992; Lelandais et al., 1998).

7.3.1. NCS mutants

Maternally inherited, abnormal growth mutations (such as maize NCS) usually reflect deletions within functional mitochondrial genes coding for subunits of respiratory complexes or ribosomal proteins (reviewed by Newton et al., 2004). These mutations affect the plant throughout its life cycle and are lethal during some stages, particularly seed development. Thus, most NCS-type mutant plants are heteroplasmic for the mitochondrial mutations, *i.e.* they contain a mixture of mutant and normal mitochondrial genomes. Deletions in ribosomal protein genes tend to cause the most severe defects. Deletions that interrupt individual genes encoding subunits of respiratory complexes usually cause less severe phenotypic defects than those affecting components of the translation apparatus. However, all of them that have been studied in maize to date cause kernel abortion. The repeats involved in the initial recombination events that gave rise to the NCS mutants in maize have been determined to be very small—16 bp for NCS2 (Marienfield and Newton, 1994), 12 bp for NCS3 (Hunt and Newton, 1991), 16 bp in NCS4 (Newton et al., 1996), 6 bp for NCS5 (Newton et al., 1990) and 31 bp for NCS6 (Lauer et al., 1990). It has been proposed that a combination of rare recombination events between small repeats, as described for NCS, and high frequency recombination across large repeats, can result in a deletion of the part of the mitochondrial genome that lies between the two repeats (Small et al., 1989; Fauron et al., 1995a). Consistent with the double recombination model, the site of the NCS2 mutation within the *nad4* gene and the site of the NCS3 mutation within the *rps3* gene are each within 1 kb of different copies of a large 11 kb repeat.

7.3.2. Protoplast culture

Although all of the maize NCS mutants studied so far arose spontaneously in the field under the influence of a specific nuclear background (WF9), protoplast culture has been the source of mtDNA deletion mutants in *Nicotiana sylvestris*. Mutants designated CMSI and CMSII, due to their partial male sterility (Li et al., 1998; Chétrit et al., 1992), carry different large deletions involving *nad7* (Pla et al., 1995; Lelandais et al., 1998) and are indistinguishable phenotypically. They do not show the leaf variegation seen in the maize NCS mutants. The CMSII mitochondrial genome results from recombination between 212-bp direct repeats, followed by differential amplification of one of the resulting subgenomes (Lelandais et al.; 1998). In the progenitor mitochondrial genome, the parental and both recombination products can detected by PCR. In the mutant, the parental genome and one of the two recombinant subgenomes are detected at only very low, substoichiometric levels.

7.3.3. Cytoplasmic male sterility

As mentioned in Section 3.4, cytoplasmic male sterility (CMS) in plants is often caused by the expression of chimeric genes (reviewed by Schnable and Wise, 1998; Chase and Gabay-Laughnan, Chapter 23 of this volume). Rearrangement mutations can lead to the loss of CMS-associated sequences from the mtDNA and recovery of male fertility, a process known as cytoplasmic reversion (reviewed by Newton et al., 2004). The events giving rise to CMS reversion via loss of the CMS-associated chimeric regions are probably similar to those that give rise to NCS mutations. In such cases, it is not sufficient to have a recombinant mutant mitochondrial genome arise; the rearranged genome must also be amplified to become predominant. Indeed, CMS-revertant mitochondrial genomes are usually homoplasmic.

7.3.4. Genome shifting

Unusual recombination events among low-level (substoichiometric) mitochondrial subgenomes, followed by the differential amplification of a recombined genome, was proposed to account for the difference in mitochondrial genotypes between modern and ancestral maize cultivars, and to be important generally in plant mitochondrial genome evolution (Small et al., 1989). Rapid shifts in the stoichiometry of mitochondrial subgenomes can in turn be associated with altered phenotypes. The shifting is seen as a rapid change in the copy number of subgenomic molecules, within a generation. It has the potential to affect the plant by modifying the expression of genes located on the shifted molecule. This phenomenon has been extensively studied in beans, where shifting has been shown to be related to the spontaneous reversion to fertility of cytoplasmic-male-sterile plants (Janska et al., 1998).

Shifts in subgenome stoichiometry can also explain the differences in the mtDNA restriction-enzyme profiles between green plants and calli of tobacco (Kanazawa et al., 1994). In the dedifferentiated callus cells, the preferentially amplified mtDNA was shown to result from recombination mediated through a 9-nucleotide repeated element. Although a nuclear gene (*chm*) has been implicated in the shifting phenomenon in Arabidopsis (Abdelnoor et al., 2003), it is not yet known how the selection of subgenomes for amplification or reduction occurs or how the segregation of plant mitochondrial genomes is controlled.

7.4. Organization of the genome in vivo: pulse-field electrophoresis and electron microscopy studies

Although mapping studies and complete sequence analyses have lead to a circular representation of plant mitochondrial genomes, the actual *in vivo* physical structure of mtDNA molecules is not known (Bendich, 1993; Backert et al., 1997; Senthilkumar and Narayanan, 1999). Pulse-field gel electrophoresis studies revealed the presence of poly-disperse linear mtDNA molecules with a wide range of sizes but no specific discrete sizes (Backert et al., 1996; Bendich, 1993, 1996; Oldenburg and Bendich, 2001). However analysis of liverwort mtDNA showed genome-sized and multimeric linear DNAs as well as branched forms with circular head-to-tail permutations of the sequence (Oldenburg and Bendich, 1998, 2001).

When plant mitochondrial genomes are viewed under the electron microscope, an array of various DNA structures is observed, including some rare circular molecules and large arrays of long linear molecules, branched linear molecules, branched circular molecules and complex rosettes. Again, no discrete size classes of molecules have been observed (Scissum-Gunn et al., 1998). Rolling-circle replication of the subgenomic circles could give rise to sigma and long linear molecules of continuous size distribution. Recombination among these molecules can give rise to branched linear molecules, or to linear or circular molecules of variable length.

8. CONCLUSION

The mitochondria of plants share functions with mitochondria of other eukaryotes, and thus share many of the same genes. However, the organization, structure, expression and evolutionary dynamics of plant mitochondrial genomes are unusual when compared with those other organisms. Structurally, plant mitochondrial genomes are very diverse relative to their fungal and animal counterparts, both of which tend to be quite uniform in organization and gene content. As we expand our view into the protists, however, the existence of structural diversity falls much more within the norm. For example, the A*moebidium parasiticum* mitochondrial genome consists of several hundred linear chromosomes that total more than 200 kb and share elaborate terminal-specific sequence patterns (Burger et al., 2003).

Multipartite mtDNA is also found in some *Cnidaria* (Warrior and Gall, 1985), the green alga *Polytomella parva* (Fan and Lee, 2002) and *Globodera*, a potato cyst nematode (Armstrong et al., 1999).

Plant mitochondrial genomes have the capacity to rearrange readily via their large repeats. While there is evidence that the genomes do recombine via such repeats, evidence for the existence of all the predicted subgenomic molecules that could be created by this recombination remains slim, and it is not at all certain whether such subgenomes represent physical reality. It is known that nuclear genes influence mitochondrial genome structure, as well as gene expression, but the mechanisms underlying that control remain unclear.

As sequence data have become more abundant, it has become clear that, although the origins of plant mitochondrial genes are relatively straightforward to determine, the origins of most of the rest of the mitochondrial genomes of the sequenced taxa are enigmatic. More than half of each of the genomes, and more than 80% of the maize mitochondrial genome, consists of sequences that have no corresponding sequences in any sequence database. This includes the sequences of their relatives within the same families or super-families (e.g. within *Brassicaceae* or within grasses). Plant mitochondrial genomes were once thought of as uninteresting because they consisted of large stretches of DNA with few genes. Those large stretches turn out to be curious indeed, providing fertile ground for further investigation.

9. ACKNOWLEDGEMENTS

The authors thank Suman Kanuganti, Leah Westgate and Mike Gibson for assistance with figure generation and Patrick Minx for stimulating discussions. The authors' research on *Zea* mitochondrial genomes is supported by the NSF Plant Genome Research Program (DBI-0110168).

10. REFERENCES

Abdelnoor, R.V., Yule, R., Elo, A., Christensen, A.C., Meyer-Gauen, G., & Mackenzie, S.A. (2003). Substoichiometric shifting in the plant mitochondrial genome is influenced by a gene homologous to *MutS*. *Proc. Natl. Acad. Sci. USA, 100*, 5968-5973.

Adams, K.L., Daley, D.O., Qiu, Y.L., Whelan, J., & Palmer, J.D. (2000). Repeated, recent and diverse transfers of a mitochondrial gene to the nucleus in flowering plants. *Nature, 408*, 354-357.

Adams, K.L., & Palmer, J.D. (2003). Evolution of mitochondrial gene content: gene loss and transfer to the nucleus. *Mol. Phylogenet. Evol., 29*, 380-395.

Adams, K.L., Qiu, Y.L., Stoutemyer, M., & Palmer, J.D. (2002). Punctuated evolution of mitochondrial gene content: high and variable rates of mitochondrial gene loss and transfer to the nucleus during angiosperm evolution. *Proc. Natl. Acad. Sci. USA, 99*, 9905-9912.

Altschul, S.F., Gish, W., Miller, W., Myers, E.W., & Lipman, D.J. (1990). Basic local alignment search tool. *J. Mol. Biol., 215*, 403-410

Anderson, S., Bankier, A.T., Barrell, B.J., de Bruijn, M.H., Coulson, A.R., Drouin, J., *et al.* (1981). Sequence and organization of the human mitochondrial genome (1981). Sequence and organization of the human mitochondrial genome. *Nature, 290*, 457-465.

Armstrong, M.R., Blok, V.C., & Phillips, M.S. (2000). A multipartite mitochondrial genome in the potato cyst nematode Globodera pallida. *Genetics, 154*, 181-192.

Backert, S., Lurz, R., & Borner, T. (1996). Electron microscopic investigation of mitochondrial DNA from *Chenopodium album* (L.). *Curr. Genet., 29*, 427-436.

Backert, S., Nielsen, B.L., & Börner, T. (1997). The mystery of the rings: structure and replication of mitochondrial genomes from higher plants. *Trends Plant Sci., 2*, 477-484.

Bendich, A.J. (1993). Reaching for the ring: the study of mitochondrial genome structure. *Curr. Genet., 24*, 279-290.

Bendich, A.J. (1996). Structural analysis of mitochondrial DNA molecules from fungi and plants using moving pictures and pulsed-field gel electrophoresis. *J. Mol. Biol., 255*, 564-588.

Benne, R., Van den Burg, J., Brakenhoff, J.P., Sloof, P., Van Boom, J.H., & Tromp, M.C. (1986). Major transcript of the frameshifted *coxII* gene from trypanosome mitochondria contains four nucleotides that are not encoded in the DNA. *Cell, 46*, 819-826.

Bogsch, E.G., Sargent, F., Stanley, N.R., Berks, B.C., Robinson, C., & Palmer, T. (1998). An essential component of a novel bacterial protein export system with homologues in plastids and mitochondria. *J. Biol. Chem., 273*, 18003-18006.

Bonen, L., & Vogel, J. (2001). The ins and outs of group II introns. *Trends Genet., 17*, 322-331.

Bonnard, G., & Grienenberger, J.M. (1995). A gene proposed to encode a transmembrane domain of an ABC transporter is expressed in wheat mitochondria. *Mol. Gen. Genet., 246*, 91-99.

Burger, G., Gray, M.W., & Lang, B.F. (2003). Mitochondrial genomes: anything goes. *Trends Genet., 19*, 709-716.

Chang, C.C., Sheen, J., Bligny, M., Niwa, Y., Lerbs-Mache, S., & Stern, D.B. (1999). Functional analysis of two maize cDNAs encoding T7-like RNA polymerases. *Plant Cell, 11*, 911-926.

Chétrit, P., Rios, R., De Paepe, R., Vitart, V., Gutierres, S., & Vedel, F. (1992). Cytoplasmic male sterility is associated with large deletions in the mitochondrial DNA of two *Nicotiana sylvestris* protoclones. *Curr. Genet., 21*, 131-137.

Cho, Y., & Palmer, J.D. (1999). Multiple acquisitions via horizontal transfer of a group I intron in the mitochondrial *cox*1 gene during evolution of the *Araceae* family. *Mol. Biol. Evol., 16*, 1155-1165.

Cho, Y., Qiu, Y.L., Kuhlman, P., & Palmer, J.D. (1998). Explosive invasion of plant mitochondria by a group I intron. *Proc. Natl. Acad. Sci. USA, 95*, 14244-14249.

Clifton, S.W., Minx, P., Fauron, C.M.-R., Gibson, M., Allen, J.O., Sun, M., *et al.* (2004). Sequence and comparative analysis of the maize NB mitochondrial genome, Submitted Plant Physiology

Conway, D.J., Fanello, C., Lloyd, J.M., Al-Joubori, B.M., Baloch, A.H., Somanath, S.D., *et al.* (2000). Origin of *Plasmodium falciparum* malaria is traced by mitochondrial DNA. Mol. *Biochem. Parasitol., 111*, 163-171.

Covello, P.S., & Gray, M.W. (1989). RNA editing in plant mitochondria. Nature, 341, 662-666.

Dewey, R.E., Levings, C.S., 3rd, and Timothy, D.H. (1986). Novel recombinations in the maize mitochondrial genome produce a unique transcriptional unit in the Texas male-sterile cytoplasm. *Cell, 44*, 439-449.

Dewey, R.E., Timothy, D.H., & Levings, C.S., 3rd (1991). Chimeric mitochondrial genes expressed in the C male-sterile cytoplasm of maize. *Curr. Genet., 20*, 475-482.

Fan, J., & Lee, R.W. (2002). Mitochondrial genome of the colorless green alga *Polytomella parva*: two linear DNA molecules with homologous inverted repeat termini. *Mol. Biol. Evol., 19*, 999-1007.

Farré, J.C., Leon, G., Jordana, X., & Araya, A. (2001). *cis* recognition elements in plant mitochondrion RNA editing. *Mol. Cell. Biol., 21*, 6731-6737.

Fauron, C., Casper, M., Gao, Y., & Moore, B. (1995a). The maize mitochondrial genome: dynamic, yet functional. *Trends Genet., 11*, 228-235.

Fauron, C., Moore, B., & Casper, M. (1995b). Maize as a model of higher plant mitochondrial genome plasticity. *Plant Sci., 112*, 11-32.

Fauron, C.M., Havlik, M., & Brettell, R.I. (1990). The mitochondrial genome organization of a maize fertile CMS-T revertant line is generated through recombination between two sets of repeats. *Genetics, 124*, 423-428.

Feagin, J.E. (2000). Mitochondrial genome diversity in parasites. *Int. J. Parasitol., 30*, 371-390.

Fey, J., Weil, J.H., Tomita, K., Cosset, A., Dietrich, A., Small, I., *et al.* (2002). Role of editing in plant mitochondrial transfer RNAs. *Gene, 286*, 21-24.

Forget, L., Ustinova, J., Wang, Z., Huss, V.A., & Lang, B.F. (2002). *Hyaloraphidium curvatum*: a linear mitochondrial genome, tRNA editing, and an evolutionary link to lower fungi. *Mol. Biol. Evol., 19*, 310-319.

Gaut, B.S. (2002). Evolutionary dynamics of grass genomes. *New Phytol., 154*, 15-28.

Giege, P., & Brennicke, A. (1999). RNA editing in Arabidopsis mitochondria effects 441 C to U changes in ORFs. *Proc. Natl. Acad. Sci. USA, 96*, 15324-15329.

Giege, P., & Brennicke, A. (2001). From gene to protein in higher plant mitochondria. *C. R. Acad. Sci. III, 324*, 209-217.

Gray, M.W., Burger, G., & Lang, B.F. (1999). Mitochondrial evolution. *Science, 283*, 1476-1481.

Gualberto, J.M., Lamattina, L., Bonnard, G., Weil, J.H., & Grienenberger, J.M. (1989). RNA editing in wheat mitochondria results in the conservation of protein sequences. *Nature, 341*, 660-662.

Handa, H. (2003). The complete nucleotide sequence and RNA editing content of the mitochondrial genome of rapeseed (*Brassica napus* L.): comparative analysis of the mitochondrial genomes of rapeseed and *Arabidopsis thaliana*. *Nucleic Acids Res., 31*, 5907-5916.

Hiesel, R., Wissinger, B., Schuster, W., & Brennicke, A. (1989). RNA editing in plant mitochondria. *Science, 246*, 1632-1634.

Hunt, M.D., & Newton, K.J. (1991). The NCS3 mutation: genetic evidence for the expression of ribosomal protein genes in *Zea mays* mitochondria. *EMBO J., 10*, 1045-1052.

Ikeda, T.M., & Gray, M.W. (1999). Identification and characterization of T3/T7 bacteriophage-like RNA polymerase sequences in wheat. *Plant Mol. Biol., 40*, 567-578.

Janska, H., Sarria, R., Woloszynska, M., Arrieta-Montiel, M., & Mackenzie, S.A. (1998). Stoichiometric shifts in the common bean mitochondrial genome leading to male sterility and spontaneous reversion to fertility. *Plant Cell, 10*, 1163-1180.

Kanazawa, A., Tsutsumi, N., & Hirai, A. (1994). Reversible changes in the composition of the population of mtDNAs during dedifferentiation and regeneration in tobacco. *Genetics, 138*, 865-870.

Karcher, D., & Bock, R. (2002). The amino acid sequence of a plastid protein is developmentally regulated by RNA editing. *J. Biol. Chem., 277*, 5570-5574.

Khazi, F.R., Edmondson, A.C., & Nielsen, B.L. (2003). An Arabidopsis homologue of bacterial *RecA* that complements an *E. coli recA* deletion is targeted to plant mitochondria. *Mol. Genet. Genom., 269*, 454-463.

Kubo, N., & Kadowaki, K. (1997). Involvement of 5' flanking sequence for specifying RNA editing sites in plant mitochondria. *FEBS Lett., 413*, 40-44.

Kubo, T., Nishizawa, S., Sugawara, A., Itchoda, N., Estiati, A., & Mikami, T. (2000). The complete nucleotide sequence of the mitochondrial genome of sugar beet (*Beta vulgaris* L.) reveals a novel gene for tRNA(Cys)(GCA). *Nucleic Acids Res., 28*, 2571-2576.

Lang, B.F., Burger, G., O'Kelly, C.J., Cedergren, R., Golding, G.B., Lemieux, C., *et al.* (1997). An ancestral mitochondrial DNA resembling a eubacterial genome in miniature. *Nature, 387*, 493-497.

Lauer, M., Knudsen, C., Newton, K.J., Gabay-Laughnan, S., & Laughnan, J.R. (1990). A partially deleted mitochondrial cytochrome oxidase gene in the NCS6 abnormal growth mutant of maize. *New Biol., 2*, 179-186.

Lazowska, J., Jacq, C., & Slonimski, P.P. (1980). Sequence of introns and flanking exons in wild-type and *box*3 mutants of cytochrome b reveals an interlaced splicing protein coded by an intron. *Cell, 22*, 333-348.

Lelandais, C., Albert, B., Gutierres, S., De Paepe, R., Godelle, B., Vedel, F., *et al.* (1998). Organization and expression of the mitochondrial genome in the *Nicotiana sylvestris* CMSII mutant. *Genetics, 150*, 873-882.

Li, X.Q., Jean, M., Landry, B.S., & Brown, G.G. (1998). Restorer genes for different forms of Brassica cytoplasmic male sterility map to a single nuclear locus that modifies transcripts of several mitochondrial genes. *Proc. Natl. Acad. Sci. USA, 95*, 10032-10037.

Lilly, J.W., & Havey, M.J. (2001). Small, repetitive DNAs contribute significantly to the expanded mitochondrial genome of cucumber. *Genetics, 159*, 317-328.

Lippok, B., Brennicke, A., & Wissinger, B. (1994). Differential RNA editing in closely related introns in *Oenothera* mitochondria. *Mol. Gen. Genet., 243*, 39-46.

Lu, B., & Hanson, M.R. (1996). Fully edited and partially edited *nad9* transcripts differ in size and both are associated with polysomes in potato mitochondria. *Nucleic Acids Res., 24*, 1369-1374.

Maier, R.M., Neckermann, K., Igloi, G.L., & Kossel, H. (1995). Complete sequence of the maize chloroplast genome: gene content, hotspots of divergence and fine tuning of genetic information by transcript editing. *J. Mol. Biol., 251*, 614-628.

Maréchal-Drouard , L., Weil, J.H., & Dietrich, A. (1993). Transfer RNAs and transfer RNA genes in plants. *Annu. Rev. Plant Physiol. Plant Mol. Biol., 44*, 13-32.

Marienfeld, J.R., & Newton, K.J. (1994). The maize NCS2 abnormal growth mutant has a chimeric *nad*4-*nad*7 mitochondrial gene and is associated with reduced complex I function. *Genetics, 138*, 855-863.

Marienfeld, J.R., Unseld, M., & Brennicke, A. (1999). The mitochondrial genome of *Arabidopsis* is composed of both native and immigrant information. *Trends Plant Sci., 4*, 495-502.

Michel, F., Umesono, K., & Ozeki, H. (1989). Comparative and functional anatomy of group II catalytic introns–a review. *Gene, 82*, 5-30.

Newton, K. J., Gabay-Laughnan, S., & De Paepe, R. (2004). Mitochondrial mutations in plants. In D. Day, H. Millar, & J. Whelan, (Eds.), *Advances in photosynthesis and respiration (vol. 17) Plant mitochondria from genome to function* (in press). Dordrecht: Kluwer.

Newton, K.J., Knudsen, C., Gabay-Laughnan, S., & Laughnan, J.R. (1990). An abnormal growth mutant in maize has a defective mitochondrial cytochrome oxidase gene. *Plant Cell, 2*, 107-113.

Newton, K.J., Mariano, J.M., Gibson, C.M., Kuzmin, E., & Gabay-Laughnan, S. (1996). Involvement of S2 episomal sequences in the generation of NCS4 deletion mutation in maize mitochondria. *Dev. Genet., 19*, 277-286.

Nosek, J., & Tomaska, L. (2003). Mitochondrial genome diversity: evolution of the molecular architecture and replication strategy. *Curr. Genet., 44*, 73-84.

Notsu, Y., Masood, S., Nishikawa, T., Kubo, N., Akiduki, G., Nakazono, M., *et al.* (2002). The complete sequence of the rice (*Oryza sativa* L.) mitochondrial genome: frequent DNA sequence acquisition and loss during the evolution of flowering plants. *Mol. Genet. Genom., 268*, 434-445.

Oda, K., Yamato, K., Ohta, E., Nakamura, Y., Takemura, M., Nozato, N., *et al.* (1992). Gene organization deduced from the complete sequence of liverwort *Marchantia polymorpha* mitochondrial DNA. A primitive form of plant mitochondrial genome. *J. Mol. Biol., 223*, 1-7.

Oda, K., Yamato, K., Ohta, E., Nakamura, Y., Takemura, M., Nozato, N., *et al.* (1992). Transfer RNA genes in the mitochondrial genome from a liverwort, *Marchantia polymorpha*: the absence of chloroplast-like tRNAs. *Nucleic Acids Res., 20*, 3773-3777.

Ohta, E., Oda, K., Yamato, K., Nakamura, Y., Takemura, M., Nozato, N., *et al.* (1993). Group I introns in the liverwort mitochondrial genome: the gene coding for subunit 1 of cytochrome oxidase shares five intron positions with its fungal counterparts. *Nucleic Acids Res., 21*, 1297-1305.

Oldenburg, D.J., & Bendich, A.J. (1998). The structure of mitochondrial DNA from the liverwort, *Marchantia polymorpha*. *J. Mol. Biol., 276*, 745-758.

Oldenburg, D.J., & Bendich, A.J. (2001). Mitochondrial DNA from the liverwort *Marchantia polymorpha*: circularly permuted linear molecules, head-to-tail concatemers, and a 5' protein. *J. Mol. Biol., 310*, 549-562.

Oro, A., Newton, K.J., & Walbot, V. (1985). Molecular analysis of the inheritance and stability of the mitochondrial genome of an inbred line of maize. *Theor. Appl. Genet., 70*, 287-293.

Ostheimer, G.J., Williams-Carrier, R., Belcher, S., Osborne, E., Gierke, J., & Barkan, A. (2003). Group II intron splicing factors derived by diversification of an ancient RNA-binding domain. *EMBO J., 22*, 3919-3929.

Palmer, J.D. (1990). Contrasting modes and tempos of genome evolution in land plant organelles. *Trends Genet., 6*, 115-120.

Palmer, J.D., & Shields, C.R. (1984). Tripartite structure of the *Brassica campestris* mitochondrial genome. *Nature, 307*, 437-440.

Parsons, J.D. (1995). Miropeats: graphical DNA sequence comparisons. *Comput. Appl. Biosci., 11*, 615-619.

Phreaner, C.G., Williams, M.A., & Mulligan, R.M. (1996). Incomplete editing of *rps*12 transcripts results in the synthesis of polymorphic polypeptides in plant mitochondria. *Plant Cell, 8*, 107-117.

Pla, M., Mathieu, C., De Paepe, R., Cherit, P., & Vedel, F. (1995). Deletion of the last two exons of the mitochondrial *nad*7 gene results in lack of the NAD7 polypeptide in a *Nicotiana sylvestris* CMS mutant. *Mol. Gen. Genet., 248*, 79-88.

Qin, P.Z., & Pyle, A.M. (1998). The architectural organization and mechanistic function of group II intron structural elements. *Curr. Opin. Struct. Biol., 8*, 301-308

Schnable, P.S., & Wise, R.P. (1998). The molecular basis of cytoplasmic male sterility and fertility restoration. *Trends Plant Sci., 3*, 175-180.

Schwartz, S., Elnitski, L., Li, M., Weirauch, M., Riemer, C., Smit, A., *et al.* (2003). MultiPipMaker and supporting tools: Alignments and analysis of multiple genomic DNA sequences. *Nucleic Acids Res., 31*, 3518-3524.

Schwartz, S., Zhang, Z., Frazer, K.A., Smit, A., Riemer, C., Bouck, J., *et al.* (2000). PipMaker–a web server for aligning two genomic DNA sequences. *Genome Res., 10*, 577-586.

Scissum-Gunn, K.D., Gandhi, M., Backert, S., & Nielsen, B.L. (1998). Separation of different conformations of plant mitochondrial DNA molecules by field inversion gel electrophoresis. *Plant Mol. Biol. Rep., 16*, 219-229.

Selosse, M., Albert, B., & Godelle, B. (2001). Reducing the genome size of organelles favours gene transfer to the nucleus. *Trends Ecol. Evol., 16*, 135-141.

Senthilkumar, P., & Narayanan, K. (1999). Analysis of rice mitochondrial genome organization using pulsed-field gel electrophoresis. *J. Biosci., 24*, 215-222.

Skovgaard, M., Jensen, L.J., Brunak, S., Ussery, D., & Krogh, A. (2001). On the total number of genes and their length distribution in complete microbial genomes. *Trends Genet., 17*, 425-428.

Small, I., Suffolk, R., & Leaver, C.J. (1989). Evolution of plant mitochondrial genomes via substoichiometric intermediates. *Cell, 58*, 69-76.

Unseld, M., Marienfeld, J.R., Brandt, P., & Brennicke, A. (1997). The mitochondrial genome of *Arabidopsis thaliana* contains 57 genes in 366,924 nucleotides. *Nature Genet., 15*, 57-61.

Wahleithner, J.A., MacFarlane, J.L., & Wolstenholme, D.R. (1990). A sequence encoding a maturase-related protein in a group II intron of a plant mitochondrial nad1 gene. *Proc. Natl. Acad. Sci. USA, 87*, 548-552.

Ward, B.L., Anderson, R.S., & Bendich, A.J. (1981). The mitochondrial genome is large and variable in a family of plants (Cucurbitaceae). *Cell, 25*, 793-803.

Warrior, R., & Gall, J. (1985). The mitochondrial DNA of *Hydra attenuata* and *Hydra littoralis* consists of two linear molecules. *Arch. Sci. Geneve, 38*, 439-445.

Williams, M.A., Kutcher, B.M., & Mulligan, R.M. (1998). Editing site recognition in plant mitochondria: the importance of 5'-flanking sequences. *Plant Mol. Biol., 36*, 229-237.

Zanlungo, S., Quinones, V., Moenne, A., Holuigue, L., & Jordana, X. (1995). Splicing and editing of *rps*10 transcripts in potato mitochondria. *Curr. Genet., 27*, 565-571.

Zimmerly, S., Hausner, G., & Wu, X. (2001). Phylogenetic relationships among group II intron ORFs. *Nucleic Acids Res., 29*, 1238-1250.

CHAPTER 7

PROTEOMICS, BIOINFORMATICS AND GENOMICS APPLIED TO PLANT ORGANELLES

C. COLAS DES FRANCS-SMALL[1], B. SZUREK[2] AND I. SMALL[2]

[1]*Institut de Biotechnologie des Plantes (UMR-CNRS 8618), Université Paris-Sud, Bât 630, 91405 Orsay CEDEX, France*
[2]*Unité de Recherche en Génomique Végétale (INRA/CNRS/UEVE), 2 rue Gaston Crémieux, 91057 Evry CEDEX, France*

Abstract. The importance of mitochondria and plastids to plant cell function has stimulated considerable research in organelle biogenesis over several decades. The basic functions and composition of these organelles are now reasonably well understood, but much remains to be discovered concerning the regulation of organelle activity in different physiological states. Coordination between the nuclear, plastid and mitochondrial compartments is still very poorly understood, as is the assembly of the many multi-component protein complexes that characterize the internal mitochondrial and plastid membranes. Progress in these areas will in many cases be dependent on techniques that can follow many genes or proteins simultaneously in order to capture the complexity of the system. In this review, we discuss the attempts to apply the latest "omics" approaches to plant organelles, using examples from the literature on plants wherever possible, but with occasional examples from other organisms when new techniques offer special promise.

1. INTRODUCTION

The completion of the first plant genome sequences and the increasing amounts of other gene and cDNA sequence data has opened the functional genomics era, but the assignment of a function to each of the 25,000+ genes is a formidable task. Sequencing often gives valuable indications about gene function by revealing similarities to other genes of known function, but is rarely sufficient by itself. Large-scale transcriptome data give valuable inferences about when, where and at what level different genes are expressed, but mRNA quantities do not always correlate with protein abundance and activity. Studies on proteins are essential for a complete understanding of the molecular biology and biochemistry of complex systems. With

H. Daniell and C.D. Chase (eds.), Molecular Biology and Biotechnology of Plant Organelles, 179—209.

regard to organelles, the relevance of proteomics, as opposed to the nucleotide sequencing normally associated with genomics, is especially clear: the vast majority of organelle proteins are encoded by the nucleus, which means that that their proteomes contain proteins of both nuclear and organellar origins. This dual origin poses a central problem in organellar genomics studies – how does one identify which nuclear-encoded proteins are present inside organelles? Much past and current research in proteomics (mass identification of proteins in complex samples), bioinformatics (targeting predictions) and functional genomics (experimental verification of protein targeting) is aimed at solving this crucial question.

2. PROTEOMICS

Proteomics is often associated with two-dimensional iso-electric focusing/SDS electrophoresis (2D PAGE) and identification using mass spectrometry, but in fact goes far beyond these techniques. The use of complementary separation techniques such as blue-native PAGE (BN PAGE) and reverse-phase chromatography (RP-HPLC), which have fewer limitations than classical 2D analysis, is very useful for analysing proteins of high molecular weight, extreme pI or hydrophobicity. The array of proteomics applications varies from straightforward identification of proteins to characterization of post-translational modifications and protein-protein interactions. In addition to these so called "bottom up" analyses, the emergence of "top down" approaches, i.e. straightforward analysis of a sample by Fourier-transform mass spectroscopy (FTMS), helps to characterize the structure of proteins and locate possible modifications.

2.1. Mitochondrial proteomics

It is predicted that mitochondria synthesize 2 to 5% of the proteins required for their function, whilst the remaining 95 to 98% of proteins are encoded by the nuclear genome and targeted to the mitochondria as protein precursors (Gray et al., 1999). In the last 20 years, a number of reports have presented two-dimensional gel profiles of mitochondrial proteins from pea (Humphrey-Smith et al., 1992), potato (Colas des Francs-Small et al., 1992), *Arabidopsis* (Davy de Virville et al., 1998) and maize (Dunbar et al., 1997). Most of these studies reported the analysis of changes in protein patterns between different tissues, physiological conditions or developmental stages. The techniques used to identify the proteins of interest were immunological recognition by cross-reacting antibodies, or N-terminal Edman sequencing, which required considerable material and was extremely costly (but has the advantage of determining the N-terminus of the mature protein when the presequence is cleaved after import into the mitochondria). The application of mass spectroscopy to protein identification has allowed the emergence of high-throughput studies, because of its extreme sensitivity, rapidity and reduced cost; though other complementary techniques are still used. More systematic mitochondrial proteomics studies of various plant species are now being published, and we will describe the most interesting breakthroughs.

2.1.1. Arabidopsis mitochondrial proteome

After the completion of the genome sequence (The Arabidopsis Genome Initiative, 2000), *Arabidopsis thaliana* became the obvious organism for plant proteomics projects. However, from a biochemist's point of view, the plant is difficult to work with; *Arabidopsis* organelle purification in quantity is difficult, as they are particularly small and fragile, often causing significant contamination of the sub-cellular fractions. *Arabidopsis* suspension cultures provide a solution to this problem, because they are quick to grow and non-photosynthetic, they are an abundant and reproducible source of mitochondria. The first published *Arabidopsis* proteomics studies employed cell suspensions (Kruft et al., 2001; Millar et al., 2001). A twin Percoll gradient density separation technique provides organelle fractions largely free of contamination by cytosol, peroxisomes, plastids and other membranes (Day et al., 1985; Kruft et al., 2001; Millar et al., 2001). The purity and integrity of the preparations can be verified by measuring enzymatic activities or pigments linked with potentially contaminating compartments (Neuburger, 1985; Millar et al., 2001), or by BN PAGE (Kruft et al., 2001). The analysis of 2D cell culture mitochondrial protein patterns revealed about 650 protein spots in a pI range of 3 to 10 (Kruft et al., 2001). This number definitely does not represent the full set of mitochondrial proteins in a plant cell. Estimates of the number of mitochondrial proteins in *Arabidopsis* range from 1000-3000, mostly based on bioinformatics predictions (Table 1).

Table 1. *Some estimates of the total numbers of organelle proteins in plants*

Reference	*Mitochondria*	*Plastids*	*Method*
AGI, 2000	2897	3574	bioinformatics
Abdallah et al., 2000		1900-2500	bioinformatics
Kruft et al., 2001	650		proteomics
Werhahn and Braun, 2002	800		proteomics
Leister, 2003	3135	3130	bioinformatics
Heazlewood et al., 2004	>416		proteomics
Heazlewood et al., 2004	3000-4500		bioinformatics

The lower number of proteins observed experimentally is due to the relatively low abundance of many proteins. Sensitive staining methods are not the only way to circumvent this problem. The fractionation of the mitochondrial proteome into sub-proteomes (Cordewell et al., 2000; Jung et al., 2000) is a very efficient way of enriching for low-abundance proteins (and also gives valuable information on sub-organellar compartmentation). Millar et al. (2001) separated their organelles into soluble, membrane and integral membrane fractions, which allowed them to locate 165 protein spots to one of these compartments with certainty. Of course, certain proteins are abundant in several compartments, due to the cross-contamination of the fractions, which compromises their sub-proteomic localization. In order to increase the number of polypeptides revealed, Kruft et al. (2001) investigated various solubilization procedures and increased the resolution by loading several gels with lower range pH gradients in the IEF stage. This latter technique allowed them to increase the number of protein spots revealed by 10%. Another problem is the constraints of conventional 2D PAGE electrophoresis: very hydrophobic or basic proteins cannot enter IEF gels, thus restricting the number of spots on gels. Other techniques like BN PAGE or SDS PAGE are therefore necessary under certain circumstances. Werhahn and Braun (2002) used BN PAGE as the first dimension of their 3D gel electrophoresis technique. The subproteomes analysed in this case were mitochondrial complexes such as ATP synthase, the cytochrome *c* reductase complex, and the preprotein translocase of the outer membrane (the TOM complex). Each complex was resolved by BN PAGE, and after electroelution to release the Coomassie blue fixed to the proteins, the complexes were then separated into their components by subsequent 2D PAGE analysis. This procedure allowed an excellent resolution of the components of each complex, including very hydrophobic ones. Differential solubilization (with sodium carbonate) of integral membrane complexes such as carrier proteins followed by SDS PAGE allowed Millar and Heazlewood (2003) to analyse this protein family in *Arabidopsis* cells. This approach identified 6 carriers (including the adenine nucleotide transporter, dicarboxylate/tricarboxylate carrier, phosphate carrier, uncoupling protein). Another interesting example is the proteomic identification of divalent metal cation-binding proteins by Herald et al. (2003), who analysed mobility shifts in the presence of divalent cations during 2D diagonal SDS PAGE. This approach allowed the identification of 23 protein spots, including the succinyl-CoA ligase □ subunit, Mn-superoxide dismutase and the Rieske iron-sulfur protein of the *b/c1* complex. This approach has broad applications for the identification of subproteomes based on metal interactions of polypeptides.

To overcome the clear limitations of the techniques used in the proteome studies described above, Heazlewood et al. (2004) analysed the *Arabidopsis* cell suspension proteome by off-line MUDPIT (multidimensional protein identification technology). This technique has the great advantage of not requiring gels: the proteins are solubilized, trypsin digested and the peptides separated by HPLC and identified by MS/MS. Overall, the authors were able to identify 390 non-redundant proteins, of which some were very basic (up to pH 11), very hydrophobic, small (6 kDa) or extremely large (428 kDa). Added to the previous data available, a total set of 416 *Arabidopsis* mitochondrial proteins has been described so far. In addition to the well-defined components of mitochondria (electron transport chain, TCA cycle, metabolic carriers and import apparatus), this approach allowed the identification of low abundance proteins with potentially key roles in mitochondrial function: cellular signalling components (protein kinases); proteins involved in the regulation of protein complex assembly (assembly factors, metalloproteases); DNA synthesis, transcription and RNA processing (transcription factors, RNA helicases, PPR proteins, etc.); carriers (ion transporter, carnitine transporter); alternative oxidase; etc. A subset of 71 proteins could not be assigned any function. The full set of experimentally identified proteins is available online at the *Arabidopsis* Mitochondrial Protein Database (see Table 2).

Table 2. *The major and best-known public databases dealing with plant organellar proteins*

Database	*URL*	*Comment*
MitoP	http://mips.gsf.de/proj/medgen/mitop/	Primarily animal and yeast proteins, *Arabidopsis* is coming (Scharfe et al., 2000)
GOBASE	http://megasun.bch.umontreal.ca/gobase/	Comprehensive organelle genome database (O'Brien et al., 2003)
AMPD	http://www.mitoz.bcs.uwa.edu.au/apmdb/Search-Page.php	*Arabidopsis* mitochondrial proteomics and targeting predictions (Heazlewood et al., 2004)
PPD	http://cbsusrv01.tc.cornell.edu/users/ppdb/	Plant plastid proteome database (Friso et al., 2004)
AMPP	http://www.gartenbau.uni-hannover.de/genetik/AMPP	2D and 3D gel analyses of *Arabidopsis* mitochondria (Kruft et al., 2001)
MitoDat	http://www.lecb.ncifcrf.gov/mitoDat/	Primarily aimed at human genetic diseases, some plant proteins (Lemkin et al., 1996)

Mitochondria isolated from suspension cell cultures present valuable starting material for studying most mitochondria's basic functions - like respiration, the citric acid cycle, amino acid and nucleotide metabolism, protein biosynthesis, mitochondrial assembly, molecular transport, protection against reactive oxygen species. etc. The extension of this research into whole plants and into specific differentiated plant tissues has the potential to identify other proteins associated with metabolic pathways that do not occur in cell culture.

2.1.2. Rice mitochondrial proteome

Another model plant is rice, as this monocot is a diploid and has considerable conserved synteny with wheat and maize; the sequencing of its genome is in progress (Yu et al., 2002), and numerous ESTs and cDNA libraries are available. A major interest in studying the rice proteome lies in the fact that differences in mitochondrial molecular function between monocots and dicots have been highlighted (Fey and Maréchal-Drouard, 1999; Thelen et al., 1999; Considine et al.,

2002). A first analysis of the rice mitochondrial proteome has recently been published (Heazlewood et al., 2003a), using the etiolated shoot as a source of organelles. This study was done using three different separation techniques, conventional 2D PAGE electrophoresis, BN/SDS PAGE and reverse-phase C18 liquid chromatography. The latter two techniques give access to very high molecular weight, very extreme pI or very hydrophobic proteins. They also allowed the identification of proteins, which, because of their physical properties, could not enter 2D gels. A total of 136 non-redundant rice mitochondrial proteins were identified, including a new set of 23 proteins of unknown function. These proteins represent all basic mitochondrial functions (electron transport, carbon metabolism, protein and metabolite transport and other various functions). This was the first direct identification of two members of the huge pentatricopeptide repeat (PPR) protein family in a proteomics study. PPR proteins have been shown to be involved in nuclear restoration of cytoplasmic male sterility in petunia, radish and rice (Bentolila et al., 2002; Brown et al., 2003; Desloire et al., 2003; Kazama and Toriyama, 2003; Koizuka et al., 2003) and belong to a large family of low abundance proteins likely to be involved in post-transcriptional processes in plant organelles (Small and Peeters, 2000). They make good indicators of the sensitivity of the approach employed.

2.1.3. Pea mitochondrial proteome

Studying mitochondria from green tissues is very important for investigating the role of mitochondria during photosynthesis and other plant-specific metabolic functions such as photorespiration. Kruft et al. (2001) purified mitochondria from green tissues of *Arabidopsis*, but judged their preparations too highly contaminated (only 90% pure) to perform good proteomics studies. The preparation of pure mitochondrial fractions from green tissues is much easier from other plant species like pea. Bardel et al. (2002) performed a very interesting comparative study of pea mitochondrial proteins from green and etiolated leaves, roots and seeds. These authors analysed the soluble fraction of pea leaf mitochondrial proteins (433 protein spots on Coomassie blue stained gels), and were able to identify minor proteins by a gel filtration step prior to classical SDS PAGE. This three-dimensional analysis provides valuable information on the oligomeric status of proteins, the composition of multisubunit complexes and protein-protein interactions.

The comparison of the protein patterns from various tissues revealed a pool of polypeptides common to all tissues tested that corresponds to proteins involved in basic mitochondrial metabolism, such as enzymes of the TCA cycle, Hsp60 and Hsp70, MnSOD etc. Other proteins were overexpressed in green tissues (the subunits of the glycine cleavage complex and serine hydroxymethyl transferase) or in etiolated tissues (some ATP synthase subunits, the E1□ pyruvate dehydrogenase subunit, etc.), roots (formate dehydrogenase, the E1□ pyruvate dehydrogenase subunit, etc.) and seeds (formate dehydrogenase, HSP 22 and a possibly mitochondrial late embryo-abundant protein), highlighting the physiological specificities of the organelle in different organs. The observation that aldehyde dehydrogenase is a protein family (9 isoforms identified in pea) led to some hypotheses on the detoxifying role of this enzyme in various cellular compartments (formaldehyde in leaves, acetaldehyde in pollen).

2.1.4. Comparison of the mitochondrial proteomes from various species

Heazlewood et al. (2003a) compared the proteins identified in the rice, *Arabidopsis* and pea proteomics studies described above. The principal observations are that 71 proteins have been found in the *Arabidopsis* and rice mitochondrial proteomes, and 28 additional proteins are orthologs of well-known mitochondrial proteins characterized in other eukaryotic systems. Out of the 48 soluble proteins identified in pea, only 28 were common to the rice and pea proteomes. The physiological functions of mitochondria in different plants are unlikely to be very dissimilar, so with more complete surveys of the proteomes of mitochondria in several plants the set of common proteins will certainly grow. The comparison of mitochondrial proteomes from different species, and especially between monocots and dicots, should allow us to better understand the crucial role and evolution of these organelles in the plant cell. This goal will be attained more easily with the creation of databases where all the groups working in this field can confront their data.

These sorts of comparisons have been widened to include human and yeast mitochondrial proteomes and that of *Rickettsia* (the nearest living relative of the mitochondrial progenitor), revealing a common pool of proteins which correspond to the central bioenergetic and metabolic functions; these proteins have been conserved throughout evolution, as opposed to those involved in mitochondrial-cellular interactions (signalling, transport) and the regulation of the mitochondrial genome, which differ across kingdoms (Heazlewood et al., 2003b; Heazlewood et al., 2004).

2.2. Chloroplast proteomics

The total plant proteome of higher plants is estimated to be over 25,000 proteins (Wortman et al., 2003) and chloroplasts contain 10% - 15% of these proteins (The Arabidopsis Genome Initiative, 2000), demonstrating the importance of these organelles in the plant cell. The chloroplast genome encodes about 120 proteins (Shimada and Sugiura, 1991). Thus the vast majority of the proteins found in chloroplasts are nuclear encoded, translated in the cytosol and imported into the organelle. When inside the chloroplast, four pathways are possible to target the protein into the thylakoid or the thylakoid lumen (Keegstra and Cline, 1999). Chloroplasts are highly compartmentalized (double-layer envelope, soluble stroma, thylakoid membranes, lumen) and contain a major protein, ribulose bisphosphate carboxylase (RuBisCo), which represents about 50% of the leaf soluble proteins and 80% of stromal proteins. Therefore, the envelope, thylakoid membrane and lumen proteins are in relatively small quantities compared to RuBisCo, and many lumenal proteins are present at concentrations 10,000-fold lower than proteins of the photosynthetic apparatus. So, the proteomics study of the chloroplast as a whole is almost impossible. Even more so than for mitochondria, further purification of the organelle into subfractions or protein complexes is an effective strategy to obtain a more in depth insight of chloroplast biogenesis (van Wijk, 2000; van Wijk, 2001).

2.2.1. Lumen and peripheral thylakoid proteome

The isolation of intact chloroplasts is relatively easy from pea and spinach leaves, but more difficult from the model plant *Arabidopsis* due to the small size of the leaves and the small amount of material. However, protein identification is easier from the latter, due to the availability of the *Arabidopsis* genome sequence. Two groups have recently published papers on the proteomics of the thylakoid lumen and peripheral proteins in pea, spinach or *Arabidopsis* (Kieselbach et al., 2000; Peltier et al., 2000; Peltier et al., 2002; Schubert et al., 2002).

The group of van Wijk systematically analysed the soluble and peripheral proteins in the thylakoids of pea (Peltier et al., 2000) and *Arabidopsis* (Peltier et al., 2002), using classical 2D PAGE electrophoresis, followed by identification by MS and N-terminal Edman sequencing. The purity of the fractions was checked by immunoblotting, using antibodies recognising compartment-specific proteins. Sixty-one lumenal and peripheral proteins of thylakoids were identified in pea and 81 in *Arabidopsis* (photosynthetic electron transport, carbon metabolism, DNA binding and transcription, oxygen radical scavenging, protein assembly...). Kieselbach et al. (Kieselbach et al., 2000; Schubert et al., 2002) studied the proteome of *Arabidopsis* and spinach lumen, and estimated that this compartment contains about 80 proteins. They identified 36 proteins in *Arabidopsis*. These studies show that the protein contents of lumen are very conserved between species (pea, spinach and *Arabidopsis*) and suggest that the narrow luminal space of the thylakoid is densely packed with proteins. In addition to the accumulation of protons necessary for ATP synthesis and the equilibration of ion currents through the thylakoid membrane, the annotation of novel lumen proteins highlights new functions for this compartment: assistance in the folding (isomerases) and proteolysis of thylakoid proteins, as well as protection against oxidative stress (m-type thioredoxins, 2-Cys peroxiredoxins).

2.2.2. Integral membrane proteome

Conventional proteomics techniques such as 2D IEF/SDS PAGE are unable to resolve all the membrane proteins, and especially the integral membrane proteins, because they are very difficult to solubilize. Organic solvent fractionation and RP-HPLC are powerful methods to study hydrophobic proteins, and both have given interesting results in membrane proteomics. Joyard and colleagues used organic solvent fractionation followed by SDS PAGE to purify and identify spinach integral chloroplast envelope and thylakoid proteins (Ferro et al., 2000). This work shows that chloroform/methanol treatments only solubilize the very hydrophobic fraction of the membrane proteins (the integral membrane proteins) and are extremely efficient at eliminating hydrophilic proteins (for example, no RuBisCo detected in the envelope fractions); the samples generated only contain a small number of proteins, which can thus be separated by classical 1-D SDS PAGE and identified by MS/MS. This is an effective way of studying integral membrane proteins (amongst others, transporters such as the malate-oxoglutarate and the phosphate-triose-phosphate translocators of the inner membrane of the envelope; components of the light-harvesting complexes (LHCs) in the thylakoid membranes).

Thylakoids are the major membrane network in chloroplasts and are the site for electron transfer reactions from water to NADP+ coupled to ATP synthesis. These photosynthetic reactions are performed by multi-subunit complexes (photosystems I (PSI) and II (PSII), cytochrome b6f complex and ATP-synthase). Each photosystem has an associated light harvesting complex (LHCI and LHCII), which functions to deliver the excitation energy deriving from light absorption to the reaction centers. These LHCs are nuclear encoded, membrane-embedded proteins. LHCII is the most abundant and represents 50% of the total chlorophyll in the thylakoid membrane as opposed to LHCI, which accounts only for 20% of the chlorophyll. Regarding the supramolecular organization, PSI seems to occur as monomeric pigment-proteins whereas PSII is believed to occur as oligomers *in situ*. Zolla et al. (2002, 2003) studied the antennas of both photosystems by RP-HPLC-ESI-MS. As the protein components of the PSI and PSII antenna systems display simple electrophoretic patterns (discrete bands between 20 and 25 kDa) by SDS PAGE, the elution times among the four antenna types of PSI and PSII in RP-HPLC differ significantly. These studies allowed the authors to determine the intact molecular masses of the protein components of LHCI and LHCII and identify different protein types. In PSI, Lhca1 and Lhca2 proteins are present in 2 or 3 isoforms, and in most plant species, Lhca2 and Lhca4 proteins are the most abundant PSI antenna proteins. In PSII, several isoforms were found for Lhcb1, Lhcb3 and Lhcb4 in most of the 14 species studied. The large amount of protein heterogeneity (due to the presence of multiple genes and to post-translational modifications) observed in PSII suggest that plants may vary PSII antenna proteins in response to light and various environmental stresses.

Envelope membranes are the site of transport of ions, metabolites, proteins, and information between plastids and surrounding cellular compartments. They catalyse the biosynthesis of a wide variety of specific plastid constituents and are functionally essential for the plant cell. Despite this complexity, only a fraction of envelope proteins have been purified or characterized. In order to characterize new envelope proteins in spinach and *Arabidopsis*, Ferro et al. (2002, 2003) used various solubilization techniques followed by SDS PAGE, digestion and MS/MS combined with the use of prediction programs for protein targeting to the envelope and protein fusions with green fluorescent protein (GFP). The authors were able to characterize minor as well as major proteins: about a third of the 54 envelope proteins identified in spinach and the 89 in *Arabidopsis* were likely to be transporters, as they share some common features like being highly hydrophobic, very basic, located in the chloroplast and homologous to transport systems; sugar, phosphate, amino acid and taurocholate transporters were identified. The high homology between components of the chloroplast envelope protein import machinery identified in this study and some of *Synechocytis* provides further evidence for the use of cyanobacterial ancestor genes to build up this machinery. The identification of many enzymes of the lipid metabolism (membrane lipid biosynthesis, fatty acid desaturases) and oxyradical scavenging and antioxidant enzymes, as well as the identification of new transporters, give new insights into the function of the chloroplast envelope, and especially its role in the metabolite and ion exchanges between the chloroplast and cytosol. Froehlich et al. (2003), using either SDS PAGE followed by in-gel digestion and analysis by LC-MS/MS and/or off-line MUDPIT, overall identified 392 non-redundant envelope proteins. A third were identified by both techniques, but 28% were identified by SDS PAGE MS/MS only and 38% by MUDPIT only, showing once more the advantage of using complementary techniques. The next major challenge will be to determine how these proteins are organised into functional complexes.

2.3. Protein complexes

Many mitochondrial and plastid proteins are present *in vivo* in large complexes and very often require the use of specialised techniques such as RP-HPLC, 2D BN/SDS PAGE and 2D BN/BN PAGE followed by digestion and MS/MS.

As an example, the proteins of the 30S and 50S ribosomal subunits in spinach chloroplasts were identified by a combination of 2D electrophoresis, RP-HPLC, MS and Edman sequencing (Yamaguchi and Subramanian, 2000; Yamaguchi et al., 2000; Yamaguchi and Subramanian, 2003). It was concluded that the spinach plastid ribosome comprises 59 proteins, of which 53 are *E. coli* orthologues and 6 are specific to higher plant chloroplasts (PSRPs). A number of these proteins were shown to be post-translationally modified. These PSRPs are thought to be involved in chloroplast translational regulation. Another example is the analysis of the 350 kD ClpP protease complex by BN PAGE and MS (Peltier et al., 2001; Peltier et al., 2003). Clp proteins degrade misfolded or unassembled proteins in an ATP-dependent manner. The isolated Clp complex contained 10 proteins. Several truncations and errors in intron and exon predictions of the annotated Clp genes were corrected using MS data and by matching genomic sequences with cDNA sequences.

More global studies have been performed on mitochondrial complexes: the subunit composition of several of them, such as respiratory complexes (Jänsch et al., 1996), import complexes (Jänsch et al., 1998) etc. were resolved using BN PAGE. Furthermore, the arrangement of mitochondrial respiratory and other complexes into supercomplexes is another very interesting aspect of proteomics, as it will probably change our understanding of mitochondrial biochemistry and physiology. In the mitochondrion, which has a high concentration of proteins and substrates, such structures offer the advantage of optimising and regulating metabolic processes. Supercomplexes have been described in yeast and mammalian mitochondria, and Eubel et al. (2003) have now isolated the respiratory supercomplexes from plant (potato, bean, *Arabidopsis* and barley) mitochondria. This was achieved by 2D BN/SDS PAGE and 2D BN/BN PAGE using two different non-ionic detergents (digitonin and dodecylmaltoside). Various supercomplexes containing either different combinations of complex I and III or a dimer of ATP synthase were isolated, but complex II, IV and AOX do not seem to take part in any supercomplex. This work opens a new and more dynamic dimension of mitochondrial proteomics, and will without doubt also be applied to chloroplast envelope and thylakoid membrane complexes.

2.4. Perspectives in proteomics

2.4.1. "Bottom up" or "top down" MS?

Most of the studies described in this review have been done by "bottom up" approaches, i.e. purification of the protein followed by its cleavage (trypsin) and peptide mass fingerprinting, generally allowing the identification of the precursor. MS/MS of individual peptides can provide more specific information for protein identification, which is important for proteins of low purity or from a non-sequenced organism. Nevertheless, the information is only partial, and protein modifications due to editing, alternative splicing, signal peptide cleavage and post-translational modifications have to be specifically searched for, and their analysis is time and sample consuming.

The "top down" approach does not require any purification or digestion step; the sample is introduced directly into the Fourrier-transform (FT) MS instrument, using electrospray ionisation (ESI). Even low-abundance proteins can be analysed, and the precise masses of all the proteins present in the sample can be determined. Subsequent fragmentation of the molecular ions of an individual protein allows the identification of the protein as well as the characterisation of the possible modifications or sequence alterations. Testing this approach on *Arabidopsis* chloroplast proteins, Zabrouskov et al. (2003) measured accurate mass values for 22 proteins, and identified and characterized 7 proteins from the 3 soluble compartments (thylakoid periphery, thylakoid lumen and stroma). This study shows that although "bottom up" analyses are still a better choice for the first identifications of precursor proteins from a genome like that of *Arabidopsis* (97 proteins identified so far), the "top down" approach shows unique capabilities for the characterization of a complex eukaryotic proteome.

2.4.2. Protein families

Proteomics can also help analyse specific protein families. A good example is that of the small multifunctional redox-active proteins named thioredoxins. In chloroplasts, 4 types of thioredoxins coexist (Lemaire et al., 2003) and play central roles in regulating enzyme activities. In order to trap potential targets of spinach chloroplast stroma thioredoxins, Balmer et al. (2003) bound modified thioredoxins to affinity columns. This approach led to the identification of 15 potential new processes in which thioredoxins are involved (vitamin biosynthesis, tetrapyrrole biosynthesis, protein assembly or degradation, glycolysis, starch degradation, DNA replication/transcription) or already known to play a role (Calvin cycle, nitrogen metabolism, stress-response, etc.). This work shows the power of such proteomic approaches to identify partners or targets of known proteins, in order to understand their physiological roles.

2.4.3. Towards the proteomics of post-translational modifications

One of the major observations from all these proteomics studies (Peltier et al., 2000; Kruft et al., 2001; Millar et al., 2001; Bardel et al., 2002; Heazlewood et al., 2004) is that among the proteins analysed, a large proportion seem to be post-translationally modified. Analysing the proteins from PSII-enriched thylakoid membranes (grana) from pea and spinach by RP-HPLC coupled with MS, Gomez et al. (2002) detected about 40 post-translationally modified gene products (mostly by palmitoylation and phosphorylation). Palmitoylation of PsbA is believed to play a role in folding, assembly or targeting of newly synthesized PsbA protein to functional PSII in granal stacks, and N-terminal phosphorylation appears to modulate the turnover of the protein with respect to photodamage (Koivuniemi et al., 1995). The N□-acetylation of several tryptic peptides from thylakoid or envelope proteins was observed in several cases (Gomez et al., 2002; Ferro et al., 2003). This modification is believed to happen after removal of the transit peptide.

In plants, protein phosphorylation is not only involved in the regulation of important cellular phenomena such as cell division and plant development, but it is also key response to environmental stimuli such as light (Allen, 1992). Precise characterisation of the phosphorylome, i.e. the subset of proteins in the proteome that becomes modified in vivo by phosphorylation, will be essential to fully understand how proteins are activated or inhibited, encouraged to interact with other components in the cell or selected for degradation. Protein phosphorylation is a transient and reversible event, so this phosphorylome will change according to environmental conditions and development. The most common method for analyzing the phosphoproteome involves the use of radioactive labelling either *in vivo* or *in vitro*. This approach is limited by the presence of pre-existing bound phosphate before labelling. The purification of phosphorylated proteins is possible thanks to affinity chromatography techniques like immobilized metal affinity chromatography (IMAC), allowing the analysis of proteins that became phosphorylated *in vivo*.

In chloroplasts, it is well established that reversible protein phosphorylation plays a direct role in the conversion of light energy to chemical energy (Allen, 1992). Studies in spinach and pea using ^{32}P labelling and phospho-amino acid antibodies showed that a number of chloroplast proteins are phosphorylated: LHCII and PSII polypeptides, the PsbA and PsbD reaction center proteins, chlorophyll-binding protein PsbC and peripheral polypeptide PsbH. Vener et al. (2001) used MS to characterize the major chloroplast thylakoid phosphoproteins in *Arabidopsis*. Thylakoids were treated with trypsin, the phosphopeptides purified by IMAC, identified by MS (PsbA, PsbD, PsbC and PsbH polypeptides of PSII, the chlorophyll-binding proteins LHCII) and their phosphorylation sites determined (all phosphorylated on threonines). This highly sensitive technique was used to analyze complex mixtures of phosphoproteins modified *in vivo*, and the phosphorylation states of the principal thylakoid proteins were quantified in various physiological conditions (day/night, continuous light, etc.). This study shows that reversible regulatory protein phosphorylation in the *Arabidopsis* photosynthetic membranes is faster during stress than during the normal light-dark cycles. Further application of

MS should help understand the role of reversible phosphorylation in photosynthetic regulation and reveal additional components that are affected by this modification.

In plant mitochondria, this type of approach has not yet been used: the first mitochondrial phosphoproteomics study was published by Bykova et al. (2003a), showing that more mitochondrial proteins can be modified and regulated by phosphorylation than previously thought (Sommarin et al., 1990). These authors performed in organello phosphorylation in the presence of radioactive ATP, and identified the labelled protein spots from 2D gel analysis by mass spectrometry. They were able to find the already known phosphorylated mitochondrial proteins (of which pyruvate dehydrogenase (PDH), whose inactivation by phosphorylation in light has been extensively described (Tovar-Mendez et al., 2003)), and could identify 14 new phosphoproteins, all house-keeping proteins also present in mammalian and fungal mitochondria (TCA cycle or associated reactions, subunits of the respiratory complexes, ATP synthesis, heat shock proteins, defense against oxidative stress).

The major protein found in pea root and seed mitochondria is formate dehydrogenase, an enzyme abundant in non-green tissues and scarce in photosynthetic tissues (Colas des Francs-Small et al., 1993), but that can accumulate in leaves in response to stress (Hourton-Cabassa et al., 1998). This protein is present in the mitochondria in multiple forms, suggesting post-translational modifications, with 4 isoforms in the pea root proteome and two in the leaf proteome. The potato tuber enzyme has recently been shown to undergo phosphorylation (Bykova et al., 2003b), but the physiological effect of this modification is not known. Two threonine phosphorylation sites were mapped.

The phosphoprotein pattern of *Arabidopsis* mitochondria is not published yet, but it is similar to that of potato mitochondria (Eubel and Braun, unpublished data), suggesting that phosphorylation processes are conserved between plant species. In silico analyses show that plant mitochondria have considerably more proteins potentially involved in signal transduction (kinases, phosphatases, etc.) than their animal and fungal counterparts (Heazlewood et al., 2004). Some targets of phosphorylation are now known, but the signalling factors and pathways remain to be elucidated.

3. BIOINFORMATICS

No high-throughput experimental approach can be successfully applied without considerable computing support (databases, scripts for data formatting and analysis, statistics, etc.), and organellar proteomics/genomics is no exception. Several well-maintained databases treating organelles have been established, including some specialising in plant mitochondria or plastids (Table 2). Most of the basic bioinformatics behind these databases are applications of standard tools widely used throughout molecular biology and shall not be considered further here. One field of bioinformatics is however, of particular interest to scientists working on organelles – the prediction of protein targeting. As we saw in the previous section, proteomics approaches have provided copious and invaluable data for constructing lists of the proteins present in organelles, a pre-requisite for full understanding of organellar biochemistry and physiology. However, the current proteomics data is far from complete for a number of reasons (but primarily the relatively low abundance of many proteins). Most estimates predict the total complement of organellar proteins to be almost an order of magnitude higher than the number of identified proteins in the proteomics studies published so far (Table 1). An alternative method is to identify organellar proteins from their characteristic N-terminal targeting sequences via gene or cDNA sequences. In principle, for species such as *Arabidopsis* and rice, the complete genome sequence offers the possibility of quickly and easily dressing complete lists of mitochondrial and plastid proteins without any complications from protein expression levels or patterns. For the majority of species, lacking complete genome sequences, large EST or cDNA collections provide a sequence resource that can be used in a similar fashion. Though such collections will suffer from representational bias due to differing expression levels, the coverage is still often much greater than in proteomics studies. In practice, unfortunately, carefully undertaken proteomics studies can guarantee with a high degree of certainty, the correct identification and localization of organellar proteins, prediction of targeting sequences is a much less exact science.

3.1. Prediction of N-terminal targeting sequences

A large minority (estimates vary from 25-40%) of proteins synthesized in the cytosol contain cleavable N-terminal signals for targeting to different compartments. These signals must be recognised with a high degree of specificity by the import machinery located on the membranes of the corresponding organelles. As transfer of these presequences to other unrelated polypeptides generally leads to targeting of the passenger protein to the relevant organelle, they must contain all the necessary targeting information and thus in theory should be identifiable. Despite an enormous amount of work and an extensive literature on the subject, we are still far from understanding the exact nature of these signals and the recognition process *in vivo.* Protein import into mitochondria and plastids shares several features (ATP-driven passage of unfolded precursor proteins through outer membrane and inner membrane complexes, involvement of Hsp70-type chaperones, reliance on an N-terminal targeting presequence that is usually cleaved after import, etc.). The targeting sequences for the two organelles also have similarities; they are both rich in arginine, serine and leucine and generally lack acidic amino acids (Glaser et al., 1998). So it is not surpring that it has been difficult in the past to clearly distinguish the two types, and indeed some proteins are imported equally well into both organelles (Peeters and Small, 2001). These similarities have complicated the development of efficient prediction tools, but the difficulty of the challenge has clearly stimulated researchers who have made numerous attempts to solve the problem. The approaches employed can be loosely grouped into three categories:

- expert systems – tools which apply strict rules developed by biologists based on known characteristics of targeting sequences

- statistical approaches – tools which categorize sequences based on explicit statistical parameters developed from analysis of known targeted proteins (typically amino acid composition)
- machine-learning approaches – tools which categorize sequences based on hidden parameters developed after training with sequences of known targeted proteins.

The first two approaches are quick and efficient when the characteristics defining the sequences to be classified are simple and obvious. They have an additional advantage in that the parameters used are explicit; one knows why a particular protein is classed as it is. The third approach is more computationally intensive and suffers from the fact that the classification is often relatively opaque; the reasons why one protein is classed one way and a second another way are often obscure. This unavoidable obscurity has provoked attempts to derive clearly understandable rules from the machine-learning results (Schein et al., 2001; Bannai et al., 2002). Nevertheless, machine-learning approaches are demonstrably more accurate when applied to the identification of mitochondrial and plastid targeting sequences (Heazlewood et al., 2004; Small et al., 2004), and the most widely used tool (TargetP) employs neural networks in a classical application of machine-learning techniques (Emanuelsson et al., 2000). Table 3 lists the best known tools that are publicly available for prediction of mitochondrial and plastid proteins.

Table 3. *Widely used targeting prediction tools*

Prediction tool	*Approach*	*URL*	*Comments*
ChloroP	Machine learning	www.cbs.dtu.dk/ services/ChloroP/	Plastid targeting only; superceded by TargetP (Emanuelsson et al., 1999)
TargetP	Machine learning	www.cbs.dtu.dk/ services/TargetP/	(Emanuelsson et al., 2000)
Predotar	Machine learning	genoplante-info.infobiogen.fr/ predotar/	(Small et al., 2004)
SubLoc	Machine learning	www.bioinfo.tsinghua. edu.cn/ SubLoc/eu_predict.htm	Mitochondrial targeting only (Hua and Sun, 2001)
PSORT	Expert system	psort.nibb.ac.jp/form.html	(Nakai and Horton, 1999)
Mitoprot	Expert system	ihg.gsf.de/ihg/mitoprot. html	Mitochondrial targeting only (Claros, 1995; Claros and Vincens, 1996)
iPSORT	Expert system	www.hypothesiscreator. net/ iPSORT/	(Bannai et al., 2002)
PCLR	Statistics	apicoplast.cis.upenn.edu/ pclr/	Plastid targeting only Schein et al., 2001)

3.2. Prediction accuracy

Several recent studies have compared prediction programs with experimentally determined lists of *Arabidopsis* organellar proteins (Peltier et al., 2002; Friso et al., 2004; Heazlewood et al., 2004; Small et al., 2004) (Table 4). These comparisons allow the estimation of the 'sensitivity' of these prediction tools (the percentage of true organellar proteins correctly predicted as organellar). At first sight the results are surprisingly bad, particularly for mitochondrial proteins. Only 30-50% of the experimentally observed mitochondrial proteins are predicted as mitochondrial by the different programs (against 55-75% of experimentally observed plastid proteins). A second important parameter is the 'specificity'; the proportion of predicted organellar proteins that are truly organellar. This gives an estimate of the number of 'false positive' predictions by each program. Unsurprisingly, there is a trade-off between sensitivity and specificity; programs with a high sensitivity tend to have a low specificity and vice versa.

The specificity parameter is extremely difficult to measure experimentally (it would require experimental verification of a large proportion of all predicted organellar proteins). However, one can estimate the relative specificity of each tool by comparing the proportions of predicted proteins that are in the experimentally determined sets (Table 4). By making a plausible assumption about the total number of organellar proteins, one can then calculate the specificity of each program, and even though the absolute figures are almost certainly inexact, the relative tendencies are valid. This calculation can be used to derive an estimate of the upper limits of the organelle proteomes. For example, many current estimates place the plastid proteome at over 3000 proteins (Table 1), but using this figure implies a specificity for Predotar of over 100%, which is impossible. Assuming that Predotar is equally good at predicting plastid proteins in the experimental set as it is those that are not, and the maximum specificity of 88% determined from test sets (Small et al., 2004), the calculation would estimate the number of Arabidopsis plastid proteins at 2450.

A little-appreciated fact is that with their default settings, most of these prediction programs predict huge numbers of false positives when used on whole proteome sets; 70-75% of predicted *Arabidopsis* mitochondrial proteins and up to two thirds of predicted *Arabidopsis* plastid proteins are likely to be wrong. Only Predotar (Small et al., 2004) manages an estimated specificity of over 50%, at the expense of a noticeably lower sensitivity than the other tools. For genomic analyses therefore, one should be very cautious when interpreting targeting predictions. False positives can be avoided to some extent by raising the stringency of the predictions (increasing the threshold at which the predictive score is taken as significant), combining predictions from several different tools (Heazlewood et al., 2004), or by taking into account data from other sources (sequence homology with known organellar proteins, localization data from other organisms, expression data etc.). Combining multiple types of data is particularly effective when coupled with expert knowledge capable of curating and interpreting this (often inaccurate) raw information. Several excellent papers show the power of these combined approaches; a common theme is the use of homology to bacterial proteins (from

Synechocystis or *Rickettsia*) to filter the targeting prediction results (Abdallah et al., 2000; Martin et al., 2002; Elo et al., 2003; Richly et al., 2003a).

Table 4. *A comparison of organelle targeting predictions using the Arabidopsis proteome.*

Mitochondrial targeting predictions

Program	*TargetP*	*Predotar (stringent)*	*Predotar (relaxed)*	*Mitoprot*	*iPSORT*	*SubLoc*
No. predicted	3101	1098	2375	3794	4491	3554
No. in experimental set predicted	157	127	165	198	188	147
% sensitivity	38.48%	31.13%	40.44%	48.53%	46.08%	36.03%
% of predicted proteins in experimental set	5.06%	11.57%	6.95%	5.22%	4.19%	4.14%
Estimated % specificity	31.02%	70.87%	42.57%	31.98%	25.65%	25.34%

Plastid targeting predictions

Program	*TargetP*	*Predotar (stringent)*	*Predotar (relaxed)*	*PCLR*	*iPSORT*
No. predicted	4251	1586	2550	6001	3143
No. in experimental set predicted	327	261	293	332	258
% sensitivity	72.03%	57.49%	64.54%	73.13%	56.83%
% of predicted proteins in experimental set	7.69%	16.46%	11.49%	5.53%	8.21%
Estimated % specificity	42.36%	90.62%	63.27%	30.46%	45.20%

Each program was used to predict targeting of the approximately 28000 proteins from the annotated Arabidopsis genome sequence. TargetP v1.01, Predotar v1.03 and PCLR results are on TIGR release 4 annotations; Mitoprot, iPSORT and SubLoc results are from AMPD (Table 2) and are based on TIGR release 3 annotations. The predictions were compared to the experimental datasets collected in AMPD (408 nuclear encoded mitochondrial proteins) and PPD (454 nuclear encoded plastid proteins) (Table 2). As the default parameters for Predotar make it more stringent than the other programs, the results with "relaxed" settings are also included (i.e. proteins predicted as only "possibly" organellar are included). The sensitivity was calculated as the percentage of the experimental set whose localization was correctly predicted. The specificity was calculated as the proportion of predicted proteins that were found in the experimental set divided by the ratio of the size of the experimental set to the estimated organelle proteome size (2000 proteins for both mitochondria and plastids).

3.3. Suborganellar compartments

Organelles comprise several membranes and compartments with distinct protein complements. Ideally, one would like to predict suborganellar localization, but currently it is difficult. Integral membrane proteins can be identified with considerable success by using tools for the prediction of transmembrane helices (Schwacke et al., 2003) and this approach has been coupled with targeting sequence predictions for the identification of mitochondrial and plastid transporters (Ferro et al., 2002; Koo and Ohlrogge, 2002; Peltier et al., 2002; Millar and Heazlewood, 2003). The only suborganellar compartment for which targeting signals have been clearly identified is the thylakoid lumen of plastids. Proteins reach the lumenal space by import from the stroma via the Sec or TAT pathways derived from bacterial protein secretion systems (Keegstra and Cline, 1999). Lumenal proteins have been predicted by combining transit peptide predictions (using TargetP) with the use of SignalP to look for bacterial secretion signals (Peltier et al., 2002; Schubert et al., 2002). Recently, a neural network-based predictor (LumenP) has been specifically developed for the identification of lumenal proteins (Westerlund et al., 2003). LumenP employs networks specifically trained on both Sec and TAT pathway precursors and, coupled with TargetP makes an effective tool for screening plant genomes for thylakoid lumen proteins.

3.4. Cleavage sites

Many, but not all, targeting sequences are cleaved after import. Cleavage of the N-terminal signals of mitochondrial and plastid precursors occurs via one or more specific proteases that recognise poorly understood characteristic features of the presequences (Glaser et al., 1998; Keegstra and Cline, 1999). Lumenal targeting signals are cleaved by yet more specific proteases (Gomez et al., 2003). Prediction of cleavage sites is useful information that is not always provided by targeting prediction tools. MitoProt (Claros and Vincens, 1996), ChloroP (Emanuelsson et al., 1999) and TargetP (Emanuelsson et al., 2000) can be used to predict cleavage sites in mitochondrial and plastid precursors, based on recognition of cleavage site motifs originally described by Gavel and von Heijne (1990a, 1990b). SignalP predicts cleavage sites in many but not all thylakoid membrane and lumenal proteins (Gomez et al., 2003).

Unfortunately, the predictions are not generally highly reliable, for a number of reasons. A major difficulty is that following cleavage by the specific presequence protease, proteins are often clipped by one or more aminopeptidases (Emanuelsson et al., 1999). Such N-terminal trimming is difficult or impossible to predict and probably depends more on structural accessibility than primary sequence. Extensive N-terminal sequence data from proteomics studies will be needed to make much improvement in this area. This information will become increasingly important: often it is essential for functional or comparative studies to know where the N-terminus of the mature protein is. For example, many organellar proteins are unlikely to fold or function correctly while the presequence is still attached. Thus for heterologous expression experiments (e.g. yeast two-hybrid assays or overexpression in *E. coli* for structural studies), it would be preferable to express the mature protein without the targeting information.

3.5. Limitations to bioinformatics predictions

It appears that the prediction tools fare much worse than expected when confronted with real data. Most of the original papers describing these tools claimed sensitivities and specificities in the range of 70-80%, but, as shown in Table 4, the predictions on *Arabidopsis* proteins are not nearly as accurate. There are many possible reasons for this discrepancy. Mitochondria are particularly badly served; all the prediction tools predict a much higher proportion of plastid proteins, and with fewer false positives. A major explanation for this is that many mitochondrial proteins lack classical N-terminal targeting sequences (e.g. outer membrane and intermembrane space proteins, many inner membrane transporters, many small components of inner membrane complexes). These proteins figure prominently in the proteomics studies but are invisible to the prediction programs (20% of the experimentally determined mitochondrial proteins are not predicted as mitochondrial by any of the programs used in Table 4). A much smaller proportion of plastid proteins lack canonical targeting sequences (Miras et al., 2002). The mitochondrial predictions are probably better considered to be matrix targeting predictions.

Other problems with predictions include doubts over gene models; large-scale transcriptome and cDNA analyses suggest that 30% of gene models for which experimental cDNA data is unavailable are incorrect (Yamada et al., 2003). The identification of the correct N-terminal methionine is often crucial for correct predictions, and so one can assume that a significant number of predicted localizations of *Arabidopsis* proteins are wrong because the wrong peptide sequence has been presented to the prediction tool. Finally, an increasing number of so-called 'dual-targeted' proteins are found to be present in both organelles (Small et al., 1998; Peeters and Small, 2001; Silva-Filho, 2003). This can occur via alternative targeting sequences (differential transcription or translation starts, alternative splicing) or via 'ambiguous' targeting sequences that are equally effectively recognized by both sets of import machinery. Either way, current prediction programs are incapable of automatically detecting such dual-targeting.

3.6. Perspectives in bioinformatics

In the future, targeting predictions should continue to improve up to a certain point. The massive increase in the numbers of characterized organellar proteins (from proteomics studies and comparative genomics) will provide much more complete training sets for machine-learning approaches which should benefit greatly in the immediate future. This influx of data should even permit the constitution of training sets specific to particular organisms or groups of organisms. The proteomics data should also help with the improvement of the recognition of cleavage sites by providing information on N-termini. However, the problems associated with proteins that lack canonical targeting sequences (a significant proportion of mitochondrial proteins) will not be solved easily, and the outlook for automatic identification of these is bleak.

4. FUNCTIONAL GENOMICS

Functional genomics is beginning to revolutionize our understanding of plant biology, and the field of plant organelle biology will also benefit from the new opportunities offered by the increasing availability of genome sequence data and other genomic resources (Ausubel and Benfey, 2002; Holtorf et al., 2002; Tabata, 2002; Hilson et al., 2003). Technological developments in *Arabidopsis thaliana* genomics have recently stimulated the initiation of projects aimed at systematically identifying the functions of plastid and mitochondrial proteins. Moreover, recombinational cloning has provided the technological breakthrough that permits high-throughput functional analysis and is starting to be used on a large scale in functional genomics projects, such as systematic reverse genetics approaches and clone-based proteomics. Coupled with bioinformatics and transcriptomics approaches, they provide us with valuable tools to decipher the roles of organelles in the biology of a plant cell.

4.1. Functional genomics of nuclear encoded organellar proteins

Most proteins required for the function of plastids and mitochondria are encoded by the nuclear genome and are targeted to these organelles as precursors. Our knowledge of the number and identity of these nuclear-encoded proteins has greatly increased over the last decade thanks to large-scale protein sequencing and the development of bioinformatics programs for uncovering candidate organellar proteins from large-scale genome sequencing projects. However, experimental evidence is needed to confirm and complement the data obtained by both of these approaches on a genome-wide scale. Furthermore, classical proteomics and bioinformatics provide sparse indications as to the <u>function</u> of the identified proteins.

4.1.1. Systematic subcellular localization

In addition to the proteomics and bioinformatics approaches already discussed, subcellular localization of proteins can be ascertained by several methods, the most widely used are immunolocalization, fluorescence tagging or *in vitro* import into isolated organelles. To date, most such studies in plants have only been carried out on the scale of small protein families (for examples, Giglione et al., 2000; Vermel et al., 2002). Very large-scale analyses of protein localization have so far only been performed in *S. cerevisiae*. Kumar et al. (2002) used high-throughput immunolocalization of epitope-tagged proteins obtained by directed topoisomerase I-mediated cloning strategies and genome-wide transposon mutagenesis, demonstrating experimental data for 55% of the yeast proteome. More recently, Huh et al. (2003) provided localization information for 75% of the yeast proteome based on the systematic analysis of a collection of yeast strains expressing full-length chromosomally tagged GFP fusion proteins. In this study, more than 500 mitochondrially targeted fusions were identified - entirely consistent with data from previous genetic and bioinformatics studies.

However, very few studies have been undertaken on a large scale in plants. Two major studies employing random cDNA fusions to green fluorescent protein (GFP) have been described, but the first used GFP::cDNA fusions (Cutler et al., 2000) and thus missed proteins targeted by their N-terminus (the vast majority of plastid and mitochondrial proteins). The second employed both GFP::cDNA and cDNA::GFP fusions expressed from a tobacco mosaic virus vector (Escobar et al., 2003). In this study, 20% of cDNA::GFP fusions expressed fluorescence ; 5% of these exhibited plastid localization and 6.8% localization in 'blobs', a descriptive term covering mitochondria, peroxisomes, small vesicles etc. This highlights one of the problems with studying targeting *in vivo*, i.e. the identification of the compartment in which the protein is localized. The large yeast study used DAPI or MitoTracker staining or mating to strains expressing RFP-tagged reference proteins to double-check mitochondrial localization (Huh et al., 2003).

An alternative and very elegant way of ensuring the correct identification of the targeted compartment has been described by Ozawa et al. (2003). They have developed a method based on the reconstitution of split GFP by protein splicing with the intein (protein splicing element) from DnaE. The C-terminal parts of the intein and GFP were targeted to mitochondria of mammalian cells using a known targeting sequence from COX8, and the cells then co-transfected with cDNAs fused to the sequences encoding the N-terminal halves of GFP and the DnaE intein. If the fused cDNA encodes a mitochondrial targeting sequence, the split inteins can reconstitute a functional module and splice the split GFP moieties to give fluorescence. Fluorescing cells were collected by fluorescence-activated cell sorting. About 250 cDNAs were isolated and identified from the cells, nine of which correspond to new mitochondrial transcripts. This method holds promise for directed recovery of cDNAs encoding targeted proteins.

So far, the majority of large-scale localization studies have been undertaken in yeast, owing primarily to the fidelity of homologous recombination in *Saccharomyces cerevisiae* and the concomitant ease with which integrated reporter gene fusions can be generated. Systematic approaches dedicated to the analyses of other eukaryotic proteomes will probably rely on the use of clone-based functional approaches (see below). For example, one hundred human cDNAs have been cloned as N- and C-terminal gene fusions to spectral variants of GFP as a means of examining the subcellular localization of these proteins in living cells (Simpson et al., 2000). Such projects have become feasible thanks to the development of efficient high-throughput cloning technologies that allow parallel handling of many ORFs or genes, one of most important recent breakthroughs in the field of functional genomics.

4.1.2. High throughput recombinational cloning and cloned-based proteomics
Recombinational cloning is based on the use of recombinases to insert ORFs from a plasmid or a PCR product into a vector. These new high-throughput cloning technologies offer the possibility of handling hundreds or thousands of genes in parallel, leading potentially to the establishment of full ORF complements representing all the proteins expressed in a given organism ('ORFeomes', Reboul et al., 2003). Such collections of ORFs enable the development of clone-based systematic functional assays, as exemplified by the pioneering work on *Caenorhabditis elegans* showing the power of ORFeome-based approaches for large-scale protein-protein interaction studies and RNAi knock-down studies (Walhout et al., 2002; Li et al., 2004). Proteins can be synthesised in any species and cell types, provided the appropriate expression cassettes and transformation procedures exist. Proteins can also be expressed as a translational fusion with any chosen peptide moieties (e.g. purification or marker tags) by cloning its ORF in frame with the desired coding sequence, and are therefore amenable to a wide range of analytical methods. These include structural genomics techniques (including crystallography and NMR), phenotypic (knock-in and knock-down) analyses, protein subcellular localization studies (using fluorescent-tag fusions and epitope tagging), molecular interaction mapping (using yeast-n-hybrid, tandem affinity purification/MS and plasmon resonance techniques) and biochemical genomics (based on tagged protein purification, protein chips and antibody libraries).

Concerning plants, several *Arabidopsis* ORFeome-type projects are underway and a quasi-complete set of ORFs is expected within a few years (Hilson et al., 2003; Yamada et al., 2003). The availibility of such resources will revolutionize plant functional genomics. Combining bioinformatics prediction models and different sources of experimental data on protein organellar localization, one can envisage the construction of sub-ORFeomes that would enable the development of specific functional genomics projects for plastids and mitochondria.

4.2. Organelle-related transcriptomics

Since the completion of the *Arabidopsis* genome sequence, transcriptome studies constitute the initial set of experiments that have quickly been performed on a genome-wide scale, leading to a plethora of expression data, some of them being focused on organelle-related nuclear genes and/or chloroplast and mitochondrial genomes. Most of the transcriptome studies of organelle-related nuclear genes performed so far are included in functional analysis projects of organelle sub-proteomes, structural components or gene families related to organelle functions, and largely rely on the analysis of publicly available resources. In their study dedicated to *Arabidopsis* plastid inner envelope membrane proteins, Koo and Ohlrogge (2002) report the analysis of the expression profiles of about 500 predicted candidates using digital northern analysis of expressed sequence tags (ESTs). This approach, which consists of the comparison of EST frequencies between pools of tissue-specific cDNA libraries, should be viewed as an initial step to find possible candidates for differentially expressed genes from a large collection of data.

Surprisingly, only a set of 21 genes displayed tissue-specific expression, suggesting that the envelope proteome candidates are relatively constitutive in terms of transcript abundance across the different tissues analyzed (flowers, roots and seeds), in contrast to stroma and thylakoid proteins. A comparable trend could be observed by analyzing gene expression patterns using publicly available cDNA microarrays. In the study of a family of 45 *Arabidopsis* mitochondrial carrier proteins, Millar and Heazlewood (2003) followed a similar approach. Analyses of microarray databases revealed differential expression profiles of the more highly expressed members in different plant tissues and in response to various plant hormones and environmental stresses. In a transcriptome and proteome study of the *Arabidopsis* mitochondrial import apparatus, Lister et al. (2004) report on a comprehensive expression analysis of 31 genes encoding the import components in all organs and cell culture.

Comparing real-time RT-PCR and microarray experimental data with EST numbers and massively parallel signature sequence data, the authors show that though many genes are present in small multi-gene families, often only one member is prominently expressed. The response of the import apparatus to mitochondrial dysfunction was investigated using Affymetrix microarrays, and revealed an increase in the expression of genes encoding import components and proteins involved in mitochondrial biogenesis after rotenone or antimycin A treatment. It is interesting that this was particularly true for minor isoforms. Even more ambitious projects are being undertaken to study chloroplast biogenesis. Kurth et al. (2002) generated a DNA array of 1827 gene-sequence-tags (GSTs) by spotting PCR products amplified from genomic DNA on nylon membranes. This GST array was employed to monitor the mRNA expression level of each of the 1827 corresponding genes under different genetic and environmental conditions, such as seedling growth in dark *versus* light. Extending this pioneer approach, Richly et al. (2003b) provide a comprehensive study of the nuclear chloroplast transcriptome by following the differential expression analysis of 3292 genes in wild-type plants grown under 16 environmental conditions and in several mutants with defects in chloroplast function and/or plastid-to-nucleus signalling.

Interestingly, from the 35 genetic and environmental different conditions that were analyzed, a well-organised structure emerged, in which only three classes of nuclear chloroplast transcriptome response are apparent. The two main responses induced by most of the apparently diverse and unrelated conditions tested are characterised by either preferential up-regulation or down-regulation of the nuclear expression profiles which are typical of, or opposite to, those of *GUN* (genomes uncoupled) mutants that affect plastid signalling. These data suggest the presence of a transcriptional master-switch regulating in a binary mode, the induction or repression of the same large set of nuclear genes that are relevant to plastid functions. The use of a nuclear chloroplast transcriptome DNA-array can also be helpful for characterizing photosynthetic and other mutants (Maiwald et al., 2003; Pesaresi et al., 2003).

4.2.1. Organelle transcriptomics and post-transcriptomics

Because of the small size and low gene content of organelle genomes, analysis of transcriptional and post-transcriptional activities in plastids and mitochondria might appear straightforward. The evaluation of array-based expression profiles is, however, particularly challenging, since organelle genes are organised in polycistronic transcription units that are subsequently subjected to multiple post-transcriptional processes leading to complex sets of overlapping transcripts of differing stability and hence relative stationary concentrations. A thorough study of the reliability and limitations of DNA-array technology for the examination of organelle gene expression profiles has been performed by Legen et al. (2002). They reported on the establishment of a macroarray-based assay to evaluate transcription rate and transcript levels of 118 genes and 11 ORFs from the tobacco plastid chromosome. This approach was successfully applied to a comparative study of transcriptome profiles of leaves from wild-type plants *versus* leaves from transplastomic mutants deficient in plastid-encoded RNA polymerase (PEP). Interestingly, using run-on transcripts from wild-type and PEP-deficient plastids as probe, the authors showed that the nuclear-encoded RNA polymerase (NEP) alone is able to transcribe the entire genetic information of the tobacco plastid chromosome.

In a landmark study on the unicellular green alga *Chlamydomonas reinhardtii*, the Stern group conducted a global gene expression analysis of both chloroplast and mitochondrial transcriptomes under a variety of environmental conditions (Lilly et al., 2002). As revealed by RNA gel blots, DNA microarray analyses and run-on assays, the results show that the most global effects on organellar RNA abundance are observed in response to sulphate and phosphate limitation. Interestingly, while removal of sulphate from the culture medium results in decreases in transcription rates, phosphate limitation conversely results in a two fold to three fold increase in RNA abundance because of increased RNA stability. How changes in transcript levels influence the organellar proteomes under stress conditions will be the next challenge in the future. More recently, Nakamura et al. (2003) reported the construction of a tobacco plastid genomic microarray consisting of 220 DNA fragments covering the whole genome sequence, i.e. with all genes and all intergenic regions. The reliability of the microarray was evaluated by comparing the plastid RNA levels in light- or dark-grown tobacco seedlings. Interestingly, unexpected signals were found in several intergenic regions, suggesting the existence of as yet uncharacterised transcripts in tobacco plastids.

4.3. Systematic reverse genetics

Transcriptome studies give valuable information on gene expression, but do not provide direct data on gene function. Studying mutants defective in specific genes is an alternative approach which provides complementary, more direct information. Numerous forward genetic screens have been employed to identify mutants affected in photosynthesis because of the ease with which pigment defects or fluorescence defects can be screened (for reviews see Leister and Schneider, 2003; Leister, 2003). *Chlamydomonas reinhardtii* is a valuable and much studied model for the genetics of photosynthesis (Dent et al., 2001). However, non-photosynthetic functions of plastids and mitochondria are much more difficult to study on a large scale by genetic approaches because of the difficulty in devising effective screens for identifying mutants. Reverse genetics is an increasingly attractive alternative. Unfortunately, knockout of nuclear genes by homologous recombination is infeasible currently in flowering plants. Consequently, functional analyses at present rely on large collections of insertion mutants (containing randomly inserted transgenes or transposons) that can be screened for mutations in genes of interest. Thanks to the systematic sequencing of insertion sites and the resulting databases, reverse genetics has become an extremely powerful tool in functional genomics. Current studies focus on the analysis of a limited number of genes (Leister, 2003), but larger scale projects are underway (Koo and Ohlrogge, 2002).

However, insertional mutagenesis programs have some limitations that make it difficult or impossible to obtain or study insertion mutants for certain types of genes. Many mutations in genes involved in organelle biogenesis will give an embryo- or seedling-lethal phenotype when homozygous (Budziszewski et al., 2001). To circumvent this limitation, efforts have been made to develop alternative, less drastic methods of mutagenesis for reverse genetics in plants. One of them is the relatively new technique of post-transcriptional gene silencing (PTGS) triggered by RNA interference (RNAi), which relies on the expression of a transgene specifically leading to the reduction in expression of the corresponding target gene. Methods exist now to carry out such studies in a systematic way (Waterhouse and Helliwell, 2003). TILLING, which relies on identifying plants carrying EMS-induced point mutations in the region of interest (Till et al., 2003), is another new reverse genetics technique offering a valuable alternative to insertional mutagenesis.

5. CONCLUSION

A significant fraction of nuclear genes in plants (probably 15-25%) code for proteins targeted to mitochondria or plastids, or is otherwise implicated in biogenesis or regulation of these organelles. Given the thousands of genes and gene products involved, researchers in this field are obviously excited by the prospects of modern high-throughput systematic approaches.

This review has tried to cover most of these new developments, ranging from the latest advances in techniques traditionally associated with organelle studies (proteomics, genetics) to newly established methods being applied to organelles (microarrays, bioinformatics). In passing, we have also touched on emerging technologies that we see as important for the future (clone-based proteomics or "ORFeomics", systematic RNAi) but that have not yet been applied to plant organelle studies. The field is at a very interesting stage, and a huge influx of data can be expected in the next few years. Our knowledge and understanding of organelle biogenesis should grow immensely over this period, provided we all brush up our computer skills; analysis of all this data will be impossible by hand.

6. ACKNOWLEDGEMENTS

We would like to offer sincere thanks to our many colleagues who helped shape this review by their comments and by giving us access to unpublished data. This review covers such a large scope that unavoidably some important findings have been omitted or at least shamefully neglected; our apologies to all those who would have chosen to emphasize different areas of research.

7. REFERENCES

Abdallah, F., Salamini, F., & Leister, D. (2000). A prediction of the size and evolutionary origin of the proteome of chloroplasts of *Arabidopsis*. *Trends Plant Sci.*, *5*, 141-142.

Allen, J. F. (1992). Protein phosphorylation in regulation of photosynthesis. *Biochim. Biophys. Acta*, *1098(3)*, 275-335.

Ausubel, F., & Benfey, P. (2002). *Arabidopsis* functional genomics. *Plant Physiol.*, *129*, 393.

Balmer, Y., Koller, A., del Val, G., Manieri, W., Schurmann, P., & Buchanan, B. B. (2003). Proteomics gives insight into the regulatory function of chloroplast thioredoxins. *Proc. Natl. Acad. Sc.i USA*, *100*, 370-375.

Bannai, H., Tamada, Y., Maruyama, O., Nakai, K., & Miyano, S. (2002). Extensive feature detection of N-terminal protein sorting signals. *Bioinformatics*, *18*, 298-305.

Bardel, J., Louwagie, M., Jaquinod, M., Jourdain, A., Luche, S., Rabilloud, T., et al. (2002). A survey of the plant mitochondrial proteome in relation to development. *Proteomics*, *2*, 880-898.

Bentolila, S., Alfonso, A. A., & Hanson, M. R. (2002). A pentatricopeptide repeat-containing gene restores fertility to cytoplasmic male-sterile plants. *Proc. Natl. Acad. Sci. USA*, *99*, 10887-10892.

Brown, G. G., Formanova, N., Jin, H., Wargachuk, R., Dendy, C., Patil, P., et al. (2003). The radish *Rfo* restorer gene of Ogura cytoplasmic male sterility encodes a protein with multiple pentatricopeptide repeats. *Plant J.*, *35*, 262-272.

Budziszewski, G. J., Lewis, S. P., Glover, L. W., Reineke, J., Jones, G., Ziemnik, L. S., et al. (2001). *Arabidopsis* genes essential for seedling viability: isolation of insertional mutants and molecular cloning. *Genetics*, *159*, 1765-1778.

Bykova, N. V., Egsgaard, H., & Moller, I. M. (2003a). Identification of 14 new phosphoproteins involved in important plant mitochondrial processes. *FEBS Lett.*, *540*, 141-146.

Bykova, N. V., Stensballe, A., Egsgaard, H., Jensen, O. N., & Moller, I. M. (2003b). Phosphorylation of formate dehydrogenase in potato tuber mitochondria. *J. Biol. Chem.*, *278*, 26021-30.

Claros, M. G., & Vincens, P. (1996). Computational method to predict mitochondrially imported proteins and their targeting sequences. *Eur. J. Biochem.*, *241*, 779-786.

Colas des Francs-Small, C., Ambard-Bretteville, F., Darpas, A., Sallantin, M., Huet, J. C., Pernollet, J. C., et al. (1992). Variation of the polypeptide composition of mitochondria isolated from different potato tissues. *Plant Physiol.*, *98*, 273-278.

Colas des Francs-Small, C., Ambard-Bretteville, F., Small, I. D., & Remy, R. (1993). Identification of a major soluble protein in mitochondria from nonphotosynthetic tissues as NAD-dependent formate dehydrogenase. *Plant Physiol.*, *102*, 1171-1177.

Considine, M. J., Holtzapffel, R. C., Day, D. A., Whelan, J., & Millar, A. H. (2002). Molecular distinction between alternative oxidase from monocots and dicots. *Plant Physiol.*, *129*, 949-953.

Cordewell, S. J., Nouwens, A. S., Verrills, N. M., Basseal, D. J., & Walsh, B. J. (2000). Subproteomics based upon protein cellular location and relative solubilities in conjunction with composite two-dimensional electrophoresis. *Electrophoresis*, *21*, 1094-1103.

Cutler, S. R., Ehrhardt, D. W., Griffitts, J. S., & Somerville, C. R. (2000). Random GFP::cDNA fusions enable visualization of subcellular structures in cells of *Arabidopsis* at a high frequency. *Proc. Natl. Acad. Sci. USA*, *97*, 3718-3723.

Davy de Virville, J., Alin, M. F., Aaron, Y., Rémy, R., Guillot-Salomon, T., & Cantrel, C. (1998). Changes in functional properties of mitochondria during growth cycle of *Arabidopsis thaliana* cell suspension cultures. *Plant Physiol. Biochem.*, *36*, 347-356.

Day, D., Neuburger, M., & Douce, R. (1985). Biochemical characterization of chlorophyll-free mitochondria from pea leaves. *Austr. J. Plant Physiol.*, *12*, 219-228.

Dent, R. M., Han, M., & Niyogi, K. K. (2001). Functional genomics of plant photosynthesis in the fast lane using *Chlamydomonas reinhardtii*. *Trends Plant Sci.*, *6*, 364-371.

Desloire, S., Gherbi, H., Laloui, W., Marhadour, S., Clouet, V., Cattolico, L., et al. (2003). Identification of the fertility restoration locus, *Rfo*, in radish, as a member of the pentatricopeptide-repeat protein family. *EMBO Rep.*, *4*, 588-594.

Dunbar, B., Elthon, T., Osterman, J., Whitaker, B., & Wilson, S. (1997). Identification of plant mitochondrial proteins: a procedure linking two-dimensional gel electrophoresis to protein sequencing from PVDF membranes using a fastblot cycle. *Plant Mol. Biol. Rep.*, *15*, 46-61.

Elo, A., Lyznik, A., Gonzalez, D. O., Kachman, S. D., & Mackenzie, S. A. (2003). Nuclear genes that encode mitochondrial proteins for DNA and RNA metabolism are clustered in the *Arabidopsis* genome. *Plant Cell*, *15*, 1619-1631.

Emanuelsson, O., Nielsen, H., Brunak, S., & von Heijne, G. (2000). Predicting subcellular localization of proteins based on their N-terminal amino acid sequence. *J. Mol. Biol.*, *300*, 1005-1016.

Emanuelsson, O., Nielsen, H., & von Heijne, G. (1999). ChloroP, a neural network-based method for predicting chloroplast transit peptides and their cleavage sites. *Protein Sci.*, *8*, 978-984.

Escobar, N. M., Haupt, S., Thow, G., Boevink, P., Chapman, S., & Oparka, K. (2003). High-throughput viral expression of cDNA-green fluorescent protein fusions reveals novel subcellular addresses and identifies unique proteins that interact with plasmodesmata. *Plant Cell*, *15*, 1507-1523.

Eubel, H., Jänsch, L., & Braun, H. P. (2003). New insights into the respiratory chain of plant mitochondria. Supercomplexes and a unique composition of complex II. *Plant Physiol.*, *133*, 274-286.

Ferro, M., Salvi, D., Brugiere, S., Miras, S., Kowalski, S., Louwagie, M., et al. (2003). Proteomics of the chloroplast envelope membranes from *Arabidopsis thaliana*. *Mol. Cell. Proteomics*, *2*, 325-345.

Ferro, M., Salvi, D., Riviere-Rolland, H., Vermat, T., Seigneurin-Berny, D., Grunwald, D., et al. (2002). Integral membrane proteins of the chloroplast envelope: identification and subcellular localization of new transporters. *Proc. Natl. Acad. Sci. USA*, *99*, 11487-11492.

Ferro, M., Seigneurin-Berny, D., Rolland, N., Chapel, A., Salvi, D., Garin, J., et al. (2000). Organic solvent extraction as a versatile procedure to identify hydrophobic chloroplast membrane proteins. *Electrophoresis*, *21*, 3517-35126.

Fey, J., & Maréchal-Drouard, L. (1999). Compilation and analysis of plant mitochondrial promoter sequences: An illustration of a divergent evolution between monocot and dicot mitochondria. *Biochem. Biophys. Res. Commun.*, *256*, 409-414.

Friso, G., Giacomelli, L., Ytterberg, A. J., Peltier, J. B., Rudella, A., Sun, Q., et al. (2004). In-depth analysis of the thylakoid membrane proteome of *Arabidopsis thaliana* chloroplasts; new proteins, functions and a plastid proteome database. *Plant Cell, 16*, 478-499.

Froehlich, J. E., Wilkerson, C. G., Ray, W. K., McAndrew, R. S., Osteryoung, K. W., Gage, D. A., et al. (2003). Proteomic study of the *Arabidopsis thaliana* chloroplastic envelope membrane utilizing alternatives to traditional two-dimensional electrophoresis. *J. Proteome Res.*, *2*, 413-425.

Gavel, Y., & von Heijne, G. (1990a). A conserved cleavage-site motif in chloroplast transit peptides. *FEBS Lett.*, *261*, 455-458.

Gavel, Y., & von Heijne, G. (1990b). Cleavage-site motifs in mitochondrial targeting peptides. *Protein Eng.*, *4*, 33-37.

Giglione, C., Serero, A., Pierre, M., Boisson, B., & Meinnel, T. (2000). Identification of eukaryotic peptide deformylases reveals universality of N-terminal protein processing mechanisms. *EMBO J.*, *19*, 5916-5929.

Glaser, E., Sjoling, S., Tanudji, M., & Whelan, J. (1998). Mitochondrial protein import in plants. Signals, sorting, targeting, processing and regulation. *Plant Mol. Biol.*, *38*, 311-338.

Gomez, S. M., Bil', K. Y., Aguilera, R., Nishio, J. N., Faull, K. F., & Whitelegge, J. P. (2003). Transit Peptide cleavage sites of integral thylakoid membrane proteins. *Mol. Cell. Proteomics*, *2*, 1068-1085.

Gomez, S. M., Nishio, J. N., Faull, K. F., & Whitelegge, J. P. (2002). The chloroplast grana proteome defined by intact mass measurements from liquid chromatography mass spectrometry. *Mol. Cell. Proteomics*, *1*, 46-59.

Gray, M. W., Burger, G., & Lang, B. F. (1999). Mitochondrial evolution. *Science*, *283(5407)*, 1476-1481.

Heazlewood, J. L., Howell, K. A., Whelan, J., & Millar, A. H. (2003a). Towards an analysis of the rice mitochondrial proteome. *Plant Physiol.*, *132*, 230-42.

Heazlewood, J. L., Millar, A. H., Day, D. A., & Whelan, J. (2003b). What makes a mitochondrion? *Genome Biol*, *4*, 218.

Heazlewood, J. L., Tonti-Filippini, J. S., Gout, A., Day, D. A., Whelan, J., & Millar, A. H. (2004). Experimental analysis of the *Arabidopsis* mitochondrial proteome highlights signalling and regulatory components, provides assessment of targeting prediction programs and points to plant-specific mitochondrial proteins. *Plant Cell, 16*, 241-256.

Herald, V. L., Heazlewood, J. L., Day, D. A., & Millar, A. H. (2003). Proteomic identification of divalent metal cation binding proteins in plant mitochondria. *FEBS Lett., 537*, 96-100.

Hilson, P., Small, I., & Kuiper, M. T. (2003). European consortia building integrated resources for *Arabidopsis* functional genomics. *Curr. Opin. Plant Biol., 6*, 426-429.

Holtorf, H., Guitton, M. C., & Reski, R. (2002). Plant functional genomics. *Naturwissenschaften, 89*, 235-249.

Hourton-Cabassa, C., Ambard-Bretteville, F., Moreau, F., Davy de Virville, J., Rémy, R., & Colas des Francs-Small, C. (1998). Stress induction of mitochondrial formate dehydrogenase in potato leaves. *Plant Physiol., 116*, 627-635.

Huh, W. K., Falvo, J. V., Gerke, L. C., Carroll, A. S., Howson, R. W., Weissman, J. S., et al. (2003). Global analysis of protein localization in budding yeast. *Nature, 425*, 686-691.

Humphrey-Smith, I., Colas des Francs-Small, C., Ambart-Bretteville, F., & Rémy, R. (1992). Tissue-specific variation of pea mitochondrial polypeptides detected by computerized image analysis of two-dimensional electrophoresis gels. *Electrophoresis, 13*, 168-172.

Jänsch, L., Kruft, V., Schmitz, U. K., & Braun, H. P. (1996). New insights into the composition, molecular mass and stoichiometry of the protein complexes of plant mitochondria. *Plant J., 9*, 357-368.

Jänsch, L., Kruft, V., Schmitz, U. K., & Braun, H. P. (1998). Unique composition of the preprotein translocase of the outer mitochondrial membrane from plants. *J. Biol. Chem., 273*, 17251-17257.

Jung, E., Heller, M., Sanchez, J. C., & Hochstrasser, D. F. (2000). Proteomics meets cell biology: the establishment of subcellular proteomes. *Electrophoresis, 21*, 3369-3377.

Kazama, T., & Toriyama, K. (2003). A pentatricopeptide repeat-containing gene that promotes the processing of aberrant *atp6* RNA of cytoplasmic male-sterile rice. *FEBS Lett., 544*, 99-102.

Keegstra, K., & Cline, K. (1999). Protein import and routing systems of chloroplasts. *Plant Cell, 11*, 557-570.

Kieselbach, T., Bystedt, M., Hynds, P., Robinson, C., & Schroder, W. P. (2000). A peroxidase homologue and novel plastocyanin located by proteomics to the *Arabidopsis* chloroplast thylakoid lumen. *FEBS Lett., 480*, 271-276.

Koivuniemi, A., Aro, E. M., & Andersson, B. (1995). Degradation of the D1- and D2-proteins of photosystem II in higher plants is regulated by reversible phosphorylation. *Biochemistry, 34*, 16022-16029.

Koizuka, N., Imai, R., Fujimoto, H., Hayakawa, T., Kimura, Y., Kohno-Murase, J., et al. (2003). Genetic characterization of a pentatricopeptide repeat protein gene, *orf687*, that restores fertility in the cytoplasmic male-sterile Kosena radish. *Plant J., 34*, 407-415.

Koo, A. J., & Ohlrogge, J. B. (2002). The predicted candidates of *Arabidopsis* plastid inner envelope membrane proteins and their expression profiles. *Plant Physiol., 130*, 823-836.

Kruft, V., Eubel, H., Jänsch, L., Werhahn, W., & Braun, H. P. (2001). Proteomic approach to identify novel mitochondrial proteins in *Arabidopsis*. *Plant Physiol., 127*, 1694-1710.

Kumar, A., Agarwal, S., Heyman, J. A., Matson, S., Heidtman, M., Piccirillo, S., et al. (2002). Subcellular localization of the yeast proteome. *Genes Dev., 16*, 707-719.

Kurth, J., Varotto, C., Pesaresi, P., Biehl, A., Richly, E., Salamini, F., et al. (2002). Gene-sequence-tag expression analyses of 1,800 genes related to chloroplast functions. *Planta, 215*, 101-109.

Legen, J., Kemp, S., Krause, K., Profanter, B., Herrmann, R. G., & Maier, R. M. (2002). Comparative analysis of plastid transcription profiles of entire plastid chromosomes from tobacco attributed to wild-type and PEP-deficient transcription machineries. *Plant J., 31*, 171-188.

Leister, D. (2003). Chloroplast research in the genomic age. *Trends Genet., 19(1)*, 47-56.

Leister, D., & Schneider, A. (2003). From genes to photosynthesis in *Arabidopsis thaliana*. *Int. Rev. Cytol., 228*, 31-83.

Lemaire, S. D., Collin, V., Keryer, E., Quesada, A., & Miginiac-Maslow, M. (2003). Characterization of thioredoxin y, a new type of thioredoxin identified in the genome of *Chlamydomonas reinhardtii*. *FEBS Lett., 543*, 87-92.

Li, S., Armstrong, C. M., Bertin, N., Ge, H., Milstein, S., Boxem, M., et al. (2004). A map of the interactome network of the metazoan *C. elegans*. *Science, 303*, 540-543.

Lilly, J. W., Maul, J. E., & Stern, D. B. (2002). The *Chlamydomonas reinhardtii* organellar genomes respond transcriptionally and post-transcriptionally to abiotic stimuli. *Plant Cell, 14*, 2681-2706.

Lister, R., Chew, O., Lee, M. N., Heazlewood, J. L., Clifton, R., Parker, K. L., et al. (2004). A transcriptomic and proteomic characterization of the *Arabidopsis* mitochondrial protein import apparatus and its response to mitochondrial dysfunction. *Plant Physiol., 134*, 777-789.

Maiwald, D., Dietzmann, A., Jahns, P., Pesaresi, P., Joliot, P., Joliot, A., et al. (2003). Knock-out of the genes coding for the Rieske protein and the ATP-synthase delta-subunit of *Arabidopsis*. Effects on photosynthesis, thylakoid protein composition, and nuclear chloroplast gene expression. *Plant Physiol., 133*, 191-202.

Martin, W., Rujan, T., Richly, E., Hansen, A., Cornelsen, S., Lins, T., et al. (2002). Evolutionary analysis of *Arabidopsis*, cyanobacterial, and chloroplast genomes reveals plastid phylogeny and thousands of cyanobacterial genes in the nucleus. *Proc. Natl. Acad. Sc.i USA, 99*, 12246-12251.

Millar, A. H., & Heazlewood, J. L. (2003). Genomic and proteomic analysis of mitochondrial carrier proteins in *Arabidopsis*. *Plant Physiol., 131*, 443-453.

Millar, A. H., Sweetlove, L. J., Giege, P., & Leaver, C. J. (2001). Analysis of the *Arabidopsis* mitochondrial proteome. *Plant Physiol., 127*, 1711-1727.

Miras, S., Salvi, D., Ferro, M., Grunwald, D., Garin, J., Joyard, J., et al. (2002). Non-canonical transit peptide for import into the chloroplast. *J. Biol. Chem., 277*, 47770-47778.

Nakamura, T., Furuhashi, Y., Hasegawa, K., Hashimoto, H., Watanabe, K., Obokata, J., et al. (2003). Array-based analysis on tobacco plastid transcripts: preparation of a genomic microarray containing all genes and all intergenic regions. *Plant Cell Physiol., 44*, 861-867.

Neuburger, M. (1985). Preparation of plant mitochondria, criteria for assessment of mitochondrial integrity and purity, survival *in vitro*. In R. Douce, & D. Day (Eds.), *Higher Plant Respiration* (pp. 7-24). Berlin: Springer-Verlag.

Ozawa, T., Sako, Y., Sato, M., Kitamura, T., & Umezawa, Y. (2003). A genetic approach to identifying mitochondrial proteins. *Nat. Biotechnol., 21*, 287-293.

Peeters, N., & Small, I. (2001). Dual targeting to mitochondria and chloroplasts. *Biochim. Biophys. Acta, 1541*, 54-63.

Peltier, J. B., Emanuelsson, O., Kalume, D. E., Ytterberg, J., Friso, G., Rudella, A., et al. (2002). Central functions of the lumenal and peripheral thylakoid proteome of *Arabidopsis* determined by experimentation and genome-wide prediction. *Plant Cell, 14*, 211-236.

Peltier, J. B., Friso, G., Kalume, D. E., Roepstorff, P., Nilsson, F., Adamska, I., et al. (2000). Proteomics of the chloroplast: systematic identification and targeting analysis of lumenal and peripheral thylakoid proteins. *Plant Cell, 12*, 319-341.

Peltier, J. B., Ripoll, D. R., Friso, G., Rudella, A., Cai, Y., Ytterberg, J., et al. (2003). Clp protease complexes from photosynthetic and non-photosynthetic plastids and mitochondria of plants, their predicted 3-D structures and functional implications. *J. Biol. Chem.*, 279, 4768-4781.

Peltier, J. B., Ytterberg, J., Liberles, D. A., Roepstorff, P., & van Wijk, K. J. (2001). Identification of a 350-kDa ClpP protease complex with 10 different Clp isoforms in chloroplasts of *Arabidopsis thaliana*. *J. Biol. Chem., 276*, 16318-16327.

Pesaresi, P., Varotto, C., Richly, E., Lessnick, A., Salamini, F., & Leister, D. (2003). Protein-protein and protein-function relationships in *Arabidopsis* photosystem I: cluster analysis of PSI polypeptide levels and photosynthetic parameters in PSI mutants. *J. Plant. Physiol., 160*, 17-22.

Reboul, J., Vaglio, P., Rual, J. F., Lamesch, P., Martinez, M., Armstrong, C. M., et al. (2003). *C. elegans* ORFeome version 1.1: experimental verification of the genome annotation and resource for proteome-scale protein expression. *Nat. Genet., 34*, 35-41.

Richly, E., Chinnery, P. F., & Leister, D. (2003a). Evolutionary diversification of mitochondrial proteomes: implications for human disease. *Trends Genet., 19*, 356-362.

Richly, E., Dietzmann, A., Biehl, A., Kurth, J., Laloi, C., Apel, K., et al. (2003b). Covariations in the nuclear chloroplast transcriptome reveal a regulatory master-switch. *EMBO Rep., 4*, 491-498.

Schein, A. I., Kissinger, J. C., & Ungar, L. H. (2001). Chloroplast transit peptide prediction: a peek inside the black box. *Nucleic Acids Res., 29*, E82.

Schubert, M., Petersson, U. A., Haas, B. J., Funk, C., Schroder, W. P., & Kieselbach, T. (2002). Proteome map of the chloroplast lumen of *Arabidopsis thaliana*. *J. Biol. Chem., 277*, 8354-8365.

Schwacke, R., Schneider, A., van, d. G. E., Fischer, K., Catoni, E., Desimone, M., et al. (2003). ARAMEMNON, a novel database for *Arabidopsis* integral membrane proteins. *Plant Physiol., 131*, 16-26.

Shimada, H., & Sugiura, M. (1991). Fine structural features of the chloroplast genome: comparison of the sequenced chloroplast genomes. *Nucleic Acids Res.*, *19*, 983-995.
Silva-Filho, M. C. (2003). One ticket for multiple destinations: dual targeting of proteins to distinct subcellular locations. *Curr. Opin. Plant Biol.*, *6*, 589-595.
Simpson, J. C., Wellenreuther, R., Poustka, A., Pepperkok, R., & Wiemann, S. (2000). Systematic subcellular localization of novel proteins identified by large-scale cDNA sequencing. *EMBO Rep.*, *1*, 287-292.
Small, I., Peeters, N., Legeai, F., & Lurin, C. (2004). Predotar: a tool for rapidly screening proteomes for N-terminal targeting sequences. *Proteomics*, *4*, 1581-1590.
Small, I., Wintz, H., Akashi, K., & Mireau, H. (1998). Two birds with one stone: genes that encode products targeted to two or more compartments. *Plant Mol. Biol.*, *38*, 265-277.
Small, I. D., & Peeters, N. (2000). The PPR motif - a TPR-related motif prevalent in plant organellar proteins. *Trends Biochem. Sci.*, *25*, 46-47.
Sommarin, M., Petit, P. X., & Moller, I. M. (1990). Endogenous protein phosphorylation in purified plant mitochondria. *Biochim. Biophys. Acta*, *1052*, 195-203.
Tabata, S. (2002). Impact of genomics approaches on plant genetics and physiology. *J. Plant Res.*, *115*, 271-275.
The Arabidopsis Genome Initiative. (2000). Analysis of the genome sequence of the flowering plant *Arabidopsis thaliana*. *Nature*, *408*, 796-815.
Thelen, J. J., Muszynski, M. G., David, N. R., Luethy, M. H., Elthon, T. E., Miernyk, J. A., et al. (1999). The dihydrolipoamide S-acetyltransferase subunit of the mitochondrial pyruvate dehydrogenase complex from maize contains a single lipoyl domain. *J. Biol. Chem.*, *274*, 21769-21775.
Till, B. J., Reynolds, S. H., Greene, E. A., Codomo, C. A., Enns, L. C., Johnson, J. E., et al. (2003). Large-scale discovery of induced point mutations with high-throughput TILLING. *Genome Res.*, *13*, 524-530.
Tovar-Mendez, A., Miernyk, J. A., & Randall, D. D. (2003). Regulation of pyruvate dehydrogenase complex activity in plant cells. *Eur. J. Biochem.*, *270*, 1043-1049.
van Wijk, K. J. (2000). Proteomics of the chloroplast: experimentation and prediction. *Trends Plant Sci.*, *5*, 420-425.
van Wijk, K. J. (2001). Challenges and prospects of plant proteomics. *Plant Physiol.*, *126(2)*, 501-508.
Vener, A. V., Harms, A., Sussman, M. R., & Vierstra, R. D. (2001). Mass spectrometric resolution of reversible protein phosphorylation in photosynthetic membranes of *Arabidopsis thaliana*. *J. Biol. Chem.*, *276*, 6959-6966.
Vermel, M., Guermann, B., Delage, L., Grienenberger, J. M., Maréchal-Drouard, L., & Gualberto, J. M. (2002). A family of RRM-type RNA-binding proteins specific to plant mitochondria. *Pro. Natl. Acad. Sci. USA*, *99*, 5866-5871.
Walhout, A. J., Reboul, J., Shtanko, O., Bertin, N., Vaglio, P., Ge, H., et al. (2002). Integrating interactome, phenome, and transcriptome mapping data for the *C. elegans* germline. *Curr. Biol.*, *12*, 1952-1958.
Waterhouse, P. M., & Helliwell, C. A. (2003). Exploring plant genomes by RNA-induced gene silencing. *Nat. Rev. Genet.*, *4*, 29-38.
Werhahn, W., & Braun, H. P. (2002). Biochemical dissection of the mitochondrial proteome from *Arabidopsis thaliana* by three-dimensional gel electrophoresis. *Electrophoresis*, *23*, 640-646.
Westerlund, I., von Heijne, G., & Emanuelsson, O. (2003). LumenP--a neural network predictor for protein localization in the thylakoid lumen. *Protein Sci.*, *12*, 2360-2366.
Wortman, J. R., Haas, B. J., Hannick, L. I., Smith, R. K. J., Maiti, R., Ronning, C. M., et al. (2003). Annotation of the *Arabidopsis* genome. *Plant Physiol.*, *132*, 461-468.
Yamada, K., Lim, J., Dale, J. M., Chen, H., Shinn, P., Palm, C. J., et al. (2003). Empirical analysis of transcriptional activity in the *Arabidopsis* genome. *Science*, *302*, 842-846.
Yamaguchi, K., & Subramanian, A. R. (2000). The plastid ribosomal proteins. Identification of all the proteins in the 50 S subunit of an organelle ribosome (chloroplast). *J. Biol. Chem.*, *275*, 28466-28482.
Yamaguchi, K., & Subramanian, A. R. (2003). Proteomic identification of all plastid-specific ribosomal proteins in higher plant chloroplast 30S ribosomal subunit. *Eur. J. Biochem.*, *270*, 190-205.
Yamaguchi, K., von Knoblauch, K., & Subramanian, A. R. (2000). The plastid ribosomal proteins. Identification of all the proteins in the 30 S subunit of an organelle ribosome (chloroplast). *J. Biol. Chem.*, *275*, 28455-28465.

Yu, J., Hu, S., Wang, J., Wong, G. K., Li, S., Liu, B., et al. (2002). A draft sequence of the rice genome (*Oryza sativa* L. ssp. indica). *Science*, *296*, 79-92.

Zabrouskov, V., Giacomelli, L., van Wijk, K. J., & McLafferty, F. W. (2003). A new approach for plant proteomics: characterization of chloroplast proteins of *Arabidopsis thaliana* by top down mass spectrometry. *Mol. Cell. Proteomics*, *2*, 1253-1260.

Zolla, L., Rinalducci, S., Timperio, A. M., & Huber, C. G. (2002). Proteomics of light-harvesting proteins in different plant species. Analysis and comparison by liquid chromatography-electrospray ionization mass spectrometry. Photosystem I. *Plant Physiol.*, *130*, 1938-1950.

Zolla, L., Timperio, A. M., Walcher, W., & Huber, C. G. (2003). Proteomics of light-harvesting proteins in different plant species. Analysis and comparison by liquid chromatography-electrospray ionization mass spectrometry. Photosystem II. *Plant Physiol.*, *131*, 198-214.

Section 2

Organelle Gene Expression and Signaling

Section 2

CHAPTER 8

THE TRANSCRIPTION OF PLANT ORGANELLE GENOMES

ANDREAS WEIHE

Institute of Biology, Genetics, Humboldt University Berlin, D-10115 Berlin, Germany

Abstract. Mitochondria and plastids possess their own genomes and transcription machineries. Although both organelles preserve features of eubacterial genomes, they have acquired during their evolution specialized components for gene expression, which are encoded in the nucleus. Plant mitochondria contain at least one nuclear-encoded, phage-type RNA polymerase that occurs widely throughout the eukaryotic lineage. Mitochondrial promoters transcribed by this RNA polymerase are rather variable, and only a certain subset of them can be distinguished by conserved tetra- and nonanucleotide sequence motifs in monocot and dicot plants, respectively. Plastids harbour at least two types of different RNA polymerases. One is a plastid-encoded eubacterial-type polymerase resembling the *E. coli* enzyme, composed of four core subunits (□, □, □' and □'') and one of several nuclear-encoded sigma factors. This holoenzyme is the plastid-encoded RNA polymerase (PEP). The other plastid polymerase is a nuclear-encoded single-subunit enzyme of the same type as the mitochondrial RNA polymerase, i.e. a phage-type enzyme, and has been named NEP (nuclear-encoded polymerase). The development of photosynthetically active chloroplasts requires both PEP and NEP. While PEP recognizes □70-like promoters with conserved –10 (TATAAT) and – 35 (TTGACA) elements, NEP recognizes promoter sequences unique to this enzyme. Two structurally different classes of plastid promoters engaged in transcription by nuclear-encoded RNA polymerases have been dissected in the plastid genome. Whereas most NEP promoters are characterized by a conserved YRTA motif and share homolgy with certain mitochondrial transcription initiation sites, a second class of NEP promoters, lacking the YRTA motif, contains a conserved sequence element downstream from the initiation site which is sufficient to support specific transcription. The organelle phage-type RNA polymerases are encoded by a small family of nuclear *RpoT* genes that consists of three members in dicot plants and only two *RpoT* genes in monocots. Aside from *RpoTm* encoding the mitochondrial RNA polymerase, a second RpoT enzyme (RpoTp) is imported into plastids both in monocot and dicot plants. Immunoblot analyses as well as studies on transgenic plants indicate that RpoTp indeed specifies a chloroplast-localized NEP. A third *RpoT* gene found exclusively in dicots (*RpoTmp*) has been shown to encode a protein that is dually targeted into both mitochondria and plastids. The principal specificity factors that interact with the plant RpoT polymerases to confer promoter-specific binding and transcription specificity have not been identified yet. It is tempting to assume that mitochondrial RpoT polymerase and plastid NEP promoter recognition is dependent on mitochondrial and plastid homologs of yeast and mammalian mitochondrial specificity factors.

H. Daniell and C.D. Chase (eds.), Molecular Biology and Biotechnology of Plant Organelles, 213—237. © 2004 *Springer. Printed in the Netherlands.*

1. INTRODUCTION

Plastids and mitochondria, acquired through endosymbiosis from ancestral free-living cyanobacteria and □-proteobacteria, respectively, maintained many prokaryotic features of their gene expression systems. Although most endosymbiont genes were subsequently transferred to the nucleus, a subset has been retained within each organelle. The presence of these genes necessitates a complete organellar transcription and translation machinery for gene expression. Recent progress has revealed a surprising complexity of the genetic systems of mitochondria and plastids. Thus, there is increasing evidence that certain genes were recruited to fulfill new functions in the coordinated genetic systems of the organelles. Both unexpected parallels and striking differences between mitochondria and plastids have become evident. The recent discovery of the *RpoT* gene family encoding phage-type single-subunit RNA polymerases (RNAPs) was an important step in our understanding of the transcriptional machineries of mitochondria and plastids. The RNAPs, the central components of the transcription apparatus of the two extranuclear organelles, fall into two classes: one eubacterial-type multi-subunit enzyme which is active in plastids, and a single-subunit core RNAP, related to RNAPs of bacteriophages T3 and T7. Phage-type enzymes are found in both organelles. The *E.coli*-like enzyme (PEP, plastid-encoded polymerase) is present exclusively in plastids. The phage-type polymerases are found in both plastids and mitochondria. Thus, plastids need at least two different types of RNAPs for the transcription of their genes: one eubacterial multi-subunit RNAP (PEP) and one nuclear-encoded polymerase (NEP) whose nuclear gene arose by duplication of the gene encoding the mitochondrial RNAP. Mitochondria, instead, rely solely on the activity of phage-type enzymes. Promoters recognized by PEP are of the □^{70}-type. NEP recognizes non-sigma-like promoters of two types, one of which (with a conserved YRTA core) resembles sequence elements of mitochondrial promoters.

This review is limited to an assessment of data on organellar RNA polymerases, their cofactors and promoters in higher plant mitochondria and plastids. Aspects of regulation and of posttranscriptional steps of gene expression as well as data on algae and lower plants are dealt with only if closely related to the topic of this article. For more details see the reviews by Sugita and Sugiura (1996), Hess and Börner (1999), Allison (2000) and Liere and Maliga (2001).

2. MITOCHONDRIAL TRANSCRIPTION

2.1. The mitochondrial RNA polymerase

In yeast and other fungi as well as in mammals, mitochondrial transcription is known to involve a nuclear-encoded phage-type RNA core enzyme (reviewed in Tracy and Stern, 1995; Hess and Börner, 1999). The primitive protist *Reclinomonas americana* is the only known organism that has retained the ancestral bacterial RNAP genes in its mitochondrial genome; no trace of the bacterial-type RNAP is left in the mitochondrial or nuclear genome of plants (Lang et al., 1997). In the mitochondrial genomes sequenced so far, no open reading frames were found that encode homologs of known RNAP subunits. Thus, the mitochondrial RNAP of plants, including algae, is encoded in the nucleus. Isolation of the *RpoT* gene family in a number of plant species with direct evidence in *Chenopodium album* (Weihe et al., 1997), *Arabidopsis thaliana* (Hedtke et al., 1997, 1999, 2000), maize (Young et al., 1998; Chang et al., 1999), wheat (Ikeda and Gray, 1999a), *Nicotiana tabacum* (Hedtke et al., 2002), *Nicotiana sylvestris* (Kobayyashi et al., 2001a, 2002) and barley (Emanuel et al., 2004), and demonstration of subcellular localization of the *RpoT* gene products, has led to the identification of plant mitochondrial and plastid RNAPs (RpoTm and RpoTp, respectively). A third *RpoT* gene found in dicots, but not in monocots (*RpoTmp*), has been shown to encode a protein targeted both into mitochondria and plastids (Hedtke et al., 2000, 2002; Kobayashi et al., 2001b). Interestingly, aside from dicots, two RpoT polymerases with bifunctional transitpeptides have also been identified in the moss *Physcomitrella patens* (Richter et al., 2002). Phylogenetically, the phage-type RNAP genes in the genome of plants arose by independent duplication events in the plant lineage and subsequently became principal components of the respective organelle transcription system.

Although no direct evidence has been provided yet that the RpoT polymerases function in mitochondrial transcription, we refer here to RpoTm (and RpoTmp) as mitochondrial RNAPs. In Arabidopsis, *RpoTm* and *RpoTmp* encode proteins with predicted molecular weights of 111 and 115 kDa, respectively, including the transit peptide. The cleavage sites for the transit peptides have not been determined yet. The overall degree of sequence similarity between the plant, the yeast and the animal enzymes increases from the amino- to the carboxy-terminus, consistent with the catalytic activity of the C-terminus, as shown for phage T7 RNAP (Sousa et al., 1993); all amino acid residues and known motifs reported to be essential for catalytic function (McAllister and Raskin, 1993) are conserved in the plant RpoT polymerases. According to the number of *RpoT* genes encoded in the nucleus, mitochondria of dicots would harbour two different catalytic subunits of phage-type RNAP, whereas in monocots only one mitochondrial RNAP would be found. Thus far, it is not understood why transcription of dicot mitochondrial genomes requires two polymerases, while transcription of monocot mitochondrial genomes requires only one. It is also completely unknown whether the two RNAPs have different preferences for certain promoters or sequence motifs.

2.2. Cofactors of the mitochondrial RNA polymerase

In contrast to the RNAPs of bacteriophages, none of the plant, animal or fungal RpoT polymerases are capable of transcriptional initiation from their cognate promoters *in vitro*, but rather require specificity factors. Such factors conferring promoter recognition have been identified in yeast and mammalian mitochondria as mtTFA and mtTFB. In yeast, the 43 kDa mtTFB (Schinkel et al., 1987), also named Mtf1p (Jang and Jaehning, 1991), was shown to play an essential role for initiation of transcription (Schinkel et al., 1987; Lisowsky and Michaelis, 1988; Jang and Jaehning, 1991). It does not bind promoter elements on its own, forming instead a holoenzyme with the core polymerase (Mangus et al., 1994). A second factor, mtTFA (also designated Mtf2), is a small protein of 19 kDa that binds DNA upstream of the transcription initiation site via two HMG boxes and enhances mitochondrial transcription *in vitro* (Diffley and Stillman, 1991; Fischer et al., 1992). This DNA-binding protein may also facilitate the interaction of other transacting factors (reviewed in Tracy and Stern, 1995; Hess and Börner, 1999). Only recently, genes encoding mtTFB homologs have been identified in mammals (McCulloch et al., 2002; McCulloch and Shadel, 2003; Falkenberg et al., 2003; Rantanen et al., 2003). Two yeast mtTFB homologs have been identified in humans as well as in mouse. These proteins, mtTFB1 and mtTFB2 (or TFB1M and TFB2M) confer specific transcription in collaboration with mtTFA and are homologous to N6 adenine RNA methyltransferases.

The recent determination of the crystal structure of the yeast mtTFB supports the relation of this protein to the family of the rRNA methyltransferases (Schubot et al., 2001). The dual-function nature of the human mtTFB1 was recently demonstrated by its ability to methylate a conserved stem-loop in bacterial 16S rRNA (Seidel-Rogol et al., 2003). TFB2M was shown to be at least one order of magnitude more active in promoting transcription than TFB1M (Falkenberg et al., 2003). A model has been proposed in which mtTFA demarcates mitochondrial promoter locations and mtTFB proteins bridge an interaction between the C-terminal tail of mtTFA and the mitochondrial RNAP to facilitate specific initiation of transcription (McCulloch and Shadel, 2003).

Homologs of mtTFB (and mtTFA) would be good candidates for mitochondrial specificity factors in plants. Indeed, in the completely sequenced genome of Arabidopsis several methyltransferase-like open reading frames can be found, and two of them (at5g66360 and at2g47420) encode predicted mitochondrial transit peptides. For at5g66360 mitochondrial targeting could be demonstrated experimentally, and both factors exhibit unspecific DNA-binding activity *in vitro* (Kühn and Weihe, unpublished), as shown for human mtTFB. Studies to analyze whether these plant homologs of mtTFB are capable of conferring promoter recognition *in vitro* are in progress.

Potential cofactors of the mitochondrial RNAP were also found by analyzing proteins binding to promoters in mitochondrial lysates used for *in vitro* transcription assays. In wheat, a 67 kDa protein was purified that stimulates *in vitro* transcriptional activity from a wheat *cox2* promoter (Ikeda and Gray, 1999b). In pea, a 43 kDa protein was isolated from mitochondria, based on its promoter-binding properties. This protein shows high similarity to isovaleryl CoA dehydrogenases from other organisms (Däschner et al., 2001). However, promoter activation by this protein in an *in vitro* assay remains to be demonstrated. It is intriguing to speculate that also in this case a protein with dual function, like the mtTFB, would have been acquired by the mitochondrial transcription machinery.

Recently, one of the five identified sigma factors (see section of maize, ZmSig2B) was shown to be localized both in plastids and mitochondria (Beardslee et al., 2000). The plant sigma-like proteins have been presumed or demonstrated to associate with the eubacteria-like RNAP of chloroplasts (see section *3.3*). A putative role for ZmSig2B in mitochondrial transcription was supported by its presence in a maize mitochondrial transcriptionally active extract. These findings are quite surprising, since no eubacterial-like RNAP activity has been demonstrated in plant mitochondria. It remains to be investigated whether and how (probably modified) sigma factors can play a role as components of the phage-type transcription machinery.

2.3. Mitochondrial promoters

Plant mitochondria encode about 50 - 60 genes in genomes of 200 - 2400 kb in size, suggesting that several promoters will be required to transcribe these genes (Unseld et al., 1997; Dombrowski et al., 1998). Indeed, individual transcription units have been found and numerous transcription initiation sites have been mapped in several plant species. One class of plant mitochondrial promoters has been identified which is characterized by a consensus sequence motif that is distinct between monocotyledonous and dicotyledonous plants. At least in dicots all the other promoters lack common structures or sequence motifs. It has been suggested that there might exist still hidden similarities between several transcription initiation sites of this divergent class of mitochondrial promoters (Binder and Brennicke, 2002). Whereas nothing more is known on these "non-consensus" promoters, mitochondrial consensus promoters have been intensively studied in dicot and in monocot plants. Both consensus- and non-consensus-type promoters can be found at transcriptional start sites of all types of RNA: mRNA, rRNA and tRNA. Extensive use of *in vitro* capping techniques, *in vitro* transcription assays and sequence analysis allowed the identification of transcription initiation sites and promoter motifs in *Zea mays* (Mulligan et al., 1988, 1991; Rapp and Stern, 1992; Rapp et al., 1993; Caoile and Stern, 1997), *Zea perennis* (Newton et al., 1995), *Glycine max* (Brown et al., 1991), *Pisum sativum* (Binder et al., 1995; Giese et al., 1997; Dombrowski et al., 1999; Hoffmann and Binder, 2002), *Triticum aestivum* (Covello and Gray, 1991), *Nicotiana* (Lelandais et al., 1996; Edqvist and Bergman, 2002), *Solanum tuberosum* (Binder et al., 1994; Lizama et al., 1994; Giese et al., 1997; Remacle and Marechal-Drouard, 1996), *Sorghum* (Yan and Pring, 1997) and *Oenothera berteriana* (Binder and Brennicke, 1993).

Among the mitochondrial consensus promoters of dicots, a conserved nonanucleotide (CRTAaGaGA, transcription initiation site underlined) shows considerable sequence identity between different genes as well as between different species. Mutational analyses have shown that the required promoter regions extend beyond the obviously conserved regions (Hoffmann and Binder, 2002). From a comparison of 11 unambiguously identified promoters from several plant species including pea, soybean, potato and *Oenothera* (Dombrowski et al., 1999), an extended consensus sequence was deduced (AAAATATCA**TAAG**A**GA**AG, 100% conserved positions in bold, transcription initiation site underlined) that is composed of three parts: the conserved nonanucleotide motif from –7 to +2, containing the transcription initiation site; the less well-conserved AT-box, consisting of predominantly adenosine and thymidine bases located through positions –14 to –8; and the positions +3 and +4, where mainly purines are found. In the AT-rich region certain nucleotide identities are required for efficient *in vitro* transcription initiation; an exchange from A to T between the different DNA strands in a given promoter can almost completely abolish recognition by the initiation complex. Extension of the conserved nonanucleotide motif into the transcribed region, found to be important for promoter activation, will cause the transcripts transcribed from this conserved type of promoter in dicot plants to begin with the same nucleotides, mostly GA (Hoffmann and Binder, 2002).

In monocots, the core promoter was identified as a CRTA tetranucleotide motif just upstream of the first transcribed nucleotide. Functional studies of the maize *atp1* and *cox3* promoters confirmed the substantial role of this core motif in monocots (Rapp et al., 1993; Caiole and Stern, 1997). The sites of transcription initiation seem to be less conserved than in dicots. Determination of transcript termini in *Sorghum* revealed also, as a variant of the CRTA motif, degenerated YRTA, AATA and CTTA sequences (Yan and Pring, 1997). Most consensus-type promoters in monocots have a small upstream element in common, which resides about 10 nucleotides further upstream and contains an AT-rich region of six nucleotides (Rapp et al., 1993; Tracy and Stern, 1995). Functional investigation of the central and upstream elements of the maize mitochondrial *atp1* promoter showed that full activity required 26 bp and an unaltered spacing between the upstream and the core elements (Caoile and Stern, 1997). The presence of more than one promoter and multiple transcription initiation sites has been described more frequently for monocots than for dicots. Three initiation sites were reported for the *Z. mays cox2* and *cob* genes, and six for the *atp9* gene. An example of monocot promoters that considerably deviate from the majority of other mitochondrial promoters is the *cox2* conditional promoter (cpc) from *Z. perennis.* This promoter does not function in the *Z. mays in vitro* system and requires the presence of a dominant nuclear allele (MCT) for activity *in vivo* (Newton et al., 1995).

3. PLASTID TRANSCRIPTION

3.1. The plastid-encoded eubacterial RNA polymerase (PEP)

Genes encoding subunits of a bacterial-like RNA polymerase with homology to the *E. coli* RNAP subunits □, □ and □' were identified in the chloroplast genomes of all non-parasitic plants and algae investigated, including *N. tabacum*, *Marchantia polymorpha*, *Spinacia oleracea*, *Z. mays*, *Oryza sativa*, *P. sativum*, *Euglena gracilis*, *Hordeum vulgare*, *Pinus thunbergii*, *Chlorella vulgaris*, *Odontella sinensis*, *Cyanophora paradoxa*, and *Porphyra purpurea.* (For detailed information see Howe, 1996; Hess and Börner, 1999). The *rpoA* gene encodes the 38 kDa □ subunit, *rpoB* the 120 kDa □ subunit, *rpoC1* the 85 kDa □' subunit, and *rpoC2* the 85 kDa □'' subunit. The subunit structure of the core RNAP complex is homologous to the multisubunit RNAPs of cyanobacteria. The eubacterial *rpoC* gene is split into the *rpoC1* (encoding the amino-terminal domain) and *rpoC2* (encoding the carboxy-terminal domain) genes in plastids as in cyanobacteria (Bergsland and Haselkorn, 1991). The gene organization is conserved between bacteria and plastids as well:

rpoB, *rpoC1*, and *rpoC2* are part of one operon, whereas *rpoA* belongs, together with genes for ribosomal proteins, to another operon (Sugiura, 1992). In *Physcomitrella patens*, and probably in all mosses, *rpoA* is absent from the plastid genome and transferred to the nucleus (Sugiura et al., 2003). The nuclear *rpoA* gene of *P. patens* encodes an N-terminal extension functioning as a transit peptide for targeting the protein to the plastids. Part of the *rpoC2* gene was found to be deleted in several CMS (cytoplasmic male-sterile) lines of *Sorghum*, providing a possible basis for cytoplasmic male sterility in this species (Chen et al., 1995). All functional *rpo* genes are lacking in the plastid genome of the non-green parasitic plant *Epifagus virginiana* (Morden et al., 1991).

3.2. Different forms of PEP

The bacterial-like RNAP of plastids, or PEP, exists in more than one form. Soluble and DNA-bound polymerase activity has been described, as well as rifampicin-sensitive and insensitive forms. Soluble and DNA-bound (also named TAC, transcriptionally active chromosome) enzymes have been observed in chloroplasts of higher plants and in *Euglena* (Hallick et al., 1976; Briat et al., 1979; Greenberg et al., 1984; Reis and Link, 1985). Both soluble polymerase and TAC seem to contain at least the □ and □ PEP subunits (Suck et al., 1996; Liere and Maliga, 2001). The complete PEP is thought to be an integral part of the TAC complex, which could be considered as a functional supercomplex with transcriptional, posttranscriptional and, probably, translational activity. It is not known whether the nuclear-encoded plastid polymerase (NEP) is part of the complex. Two forms of PEP, differing in their susceptibility to rifampicin, have been isolated from mustard plastids and called the "A" and the "B enzyme" (PEP-A and PEP-B) (Pfannschmidt and Link, 1994). PEP-A, found to be rifampicin-sensitive, was abundant in chloroplasts from green leaves, whereas the rifampicin-insensitive B enzyme was isolated from etioplasts. The B polymerase, identified as PEP, contained the four subunits □, □, □' and □", whereas the A enzyme consisted of at least 13 polypeptides. However, both forms of the polymerase recognize the same set of promoters, suggesting recruitment of different additional components in etioplasts and chloroplasts, respectively (Pfannschmidt and Link, 1997). Although the two forms of PEP share the same catalytic core, only the simpler B enzyme is, like the *E. coli* RNAP, rifampicin-sensitive. The more complex PEP-A is rifampicin-resistant *in vitro*, probably because the rifampicin target site in the □ subunit is inaccessible. One of the additional components, associated with PEP-A, is a heterotrimeric kinase (PTK, plastid transcription kinase), which is related to the catalytic □ subunit of nucleocytoplasmic casein kinase (CK2) and subject to regulation by the redox-reactive reagent glutathione *in vitro* (Baginsky et al., 1997, 1999; Link, 2003).

3.3. PEP specificity factors: sigma-like proteins

Beyond the plastid-encoded subunits, PEP is thought to require a □^{70}-like specificity factor (Burton et al., 1981) that mediates transcription initiation downstream of promoter sequences resembling bacterial –10/–35-promoters. As in bacteria, several sigma factors seem to be involved in the transcription of the plastid genome. In the green alga *Chlamydomonas reinhardtii* sigma-like activities were described by Surzycki and Shellenbarger (1976). For higher plants, sigma factors were analyzed first in spinach (Lerbs et al., 1983) and mustard (Bülow and Link, 1988). From mustard three different sigma-like factors (SLFs) were purified (Tiller and Link, 1993). Using antiserum against cyanobacterial sigma factors, immunological evidence for SLFs was obtained in chloroplast RNAP preparations from *Z. mays*, *O. sativa*, *C. reinhardtii* and *Cyanidium caldarum* (Troxler et al., 1994). During the last years, genes encoding SLFs were identified within the nuclear genome of numerous species and have been shown to constitute a gene family in cyanobacteria (Brahamsha and Haselkorn, 1992), red algae (Oikawa et al., 1998) and Streptophytes (Isono et al., 1997; Tanaka et al., 1997; Tan and Troxler, 1999; Fujiwara et al., 2000; Oikawa et al., 2000). Land plants encode up to six different SLFs, which have been named differently by the numerous groups investigating SLFs. To avoid confusion, a common nomenclature has been proposed (http://sfns.u-shizuoka-ken.ac.jp/pctech/sigma/proposal/proposal.html) which is based on a phylogenetic tree using the amino sequence data registered in GenBank. Based on the topology of the tree, six subfamilies of plant SLFs can be distinguished, proposed to be named Sig1 through Sig6. Members of all six subfamilies were identified in Arabidopsis (Allison, 2000; Fujiwara et al., 2000).

The C-terminal part of the sigma factors is conserved between plants and highly conserved prokaryotic sigma subregions 2.1 to 4.2. The N-terminal subregion is not conserved in plants when compared to bacteria. It was shown that the N-terminal parts of Sig1, Sig2, and Sig3 from Arabidopsis are not functional in an *E. coli* transcriptional system, suggesting plant-specific functions of the non-conserved amino terminus (Mache et al., 2002). The Sig1 gene has been suggested to encode the principal chloroplast SLF, because it is highly expressed during chloroplast biogenesis and coincides with the highest levels of PEP activity (Allison, 2000).

Consistent with the presumed interaction of plant sigma factors with the plastid-encoded RNAP, import into chloroplasts has been shown for some of these factors (Isono et al., 1997; Lahiri et al., 1999; Tan and Troxler, 1999; Fujiwara et al., 2000). One exception is the Sig2B factor from maize, which was shown to co-purify with both plastids and mitochondria (Beardslee et al., 2000) and might play, aside its function in the plastids, a role in mitochondrial transcription.

The conservation of the primary structure among eubacterial and plant sigma factors is sufficiently high to permit the reconstitution of a functional holoenzyme with the *E. coli* core RNA polymerase *in vitro* (Kestermann et al., 1998; Hakimi et al., 2000; Beardslee et al., 2000; Homann and Link, 2003) and *in vivo* (Hakimi et al., 2000). Plant SLFs show various types of regulation such as tissue- or organ-specific expression (Isono et al., 1997; Tozawa et al., 1998; Lahiri et al., 1999; Tan and Troxler, 1999; Lahiri and Allison, 2000), light-dependent expression (Liu and Troxler 1996; Isono et al., 1997; Tanaka et al., 1997; Oikawa et al., 1998; Tozawa et al., 1998; Morikawa et al. 1999), circadian clock-controlled expression (Kanamaru et al. 1999; Morikawa et al. 1999), or plastid development-dependent expression (Lahiri and Allison, 2000).

These data suggest that plastid SLFs regulate different sets of plastid genes, and therefore the nucleus exerts a level of control over chloroplast gene expression in response to various environmental and endogenous signals. This model has derived some support from the phenotypes of Arabidopsis insertion mutants lacking one sigma factor; such mutants show preferential loss of transcription from different gene subsets and, in one case, are lethal due to a developmental defect (Hanaoka et al., 2003; Yao et al., 2003).

Phosphorylation of SLFs seems to play an important role for the regulation of PEP activity (reviewed in Link, 1996). *In vitro* studies in mustard have shown that SLFs in their active (non-phosphorylated) form bind relatively loosely to the PEP core enzyme, whereas the phosphorylated SLFs bind tightly to the RNAP. In the latter case, subsequent release of the SLFs is hindered resulting in inefficient recycling of SLFs and hence reduced transcription activity (Tiller et al., 1991; Tiller and Link, 1993). A central component of this regulatory mechanism is the plastid transcription kinase PTK, proposed to be part of the signaling pathway controlling PEP activity (Baginsky et al., 1999).

3.4. The plastid nuclear-encoded RNA polymerase (NEP)

Several lines of evidence have suggested the existence of a fully nuclear-encoded transcriptional activity in plastids. The earliest reports came from data on synthesis of RNA in ribosome-deficient plastids of heat-bleached rye leaves and in white tissue of the barley mutant *albostrians* (Bünger and Feierabend, 1980; Siemenroth et al., 1981). Since plastid-encoded proteins are not synthesized in these leaves, RNA synthesis in the plastids was taken as evidence for the nuclear localization of genes for a plastid RNAP (NEP). The transcription of distinct sets of genes was demonstrated in ribosome-free plastids of *albostrians* barley (Hess et al., 1993), *iojap* maize (Han et al., 1993) and heat-bleached rye and barley seedlings (Falk et al., 1993; Hess et al., 1993). Further support for the existence of NEP came from the genome analysis of the plastids of the achlorophyllous parasitic plant *Epifagus virginiana*, possessing a reduced plastid genome of about 70 kb (Wolfe et al., 1992) that lacks the *rpoB*, *rpoC1* and *rpoC2* genes, while *rpoA* is present as a pseudogene (Morden et al., 1991). Nevertheless, transcription of a subset of plastid genes is maintained (DePamphilis and Palmer, 1990; Ems et al., 1995). A third line of evidence has emerged from biochemical studies on chloroplast RNAP preparations from spinach. Using an antibody-linked polymerase assay, Lerbs et al. (1985) detected chloroplast RNAP activity among polypeptides translated from purified polyA+-mRNA, suggesting a nuclear origin of the plastid RNAP. Further purification of plastid RNAP identified a single 110 kDa protein with enhanced activity on supercoiled versus linear templates. The purified protein also recognized the T7 promoter. These findings led to the assumption that a nuclear-encoded phage-type RNAP exists in plastids (Lerbs-Mache, 1993). The existence of NEP has also been established by targeted deletion of *rpo* genes from the tobacco plastid genome (Allison et al., 1996; Serino and Maliga, 1998; De Santis-Maciossek et al., 1999; Krause et al., 2000). The obtained transgenic plants, although lacking PEP, still transcribe a subset of plastid genes. Finally, the pattern of plastid NEP drug resistance was shown to be different from bacterial-type RNA polymerases and analogous to the phage T7 RNA polymerase, again pointing to phage-type RNAP as candidates for NEP activity (Liere and Maliga, 1999).

In both mocot and dicot plants, one of the *RpoT* genes (see section *2.1*), *RpoTp*, encodes a phage-type RNAP that is imported into the plastids (Hedtke *et al.*, 1997; Chang *et al.*, 1999; Ikeda and Gray, 1999a; Hedtke et al., 2002; Kobayashi et al., 2002). Using an antibody raised against the C-terminus of RpoTp in an antibody-linked polymerase assay, Chang et al. (1999) showed that the maize *RpoTp* indeed specifies a chloroplast-localized enzyme. Direct evidence that RpoTp is a catalytic subunit of NEP comes from the analysis of plastid transcription in mutant tobacco plants overexpressing *N. sylvestris* and *A. thaliana* RpoTp (Liere et al., 2004). Primer extension assays revealed that transcription from typical type-I NEP promoters as e.g. P*atpB*-289 is enhanced in comparison to the PEP promoters. Support for the identity of RpoTp and NEP comes also from the analysis of transcription profiles in a barley leaf gradient, in which the pattern of mRNA accumulation of NEP-transcribed plastid genes closely followed that of *RpoTp* transcripts (Emanuel et al., 2004).

Another *RpoT* gene found exclusively in dicots (*RpoTmp*) has been shown to encode a gene product dually targeted into both mitochondria and plastids (Hedtke et al., 2000, 2002; Kobayashi et al., 2001b). It is not known why dicots have two plastid targeted phage-type RNA polymerases and if they fulfill distinct roles in plastid gene expression. Only recently, analysis of an Arabidopsis activation-tagged T-DNA insertion line has provided evidence for a specific role of RpoTmp in plastid transcription. The mutation in the *RpoTmp* gene affected the light-induced accumulation of several plastid mRNAs and proteins and resulted in a lower photosynthetic efficiency. In contrast to these alterations in plastid gene expression, no major effect of the *RpoTmp* mutation on the accumulation of examined mitochondrial gene transcripts and proteins was observed (Baba et al., 2004).

Analysis of the spinach *rrn16* PC promoter (see section *3.6.3.*) suggests that there might exist an additional nuclear-encoded RNAP in plastids that is different from RpoTp or RpoTmp. It was shown that an RpoT antibody from maize repressed a biochemically purified NEP transcription activity, named NEP-1, from the unique *rrn16* PC promoter of spinach *in vitro* (Bligny et al., 2000). Further biochemical analyses revealed a second NEP transcription activity, NEP-2, as well recognizing the PC promoter *in vitro*. However, the maize RpoTp antibody did not inhibit NEP-2 transcription. Although both enzymes seemed to recognize the T7 promoter *in vitro* it remained uncertain as to whether NEP-1 or NEP-2 are phage-type RNA polymerases or even made of the same core enzyme (i.e. RpoTp or RpoTmp).

3.5. NEP cofactors

While the genuine phage-encoded RNAPs do not require any cofactors for either promoter recognition or processivity, one should expect that the plastid phage-type NEP, like the mitochondrial RNAP, does need such factors. As to the NEP subunit composition, however, one can only speculate based on information about related mitochondrial RNAP cofactors. Since the genes for the catalytic subunits of the plastid and mitochondrial phage-type RNAPs are closely related, one can expect that the organellar specificity factors are encoded by related genes as well. Candidate genes are those encoding homologs of the mitochondrial specificity factor mtTFB and RNA methyltransferases. Several methyltransferase-like proteins with predicted plastid transit peptides were identified in Arabidopsis, and two of them (at1g01860 and at1g65760) have been shown to be imported into plastids (Kuhla and Liere, unpublished). Their possible involvement with NEP and promoter recognition remains to be determined.

Another type of factor that may play a role in NEP transcription is the plastid ribosomal protein RPL4, encoded by the nuclear *Rpl4* gene. RPL4 was found to co-purify with a T7-like transcription complex in spinach (Trifa et al., 1998). In prokaryotes, the ribosomal protein L4 fulfills extraribosomal functions in transcriptional regulation (Zengel et al., 1990). The spinach and Arabidobsis *Rpl4* genes exhibit a remarkable 3' extension that resembles highly acidic C-terminal ends of some transcription factors. Like the case of methyltransferase-related proteins, a function of RPL4 in plastid transcription has yet to be demonstrated.

In the case of the NEP-2 activity reported to transcribe the spinach *rrn* operon, CDF2 was identified as a transcription factor that had previously been shown to recruit PEP to the *rrn* promoter to repress transcription. rDNA transcriptional regulation, at least in spinach, seems to be based on the interaction of CDF2 with two different transcriptional systems (Bligny et al., 2002).

3.6. Plastid promoters

3.6.1. PEP promoters

Coinciding with the existence of a bacterial-like RNAP in plastids, –10 (TAtaaT) and –35 (TTGaca) □70-type promoters were found to precede most of the transcription units in these organelles (Gruissem and Zurawski, 1985; Strittmatter et al., 1985; Mullet, 1988; Gruissem and Tonkyn, 1993). This promoter type, found in all higher plants and algae, has also been named consensus type (CT) promoter (Kapoor et al., 1997). CT promoters function correctly in *E. coli* as shown for the *Z. mays rbcL* (Gatenby et al., 1982) and *atpB* (Bradley and Gatenby, 1985) as well as the *S. oleracea psbA* promoter (Boyer and Mullet, 1986).

Several sigma-type promoters contain additional cis-elements of regulatory significance. Detailed *in vitro* characterization of the *psbA* promoter in mustard (Eisermann et al., 1990) revealed a TATATA element between the –10 and –35 hexamers resembling the TATA box of nuclear genes transcribed by RNAP II (Link, 1994). The TATATA element and the –10 region were sufficient to obtain basic transcription levels in plastid extracts from both dark- and light-grown plants. However, the –35 region was indispensable for enhanced transcription rates, characteristic of chloroplasts from light-grown plants (Eisermann et al., 1990). The barley *psbA* promoter also has the TATA-box element that, as in mustard, is necessary for full promoter activity. But unlike in mustard, the –35 region of barley is absolutely required for transcription *in vitro* (Kim et al., 1999a). In another monocotyledonous plant, wheat, the *psbA* promoter possesses, in addition to the TATA-box, a TGn motif upstream of the –10 region (extended –10 sequence, Bown et al., 1997; Satoh et al., 1999). The TATA element in wheat does not seem to be important. Constitutive transcription was shown to be dependent from the –10 and –35 regions when a PEP prepared from the base of the leaf (young chloroplasts) was used in the *in vitro* assay, whereas transcription by PEP from the leaf tip (mature chloroplasts) was dependent only on the –10 region. It was proposed that the two PEP types use different sigma factors and that the extended –10 region may play a role in recognition by the PEP holoenzyme in mature chloroplasts (Satoh et al., 1999).

Sequence motifs, upstream of the –35 region, that have regulatory significance were identified in the *rbcL* and the *psbD* promoters. The *rbcL* promoter has been mapped in tobacco (Shinozaki and Sugiura, 1982), maize, spinach, pea (Mullet et al., 1985), barley (Reinbothe et al., 1993) and Arabidopsis (Isono et al., 1997). It was shown that both the –10 and –35 sequences and their spacing are important for the *rbcL* promoter strength (Gruissem and Zurawski, 1985; Hanley-Bowdoin et al., 1985). Using chimeric *uidA* genes, it was confirmed that the *rbcL* core promoter is sufficient to support wild-type rates of transcription (Shiina et al., 1998). Sequences upstream of the core promoter to position –102 were proposed to provide a binding site for the chloroplast DNA-binding factor 1 (CDF1) in maize, and segments of the CDF1 binding site are conserved between maize, spinach, pea and tobacco (Lam et al., 1988). The rates of *rbcL* transcription were significantly reduced in dark-adapted plants, resulting in lower steady state levels of *uidA* mRNA in the dark, unless the chimeric constructs contained the *rbcL* 5' UTR. Thus, accumulation of *rbcL* mRNA in a light-independent manner seems to be due to stabilization of the mRNA by its 5' UTR in the dark (Shiina et al., 1998).

The *psbD-psbC* genes, unlike most photosynthesis genes, are highly expressed in mature chloroplasts due to activation of the blue-light responsive promoter (BLRP; Sexton et al., 1990). The *psbD* promoter, studied in a variety of species (Christopher et al., 1992), including barley (Kim and Mullet, 1995; Kim et al., 1999a), tobacco (Allison and Maliga, 1995), wheat (Wada et al., 1994), rice (To et al., 1996) and Arabidopsis (Hoffer and Christopher, 1997), has relatively closely spaced (15 nucleotides), poorly conserved –10/–35 regions. In a transgenic approach (Allison and Maliga, 1995) the architecture of the *psbD* BLRP was studied *in vivo*. The 107 bp region studied comprised the –10/–35 regions, the AAG-box and part of the PGT-box. These sequences directly upstream of the promoter core were found to be responsible for light-induced 150-fold transcript accumulation. In rice, wheat and barley the transcription from the *psbD* promoter depends solely on the –10 element (To et al., 1996; Satoh et al., 1997; Kim et al., 1999a). The nuclear-encoded AAG-binding factor (AGF) was found to bind to the AAG-box of the barley promoter *in vitro* (Kim and Mullet, 1995). The role of the PGT-binding factor (PGTF), having activity regulated by an ADP-dependent kinase, was also investigated (Kim et al., 1999b). Based on these *in vitro* studies, it is assumed that binding of PEP is mediated by AGF constitutively binding to the AAG-box. Light-induced transcription is initiated by PGTF binding to the PGT-box. The lack of transcription in the dark is thought to be caused by phosphorylation of PGTF resulting in detachment from the PGT element. In Arabidopsis, the DET1 gene product seems to regulate the activity of these DNA-binding complexes via a photosensoric pathway (Christopher and Hoffer, 1998).

3.6.2. NEP promoters

Direct evidence for the existence of NEP promoters was provided by experimental systems in which the transcriptional activity of PEP had been eliminated. These included the ribosome-deficient plastids of *albostrians* barley and *iojap* maize mutants (Hübschmann and Börner, 1998; Silhavy and Maliga, 1998a) and tobacco □rpo plants lacking the chloroplast-encoded RNAP (Allison et al., 1996; Hajdukiewicz et al., 1997). Other systems in which NEP promoters have been characterized are a photosynthetically inactive tobacco suspension culture and an embryogenic culture of rice (Kapoor et al. 1997; Silhavy and Maliga, 1998b).

While photosynthesis genes and operons have PEP promoters, most house-keeping genes possess promoters for both PEP and NEP. Only a few plastid genes are transcribed exclusively from NEP promoters: the *rpoB* operon encoding three subunits of PEP, *clpP* encoding the proteolytic subunit of the Clp protease in monocots, and *accD* encoding a subunit of the acetyl-CoA carboxylase in dicots (Silhavy and Maliga, 1998a; Hajdukiewicz et al., 1997). Promoters utilized by NEP do not belong to the eubacterial –10/–35 □70 type and were therefore occasionally named NC (nonconsensus) type II promoters (Kapoor et al., 1997).

Primer extension and *in vitro* capping of plastid transcripts led to the identification of NEP transcription initiation sites. Several sequence motifs have been identified around the transcription start sites (Hajdukiewicz et al., 1997; Hübschmann and Börner, 1998; Silhavy and Maliga, 1998a, b) and used to distinguish three classes of NEP promoters (Weihe and Börner, 1999). Most strikingly, an AT-rich region of about 15 nucleotides that is conserved between monocots and dicots can be dissected immediately upstream of the transcription initiation site. This sequence, –15 to +5 upstream of the transcription initiation site, is characterized by a highly conserved YRT core motif and flanked by less conserved AT-rich sequences. This region, identified in all but one of the NEP promoters investigated so far (see below), was designated as YRT-box, and NEP promoters possessing this box were designated class I promoters. A second characteristic element of about 10 nucleotides, 18 to 20 nucleotides upstream of the YRT-box, was found in a subset of class I promoters and named the GAA-box. Promoters carrying the YRT-box and the GAA-box were classed as Ib NEP promoters. In contrast, class Ia promoters possess the YRT-box, but lack the GAA-box.

A precise dissection of promoter architecture through *in vitro* deletion and point mutagenesis has been documented for the *PrpoB* and the *PaccD* (Liere and Maliga, 1999) and *PatpB* (Kapoor and Sugiura, 1999) promoters in tobacco. The core promoter element of class Ia (*rpoB* and *accD*) in these *in vitro* studies was identified as being identical to the characteristic YRT-motif. It was further demonstrated that the NEP activity in the tobacco plastid extract faithfully recognized the maize *rpoB* promoter containing the same functionally important nucleotides as revealed for the tobacco *rpoB* promoter. This is an indication that the NEP-dependent transcription system is conserved between monocots and dicots (Liere and Maliga, 1999). In similar *in vitro* transcription experiments, a functional role of the GAA-box in a class Ib promoter, PatpB–289, could be demonstrated (Kapoor and Sugiura, 1999).

The existence of a further class of NEP promoters (class II) that lack the YRT-motif and differ completely in sequence and organization from the class I promoters is represented so far by a single example. Using transplastomic plants expressing a *uidA* reporter gene from one of the two NEP promoters of the *clpP* gene in tobacco plastids, it was demonstrated that a sequence downstream from the initiation site (–5 to +25) was sufficient to support specific transcription initiation (Sriraman et al., 1998a). The critical motif was found to be conserved among dicots, monocots, conifers and liverworts, an indication that NEP-dependent transcription appeared early in evolution. Although the motif is also present in maize, the sequence is not used as a promoter, suggesting that an auxiliary factor necessary for transcription initiation is absent in the monocot plant. It is tempting to assume that the two promoter classes are recognized by different RpoT-type enzymes and/or specificity factors, but no experimental data on promoter usage or transcription factors involved in specific promoter recognition are available yet.

3.6.3. The rrn16 promoter and other multiple promoters

The DNA region preceding the *rrn16* gene is highly conserved between different plant species. However, promoters and transcription initiation sites differ substantially between the species. In tobacco, two promoters (P1 and P2) were mapped and shown to be active *in vivo* (Vera and Sugiura, 1995; Allison et al., 1996). The tobacco P1 promoter at –116 is of the –10/–30 σ70 type, whereas P2 at –62/–64 is a NEP promoter. Transcripts from both P1 and P2 promoters were detected in RNA from green chloroplasts. However, the steady-state level of transcripts from the tobacco P2 promoter is enhanced in proplastids of heterotrophic cells, and it is the sole *rrn16* promoter active in plants deficient in one of the plastid *rpo* genes (Δrpo plants), i.e. plants lacking PEP (Allison et al., 1996; Serino and Maliga, 1998). In a recent study, sequence elements important for the tobacco *rrnP1* activity were identified (Suzuki et al., 2004). *In vivo* dissection of the *rrn* operon promoter indicates that sequences upstream of the conserved –35 box are important for promoter function. A more detailed *in vitro* dissection identified RUA (rRNA upstream activator), a conserved 6-bp sequence directly upstream of the –35 core promoter element responsible for enhanced transcription from the *PrrnP1* promoter core. Furthermore, it was shown that the –35 hexamer, but not the –10 element, is crucial for promoter activity.

In pea and maize, no NEP promoters directly upstream of the *rrn* operon have been found (Sun et al., 1989; Silhavy and Maliga, 1998a). In spinach, two different transcription initiation sites were mapped *in vitro* using the *E. coli* RNAP (Lescure et al., 1985). Although also named P1 and P2, they are not equivalent to P1 and P2 in tobacco. Analysis of primary transcripts revealed, however, that these sequences are not utilized as promoters *in vivo*. Instead, transcription is initiated at a third site, called PC, located between P1 and P2 (Baeza et al., 1991; Iratni et al., 1997). The PC promoter has no similarity to the –10/–35 □70 type promoters, nor to the *rrn16* NEP promoter (P2) in tobacco or other NEP promoters. The spinach PC promoter seems to be recognized by the mustard PEP *in vitro* (Pfannschmidt and Link, 1997). The promoter is active in chloroplasts, but not in roots, where transcription starts from a more upstream promoter preceding the *trnV* gene that in this case is cotranscribed with the *rrn16* gene (Iratni et al., 1997). A protein factor, CDF2, was shown to be involved with the activation of PC and is supposed to mediate the silencing of P1 and P2 (Iratni et al., 1994). In Arabidopsis, active transcription was found from both the PC promoter (identical to spinach) and the P1 promoter (identical to tobacco). Studies of promoters in heterologous plastids indicate the existence of transcription factors that are specific for the respective promoter and species (Sriraman et al., 1998b).

The *rrn* operon is not the only case of multiple promoters. It seems rather common that plastid genes and operons are transcribed from two or even more promoters. Thus, the *psbK-psbI-psbD-psbC* operon of *Hordeum vulgare* is transcribed from at least three different promoters (Sexton et al., 1990; Christopher et al., 1992), the tobacco *clpP* gene has four transcription initiation sites, at –35, –95, –173 and –511 relative to the coding region (Hajdukiewicz et al., 1997; Sriraman et al., 1998a). The *rpl32* gene possesses two initiation sites at –1101 and –1030 in *N. tabacum* (Vera et al., 1992, 1996), and the *atpB* gene of *Z. mays* is transcribed from –298 and –601 (Silhavy and Maliga, 1998a).

3.6.4. Exceptional promoters: tRNA transcription

While most plastid tRNA genes are transcribed by PEP from upstream □70-type promoters, a few tRNA genes do not seem to require 5' upstream promoter elements for their expression. Instead, internal promoters were suggested for the spinach tRNA genes *trnR1* and *trnS1* as well as *trnS* and *trnQ* (Gruissem et al., 1986; Neuhaus et al., 1989, 1990) and *trnH* (Nickelsen and Link, 1990) from mustard. The best characterized case, the spinach *trnS* gene (Wu et al., 1997) contains internal sequence elements that resemble blocks A and B of RNAP III-transcribed nuclear tRNA genes (Galli et al., 1981). The *in vitro* transcription start site was mapped 12 nucleotides upstream of the mature tRNA coding region. The coding region without any upstream sequences, but containing blocks A and B, yielded a basal level (8%) of transcription. To restore wild type promoter activity, an AT-rich region between –31 and –11 upstream of the transcription start site was necessary. The architecture of this promoter resembles that of yeast tRNA genes (Geiduschek et al., 1995) and suggests that the plastid tRNA genes may be transcribed by an RNAP III-like enzyme. However, no such enzyme activity has been described so far in plastids. On the other hand, it cannot be excluded that these tRNA genes are transcribed by NEP (or PEP) associated with specialized transcription factors.

3.7. The specific roles of PEP and NEP

The observation that plastid genes are transcribed by at least two different types of RNAPs, the eubacterial-type PEP and the phage type NEP, raises the question as to why plastids need these different polymerases and what is their specific function in plastid gene expression.

When, in evolution, the nuclear-encoded NEP was acquired by the plastid gene expression system, certain genes driven by PEP promoters became transcribed by NEP. This change can be considered as a critical step of the nucleus taking control of the transcription of plastid genes, thereby integrating the photosynthetic organelle into the developmental network of the multicellular plant (Liere and Maliga, 2001). According to the distribution of NEP and PEP promoters, plastid genes could be classified into PEP-transcribed genes (mostly photosynthesis genes), NEP-transcribed genes (mostly housekeeping genes, including the genes for the core subunits of PEP) and genes that have both PEP and NEP promoters (Hajduiewicz et al., 1997; Hübschmann and Börner, 1998; Silhavy and Maliga, 1998a; Sriraman et al., 1998a; Hess and Börner, 1999; Liere and Maliga, 1999). Based on these data, it has been proposed that NEP activity is, in first line, essential in non-green tissues to initiate the activity of PEP by transcribing the plastid *rpo* genes, whereas PEP plays a major role in green tissue, in particular in the biogenesis and maintenance of the photosynthetic apparatus. However, the situation appears to be more complex. NEP is present in all plastid types, as there are essential genes expressed from a single NEP promoter, e.g. the class I-NEP promoter of *clpP* in monocots, which is expressed also in mature chloroplasts. On the other hand, the *rrn* operon in monocots is transcribed only from a PEP promoter (Silhavy and Maliga, 1998a,b). Transplastomic tobacco plants that lack PEP activity because of targeted mutation of one of the relevant *rpo* genes are able to transcribe all plastid genes including those that are reported to have only PEP promoters (Krause et al., 2000; Legen et al., 2002). All these data argue against a hierarchical role of the two transcription systems and rather suggest a parallel function of NEP and PEP.

4. OUTLOOK

Important steps towards unravelling the amazingly complex transcriptional machinery of mitochondria and plastids have been made, aside the dissection of promoters, through the identification and characterization of the main components of organelle gene expression – the different RNA polymerases involved in mitochondrial and plastid transcription and some of their cofactors, as the family of sigma proteins functioning in plastid transcription. An upcoming challenge is to assign specific functions to the different RNAPs, the sigma factors and putative cofactors of both the mitochondrial and plastid phage-type RNAPs, which have still to be unambiguously identified. The identification of further sites of regulation needs to be considerably extended to fully understand organelle gene expression. Studies of the components of the transcriptional apparatus will have to be complemented by studies of the regulation of mitochondrial and chloroplast gene activities (as well as posttranscriptional regulatory processes) in different tissues, developmental stages and physiological regimes. This will also shed light on the cross talk between the coordinated genetic systems of the organelles within the regulatory grid of the plant cell.

Complementation studies, including creation of chloroplast vectors that can function in green and non-green tissues, recently helped to achieve plastid transformation in cotton and carrot (Kumar et al., 2004 a,b). Both transgenes (nptII, aphA6) were transcribed by the full-length plastid Prrn promoter containing binding sites for both nuclear-encoded and plastid-encoded RNA polymerase and was expected to function both in proplastids and mature chloroplasts. Translation of the aphA-6 gene was regulated by the T7 gene 10 5'UTR capable of efficient translation in the dark, in proplastids present in non-green tissues. The T7 gene 10 5' UTR and rps16 3' UTR facilitated 74.8% transgene (*badh*) expression in non-green edible parts (carrots) containing chromoplasts (grown under the ground in the dark) and 48% in proplastids (in cultured cells) when compared to chloroplasts (100%) in leaves and conferred salt tolerance, as high as in halophytes (Kumar et al., 2004). The nptII gene in the cotton plastid transformation vector was driven by the psbA 5' and psbA 3' UTRs, which were shown to be responsible for light regulated expression of transgenes integrated into the plastid genome (Dhingra et al., 2004; Devine and Daniell, 2004; Daniell et al., 2004; Fernandez San Millan et al., 2003). Therefore, a combination of both aphA-6 and aphA-2 genes, driven by regulatory signals in the light and in the dark, in both proplastids and chloroplasts, provided continuous protection for transformed plastids/chloroplasts from the selectable agent. These approaches helped to achieve a high frequency plastid transformation of a recalcitrant, but very important crop. This provides one example for the power of knowledge obtained from basic studies to achieve major goals in biotechnology. For more such examples, see chapter 16 of this volume.

5. REFERENCES

Allison, L.A. (2000). The role of sigma factors in plastid transcription. *Biochimie, 82*, 537-548.

Allison, L.A., & Maliga, P. (1995). Light-responsive and transcription-enhancing elements regulate the plastid *psbD* core promoter. *EMBO J., 14*, 3721-3730.

Allison, L.A., Simon, L.D., & Maliga, P. (1996). Deletion of *rpoB* reveals a second distinct transcription system in plastids of higher plants. *EMBO J., 15*, 2802-2809.

Baba, K., Schmidt, J., Espinosa-Ruiz, A., Villarejo, A., Shiina, T., Gardestrom, P., Sane, A.P., & Bhalerao, R.P. (2004). Organellar gene transcription and early seedling development are affected in the rpoT;2 mutant of Arabidopsis. *Plant J., 38*, 38-48.

Baeza, L., Bertrand, A., Mache, R., & Lerbs-Mache, S. (1991). Characterization of a protein-binding sequence in the promoter region of the 16S rRNA gene of thze spinach chloroplast genome. *Nucleic Acids Res., 19*, 3577-3581.

Baginski, S., Tiller, K., Pfannschmidt, T., & Link, G. (1999). PTK, the chloroplast RNA polymerase-associated protein kinase from mustard (*Sinapis alba* L.) mediates redox control of plastid *in vitro* transcription. *Plant Mol. Biol., 39*, 1013-1023.

Baginsky, S., Tiller, K., & Link, G. (1997). Transcription factor phosphorylation by a protein kinase associated with chloroplast RNA polymerase from mustard (*Sinapis alba*). *Plant Mol. Biol., 34*, 181-189.

Beardslee, T.A., Roy-Chowdhury, S., Jaiswal, P., Buhot, L., Lerbs-Mache, S., Stern, D.B., & Allison, L.A. (2002). A nucleus-*encoded maize protein with sigma factor activity accumulates in mitochondria and chloroplasts. Plant J., 31*, 1-13.

Bergsland, K.J., & Haselkorn R. (1991). Evolutionary relationships among eubacteria, cyanobacteria, and chloroplasts: Evidence from the *rpoC1* gene of Anabaena sp. strain PCC7120. *J. Bacteriol., 173*, 3446-3455.

Binder, S., & Brennicke, A. (1993). Transcription initiation sites in mitochondria of *Oenothera berteriana*. *J. Biol. Chem., 268*, 7849-7855.

Binder, S., & Brennicke, A. (2002). Gene expression in plant mitochondria: transcriptional and post-transcriptional control. *Phil. Trans. R. Soc. Lond. B, 358*, 181-189.

Binder, S., Hatzack, F., & Brennicke, A. (1995). A novel pea mitochondrial *in vitro* transcription system recognizes homologous and heterologous mRNA and tRNA promoters. *J. Biol. Chem., 270*, 22182-22189.

Bligny, M., Courtois, F., Thaminy, S., Chang, C.C., Lagrange, T., Baruah-Wolff, J., Stern, D.B., & Lerbs-Mache, S. (2000) Regulation of plastid rDNA transcription by interaction of CDF2 with two different RNA polymerases. *EMBO J., 19*, 1851-1860.

Bown, J., Barne, K., Minchin, S., & Busby, S. (1997). Extended –10 promoters. *Nucleic Acids Mol. Biol., 11*, 41-52.

Boyer, S.K., & Mullet, J.E. (1988). Sequence and transcript map of barley chloroplast *psbA* gene. *Nucleic Acids Res., 16*, 8184.

Bradley, D., & Gatenby, A.A. (1985). Mutational analysis of the maize chloroplast ATPase-beta subunit gene promoter: the isolation of promoter mutants in *E. coli* and their characterization in a chloroplast *in vitro* transcription system. *EMBO J., 4*, 3641-3648.

Brahamsha, B., & Haselkorn, R. (1992). Identification of multiple RNA polymerase sigma factors in the cyanobacterium *Anabaena* sp. Strain PCC 7120: cloning, expression, and inactivation of the *sigB* and *sigC* genes. *J. Bacteriol., 174*, 7273-7282.

Briat, J.F., Laulhere, J.P., & Mache, R. (1979). Transcription activity of a DNA-protein complex isolated from spinach plastids. *Eur. J. Biochem., 98*, 285-292.

Brown, G.G., Auchincloss, A.H., Covello, P.S., Gray, M.W., Menassa, R., & Singh, M. (1991). Characterization of transcription initiation sites on the soybean mitochondrial genome allows identification of a transcription- associated sequence motif. *Mol. Gen. Genet., 228*, 345-355.

Bülow, S., & Link, G. (1988). Sigma-like activity from mustard (*Sinapis alba* L.) chloroplasts conferring DNA-binding and transcription specificity to *E. coli* core RNA polymerase. *Plant Mol. Biol., 10*, 349-357.

Bünger, W., & Feierabend, J. (1980). Capacity for RNA synthesis in 70S ribosome-deficient plastids of heat-bleached rye leaves. *Planta, 149*, 163-169.

Burton, Z., Burgess, R.R., Lyn, J., Moore, S., Holder, S., & Gross, C. (1981). The nucleotide sequence of the cloned *rpoD* gene for the RNA polymerase sigma subunit for *E. coli* K12. *Nucleic Acids Res., 9*, 2889-2903.

Caoile, A.G.F.S., & Stern, D.B. (1997). A conserved core element is functionally important for maize mitochondrial promoter activity *in vitro*. *Nucleic Acids Res., 25*, 4055-4060.

Chang, C.C., Sheen, J., Bligny, M., Niwa, Y., Lerbs-Mache, S., & Stern, D.B. (1999). Functional analysis of two maize cDNAs encoding T7-like RNA polymerases. *Plant Cell, 11*, 911-926.

Chen, Z., Schertz, K.F., Mullet, J.E., DuBell, A., & Hart, G.E. (1995). Characterization and expression of *rpoC2* in CMS and fertile lines of sorghum. *Plant. Mol. Biol., 28*, 799-809.

Christopher, D.A., & Hoffer, D.H. (1998). DETI represses a chloroplast blue light responsive promoter in a developmental and tissue-specific manner in *Arabidopsis thaliana*. *Plant J., 14*, 1-11.

Christopher, D.A., Kim, M., & Mullet, J.E. (1992). A novel light-regulated promoter is conserved in cereal and dicot chloroplasts. *Plant Cell, 4*, 785-798.

Covello, P.S., & Gray, M.W. (1991). Sequence analysis of wheat mitochondrial transcripts capped *in vitro*: Definitive identification of transcription initiation sites. *Curr. Genet., 20*, 245-252.

Daniell, D., Carmona-Sanchez, O., & Burns, B. B. (2004). Chloroplast derived antibodies, biopharmaceuticals and edible vaccines. In R. Fischer & S. Schillberg (Eds.) *Molecular Farming* (pp.113-133). Weinheim: WILEY-VCH Verlag.

Däschner, K., Couée, I, & Binder, S. (2001). The nitochobdrial isovaleryl-coenzyme A dehydrogenease form *Arabidopsis thaliana* oxidizes intermediates of leucine and valine catabolism. *Plant Physiol., 126*, 601-612.

De Santis-Maciossek, G., Kofer, W., Bock, A., Schoch, S., Maier, R.M., Wanner, G., Rüdiger, W., Koop, H.U., & Herrmann, R.G. (1999). Targeted disruption of the plastid RNA polymerase genes *rpoA*, *B* and *C1*: molecular biology, biochemistry and ultrastructure. *Plant J., 18*, 477-489.

DePamphilis, C.W., & Palmer, J.D. (1990). Loss of photosynthetic and chlororespiratory genes from the plastid genome of a parasitic flowering plant. *Nature, 348*, 337-339.

Devine, A. L., & Daniell, H. (2004). Chloroplast genetic engineering. In S. Moller (Ed.), *Plastids* (pp. 283-320). United Kingdom: Blackwell Publishers.

Dhingra, A., Portis, A.R. & Daniell, H. (2004). Enhanced translation of a chloroplast expressed *RbcS* gene restores SSU levels and photosynthesis in nuclear antisense *RbcS* plants. *Proc. Natl. Acad. Sci.*, U.S.A.101: 6315-6320.

Diffley, J.F.X., & Stillmann, B. (1991). A close relative of the nuclear chromosomal high-mobility group protein HMG1 in yeast mitochondria. *J. Biol. Chem., 267*, 3368-3374.

Dombrowski, S., Hoffmann, M., Guha, C., & Binder, S. (1999). Continuous primary sequence requirements in the 18-nucleotide promoter of dicot plant mitochondria. *J. Biol. Chem., 274*, 10094-10099.

Dombrowski, S., Hoffmann, M., Kuhn, J., Brennicke, A., & Binder, S. (1998). On mitochondrial promoters in *Arabidopsis thaliana* and other flowering plants. In M. Møller, P. Gardestrom , K. Glimelius and E. Glaser (Eds.), *Plant mitochondria: from gene to function.* (pp. 165-170). Leiden: Backhuys Publishers.

Eisermann, A., Tiller, K., & Link, G. (1990). *In vitro* transcription and DNA binding characteristics of chloroplast & etioplast extracts from mustard (*Sinapis alba*) indicate differential usage of the *psbA* promoter. *EMBO J., 9*, 3981-3987.

Emanuel, C., Weihe, A., Graner, A., Hess, W.R., & Börner, T. (2004). Chloroplast development affects expression of phage-type RNA polymerases in barley leaves. *Plant J., 38*, 460-472.

Ems, S., Morden, C.W., Dixon, C., Wolfe, K.H., DePamphilis, C.W., & Palmer, J.D. (1995). Transcription, splicing and editing of plastid RNAs in the nonphotosynthetic plant *Epifagus virginiana*. *Plant Mol. Biol., 29*, 721-733.

Falk, J., Schmidt, A., & Krupinska, K. (1993) Characterization of plastid DNA transcription in ribosome deficient plastids of heat-bleached barley leaves. *J. Plant Physiol., 141*, 176-181.

Falkenberg, M., Gaspari, M., Rantanen, A., Trifunovic, A., Larsson, N. G., & Gustafsson (2003). Mitochondrial transcription factors B1 and B2 activate transcription of human mtDNA. *Nat. Genet., 31*, 289-294.

Fernandez-San Millan, A., Mingeo-Castel, A. M., Miller, M., & Daniell, H. (2003). A chloroplast transgenic approach to hyper-express and purify human serum albumin, a protein highly susceptible to proteolytic degradation. *Plant Biotechnology Journal 1*, 71-79.

Fischer, R.P., Lisowsky, T., Parisi, M.A., & Clayton, D.A. (1992) DNA wrapping and bending by a mitochondrial high-mobility group-like transcriptional activator protein. *J. Biol. Chem., 267*, 3358-3367.

Fujiwara, M., Nagashima, A., Kanamaru, K., Tanaka, K., & Takahashi, H. (2000). Three new nuclear genes, *sigD*, *sigE* and *sigF*, encoding putative plastid RNA polymerase sigma factors in *Arabidopsis thaliana*. *FEBS Lett., 481*, 47-52.

Galli, G., Hofstetter, H., & Birnstiel, M.L. (1981). Two conserved sequence blocks within eukaryotic tRNA genes are major promoter elements. *Nature, 294*, 626-631.

Gatenby, A.A., Castleton, J.A., & Saul, M.W. (1981). Expression in *E. coli* of *Z. mays* and wheat chloroplast genes for large subunit of ribulose biphosphate carboxylase. *Nature, 291*, 117-121.

Geiduschek, E.P., Bardeleben, C., Joazeiro, C.A., Kassavetis, G.A., & Whitehall, S. (1995). Yeast RNA polymerase III: transcription complexes and RNA synthesis. *Braz. J. Med. Biol. Res., 28*, 147-159.

Giese, A., Thalheim, C., Brennicke, A., & Binder, S. (1996). Correlation of nonanucleotide motifs with transcript initiation of 18S rRNA genes in mitochondria of pea, potato and *Arabidopsis*. *Mol. Gen. Genet., 252*, 429-436.

Gray, M.W. (1993). Origin and evolution of organelle genomes. *Curr. Opin. Genet. Dev., 3,* 884-890.

Greenberg, B.M., Narita, J.O., DeLuva-Flaherty, , C., Gruissem, W., Rushlow, K.A., & Hallick, R.B. (1984). Evidence for two RNA polymerase activities in *Euglena gracilis* chloroplasts. *J. Biol. Chem., 259*, 14880-14887.

Gruissem, W., & Tonkyn, J.C. (1993).Control mechanisms of plastid gene expression. *Crit. Rev. Plant Sci., 12*, 19-55.

Gruissem, W., & Zurawski, G. (1985) Analysis of promoter regions for the spinach chloroplast *rbcL*, *atpB* and *psbA* genes. *EMBO J., 4*, 3375-3383.

Gruissem, W., Elsner-Menzel, C., Latshaw, S., Narita, J.O., Schaffer, M.A., & Zurawski, G. (1986). A subpopulation of spinach chloroplast tRNA genes does not require upstream promoter elements for transcription. *Nucleic Acids Res., 14*, 7541-7556.

Hajdukiewicz, P.T., Allison, L.A., & Maliga, P. (1997). The two RNA polymerases encoded by the nuclear and the plastid compartments transcribe distinct groups of genes in tobacco plastids. *EMBO J., 16*, 4041-4048.

Hakimi, M.A., Privat, I., Valay, J.-G., & Lerbs-Mache, S. (2000). Evolutionary conservation of C-terminal domains of primary sigma^{70}-type transcription factors between plants and bacteria. *J. Biol. Chem., 275*, 9215-9221.

Hallick, R.B., Lipper, C., Richards, O.C., & Rutter, W.J. (1976).Isolation of transcriptionally active chromosome from chloroplasts of *Euglena gracilis*. *Biochemistry, 15*, 3039-3045.

Han, C.D., Patrie, W., Polacco, M., & Coe, E.H. (1993). Aberrations in plastid transcripts and deficiency of plastid DNA in striped and albino mutants of maize. *Planta, 191*, 552-563.

Hanaoka, M., Kanamaru, K., Takahashi, H., & Tanaka, K. (2003). Molecular genetic analysis of chloroplast gene promoters dependent on SIG2, a nucleus-encoded sigma factor for the plastid-encoded RNA polymerase, in *Arabidopsis thaliana*. *Nucleic Acids Res., 31*, 7090-7098.

Hanley-Bowdoin, L., Orozco, E.M. Jr, & Chua, N.H. (1985). *In vitro* synthesis and processing of a maize chloroplast transcript encoded by the ribulose 1,5-bisphosphate carboxylase large subunit gene. *Mol. Cell Biol., 5*, 2733-2745.

Hedtke, B., Börner, T., & Weihe, A. (1997). Mitochondrial and chloroplast phage-type RNA polymerases in *Arabidopsis*. *Science, 277*, 809-811.

Hedtke, B., Börner, T., & Weihe, A. (2000). One RNA polymerase serving two genomes. *EMBO Rep., 1*, 435-440.

Hedtke, B., Legen, J., Weihe, A., Herrmann, R.G., & Börner, T. (2002). Six active phage-type RNA polymerase genes in *Nicotiana tabacum*. *Plant J., 30*, 625-637.

Hedtke, B., Meixner, M., Gillandt, S., Richter, E., Börner, T., & Weihe, A. (1999). Green fluorescent protein as a marker to investigate targeting of organellar RNA polymerases of higher plants *in vivo*. *Plant J., 17*, 557-561.

Hess, W., & Börner, T. (1999). Organellar RNA polymerases of higher plants. *Int. Rev. Cytol., 190*, 1-59.

Hess, W.R., Prombona, A., Fieder, B., Subramanian, A.R., & Börner, T. (1993). Chloroplast *rps15* and the *rpoB/C1/C2* gene cluster are strongly transcribed in ribosome-deficient plastids: evidence for a functioning non-chloroplast-encoded RNA polymerase. *EMBO J., 12*, 563-571.

Hoffer, P.A., & Christopher, D.A. (1997). Structure and blue light-responsive transcription of a chloroplast *psbD* promoter from *Arabidopsis thaliana*. *Plant Physiol., 115*, 213-222.

Hoffmann, M., & Binder, S. (2002). Functional importance of nucleotide identities within the pea *atp9* mitochondrial promoter sequence. *J. Mol. Biol., 320*, 943-950.

Homann, A., & Link, G. (2003). DNA-binding and transcription characteristics of three cloned sigma factors from mustard (*Sinapis alba* L.) suggest overlapping and distinct roles in plastid gene expression. *Eur. J. Biochem., 270*, 1288-1300.

Howe, C.J. (1996). RNA polymerases and plastid evolution. *Trends Plant Sci., 1*, 323-324.

Hübschmann, T., & Börner, T. (1998). Characterisation of transcript initiation sites in ribosome-deficient barley plastids. *Plant. Mol. Biol., 36*, 493-496.

Ikeda, T.M., & Gray, M.W. (1999a). Identification and characterization of T3/T7 bacteriophage-like RNA polymerase sequences in wheat. *Plant. Mol. Biol., 40*, 567-578.

Ikeda, T.M., & Gray, M.W. (1999b). Characterization of a DNA-binding protein implicated in transcription in wheat mitochondria. *Mol. Cell. Biol., 19*, 8113-8122.

Iratni, R., Baeza, L., Andreeva, A., Mache, R., & Lerbs-Mache, S. (1994). Regulation of rDNA transcription in chloroplasts: Promoter exclusion by constitutive repression. *Genes Dev., 8*, 2928-2938.

Iratni, R., Diederich, L., Harrak, H., Bligny, M., & Lerbs-Mache, S. (1997). Organ-specific transcription of the rrn operon in spinach plastids. *J. Biol. Chem., 272*, 13676-13682.

Isono, K., Niwa, Y., Satoh, K., & Kobayashi, H. (1997). Evidence for transcriptional regulation of plastid photosynthesis genes in *Arabidopsis thaliana* roots. *Plant Physiol., 114*, 623-630.

Isono, K., Shimizu, M., Yoshimoto, K., Niwa, Y., Satoh, K., Yokota, A., & Kobayashi, H. (1997). Leaf-specifically expressed genes for polypeptides destined for chloroplasts with domains of sigma70-factors of bacterial RNA polymerases in *Arabidopsis thaliana. Proc. Natl. Acad. Sci. USA, 94*, 14948-14953.

Jaehning, J.A. (1993). Mitochondrial transcription: is a pattern emerging? *Mol. Microbiol., 8*, 1-4.

Jang, S.H., & Jaehning, J.A. (1991). The yeast mitochondrial RNA polymerase specificity factor, MTF1, is similar to bacterial sigma factors. *J. Biol. Chem., 266*, 22671-22677.

Kanamaru, K., Fujiwara, M., Seki, M., Katagiri, T., Nakamura, M., Mochizuki, N., et al. (1999). Plastidic RNA polymerase sigma factors in *Arabidopsis*. *Plant Cell Physiol., 40*, 832-842.

Kapoor, S., & Sugiura, M. (1999). Identification of two essential sequence elements in the nonconsensus Type II P*atpB*-290 plastid promoter by using plastid transcription extracts from cultured tobacco BY-2 cells. *Plant Cell, 11*, 1799-1810.

Kapoor, S., Suzuki, J.Y., & Sugiura, M. (1997). Identification and functional significance of a new class of non-consensus-type plastid promoter. *Plant J., 11*, 327-337.

Kestermann, M., Neukirchen, S., Kloppstech, K., & Link, G. (1998). Sequence and expression characteristics of a nuclear-encoded chloroplast sigma factor from mustard (*Sinapis alba*). *Nucleic Acids Res., 26*, 2747-2753.

Kim, M., & Mullet, J.E. (1995). Identification of a sequence-specific DNA binding factor required for transcription of the barley chloroplast blue light-responsive *psbD-psbC* promoter. *Plant Cell, 7*, 1445-1457.

Kim, M., Christopher, D.A., & Mullet, J.E. (1999b) ADP-Dependent Phosphorylation Regulates Association of a DNA-Binding Complex with the Barley Chloroplast *psbD* Blue-Light-Responsive Promoter. *Plant Physiol., 119*, 663-670.

Kim, M., Thum, K.E., Morishige, D.T., & Mullet, J.E. (1999a). Detailed architecture of the barley chloroplast *psbD-psbC* blue light-responsive promoter. *J. Biol. Chem., 274*, 4684-4692.

Kobayashi, Y., Dokiya, Y., Sugiura, M., Niwa, Y., & Sugita, M. (2001a). Genomic organization and organ-specific expression of a nuclear gene encoding phage-type RNA polymerase in *Nicotiana sylvestris, Gene, 279*, 33-40.

Kobayashi, Y., Dokiya, Y., & Sugita, M. (2001b). Dual targeting of phage-type RNA polymerase to both mitochondria and plastids is due to alternative translation initiation in single transcripts. *Biochem. Biophys. Res. Commun., 289*, 1106-1113.

Kobayashi, Y., Dokiya, Y., Kumazawa, Y., & Sugita, M. (2002). Non-AUG translation initiation of mRNA encoding plastid-targeted phage-type RNA polymerase in *Nicotiana sylvestris. Biochem. Biophys. Res. Commun.*, 299, 57-61.

Krause, K., Maier, R.M., Kofer, W., Krupinska, K., & Herrmann, R.G. (2000). Disruption of plastid-encoded RNA polymerase genes in tobacco: expression of only a distinct set of genes is not based on selective transcription of the plastid chromosome. *Mol. Gen. Genet., 263*, 1022-1030.

Kumar, S., Dhingra, A. & Daniell, H. (2004a). Plastid expressed *betaine aldehyde dehydrogenase* gene in carrot cultured cells, roots and leaves confers enhanced salt tolerance. *Plant Physiol* in press.

Kumar, S., Dhingra, A., & Daniell, H. (2004b). Manipulation of gene expression facilitates cotton plastid transformation of cotton by somatic embryogenesis & maternal inheritance of transgenes. *Plant Mol. Biol.* in press.

Lahiri, S.D., & Allison, L.A. (2000). Complimentary expression of two plastid-localized sigma-like factors in maize. *Plant Physiol., 123*, 883-894.

Lahiri, S.D., Yao, J., McCumbers, C. & Allison, L.A. (1999). Tissue-specific and light-dependent expression within a family of nuclear encoded sigma-like factors from *Zea mays. Mol. Cell. Biol. Res. Commun., 1*, 14-20.

Lam, E., Hanley-Bowdoin, L., & Chua, N.H. (1988). Characterization of a chloroplast sequence-specific DNA binding factor. *J. Biol. Chem., 263*, 8288-8293.

Lang, B.F., Burger, G., O'Kelly, C.J., Cedergren, R., Golding, G.B., Lemieux, C., et al., (1997). An ancestral mitochondrial DNA resembling a eubacterial genome in miniature. *Nature, 387*, 493-497.

Legen, J., Kemp, S., Krause, K., Profanter, B., Herrmann, R.G., & Maier, R.M. (2002). Comparative analysis of plastid transcription profiles of entire plastid chromosomes from tobacco attributed to wild-type and PEP-deficient transcription machineries. *Plant J., 2*, 171-188.

Lelandais, C., Guiterres, S., Mathieu, C., Vedel, F., Remacle, C., Marechal-Drouard, L., et al. (1996). A promoter element active in run-off transcription controls the expression of two cistrons of *nad* and *rps* genes in *Nicotiana sylvestris* mitochondria. *Nucleic Acids Res., 24*, 4798-4804.

Lerbs, S., Bräutigam, E., & Mache, R. (1983). DNA-dependent RNA polymerase of spinach chloroplasts: characterization of alpha-like and sigma-like polypeptides. *Mol. Gen. Genet., 211*, 458-464.

Lerbs, S., Bräutigam, E., & Parthier, B. (1985). Polypeptides in DNA-dependent RNA polymerase of spinach chloroplasts: Characterization by antibody-linked polymerase assay and determination of sites of synthesis. *EMBO J., 4*, 1661-1666.

Lerbs-Mache, S. (1993). The 110-kDa polypeptide of spinach plastid DNA-dependent RNA polymerase: single-subunit enzyme or catalytic core of multimeric enzyme complexes? *Proc. Natl. Acad. Sci. USA, 90*, 5509-5513.

Lescure, A.-M., Bisanz-Seyer, C., Pesey, H., & Mache, R. (1985). *In vitro* transcription initiation of the spinach chloroplast 16S rRNA at two tandem promoters. *Nucleic Acids Res., 13*, 8787-8796.

Liere, K., & Maliga, P. (1999). *In vitro* characterization of the tobacco *rpoB* promoter reveals a core sequence motif conserved between phage-type plastid and plant mitochondrial promoters. *EMBO J., 18*, 249-257.

Liere, K., & Maliga, P. (2001). Plastid RNA Polymerases. In E. -M. Aro and B. Andersson (Eds.), *Regulation of Photosynthesis*. (pp. 29-49). Dordrecht: Kluwer Academic Publishers.

Liere, K., Kaden, D., Maliga, P., & Börner, T. (2004). Over expression of phage-type RNA polymerase RpoTp in tobacco demonstrates its role in chloroplast transcription by recognizing a distinct promoter type. *Nucleic Acids Res., 32*, 1159-1165.

Link, G. (1994). Plastid differentiation: Organelle promoters and transcription factors. In L. Nover (Ed.), *Results and Problems in Cell Differentiation: Vol. 20, Plant Promoters and Transcription Factors* (pp. 65-85). Berlin: Springer-Verlag.

Link, G. (1996). Green Life: Control of chloroplast gene transcription. *Bioessays, 18*, 465-471.

Link, G. (2003). Redox regulation of chloroplast transcription. *Antioxid. Redox Signal., 5*, 79-87.

Link, G., & Homann, A. (2003). DNA-binding and transcription characteristics of three cloned sigma factors from mustard (*Sinapsis alba*) suggest overlapping and distinct roles in plastid gene expression. *Eur. J. Biochem., 270*, 1288-1300.

Lisowski, T., & Michaelis, G. (1988). A nuclear gene essential for mitochondrial replication suppresses a defect of mitochondrial transcription in *Saccharomyces cerevisiae*. *Mol. Gen. Genet., 214*, 218-223.

Liu, B., & Troxler, R.F. (1996). Molecular characterization of a positively photoregulated nuclear gene for a chloroplast RNA polymerase sigma-factor in *Cyanidium caldarium*. *Proc. Natl. Acad. Sci. USA, 93*, 3313-3318.

Lizama, L., Holuigue, L., & Jordana, X. (1994). Transcription initiation sites for the potato mitochondrial gene coding for subunit 9 of ATP synthase (*atp9*). *FEBS Lett., 349*, 243-248.

Mache, R., Cottet, A., Imberty, A., Hakimi, A.M., & Lerbs-Mache, S. (2002). The plant sigma factors: structure and phylogenetic origin. *Genome Lett., 1*, 71-76.

Mangus, D.A., Jang, S.H., & Jaehning, J.A. (1994). Release of the yeast mitochondrial RNA polymerase specificity factor from transcription complexes. *J. Biol. Chem., 269*, 26568-26574.

McAllister, W.T., & Raskin, C.A. (1993). The phage RNA polymerases are related to DNA polymerases and reverse transcriptases. *Mol. Microbiol., 10*, 1-6.

McCulloch, V., & Shadel, G.S. (2003). Human mitochondrial transcription factor B1 interacts with the C-terminal activation region of h-mtTFA and stimulates transcription independently of its RNA methyltransferase activity. *Mol. Cell. Biol., 23*, 5816-5824.

McCulloch, V., Seidel-Rogol, B.L., & Shadel, G.S. (2002) A human mitochondrial transcription factor is related to RNA adenine methyltransferases and binds S-adenosylmethionine. *Mol. Cell. Biol., 22*, 1116-1125.

Morden, C.W., Wolfe, K.H., dePamphilis, C.W., & Palmer, J.D. (1991). Plastid translation and transcription genes in a non-photosynthetic plant: intact, missing and pseudo genes. *EMBO J., 10*, 3281-3288.

Morikawa, K., Ito, S., Tsunoyama, Y., Nakahira, Y., Shiina, T., & Toyoshima, Y. (1999). Circadian-regulated expression of a nuclear-encoded plastid sigma factor gene (*sigA*) in wheat seedlings. *FEBS Lett.*, 451, 275-278.

Mullet, J.E. (1988). Chloroplast development and gene expression. *Annu. Rev. Plant Physiol. Plant Mol. Biol., 39*, 475-502.

Mullet, J.E., Orozco, E.M., & Chua, N.H. (1985). Multiple transcripts for higher plant *rbcL* and *atpB* genes and localization of the transcription initiation site of the *rbcL* gene. *Plant Mol. Biol., 4*, 3954

Mulligan, R.M., Lau, G.T., & Walbot, V. (1988). Numerous transcription initiation sites exist for the maize mitochondrial genes for subunit 9 of the ATP synthase and subunit 3 of cytochrome oxidase. *Proc. Natl. Acad. Sci. USA, 85*, 7998-8002.

Mulligan, R.M., Leon, P., & Walbot, V. (1991). Transcriptional and post-transcriptional regulation of maize mitochondrial gene expression. *Mol. Cell. Biol., 11*, 533-543.

Neuhaus, H., Pfannschmidt, T., & Link, G. (1990). Nucleotide sequence of the chloroplast *psbI* and *trnS*-GCU genes from mustard (*Sinapis alba* L.). *Nucleic Acids Res., 18*, 368.

Neuhaus, H., Scholz, A., & Link, G. (1989). Structure and expression of a split chloroplast gene from mustard (*Sinapis alba* L.): Ribosomal protein rps16 reveals unusual transcriptional features and complex RNA maturation. *Curr. Genet., 15*, 63-70.

Newton, K.J., Winberg, B., Yamato, K., Lupold, S., & Stern, D.B. (1995). Evidence for a novel mitochondrial promoter preceding the *cox2* gene of perennial teosintes. *EMBO J., 14*, 585-593.

Nickelsen, J., & Link, G. (1990). Nucleotide sequence of the mustard chloroplast genes *trnH* and *rps19*. *Nucleic Acids Res., 18*, 1051.

Oikawa, K., Fujiwara, M., Nakazato, E., Tanaka, K., & Takahashi, H. (2000). Characterization of two plastid sigma factors, SigA1 and SigA2, that mainly function in matured chloroplasts in *Nicotiana tabacum*. *Gene, 261*, 221-228.

Oikawa, K., Tanaka, K., & Takahashi, H. (1998) Two types of differentially photo-regulated nuclear genes that encode sigma factors for chloroplast RNA polymerase in the red alga *Cyanidium caldarium* strain RK-1. *Gene, 210*, 277-285.

Pfannschmidt, T., & Link, G. (1994). Separation of two classes of plastid DNA-dependent RNA polymerase that are differentially expressed in mustard (*Sinapis alba* L.) seedlings. *Plant Mol. Biol., 25*, 69-81.

Pfannschmidt, T., & Link, G. (1997). The A and B forms of plastid DNA-dependent RNA polymerase from mustard (*Sinapis alba* L.) transcribe the same genes in a different developmental context. *Mol. Gen. Genet., 257*, 35-44.

Rantanen, A., Gaspari, M., Falkenberg, M., Gustafsson, C.M., & Larsson, N. G. (2003) Characterization of the mouse genes for mitochondrial transcription factors B1 and B2 . *Mamm. Genome, 14*, 1-6.

Rapp, W.D., & Stern, D.B. (1992). A conserved 11 nucleotide sequence contains an essential promoter element of the maize mitochondrial *atp1* gene. *EMBO J., 11*, 1065-1073.

Rapp, W.D., Lupold, D.S., Mack, S., & Stern, D.B. (1993). Architecture of the maize mitochondrial *atp1* promoter as determined by linker-scanning and point mutagenesis. *Mol. Cell. Biol., 13*, 7232-7238.

Reinbothe, S., Reinbothe, C., Heintzen, C., Seidenbecher, C., and Parthier, B. (1993). A methyl jasmonate-induced shift in the length of the 5' untranslated region impairs translation of the plastid *rbcL* transcript in barley. *EMBO J., 12*, 1505-1512.

Reis, T., & Link, G. (1985). Characterization of transcriptionally active DNA-protein complexes from chloroplasts and etioplast of mustard *(Sinapis alba* L.). *Eur. J. Biochem., 148*, 207-212.

Remacle, C., & Marechal Drouard, L. (1996) Characterization of the potato mitochondrial transcription unit containing 'native' *trnS* (GCU), *trnF* (GAA) and *trnP* (UGG). *Plant Mol. Biol., 30*, 553-563.

Richter, U., Kiessling, J., Hedtke, B., Decker, E., Reski, R., Börner, T., & Weihe, A. (2002) Two *RpoT* genes of *Physcomitrella* patens encode phage-type RNA polymerases with dual targeting to mitochondria and plastids. *Gene, 290*, 95-105.

Satoh, J., Baba, K., Nakahira, Y., Shiina, T., & Toyoshima, Y. (1997). Characterization of dynamics of the *psbD* light-induced transcription in mature wheat chloroplasts. *Plant. Mol. Biol., 33*, 267-278.

Satoh, J., Baba, K., Nakahira, Y., Shiina, T., & Toyoshima, Y. (1999). Developmental stage-specific multi-subunit plastid RNA polymerase (PEP) in wheat. *Plant J., 18*, 407-415.

Schinkel, A.H., Groot Koerkamp, M.J.A., Touw, E.P.W., & Tabak, H.F. (1987). Specificity factor of yeast mitochondrial RNA polymerase. Purification and interaction with core RNA polymerase. *J. Biol. Chem., 262*, 12785-12791.

Schubot, F.D., Chen, C.J., Rose, J.P., Dailey, T.A., Dailey, H.A., & Wang, B.C. (2001). Crystal structure of the transcription factor sc-mtTFB offers insights into mitochondrial transcription. *Protein Sci., 10*, 1980-1988.

Seidel-Rogol, B.L., McCulloch, V., & Shadel, G.S. (2003). Human mitochondrial transcription factor B1 methylates ribosomal RNA at a conserved stem-loop. *Nat. Genet., 33*, 23-24.

Serino, G., & Maliga, P. (1998). RNA polymerase subunits encoded by the plastid *rpo* genes are not shared with the nucleus-encoded plastid enzyme. *Plant Physiol., 117*, 1165-1170.

Sexton, T.B., Jones, J.T., & Mullet, J.E. (1990). Sequence and transcriptional analysis of the barley ctDNA region upstream of psbD-psbC encoding *trnK*(UUU), *rsp16*, *trnQ*(UUG), *psbK*, *psbI* and *trnS*(GCU). *Curr. Genet., 17*, 445-454.

Shiina, T., Allison, L., & Maliga, P. (1998). *rbcL* transcript levels in tobacco plastids are independent of light: reduced dark transcription rate is compensated by increased mRNA stability. *Plant Cell, 10*, 1713-1722.

Shinozaki, K., & Sugiura, M. (1982). The nucleotide sequence of the tobacco chloroplast gene for the large subunit of ribulose-1,5-biphosphate carboxylase/oxygenase. *Gene, 20*, 91-102.

Siemenroth, A., Wollgiehn, R., Neumann, D., & Börner, T. (1981). Synthesis of ribosomal RNA in ribosome-deficient plastids of the mutant *"albostrians"* of *Hordeum vulgare* L. *Planta, 153*, 547-555.

Silhavy, D., & Maliga, P. (1998a). Mapping of promoters for the nucleus-encoded plastid RNA polymerase (NEP) in the iojap maize mutant. *Curr. Genet., 33*, 340-344.

Silhavy, D., & Maliga, P. (1998b) Plastid promoter utilization in a rice embryogenic cell culture. *Curr. Genet., 34*, 67-70.

Sousa, R., Chung, Y.J., Rose, J.P., & Wang, B.C. (1993). Crystal structure of bacteriophage T7 RNA polymerase at 3.3 Å resolution. *Nature, 364*, 593-599.

Sriraman, P., Silhavy, D., & Maliga, P. (1998a). The phage-type *PclpP–53* plastid promoter comprises sequences downstream of the transcription initiation site. *Nucleic Acids Res., 26*, 4874-4879.

Sriraman, P., Silhavy, D., & Maliga, P. (1998b). Transcription from heterologous rRNA operon promoters in chloroplasts reveals requirement for specific activating factors. *Plant Physiol., 117*, 1495-1499.

Strittmatter, G., Godzicka-Josefiak, A., & Kössel, H. (1985). Identification of an rRNA operon promoter from *Zea mays* chloroplast which excludes the proximal tRNAVal from the primary transcript. *EMBO J., 4*, 599-604.

Suck, R, Zeltz, P., Falk, J., Acker, A., Kössel, H., & Krupinska, K. (1996). Transcriptionally active chromosomes (TACs) of barley chloroplasts contain the a-subunit of plastome-encoded RNA polymerase. *Curr. Genet., 30*, 515-521.

Sugita, M., & Sugiura, M. (1996). Regulation of gene expression in chloroplasts of higher plants. *Plant Mol. Biol., 32*, 315-326.

Sugiura, C., Kobayashi, Y., Aoki, S., Sugita, C., & Sugita, M. (2003). Complete chloroplast DNA sequence of the moss *Physcomitrella patens*: evidence for the loss and relocation of *rpoA* from the chloroplast to the nucleus. *Nucleic Acids. Res., 31*, 5324-5331.

Sugiura, M. (1992). The chloroplast genome. *Plant Mol. Biol., 9*, 5650-5659.

Sun, E., Wu, B.W., & Tewari, K.K. (1989). *In vitro* analysis of the pea chloroplast 16S rRNA gene promoter. *Mol. Cell. Biol., 9*, 5650-5659.

Surzycki, S.J., & Shellenbarger, D.L. (1976). Purification and characterization of a putative sigma factor from *Chlamydomonas reinhardtii*. *Proc. Natl. Acad. Sci. USA, 73*, 3961-3965.

Suzuki, J.Y., Sriraman, P., Svab, Z., & Maliga, P. (2004). Unique architecture of the plastid ribosomal RNA operon promoter recognized by the multisubunit RNA polymerase in tobacco and other higher plants. *Plant Cell, 15*, 195-205.

Tan, S., & Troxler, R.F. (1999). Characterization of two chloroplast RNA polymerase sigma factors from *Zea mays*: Photoregulation and differential expression. *Proc. Natl. Acad. Sci. USA, 96*, 5316-5321.

Tanaka, K., Tozawa, Y., Mochizuki, N., Shinozaki, K., Nagatani, A., Wakasa, K., & Takahashi, H. (1997). Characterization of three cDNA species encoding plastid RNA polymerase sigma factors in *Arabidopsis thaliana*: evidence for the sigma heterogenity in higher plant plastid. *FEBS Lett., 413*, 309-313.

Tiller, K., & Link, G. (1993). Sigma-like transcription factors from mustard (*Sinapis alba* L.) etioplasts are similar in size to, but functionally distinct from, their chloroplast counterparts. *Plant Mol. Biol. 21*, 503-513.

Tiller, K., Eisermann, A., & Link, G. (1991). The chloroplast transcription apparatus from mustard (*Sinapis alba* L.). Evidence for three different transcription factors which resemble bacterial sigma factors. *Eur. J. Biochem., 198*, 93-99.

To, K.Y., Cheng, M.C., Suen, D.F., Mon, D.P., Chen, L.F.O., & Chen, S.C.G. (1996). Characterization of the light-responsive promoter of rice chloroplast *psbD-C* operon and the sequence-specific DNA binding factor. *Plant Cell Physiol., 37*, 660-666.

Tozawa, Y., Tanaka, K., Takahashi, H., & Wakasa, K. (1998). Nuclear encoding of a plastid sigma factor in rice and its tissue- and light-dependent expression. *Nucleic Acids Res., 26*, 415-419.

Tracy, R.L., & Stern, D.B. (1995). Mitochondrial transcription initiation: promoter structures & RNA polymerases. *Curr. Genet., 28*, 205-216.

Trifa, Y., Privat, I., Gagnon, J., Baeza, L., & Lerbs-Mache, S. (1998). The nuclear *RPL4* gene encodes a chloroplast protein that co-purifies with the T7-like transcription complex as well as plastid ribosomes. *J. Biol. Chem., 273*, 3980-3985.

Troxler, R.F., Zhang, F., Hu, J., & Bogorad, L. (1994). Evidence that sigma factors are components of chloroplast RNA polymerase. *Plant Physiol., 104*, 753-759.

Unseld, M., Marienfeld, J.R., Brandt, P., & Brennicke A. (1997). The mitochondrial genome of Arabidopsis thaliana contains 57 genes in 366,924 nucleotides. *Nature Genet., 15*, 57-61.

Vera, A., & Sugiura, M. (1995). Chloroplast rRNA transcription from structurally different tandem promoters. An additional novel-type promoter. *Curr. Genet., 27*, 280-284.

Vera, A., Hirose, T., & Sugiura, M. (1996). A ribosomal protein gene (*rpl32*) from tobacco chloroplast DNA is transcribed from alternative promoters: Similarities in promoter region organization of plastid housekeeping genes. *Mol. Gen. Genet., 251*, 518-525.

Wada, T., Tunoyama, Y., Shiina, T., & Toyoshima, Y. (1994). *In vitro* analysis of light-induced transcription in the wheat *psbD/C* gene cluster using plastid extracts from dark-grown and short-term-illuminated seedlings. *Plant Physiol., 104*, 1259-1267.

Weihe, A., & Börner, T. (1999). Transcription and the architecture of promoters in chloroplasts. *Trends Plant Sci., 5*, 169-170.

Weihe, A., Hedtke, B., & Börner, T. (1997). Cloning and characterization of a cDNA encoding a bacteriophage-type RNA polymerase from the higher plant *Chenopodium album*. *Nucleic Acids Res., 25*, 2319-2325.

Wolfe, K.H., Morden, C.W., Ems, S.C., & Palmer, J.D. (1992). Rapid evolution of the plastid translational apparatus in a nonphotosynthetic plant: Loss or accelerated sequence evolution of tRNA and ribosomal protein genes. *J. Mol. Evol., 35*, 304-317.

Wu, C.Y., Lin, C.H., & Chen, L.J. (1997). Identification of the transcription start site for the spinach chloroplast serine tRNA gene. *FEBS Lett., 418*, 157-161.

Yan, B, & Pring, D.R. (1997).Transcriptional initiation sites in sorghum mitochondrial DNA indicate conserved and variable features. *Curr. Genet., 32*, 287-295.

Yao, J., Roy-Chowdhury, S., & Allison, L.A. (2003). AtSig5 is an essential nucleus-encoded *Arabidopsis* sigma-like factor. *Plant Physiol., 132*, 739-747

Young, D.A., Allen, R.L., Harvey, A.J., & Lonsdale, D.M. (1998). Characterization of a gene encoding a single-subunit bacteriophage-type RNA polymerase from maize which is alternatively spliced. *Mol. Gen. Genet., 260*, 30-37.

Zengel, J.M., & Lindahl, L. (1990). Ribosomal protein L4 stimulates *in vitro* termination of transcription at a NusA-dependent terminator in the S10 operon leader. *Proc. Natl. Acad. Sci. USA, 87*, 2675-2679.

CHAPTER 9

RNA EDITING IN PLANT ORGANELLES

R. MICHAEL MULLIGAN

The Department of Developmental and Cell Biology, University of California, Irvine, California, 92697-2300, USA

Abstract. RNA editing is a post-transcriptional process that occurs in the organelles of vascular plants and changes the coding information in mRNAs. In higher plants, specific cytidine residues are converted to uridine residues in chloroplasts or in mitochondrial transcripts, and this process frequently re-specifies the codon to direct the incorporation of a non-synonymous amino acid residue. The amino acid incorporated by the edited codon is typically the evolutionarily conserved amino acid at that position, and the unedited codon would direct the incorporation of a radical amino acid substitution. In several instances, expression of an unedited version of a transcript that requires editing results in a mutant phenotype; thus, demonstrating a genetic requirement for RNA editing in gene expression. Analysis of RNA editing in transgenic chloroplasts and development of *in vitro* editing systems suggest that RNA editing site recognition may occur by specific *trans*-acting protein factors. In addition, recent results demonstrate that a single *trans*-acting factor may recognize several editing sites with sequence similarity and editing site "clusters" may exist.

1. INTRODUCTION

In the 1950s, the central dogma of molecular biology established DNA as the informational molecule in biology, and RNA was thought to be a transient form of the information that was produced when genetic information needed to be accessed to produce a polypeptide. A series of modifications to the central dogma occurred over the next 50 years, and surprising developments in the way genes are expressed have continued to fascinate scientists and students alike. Reverse transcriptase was discovered, and then DNA could be produced from RNA. Viral RNA genomes and RNA dependent RNA polymerases showed that RNA could be the genetic material, and DNA was not essential. The discovery of introns in genes' sequences demonstrated that the genetic information need not be contiguous, and alternative splicing could produce different proteins from a single gene. In 1985, it was reported that genetic information could be modified at the level of RNA by a correction

H. Daniell and C.D. Chase (eds.), Molecular Biology and Biotechnology of Plant Organelles, 239—260.

process that inserted or deleted nucleotides from transcripts, a process known as RNA editing (Benne et al., 1986).

2. TYPES OF RNA EDITING

RNA editing is a co- or post-transcriptional process that modifies the sequence of an RNA through nucleotide insertion, deletion or modification to make it different from the DNA that encoded the RNA (Smith et al., 1997). Editing may be divided into two types of RNA modifications: insertion/deletion editing and nucleotide modification editing (Table 1).

Insertion/deletion editing causes the RNA to differ from the encoding genomic sequence by a different number of nucleotides, and changes the reading frame of an mRNA. Insertion/deletion editing was first discovered in Trypanosome kinetoplasts, specialized mitochondria, and has since been recognized in *Physarum* mitochondria and in several viruses. Some kinetoplast transcripts are subjected to very extensive RNA editing; over 50% of the *cox3* mRNA sequence in *Trypansoma brucei* is created by post-transcriptional uridine (U) insertion (Feagin et al., 1988). The U insertion process is directed by an overlapping series of small antisense guide RNAs (gRNAs) that anneal to the pre-mRNA and act as a template for the insertion of the correct number of Us (Simpson et al., 2003). The structural gene for the kinetoplast protein lacks substantial genetic information that is borne elsewhere in the genome as guide RNA genes. Thus, gRNA genes actually encode portions of the genetic information for their respective mRNA. Extended laboratory culture of trypanosomes has been shown to result in the loss of some gRNA genes and the inability to express a functional mRNA and protein (Thiemann et al., 1994). Thus, the idea of the "gene" must be extended in this system to include not only the information for the expression of the pre-mRNA, but also the constellation of gRNAs genes that provide the genetic information for the editing process.

Insertion editing in viral transcripts appears to result from the displacement of the nascent RNA on the DNA template in the transcription complex and is usually detected as insertion of guanidine (G) (Vidal et al., 1990). Editing tends to occur in a region where the nascent RNA may slide one or more nucleotides relative to the DNA and retain complementary base pairing with the template DNA. Thus, a specific nucleotide may be transcribed two or more times, and multiple transcripts may arise that differ by the insertion of one or more Gs (Vidal et al., 1990).

Base modification editing is widely distributed, and in some cases results in the production of proteins with distinct functions or properties (Table 1b). The mechanisms of base modification editing differ and incidences of base modification editing appear not to be related in origin, however, the mechanisms frequently share similarities. Base modification editing of structural RNAs such as tRNAs and rRNAs has been known for decades and modification reactions include methylation, acetylation, deamination, rearrangements, and other chemical reactions (Gott and Emeson, 2000). The mechanisms of such reactions are becoming well understood with the identification of genes involved in these processes. The best understood

Table 1. *Types and occurrence of RNA editing*

a. Insertion/Deletion Editing

Nucleotide Changes	*Organisms*	*Genes*	*Mechanism*
U insertion	*Trypanosoma* kinetoplastids	mRNAs (mitochondrial)	gRNA directed-cleavage, U insertion, RNA ligation
C insertion, dinucleotide insertion	*Physarum*	mRNAs, rRNAs, tRNAs (mitochondrial)	unknown
G insertion	Paramyxoviruses	P mRNA (viral)	polymerase stuttering

b. Nucleotide Modification Editing

Nucleotide Changes	*Organisms*	*Genes*	*Mechanism*
C-to-U	Land Plants	mRNAs, tRNAs (chloroplast, mitochondrial)	C deamination
	Physarum	mRNAs (mitochondrial)	unknown
	Mammals	*apoB* mRNA	C deamination
U-to-C	Land Plants	mRNAs (chloroplast, mitochondrial)	unknown
A-to-I	Vertebrates	*GluRB*, related genes	dsRNA dependent A deamination
	Viral genomes	Viral genes	
	Drosophila	*4f-rnp* mRNAs	
several	*Acanthomoeba*	tRNAs (mitochondrial)	unknown

forms of base modification editing in gene expression are the cytidine-to-uridine (C-to-U) conversions of apolipoprotein B (*apoB*) mRNAs (Anant and Davidson, 2002; Wang et al., 2003; Wedekind et al., 2003), and adenosine-to-inosine (A-to-I) conversion of mRNAs for several channel proteins and membrane receptors in neuronal tissues (Bass, 2002; Maas et al., 2003; Saunders and Barber, 2003; Schaub and Keller, 2002; Seeburg and Hartner, 2003). These reactions are catalyzed by RNA-specific cytidine or adenosine deaminases that are mechanistically related to nucleotide deaminases that function in the biosynthesis of nucleotides. A-to-I editing is directed by one of several enzymes in the family of adenosine deaminases acting on RNA (ADARs) and causes adenosine deamination to inosine. RNA substrates for A-to-I editing typically form imperfect double-stranded substrates, and the recombinant enzyme will recognize and deaminate adenosines in double-stranded RNAs *in vitro*. Inosine base pairs like guanine, and inosine residues in mRNAs are translated by the ribosome as a guanine. Thus, A-to-I editing effectively re-specifies the edited codons by an A-to-G transition.

C-to-U editing of mammalian *apoB* mRNAs is directed by Apobec, an RNA specific cytidine deaminase. The mechanism of C-to-U editing by Apobec is especially relevant to consider because the mechanism of C-to-U editing in plant organelles may occur through similar reactions. Apobec is the catalytic subunit of the editing complex that catalyzes a site-specific deamination of a single C residue to produce a stop codon in apolipoprotein B mRNAs. Apobec has a substantial degree of amino acid identity and protein structural similarity with cytidine deaminases, and appears to be a member of the cytidine deaminase family (Bhattacharya et al., 1994). The *apoB* mRNAs are partially edited in intestinal enterocytes, and these cells produce two polypeptides from the *apoB* gene. ApoB100 is produced by unedited *apoB* mRNAs and forms the polypeptide component of the chylomicrons that carry triglycerides. ApoB48 is produced from edited transcripts and produces a truncated polypeptide as a result of stop codon formation at the editing site. ApoB48 forms the protein component of low-density lipoprotein (LDL) that carries cholesterol in the blood stream. Thus, the regulation of editing of *apoB* mRNAs is critical in cardiovascular health.

The *apoB* mRNAs are edited by a single C-to-U conversion that converts a CAA^{glu} to a UAA^{ter} codon within a 14,000 nucleotide transcript; thus, this reaction is extremely site specific. The *cis*-acting elements include an AU-rich context with an 11 nucleotide motif immediately downstream of the edited C (Anant and Davidson, 2000). *Trans*-acting factors include Apobec, the RNA-specific cytidine deaminase, and an accessory protein known as ACF (Apobec complementation factor), that acts as an RNA recognition component of the complex (Blanc et al., 2001; Henderson et al., 2001).

A homologue of Apobec has recently been shown to play an important role in immunology in several important steps that occur in immune B cells. An Apobec-like protein known as activation-induced cytidine deaminase (AID) is required for class switch recombination, somatic hypermutation, and gene conversion (Okazaki et al., 2003). The AID polypeptide includes the entire cytidine deaminase active site region with high sequence identity to Apobec. In addition, recombinant AID exhibits cytidine deaminase activity in an *in vitro* assay and sensitivity to zinc chelation, as expected for a member of the cytidine deaminase family (Muramatsu et al., 1999). AID appears to catalyze C deamination to U on single-stranded DNA templates and may be involved in the introduction of untemplated point mutations in the variable region of immunoglobulin genes (Okazaki et al., 2003).

3. EDITING IN PLANT RNAS

RNA editing was discovered in plant mitochondria a few years after editing was reported in trypanosomes. Three groups working on various aspects of gene expression in plant mitochondria demonstrated that C-to-U changes occurred in some positions when genomic and cDNA sequences were compared (Covello and

Gray, 1989; Gualberto et al., 1989; Hiesel et al., 1989). These reports resolved an enigma in plant mitochondrial gene expression; arginine codons (CGG) were sometimes detected in gene sequences at positions where tryptophan residues were highly conserved, and the possibility that CGG coded for tryptophan was suggested (Fox and Leaver, 1981). Analysis of cDNA sequences revealed that the unexpected CGG codons had been converted into UGG^{trp} codons by C-to-U editing. Thus, RNA editing in plant organelles was established as a sequence correction mechanism that re-specified genetic information by making C-to-U changes in the RNAs (Gualberto et al., 1989).

Several higher plant chloroplast genomes have been sequenced and analyzed for editing, and generally have about 30 C-to-U editing sites (Kugita et al., 2003a; Kugita et al., 2003b; Maier et al., 1995; Sugiura, 1995). All of the editing sites described for chloroplasts from vascular plants are C-to-U editing sites, and no U-to-C (reverse) edits have been identified. U-to-C editing is prevalent in the chloroplast of some non-vascular plants such as the hornwort, *Anthoceros*, with roughly half of the ~900 editing events specified by that chloroplast genome being reverse edits (Kugita et al., 2003a).

The complete *Arabidopsis* and rice mitochondrial genomes have been sequenced and analyzed for RNA editing, and these genomes encode 441 and 491 C-to-U editing sites, respectively (Giege and Brennicke, 1999; Notsu et al., 2002). Three reverse edits (U-to-C) were reported in wheat and *Oenothera* mitochondria soon after the discovery of editing (Gualberto et al., 1990; Hiesel et al., 1990; Schuster et al., 1990); however, additional U-to-C editing sites have not been described in higher plants, even after the complete sequence analysis of rice and *Arabidopsis* mitochondrial genomes.

4. WHAT IS THE FUNCTION OF RNA EDITING?

Conversion of C to U generally causes a radical change in the amino acid specified by a codon, and would be predicted to perturb the structure and function of a protein. For example, the three most common amino acid changes directed by editing in the *Arabidopsis* mitochondrial genome result in drastic changes of the hydrophobicity of the amino side chain: serine to leucine (93 sites); proline to leucine (80 sites); and serine to phenyalanine (47 sites) (Giege and Brennicke, 1999). Moreover, almost 90% of the editing sites in the *Arabidopsis* mitochondrial genome are in the first two positions of the codon that typically convert the amino acid specified by the codon to an evolutionarily conserved residue. An example of the evolutionarily conserved nature of the non-synonymous amino acid changes directed by editing is shown in Table 2. Thus, the primary function of RNA editing is a genetic correction mechanism that corrects mutations in the genomic sequence such that the mRNA codes for the correct amino acid sequence.

Table 2. *Editing restores the evolutionary conserved amino acid residue in maize mitochondrial rps12 RNAs*

Rps12 gene	*Amino acid residue*					
	24	66	73	90	95	97
Zea mays mt genomic	S	H	S	S	S	R
Zea mays mt edited	L	Y	L	L	F	C
Acanthomoeba castellani mt	L	Y	L	L	Y	L
Drosophila melanogaster mt	L	Y	L	V	L	A
Nicotiana tabacum ct	L	Y	L	L	Y	I
Chlamydomnas reinhardtii ct	L	Y	L	L	Y	I
Marchantia polymorpha ct	L	Y	L	L	Y	I
Escherichia coli	L	Y	L	L	Y	T

Amino acid residues encoded by unedited and edited maize mitochondrial (mt) transcripts are compared to amino acid residues in Rps12 *polypeptides encoded by other mt , chloroplast (ct) and bacterial genetic systems.*

The most conclusive example of the requirement of editing for protein function was shown by the requirement of RNA editing in the expression of *psbF*, a photosystem II polypeptide (Bock et al., 1994). The spinach *psbF* transcript is edited at a single C to cause a serine-to-phenylalanine conversion, but tobacco *psbF* mRNAs encode a phenylalanine residue and do not require editing at that site. The spinach editing site was introduced into the tobacco chloroplast *psbF* gene, and the heterologous editing site could not be edited by the tobacco chloroplast editing machinery, thus creating a phenyalanine-to-serine mutation in the PsbF polypeptide. These transgenic plants exhibited a photosystem II-deficient phenotype with retarded growth and pale green phenotype (Bock et al., 1994). Thus, failure to convert the editing site in the transgenic tobacco chloroplasts resulted in a mutant phenotype and this emphasizes the genetic requirement for editing.

The requirement for editing in gene expression has also been demonstrated by the introduction of a proline-to-leucine mutation in the *Chlamydomonas petB* gene (Zito et al., 1997). A proline-to-leucine change is corrected by editing in several higher plant plastid systems, but is not corrected in *Chlamydomonas* because the *Chlamydomonas* chloroplast genome normally codes for the critical leucine residue. The *Chlamydomonas petB* mutant strains could not grow phototrophically, and exhibited blocked electron transport properties expected from a nonfunctional cytochrome b_6/f complex. In the "unedited" *Chlamydomonas* strains, the mutant phenotype resulted from the inability to assemble the cytochrome b_6/f complex. Comparison of the "unedited" *petB* strain with a characterized heme-attachment mutant of cytochrome b_6 suggested that the primary affect was on heme assembly with apocytochrome b_6, consequently preventing assembly of the cytochrome b_6/f complex (Zito et al., 1997). These results nicely illustrate a genetic requirement for

RNA editing, and editing apparently functions as a genetic correction mechanism to correct nucleotide changes that would otherwise result in a mutant phenotype.

Several investigations suggest that editing may have a function in cytoplasmic male sterility. Cytoplasmic male sterility (CMS) occurs when pollen fails to develop normally. This trait is frequently associated with the presence of a novel chimeric ORF in the mitochondrial genome that may somehow disrupt mitochondrial functions (Conley and Hanson, 1995; Hanson et al., 1999; Chapters 22 and 23 in this volume). In tobacco plants, mitochondria carrying "edited" and "unedited" Atp9 proteins were created by nuclear transformation and protein targeting to the mitochondrion (Hernould et al., 1993). Transgenic plants expressing the unedited *atp9* gene exhibited either fertile, semi-fertile, or male-sterile phenotypes, while control plants, which expressed the edited *atp9* gene or the selectable marker alone, were all male-fertile. Subsequent studies demonstrated that antisense inhibition dramatically reduced the abundance of the nuclear-encoded *atp9* transcripts and restored male fertility to the male-sterile plants (Zabaleta et al., 1996).

These data suggest that mitochondrial dysfunction may result from the presence of unedited translation products within an organelle. Many of the mitochondrial and chloroplast genes are members of multisubunit complexes that participate in complicated electron transport or ATP synthesis reactions. Point mutations produced by the translation of unedited RNAs could produce polypeptides that initiate assembly, but fail to produce functional complexes, as would appear to be the case with the "unedited" *petB* mutant in *Chlamydomonas* (Zito et al., 1997). If unedited translation products do interfere with the assembly of multisubunit complexes, this process could act like a dominant negative allele through interference with the assembly of functional organellar complexes.

5. INCOMPLETE EDITING AND POLYMORPHISM IN GENE EXPRESSION

Editing site conversion is typically highly efficient in chloroplasts, and most cDNAs are completely or nearly completely edited. A comprehensive analysis of the efficiency of editing site conversion in maize plastids was evaluated at all 27 maize plastid editing sites in 10 different tissues that included chloroplasts, etioplasts, and amyloplasts (Peeters and Hanson, 2002). The extent of RNA editing differed among transcripts in the same tissue, and between the same site in different tissues. The editing site efficiency of *ndhB* RNAs was least at 8% and 1% in roots and callus plastids, respectively. In contrast, editing efficiency of some sites was unaffected by developmental stage and these sites were edited with 80 to 100% efficiency. The editing status of all plastid editing sites in young green leaves was generally high, in the 80 to 100% range, suggesting very efficient RNA editing in young green leaves.

The magnitude and developmental changes associated with C-to-U editing in maize mitochondria was evaluated with ~400 cDNA clones for *rps12*, *rps13*, *nad3*, spliced *cox2*, and *atp9* transcripts (Grosskopf and Mulligan, 1996). The *atp9* and spliced *cox2* transcripts were essentially 100% edited in all samples. The *rps12* and *rps13* transcripts exhibit a substantial level of heterogeneity, with only 56% and

83% of the editing sites converted, respectively (Lu et al., 1996; Phreaner et al., 1996; Williams et al., 1998b). In addition, *nad3* mRNAs were incompletely edited with a range of values between 50 to 85% under various environmental and developmental conditions (Grosskopf and Mulligan, 1996). Thus, editing site conversion for some RNAs is incomplete, and the transcripts represent a heterogeneous mixture of sequences and corresponding genetic information. These results suggested that editing status changed during development, although no single form of the incompletely edited transcripts was especially prevalent, indicating that no specific isotype was preferentially expressed.

Incompletely edited transcripts would encode polypeptides with radical amino acid substitutions that are likely to be non-functional and may even be deleterious. The amino acid sequence of the wheat Atp9 polypeptide was shown to reflect the fully edited mRNA sequence, and the corresponding region of the RNA exhibits little to no heterogeneity from incomplete C-to-U editing (Begu et al., 1990). In potato mitochondria, *nad9* transcripts are represented by two size classes: larger mRNAs that are incompletely edited and more abundant, smaller mRNAs that are fully edited (Lu and Hanson, 1996). Both large and small mRNAs are incorporated into mitochondrial polysomes. Amino acid sequence analysis of the NADH dehydrogenase subunit 9 polypeptide (Nad9) from potato demonstrated that the subunit isolated from purified complex 1 reflects the fully edited translation product (Grohmann et al., 1994), although the degree of incomplete editing would probably be below the level of detection in protein sequence analyses. In addition, a single homogenous form of the ATP synthase subunit 6 polypeptide (Atp6) accumulated, despite the presence of incompletely edited *atp6* mRNAs in petunia mitochondria (Lu and Hanson, 1994). Thus, analysis of polypeptides from assembled the ATP synthase or NADH dehydrogenase complex indicated that only edited translation products accumulated in mature, functional complexes.

Detection of unedited translation products was first made in maize mitochondria with antibodies to the ribosomal protein S12 polypeptide (Rps12) (Phreaner et al., 1996). In maize, *rps12* mRNAs have six editing sites that specify three amino changes in an eight-codon region, and each editing site is approximately 50% converted. Thus, the Rps12 polypeptide was an ideal product to analyze whether incompletely edited RNAs were translated in plant mitochondria. Epitope-specific antibodies were produced from synthetic peptides that represented the fully edited and fully unedited translation products, and were shown to discriminate between edited and unedited translation products (Phreaner et al., 1996). Edited *rps12* translation products were detected in intact mitochondria, and were shown to accumulate in a ribosomal fraction, but were depleted from the post-ribosomal supernatant (Lu et al., 1996; Phreaner et al., 1996). Thus, the edited translation product is assembled into ribosomes and provides the functional polypeptide for the small subunit of the ribosome. However, unedited *rps12* translation products were also detected in maize and petunia mitochondria, conclusively demonstrating that unedited transcripts are and can be translated by this system (Lu et al., 1996; Phreaner et al., 1996). The unedited Rps12 polypeptides accumulated in the post-ribosomal supernatant, and were not detected in the ribosomal fraction in maize mitochondria. Unedited Rps12 polypeptides were detected in petunia ribosomes, and this raises interesting questions about how these polypeptides affected ribosomal functions of these subunits. Thus, the translation of incompletely edited *rps12* transcripts does occur and apparently represents an infidelity in the expression of mitochondrial genes.

A comparable study on the translation of *rps13* mRNAs in maize mitochondria detected fully edited Rps13 translation products, but unedited translation products were not detected (Williams et al., 1998b). In this case, the transcript pool exhibited about 80% editing site conversion, a much higher degree of editing than existed in the *rps12* example. In addition, the sensitivity of the antibodies was moderate, and unedited translation products could have gone undetected.

Incompletely edited transcripts appear to at least initially engage in translation, but typically the encoded proteins fail to accumulate. The incompletely edited transcripts potentially encode aberrant amino acid substitutions. Perhaps the abnormal polypeptides fail to assemble properly, and this leads to termination of translation. Alternatively, the incompletely edited translation products are unstable, and proteolytic degradation destroys the individual polypeptides or entire protein complexes.

Translation of incompletely edited transcripts does result in polymorphism in gene expression, at least in the case of *rps12*, and represents inefficiency in the system. These results indicate that the mitochondrial translation apparatus is insensitive to the editing status of mRNAs, and that the mitochondrial translation system apparently does not have a mechanism to regulate translation initiation on partially edited transcripts or to scan for incomplete editing during translation.

6. EDITING IN THE TEMPORAL PROCESSES OF GENE EXPRESSION

A number of processes in organellar gene expression may occur before final maturation of an mRNA. These include the cleavage of polycistronic transcripts, 5' processing and exonucleolytic trimming of the 3' terminus to a stem loop (reviewed in Chapter 10 of the volume); intron splicing (reviewed in Chapter 11); and incorporation into polyribosomes (reviewed in Chapter 12). The relationship of RNA editing to these other processes has been the topic of numerous studies, both as the temporal relationship of editing amidst these other steps and as a potential regulator of various steps in gene expression.

Early studies performed with plant mitochondria compared the editing status of unspliced pre-mRNAs and spliced mRNAs. The pool of pre-mRNAs is composed of nascent RNAs, while the fully spliced RNAs reflect a mature pool of transcripts. Unspliced cytochrome c oxidase subunit 2 (*cox2*) pre-mRNAs were analyzed from maize and petunia mitochondria (Sutton et al., 1991; Yang and Mulligan, 1991). Maize *cox2* RNAs have 18 editing sites, and unspliced *cox2* cDNAs exhibit forms ranging from completely edited to fully unedited transcripts. By contrast, spliced cDNAs indicated that the mature mRNAs were completely edited. Several important observations came out of these results. First, since unspliced transcripts exhibited a range of editing site conversion, splicing prior to editing was not required and editing could be complete in an unspliced transcript. The conversion of editing sites in the unspliced transcripts appeared to be random with no evidence of a processive action or directional bias that would exist if the editing complex "scanned" a transcript. Finally, since completely unedited transcripts were well represented in the unspliced cDNA pool, editing probably continues well after completion of transcription.

An interesting contingency of splicing prior to editing may occur in the wheat *nad7* transcript. An editing site that is three nucleotides downstream of a 5' splice site is fully edited in spliced transcripts, but never edited in the unspliced transcripts (Carrillo and Bonen, 1997). Three additional editing sites in the downstream exon exhibit more than 50% editing in the unspliced mRNAs, and editing sites in other exons were edited at over 50% in these transcripts. Thus, the editing of the site immediately downstream of the intron behaved very differently than any other editing site in the pre-mRNAs. A requirement for splicing prior to editing may be explained in this example if editing site recognition sequences are present in the upstream exon and splicing brings the recognition region into proximity of the editing site. Taken together, these studies demonstrate that splicing is a post-transcriptional process that is generally not required prior to editing, and that both spliced and unspliced RNAs may be substrates for the editing process.

Comparable results were obtained in a chloroplast editing system with polycistronic transcripts that contain introns (Freyer et al., 1993). The *psb* operon expresses a polycistronic transcript that is cleaved to monocistronic *psbB*, *psbH*, *petB*, and *petD* mRNAs. In addition, *petB* and *petD* transcripts each have a single intron that must be spliced. In this chloroplast system, editing was shown to be an early step in mRNA processing with substantial editing detected in unspliced polycistronic transcripts. Thus, editing can precede splicing or cleavage reactions, and appears to be independent of these processes.

A novel mechanism of post-transcriptional control involving polycistronic RNA cleavage and editing occurs in the chloroplast *ndhD*/*psaC* operon (Del Campo et al., 2002). The *psaC* mRNAs accumulate to much higher levels than the six *ndh* gene transcripts expressed from the same operon. RNA editing creates the *ndhD* start codon and modifies an additional downstream codon. The downstream editing sites are rapidly converted and are edited in both polycistronic and unspliced RNAs; however, the *ndhD* start codon is incompletely edited. Functional *ndhD* and *psaC* transcripts are produced from polycistronic transcripts by two mechanisms. Intergenic cleavage produces both *psaC* and fully edited *ndhD* transcripts, and downstream cleavage in the *ndhD* coding sequence occurs to produce *psaC* transcripts from transcripts with unedited *ndhD* start codons. Thus, monocistronic *ndhD* transcripts have fully edited start codons, as required for translation of the mRNA, and additional *psaC* transcripts are produced by cleavage in the *ndhD* coding sequence of unedited *ndhD* transcripts.

7. EDITING SITE RECOGNITION

The plant organellar editing complex must specifically recognize ~30 editing sites in chloroplasts and hundreds of editing sites in plant mitochondria. The *cis*-elements required for editing site recognition in chloroplasts have been carefully analyzed by insertion of transgenes into tobacco chloroplast genomes by the Maliga, Bock and Hanson laboratories. Analysis of three editing sites in transgenic tobacco chloroplasts by 5' and 3' deletion lead to the broad conclusion that recognition elements exist in the 5' flanking region with some sequence requirements in the 3' region (Bock et al., 1997; Bock et al., 1996; Chateigner-Boutin and Hanson, 2002; Chateigner-Boutin and Hanson, 2003; Chaudhuri et al., 1995; Chaudhuri and Maliga, 1996; Hermann and Bock, 1999; Reed and Hanson, 1997; Reed et al., 2001b). Typically 20 to 80 nucleotides of 5' flanking sequence are sufficient for editing of a transgene sequence. In the minimal example, the start codon for *psbL* requires as little as 16 nucleotides of 5' flanking and 6 nucleotides of 3' sequence to specify editing site conversion (Chaudhuri and Maliga, 1996). Only a few editing sites have been carefully studied, and some editing sites may require additional 3' sequence.

The *cis*-elements required for editing site recognition in plant mitochondria appear to be similar to those described for chloroplast systems. *In vivo* analysis of transgene editing has not been possible because mitochondrial transformation has not been achieved in plants; however, plant mitochondria have an active genomic recombination system that has created 5' and 3' break points near several editing sites (Handa et al., 1995; Kubo and Kadowaki, 1997; Mulligan et al., 1999; Williams et al., 1998a). These recombinant editing sites are natural substitution mutations, and demonstrate that 5' flanking sequences are required to specify editing sites. Recombination breakpoints a few nucleotides downstream of an editing site have little to no effect on the extent of editing (Mulligan et al., 1999; Williams et al., 1998a). In one example, the recognition sequences can be concluded to exist within the 5' 38 nucleotides of the editing site (Handa et al., 1995).

A confounding aspect in the understanding of editing site recognition has been how ~30 to hundreds of editing sites can be specifically recognized in chloroplasts and mitochondria. Recent results from Suguira's group suggest that protein factors may be responsible for editing site recognition (Hirose and Sugiura, 2001; Miyamoto et al., 2002; Sasaki et al., 2003; Tsudzuki et al., 2001). RNA cross-linking experiments with a tobacco chloroplast extract capable of *in vitro* RNA editing indicated that different polypeptides associate with *petB* and *psbL* RNAs. RNA competition and cross linking reactions suggest the involvement of a 70 kD polypeptide with *petB* RNA, and a 56 kD polypeptide cross links specifically to *psbE* transcripts.

These distinct protein factors appear to be involved in editing site recognition. Could each editing site be identified by a separate *trans*-acting protein factor? Thirty distinct proteins might be conceivable in the chloroplast system, but would a cell make many hundreds of proteins to individually recognize all the mitochondrial editing sites?

A major breakthrough on editing site recognition came from the Hanson lab through a comprehensive analysis of all plastid editing sites in transgenic chloroplasts that over express *rpoB* or *ndhF* editing substrates (Chateigner-Boutin and Hanson, 2002). Over-expression of an editing substrate was known to compete with and decrease the editing frequency of the endogenous homologous editing site (Chaudhuri et al., 1995; Reed and Hanson, 1997; Reed et al., 2001a). The Hanson lab demonstrated that additional editing sites of the chloroplast also experience the competition effect, and that these editing sites share 5' sequence identity with the over-expressed substrate. This result suggested that a common *trans*-acting factor recognizes similar RNA sequence motifs at these editing sites. Conserved nucleotides identified in the 5' flanking sequences may represent recognition features for that group of editing sites. Inspection of sequences surrounding other chloroplast editing sites indicated that they could be grouped into clusters that include conserved nucleotides in the 5' flanking region of the editing site.

Further experimental support for editing site clusters comes from the analysis of developmental variation in chloroplast editing site conversion. Individual members of an editing site cluster tend to be edited to a similar extent in various tissues (Chateigner-Boutin and Hanson, 2003). Editing site conversion varies in a similar manner for members of the plastid *ndh* editing site cluster. All members of this cluster exhibit low editing site conversion in root plastids compared to leaf chloroplasts. This suggests that these cluster-specific factors are developmentally regulated and may limit editing site conversion in root tissue.

Taken together, these two studies support a model in which a single *trans*-factor may be capable of recognizing several editing sites through the recognition of critical nucleotides in the 5' region. Thus, rather than thirty-something plastid editing factors, perhaps a dozen *trans*-acting factors may be capable of recognizing the all of the chloroplast editing sites.

8. CATALYTIC MECHANISMS OF PLANT RNA EDITING

Enzymatic reactions potentially responsible for plant organelle RNA editing are summarized in Table 3. C and U are interconverted as free nucleotides and nucleosides by biological systems in pyrimidine biosynthetic and salvage pathways. The interconversion of cytidine and uridine may be mechanistically similar to editing reactions that might occur during the editing of RNAs. Two enzymes are generally involved in nucleotide metabolism. Cytidine deaminase catalyzes the irreversible conversion of C to U through a hydrolytic deamination reaction, and CTP synthase utilizes ammonia and ATP in the irreversible conversion of U to C. In principle, a transamination reaction could reversibly convert C to U, utilizing amino donors and acceptors such as aspartate and oxaloacetate. Transaminases are common in the metabolism of amino acids and in intermediary metabolism, but are not known to be involved in nucleotide metabolism.

The mechanism of C-to-U editing in higher plant mitochondria and chloroplasts has been studied *in organello* (Rajasekhar and Mulligan, 1993; Yu and Schuster, 1995) and *in vitro* (Blanc et al., 1995; Hayes, 2004; Hirose and Sugiura,

2001; Miyamoto et al., 2002). Radio-labeled nucleic acids have been used to show that the editing reaction occurs with retention of the □-phosphate and the ring of the nucleotide base (Rajasekhar and Mulligan, 1993; Yu and Schuster, 1995) and therefore appears to occur by a base deamination reaction rather than nucleotide insertion/deletion or other reactions involving removal of the base. *In vitro* editing systems have been established with both chloroplast and mitochondrial extracts (Hirose and Sugiura, 2001; Takenada and Brennicke, 2003), and these reactions do not require additional substrates such as amino acceptors for a transamination reaction. Thus, higher plant editing systems appear to utilize a cytidine deaminase-like reaction in the conversion of Cs to Us in edited RNAs.

Table 3. *Possible enzymatic reactions involved in plant RNA editing*

Protoypic enzyme	*Reaction*	*Catalytic mechanism*	*Reversible*	*Comment*
Cytidine deaminase	CMP+ water > UMP + NH3	hydrolytic deamination	No	Zn (or Fe) metallo-protein
CTP synthetase	UTP + NH3 + ATP > CTP + ADP + Pi	ATP-driven amination	No	
CTP synthase	UTP + gln + ATP > CTP + glu ADP + Pi	ATP-driven amination	No	
Amino-transferase	□-amine + oxaloacetate > □-keto- + asp	Trans-amination	Yes	Pyridoxyl-phosphate dependent

Cytidine deaminase is a well-characterized zinc metalloprotein that performs an irreversible hydrolytic deamination reaction. The structure of the protein and the enzyme complexed with substrate and inhibitors has been studied by X-ray crystallography (Betts et al., 1994). The mammalian editing enzyme, Apobec, shares many critical features with the cytidine deaminase although the architecture of the editing enzyme is modified to accept polyribonucleotide substrates (Navaratnam et al., 1998). Cytidine deaminase and the editing deaminases are inhibited by removal of the zinc ion through metal chelators such as 1,10 *o*-phenanthroline. The only empirical evidence against the involvement of an editing deaminase in higher plant organelle editing is the report that the pea mitochondrial extract is not inhibited by *o*-phenanthroline (Takenaka and Brennicke, 2003). An extensive characterization of the *in vitro* editing reaction has not been reported, and substantial surprises may yet be revealed.

The *Arabidopsis* genome includes nine cytidine deaminase-like genes that were initially characterized in an attempt to identify an editing deaminase (Faivre-Nitschke et al., 1999). Additional searches of *Arabidopsis* and other plant databases with short segments of highly conserved amino acid residues from cytidine deaminases do not reveal compelling candidates for an editing deaminase. Thus, the identity of the catalytic subunit of the editing complex remains a mystery, and this mystery may not be resolved until the components of the editing complex are cloned and characterized.

U-to-C conversion is a prevalent form of editing in some non-vascular chloroplast systems, and these plastids also exhibit extensive C-to-U editing. In addition, there are a few early reports of U-to-C editing in higher plant mitochondria; however comprehensive analysis of the completely sequenced *Arabidopsis* and rice mitochondrial genomes failed to identify a single U-to-C editing site in these organelles (Giege and Brennicke, 1999; Notsu et al., 2002). A cytidine deaminase-like enzyme is a very unlikely candidate for the reverse editing reaction because the reaction in that direction is highly unfavorable.

A transamination reaction could reversibly transfer an amine to uridine, and these are common reactions in intermediary metabolism utilizing aspartate/oxaloacetate or glutamate/□-ketoglutarate as amino donors and acceptors. Since, amino acceptors such as oxaloacetate would be required for a transamination mechanism, and these are not required in any plant organelle editing system reported (Hirose and Sugiura, 2001; Miyamoto et al., 2002; Miyamoto et al., 2004; Takenaka and Brennicke, 2003) transamination is an unlikely candidate mechanism for higher plant C-to-U editing.

UTP may be converted to CTP by CTP synthase or CTP synthetase, and ATP is used to drive the amination reaction with ammonia or glutamine as the source of the amine. Although these enzymes are viable candidates for U-to-C editing, this reaction is unfavorable in the C-to-U direction, and is unlikely to be utilized by higher plant organelles in C-to-U editing. In summary, the only reversible mechanism for C-to-U interconversion would seem to be transamination, and higher plant C-to-U editing is clearly not dependent on an amino acceptor; thus, it may be that U-to-C and C-to-U editing occur by distinct mechanisms.

9. *TRANS*-ACTING FACTORS IN RNA EDITING

The editing apparatus would need to have two basic capacities: the ability to specifically recognize an editing site, presumably by binding directly to critical nucleotides in the near 5' flanking region of the editing site; and a catalytic function that is able to convert the critical cytidine residue to a uridine. These capacities could exist in a single enzyme, or in separate macromolecules that form a complex. For example, a *trans*-acting factor could be responsible for recognizing editing sites, and a single catalytic subunit could provide the deamination function for any editing site recognized by the trans-acting factor.

The development of extracts and assays for *in vitro* analysis of RNA editing will allow the rapid identification and characterization of the macromolecules that are involved. Considerable progress has been made in the identification of trans-acting proteins that specifically interact with edited transcripts. Sugiura's group developed a tobacco chloroplast extract and editing assay that allowed them to perform binding and competition assays (Hirose and

Sugiura, 2001). A comprehensive analysis of the cis-acting elements for *psbE* and *petB* editing sites was performed by scanning mutagenesis, with analysis of the mutated RNAs as substrates and competitors. These results established that critical nucleotides exist in the –15 to –6 region of the *psbE* editing site and in the –20 to –6 region of the *petB* editing site (Miyamoto et al., 2002). The critical regions identified for editing site recognition in this study are similar to those identified through *in vivo* studies with transgenic tobacco chloroplasts (Bock et al., 1997; Bock et al., 1996; Chaudhuri and Maliga, 1996; Reed et al., 2001b).

RNA cross-linking experiments identified distinct 56 KD and 70 kD proteins that specifically bind to *psbE* and *petB* mRNAs, respectively (Miyamoto et al., 2002). These proteins specifically bind to ~10 nucleotide sequence elements located five nucleotides upstream of the edited nucleotide, and may represent specificity factors that are responsible for editing site recognition (Miyamoto et al., 2004). Additional competition and cross-linking studies showed that the *trans*-acting proteins interact strongly with the upstream sequence element as well as weakly with the nucleotide immediately surrounding the editing site. Thus, the specificity factor may form a stable interaction with the upstream *cis*-element, but interact weakly with the editing site. Since the same polypeptide is involved in binding both the upstream element and the editing site, these site-specific factors may also possess catalytic activity for C-to-U conversion (Miyamoto et al., 2004).

10. ROLE OF AN RNA HELICASE IN ORGANELLE EDITING

In vitro editing assays have been performed with chloroplast extracts from pea and tobacco and with mitochondrial extracts from pea. Tobacco chloroplast extracts support editing without the addition of ATP, but are strongly stimulated by the addition of ATP (Hirose and Sugiura, 2001). The pea mitochondrial system was reported to exhibit a broad nucleotide specificity (Takenaka and Brennicke, 2003). The broad nucleotide specificity of the *in vitro* editing reaction is a potentially important observation. Most enzyme systems utilize ATP or GTP for phosphorylation; the ability to substitute other NTPs and dNTPs is rather unusual. RNA helicases are known to have a broad nucleotide specificity (Claude et al., 1991; Du et al., 2002; Lin and Kim, 1999; Shuman, 1993; Zhang and Grosse, 1994) that is similar to the specificity detected in the pea mitochondrial system.

RNA helicases are involved in many processes and multicomponent complexes within the cell (Delagoutte and von Hippel, 2003). The role of RNA helicases in multiprotein complexes is becoming better understood, and probably involves not only the unwinding of duplex RNA, but processive motor movement on nucleic acids, disruption of RNA protein complexes, and RNA remodeling for RNA folding. As an example of the numerous roles that RNA helicases play in a complicated reaction such as splicing, eight different RNA helicases are required in essential steps of pre-mRNA splicing: spliceosome assembly, lariat formation, exon ligation, release of the mature mRNA, and recycling of the snRNPs (Silverman et al., 2003). RNA helicases have been shown to be involved in translation, mRNA export, mRNA turnover, and RNA interference. Thus, it is not surprising, and is perhaps predictable, that RNA helicase might be involved in RNA editing.

11. THE MOLECULAR BUREAUCRAT MODEL FOR THE ORIGIN OF RNA EDITING

RNA editing in plant organelles is a curious process in gene expression. Editing in plant organelles is a post-transcriptional process required to correct mutations present in the DNA, and there is no apparent advantage associated with this particular mechanism of dealing with mutations. By contrast, editing of apolipoprotein B or glutamate receptors creates two or more forms of these polypeptides that are expressed in various cells and have distinct functions or properties (Davidson and Shelness, 2000; Schaub and Keller, 2002). In the apolipoprotein and glutamate receptor editing systems, RNA editing plays a critical role in the tissue specificity of gene expression and affects the function of the products; thus, editing plays a clear role in affecting the function of the products.

Extensive analyses have been made in taxa representing the major lineages in angiosperm evolution (Table 4). RNA editing has not been detected in prokaryotes or eukaryotic algae. RNA editing occurs, however, in chloroplasts and mitochondria of some bryophytes (Freyer et al., 1997; Pruchner et al., 2002), which are non-vascular plants that include mosses, hornworts, and liverworts. The mitochondria and chloroplasts of all higher plants examined exhibit C-to-U editing, and U-to-C editing has been rarely detected in some angiosperm mitochondria.

Table 4. *Taxonomic distribution of RNA editing processes*

Taxa	*Distribution*	*C-to-U*	*U-to-C*
Prokaryotes	None	Not detected	Not detected
Alga	None	Not detected	Not detected
Bryophyta*	Some	Chloroplasts Mitochondria	Chloroplasts Mitochondria
Pteridophyta	All	Chloroplasts Mitochondria	Not Detected
Spermatophyta	All	Chloroplasts Mitochondria	Mitochondria**

**RNA editing is absent in chloroplast and mitochondria of Marchantia, but prevalent in other representatives; **U-to-C editing has only rarely been detected in angiosperm mitochondria.*

Editing has not been detected in the prokaryotic progenitors of chloroplasts and mitochondria; however, it is widely distributed in land plants. Thus, RNA editing is apparently a derived characteristic that arose at some point after the divergence of land plants. In the more primitive groups studied, the distribution of RNA editing is sporadic, present in most representatives, but apparently absent in the liverwort *Marchantia*. In addition, U-to-C editing is prevalent in the chloroplast of the hornwort, *Anthoceros*, but has not been detected in higher plant chloroplasts. The scattered distribution of editing and types of editing in the Bryopyta could reflect multiple originations of RNA editing during the evolution of land plants. The mechanisms driving C-to-U and U-to-C editing may be quite different. The prevalence of U-to-C editing in the hornworts may reflect a process that was involved in early forms of editing and that was subsequently lost in higher plant chloroplasts. The elucidation of the editing machinery in these various forms of plant organelle editing will surely result in an interesting perspective on the evolution of RNA editing.

Several questions frequently arise about the existence of RNA editing, the simplest among them being: Why? This extra step in organelle gene expression seems to be an unnecessary process to correct mistakes in the genomic DNA sequence. Why wouldn't the genome just encode the correct genetic information and eliminate this unnecessary post-transcriptional correction? The origin of RNA editing was the topic of a particularly insightful perspective that was published soon after the discovery of editing in plants (Covello and Gray, 1993). Basically the model for the origin of editing was described by these investigators in three steps: 1) the appearance of an RNA editing activity; 2) the occurrence of organelle genome mutations that could be corrected by editing and the fixation of these mutations by genetic drift; and 3) maintenance of RNA editing activity by natural selection.

The appearance of an editing activity might occur through modification of the specificity in an existing enzyme system. This is relatively easy to envision with nucleotide modification editing arising as an enzyme involved with nucleotide metabolism acquires the ability to accept RNA as a substrate. Apobec, the mammalian C-to-U editing enzyme for *apoB* mRNAs, bears such similarity to cytidine deaminase that it is considered a member of the cytidine deaminase family (Navaratnam et al., 1995). Further characterization of the editing apparatus in chloroplasts and plant mitochondria will certainly reveal interesting relationships of the editing machinery with other cellular processes, and will help elucidate the probable origin and evolution of the editing process.

The second proposed step in the origin of editing refers to mutations at "editable nucleotides" is especially insightful in view of the recent recognition of editing site clusters. T-to-C transitions could occur, and be corrected at the RNA level by the editing apparatus. Furthermore, if a specific set of *trans*-acting factors were present in the organelle that could bind and edit a family of closely related RNA sequences, then a T-to-C mutation in that RNA context may immediately be correctable by C-to-U editing. Thus, some Cs may exist in a context of "editable" sites, and mutations in this context could immediately be corrected by editing. In this scenario, editing sites would be expected to arise relatively rapidly within the constraints of editing site recognition by the *trans*-acting factors. In addition, the editing sites would be expected to experience transition mutations at a higher frequency than non-editable sites, being created and destroyed by T-to-C and C-to-T transitions.

The ability to make facile changes at RNA editing sites is a unique feature of these systems that may facilitate more complicated changes in protein evolution.

The ability to express incompletely edited RNAs affords this system the ability to constantly test many different combinations of amino acid changes in the polypeptide. The system may access two amino acid conditions at an edited codon by leaving a C, or editing to a U, and this allows an intrinsic flexibility in evolution. Two amino acid conditions are rapidly accessed at each editing site. Thus, if a mutation arose at any site in a polypeptide, compensatory changes at each editing site in the amino acid sequence could be tested for fitness and function. In addition, all permutations of the editing sites could be accessed by the system, thus offering immediate access to a host of changes.

The impacts of RNA editing on gene evolution have been demonstrated in a study of the mitochondrial *coxI* gene in gymnosperms. A high rate of non-synonymous substitutions occurred in the *cox1* genes with a 98% frequency of C-to-T transitions (Lu et al., 1998). All of the non-synonymous T-to-C changes in genomic DNA sequences were eliminated in the mRNA sequences by RNA editing, and resulted in nearly identical Cox1 amino acid sequences in these species. In addition, the number and location of editing sites varied widely in the group of gymnosperms examined. While most species had about 30 editing sites in the *cox1* transcript, no editing occurred in *Gingko biloba* or *Larix sibirica* transcripts. The elimination of the editing sites in the *cox1* genes of *G. biloba* and *L. sibirica* may have resulted form integration of the edited mitochondrial *coxI* sequence through reverse transcription and insertion into the mitochondrial genome. The authors demonstrated that RNA editing can accelerate the divergence of DNA sequences among species.

Finally, the last step in the origin of the editing apparatus is the maintenance of the editing apparatus by selection. A number of editing sites may accumulate in the genome. It is not probable for all of these editing sites to be simultaneously destroyed by mutation or other molecular process, and the editing apparatus becomes a required part of the gene expression machinery. So, the editing apparatus is a molecular bureaucrat. It created a job for itself, and now it cannot be removed from the administration of gene expression!

12. REFERENCES

Anant, S., & Davidson, N. O. (2000). An AU-rich sequence element (UUUN[A/U]U) downstream of the edited C in apolipoprotein B mRNA is a high-affinity binding site for Apobec-1: binding of Apobec-1 to this motif in the 3' untranslated region of c-myc increases mRNA stability. *Mol. Cell. Biol., 20*, 1982-1992.

Anant, S., & Davidson, N. O. (2002). Identification and regulation of protein components of the apolipoprotein B mRNA editing enzyme. A complex event. *Trends Cardiovasc. Med., 12*, 311-317.

Bass, B. L. (2002). RNA editing by adenosine deaminases that act on RNA. *Annu. Rev. Biochem., 71*, 817-846.

Begu, D., Graves, P. V., Domec, C., Arselin, G., Litvak, S., & Araya, A. (1990). RNA editing of wheat mitochondrial ATP synthase subunit 9: direct protein and cDNA sequencing. *Plant Cell, 2*, 1283-1290.

Benne, R., Van den Burg, J., Brakenhoff, J. P., Sloof, P., Van Boom, J. H., & Tromp, M. C. (1986). Major transcript of the frameshifted coxII gene from trypanosome mitochondria contains four nucleotides that are not encoded in the DNA. *Cell, 46*, 819-826.

Betts, L., Xiang, S., Short, S. A., Wolfenden, R., & Carter, C. W., Jr. (1994). Cytidine deaminase. The 2.3 A crystal structure of an enzyme: transition-state analog complex. *J. Mol. Biol., 235*, 635-656.

Bhattacharya, S., Navaratnam, N., Morrison, J. R., Scott, J., & Taylor, W. R. (1994). Cytosine nucleoside/nucleotide deaminases and apolipoprotein B mRNA editing. *Trends Biochem. Sci., 19*, 105-106.

Blanc, V., Henderson, J. O., Kennedy, S., & Davidson, N. O. (2001). Mutagenesis of apobec-1 complementation factor reveals distinct domains that modulate RNA binding, protein-protein interaction with apobec-1, and complementation of C-to-U RNA-editing activity. *J. Biol. Chem., 276*, 46386-46393.

Blanc, V., Litvak, S., & Araya, A. (1995). RNA editing in wheat mitochondria proceeds by a deamination mechanism. *FEBS Lett., 373*, 56-60.

Bock, R., Hermann, M., and Fuchs, M. (1997). Identification of critical nucleotide positions for plastid RNA editing site recognition. *RNA, 3*, 1194-1200.

Bock, R., Hermann, M., & Kossel, H. (1996). *In vivo* dissection of cis-acting determinants for plastid RNA editing. *EMBO J., 15*, 5052-5059.

Bock, R., Kossel, H., & Maliga, P. (1994). Introduction of a heterologous editing site into the tobacco plastid genome: the lack of RNA editing leads to a mutant phenotype. *EMBO J., 13*, 4623-4628.

Carrillo, C., & Bonen, L. (1997). RNA editing status of *nad7* intron domains in wheat mitochondria. *Nucleic Acids Res., 25*, 403-409.

Chateigner-Boutin, A. L., & Hanson, M. R. (2002). Cross-competition in transgenic chloroplasts expressing single editing sites reveals shared cis elements. *Mol. Cell. Biol., 22*, 8448-8456.

Chateigner-Boutin, A. L., & Hanson, M. R. (2003). Developmental co-variation of RNA editing extent of plastid editing sites exhibiting similar cis-elements. *Nucleic Acids Res., 31*, 2586-2594.

Chaudhuri, S., Carrer, H., & Maliga, P. (1995). Site-specific factor involved in the editing of the psbL mRNA in tobacco plastids. *EMBO J., 14*, 2951-2957.

Chaudhuri, S., & Maliga, P. (1996). Sequences directing C-to-U editing of the plastid *psbL* mRNA are located within a 22 nucleotide segment spanning the editing site. *EMBO J., 15*, 5958-5964.

Claude, A., Arenas, J., & Hurwitz, J. (1991). The isolation and characterization of an RNA helicase from nuclear extracts of HeLa cells. *J. Biol. Chem., 266*, 10358-10367.

Conley, C. A., & Hanson, M. R. (1995). How do alterations in plant mitochondrial genomes disrupt pollen development? *J. Bioenerg. Biomembr., 27*, 447-457.

Covello, P. S., & Gray, M. W. (1989). RNA editing in plant mitochondria. *Nature, 341*, 662-666.

Covello, P. S., & Gray, M. W. (1993). On the evolution of RNA editing. *Trends Genet., 9*, 265-268.

Davidson, N. O., & Shelness, G. S. (2000). APOLIPOPROTEIN B: mRNA editing, lipoprotein assembly, and pre-secretory degradation. *Annu, Rev. Nutr., 20*, 169-193.

Del Campo, E. M., Sabater, B., & Martin, M. (2002). Post-transcriptional control of chloroplast gene expression. Accumulation of stable *psaC* mRNA is due to downstream RNA cleavages in the *ndhD* gene. *J. Biol. Chem., 277*, 36457-36464.

Delagoutte, E., & von Hippel, P. H. (2003). Helicase mechanisms and the coupling of helicases within macromolecular machines. Part II: Integration of helicases into cellular processes. *Q. Rev. Biophys., 36*, 1-69.

Du, M. X., Johnson, R. B., Sun, X. L., Staschke, K. A., Colacino, J., & Wang, Q. M. (2002). Comparative characterization of two DEAD-box RNA helicases in superfamily II: human translation-initiation factor 4A and hepatitis C virus non-structural protein 3 (NS3) helicase. *Biochem. J., 363*, 147-155.

Faivre-Nitschke, S. E., Grienenberger, J. M., & Gualberto, J. M. (1999). A prokaryotic-type cytidine deaminase from *Arabidopsis thaliana* gene expression and functional characterization. *Eur. J., Biochem., 263*, 896-903.

Feagin, J. E., Abraham, J. M., & Stuart, K. (1988). Extensive editing of the cytochrome c oxidase III transcript in *Trypanosoma brucei*. *Cell, 53*, 413-422.

Fox, T. D., & Leaver, C. J. (1981). The *Zea mays* mitochondrial gene coding cytochrome oxidase subunit II has an intervening sequence and does not contain TGA codons. *Cell, 26*, 315-323.

Freyer, R., Hoch, B., Neckermann, K., Maier, R. M., & Kossel, H. (1993). RNA editing in maize chloroplasts is a processing step independent of splicing and cleavage to monocistronic mRNAs. *Plant J., 4*, 621-629.

Freyer, R., Kiefer-Meyer, M. C., & Kossel, H. (1997). Occurrence of plastid RNA editing in all major lineages of land plants. *Proc. Natl. Acad. Sci. USA, 94*, 6285-6290.

Giege, P., & Brennicke, A. (1999). RNA editing in Arabidopsis mitochondria effects 441 C-to-U changes in ORFs. *Proc. Natl. Acad. Sci. USA, 96*, 15324-15329.

Gott, J. M., & Emeson, R. B. (2000). Functions and mechanisms of RNA editing. *Annu. Rev. Genet., 34*, 499-531.

Grohmann, L., Thieck, O., Herz, U., Schroder, W., & Brennicke, A. (1994). Translation of *nad9* mRNAs in mitochondria from *Solanum tuberosum* is restricted to completely edited transcripts. *Nucleic Acids Res., 22*, 3304-3311.

Grosskopf, D., & Mulligan, R. M. (1996). Developmental- and tissue-specificity of RNA editing in mitochondria of suspension-cultured maize cells and seedlings. *Curr. Genet., 29*, 556-563.

Gualberto, J. M., Lamattina, L., Bonnard, G., Weil, J. H., & Grienenberger, J. M. (1989). RNA editing in wheat mitochondria results in the conservation of protein sequences. *Nature, 341*, 660-662.

Gualberto, J. M., Weil, J. H., & Grienenberger, J. M. (1990). Editing of the wheat *coxIII* transcript: evidence for twelve C-to-U and one U-to-C conversions and for sequence similarities around editing sites. *Nucleic Acids Res., 18*, 3771-3776.

Handa, H., Gualberto, J. M., & Grienenberger, J. M. (1995). Characterization of the mitochondrial *orfB* gene and its derivative, *orf224*, a chimeric open reading frame specific to one mitochondrial genome of the "Polima" male-sterile cytoplasm in rapeseed (Brassica napus L.). *Curr. Genet., 28*, 546-552.

Hanson, M. R., Wilson, R. K., Bentolila, S., Kohler, R. H., & Chen, H. C. (1999). Mitochondrial gene organization and expression in petunia male fertile and sterile plants. *J. Hered., 90*, 362-368.

Hayes, M. L., Hegeman-Crim, C. E., Chateigner-Boutin, A.L. & Hanson, M.R (2004). *In vitro* analysis of RNA substrate requirements for plastid RNA editing. Annual Meeting of the Canadian Society of Plant Physiologists Abstract.

Henderson, J. O., Blanc, V., & Davidson, N. O. (2001). Isolation, characterization and developmental regulation of the human apobec-1 complementation factor (ACF) gene. *Biochim. Biophys. Acta, 1522*, 22-30.

Hermann, M., & Bock, R. (1999). Transfer of plastid RNA-editing activity to novel sites suggests a critical role for spacing in editing-site recognition. *Proc. Natl. Acad. Sci. USA, 96*, 4856-4861.

Hernould, M., Suharsono, S., Litvak, S., Araya, A., & Mouras, A. (1993). Male-sterility induction in transgenic tobacco plants with an unedited *atp9* mitochondrial gene from wheat. *Proc. Natl. Acad. Sci. USA, 90*, 2370-2374.

Hiesel, R., Wissinger, B., & Brennicke, A. (1990). Cytochrome oxidase subunit II mRNAs in Oenothera mitochondria are edited at 24 sites. *Curr. Genet., 18*, 371-375.

Hiesel, R., Wissinger, B., Schuster, W., & Brennicke, A. (1989). RNA editing in plant mitochondria. *Science, 246*, 1632-1634.

Hirose, T., & Sugiura, M. (2001). Involvement of a site-specific trans-acting factor and a common RNA-binding protein in the editing of chloroplast mRNAs: development of a chloroplast *in vitro* RNA editing system. *EMBO J., 20*, 1144-1152.

Kubo, N., & Kadowaki, K. (1997). Involvement of 5' flanking sequence for specifying RNA editing sites in plant mitochondria. *FEBS Lett., 413*, 40-44.

Kugita, M., Kaneko, A., Yamamoto, Y., Takeya, Y., Matsumoto, T., & Yoshinaga, K. (2003a). The complete nucleotide sequence of the hornwort (Anthoceros formosae) chloroplast genome: insight into the earliest land plants. *Nucleic Acids Res., 31*, 716-721.

Kugita, M., Yamamoto, Y., Fujikawa, T., Matsumoto, T., & Yoshinaga, K. (2003b). RNA editing in hornwort chloroplasts makes more than half the genes functional. *Nucleic Acids Res., 31*, 2417-2423.

Lin, C., & Kim, J. L. (1999). Structure-based mutagenesis study of hepatitis C virus NS3 helicase. *J. Virol., 73*, 8798-8807.

Lu, B., & Hanson, M. R. (1994). A single homogeneous form of ATP6 protein accumulates in petunia mitochondria despite the presence of differentially edited *atp6* transcripts. *Plant Cell, 6*, 1955-1968.

Lu, B., & Hanson, M. R. (1996). Fully edited and partially edited *nad9* transcripts differ in size and both are associated with polysomes in potato mitochondria. *Nucleic Acids Res., 24*, 1369-1374.

Lu, B., Wilson, R. K., Phreaner, C. G., Mulligan, R. M., & Hanson, M. R. (1996). Protein polymorphism generated by differential RNA editing of a plant mitochondrial *rps12* gene. *Mol. Cell. Biol., 16*, 1543-1549.

Lu, M. Z., Szmidt, A. E., & Wang, X. R. (1998). RNA editing in gymnosperms and its impact on the evolution of the mitochondrial *coxI* gene. *Plant Mol.Biol., 37*, 225-234.

Maas, S., Rich, A., & Nishikura, K. (2003). A-to-I RNA editing: recent news and residual mysteries. *J. Biol. Chem., 278*, 1391-1394.

Maier, R. M., Neckermann, K., Igloi, G. L., & Kossel, H. (1995). Complete sequence of the maize chloroplast genome: gene content, hotspots of divergence and fine tuning of genetic information by transcript editing. *J. Mol. Biol., 251*, 614-628.

Miyamoto, T., Obokata, J., & Sugiura, M. (2002). Recognition of RNA editing sites is directed by unique proteins in chloroplasts: biochemical identification of cis-acting elements and trans-acting factors involved in RNA editing in tobacco and pea chloroplasts. *Mol. Cell. Biol., 22*, 6726-6734.

Miyamoto, T., Obokata, J., & Sugiura, M. (2004). A site-specific factor interacts directly with its cognate RNA editing site in chloroplast transcripts. *Proc. Natl. Acad. Sci. USA, 101*, 48-52.

Mulligan, R. M., Williams, M. A., & Shanahan, M. T. (1999). RNA editing site recognition in higher plant mitochondria. *J. Hered., 90*, 338-344.

Muramatsu, M., Sankaranand, V. S., Anant, S., Sugai, M., Kinoshita, K., Davidson, N. O., & Honjo, T. (1999). Specific expression of activation-induced cytidine deaminase (AID), a novel member of the RNA-editing deaminase family in germinal center B cells. *J. Biol. Chem., 274*, 18470-18476.

Navaratnam, N., Bhattacharya, S., Fujino, T., Patel, D., Jarmuz, A. L., & Scott, J. (1995). Evolutionary origins of *apoB* mRNA editing: catalysis by a cytidine deaminase that has acquired a novel RNA-binding motif at its active site. *Cell, 81*, 187-195.

Navaratnam, N., Fujino, T., Bayliss, J., Jarmuz, A., How, A., Richardson, N., Somasekaram, A., Bhattacharya, S., Carter, C., & Scott, J. (1998). *Escherichia coli* cytidine deaminase provides a molecular model for *apoB* RNA editing and a mechanism for RNA substrate recognition. *J. Mol. Biol., 275*, 695-714.

Notsu, Y., Masood, S., Nishikawa, T., Kubo, N., Akiduki, G., Nakazono, M., Hirai, A., & Kadowaki, K. (2002). The complete sequence of the rice (*Oryza sativa* L.) mitochondrial genome: frequent DNA sequence acquisition and loss during the evolution of flowering plants. *Mol. Genet. Genom., 268*, 434-445.

Okazaki, I., Yoshikawa, K., Kinoshita, K., Muramatsu, M., Nagaoka, H., & Honjo, T. (2003). Activation-induced cytidine deaminase links class switch recombination and somatic hypermutation. *Ann. NY Acad. Sci., 987*, 1-8.

Peeters, N. M., & Hanson, M. R. (2002). Transcript abundance supercedes editing efficiency as a factor in developmental variation of chloroplast gene expression. *RNA, 8*, 497-511.

Phreaner, C. G., Williams, M. A., & Mulligan, R. M. (1996). Incomplete editing of *rps12* transcripts results in the synthesis of polymorphic polypeptides in plant mitochondria. *Plant Cell, 8*, 107-117.

Pruchner, D., Beckert, S., Muhle, H., & Knoop, V. (2002). Divergent intron conservation in the mitochondrial *nad2* gene: signatures for the three bryophyte classes (mosses, liverworts, and hornworts) and the lycophytes. *J. Mol. Evol., 55*, 265-271.

Rajasekhar, V. K., & Mulligan, R. M. (1993). RNA Editing in Plant Mitochondria: [alpha]-Phosphate Is Retained during C-to-U Conversion in mRNAs. *Plant Cell, 5*, 1843-1852.

Reed, M. L., & Hanson, M. R. (1997). A heterologous maize *rpoB* editing site is recognized by transgenic tobacco chloroplasts. *Mol. Cell. Biol., 17*, 6948-6952.

Reed, M. L., Lyi, S. M., & Hanson, M. R. (2001a). Edited transcripts compete with unedited mRNAs for trans-acting editing factors in higher plant chloroplasts. *Gene, 272*, 165-171.

Reed, M. L., Peeters, N. M., & Hanson, M. R. (2001b). A single alteration 20 nt 5' to an editing target inhibits chloroplast RNA editing *in vivo*. *Nucleic Acids Res., 29*, 1507-1513.

Sasaki, T., Yukawa, Y., Miyamoto, T., Obokata, J., & Sugiura, M. (2003). Identification of RNA editing sites in chloroplast transcripts from the maternal and paternal progenitors of tobacco (*Nicotiana tabacum*): comparative analysis shows the involvement of distinct trans-factors for *ndhB* editing. *Mol. Biol. Evol., 20*, 1028-1035.

Saunders, L. R., & Barber, G. N. (2003). The dsRNA binding protein family: critical roles, diverse cellular functions. *FASEB J., 17*, 961-983.

Schaub, M., & Keller, W. (2002). RNA editing by adenosine deaminases generates RNA and protein diversity. *Biochimie, 84*, 791-803.

Schuster, W., Hiesel, R., Wissinger, B., & Brennicke, A. (1990). RNA editing in the cytochrome b locus of the higher plant *Oenothera berteriana* includes a U-to-C transition. *Mol. Cell. Biol., 10*, 2428-2431.

Seeburg, P. H., & Hartner, J. (2003). Regulation of ion channel/neurotransmitter receptor function by RNA editing. *Curr. Opin . Neurobiol., 13*, 279-283.

Shuman, S. (1993). Vaccinia virus RNA helicase. Directionality and substrate specificity. *J. Biol. Chem., 268*, 11798-11802.

Silverman, E., Edwalds-Gilbert, G., & Lin, R. J. (2003). DExD/H-box proteins and their partners: helping RNA helicases unwind. *Gene, 312*, 1-16.

Simpson, L., Sbicego, S., & Aphasizhev, R. (2003). Uridine insertion/deletion RNA editing in trypanosome mitochondria: a complex business. *RNA, 9*, 265-276.

Smith, H. C., Gott, J. M., & Hanson, M. R. (1997). A guide to RNA editing. *RNA, 3*, 1105-1123.

Sugiura, M. (1995). The chloroplast genome. *Essays Biochem., 30*, 49-57.

Sutton, C. A., Conklin, P. L., Pruitt, K. D., & Hanson, M. R. (1991). Editing of pre-mRNAs can occur before cis- and trans-splicing in Petunia mitochondria. *Mol. Cell. Biol., 11*, 4274-4277.

Takenaka, M., & Brennicke, A. (2003). *In vitro* RNA editing in pea mitochondria requires NTP or dNTP, suggesting involvement of an RNA helicase. *J. Biol. Chem., 278*, 47526-47533.

Thiemann, O. H., Maslov, D. A., and Simpson, L. (1994). Disruption of RNA editing in *Leishmania tarentolae* by the loss of minicircle-encoded guide RNA genes. *EMBO J., 13*, 5689-5700.

Tsudzuki, T., Wakasugi, T., & Sugiura, M. (2001). Comparative analysis of RNA editing sites in higher plant chloroplasts. *J. Mol. Evol., 53*, 327-332.

Vidal, S., Curran, J., & Kolakofsky, D. (1990). A stuttering model for paramyxovirus P mRNA editing. *EMBO J., 9*, 2017-2022.

Wang, A. B., Liu, D. P., & Liang, C. C. (2003). Regulation of human apolipoprotein B gene expression at multiple levels. *Exp. Cell Res., 290*, 1-12.

Wedekind, J. E., Dance, G. S., Sowden, M. P., & Smith, H. C. (2003). Messenger RNA editing in mammals: new members of the APOBEC family seeking roles in the family business. *Trends Genet., 19*, 207-216.

Williams, M. A., Kutcher, B. M., & Mulligan, R. M. (1998a). Editing site recognition in plant mitochondria: the importance of 5'-flanking sequences. *Plant Mol. Biol., 36*, 229-237.

Williams, M. A., Tallakson, W. A., Phreaner, C. G., & Mulligan, R. M. (1998b). Editing and translation of ribosomal protein S13 transcripts: unedited translation products are not detectable in maize mitochondria. *Curr. Genet., 34*, 221-226.

Yang, A. J., & Mulligan, R. M. (1991). RNA editing intermediates of *cox2* transcripts in maize mitochondria. *Mol. Cell. Biol., 11*, 4278-4281.

Yu, W., & Schuster, W. (1995). Evidence for a site-specific cytidine deamination reaction involved in C to U RNA editing of plant mitochondria. *J. Biol. Chem., 270*, 18227-18233.

Zabaleta, E., Mouras, A., Hernould, M., Suharsono, & Araya, A. (1996). Transgenic male-sterile plant induced by an unedited *atp9* gene is restored to fertility by inhibiting its expression with antisense RNA. *Proc. Natl. Acad. Sci. USA, 93*, 11259-11263.

Zhang, S., & Grosse, F. (1994). Nuclear DNA helicase II unwinds both DNA and RNA. *Biochemistry, 33*, 3906-3912.

Zito, F., Kuras, R., Choquet, Y., Kossel, H., & Wollman, F. A. (1997). Mutations of cytochrome b6 in *Chlamydomonas reinhardtii* disclose the functional significance for a proline to leucine conversion by *petB* editing in maize and tobacco. *Plant Mol. Biol., 33*, 79-86.

CHAPTER 10

PLASTID AND PLANT MITOCHONDRIAL RNA PROCESSING AND RNA STABILITY

A. MARCHFELDER AND S. BINDER

Molekulare Botanik, Universität Ulm, Albert-Einstein-Allee 11, 89069 Ulm, Germany

Abstract. In recent years considerable progress has been made in the analysis of posttranscriptional processes in both mitochondria and chloroplasts. Processes such as RNA editing, intron splicing, excision of individual mRNAs from large precursor transcripts, endonucleolytic cleavages in the 5' and 3' untranslated regions (UTRs), polyadenylation as well as translation have been described and functionally investigated. All these processes can influence RNA stability and contribute to posttranscriptional gene regulation in a complex scaffold of regulatory levels including various interactions of the different mechanisms. However, processes that directly interfere with RNA stability have the strongest influence on gene expression and thus are the major checkpoints of posttranscriptional gene regulation. In chloroplasts and mitochondria of higher plants 3' end trimming interferes with RNA decay, and polyadenylation of RNAs enhances degradation. In *Chlamydomonas reinhardtii* and also in higher plant chloroplasts *cis*-elements in the 5' UTRs are have been implicated to be important for gene regulation by RNA stability. Beside these processes the excision of translatable mRNAs from large precursor molecules may also represent an important level to regulate gene expression posttranscriptionally. In this chapter we will summarize recent developments in the studies of mRNA processing and stability and their impact on gene expression. We will also discuss progress that has been made in the analysis of tRNA and rRNA maturation, which may also have a regulatory influence on overall organellar gene activity.

1. INTRODUCTION

Mitochondria and plastids have developed from free living bacteria-like organisms and as a consequence of that still contain their own DNA, which encodes genes essential for the eukaryotic cell (Hoffmann *et al.*, 2001). Inherent is the requirement of the cell to realize this genetic information. The integration of the formerly free living organisms and their development to the present day organelles changed their environment from most likely variable conditions to the relatively stable surroundings of the eukaryotic cell. However, these cells must differentiate and fulfill diverse functions in the organism and they are also exposed to changing environmental conditions, which require physiological and biochemical changes at

H. Daniell & C. Chase (eds.), Molecular Biology and Biotechnology of Plant Organelles, 261—294.

intracellular levels. Organellar gene expression also has to respond to these endogenous and exogenous changes. The embedding of the organelles in the cellular context may have reduced the requirement for fast and dramatic adaptions of organellar gene expression to exogenous stimuli but it has added new requirements in terms of coordination with nuclear gene expression since many components of organellar genetic systems are encoded in the nucleus, particularly the high molecular weight respiratory and photosynthetic complexes are composed of organelle- and nucleus-encoded subunits.

Gene expression is predominately regulated at the level of transcription initiation. Distinct transcription factors control specific initiation events in response to exo- and endogenous stimuli. Although enormous progress has been made towards elucidating the transcription initiation processes in organelles by analyzing promoters and transacting factors and although light regulated promoters have been characterized in detail (Allison and Maliga, 1995; Christopher *et al.*, 1992), the overall contribution of the transcription initiation process to the regulation of organellar gene expression seems to be of minor relevance since most of the transcription rates remain constant under varying endo- and exogenous conditions (Stern *et al.*, 1997).

In the mid eighties the first complete sequences of plastid DNAs from tobacco and liverwort were reported (Ohyama, 1996; Sugiura, 1995). Comparison with partial sequences and complete plastid sequences from many other species revealed that gene arrangements and transcription units are extremely conserved. Although most of these transcription units have been rearranged after the initial endosymbiosis, some of them still resemble operons in cyanobacteria (details see Chapter 5). Most of the plastid-encoded genes are transcribed as parts of polycistronic precursor molecules, either by the multisubunit RNA polymerase (PEP) similar to bacterial RNA polymerases or the phage-type single subunit polymerase (NEP). While PEP is - with the exception of the sigma factors - encoded in the plastid DNA, the latter enzyme is encoded in the nucleus (details see Chapter 14).

Pioneering work on plastid gene expression regulation performed in the end eighties indicated that besides the modulation of transcriptional rates, changes in the stability of mRNAs are the major parameters determining gene expression. This comparative study of promoter activities and transcript abundances during light induced chloroplast development and plastid differentiation revealed significant changes of relative transcript quantities upon constant relative transcription initiation frequencies. The discrepancy suggested that the steady state levels of several RNAs are controlled rather by changes in the stability than by variation in transcription initiation rates (Deng and Gruissem, 1987; Deng *et al.*, 1989; Gruissem, 1989). These studies indicated for the first time that posttranscriptional processes are the major players in regulation of gene expression in chloroplasts.

Chloroplasts are the dominant organelles in vegetative tissues of an adult plant, whereas mitochondria are rather underrepresented in this developmental phase. Nonetheless their function is essential for the survival of a plant, and last but not least they are the predominant energy source in certain developmental phases such as germination. Studies of cytoplasmic male sterility (CMS) established the importance of full mitochondrial activity in generative plant tissues (details see Chapter 25). This trait is manifested by the expression of chimeric open reading frames created by recombination, whose expression leads to abnormal pollen development (Budar and Pelletier, 2001). Interestingly, the restoration of CMS by respective restorer of fertility (*rf*) genes is often achieved on the posttranscriptional level (for details see below).

One of the hallmarks of the molecular biological studies of plant mitochondria was the presentation of complete sequences of the mitochondrial DNAs from liverwort (*Marchantia polymorpha*) and *Arabidopsis thaliana* (Oda *et al.*, 1992; Unseld *et al.*, 1997), providing fundamental information for the analysis of gene expression. The distribution of the genes suggests that most of them could be expressed in multipartite transcription units accompanied by extensive processing of respective precursor RNAs (Dombrowski *et al.*, 1998). A comprehensive analysis of the relative contribution of transcriptional and posttranscriptional processes to the steady state mRNA levels was later performed in *A. thaliana* (Giegé *et al.*, 2000). This study revealed high variations in the transcriptional rates among the genes for different subunits of the same complex, which appear to be counterbalanced by posttranscriptional activities resulting in relatively homogenous steady state mRNA levels. This analysis thus confirmed other reports that had suggested posttranscriptional processes to be important for regulation of gene expression in plant mitochondria (see below).

Despite the progress that has been achieved in recent years many more detailed functional studies are required to sketch a reasonable comprehensive picture of the posttranscriptional processes involved in organellar gene regulation. The complete nuclear sequences of the model organisms *A. thaliana* and rice and the availability of respective mutant collections have already started to enhance the analysis of posttranscriptional processes in organelles. (Alonso *et al.*, 2003; A.G.I., 2000; Goff, 2002; Yu, 2002).

2. MESSENGER RNA PROCESSING AND STABILITY

2.1. Disassembly of organellar polycistronic precursor RNAs and generation of translatable transcripts

As mentioned above, most plastid encoded genes are transcribed within polycistronic precursor RNAs that undergo extensive processing to generate mono- or dicistronic translatable transcripts. This is reflected by the presence of highly complex transcription patterns often additionally complicated by multiple transcription initiation sites. The *psb*B operon encoding *psb*B, *psb*T, *psb*H *pet*B and *pet*D is one of the best characterized plastid transcription unit and a paradigm for processing of a typical polycistronic transcription unit (Barkan, 1988; Westhoff and Herrmann, 1988). The *in vivo* importance of the disassembly of the multicistronic precursors into certain mRNAs is still not fully understood as is the identity of the components necessary for these processing events. Nevertheless the generation of translatable mRNAs by endonucleolytic cleavage offers the first opportunity for posttranscriptional regulation of gene expression (Fig. 1).

Recent analyses of various mutants now shed light on the identity of some of these components operating in this early stage of posttranscriptional processing. But often the multiple molecular phenotypes of the mutants or concomitant indirect effects hamper the unambiguous and clear assignment of a function to the individual protein affected in the respective mutant. For instance the pentatricopeptide repeat (PPR) protein CRP1 with similarity to mitochondrial and plastidal translation regulators is required for the translation of *pet*A and *pet*D mRNAs, but independent from that also for the processing of *pet*D transcripts (Barkan *et al.*, 1994; Fisk *et al.*, 1999; Small and Peeters, 2000). Similarly a mutation in a protein encoded by the

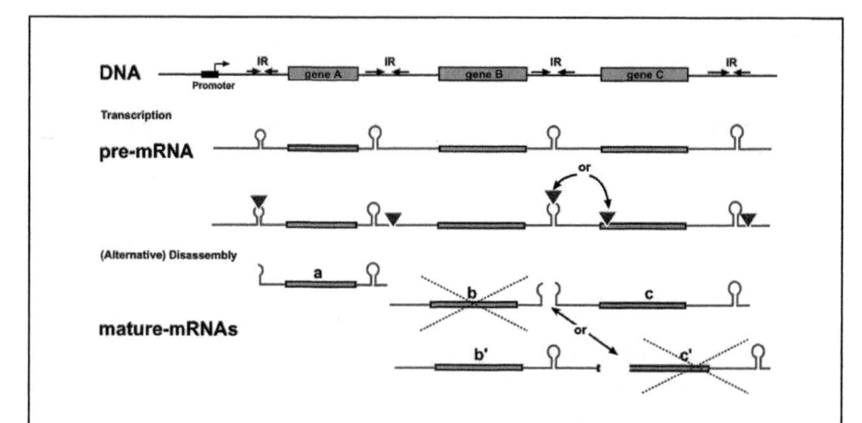

Figure 1. Generation of translatable transcripts from multicistronic precursor RNAs in plastids. *After or possibly also during transcription large pre-mRNAs are disassembled most likely by endonucleolytic processing indicated by dark gray triangles. The endonucleolytic cleavage can directly generate translatable mRNA (transcript a). It may destroy 5' stem-loops preventing translation (transcript a) or may initiate additional 3' trimming (transcripts b' and c; 3' trimming see Fig. 2). Alternative endonucleolytic cuts may generate distinct intact transcripts (b' and c) on the expense of others (b and c'), which cannot be translated or which are not stable (crossed out). If these alternative processing events occur with different efficiencies the relative abundances of mRNAs of a transcription unit may shift.*

*hcf*107 locus causes miscellaneous molecular phenotypes (Felder *et al.*, 2001). It is required for processing at a site 45 nucleotides upstream of the *psb*H reading frame and consequently in the respective *hcf*107 mutant no transcripts with *psb*H as leading cistron accumulate and no *psb*H gene product is synthesized. It was speculated that the endonucleolytic processing might unfold or destroy a secondary structure that prevents translation of *psb*H mRNAs with 5' ends extending upstream of this cleavage site (Fig. 1). This would nicely fit with a function the HCF107 protein alternatively as a site-specific endoribonucleases, an auxillary factor for such a nuclease or a stabilizing factor of ~45 processed RNAs. However, these putative functions do not account for the observation that another protein, the *psb*B encoded CP47, is not synthesized in the mutant but in this case without any apparent changes of the *psb*B transcripts. Therefore it is hypothesized that HCF107 might be part of a multiprotein complex fulfilling different functions such as endonucleolytic cleavage, transcript stabilization and translation activation (Felder *et al.*, 2001).

Analysis of two allelic EMS and T-DNA insertion mutants named *hcf*152, which are deficient in the accumulation of the cytochrome b_6f complex, identified an 80 kDa PPR protein that shows structural similarity to the above mentioned CRP1. This protein was found to interfere with the efficient accumulation of spliced *pet*B RNAs. In addition the accumulation of transcripts cleaved between the *psb*H and *pet*B genes is affected at least in the T-DNA insertion mutant. The reduced steady state level of spliced *pet*B mRNAs in the mutants implies that the HCF152 protein might be required for the splicing event; however, this defect should be accompanied by reduced levels of the excised intron, which was not found to be the case. It seems thus more plausible that this PPR protein is a specific stabilizing factor for processed and spliced transcripts, or a translation factor, whose absence in the mutants results in a decay of spliced *pet*B RNAs. In addition the lack of cleavage between *psb*H and *pet*B in *hcf*152-1 indicates an additional role of this protein in endonucleolytic cleavage since so far no evidence has been found for an assembly with other proteins (Meierhoff *et al.*, 2003). The study of this 80 kDa PPR protein, which was also found to be an RNA binding protein prefentially interacting with exon-intron junctions in *pet*B RNAs (Nakamura *et al.*, 2003), as well as of CRP1 and the *hcf*107 gene product demonstrates the benefits of genetic approaches, but also clearly shows the difficulties of unambiguous assignments of protein function(s) in plastid transcript maturation.

A member of the plant combinatorial and modular protein (PCMP) family (Aubourg *et al.*, 2000), CRR2, which is functionally related to PPR proteins, is reported to be involved in processing of *ndh*B mRNAs in *A. thaliana* plastids. In respective allelic mutants, which do not accumulate NADH dehydrogenase complexes, endonucleolytic cleavage at a site 12 nucleotides upstream of the translation start codon is most likely impaired (Hashimoto *et al.*, 2003). But as in other genetic studies the exact function of the identified protein remains somewhat unclear.

Another nice example for the regulatory role of the disassembly of large precursor RNAs has recently been described for the *ndh*H-D operon in leek (*Allium porrum*)(Del Campo *et al.*, 2002). Intergenic cleavages, intron splicing and RNA editing are required to generate multiple overlapping transcripts, of which the *psa*C mRNA accumulates to steady state levels estimated to be around two orders of magnitude higher than the six *ndh* RNAs. Investigation of *in vivo* RNAs indicates that the state of the editing site restoring *ndh*D AUG from ACG and two alternative endonucleolytic cleavage scenarios may be responsible for different amounts of *psa*C and *ndh*D transcripts (Fig. 1). When cleavage occurs in the *psa*C-*ndh*D intergenic region, three different *ndh*D mRNAs with 5' ends at position –56, –45, and –31 upstream of the translation start codon are released. Processing sites further downstream are used to generate stable *psa*C transcripts with 3' UTRs extending beyond the *ndh*D start codon, which remained unedited in most of these RNAs.

Thus RNA editing may direct endonucleolytic cleavage generating *psa*C mRNAs at the expense of *ndh*D (Del Campo *et al.*, 2002).

The transcription modes of mitochondrially and plastid-encoded genes resemble each other in many respects. For instance many mitochondrial genes are, as in plastids, cotranscribed within large transcription units and the primary precursor RNA is intensively processed as indicated by complicated transcript patterns (Fig. 1). In contrast to plastids the gene order and the composition of the mitochondrial transcription units are highly variable between different plant species (Handa, 2003; Kubo *et al.*, 2000; Notsu *et al.*, 2002; Unseld *et al.*, 1997). Only a few transcription units for instance the 18S-5S rRNAs, the *rps*12-*nad*3 and the *rpl*5-*rps*14-*cob* are conserved in many different plant species, the latter, for example, being completely maintained in pea, rape seed, potato and rice but found partially dismantled in *Oenothera* and *Vicia faba*. (Aubert *et al.*, 1992; Brandt *et al.*, 1993; Hoffmann *et al.*, 1999; Quiñones *et al.*, 1996; Schuster, 1993; Schuster *et al.*, 1990; Wahleithner and Wolstenholme, 1988; Ye *et al.*, 1993).

Monocistronic genes can also be represented by many different steady state RNAs, which partially originate from different transcription initiation sites but are also generated by multiple processing events in the 5' and 3' UTRs (Hoffmann *et al.*, 2001). Current studies of the steady state RNAs in *A. thaliana* mitochondria even suggest that most major RNAs contain posttranscriptionally generated 5' ends (J. Forner, B. Weber and S. Binder, unpublished results). This is similar to the above discussed *psb*H 5' processing, which is crucial for the respective mRNAs to become a valid template for translation. It is thus tempting to speculate that these 5' processing events in mitochondria might have a regulatory influence on translation and thus on gene expression.

Altered transcript profiles most likely by different endonucleolytic cleavages or specific destabilization of certain mitochondrial RNAs have also been observed as an effect of restoration of cytoplasmic male sterility (CMS). Some genes encoding PPR proteins have been identified as restorer of fertility (*rf*) genes. For instance in restored petunia plants the abundance of a distinct *pcf* transcript is substantially reduced (Bentolila *et al.*, 2002; Nivison and Hanson, 1989). Likewise a PPR protein with 18 repeats is identified as restorer of fertility gene 1 (*rf*-1) in rice that seems to be responsible for processing of a cytoplasmic male sterility (CMS) related aberrant *atp*6 RNA (Kazama and Toriyama, 2003). Both PPR proteins thus seem to be involved in RNA metabolism, although their exact modes of action remain to be investigated. A direct correlation of such PRR proteins with RNA metabolism, however, seems not to be obligatory, since another member of this gene family restores fertility in the Kosena CMS radish, and lowers the level of the CMS-causing *orf*125 gene product without any effect on the respective mRNA (Koizuka *et al.*, 2003).

In summary these studies show the importance of the detailed dissection of precursor RNAs and the endonucleolytic processing of monocistronic transcripts in plant organelles. The genetic approaches studying this feature in chloroplasts are promising start points for these investigations. But much more detailed studies, especially in mitochondria, are necessary to unravel the mechanisms of the various complex processing events to determine the identity and exact function(s) of the proteins involved and their contributions to regulation of gene expression in plastids and mitochondria.

2.2. Cis *elements in the 5' UTRs in chloroplast mRNAs are important for RNA stability*

The importance of the 5' UTRs for RNA stability of plastid transcripts is by far best explored in *C. reinhardtii*. This unicellular alga is amenable to both biochemical as well as genetic approaches, which in recent years revealed *cis*-elements in these

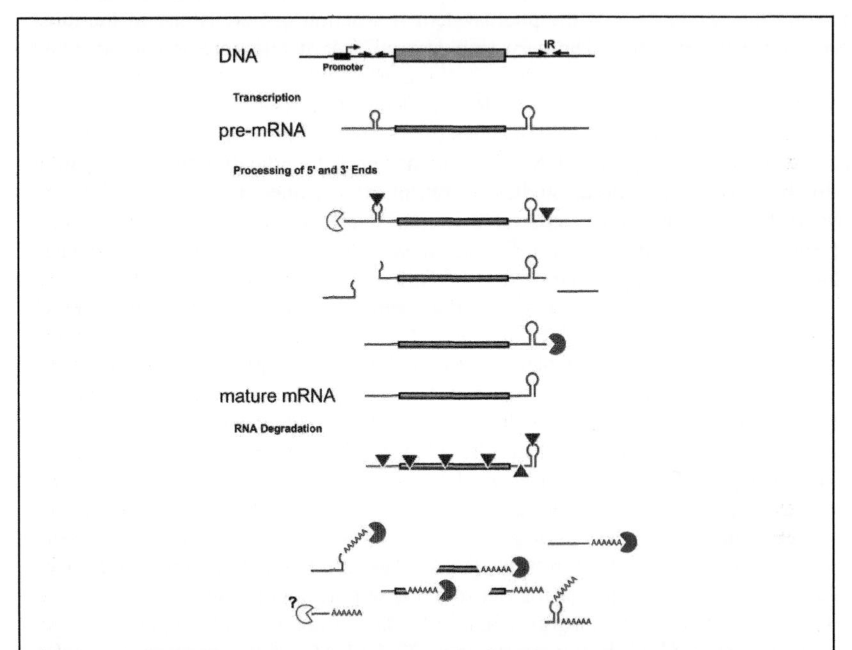

Figure 2. Processing and degradation of mRNAs in chloroplasts. *A gene is transcribed with large 5' and 3' UTRs containing stem-loop structures. Endonucleolytic (dark gray triangle) cleavage within a secondary structure or 5' exonucleolytic (white pacman) processing occurs and may be required for subsequent translation. 3' trimming is initiated by an endonucleolytic cut downstream of a stem-loop structure (encoded by an inverted repeat (IR), indicated by black arrows) followed by 3' exonucleolytic (gray pacman) removal of sequences downstream of the stem-loop. Degradation of mRNA is most likely initiated by endonucleolytic cuts within the coding region and upstream or within the 3' stem-loop structure. Cleavage products are then polyadenylated followed by enhanced degradation by 3' exonucleolytic activity or may be also by 5' exonucleases (white pacman with question mark).*

parts of the RNAs and identified several RNA binding proteins necessary for mRNA stability (Nickelsen, 2003; Rochaix, 1996). Genetic approaches particularly provide evidence for the existence of a large number of nuclear encoded proteins necessary for the stability of specific mRNAs (Barkan and Goldschmidt-Clermont, 2000). The recent publication of the complete sequence of the chloroplast genome as well as of a comprehensive transcriptome analysis further provides valuable general background information for using this model organism in the analysis of chloroplasts gene expression (Lilly *et al.*, 2002; Maul *et al.*, 2002). The 5' part of an mRNA, however, plays also an important role in translation initiation and this process can also influence the stability of an RNA. This section will focus mainly on those processes that directly affect RNA stability.

In the context of mRNA stability mediated by 5' UTRs the *psb*D gene, encoding the D2 protein of the photosystem II reaction center, is one of the most intensively studied genes. The very abundant mRNA of this gene is cotranscribed with exon 2 of *psa*A (Rochaix, 1996). Initial evidence that the stability of this RNA may be determined through its 5' UTR came from studies of a nuclear mutant with impaired photosystem II activity (Kuchka *et al.*, 1989). In this *nac*2 mutant *psb*D is transcribed, but the respective RNA does not accumulate indicating that stabilization of this transcript is impaired. Studies of transplastomic lines, in which the *psb*D 5' UTR is fused to the *aad*A reporter gene, subsequently revealed that the leader sequence contains a target site for the nuclear encoded NAC2 factor, whose absence in the mutant specifically destabilizes *psb*D mRNA (Nickelsen *et al.*, 1994).

The *nac*2 gene product is a stroma localized tetratricopeptide repeat (TPR) protein, with a predicted molecular mass of about 140 kDa. The NAC2 protein seems to be part of a high molecular weight protein complex, which transiently interacts with the so-called PRB2 *cis*-element, one of several identified target sites in the 5' UTR of this RNA (Boudreau *et al.*, 2000; Nickelsen *et al.*, 1999). The binding of NAC2 probably protects the RNA directly or indirectly from 5' to 3' exonucleolytic digestion (Drager *et al.*, 1998; Drager *et al.*, 1999; Nickelsen *et al.*, 1999), explaining the specific destabilization of *psb*D mRNA in the mutant (Fig. 2). But it also may guide the binding of a "poly(U) tract-sensitive" 40 kDa protein, which in turn releases NAC2 from the RNA (Ossenbühl and Nickelsen, 2000). This protein is most likely required for ribosomal assembly (Nickelsen, 2003). Another NAC2 dependent RNA binding protein of 47 kDa is found to bind only to the longer –74, unprocessed 5' UTR but not to the –47 5' UTR of the predominant *psb*D transcript. This strengthens the potential interaction of NAC2 with other proteins although the exact function(s) of this 47 kDa protein still remains unclear (Nickelsen, 2003; Ossenbühl and Nickelsen, 2000).

Further evidence for the function of the 5' UTR in mRNA stabilization was obtained through the analysis of *rbc*L transcripts. In respective transplastomic *C. reinhardtii* mutants the 5' leader sequence of this RNA was fused to a reporter gene and the generated chimeric RNA was strongly destabilized upon illumination (Salvador *et al.*, 1993). Futhermore, it has been observed that the redox state of the cell controls stability of the *rbc*L transcript via the 5' UTR, indicating that the cellular redox is poised to be an important parameter for the regulation of gene expression via the 5' UTR (Salvador and Klein, 1999).

A different TPR protein (MBB1) is required for stabilizing the mRNAs of the *psb*B/*psb*T/*psb*H transcription unit. In analogy to NAC2, the target site of MBB1 is located within the 5' UTR and the protein is found in a high molecular weight complex, which is associated with mRNA. The fact that an *mbb*1 homologue is found in *A. thaliana* suggests that a specific stability factor binding to the 5' UTR(s) of RNA(s) plays an important role also in the stabilization of at least one mRNA in chloroplasts of higher plants (Vaistij *et al.*, 2000a; Vaistij *et al.*, 2000b). This conservation between an alga and a flowering plant highlights the importance of this regulation via the 5' UTR in plastids.

Besides these *trans*-acting factors, *cis*-elements required for 5' UTR mediated RNA stability have been the subject of many studies in *C. reinhardtii*. In *psb*D mRNA, in addition to the above mentioned PRB2 located around position –30, other target sites for *trans*-acting factors have been characterized. For instance, yet another *cis*-element important for RNA stability resides within the first twelve nucleotides of the long unprocessed –74 precursor *psb*D RNA. In addition, the 5' UTR of this transcript contains two further elements, a potential Shine-Dalgarno sequence, designated PRB1, and a U-tract spacer between the PRB1 and PRB2. Both PRB1 and the oligo(U) sequence were found to be important for translation (Nickelsen *et al.*, 1999; Ossenbühl and Nickelsen, 2000).

In the 5' UTR of other mRNAs, similar *cis*-elements were identified and changes within these target sites can accelerate degradation more than 50-fold, as seen with *rbc*L transcripts (Salvador *et al.*, 2004). This leader contains an RNA stabilizing element between positions +38 and +47 (Anthonisen *et al.*, 2001), which is consistent with previous mapping results locating crucial sequences between positions +21 and +41 (positions are in this case given with respect to the first transcribed nucleotide +1, (Anthonisen *et al.*, 2001; Singh *et al.*, 2001)). Similarly sequences between +31 and +42 are crucial for *atp*B mRNA. Interestingly all of these sequences are part of potential stem-loop structures that might have some function in the stabilization process (Anthonisen *et al.*, 2001).
Three *cis*-acting sequence elements have also been mapped in the *pet*D 5' UTR. While two of them, elements II and III, located in the center and in the 3' part of the leader, respectively, are essential exclusively for translation, the most 5' located element I is found to be important for both translation and RNA stability (Higgs *et al.*, 1999).

While common primary sequence motives are not apparent in this relatively small set of target site sequences, some of them form part of potential stem-loop structures, whose biological significance is, however still unclear at present (Anthonisen *et al.*, 2001; Higgs *et al.*, 1999; Nickelsen *et al.*, 1999). The investigation of the 5' UTRs of chloroplast mRNAs in *C. reinhardtii* furthermore revealed multiple 5' ends for individual loci. In *psb*D and *psb*A transcripts two different 5' termini are detected, respectively. The generation of the 5' shorter transcripts could somehow interfere with translatability (Fig. 2), but the exact functions of these processing events are still unclear (Barkan and Goldschmidt-Clermont, 2000).

Much less is known about the role of 5' UTRs in mRNA stability in the chloroplasts of vascular plants. A recent study used transplastomic tobacco lines expressing chimeric RNAs composed of differently mutagenized 5' UTRs of *psb*A with the *uid*A reading frame as reporter. Mutations within a potential stem-loop structure were found to decrease the life time of the RNA up to 3-fold, while translation efficiency is concomitantly reduced up to 6-fold (Zou *et al.*, 2003). A similar result was obtained by the *in vivo* promoter analysis of the *rbc*L gene in tobacco. Here it was found that chimeric transcripts of the *rbc*L 5' region and *uid*A coding region accumulated to higher levels when a putative stem-loop structure in the 5' UTR is present in the promoter construct (Fig. 2) (Shiina *et al.*, 1998). In addition, the stability of the *rbc*L transcript is increased in the dark, which compensates for reduced transcriptional rates, an observation confirmed in the above mentioned report (Zou *et al.*, 2003). Apart from transplastomic *in vivo* approaches, biochemical studies have found and partially identified proteins binding to the 5' UTR of *psb*A transcripts in spinach and *A. thaliana* (Alexander *et al.*, 1998; Shen *et al.*, 2001).

Both studies detected a 43 kDa protein, which was in spinach identified as a homologue of the *E. coli* S1 ribosomal protein. This protein binds together with an unknown 47 kDa protein to a region ranging from –40 to –9 (with respect to the ATG) covering a putative Shine-Dalgarno motif. The exact function of these proteins in either RNA stability and/or translation remains unclear as those identified in *A. thaliana*, where in addition a 30 kDa 5' binding protein was found. The protein-RNA interaction in *A. thaliana* is reported to depend on a potential stem-loop structure and occurs in response to the cellular redox state (see below). The few reports of 5' UTR-dependent RNA stability in chloroplasts of land plants suggest that these plants may have retained similar mechanisms as those described for *C. reinhardtii*.

2.3. 3' processing and RNA stability of plastid mRNAs

When it was first established that posttranscriptional control is the major level of gene regulation in chloroplasts, 3' inverted repeats (IRs) were suggested to be potentially important *cis* elements determining the rates of degradation. The underlying signatures of inverted sequences are frequently observed in the 3' non-coding regions of many plastid genes and can be predicted to fold into stem-loop structures in the RNA (Fig. 2). Indeed it was found that such structures do not terminate transcription in *in vitro* and in *run-on* experiments (Gruissem, 1989; Stern and Gruissem, 1987).

Instead, *in vitro* experiments with chloroplast protein lysates demonstrated that these 3' IR precursors are substrates for rapid and faithful generation of 3' ends identical to those found *in vivo*. These 3' termini are located immediately downstream of the base paired nucleotides and it was consequently suggested that the stem loop is a sterical barrier preventing progression of a 3' to 5' exoribonuclease into coding regions (Fig. 2). In subsequent studies a number of proteins were found to be involved in the 3' trimming process either by binding and stabilizing the secondary structure or by acting as exo- or endoribonucleases (Nickelsen, 2003).

Among the several initially identified proteins a 28 kDa RNA binding protein (28 RNP) is one of the best investigated. This polypeptide contains two RNA binding domains and interacts with precursor and mature RNAs from different genes, but does not exhibit endo- or exoribonuclease activity (Schuster and Gruissem, 1991). The 28 RNP and possibly other polypeptides were later found to be crucial for the decision on whether an mRNA is degraded (in the absence of the 28 RNP) or correctly processed (in the presence of the 28 RNP). Thus this protein, which can be phosphorylated by chloroplast protein extracts, seems to regulate the activity of a high molecular weight protein complex, the chloroplast degradosome. This complex consists of a 100 kDa protein with strong similarity to bacterial polynucleotide phosphorylase (PNPase, called cpPNPase or 100 RNP in chloroplasts), a p67 protein has been suggested to function as an endoribonuclease and a 33 kD protein (33 RNP) (Hayes *et al.*, 1996; Lisitsky and Schuster, 1995). A later independent study of the composition of this high molecular weight complex confirmed the presence of the cpPNPase, but failed to detect any other protein as an integral part of the complex (Baginsky *et al.*, 2001).

Consequently it was suggested that the chloroplast degradosome may be a PNPase homopolymer and thus very different from its *E. coli* counterpart, which in addition contains a DEAD box RNA helicase, an enolase and the endoribonuclease RNase E as major components (Carpousis, 2002). The p67, a protein that crossreacts with an antibody directed against the bacterial RNase E, and was identified as a GroEL homologue, was not detected in the chloroplast complex. Likewise the *in silico* identification of a nuclear encoded *A. thaliana* RNase E homologue lacking the C-terminal part, which in *E. coli* is responsible for protein interactions and serves as a platform for complex formation, provides a further albeit indirect argument for the absence or at least stable incorporation of an RNase E like protein in the plastid degradosome in higher plants (Baginsky *et al.*, 2001; Vanzo *et al.*, 1998).

Endonucleolytic cleavage appears to be crucial in determining the fate of a stem-loop stabilized RNA in plastids (Fig. 2). While an endonucleolytic cut downstream of the inverted repeat most likely initiates correct 3' processing, a cut upstream of the stem-loop is suggested to initiate complete degradation of the transcript. The identity of the endoribonuclease(s) responsible for these cleavages is unknown.

One endonucleolytic protein, CSP41, initially detected as an RNA binding protein, has been unambiguously identified (Chen and Stern, 1991a; Yang *et al.*, 1996; Yang and Stern, 1997). CSP41 is a member of the huge short chain dehydrogenase/reductase super family (Bollenbach and Stern, 2003). This protein appears to be a rather unspecific endonuclease that needs divalent metal ions for catalytic activity and most likely also for substrate recognition and/or binding, and which cleaves preferentially in the 3' stem-loop structures. Intact fully base paired stems were found to be optimal for substrate recognition, while the exchange of nucleotide identities at the cleavage site has no effect on cleavage site specificity, corroborating that this protein has no sequence specificity.

The suggestion that the most important function of CSP41 is the degradation of a stem-loop stabilized mRNA has now also been confirmed *in vivo*

(Bollenbach *et al.*, 2003). In transgenic tobacco lines, the expression of this enzyme has been knocked out by antisense RNA. This promotes a two- to sevenfold reduction of the degradation rates of *rbc*L, *psb*A and *pet*D mRNAs, which surprisingly is compensated by reduced transcriptional rates of these loci. This leads to similar accumulation of *rbc*L and *psb*A transcript levels in antisense and wild-type (wt) plants, while the accumulation of *pet*D decreases by 25% in antisense plants. Data bank searches have in addition revealed, at least in *A. thaliana,* two different forms of this enzyme. Beside CSP41a (the original CSP41), which is knocked out in the antisense approach in tobacco, the second predicted CSP41b shares 35% identical amino acids with the former.

Two other chloroplast proteins, EndoC2 from spinach and p54 from mustard, exhibit endonucleolytic activity (Chen and Stern, 1991b; Liere and Link, 1997; Nickelsen and Link, 1993). In contrast to CSP41a, p54 is involved in the 3' end generation of *trn*K and *rps*16 transcripts, which lack any obvious stem-loop structures. Interestingly, *in vitro* studies suggest this protein to be regulated by redox state and phosphorylation. This endonucleolytic cleavage of RNA was also observed in plastids of dark-grown spinach. Here most likely (a) constitutively expressed endonuclease(s) is (are) probably blocked by a 41 kDa protein in plants grown under 8/16 h light/dark cycle, while an 48 h dark application of these plants releases the endonucleolytic activity initiating mRNA degradation (Fig. 2). But as mentioned above the identity of the(se) endonuclease(s) as well as the inhibiting protein is unknown (Baginsky *et al.*, 2002).

Among the many chloroplast RNPs, some may function as general binding proteins stabilizing RNA. For instance the highly abundant stroma proteins CP28, 29A, 29B, 31 and 33 from tobacco together stabilize *psb*A mRNA (Li and Sugiura, 1990; Nakamura *et al.*, 2001; Ye *et al.*, 1991). These proteins each contain two RNA binding domains and form high molecular weight complexes either dependent on (CP29A, CP29B, CP28B) or independent of (CP31, CP33) intact RNA. For instance, ribosome-free *psb*A mRNA was found to coprecipitate with antibodies against CP29A, CP28 and CP31 but not with antibodies against CP33. Depletion of these chloroplast proteins from a protein extract reduced stability of a synthetic *psb*A mRNA *in vitro,* and the degradation enhancing effect was reversed by readdition of the proteins to the depleted extracts. Furthermore, at least one of these proteins (CP31) also participates most likely as a general factor in RNA editing (Hirose and Sugiura, 2001). It is thus proposed that these proteins bind to nascent RNA thereby protecting it. Additionally the polypeptides might also serve as platforms to recruit other polypeptides, which then process RNA in different manners (Nakamura *et al.*, 2001).

The function of 3' inverted repeats in mRNA stability has also been analyzed in *C. reinhardtii. In vitro* and *in vivo* analyses revealed that the deletion of such a structure reduces the steady state amount of *atp*B mRNA to 20 to 35 % of wt levels without affecting transcriptional rates. Furthermore this deletion leads to the generation of *atp*B RNA of heterogeneous sizes indicating that the secondary structure is required not only for stability but also for correct 3' end formation (Stern *et al.*, 1991). The latter seems to be a two-step mechanism with an initial endonucleolytic cut ten nucleotides downstream of the stem-loop followed by the 3' exonucleolytic removal of the trailer nucleotides to generate in a synergistic manner the mature 3' end, which seems to be required for translation (Drager *et al.*, 1996; Hicks *et al.*, 2002; Stern and Kindle, 1993). This is similar to the above summarized 3' end formation process later suggested for chloroplasts in vascular plants (Fig. 2). In contrast to the latter, none of the 3' processing components are known so far in the alga, although a recessive nuclear mutation was described that suppresses the enhanced *atp*B degradation caused by the 3' IR deletion. This mutation also affects 3' end formation of *psb*I and *cem*A containing mRNAs of the *atp*A gene cluster (Levy *et al.*, 1997; Levy *et al.*, 1999). In addition to these reports suggesting a function of the 3' IR as a stability element and also 3' processing signals in *C. reinhardtii*, contradictory results were observed by the *in vivo* analysis of *rbc*L and *psa*B 3' IRs. Here the structures were found to be dispensable for stability, but not for transcription termination, which has, however, not been found to be efficient in another report (Blowers *et al.*, 1993; Rott *et al.*, 1996).

2.4. Polyadenylation enhances degradation of RNA in plastids

The decision whether RNA is processed and stabilized or whether it becomes degraded depends on structures or modifications of the RNA molecule. As mentioned above the removal of a 3' stem-loop structure from an mRNA by an endonucleolytic cut initiates degradation, which can be enhanced by the addition of poly(A) tails, which substantially diminish the stability of an RNA molecule (Fig. 2).

Plastid poly(A) RNAs were initially detected almost 30 years ago in maize (Haff and Bogorad, 1976). It has been estimated that about 6% of plastid RNAs are polyadenylated and that the poly(A) tracts have an average length of about 45 nucleotides. Interestingly two years later poly(A) as well as poly(G) polymerase activities were detected in wheat chloroplasts (Burkard and Keller, 1974). However, it took more than two decades that, stimulated by the analysis of polyadenylation and its function in mRNA decay in *E. coli*, the importance of poly(A) sequences in the degradation of plastid RNAs was further explored. Applying a 3' RACE approach with an oligo(dT) adapter primer, potential 3' poly(A) tracts were rediscovered in various plastid mRNAs (Kudla *et al.*, 1996; Lisitsky *et al.*, 1996).

These non-encoded nucleotide stretches of predominantly adenosines but also containing guanosines and to a minor extent cytidines and uridines were found at 3' ends, generated by endonucleolytic cleavages primarily within the coding regions but also at the mature 3' ends. In contrast to the function of the poly(A) tails in nuclear-cytoplasmic transcripts known to support stability, enhance translation efficiency and required for the export from the nucleus, the poly(A) stretches in plastids were suggested to destabilize the RNA similar to their function in bacteria (Carpousis *et al.*, 1999; Wahle and Ruegsegger, 1999). This was concluded from *in vitro* studies revealing polyadenylated RNA to be the primary target of exoribonucleolytic digestion in chloroplast protein lysates and more specifically with cpPNPase. This enzyme was also found to have enhanced binding affinity towards poly(A) RNA. The prominent role of polyadenylation for plastid RNA degradation was further substantiated by the observation that the polyadenylation blocking compound cordycepin inhibits chloroplast mRNA decay (Lisitsky *et al.*, 1997a; Lisitsky *et al.*, 1997b; Schuster *et al.*, 1999).

The degradation promoting effect of plastid poly(A) sequences thus seems to be experimentally established. It is, however, still unclear how long these non-encoded extensions actually are *in vivo* and furthermore which enzyme catalyzes the addition of the non-encoded nucleotides. In *E. coli* a poly(A) polymerase was identified as the primary enzyme catalyzing the addition of non-encoded nucleotides, predominantly adenosines; a minor adenosine adding enzyme is the PNPase. For chloroplasts, however, it was suggested that the PNPase homologue is the major if not sole poly(A) polymerizing protein. Whether cpPNPase exhibits exoribonuclease or a poly(A) polymerase activity is seen at least *in vitro* depending on the concentration of ADP, which enhances poly(A) polymerization and P_i, which favors degradation (Yehudai-Resheff *et al.*, 2001).

These apparently counteracting activities are assigned to different domains of the enzyme. The enzyme is composed of two RNase PH domains, separated by an □-helical linker, and two RNA binding domains in the C-terminal part. The N-terminal RNase PH domain harbors the degradation, but cannot polymerize polynucleotides. The C-terminal domain was found to have only limited exonuclease activity, which is, however, enhanced when tested with polyadenylated RNAs. Thus this enzyme first adds adenosines to generate its favorite substrate for degradation (Yehudai-Resheff *et al.*, 2003). The plastid PNPase could thus very well perform both functions, but on the other hand in one of the initial studies, a poly(A) polymerase activity separate from cpPNPase has been reported (Lisitsky *et al.*, 1996). Similarly an extra poly(A) adding enzyme is suggested by an investigation of steady state RNA from transgenic *A. thaliana* plants with reduced PNPase activity. Here a 3' RACE analysis revealed accumulation of polyadenylated *psb*A, *rbc*L and *rps*14 RNAs in the PNPase deficient plants being much higher than in PNPase overexpressing or wild type plants (Walter *et al.*, 2002).

Consistent with this observation, several genes encoding proteins with substantial similarity to poly(A) polymerases or/and to the closely related tRNA terminal nucleotidyl transferase are *in silico* found in the *A. thaliana* nuclear genome, which are predicted to be targeted to plastids or mitochondria (A.G.I, 2000). Thus the role of cpPNPase in plastid polyadenylation remains somehow unclear and needs further investigation. Also controversial are the reports concerning the importance of cpPNPase for degradation. Since in both PNPase-deficient and PNPase-overexpressing plants the steady state levels of certain RNAs were found to be almost unchanged, it was suggested that this enzyme is dispensable for degradation. However this contradicts the increased accumulation of poly(A) RNA in the PNPase reduced plants in the same study (Walter *et al.*, 2002).

Even with these uncertainties the cpPNPase together with the CSP41 is so far the only unambiguously identified and thus best characterized enzyme of the land plant chloroplast 3’ processing/degradation machinery. Beside its function in 3' trimming of mRNAs, which is consistently observed both *in vivo* and *in vitro*, it appears also to be directly or indirectly involved in the 3' processing of the 23 S rRNA and in the degradation of two tRNA species (Walter *et al.*, 2002).

Considering all the data obtained in recent years a tremendous advance has been made in understanding the mechanisms of RNA 3' processing, stability and degradation in plastids of higher plants, but still many components are not identified and more important it is still far from clear how the interplay of the different proteins and their various functions is orchestrated.

2.5. mRNA 3' processing, stability and polyadenylation in plant mitochondria

In plant mitochondria, the processing of mRNAs in terms of posttranscriptional 5' and 3' end generation has long been concluded from descriptive data. Multiple termini were observed for many mitochondrial transcripts, both derived from poly- or monocistronic transcription units. The identification of transcription initiation sites provided circumstantial proof that many of the 5' ends result from posttranscriptional processing. This indirect evidence, as well as the different mapping techniques employed, bear some ambiguities, since termini artificially generated during preparation of the RNA for the mapping experiments can not be completely ruled out. Presently only little is known about the processing mechanisms and components involved.

As mentioned above, the complete mitochondrial genomes of the moss *Marchantia polymorpha* and four different seed plants are now sequenced (Handa, 2003; Kubo *et al.*, 2000; Notsu *et al.*, 2002; Oda *et al.*, 1992; Unseld *et al.*, 1997). They encode more or less the same set of about 54 to 59 genes, which are apparently randomly distributed across the genomes, and their orders differ significantly with the exception of the few linked genes mentioned above. The complete genomes are a valuable basis to functionally study gene expression, but so far no complete transcript map is available for any of these genomes.

A rough calculation considering the distance between individual genes in *A. thaliana* suggests that about 40 genes, with spacers of less than three kb or less at an average distance of eight kb, are potentially transcribed as clusters with two or more reading frames each (Dombrowski *et al.*, 1998). Thus in many cases genes may be transcribed within larger precursors, which are disassembled into multiple translatable RNAs (see references in Hoffmann *et al.*, 2001). We are currently establishing a complete transcript map for *A. thaliana* mitochondria. The analysis of ten genes detected many different 5' ends, but generally only a single major 3' terminus, independent from the organization in mono- or oligocistronic transcription units (J. Forner and S. Binder, unpublished results). Presently the origin, function or relevance of these multiple 5' termini is unclear. They may just represent different intermediates on the way to the mature 5' end, which is required for efficient translation or which is the final product of a 5' stabilization/processing process, analogous to the situation found in *C. reinhardtii* or higher plant chloroplasts (see above).

Clearer, though far from fully understood, is the formation of 3' termini of mitochondrial transcripts. These ends are either generated directly by transcription termination or secondary products of posttranscriptional processes. Nothing is known about the former process and it might be conceivable that in plant mitochondria no specific mechanism exists and termination just occurs randomly. In animal mitochondria an active transcription termination process has been detected with a specific termination factor causing the RNA polymerase to fall off the template downstream of the rRNA genes (Fernandez-Silva *et al.*, 1997; Fernandez-Silva *et al.*, 2001; Loguercio Polosa *et al.*, 1999; Roberti *et al.*, 2003). Several homologues of these factors can be identified in the *A. thaliana* nuclear genome but their function has not been experimentally addressed (Forner and Binder, unpublished results).

The 3' termini of several plant mitochondrial mRNAs are found just at the ends of inverted repeats that potentially fold into stem-loop structures in the RNA (Schuster *et al.*, 1986). It has been speculated that this close spatial relationship originates from a function of the IRs as transcription terminators or as signals for 3' end processing. In one of the initial functional *in vitro* studies no indication was found that would be consistent with an operation of mitochondrial stem loops in transcription termination (Dombrowski *et al.*, 1997). In this mitochondrial *in vitro* transcription system, RNA elongation runs unimpeded through the IR. Further *in vitro* processing analysis strongly suggested that at least the pea *atp*9 stem loop functions as a processing signal and stabilizing element (Dombrowski and Binder, 1998). These assays thus indicated that mitochondrial 3' mRNA ends can indeed be generated posttranscriptionally and that stem-loop structures might be *cis*-elements necessary for correct processing and increased stability. Similar results have also been obtained through the analysis of RNAs of a chimeric gene implicated in CMS in *Brassica* cybrids (Bellaoui *et al.*, 1997).

The *in vivo* accumulation of the CMS-related *orf*138 transcripts correlates with the presence (in sterile) or absence (in fertile) of a secondary structure in the 3' region of the RNA. *In vitro* decay assays corroborated that this structure increases RNA stability. This function of potential stem loop structures appears similar to the situation in chloroplasts. However, although the presence of stem-loop structures seems to generally mediate RNA stability as they are also found in other transcripts (Kaleikau *et al.*, 1992; Saalaoui *et al.*, 1990), we must assume that unlike chloroplasts' distinct structures and mechanisms that might be involved in 3' end formation, mitochondrial stem-loops are found only in a minority of the mRNAs. Thus other *cis*-elements or less obvious higher order structures may determine 3' end formation and RNA stability here, further separating mitochondrial processes from those in plastids.

Similarities between the two organelles exist in the 3' polyadenylation of mRNAs and its most obvious function. In mitochondria, this process was initially detected in the molecular investigation of PET-1 CMS in sunflower, which is caused by the expression of a 15 kDa protein encoded by *orf*522. This gene is cotranscribed with the upstream *atp*1 gene in a 3.0 kb transcript. The steady-state level of this RNA is tissue- and cell-specifically decreased in the nuclear genetic background of a restorer gene, whose presence correlates with the substantially reduced accumulation of this protein in male florets (Monéger *et al.*, 1994; Smart *et al.*, 1994). 3' RACE experiments using an oligo(dT) adapter primer detected a specifically higher level of polyadenylated *orf*522 mRNAs in male florets. This reverse correlation of a generally reduced level of *atp*1-*orf*522 mRNA with a higher level of polyadenylated RNA from these genes strongly suggested that polyadenylation enhances degradation in plant mitochondria (Gagliardi and Leaver, 1999). *In vitro* studies indeed revealed two distinct ribonuclease activities, one of which preferentially degrades transcripts with poly(A) tails. These non-encoded poly(A) extensions have also been found to accelerate transcript degradation in maize and potato mitochondria (Gagliardi *et al.*, 2001; Lupold *et al.*, 1999).

The degradation promoting effect of poly(A) tails seems to be much weaker in RNAs with terminal stem-loops compared to RNAs without such structures. Respective *in vitro* studies addressing this problem suggested that several cycles of polyadenylation and partial digestion are necessary to overcome the relatively stable stem-loop structure and to completely degrade the RNA (Kuhn *et al.*, 2001). However, the stem-loop structures seem to be not a general hindrance in 3' to 5' degradation processivity, since exonuclease activity in potato proceeds unimpeded through such structures present in synthetic potato *atp*9 and maize *rps*2 RNA substrates (Gagliardi *et al.*, 2001). The contradicting results may stem from the differing systems used. The stem-loop structure in potato for instance is more than 60 nucleotides upstream of the mature 3' end and may have a different function (Lizama *et al.*, 1994). Furthermore, additional proteins supporting the stabilizing effect of the secondary structure may be required and the relatively crude extracts used in these experiments may significantly vary in their composition of such auxiliary proteins.

As in the case of plastids, the actual lengths of the mitochondrial 3' poly(A) tails are unknown. Extensions of ten (optimally 20 to 30) adenosines are sufficient to enhance degradation *in vitro* (Gagliardi *et al.*, 2001; Kuhn *et al.*, 2001). RT-PCR analysis of *in vitro* 5' to 3' end ligated RNAs (Kuhn and Binder, 2002) allows a more realistic estimate of the number of posttranscriptionally added untemplated nucleotides than oligo(dT) primed RT-PCR. This approach detects homopolymeric adenosine stretches up to 25 nucleotides in *A. thaliana atp*1 transcripts (J. Forner and S. Binder, unpublished results). Interestingly, a series of very short non-encoded extensions have been found in maize and pea (Kuhn *et al.*, 2001; Williams *et al.*, 2000). These are composed mainly of adenosines and cytidines and some of them even perfectly match the CCA extensions posttranscriptionally added to tRNA 3' ends. This may suggest two different types of 3' terminal extensions that could be polymerized by different enzymes. As mentioned above, several genes for putative mitochondrial poly(A) polymerases and/or ATP(CTP):tRNA nucleotidyltransferases, closely related proteins indistinguishable on the level of amino acid sequences, are found in the *A. thaliana* nuclear genome, which may polymerize different types of extensions. The function of the short non-encoded extensions is unclear, since three adenosines have no influence on RNA stability at least *in vitro* (Kuhn *et al.*, 2001).

Beside the obscurities in terms of the *cis*-elements necessary and the mechanisms underlying 3' processing, RNA stability and degradation of the *trans*-acting components involved in these processes in mitochondria are almost completely unknown. An *A. thaliana* protein homologous to the *suv*3 RNA helicase from yeast has been identified *in silico* and shown to be likewise located in mitochondria (Gagliardi *et al.*, 1999). In *Saccharomyces* this protein is part of the mitochondrial degradosome, and a potential analog in plants would thus suggest participation in RNA processing or degradation (Dziembowski *et al.*, 2003). On the other hand the human homologue of this protein has surprisingly been found to accept DNA rather than RNA as substrate (Minczuk *et al.*, 2002). Thus the exact function of this protein in plant mitochondria remains so far unclear and requires experimental analysis.

The identification of *trans*-acting factors will definitively further benefit from the complete nuclear genome sequences from *A. thaliana* and rice as well as from the various proteomic approaches to identify mitochondrial proteins (Alonso *et al.*, 2003; Arabidopsis, 2000; Elo *et al.*, 2003; Goff, 2002; Heazlewood *et al.*, 2003; Heazlewood *et al.*, 2004; Kruft *et al.*, 2001; Millar *et al.*, 2001; Sickmann *et al.*, 2003; Yu, 2002). Genes encoding potential components can readily be identified *in silico* and their functions in mitochondrial RNA metabolism can than be addressed by reverse genetic approaches.

For instance, several other putative DEAD box RNA helicases have been found through similarity searches and their mitochondrial localization has already been experimentally confirmed by expressing respective GFP fusion proteins in tobacco protoplasts (S. Schmidt-Gattung and S. Binder, unpublished results). Likewise genes encoding PNPase-like and RNase 2-like enzymes have been identified, and the respective GFP fusion proteins are found to be imported into mitochondria, which is consistent with the above mentioned two different ribonuclease activities (S. Schmidt-Gattung, S. Binder, D. Gagliardi and J. Gualberto, unpublished results). These proteins may have crucial functions in plant mitochondrial RNA metabolism and might be assisted by one or several RNA binding proteins with so far unidentified functions or by the PPR proteins (Small and Peeters, 2000; Vermel *et al.*, 2002). The latter are encoded by one of the largest plant gene families with more than 450 members in *A. thaliana*, many of which are predicted to be imported into mitochondria.

Their anticipated participation in mitochondrial RNA metabolism has recently been substantiated by the identification of members of this protein family as restorers of fertility of CMS in petunia and rice. In the first species CMS correlates with the presence of the *pcf* chimeric reading frame cotranscribed with the downstream-located *nad*3 and *rps*12 genes (Nivison and Hanson, 1989).

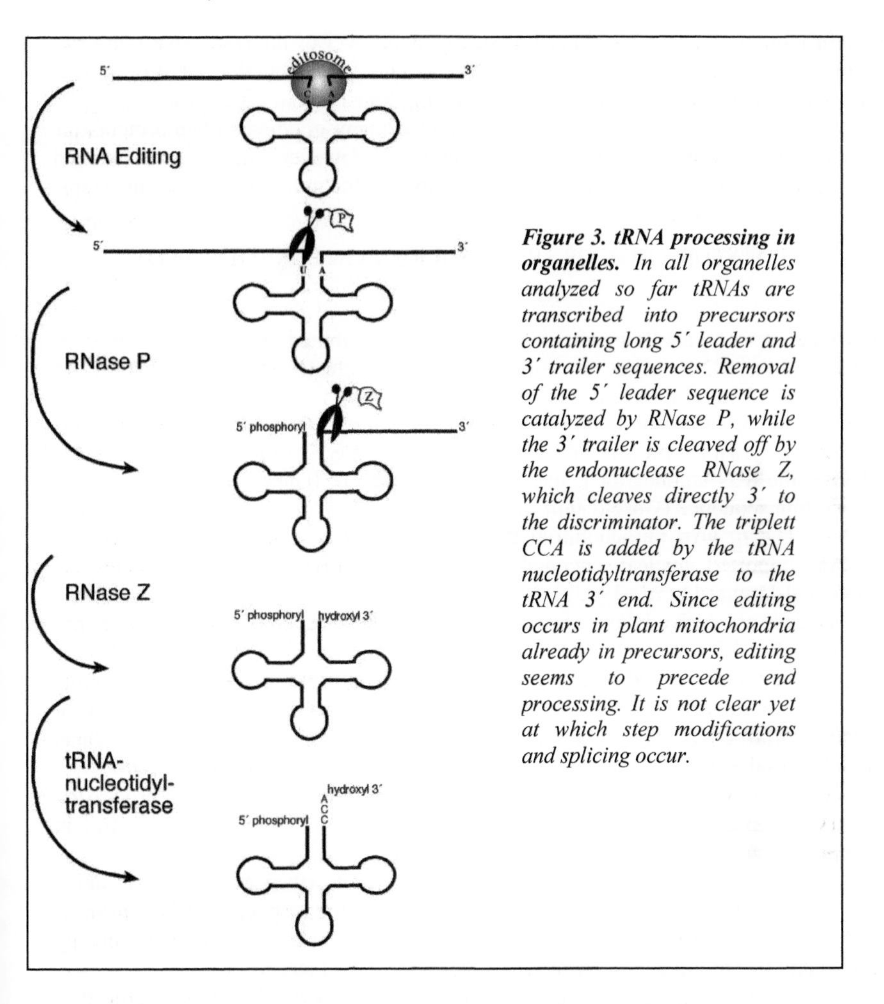

Figure 3. tRNA processing in organelles. *In all organelles analyzed so far tRNAs are transcribed into precursors containing long 5′ leader and 3′ trailer sequences. Removal of the 5′ leader sequence is catalyzed by RNase P, while the 3′ trailer is cleaved off by the endonuclease RNase Z, which cleaves directly 3′ to the discriminator. The triplett CCA is added by the tRNA nucleotidyltransferase to the tRNA 3′ end. Since editing occurs in plant mitochondria already in precursors, editing seems to precede end processing. It is not clear yet at which step modifications and splicing occur.*

Restoration is accompanied by an altered *pcf* transcript profile and substantially reduced levels of the PCF protein. A 592 amino acids PPR protein with 14 repeats has now been found to restore fertility upon transformation into CMS plants (Bentolila *et al.*, 2002). The restoration of fertility by transformation is indeed correlated with a ten-times lower level of the PCF protein; it is, however, unclear whether it is also accompanied by an altered transcript profile, which is always observed in crossings of CMS plants with restorer lines. Another PPR protein of 18 repeats was found to restore ms-bo CMS in rice by processing the CMS related aberrant *atp*6 RNA, indicating a direct or indirect interference with the metabolism of this RNA (Kazama and Toriyama, 2003). A third PPR protein has been found to restore Kosena and Ogura CMS in radish by reducing the CMS-associated ORF125 protein without altering the pattern or steady-state level of the respective transcript (Brown *et al.*, 2003; Koizuka *et al.*, 2003). The exact function of this protein remains unclear; it may be involved in regulating translational efficiency.

Thus, recent developments have yielded many good candidate proteins for components of mitochondrial RNA metabolism; however, much more detailed studies are required to define their exact functions in the individual processing steps.

3. MATURATION OF TRANSFER RNAS IN PLASTIDS AND MITOCHONDRIA

tRNA genes in organelle genomes are generally distributed around the genome. In mitochondria, tRNA genes are either encoded in clusters with each other, are physically linked with rRNA or mRNA genes, or found as solitary genes flanked by extensive stretches of non-coding DNA. In *A. thaliana*, for instance, the tRNA genes for serine, tyrosine, proline and cysteine are clustered within a region of 2 kb, whereas the tRNA genes for lysine and methionine are flanked by more than 5 kb non-coding sequence (Unseld *et al.*, 1997).

The *in vivo* mitochondrial tRNA population is a mix of chloroplast-like tRNAs, imported nuclear encoded tRNAs and "native" mitochondrial tRNAs (Maréchal-Drouard *et al.*, 1993b). In potato, for instance, 16% of the mitochondrial tRNA genes are derived from plastid-like genes, which have been incorporated into the mitochondrial genome; 36% are imported from the nuclear-cytoplasmic compartment and only 48% are "native" mitochondrial tRNAs (Maréchal-Drouard *et al.*, 1990). In addition, tRNA-like structures (t-elements) have been found in wheat mitochondrial DNA, which are recognized and processed efficiently by wheat mitochondrial extracts (Hanic-Joyce *et al.*, 1990). Since these t-elements are often encoded close to active genes they might function as punctuation for the processing of larger transcripts, analogous to the processing of the interspersed tRNAs in animal mitochondrial transcripts (Ojala *et al.*, 1981).

In chloroplasts tRNA genes appear to be transcribed as monomeric units, with putative promotor elements located upstream of the respective tRNA gene, as well as into polymeric precursors, containing other tRNAs, rRNAs, open reading frames and/or protein coding sequences , and references therein).

Organellar tRNA molecules are generally transcribed as longer precursor RNAs, which have to undergo a variety of processing steps to yield the functional molecules (Martin, 1995) (Fig. 3). Additional 5′ leader and 3′ trailer sequences have to be removed, the CCA has to be added and numerous base modification have to be introduced. In mitochondria of lower plants, very few examples of tRNA genes with intervening sequences have been reported, while in chloroplasts several tRNA precursors contain either group I or group II introns in lower as well as in flowering plants (see Chapter 11). Some mitochondrial tRNAs have to be edited (see also Chapter 9), and histidine tRNAs have to be posttranscriptionally completed with an additional 5′ nucleotide ("minus 1") (Dirheimer *et al.*, 1995). Organellar tRNAs contain less modified bases than cytosolic tRNAs, but so far little information is available about the enzymes responsible for modification of tRNAs in organelles (Björk and Kohli, 1990).

In organelles, processing of tRNAs has been comparatively well studied in comparison to other RNA maturation events. Protein extracts, which are competent in 5′ and 3′ tRNA processing as well as CCA addition, have been isolated from mitochondria and chloroplasts, respectively. All tRNA processing activities analyzed so far resemble those in their eukaryotic counterparts better than their analogs in their bacterial predecessors.

Removal of the 5′ extension of a pre-tRNA is catalyzed by the ubiquitous endoribonuclease RNase P (Fig. 3, EC 3.1.26.6; for review see Frank and Pace, 1998). RNase P is usually composed of RNA and protein subunits, with the RNA subunit contributing the catalytical activity of the complex. RNase P-like activities have been observed in mitochondrial protein extracts from wheat, potato and *Oenothera berteriana* (Hanic-Joyce and Gray, 1990; Marchfelder and Brennicke, 1994; Marchfelder *et al.*, 1990). In chloroplast extracts, similar activities have been detected in *Euglena gracilis*, spinach, tobacco and pea (Greenberg and Hallick, 1986; Gruissem *et al.*, 1983; Marion-Poll *et al.*, 1988; Yamaguchi-Shinozaki *et al.*, 1987). Interestingly, the plastid RNase P from higher plants does not seem to contain an RNA subunit (Gegenheimer, 1995). Density gradient experiments and micrococcal nuclease digests suggest that the plastid RNase P consists solely of protein(s). For a final determination of its subunit composition the plastid RNase P enzyme has to be isolated. In contrast, an RNA component is present in plastid RNase P from primitive algae, the gene for which has been isolated. It is encoded in the respective plastid genomes and shows distinct similarity to the RNase P RNA of *Cyanophora*. These genes are so far the only ones identified for plant organellar RNase P moieties (Cordier and Schön, 1999).

Like the 5′ processing activity, cleavage 3' of tRNAs in larger precursor RNA has been observed in mitochondrial extracts from several plants (Hanic-Joyce and Gray, 1990; Kunzmann *et al.*, 1998) and in chloroplast extracts from *Euglena gracilis*, spinach, tobacco, pea and wheat (Greenberg and Hallick, 1986; Gruissem *et al.*, 1983; Marion-Poll *et al.*, 1988; Oommen *et al.*, 1992; Yamaguchi-Shinozaki *et al.*, 1987). In all cases, an endonuclease, termed RNase Z (Fig. 3), has been found to process all of the pre-tRNAs tested (EC 3.1.26.11; for a review on tRNA 3′ end processing see Mörl and Marchfelder, 2001). Whereas in mitochondria this endonuclease processes the pre-tRNA directly 3′ to the discriminator, which is the nucleotide 5′ to the CCA triplet and serves as an identity element in many tRNAs (Hanic-Joyce and Gray, 1990; Kunzmann *et al.*, 1998), the RNase Z cleavage in chloroplast tRNAs occurs one nucleotide further downstream (Oommen *et al.*, 1992).

Thus, organellar tRNA 3′ processing seems to resemble rather the endonucleolytic eukaryotic 3′ end generation than the bacterial 3′ maturation pathway. Only recently the first RNase Z protein and gene sequences were identified in plants (Schiffer *et al.*, 2002). The knockout of RNase Z activity by RNAi in *Drosophila melanogaster* confirmed that this protein indeed catalyzes tRNA 3′ processing *in vivo* (Dubrovsky *et al.*, 2004). In *A. thaliana*, four RNase Z genes have been found, AthZ1-4 (S. Binder and A. Marchfelder, unpublished results). Two of these proteins (AthZ2 and 4) are predicted to be routed to chloroplasts while one is *in silico* targeted to mitochondria (AthZ3). These N-terminal pre-sequence evaluations by several sorting programs still have to be verified experimentally, but it seems at least from these estimations that for all plant cell compartments RNase Z activities can be accounted for in the nuclear genome.

tRNA molecules are also subject to RNA editing (Fig. 3; see also chapter 9). So far three tRNAs ($tRNA^{Phe}$, $tRNA^{Cys}$, $tRNA^{His}$) - all "native" tRNAs - have been found to require editing to become functional tRNAs (Binder *et al.*, 1994; Maréchal-Drouard *et al.*, 1996b; Maréchal-Drouard *et al.*, 1993a). Edited nucleotides are already found in the 5' and 3' extended precursor molecules, indicating that editing can precede endonucleolytic 5′ and 3′ end processing. *In vitro* processing experiments with unedited mitochondrial $tRNA^{Phe}$ precursors from pea and unedited mitochondrial $tRNA^{His}$ from larch showed that editing is actually a prerequisite for 5' and 3' end cleavage in some tRNA transcripts (Marchfelder *et al.*, 1996; Maréchal-Drouard *et al.*, 1996a). In contrast, *in vitro* processing, CCA addition and aminoacylation of pre-$tRNA^{Cys}$ does not require editing. These data suggest that the order of processing (editing and end processing) is dependent on whether editing generates more stable folding in the pre-tRNA, which is a better substrate for the end processing enzymes. In the case of $tRNA^{Phe}$, the editing site resides in the acceptor stem, and structural probing experiments with lead ions showed that the unedited pre-$tRNA^{Phe}$ does not fold into the typical tRNA structure (Marchfelder *et al.*, 1996). In *Oenothera* and potato mitochondria, $tRNA^{Cys}$ is edited in the anticodon stem, where a C-U mismatch is changed to a U-U mismatch, which does not change the precursor stability (Binder *et al.*, 1994).

The 3′ terminal CCA sequence, which is part of every tRNA, has to be added by the ATP(CTP):tRNA nucleotidyltransferase (Fig. 3; E.C. 2.7.7.25). Mitochondrial tRNA nucleotidyltransferases have been characterized in extracts from lupine and wheat (Shanmugam *et al.*, 1996; Vicaretti and Joyce, 1998) and a plastid transferase was described from wheat (Oommen *et al.*, 1992). The lupine mitochondrial transferase seems to be encoded by a nuclear locus which also encodes the cytoplasmic form of that enzyme (Shanmugam *et al.*, 1996). As already mentioned above, several genes encoding potential organellar CCA transferases or poly(A) polymerases are found in the *A. thaliana* nuclear genome (Arabidopsis, 2000).

Despite their large genome size, plant mitochondria do not encode a full set of tRNAs and thus have to import some of their tRNAs required for translation (for a review see Schneider and Maréchal-Drouard, 2000). As yet it is not clear whether the tRNAs are imported as precursor molecules or as mature tRNAs. Recently Maréchal-Drouard and coworkers showed that at least *in vitro* tRNAs can be imported into mitochondria in their mature form (Delage *et al.*, 2003). *In vitro* processing experiments showed that plant mitochondrial extracts are capable of processing not only native mitochondrial pre-tRNAs, but also plastid, nuclear and heterologous pre-tRNAs from different origins (Kunzmann *et al.*, 1998; Marchfelder, 1995). Thus even if the tRNAs are imported as precursors they can be matured by the mitochondrial processing activities.

Organellar tRNAs contain fewer modified nucleotides than their cytosolic counterparts. So far only a spinach chloroplast extract active in pseudourilydation has been described (Greenberg and Hallick, 1986). Although organellar tRNAs contain few modifications, these can nevertheless be very important for the tRNA function. In one investigated example, codon recognition of a tRNA is switched by a modification. In potato mitochondria the gene for $tRNA^{Ile}$ encodes an anticodon specific for methionine (CAT), but the tRNA is aminoacylated with isoleucine (Weber *et al.*, 1990). RNA sequencing revealed a lysidine modification in the first postition of the anticodon, which allows base pairing to the isoleucine specific codons. A similar situation has been found in spinach chloroplasts (Francis and Dudock, 1982).

All histidine tRNAs require an extra nucleotide at their 5′ end and form an eight base pair acceptor stem instead of the usual seven base pair acceptor stem. Organellar histidine tRNAs also adhere to this rule and contain a G as an extra pairing nucleotide at the 5′ end. In chloroplasts this extra G is already encoded in the gene for $tRNA^{His}$ and the plastid RNase P processes the pre-tRNA without removing this G which remains part of the tRNA (Burkard and Söll, 1988). In mitochondria the G is also encoded in the tRNA gene, but so far it has not been investigated whether the mitochondrial RNase P processes the pre-tRNA 5′ retaining this G or whether it is removed and subsequently re-added.

4. PROCESSING OF ORGANELLAR RIBOSOMAL RNAS

The genes for ribosomal RNAs in chloroplasts are located as duplicates on the plastid genome and are transcribed into multicistronic RNAs containing the 16S rRNA, the gene for $tRNA^{Ile}$ and $tRNA^{Ala}$, the 23S rRNA, the 4.5S rRNA and the 5S rRNA. The primary processing sites in the initial precursor transcript of about 7200 nucleotides have not been yet identified (Hartley and Head, 1979). The only information comes from *Euglena gracilis bacillaris*, where the primary transcript is processed by several endonucleolytic cuts in the spacer regions between the tRNAs and the rRNAs (McGarvey and Helling, 1989).

Recently it was reported that the nuclear-encoded chloroplast protein DAL is required for the maturation of plastid ribosomal RNAs (Bisanz *et al.*, 2003). Processing intermediates of the 16S rRNA accumulate in the *dal*1-2 mutant. Thus DAL either has a role in organizing the overall structure of either the 30S (or both the 30S and the 50S) subunit and/or is involved directly in 16S rRNA processing.

In higher plant mitochondria the situation is slightly different. The genes for the 18S and 5S rRNAs are tightly linked and are most likely cotranscribed in all species investigated (Coulthart *et al.*, 1990; Handa, 2003; Kubo *et al.*, 2000; Notsu *et al.*, 2002; Unseld *et al.*, 1997). Evidence for such cotranscription of the 18S and 5S rRNA genes was at least found in maize. Five potential processing intermediates were identified, suggesting that several exo- and endonucleolytic processing events are required for the generation of mature rRNAs (Maloney and Walbot, 1990).

In wheat mitochondria the gene for $tRNA^{fMet}$ is located just upstream from the 18S rRNA gene with only one base pair separating the 3' end of the $tRNA^{fMet}$

from the 5' terminus of the 18S rRNA. The 5′ end of the latter is thus generated by the endonucleolytic tRNA 3′ processing of tRNA^{fMet} (Hanic-Joyce and Gray, 1990). In addition a t-element (a tRNA-like element; see previous section), which is efficiently processed like a genuine tRNA in *in vitro* extracts, is located downstream of the 5S rRNA gene (Hanic-Joyce *et al.*, 1990). Thus processing of the 5' end of the t-element could generate the 3′end of 5S rRNA.

For the 26S rRNA gene, which is separately transcribed as a monocistronic precursor, several transcripts are also observed, suggesting that the mature 26S rRNA is also processed from the initially longer primary transcripts (Handa, 2003; Kubo *et al.*, 2000; Maloney *et al.*, 1989; Mulligan *et al.*, 1988; Notsu *et al.*, 2002; Unseld *et al.*, 1997). To further explore whether 5' and 3' ends are generated endo- and/or exonucleolytically and which proteins are required for this maturation in both chloroplasts and mitochondria, respective *in vitro* processing systems are required.

So far only one organellar rRNA modification enzyme has been described in *A. thaliana*, a plastid 16S rRNA methylase. Mutation of the methylase gene *pcf1* results in chilling sensitive plants, showing that the gene is essential for the low temperature development of chloroplasts (Tokuhisa *et al.*, 1998).

5. RNA STABILITY VERSUS RNA DEGRADATION: REGULATION BY PHOSPHORYLATION AND/OR REDOX STATE?

The destiny of an organellar RNA, *i.e.* processing and stabilization or degradation, is determined by the balance between different processes engaging a large number of proteins. To guarantee adequate regulation on the posttranscriptional level, the involved proteins and the respective processes have to be ingeniously coordinated and governed. How this is achieved will be one of the most interesting issues in the future, but connected with and dependent on the full characterization of the organellar posttranscriptional machineries.

Protein phosphorylation and the redox state of the organelle (and the cell) are good candidates for processes and/or internal parameters with the potential to coordinate and integrate not only different processing mechanisms but also the various levels of gene regulation. The investigation of the redox state as potential trigger of regulation particularly has attracted high attention. Its influence on translation in *C. reinhardtii*, on transcription initiation in chloroplasts of higher plants and transcription regulation of nuclear encoded plastid proteins has recently been reviewed (Barnes and Mayfield, 2003; Link, 2003; Pfannschmidt, 2003; Pfannschmidt *et al.*, 2003). Although promising progress has been made in this field, the overall picture remains patchy, especially since data on the role of protein phosphorylation and the redox state control of processes determining RNA stability are so few. So far only the phosphorylation of two higher chloroplast proteins, the 28 RNP from spinach and the p54 from mustard, has been reported. As discussed above it was speculated that the 28 RNP has a regulatory role in directing the chloroplast degradosome action on RNA. The binding affinity of this protein has been found to be three- to fourfold reduced upon phosphorylation; however, no experimental evidence has so far been presented that would support a direct role of this phosphorylation in the control of the degradosome action (Hayes *et al.*, 1996; Lisitsky and Schuster, 1995).

In the case of the p54 protein, phosphorylation was found to enhance both RNA binding and processing activities. In addition, this protein can, at least *in vitro,* be activated by the oxidized form of glutathione (GSSG) but not by various other reductants. Interestingly the strongest activation of this protein is observed when phosphorylation occurs prior to oxidation. But as in the case of the spinach 28 RNP, the exact function of p54 phosphorylation and oxidation *in vivo* is as yet unclear. Particularly the *in vivo* importance of the oxidation observed *in vitro* only at a glutathione concentration ten times higher than those measured in isolated chloroplasts needs further experimentation (Liere and Link, 1997). The binding of two *A. thaliana* proteins to the 5' UTR of *psb*A occurs in a redox-dependent manner; however, the identity of the polypeptides is still unclear (Shen *et al.*, 2001).

In vivo evidence for an influence of the redox poised on RNA degradation has been found by the investigation of transplastomic *C. reinhardtii* lines expressing a chimeric transcript composed of the 5' UTR of *rbc*L, the □-GUS reading frame and the *psa*B 3' UTR (MU7; Salvador and Klein, 1999). The addition of DCMU (3-(3,4-dichlorophenyl)-1,1-dimethylurea), blocking photosynthetic electron transport, prevents the light-dependent degradation of this transcript usually observed in light phases when these algae are grown in 12-h dark/12-h light cycles. The effect can be partially abolished by the additional application of DTT (dithiothreitol). Together with the observation that these compounds have the same effect on endogenous transcripts, these finding strongly suggest that via the 5' UTR the redox state influences the half-life time of at least several transcripts in *C. reinhardtii*.

In contrast, only very few experimental data supporting a regulatory role of protein phosphorylation and redox state in mitochondrial gene expression are available. At least protein synthesis in isolated pea mitochondria seems to require electron transport through succinate:ubiquinon oxidoreductase (complex II; Escobar Galvis *et al.*, 1998). This report, together with the finding that a variety of electron transport inhibitors reduce UTP incorporation in RNA and thus decrease RNA synthesis, represent basic data suggesting an influence of the redox poise on mitochondrial gene expression at two different levels (Wilson *et al.*, 1996).

Besides phosphorylation and redox level, Mg^{2+} concentrations may have a regulatory function in chloroplasts. At least *in vitro*, the half-lives of *psb*A mRNA, *rbc*L mRNA, 16S rRNA and $tRNA^{His}$ are enhanced with increasing Mg^{2+} concentrations (Horlitz and Klaff, 2000). Since the Mg^{2+} concentrations were indeed found to increase from 3 mM in chloroplasts of young leaves to 10 mM in adult plants, it was suggested that the concentration of this divalent cation might play a role in regulating RNA stability in a developmental manner. In this process the Mg^{2+}-dependent enzyme CSP41a might have relevant role (Bollenbach *et al.*, 2003). The comparison of degradation rates in chloroplast lysates from wt and CSP41a antisense plants measured at low (< 1 mM free Mg^{2+}) and relatively high (4 mM Mg^{2+}) concentrations revealed a threefold lower decay rate under the latter condition in the antisense lysate, while no differences are observed under low Mg^{2+}

concentration. This suggests the CSP41a protein may be part of the Mg^{2+} dependent control mechanism.

6. ACKNOWLEDGEMENTS

We thank Axel Brennicke for helpful comments on the manuscript. We apologize if we did not mention all relevant publications due to space limits and because of accidental oversight!

7. REFERENCES

Alexander, C., Faber, N. & Klaff, P. (1998). Characterization of protein-binding to the spinach chloroplast *psbA* mRNA 5' untranslated region. *Nucleic Acids Res.*, *26*, 2265-2272.

Allison, L.A. & Maliga, P. (1995). Light-responsive and transcription-enhancing elements regulate the plastid *psbD* core promoter. *EMBO J.*, *14*, 3721-3730.

Alonso, J.M. *et al.* (2003). Genome-wide insertional mutagenesis of *Arabidopsis thaliana*. *Science*, *301*, 653-657.

Anthonisen, I.L., Salvador, M.L. & Klein, U. (2001). Specific sequence elements in the 5' untranslated regions of *rbcL* and *atpB* gene mRNas stabilize transcripts in the chloroplast of *Chlamydomonas reinhardtii*. *RNA*, *7*, 1024-1033.

The Arabidopsis Genome Initiative (A.G.I.). (2000). Analysis of the genome sequence of the flowering plant *Arabidopsis thaliana*. *Nature*, *408*, 796-815.

Aubert, D., Bisanz-Seyer, C. & Herzog, M. (1992). Mitochondrial *rps*14 is a transcribed and edited pseudogene in *Arabidopsis thaliana*. *Plant Mol. Biol.*, *20*, 1169-1174.

Aubourg, S., Boudet, N., Kreis, M. & Lecharny, A. (2000). In *Arabidopsis thaliana*, 1% of the genome codes for a novel protein family unique to plants. *Plant Mol. Biol.*, *42*, 603-313.

Baginsky, S., Gruissem, W., Shteiman-Kotler, A., Liveanu, V., Yehudai-Resheff, S., Bellaoui, M. *et al.* (2002). Endonucleolytic activation directs dark-induced chloroplast mRNA degradation. *Nucleic Acids Res.*, *30*, 4527-4533.

Baginsky, S., Shteiman-Kotler, A., Liveanu, V., Yehudai-Resheff, S., Bellaoui, M., Settlage, R.E. *et al.* (2001). Chloroplast PNPase exists as a homo-multimer enzyme complex that is distinct from the Escherichia coli degradosome. *RNA*, *7*, 1464-1475.

Barkan, A. (1988). Proteins encoded by a complex chloroplast transcription unit are each translated from both monocistronic and polycistronic mRNAs. *EMBO J.*, *7*, 2637-2644.

Barkan, A. & Goldschmidt-Clermont, M. (2000). Participation of nuclear genes in chloroplast gene expression. *Biochimie*, *82*, 559-572.

Barkan, A., Walker, M., Nolasco, M. & Johnson, D. (1994). A nuclear mutation in maize blocks the processing and translation of several chloroplast mRNAs and provides evidence for the differential translation of alternative mRNA forms. *EMBO J.*, *13*, 3170-3181.

Barnes, D. & Mayfield, S.P. (2003). Redox control of posttranscriptional processes in the chloroplast. *Antioxid. Redox Signal.*, *5*, 89-94.

Bellaoui, M., Pelletier, G. & Budar, F. (1997). The steady-state level of mRNA from the *Ogura* cytoplasmic male sterility locus in *Brassica* cybrids is determined post-transcriptionally by its 3' region. *EMBO J.*, *16*, 5057-5068.

Bentolila, S., Alfonso, A.A. & Hanson, M.R. (2002). A pentatricopeptide repeat-containing gene restores fertility to cytoplasmic male-sterile plants. *Proc. Natl. Acad. Sci. U S A*, *99*, 10887-10892.

Binder, S., Marchfelder, A. & Brennicke, A. (1994). RNA editing of tRNA(Phe) and tRNA(Cys) in mitochondria of *Oenothera berteriana* is initiated in precursor molecules. *Mol. Gen. Genet.*, *244*, 67-74.

Bisanz, C., Bégot, L., Carol, P., Perez, P., Bligny, M., Pesey, H. *et al.* (2003). The *Arabidopsis* nuclear DAL gene encodes a chloroplast protein which is required for the maturation of the plastid ribosomal RNAs and is essential for chloroplast differentiation. *Plant Mol. Biol.*, *51*, 651-663.

Björk, G.R. & Kohli, J. (1990). Synthesis and function of modified nucleosides in tRNA. In N, G.C. and K.C., K. (eds.), *Chromatography and Modifications of Nucleosides*. Elsevier, Amsterdam, pp. 813-867.

Blowers, A.D., Klein, U., Ellmore, G.S. & Bogorad, L. (1993). Functional *in vivo* analyses of the 3' flanking sequences of the *Chlamydomonas* chloroplast *rbcL* and *psaB* genes. *Mol. Gen. Genet.*, *238*, 339-349.

Bollenbach, T.J. & Stern, D.B. (2003). Secondary structures common to chloroplast mRNA 3'-untranslated regions direct cleavage by CSP41, an endoribonuclease belonging to the short chain dehydrogenase/reductase superfamily. *J. Biol. Chem.*, *278*, 25832-25838.

Bollenbach, T.J., Tatman, D.A. & Stern, D.B. (2003). CSP41a, a multifunctional RNA-binding protein, initiates mRNA turnover in tobacco chloroplasts. *Plant J.*, *36*, 842-852.

Boudreau, E., Nickelsen, J., Lemaire, S.D., Ossenbuhl, F. & Rochaix, J.D. (2000). The Nac2 gene of *Chlamydomonas* encodes a chloroplast TPR-like protein involved in *psbD* mRNA stability. *EMBO J.*, *19*, 3366-3376.

Brandt, P., Unseld, M., Eckert-Ossenkopp, U. & Brennicke, A. (1993). An *rps*14 pseudogene is transcribed and edited in *Arabidopsis* mitochondria. *Curr. Genet.*, *24*, 330-336.

Brown, G.G., Formanova, N., Jin, H., Wargachuk, R., Dendy, C., Patil, P. *et al.* (2003). The radish Rfo restorer gene of Ogura cytoplasmic male sterility encodes a protein with multiple pentatricopeptide repeats. *Plant J.*, *35*, 262-272.

Budar, F. & Pelletier, G. (2001). Male sterility in plants: occurrence, determinism, significance and use. *C. R. Acad. Sci. III*, *324*, 543-550.

Burkard, G. & Keller, E.B. (1974). Poly(A) polymerase and poly(G) polymerase in wheat chloroplasts. *Proc. Natl. Acad. Sci. U S A*, *71*, 389-393.

Burkard, U. & Söll, D. (1988). The 5'-terminal guanylate of chloroplast histidine tRNA is encoded in its gene. *J. Biol. Chem.*, *263*, 9578-9581.

Carpousis, A.J. (2002). The *Escherichia coli* RNA degradosome: structure, function and relationship in other ribonucleolytic multienzyme complexes. *Biochem. Soc. Trans.*, *30*, 150-155.

Carpousis, A.J., Vanzo, N.F. & Raynal, L.C. (1999). mRNA degradation. A tale of poly(A) and multiprotein machines. *Trends Genet.*, *15*, 24-28.

Chen, H.C. & Stern, D.B. (1991a). Specific binding of chloroplast proteins in vitro to the 3' untranslated region of spinach chloroplast petD mRNA. *Mol. Cell. Biol.*, *11*, 4380-4388.

Chen, H.C. & Stern, D.B. (1991b). Specific ribonuclease activities in spinach chloroplasts promote mRNA maturation and degradation. *J. Biol. Chem.*, *266*, 24205-24211.

Christopher, D.A., Kim, M. & Mullet, J.E. (1992). A novel light-regulated promoter is conserved in cereal and dicot chloroplasts. *Plant Cell*, *4*, 785-798.

Cordier, A. & Schön, A. (1999). Cyanelle RNase P: RNA structure analysis and holoenzyme properties of an organellar ribonucleoprotein enzyme. *J. Mol. Biol.*, *289*, 9-20.

Coulthart, M.B., Huh, G.S. & Gray, M.W. (1990). Physical organization of the 18S and 5S ribosomal RNA genes in the mitochondrial genome of rye (*Secale cereale* L.). *Curr. Genet.*, *17*, 339-346.

Del Campo, E.M., Sabater, B. & Martin, M. (2002). Post-transcriptional control of chloroplast gene expression. Accumulation of stable *psa*C mRNA is due to downstream RNA cleavages in the *ndh*D gene. *J. Biol. Chem.*, *277*, 36457-36464.

Delage, L., Dietrich, A., Cosset, A. & Maréchal-Drouard, L. (2003). *In vitro* import of a nuclearly encoded tRNA into mitochondria of *Solanum tuberosum*. *Mol. Cell. Biol.*, *23*, 4000-4012.

Deng, X.W. & Gruissem, W. (1987). Control of plastid gene expression during development: the limited role of transcriptional regulation. *Cell*, *49*, 379-387.

Deng, X.W., Tonkyn, J.C., Peter, G.F., Thornber, J.P. & Gruissem, W. (1989). Post-transcriptional control of plastid mRNA accumulation during adaptation of chloroplasts to different light quality environments: Control of plastid gene expression during development: the limited role of transcriptional regulation. *Plant Cell*, *1*, 645-654.

Dirheimer, G., Keith, G., Dumas, P. & Westhof, E. (1995). Primary, secondary and tertiary structures of tRNAs. In Söll, D. & A., R. (eds.), *tRNA: structure, biosynthesis and function*. ASM, Washington DC, pp. 93-126.

Dombrowski, S. & Binder, S. (1998). 3' Inverted repeats in plant mitochondrial mRNAs act as processing and stabilizing elements but do not terminate transcription. In Møller, I.M., Gardeström, P., Glimelius, K. & Glaser, E. (eds.), *Plant Mitochondria: From Gene to Function*. Backhuys Publications, Leiden Netherlands, pp. 9-12.

Dombrowski, S., Brennicke, A. & Binder, S. (1997). 3'-Inverted repeats in plant mitochondrial mRNAs are processing signals rather than transcription terminators. *EMBO J.*, *16*, 5069-5076.
Dombrowski, S., Hoffmann, M., Kuhn, J., Brennicke, A. & Binder, S. (1998). On mitochondrial promoters in *Arabidopsis thaliana* and other flowering plants. In Møller, I.M., Gardeström, P., Glimelius, K. & Glaser, E. (eds.), *Plant Mitochondria: From Gene to Function*. Backhuys Publications, Leiden Netherlands, pp. 165-170.
Drager, R.G., Girard-Bascou, J., Choquet, Y., Kindle, K.L. & Stern, D.B. (1998). In vivo evidence for 5'-->3' exoribonuclease degradation of an unstable chloroplast mRNA. *Plant J.*, *13*, 85-96.
Drager, R.G., Higgs, D.C., Kindle, K.L. & Stern, D.B. (1999). 5' to 3' exoribonucleolytic activity is a normal component of chloroplast mRNA decay pathways. *Plant J.*, *19*, 521-531.
Drager, R.G., Zeidler, M., Simpson, C.L. & Stern, D.B. (1996). A chloroplast transcript lacking the 3' inverted repeat is degraded by 3' to 5' exoribonuclease activity. *RNA*, *2*, 652-663.
Dubrovsky, E.B., Dubrovskaya, V.A., Levinger, L., Schiffer, S. & Marchfelder, A. (2004). *Drosophila* RNase Z processes mitochondrial and nuclear pre-tRNA 3' ends in vivo. *Nucleic Acids Res.*, *32*, 255-262.
Dziembowski, A., Piwowarski, J., Hoser, R., Minczuk, M., Dmochowska, A., Siep, M. *et al.* (2003). The yeast mitochondrial degradosome. Its composition, interplay between RNA helicase and RNase activities and the role in mitochondrial RNA metabolism. *J. Biol. Chem.*, *278*, 1603-1611.
Elo, A., Lyznik, A., Gonzalez, D.O., Kachman, S.D. & Mackenzie, S.A. (2003). Nuclear genes that encode mitochondrial proteins for DNA and RNA metabolism are clustered in the *Arabidopsis* genome. *Plant Cell*, *15*, 1619-1631.
Escobar Galvis, M.L., Allen, J.F. & Håkansson, G. (1998). Protein synthesis by isolated pea mitochondria is dependent on the activity of respiratory complex II. *Curr. Genet.*, *33*, 320-329.
Felder, S., Meierhoff, K., Sane, A.P., Meurer, J., Driemel, C., Plucken, H. *et al.* (2001). The nucleus-encoded HCF107 gene of *Arabidopsis* provides a link between intercistronic RNA processing and the accumulation of translation-competent *psb*H transcripts in chloroplasts. *Plant Cell*, *13*, 2127-2141.
Fernandez-Silva, P., Martinez-Azorin, F., Micol, V. & Attardi, G. (1997). The human mitochondrial transcription termination factor (mTERF) is a multizipper protein but binds to DNA as a monomer, with evidence pointing to intramolecular leucine zipper interactions. *EMBO J.*, *16*, 1066-1079.
Fernandez-Silva, P., Polosa, P.L., Roberti, M., Di Ponzio, B., Gadaleta, M.N., Montoya, J. *et al.* (2001). Sea urchin mtDBP is a two-faced transcription termination factor with a biased polarity depending on the RNA polymerase. *Nucleic Acids Res.*, *29*, 4736-4743.
Fisk, D.G., Walker, M.B. & Barkan, A. (1999). Molecular cloning of the maize gene *crp1* reveals similarity between regulators of mitochondrial and chloroplast gene expression. *EMBO J.*, *18*, 2621-2630.
Francis, M.A. & Dudock, B.S. (1982). Nucleotide sequence of a spinach chloroplast isoleucine tRNA. *J. Biol. Chem.*, *257*, 11195-11198.
Frank, D.N. & Pace, N.R. (1998). Ribonuclease P: unity and diversity in a tRNA processing ribozyme. *Annu. Rev. Biochem.*, *67*, 153-180.
Gagliardi, D., Kuhn, J., Spadinger, U., Brennicke, A., Leaver, C.J. & Binder, S. (1999). An RNA helicase (AtSUV3) is present in *Arabidopsis thaliana* mitochondria. *FEBS Lett.*, *458*, 337-342.
Gagliardi, D. & Leaver, C.J. (1999). Polyadenylation accelerates the degradation of the mitochondrial mRNA associated with cytoplasmic male sterility in sunflower. *EMBO J.*, *18*, 3757-3766.
Gagliardi, D., Perrin, R., Marechal-Drouard, L., Grienenberger, J.M. & Leaver, C.J. (2001). Plant mitochondrial polyadenylated mRNAs are degraded by a 3'- to 5'exoribonuclease activity, which proceeds unimpeded by stable secondary structures. *J. Biol. Chem.*, *276*, 43541-43547.
Gegenheimer, P. (1995). Structure, mechanism and evolution of chloroplast transfer RNA processing systems. *Mol. Biol. Rep.*, *22*, 147-150.
Giegé, P., Hoffmann, M., Binder, S. & Brennicke, A. (2000). RNA degradation buffers asymmetries of transcription in *Arabidopsis* mitochondria. *EMBO Rep.*, *1*, 164-170.
Goff, S.A.*et al.* (2002). A draft sequence of the rice genome (*Oryza sativa* L. ssp. japonica). *Science*, *296*, 92-100.
Greenberg, B.M. & Hallick, R.B. (1986). Accurate transcription and processing of 19 *Euglena* chloroplast tRNAs in a *Euglena* soluble extract. *Plant Mol. Biol.*, *6*, 89-100.
Gruissem, W. (1989). Chloroplast gene expression: how plants turn their plastids on. *Cell*, *56*, 161-170.
Gruissem, W., Greenberg, B.M., Zurawski, G., Prescott, D.M. & Hallick, R.B. (1983). Biosynthesis of chloroplast transfer RNA in a spinach chloroplast transcription system. *Cell*, *35*, 815-828.

Haff, L.A. & Bogorad, L. (1976). Poly(adenylic acid)-containing RNA from plastids of maize. *Biochemistry*, *15*, 4110-4115.

Handa, H. (2003). The complete nucleotide sequence and RNA editing content of the mitochondrial genome of rapeseed (*Brassica napus* L.): comparative analysis of the mitochondrial genomes of rapeseed and *Arabidopsis thaliana*. *Nucleic Acids Res.*, *31*, 5907-5916.

Hanic-Joyce, P.J. & Gray, M.W. (1990). Processing of transfer RNA precursors in a wheat mitochondrial extract. *J. Biol. Chem.*, *265*, 13782-13791.

Hanic-Joyce, P.J., Spencer, D.F. & Gray, M.W. (1990). *In vitro* processing of transcripts containing novel tRNA-like sequences ('t-elements') encoded by wheat mitochondrial DNA. *Plant Mol. Biol.*, *15*, 551-559.

Hartley, M.R. & Head, C. (1979). The synthesis of chloroplast high-molecular-weight ribosomal ribonucleic acid in spinach. *Eur. J. Biochem.*, *96*, 301-309.

Hashimoto, M., Endo, T., Peltier, G., Tasaka, M. & Shikanai, T. (2003). A nucleus-encoded factor, CRR2, is essential for the expression of chloroplast *ndh*B in *Arabidopsis*. *Plant J.*, *36*, 541-549.

Hayes, R., Kudla, J., Schuster, G., Gabay, L., Maliga, P. & Gruissem, W. (1996). Chloroplast mRNA 3'-end processing by a high molecular weight protein complex is regulated by nuclear encoded RNA binding proteins. *EMBO J.*, *15*, 1132-1141.

Heazlewood, J.L., Howell, K.A., Whelan, J. & Millar, A.H. (2003). Towards an analysis of the rice mitochondrial proteome. *Plant Physiol.*, *132*, 230-242.

Heazlewood, J.L., Tonti-Filippini, J.S., Gout, A.M., Day, D.A., Whelan, J. & Millar, A.H. (2004). Experimental analysis of the *Arabidopsis* mitochondrial proteome highlights signaling and regulatory components, provides assessment of targeting prediction programs, and indicates plant-specific mitochondrial proteins. *Plant Cell*, *16*, 241-256.

Hicks, A., Drager, R.G., Higgs, D.C. & Stern, D.B. (2002). An mRNA 3' processing site targets downstream sequences for rapid degradation in *Chlamydomonas* chloroplasts. *J. Biol. Chem.*, *277*, 3325-3333.

Higgs, D.C., Shapiro, R.S., Kindle, K.L. & Stern, D.B. (1999). Small *cis*-acting sequences that specify secondary structures in a chloroplast mRNA are essential for RNA stability and translation. *Mol. Cell. Biol.*, *19*, 8479-8491.

Hirose, T. & Sugiura, M. (2001). Involvement of a site-specific trans-acting factor and a common RNA-binding protein in the editing of chloroplast mRNAs: development of a chloroplast *in vitro* RNA editing system. *EMBO J.*, *20*, 1144-1152.

Hoffmann, M., Dombrowski, S., Guha, C. & Binder, S. (1999). Cotranscription of the *rpl5-rps14-cob* gene cluster in pea mitochondria. *Mol. Gen. Genet.*, *261*, 537-545.

Hoffmann, M., Kuhn, J., Däschner, K. & Binder, S. (2001). The RNA world of plant mitochondria. *Prog. Nucleic Acid Res. Mol. Biol.*, *70*, 119-154.

Horlitz, M. & Klaff, P. (2000). Gene-specific trans-regulatory functions of magnesium for chloroplast mRNA stability in higher plants. *J. Biol. Chem.*, *275*, 35638-35645.

Kaleikau, E.K., Andre, C.P. & Walbot, V. (1992). Structure and expression of the rice mitochondrial apocytochrome b gene (*cob-1*) and pseudogene (*cob-2*). *Curr. Genet.*, *22*, 463-470.

Kazama, T. & Toriyama, K. (2003). A pentatricopeptide repeat-containing gene that promotes the processing of aberrant *atp*6 RNA of cytoplasmic male-sterile rice. *FEBS Let.*, *544*, 99-102.

Koizuka, N., Imai, R., Fujimoto, H., Hayakawa, T., Kimura, Y., Kohno-Murase, J. *et al.* (2003). Genetic characterization of a pentatricopeptide repeat protein gene, orf687, that restores fertility in the cytoplasmic male-sterile Kosena radish. *Plant J.*, *34*, 407-415.

Kruft, V., Eubel, H., Jansch, L., Werhahn, W. & Braun, H.P. (2001). Proteomic approach to identify novel mitochondrial proteins in *Arabidopsis*. *Plant Physiol.*, *127*, 1694-1710.

Kubo, T., Nishizawa, S., Sugawara, A., Itchoda, N., Estiati, A. & Mikami, T. (2000). The complete nucleotide sequence of the mitochondrial genome of sugar beet (*Beta vulgaris* L.) reveals a novel gene for tRNA(Cys)(GCA). *Nucleic Acids Res.*, *28*, 2571-2576.

Kuchka, M.R., Goldschmidt-Clermont, M., van Dillewijn, J. & Rochaix, J.D. (1989). Mutation at the *Chlamydomonas nuclear* NAC2 locus specifically affects stability of the chloroplast *psbD* transcript encoding polypeptide D2 of PS II. *Cell*, *58*, 869-876.

Kudla, J., Hayes, R. & Gruissem, W. (1996). Polyadenylation accelerates degradation of chloroplast mRNA. *EMBO J.*, *15*, 7137-7146.

Kuhn, J. & Binder, S. (2002). RT-PCR analysis of 5' to 3'-end-ligated mRNAs identifies the extremities of *cox2* transcripts in pea mitochondria. *Nucleic Acids Res.*, *30*, 439-446.

Kuhn, J., Tengler, U. and Binder, S. (2001). Transcript lifetime is balanced between stabilizing stem-loop structures and degradation-promoting polyadenylation in plant mitochondria. *Mol. Cell. Biol.*, *21*, 731-742.

Kunzmann, A., Brennicke, A. & Marchfelder, A. (1998). 5' end maturation and RNA editing have to precede tRNA 3' processing in plant mitochondria. *Proc. Natl. Acad. Sci. U S A*, *95*, 108-113.

Levy, H., Kindle, K.L. & Stern, D.B. (1997). A nuclear mutation that affects the 3' processing of several mRNAs in *Chlamydomonas* chloroplasts. *Plant Cell*, *9*, 825-836.

Levy, H., Kindle, K.L. & Stern, D.B. (1999). Target and specificity of a nuclear gene product that participates in mRNA 3'-end formation in *Chlamydomonas* chloroplasts. *J. Biol. Chem.*, *274*, 35955-35962.

Li, Y.Q. & Sugiura, M. (1990). Three distinct ribonucleoproteins from tobacco chloroplasts: each contains a unique amino terminal acidic domain and two ribonucleoprotein consensus motifs. *EMBO J.*, *9*, 3059-3066.

Liere, K. & Link, G. (1997). Chloroplast endoribonuclease p54 involved in RNA 3'-end processing is regulated by phosphorylation and redox state. *Nucleic Acids Res.*, *25*, 2403-2408.

Lilly, J.W., Maul, J.E. & Stern, D.B. (2002). The *Chlamydomonas reinhardtii* organellar genomes respond transcriptionally and post-transcriptionally to abiotic stimuli. *Plant Cell*, *14*, 2681-2706.

Link, G. (2003). Redox regulation of chloroplast transcription. *Antioxid. Redox Signal.*, *5*, 79-87.

Lisitsky, I., Klaff, P. & Schuster, G. (1996). Addition of destabilizing poly (A)-rich sequences to endonuclease cleavage sites during the degradation of chloroplast mRNA. *Proc. Natl. Acad. Sci. U S A*, *93*, 13398-13403.

Lisitsky, I., Klaff, P. & Schuster, G. (1997a). Blocking polyadenylation of mRNA in the chloroplast inhibits its degradation. *Plant J.*, *12*, 1173-1178.

Lisitsky, I., Kotler, A. & Schuster, G. (1997b). The mechanism of preferential degradation of polyadenylated RNA in the chloroplast. The exoribonuclease 100RNP/polynucleotide phosphorylase displays high binding affinity for poly(A) sequence. *J. Biol. Chem.*, *272*, 17648-17653.

Lisitsky, I. and Schuster, G. (1995). Phosphorylation of a chloroplast RNA-binding protein changes its affinity to RNA. *Nucleic Acids Res.*, *23*, 2506-2511.

Lizama, L., Holuigue, L. & Jordana, X. (1994). Transcription initiation sites for the potato mitochondrial gene coding for subunit 9 of ATP synthase (*atp9*). *FEBS Let.*, *349*, 243-248.

Loguercio Polosa, P., Roberti, M., Musicco, C., Gadaleta, M.N., Quagliariello, E. & Cantatore, P. (1999). Cloning and characterisation of mtDBP, a DNA-binding protein which binds two distinct regions of sea urchin mitochondrial DNA. *Nucleic Acids Res.*, *27*, 1890-1899.

Lupold, D.S., Caoile, A.G. & Stern, D.B. (1999). Polyadenylation occurs at multiple sites in maize mitochondrial *cox2* mRNA and is independent of editing status. *Plant Cell*, *11*, 1565-1578.

Maloney, A.P., Traynor, P.L., Levings, C.S. & Walbot, V. (1989). Identification in maize mitochondrial 26S rRNA of a short 5'-end sequence possibly involved in transcription initiation and processing. *Curr. Genet.*, *15*, 207-212.

Maloney, A.P. & Walbot, V. (1990). Structural analysis of mature and dicistronic transcripts from the 18 S and 5 S ribosomal RNA genes of maize mitochondria. *J. Mol. Biol.*, *213*, 633-649.

Marchfelder, A. (1995). Plant mitochondrial RNase P. *Mol. Biol. Rep.*, *22*, 151-156.

Marchfelder, A. & Brennicke, A. (1994). Characterization and partial purification of tRNA processing activities from potato mitochondria. *Plant Physiol.*, *105*, 1247-1254.

Marchfelder, A., Brennicke, A. & Binder, S. (1996). RNA editing is required for efficient excision of $tRNA^{Phe}$ from precursors in plant mitochondria. *J. Biol. Chem.*, *271*, 1898-1903.

Marchfelder, A., Schuster, W. & Brennicke, A. (1990). *In vitro* processing of mitochondrial and plastid derived tRNA precursors in a plant mitochondrial extract. *Nucleic Acids Res.*, *18*, 1401-1406.

Maréchal-Drouard, L., Cosset, A., Remacle, C., Ramamonjisoa, D. & Dietrich, A. (1996a). A single editing event is a prerequisite for efficient processing of potato mitochondrial phenylalanine tRNA. *Mol. Cell. Biol.*, *16*, 3504-3510.

Maréchal-Drouard, L., Guillemaut, P., Cosset, A., Arbogast, M., Weber, F., Weil, J.H. *et al.* (1990). Transfer RNAs of potato (*Solanum tuberosum*) mitochondria have different genetic origins. *Nucleic Acids Res.*, *18*, 3689-3696.

Maréchal-Drouard, L., Kumar, R., Remacle, C. & Small, I. (1996b). RNA editing of larch mitochondrial tRNA(His) precursors is a prerequisite for processing. *Nucleic Acids Res.*, *24*, 3229-3234.

Maréchal-Drouard, L., Ramamonjisoa, D., Cosset, A., Weil, J.H. & Dietrich, A. (1993a). Editing corrects mispairing in the acceptor stem of bean and potato mitochondrial phenylalanine transfer RNAs. *Nucleic Acids Res.*, *21*, 4909-4914.

Maréchal-Drouard, L., Weil, J.H. & Dietrich, A. (1993b). Transfer RNA and transfer genes in plants. *Annu. Rev. Plant Physiol. Plant Mol. Biol.*, *44*, 13-32.

Marion-Poll, A., Hibbert, C.S., Radebaugh, C.A. & Hallick, R.H. (1988). Processing of mono-, di-, and tricistronic transfer RNA precursors in a spinach or pea chloroplast soluble extract. *Plant Mol. Biol.*, *11*.

Martin, N.C. (1995). Organellar tRNAs: biosynthesis and function. In Söll, D. and RajBhandary, A. (eds.), *tRNA: structure, biosynthesis and function*. ASM, Washington DC, pp. 127-140.

Maul, J.E., Lilly, J.W., Cui, L., dePamphilis, C.W., Miller, W., Harris, E.H. *et al.* (2002) The *Chlamydomonas reinhardtii* plastid chromosome: islands of genes in a sea of repeats. *Plant Cell*, *14*, 2659-2679.

McGarvey, P. & Helling, R.B. (1989). Processing of chloroplast ribosomal RNA transcripts in *Euglena gracilis bacillaris*. *Curr. Genet.*, *15*, 363-370.

Meierhoff, K., Felder, S., Nakamura, T., Bechtold, N. & Schuster, G. (2003). HCF152, an *Arabidopsis* RNA binding pentatricopeptide repeat protein involved in the processing of chloroplast *psb*B-*psb*T-*psb*H-*pet*B-*pet*D RNAs. *Plant Cell*, *15*, 1480-1495.

Millar, A.H., Sweetlove, L.J., Giegé, P. & Leaver, C.J. (2001). Analysis of the *Arabidopsis* mitochondrial proteome. *Plant Physiol.*, *127*, 1711-1727.

Minczuk, M., Piwowarski, J., Papworth, M.A., Awiszus, K., Schalinski, S., Dziembowski, A. *et al.* (2002). Localisation of the human hSuv3p helicase in the mitochondrial matrix and its preferential unwinding of dsDNA. *Nucleic Acids Res.*, *30*, 5074-5086.

Monéger, F., Smart, C.J. & Leaver, C.J. (1994). Nuclear restoration of cytoplasmic male sterility in sunflower is associated with the tissue-specific regulation of a novel mitochondrial gene. *EMBO J.*, *13*, 8-17.

Mörl, M. & Marchfelder, A. (2001). The final cut. The importance of tRNA 3'-processing. *EMBO Rep.*, *2*, 17-20.

Mulligan, R.M., Maloney, A.P. & Walbot, V. (1988). RNA processing and multiple transcription initiation sites result in transcript size heterogeneity in maize mitochondria. *Mol. Gen. Genet.*, *211*, 373-380.

Nakamura, T., Meierhoff, K., Westhoff, P. & Schuster, G. (2003). RNA-binding properties of HCF152, an *Arabidopsis* PPR protein involved in the processing of chloroplast RNA. *Eur. J. Biochem.*, *270*, 4070-4081.

Nakamura, T., Ohta, M., Sugiura, M. & Sugita, M. (2001). Chloroplast ribonucleoproteins function as a stabilizing factor of ribosome-free mRNAs in the stroma. *J. Biol. Chem.*, *276*, 147-152.

Nickelsen, J. (2003). Chloroplast RNA-binding proteins. *Curr. Genet.*, *43*, 392-399.

Nickelsen, J., Fleischmann, M., Boudreau, E., Rahire, M. & Rochaix, J.D. (1999). Identification of *cis*-acting RNA leader elements required for chloroplast *psbD* gene expression in *Chlamydomonas*. *Plant Cell*, *11*, 957-970.

Nickelsen, J. & Link, G. (1993). The 54 kDa RNA-binding protein from mustard chloroplasts mediates endonucleolytic transcript 3' end formation *in vitro*. *Plant J.*, *3*, 537-544.

Nickelsen, J., van Dillewijn, J., Rahire, M. & Rochaix, J.D. (1994). Determinants for stability of the chloroplast *psbD* RNA are located within its short leader region in *Chlamydomonas reinhardtii*. *EMBO J.*, *13*, 3182-3191.

Nivison, H.T. & Hanson, M.R. (1989). Identification of a mitochondrial protein associated with cytoplasmic male sterility in petunia. *Plant Cell*, *1*, 1121-1130.

Notsu, Y., Masood, S., Nishikawa, T., Kubo, N., Akiduki, G., Nakazono, M. *et al.* (2002). The complete sequence of the rice (*Oryza sativa* L.) mitochondrial genome: frequent DNA sequence acquisition and loss during the evolution of flowering plants. *Mol. Genet. Genomics*, *268*, 434-445.

Oda, K., Yamato, K., Ohta, E., Nakamura, Y., Takemura, M., Nozato, N. *et al.* (1992). Gene organization deduced from the complete sequence of liverwort *Marchantia polymorpha* mitochondrial DNA. A primitive form of plant mitochondrial genome. *J. Mol. Biol.*, *223*, 1-7.

Ohyama, K. (1996). Chloroplast and mitochondrial genomes from a liverwort, *Marchantia polymorpha*-gene organization and molecular evolution. *Biosci. Biotechnol. Biochem.*, *60*, 16-24.

Ojala, D., Montoya, J. & Attardi, G. (1981). tRNA punctuation model of RNA processing in human mitochondria. *Nature*, *290*, 470-474.

Oommen, A., Li, X.Q. & Gegenheimer, P. (1992). Cleavage specificity of chloroplast and nuclear tRNA 3'-processing nucleases. *Mol. Cell. Biol.*, *12*, 865-875.

Ossenbühl, F. & Nickelsen, J. (2000). *cis*- and *trans*-Acting determinants for translation of *psbD* mRNA in *Chlamydomonas reinhardtii*. *Mol. Cell. Biol.*, *20*, 8134-8142.

Pfannschmidt, T. (2003). Chloroplast redox signals: how photosynthesis controls its own genes. *Trends Plant Sci.*, *8*, 33-41.

Pfannschmidt, T., Schutze, K., Fey, V., Sherameti, I. & Oelmuller, R. (2003). Chloroplast redox control of nuclear gene expression--a new class of plastid signals in interorganellar communication. *Antioxid. Redox Signal.*, *5*, 95-101.

Quiñones, V., Zanlungo, S., Moenne, A., Gómez, I., Holuigue, L., Litvak, S. *et al.* (1996). The *rpl5-rps14-cob* gene arrangement in *Solanum tuberosum*: *rps14* is a transcribed and unedited pseudogene. *Plant Mol. Biol.*, *31*, 937-943.

Roberti, M., Polosa, P.L., Bruni, F., Musicco, C., Gadaleta, M.N. & Cantatore, P. (2003). DmTTF, a novel mitochondrial transcription termination factor that recognises two sequences of *Drosophila melanogaster* mitochondrial DNA. *Nucleic Acids Res.*, *31*, 1597-1604.

Rochaix, J.D. (1996). Post-transcriptional regulation of chloroplast gene expression in *Chlamydomonas reinhardtii*. *Plant Mol. Biol.*, *32*, 327-341.

Rott, R., Drager, R.G., Stern, D.B. & Schuster, G. (1996). The 3' untranslated regions of chloroplast genes in Chlamydomonas reinhardtii do not serve as efficient transcriptional terminators. *Mol. Gen. Genet.*, *252*, 676-683.

Saalaoui, E., Litvak, S. & Araya, A. (1990). The apocytochrome from alloplasmic line wheat (T. aestivum, cytoplasm-T. timopheevi) exists in two differently expressed forms. *Plant Sci.*, *66*, 237-246.

Salvador, M.L. & Klein, U. (1999). The redox state regulates RNA degradation in the chloroplast of *Chlamydomonas reinhardtii*. *Plant Physiol.*, *121*, 1367-1374.

Salvador, M.L., Klein, U. & Bogorad, L. (1993). 5' sequences are important positive and negative determinants of the longevity of *Chlamydomonas* chloroplast gene transcripts. *Proc. Natl. Acad. Sci. U S A*, *90*, 1556-1560.

Salvador, M.L., Suay, L., Anthonisen, I.L. & Klein, U. (2004). Changes in the 5'-untranslated region of the *rbcL* gene accelerate transcript degradation more than 50-fold in the chloroplast of *Chlamydomonas reinhardtii*. *Curr. Genet.*, in press.

Schiffer, S., Rösch, S. & Marchfelder, A. (2002). Assigning a function to a conserved group of proteins: the tRNA 3'-processing enzymes. *EMBO J.*, *21*, 2769-2777.

Schneider, A. & Maréchal-Drouard, L. (2000). Mitochondrial tRNA import: are there distinct mechanisms? *Trends Cell Biol.*, *10*, 509-913.

Schuster, G. & Gruissem, W. (1991). Chloroplast mRNA 3' end processing requires a nuclear-encoded RNA-binding protein. *EMBO J.*, *10*, 1493-1502.

Schuster, G., Lisitsky, I. & Klaff, P. (1999). Polyadenylation and degradation of mRNA in the chloroplast. *Plant Physiol.*, **120**, 937-944.

Schuster, W. (1993) Ribosomal protein gene *rpl5* is cotranscribed with the *nad3* gene in *Oenothera* mitochondria. *Mol. Gen. Genet.*, *240*, 445-449.

Schuster, W., Hiesel, R., Isaac, P.G., Leaver, C.J. & Brennicke, A. (1986). Transcript termini of messenger RNAs in higher plant mitochondria. *Nucleic Acids Res.*, *14*, 5943-5954.

Schuster, W., Unseld, M., Wissinger, B. & Brennicke, A. (1990). Ribosomal protein S14 transcripts are edited in *Oenothera* mitochondria. *Nucleic Acids Res.*, *18*, 229-233.

Shanmugam, K., Hanic-Joyce, P.J. & Joyce, P.B. (1996). Purification and characterization of a tRNA nucleotidyltransferase from *Lupinus albus* and functional complementation of a yeast mutation by corresponding cDNA. *Plant Mol. Biol.*, *30*, 281-295.

Shen, Y., Danon, A. & Christopher, D.A. (2001). RNA binding-proteins interact specifically with the *Arabidopsis* chloroplast *psbA* mRNA 5' untranslated region in a redox-dependent manner. *Plant Cell Physiol.*, *42*, 1071-1078.

Shiina, T., Allison, L. & Maliga, P. (1998). *rbcL* Transcript levels in tobacco plastids are independent of light: reduced dark transcription rate is compensated by increased mRNA stability. *Plant Cell*, *10*, 1713-1722.

Sickmann, A., Reinders, J., Wagner, Y., Joppich, C., Zahedi, R., Meyer, H.E. *et al.* (2003). The proteome of *Saccharomyces cerevisiae* mitochondria. *Proc. Natl. Acad. Sci. U S A*, *100*, 13207-13212.

Singh, M., Boutanaev, A., Zucchi, P. & Bogorad, L. (2001). Gene elements that affect the longevity of *rbcL* sequence-containing transcripts in *Chlamydomonas* reinhardtii chloroplasts. *Proc. Natl. Acad. Sci. U S A*, *98*, 2289-2294.

Small, I.D. & Peeters, N. (2000). The PPR motif - a TPR-related motif prevalent in plant organellar proteins. *Trends Biochem. Sci.*, *25*, 46-47.

Smart, C.J., Monéger, F. & Leaver, C.J. (1994). Cell-specific regulation of gene expression in mitochondria during anther development in sunflower. *Plant Cell*, *6*, 811-825.

Stern, D.B. & Gruissem, W. (1987).Control of plastid gene expression: 3' inverted repeats act as mRNA processing and stabilizing elements, but do not terminate transcription. *Cell*, *51*, 1145-1157.

Stern, D.B., Higgs, D.C. and Yang, J. (1997). Transcription and translation in chloroplasts. *Trends Plant Sci.*, *2*, 308-315.

Stern, D.B. & Kindle, K.L. (1993). 3'end maturation of the Chlamydomonas reinhardtii chloroplast atpB mRNA is a two-step process. *Mol. Cell. Biol.*, *13*, 2277-2285.

Stern, D.B., Radwanski, E.R. & Kindle, K.L. (1991). A 3' stem/loop structure of the *Chlamydomonas* chloroplast *atpB* gene regulates mRNA accumulation *in vivo*. *Plant Cell*, ., 285-297.

Sugiura, M. (1995). The chloroplast genome. *Essays Biochem.*, *30*, 49-57.

Tokuhisa, J.G., Vijayan, P., Feldmann, K.A. & Browse, J.A. (1998) Chloroplast development at low temperatures requires a homolog of DIM1, a yeast gene encoding the 18S rRNA dimethylase. *Plant Cell*, *10*, 699-711.

Unseld, M., Marienfeld, J.R., Brandt, P. & Brennicke, A. (1997). The mitochondrial genome of *Arabidopsis thaliana* contains 57 genes in 366,924 nucleotides. *Nat. Genet.*, *15*, 57-61.

Vaistij, F.E., Boudreau, E., Lemaire, S.D., Goldschmidt-Clermont, M., Rochaix, J.D. & Wostrikoff, K. (2000a). Characterization of Mbb1, a nucleus-encoded tetratricopeptide-like repeat protein required for expression of the chloroplast *psbB/psbT/psbH* gene cluster in *Chlamydomonas reinhardtii*. *Proc. Natl. Acad. Sci. U S A*, *97*, 14813-14818.

Vaistij, F.E., Goldschmidt-Clermont, M., Wostrikoff, K. & Rochaix, J.D. (2000b). Stability determinants in the chloroplast *psbB/T/H* mRNAs of *Chlamydomonas reinhardtii*. *Plant J.*, *21*, 469-482.

Vanzo, N.F., Li, Y.S., Py, B., Blum, E., Higgins, C.F., Raynal. *et al.* (1998). Ribonuclease E organizes the protein interactions in the *Escherichia coli* RNA degradosome. *Genes Dev.*, *12*, 2770-2781.

Vermel, M., Guermann, B., Delage, L., Grienenberger, J.M., Marechal-Drouard, L. & Gualberto, J.M. (2002). A family of RRM-type RNA-binding proteins specific to plant mitochondria. *Proc. Natl. Acad. Sci. U S A*, *99*, 5866-5871.

Vicaretti, R. & Joyce, P.B. (1998). Comparison of mitochondrial and cytosolic tRNA nucleotidyltransferases from *Triticum aestivum*. *Can. J. Bot.*, *77*, 230-239.

Wahle, E. and Ruegsegger, U. (1999). 3'-End processing of pre-mRNA in eukaryotes. *FEMS Microbiol. Rev.*, *23*, 277-295.

Wahleithner, J.A. & Wolstenholme, D.R. (1988). Ribosomal protein S14 genes in broad bean mitochondrial DNA. *Nucleic Acids Res.*, *16*, 6897-6913.

Walter, M., Kilian, J. & Kudla, J. (2002). PNPase activity determines the efficiency of mRNA 3'-end processing, the degradation of tRNA and the extent of polyadenylation in chloroplasts. *EMBO J.*, *21*, 6905-6914.

Weber, F., Dietrich, A., Weil, J.H. & Maréchal-Drouard, L. (1990). A potato mitochondrial isoleucine tRNA is coded for by a mitochondrial gene possessing a methionine anticodon. *Nucleic Acids Res.*, **18**, 5027-5030.

Westhoff, P. & Herrmann, R.G. (1988) Complex RNA maturation in chloroplasts. The *psbB* operon from spinach. *Eur. J. Biochem.*, *171*, 551-564.

Williams, M.A., Johzuka, Y. & Mulligan, R.M. (2000). Addition of non-genomically encoded nucleotides to the 3'-terminus of maize mitochondrial mRNAs: truncated *rps12* mRNAs frequently terminate with CCA. *Nucleic Acids Res.*, *28*, 4444-4451.

Wilson, S.B., Davidson, G.S., Thomson, L.M. & Pearson, C.K. (1996). Redox control of RNA synthesis in potato mitochondria. *Eur. J. Biochem.*, *242*, 81-85.

Yamaguchi-Shinozaki, K., Shinozaki, K. & Sugiura, M. (1987). Processing of precursor tRNAs in a chloroplast lysate. *FEBS Let.*, *215*, 132-136.

Yang, J., Schuster, G. & Stern, D.B. (1996). CSP41, a sequence-specific chloroplast mRNA binding protein, is an endoribonuclease. *Plant Cell*, *8*, 1409-1420.

Yang, J. & Stern, D.B. (1997). The spinach chloroplast endoribonuclease CSP41 cleaves the 3'-untranslated region of *petD* mRNA primarily within its terminal stem-loop structure. *J. Biol. Chem.*, *272*, 12874-12880.

Ye, F., Bernhardt, J. & Abel, W.O. (1993). Genes for ribosomal proteins S3, L16, L5 and S14 are clustered in the mitochondrial genome of *Brassica napus* L. *Curr. Genet.*, *24*, 323-329.

Ye, L.H., Li, Y.Q., Fukami-Kobayashi, K., Go, M., Konishi, T., Watanabe, A. & Sugiura, M. (1991). Diversity of a ribonucleoprotein family in tobacco chloroplasts: two new chloroplast ribonucleoproteins and a phylogenetic tree of ten chloroplast RNA-binding domains. *Nucleic Acids Res.*, *19*, 6485-6490.

Yehudai-Resheff, S., Hirsh, M. & Schuster, G. (2001). Polynucleotide phosphorylase functions as both an exonuclease and a poly(A) polymerase in spinach chloroplasts. *Mol. Cell. Biol.*, *21*, 5408-5416.

Yehudai-Resheff, S., Portnoy, V., Yogev, S., Adir, N. & Schuster, G. (2003). Domain analysis of the chloroplast polynucleotide phosphorylase reveals discrete functions in RNA degradation, polyadenylation, and sequence homology with exosome proteins. *Plant Cell*, *15*, 2003-2019.

Yu, J. *et al.* (2002). A draft sequence of the rice genome (*Oryza sativa* L. ssp. indica). *Science*, *296*, 79-92.

Zou, Z., Eibl, C. & Koop, H.U. (2003). The stem-loop region of the tobacco *psbA* 5'UTR is an important determinant of mRNA stability and translation efficiency. *Mol. Genet. Genomics*, *269*, 340-349.

CHAPTER 11

INTRON SPLICING IN PLANT ORGANELLES

ALICE BARKAN

Department of Biology and Institute of Molecular Biology, University of Oregon, Eugene, OR, 97403, USA

Abstract. The mitochondrial and chloroplast genomes of plants and algae are rich in intervening sequences. With the exception of the group III introns in the chloroplasts of euglenoid protists, plant organellar introns all belong to either the group I or group II intron ribozyme families. Members of the group II intron class are particularly diverse and plentiful in plant organelles and provide fertile ground for the identification of novel modes of protein-RNA cooperation and coevolution. This review summarizes recent progress in understanding splicing mechanisms, *trans*-acting splicing factors, and the regulation of splicing in the organelles of plants.

1. INTRODUCTION

Many aspects of organellar gene organization and expression in plants and algae reflect the kinship between the organelles and their bacterial progenitors. In this context, it is striking that introns are plentiful in plant organellar genomes but scarce in bacterial genomes. The presence of introns in genes essential for both energy metabolism (photosynthesis or respiration) and organellar gene expression necessitates mechanisms to ensure faithful and efficient intron splicing, and provides opportunities for regulation by environmental and developmental cues. It is now well established that the splicing of some chloroplast introns is subject to regulation, and that proteins of both nuclear and organellar origin play essential roles in organellar RNA splicing. This review summarizes recent advances in understanding splicing mechanisms, regulation, and *trans*-acting factors in plant organelles. Detailed information on intron diversity and phylogenetic distribution can be obtained elsewhere (Bonen and Vogel, 2001; Malek and Knoop, 1998; Palmer *et al.*, 2000; Turmel *et al.*, 2003). Plant organellar intron sequences, secondary structures, and related information can be accessed through several on-line databases: FUGOID (Li and Herrin, 2002) (http://wnt.cc.utexas.edu /~ifmr530/introndata/main.htm), GOBASE (http://megasun.bch.umontreal.ca /gobase/) (O'Brien *et al.*, 2003), and the Comparative RNA Web site (http://www.rna.icmb.utexas.edu/) (Cannone *et al.*, 2002).

H. Daniell and C. D. Chase (eds.), Molecular Biology and Biotechnology of Plant Organelles, 295—322.

2. TYPES OF ORGANELLAR INTRONS IN PLANTS AND THEIR PHYLOGENETIC DISTRIBUTION

2.1. Intron Classes and Splicing Mechanisms

Plant organellar introns fall into three categories - group I, group II, and group III. Each class is defined by conserved features of primary sequence and predicted secondary structure (Bonen and Vogel, 2001; Copertino and Hallick, 1993; Lambowitz *et al.*, 1999; Michel and Dujon, 1983; Michel *et al.*, 1989). Group I and group II introns have a broad phylogenetic distribution, with representatives in bacteria, fungal mitochondria, protists and plant organelles. All of the organellar introns in plants and algae fall into the group I or group II categories. In contrast, group III introns have been found only in the chloroplasts of euglenoid protists. As discussed below, group II and group III introns share key features and likely arose from a common ancestor.

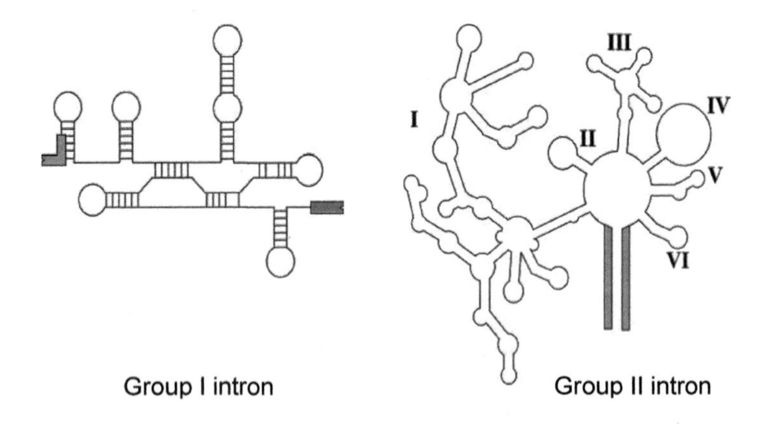

Figure 1*. Secondary structure of a canonical group I and group II intron. Exons are shown as gray bars, and introns as solid lines. The six group II intron domains are indicated.*

Both group I and group II introns are considered to be ribozymes because some introns in each class exhibit self-splicing activity *in vitro*. However, these two intron classes differ fundamentally in structure (see Figure 1) and catalytic mechanism. The catalytic core of a group I intron consists of two helical domains that pack to form a compact catalytic center (Golden *et al.*, 1998; Michel and Westhof, 1990). Non-conserved peripheral domains define four group I intron subgroups (Michel and Westhof, 1990) and serve to stabilize the intron's three-dimensional structure (Lehnert *et al.*, 1996; Saldanha *et al.*, 1993; Westhof, 2002).

Group II introns are larger, more structurally complex and more evolutionarily malleable than group I introns. The group II intron secondary structure is typically depicted as six helical domains emanating from a central core (Bonen and Vogel, 2001; Michel and Ferat, 1995; Michel *et al.*, 1989; Qin and Pyle, 1998) (see Figure 1). *In vitro* analyses of self-splicing fungal group II introns have assigned specific roles in splicing to each domain (Qin and Pyle, 1998). High resolution structural information is available only for domains 5 and 6 (Sigel *et al.*, 2004; Zhang and Doudna, 2002), but biochemical, phylogenetic, and genetic approaches have provided evidence for an array of tertiary contacts that create a compact catalytic center involving primarily domains 1 and 5 (Bonen and Vogel, 2001; Qin and Pyle, 1998). Group II introns are classified as belonging to either subgroup IIA or IIB, based on several differences in their predicted structures (Bonen and Vogel, 2001; Michel *et al.*, 1989; Toor *et al.*, 2001).

The splicing of both group I and group II introns consists of two consecutive *trans*-esterification reactions involving first the 5', and then the 3' splice junctions, but the reactions otherwise differ. Group I splicing is initiated by an exogenous guanosine cofactor that attacks the 5' splice junction; the liberated 3' hydroxyl group of the 5' exon then attacks the 3' splice junction, resulting in exon ligation and intron release. Group II splicing is generally initiated by the 2' hydroxyl group of a "bulged" adenosine residue in the domain 6 helix, which attacks the 5' splice junction. The covalent attachment of the 2'hydroxyl group to the 5' intron residue results in a branched structure referred to as a lariat, and the attacking adenosine is called the "branchpoint". The 3'hydroxyl of the cleaved 5' exon then attacks the 3' splice junction, yielding ligated exons and an excised intron lariat.

These canonical splicing mechanisms appear to apply to many organellar introns in plants: where excised intron lariats have been sought, they have generally been found (Carrillo *et al.*, 2001; Kim and Hollingsworth, 1993; Vogel and Boerner, 2002). The site-directed mutation of predicted functional elements in several plant organellar group I and group II introns disrupted splicing in engineered *C. reinhardtii* chloroplasts (Hollander and Kuck, 1999a; Lee and Herrin, 2003), during *in vitro* self-splicing (Lee and Herrin, 2003), in engineered *E. coli* cells (Hollander and Kuck, 1999b), or during transient expression in electroporated wheat mitochondria (Farre and Araya, 2002).

An alternative, "hydrolytic" pathway for group II splicing *in vitro* uses water or hydroxide as the nucleophile in the first step of the reaction, releasing the excised intron as a linear molecule instead of as a lariat (Daniels *et al.*, 1996). The relevance of this pathway to natural group II introns was shown through the use of an innovative PCR-based method with barley chloroplast introns (Vogel and Boerner, 2002; Vogel *et al.*, 1997a). Most of the excised group II introns were detected as lariats and thus splice with the classic branching pathway; a notable exception, however, was the *trn*V intron, which lacks a predicted bulged branchpoint adenosine in all land plant chloroplasts, and was detected only in a linear form. This provided strong evidence that the *trn*V intron excises via a hydrolytic pathway *in vivo*.

Similarities between the chemistry of group II splicing and nuclear spliceosomal splicing, and structural similarities between specific snRNAs and group II intron domains have fueled speculation that the snRNAs in the spliceosome represent "escaped" domains of an ancestral, self-splicing group II intron (Cech, 1986; Hetzer *et al.*, 1997; Shukla and Padgett, 2002; Valadkhan and Manley, 2001; Villa *et al.*, 2002). The remarkable evolutionary lability of group II introns evident in plant organelles lends credence to this idea. Many group II introns in plants differ markedly from the canonical group II intron structural framework. Some lack elements that are considered to be essential to the formation of the catalytic center, and some have become fragmented during the course of evolution such that they are transcribed in pieces that must then be "*trans*-spliced" (reviewed in Bonen and Vogel, 2001; Malek and Knoop, 1998). Examples of group II *trans*-splicing occur in the *C. reinhardtii* chloroplast *psa*A mRNA (Choquet *et al.*, 1988; Herrin and Schmidt, 1988; Kuck *et al.*, 1987), the land plant chloroplast *rps*12 mRNA (Fromm *et al.*, 1986; Zaita *et al.*, 1987), and in the angiosperm mitochondrial *nad*1, *nad*2, and *nad*5 mRNAs (Binder *et al.*, 1992; Chapdelaine and Bonen, 1991; Knoop *et al.*, 1997; Knoop *et al.*, 1991). The *trans*-spliced intron fragments represent structural modules that assemble with one another via RNA-RNA interactions to recreate an intact group II intron structure. The efficiency and fidelity of this complex recognition process seem likely to be aided by protein cofactors, and there is genetic evidence for the participation of many proteins in the *trans*-splicing of the *psa*A introns in *C. reinhardtii* chloroplasts (Goldschmidt-Clermont *et al.*, 1990).

The sequences of some plant organellar group II introns are modified post-transcriptionally by RNA editing (*e.g.* Binder and Brennicke, 2003; Bonen and Vogel, 2001; Carrillo *et al.*, 2001; Kugita *et al.*, 2003b), and it has been suggested that editing may be necessary to restore function to the highly divergent domains 5 and 6 found in many mitochondrial introns in plants (Bonen and Vogel, 2001). However, the unedited forms of several of these were detected as excised introns, showing that their editing is not a prerequisite for splicing (Carrillo and Bonen, 1997; Carrillo *et al.*, 2001). Nonetheless, most organellar introns have not been sequenced at the RNA level to determine whether they are edited, so caution should be used in assuming that functional intron RNAs have a nucleotide sequence corresponding to that of their DNA template.

The group III introns provide a special case and may represent an extreme example of group II intron degeneration. Group III introns have been found only in the chloroplasts of euglenoid protists and are the smallest of the organellar introns, at ~100 nt (reviewed in Copertino and Hallick, 1993). They are characterized by splice junction consensus sequences that resemble the group II consensus, by an A/U rich nucleotide sequence and by a conserved structure near the 3'splice junction that resembles domain VI of group II introns. On the basis of these properties, it has been proposed that group III introns are highly degenerate group II introns (Christopher and Hallick, 1989). In accordance with this hypothesis, group III introns are excised as lariats (Copertino *et al.*, 1994), and one group III intron harbors an open reading frame encoding a protein with similarity to group II intron maturases (see below) (Copertino *et al.*, 1994; Mohr *et al.*, 1993). The alternative hypothesis, that group III introns represent the ancestral state in euglenoid chloroplasts with additional group II domains acquired subsequently, cannot be eliminated (Doetsch *et al.*, 1998).

Euglenoid chloroplasts also contain the most highly reduced, unambiguous group II introns identified to date. Despite their small size (e.g. ~300 nucleotides), these conform to the canonical six domain secondary structure; however, motifs that mediate important tertiary interactions in self-splicing group II introns are often lacking (Copertino and Hallick, 1993; Thompson *et al.*, 1997). Euglenoid plastids also house many "twintrons" – introns within introns (Copertino *et al.*, 1991; Copertino and Hallick, 1993; Copertino *et al.*, 1992). The individual introns in twintrons fall into either the group II or group III class and are excised in sequential splicing reactions, such that removal of the inner intron reconstructs an intact outer intron, which is then excised. A transformation system was recently reported for *Euglena gracilis* chloroplasts and will be valuable for dissecting the *cis*-acting requirements of the diverse intron types found in these organelles (Doetsch *et al.*, 2001).

2.2. Intron Distribution

The basic set of introns present in land plant chloroplasts was established prior to the divergence of vascular and non-vascular plants, as it is shared by bryophytes, gymnosperms, angiosperms and their closest algal relatives, the charophyta (Kugita *et al.*, 2003a; Sugiura *et al.*, 2003; Turmel *et al.*, 2002; Wakasugi *et al.*, 2001). This conservation of intron content parallels the overall conservation of chloroplast genome organization in the land plants and charophyta (see Chapter 5 in this volume). Chloroplasts in land plants generally have between 17 and 20 group II introns, which interrupt tRNA and protein-coding sequences and include a single *trans*-spliced intron, in the *rps*12 gene. The small inter-species variation in group II intron content reflects the complete loss of several intron-containing genes in specific lineages, (e.g. two intron-containing *ndh* genes are absent in black pine chloroplasts (Wakasugi *et al.*, 1994)) or the presence of intron-less alleles in some lineages (e.g. the monocot grasses lack introns in the *clp*P and *rpo*C1 genes that are found in dicots and in the common ancestor). Thus, maize and rice chloroplasts have seventeen group II introns, whereas *Arabidopsis* and tobacco chloroplasts have twenty. Nonphotosynthetic angiosperms with highly reduced chloroplast genomes show a corresponding reduction in intron content (Wolfe *et al.*, 1992), but the six retained introns in *Epifagus virgniana* are found at conserved locations and are spliced (Ems *et al.*, 1995).

Land plants and charophyta (together called the streptophyta) contain a single group I intron in their chloroplast genomes, in a gene for *trnL(UAA)*. This intron is considered to be the most ancient of any intervening sequence, represented even in cyanobacteria (Xu *et al.*, 1990) and in the chloroplasts of all algal lineages (Simon *et al.*, 2003). The *trnL* intron was subject to frequent loss in the green and red algae, and is lacking, for example, in the *C. reinhardtii* chloroplast genome.

Introns are notably absent in the chloroplast genome of *M. viride,* a green biflagellate representing the earliest divergence within the streptophyta (Lemieux *et al.*, 2000). The chloroplast gene organization of *M. viride* is otherwise very similar to that in land plants and in the charophyta. This supports the idea that the acquisition of chloroplast introns within and outside the streptophyta were independent events. Accordingly, the chloroplast genome of the chlorophyte *C. reinhardtii* harbours five group I introns (four in the *psb*A gene and one in the 23S-like rRNA gene), and only two group II introns (both in the *psa*A gene and *trans*-spliced) (Maul *et al.*, 2002), none of which are conserved in land plants. *Euglena gracilis*, a photosynthetic protist, houses the most intron rich chloroplast genome described to date, with at least 155 introns, accounting for an astonishing 39% of its chloroplast DNA content. These introns all fall into the group II and group III categories (Hallick *et al.*, 1993).

Plant mitochondrial intron content is substantially less conserved than chloroplast intron content, reflecting the more dynamic nature of mitochondrial genome structure (Palmer *et al.*, 2000). Angiosperm mitochondria share a set of 23 group II introns (with sporadic loss of individual introns), all in protein coding genes (Kubo *et al.*, 2000; Notsu *et al.*, 2002; Unseld *et al.*, 1997). This set includes five *trans*-spliced introns, which are represented by *cis*-arranged homologs in more ancient land plants (Malek and Knoop, 1998). Most of the introns in angiosperm mitochondria were acquired much more recently than those in chloroplasts, having become incorporated after the emergence of land plants (Turmel *et al.*, 2003). For example, of the 25 group II introns in *Marchantia* mitochondria (Oda *et al.*, 1992a; Oda *et al.*, 1992b), only one is at a position that is conserved in the angiosperms. Four of the group II introns present in the mitochondria of the last common ancestor of green algae and land plants continue to be represented in vascular and non-vascular plants: two are found in *Marchantia* and a different two are represented in angiosperms (Turmel *et al.*, 2003). Thus, most mitochondrial introns appear to have been acquired independently in the bryophytes and angiosperms, with mitochondrial intron loss occurring frequently in the land plant lineage.

Only one group I intron has been found in the mitochondria of vascular plants, residing in the *cox*1 gene. This intron has a patchy phylogenetic distribution that has been attributed to numerous independent horizontal transfer events during recent angiosperm evolution (Palmer *et al.*, 2000). The *Marchantia* mitochondrial genome has seven group I introns (Ohta *et al.*, 1993), none of which appear to have been inherited vertically from the common ancestor into the angiosperm lineage.

Clearly, introns have come and gone frequently during the evolution of plant organelles and have moved from one site to another within an organellar genome. This fluidity is thought to result from the fact that many group I and group II introns are also mobile genetic elements. They exhibit two modes of transposition: intron "homing" in which an intron is transmitted to an intronless allele of the same gene, and transposition to ectopic (albeit loosely related) sites. Both types of movement involve the participation of proteins encoded by open reading frames (ORFs) within the introns themselves. These intron-encoded proteins are sometimes referred to as "maturases", because some are dual function proteins that also promote the splicing of their host intron (see section 3.3).

The mechanism of group II intron mobility has been studied in fungi and bacteria, and involves a variety of retrotransposition mechanisms that require both the maturase protein and the intron RNA (Bonen and Vogel, 2001; Lambowitz *et al.*, 1999). Functional group II intron maturases are characterized by a reverse-transcriptase domain involved in intron homing and retrotransposition, an endonuclease domain involved in retrotransposition, and a "domain X" implicated in RNA binding and splicing. Most of the group II introns in plant organelles do not encode a maturase, although there is evidence that all group II introns arose from maturase-encoding ancestors (Toor *et al.*, 2001). A single maturase-like ORF, called *mat*K, survives in the chloroplasts of land plants. *Mat*K lacks most of the reverse-transcriptase domain and is presumed to be incapable of promoting intron mobility, but there is increasing evidence that it does facilitate splicing (see section 3.3.2). Angiosperm mitochondria also harbor a single ORF related to group II maturases, in the *nad*1 pre-mRNA (Unseld *et al.*, 1997); this protein is called Mat-r and includes both a reverse-transcriptase domain and domain X. The editing of the Mat-r – encoding RNA serves to increase the similarity between both of these domains and their counterparts in functional group II maturases in bacteria and fungi (Begu *et al.*, 1998; Bonen and Vogel, 2001; Thomson *et al.*, 1994), implying strong selective pressure to maintain protein function. Thus, it can be anticipated that Mat-r promotes both splicing and intron mobility.

Group I intron ORFs encode site-specific DNA endonucleases that fall into several classes (Lambowitz *et al.*, 1999). Group I intron homing is initiated when the endonuclease cleaves the DNA of an intronless allele, whereas movement to ectopic sites involves a reverse-splicing mechanism. Three of the five group I introns in the *C. reinhardtii* chloroplast (two in *psb*A and the 23S rRNA intron) encode proteins that are related to known homing endonucleases (Holloway *et al.*, 1999; Rochaix *et al.*, 1985), and all three of these proteins have been shown to promote the insertion of their host intron into intronless alleles *in vivo* (Duerrenberger and Rochaix, 1991; Odom *et al.*, 2001).

3. SPLICING FACTORS IN PLANT ORGANELLES

3.1. Proteins are Required for Organellar Splicing

Despite their designation as "ribozymes", many introns with characteristic group I or group II intron features have acquired an obligate dependence upon *trans*-acting factors for their splicing. In fact, none of the ~40 introns in the organelles of vascular plants have been reported to self-splice. In some cases the loss of self-splicing activity is to be expected because of the clear absence of essential intron elements. However, most of the ~20 group II introns in land plant chloroplasts appear to be "intact" but nonetheless fail to self-splice under the standard conditions that elicit the self-splicing of model group II introns from other organisms.

Only three examples of self-splicing group II introns have been reported in plant-related lineages, all in protist or algal organelles (Costa *et al.*, 1997; Kuck *et al.*, 1990; Sheveleva and Hallick, 2004). In contrast, group I introns seem less subject to evolutionary degeneration, and most have retained self-splicing activity. Self-splicing has been detected for all five of the group I introns in *C. reinhardtii* chloroplasts (Deshpande *et al.*, 1997; Herrin *et al.*, 1991; Herrin *et al.*, 1990) and for chloroplast group I introns in other algae (Kapoor *et al.*, 1997; Simon *et al.*, 2003). The group I intron in land plant chloroplast *trn*L provides an exception, in that it fails to self-splice (Simon *et al.*, 2003).

Importantly, even where self-splicing activity has been detected, the conditions required to elicit this activity are distinctly non-physiological, involving elevated [Mg^{2+}] (25-100 mM), monovalent salt (e.g. 500 mM NH_4Cl) and temperature (42-55°C). Furthermore, a self-splicing group I intron from *C. reinhardtii* chloroplasts and a self-splicing group II intron from *S. obliquus* mitochondria were more tolerant of mutations in core elements when expressed in *C. reinhardtii* chloroplasts than during self-splicing *in vitro* (Hollander and Kuck, 1998; Lee and Herrin, 2003), indicating that factors present *in vivo* optimize the splicing environment. Taken together, these observations indicate that accessory factors promote the splicing of virtually all group I and group II introns *in vivo*. Indeed, as discussed below, genetic studies have proven the involvement of proteins in the splicing of 16 of the 17 group II introns in the maize chloroplast, and both of the group II introns in the *C. reinhardtii* chloroplast.

Protein cofactors for group I and group II splicing originated in two ways: some are encoded within the introns themselves and have an ancient relationship with their host intron, whereas others were recruited more recently from their host genome (i.e. the nuclear genome in the case of eukaryotic cells) (reviewed in Lambowitz *et al.*, 1999). The intron-encoded splicing factors are called "maturases", and, as discussed above, they often serve dual roles by promoting both the splicing of their host intron and intron mobility. Host-encoded protein cofactors arose through the modification of genes that evolved in other contexts; these are diverse in origin and, presumably, mechanism. The coevolution of group I and group II introns with host-encoded splicing factors is an example of a broader class of evolutionary processes involving the incorporation of catalytic RNAs into catalytic ribonucleoprotein particles (RNPs) within which the protein and RNA components cooperate. Particularly ancient and complex examples are the ribosome and, presumably, the spliceosome. The study of nuclear-encoded proteins that facilitate the splicing of organellar introns will therefore bear not only on our understanding of nuclear-organellar crosstalk, but will also elucidate the general mechanisms underlying the cooperation between protein and RNA molecules in catalytic RNPs.

Several host factors for group I splicing in fungal mitochondria have been studied in detail (Bassi *et al.*, 2002; Lambowitz *et al.*, 1999; Weeks, 1997), but fungal systems have been less useful for the identification of host factors for group II splicing. Plant organelles harbor a particularly large and diverse set of group II introns (Bonen and Vogel, 2001). This diversity is likely to be reflected by a wealth of nuclear-encoded splicing factors and associated mechanisms. Genetic approaches for the identification of nuclear genes involved in organellar gene expression are well-established in model algal and vascular plant species and can be used to identify nuclear genes required for organellar splicing (Barkan and Goldschmidt-Clermont, 2000; Leister and Schneider, 2003; Rochaix, 2002; Stern *et al.*, 2004). Thus, plants have emerged as the primary system for the identification of host factors for group II intron splicing.

3.2. Possible Roles of Organellar Splicing Factors

The catalytic activity of group I and group II introns is dependent upon their acquisition of a properly structured active site. The folding of group I and group II introns, like that of other large and highly structured RNAs, is problematic because numerous non-native conformations are similar in stability to the native structure, so the RNAs are easily trapped in an inactive conformation (reviewed in Herschlag, 1995; Thirumalai and Woodson, 2000; Treiber and Williamson, 2001; Weeks, 1997). Furthermore, the tertiary interactions that establish the intron's three-dimensional architecture can be weak (Swisher *et al.*, 2002). Thus, the productive folding of group I and group II introns is considered to be their Achilles heel, and it is assumed that many proteins that facilitate group I and group II intron splicing do so by enhancing the yield of folded, active intron molecules. Proteins could promote intron folding in two fundamentally different ways. High-affinity, sequence-specific interactions could guide intron folding by stabilizing an otherwise transient tertiary structure, or by precluding a competing, non-productive folding pathway. Alternatively, proteins could resolve misfolded RNA structures through the active disruption of RNA helices (Lorsch, 2002), or through non-specific, low affinity binding to unstructured RNA. Proteins that promote productive RNA folding by resolving misfolded RNAs have been dubbed "RNA chaperones" to draw attention to the conceptual similarity between their action and that of protein chaperones like Hsp70 and Hsp60 (Herschlag, 1995).

There is ample precedent from non-plant systems for proteins that facilitate "self- splicing" via high affinity, sequence-specific interactions with an intron: three fungal host factors for group I splicing, fungal group I maturases, and a bacterial group II maturase employ variations on this theme (Ho and Waring, 1999; Lambowitz *et al.*, 1999; Matsuura *et al.*, 2001; Solem *et al.*, 2002; Wank *et al.*, 1999; Weeks, 1997; Weeks and Cech, 1996). By extrapolation, it seems likely that some of the maturase-like molecules encoded in plant organelles function in this way. Results summarized below indicate that CRS1, a nuclear-encoded group II splicing factor in maize chloroplasts, also promotes splicing in this general manner.

The RNA chaperone concept is attractive, but examples of RNA chaperone activity in a native context remain scarce. Several *E. coli* proteins that bind RNA non-specifically promote the folding of a phage group I intron *in vitro* or when over-expressed *in vivo* (reviewed in Mayer *et al.*, 2002). The most compelling evidence for the involvement of an RNA chaperone in RNA folding in its natural environment involves CYT19, a putative RNA helicase in *Neurospora* that promotes the splicing of a mitochondrial group I intron by destabilizing non-native RNA structures (Mohr *et al.*, 2002). The involvement of RNA helicases in plant organellar splicing seems likely, but has not been demonstrated. Ribosome transit through exon sequences promotes the productive folding of a phage group I intron by preventing the exons from interacting inappropriately with intron sequences (Semrad and Schroeder, 1998); a similar mechanism might account for some of the splicing defects observed in angiosperm plastids lacking ribosomes (Jenkins *et al.*, 1997; Vogel *et al.*, 1999) (see section 3.3.2).

All of the group I and group II splicing factors that have been studied in detail act by promoting intron folding, but few protein/intron partners have been analysed, and it is likely that study of the diverse introns found in plant organelles will reveal additional mechanisms. In the special case of the *trans*-spliced group II introns, proteins are likely to assist in the assembly of the intron fragments. Proteins could serve to recycle splicing factors by disrupting protein/RNA interactions after intron excision (Jankowsky *et al.*, 2001; Staley and Guthrie, 1998), or they might facilitate substrate (*i.e.* splice-junction) recognition, in analogy to the protein component of the RNAse P ribozyme (Hsieh *et al.*, 2004). Proteins could, in principle, even contribute to catalysis itself. There are no documented examples in which both protein and RNA contribute functional groups to the same active site, but such a function is being considered for CRS2, a chloroplast group II splicing factor (see section 3.4). It is also an attractive possibility for plant mitochondrial introns with non-canonical sequences in "catalytically-essential" domain 5 (Carrillo *et al.*, 2001).

3.3. Intron-Encoded Maturases in Plant Organelles

3.3.1. Group I Maturases

All of the established group I intron maturases have been found in fungi and are related to the LAGLIDADG class of homing endonucleases (Lambowitz *et al.*, 1999). Some of these proteins are bifunctional and promote both intron homing and splicing, whereas others have lost their DNA endonuclease function and are now specialized splicing factors. The I-*Ani*I maturase from *Aspergillus nidulans* mitochondria is an example of a bifunctional maturase (Ho *et al.*, 1997), and is the only group I maturase whose role in splicing has been studied in depth. I-*Ani*I binds specifically to its host intron *in vitro*, promoting intron folding and stabilization of the active tertiary structure (Ho *et al.*, 1997; Ho and Waring, 1999; Solem *et al.*, 2002). Several algal group I introns encode DNA endonucleases that participate in intron homing (see above), but there is as yet no evidence that these proteins also participate in splicing. For example, the group I intron in the *C. reinhardtii* 23S rRNA gene encodes a homing endonuclease (Duerrenberger and Rochaix, 1991), but deletion of this ORF does not disrupt 23S rRNA splicing (Thompson and Herrin, 1991). Therefore, either this protein does not participate in splicing, or its role can be assumed by other proteins (*e.g.* perhaps by one of the group I intron-encoded proteins in the *psb*A gene in the same organelle (Holloway *et al.*, 1999)). Conversely, deletion of all four of the group I introns in *psb*A did not cause a growth defect (Johanningmeier and Heiss, 1993), indicating that the splicing of the sole remaining group I intron, in the 23S-like rRNA gene, is not dependent upon the *psb*A intron ORFs. Deletion of the ORFs in *psb*A introns 2 and 3 also did not disrupt splicing (D. Herrin, personal communication). Thus, the available data suggest that none of the group I intron ORFs in the *C. reinhardtii* chloroplast function in splicing, although the possibility that two or more of these ORFs are functionally-redundant cannot be eliminated until they are deleted simultaneously.

3.3.2. Group II Maturases

Predicted maturase ORFs are rare among the group II introns in plant organelles: for example, they are absent in the *C. reinhardtii* chloroplast genome, and they are found in only two of the ~40 group II introns in vascular plant organelles. MatK resides in the *trn*K intron of land plant chloroplasts (Neuhaus and Link, 1987) and includes a small remnant of the reverse transcriptase domain fused to domain X, which is responsible for the splicing activity of fungal maturases (Mohr *et al.*, 1993). Mat-r is encoded in a *nad*1 intron in angiosperm mitochondria and includes both domain X and a reverse-transcriptase domain. Interestingly, an internal promoter in the wheat *mat-r* ORF drives transcription of only the domain X region (Farre and Araya, 1999), suggesting that domain X may sometimes be expressed as an independent entity in *mat-r*, as in *mat*K.

There is no proof that MatK or Mat-r function in splicing, but there is accumulating evidence that they might do so. For example, the domain X-encoding sequences in both genes are modified by RNA editing, resulting in increased amino acid conservation with maturases in other species (Begu *et al.*, 1998; Thomson *et al.*, 1994; Vogel *et al.*, 1997b). MatK has been shown to bind RNA *in vitro* (Liere and Link, 1995), and its absence in barley and maize mutants lacking chloroplast ribosomes is correlated with the failure to splice the *trn*K intron (Vogel *et al.*, 1997b) as well as other introns in subgroup IIA (Jenkins *et al.*, 1997; Vogel *et al.*, 1999). An intriguing implication of those genetic data is that MatK may be a general group IIA splicing factor, in contrast to canonical group II maturases, which are specific for their host intron (Lambowitz *et al.*, 1999). However, the mutant plastids used for those experiments lack *all* plastid-encoded proteins and also have an aberrant nucleus-derived proteome, so the contribution of MatK's absence to the splicing defects is difficult to assess.

The most compelling evidence that MatK promotes the splicing of multiple group II introns arose from the sequence of the plastid genome of the non-photosynthetic angiosperm *Epifagus virginiana*. The Epifagus plastid genome has retained a subset of the genes involved in chloroplast gene expression and four genes that function in cellular processes other than photosynthesis (Wolfe *et al.*, 1992). The *trn*K gene has been lost from *Epifagus*, but *mat*K has been maintained as a freestanding ORF. It has been proposed that the retention of *matK* reflects an essential role for MatK in the splicing of one or more of the six retained group II introns (Wolfe *et al.*, 1992), all of which are accurately spliced *in vivo* (Ems *et al.*, 1995). An analogous situation exists in euglenoid plastids: a group III intron in the *E. gracilis psb*C gene houses an ORF resembling a group II maturase; this ORF has been maintained in a free-standing context in a derived nonphotosynthetic euglenoid species, suggesting that it too plays an essential role in the splicing of retained plastid introns (Copertino *et al.*, 1994).

Technology for engineering the tobacco chloroplast genome is well-established (Daniell *et al.*, 2002, 2004; Devine and Daniell 2004; see also chapter 16 of this book), so it would seem to be straightforward to definitively test the role of tobacco MatK by deleting its ORF. In fact, this approach has been attempted, but homoplastomic *mat*K knockouts could not be recovered, even when an intronless *trn*K allele replaced the intron-containing allele and when heteroplastomic plantlets were grown in the absence of selective antibiotic (R. Maier, personal communication). This strongly suggests that MatK is required for the expression of one or more of the cell-essential plastid ORFs in tobacco, in addition to any role it might play in the splicing of the *trn*K intron. Group II introns in tobacco are found in several plastid tRNA and ribosomal protein genes, as well as in the cell-essential *clp*P gene (Kuroda and Maliga, 2003). A role for MatK in the splicing of any or all of these introns could account for the cell-essential nature of MatK itself.

3.4. Nuclear-Encoded Splicing Factors in Plant Organelles

Genetic screens have been quite successful at identifying nuclear-encoded group II intron splicing factors in *C. reinhardtii* and maize chloroplasts. A particularly intriguing and complex example involves the *C. reinhardtii* chloroplast *psa*A gene, which includes two group II introns that are transcribed in pieces and spliced in *trans*. In fact, the *psa*A mRNA is assembled from at least four independently transcribed RNA fragments (Choquet *et al.*, 1988; Herrin and Schmidt, 1988; Kuck *et al.*, 1987). Intron 2 is transcribed in two segments, together with the flanking exons. Intron 1 consists of at least three pieces: 5' and 3' intron fragments that are cotranscribed with flanking exon sequences and an internal intron fragment that is independently transcribed from the chloroplast *tsc*A locus (Goldschmidt-Clermont *et al.*, 1991; Roitgrund and Mets, 1990). The *tsc*A RNA is proposed to bridge the 5' and 3' intron fragments, but domain 1 of this composite intron appears to lack critical elements, suggesting that additional, independently transcribed intron fragments remain to be discovered (Turmel *et al.*, 1995).

Mutations that disrupt the *trans*-splicing of the *C. reinhardtii psa*A mRNA were identified by RNA gel blot analysis of mutants lacking photosystem I, and define at least fourteen nuclear genes (Goldschmidt-Clermont *et al.*, 1990; Herrin and Schmidt, 1988). Several loci are required for the splicing of both introns, but most function in the splicing of either intron 1 or intron 2, demonstrating that the two introns splice independently of one another. One locus is required for the processing of the *tsc*A RNA and is likely to influence splicing only indirectly (Hahn *et al.*, 1998).

The molecular cloning of two of these genes has been reported. *Raa*2 (*RNA maturation of psaA*, formerly called *Maa*2) is required for the splicing of *psa*A intron 2 and resembles pseudouridine synthases (Perron *et al.*, 1999). Site-directed mutagenesis of conserved residues that are essential for the catalytic activity of related bacterial enzymes did not prevent Raa2 from complementing the splicing defect in *raa*2 mutants. Thus, Raa2's role in splicing appears not to involve pseudouridine synthase activity. Interestingly, analogous observations have been made for two pseudouridine synthases in *E. coli*, where mutant variants lacking essential active site residues retained the ability to complement growth defects caused by null alleles. Those authors suggested that the bacterial pseudouridine synthases are bifunctional, with independent RNA chaperone and pseudouridine synthase activity (Gutgsell *et al.*, 2001; Gutgsell *et al.*, 2000). The same theme may hold true for Raa2. Raa2 is found in the chloroplast, where it is associated through ionic interactions with a low-density membrane fraction (Perron *et al.*, 1999) that has been proposed to be an assembly site for thylakoid membrane complexes (Zerges and Rochaix, 1998). Raa2 is found in a multiprotein complex that also contains Raa1, another nuclear-encoded protein required for the splicing of *psa*A intron 2 (Perron *et al.*, 2004).

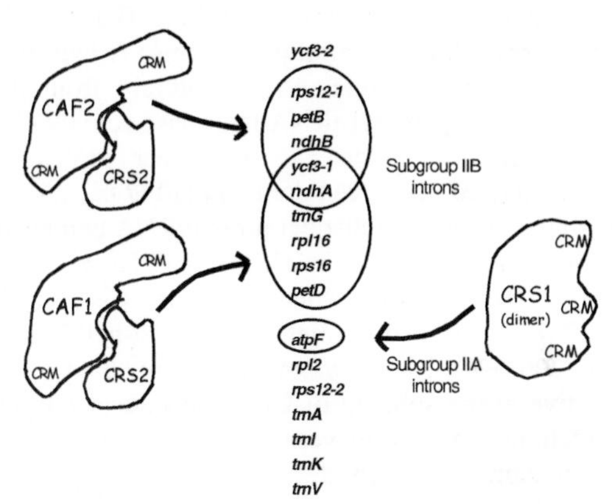

Figure 2. *Nuclear-encoded splicing factors and their intron targets in maize chloroplasts. CRS2-CAF complexes and the CRM domains in CAF1, CAF2, and CRS1 are shown. Recombinant CRS1 is dimeric. All seven subgroup IIA introns fail to splice in mutant plastids lacking ribosomes, implicating a plastid translation product in their splicing. Results are summarized from Jenkins* et al., *1997, Ostheimer* et al., *2003, and Vogel* et al., *1999.*

The *Raa*3 gene is required for the splicing of *psa*A intron 1 and encodes a large protein with limited similarity to pyridoxamine 5'-phosphate oxidases (Rivier *et al.*, 2001); surprisingly, the N-terminal ~1/3 of the protein appears to be dispensable for its splicing function. Raa3 is localized to the chloroplast stroma, where it is found in a ribonuclease-sensitive complex of ~1700 kDa. The *tsc*A and *psa*A exon 1 precursor RNAs are found in this same complex, as shown by their cofractionation during size exclusion chromatography and by the coordinate reduction in the size of the Raa3 and *psa*A exon 1-precursor particles in a mutant strain that was deleted for *tsc*A. It will be interesting to learn how the splicing of the two *psa*A introns is orchestrated, given that the complexes involved in the splicing of introns 1 and 2 (containing Raa3/*tsc*A and Raa1/Raa2, respectively) localize to different chloroplast subcompartments.

An analogous genetic approach in maize has led to the identification of four nuclear-encoded proteins that are required for the splicing of various subsets of the group II introns in the chloroplast, and which form three protein complexes that are bound to their cognate intron RNAs *in vivo*. This work began with a screen for chloroplast splicing defects among a collection of transposon-induced non-photosynthetic maize mutants. The initial screen yielded mutations in two nuclear genes, chloroplast RNA splicing 1 and 2 (*crs*1 and *crs*2) (Jenkins *et al.*, 1997). *Crs*1 mutants lack the chloroplast ATP synthase due to a specific defect in the splicing of the subgroup IIA intron in the chloroplast *atp*F gene. *Crs*2 mutants fail to splice nine of the ten chloroplast introns in subgroup IIB, but splice the seven subgroup IIA introns normally (Jenkins *et al.*, 1997; Vogel *et al.*, 1999). These intron specificities are summarized in Figure 2. As noted above, a plastid translation product(s) (potentially MatK) is required for the splicing of all of the subgroup IIA introns. Taken together, these results show that sixteen of the seventeen group II introns in the maize chloroplast rely on proteins for their splicing *in vivo* (Jenkins *et al.*, 1997; Vogel *et al.*, 1999); the splicing of the *ycf3*-2 intron is not disrupted in any of the mutant backgrounds analysed to date (Ostheimer *et al.*, 2003) and is the only candidate for a truly self-splicing group II intron in the maize chloroplast.

Both *crs*1 and *crs*2 were cloned by transposon-tagging and proved to be unrelated to one another (Jenkins and Barkan, 2001; Till *et al.*, 2001). CRS2 strongly resembles bacterial peptidyl-tRNA hydrolases, which recycle tRNAs when peptidyl-tRNAs are released prematurely from the ribosome. However, CRS2 appears to lack peptidyl-tRNA hydrolase activity (Jenkins and Barkan, 2001). CRS1 was initially recognized as a pioneer protein harboring three copies of an uncharacterised conserved domain (Till *et al.*, 2001). CRS1 and CRS2 reside in the chloroplast stroma, in distinct protein/RNA complexes of 500-700 kDa (Jenkins and Barkan, 2001; Ostheimer *et al.*, 2003; Till *et al.*, 2001).

Two additional splicing factors, CRS2 Associated Factors 1 and 2 (CAF1 and CAF2), surfaced through a yeast two-hybrid screen with CRS2 bait (Ostheimer *et al.*, 2003). *Caf1* and *caf2* mutants were obtained in reverse genetic screens and are defective in the splicing of CRS2-dependent introns, showing that the CRS2-CAF interactions detected in yeast are physiologically relevant. Interestingly, *caf1* and *caf2* mutants are each defective in the splicing of different intron subsets, as summarized in Figure 2. Coimmunoprecipitation experiments showed further that CRS2-CAF complexes reside in the chloroplast stroma, where they are tightly bound to their cognate intron RNA targets (Ostheimer *et al.*, 2003). Analogously, anti-CRS1 antibodies specifically coimmunoprecipitated the *atp*F intron from chloroplast extract (Ostheimer *et al.*, 2003).

Taken together, these experiments show that CRS1, CAF1-CRS2, and CAF2-CRS2 complexes facilitate the splicing of specific intron subsets in the chloroplast, and that the genetically-defined intron specificities of each protein reflect the specific association between that protein and the corresponding introns *in vivo*. Whereas it appeared at first that the CRS1 and CRS2 trails would lead toward distinct mechanisms of protein-assisted splicing, the two stories converged with the realization that the CRS2-associated factors CAF1 and CAF2 are both related to CRS1 (Ostheimer *et al.*, 2003). The similarity between CAF1, CAF2, and CRS1 is confined to 10 kDa segments corresponding to the ancient domain originally recognized in CRS1. These domains align with the product of a conserved prokaryotic open reading frame (Ostheimer *et al.*, 2002; Ostheimer *et al.*, 2003; Till *et al.*, 2001) whose *E. coli* representative, YhbY, is specifically bound to pre-50S ribosomal subunits *in vivo* (T. Kawamura and A. Barkan, in preparation). Thus, the domain was named the Chloroplast RNA Splicing and Ribosome Maturation domain (CRM) (Ostheimer *et al.*, 2003) to reflect what is known about the four characterized proteins in the family: CRS1, CAF1, CAF2, and YhbY.

It is now clear that CRM domains are previously unrecognised RNA binding domains. The crystal structure of YhbY showed structural similarity with a known RNA binding protein, and the core structure-determining residues are conserved in the plant CRM domains, suggesting they adopt a similar structure (Ostheimer *et al.*, 2002). Biochemical data have confirmed that CRM domains bind RNA: recombinant CRS1, CAF1, and CAF2 bind RNA with high affinity *in vitro,* as do isolated CRM domain from CRS1 and CAF2 (O. Ostersetzer, K. Watkins, and A. Barkan, manuscripts in preparation). CRM domains are found in a plant-specific protein family represented by 16 members in *Arabidopsis*. Thus, it appears that an ancestral pre-ribosome binding protein was recruited as a malleable RNA binding module during the evolution of plant genomes, giving rise to a family of intron-specific group II intron splicing factors and other presumed RNA binding proteins. Some of the uncharacterised members of the CRM family are closely related to either CRS1 or to the CAFs, and are predicted to be targeted to mitochondria or chloroplasts. It seems likely that some of these will also prove to be organellar group II intron spicing factors.

The following working model has emerged for the biochemical roles in splicing of CRS1, CAF1, CAF2, and CRS2.

(i)The CRM domain proteins CRS1, CAF1, and CAF2 are intron-specific binding proteins that guide intron folding. Interactions between recombinant CRS1 and the *atp*F intron have been studied in depth (Ostersetzer *et al.*, 2004). CRS1 binds *in vitro* with high affinity and specificity to the *atp*F intron through the recognition of non-conserved sequence elements in intron domains 1 and 4. Hydroxyl-radical footprinting assays showed that CRS1 binding causes a profound change in intron structure, promoting the internalisation of intron elements expected to be at the core of the functional ribozyme. These results lead to a picture in which CRS1 promotes the folding of its group II intron target through tight and specific interactions with two peripheral intron segments (Ostersetzer *et al.*, 2004). It seems reasonable to propose that the related proteins CAF1 and CAF2 might act similarly.

(ii) CRS2 is recruited to specific introns via interaction with CAF1 or CAF2. Recombinant CAF1 and CAF2 bind intron RNA with high affinity *in vitro*, whereas recombinant CRS2 does not (Barkan lab, unpublished observations). CRS2 might nonetheless contribute to the RNA binding activity of the CRS2/CAF complexes through a C-terminal appendage that resembles an RNA binding motif (Jenkins and Barkan, 2001).

(iii) CRS2's conserved peptidyl-tRNA hydrolase active site may contribute functional groups during splicing catalysis. The CRS2 crystal structure strongly resembles that of bacterial peptidyl-tRNA hydrolases (G. Ostheimer, A.B., and B. Matthews, in preparation), and CRS2 has maintained all amino acid residues that are known to be essential for catalysis by peptidyl-tRNA hydrolases. Nonetheless, CRS2 has failed to exhibit any hint of peptidyl-tRNA hydrolase activity in *in vivo* complementation assays (Jenkins and Barkan, 2001) or in *in vitro* assays (N. Henderson and A.B., unpublished results). Furthermore, a second chloroplast-localized peptidyl-tRNA hydrolase-like molecule is predicted from maize and Arabidopsis genome sequence, suggesting that CRS2 need not supply a peptidyl-tRNA hydrolase function *in vivo.* If CRS2 does not function as a peptidyl-tRNA hydrolase, why has its peptidyl-tRNA hydrolase-like active site been so highly conserved? One intriguing possibility is that the ancestral active site, which cleaves the ester bond linking tRNA to nascent peptide, has been modified such that it now participates in a chemical step of splicing.

Predicted orthologs of these maize chloroplast splicing factors can be identified from rice and *Arabidopsis* genome sequence data, and these proteins presumably play similar roles. However, there is no evidence for the participation of related proteins in the splicing of organellar introns in algae. As the nuclear genome sequences of various "lower" plants become available, it will be interesting to see whether the emergence of *crs*1-, *crs*2- and *caf*-like genes coincides with the appearance of the chloroplast genome organization that is characteristic of land plants.

The *Arabidopsis* protein HCF152 is necessary for the accumulation of spliced *pet*B RNA but may not be necessary for splicing itself, since the excised *pet*B intron accumulates normally in *hcf*152 mutants (Meierhoff *et al.*, 2003). However, a role for HCF152 in a late step in splicing remains possible: mutation of intron residues mediating the □-□' interaction in an algal mitochondrial group II intron reduced exon ligation and stabilized the excised intron (Hollander and Kuck, 1999a); the *hcf*152 mutant phenotype is consistent with a similar splicing defect. In any case, mutations that unambiguously disrupt the splicing of chloroplast introns have not surfaced in Arabidopsis despite intensive use of this organism for the genetic analysis of chloroplast biogenesis. This may reflect species-specific features of chloroplast gene and intron content. There is accumulating evidence that plastid translation is essential for cellular viability in *Arabidopsis* but not in maize (reviewed in Stern *et al.*, 2004); this difference correlates with the presence of two cell-essential ORFs in the *Arabidopsis* chloroplast genome (Drescher *et al.*, 2000) and the absence of those ORFs in maize (Maier *et al.*, 1995). Null mutations in the *Arabidopsis crs*2, *caf*1 and *caf*2 orthologs may not be recoverable, as these genes are required for the biogenesis of the plastid translation machinery due to their roles in tRNA and ribosomal protein mRNA splicing (Jenkins *et al.*, 1997; Ostheimer *et al.*, 2003). The maintenance of two group II introns in the *Arabidopsis* chloroplast *clp*P gene may present an additional barrier to the recovery of chloroplast splicing mutants, since *clp*P expression is essential for shoot development in *Arabidopsis* (Kuroda and Maliga, 2003). The maize chloroplast *clp*P gene lacks introns (Maier *et al.*, 1995) and may not be cell-essential in any case (Cahoon *et al.*, 2003). These organism-specific "quirks" highlight the importance of using multiple model organisms for the study of organellar splicing and, indeed, all biological processes.

3.5. Strategies for the Identification of Additional Organellar Splicing Factors

It is likely that many splicing factors remain to be identified even in the relatively well-developed maize and *C. reinhardtii* chloroplast systems. Many of the nuclear genes identified in genetic screens as being required for the *trans*-splicing of the *C. reinhardtii* chloroplast *psa*A mRNA (Goldschmidt-Clermont *et al.*, 1990) have not been cloned. Similarly, additional nuclear mutations that disrupt maize chloroplast splicing have been recovered, and these define two unidentified nuclear genes (Barkan lab, unpublished). It is unlikely that genetic screens in maize and *C. reinhardtii* have been done to saturation, so still more genes are likely to emerge through this traditional approach.

The identification of plant mitochondrial splicing factors presents additional challenges and is just beginning. A mutation in the nuclear gene *nms*1 in

Nicotiana sylvestris disrupts the splicing of a group II intron in the mitochondrial *nad*4 gene (Brangeon *et al.*, 2000) and is the sole example of a mitochondrial splicing mutant in plants. *Nms*1 mutant plants are defective in both vegetative and reproductive development, as might be anticipated from a defect in respiration. The description of this mutant provides clues about the gross phenotype to be anticipated from mitochondrial splicing defects, and could open the door to the use of genetic screens for mitochondrial splicing factors in more genetically tractable plants. However, the essential role of respiration in most plant cells is likely to complicate such approaches.

The *C. reinhardtii* chloroplast provides a promising venue for the genetic identification of plant organellar group I intron splicing factors because it houses five group I introns (in contrast to the lone group I intron in land plant chloroplasts) and sophisticated genetic screens can be applied. The nuclear mutant *ac20* over-accumulates the unspliced form of the plastid 23S-like rRNA and `may define a gene for a group I splicing factor (Herrin *et al.*, 1990). However, replacement of the 23S-like rRNA gene with an intronless allele did not alleviate the defect in chloroplast ribosome maturation in *ac20* mutant cells (Holloway and Herrin, 1998). Thus, either the splicing defect in this mutant is a secondary effect of a defect in rRNA processing, or the Ac20 gene product is bifunctional, promoting splicing and ribosome maturation independently.

It may be difficult to recover mutants in *C. reinhardtii* with defects in plastid group I intron splicing because the presence of cell-essential ORFs in its chloroplast genome (Boudreau *et al.*, 1997) makes plastid translation (and thus the expression of its 23S-like rRNA) essential. In an alternative approach that may circumvent this problem, Li *et al.* identified nuclear mutations that suppress the splicing defects caused by engineered mutations in core helices of the 23S-like rRNA intron (Li *et al.*, 2002). These mutations mapped to two genes, and also suppressed the splicing defects caused by similar mutations in a group I intron in the *psb*A gene, leading the authors to propose that the nuclear genes encode general group I splicing factors. It will be interesting to learn the identity of these genes, and whether their wild-type alleles promote the splicing of unmodified group I introns in the *C. reinhardtii* chloroplast.

Genetic approaches are quite powerful but they are unlikely to identify the complete set of organellar splicing factors, so biochemical and candidate gene strategies can be anticipated to play increasingly important roles. In cases where one splicing factor has been identified for a particular intron, this protein provides a handle for the identification of interacting proteins through yeast two-hybrid screens or through the purification of native complexes by co-immunoprecipitation. However, where no factors are known, more generalized biochemical fishing is required. For example, two proteins in *C. reinhardtii* chloroplast extract were detected that bind *in vitro* to a group II intron from *Scenedesmus obliquus* mitochondria (Bunse *et al.*, 2001). In principle, splicing factors could be identified via activity-based assays, through the purification of splicing-active fractions from organellar extracts. However, there are as yet no reports of plant organellar extracts that exhibit splicing activity.

As the number of identified group I and group II splicing factors increases, a "candidate-gene" approach will become increasingly useful for the identification of additional factors. Many excellent candidates can now be predicted in vascular plants. These include a set of nuclear-encoded group II maturase-like proteins that are predicted to be organelle-targeted (Mohr and Lambowitz, 2003). Vascular plant genomes also encode predicted mitochondrial paralogs of the maize chloroplast splicing factors CRS2, CAF1, and CAF2. Thus, it is quite possible that CRS2/CAF-like complexes promote group II intron splicing in plant mitochondria. Additional candidates include predicted chloroplast-targeted CRS1 paralogs in plants (five such paralogs are encoded by the *Arabidopsis* genome), algal homologs of plant splicing factors (and *vice versa*), and homologs of fungal mitochondrial splicing factors. Reverse-genetic and biochemical tests of the roles of such candidates are certain to be reported in the coming years.

4. THE REGULATION OF ORGANELLAR RNA SPLICING IN PLANTS

RNA splicing is an essential step in the expression of intron-containing genes and, as such, is a potential regulatory step in organellar gene expression. The possibility that organellar RNA splicing might be exploited for gene regulation is appealing given that unspliced organellar transcripts typically accumulate to high levels, suggesting that splicing is rate-limiting for the production of mature RNAs. The basis for the inefficiency of organellar splicing is unknown. The rate of splicing may simply be slow relative to the rate of transcription. However, it is also possible that the extent of splicing is limited by the length of a kinetic window following transcription during which a nascent intron RNA can most easily assume its active conformation. In this case, much of the unspliced RNA that accumulates in organelles may be incompetent for splicing, representing a dead-end product that is ultimately degraded.

There have been few systematic investigations of the regulation of plant organellar splicing, but the available data support the notion that splicing in chloroplasts *is* regulated and that this regulation makes an important contribution to the overall regulation of chloroplast gene expression. In maize, the group II introns in the plastid *atp*F, *pet*D, *pet*B and *rpl*16 genes are predominantly unspliced in roots and in the leaf meristem, but are predominantly spliced in etiolated and green leaf blades (Barkan, 1989). Similarly, most *ycf*3 transcripts are unspliced in maize endosperm (McCullough *et al.*, 1992), whereas they are predominantly spliced in the leaf (Jenkins *et al.*, 1997; McCullough *et al.*, 1992). In mustard chloroplasts, unspliced *trn*G transcripts peak in abundance ~48 hours after sowing and then drop precipitously, in concert with an increase in the abundance of the spliced tRNA (Liere and Link, 1994).

In contrast, exposure to light had no affect on the proportion of spliced *trn*G transcripts in mustard (Liere and Link, 1994), spliced *atp*F, *rpl*16, *pet*B, or *pet*D transcripts in maize leaves (Barkan, 1989), or spliced *psb*A transcripts in *Euglena* chloroplasts (Hollingsworth *et al.*, 1984). In *C. reinhardtii*, however, light activates the splicing of the four group I introns in the chloroplast *psb*A gene, and does so in a manner that is mediated by photosynthetic electron transport (Deshpande *et al.*, 1997). This effect is specific for the *psb*A introns: the splicing of the group I intron in the 23S rRNA gene was not influenced by light (Deshpande *et al.*, 1997).

These examples suggest that the splicing of some chloroplast introns is regulated by developmental signals in plants and by light in green algae. However, these conclusions are based on steady-state levels of spliced versus unspliced RNAs, and changes in this ratio could potentially reflect changes in the relative stabilities of the precursor and product forms rather than the regulation of splicing rate. Furthermore, a change in the abundance of a spliced mRNA may not always influence the rate of synthesis of its translation product because some chloroplast mRNAs may be in excess of the amount needed for maximal translation (Eberhard *et al.*, 2002). Evidence that the light-induced increase in *psb*A splicing in *C. reinhardtii* chloroplasts is important for optimal photoautotrophic growth was provided by the analysis of destabilizing mutations in core intron elements (Lee and Herrin, 2003); one mutant strain exhibited a 45% reduction in the level of spliced *psb*A RNA, and this was reflected by small but significant decreases in the rates of *psb*A synthesis and photoautotrophic growth.

A variety of mechanisms could account for the changes in the ratio of spliced to unspliced transcripts that have been observed in chloroplasts. One possibility is that the splicing rate truly changes, and does so as a consequence of changes in the abundance or activities of splicing factors. Indeed, the abundance of the mRNA for the maize chloroplast splicing factor CRS1 roughly parallels the differences in the proportion of spliced *atp*F mRNA observed in various seedling tissues (Till *et al.*, 2001). However, the more informative experiment, not yet reported, is to determine whether the proportion of spliced RNA changes when the abundance of each splicing factor is increased and decreased in small increments. Developmentally-regulated changes in chloroplast Mg^{2+} concentration may also contribute to the developmental regulation of plastid splicing. Mg^{2+} is critical for group II intron folding and catalysis (Pyle, 2002) and mutations in a gene for a mitochondrial Mg^{2+} transporter disrupt the splicing of group II introns in yeast mitochondria (Bui *et al.*, 1999; Wiesenberger *et al.*, 1992). The concentration of free Mg^{2+} rises from ~3 mM to ~10 mM during the maturation of young to mature chloroplasts in spinach (Horlitz and Klaff, 2000), and this change may well impact the rates of both group I and group II intron splicing.

Introns presumably start to fold *in vivo* while still being transcribed. Cotranscriptional folding may enhance the yield of productively folded introns by reducing the number of potential partners available to nascent intron segments. If so, the transcription elongation rate could influence the yield of productively folded introns by affecting the length of the kinetic window within which competing, non-native partners are excluded. A chloroplast-encoded bacterial-type RNA polymerase provides the bulk of the transcription activity in mature chloroplasts, whereas transcription is catalyzed by a nuclear-encoded phage-like RNA polymerase in proplastids and non-leaf plastids (Allison, 2000). Considering that T7 RNA polymerase elongates ~eight-fold faster than bacterial RNA polymerase (Iost *et al.*, 1992), these two plastid polymerases may differ substantially in their elongation rates. If so, the more rapid transcription elongation rate of the phage-like polymerase may reduce the yield of productively folded introns in plastids where this polymerase predominates (*e.g.* in proplastids and root plastids), resulting in an increase in the proportion of unspliced transcripts in these plastid forms.

An increase in the ratio of spliced to unspliced transcripts could also result from an increase in the relative stability of the spliced versus the unspliced form. Such a mechanism would require a means for distinguishing spliced RNAs from their unspliced precursors. The act of translation itself could serve this purpose, since translation can modulate the stabilities of chloroplast mRNAs (Barkan, 1993; Chen *et al.*, 1993; Klaff and Gruissem, 1991; Mayfield *et al.*, 1994; Monde *et al.*, 2000). Unspliced mRNAs cannot be fully translated and therefore may be inherently unstable. Spliced mRNAs may then become stabilized in those plastid types containing an active translation machinery (i.e. in chloroplasts and etioplasts). As a consequence, the ratio of spliced to unspliced mRNAs could be higher in chloroplasts and etioplasts than in proplastids and root plastids even in the absence of changes in splicing rates.

However, this mechanism is unlikely to account fully for the increase in the proportion of spliced *pet*B and *pet*D transcripts observed during the proplastid to chloroplast transition in maize, because these transcripts are spliced to a greater extent in mutant leaf plastids lacking ribosomes than they are in wild-type roots and leaf meristem (Barkan, 1989; Jenkins *et al.*, 1997). The *Arabidopsis* protein HCF152 is required for the accumulation of spliced but not unspliced *pet*B transcripts, but it does not influence the abundance of the excised intron, suggesting that it stabilizes the spliced mRNA (Meierhoff *et al.*, 2003). An increase in the abundance or activity of this protein during leaf development could provide another means to increase the proportion of spliced *petB* transcripts.

Studies addressing the regulation of splicing in plant mitochondria have not been reported, and analyses of the regulation of chloroplast splicing are far from comprehensive. A more complete understanding of regulatory mechanisms involving organellar introns can be anticipated as these descriptive studies are completed and as more splicing factors are identified. A variety of stable and transient transformation systems have been developed for the expression of modified introns in plant and algal organelles (Bock and Maliga, 1995; Doetsch *et al.*, 2001; Farre and Araya, 2002; Hollander and Kuck, 1999a; Lee and Herrin, 2003). These will provide an important complement to the work with *trans*-acting splicing factors, by allowing the dissection of the *cis*-acting requirements for both regulated and constitutive splicing reactions in the organelles of plants.

5. ACKNOWLEDGEMENTS

I would like to thank Christian Schmitz-Linneweber, Kenny Watkins, and David Herrin for comments on the manuscript, and Rainer Maier and David Herrin for communicating unpublished data.

6. REFERENCES

Allison, L. A. (2000). The role of sigma factors in plastid transcription. *Biochimie, 82*(6-7), 537-548.

Barkan, A. (1989). Tissue-dependent plastid RNA splicing in maize: Transcripts from four plastid genes are predominantly unspliced in leaf meristems and roots. *Plant Cell, 1*, 437-445.

Barkan, A. (1993). Nuclear mutants of maize with defects in chloroplast polysome assembly have altered RNA metabolism. *Plant Cell, 5*, 389-402.

Barkan, A., & Goldschmidt-Clermont, M. (2000). Participation of nuclear genes in chloroplast gene expression. *Biochimie, 82*, 559-572.

Bassi, G., Oliveira, D. , White, M., & Weeks, K. (2002). Recruitment of intron-encoded and co-opted proteins in splicing of the bI3 group I intron RNA. *Proc. Natl. Acad. Sci. USA, 99*, 128-133.

Begu, D., Mercado, A., Farre, J. C., Moenne, A., Holuigue, L., Araya, A., & Jordana, X. (1998). Editing status of mat-r transcripts in mitochondria from two plant species: C-to-U changes occur in putative functional RT and maturase domains. *Curr Genet, 33*(6), 420-428.

Binder, S., & Brennicke, A. (2003). Gene expression in plant mitochondria: transcriptional and post-transcriptional control. *Philos Trans R Soc Lond B Biol Sci, 358*, 181-188.

Binder, S., Marchfelder, A., Brennicke, A., & Wissinger, B. (1992). RNA editing in trans-splicing intron sequences of nad2 mRNAs in Oenothera mitochondria. *J Biol Chem, 267*(11), 7615-7623.

Bock, R., & Maliga, P. (1995). Correct splicing of a group II intron from a chimeric reporter gene transcript in tobacco plastids. *Nucl. Acids Res., 23*, 2544-2547.

Bonen, L., & Vogel, J. (2001). The ins and outs of group II introns. *Trends in Genet, 17*, 322-331.

Boudreau, E., Turmel, M., Goldschmidt-Clermont, M., Rochaix, J. D., Sivan, S., Michaels, A., & Leu, S. (1997). A large open reading frame (orf1995) in the chloroplast DNA of Chlamydomonas reinhardtii encodes an essential protein. *Mol Gen Genet, 253*(5), 649-653.

Brangeon, J., Sabar, M., Gutierres, S., Combettes, B., Bove, J., Gendy, C., *et al.* (2000). Defective splicing of the first nad4 intron is associated with lack of several complex I subunits in the Nicotiana sylvestris NMS1 nuclear mutant. *Plant J, 21*(3), 269-280.

Bui, D., Gregan, J., Jarosch, E., Ragnini, A., & Schweyen, R. (1999). The bacterial magnesium transporter CorA can functionally substitute for its putative homologue Mrs2p in the yeast inner mitochondrial membrane. *J. Biol. Chem., 274*, 20438-20443.

Bunse, A., Nickelsen, J., & Kuck, U. (2001). Intron-specific RNA binding proteins in the chloroplast of the green alga *Chlamyhdomonas reinhardtii*. *Biochim Biophys Acta, 1519*, 46-54.

Cahoon, A., Cunningham, K., & Stern, D. (2003). The plastid *clp*P gene may not be essential for plant cell viability. *Plant Cell Physiol, 44*, 93-95.

Cannone, J., Subramanian, S., Schnare, M., Collett, J., D'Souza, L., Du, Y., Feng, B., Lin, N., Madabusi, L., Muller, K., Pande, N., Shang, Z., Yu, N., & Gutell, R. (2002). The Comparative RNA Web (CRW) Site: An Online Database of Comparative Sequence and Structure Information for Ribosomal, Intron, and other RNAs. *BioMed Central Bioinformatics, 3*, 2.

Carrillo, C., & Bonen, L. (1997). RNA editing status of *nad7* intron domains in wheat mitochondria. *Nucleic Acids Res, 25*(2), 403-409.

Carrillo, C., Chapdelaine, Y., & Bonen, L. (2001). Variation in sequence and RNA editing within core domains of mitochondrial group II introns among plants. *Mol Gen Genet, 264*(5), 595-603.

Cech, T. R. (1986). The generality of self-splicing RNA: relationship to nuclear mRNA splicing. *Cell, 44*, 207-210.

Chapdelaine, Y., & Bonen, L. (1991). The wheat mitochondrial gene for subunit I of the NADH dehydrogenase complex: a trans-splicing model for this gene-in-pieces. *Cell, 65*(3), 465-472.

Chen, X., Kindle, K., & Stern, D. (1993). Initiation codon mutations in the *Chlamydomonas* chloroplast *pet*D gene result in temperature-sensitive photosynthetic growth. *EMBO J., 12*, 3627-3635.

Choquet, Y., Goldschmidt-Clermont, M., Girard-Bascou, J., Kueck, U., Bennoun, P., & Rochaix, J.-D. (1988). Mutant phenotypes support a *trans*-splicing mechanism for the expression of the tripartite *psa*A gene in the C. reinhardtii chloroplast. *Cell, 52*, 903-913.

Christopher, D. A., & Hallick, R. B. (1989). *Euglena gracilis* chloroplast ribosomal protein operon: a new chloroplast gene for ribosomal protein L5 and description of a novel organelle intron category designated group III. *Nucl. Acids Res., 17*, 7591-7608.

Copertino, D. W., Christopher, D. A., & Hallick, R. B. (1991). A mixed group II/group III twintron in the *Euglena gracilis* chloroplast ribosomal protein S3 gene: evidence for intron insertion during gene evolution. *Nucl. Acids Res., 19*, 6491-6497.

Copertino, D. W., Hall, E. T., Hook, F. W. V., Jenkins, K. P., & Hallick, R. B. (1994). A group III twintron encoding a maturase-like gene excises through lariat intermediates. *Nucl. Acids Res., 22*, 1029-1036.

Copertino, D. W., & Hallick, R. B. (1993). Group II and group III introns of twintrons: potential relationships with nuclear pre-mRNA introns. *Trends Biochem Sci, 18*, 467-471.

Copertino, D. W., Shigeoka, S., & Hallick, R. B. (1992). Chloroplast group III twintron excision utilizing multiple 5' and 3' splice sites. *EMBO J., 11*, 5041-5050.

Costa, M., Fontaine, J. M., Loiseaux-de Goer, S., & Michel, F. (1997). A group II self-splicing intron from the brown alga Pylaiella littoralis is active at unusually low magnesium concentrations and forms populations of molecules with a uniform conformation. *J Mol Biol, 274*(3), 353-364.

Daniell, H., Khan, M. S., & Allison, L. (2002). Milestones in chloroplast genetic engineering: an environmentally friendly era in biotechnology. *Trends Plant Sci, 7*(2), 84-91.

Daniell, D., Carmona-Sanchez, O., & Burns, B. B. (2004). Chloroplast derived antibodies, biopharmaceuticals and edible vaccines. In R. Fischer & S. Schillberg (Eds.) *Molecular Farming* (pp.113-133). Weinheim: WILEY-VCH Verlag.

Daniels, D., Michels, W. J., & Pyle, A. M. (1996). Two competing pathways for self-splicing by group II introns; a quantitative analysis of in-vitro reaction rates and products. *J. Mol. Biol., 256*, 31-49.

Deshpande, N. N., Bao, Y., & Herrin, D. L. (1997). Evidence for light/redox-regulated splicing of *psb*A pre-RNAs in *Chlamydomonas* chloroplasts. *RNA, 3*, 37-48.

Devine, A. L., & Daniell, H. (2004). Chloroplast genetic engineering. In S. Moller (Ed.), *Plastids* (pp. 283-320). United Kingdom: Blackwell Publishers.

Doetsch, N., Favreau, M., Kuscuoglu, N., Thompson, M., & Hallick, R. (2001). Chloroplast transformation in *Euglena gracilis:* splicing of a group III twintron transcribed from a transgenic *psb*K operon. *Curr. Genet., 39*, 49-60.

Doetsch, N., Thompson, M., & Hallick, R. (1998). A maturase-encoding group III twintron is conserved in deeply rooted euglenoid species: are group III introns the chicken or the egg? *Mol. Biol. Evol., 15*, 76-86.

Drescher, A., Ruf, S., Calsa, T., Carrer, H., & Bock, R. (2000). The two largest chloroplast genome-encoded open reading frames of higher plants are essential genes. *Plant J, 22*, 97-104.

Duerrenberger, F., & Rochaix, J.-D. (1991). Chloroplast ribosomoal intron of *Chlamydomonas reinhardtii: in vitro* self-splicing, DNA endonuclease activity and *in vivo* mobility. *EMBO J., 10*, 3495-3501.

Eberhard, S., Drapier, D., & Wollman, F. (2002). Searching limiting steps in the expression of chloroplast-encoded proteins: relations between gene copy number, transcription, transcript abundance and translation rate in the chloroplast of *Chlamydomonas reinhardtii*. *Plant J., 31*, 149-160.

Ems, S. C., Morden, C. W., Dixon, C. K., Wolfe, K. H., dePamphilis, C. W., & Palmer, J. D. (1995). Transcription, splicing and editing of plastid RNAs in the nonphotosynthetic plant *Epifagus virginiana*. *Plant Mol. Biol., 29*, 721-733.

Farre, J., & Araya, A. (1999). The mat-r open reading frame is transcribed from a non-canonical promoter and contains an internal promoter to co-transcribe exons nad1e and nad5III in wheat mitochondria. *Plant Mol Biol, 40*(6), 959-967.

Farre, J., & Araya, A. (2002). RNA splicing in higher plant mitochondria: determination of functional elements in a group II intron from a chimeric *cox* II gene in electroporated wheat mitochondria. *Plant J, 29*, 203-213.

Fromm, H., Edelman, M., Koller, B., Goloubinoff, P., & Galun, E. (1986). The enigma of the gene coding for ribosomal protein S12 in the chloroplast of *Nicotiana*. *Nucleic Acids Res., 14*, 883-898.

Golden, B. L., Gooding, A. R., Podell, E. R., & Cech, T. R. (1998). A preorganized active site in the crystal structure of the *Tetrahymena* ribozyme. *Science, 282*(5387), 259-264.

Goldschmidt-Clermont, M., Choquet, Y., Girard-Bascou, J., Michel, F., Schirmer-Rahire, M., & Rochaix, J.-D. (1991). A small chloroplast RNA may be required for *trans*-splicing in Chlamydomonas reinhardtii. *Cell, 65*, 135-143.

Goldschmidt-Clermont, M., Girard-Bascou, J., Choquet, Y., & Rochaix, J.-D. (1990). *Trans*-splicing mutants of *Chlamydomonas reinhardtii*. *Mol. Gen. Genet., 223*, 417-425.

Gutgsell, N., Campo, M. D., Raychaudhuri, S., & Ofengand, J. (2001). A second function for pseudouridine synthases: A point mutant of RluD unable to form pseudouridines 1911, 1915, and

1917 in Escherichia coli 23S ribosomal RNA restores normal growth to an RluD-minus strain. *RNA, 7*, 990-998.

Gutgsell, N., Englund, N., Niu, L., Kaya, Y., Lane, B., & Ofengand, J. (2000). Deletion of the *Escherichia coli* pseudouridine synthase gene *tru*B blocks formation of pseudouridine 55 in tRNA in vivo, does not affect exponential growth, but confers a strong selective disadvantage in competition with wild-type cells. *RNA, 6*(12), 1870-1881.

Hahn, D., Nickelsen, J., Hackert, A., & Kuck, U. (1998). A single nuclear locus is involved in both chloroplast RNA trans-splicing and 3' end processing. *Plant J., 15*, 575-581.

Hallick, R. B., Hong, L., Drager, R. G., Favreau, M. R., Monfort, A., Orsat, B., Spielmann, A., & Stutz, E. (1993). Complete sequence of *Euglena gracilis* chloroplast DNA. *Nucl. Acids Res., 21*, 3537-3544.

Herrin, D. L., Bao, Y., Thompson, A. J., & Chen, Y.-F. (1991). Self-splicing of the Chlamydomonas chloroplast *psb*A introns. *Plant Cell, 3*, 10951107.

Herrin, D. L., Chen, Y.-F., & Schmidt, G. W. (1990). RNA splicing in *Chlamydomonas* chloroplasts: self-splicing of 23S preRNA. *J. Biol. Chem., 265*, 21134-21140.

Herrin, D. L., & Schmidt, G. W. (1988). *trans*-Splicing of transcripts for the chloroplast *psa*A1 gene. *J. Biol. Chem., 263*, 14601-14604.

Herschlag, D. (1995). RNA chaperones and the RNA folding problem. *J. Biol. Chem., 270*, 20871-20874.

Hetzer, M., Wurzer, G., Schweyen, R., & Mueller, M. (1997). Trans-activation of group II intron splicing by nuclear U5 snRNA. *Nature, 386*, 417-420.

Ho, Y., Kim, S. J., & Waring, R. B. (1997). A protein encoded by a group I intron in *Aspergillus nidulans* directly assists RNA splicing and is a DNA endonuclease. *Proc. Natl. Acad. Sci. USA, 94*, 8994-8999.

Ho, Y., & Waring, R. (1999). The maturase encoded by a group I intron from *Aspergillus nidulans* stabilizes RNA tertiary structure and promotes rapid splicing. *J. Mol. Biol., 292*, 987-1001.

Hollander, V., & Kuck, U. (1998). Splicing of the mitochondrial group-II intron rI1: conserved intron-exon interactions diminish splicing efficiency. *Curr Genet, 33*, 117-123.

Hollander, V., & Kuck, U. (1999a). Group II intron splicing in chloroplasts: identification of mutations determining intron stability and fate of exon RNA. *Nucleic Acids Res, 27*(11), 2345-2353.

Hollander, V., & Kuck, U. (1999b). Group II intron splicing in *Escherichia coli:* phenotypes of *cis*-acting mutations resemble splicing defects observed in organelle RNA processing. *Nucl. Acids Res., 27*, 2339-2344.

Hollingsworth, M. J., Johanningmeir, U., Karabin, G. D., Stiegler, G. L., & Hallick, R. B. (1984). Detection of multiple, unspliced precursor mRNA transcripts for the Mr 32,000 thylakoid membrane protein from *Euglena gracilis* chloroplasts. *Nucl. Acids Res., 12*, 2001-2017.

Holloway, S., Deshpande, N., & Herrin, D. (1999). The catalytic group-I introns of the *psb*A gene of *Chlamydomonas reinhardtii*: core structures, ORFs and evolutionary implications. *Curr Genet, 36*, 69-78.

Holloway, S., & Herrin, D. (1998). Processing of a composite large subunit rRNA. Studies with chlamydomonas mutants deficient in maturation of the 23s-like rRNA. *Plant Cell, 10*, 1193-1206.

Horlitz, M., & Klaff, P. (2000). Gene-specific *trans*-regulatory functions of magnesium for chloroplast mRNA stability in higher plants. *J. Biol. Chem., 275*, 35638-35645.

Hsieh, J., Andrews, A. J., & Fierke, C. A. (2004). Roles of protein subunits in RNA-protein complexes: lessons from ribonuclease P. *Biopolymers, 73*(1), 79-89.

Iost, I., Guillerez, J., & Dreyfus, M. (1992). Bacteriophage T7 RNA polymerase travels far ahead of ribosomes *in vivo*. *J. Bacteriol., 174*, 619-622.

Jankowsky, E., Gross, C. H., Shuman, S., & Pyle, A. M. (2001). Active disruption of an RNA-protein interaction by a DExH/D RNA helicase. *Science, 291*(5501), 121-125.

Jenkins, B., & Barkan, A. (2001). Recruitment of a peptidyl-tRNA hydrolase as a facilitator of group II intron splicing in chloroplasts. *EMBO J, 20*, 872-879.

Jenkins, B., Kulhanek, D., & Barkan, A. (1997). Nuclear mutations that block group II RNA splicing in maize chloroplasts reveal several intron classes with distinct requirements for splicing factors. *Plant Cell, 9*, 283-296.

Johanningmeier, U., & Heiss, S. (1993). Construction of a *Chlamydomonas reinhardtii* mutant with an intronless *psb*A gene. *Plant Mol. Biol., 22*, 91-99.

Kapoor, M., Nagai, T., Wakasugi, T., Yoshinaga, K., & Sugiura, M. (1997). Organization of chloroplast ribosomal RNA genes and in vitro self-splicing activity of the large subunit rRNA intron from the green alga *Chlorella vulgaris* C-27. *Curr. Genet., 31*, 503-510.

Kim, J.-K., & Hollingsworth, M. J. (1993). Splicing of group II introns in spinach chloroplasts (*in vivo*): analysis of lariat formation. *Curr Genet, 23*, 175-180.

Klaff, P., & Gruissem, W. (1991). Changes in chloroplast mRNA stability during leaf development. *Plant Cell, 3*, 517-529.

Knoop, V., Altwasser, M., & Brennicke, A. (1997). A tripartite group II intron in mitochondria of an angiosperm plant. *Mol Gen Genet, 255*(3), 269-276.

Knoop, V., Schuster, W., Wissinger, B., & Brennicke, A. (1991). Trans splicing integrates an exon of 22 nucleotides into the *nad5* mRNA in higher plant mitochondria. *EMBO J, 10*(11), 3483-3493.

Kubo, T., Nishizawa, S., Sugawara, A., Itchoda, N., Estiati, A., & Mikami, T. (2000). The complete nucleotide sequence of the mitochondrial genome of sugar beet (*Beta vulgaris* L.) reveals a novel gene for tRNA(Cys)(GCA). *Nucleic Acids Res, 28*(13), 2571-2576.

Kuck, U., Choquet, Y., Schneider, M., Dron, M., & Bennoun, P. (1987). Structural and transcriptional analysis of two homologous genes for the P700 chlorophyll a-apoproteins in *Chlamydomonas reinhardtii*: evidence for *in vivo trans* splicing. *EMBO J, 6*, 2185-2195.

Kuck, U., Godehardt, I., & Schmidt, U. (1990). A self-splicing group-II intron in the mitochondrial large subunit rRNA (LSUrRNA) gene of the eukaryotic alga *Scenedesmus obliquus*. *Nucl. Acids Res., 18*, 2691-2697.

Kugita, M., Kaneko, A., Yamamoto, Y., Takeya, Y., Matsumoto, T., & Yoshinaga, K. (2003a). The complete nucleotide sequence of the hornwort (*Anthoceros formosae*) chloroplast genome: insight into the earliest land plants. *Nucleic Acids Res, 31*(2), 716-721.

Kugita, M., Yamamoto, Y., Fujikawa, T., Matsumoto, T., & Yoshinaga, K. (2003b). RNA editing in hornwort chloroplasts makes more than half the genes functional. *Nucleic Acids Res, 31*(9), 2417-2423.

Kuroda, H., & Maliga, P. (2003). The plastid *clp*P1 protease gene is essential for plant development. *Nature, 86-89.*

Lambowitz, A., Caprara, M., Zimmerly, S., & Perlman, P. (1999). Group I and Group II Ribozymes as RNPs: Clues to the Past and Guides to the Future. In R. Gesteland, T. Cech & J. Atkins (Eds.), *The RNA World* (second ed., pp. 451-485). Cold Spring Harbor: Cold Spring Harbor Laboratory Press.

Lee, J., & Herrin, D. (2003). Mutagenesis of a light-regulated *psb*A intron reveals the importance of efficient splicing for photosynthetic growth. *Nucl. Acids Res., 31*, 4361-4372.

Lehnert, V., Jaeger, L., Michel, F., & Westhof, E. (1996). New loop-loop tertiary interactions in self-splicing introns of subgroup IC and ID: a complete 3D model of the *Tetrahymena thermophila* ribozyme. *Chem Biol, 3*(12), 993-1009.

Leister, D., & Schneider, A. (2003). From genes to photosynthesis in *Arabidopsis thaliana*. *Int Rev Cytol, 228*, 31-83.

Lemieux, C., Otis, C., & Turmel, M. (2000). Ancestral chloroplast genome in *Mesostigma viride* reveals an early branch of green plant evolution. *Nature, 403*, 649-652.

Li, F., & Herrin, D. (2002). FUGOID: functional genomics of organellar introns database. *Nucl. Acids Res., 30*, 385-386.

Li, F., Holloway, S., Lee, J., & Herrin, D. (2002). Nuclear genes that promote splicing of group I introns in the chloroplast 23S rRNA and *psb*A genes in *Chlamydomonas reinhardtii*. *Plant J., 32*, 467-480.

Liere, K., & Link, G. (1994). Structure and expression characteristics of the chloroplast DNA region containing the split gene for tRNA-gly (UCC) from mustard. *Curr. Genet., 26*, 557-563.

Liere, K., & Link, G. (1995). RNA-binding activity of the matK protein encoded by the chloroplast *trn*k intron from mustard. *Nucl. Acids Res., 23*, 917-921.

Lorsch, J. R. (2002). RNA chaperones exist and DEAD box proteins get a life. *Cell, 109*(7), 797-800.

Maier, R. M., Neckermann, K., Igloi, G. L., & Koessel, H. (1995). Complete sequence of the maize chloroplast genome: Gene content, hotspots of divergence and fine tuning of genetic information by transcript editing. *J. Molec. Biol., 251*, 614-628.

Malek, O., & Knoop, V. (1998). Trans-splicing group II introns in plant mitochondria: the complete set of *cis*-arranged homologs in ferns, fern allies, and a hornwort. *RNA, 4*(12), 1599-1609.

Matsuura, M., Noah, J., & Lambowitz, A. (2001). Mechanism of maturase-promoted group II intron splicing. *EMBO J, 20*, 7259-7270.

Maul, J. E., Lilly, J. W., Cui, L., dePamphilis, C. W., Miller, W., Harris, E. H., & Stern, D. B. (2002). The *Chlamydomonas reinhardtii* plastid chromosome: islands of genes in a sea of repeats. *Plant Cell, 14*(11), 2659-2679.
Mayer, O., Waldsich, C., Grossberger, R., & Schroeder, R. (2002). Folding of the td pre-RNA with the help of the RNA chaperone StpA. *Biochem Soc Trans, 30*(Pt 6), 1175-1180.
Mayfield, S. P., Cohen, A., Danon, A., & Yohn, C. B. (1994). Translation of the *psb*A mRNA of *Chlamydomonas reinhardtii* requires a structured RNA element contained within the 5' untranslated region. *J. Cell Biol., 127*, 1537-1545.
McCullough, A., Kangasjarvi, J., Gengenback, B., & Jones, R. (1992). Plastid DNA in developing maize endosperm. *Plant Physiol., 100*, 958-964.
Meierhoff, K., Felder, S., Nakamura, T., Bechtold, N., & Schuster, G. (2003). HCF152, an *Arabidopsis* RNA binding pentatricopeptide repeat protein involved in the processing of chloroplast *psb*B-*psb*T-*psb*H-*pet*B-*pet*D RNAs. *Plant Cell, 15*, 1480-1495.
Michel, F., & Dujon, B. (1983). Conservation of RNA secondary structures in two intron families including mitochondrial-, chloroplast- and nuclear-encoded members. *EMBO J., 2*, 33-38.
Michel, F., & Ferat, J.-L. (1995). Structure and actvities of group II introns. *Ann. Rev. Biochem., 64*, 435-461.
Michel, F., Umesono, K., & Ozeki, H. (1989). Comparative and functional anatomy of group II catalytic introns - A review. *Gene, 82*, 5-30.
Michel, F., & Westhof, E. (1990). Modelling of the three-dimensional architecture of group I catalytic introns based on comparative sequence analysis. *J. Mol. biol., 216*, 585-610.
Mohr, G., & Lambowitz, A. (2003). Putative proteins related to group II intron reverse transcriptase/maturases are encoded by nuclear genes in higher plants. *Nucl. Acids Res., 31*, 647-652.
Mohr, G., Perlman, P., & Lambowitz, A. (1993). Evolutionary relationships among group II intron-encoded proteins and identification of a conserved domain that may be related to maturase function. *Nucl. Acids Res., 21*, 4991-4997.
Mohr, S., Stryker, J., & Lambowitz, A. (2002). A DEAD-box protein functions as an ATP-dependent RNA chaperone in group I intron splicing. *Cell, 109*, 769-779.
Monde, R., Schuster, G., & Stern, D. (2000). Processing and degradation of chloroplast mRNA. *Biochimie, 82*, 573-582.
Neuhaus, H., & Link, G. (1987). The chloroplast tRNA(Lys)(UUU) gene from mustard (*Sinapis alba*) contains a class II intron potentially encoding for a maturase-related polypeptide. *Curr. Genet., 11*, 251-257.
Notsu, Y., Masood, S., Nishikawa, T., Kubo, N., Akiduki, G., Nakazono, M., *et al.* (2002). The complete sequence of the rice (*Oryza sativa* L.) mitochondrial genome: frequent DNA sequence acquisition and loss during the evolution of flowering plants. *Mol Genet Genomics, 268*(4), 434-445.
O'Brien, E., Badidi, E., Barbasiewicz, A., deSousa, C., Lang, B., & Burger, G. (2003). GOBASE - a database of mitochondrial and chloroplast information. *Nucl. Acids Res., 31*, 176-178.
Oda, K., Kohchi, T., & Ohyama, K. (1992a). Mitochondrial DNA of *Marchantia polymorpha* as a single circular form with no incorporation of foreign DNA. *Biosci Biotechnol Biochem, 56*(1), 132-135.
Oda, K., Yamato, K., Ohta, E., Nakamura, Y., Takemura, M., Nozato, N., Akashi, K., Kanegae, T., Ogura, Y., Kohchi, T., & *et al.* (1992b). Gene organization deduced from the complete sequence of liverwort Marchantia polymorpha mitochondrial DNA. A primitive form of plant mitochondrial genome. *J Mol Biol, 223*(1), 1-7.
Odom, O. W., Holloway, S. P., Deshpande, N. N., Lee, J., & Herrin, D. L. (2001). Mobile self-splicing group I introns from the *psb*A gene of *Chlamydomonas reinhardtii*: highly efficient homing of an exogenous intron containing its own promoter. *Mol Cell Biol, 21*(10), 3472-3481.
Ohta, E., Oda, K., Yamato, K., Nakamura, Y., Takemura, M., Nozato, N., *et al.* (1993). Group I introns in the liverwort mitochondrial genome: the gene coding for subunit 1 of cytochrome oxidase shares five intron positions with its fungal counterparts. *Nucleic Acids Res, 21*(5), 1297-1305.
Ostersetzer, O., Watkins, K., Cooke, A., & Barkan, A. (in review). CRS1, a host factor for group II intron splicing, promotes intron folding by binding to non-conserved elements in domains 1 and 4.
Ostheimer, G., Barkan, A., & Matthews, B. (2002). Crystal structure of *E. coli* YhbY: a representative of a novel class of RNA binding proteins. *Structure, 10*, 1593-1601.

Ostheimer, G., Williams-Carrier, R., Belcher, S., Osborne, E., Gierke, J., & Barkan, A. (2003). Group II intron splicing factors derived by diversification of an ancient RNA binding module. *EMBO J, 22*, 3919-3929.

Palmer, J., Adams, K., Cho, Y., Parkinson, C., Qiu, Y., & Song, K. (2000). Dynamic evolution of plant mitochondrial genomes: Mobile genes and introns and highly variable mutation rates. *Proc. Natl. Acad. Sci. USA, 97*, 6960-6966.

Perron, K., Goldschmidt-Clermont, M., & Rochaix, J.-D. (1999). A factor related to pseudouridine synthases is required for chloroplast group II intron *trans*-splicing in *Chlamydomonas reinhardtii. EMBO J, 18*, 6481-6490.

Perron, K., Goldschmidt-Clermont, M., & Rochaix, J.-D. (2004) A multiprotein complex involved in chloroplast group II intron splicing. *RNA, 10*, 704-711.

Pyle, A. (2002). Metal ions in the structure and function of RNA. *J. Biol. Inorg. Chem., 7*, 679-690.

Qin, P. Z., & Pyle, A. M. (1998). The architectural organization and mechanistic function of group II intron structural elements. *Curr. Opin. Struct. Biol., 8*, 301-308.

Rivier, C., Goldschmidt-Clermont, M., & Rochaix, J. (2001). Identification of an RNA-protein complex involved in chloroplast group II intron *trans*-splicing in *Chlamydomonas reinhardtii. EMBO J, 20*, 1765-1773.

Rochaix, J. (2002). *Chlamydomonas*, a model system for studying the assembly and dynamics of photosynthetic complexes. *FEBS Lett, 529*(1), 34-38.

Rochaix, J., Rahire, M., & Michel, F. (1985). The chloroplast ribosomal intron of *Chlamydomonas reinhardii* codes for a polypeptide related to mitochondrial maturases. *Nucl. Acids. Res., 13*, 975-983.

Roitgrund, C., & Mets, L. J. (1990). Localization of two novel chloroplast genome functions: *trans*-splicing of RNA and protochlorophyllide reduction. *Curr. Genet., 17*, 147-153.

Saldanha, R., Mohr, G., Belfort, M., & Lambowitz, A. M. (1993). Group I and group II introns. *FASEB J., 7*, 15-24.

Semrad, K., & Schroeder, R. (1998). A ribosomal function is necessary for efficient splicing of the T4 phage thymidilate synthase intron *in vivo*. *Genes Dev., 12*, 1327-1337.

Sheveleva, E., & Hallick, R. (2004). Recent horizontal intron transfer to a chloroplast genome. *Nucl. Acids Res., 32*, 803-810.

Shukla, G., & Padgett, R. (2002). A catalytically active group II intron domain 5 can function in the U12-dependent spliceosome. *Molec. Cell, 9*, 1145-1150.

Sigel, R. K., Sashital, D. G., Abramovitz, D. L., Palmer, A. G., Butcher, S. E., & Pyle, A. M. (2004). Solution structure of domain 5 of a group II intron ribozyme reveals a new RNA motif. *Nat Struct Mol Biol, 11*(2), 187-192.

Simon, D., Fewer, D., Friedl, T., & Bhattacharya, D. (2003). Phylogeny and self-splicing ability of the plastid tRNA-Leu group I Intron. *J Mol Evol, 57*(6), 710-720.

Solem, A., Chatterjee, P., & Caprara, M. (2002). A novel mechanism for protein-assisted group I intron splicing. *RNA, 8*, 412-425.

Staley, J. P., & Guthrie, C. (1998). Mechanical devices of the spliceosome: motors, clocks, springs, and things. *Cell, 92*(3), 315-326.

Stern, D., Hanson, M., & Barkan, A. (2004). Genetics and genomics of chloroplast biogenesis: maize emerges as a model system. *Trends Plant Sci., in press.*

Sugiura, C., Kobayashi, Y., Aoki, S., Sugita, C., & Sugita, M. (2003). Complete chloroplast DNA sequence of the moss *Physcomitrella patens*: evidence for the loss and relocation of rpoA from the chloroplast to the nucleus. *Nucleic Acids Res, 31*(18), 5324-5331.

Swisher, J., Su, L., Brenowitz, M., Anderson, V., & Pyle, A. (2002). Productive folding to the native state by a group II intron ribozyme. *J. Mol. Biol., 315*, 297-310.

Thirumalai, D., & Woodson, S. (2000). Maximizing RNA folding rates: A balancing act. *RNA, 6*, 790-794.

Thompson, A., & Herrin, D. (1991). *In vitro* self-splicing reactions of the chloroplast group I intron Cr.LSU from *Chlamdomonas reinhardtii* and *in vivo* manipulation via gene-replacement. *Nucl. Acids Res., 19*, 6611-6618.

Thompson, M., Zhang, L., Hong, L., & Hallick, R. (1997). Extensive structural conservation exists among several homologs of two *Euglena* chloroplast group II introns. *Mo. Gen. Genet., 234*, 45-54.

Thomson, M. C., Macfarlane, J. L., Beagley, C. T., & Wolstenholme, D. R. (1994). RNA editing of mat-r transcripts in maize and soybean increases similarity of the encoded protein to fungal and

bryophyte group II intron maturases: evidence that mat-r encodes a functional protein. *Nucleic Acids Res, 22*(25), 5745-5752.
Till, B., Schmitz-Linneweber, C., Williams-Carrier, R., & Barkan, A. (2001). CRS1 is a novel group II intron splicing factor that was derived from a domain of ancient origin. *RNA, 7*, 1227-1238.
Toor, N., Hausner, G., & Zimmerly, S. (2001). Coevolution of group II intron RNA structures with their intron-encoded reverse transcriptases. *RNA, 7*, 1142-1152.
Treiber, D. K., & Williamson, J. R. (2001). Beyond kinetic traps in RNA folding. *Curr Opin Struct Biol, 11*(3), 309-314.
Turmel, M., Choquet, Y., Goldschmidt-Clermont, M., Rochaix, J. D., Otis, C., & Lemieux, C. (1995). The trans-spliced intron 1 in the *psa*A gene of the *Chlamydomonas* chloroplast: a comparative analysis. *Curr Genet, 27*(3), 270-279.
Turmel, M., Otis, C., & Lemieux, C. (2002). The chloroplast and mitochondrial genome sequences of the charophyte Chaetosphaeridium globosum: insights into the timing of the events that restructured organelle DNAs within the green algal lineage that led to land plants. *Proc Natl Acad Sci U S A, 99*, 11275-11280.
Turmel, M., Otis, C., & Lemieux, C. (2003). The mitochondrial genome of *Chara vulgaris:*Insights into the mitochondrial DNA architecture of the last common ancestor of green algae and land plants. *Plant Cell, 15*, 1888-1903.
Unseld, M., Marenfeld, J., Brandt, P., & Brennicke, A. (1997). The mitochondrial genome of *Arabidopsis thaliana* contains 57 genes in 366,924 nucleotides. *Nat. Genet., 15*, 57-61.
Valadkhan, S., & Manley, J. (2001). Splicing-related catalysis by protein-free snRNAs. *Nature, 413*, 701-707.
Villa, T., Pleiss, J., & Guthrie, C. (2002). Spliceosomal snRNAs: Mg^{2+} - dependent chemistry at the catalytic core? *Cell, 109*, 149-152.
Vogel, J., & Boerner, T. (2002). Lariat formation and a hydrolytic pathway in plant chloroplast group II intron splicing. *EMBO J, 21*, 3794-3803.
Vogel, J., Boerner, T., & Hess, W. (1999). Comparative analysis of splicing of the complete set of chloroplast group II introns in three higher plant mutants. *Nucl. Acids Res., 27*, 3866-3874.
Vogel, J., Hess, W., & Boerner, T. (1997a). Precise branch point mapping and quantification of splicing intermediates. *Nucleic Acids Res., 25*, 2030-2031.
Vogel, J., Hubschmann, T., Borner, T., & Hess, W. (1997b). Splicing and intron-internal RNA editing of *trn*K-*mat*K transcripts in barley plastids: support for MatK as an essential splice factor. *J Mol Biol, 270*, 179-187.
Wakasugi, T., Tsudzuki, J., Ito, S., Nakashima, K., Tsudzuki, T., & Sugiura, M. (1994). Loss of all *ndh* genes as determined by sequencing the entire chloroplast genome of the black pine *Pinus thunbergii*. *Proc Natl Acad Sci USA, 91*, 9794-9798.
Wakasugi, T., Tsudzuki, T., & Sugiura, M. (2001). The genomics of land plant chloroplasts: Gene content and alteration of genomic information by RNA editing. *Photosynthesis Res., 70*, 107-118.
Wank, H., SanFilippo, J., Singh, R., Matsuura, M., & Lambowitz, A. (1999). A reverse transcriptase/maturase promotes splicing by binding at its own coding segment in a group II intron RNA. *Mol. Cell, 4*, 239-250.
Weeks, K. M. (1997). Protein-facilitated RNA folding. *Curr. Opin. Struct. Biol., 7*, 336-342.
Weeks, K. M., & Cech, T. R. (1996). Assembly of a ribonucleoprotein catalyst by tertiary structure capture. *Science, 271*, 345-348.
Westhof, E. (2002). Group I introns and RNA folding. *Biochem Soc Trans, 30*(Pt 6), 1149-1152.
Wiesenberger, G., Waldherr, M., & Schweyen, R. (1992). The nuclear gene mrs2 is essential for the excision of group II introns from yeast mitochondrial transcripts *in vivo*. *J. Biol. Chem., 267*, 6963-6969.
Wolfe, K. H., Morden, C. W., & Palmer, J. D. (1992). Function and evolution of a minimal plastid genome from a nonphotosynthetic parasitic plant. *Proc. Natl. Acad. Sci. USA, 89*, 10648-10652.
Xu, M.-Q., Kaathe, S., Goodrich-Blair, H., Nierzwicki-Bauer, S., & Shub, D. (1990). Bacterial origin of a chloroplast intron: Conserved self-splicing group I introns in cyanobacteria. *Science, 250*, 1566-1569.
Zaita, N., Torazawa, K., Shinozaki, K., & Sugiura, M. (1987). *Trans*-splicing in vivo: joining of transcripts from the "divided" gene for ribosomal protein S12 in the chloroplasts of tobacco. *FEBS Lett, 210*, 153-156.

Zerges, W., & Rochaix, J. D. (1998). Low density membranes are associated with RNA-binding proteins and thylakoids in the chloroplast of Chlamydomonas reinhardtii. *J. Cell Biol., 140*, 101-110.

Zhang, L., & Doudna, J. (2002). Structural insights into Group II intron catalysis and branch-site selection. *Science, 295*, 2084-2088.

CHAPTER 12

TRANSLATIONAL MACHINERY IN PLANT ORGANELLES

LINDA BONEN

Biology Department, University of Ottawa, Ottawa, Canada K1N 6N5

Abstract. Chloroplasts and mitochondria have their own distinctive translation systems to direct synthesis of proteins encoded by the organellar genomes. The machinery, which includes ribosomal RNAs, ribosomal proteins, transfer RNAs, translation factors and aminoacyl-tRNA synthetases, is partly specified by organellar genes and partly by nuclear genes. The latter give rise to proteins synthesized in the cytoplasm (by the third distinct translation system in plant cells) and then imported into the appropriate organelle. The chloroplast and mitochondrial translation systems in flowering plants each are comprised of more than 100 different components, so that their complexity in fact exceeds the number of different protein-coding genes that they decode. The organellar systems retain traces of their endosymbiotic bacterial ancestry, the chloroplast to a greater extent than the mitochondrion; however, they have acquired a number of novel features and unique complexities. Of particular note is the mosaic composition of the organellar translation apparatus, arising in part from the recent intracellular movement of genetic information, as well as the sharing of certain nuclear gene products. More specifically, ribosomal protein genes have migrated from the organelle to the nucleus, chloroplast transfer RNA gene copies have moved to the mitochondrion, cytosol-type transfer RNAs are imported for use in the mitochondrion, and nuclear-encoded aminoacyl-tRNA synthetases are dual targeted to both the chloroplast and mitochondrion. In addition, there is evidence for intercellular plant-to-plant horizontal movement of mitochondrial ribosomal protein genes during recent evolution.

1. INTRODUCTION

1.1. Overview of translation in plant chloroplasts and mitochondria

The basic process of translation, whereby information in the nucleotide sequence of the messenger RNA is converted into protein is fundamentally the same among all life-forms. The ribosomes, comprised of the large subunit (LSU) and small subunit (SSU) with their ribosomal RNAs (rRNAs) and ribosomal proteins, ensure the correct positioning of the mRNA codons to base-pair with anticodons of the transfer RNAs (tRNAs), which carry the amino acids.

H. Daniell and C. D. Chase (eds.), Molecular Biology and Biotechnology of Plant Organelles, 323—345.

The mRNAs are read triplet by triplet in a 5' to 3' direction to generate the polypeptide chain, and this decoding property is provided by the small subunit. The large subunit contains the peptidyltransferase centre for peptide bond formation, and this reaction is catalyzed by RNA rather than proteins, so that the ribosome is a ribozyme (reviewed in Steitz and Moore, 2003). The tRNA binding sites in the ribosome are the A site (for aminoacyl-tRNA), P site (for peptidyl-tRNA) and E site (for the exiting deacylated tRNA). Accurate and efficient progression of these steps is mediated by translation factors, namely initiation factors (IF), elongation factors (EF) and release factors (RF). In bacteria, the 30S subunit forms a pre-initiation complex with mRNA. It is mediated by IF-1, IF-2, IF-3, tRNA-fMet and GTP, with IF-2 being the GTP binding protein which brings fMet-tRNA to the 30S subunit. IF-3 is the ribosome dissociation factor, which prevents the subunits from associating prematurely. The elongation factor EF-Tu delivers aminoacyl-tRNAs to the ribosome, and leaves when GTP is hydrolyzed. EF-G drives translocation of the tRNAs and mRNA after peptide bond formation, again with the hydrolysis of GTP. EF-Ts acts as the GTP exchange factor to prepare EF-Tu for the next round of elongation. The release factors terminate translation to liberate the newly synthesized protein. RF-1 decodes UAA and UAG, RF-2 decodes UAA and UGA, RF-3 releases the polypeptide chain and RRF, the ribosome recycling factor, dissociates the ribosome from the mRNA, which is particularly important for polycistronic mRNAs. The "charging" of tRNAs with the correct amino acids (by their ligation to the 3' ends of the tRNAs) is a crucial facet in the decoding process and is executed by the set of aminoacyl-tRNA synthetases.

Plant mitochondria and chloroplasts both exhibit certain bacterial-like features of protein synthesis, the chloroplast to a greater extent (as will be discussed below), and both use the standard genetic code, unlike certain other organellar systems such as animal or yeast mitochondria. However, there was a period of time in the 1980's when it was believed that there were several deviations from the conventional code in plant organelles. It is now known, however, that C-to-U type RNA editing "corrects" such codons to the standard genetic code.

Chloroplast messenger RNAs are typically synthesized as polycistronic transcripts, often with subsequent cleavage to monocistronic form (unlike in bacteria). In both chloroplasts and mitochondria, RNA processing (including splicing and RNA editing) plays an important role in generating mature mRNAs. Recognition of the initiation codon in the mRNA by the translational machinery is not yet fully understood. In chloroplasts, a bacterial-type "Shine-Dalgarno" base-pairing between the 3' end of the SSU rRNA and sequences preceding the initiation codon of the mRNA appears not to be universally used. In plant mitochondria, where the genomes are large and recombinogenic, relatively fewer genes are co-transcribed, and homologous genes from closely-related plants can have unrelated flanking (regulatory) sequences.

Plant organellar mRNAs are relatively long-lived compared to bacterial ones, although they lack eukaryotic-type 5' caps or polyA tails for stability. In fact, polyadenylation is involved in a bacterial-type RNA turnover pathway. Because translation occurs in the same compartment as transcription and RNA processing, this poses logistical problems such as the translation of unedited or unspliced transcripts and raises questions about temporal and spatial segregation of (membrane-associated) gene expression events in organelles.

This chapter will highlight novel aspects of translation and its machinery in plant chloroplasts and mitochondria, from both evolutionary and mechanistic perspectives. With respect to chloroplast translation, there are excellent reviews dealing with aspects of this topic (cf. Zerges 2000; Sugiura et al.1998). In contrast, relatively little is known about the specific nature of plant mitochondrial translation and its machinery, and so it has been discussed only briefly in recent gene expression reviews (.Binder and Brennicke 2003; Hoffmann et al. 2001, Kumar 1995).

1.2. Evolutionary origin of plant organellar translational machinery

The early discovery that chloroplasts and mitochondria have their own ribosomes, with sizes and properties uniquely distinctive from those in the cytosol, and hence their own protein synthesis systems, has played an important role in shaping our thinking over the years, both about the activities occurring within organelles, and indeed the evolutionary origin of these organelles (reviewed in Gray and Doolittle 1982; Lang et al. 1999). The bacterial-like features of the chloroplast and mitochondrial translation systems (cf. ribosomal RNA sequence similarity, fMet-tRNA as initiator tRNA, and antibiotic sensitivities, just to name a few) were instrumental in arriving at the widely accepted view that chloroplasts arose from a photosynthetic cyanobacterial-type lineage and mitochondria from a respiring □-proteobacterial-type lineage. Indeed the early sequence analysis of SSU ribosomal RNA (Bonen and Doolittle 1976; Gray and Doolittle 1982) provided compelling evidence for the endosymbiotic hypothesis. The plant mitochondrial rRNA data provided key information about the evolutionary origin of these organelles because the rate of nucleotide substitution is much lower in plant mitochondria than in fungal or animal mitochondria (Palmer and Herbon 1988). Thus their bacterial heritage is less obscured.

The early stage of endosymbiotic association must have involved massive transfer of genes to nucleus, with others that were no longer needed being lost (or perhaps recruited for new non-organellar functions). This is strongly supported by genomic sequence data (Adams and Palmer 2003; Timmis et al. 2004). Free-living cyanobacteria and □-proteobacteria typically contain several thousand protein-coding genes, whereas chloroplasts and mitochondria of flowering plants encode about 75 and 30, respectively. Because the ability of organelles to synthesis their own proteins is now only semi-autonomous, with some genes being in the nucleus and others retained in the organelle, this necessitates tight coordination of biosynthesis and assembly of the translational machinery. Notably, a large fraction of the genes in the organelles are devoted to translation. For example, of the 100 or so chloroplast genes in flowering plants, 30 are tRNA genes and ~20 are ribosomal protein genes. The plant mitochondrial gene set is smaller (about 60 genes in total), but approximately the same proportion is dedicated to the translational process, namely, about 20 tRNA genes and 5 to10 ribosomal protein genes in flowering plant mitochondria.

In addition to gene transfer events early in organellar evolution, successful movement of genes from the organelle to the nucleus is still ongoing in the plant lineage. The chloroplast systems are relatively conservative and stable in their gene content (and even gene order) among land plants, but lateral transfer from the mitochondrion to the nucleus appears to occur with remarkable frequency and in a punctuated fashion over evolutionary time (Adams et al. 2002b). Ribosomal protein genes appear particularly prone to movement, perhaps in part because they are typically hydrophilic (and many are quite short). Various stages of transfer have been documented, including "transition states" where potentially functional genes are found in both the nucleus and organelle, and cases of remnant pseudogenes not yet lost from the organelle. Remarkably, intercellular movement of mitochondrial genes between different plants appears to occur with the "recapture" of ribosomal protein genes previously lost from the mitochondrion, and even with the creation of chimeric "half-monocot, half-dicot" genes (Bergthorsson et al. 2003). Movement of copies of individual tRNA genes from the chloroplast to the mitochondrion has also occurred, and very recently, been based on plant-specific variation. Moreover, even though plant mitochondrial genomes are very large, they surprisingly do not encode a full set of tRNAs; some cytosol-type ones are imported into the organelle, and plant-specific distributions again are suggestive of bursts of genetic fluidity. This take-over by "alien" tRNAs must be accompanied by the appropriate aminoacyl-tRNA synthetases to be successful. Such events contribute to the mosaic nature of the plant mitochondrial translation machinery and raise intriguing questions as to how the fidelity and efficiency of such an essential fundamental cellular process is maintained.

2. MACHINERY FOR ORGANELLAR PROTEIN SYNTHESIS

2.1. Chloroplast translational machinery

Our initial knowledge about the nature of the chloroplast protein synthesis machinery dates back to the isolation of ribosomes and tRNAs in the 1970's and the discovery that they show particularly strong similarities in composition and mode of function to counterparts in cyanobacteria (reviewed in Gray and Doolittle 1982). Bacterial model systems and unicellular green algal plastid ones (such as Chlamydomonas and Euglena) have proven valuable for gaining insight into the processes in flowering plants (reviewed in Zerges 2000; Sugiura et al. 1998) and various approaches have included biochemical analysis (such as heterologous reconstitution studies with chloroplast/bacterial components), molecular genetic studies (such as complementation of *E. coli* mutants), and *in vitro* translation systems in tobacco (Hirose and Sugiura 1996). The availability of complete nuclear genome sequences enables powerful comparative analysis exploiting predictions based on the cyanobacterial origin of plastids and large-scale proteomic analysis of chloroplasts from flowering plants is also being carried out (cf. http://cbsusrv01.tc.cornell.edu /users/ppdb/).

2.1.1. Chloroplast ribosomes

The chloroplast 70S ribosomes are comprised of the 50S large subunit and the 30S small subunit, as in bacteria. In land plants, the chloroplast LSU rRNA is split into two pieces, designated as 23S and 4.5S rRNA, the latter corresponding to the 3' end of *E. coli* LSU (23S) rRNA. The large subunit also contains the 5S rRNA and 33 ribosomal proteins. The precision of the latter number is based on detailed analysis of the spinach chloroplast ribosome (Yamaguchi and Subramanian 2000). The chloroplast large subunit lacks homologues of *E.coli* L25 and L30 proteins, which incidentally are not universally present among eubacteria (Lecompte et al. 2002) and it has been proposed that L25 function may have been taken over by L22. On the other hand, the chloroplast contains two "plastid-specific" ones designated PSRP-5 and PSRP-6 which show no detectable homology to any proteins from other organisms.

The small subunit of the chloroplast ribosome contains the 16S rRNA and 25 ribosomal proteins (Yamaguchi et al. 2000) and includes homologues for all 21 ribosomal proteins found in the *E.coli* small subunit. Three of the four extra ones (PSRP-1, PSRP-2, PSRP-3) are related to cyanobacterial proteins, and PSRP-2 has RNA binding (RRM) domains, with a proposed possible role in recruiting stored mRNAs for active protein synthesis (Yamaguchi and Subramanian, 2003). This would be analogous to the role of translational activators in yeast mitochondria that deliver mRNAs to the translation system on the inner mitochondrial membrane (Naithani et al. 2003). Notably, PSRP-2, PSRP-3, and PSRP-5 were among sixteen translation-type proteins identified in recent Arabidopsis proteomic analysis of the chloroplast thylakoid membrane fraction (Friso et al. 2004).

Chloroplast ribosomes also resemble bacterial ones in their sensitivities to antibiotics which inhibit protein synthesis (reviewed in Sugiura et al. 1998). Moreover, the use of translational inhibitors (such as spectinomycin) has demonstrated the key role of plastid protein synthesis in tobacco cell division and development (cf. Ahlert et al. 2003). Nuclear mutants defective in protein synthesis are also providing insight into the nature of the machinery and its biosynthesis. For example, an Arabidopsis mutant (DAL) exhibits high accumulation of precursor 16S and 23S rRNA species, and this lack of rRNA processing is correlated with defects in chloroplast differentiation (Bisanz et al. 2003). Among a collection of maize nuclear mutants with defects in polysome assembly (Barkan 1993), a PPR-type (pentatricopeptide repeat) protein gene (*ppr2*) has been identified and may play a role in the synthesis or assembly of the chloroplast translational apparatus (Williams and Barkan 2003). Chloroplast ribosome-deficient mutants (such as barley albostrians) have also been used to assess the contribution of plastid protein synthesis to RNA processing events, one example being the maturation of intron-containing tRNAs (Vogel et al.1999).

2.1.2. Chloroplast translation factors

The chloroplast translation factors for initiation, elongation and termination resemble those in bacteria, and insight into their roles has benefited from studies using unicellular model systems (reviewed in Sugiura et al. 1998). For example, IF-2 (the GTP binding protein which brings fMet-tRNA to the 30S subunit) from the common bean was recently demonstrated to complement an *E.coli* mutant (Campos et al. 2001). Translation factors may have multiple functions in the chloroplast; for example, the peptide release factor RF-2 (for UGA triplets, which are relatively frequent in flowering plant chloroplasts) appears to play a role in mRNA stability in Arabidopsis (Meurer et al. 2002). Similarly, the ribosome recycling factor RRF is present in the plastid 70S ribosome in a stoichiometry of approximately one (so much higher than for *E.coli*), suggesting that it may have additional regulatory functions (Yamaguchi and Subramanian, 2000). Incidentally, this protein is abundantly represented in the thylakoid membrane fraction (Friso et al. 2004), consistent with membrane-associated translation.

2.1.3. Chloroplast transfer RNAs

The chloroplasts of flowering plants have a set of 30 different tRNAs which are encoded in the chloroplast genome, and as in bacteria, fMet-tRNA acts as the initiator tRNA. This set is adequate for translation of the 61 sense codons used in chloroplast coding sequences, even without tRNAs to recognize CUPy (Leu), CCPy (Pro), GCPy (Ala) or CGC/G (Arg), assuming that "two-out-of-three" or "U:N wobble" recognition mechanisms are used (reviewed in Sugiura et al.1998). Additionally, some tRNAs may be imported. Indeed this has been proposed for non-photosynthetic parasitic plants such as *Epifagus viriginia*, which have "degenerate" chloroplast genomes lacking certain tRNA genes (Lohan and Wolfe 1998).

The mechanisms of biosynthesis and post-transcriptional modification of chloroplast tRNAs show strong similarities to the (cyano)bacterial process, but one notable difference is the nature of RNAse P, which generates the mature 5' end of tRNAs. It typically has ribozymic function, but in flowering plant chloroplasts, the enzymatic activity appears to be solely protein-based (Thomas et al. 2000).

2.2. Mitochondrial translational machinery

Our knowledge of the precise nature and mode of action of the plant mitochondrial translational machinery has lagged behind that of the chloroplast, partly because of the lack of unicellular model systems or *in vitro* translation systems. It is notable that very early *in organello* studies revealed differences in mitochondrial protein synthesis in normal maize vs. cytoplasmic male sterile mutants (Forde et al.1978). In addition, although plant mitochondrial translation appears more bacteria-like than in animals or fungi, it is less so than the chloroplast, so comparative analysis is somewhat less powerful. Moreover, the mosaic nature of the machinery - due to recent gene transfer (or gene replacement) events - adds another level of complexity. Large-scale proteomic analysis of Arabidopsis mitochondrial proteins is under way, and a recent study revealed 15 polypeptides in the "protein synthesis" category

(Heazlewood et al. 2004). Data are available at sites such as http://www.mitoz.bcs.uwa.edu.au/apmdb/APMDB_Database.php.

2.2.1. Mitochondrial ribosomes

Plant mitochondrial (78S-80S) ribosomes are larger than their bacterial 70S counterparts, as are their LSU and SSU rRNAs (cf. 26S and 18S rRNAs in plant mitochondria vs. 23S and 16S rRNAs in bacteria/plastids). Plant mitoribosomes also clearly contain a mitochondrial-encoded 5S rRNA, unlike in yeast and animals where a cytosol-type 5S rRNA may be imported (cf. Magalhaes et al. 1998). Two-dimensional gel electrophoretic analysis of mitochondrial ribosomes from potato suggests that the large and small subunits contain at least 33 and 35 proteins, respectively (Pinel et al. 1986). For broad bean mitoribosomes about 80 species in total were observed (Maffey et al. 1997).

This is similar to the number (namely, 79) in mammalian mitoribosomes (reviewed in O'Brien 2002), but the latter have no detectable homologues to 12 E.coli ones. In contrast, homologues to virtually all 54 of the *E.coli* ribosomal protein genes are present either in the organelle or nucleus in flowering plants (see below). The yeast mitoribosome also contains ~80 ribosomal proteins (Graack and Wittmann-Leopold 1998), though all have yet to be characterized (Saveanu et al. 2001). In Arabidopsis mitochondrial proteomic analysis, L3, L7/L12 and L25 were among the fifteen "protein-synthesis" type ones characterized (Heazlewood et al. 2004).

Plant mitochondrial ribosomal proteins show a complex evolutionary history, with regard to sequence origin and gene location. For example the L2 protein, which is regarded as one of the most "universal" and conservative ribosomal proteins in nature, in rice mitochondria has an internal insertion that doubles its length relative to bacterial (or chloroplast) homologues (Kubo et al. 1996). More remarkably, in some plants (such as Arabidopsis) it is fractured into two parts, with one in the nucleus and the other in the mitochondrion (Adams et al. 2001). This necessitates exquisite co-ordination for their assembly into the mitoribosome. The mitochondrial S2 has long carboxy-terminal extensions (differing in sequence among cereals), which are proteolytically cleaved prior to assembly in the ribosomal subunit (Perrotta et al. 2002). Incidentally, the mitochondrial *rps2* and *rps11* genes appear to have been "recaptured" through plant-to-plant lateral gene transfer in several lineages (Berthorsson et al. 2003).

This study also revealed a curious chimeric half-dicot, half-moncot structure for the mitochondrial S11 protein in *Sanguinaria* (bloodroot), again attributed to horizontal gene flow. The nuclear-encoded mitochondrial S19 protein in Arabidopsis has an extra RNA-binding domain, originally proposed to be a functional substitute for S13 (whose gene could not be found in either the mitochondrial or nuclear genome) (Sanchez et al. 1996). More recent studies have demonstrated, however, that a duplicated (and divergent) copy of the nuclear-encoded chloroplast S13 is imported into the mitochondrion in Arabidopsis (Adams et al. 2002a; Mollier et al. 2002).

The essential nature of mitochondrial ribosomal proteins in translation is also indirectly illustrated by mutants, such as non-chromosomal stripe (NCS3) in maize (Hunt and Newton, 1991) and maternal distorted leaf (MDL) mutants in Arabidopsis (Sakamoto et al. 1996). Both show deletion/rearrangements within the *rps3/rpl16* locus, correlated with impaired mitochondrial protein synthesis. Similarly, mutations in Arabidopsis nuclear genes for mitochondrial ribosomal proteins show severe effects. For example, a transposon-induced knockout of the *rps16* gene causes an embryo-defective lethal phenotype (Tsugeki et al. 1996) and a *rpl14* mutant (HUELLENLOS) shows defective ovule development (Skinner et al. 2001).

2.2.2. Mitochondrial translation factors

Recent mitochondrial proteomic studies in Arabidopsis have identified bacterial-like translation factors and associated machinery, all of which are encoded by nuclear genes. They include mitochondrial elongation factors EF-Tu and EF-Ts, a putative ribosome recycling factor, tRNA adenylyltransferase and methionyl-tRNA synthetase (Heazlewood et al. 2004). In addition, bioinformatic analysis of the Arabidopsis nuclear genome yields candidates for other bacterial-type mitochondrial translational components, such as initiation factors (IF-2 and IF-3) and elongation factor EF-G. Interestingly, there is no detectable homologue for a mitochondrial-type IF-1 in the Arabidopsis nuclear genome, and import experiments have suggested that the nuclear-encoded chloroplast IF-1 protein is not dual targeted to the mitochondrion (Millen et al. 2001).

2.2.3. Mitochondrial transfer RNAs

The set of transfer RNAs in flowering plant mitochondria has the distinction of being comprised of three different categories: native tRNAs derived from the □-proteobacterial-type ancestor, chloroplast-like tRNAs derived from promiscuous chloroplast DNA integrated into the mitochondrial genome, and cytosol-type tRNAs, which are nuclear-encoded and imported into the mitochondrion. Notably, the specific composition of the tRNA set differs among plants (reviewed in Small et al. 1999; Dietrich et al. 1996).

This suggests an ongoing "capture-and-loss" of alien tRNA species for use in mitochondrial protein synthesis. Interestingly, a tRNA-Phe(GAA) gene copy has been converted into a tRNA-Tyr(GUA) gene in Arabidopsis mitochondria, and a novel tRNA-Cys(GCA) of unknown origin is present in the sugar beet mitochondrial genome (Kubo et al. 2000). Nevertheless, each set is adequate for translation of all 61 sense codons used in mitochondrial protein coding sequences. As in bacteria and chloroplasts, fMet-tRNA is the initiator tRNA.

A comprehensive study of the tRNA population present in potato mitochondria identified 31 distinct tRNA species, of which 20 are encoded in the mitochondrial genome (with 15 native and 5 chloroplast-like species) and eleven species are imported cytosol-type (Marechal-Drouard et al.1990). Similar studies in wheat mitochondria have revealed 14 distinct imported tRNA species (Glover et al. 2001), in addition to 9 native and 6 chloroplast-like ones encoded within the mitochondrion (Joyce and Gray 1989).

This remarkable plasticity raises intriguing questions as to how tRNAs are imported into the mitochondrion, and how the appropriate machinery is provided for their biosynthesis/maturation (which can include C-to-U editing), as well as their correct charging by aminoacyl-tRNA synthetases (reviewed in Small et al. 1999; Dietrich et al. 1996).

3. GENES FOR ORGANELLAR TRANSLATIONAL MACHINERY

3.1. Chloroplast genes for translational machinery

Certain components of the chloroplast translational machinery are encoded within the organelle whereas others are specified by nuclear genes. The chloroplast genomes have been completely sequenced for a number of flowering plants, and information is available at websites http://megasun.bch.umontreal. ca/gobase/ and www.ncbi.nlm.nih.gov/genomes/ORGANELLES/organelles.html. Table 1 gives a summary of translational machinery components which are encoded in the chloroplasts of two flowering plants, Arabidopsis [NC_000932] and rice [NC_001320], as well as the bryophyte *Marchantia polymorpha* [X04465].The mitochondrial-encoded translational machinery is discussed below.

Table 1. *Translational machinery encoded in plant organellar genomes*

		Chloroplast			*Mitochondria*	
	Arabidopsis	**Rice**	**Marchantia**	**Arabidopsis**	**Rice**	**Marchantia**
Ribosomal RNAs LSU rRNA	+	+	+	+	+	+
SSU rRNA	+	+	+	+	+	+
5S rRNA	+	+	+	+	+	+
4.5S rRNA	+	+	+			
Transfer RNAs tRNA-Ala (UGC)	+	+	+			+
tRNA-Arg (UCU)	+	+	+			+
tRNA-Arg (ACG)	+	+	+			+
tRNA-Arg (CCG)			+			
tRNA-Arg (UCG)						+
tRNA-Asn (GUU)	+	+	+	+ (c)	+ (c)	+
tRNA-Asp (GUC)	+	+	+	+ (c)	+ (m)	+
tRNA-Cys (GCA)	+	+	+	+ (m)	+ (c)	+
tRNA-Gln (UUG)	+	+	+	+ (m)	+ (m)	+
tRNA-Glu (UUC)	+	+	+	+ (m)	+ (m)	+
tRNA-Gly (GCC)	+	+	+	+ (m)		+
tRNA-Gly (UCC)	+	+	+			+
tRNA-His (GUG)	+	+	+	+ (c)	+ (c)	+
tRNA-Ile (CAU)	+	+	+	+ (m)	+ (m)	+
tRNA-Ile (GAU)	+	+	+			
tRNA-Leu (UAA)	+	+	+			+
tRNA-Leu (CAA)	+	+	+			+
tRNA-Leu (UAG)	+	+	+			+
tRNA-Lys (UUU)	+	+	+	+ (m)	+ (m)	+
tRNA-Met (CAU)	+	+	+	+ (c)	+ (m), + (c)	+
tRNA-fMet (CAU)	+	+	+	+ (m)	+ (m)	+
tRNA-Phe (GAA)	+	+	+		+ (c)	+
tRNA-Pro (UGG)	+	+	+	+ (m)	+ (m)	+
tRNA-Pro (GGG)			+			
tRNA-Ser (GCU)	+	+	+	+ (m)	+ (m)	+
tRNA-Ser (UGA)	+	+	+	+ (m)	+ (m)	+

Table 1 (cont.)

Table 1 (cont.)

	Chloroplast			Mitochondria		
	Arabidopsis	Rice	Marchantia	Arabidopsis	Rice	Marchantia
tRNA-Ser (GGA)	+	+	+	+ ©	+ ©	
tRNA-Thr (GGU)	+	+	+			+
tRNA-Thr (UGU)	+	+	+			
tRNA-Trp (CCA)	+	+	+	+ ©	+ ©	+
tRNA-Tyr (GUA)	+	+	+	+ (m)	+ (m)	+
tRNA-Val (UAC)	+	+	+			+
tRNA-Val (GAC)	+	+	+			
Ribosomal proteins S1					+	+
S2	+	+	+		+	+
S3	+	+	+	+	+	+
S4	+	+	+	+	+	+
S7	+	+	+	+	+	+
S8	+	+	+			+
S10						+
S11	+	+	+		□	+
S12	+	+	+	+	+	+
S13					+	+
S14	+	+	+	□	□	+
S15	+	+	+			
S16	+	+	+			
S18	+	+	+			
S19	+	+	+	□	+	+
L2	+	+	+	+ (5' part)	+	+
L5				+	+	+
L6						+
L14	+	+	+			
L16	+	+	+	+	+	+
L20	+	+	+			
L21			+			
L22	+	+	+			
L23	+	+	+			
L32	+	+	+			
L33	+	+	+			
L36	+	+	+			
Translation factors IF-1	□	+	+			

The chloroplast ribosomal RNA genes have an operon-type organization (16SrRNA – tRNA-Ile – tRNA-Ala – 23SrRNA – 4.5SrRNA – 5SrRNA) identical to that typically found in cyanobacteria. This cluster is usually located on the inverted repeat in the chloroplast genome, a notable exception being in legumes which lack an inverted repeat. The ribosomal RNAs undergo post-transcriptional modifications, including several key ones near the peptidyltransferase center. It is as yet unknown if small guide RNAs are involved, but a recent search for non-coding RNAs in Arabidopsis revealed seven chloroplast-encoded ones (Marker et al. 2002), including a 55-nt one (Ath-243), which intriguingly maps between the 4.5S rRNA and 5S rRNA genes.

A set of 20-21 ribosomal protein genes is typically present in plant chloroplast genomes with occasional lineage-specific differences. For example, the gene for L22 is in the nucleus in legumes (Gantt et al. 1991) and a cytosol-type L23 protein replaces the chloroplast homologue in spinach (Bubunenko et al. 1994; Yamaguchi and Subramanian 2000). Ten of the ribosomal protein genes are physically clustered in the chloroplast genome (namely those for L23, L2, S19, L22, S3, L16, L14, S8, L36, and S11) and this order resembles the *Synechococcus* "L3 cluster". It appears to be an ancient fusion of the *E. coli* S10-spc-□ operons (reviewed in Sugiura et al.1998), although some are missing because relocated elsewhere in the chloroplast genome or transferred to the nucleus. Four of the chloroplast ribosomal protein genes typically have group II (ribozymic-family) introns (namely, *rps12*, *rps16*, *rpl2*, *rpl16*), and notably the *rps12* gene has undergone DNA rearrangements, so that dispersed segments are independently transcribed and the mRNA generated through trans-splicing.

The set of 30 different tRNA genes vary in organization, with some being clustered and others dispersed. Six of them contain group I or group II introns, and the tRNA-Lys(UUU) intron encodes a RNA maturase (matK) involved in splicing. Incidentally, the group I intron in the chloroplast tRNA-Leu(UAA) gene prompted the discovery of introns at the homologous site in (cyano)bacteria.

All of the chloroplast translation factors, except for IF-1, are nuclear-encoded (and imported), and even the IF-1 (*infA*) gene is not universally present in the chloroplast genomes of flowering plants. Rather, it has undergone multiple independent transfers to the nucleus during angiosperm evolution and holds the distinction of being "by far the most mobile chloroplast gene known in plants" (Millen et al. 2001). Interestingly, when present in the chloroplast, the *infA* gene is located between S8 and L36 within the large "L3 cluster" of ribosomal protein genes, and in some plants (such as Arabidopsis) there is a remnant pseudogene.

3.2. Mitochondrial genes for translational machinery

The set of translational machinery components encoded within the mitochondrion in flowering plants is somewhat smaller than for the chloroplast, and Table 1 gives a summary for Arabidopsis [NC_001284], rice [AB076665, AB076666] and *Marchantia polymorpha* [M68929]. Also unlike the chloroplast, where gene order is well-conserved during evolution, plant mitochondrial genomes are highly recombinogenic and gene organization (and even gene presence) can differ among closely-related plants. Hence, traces of ancestral operon-type organization have been eroded. One clear example being the separation of the LSU rRNA gene from the SSU rRNA gene and the atypical linkage of SSU rRNA – 5S rRNA genes (Bonen and Gray, 1980). In contrast, *Marchantia* mitochondria retains the bacterial-type gene order of LSU rRNA-5S rRNA-SSU rRNA. The transfer RNA genes are typically dispersed in the mitochondrial genome, sometimes co-transcribed with other genes, and pseudo-tRNA gene copies are also found in spacer regions. There are both native and chloroplast-like tRNA genes, and they are designated by (m) and © respectively in Table 1. The latter sometimes retain flanking homology to the chloroplast progenitor sequence (suggestive of a DNA mode of transfer from the chloroplast to the mitochondrion), whereas the absence in other cases is compatible with RNA-mediated transfer.

The mitochondrial rRNAs undergo several important post-transcriptional modifications (pseudouridylation and methylation), including key ones near the peptidyltransferase center, and homologues of the yeast mitochondrial modifying enzymes (cf. Pintard et al. 2002) are present in the Arabidopsis nuclear genome. It is unknown whether small guide RNAs are involved in organellar rRNA maturation, but Rnomics studies may be useful in uncovering candidates. For example, eight mitochondrial-encoded small non-coding RNAs were recently identified in Arabidopsis (Marker et al. 2002).

A total of fourteen ribosomal protein genes have been found in the mitochondrial genomes of various flowering plants, although usually only a subset is present in any given plant (Adams et al. 2002b), and these fourteen are in turn a subset of the 16 in the Marchantia mitochondrial genome (Takemura et al. 1992). It should be noted that S10 (although absent from Table 1) is present in some angiosperm mitochondrial genomes (cf. Adams et al. 2000). Traces of bacterial-like gene order are sometimes seen, for example, the *rps3-rpl16-rpl5-rps14* gene linkage in Brassica, but DNA rearrangements have resulted in *rps3-rpl16* being physically distant from *rpl5-□rps14* in its relative, Arabidopsis. Ribosomal protein genes are often co-transcribed with other genes, and there are instances where ribosomal protein pseudogenes appear to provide regulatory sequences for co-transcribed functional genes (Subramanian et al. 2001). RNA maturation can include splicing (cf. group II introns in *rpl2*, *rps3* and *rps10*) as well as the expected C-to-U editing within protein coding sequences. Interestingly, ribosomal protein mRNAs appear somewhat less heavily edited (cf. < 10 sites per kb) than certain other categories (cf. ~ 20 sites per kb for NADH dehydrogenase mRNAs) in Arabidopsis (Giege and Brennicke 1999), and ribosomal protein mRNAs sometimes lack predicted editing, such as *rps1* in Oenothera (Mundel and Schuster 1996). Such features might, in some cases, reflect a "transition state" of reduced functional constraint due to the presence of gene copies in both the organelle and nucleus.

As discussed above, successful gene transfer to the nucleus is still going on in flowering plants, and a large-scale Southern hybridization survey of 280 genera of angiosperms suggests multiple independent losses, with a remarkable inferred frequency of 6 to 42 times for the 14 ribosomal protein genes (Adams et al. 2002b). In certain plants, such as *Lachnocaulon* (bog button), no ribosomal protein homologous sequences were detected in the mitochondrion, whereas in others, certain truncated (or otherwise inactivated) pseudogenes have been found (Table 1). Interestingly, in the case of wheat L5, potentially functional copies are present in both the nucleus and the mitochondrion, suggestive of a "transition state" (Sandoval et al. 2004). Certain ribosomal protein genes although apparently functional (cf. wheat S13) can exhibit relatively low steady state levels of mRNAs (Bonen, 1987; Takvorian et al. 1997). Moreover, as mentioned above, gene transfer events can result in non-native ribosomal proteins being used in plant mitochondrial protein synthesis, some examples being "chimeric monocot-dicot" S11, "recaptured" S2, and "alien" S7 proteins (reviewed in Adams and Palmer 2003; Bergthorsson et al. 2003).

3.3. Nuclear genes for organellar translational machinery

3.3.1. Nuclear-encoded proteins for organellar translation

In contrast to the situation for organelle-encoded gene products involved in protein synthesis, our knowledge about nuclear-encoded translational machinery is incomplete. Bioinformatic analysis of completely-sequenced plant nuclear genomes and large-scale proteomic studies are under way. Such analysis is assisted by targeting algorithms (such as SignalP, Predotar, Psort) accompanied by functional tests, such as *in vivo* GFP-fusion import (cf. Peeters and Small 2001; Timmis et al. 2004). Table 2 gives a summary of candidate Arabidopsis chloroplast and mitochondrial ribosomal proteins, which are homologous to the 54 in *E. coli* ribosomes. Notably, the chloroplast set lacks L25 and L30 (Yamaguchi and Subramanian 2000), and for seven of the nine missing mitochondrial ones, there are nuclear-encoded chloroplast homologues, raising the opportunity for dual targeting (or recruitment of duplicated gene copies). Identification is complicated by the presence of multi-gene families, members of which may have been recruited for non-organellar functions (cf. Timmis et al. 2004) or show tissue-specific differences, as for rice mitochondrial L11 (Handa et al. 2001).

Typically, the nuclear-encoded organelle-destined ribosomal protein sequence is preceded by an amino-terminal targeting sequence, either fused to the coding region or linked by an intron in the gene. For example, all of the nuclear-encoded chloroplast ribosomal proteins in spinach have N-terminal transit peptide signals (Yamaguchi and Subramanian 2000). The lower proportion of SSU ones being nuclear-encoded (Table 2) has been suggested to be related to a bias in retention in the organelle of those involved in early stages of ribosome assembly and direct interactions with rRNAs. In some cases, the regulatory sequences have been recruited from duplications of pre-existing genes (or parts thereof). For example, in Arabidopsis nucleus, the chloroplast S9 and L12 have similar transit peptides (Arimura et al.1999), as do the nuclear-located mitochondrial L5 and L4 in wheat (Sandoval, et al. 2004).

Table 2. *Candidate bacterial-type organellar ribosomal proteins encoded in Arabidopsis nuclear DNA*

LSU	*Chloroplast*	*Mitochondria*	*SSU*	*Chloroplast*	*Mitochondria*
L1*	NP_191908	NP_181799	**S1**	NP_850903	?
L2*	chl	mt (5'), NP_566007 (3')	**S2***	chl	AAG03026
L3*	NP_181831	NP_566579	**S3***	chl	mt
L4*	NP_563786	NP_565463	**S4***	chl	mt
L5*	NP_192040	mt	**S5***	NP_180936	NP_564842
L6*	NP_172011	NP_565438	**S6**	NP_176632	BAA94995
L7/L2*	NP_189421	NP_564986	**S7***	chl	mt
L9	NP_190075	NP_200119	**S8***	chl	CAB79701
L10*	NP_196855	NP_187843	**S9***	NP_177635	NP_190477
L11*	NP_174575	NP_195274	**S10***	NP_187919	P42797
L13*	NP_177984	NP_186828	**S11***	chl	AAG50732
L14*	chl	AAL60452	**S12***	chl	mt
L15*	NP_189221	NP_201272	**S13***	NP_568299	AAG51625
L16	chl	mt	**S14***	chl	CAB40383
L17	NP_190989	NP_568216	**S15***	chl	AAF27115
L18*	NP_566655	NP_198134	**S16**	chl	NP_195188
L19	NP_568677	NP_564213	**S17***	NP_178103	NP_175365
L20	chl	NP_173118	**S18**	chl	NP_172201

Table 2 (cont.)

Table 2 (cont.)

LSU	Chloroplast	Mitochondria	SSU	Chloroplast	Mitochondria
L21	NP_174808	NP_567861	S19*	chl	NP_568681
L22*	chl	NP_567805	S20	NP_188137	?
L23*	chl	NP_195698	S21	NP_566809	?
L24*	NP_200271	BAA97247			
L25	-	?			
L27	NP_198911	AAC64229			
L28	NP_565765	?			
L29*	NP_201325	NP_172261			
L30	-	NP_568821			
L31	NP_565109	?			
L32	chl	?			
L33	chl	NP_187283			
L34	NP_174202	?			
L35	NP_850047	?			
L36	chl	NP_197518			

* universally present in eubacteria, archaea and eukaryotic cytosol ribosomes (Lecompte et al. 2002). Compiled by G. Voros and L. Bonen.

The organelle-destined proteins do not always possess the expected amino-terminal targeting signals; for example, the mitochondrial-type S10 in the nucleus of spinach, maize, and Oxalis (all apparently derived from independent transfer events) have none, in contrast to those of carrot and Fuchsia, which can be traced back to duplicated HSP22 and HSP70 pre-sequences, respectively (Adams et al. 2000). The nuclear-located mitochondrial *rps14* gene illustrates yet another clever strategy, namely "hitch-hiking" within an intron in the *sdh2* gene, and expression through alternative splicing (Figueroa et al. 1999; Kubo et al. 1999). After import, the fused N-terminal SDH2 stretch is removed by proteolytic cleavage (Figueroa et al. 2000). As mentioned above, L2 gene structure is very unusual in some plants, with one part in the nucleus and the other in the mitochondrion (as in Arabidopsis) or with both parts separately encoded in the nucleus (cf. soybean and Medicago) (Adams et al. 2001).

Nuclear-encoded translational proteins can also be shared between compartments in plant cells, either directly (via dual targeting) or indirectly (through recruitment of duplicated copies). For example, the Arabidopsis chloroplast L21 is nuclear-encoded, but phylogenetic analysis indicates that it is of mitochondrial origin (Gallois et al. 2001). The spinach chloroplast L23 gene has been functionally replaced by a cytosol-type L23 counterpart (Bubunenko et al. 1994), and the mitochondrial S8 gene has been lost, with the cytosol homologue (named S15e) being targeted to the mitochondrion (Adams et al. 2002a). A duplicated (and divergent) copy of the nuclear gene for chloroplast S13 has substituted for mitochondrial S13 in Arabidopsis (Adams et al. 2002a; Mollier et al. 2002). Similarly there are numerous cases of tRNA machinery being shared. For example, certain aminoacyl-tRNA synthetases are targeted to both the mitochondrion and chloroplast (such as methionine-tRNA synthetase in Arabidopsis). While others are shared by both the cytosolic and organellar compartments (such as alanyl-tRNA synthetase) (reviewed in Peeters and Small 2001; Small et al. 1999). The presence of such "non-native" enzymes in the mitochondrion correlates well with the chimeric origin of tRNAs (chloroplast-like and cytosol-imported), which must be correctly "charged" with the appropriate amino acids to be used in the mitochondrion.

3.3.2. Nuclear-encoded transfer RNAs for organellar translation

Approximately one-third of the different tRNAs used in mitochondrial protein synthesis are imported cytosol-type ones (see above), and plant-specific differences suggest that functional substitution is a dynamic, ongoing process (reviewed in Dietrich et al. 1996; Small et al. 1999). Although the mechanism of import is not yet well understood, a protein component on the mitochondrial surface (but not the cytosolic fraction) is required, as well as ATP and a membrane potential, based on *in vitro* import studies with potato mitochondria (Delage et al. 2003). It is worth noting that in many unicellular eukaryotes (such as Chlamydomonas, Euglena, and trypanosomes), tRNAs are imported into the mitochondrion. Even the large mitochondrial genome in Marchantia lacks a full set, with at least three different tRNAs being imported, namely, tRNA-Ile(AAU), tRNA-Thr(AGU), and tRNA-Val(AAC), the latter co-existing with a mitochondrial-encoded tRNA-Val(UAC) species (Akashi et al. 1998). Also as mentioned above, import of tRNAs into the chloroplast is predicted for certain non-photosynthetic parasitic plants, which have incomplete gene sets in their atrophied plastid genomes (Lohan and Wolfe 1998).

4. TRANSLATION INITIATION IN PLANT ORGANELLES

4.1. Initiation of translation in chloroplasts

The mechanism whereby the initiation codons of chloroplast mRNAs are recognized by the translational machinery is under investigation and appears to differ somewhat from bacteria. In *E. coli*, a key step in the formation of the translation initiation complex (whereby the 30S subunit is correctly positioned at the initiation codon of the mRNA) is base-pairing interactions between a pyrimidine-rich stretch near the 3' end of the SSU (16S) rRNA and a purine-rich Shine-Dalgarno (SD) motif (GGAGG) located slightly upstream (approximately -5 to -9 nt) of the initiation codon in the mRNA. The S1 ribosomal protein is also important for mRNA recognition and binding to the 30S subunit in *E. coli*. As shown in Figure 1, plant chloroplast 16S rRNAs have a conserved anti-SD sequence at their 3' termini, as do cyanobacteria and □-proteobacteria.

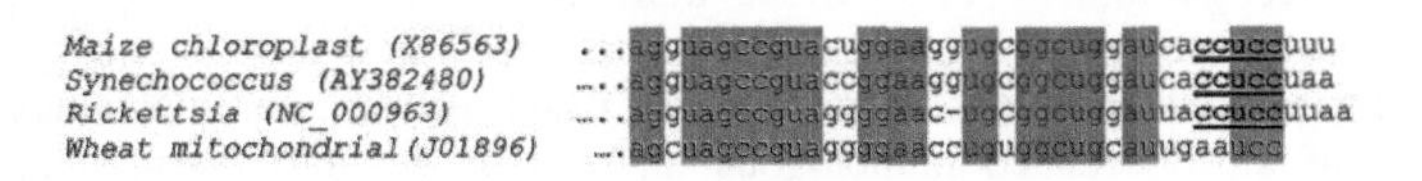

Maize chloroplast (X86563) ...agguagccguacuggaaggugcggcuggaucaccuccuuu
Synechococcus (AY382480) ...agguagccguaccggaaggugcggcuggaucaccuccuaa
Rickettsia (NC_000963) ...agguagccguaggggaac-ugcggcuggauuaccuccuuaa
Wheat mitochondrial(J01896) ...agcuagccguaggggaaccuguggcugcauugaaucc

Figure 1. Alignment of the 3' terminal sequences of SSU ribosomal RNAs from maize chloroplast, *Synechococcus* (cyanobacteria), *Rickettsia* (□-proteobacteria) and wheat mitochondria. The precise 3' end of the wheat mitochondrial 18S rRNA was experimentally determined by RNA sequencing (Schnare and Gray 1982). Identical positions are shaded and the anti-Shine-Dalgarno sequence is underlined in bold.

About two-thirds of the chloroplast protein-coding genes have SD-like purine-rich sequences in the 5' UTRs of their mRNAs, but not all are at the expected position (reviewed in Zerges 2000; Sugiura et al. 1998). This may influence their role. For example i*n vitro* analysis in tobacco chloroplasts suggests a bacterial-type interaction when the SD-like sequences are at the conventional location (as for *rps14* and *rbcL*), but not if positioned either farther away (*rps12*) or too close to the initiation codon (*petB*) (Hirose and Sugiura, 2004). Interestingly, a conventional SD-like sequence may also act as a negative regulatory element in the case of tobacco chloroplast *rps2* mRNA (Plader and Sugiura, 2003). A role for the SD-like sequences in light-regulated translation (cf. *psbA*) has been of particular interest (cf. Kim and Mullet 1994; Hirose and Sugiura 1996). Thus it appears that chloroplast 5' UTRs contain various cis-elements which potentially interact with specific (nuclear-encoded) proteins for translation initiation and regulation.

In Chlamydomonas chloroplast, as in bacteria, there is experimental support for an extended codon-anticodon interaction between A-37 of tRNA-fMet and U at position -1 relative to the initiation codon of the mRNA (Esposito et al. 2003), and the initiation codon context and higher order RNA structure is expected to also be important for specificity or fidelity of translation in plants. Atypical initiation codons are occasionally used, such as GUG for *rps19* and *psbC*, or UUG for *infA* (reviewed in Sugiura et al. 1998), and the initiation codons of several other genes (namely *ndhD*, *psbL*, and *rpl2*) are created by editing of ACG to AUG (reviewed in Maier et al. 1996), posing another potential regulatory control point.

In *E. coli*, the S1 protein plays an important role in initial mRNA alignment in the ribosome, through RNA binding repeats in a carboxy-terminal domain. However, the Chlamydomonas chloroplast S1 protein (which has a shorter extension, as do those in flowering plants) appears to have different RNA-binding properties (Shteiman-Kotler and Schuster 2000). Moreover, it has been proposed that other ribosomal proteins (such as S7) may play an important role in translation initiation or regulation in the Chlamydomonas chloroplast (cf. Fargo et al. 2001). It is worth noting that flowering plants and Chlamydomonas show certain differences in their chloroplast ribosomal machinery, for example the latter possess a novel S1-domain type protein, as well as larger S2, S3 and S5 proteins, compared to chloroplasts in flowering plants (Yamaguchi et al. 2002).

4.2. Initiation of translation in plant mitochondria

The mode of recognition of the correct translation initiation site in plant mitochondria is as yet unknown. The SSU rRNA does not possess the bacterial-type anti-Shine-Dalgarno sequence at its 3' end (Figure 1), and sequence variation among the 5' UTRs of closely-related homologous mRNAs brings into question any conserved role for the sequence preceding initiation codons (cf. Boer et al. 1985). On the other hand, the 5' UTRs of mRNAs for different genes sometimes share blocks of sequence similarity, which are derived from duplication/rearrangement events (Pring et al. 1992). For example, a purine-rich loose consensus sequence and pyrimidine-rich stretches farther upstream, are shared by the 5' UTRs of *cox2*, *atp6* and *orf25*, raising the possibility that they serve a role in translation initiation or regulation. RNA editing sites are also occasionally observed in the 5' UTRs of mRNAs, and thus are potential candidates for translation regulation. For example, there is an editing site within the 3 nt spacer between the *rpl2* and *rps19* coding sequences in rice mitochondria (Kubo et al. 1996).

As in the chloroplast, GUG initiators are occasionally used (cf. *rpl16*) and RNA editing sometimes creates initiation codons (reviewed in Maier et al. 1996). The S1 protein, which in *E.coli* assists in aligning the mRNA in the ribosome for initiation (see above), lacks this C-terminal repeated domain in plant mitochondria (cf. Mundel and Schuster 1996). By analogy to yeast mitochondria and plant chloroplasts, it is anticipated that gene-specific translational activators may be important in membrane-associated translation (cf. Naithani et al. 2003; Zerges 2000).

5. TRANSLATION AND RNA MATURATION IN PLANT ORGANELLES

The co-existence of transcription, RNA processing and translation in the same compartment raises questions about their temporal and spatial relationships. The issue of whether unedited (or partially edited) transcripts undergo translation has been addressed using antibodies specific to S12 proteins derived from edited vs. unedited mRNAs in maize and petunia mitochondria (Phreaner et al. 1996; Lu et al. 1996). Polymorphic S12 proteins were observed, although only "edited-type" proteins appeared to be assembled into functioning ribosomes. Recent *in vivo* studies in tobacco chloroplast revealed that unedited *ndhD* transcripts (lacking an AUG start codon) were associated with polysomes, raising the additional possibility that ACG might act as an initiation codon (Zandueta-Criado and Bock 2004). In contrast, unedited-type S13 translation products (as monitored by editing-state-specific antibodies) were not detected in maize mitochondria, but this might reflect rapid turnover of polypeptides rather than exclusion of transcripts from translation (Williams et al. 1998). Studies using chloroplast ribosome-deficient mutants (barley albostrians) also suggest that RNA editing is not dependent on translation (reviewed in Maier et al. 1996).

6. MULTI-FUNCTIONAL POTENTIAL OF ORGANELLAR TRANSLATION MACHINERY

It has long been appreciated that translational machinery components (ribosomal proteins in particular) are adept at performing multiple functions (Wool 1996). In plant organelles, these proteins are well placed for "moonlighting" roles as RNA binding proteins in events such as editing, splicing or RNA stability. One particularly pertinent example is the recruitment of tyrosyl-tRNA synthetase for group I intron splicing in Neurospora mitochondria (and leucyl-tRNA synthetase in yeast) (cf. Myers et al. 2002). In addition, a tRNA modifying-type enzyme (with cytidine deaminase activity) is often cited as a candidate for providing C-to-U type editing. Other examples include a role in the feedback control of transcription by the chloroplast L4 ribosomal protein, based on its ability to functionally substitute for the *E.coli* L4 homologue in NusA-dependent attenuation of the "S10" operon (Trifa and Lerbs-Mache 2000).

Variation in the stoichiometries of certain mitochondrial ribosomal proteins during development (cf. L29 and L21 levels elevated relative to L2) suggests that different subpopulations of ribosomes may have different functions or certain ribosomal proteins may be co-opted for other cellular functions (Zhao et al.1999). With respect to the "plastid-specific" PSRP chloroplast ribosomal proteins, various extra roles have been proposed, including RNA processing/ stabilization of transcripts or acting as translator activators in delivering mRNAs to (thylakoid-membrane-associated) translational machinery (Yamaguchi and Subramanian, 2003). Similarly, ribosomal proteins with extra RNA binding domains (such as Arabidopsis mitochondrial S19) are candidates for participating in RNA metabolism events.

7. CONCLUDING REMARKS

The translational machinery in plant organelles is revealing a fascinating mixture of ancestral prokaryotic-like features, as well as idiosyncratic recently-acquired traits. The former are particularly well illustrated by the chloroplast system, whereas there is a remarkably mosaic composition of the plant mitochondrial translation machinery. It seems rather paradoxical that one of the most conservative and fundamental life processes, namely protein synthesis, is able, in plant organelles, to tolerate the switching of components with such apparent ease. For example, chloroplast-like or cytosol tRNAs can replace native mitochondrial ones, and core ribosomal proteins can be lost or replaced by functional substitutes. As yet little is known about crosstalk between the chloroplast and mitochondrion, but the sharing of certain translational machinery components by the two organelles (whether through lateral transfer or dual targeting of nuclear gene products) sets the stage for co-ordinated regulation. Indeed in Chlamydomonas there is intriguing genetic evidence of inter-dependence between chloroplast and mitochondrial protein synthesis (cf. Bennoun and Delosme 1999). It will be interesting to see what surprises await in the further uncovering of translation mechanisms used by plant organelles to ensure that proteins essential for photosynthetic and respiratory function are properly synthesized.

An increased understanding of the mechanisms which lead to efficient protein synthesis in plant organelles also has important practical applications, one example being in the design of effective transformation systems to enable hyper-expression of foreign proteins such as vaccines or biopharmaceuticals in the chloroplasts of transgenic plants (Fernandez San-Millan et al., 2003; Molina et al., 2004; Daniell et al., 2004; Chebolu and Daniell, 2004, also see chapter 16). Similarly, enhanced translation resulted in hyper-expression of the CRY insecticidal protein (up to 46% of total leaf protein) and conferred the highest level of tolerance against insects reported so far (DeCosa et al., 2001). Enhanced translation of the *badh* gene in transgenic chloroplasts also resulted in very high levels of salt tolerance (up to 400 mM) and accumulated as much betaine as in halophytes (Kumar et al., 2004). Enhanced translation of a plastid-relocated copy of *rbcS* in tobacco antisense mutant led to the restoration of Rubisco assembly and photosynthesis (Dhingra et al. 2004), and such work has exciting implications for improving crop productivity.

8. REFERENCES

Adams, K.L., Palmer, J.D. (2003) Evolution of mitochondrial gene content: gene loss and transfer to the nucleus. *Mol. Phylogenet. Evol. 29*, 380-395.

Adams, K.L., Daley, D.O., Qiu, Y.L., Whelan, J., & Palmer, J.D. (2000) Repeated, recent and diverse transfers of a mitochondrial gene to the nucleus in flowering plants. *Nature 408*, 354-357.

Adams, K.L., Ong, H.C., & Palmer, J.D. (2001) Mitochondrial gene transfer in pieces: fission of the ribosomal protein gene *rpl2* and partial or complete gene transfer to the nucleus. *Mol. Biol. Evol. 18*, 2289-2297.

Adams, K.L., Daley, D.O., Whelan, J., & Palmer, J.D. (2002a) Genes for two mitochondrial ribosomal proteins in flowering plants are derived from their chloroplast or cytosolic counterparts. *Plant Cell 14*, 931-943.

Adams, K.L., Qiu, Y.L., Stoutemyer, M., & Palmer, J.D. (2002b) Punctuated evolution of mitochondrial gene content: high and variable rates of mitochondrial gene loss and transfer to the nucleus during angiosperm evolution. *Proc. Natl. Acad. Sci. USA 99*, 9905-9912.

Ahlert, D., Ruf, S., & Bock, R. (2003) Plastid protein synthesis is required for plant development in tobacco. Proc. Natl. Acad. Sci. USA 100, 15730-15735.

Akashi, K., Takenaka, M., Yamaoka, S., Suyama, Y., Fukuzawa, H., & Ohyama, K. (1998) Coexistence of nuclear DNA-encoded tRNA-Val(AAC) and mitochondrial DNA-encoded tRNA-Val(UAC) in mitochondria of a liverwort *Marchantia polymorpha*. *Nucl. Acids Res. 26*, 2168-2172.

Arimura, S., Takusagawa, S. Hatano, S., Nakazono, M., Hirai, A., & Tsutsumi, N. (1999) A novel plant nuclear gene encoding chloroplast ribosomal protein S9 has a transit peptide related to that of rice chloroplast ribosomal protein L12. *FEBS Lett. 450*, 231-234.

Barkan, A. (1993) Nuclear mutants of maize with defects in chloroplast polysome assembly have altered chloroplast RNA metabolism. *Plant Cell 5*, 389-402.

Bennoun, P., & Delosme, M. (1999) Chloroplast suppressors that act on a mitochondrial mutation in *Chlamydomonas reinhardtii*. *Mol. Gen. Genet. 262*, 85-89.

Bergthorsson, U., Adams, K.L., Thomason, B., & Palmer, J.D. (2003) Widespread horizontal transfer of mitochondrial genes in flowering plants. *Nature 424*, 197-201.

Binder, S., & Brennicke, A. (2003) Gene expression in plant mitochondria: transcriptional and post-transcriptional control. *Phil. Trans. R. Soc. Lond. B 358*, 181-189.

Bisanz, C., Begot, L., Carol, P., Perez, P., Bligny, M., Pesey, H., Gallois, J.L., Lerbs-Mache, S., & Mache, R. (2003) The Arabidopsis nuclear DAL gene encodes a chloroplast protein which is required for the maturation of the plastid ribosomal RNAs and is essential for chloroplast differentiation. *Plant Mol. Biol. 51*, 651-663.

Boer, P.H., McIntosh, J.E., Gray, M.W., & Bonen, L. (1985) The wheat mitochondrial gene for apocytochrome b: absence of a prokaryotic ribosome binding site. *Nucl. Acids Res. 13*, 2281-2292.

Bonen, L. (1987) The mitochondrial S13 ribosomal protein gene is silent in wheat embryos and seedlings. *Nucl. Acids Res. 15*, 10393-10404.

Bonen, L., & Doolittle, W.F. (1976) Partial sequences of 16S rRNA and the phylogeny of blue-green algae and chloroplasts. *Nature 261*, 669-673.

Bonen, L., & Gray M.W. (1980) The genes for wheat mitochondrial ribosomal RNA and tRNA: evidence for an unusual arrangement. *Nucl. Acids Res. 8*, 319-335.

Bubunenko, M.G., Schmidt, J., & Subramanian, A.R. (1994) Protein substitution in chloroplast ribosome evolution: a eukaryotic cytosolic protein has replaced its organelle homologue (L23) in spinach. *J. Mol. Biol. 240*, 28-41.

Campos, F., Garcia-Gomez, B.I., Solorzano, R.M., Salazar, E., Estevez, J., Leon, P., Alvarez-Buylla, E.R., & Covarrubias, A.A. (2001) A cDNA for nuclear-encoded chloroplast translational initiation factor 2 from a higher plant is able to complement an *infB Escherichia coli* null mutant. *J. Biol. Chem. 276,* 28388-28394.

Chebolu, S & Daniell, H. (2004). Chloroplast derived vaccine antigens and biopharmaceuticals: expression, folding, assembly and functionality. Current Trends in Microbiology and Immunology, in press.

Daniell, D., Carmona-Sanchez, O., & Burns, B. B. (2004). Chloroplast derived antibodies, biopharmaceuticals and edible vaccines. In R. Fischer & S. Schillberg (Eds.) *Molecular Farming* (pp.113-133). Weinheim: WILEY-VCH Verlag.

DeCosa, B., Moar, W., Lee, S. B., Miller, M., & Daniell, H. (2001). Overexpression of the *Bt* cry2Aa2 operon in chloroplasts leads to formation of insecticidal crystals. *Nat. Biotechnol., 19,* 71-74.

Delage, L., Dietrich, A., Cosset, A., & Marechal-Drouard, L. (2003) In vitro import of a nuclearly encoded tRNA into mitochondria of *Solanum tuberosum. Mol. Cell. Biol. 23,* 4000-4012.

Dhingra, A., Portis, A.R. & Daniell, H. (2004). Enhanced translation of a chloroplast expressed *RbcS* gene restores SSU levels and photosynthesis in nuclear antisense *RbcS* plants. *Proc. Natl. Acad. Sci.,* U.S.A.101: 6315-6320.

Dietrich, A., Small, I., Cosset, A., Weil, J.H., & Marechal-Drouard, L. (1996) Editing and import: strategies for providing plant mitochondria with a complete set of functional transfer RNAs. *Biochimie 78,* 518-529.

Esposito,D., Fey, J.P., Eberhard, S., Hicks, A.J., & Stern D.B. (2003) *In vivo* evidence for the prokaryotic model of extended codon-anticodon interaction in translation initiation. *EMBO J. 22,* 651-656.

Fargo, D.C., Boynton, J.E., & Gillham, N.W. (2001) Chloroplast ribosomal protein S7 of Chlamydomonas binds to chloroplast mRNA leader sequences and may be involved in translation initiation. *Plant Cell 13,* 207-218.

Fernandez-San Millan, A., Mingeo-Castel, A. M., Miller, M., & Daniell, H. (2003). A chloroplast transgenic approach to hyper-express and purify human serum albumin, a protein highly susceptible to proteolytic degradation. *Plant Biotechnology Journal 1,* 71-79.

Figueroa, P., Holuigue, L., Araya, A., & Jordana, X. (2000) The nuclear-encoded SDH2-RPS14 precursor is proteolytically processed between SDH2 and RPS14 to generate maize mitochondrial RPS14. Biochem. Biophys. *Res. Comm. 271,* 380-385.

Figueroa, P., Gomez, I., Holuigue, L., Araya, A., & Jordana, X. (1999) Transfer of *rps14* from the mitochondrion to the nucleus in maize implied integration within a gene encoding the iron-sulphur subunit of succinate dehydrogenase and expression by alternative splicing. *Plant J. 18,* 601-609.

Forde, B.G., Oliver, R.J., & Leaver, C.J. (1978) Variation in mitochondrial translation products associated with male-sterile cytoplasms in maize. *Proc. Natl. Acad. Sci. USA 75,* 3841-3845.

Friso, G., Giacomelli, L., Ytterberg, A.J., Peltier, J.B., Rudella, A., Sun, Q., & van Wijk, K.J. (2004) In-depth analysis of the thylakoid membrane proteome of *Arabidopsis thaliana* chloroplasts: new proteins, new functions, and a plastid proteome database. *Plant Cell 16,* 478-499.

Gallois, J.L., Achard, P., Green, G., & Mache, R. (2001) The *Arabidopsis* chloroplast ribosomal protein L21 is encoded by a nuclear gene of mitochondrial origin. *Gene 274,* 179-185.

Gantt, J.S., Baldauf, S.L., Calie, P.J., Weeden, N.F., & Palmer, J.D. (1991) Transfer of *rpl22* to the nucleus greatly preceded its loss from the chloroplast and involved the gain of an intron. *EMBO J. 10,* 3073-3078.

Giege, P., & Brennicke, A. (1999) RNA editing in *Arabidopsis* mitochondria effects 441 C to U changes in ORFs. *Proc. Natl. Acad. Sci. USA 96,* 15324-15329.

Glover, K.E., Spencer, D.F., & Gray, M.W. (2001) Identification and structural characterization of nucleus-encoded transfer RNAs imported into wheat mitochondria. *J. Biol. Chem. 276,* 639-648.

Graack, H.R., & Wittmann-Liebold, B. (1998) Mitochondrial ribosomal proteins (MRPs) of yeast. *Biochem. J. 329,* 433-448.

Gray, M.W., & Doolittle, W.F. (1982) Has the enodsymbiont hypothesis been proven? *Microbiol. Rev. 46,* 1-42.

Handa, H., Kobayashi-Uehara, A., & Murayama, S. (2001) Characterization of a wheat cDNA encoding mitochondrial ribosomal protein L11: qualitative and quantitative tissue-specific differences in its expression. *Mol. Gen. Genomics 265,* 569-575.

Heazlewood, J.L., Tonti-Filippini, J.S., Gout, A.M., Day, D.A., Whelan, J., & Millar, A.H. (2004) Experimental analysis of the Arabidopsis mitochondrial proteome highlights signaling and regulatory components, provides assessment of targeting prediction programs, and indicates plant-specific mitochondrial proteins. *Plant Cell 16,* 241-256.

Hirose, T., & Sugiura, M. (1996) Cis-acting elements and trans-acting factors for accurate translation of chloroplast *psbA* mRNAs: development of an *in vitro* translation system from tobacco chloroplasts. *EMBO J. 15*, 1687-1695.

Hirose, T., & Sugiura, M. (2004) Functional Shine-Dalgarno-like sequences for translational initiation of chloroplast mRNAs. *Plant Cell Physiol. 45,* 114-117.

Hoffmann, M., Kuhn, J., Daschner, K., & Binder, S. (2001) The RNA world of plant mitochondria. *Prog. Nucl. Acid Res. Mol. Biol.* ***70,*** 119-154.

Hunt, M.D., & Newton, K.J. (1991) The NCS3 mutation: genetic evidence for the expression of ribosomal protein genes in *Zea mays* mitochondria. *EMBO J. 10,* 1045-1052.

Joyce, P.B.M., & Gray, M.W. (1989) Chloroplast-like transfer RNA genes expressed in wheat mitochondria. *Nucl. Acids Res. 17,* 5461-5476.

Kim, J., & Mullet, J.E. (1994) Ribosome-binding sites on chloroplast *rbcL* and *psbA* mRNAs and light-induced initiation of D1 translation. *Plant Mol. Biol. 25*, 437-448.

Kubo, N., Ozawa, K., Hino, T., & Kadowaki, K. (1996) A ribosomal protein L2 gene is transcribed, spliced and edited at one site in rice mitochondria. *Plant Mol. Biol. 31*, 853-862.

Kubo, N., Harada, K., Hirai, A., & Kadowaki, K. (1999) A single nuclear transcript encoding mitochondrial RPS14 and SDHB of rice is processed by alternative splicing: common use of the same mitochondrial targeting signal for different proteins. *Proc. Natl. Acad. Sci. USA* 96, 9207-9211.

Kubo, T., Nishizawa, S., Sugawara, A., Itchoda, N., Estiati, A., & Mikami, T. (2000) The complete nucleotide sequence of the mitochondrial genome of sugar beet (Beta vulgaris L.) reveals a novel gene for tRNA-Cys (GCA). *Nucl. Acids Res. 28,* 2571-2576.

Kumar, R. (1995) Mitochondrial ribosomes and their proteins. In Levings, C.S. and Vasil, I.K. (Eds.) *The Molecular Biology of Plant Mitochondria* (pp.131-184). Hingham, MA: Kluwer Academic.

Kumar, S., Dhingra, A. & Daniell, H. (2004). Plastid expressed *betaine aldehyde dehydrogenase* gene in carrot cultured cells, roots and leaves confers enhanced salt tolerance. *Plant Physiol* in press.

Lang, B.F., Gray, M.W., Burger, G. (1999) Mitochondrial genome evolution and the origin of eukaryotes. *Ann. Rev. Genet. 33*, 351-397.

Lecompte, O., Ripp, R., Thierry, J.C., Moras, D., & Poch, O. (2002) Comparative analysis of ribosomal proteins in complete genomes : an example of reductive evolution at the domain scale. *Nucl. Acids Res. 30,* 5382-5390.

Lohan, A.J., & Wolfe, K.H. (1998) A subset of conserved tRNA genes in plastid DNA of nongreen plants. *Genetics 150,* 425-433.

Lu, B., Wilson, R.K., Phreaner, C.G., Mulligan, R.M., & Hanson, M.R. (1996) Protein polymorphism generated by differential RNA editing of a plant mitochondrial *rps12* gene. *Mol. Cell. Biol. 16,* 1543-1549.

Maffey, L., Degand, & H, Boutry, M. (1997) Partial purification of mitochondrial ribosomes from broad bean and identification of proteins encoded by the mitochondrial genome. *Mol. Gen. Genet. 254,* 365-371.

Magalhaes, P.J., Andreu, A.L., & Schon, E.A. (1998) Evidence for the presence of 5S rRNA in mammalian mitochondria. *Mol. Biol. Cell 9,* 2375-2382.

Maier, R.M., Zeltz, P., Kossel, H., Bonnard, G., Gualberto, J.M., & Grienenberger, J.M. (1996) RNA editing in plant mitochondria and chloroplasts. *Plant Mol. Biol. 32,* 343-365.

Marechal-Drouard, L., Guillemaut, P., Cosset, A., Arbogast, M., Weber, F., Weil, J.H., & Dietrich, A. (1990) Transfer RNAs of potato (Solanum tuberosum) mitochondria have different genetic origins. *Nucl. Acids Res. 18,* 3689-3696.

Marker, C., Zemann, A., Terhorst, T., Kiefmann, M., Kastenmayer, J.P., Green, P., Bachellerie, J.P., Brosius, J., & Huttenhofer, A. (2002) Experimental RNomics: identification of 140 candidates for small non-messenger RNAs in the plant *Arabidopsis thaliana. Curr. Biol. 12,* 2002-2013.

Meurer, J., Lezhneva, L., Amann, K., Godel, M., Bezhani. S., Sherameti, I., & Oelmuller, R. (2002) A peptide chain release factor 2 affects the stability of UGA-containing transcripts in *Arabidopsis* chloroplasts. *Plant Cell 14,* 3255-3269.

Millen, R.S., Olmstead, R.G., Adams, K.L., Palmer, J.D., Lao, N.T., Heggie, L., Kavanagh, T.A., Hibberd, J.M., Gray, J.C., Morden, C.W., Calie, P.J., Jermiin, L.S., & Wolfe, K.H. (2001) Many parallel losses of *infA* from chloroplast DNA during angiosperm evolution with multiple independent transfers to the nucleus. *Plant Cell 13,* 645-658.

Mollier, P., Hoffmann, B., Debast, C., & Small, I. (2002) The gene encoding *Arabidopsis thaliana* mitochondrial ribosomal protein S13 is a recent duplication of the gene encoding plastid S13. *Curr. Genet. 40*, 405-409.

Molina, A., Herva-Stubbs, S., Daniell, H., Mingo-Castel, A. M., & Veramendi, J. (2004). High yield expression of a viral peptide animal vaccine in transgenic tobacco chloroplasts. *Plant Biotechnol. Journal, 2*, 141-153.

Mundel, C., & Schuster, W. (1996) Loss of RNA editing of rps1 sequences in Oenothera mitochondria. *Curr. Genet. 30*, 455-460.

Myers, C.A., Kuhla, B., Cusack, S.,& Lambowitz, A.M. (2002) tRNA-like recognition of group I introns by a tyrosyl-tRNA synthetase. *Proc. Natl. Acad. Sci. USA 99*, 2630-2635.

Naithani, S., Saracco, S.A., Butler, C.A., & Fox, T.D. (2003) Interactions among COX1, COX2, and COX3 mRNA-specific translational activator proteins on the inner surface of the mitochondrial inner membrane of *Saccharomyces cerevisiae*. *Mol. Biol. Cell 14*, 324-333.

O'Brien, T.W. (2002) Evolution of a protein-rich mitochondrial ribosome: implications for human genetic disease. *Gene 286*, 73-79.

Palmer, J.D., & Herbon, L.A. (1988) Plant mitochondrial DNA evolves rapidly in structure, but slowly in sequence. *J. Mol. Evol. 28*, 87-97.

Peeters, N., & Small, I. (2001) Dual targeting to mitochondria and chloroplasts. *Biochim. Biophys. Acta* 1541, 54-63.

Perrotta, G., Grienenberger, J.M., & Gualberto, J.M. (2002) Plant mitochondrial *rps2* genes code for proteins with a C-termimal extension that is processed. *Plant Mol. Biol. 50*, 523-533.

Phreaner, C.G., Williams, M.A., & Mulligan, R.M. (1996) Incomplete editing of *rps12* transcripts results in the synthesis of polymorphic polypeptides in plant mitochondria. *Plant Cell 8*, 107-117.

Pinel, C., Douce, R., & Mache, R. (1986) A study of mitochondrial ribosomes from the higher plant *Solanum tuberosum*. *Mol. Biol. Rep. 11*, 93-97.

Pintard, L. Bujnicki, J.M., Lapeyre, B., & Bonnerot, C. (2002) MRM2 encodes a novel yeast mitochondrial 21S rRNA methyltransferase. *EMBO J. 21*, 1139-1147.

Plader, W., & Sugiura, M. (2003) The Shine-Dalgarno-like sequence is a negative regulatory element for translation of tobacco chloroplast *rps2* mRNA: an additional mechanism for translational control in chloroplasts. *Plant J. 34*, 377-382.

Pring, D.R., Mullen, J.A., & Kempken, F. (1992) Conserved sequence blocks 5' to start codons of plant mitochondrial genes. *Plant Mol. Biol. 19*, 313-317.

Sakamoto, W., Kondo, H., Murata, M., & Motoyoshi, F. (1996) Altered mitochondrial gene expression in a maternal distorted leaf mutant of Arabidopsis induced by *chloroplast mutator*. *Plant Cell 8*, 1377-1390.

Sanchez, H., Fester, T., Kloska, S., Schroder, W., & Schuster, W. (1996) Transfer of *rps19* to the nucleus involves the gain of an RNP-binding motif which may functionally replace RPS13 in *Arabidopsis* mitochondria. *EMBO J. 15*, 2138-2149.

Sandoval, P., Leon, G., Gomez, I., Carmona, R., Figueroa, P., Holuigue, L., Araya, A., & Jordana, X. (2004) Transfer of *RPS14* and *RPL5* from the mitochondrion to the nucleus in grasses. Gene 324, 139-147.

Saveanu, C., Fromont-Racine, M., Harington, A., Ricard, F., Namane, A., & Jacquier, A. (2001) Identification of 12 new yeast mitochondrial proteins including 6 that have no prokaryotic homologues. *J. Biol. Chem. 276*, 15861-15867.

Schnare, M.N., & Gray, M.W. (1982) 3'terminal sequence of wheat mitochondrial 18S ribosomal RNA: further evidence of a eubacterial evolutionary origin. *Nucl. Acids Res. 10*, 3921-3932.

Shteiman-Koller, A., & Schuster, G. (2000) RNA-binding characteristics of the chloroplast S1-like ribosomal protein CS1. *Nucl. Acids Res. 28*, 3310-3315.

Skinner, D.J., Baker, S.C., Meister, R.J., Broadhvest, J., Schneitz, K., & Gasser, C.S. (2001) The *Arabidopsis* HUELLENLOS gene, which is essential for normal ovule development, encodes a mitochondrial protein. *Plant Cell 13*, 2719-2730.

Small, I. Akashi, K., Chapron, A., Dietrich, A., Duchene, A.M., Lancelin, D., Marechal-Drouard, L., Menard, B., Mireau, H., Moudden, Y., Ovesna, J., Peeters, N., Sakamoto, W., Souciet, G., & Wintz, H. (1999) The strange evolutionary history of plant mitochondrial tRNAs and their aminoacyl-tRNA synthetases. *J. Hered. 90*, 333-337.

Steitz, T.A., & Moore, P.B. (2003) RNA, the first macromolecular catalyst: the ribosome is a ribozyme. *Trends Biochem. Sci. 28*, 411- 418.

Subramanian, S., Fallahi, M., & Bonen, L. (2001) Truncated and dispersed *rpl2* and *rps19* pseudogenes are co-transcribed with neighbouring downstream genes in wheat mitochondria. *Curr. Genet. 39*, 264-272.

Sugiura, M., Hirose, T., & Sugita, M. (1998) Evolution and mechanism of translation in chloroplasts. *Ann. Rev. Genet. 32*, 437-59.

Takemura, M., Oda, K., Yamato, K., Ohta, E., Nakamura, Y., Nozato, N., Akashi, K., & Ohyama, K. (1992) Gene clusters for ribosomal proteins in the mitochondrial genome of a liverwort, *Marchantia polymorpha*. *Nucl. Acids Res. 20*, 3199-3205.

Takvorian, A., Coville, J.L., Haouazine-Takvorian, N., Rode, A., & Hartmann, C. (1997) The wheat mitochondrial *rps13* gene: RNA editing and co-transcription with the *atp6* gene. *Curr. Genet. 31*, 497-502.

Timmis, J.N., Ayliffe, M.A., Huang, C.Y., & Martin, W. (2004) Endosymbiotic gene transfer: organelle genomes forge eukaryotic chromosomes. *Nat. Rev. Genet. 5*, 123-136.

Thomas, B.C., Li, X., & Gegenheimer, P. (2000) Chloroplast ribonuclease P does not utilize the ribozyme-type pre-tRNA cleavage mechanism. *RNA 6*, 545-553.

Trifa, Y., & Lerbs-Mache, S. (2000) Extra-ribosomal function(s) of the plastid ribosomal protein L4 in the expression of ribosomal components in spinach. *Mol. Gen. Genet. 263*, 642-647.

Tsugeki, T.R., Kochieva, E.Z., & Fedoroff, N.V. (1996) A transposon insertion in the Arabidopsis SSR16 gene causes an embryo-defective lethal mutation. *Plant J. 10*, 479-489.

Vogel, J., Borner, T., & Hess, W.R. (1999) Comparative analysis of splicing in the complete set of chloroplast group II introns in three higher plant mutants. *Nucl. Acids Res. 27*, 3866-3874.

Williams, P.M., & Barkan, A. (2003) A chloroplast-localized PPR protein required for plastid ribosome accumulation. *Plant J. 36*, 675-686.

Williams, M.A., Tallakson, W. A., Phreaner, C.G., & Mulligan, R.M. (1998) Editing and translation of ribosomal protein S13 transcripts: unedited translation products are not detectable in maize mitochondria. *Curr. Genet. 34*, 221-226.

Wool, I.G. (1996) Extraribosomal functions of ribosomal proteins. *Trends Biochem. Sci. 21*, 164-165.

Yamaguchi, K., & Subramanian, A.R. (2000) The plastid ribosomal proteins: Identification of all the proteins in the 50S subunit of an organelle ribosome (chloroplast) *J. Biol. Chem. 275*, 28466-28482.

Yamaguchi, K., & Subramanian, A.R. (2003) Proteomic identification of all plastid-specific ribosomal proteins in higher plant chloroplast 30S ribosomal subunit. *Eur. J. Biochem. 270*, 190-205.

Yamaguchi, K., von Knoblauch, K., & Subramanian, A.R. (2000) The plastid ribosomal proteins: Identification of all the proteins in the 30S subunit of an organelle ribosome (chloroplast) *J. Biol. Chem. 275*, 28455-28465.

Yamaguchi, K., Prieto, S., Beligni, M.V., Haynes, P.A., McDonald, W.H., Yates, J.R., & Mayfield, S.P. (2002) Proteomic characterization of the small subunit of Chlamydomonas reinhardtii chloroplast ribosome: identification of a novel S1 domain-containing protein and unusually large orthologs of bacterial S2, S3 and S5. *Plant Cell 14*, 2957-2974.

Zandueta-Criado, A., & Bock, R. (2004) Surprising features of plastid *ndhD* transcripts: addition of non-encoded nucleotides and polysome association of mRNAs with an unedited start codon. *Nucl. Acids Res. 32*, 542-550.

Zerges, W. (2000) Translation in chloroplasts. *Biochimie 82*, 583-601.

Zhao, Y.Y., Xu, T., Zucchi, P., & Bogorad, L. (1999) Subpopulations of chloroplast ribosomes change during photoregulated development of *Zea mays* leaves: ribosomal proteins L2, L21 and L29. *Proc. Natl. Acad. Sci. USA 96*, 8997-9002.

CHAPTER 13

REGULATION OF TRANSLATION IN CHLOROPLASTS

By Light and for Protein Complex Assembly

WILLIAM ZERGES

Biology Department, Concordia University, Montreal, Canada

Abstract Translation is an important regulated step in the expression of chloroplast genomes. This chapter covers our understanding of how and why translation is regulated by light during chloroplast differentiation and in mature chloroplasts. Experimental approaches and results are emphasized. Light is perceived through its effects on specific intermediates and products of photosynthesis (e.g., the redox states of electron carriers, the stromal ADP and ATP concentrations, and the electrochemical proton gradient across thylakoid membranes) rather than by specialized light receptors. Some of the translational regulatory factors and molecular mechanisms by which these factors respond to light-dependent biochemical signals are being elucidated. Translational regulation is imposed at the phases of initiation and elongation and at the transition between these phases. Translation of certain chloroplast mRNAs has direct roles in the assembly and the repair of the multi-subunit integral membrane complexes of the photosynthetic apparatus. In at least one case, translation occurs concurrently with assembly of the nascent polypeptide into a complex. A hierarchical system involving translational control determines the order in which chloroplast mRNAs are translated to ensure the production of subunits in the correct stoichiometry and possibly for their proper folding and insertion into an assembling complex. Chloroplast mRNA-specific translational regulators might impose this regulation. Current debates address where in chloroplasts mRNAs are translated and complexes are assembled and how processes involving translation of chloroplast mRNAs could constrain chloroplast genome evolution.

1. INTRODUCTION

Translation is arguably the most important regulated step in the expression of chloroplast genomes. Many more examples have been described of translational regulation, with effects of greater magnitudes than at other levels of gene expression. This probably

H. Daniell and C.D. Chase (eds.), Molecular Biology and Biotechnology of Plant Organelles, 347—383.

reflects the effectiveness of translational control in determining the levels of proteins encoded by mRNAs with the long half-lives, which in chloroplasts are on the order of several hours (Kim et al., 1993; Klaff and Gruissem, 1991). The importance of translational control is also supported by the finding that mRNA levels do not generally limit gene expression in the chloroplast of the unicellular green alga *Chlamydomonas reinhardtii* (Eberhard et al., 2002; Goodenough, 1971; Hosler et al., 1989).

Most chloroplast proteins are encoded by nuclear genes, synthesized by cytosolic ribosomes and targeted to the appropriate chloroplast compartment via N-terminal targeting signals, which are subsequently removed by proteolytic cleavage (reviewed by Adam in Chapter 15 of this volume). Chloroplast genes are expressed within the organelle by a genetic system that most closely resembles those of prokaryotes. (See Chapter 12 for more details). This chapter covers regulation of translation in chloroplasts by light and for the assembly of the complexes of the photosynthetic apparatus. Recent reviews have covered these and other themes (Barkan and Goldschmidt-Clermont, 2000; Bruick and Mayfield, 1999; Choquet and Vallon, 2000; Choquet and Wollman, 2002; Choquet et al., 2001; Danon, 1997; Gillham, 1994; Goldschmidt-Clermont, 1998; Harris et al., 1994; Hauser et al., 1998; Mayfield, 1990; Pfannschmidt, 2003; Pfannschmidt et al., 2003; Stern, 1997; Sugiura et al., 1998; Wollman et al., 1999; Zerges, 2000, 2002).

1.1. Experimental organisms

Studies of translational control in chloroplasts use barley, tobacco, maize, pea, spinach, and Arabidopsis, exploit advantages of each of these systems, and reveal information relevant to the physiology and development of the chloroplasts of vascular plants. General translation factors have been extensively studied in the alga *Euglena gracilis* (Betts and Spremulli, 1994; Koo and Spremulli, 1994a, b; Yu and Spremulli, 1998). The most versatile experimental system is *C. reinhardtii* because it is amenable to multiple experimental approaches (Harris, 2001; Rochaix et al., 1998). Its ability to use exogenous reduced carbon sources, its short life cycle, and the use of microbial genetic techniques have allowed the isolation of photosynthesis mutants. Studies using *C. reinhardtii* have identified nuclear genes that function in chloroplast mRNA translation. Large quantities of chloroplasts can be isolated for biochemical studies. Routine transformation of the chloroplast genome by homologous recombination has allowed the use of site-directed mutagenesis and chimeric reporter genes in studies of *cis*-acting translational regulatory elements (Boynton et al., 1988; Rochaix, 1996). The recently determined *C. reinhardtii* nuclear genome sequence will allow rapid identification of translational regulators identified through genetic approaches and homologues of known translational regulators from other systems. Phylogenetically, *C. reinhardtii* belongs to the "green" plant lineage and is within a sister clade to the vascular plants. Thus it is more closely related to agricultural plants than are most other algae (Bhattacharya and Medlin, 1998).

1.2. Translational activation is more common in chloroplasts than in other systems

While most examples of translation control of mRNAs in the nuclear-cytosolic compartments or bacteria involve repression of the basal translation initiation rate (reviewed by Draper et al., 1998; Gold, 1988; Mathews et al., 1996), regulation of translation initiation on chloroplast mRNAs involves both activation and repression. The positive roles of most *cis*-acting elements suggest that regulation of translation occurs though modulating molecular events that inherently promote translation (reviewed by Zerges, 2000). Similarly, most *trans*-acting translational factors appear to activate translation (reviewed by Zerges, 2000). Repressive *cis*-acting translational control elements have also been reported (Hirose and Sugiura, 1997; Plader and Sugiura, 2003; Reinbothe et al., 1993). A long-awaited *in vitro* translation system for chloroplasts (of tobacco) was developed in 1996 and is beginning to be used to address these molecular mechanisms (Hirose et al., 1998; Hirose and Sugiura, 1996, 1997). Due to difficulties in the development and general use of this system, however, most studies have examined translation *in vivo*, in isolated chloroplasts, in crude chloroplast lysates or the analysis of mRNA distributions on polyribosomes.

2. LIGHT REGULATES TRANSLATION IN MATURE CHLOROPLASTS

While light is the primary energy source for plants, it also serves as a signal for a variety of regulatory responses, many of which involve gene expression. Regulation of nuclear genes by light occurs primarily at the level of transcription (Nagy and Schafer, 2000; Wang and Deng, 2003). However, in chloroplasts many more examples of light-regulation of translation have been described than examples of light-regulated transcription (Kim et al., 1999; Pfannschmidt et al., 1999; Pfannschmidt et al., 2001), RNA processing (Deshpande et al., 1997; Lee and Herrin, 2003), mRNA stability (Hayes et al., 1999) or protein stability (Kim et al., 1994a; Mullet et al., 1990). This section describes the physiological functions of translational regulation by light in chloroplasts, the experimental evidence for this regulation, how light is perceived, and the known regulatory factors and mechanisms involved. Separate coverage is given to differentiating and mature chloroplasts because the light receptors, regulatory mechanisms, and physiological functions of this regulation differ in most respects.

2.1. Physiological functions of translational regulation by light in mature chloroplasts

Plants and algae use a variety of mechanisms to optimize photosynthesis and minimize light-induced damage to the photosynthetic apparatus and the generation of free radicals by damaged photosystems. Many of these responses occur at the translational level within chloroplasts to control the synthesis of the polypeptide subunits for the assembly or the repair of the complexes of the photosynthetic apparatus. Chloroplast biogenesis occurs during the day (Adamska et al., 1991; Beator and Kloppstech, 1993, 1996; Kreps and Kay, 1997; Nakahira et al., 1998; Oelmuller et al., 1995) and requires polypeptides encoded by nuclear and chloroplast genes (reviewed by Goldschmidt-Clermont, 1998; Rochaix, 1996; Zerges, 2000). Light might act as a common signal that is perceived by both compartments and induces expression of genes whose products are required for the assembly of the photosynthetic apparatus (Lee and Herrin, 2002). Moreover, photochemical reactions within photosystem I (PSI) and photosystem II (PSII) damage certain protein subunits. The D1 subunit of PSII is particularly sensitive and often damaged in the light (Ohad et al., 1990). Damaged D1 is removed by proteolysis (see Chapter 15) and replaced by a newly synthesized protein (reviewed by Adam, 2000; Aro et al., 1993). Translation of *psbA* mRNA (which encodes D1) is stimulated by light in all plants and algae (that have been examined to date) to provide new protein for this repair process (reviewed by Bruick and Mayfield, 1999; Zerges, 2000, 2002). A direct role of *psbA* translation in the assembly of nascent D1 into PSII during its repair is described later in this chapter.

2.2. Light-dependent translation of mRNAs in mature chloroplasts has been revealed by pulse-labeling with radioisotopes

Most examples of translational regulation by light were revealed by light-dependent changes in protein synthesis rates that occur in the absence of corresponding alterations of the abundance of the mRNAs (Herrin et al., 1986; Lee and Herrin, 2002; Kettunen et al., 1997; Malnoe et al., 1988). Protein synthesis rates are measured by isotope pulse-labeling experiments in which isolated chloroplasts or Arabidopsis hypocotyls (Meurer et al., 1996) are allowed to take up [^{35}S]methionine and incorporate it into newly synthesized proteins during a specific pulse period, typically 5-45 minutes. Experiments involving *C. reinhardtii* use whole cells and label with [^{35}S]sulfate or [^{14}C]acetate. Proteins are extracted and fractionated by denaturing polyacrylamide gel electrophoresis and the proteins that were synthesized during the pulse are revealed by autoradiography. The high background due to protein synthesis in the cytoplasm can be eliminated by cycloheximide treatment or by using isolated chloroplasts.

Using this approach, pea chloroplasts were found to induce D1 synthesis after a shift from dark to light growth conditions (Kettunen et al., 1997). The reverse shift from light to dark conditions is followed by a decrease in D1 synthesis by over 10 fold in the aquatic angiosperm *Spirodela* (Fromm et al., 1985). In *C. reinhardtii*, a transition from dark to light conditions induces the synthesis of the PSII subunits D1 and D2 (encoded by *psbA* and *psbD*, respectively) and the large subunit of ribuolose bisphosphate carboxylase-oxygenase (LSU of rubisco) which is encoded by the chloroplast *rbcL* gene (Herrin et al., 1986; Malnoe et al., 1988). Light regulation of other chloroplast genes might also occur but have not been detected because only the major proteins (D1, D2 and LSU) are readily detected by pulse-labeling experiments.

The results of isotopic pulse-labeling experiments can be misleading. Although pulse-labeling of a newly synthesized protein can reveal the rate at which the mRNA encoding is translated, and the studies described here controlled for effects at the levels of transcription and mRNA stability with RNA-gel blot experiments, the amount of isotope-labeled protein also can be determined by other factors. As described below, comparisons of isotope-labeled protein levels in short pulses with etioplasts and differentiating chloroplasts revealed differences due to rapid protein degradation and not translational control. The amount of isotope-labeled protein could also be affected by other processes involving the mRNA that must precede translation (e.g., editing of the initiation codon (Hirose and Sugiura, 1997; Neckermann et al., 1994), 5' terminal processing (Rochaix, 1996), or intra-organellar localization) and which were not monitored by the RNA-gel-blot control experiments used in these studies.

2.3. Other approaches confirm that translation is regulated by light in mature chloroplasts

A study of *C. reinhardtii* cells cultured in a 12:12 dark-light regime revealed by transmission electron microscopy that chloroplast ribosomes are recruited to thylakoid membranes within the first 10 min of the light phase and released to the stroma within the first 10 min of the dark phase (Chua et al., 1976). Analyses of subfractions of pea chloroplasts that had been shifted from dark to light conditions revealed the recruitment of the *psbA* mRNA to thylakoid membranes (Kettunen et al., 1997). Light-dependent loading of certain chloroplast mRNAs on poly-ribosomes revealed that translational regulation underlies changes in protein accumulation. For example, a transition from dark to light conditions results in an increase in the proportion of *psbA* mRNAs associated with poly-ribosomes in *C. reinhardtii* (Herrin, D. et al., 1986; Trebitsh et al., 2000) and barley (Kim and Mullet, 1994). These results also show that the regulation occurs at translation initiation steps because inhibition of elongation or termination steps in the dark should maintain an mRNA on poly-ribosomes. Regulation of the initiation of *psbA* mRNA translation in tobacco chloroplasts is also supported by *in vivo* light regulation of chimeric reporter genes with the *psbA* 5' untranslated region (UTR) fused to a heterologous reporter gene (Staub and Maliga, 1993). Although *C. reinhardtii* is much more amenable to chloroplast transformation than tobacco, it is not yet known whether the *psbA* 5' UTR can confer light-regulated translation upon a heterologous reporter mRNA in *C. reinhardtii* chloroplast transformants.

2.4. A role of photosynthesis in light perception and signaling to translational regulatory factors

A recent study used pulse-labeling to measure protein synthesis rates in the *C. reinhardtii* chloroplast throughout the diurnal (12:12 hour dark-light) cycle (Lee and Herrin, 2002). The proteins encoded by *psbA*, *psbD*, and *rbcL* were radiolabeled at higher levels in the light phase than in the dark phase. The absence of corresponding changes in the mRNAs led these authors to propose that translation of these mRNAs is regulated by light. Radio-labeling of the chloroplast elongation factor ef-Tu during these pulse periods declined late in the dark phase while the *tufA* mRNA accumulated, suggesting that its translation is repressed (Lee and Herrin, 2002). *C. reinhardtii* divides early in each light phase (Harris, 2001) and exhibits circadian regulation of the transcription of at least a few genes (Jacobshagen and Johnson, 1994; Jacobshagen et al., 1996; Mittag and Wagner, 2003; Savard et al., 1996). However, the diurnal regulation revealed by Lee and Herrin (2002) is probably not mediated primarily by a circadian clock or the cell cycle for two reasons. Firstly, when light was maintained during a subjective dark phase, during which a circadian or cell cycle control should continue to function, radio-labeling of these proteins only declined slightly. Secondly, protein synthesis could be induced during the dark phase by an exogenous reduced carbon source and this induction was blocked by dissipation of the trans-thylakoid membrane electrochemical proton gradient by a proton ionophore (Lee and Herrin, 2002; Michaels and Herrin, 1990).

These findings, and others described below, reveal that the perception of light and transmission of this signal to translational regulatory factors is mediated by photosynthesis (Figure 1) (Bruick and Mayfield, 1999; Pfannschmidt, 2003; Pfannschmidt et al., 2003; Zerges, 2000, 2002). This differs from the specialized light receptors and signal transduction pathways that regulate transcription of some nuclear genes (Fankhauser and Staiger, 2002; Lin, 2002; Nagy and Schafer, 2000) and the chloroplast *psbC-psbD* operon (Christopher and Hoffer, 1998; Christopher et al., 1992; Christopher and Mullet, 1994; Kim et al., 1999; Sexton et al., 1990). While photosynthesis mediates the light-regulation of nuclear and chloroplast genes (reviewed by Pfannschmidt, 2003; Pfannschmidt et al., 2003), only examples of translational regulation in chloroplasts are covered here.

In photosynthesis light-driven charge separation reactions within PSI and PSII generate high-energy electrons that drive a series of redox reactions involving plastoquinone, the cytochrome *b6/f* complex, plastocyanin and ferredoxin to generate the electrochemical proton gradient across thylakoid membranes and NADPH (Figure 1). The ATP sythase complex uses this proton gradient for synthesis of ATP from ADP and phosphate. ATP and NADPH are used for the carbon fixation in light-independent reactions. The current model proposes that light intensity is monitored through specific biochemical cues or "sensors" within photosynthesis (Allen, 1993a, b; Allen et al., 1995; Allen and Raven, 1996; Bruick and Mayfield, 1999; Pfannschmidt, 2003; Pfannschmidt et al., 2003; Zerges, 2000, 2002). The stromal ADP/ATP ratio, the magnitude of the electrochemical proton gradient across thylakoid membranes, and the redox states of specific electron carriers in the photosynthetic electron transport chain are modulated by light and monitored by genetic regulatory factors (Figure 1)(Allen, 1993b; Pfannschmidt, 2003; Pfannschmidt et al., 2003). These regulatory factors respond to these cues by controlling the expression of specific chloroplast genes at the levels of transcription (Allen and Pfannschmidt, 2000; Krause et al., 2000; Legen et al., 2002; Pfannschmidt et al., 1999; Tullberg et al., 2000), RNA splicing (Deshpande et al., 1997), mRNA stability (Salvador and Klein, 1999) and translation (Bruick and Mayfield, 1999; Danon, 1997; Zerges, 2000, 2002).

A growing body of evidence supports this mode of light perception and signal transduction. We know that light is perceived through photosynthesis because inhibition of specific steps of the photosynthetic electron transport chain or dissipation of the electrochemical proton gradient also inhibit light-dependent *psbA* translation in *C. reinhardtii* (Trebitsh and Danon, 2001). Chloroplasts are known to house the light receptors and signal transduction pathways because light activates *psbA* translation in isolated chloroplasts (Kettunen et al., 1997; Taniguchi et al., 1993; Trebitsh and Danon, 2001). However, the signals involved are complex and what we know about them is described below.

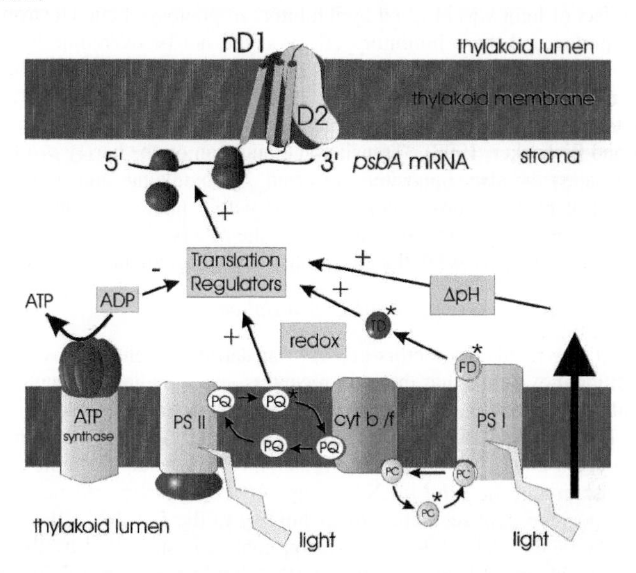

Figure 1. *Light regulates translation of the chloroplast psbA mRNA through photosynthesis. Light regulates psbA translation through its effects on stromal [ADP], the redox states of plastoquinone (PQ) and thioredoxin (TD) (reduced states are indicated by asterisks), and the electrochemical proton gradient (large arrow). The four major photosynthesis complexes that mediate these effects are shown in the lower thylakoid membrane cross-section. In the stroma translational regulatory factors monitor these sensors and regulate the initiation and elongation phases of psbA mRNA translation. The nascent D1 protein (nD1) is assembled into the PSII subcomplex. (Abbreviations: ferredoxin, FD; photosystem II, PSII; cytochrome b6/f complex, cyt b6/f; photosystem I, PSI; plastocyanin, PC) (see Color Plate 2).*

2.5. Light regulates the elongation phase of translation

In barley chloroplasts, light stimulates the elongation phase of *psbA* mRNA translation. This was revealed by comparisons of the level of protein synthesis in isolated chloroplasts in the dark and in the light by isotopic pulse-labeling experiments (Edhofer et al., 1998; Muhlbauer and Eichacker, 1998). Light stimulated [^{35}S]methionine incorporation into D1 protein by several fold. This induction might occur at the level of translation elongation because it was not affected by an inhibitor of translation initiation, lincomysin. This effect could partially reflect the increased ATP levels generated by photosynthesis because exogenous ATP stimulated elongation of *psbA* mRNA translation by 5-fold in chloroplast lysates from dark grown plants (Edhofer et al., 1998). However, the electrochemical proton gradient appears to be the major sensor because the stimulatory effect of light was blocked by inhibitors of photosynthetic electron transport, a proton ionophore, and these inhibitory effects could not be overcome by addition of exogenous ATP (Muhlbauer and Eichacker, 1998; Zhang et al., 2000). These authors propose that the rate of D1 synthesis into the thylakoid membrane is regulated by a pH-sensitive factor that regulates translation elongation by ribosomes on the *psbA* mRNA (Muhlbauer and Eichacker, 1998). Translation elongation on the barley *psbA* mRNA (in chloroplast lysates) is also oppositely affected by oxidizing and reducing agents suggesting control by the redox states of sensor within photosynthesis (Zhang et al., 2000). Further research is required to determine the relative importance and molecular mechanisms of regulation by ATP, the electrochemical pH gradient, and redox factors.

2.6. Translational regulators of psbA mRNA translation

Studies of candidate regulatory factors of *psbA* translation by light in *C. reinhardtii* have provided additional evidence that the light receptors and signaling molecules are within the photosynthesis apparatus and, thus, located entirely within the chloroplast (Section 2.4, Figure 1, and reviewed by (Bruick and Mayfield, 1999; Pfannschmidt, 2003; Pfannschmidt et al., 2003). A role of a multi-subunit RNA-binding complex in the light-dependent translation of the *psbA* mRNA is supported by genetic evidence (Yohn et al., 1998) and the positive light-regulation of its binding to the 5' UTR of the *psbA* mRNA (Danon and Mayfield, 1991, 1994a, b). Redox control is supported by the abilities of reduced thioredoxin or DTT to activate the RNA-binding activity of this complex *in vitro* (Danon, 1997). This complex also appears to be repressed by another mechanism in the dark because these reducing agents cannot activate the complex isolated from dark-grown cells (Trebitsh et al., 2001). This primary repression could be mediated by the redox state of the plastoquinone pool, ADP-dependent phosphorylation of the RB60 subunit of the complex, or both mechanisms (Danon and Mayfield, 1994a) (reviewed by (Pfannschmidt, 2003; Pfannschmidt et al., 2003). RB60 is a homologue of protein disulfide isomerase, an enzyme that catalyzes the isomerization of protein disulfides and is itself under redox control (Kim and Mayfield, 1997). RB60 has been proposed to carry out the redox control of the RNA-binding activity of another protein of the complex, RB47, by reducing a disulfide bond that bridges cysteine residues in two of its RNA-binding domains, thereby relieving steric inhibition to RNA binding (Fong et al., 2000). In the absence of photosynthesis in the dark higher ADP pool levels and/or oxidation of the plastoquinone pool could result in inactivation of this complex (Danon and Mayfield, 1994a). The reverse process in the light would "prime" RB60 to sense the redox state thioredoxin and modulate the regulatory activity of the complex in response to varying light intensity (Bruick and Mayfield, 1999). The *psbA* mRNA might be targeted to thylakoid membranes by a 63 kDa protein that binds preferentially to the *psbA* 5' untranslated region and is associated with thylakoid membranes in *C. reinhardtii* (Ossenbuhl et al., 2002). Proteins that bind to the 5' untranslated leader of the *psbA* mRNA have been identified in chloroplasts of vascular plants and they could be involved in light regulation of *psbA* translation (Alexander et al., 1998; Klaff et al., 1997; Nickelsen, 2003; Shen et al., 2001).

2.7. Other translational factors might mediate regulation by light.

The complex with RB47 and RB60 has been reported to bind specifically to the *psbA* 5' UTR (Danon and Mayfield, 1991), and yet light activates the translation of many, if not most, of the chloroplast mRNAs that encode subunits of the photosynthesis apparatus in *C. reinhardtii* (Lee and Herrin, 2002; Malnoe et al., 1988; Trebitsh and Danon, 2001; Trebitsh et al., 2000). Chloroplast RNA-binding proteins that could be involved in general light regulation of translation have been identified, but they have not been characterized at the molecular level (Hauser et al., 1996, reviewed by Nickelsen, 2003). The RNA-binding activity of a 46 kDa protein associated with chloroplast membranes is dramatically stimulated within 5 to 10 minutes following a shift from dark to light conditions (Zerges et al., 2002). This light-regulation is probably mediated by the effect of photosynthesis on ADP pools because 1) it is prevented by two inhibitors that have the common effect of blocking ATP synthesis and 2) ADP inhibits the RNA-binding activity *in vitro* with a Ki of 50 nM (Zerges et al., 2002). An intermediate in chlorophyll biosynthesis and known light-signaling molecule, Mg-protoporphryin IX, might also be involved in this regulation because it inhibits the RNA-binding activities of two proteins *in vitro*.

2.8. Light regulates translation of the chloroplast chlL mRNA.

Translational regulation of the chloroplast *chlL* mRNA in *C. reinhardtii* chloroplasts is repressed by light through photosynthesis. CHLL is one of three chloroplast gene products that are required for light-independent chlorophyll synthesis (Timko, 1998). In the dark, CHLL accumulates and its mRNA is associated with polysomes (Cahoon and Timko, 2000). Following a shift from dark to light growth conditions, CHLL becomes undetectable and the *chlL* mRNA shifts from large polysomes to the free mRNA pool.

As in the translational regulation involving the *psbA* mRNA by light, the light signal is perceived and transmitted through photosynthesis because inhibitors of specific steps of photosynthesis can induce CHLL accumulation in the light. These studies also revealed that the sensors could be the ADP/ATP ratio or the redox state(s) of an electron acceptor located downstream of the cytochrome *b6/f* complex. The translational regulatory factors involved could be encoded by the nuclear genes *y-1*, *y-5*, *y-6*, *y-7*, *y-8*, and *y-10* because they are required for CHLL accumulation in the dark. In particular, a *y-1* gene product is required for *chlL* translation because the *y-1* mutant fails to load the *chlL* mRNA on polysomes in the dark (Cahoon and Timko, 2000).

2.9. Oxidative stress caused by high light intensity represses translation of the rbcL mRNA in C. reinhardtii chloroplasts.

In *C. reinhardtii*, light of an intensity high enough to cause the photosynthetic electron transport chain to generate reactive oxygen species represses translation of the *rbcL* mRNA for several hours (Shapira et al., 1997). This response appears to be cued to the alga's antioxidant protective response with glutathione acting as the sensor (Irihimovitch and Shapira, 2000). This regulation could be imposed by autoregulatory feedback repression involving the *rbcL* product, LSU, which binds to RNA following its isolation from cells under oxidative stress, but not following isolation from non-stressed cells (Yosef et al., 2003). Binding might involve a putative RNA-binding domain in the N-terminal region of the protein that becomes accessible following partial unfolding of the protein in oxidizing conditions.

3. TRANSLATIONAL REGULATION DURING LIGHT-DEPENDENT CHLOROPLAST DIFFERENTIATION

Chloroplasts differentiate from proplastids of the apical meristem during photomorphogenesis in vascular plants. However, due to difficulties involved in isolating proplastids, chloroplast differentiation has been studied primarily in angiosperm seedlings that have germinated in the dark and in which chloroplasts do not differentiate because chlorophyll is synthesized only by a light-dependent pathway (Reinbothe and Reinbothe, 1996). Consequently, mature thylakoid membranes are not assembled (Vothknecht and Westhoff, 2001). Proplastids in leaves and stems differentiate to etioplasts, a specialized plastid type that is poised to initiate thylakoid biogenesis in response to light. Other studies have used the *y-1* mutant of *C. reinhardtii* which lacks the light-independent pathway (Cahoon and Timko, 2000; Timko, 1998). These mutants undergo an analogous chloroplast differentiation following a shift from dark to light growth conditions (Malnoe et al., 1988; Ohad, 1974; Ohad et al., 1967a, b). Separate coverage is given here to translational control in chloroplast differentiation because the known light receptors and underlying regulatory mechanisms differ from those in mature chloroplasts.

Illumination has multiple effects on the differentiation of etioplasts to chloroplasts. The accumulation of chlorophyll is accompanied by the assembly of the network of thylakoid membranes, the accumulation of the chlorophyll binding proteins, and their assembly into the photosystem complexes (Malnoe et al., 1988; Timko, 1998; Vothknecht and Westhoff, 2001). Within the first minutes of chloroplast differentiation pulse-labeling experiments have detected newly synthesized chlorophyll binding apoproteins of the PSI reaction center (PsaAp and PsaBp), the D1 subunit of PSII, and the large subunit of rubisco (LSU, encoded by *RbcL*) in barley (Gamble and Mullet, 1989; Kim et al., 1994a; Klein and Mullet, 1986), amaranth (Berry et al., 1986), and the *y1* mutant of *C. reinhardtii* (Malnoe et al., 1988). Translational control was attributed to these effects because alterations in the levels of the relevant mRNAs were not detected (Klein and Mullet, 1986; Klein and Mullet, 1987; Malnoe et al., 1988).

3.1. Mechanisms of light perception and regulation during chloroplast differentiation

The mechanisms by which light regulates chloroplast gene expression in differentiating chloroplasts are also complex and occur at multiple levels. Evidence has been reported for regulation of translation initiation (Kim and Mullet, 1994), translation elongation (Kim et al., 1994b), and by post-translational stabilization of the apoproteins by chlorophyll. Light probably does not signal these changes in gene expression through photosynthesis (as described above) because these effects occur within the first minutes of chloroplast differentiation and before the assembly of the photosynthetic electron transport chain (Klein and Mullet, 1986; Klein and Mullet, 1987).

3.2. Light receptors and signaling molecules

As in mature chloroplasts, gene regulation in differentiating chloroplasts is not cued to specialized light receptors and signal transduction pathways, but rather to a light-dependent step in chlorophyll synthesis. This is supported by the similar time courses for the accumulation of chlorophyll and chlorophyll-binding proteins during chloroplast differentiation (Klein and Mullet, 1986), the analyses of protein accumulation in barley mutants defective in various steps of chlorophyll synthesis (Herrin et al., 1992; Klein et al., 1988a), and the effects of directly manipulating chlorophyll synthesis on protein synthesis in chloroplast lysates (Eichacker et al., 1990). Moreover, a role of phytochrome, the primary light receptor involved in activation of nuclear gene expression by light, was excluded by the inability of a pulse of far-red light to reverse the light induction of apoprotein accumulation in barley chloroplasts (Klein et al., 1988a). The primary candidate light receptor is the light-dependent protochlorophylide-oxidoreductase and the sensors could be chlorophyll or chlorophyll intermediates (upstream from this step) that accumulate in the dark and are known signaling molecules in the regulation of gene expression in the chloroplast (Johanningmeier, 1988; Johanningmeier and Howell, 1984) and the nucleus (Kropat et al., 1997, 2000). In barley chloroplasts polyribosomes are associated with Mg-chelatase (Kannangara et al., 1997), an enzyme of chlorophyll biosynthesis that precedes these regulatory intermediates and is itself a component of a signal transduction pathway that informs the nucleus of the developmental state of the chloroplast (Brown et al., 2001; Jarvis, 2001; Mochizuki et al., 2001). This association could reflect a role of Mg-cheletase in translational regulation during chloroplast differentiation (Kannangara et al., 1997).

3.3. Physiological functions of regulation by chlorophyll biosynthesis

Coordination of the synthesis of chlorophyll and the accumulation of apoproteins that bind this pigment could ensure that these apoproteins are synthesized only when chlorophyll is available. This regulation might also prevent the generation of free radicals by free chlorophyll or photosystems deficient in chlorophyll (Mullet et al., 1990). It could also reflect requirements for chlorophyll binding to the apoproteins during their synthesis to promote their proper folding or before the amino acid residues involved become inaccessible within large multi-subunit integral membrane complexes.

3.4. Examples of translational control during chloroplast differentiation

Early activation of *psbA* translation initiation was revealed by an increase in ribosomes bound to the initiation codon following illumination of 8-day-old barley seedlings that was revealed by results of primer-extension inhibition ("ribosome toe-printing") experiments using isolated poly-ribosomes from etioplasts and differentiating chloroplasts (Kim and Mullet, 1994). Translational regulation of *psbA* and *rbcL* mRNAs is also suggested by their recruitment to membranes during barley chloroplast differentiation (Klein et al., 1988b).

A recent study observed a drastic induction of *rbcL* mRNA translation during chloroplast differentiation from etioplasts of 8-day dark grown barley seedlings from comparisons of LSU synthesis rates by pulse-labeling and analyses of the distribution of the mRNA on polysomes in etioplasts and differentiating chloroplasts (Kim and Mullet, 2003). The primary block in the dark occurs after initiation because no increase in the number of ribosomes at the translation initiation region of *rbcL* mRNAs on polysomes was observed with primer-extension inhibition experiments (Kim and Mullet, 2003). Moreover, in the presence of an inhibitor of initiation, lincomysin, the number of ribosomes at the initiation codon decreased during chloroplast differentiation, but not when seedlings were maintained in the dark. Thus, ribosomes from illuminated plants appear to leave the initiation region and enter the elongation phase while those from dark-grown plants do not. When elongation was inhibited by chloramphenicol, the number of initiating ribosomes increased during differentiation, revealing that additional rounds of initiation occurred. The authors proposed that translation initiation complexes assemble in both etioplasts and differentiating chloroplasts and an unknown mechanism prevents these ribosomes from entering the elongation phase in etioplasts and that translational activation in response to light results from a release of this block, thereby allowing the ribosomes to clear the initiation region and translate the coding sequence (Kim and Mullet, 2003). This resembles the "entrapment" mechanism that occurs in the auto-regulatory feedback translational repression involving the *Escherichia coli* mRNAs of the alpha operon and the *rpsO* gene (reviewed by Draper et al., 1998; see also Ehresmann et al., 1995; Schlax et al., 2001).

3.5. Chlorophyll stabilizes apoproteins encoded by chloroplast mRNAs

Regulation of translation is not the only process underlying the accumulation of chloroplast genome-encoded proteins during chloroplast differentiation. Results of several studies indicate that chlorophyll stabilizes apoproteins following their synthesis. This process is covered here because translation was originally proposed to underlie this regulation and to show the difficulties involved in interpretation of the results of pulse-labeling experiments. During chloroplast differentiation, the slight increases in polysome association of mRNAs encoding chlorophyll-binding proteins of PSI and PSII cannot account for the drastic increases in the radio-labeling of these proteins in pulse-labeling experiments (Berry et al., 1990; Klein et al., 1988b). Thus, the regulation is imposed primarily after translation initiation. Although studies in barley have characterized D1 translation intermediates due to ribosome pausing at specific sites on the coding region of the *psbA* mRNA, induction probably does not result from release of these ribosomes because the pausing frequency *increases* during chloroplast differentiation (Kim et al., 1994b; Kim and Mullet, 1994). Rather, pausing could function to facilitate binding of chlorophyll to nascent D1 polypeptides (Kim et al., 1994b; Kim and Mullet, 1994). Evidence against a role of chlorophyll in the regulation of translation elongation on the *psaA* and *psbA* mRNAs was also provided by translation runoff assays in lysates of barley etioplasts in which exogenous chlorophyll did not stimulate elongation (Kim et al., 1994a).

Newly synthesized chlorophyll is thought to stabilize the D1 apoprotein because discrete unstable D1 intermediates have been detected in barley etioplasts (Kim et al., 1994a; Kim and Mullet, 1994; Mullet et al., 1990). Moreover, chlorophyll greatly stabilizes the D1 that is produced in translation runoff reactions in barley etioplast lysates (Kim et al., 1994a). The nuclear gene *viridis-115* in barley is required for the accumulation of D1 and CP47 and thus its product could be involved in the stabilization of these apoproteins by chlorophyll (Gamble and Mullet, 1989; Kim et al., 1994c).

3.6. Do mechanisms of light regulation vary with the etioplast age?

The relative degrees to which translational control and apoprotein stabilization by chlorophyll contribute to the accumulation of photosynthesis proteins during chloroplast differentiation are unclear. Moreover, the mechanisms of light-induced induction of gene expression might vary with the developmental stage of the etioplast, with translational control being more important after eight days of seedling differentiation in the dark than after 4.5 days. The studies that found that chlorophyll stabilizes apoproteins used etioplasts from 4.5-day dark-grown seedlings and the evidence for translational control was found when etioplasts from 8-day dark-grown seedlings were used (as described above). Drastically stronger induction of *rbcL* mRNA translation was observed during the differentiation of chloroplasts from etioplasts from 8-day dark-grown seedlings (Kim and Mullet, 2003) than with etioplasts from 4.5-day dark-grown seedlings (Klein and Mullet, 1986; Klein and Mullet, 1987). Moreover, the increase in the number of initiating ribosomes on the *psbA* mRNA was detected during differentiation of etioplasts from 8-day dark-grown seedlings (Kim and Mullet, 1994) but not when 4.5-day dark-grown seedlings were used (Kim et al., 1994a). Future research could address the light receptors and signal transduction pathways that regulate these translational responses in etioplasts from 8-day dark-grown seedlings using the approaches that have been used for 4.5-day dark-grown seedlings.

4. TRANSLATION IN REPAIR AND ASSEMBLY OF PHOTOSYNTHETIC COMPLEXES

4.1. Cotranslational insertion of D1 into thylakoid membranes and PSII

The synthesis of certain chloroplast genome-encoded proteins of the photosynthetic electron transport chain is coupled to their insertion into thylakoid membranes, their binding to chlorophyll, and their assembly into the complex in which they function. During the damage-and-repair cycle undergone by PSII (as described above) polyribosomes with *psbA* mRNA associate with thylakoid membranes in chloroplasts of barley (Klein et al., 1988b) and *C. reinhardtii* (Herrin and Michaels, 1985). In maize chloroplasts D1 synthesis requires the SecY component of a pathway that targets proteins to thylakoid membranes (Roy and Barkan, 1998). Pulse-chase studies in spinach chloroplasts detected no free newly synthesized D1 protein, but they did reveal nascent D1 intermediates assembled in PSII core complexes (Zhang et al., 1999). Moreover, the D2 subunit of the PSII reaction center appears to function as a receptor or stabilizing factor for nascent D1 as it is synthesized and inserted into the complex (Ohad et al., 1990; van Wijk et al., 1997; Zhang et al., 1999). Nascent D1 intermediates were co-immunoprecipitated with antibodies against D2 (Zhang et al., 1999). Additional support for close contacts between nascent elongating D1 protein with cysteine residues in D2 was provided by cross-linking experiments (Zhang et al., 1999, 2000). Formation of disulfide bonds between D1 and D2 are required for stability of the early intermediate D1-D2 dimer because a thio-modifying agent and a strong oxidant prevented assembly of this intermediate (Zhang et al., 2000). Obligate coupling of *psbA* translation and D1 incorporation into PSII is also supported by the abilities of a thio-modifying agent and a strong oxidant to block synthesis *and* assembly of D1 (Zhang et al., 2000). Additional evidence for obligate co-translational assembly of subunits encoded by the chloroplast genome is supported by their apparent inability to assemble after their synthesis. For example, *in vitro* assembly of a subunit into a thylakoid membrane complex has been described only using crude thylakoid membranes and run-off translation of mRNAs that had been engaged by ribosomes and associated with membranes *in vivo* (Minami et al., 1986) or when the subunit is synthesized by a cell-free chloroplast translation system (Nilsson et al., 1999; Rohl and van Wijk, 2001). In contrast, three nuclear genome-encoded subunits can assemble into complexes *in vitro* (Bertsch and Soll, 1995; Minai et al., 1996; Minai et al., 2001; Reinsberg et al., 2001).

4.2. Translation and assembly of thylakoid membrane complexes of photosynthesis

A model for the *de novo* assembly of the photosynthesis complexes in *C. reinhardtii* termed "control by epistasy of synthesis" (CES) proposes that a hierarchical system involving translational regulation ensures that subunits are synthesized and inserted into an assembling complex in a specific order (Figure 2) (Choquet and Vallon, 2000; Choquet and Wollman, 2002; Choquet et al., 2001; Wollman et al., 1999; Zerges, 2002). This system could ensure the production of subunits in the correct stoichiometry and, in some cases (e.g., D1) ensure that a subunit is synthesized only when the partner subunits it requires as an assembly scaffold are available. Subunits synthesized early in the pathway could have two functions. Firstly, their presence could signal (directly or indirectly) to regulators to activate translation of the mRNAs encoding subunits immediately downstream in the pathway. Candidate regulators are discussed in the next section. In the absence of an upstream subunit the translation of the mRNA(s) encoding the downstream subunit(s) would be repressed. Secondly, upstream subunits serve as a receptor or scaffold for the cotranslational assembly of the downstream subunit as described above for the PSII damage-repair cycle. Such subunits are often referred to as "assembly partners".

CES was first based upon results of the analyses of the chloroplast proteins synthesized in *C. reinhardtii* mutants deficient for a single subunit of a complex (de Vitry et al., 1989; Girard-Bascou et al., 1987; Kuras and Wollman, 1994; Morais et al., 1998; Rochaix and Erickson, 1988; Stampacchia et al., 1997). These studies revealed that certain subunits are required for the synthesis or immediate stabilization of one or more of their partner subunits. In none of these cases could alterations in the level of the mRNA of a gene explain its repression and therefore translational control was implicated.

4.3. Evidence supporting the CES pathway in PSII biogenesis.

The evidence for CES in PSII biogenesis is covered first to maintain continuity with the previous section and because this model was first proposed to underlie PSII assembly

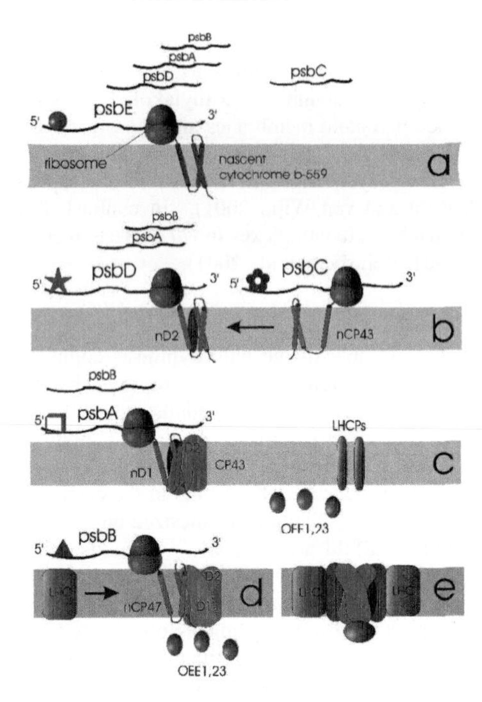

Figure 2. This hierarchical pathway proposed for PSII assembly involves translational control by chloroplast mRNA-specific regulators, which are shown as various shapes bound to the mRNAs. (a) The cytochrome b-559 □-subunit is inserted into the thylakoid membrane during translation of the psbE mRNA. (b) Activation of psbD and psbC translation produces D2 and CP43, respectively. (c) Activation of psbA translation occurs and nascent D1 (nD1) assembles with the subcomplex. Proteins encoded by nuclear genes (light harvesting proteins (LHCPs) and subunits of the oxygen-evolving complex (OEE 1, 2, and 3) are imported from the cytosol into the thylakoid lumen. (d) Activation of psbB translation occurs and nascent CP47 is assembled. (e) Association of light harvesting complexes, the oxygen evolving complex, and small subunits (not shown) forms a functional PSII.

(de Vitry et al., 1989) (Figure 2). In a null mutant for the □-subunit of cytochrome *b-559* of PSII, synthesis of the other core subunits (D1, D2, CP43 and CP47) was undetectable (Morais et al., 1998). Similarly, in a mutant lacking D2, newly synthesized D1 and CP47 were not detected (de Vitry et al., 1989; Erickson et al., 1989). In the absence of D1 protein only CP47 synthesis was undetected (Bennoun and Spierer-Herz, 1986; de Vitry et al., 1989; Erickson et al., 1989). Synthesis of CP43 requires the cytochrome *b-*

559 □-subunit but is independent of the other subunits (de Vitry et al., 1989; Morais et al., 1998). According to the CES model newly synthesized cytochrome *b-559* □-subunit activates translation of the mRNAs of *psbD* and *psbC*, encoding D2 and CP43, respectively. The □-subunit might function as an assembly partner for cotranslational insertion of D2 into the membrane and a scaffold for its assembly. The sub-complex with cytochrome *b-559* and D2 would have similar functions for D1 assembly in a manner analogous to the process described above for the PSII damage-repair cycle., the PSII subcomplex with cytochrome *b-559*, D2, and D1 activates *psbB* translation and serves as an assembly partner for its product, the CP47 subunit (de Vitry et al., 1989). Assembly of CP43, the nuclear genome-encoded small subunits, the light harvesting antenna proteins and the oxygen-evolving complex forms a functional PSII complex.

Indeed, the results of a very recent and unpublished study show that D2 is required for translation of the *psbA* mRNA because unassembled D1 inhibits its translation through the 5' UTR. In turn, autoregulatory inhibition *psbB* and *psbH* translation is exerted by their unassembled products in the absence of D1 or D2 (L. Minai, Y. Choquet, F.-A. Wollman, personal communication). The role of the *psbE* product as the upstream assembly partner remains to be established. The next section covers an example of autoregulatory translational control during the assembly dependent translational control of the *petA* mRNA, which has been extensively investigated.

4.4. *Translational control and assembly of the cytochrome b6/f complex*

Translational control is involved in the expression of cytochrome f for its assembly into the cytochrome b_6/f complex in *C. reinhardtii* in an analogous CES pathway (Figure 3). A systematic deletion study of the three chloroplast genes encoding subunits of this complex revealed that synthesis of cytochrome f depends on the presence of the other subunits, cytochrome b_6 and subunit IV (suIV) (Kuras and Wollman, 1994) Synthesis of cytochrome *b6* and SUIV is independent upon the presence of the other three chloroplast genome-encoded subunits, although suIV stability requires cytochrome *b6* (Kuras and Wollman, 1994). A defect in translation initiation on the *C. reinhardtii petA* mRNA results from the absence of cytochrome b_6 or suIV because the *petA* 5' UTR is both necessary and sufficient for these interactions, based on results of 5' UTR-swapping experiments in chloroplast transformants (Choquet et al., 1998). Analyses of the effects of site-directed mutagenesis of *petA* revealed that the initial signal for this translational repression is the exposition of a tetrapeptide motif (KQFE) within the C-terminal domain of cytochrome *f* (Kuras et al., 1995). When this domain cannot assemble into the complex, due to the absence of the upstream subunits of the CES pathway (i.e., cytochrome *b6* and suIV), its presence in the chloroplast stroma represses *petA* mRNA translation (Choquet et al., 1998; Choquet and Vallon, 2000; Choquet and Wollman, 2002; Choquet et al., 2001). Several examples of auto-regulatory feedback translational repression have been described in *E. coli*. This occurs when a polypeptide represses its own synthesis by binding to the 5' UTR of the mRNA encoding it and thereby either blocks the assembly of an initiation complex or prevents it from leaving the initiation region (Draper et al., 1998; Gold, 1988). In cytochrome *f* assembly-dependent regulation of *petA* mRNA translation, however, it seems improbable that the repressor motif interacts directly with the *petA* 5' UTR because one of its residues is embedded in the thylakoid membrane membrane (Choquet and Wollman, 2002) and, unlike the bacterial auto-regulatory proteins, cytochrome *f* is not known to bind RNA.

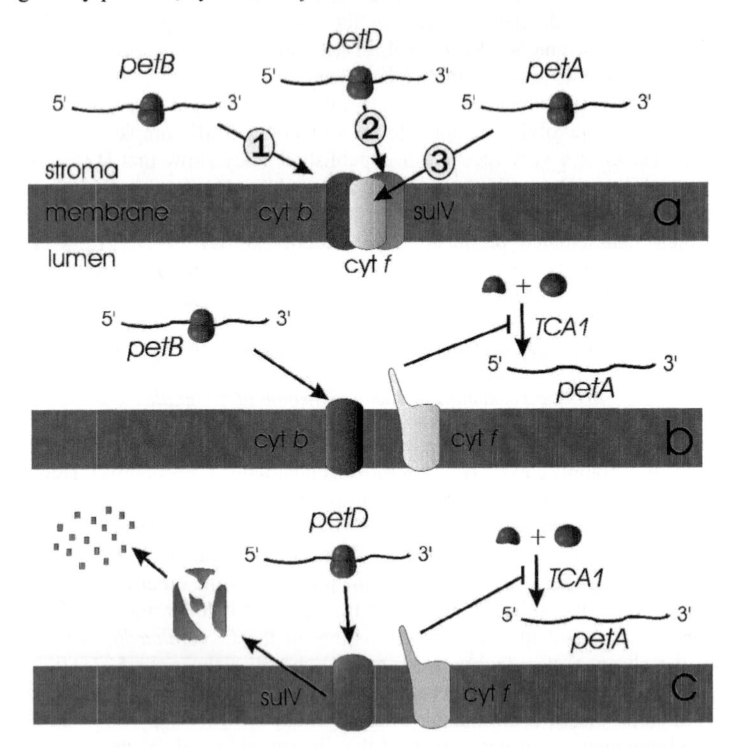

Figure 3. *Cytochrome b_6/f complex Assembly. (a) The CES pathway involves an ordered synthesis and assembly of the three subunits. (b) In the absence of sulV (in a deletion mutant for petD), cytochrome b_6 is synthesized. The free C-terminal domain of non-assembled cytochrome f represses translation of the petA mRNA. TCA1 functions in the activation of petA translation. (c) In the absence of cytochrome b_6 (in a deletion mutant for petB) sulV is synthesized and rapidly degraded. petA translation is repressed.*

4.5. Photosystem I biogenesis also involves assembly-dependent translational control.

When the major photosystem I core subunit PsaA is synthesized in excess of its partner PsaB in *C. reinhardtii* chloroplasts, translation of the *psbA* mRNA is repressed through mechanisms that operate via its 5' UTR (Girard-Bascou et al., 1987, Stampacchia et al., 1997, Takahashi et al., 1996, Wostrikoff et al., 2004). Similarly, translation of the chloroplast *psaC* mRNA s repressed via its 5' UTR when the PSI subunit it encodes is expressed in excess of PsaA.

4.6. Generality of assembly-dependent translational control of subunit synthesis

The results reviewed above reveal that autoregulatory feedback inhibition of translation appears to be a general mechanism controlling the expression of the chloroplast genome-encoded subunits of the oxidoreductase complexes of the photosynthetic electron transport system in the chloroplast of *C. reinhardtii.* Similar co-dependencies to those observed in *C. reinhardtii* have been found for cytochrome *b6/f* biogenesis in tobacco chloroplasts (Monde et al., 2000). There is also evidence for CES in biogenesis of other complexes of photosynthesis. *C. reinhardtii* mutants deficient for the ATP synthase □ subunit have decreased levels of α subunit synthesis (Drapier et al., 1992). Deficiency of the rubisco small subunit by anti-sense repression of the nuclear *rbcS* gene in tobacco (Rodermel et al., 1996) or deletion of the genes in *C. reinhardtii* (Khrebtukova and Spreitzer, 1996) also diminished translation of the chloroplast *rbcL* mRNA encoding the large subunit. Excess LSU from the nuclear-cytosolic compartments might repress translation of the *rbcL* mRNA encoding it through the autoregulatory mechanism recently proposed to occur during oxidative stress (Yosef et al., 2003). In this case, the autoregulatory feedback repression could ensure production of the two subunits in the correct stoichiometry from these different compartments (Khrebtukova and Spreitzer, 1996; Rodermel et al., 1996).

5. TRANSLATIONAL REGULATORS OF SPECIFIC CHLOROPLAST mRNAS

5.1. The nuclear TCA1 gene is required for translation of the chloroplast petA mRNA and might mediate assembly-dependent translation of the petA mRNA

A protein that might mediate the interaction between the cytochrome f repressor motif and the *petA* 5' UTR was identified by genetic analyses of the nuclear gene *TCA1* in *C. reinhardtii* (Figure 3, Wostrikoff et al., 2001). *TCA1* is specifically required for translation of the *petA* mRNA because mutants fail to synthesize cytochrome f in 5-minute pulse-labeling experiments even though the *petA* mRNA accumulates to 15-30% of the level in a wild-type strain. Other evidence for specificity is presented below. The primary defect probably occurs during translation initiation on the *petA* mRNA because the 5' UTR is necessary and sufficient for the dependence upon wild-type TCA1 function. That the *petA* translation defect in each of seven mutants results from a mutant

TCA1 allele suggests that *TCA1* encodes the only trans-acting regulator of *petA* mRNA translation. The recessive phenotypes of the seven mutant alleles and the general observation that recessive alleles have diminished function indicate that *TCA1* activates *petA* translation rather than represses it. Analyses of double mutants revealed that TCA1 is epistatic to the repressor motif in the cytochrome f C-terminal domain; deletion of the repressor motif (recall this causes constitutive *petA* translation in a wild-type genetic background) did not suppress the *petA* translation defect produced by a mutant *TCA1* allele. Because in such a negative regulatory hierarchy the epistatic gene functions downstream (Avery and Wasserman, 1992), this result supports a role of a *TCA1* product as a regulatory factor that mediates the *petA* translational repression by the unassembled repressor motif (Choquet and Wollman, 2002; Wostrikoff et al., 2001).

5.2. Other mRNA-specific trans-acting chloroplast translation factors

Other nuclear genes are required for translation of specific chloroplasts mRNAs (Table 1) (reviewed by Barkan and Goldschmidt-Clermont, 2000; Goldschmidt-Clermont, 1998). The nuclear location of these genes underlies a belief that chloroplast gene expression is controlled by the nucleus. However, this view probably exaggerates the importance of gene location. One can also view the nuclear genome as merely encoding regulatory factors that, once they are synthesized and localized, exert their control in response to diverse signals and from wherever they function in the cell. In contrast, the chloroplast does control gene expression in the nucleus (Brown et al., 2001; Jarvis, 2001; Pfannschmidt, 2003), perhaps making it the master regulatory compartment of the cell.

One class of these nuclear genes includes *TCA1* and was identified by mutations that reduce or abolish translation of the mRNAs from one or two chloroplast genes (Barkan et al., 1994; Drapier et al., 1992; Girard-Bascou et al., 1992; Kuchka et al., 1988; McCormac and Barkan, 1999; Rochaix et al., 1989; Stampacchia et al., 1997; Wostrikoff et al., 2001; Yohn et al., 1996; Yohn et al., 1998b). The second class was identified by nuclear suppressor mutations that alleviate the requirement for the *wild-type* genes of the first class (Wu and Kuchka, 1995; Zerges et al., 1997) and, in one case, also for *cis*-acting sequences within the 5' UTR of the target chloroplast mRNA (Zerges et al., 2003; Zerges et al., 1997). A third class includes nuclear genes that are required to stabilize chloroplast mRNAs, possibly via 5' terminal processing events that must precede translation. This class is not covered here since their role in translation has not been firmly established. (For reviews see Choquet and Wollman, 2002; Zerges, 2000.)

That the target chloroplast mRNAs in *C. reinhardtii* are expressed within CES pathways suggests that the nucleus-encoded translational regulators might coordinate the expression of individual subunits with their assembly into the thylakoid membrane complex in which they function. For example, each chloroplast mRNA-specific translational regulator might control when a subunit is synthesized with respect to the synthesis of other subunits in the pathway (as described above). Some of these mRNA-specific translation factors might localize their target mRNA to membranes or a hypothetical assembly apparatus where other factors activate their translation (Dauvillee et al., 2003; Zerges, 2000; Zerges and Rochaix, 1998), and thus not be translational regulators *per se* (reviewed by Darzacq et al., 2003; Johnstone and Lasko, 2001).

Several of these nuclear gene products could control translation initiation because they function via the 5' UTR of their target chloroplast mRNA. This was demonstrated by the ability of the 5' UTRs to confer dependency upon the nuclear gene to the expression of a reporter gene in chloroplast transformants (Stampacchia et al., 1997; Wostrikoff et al., 2001; Zerges and Rochaix, 1994). We should keep in mind, however, that some of these nuclear genes might be required for processes that *precede* initiation because 5' UTRs can also be involved in editing of the translation initiation codon, 5' terminal processing (reviewed by Zerges, 2000) and, at least in other systems, mRNA localization (Johnstone and Lasko, 2001). In two cases, roles of 5' UTRs are supported by point mutations within them that partially alleviate the requirement for the *trans*-acting function encoded by a nuclear gene (Rochaix et al., 1989; Stampacchia et al., 1997; Zerges and Rochaix, 1994). In another case, initiation control is suggested by reduced association of the target *psbA* mRNA with chloroplast ribosomes in the nuclear mutant (Yohn et al., 1996). Conversely, the presence of the *psbD* mRNA on polysomes in mutants of either of two nuclear genes required for its translation suggests that they function after initiation (Cohen et al., 2001; Rattanachaikunsopon et al., 1999).

The genetic approaches used to characterize the functions of these nuclear genes are subject to some limitations. The chloroplast mRNA specificity of a translational defect in some cases could result from partial loss or alteration of a general translation factor for which one chloroplast mRNA has a particularly high requirement. However, null mutations of *crp1*, *AC115* and *TAB2* do specifically affect translation of their target chloroplast mRNA thus supporting specificity of their functions (Dauvillee et al., 2003; Fisk et al., 1999; Rattanachaikunsopon et al., 1999). Specific roles of *TBC1*, *TAB1* and *TCA1* are supported by the ability of suppressor mutations that alter or replace the 5' UTR of the target chloroplast mRNA to alleviate defects in translation and photosynthesis produced by mutant alleles of these loci (Rochaix et al., 1989; Stampacchia et al., 1997; Wostrikoff et al., 2001). Genetic approaches cannot determine where a nuclear gene product functions in a biochemical pathway controlling translation. Upstream roles of nuclear genes that function in *psbA* translation were revealed by their requirement for the accumulation of the complex that binds to the *psbA* mRNA (Yohn et al., 1996; Yohn et al., 1998b). A direct role of Tab2 was revealed by its detection with the target chloroplast *psaB* mRNA in a complex purified from chloroplasts and its ability to bind the mRNA in *in vitro* RNA-protein binding assays (Dauvillee et al., 2003).

Table 1. *Chloroplast mRNAs require nuclear gene products for their translation*

Complex	*Chloroplast gene*	*Nuclear gene*	*Organism*	*Nuclear Gene Cloned*	*Reference*
ATP Sythase	*atpA* □ subunit	(F54)	Chlamydomonas		(Drapier et al., 1992)
"	*atpB* □ subunit	*atp1*	Maize		(McCarthy and Brimacombe, 1994)
Cyt. b_6/f	*petA* cytochrome *f*	*TCA1*	Chlamydomonas	√	(Wostrikoff et al., 2001)
"	*petA* *petD*	*crp1*	Maize	√	(Barkan et al., 1994; Fisk et al., 1999)
PSI	*psaB* PsaB	*TAB1*	Chlamydomonas		(Stampacchia et al., 1997)
"	"	*TAB2*	"	√	(Dauvillee et al., 2003)
PSII	*psbA* D1	(F35)	"		(Girard-Bascou et al., 1992; Yohn et al., 1996)
"	"	(hf149)	"		(Yohn et al., 1998b)
"	"	(hf233)	"		"
"	"	(hf261)	"		"
"	"	(hf859)	"		"
"	"	(hf1085)	"		"
"	"	*Vir-115*	Barley		(Kim et al., 1994c)
"	*psbC* CP43	*TBC1*	Chlamydomonas		(Rochaix et al., 1989)
"	"	*TBC2*	"	√	(Auchincloss et al., 2002)
"	"	*TBC3*	"		(Zerges et al., 1997)
"	*psbD* D2	*NAC1*	"		(Kuchka et al., 1988)
"		*AC115*	"	√	(Rattanachaikunsopon et al., 1999)
"		*sup4b*	"		(Wu and Kuchka, 1995)

Unnamed loci are indicated by the name of the mutant strain in parentheses.

5.3. Conservation of mRNA-specific translational regulators

Protein sequence similarities between some of these nuclear gene products suggest that they arose early in the evolution of plants and that their functions have been conserved. Crp1 orthologues are in *Neurospora crassa* (*cya5*) and *Saccharomyces cerevisiae* (*pet309*) where they activate translation of the mitochondrial *cox1* mRNA (Coffin et al., 1997; Manthey and McEwen, 1995; Manthey et al., 1998). Partial amino acid sequence similarity exists between Crp1 and Tbc2 (Auchincloss et al., 2002). The *C. reinhardtii* protein Tab2 has 31-46% sequence identity with orthologues in eukaryotic and prokaryotic organisms with oxygenic photosynthesis (Dauvillee et al., 2003). In other cases, genes that evolved to function in other processes and in other intracellular compartments may have been recruited to control translation, for example, in the light-dependent expression of *psbA* as described above (Kim and Mayfield, 1997; Yohn et al., 1998a).

5.4. Intraorganellar localization of the nucleus-encoded proteins

Molecular cloning of five of these nuclear genes has permitted direct analyses of their products and functions (Table 1). Membrane association of AC115 is suggested by a hydrophobic stretch of amino acids at the COOH terminus (Rattanachaikunsopon et al., 1999). crp1, Tbc2, and Tab2 are localized to the chloroplast stroma and present in large complexes (Auchincloss et al., 2002; Fisk et al., 1999). crp1, Tbc2, Nac2, Tab2 are not associated with ribosomes suggesting they transiently function with the translation machinery or indirectly through other factors (Auchincloss et al., 2002; Boudreau et al., 2000; Dauvillee et al., 2003; Fisk et al., 1999). Tab2 is the first mRNA-specific translation factor found to bind its target mRNA (Dauvillee et al., 2003).

6. WHERE DOES TRANSLATION OCCUR IN CHLOROPLASTS?

The belief that thylakoid membrane proteins are co-translationally inserted into thylakoid membranes is based on the observation of poly-ribosomes associated with thylakoid membranes by electron microscopy and co-sedimentation of polysomes and membranes on sucrose density gradients (Chua et al., 1976; Klein et al., 1988b; Margulies and Michaels, 1974; Margulies et al., 1975). However, results of other studies suggest that chloroplast mRNAs are translated and their products assembled into complexes at specialized regions of the inner envelope membrane (reviewed by Sato et al., 1999; Zerges, 2000; Zerges and Rochaix, 1998, see also Fisk et al., 1999; Joyard et al., 1998). An early study found that half of chloroplast polysomes are associated with small fragments of membrane that differ from thylakoid membranes in polypeptide composition and the amount of chlorophyll they contained (0.5% of level in thylakoid membranes) and thus resemble inner envelope membrane (Margulies and Weistrop, 1980). A set of RNA-binding proteins including RB47 (Zerges and Rochaix, 1998) and a protein that functions in splicing of the chloroplast *psaA* mRNA (Perron et al., 1999) are associated with chloroplast membranes that are physically associated with thylakoids and yet they have the lower buoyant density and pigment compositions characteristic of inner envelope membrane. Nucleoids containing the chloroplast chromosome are localized to the envelope of pea chloroplasts during a period of chloroplast differentiation suggesting that some steps of gene expression occur at the envelope (reviewed by Sato et al., 1999; see also Sato et al., 1993; Sato et al., 1998).

Much of the evidence for ribosomes on thylakoid membranes is based on their co-fractionation on sucrose buoyancy density gradients (Klein et al., 1988b; Margulies and Michaels, 1974, 1975; Muhlbauer and Eichacker, 1999). However, inner envelope-like membranes containing thylakoid polypeptides and ribosomes should have higher buoyant density than previously characterized envelope and would co-fractionate with thylakoid membranes on these gradients.

Thylakoid proteins encoded by nuclear genes are imported from the cytosol through the specific regions of the chloroplast envelope containing the import apparatus (reviewed by Cline and Henry, 1996). Synthesis of thylakoid proteins encoded by chloroplast genes at the chloroplast envelope could bind chlorophyll, which is synthesized there (reviewed by Joyard et al., 1998) and assemble with incoming polypeptide subunits from the cytosol.

If thylakoid membrane complexes are assembled in the inner envelope, how do they get to thylakoid membranes? It seems highly unlikely that polytopic membrane proteins are removed from the inner envelope membrane, transported through the stroma and inserted into thylakoid membranes. Evidence has been found for a vesicle transport system in chloroplasts (Westphal et al., 2001, 2003). Chloroplast homologues exist for two proteins that function in vesicular trafficking in the cytosol NSF (Hugueney et al., 1995) and dynamin (reviewed by Cline and Henry, 1996; see also Park et al., 1998). Genetic evidence demonstrated that the chloroplast dynamin homologue is required for thylakoid biogenesis in Arabidopsis (Park et al., 1998). Also in Arabidopsis, a protein associated with the membranes of the inner envelope and thylakoids, called VIPP1, is required for the maintenance of thylakoid membranes and for the appearance of vesicles in the stroma (Kroll et al., 2001). Extensions of the inner envelope membrane and accumulation of membrane vesicles in the stroma have been seen by electron microscopy following induction of rapid thylakoid development in *Chlamydomonas* (reviewed by Hoober et al., 1998). Intriguing evidence has been reported for occurrence of the early steps of photosystem assembly in the plasma membrane, and not thylakoid membranes, of the closest free-living relative of chloroplasts, the cyanobacterium *Synechocystis* sp. PCC 6803 (Zak et al., 2001).

7. CHLOROPLAST GENOME EVOLUTION

Most genes of the photosynthetic bacterium that founded chloroplasts by endosymbiosis 1-2 billion years ago have been deleted or translocated to the nucleus (Martin et al., 2002; reviewed by Gray in Chapter 2 of this volume). Why certain genes and gene expression systems have been maintained by all chloroplasts examined to date is the subject of an active debate (Allen, 1993a; Allen and Raven, 1996; Race et al., 1999; Zerges, 2002). The importance of rapid redox control has been proposed to retain certain genes in chloroplasts (Allen, 1993a; Allen and Raven, 1996). Readers will appreciate that certain chloroplast mRNAs are components of essential processes that are highly complex both in the number of components (e.g., regulatory factors, mRNAs encoding partner subunits, upstream subunits in CES pathways, membranes) and the temporal order of events involved. This complexity could impose two types of constraints on the translocation of the chloroplast genes encoding these mRNAs to the nuclear genome (Zerges, 2002). Physical constraints could result from a requirement that assembly and repair of the photosynthesis complexes occur within the organelle because multi-subunit complexes assembled in another compartment are most probably too large and hydrophobic to cross the chloroplast envelope membrane with existing import machinery. Thus, chloroplast genes encoding core subunits could be maintained in chloroplast genomes because their mRNAs must be translated there for assembly of their polypeptide product into a thylakoid membrane complex. These genes could also be maintained in chloroplasts by phylogenetic constraints resulting from the high improbability that the infrequent events that we believe translocate an organelle gene to the nucleus (Blanchard and Lynch, 2000) also relocate all the other components required for its translation and assembly of its polypeptide product into a complex.

8. CONCLUSIONS

Although many intriguing problems remain unresolved, our understanding of translation regulation in chloroplasts is remarkably good. The impetus for the intensive research that has provided this understanding originated with the excitement over the discoveries of genetic systems in chloroplasts and mitochondria and was facilitated by the concurrent advent of molecular biology. Researchers in the field hope that future discoveries will reveal precise molecular mechanisms of translational control and the assembly of integral membrane complexes and general principles involved in these processes. More recently we are also motivated by the great potential for agriculture and biotechnology provided by our ability to express foreign proteins in transgenic chloroplasts (see chapter 16 of this volume). The chapters in Section 3 of this book describe these emerging biotechnologies how they are enhancing the productivity of agricultural plants through the optimization of photosynthesis (Dhingra et al., 2004) and the engineering of resistance to herbicides (Daniell et al., 1998), insects (McBride et al., 1995; DeCosa et al., 2001), diseases (DeGray et al., 2001), tolerance to drought (Lee et al., 2003), salt (Kumar et al., 2004) and phytoremediation (Ruiz et al., 2003).

Chapter 16 also describes how chloroplasts can be remarkable productive factories for the expression of extremely high levels of useful heterologous proteins and biomolecules for pharmaceuticals, vaccines, and polymers for plastic substitutes (Daniell et al., 2004a,b). A 500-fold increase in translation of Human Serum Albumin was achieved when the *hsa* coding sequence was regulated by the chloroplast psbA 5' and 3' UTRs; rapid translation of *hsa* resulted in aggregates of Human Serum albumin, which protected the newly synthesized protein from proteolytic degradation (Fernandez-San Millan et al., 2003). One acre of chloroplast transgenic plants can produce up to 400 million doses of clean and fully functional anthrax vaccine, when regulated by psbA UTRs (Chebolu and Daniell, 2004). Expression of ubiC gene under the control of psbA 5' and 3' UTRS and illumination of transgenic plants under continuous light resulted in 26.5% of liquid crystal polymer; this is 250-fold higher corismate pyruvate lyase activity and 50-fold higher polymer accumulation than the best nuclear transgenic plants (chapter 16). These examples illustrate the power of gene regulation in biotechnology applications. An understanding of chloroplast gene expression at all levels is critical in these endeavors in basic and applied sciences.

9. ACKNOWLEDGEMENTS

I thank E. Freese, C. Hauser, M. Herrington, and J. Wyglinski for critical reading of the manuscript.

10. REFERENCES

Adam, Z. (2000). Chloroplast proteases: possible regulators of gene expression? *Biochimie, 82*, 647-654.

Adamska, I., Scheel, B. & Kloppstech, K. (1991) Circadian oscillations of nuclear-encoded chloroplast proteins in pea (Pisum sativum). *Plant. Mol. Biol., 17*, 1055-1065.

Alexander, C., Faber, N. & Klaff, P. (1998). Characterization of protein-binding to the spinach chloroplast *psbA* mRNA 5' untranslated region. *Nucleic Acids Res., 26*, 2265-2272.

Allen, J.F. (1993a). Control of gene expression by redox potential and the requirement for chloroplast and mitochondrial genomes. *J. Theor. Biol., 165*, 609-631.

Allen, J.F. (1993b). Redox control of transcription: sensors, response regulators, activators and repressors. *FEBS Lett., 332*, 203-207.

Allen, J.F., Alexciev, K. & Hakansson, G. (1995). Photosynthesis. Regulation by redox signalling. *Curr. Biol., 5*, 869-872.

Allen, J.F., & Pfannschmidt, T. (2000). Balancing the two photosystems: photosynthetic electron transfer governs transcription of reaction centre genes in chloroplasts. *Philos. Trans. R. Soc. Lond. B Biol. Sci., 355*, 1351-1359.

Allen, J.F., & Raven, J.A. (1996). Free-radical-induced mutation vs redox regulation: costs and benefits of genes in organelles. *J. Mol. Evol., 42*, 482-492.

Aro, E.M., Virgin, I., & Andersson, B. (1993). Photoinhibition of Photosystem II. Inactivation, protein damage and turnover. *Biochim. Biophys. Acta, 1143*, 113-134.

Auchincloss, A.H., Zerges, W., Perron, K., Girard-Bascou, J., & Rochaix, J.D. (2002). Characterization of Tbc2, a nucleus-encoded factor specifically required for translation of the chloroplast *psbC* mRNA in *Chlamydomonas reinhardtii*. *J. Cell. Biol., 157*, 953-962.

Avery, L., & Wasserman, S. (1992). Ordering gene function: the interpretation of epistasis in regulatory hierarchies. *Trends Genet., 8*, 312-316.

Barkan, A., & Goldschmidt-Clermont, M. (2000). Participation of nuclear genes in chloroplast gene expression. *Biochimie, 82*, 559-572.

Barkan, A., Walker, M., Nolasco, M., & Johnson, D. (1994). A nuclear mutation in maize blocks the processing and translation of several chloroplast mRNAs and provides evidence for the differential translation of alternative mRNA forms. *EMBO J., 13*, 3170-3181.

Beator, J., & Kloppstech, K. (1993). The Circadian Oscillator Coordinates the Synthesis of Apoproteins and Their Pigments during Chloroplast Development. *Plant Physiol., 103*, 191-196.

Beator, J., & Kloppstech, K. (1996). Significance of circadian gene expression in higher plants. *Chronobiol. Int., 13,* 319-339.

Bennoun, P., & Spierer-Herz, J. E., Girard-Bascou, J. Pierre, Y., Delosome M., & Rochaix, J-D. (1986). Characterization of photosystem II mutants of *Chlamydomonas rienhardtii* lacking the *psbA* gene. *Plant Mol. Biol., 6,* 151-160.

Berry, J.O., Breiding, D.E., & Klessig, D.F. (1990). Light-mediated control of translational initiation of ribulose-1, 5- bisphosphate carboxylase in amaranth cotyledons. *Plant Cell, 2,* 795-803.

Berry, J.O., Nikolau, B.J., Carr, J.P., & Klessig, D.F. (1986). Translational regulation of light-induced ribulose 1,5-bisphosphate carboxylase gene expression in amaranth. *Mol. Cell. Biol., 6,* 2347-2353.

Bertsch, U., & Soll, J. (1995). Functional analysis of isolated cpn10 domains and conserved amino acid residues in spinach chloroplast co-chaperonin by site-directed mutagenesis. *Plant Mol. Biol., 29,* 1039-1055.

Betts, L., & Spremulli, L.L. (1994). Analysis of the role of the Shine-Dalgarno sequence and mRNA secondary structure on the efficiency of translational initiation in the *Euglena gracilis* chloroplast *atpH* mRNA. *J. Biol. Chem., 269,* 26456-26463.

Bhattacharya, D., & Medlin, L. (1998). Algal Phylogeny and the Origin of Land Plants. *Plant Physiol., 116,* 9-15.

Blanchard, J.L., & Lynch, M. (2000). Organellar genes: why do they end up in the nucleus? *Trends Genet., 16,* 315-320.

Boudreau, E., Nickelsen, J., Lemaire, S.D., Ossenbuhl, F., & Rochaix, J.D. (2000). The *Nac2* gene of Chlamydomonas encodes a chloroplast TPR-like protein involved in *psbD* mRNA stability. *EMBO J., 13,* 3366-3376.

Boynton, J.E., Gillham, N.W., Harris, E.H., Hosler, J.P., Johnson, A.M., Jones, A.R., Randolph-Anderson, B.L., Robertson, D., Klein, T.M., Shark, K.B., et al. (1988). Chloroplast transformation in Chlamydomonas with high velocity microprojectiles. *Science, 240,* 1534-1538.

Brown, E.C., Somanchi, A., & Mayfield, S.P. (2001). Interorganellar crosstalk: new perspectives on signaling from the chloroplast to the nucleus. *Genome Biol., 2,* 102.1-102.4.

Bruick, R.K., & Mayfield, S.P. (1999). Light-activated translation of chloroplast mRNAs. *Trends Plant. Sci., 4,* 190-195.

Cahoon, A.B., & Timko, M.P. (2000). yellow-in-the-dark mutants of Chlamydomonas lack the CHLL subunit of light-independent protochlorophyllide reductase. *Plant Cell, 12,* 559-568.

Chebolu, S & Daniell, H. (2004). Chloroplast derived vaccine antigens and biopharmaceuticals: expression, folding, assembly and functionality. Current Trends in Microbiology and Immunology, in press.

Choquet, Y., Stern, D.B., Wostrikoff, K., Kuras, R., Girard-Bascou, J., & Wollman, F.A. (1998). Translation of cytochrome f is autoregulated through the 5' untranslated region of *petA* mRNA in Chlamydomonas chloroplasts. *Proc. Natl. Acad. Sci. USA, 95,* 4380-4385.

Choquet, Y., & Vallon, O. (2000). Synthesis, assembly and degradation of thylakoid membrane proteins. *Biochimie, 82,* 615-634.

Choquet, Y., & Wollman, F.A. (2002). Translational regulations as specific traits of chloroplast gene expression. *FEBS Lett., 529,* 39-42.

Choquet, Y., Wostrikoff, K., Rimbault, B., Zito, F., Girard-Bascou, J., Drapier, D., & Wollman, F.A. (2001). Assembly-controlled regulation of chloroplast gene translation. *Biochem. Soc. Trans., 29,* 421-426.

Christopher, D.A., & Hoffer, P.H. (1998). DET1 represses a chloroplast blue light-responsive promoter in a developmental and tissue-specific manner in *Arabidopsis thaliana. Plant J., 14,* 1-11.

Christopher, D.A., Kim, M., & Mullet, J.E. (1992). A novel light-regulated promoter is conserved in cereal and dicot chloroplasts. *Plant Cell, 4,* 785-798.

Christopher, D.A., & Mullet, J.E. (1994). Separate photosensory pathways co-regulate blue light/ultraviolet-A-activated *psbD-psbC* transcription and light-induced D2 and CP43 degradation in barley (*Hordeum vulgare*) chloroplasts. *Plant Physiol., 104,* 1119-1129.

Chua, N.H., G., B., P., S., & G.E., P. (1976) Periodic variations in the ratio of free to thylakoid-bound chloroplast ribosomes during the cell cycle of *Chlamydomonas reinhardtii. J. Cell Biol., 71,* 497-514.

Cline, K., & Henry, R. (1996). Import and routing of nucleus-encoded chloroplast proteins. *Annu. Rev. Cell. Dev. Biol., 12*, 1-26.

Coffin, J.W., Dhillon, R., Ritzel, R.G., & Nargang, F.E. (1997). The *Neurospora crassa* cya-5 nuclear gene encodes a protein with a region of homology to the *Saccharomyces cerevisiae* PET309 protein and is required in a post-transcriptional step for the expression of the mitochondrially encoded COXI protein. *Curr. Genet., 32*, 273-280.

Cohen, A., Yohn, C.B., & Mayfield, S. (2001). Translation of the chloroplast-encoded *psbD* is arrested post-initiation in a nuclear mutant of *Chlamydomonas reinhardtii. J. Plant Physiol., 158*, 1069-1075.

Daniell, H., Khan, M.S., & Allison, L. (2002). Milestones in chloroplast genetic engineering: an environmentally friendly era in biotechnology. *Trends Plant Sci., 7*, 84-91.

Daniell, D., Carmona-Sanchez, O., & Burns, B. B. (2004a). Chloroplast derived antibodies, biopharmaceuticals and edible vaccines. In R. Fischer & S. Schillberg (Eds.) *Molecular Farming* (pp.113-133). Weinheim: WILEY-VCH Verlag.

Daniell, H., Chebolu, S., Kumar, S., Singleton, M., Falconer, R. (2004b). Chloroplast-derived vaccine antigens and other therapeutic proteins. *Vaccine, in press*

Daniell, H., Datta, R., Varma, S., Gray, S., & Lee, S. B. (1998). Containment of herbicide resistance through genetic engineering of the chloroplast genome. *Nat. Biotechnol., 16*, 345-348.

Danon, A. (1997) Translational regulation in the chloroplast. *Plant Physiol., 115*, 1293-1298.

Danon, A., & Mayfield, S.P. (1991). Light regulated translational activators: identification of chloroplast gene specific mRNA binding proteins. *EMBO J., 10*, 3993-4001.

Danon, A., & Mayfield, S.P. (1994a). ADP-dependent phosphorylation regulates RNA-binding in vitro: implications in light-modulated translation. *EMBO J., 13*, 2227-2235.

Danon, A., & Mayfield, S.P. (1994b). Light-regulated translation of chloroplast messenger RNAs through redox potential. *Science, 266*, 1717-1719.

Darzacq, X., Powrie, E., Gu, W., Singer, R.H., & Zenklusen, D. (2003). RNA asymmetric distribution and daughter/mother differentiation in yeast. *Curr. Opin. Microbiol. 6*, 614-620.

Dauvillee, D., Stampacchia, O., Girard-Bascou, J., & Rochaix, J.D. (2003). Tab2 is a novel conserved RNA binding protein required for translation of the chloroplast *psaB* mRNA. *EMBO J., 22*, 6378-6388.

DeCosa, B., Moar, W., Lee, S. B., Miller, M., & Daniell, H. (2001). Overexpression of the *Bt* cry2Aa2 operon in chloroplasts leads to formation of insecticidal crystals. *Nat. Biotechnol., 19*, 71-74.

DeGray, G., Rajasekaran, K., Smith, F., Sanford, J., & Daniell, H. (2001). Expression of an antimicrobial peptide via the chloroplast genome to control phytopathogenic bacteria and fungi. *Plant Physiology, 127*, 852-862.

de Vitry, C., Olive, J., Drapier, D., Recouvreur, M., & Wollman, F.A. (1989). Posttranslational events leading to the assembly of photosystem II protein complex: a study using photosynthesis mutants from *Chlamydomonas reinhardtii. J. Cell Biol., 109*, 991-1006.

Dhingra, A., Portis, A.R. & Daniell, H. (2004). Enhanced translation of a chloroplast expressed *RbcS* gene restores SSU levels and photosynthesis in nuclear antisense *RbcS* plants. *Proc. Natl. Acad. Sci.*, U.S.A.101: 6315-6320.

Deshpande, N.N., Bao, Y., & Herrin, D.L. (1997). Evidence for light/redox-regulated splicing of *psbA* pre-RNAs in Chlamydomonas chloroplasts. *RNA, 3*, 37-48.

Draper, D.E., Gluick, T.C., & Schlax, P.J. (1998). Pseudoknots, RNA Folding, and Translational Regulation. In R.W. Simons and M. Grunberg-Manago (Eds.), *RNA Structure and Function* (pp. 415-436). Plainview: Cold Spring Harbor Laboratory Press.

Drapier, D., Girard-Bascou, J., & Wollman, F.A. (1992). Evidence for Nuclear Control of the Expression of the *atpA* and *atpB* Chloroplast Genes in Chlamydomonas. *Plant Cell, 4*, 283-295.

Eberhard, S., Drapier, D., & Wollman, F.A. (2002). Searching limiting steps in the expression of chloroplast-encoded proteins: relations between gene copy number, transcription, transcript abundance and translation rate in the chloroplast of *Chlamydomonas reinhardtii. Plant J., 31*, 149-160.

Edhofer, I., Muhlbauer, S.K., & Eichacker, L.A. (1998). Light regulates the rate of translation elongation of chloroplast reaction center protein D1. *Eur. J. Biochem., 257*, 78-84.

Ehresmann, C., Philippe, C., Westhof, E., Benard, L., Portier, C., & Ehresmann, B. (1995). A pseudoknot is required for efficient translational initiation and regulation of the *Escherichia coli rpsO* gene coding for ribosomal protein S15. *Biochem. Cell Biol., 73*, 1131-1140.

Eichacker, L.A., Soll, J., Lauterbach, P., Rudiger, W., Klein, R.R., & Mullet, J.E. (1990). *In vitro* synthesis of chlorophyll a in the dark triggers accumulation of chlorophyll a apoproteins in barley etioplasts. *J. Biol. Chem., 265*, 13566-13571.

Erickson, J.M., Pfister, K., Rahire, M., Togasaki, R.K., Mets, L., & Rochaix, J.D. (1989). Molecular and biophysical analysis of herbicide-resistant mutants of *Chlamydomonas reinhardtii*: structure-function relationship of the photosystem II D1 polypeptide. *Plant Cell, 1*, 361-371.

Fankhauser, C., & Staiger, D. (2002). Photoreceptors in *Arabidopsis thaliana*: light perception, signal transduction and entrainment of the endogenous clock. *Planta, 216*, 1-16.

Fernandez-San Millan, A., Mingeo-Castel, A. M., Miller, M., & Daniell, H. (2003). A chloroplast transgenic approach to hyper-express and purify human serum albumin, a protein highly susceptible to proteolytic degradation. *Plant Biotechnology Journal 1*, 71-79.

Fisk, D.G., Walker, M.B., & Barkan, A. (1999). Molecular cloning of the maize gene *crp1* reveals similarity between regulators of mitochondrial and chloroplast gene expression. *EMBO J., 18*, 2621-2630.

Fong, C.L., Lentz, A., & Mayfield, S.P. (2000). Disulfide bond formation between RNA binding domains is used to regulate mRNA binding activity of the chloroplast poly(A)-binding protein. *J. Biol. Chem., 275*, 8275-8278.

Fromm, H., Devic, M., Fluhr, R., & Edelman, M. (1985). Control of *psbA* gene expression: in mature Spirodela chloroplasts light regulation of 32-kd protein synthesis is independent of transcript level. *EMBO J., 4*, 291-295.

Gamble, P.E., & Mullet, J.E. (1989). Translation and stability of proteins encoded by the plastid *psbA* and *psbB* genes are regulated by a nuclear gene during light-induced chloroplast development in barley. *J. Biol. Chem., 264*, 7236-7243.

Gillham, N.W. (1994). *Organelle genes and genomes*. New York: Oxford University Press.

Girard-Bascou, J., Choquet, Y., Schneider, M., Delosme, M., & Dron, M. (1987). Characterization of a chloroplast mutation in the *psaA2* gene of *Chlamydomonas reinhardtii*. *Curr. Genet., 12*, 489-495.

Girard-Bascou, J., Pierre, Y., & Drapier, D. (1992). A nuclear mutation affects the synthesis of the chloroplast *psbA* gene production *Chlamydomonas reinhardtii*. *Curr. Genet., 22*, 47-52.

Gold, L. (1988). Posttranscriptional regulatory mechanisms in *Escherichia coli*. *Annu. Rev. Biochem., 57*, 199-233.

Goldschmidt-Clermont, M. (1998). Coordination of nuclear & chloroplast gene expression in plant cells. *Int. Rev. Cytol., 177*, 115-180.

Goodenough, U.W. (1971). The effects of inhibitors of RNA and protein synthesis on chloroplast structure and function in wild-type *Chlamydomonas reinhardtii*. *J. Cell Biol. 50*, 35-49.

Harris, E.H. (2001). Chlamydomonas as a Model Organism. *Annu. Rev. Plant Physiol. Plant Mol. Biol., 52*, 363-406.

Harris, E.H., Boynton, J.E., & Gillham, N.W. (1994). Chloroplast ribosomes and protein synthesis. *Microbiol. Rev., 58*, 700-754.

Hauser, C.R., Gillham, N.W., & Boynton, J.E. (1996). Translational regulation of chloroplast genes. Proteins binding to the 5'-untranslated regions of chloroplast mRNAs in *Chlamydomonas reinhardtii*. *J. Biol. Chem., 271*, 1486-1497.

Hauser, C.R., Gillham, N.W., & Boynton, J.E. (1998). Regulation of Chloroplast Translation. In J-D. Rochaix, M. Goldschmidt-Clermont and S. Merchant (Eds.), *The Molecular Biology of Chloroplasts and Mitochondria in Chlamydomonas* (pp. 197-21). Norwell, Dordrecht: Kluwer Academic Publishers.

Hayes, R., Kudla, J., & Gruissem, W. (1999). Degrading chloroplast mRNA: the role of polyadenylation. *Trends Biochem. Sci., 24*, 199-202.

Herrin, D., Michaels, A.S., & Paul, A.L. (1986). Regulation of genes encoding the large subunit of ribulose-1,5- bisphosphate carboxylase and the photosystem II polypeptides D-1 and D- 2 during the cell cycle of *Chlamydomonas reinhardtii*. *J. Cell Biol., 103*, 1837-1845.

Herrin, D.L., Battey, J.F., Greer, K., & Schmidt, G.W. (1992). Regulation of chlorophyll apoprotein expression and accumulation. Requirements for carotenoids and chlorophyll. *J. Biol. Chem., 267*, 8260-8269.

Herrin, D.L., & Michaels, A. (1985). *In vitro* synthesis and assembly of the peripheral subunits of coupling factor CF1 (alpha and beta) by thylakoid-bound ribosomes. *Arch. Biochem. Biophys., 237*, 224-236.

Hirose, T., Kusumegi, T., & Sugiura, M. (1998). Translation of tobacco chloroplast *rps14* mRNA depends on a Shine-Dalgarno-like sequence in the 5'-untranslated region but not on internal RNA editing in the coding region. *FEBS Lett., 430*, 257-260.

Hirose, T., & Sugiura, M. (1996). Cis-acting elements and trans-acting factors for accurate translation of chloroplast *psbA* mRNAs: development of an in vitro translation system from tobacco chloroplasts. *EMBO J., 15*, 1687-1695.

Hirose, T., & Sugiura, M. (1997). Both RNA editing and RNA cleavage are required for translation of tobacco chloroplast *ndhD* mRNA: a possible regulatory mechanism for the expression of a chloroplast operon consisting of functionally unrelated genes. *EMBO J., 16*, 6804-6811.

Hoober, J.K., Boyd, C.O., & L.G., P. (1991). Origin of the thylakoid membranes in *Chlamydomonas reinhardtii y-1* in the light or dark. *Plant Physiol., 96*, 1321-1328.

Hoober, J.K., Park, J.M., G.R., W., Y., K., & L.L., E. (1998). Assembly of Light-Harvesting Systems. In J-D. Rochaix, M. Goldschmidt-Clermont and S. Merchant (Eds.), *The Molecular Biology of Chloroplasts and Mitochondria in Chlamydomonas* (pp. 363-376). Norwell, Dordrecht: Kluwer Academic Publishers.

Hosler, J.P., A., W.E., Harris, E.H., Gillham, N., & Boynton, J. (1989). Relationship between gene dosage and gene expression in the chloroplast of *Chlamydomonas reinhardtii*. *Plant Physiol., 91*, 648-655.

Hugueney, P., Bouvier, F., Badillo, A., d'Harlingue, A., Kuntz, M., & Camara, B. (1995). Identification of a plastid protein involved in vesicle fusion and/or membrane protein translocation. *Proc. Natl. Acad. Sci. USA, 92*, 5630-5634.

Irihimovitch, V., & Shapira, M. (2000). Glutathione redox potential modulated by reactive oxygen species regulates translation of Rubisco large subunit in the chloroplast. *J. Biol. Chem., 275*, 16289-16295.

Jacobshagen, S., & Johnson, C.H. (1994). Circadian rhythms of gene expression in *Chlamydomonas reinhardtii*: circadian cycling of mRNA abundances of cab II, and possibly of beta-tubulin and cytochrome c. *Eur. J. Cell Biol., 64*, 142-152.

Jacobshagen, S., Kindle, K.L., & Johnson, C.H. (1996). Transcription of CABII is regulated by the biological clock in *Chlamydomonas reinhardtii*. *Plant Mol. Biol., 31*, 1173-1184.

Jarvis, P. (2001). Intracellular signalling: The chloroplast talks! *Curr. Biol., 11*, R307-310.

Johanningmeier, U. (1988). Possible control of transcript levels by chlorophyll precursors in Chlamydomonas. *Eur. J. Biochem., 177*, 417-424.

Johanningmeier, U., & Howell, S.H. (1984). Regulation of light-harvesting chlorophyll-binding protein mRNA accumulation in *Chlamydomonas reinhardtii*. Possible involvement of chlorophyll synthesis precursors. *J. Biol. Chem., 259*, 13541-13549.

Johnstone, O., & Lasko, P. (2001). Translational regulation and RNA localization in Drosophila oocytes and embryos. *Annu. Rev. Genet., 35*, 365-406.

Joyard, J., Teyssier, E., Miege, C., Berny-Seigneurin, D., Marechal, E., Block, M.A., Dorne, A.J., Rolland, N., Ajlani,, G. & Douce, R. (1998). The biochemical machinery of plastid envelope membranes. *Plant Physiol., 118*, 715-723.

Kannangara, C.G., Vothknecht, U.C., Hansson, M., & von Wettstein, D. (1997). Magnesium chelatase: association with ribosomes and mutant complementation studies identify barley subunit Xantha-G as a functional counterpart of Rhodobacter subunit BchD. *Mol. Gen. Genet., 254*, 85-92.

Kettunen, R., Pursiheimo, S., Rintamaki, E., Van Wijk, K.J., & Aro, E.M. (1997). Transcriptional and translational adjustments of *psbA* gene expression in mature chloroplasts during photoinhibition and subsequent repair of photosystem II. *Eur. J. Biochem., 247*, 441-448.

Khrebtukova, I., & Spreitzer, R.J. (1996). Elimination of the Chlamydomonas gene family that encodes the small subunit of ribulose-1,5-bisphosphate carboxylase/oxygenase. *Proc. Natl. Acad. Sci. USA, 93*, 13689-13693.

Kim, J., Eichacker, L.A., Rudiger, W., & Mullet, J.E. (1994a). Chlorophyll regulates accumulation of the plastid-encoded chlorophyll proteins P700 and D1 by increasing apoprotein stability. *Plant Physiol., 104*, 907-916.

Kim, J., Klein, P.G., & Mullet, J.E. (1994b). Synthesis and turnover of photosystem II reaction center protein D1. Ribosome pausing increases during chloroplast development. *J. Biol. Chem., 269*, 17918-17923.

Kim, J., Klein, P.G., & Mullet, J.E. (1994c). Vir-115 gene product is required to stabilize D1 translation intermediates in chloroplasts. *Plant Mol. Biol., 25*, 459-467.

Kim, J., & Mayfield, S.P. (1997). Protein disulfide isomerase as a regulator of chloroplast translational activation. *Science, 278*, 1954-1957.

Kim, J., & Mullet, J.E. (1994) Ribosome-binding sites on chloroplast *rbcL* and *psbA* mRNAs and light-induced initiation of D1 translation. *Plant Mol. Biol., 25*, 437-448.

Kim, J., & Mullet, J.E. (2003). A mechanism for light-induced translation of the *rbcL* mRNA encoding the large subunit of ribulose-1,5-bisphosphate carboxylase in barley chloroplasts. *Plant Cell Physiol., 44*, 491-499.

Kim, M., Christopher, D.A., & Mullet, J.E. (1993). Direct evidence for selective modulation of *psbA*, *rpoA*, *rbcL* and 16S RNA stability during barley chloroplast development. *Plant Mol. Biol., 22*, 447-463.

Kim, M., Thum, K.E., Morishige, D.T., & Mullet, J.E. (1999). Detailed architecture of the barley chloroplast *psbD-psbC* blue light-responsive promoter. *J. Biol. Chem., 274*, 4684-4692.

Klaff, P., & Gruissem, W. (1991). Changes in Chloroplast mRNA Stability during Leaf Development. *Plant Cell, 3*, 517-529.

Klaff, P., Mundt, S.M., & Steger, G. (1997). Complex formation of the spinach chloroplast *psbA* mRNA 5' untranslated region with proteins is dependent on the RNA structure. *RNA, 3*, 1468-1479.

Klein, R.R., Gamble, P.E., & Mullet, J.E. (1988a). Light-Dependent Accumulation of Radiolabeled Plastid-Encoded Chlorophyll a-Apoproteins Required Chlorophyll a. *Plant Physiol., 88*, 1246-1256.

Klein, R.R., Mason, H.S., & Mullet, J.E. (1988b). Light-regulated translation of chloroplast proteins. I. Transcripts of *psaA-psaB*, *psbA*, and *rbcL* are associated with polysomes in dark-grown and illuminated barley seedlings. *J. Cell Biol., 106*, 289-301.

Klein, R.R., & Mullet, J. (1986). Regulation of Chloroplast-encoded Chlorophyll-binding protein translation during higher plant chloroplast biogenesis. *J. Biol. Chem., 261*, 11138-11145.

Klein, R.R., & Mullet, J.E. (1987). Control of gene expression during higher plant chloroplast biogenesis. Protein synthesis and transcript levels of *psbA*, *psaA-psaB*, and *rbcL* in dark-grown and illuminated barley seedlings. *J. Biol. Chem., 24*, 4341-4348.

Koo, J.S., & Spremulli, L.L. (1994a). Analysis of the translational initiation region on the Euglena gracilis chloroplast ribulose-bisphosphate carboxylase/oxygenase (*rbcL*) messenger RNA. *J. Biol. Chem. 269*, 7494-7500.

Koo, J.S., & Spremulli, L.L. (1994b). Effect of the secondary structure in the Euglena gracilis chloroplast ribulose-bisphosphate carboxylase/oxygenase messenger RNA on translational initiation. *J. Biol. Chem., 269*, 7501-7508.

Krause, K., Maier, R.M., Kofer, W., Krupinska, K., & Herrmann, R.G. (2000). Disruption of plastid-encoded RNA polymerase genes in tobacco: expression of only a distinct set of genes is not based on selective transcription of the plastid chromosome. *Mol. Gen. Genet., 263*, 1022-1030.

Kreps, J.A., & Kay, S.A. (1997). Coordination of Plant Metabolism and Development by the Circadian Clock. *Plant Cell, 9*, 1235-1244.

Kroll, D., Meierhoff, K., Bechtold, N., Kinoshita, M., Westphal, S., Vothknecht, U.C., Soll, J., & Westhoff, P. (2001). VIPP1, a nuclear gene of *Arabidopsis thaliana* essential for thylakoid membrane formation. *Proc. Natl. Acad. Sci. USA, 98*, 4238-4242.

Kropat, J., Oster, U., Rudiger, W., & Beck, C.F. (1997). Chlorophyll precursors are signals of chloroplast origin involved in light induction of nuclear heat-shock genes. *Proc. Natl. Acad. Sci. USA, 94*, 14168-14172.

Kropat, J., Oster, U., Rudiger, W., & Beck, C.F. (2000). Chloroplast signaling in the light induction of nuclear HSP70 genes requires the accumulation of chlorophyll precursors and their accessibility to cytoplasm/nucleus. *Plant J., 24*, 523-531.

Kuchka, M., Mayfield, S., & Rochaix, J.D. (1988). Nuclear mutations specifically affect the synthesis and/or degradation of the chloroplast-encoded D2 polypeptide of photosystem II in *Chlamydomonas reinhardtii. EMBO J., 7*, 319-324.

Kumar, S., Dhingra, A. & Daniell, H. (2004). Plastid expressed *betaine aldehyde dehydrogenase* gene in carrot cultured cells, roots and leaves confers enhanced salt tolerance. *Plant Physiol* in press.

Kuras, R., & Wollman, F.A. (1994). The assembly of cytochrome b6/f complexes: an approach using genetic transformation of the green alga *Chlamydomonas reinhardtii. EMBO J., 13*, 1019-1027.

Kuras, R., Wollman, F.A., & Joliot, P. (1995). Conversion of cytochrome f to a soluble form *in vivo* in *Chlamydomonas reinhardtii. Biochemistry, 34*, 7468-7475.

Lee, J., & Herrin, D. (2002). Assessing the relative importance of light and the circadian clock in controlling chloroplast translation in *Chlamydomonas reinhardtii. Photosynth. Res., 72*, 295-306.

Lee, J., & Herrin, D.L. (2003). Mutagenesis of a light-regulated *psbA* intron reveals the importance of efficient splicing for photosynthetic growth. *Nucleic Acids Res., 31*, 4361-4372.

Lee, S. B., Kwon, H. B., Kwon, S. J., Park, S. C., Jeong, M. J., Han, S. E., Byun, M.O., Daniell, H. (2003). Accumulation of trehalose within transgenic chloroplasts confers drought tolerance. *Mol. Breeding, 11*, 1-13.

Legen, J., Kemp, S., Krause, K., Profanter, B., Herrmann, R.G., & Maier, R.M. (2002). Comparative analysis of plastid transcription profiles of entire plastid chromosomes from tobacco attributed to wild-type and PEP-deficient transcription machineries. *Plant J., 31*, 171-188.

Lin, C. (2002). Blue light receptors and signal transduction. *Plant Cell, 14 (Suppl)*, S207-S225.

Malnoe, P., Mayfield, S.P., & Rochaix, J.D. (1988). Comparative analysis of the biogenesis of photosystem II in the wild- type and *y-1* mutant of *Chlamydomonas reinhardtii. J. Cell Biol., 106*, 609-616.

Manthey, G.M., & McEwen, J.E. (1995). The product of the nuclear gene PET309 is required for translation of mature mRNA and stability or production of intron-containing RNAs derived from the mitochondrial COX1 locus of *Saccharomyces cerevisiae. EMBO J., 14*, 4031-4043.

Manthey, G.M., Przybyla-Zawislak, B.D., & McEwen, J.E. (1998). The *Saccharomyces cerevisiae* Pet309 protein is embedded in the mitochondrial inner membrane. *Eur. J. Biochem., 255*, 156-161.

Margulies, M.M., & Michaels, A. (1974). Ribosomes bound to chloroplast membranes in *Chlamydomonas reinhardtii. J. Cell Biol., 60*, 65-77.

Margulies, M.M., & Michaels, A. (1975). Free and membrane-bound chloroplast polyribosomes *Chlamydomonas reinhardtii. Biochim. Biophys. Acta., 402*, 297-308.

Margulies, M.M., Tiffany, H.L. & Michaels, A. (1975). Vectorial discharge of nascent polypeptides attached to chloroplast thylakoid membranes. *Biochem. Biophys. Res. Commun., 64*, 735-739.

Margulies, M.M., & Weistrop, J.S. (1980). Sub-thylakoid fractions containing ribosomes. *Biochim. Biophys. Acta., 606*, 20-33.

Martin, W., Rujan, T., Richly, E., Hansen, A., Cornelsen, S., Lins, T., Leister, D., Stoebe, B., Hasegawa, M., & Penny, D. (2002). Evolutionary analysis of Arabidopsis, cyanobacterial, and chloroplast genomes reveals plastid phylogeny and thousands of cyanobacterial genes in the nucleus. *Proc. Natl. Acad. Sci. USA, 99*, 12246-12251.

Mathews, D.H., Sonenberg, N., & Hershey, J.W.B. (1996). *Origins and targets of translational control. Translational Control*. Plainview: Cold Spring Harbor Laboratory Press.

Mayfield, S.P. (1990). Chloroplast gene regulation: interaction of the nuclear and chloroplast genomes in the expression of photosynthetic proteins. *Curr. Opin. Cell. Biol., 2*, 509-513.

McCarthy, J.E., & Brimacombe, R. (1994. Prokaryotic translation: the interactive pathway leading to initiation. *Trends Genet., 10*, 402-407.

McBride, K. E., Svab, Z., Schaaf, D. J., Hogan, P. S., Stalker, D. M., & Maliga, P. (1995). Amplification of a chimeric *Bacillus* gene in chloroplasts leads to an extraordinary level of an insecticidal protein in tobacco. *Bio/Technology 13*, 362-365.

McCormac, D.J., & Barkan, A. (1999). A nuclear gene in maize required for the translation of the chloroplast atpB/E mRNA. *Plant Cell, 11*, 1709-1716.

Meurer, J., Berger, A., & Westhoff, P. (1996). A nuclear mutant of Arabidopsis with impaired stability on distinct transcripts of the plastid *psbB*, *psbD/C*, *ndhH*, and *ndhC* operons. *Plant Cell, 8*, 1193-1207.

Michaels, A., & Herrin, D.L. (1990). Translational regulation of chloroplast gene expression during the light-dark cell cycle of Chlamydomonas: evidence for control by ATP/energy supply. *Biochem Biophys. Res. Commun., 170*, 1082-1088.

Minai, L., Cohen, Y., Chitnis, P.R., & Nechushtai, R. (1996). The precursor of PsaD assembles into the photosystem I complex in two steps. *Proc. Natl. Acad. Sci. USA, 93*, 6338-6342.

Minai, L., Fish, A., Darash-Yahana, M., Verchovsky, L., & Nechushtai, R. (2001). The Assembly of the PsaD Subunit into the Membranal Photosystem I Complex Occurs via an Exchange Mechanism. *Biochemistry, 40*, 12754-12760.

Minami, E., Shinohara, K., Kuwabara, T., & Watanabe, A. (1986). *In vitro* synthesis and assembly of photosystem II proteins of spinach chloroplasts. *Arch. Biochem. Biophys., 244*, 517-527.

Mittag, M., & Wagner, V. (2003). The circadian clock of the unicellular eukaryotic model organism *Chlamydomonas reinhardtii. Biol. Chem., 384*, 689-695.

Mochizuki, N., Brusslan, J.A., Larkin, R., Nagatani, A., & Chory, J. (2001). Arabidopsis genomes uncoupled 5 (GUN5) mutant reveals the involvement of Mg-chelatase H subunit in plastid-to-nucleus signal transduction. *Proc. Natl. Acad. Sci. USA, 98*, 2053-2058.

Monde, R.A., Zito, F., Olive, J., Wollman, F.A., & Stern, D.B. (2000). Post-transcriptional defects in tobacco chloroplast mutants lacking the cytochrome b6/f complex. *Plant J., 21*, 61-72.

Morais, F., Barber, J., & Nixon, P.J. (1998). The chloroplast-encoded alpha subunit of cytochrome b-559 is required for assembly of the photosystem two complex in both the light and the dark in *Chlamydomonas reinhardtii. J. Biol. Chem., 273*, 29315-29320.

Muhlbauer, S.K., & Eichacker, L.A. (1998). Light-dependent formation of the photosynthetic proton gradient regulates translation elongation in chloroplasts. *J. Biol. Chem., 273*, 20935-20940.

Muhlbauer, S.K., & Eichacker, L.A. (1999). The stromal protein large subunit of ribulose-1,5-bisphosphate carboxylase is translated by membrane-bound ribosomes. *Eur. J. Biochem., 261*, 784-788.

Mullet, J.E., Klein, P.G., & Klein, R.R. (1990). Chlorophyll regulates accumulation of the plastid-encoded chlorophyll apoproteins CP43 and D1 by increasing apoprotein stability. *Proc. Natl. Acad. Sci. USA 87*, 4038-4042.

Nagy, F., & Schafer, E. (2000). Nuclear and cytosolic events of light-induced, phytochrome-regulated signaling in higher plants. *EMBO J., 19*, 157-163.

Nakahira, Y., Baba, K., Yoneda, A., Shiina, T., & Toyoshima, Y. (1998). Circadian-regulated transcription of the psbD light-responsive promoter in wheat chloroplasts. *Plant Physiol., 118*, 1079-1088.

Neckermann, K., Zeltz, P., Igloi, G.L., Kossel, H., & Maier, R.M. (1994). The role of RNA editing in conservation of start codons in chloroplast genomes. *Gene, 146*, 177-182.

Nickelsen, J. (2003). Chloroplast RNA-binding proteins. *Curr. Genet. 43*, 392-399.

Nilsson, R., Brunner, J., Hoffman, N.E., & van Wijk, K.J. (1999). Interactions of ribosome nascent chain complexes of the chloroplast- encoded D1 thylakoid membrane protein with cpSRP54. *EMBO J., 18*, 733-742.

Oelmuller, R., Schneiderbauer, A., Herrmann, R.G., & Kloppstech, K. (1995). The steady-state mRNA levels for thylakoid proteins exhibit coordinate diurnal regulation. *Mol. Gen. Genet., 246*, 478-484.

Ohad, I. (1974). Biogenesis and modulation of membrane properties in greening *Chlamydomonas reinhardtii, y-1* cells. *Methods Enzymol., 32 (Part B)*, 865-871.

Ohad, I., Adir, N., Koike, H., Kyle, D.J., & Inoue, Y. (1990). Mechanism of photoinhibition *in vivo*. A reversible light-induced conformational change of reaction center II is related to an irreversible modification of the D1 protein. *J. Biol. Chem., 265*, 1972-1979.

Ohad, I., Siekevitz, P., & Palade, G.E. (1967a). Biogenesis of chloroplast membranes. I. Plastid dedifferentiation in a dark-grown algal mutant (*Chlamydomonas reinhardtii*). *J. Cell Biol., 35*, 521-552.

Ohad, I., Siekevitz, P., & Palade, G.E. (1967b). Biogenesis of chloroplast membranes. II. Plastid differentiation during greening of a dark-grown algal mutant (*Chlamydomonas reinhardtii*). *J. Cell Biol., 35*, 553-584.

Ossenbuhl, F., Hartmann, K., & Nickelsen, J. (2002). A chloroplast RNA binding protein from stromal thylakoid membranes specifically binds to the 5' untranslated region of the *psbA* mRNA. *Eur. J. Biochem., 269*, 3912-3919.

Park, J.M., Cho, J.H., Kang, S.G., Jang, H.J., Pih, K.T., Piao, H.L., Cho, M.J., & Hwang, I. (1998). A dynamin-like protein in *Arabidopsis thaliana* is involved in biogenesis of thylakoid membranes. *EMBO J., 17*, 859-867.

Perron, K., Goldschmidt-Clermont, M., & Rochaix, J.D. (1999). A factor related to pseudouridine synthases is required for chloroplast group II intron trans-splicing in *Chlamydomonas reinhardtii. EMBO J., 18*, 6481-6490.

Pfannschmidt, T. (2003). Chloroplast redox signals: how photosynthesis controls its own genes. *Trends Plant Sci., 8*, 33-41.

Pfannschmidt, T., Nilsson, A., Tullberg, A., Link, G., & Allen, J.F. (1999). Direct transcriptional control of the chloroplast genes *psbA* and *psaAB* adjusts photosynthesis to light energy distribution in plants. *IUBMB Life, 48*, 271-276.

Pfannschmidt, T., Schutze, K., Brost, M., & Oelmuller, R. (2001). A novel mechanism of nuclear photosynthesis gene regulation by redox signals from the chloroplast during photosystem stoichiometry adjustment. *J. Biol. Chem., 276*, 36125-36130.

Pfannschmidt, T., Schutze, K., Fey, V., Sherameti, I., & Oelmuller, R. (2003). Chloroplast redox control of nuclear gene expression--a new class of plastid signals in interorganellar communication. *Antioxid. Redox Signal., 5*, 95-101.

Plader, W., & Sugiura, M. (2003). The Shine-Dalgarno-like sequence is a negative regulatory element for translation of tobacco chloroplast *rps2* mRNA: an additional mechanism for translational control in chloroplasts. *Plant J., 34*, 377-382.

Race, H.L., Herrmann, R.G., & Martin, W. (1999). Why have organelles retained genomes? *Trends Genet., 15*, 364-370.

Rattanachaikunsopon, P., Rosch, C., & Kuchka, M.R. (1999). Cloning and characterization of the nuclear AC115 gene of *Chlamydomonas reinhardtii. Plant. Mol. Biol., 39*, 1-10.

Reinbothe, S., & Reinbothe, C. (1996). Regulation of Chlorophyll Biosynthesis in Angiosperms. *Plant Physiol., 111*, 1-7.

Reinbothe, S., Reinbothe, C., Heintzen, C., Seidenbecher, C., & Parthier, B. (1993). A methyl jasmonate-induced shift in the length of the 5' untranslated region impairs translation of the plastid *rbcL* transcript in barley. *EMBO J., 12*, 1505-1512.

Reinsberg, D., Ottmann, K., Booth, P.J., & Paulsen, H. (2001). Effects of chlorophyll a, chlorophyll b, and xanthophylls on the in vitro assembly kinetics of the major light-harvesting chlorophyll a/b complex, LHCIIb. *J. Mol. Biol., 308*, 59-67.

Rochaix, J.D. (1996). Post-transcriptional regulation of chloroplast gene expression in *Chlamydomonas reinhardtii. Plant Mol. Biol., 32*, 327-341.

Rochaix, J.D., & Erickson, J. (1988). Function and assembly of photosystem II: genetic and molecular analysis. *Trends Biochem. Sci ., 13*, 56-59.

Rochaix, J.D., Goldschmidt-Clermont, M., & Merchant, S. (1998). *The molecular biology of chloroplasts and mitochondria in Chlamydomonas*. Dordrecht, Boston: Kluwer Academic Publishers.

Rochaix, J.D., Kuchka, M., Mayfield, S., Schirmer-Rahire, M., Girard-Bascou, J., & Bennoun, P. (1989. Nuclear and chloroplast mutations affect the synthesis or stability of the chloroplast *psbC* gene product in *Chlamydomonas reinhardtii. EMBO J., 8*, 1013-1021.

Rodermel, S., Haley, J., Jiang, C.Z., Tsai, C.H., & Bogorad, L. (1996). A mechanism for intergenomic integration: abundance of ribulose bisphosphate carboxylase small-subunit protein influences the translation of the large-subunit mRNA. *Proc. Natl. Acad. Sci. USA, 93*, 3881-3885.

Rohl, T., & van Wijk, K.J. (2001). *In vitro* reconstitution of insertion and processing of cytochrome f in a homologous chloroplast translation system. *J. Biol. Chem., 276*, 35465-35472.

Roy, L.M., & Barkan, A. (1998). A SecY homologue is required for the elaboration of the chloroplast thylakoid membrane and for normal chloroplast gene expression. *J. Cell Biol., 141*, 385-395.

Ruiz, O. N., Hussein, H., Terry, N., & Daniell, H. (2003). Phytoremediation of organomercurial compounds via chloroplast genetic engineering. *Plant Physiol., 132*, 1-9.

Salvador, M.L., & Klein, U. (1999). The redox state regulates RNA degradation in the chloroplast of *Chlamydomonas reinhardtii. Plant Physiol ., 121*, 1367-1374.

Sato, N., Albrieux, C., Joyard, J., Douce, R., & Kuroiwa, T. (1993). Detection and characterization of a plastid envelope DNA-binding protein which may anchor plastid nucleoids. *EMBO J., 12*, 555-561.

Sato, N., Ohshima, K., Watanabe, A., Ohta, N., Nishiyama, Y., Joyard, J., & Douce, R. (1998). Molecular characterization of the PEND protein, a novel bZIP protein present in the envelope membrane that is the site of nucleoid replication in developing plastids. *Plant Cell, 10*, 859-872.

Sato, N., Rolland, N., Block, M.A., & Joyard, J. (1999). Do plastid envelope membranes play a role in the expression of the plastid genome? *Biochimie, 81*, 619-629.

Savard, F., Richard, C., & Guertin, M. (1996). The *Chlamydomonas reinhardtii* LI818 gene represents a distant relative of the *cabI/II* genes that is regulated during the cell cycle and in response to illumination. *Plant Mol. Biol., 32*, 461-473.

Schlax, P.J., Xavier, K.A., Gluick, T.C., & Draper, D.E. (2001). Translational repression of the *Escherichia coli* alpha operon mRNA: importance of an mRNA conformational switch and a ternary entrapment complex. *J. Biol. Chem., 276*, 38494-38501.

Sexton, T.B., Christopher, D.A., & Mullet, J.E. (1990). Light-induced switch in barley *psbD-psbC* promoter utilization: a novel mechanism regulating chloroplast gene expression. *EMBO J., 9*, 4485-4494.

Shapira, M., Lers, A., Heifetz, P.B., Irihimovitz, V., Osmond, C.B., Gillham, N.W., & Boynton, J.E. (1997). Differential regulation of chloroplast gene expression in *Chlamydomonas reinhardtii* during photoacclimation: light stress transiently suppresses synthesis of the Rubisco LSU protein while enhancing synthesis of the PS II D1 protein. *Plant Mol. Biol., 33*, 1001-1011.

Shen, Y., Danon, A., & Christopher, D.A. (2001). RNA binding-proteins interact specifically with the Arabidopsis chloroplast *psbA* mRNA 5' untranslated region in a redox-dependent manner. *Plant Cell Physiol., 42*, 1071-1078.

Stampacchia, O., Girard-Bascou, J., Zanasco, J.L., Zerges, W., Bennoun, P., & Rochaix, J.D. (1997). A nuclear-encoded function essential for translation of the chloroplast *psaB* mRNA in Chlamydomonas. *Plant Cell, 9*, 773-782.

Staub, J.M., & Maliga, P. (1993). Accumulation of D1 polypeptide in tobacco plastids is regulated via the untranslated region of the *psbA* mRNA. *EMBO J., 12*, 601-606.

Stern, D.B., Higgs, D.C., & Yang, J. (1997). Transcription and translation in chloroplasts. *Trends Plant Sci., 2*, 308-315.

Sugiura, M., Hirose, T., & Sugita, M. (1998). Evolution and mechanism of translation in chloroplasts. *Annu. Rev. Genet., 32*, 437-459.

Takahashi, Y., Utsumi, K., Yamamoto, Y., Hatano, A., & Satoh, K. (1996). Genetic engineering of the processing site of D1 precursor protein of photosystem II reaction center in *Chlamydomonas reinhardtii*. *Plant Cell Physiol., 37*, 161-168.

Taniguchi, M., Kuroda, H., & Satoh, K. (1993). ATP-dependent protein synthesis in isolated pea chloroplasts. Evidence for accumulation of a translation intermediate of the D1 protein. *FEBS Lett., 317*, 57-61.

Timko, M.P. (1998). Pigment biosynthesis: Chlorophylls, heme, and carotenoids. In J-D. Rochaix, M. Goldschmidt-Clermont and S. Merchant (Eds.), *The Molecular Biology of Chloroplasts and Mitochondria in Chlamydomonas* (pp. 403-420). Norwell, Dordrecht: Kluwer Academic Publishers.

Trebitsh, T., & Danon, A. (2001). Translation of chloroplast *psbA* mRNA is regulated by signals initiated by both photosystems II and I. *Proc. Natl. Acad. Sci. USA, 98*, 12289-12294.

Trebitsh, T., Levitan, A., Sofer, A., & Danon, A. (2000). Translation of chloroplast *psbA* mRNA is modulated in the light by counteracting oxidizing and reducing activities. *Mol. Cell. Biol., 20*, 1116-1123.

Trebitsh, T., Meiri, E., Ostersetzer, O., Adam, Z., & Danon, A. (2001). The Protein Disulfide Isomerase-like RB60 Is Partitioned between Stroma and Thylakoids in *Chlamydomonas reinhardtii* Chloroplasts. *J. Biol. Chem., 276*, 4564-4569.

Tullberg, A., Alexciev, K., Pfannschmidt, T., & Allen, J.F. (2000). Photosynthetic electron flow regulates transcription of the psaB gene in pea (*Pisum sativum* L.) chloroplasts through the redox state of the plastoquinone pool. *Plant Cell Physiol., 41*, 1045-1054.

van Wijk, K.J., Roobol-Boza, M., Kettunen, R., Andersson, B., & Aro, E.M. (1997). Synthesis and assembly of the D1 protein into photosystem II: processing of the C-terminus and identification of the initial assembly partners and complexes during photosystem II repair. *Biochemistry, 36*, 6178-6186.

Vothknecht, U.C., & Westhoff, P. (2001). Biogenesis and origin of thylakoid membranes. *Biochim. Biophys. Acta., 1541*, 91-101.

Wang, H., & Deng, X.W. (2003). Dissecting the phytochrome A-dependent signaling network in higher plants. *Trends Plant Sci., 8*, 172-178.

Westphal, S., Soll, J., & Vothknecht, U.C. (2001). A vesicle transport system inside chloroplasts. *FEBS Lett., 506*, 257-261.

Westphal, S., Soll, J., & Vothknecht, U.C. (2003). Evolution of chloroplast vesicle transport. *Plant Cell Physiol., 44*, 217-222.

Wollman, F.A., Minai, L., & Nechushtai, R. (1999). The biogenesis and assembly of photosynthetic proteins in thylakoid membranes. *Biochim. Biophys. Acta, 411*, 21-85.

Wostrikoff, K., Choquet, Y., Wollman, F.A., & Girard-Bascou, J. (2001) TCA1, a single nuclear-encoded translational activator specific for *petA* mRNA in *Chlamydomonas reinhardtii* chloroplast. *Genetics, 159*, 119-132.

Wostrikoff, K., Girard-Bascou, J., Wollman, F.A. and Choquet, Y. (2004) Biogenesis of PSI involves a cascade of translational autoregulation in the chloroplast of Chlamydomonas. Embo J 10, 10.

Wu, H.Y., & Kuchka, M.R. (1995). A nuclear suppressor overcomes defects in the synthesis of the chloroplast *psbD* gene product caused by mutations in two distinct nuclear genes of Chlamydomonas. *Curr. Genet., 27*, 263-269.

Yohn, C.B., Cohen, A., Danon, A., & Mayfield, S.P. (1996). Altered mRNA binding activity and decreased translational initiation in a nuclear mutant lacking translation of the chloroplast *psbA* mRNA. *Mol. Cell. Biol., 16*, 3560-3566.

Yohn, C.B., Cohen, A., Danon, A., & Mayfield, S.P. (1998a). A poly(A) binding protein functions in the chloroplast as a message- specific translation factor. *Proc. Natl. Acad. Sci. USA, 95*, 2238-2243.

Yohn, C.B., Cohen, A., Rosch, C., Kuchka, M.R., & Mayfield, S.P. (1998b). Translation of the chloroplast *psbA* mRNA requires the nuclear-encoded poly(A)-binding protein, RB47. *J. Cell Biol., 142*, 435-442.

Yosef, I., Irihimovitch, V., Knopf, J.A., Cohen, I., Dahan, I., Nahum, E., Keasar, C., & Shapira, M. (2003). RNA binding activity of ribulose-1,5-bisphosphate carboxylase/oxygenase large subunit from *Chlamydomonas reinhardtii. J. Biol. Chem.*, 279, 10148-10156.

Yu, N.J., & Spremulli, L.L. (1998). Regulation of the activity of chloroplast translational initiation factor 3 by NH2- and COOH-terminal extensions. *J. Biol. Chem., 273*, 3871-3877.

Zak, E., Norling, B., Maitra, R., Huang, F., Andersson, B., & Pakrasi, H.B. (2001). The initial steps of biogenesis of cyanobacterial photosystems occur in plasma membranes. *Proc. Natl. Acad. Sci. USA, 98*, 13443-13448.

Zerges, W. (2000). Translation in chloroplasts. *Biochimie, 82*, 583-601.

Zerges, W. (2002). Does complexity constrain organelle evolution? *Trends Plant Sci., 7*, 175-182.

Zerges, W., Auchincloss, A.H., & Rochaix, J.D. (2003). Multiple Translational Control Sequences in the 5' Leader of the Chloroplast *psbC* mRNA Interact With Nuclear Gene Products in *Chlamydomonas reinhardtii. Genetics, 163*, 895-904.

Zerges, W., Girard-Bascou, J. & Rochaix, J.D. (1997) Translation of the chloroplast *psbC* mRNA is controlled by interactions between its 5' leader and the nuclear loci TBC1 and TBC3 in *Chlamydomonas reinhardtii. Mol. Cell. Biol., 17*, 3440-3448.

Zerges, W. & Rochaix, J.D. (1994) The 5' leader of a chloroplast mRNA mediates the translational requirements for two nucleus-encoded functions in *Chlamydomonas reinhardtii. Mol. Cell. Biol., 14*, 5268-5277.

Zerges, W. and Rochaix, J.D. (1998) Low density membranes are associated with RNA-binding proteins and thylakoids in the chloroplast of *Chlamydomonas reinhardtii. J. Cell Biol., 140*, 101-110.

Zerges, W., Wang, S. and Rochaix, J.D. (2002) Light activates binding of membrane proteins to chloroplast RNAs in *Chlamydomonas reinhardtii. Plant Mol. Biol., 50*, 573-585.

Zhang, L., Paakkarinen, V., van Wijk, K.J. and Aro, E.M. (1999) Co-translational assembly of the D1 protein into photosystem II. J. *Biol. Chem., 274*, 16062-16067.

Zhang, L., Paakkarinen, V., van Wijk, K.J. and Aro, E.M. (2000) Biogenesis of the chloroplast-encoded D1 protein: regulation of translation elongation, insertion, and assembly into photosystem II. *Plant Cell, 12*, 1769-1782.

CHAPTER 14

TARGETING SIGNALS AND IMPORT MACHINERY OF PLASTIDS AND PLANT MITOCHONDRIA

ELZBIETA GLASER[1] AND JÜRGEN SOLL[2]

[1]*Department of Biochemistry and Biophysics, Stockholm University, Arrhenius Laboratories for Natural Sciences, 106 91 Stockholm, Sweden, E-Mail: e_glaser@dbb.su.se*

[2]*Department für Biologie I, Botanik, Ludwig-Maximilians-Universität München, Menzinger Straße 67, 80638 München, Germany, E-Mail: soll@uni-muenchen.de*

Abstract: Genetic information in plants is localized in three compartments: the nucleus, pastids and mitochondria. Despite both plastids and mitochondria containing their own DNA, most of the proteins of these organelles are nuclear encoded. The biogenesis of a functional organelle requires coordinate expression of both nuclear and organellar genomes and specific intracellular protein trafficking, processing and assembly machinery. Most nuclear-encoded organellar proteins are synthesized in the cytosol as precursor proteins with N-terminal signal peptides, required for sorting and targeting the protein to the correct organelle. Both plastid and plant mitochondrial signal peptides show a remarkable similarity in amino acid composition. They are rich in hydrophobic, hydroxylated and positively charged amino acid residues and deficient in acidic amino acids. However, despite great similarities of signal peptides, the plastid and mitochondrial protein targeting is specific. Yet, there is a group of proteins that are dually targeted to both organelles. Molecular chaperones and other cytosolic factors are involved in the import process: maintaining precursors in an import-competent conformation, guiding them to the organelle and assisting in transport through the membrane. Precursor proteins are transported into plastids through the translocons of the outer and the inner envelope membrane, TOC and TIC complexes, and into mitochondria through the translocases of the outer and the inner mitochondrial membrane, TOM and TIM complexes. Mitochondrial molecular chaperone Hsp70 functions as a molecular motor pulling precursor proteins into the mitochondria. The signal peptides are proteolytically cleaved off inside the organelles by highly specific organellar processing peptidases, the mitochondrial processing peptidases, that in plant mitochondria are an integral part of the cytochrome bc_1 complex of the respiratory chain, and the stromal processing peptidase, SPP, in plastids. The mature proteins, after intraorganellar sorting, are assembled by molecular chaperones into functional oligomeric protein complexes, whereas the cleaved targeting peptides, potentially destructive for biological membranes, are degraded inside the organelles by a newly identified signal peptide degrading zinc metalloprotease, both in plastids and in mitochondria.

H. Daniell and C. D. Chase (eds.), Molecular Biology and Biotechnology of Plant Organelles 385—417.

1. INTRODUCTION

Chloroplasts and mitochondria are genetically semiautonomous; they are believed to originate from endosymbiotic prokaryotic ancestors, chloroplasts from a cyanobacterium (McFadden, 2001), and mitochondria from an □-proteobacterium (Andersson and Kurland., 1999; Gray *et al.*, 1999; Lang *et al.*, 1999) that invaded primitive eukaryotic host cells and evolved to become organelles. During evolution most of the bacterial DNA has been either lost or transferred to the nucleus and only a small portion of the DNA is still present in both organelles. This residual DNA encodes 60-200 (Leister, 2003) proteins in chloroplasts and 3-67 proteins (Gray *et al.*, 1999) in mitochondria in different plant species; that corresponds to less than 1% of the total organellar protein content. The *Arabidopsis thaliana* genome has been completely sequenced and 25498 genes were annotated (Arabidopsis Genome Initiative, 2000). The recent proteome analysis of *Arabidopsis* organelles identified 80 lumenal (Peltier *et al.*, 2002; Schubert *et al.*, 2002) and 392 envelope (Froehlich *et al.*, 2003) proteins in chloroplasts and 416 (Heazlewood *et al.*, 2003) proteins in mitochondria.

Subcellular localisation programs, TargetP (cbs.dtu.dk/services/TargetP/), Predotar (inra.fr/predotar/), MitoProt (ihg.gsf.de/ihg/mitoprot.html) and ChloroP (cbs.dtu.dk/services/ChloroP/) predict a subset of 10 to 15% of the *Arabidopsis* proteome to be localized in the mitochondria and plastids. This prediction is based on the presence of an N-terminal extension, the targeting peptide in the majority of nucleus-encoded organellar proteins. Plastids are predicted to contain ~2500 to 3500 proteins (Arabidopsis Genome Initiative, 2000) and mitochondria ~3000-4500 proteins (Heazlewood *et al.*, 2003). A new, more stringent version of Predotar v1.03 (genoplante-info.infobiogen.fr/predotar/) predicts a smaller set of 1098 proteins to be localized in mitochondria. Even though the predicted numbers might be overestimated, it remains evident that a few thousand of the mitochondrial and plastid proteins are encoded by the nuclear genome, expressed on cytosolic polyribosomes and imported into the organelles. This demonstrates the importance of organellar protein trafficking in the plant cell.

The biosynthesis of functional plastids and mitochondria requires thus a coordinated expression of both the nuclear and organellar genomes and the import of a great majority of the organellar proteins from the cytosol. Most of the nuclear-encoded chloroplast and mitochondrial precursors carry an N-terminal extension, the targeting peptide (also called a transit peptide for chloroplast proteins and a presequence for mitochondrial proteins). Molecular chaperones, Hsp70 and Hsp90, 14-3-3 proteins and a series of other cytosolic factors interact with precursor proteins (either with the targeting peptide or the mature portion) or with the organellar receptors to mediate, facilitate or inhibit organellar protein import (Soll, 2002; Zhang and Glaser, 2002, Truscott *et al.*, 2003). Conventionally, the mitochondrial and chloroplast import process is viewed as post-translational, however, co-translational import of mitochondrial precursors mediated by mRNA binding to mitochondria has also been suggested for a subset of proteins (Marc *et al.*, 2002). Mitochondrial and chloroplast precursors, targeted to the organelle surface, are recognized by organellar receptors and transported across the organellar membranes via high molecular mass, oligomeric import complexes, Translocases of Outer and Inner Mitochondrial Membrane - TOM and TIM; and via the Translocon of the Outer and Inner envelope of Chloroplast - TOC and TIC, respectively. After import, targeting peptides are cleaved off in the mitochondria by the general mitochondrial processing peptidase (MPP) that in plants is integrated into the cytochrome bc_1 complex of the respiratory chain (Glaser and Dessi, 1999) and by the stromal processing peptidase (SPP) in chloroplasts (Richter and Lamppa, 1999). The mature proteins are transported within the organelle to their final destination and assembled with their partner proteins either spontaneously or upon the action of molecular chaperones. The cleaved targeting peptides, potentially harmful to biological membranes, are degraded inside the organelles by a newly identified presequence peptide-degrading zinc metalloprotease, PreP, that is dually targeted and executes its function both in mitochondria and in plastids (Ståhl *et al.*, 2002; Moberg *et al.*, 2003; Bhushan *et al.* 2003). The proteolytic events associated with precursor processing and degradation of targeting peptides are reviewed in this volume by Z. Adam.

This chapter will review our present knowledge on chloroplast and mitochondrial targeting peptides, cytosolic factors involved in the import process and the chloroplast and mitochondrial import machineries.

2. CHLOROPLAST AND MITOCHONDRIAL TARGETING SIGNALS

2.1. The targeting destination of the chloroplast and plant mitochondrial targeting peptides

The targeting of most of the chloroplast and mitochondrial proteins is mediated by cleavable N-terminal targeting peptides (Bruce, 2001). The chloroplast targeting peptide, cTP, is recognized by the receptor at the chloroplast envelope and the precursor protein is translocated into the chloroplast, followed by processing of the cTP. Proteins then are directed into the inner envelope membrane, into the thylakoid membranes, or remain in the chloroplast stroma. The inner envelope proteins, the peripheral thylakoid proteins located at the stromal site of the thylakoid membrane, and most of the integral thylakoid membrane proteins have no additional N-terminal transit peptides. In contrast, proteins located in the thylakoid lumen are targeted and translocated via a second transit peptide - the lumenal transit peptide which is located directly C-terminal of the cTP. Outer envelope proteins do not have a cleavable targeting peptide.

The mitochondrial targeting peptide, mTP, is recognised by mitochondrial import receptors and targets the precursor protein into the mitochondrial matrix, where it is processed (Sjöling and Glaser, 1998). Some inner membrane proteins require a non-cleavable, hydrophobic stop signal sequence located C-terminal of the mTP. The intermembrane space proteins and some other membrane proteins contain an additional cleavable signal peptide located directly down-stream of mTP, a so-called bipartite targeting peptide. There also exists a second group of the mitochondrial targeting signals that has been found for hydrophobic carrier proteins of the inner membrane, the internal signals (Brix *et al.*, 1999). These signals are less well characterized than mTPs; they do not require the presence of mTPs, are not processed and seem to be distributed all over the protein. Outer mitochondrial membrane proteins contain a non-cleavable targeting and sorting signal within the protein itself (Rapaport, 2003).

A group of proteins exist that are encoded by a single gene in the nucleus, translated in the cytosol and targeted to both chloroplasts and mitochondria. These proteins contain an N-terminal signal peptide referred to as a dual targeting peptide that is recognized by both mitochondrial and chloroplast receptors and has the capacity to target the precursor protein to both the mitochondrial matrix and chloroplast stroma (Peeters and Small, 2001; Silva-Filho, 2003).

2.2. The features of chloroplast and plant mitochondrial targeting peptides

The features of plant mitochondrial and chloroplast N-terminal signal peptides have been compared using statistical analysis and structural prediction methods on a data set containing 58 plant mTPs and 277 cTPs with an experimentally determined cleavage site (Zhang and Glaser, 2002). The length of organellar signal peptides varies substantially, from 18 to 107 residues for plant mTPs and from 13 to 146 for cTPs; however, cTPs are considerably longer (58 on average) than mTPs (42 on average). The great majority (>80%) of the mTPs are in the range of 20-60 residues long, whereas the majority of cTPs consist of 30-80 residues. Both plant mitochondrial and chloroplast signal peptides show a remarkable similarity in amino acid composition. They are rich in hydrophobic (Ala, Leu, Phe, Val), hydroxylated (Ser, Thr) and positively charged (Arg, Lys) amino acid residues and deficient in acidic amino acids. Serine (16-17%) and alanine (12-13%) are greatly overrepresented in mTPs and cTPs and arginine (12%) in mTPs. In comparison to mTPs from non-plant sources, plant mTPs are about 7-9 amino acid residues longer and contain about 2-5 fold more serine residues (Sjöling and Glaser, 1998). Characteristics of mTPs and cTPs are important for the import process. The mTPs interact with receptors through hydrophobic residues (Abe *et al.*, 2000). Hydroxylated residues of cTPs supply the phosphorylation motif, which is important for the interaction of cTPs with the cytosolic 14-3-3 proteins shown to form an import guidance complex with the precursors (May *et al.*, 2000). Despite great similarities between the mTPs and cTPs in amino acid composition, some differences have been found when mitochondrial and chloroplast targeting peptides were analysed using Sequence Logos (Fig. 1), a program that calculates how often each residue occurs at each position, in conjunction with the degree of sequence conservation (Schneider et al, 1990). Alanine, leucine, serine and arginine were found to be highly abundant in the N-terminal domain of mTPs, whereas alanine and serine dominate the N-terminal domain of cTPs. Positively charged residues are rarely found within the 15 N-terminal residues of cTPs, but they are abundant within the corresponding segment of mTPs (Zhang and Glaser, 2002).

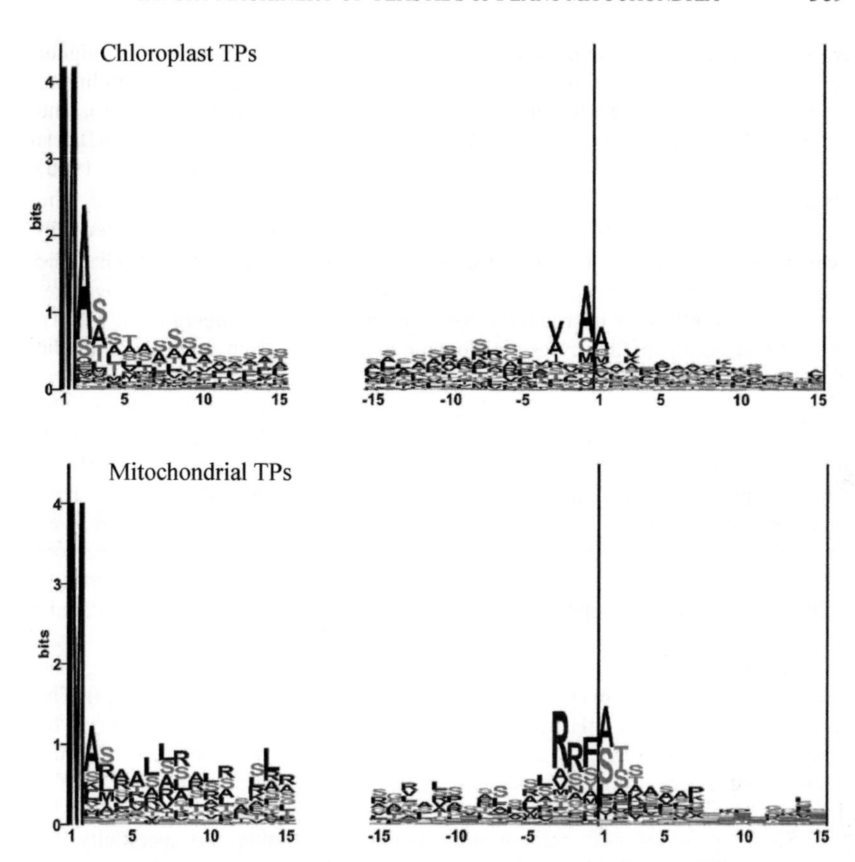

***Figure 1**. Sequence logos of chloroplast and plant mitochondrial signal peptides (from Zhang, 2001; see also Zhang and Glaser, 2002) (see Color Plate 3).*

Though signal sequences are largely unstructured in aqueous solution, a lipid environment can induce the formation of helical elements on both mTPs (MacLachlan *et al.*, 1994) and cTPs (Endo *et al.*, 1992). mTPs have the potential to form an amphiphilic □-helix (von Heijne, 1986) that is important for import. cTPs contain, in general, three distinct regions, an uncharged N-terminal domain, a central hydrophilic domain and a C-terminal domain with the potential to form an amphiphilic □-strand (von Heijne *et al.*, 1989). cTPs form random coils (von Heijne and Nishikawa, 1991) although recent reports indicate that an □-helical structure forms in a hydrophobic environment (Bruce, 2000). Most mTPs and cTPs have regions with the potential to form helical element(s); helical content is ~40% for mTPs and ~20% for cTPs. These theoretical observations have been confirmed by NMR studies. Both ferredoxin (Lencelin *et al.*, 1994) and Rubisco activase (Krimm *et al.*, 1999) from *Chlamydomonas reinhardtii*, determined in the presence of TFE, were shown to contain a helix and a random coil. The NMR structure of a higher plant cTP, *Silene pratensis* ferredoxin, is mainly unstructured with the ability to form two helical domains when introduced into micelles (Wienk *et al.*, 1999). The structures of a few relatively short mTPs, *e.g.* aldehyde dehydrogenase (ALDH) (Karlslake *et al.*, 1990) and cytochrome c oxidase subunit IV (Chupin *et al.*, 1995) were determined to form amphiphilic □-helices in membrane-like environments. The first NMR structure of a mTP from higher plants (*Nicotiana plumbaginifolia* F_1□) revealed the presence of two helices, an N-terminal amphipathic □-helix, the putative receptor binding site and a C-terminal □-helix upstream of the cleavage site, all separated by a largely unstructured long internal domain (Moberg *et al.*, 2004). The NMR structure of the cytosolic domain of rat Tom20 in complex with the presequence of ALDH revealed that the hydrophobic side of the pALDH amphipathic □-helix is located in a hydrophobic groove of the Tom20 receptor (Abe *et al.*, 2000). Also, other mTPs interacted with the same binding site of Tom20 through hydrophobic interactions (Muto *et al.*, 2001). The mechanisms underlying chloroplast preprotein recognition at the envelope are not fully understood, but both protein and lipid components of the outer envelope membrane appear to be involved (see section 3). The interactions between different cTPs and chloroplast receptors were shown to involve both conformational and electrostatic interactions (Hinnah *et al.*, 2002), as well as a phosphorylated form of a transit peptide (Sveshnikova *et al.*, 2000b).

Nearly all targeting peptides contain Hsp70 binding motifs (Zhang and Glaser, 2002); 97% plant mTPs and 82.5% cTPs contain Hsp70-binding site(s). The interaction of mTPs with mtHsp70 during the import process has been supported by biochemical studies (Zhang *et al.* 2001); mtHsp70 has an essential role in the mitochondrial import process (Okamoto *et al.*, 2002), whereas the role of the chloroplast Hsp70 in protein import into chloroplasts is not clear (Rial *et al.*, 2003).

Signal peptides also contain information determining the specificity for cleavage by MPP or SPP (Fig. 1). This information is located as a loosely conserved motif at the C-terminus of the peptides in combination with structural properties upstream of the cleavage site (Sjöling and Glaser, 1998; Zhang *et al.*, 2002). Mitochondrial precursors can be divided into two major groups. These include 38% and 42% of mTPs containing an arginine in position -2 (-2Arg) or -3 (-3Arg), respectively, relative to the cleavage site. This results in loosely conserved motifs around the cleavage site, Arg-X□Ser-Thr/Ser-Thr and Arg-X-Phe/Tyr□Ala/Ser-Thr/Ser/Ala, respectively. The third group includes mTPs containing no conserved arginine (no Arg) in the proximity of the cleavage site. In chloroplast precursors, sequence conservation around the cleavage site is lower and different from the mitochondrial motif. It has been suggested that cTPs contain a loosely conserved motif around the processing site (Val/Ile-X-Ala/Cys□Ala) and that most cTPs contain a basic residue within the 7 most C-terminal amino acids of the targeting peptide (Gavel and von Heijne, 1990). Sequence logos analysis confirmed the high abundance of Val in position –3 and Ala in position -1 (Zhang and Glaser, 2002), whereas analysis of cTPs by the neural network program, ChloroP, yielded an alternative consensus sequence, Val-Arg□Ala-Ala-Ala-Vxx (Emanuelsson *et al.*,

1999). This indicates that although SPP and MPP belong to the same pitrilysin family of metalloendopeptidases (Rawlings and Barrett, 1995), they recognise different features of the precursors.

2.3. Dual targeting peptides

Proteins that are encoded by a single gene but are targeted to both chloroplasts and mitochondria are referred to as dual targeted proteins (Peeters and Small, 2001). Since the first report of dual targeting of pea glutathione reductase (GR) in 1995 (Creissen *et al.*, 1995), twenty four dually targeted proteins have been identified and it is expected that this event may be far more common than originally thought. Dual targeting results from 'twin' or 'ambiguous' targeting peptides (Peeters and Small, 2001; Silva-Filho, 2003). Twin targeting signals may arise from multiple transcription initiation sites, alternative pre-mRNA processing or from variable post-translational modification, all resulting in multiple precursor proteins each possessing different targeting information. For example, translation of protoporphirin oxidase II (protox-II) generates two products using in-frame initiation codons for chloroplast and mitochondrial import, respectively (Watanabe et al, 2001). The 'ambiguous' signal arises from genes encoding single precursors with a targeting signal that is recognized by the import apparatus of both organelles (Small et al, 1998). The majority of proteins that belong to that group are involved in gene expression, *e.g.* aminoacyl-tRNA synthetases (MetRS, HisRS, CysRS, AspRS, GlyRS, LysRS, TrpRS), RNA polymerase, methionine aminopeptidases and a peptidyl deformylase. Other dual targeted enzymes are related to protection against oxidative stress, *e.g.* GR, ascorbate peroxidase (Chew, *et al.* 2004), or to cellular protein turnover, such as targeting peptide-degrading zinc metallopeptidase (Moberg *et al.* 2003; Bhushan *et al.* 2003).

The evidence from *in vitro* work suggests that the import pathway used by these proteins is indistinguishable from that taken by other imported precursors. Peeters and Small (2001) compared features of ambiguous targeting peptides with a large set of mitochondrial and chloroplastic targeting peptides; they concluded that dual targeting peptides contain classical features of both mTPs and cTPs but they contain fewer alanines and a greater abundance of phenylalanine and leucine, suggesting that they are more hydrophobic. Secondary structure predictions demonstrate a lower abundance of potential N-terminal helices in ambiguous targeting sequences than in classical mitochondrial targeting sequences. It appears that these sequences are intermediary in character between mitochondrial and plastid targeting sequences and contain features from both.

A study where mitochondrial and chloroplastic targeting peptides were fused in tandem showed that the position of the targeting sequence is important, and that the N-terminal portion has the greatest importance for the final location (Silva-Filho *et al*, 1997). Hedtke *et al.* (2000) reported that the dual targeted protein, RNA polymerase RpoT;2 may have a domain structure with the N-terminal region being required for chloroplastic import and the C-terminal region being important for mitochondrial targeting. Also the dual targeting peptide of the recently identified targeting peptide-degrading zinc metallopeptidase reveals a domain structure but with an opposite orientation, the N-terminal region mediates mitochondrial import while the C-terminal region is important for chloroplastic targeting (Bhushan *et al.* 2003). Furthermore, studies of the dual targeting peptide of GR suggested a triple domain structure with two domains responsible for the organellar targeting and the third N-terminal domain controlling the efficiency of organellar import (Rudhe *et al.* 2002). Overall single mutations of positive and hydrophobic residues had a greater effect on mitochondrial import in comparison to chloroplast import (Chew *et al.*, 2004). It will be of great interest to perform a high throughput investigation to identify all dual targeted proteins to mitochondria and chloroplasts in the plant cell, their determinants for import and mode of interaction with organellar receptors.

3. PROTEIN IMPORT INTO CHLOROPLASTS

3.1. Cytosolic processes

As soon as the polypeptide emerges from the ribosome it encounters a highly concentrated aqueous environment, with protein concentrations of about 300 mg/ml. Furthermore, polypeptides will start to attain native-like secondary and tertiary structure. Together, with the danger of aggregation, these conditions are not favourable for polypeptides, which have to be posttranslationally imported into organelles. Cytosolic factors are likely to exist which facilitate import by binding to precursor proteins and hence influence their tendency to aggregate, fold prematurely or become mistargeted.

During or shortly after translation most if not all preproteins associate with molecular chaperones (Beckmann *et al.*, 1990; Frydman *et al.*, 1994). This interaction is most likely nonselective and due to the unfolded polypeptide chain exposing hydrophobic amino acid stretches. *In vitro* the import of preLHCP into chloroplasts, which was expressed in *E. coli* and denatured in urea prior to the import reaction, was greatly stimulated by cytosolic factors (Waegemann *et al.*, 1993), one of which could be replaced by purified Hsp70. Other less hydrophobic preproteins such as preSSU and preFd were imported in similar amounts into chloroplasts either in the absence or presence of Hsp70 (Pilon *et al.*, 1990). PreFd maintained a loosely folded conformation for several hours upon dilution with an aqueous solution (Pilon *et al.*, 1992); this situation is very different from the *in vivo* situation where the preproteins encounter a crowded, highly concentrated environment. Therefore, a direct comparison between *in vivo* and *in vitro* requirements might be difficult.

A more selective interaction between plastid-directed preproteins and cytosolic polypeptides seems to occur in case of a protein kinase which can phosphorylate a serine or threonine residue located within the targeting signal. Abundant chloroplast preproteins like preSSU, preLHCP, preOE23 or preOE33 were experimentally shown to become phosphorylated (Waegemann *et al.*, 1996), while more are predicted from *in silico* analysis to be potential phosphorylation targets. 14-3-3 proteins bind to the phosphorylated preprotein, and together, with perhaps other factors, form a cytosolic preprotein guidance complex (May and Soll, 2000). Preproteins bound to the guidance complex are transported with 4-5 fold higher rates into chloroplasts than monomeric preproteins. Deletion of the phosphorylation site, which leads to the failure of guidance complex formation, still does not result in mistargeting to other organelles, such as mitochondria (Waegemann *et al.*, 1996). Together the data indicate that phosphorylation followed by guidance complex formation puts a preprotein on the fast lane for import and gives it priority over non phosphorylated ones. This regulatory property might be very important during developmental or environmental adaptations, *e. g.* greening, stress or light-dark switches. While phosphorylation enhances the capacity for import, other factors exist that inhibit protein import into chloroplasts *in vitro* in wheat germ lysate (Schleiff *et al.*, 2002). The nature of these factors(s) and their relevance *in vivo* is not clear. However, *in vitro* these factors strongly influence import experiments and must be considered when comparing data from different import studies.

3.2. Chloroplast protein import machinery

In this section we will try to outline the import route and the translocon components involved for a prominent chloroplast preprotein like the precursor of Rubisco small subunit (pSSU). Other pathways are likely to exist but knowledge is still very limited and so we will not deal with it.here. PSSU is recognized at the chloroplast surface by protease sensitive receptor polypeptides. Binding is GTP dependent and leads to the formation of import intermediates. Complete translocation across both membranes requires ATP, probably for stromal chaperones to pull the precursor into the organelle. Either during or shortly after translocation, the transit peptide is cleaved off by SPP and the mature form now folds, assembles and attains its native function (for review see Bauer *et al.*, 2001; Keegstra and Cline, 1999; Soll and Schleiff, 2004). The engineering of cTPs containing protein genes into the chloroplast genome and their expression showed that processing by SPP can occur independently of protein translocation (Daniell *et al.*, 1998; Dhingra *et al.*, 2004).

3.2.1. The Toc complex

Different experimental approaches were used in order to identify subunits of the translocon at the outer envelope of chloroplasts (Toc complex). Fig. 2 and Table 1 give a graphical and an index overview of the Toc and Tic components and their putative functions.Chemical crosslinking of preproteins to chloroplasts under different import conditions, *e. g.* low NTP concentrations or low temperature, which both favor the formation of early import intermediates (Perry and Keegstra, 1994; Ma *et al.*, 1996), identified Toc159 and Toc75 as potential subunits. A crosslinked precursor Toc34 product was observed especially in the absence of GTP (Kouranov *et al.*, 1997). On the other hand envelope membranes were solubilized by different detergents, such as digitonin, TritonX100 or dodecylmaltoside, and protein membrane complexes were isolated by sucrose density centrifugation (Waegemann and Soll, 1991; Kessler *et al.*, 1994; Schnell *et al.*, 1994).

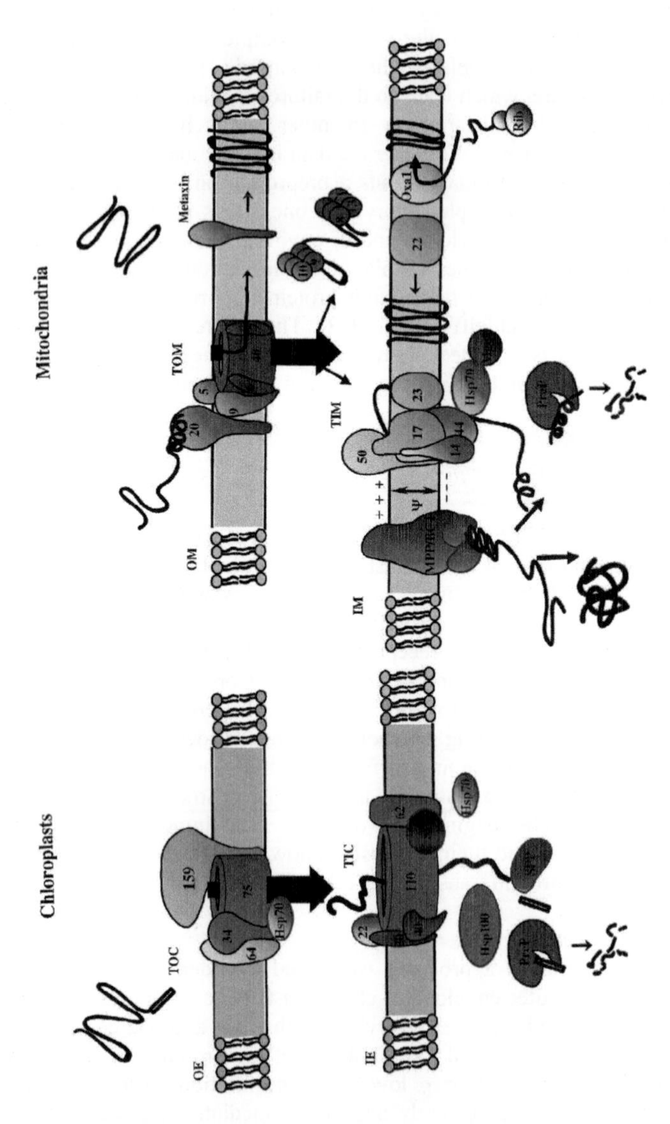

Figure 2. Graphical overview of the chloroplast and plant mitochondrial import machineries (see Color Plate 4).

These complexes contained again Toc159, Toc75 and Toc34 and, dependent on the detergent and solubilisation conditions, Toc64 and other potential Toc subunits (Sohrt and Soll, 2000; Schleiff *et al.*, 2003a). Cloning of the cDNAs for each of these subunits allowed then a detailed analysis of each protein. Since the completion of the *Arabidopsis* genome it has become clear that most Toc and Tic subunits are encoded by several genes (Jackson-Constan and Keegstra, 2001). Therefore transcriptional and translational control will play an important role in the composition of translocon complexes in different developmental stages or different plant organs (see below).

Toc75 is the most abundant outer envelope protein. Heterologously expressed Toc75 forms a cation-selective high conductance channel as determined in lipid bilayer experiments (Hinnah *et al.*, 1997). Structural analysis as well as computational analysis indicate that Toc75 forms a □-barrel-type anion channel, which is lined by 16 transmembrane □-sheets (Svesnikova *et al.*, 2000a). Calculation of the pore diameter indicates that the aqueous Toc75 channel has a minimum diameter of 15 Å and a maximal diameter of 25 Å (Hinnah *et al.*, 2002). Toc75 has a preprotein binding site at the cytosolic face of the channel, which can discriminate between the precursor form and the mature form (Ma *et al.*, 1996; Hinnah *et al.*, 1997). Surprisingly, two homologs of Toc75 are expressed in green leaves, namely Toc75III (III indicates the chromosome number in *Arabidopsis*), which represents 90% of the Toc75 isoforms and Toc75V, which represents the other 10% (Eckart *et al.*, 2002). Both subunits show clear homologies to proteins in the outer membrane of gram-negative bacteria, which seem to be involved in the biogenesis of the outer membrane of gram-negative bacteria (Bölter *et al.*, 1998; Reumann *et al.*, 1999; Heins and Soll, 1998; Genevrois *et al.*, 2003; Voulhoux *et al.*, 2003). Most intriguingly, very recent findings show that mitochondria possess related proteins in their outer membrane, which are required for outer membrane protein insertion (Gentle *et al.*, 2004; Paschen *et al.*, 2003). This ancestral channel family, which was inherited from the endosymbiont in the arising eukaryotic cell, might represent the initial translocation site to which receptors, regulators and motor proteins were added to make protein import unidirectional. To date, no loss of function mutants have been describe for Toc75III (Jackson-Constan and Keegstra, 2001), which could indicate that the gene is essential for plastid biogenesis.

Toc34 is present in a 1:1 ratio with Toc75 in the translocon (Schleiff *et al.*, 2003a). It is anchored in the outer envelope by a carboxy terminal □-helical membrane anchor while a large N-terminal domain protrudes into the cytosol (Seedorf *et al.*, 1995). This N-terminal domain houses the GTP-binding and GTPase activity of Toc34 (Kessler *et al.*, 1994). Toc34 binds precursor proteins with high affinity in its GTP-bound form. The precursor functions as a GTPase activating factor and stimulates the GTP hydrolysis of Toc34 by about 40-50 fold (Jelic *et al.*, 2002). Toc34-GDP has a much lower affinity for the preprotein, which continues its path to the next translocon subunit, most likely Toc159 (see below). Toc34 can now follow two routes which influence its activity; first,, Toc34 can be recharged with GTP and enter a new round of precursor binding, and second, Toc34 becomes phosphorylated by an outer envelope protein kinase (Jelic *et al.*, 2002).

Phosphorylation inhibits GTP binding und switches the receptor off. Dephosphorylation has to occur in order to allow GTP binding and activation of Toc34 receptor. In *Arabidopsis* two Toc34 homologes are present, atToc34 and atToc33. AtToc33 seems to be predominantly expressed in photosynthetic and meristematic tissue, while atToc34 is expressed in all tissues, but present at lower amounts (Jarvis *et al.*, 1998; Gutensohn *et al.*, 2000). Both atToc33 and atToc34 bind precursor proteins in a GTP dependent manner; however, only atToc33 can be inactivated by protein phosphorylation (Jelic *et al.*, 2003). Therefore, plant cells simultaneously contain two isoforms of the same receptor: atToc34, which seems to be present and active constitutively, and atToc33, which can be much more abundant but has more complex regulation and can be completely switched off. The nature of the protein kinase and its regulation are currently unknown. The protein Toc34 of pea seems therefore equivalent to atToc33. The plastid protein import mutant *ppi*1, which has a loss of function mutation in the atToc33 gene, exhibits a pale green phenotype and retarded chloroplast development (Jarvis *et al.*, 1998). However atToc33 function is not essential and plants recover and can grow photoautotrophically on soil. Together the *in vivo* and *in vitro* data indicate that both Toc34 isoforms have overlapping functions. However both receptors show distinct preferences for certain classes of preproteins (Gutensohn *et al.*, 2000; Jelic *et al.*, 2003).

Toc159 was identified very early as a potential translocon subunit (Waegemann and Soll, 1991; Perry and Keegstra, 1994; Ma *et al.*, 1996). It forms a prominent crosslinking product with precursor proteins (Bauer *et al.*, 2000) and was always recovered in detergent-solubilized Toc complexes (Kessler *et al.*, 1994; Schleiff *et al.*, 2003a). Toc159 is composed of three domains: an N-terminal A domain that contains many acidic amino acid residues, a central G-domain that contains a GTP binding domain with strong homologies to Toc34 (Hirsch *et al.*, 1994; Kessler *et al.*, 1994), and a carboxy-terminal M domain that is essential for targeting and anchoring the protein into the membrane (Muckel and Soll, 1996; Lee *et al.*, 2003). Toc159 is often recovered as an 86 kDa (Toc159f) proteolytic fragment which has lost the A domain, and was referred to earlier as Toc86. Toc159 is essential for life and *ppi*2 seedlings die early in development (Bauer *et al.*, 2000; Hiltbrunner *et al.*, 2001a). They are not viable on soil. Toc159- related proteins like Toc132, Toc120 and Toc90 (Bauer *et al.*, 2000) are not able to substitute for Toc159 and play its central role in the function of the Toc complex (Hiltbrunner *et al.*, 2001b). A soluble Toc159 population was observed after mechanical rupture of leaf mesophyll cells (Hiltbrunner *et al.*, 2001b). In addition, expression of an atToc159-GFP fusion protein under the strong CaMV promoter not only showed outer envelope localization but diffuse staining of the cytosol. This led to the hypothesis that Toc159 could function as a receptor in the cytosol which shuttles its cargo to the outer envelope, inserts into the membrane and releases its cargo to the translocon. There are several lines of evidence that do not support this idea; first, Toc159 behaves as an integral membrane protein and cannot be extracted by 0.1 M Na_2CO_3 or 6 M urea (Hiltbrunner *et al.*, 2001b; Becker *et al.*, 2004). Second, expression of the atToc159 M domain fused to a T7 tag showed only envelope but no cytosolic location (Lee *et al.*, 2003). Third the "cytosolic" form of Toc159 is associated with chloroplast specific galactolipids. Antibodies raised against these lipids can actually co-immunoprecipitate Toc159, indicating that Toc159 is present in low density lipid fragments and not truly soluble (Sickmann *et al.*, 2003). Furthermore it has recently been demonstrated that expression under the control of a strong promoter, as used in the above studies, can lead to mistargeting (Sickmann *et al.*, 2003).

A series of events and functional properties can be deduced from reconstitution studies using single purified components or the purified Toc complex (Schleiff *et al.*, 2003b). The Toc complex, when reconstituted into liposomes, is able to recognize and translocate proproteins into the lumen of the liposome in a GTP dependent manner. ATP and molecular chaperones play no role in this process. By reconstituting single Toc subunits into proteoliposomes it could be demonstrated that the minimal unit, which can drive preproteins across a membrane, consists of Toc75, the channel, and Toc159, which functions as a molecular motor. GTP binding and hydrolysis are both necessary and sufficient for preprotein binding and translocation (Schleiff *et al.*, 2003b). Toc34 probably functions as an initial receptor. This assumption is supported by the ratio of 4 Toc34: 4 Toc75: 1 Toc159 in the purified Toc complex (Schleiff *et al.*, 2003a). The catalytic function of Toc159 does not require a 1 : 1 ratio with other components.

The role of Toc64 is less clear. Toc64, like the peroxisomal receptor Pex5 or the mitochondrial receptor Tom70, exposes tetratricopeptide repeats on the chloroplast surface (Sohrt and Soll, 2000). Tom70 functions as a receptor for hydrophobic inner membrane localized transporters that arrive in a complex with Hsp70. The Hsp70 binds to the tetratricopeptide motif and Tom70 then calyses the transfer of the preprotein to the translocon (Wiedemann *et al.*, 2001; Young *et al.*, 2003). A similar role could be fulfilled by Toc64.

3.2.2. The Tic complex

Most likely, translocation occurs jointly between the Toc and the Tic complex, though the Tic complex has the capacity to act independently of the Toc translocon (Scott and Theg, 1996). Several Tic subunits have been identified; however, the role in import is less well defined. The Tic subunits known to date are Tic110, Tic62, Tic55, Tic40, Tic32, Tic22 und Tic20.

Tic110 is an abundant protein in the inner envelope. It was identified early on as a translocon subunit (Kessler and Blobel, 1996; Lübeck *et al.*, 1996). At the stromal side Tic110 can interact with molecular chaperones such as chaperonin 60 or hsp 93 (Kessler *et al.*, 1996; Nielsen *et al.*, 1997). If these interactions occur directly or through other Tic subunits like Tic40 (see below) remains to be established. Reconstitution of heterologously expressed Tic110 into lipid bilayers showed that it forms a cation-selective high conductance channel (Heins *et al.*, 2002). The pore diameter was estimated to be between 15-20 Å which is in good agreement with the data obtained for Toc75. There are still conflicting results about the topology and structure of Tic110, which, when resolved, should give a clearer picture on the exact role of Tic110 (Lübeck *et al.*, 1996, Jackson *et al.*, 1998; Inaba *et al.*, 2003).

The coordination of molecular chaperones at the stromal exit site of the Tic translocon could also be coordinated by Tic40 (Stahl *et al.*, 1999). Tic40 is in close proximity to Tic110 and has domains that are related to Hsp70 interacting proteins (Hip) as well as Hsp70/Hsp90 organizing proteins (Hop) in connection with a tetratricopeptide repeat motif. Tic40 is encoded by a single gene in *Arabidopsis* (Jackson-Constan and Keegstra, 2001). Tic40 knockout plants show a pale green phenotype and retarded chloroplast development but grow photoautotrophically on soil (Chou *et al.*, 2003). A role for Tic40 as a chaperone recruitment factor is further supported by the observation that Hsp93 and Tic40 can be immunoprecipitated together (Chou *et al.*, 2003).

Isolation of the Tic complex by blue native PAGE indicated that Tic110 is associated with several more proteins (Caliebe *et al.*, 1997). We have recently identified Tic62, Tic55 and Tic32 in this complex. All three subunits are redox components and could form a regulatory circuit at the level of the Tic complex. Tic55 contains a Rieske iron sulphur center and a mononuclear iron binding site, which indicates the potential for electron transfer (Caliebe *et al.*, 1997). Tic62 contains a conserved NAD(P) binding site and a carboxy-terminal repeated peptide motif, leading to interaction with stromal ferredoxin-NAD(P) reductase (Küchler *et al.*, 2002). Ferredoxin-NADP-reductase connects photosynthetic electron transfer with metabolically required reducing power. Ferredoxin-NADP-reductase is a central component in maintaning the NAD(P)/NAD(P)H ratio. Tic62 might therefore represent a link between the metabolic redox status of the chloroplast and the Tic translocon. The possibility that protein import into chloroplasts is redox-regulated was demonstrated for two isoforms of ferrodoxin. While ferrodoxin I, which is involved in photosynthetic electron transport, imported correctly both in the dark and in the light, Ferredoxin III imported correctly only in the dark. In the light Ferredoxin III was mistargeted to the interenvelope space (Hirohashi *et al.*, 2001). Ferredoxin III is thought to be involved directly in metabolic processes. Tic32 belongs to the family of short chain dehydrogenases, which also use NAD(P) as a cofactor. Tic32 binds to the N-terminal □-helical region of Tic110 (Soll, Hörmann, unpublished). Its role in import is not clear at the moment.

Tic22 is a peripheral subunit of the Tic translocon, localized to the space between the two envelope membranes (Ma *et al.*, 1996; Kouranov and Schnell, 1997). It might play a role in coordinating Tic-Toc function or in guiding preproteins across the soluble interenvelope space.

Another integral Tic subunit is Tic20 (Kouranov *et al.*, 1998). Tic20 shows some homologies to bacterial amino acid transporters and the mitochondrial import component Tim17 (Rassow *et al.*, 1999). Therefore Tic20 was proposed to form part of the import channel (Kouranov *et al.*, 1998). Experimental evidence is however lacking. Antisense plants for atTic20 showed a pale green phenotype but were viable on soil, though chloroplast ultrastructure was somewhat disturbed (Chen *et al.*, 2002). Chloroplasts isolated from Tic20 antisense plants retained the capacity to import preproteins, though at reduced levels. Together these data could indicate that more than one Tic complex exists that is similar to the situation in mitochondria. More biochemical work is needed to clearly define the function of single subunits.

Table 1. *Chloroplast translocon subunits and their putative functions. N.a. stands for not available.*

		Experimental evidence		
Translocon subunit	*Function*	*In vitro*	*In vivo*	*Phenotype*
Toc 159	Receptor, GTP motor	Reconstitution, crosslinking	tDNA	Pale, not viable on soil
Toc75III	Import channel	Crosslinking, electro-physiology	N.a.	Could be essential
Toc75V	Channel ?	Homology to Toc75III	N.a.	-
Toc64	Putative receptor	Cofractionates with Toc core complex	N.a.	-
AtToc34	Constitutive GTP dependent receptor	Crosslinking reconstitution	RNAi	No phenotype
AtToc33	Regulated GTP dependent receptor	Crosslinking, reconstitution	tDNA RNAi	Light green, slightly retarded growth, photo-autotrophic
Tic110	Import channel, chaperone binding	Crosslinking, reconstitution	N.a.	Could be essential
Tic62	Regulation	NADP/FNR binding	-	-
Tic55	Regulation	Rieske/Fe-S center	-	-
Tic40	Chaperone binding	Homology to heat shock binding protein	tDNA	Photoautotrophic, pale, retarded flowering
Tic22	?	Crosslinking	-	-
Tic20	Channel forming	Crosslinking	RNAi	Photoautotrophic, but pale, retarded chloroplast development

3.3. The role of lipids in protein import

All plastid membranes contain specific galactolipids, such as monogalactosyldiacylglycerol, digalactosyldiacylglycerol and sulphoquinovosyldiacylglycerol. Therefore, plastids expose a unique lipid surface to the cytosol. So, what roles do lipids have in protein import (Bruce, 1998; Horniak *et al.*, 1993)? Several studies have indicated that chloroplast transit sequences are largely unstructured in an aqueous environment. However, in the presence of hydrophobic solvents or artificial membranes that contain galactolipids, these sequences can adopt some □-helical structure (Chupin *et al.*, 1994; Pinnaduwage and Bruce, 1996; Wienk *et al.*, 2000), although this structure could not be described as being stable, regular secondary structure. This structural change could have a large impact on preprotein recognition by Toc receptors. Phosphorylation (Waegemann and Soll, 1996) (see above) could have a similar effect on the secondary structure of the transit sequence and might replace the lipid effect. *In vitro* binding studies using pea Toc34 and either phosphorylated or nonphosphorylated preproteins indicated that the receptor can bind the transit sequence with a high affinity in the absence of lipids (Schleiff *et al.*, 2002).

Changing the lipid composition of the outer envelope influences protein import. Phosphatidylcholine is required for the transfer of the preprotein from the receptor site to the translocation site (Kerber and Soll, 1992). Furthermore, in chloroplasts that do not contain digalactosyldiacylglycerol, import is greatly impaired, whereas preprotein binding seems to be unaltered (Chen and Li, 1998). In line with these observations is the finding that the isolated Toc complex contains predominantly digalactosyldiacylglycerol and phosphatidylcholine (Schleiff *et al.*, 2003a).

3.4. Import/insertion into the outer envelope of chloroplasts

Except for Toc75, all of the outer envelope proteins that have been identified to date have no cleavable transit sequence, and are targeted to the chloroplast surface by internal information. The transit sequence in Toc75 is bipartite, and the amino-proximal part contains the chloroplast-targeting, envelope-transfer information (Tranel *et al.*, 1995; Inoue *et al.*, 2001). The Toc75 preprotein uses the standard import pathway, but translocation is halted early by a stop-transfer signal in the preprotein. The amino-proximal part of the transit sequence is cleaved by SPP, and the intermediate form of Toc75 retreats from the Tic translocon. This form is subsequently inserted into the outer envelope and is terminally processed by an unknown peptidase (Inoue *et al.*, 2001).

Two outer envelope proteins that lack a cleavable transit sequence and that have been studied in some detail are OEP7 (outer envelope protein 7) and Toc34 (Schleiff *et al.*, 2001; Tu and Li, 2000; Salamon *et al.*, 1990; May and Soll, 1998; Lee *et al.*, 2001, Qbadou *et al.*, 2003). Both proteins contain a single transmembrane □-helix - OEP7 in the orientation amino-terminus in and carboxyl-terminus out (Salomon *et al.*, 1990), and Toc34 in the reverse orientation (Seedorf and Soll,

1995). The topology of the proteins is determined by three distinct features. First, positive charges that are clustered at one end of the transmembrane helix favor the retention of this section of the polypeptide on the cytoplasmic face of the membrane (May and Soll, 1998; Schleiff *et al.*, 2001; von Heijne, 1992) (positive-inside rule). Second, the inner leaflet of the outer envelope has a lower concentration of phosphatidylcholine and a higher concentration of phosphatidylglycerol than the outer leaflet. This lipid asymmetry, which results in an uneven charge distribution, is another important determinant of the orientation of a transmembrane □-helix (Schleiff *et al.*, 2001; Qbadou *et al.*, 2003). Third, a large hydrophilic domain, such as that present in Toc34, can overrule the positive-inside determinant and become a major topogenic signal; that is, it is retained on the cytoplasmic side of the membrane (Qbadou *et al.*, 2003). The subunits of the Toc translocon do not seem to be involved in the insertion of these proteins (Tu and Li, 2000; Schleiff *et al.*, 2001), and no other proteinaceous factors have been identified as being involved (Tsai *et al.*, 1999). OEP7 and Toc34 can spontaneously insert into protein-free liposomes in the correct topology, although this insertion occurs with a low efficiency (Schleiff *et al.*, 2001; Qbadou *et al.*, 2003). This indicates that helper proteins might facilitate outer envelope insertion (Tsai *et al.*, 1999), but that such proteins are not essential for insertion.

4. PROTEIN IMPORT INTO PLANT MITOCHONDRIA

4.1. Cytosolic factors

Plant mitochondrial import shares several conceptual similarities with the chloroplast protein import process. Chaperones of the Hsp70 class have a general role in maintaining nascent proteins in a soluble and import-competent state (Murakami *et al.*, 1988). Hsp70s may operate in concert with specific cytosolic factors enhancing the interaction between mitochondrial receptors and the precursor/chaperone complex. The mammalian mitochondrial import stimulating factor, MSF, a member of the 14-3-3 protein family (Hachiya *et al.*, 1993), was considered to have an important role in the targeting of a subset of mitochondrial precursors. Precursor binding factor, PBF, an oligomer of 50 kDa subunits, was reported to bind to the presequence (Murakami *et al.*, 1996) and to interact with Hsp70. Targeting factor (TF), a 200 kDa homo-oligomer made from the 28 kDa subunits, was shown to specifically bind to the presequence of rat ornithine aminotransferase precursor protein (Ono and Tuboi, 1990). Recently, using two-hybrid screening, Yano *et al.* (2003) identified arylhydrocarbon receptor-interacting protein (AIP) interacting with the mitochondrial receptor Tom20. Binding to Tom20 was mediated by the tetratricopeptide repeats of AIP. AIP binds specifically to mitochondrial preproteins and forms a ternary complex with Tom20 and preprotein. AIP-enhanced import of preornithine transcarbamylase and depletion of AIP by RNA interference impaired the import. Furthermore, AIP has a chaperone-like activity and prevents substrate proteins from aggregating. These results suggest that

AIP functions as a cytosolic factor that mediates preprotein import into mitochondria. There is an *Arabidopsis* homologue of AIP that shows 45% sequence similarity (27% identity) however, its function has not been investigated so far.

Surprisingly, mitochondrial precursor proteins synthesised in rabbit reticulocyte lysate are easily imported into mitochondria, whereas the same precursors synthesised in wheat germ extract (WGE) fail to be imported (Dessi *et al.*, 2003). Only a precursor that does not require the addition of extra-mitochondrial ATP for import, the F_Ad ATP synthase subunit, could be imported from WGE. Investigation of chimeric constructs of precursors with switched presequences revealed that the mature domain and not the presequence of the F_Ad precursor defines the import competence in WGE. Interestingly, the dual-targeted GR precursor synthesised in WGE was imported into chloroplasts, but not into mitochondria (Dessi *et al.*, 2003). Investigations on the composition of WGE have been carried out by several laboratories with regard to how WGE components interact with precursor proteins. Hsp70 in WGE associates with nascent precursors in an ATP-dependent manner (Miernyk et al, 1992). A series of studies using the precursor of aspartate aminotransferase showed that the protein makes a very stable complex with Hsp70, possibly due to a lack or inactivity of cytosolic factors required to release proteins from plant chaperones (Mattingly et al, 1993). Results of import experiments with an overexpressed F_1□ precursor depleted of any cytosolic factors point to the conclusion that the import incompetence of WGE-synthesised mitochondrial precursors is a result of the interaction of WGE inhibitory factors with the mature portion of precursor proteins (Dessi *et al.*, 2003). There also exists a WGE factor that inhibits chloroplast protein import; see section 3.1 (Schleiff *et al.*, 2002).

Several recent observations support the co-translational mitochondrial import of some proteins (Lithgow, 2000). In *S. cerevisiae*, co-translational import is required to transport fumarase into the mitochondria (Stein *et al.*, 1994). The 3′ untranslated region (UTR) of ATM1 is required for the localization of its mRNA to the surface of mitochondria (Corral-Debrinski *et al.*, 2000) and subsequent import. A mRNA binding protein, protein kinase A anchoring protein (AKAP121) of the mitochondrial outer membrane (Ginsberg *et al.*, 2003), contains an RNA-binding domain that was shown to specifically interact with the 3′UTR of transcripts encoding the manganese superoxide dismutase and the F_0f subunit of ATP synthase. Yeast mutants of the nascent polypeptide-associated factor and the ribosome-associated complex display a slow growth phenotype on non-fermentable carbon sources and have defects in protein targeting to mitochondria, suggesting that translation and import processes are associated (Funfschilling and Rospert, 1999; George *et al.*, 1998).

4.2. Mis-sorting of chloroplast precursors to plant mitochondria

The targeting of proteins to mitochondria and chloroplasts can be studied using both *in vitro* and *in vivo* approaches. *In vivo* approaches use an intact cellular system and obviously reflect the *in vivo* targeting capacity of a signal. However, they have several limitations arising from the fact that the investigated proteins are over-expressed at high levels, no kinetics can be assessed and passenger proteins intsead of native mature proteins are used. *In vitro* import approaches overcome these limitations but they have other disadvantages due to a lack of an intact cellular system, *e.g.* lack of cytosolic factors and/or competition with other organelles.

Mitochondrial and chloroplast protein import was demonstrated to be highly specific *in vivo* (Boutry *et al.*, 1987; Glaser *et al.*, 1998; Soll and Tien, 1998). However, whereas mis-sorting of mitochondrial proteins into chloroplasts has not been reported, several studies have reported mis-sorting of chloroplast proteins into mitochondria *in vitro* using different chloroplast precursors, such as the PsaF protein (Hugosson *et al.*, 1995), triose-3-phosphoglycerate phosphate translocator (Brink *et al.*, 1994; Silva Filho *et al.*, 1997), plastocyanin, the 33 kDa photosystem II protein (Cleary *et al.* 2001), and small subunit of Rubisco (Lister *et al.*, 2001).

In order to overcome the limitations of the *in vitro* import system and to ensure the correct specificity of targeting, a novel *in vitro* dual import system for simultaneous targeting of precursor proteins into mitochondria and chloroplasts has been developed (Rudhe *et al.*, 2002). Purified organelles are mixed, incubated with precursors and re-purified after import. This allows the determination of the targeting specificity into either organelle and the use of authentic precursors. Using the dual import system, it has been shown that the chloroplastic precursor of the small subunit of Rubisco was mis-targeted to pea mitochondria in a single import system but was imported only into chloroplasts in the dual import system, whereas the dual targeted GR precursor was targeted to both mitochondria and chloroplasts in both systems (Rudhe *et al.* 2002).

4.3. Plant mitochondrial protein import machinery

Proteins that have been targeted to or synthesized at the mitochondrial surface are transported into the organelle through oligomeric TOM and TIM complexes of the mitochondrial membranes (Truscott *et al.*, 2003). The TOM and TIM complexes were first characterized in yeast and subsequently identified in mammalian and plant systems. Notably, comparison of the homologous yeast, human and *Arabidopsis* genes shows that whereas many non-plant proteins are encoded by a single gene, they are often represented by multiple homologous gene copies in *Arabidopsis*. An overview of an up-to-date collection of annotated mitochondrial TOM and TIM genes in *Arabidopsis* is available at MPIPM (mitoz.bcs.uwa.edu.au/MPIMP/index.html) (Lister *et al.*, 2003).

4.3.1. The TOM complex

The TOM complex consists of receptor subunits that recognise and bind mitochondrial precursor proteins and a core translocase composed of a protein forming the general import pore (GIP) together with a set of small, tightly associated accessory subunits. Fig. 2 and Table 2 give a graphical and an index overview of the TOM and TIM components and their putative functions.

The Tom complex in yeast forms an oligomer of 400 kDa (Rapaport, 2002). The receptor subunits, Tom20, Tom70 and Tom22, are anchored to the mitochondrial outer membrane. Precursors with a cleavable presequence are recognized by Tom20 and subsequently delivered to Tom22 (Brix *et al.*, 1997). The interaction of mTPs with Tom20 is accomplished through hydrophobic amino acid residues of mTPs (Abe *et al.*, 2000; Muto *et al.*, 2001), whereas binding to Tom22 has a more ionic character and may involve interactions of the basic residues of mTPs with acidic residues of Tom22. Internal mitochondrial targeting signals of the hydrophobic carrier proteins bind through the Tom70 receptor (Chacinska *et al.*, 2002). Tom5 accessory subunit functions as a secondary receptor, transferring precursors from the receptor subunits to the GIP subunit, Tom40. Tom40 forms the protein-conducting cation-selective channel ~22Å wide (Hill *et al.*, 1998) that allows transfer of a single □-helix or a loop. Electron microscopy analysis revealed that the TOM complex contains up to three pore-like structures in each complex (Kunkele *et al.*, 1998). Two additional small Tom proteins, Tom6 and Tom7, tightly bind to Tom40 and affect structural organization and stability of the TOM complex (Neupert, 1997). There also exists in the outer membrane a sorting and assembly machinery (SAM) required for sorting and assembly of outer membrane proteins (Wiedemann *et al.*, 2003) that contains Tom37 (homologue of the mammalian Metaxin) and Sam50 subunits (Kozjak *et al.*, 2003).

The plant TOM complex differs in many aspects from the characterized yeast complex. It forms a smaller complex of 230 kDa (Werhahn *et al.*, 2001). A noteworthy difference is the absence of the Tom22 receptor. A subunit with a molecular mass of 9 kDa, which is structurally similar to Tom22 but lacks the cytosolic acidic receptor domain, was found (Jansch *et al.*, 1998). The 9 kDa subunit is the likely replacement for Tom22, and the loss of the acidic domain has been suggested to be crucial in preventing chloroplastic proteins from binding to the TOM complex (Mascasev *et al.*, 2000). The *Solanum tuberosum* TOM complex does not contain any counterpart to Tom70 (Jansch *et al.*, 1998; Werhahn *et al.*, 2001); however, six potential Tom70 homologues have been proposed to exist in the *Arabidopsis* genome (Lister et al, 2003). Furthermore, plant Tom20 is anchored in the outer membrane by the C-terminal domain in contrast to the N-terminal domain of yeast and mammalian Tom20 (Zhang *et al.*, 2001; Werhahn *et al.*, 2001).

Moreover, plant genome sequencing projects have revealed that almost all Tom subunits are encoded by multiple genes (Tom20:1-4, Tom9:1-2, Tom7:1-2, Tom40:1-2). Isoforms of receptor proteins may reflect different substrate specificities (Werhahn *et al.*, 2001). Expression analysis of Tom isoforms in various organs showed that although they were present in small multigene families, only one member was prominently expressed (Lister *et al.*, 2004). This was supported by comparison of real-time RT-PCR and microarray experimental data with expressed sequence tag numbers and massive parallel signature sequence data, and confirmed by mass spectrometric analysis.

Interestingly, the outer mitochondrial membrane of *Arabidopsis* contains a novel protein that displays 67% sequence identity with the Toc64 translocase of the outer envelope membrane of chloroplasts (Chew *et al.*, 2003) as well a homologue of Tom37 (Metaxin) of the SAM machinery.

4.3.2. The TIM complexes

Transport of precursors through the inner mitochondrial membrane in yeast is mediated by at least two distinct complexes, the TIM23 and the TIM22 complex (Jensen and Dunn, 2002). The TIM23 complex mediates the import of precursor proteins with N-terminal mTPs, whereas the TIM22 complex supports the import of polytopic membrane proteins with internal signals. In addition, the OXA1 translocase mediates insertion of precursor proteins from the matrix into the inner membrane (Hell *et al.*, 1998).

The TIM23 complex consists of two integral membrane proteins, Tim23 and Tim17, a peripherally attached Tim44 that constitutes an anchor for mtHsp70, and two recently identified subunits, Tim50 and Tim14 (Truscott *et al.*, 2003, 2003a). The TIM23 complex is less abundant (15 pmol Tim23/mg protein) than the TOM complex (300 pmol Tom40/mg protein) (Rehling *et al.*, 2001). Tim23 and Tim17 are homologous and span the inner membrane four times. Tim23 forms a voltage-activated cation-selective channel with a diameter of 13Å (Schwartz and Matouschek, 1999); it occurs as a dimer and contains in yeast a hydrophilic 100 amino acid residue-long segment exposed to the inter membrane space and spanning to the outer membrane (Donzeau *et al.*, 2000). Dimer formation is mediated by leucine zippers and the dimer dissociates when Tim23 binds mTPs (Bauer *et al.*, 1996). The function of Tim17 is unclear (Pfanner and Chacinska, 2002). Tim50 is anchored to the membrane and exposes a large C-terminal domain towards the inter membrane space (Geissler *et al.*, 2002). That domain binds mTPs and mediates precursor translocation. Tim14 is also anchored to the membrane but exposes its J-domain to the matrix (Mokranjac *et al.*, 2003). This suggests that Tim14 has a function similar to DnaJ, *i.e.* stimulation of the ATPase activity of mtHsp70.

Multiple homologues of the TIM23 subunits are present in the *Arabidopsis* genome: Tim23:1-3, Tim17:1-3, Tim14:1-3, Tim44:1-2 and a single Tim50. All of Tim23 and Tim17 isoenzymes are expressed but have different tissue and developmental expression profiles (Murcha *et al.*, 2003). Mass spectrometric analysis of purified mitochondria identified and verified the presence of Tim17, Tim23, Tim44, and Tim50, that correlated with the most abundant gene transcript measured by expression data (Lister *et al.*, 2004). Interestingly, the *Arabidopsis* Tim23:17 translocase displays topological differences in comparison to fungal and mammalian systems. In plants, it appears that Tim17 rather than Tim23 contains an additional C-terminal segment of 85 amino acid residues that is exposed to the inter membrane space (Murcha *et al.*, 2003). The authors suggested that this segment may function to link the outer and inner mitochondrial membrane (Murcha *et al.*, 2003).

Another translocase of the inner membrane, the TIM22 complex, mediates insertion of carrier proteins and translocases in yeast (Kerscher *et al.*, 1997, 2000). It forms in yeast a complex of 300 kDa that consists of three integral membrane

***Table 2**. Plant mitochondrial translocase subunits and their putative function. Transcriptome and proteome data from Lister* et al. *2004. Phenotype in* Arabidopsis *from TAIR database (http://www.arabidopsis.org/index.jsp). Phenotype in yeast from http://www-deletion.stanford.edu/cgi-bin/deletion/search3.pl.*
N.a. stands for not available and N.o. for not observable.

Import component	*Function*	*Detected in* Arabidopsis thaliana		*Phenotype in* Arabidopsis thaliana	*Pheno-type in yeast*
		Trans-criptome	*Proteome*		
Tom5	Secondary receptor	Yes		N.a.	Viable
Tom6	Formation and stability of TOM	Yes	Yes	N.o.	Viable
Tom7:1-2	Organisation and stability of TOM	Yes		N.a.	Viable
Tom9:1-2	Receptor	Yes	Yes	N.o.	Lethal
Tom20:1-4	Receptor	Yes	Yes	N.o.	Viable
Metaxin	Sorting, assembly factor	Yes	Yes	N.o.	Viable
Tom40:1-2	Import channel	Yes	Yes	N.o.	Lethal
Tim8	Chaperone in IMS	Yes	Yes	N.o.	Viable
Tim9	Chaperone in IMS	Yes	Yes	Yes	Lethal
Tim10	Chaperone in IMS	Yes	Yes	N.a.	Lethal
Tim13	Chaperone in IMS	Yes	Yes	N.a.	Viable
Tim14:1-3	Chaperone function, import motor			N.o.	Lethal
Tim17:1-3	Import channel ?	Yes	Yes	N.o.	Lethal
Tim23:1-3	Import channel	Yes	Yes	N.o.	Lethal
Tim50	Connection between TOM and TIM23	Yes	Yes	Yes	Lethal
Tim44:1-2	Anchor for mtHsp70, import motor	Yes	Yes	N.o.	Lethal
Tim22:1-3	Insertion pore for IM proteins	Yes		N.o.	Lethal
Oxa1	Insertion to IM	Yes	Yes	N.o.	Viable

proteins, Tim22, Tim54 and Tim18. Tim22 is a homologue of Tim23 and Tim17 and forms voltage-activated pores. Tim54 and Tim18 are needed for the formation or stability of the TIM22 complex. Purified TIM22 complex revealed a twin-pore structure of ~16 Å each that was voltage activated (Rehling *et al.*, 2003).

Homologues of the TIM22 complex, Tim22:1-3 isoenzymes but not Tim54 and Tim18, are present in *Arabidopsis*.

The function of the TIM22 translocase is dependent on a group of molecular chaperones located in the inter membrane space called small Tims, represented by Tim8, Tim9, Tim10, Tim11, Tim12 and Tim13 (Koehler *et al.*, 1999). Small Tims contain a twin zinc-finger motif, CX_3C, that binds Zn^{2+} (Lutz *et al.*, 2003). Small Tims guide unfolded, hydrophobic membrane proteins through the aqueous inter membrane space (Curran *et al.*, 2002). Mutation in *ddp*1, a close homologue of human *tim*8, causes the Mohr-Tranebjaerg syndrome, a neurodegenerative disorder characterised by progressive sensorineural deafness, cortical blindness, dystonia and paranoia. This is the first human disease caused by a defect in the mitochondrial protein import machinery (Koehler *et al.*, 1999). Homologues of the small Tims, Tim8, Tim9, Tim10 and Tim13 are found in *Arabidopsis*.

Insertion of proteins into the inner membrane may also occur from the matrix side and is mediated by the OXA1 complex. Oxal protein is a member of a conserved protein family represented in prokaryots by YidC that mediates insertion of proteins into the plasma membrane and by Alb3 in chloroplasts that mediates insertion of LHCP into the thylakoid membrane (Moore *et al.*, 2000). In mitochondria, Oxa1p has been shown to be involved in insertion of CoxI, CoxII, CoxIII and cytochrome b. *Oxa*1p gene is present in *Arabidopsis* in a single copy.

Twin-arginine targeting (Tat) protein secretion systems are found in many Archaea, bacteria, chloroplasts and mitochondria. The Tat system consists of two protein types, members of the TatA and TatC families. Interestingly, although many mitochondrially encoded TatC homologues have been identified, corresponding TatA homologues have not been found in this organelle (Yen *et al.*, 2002).

4.4. The mechanism and regulation of plant mitochondrial protein import

What factors drive and regulate import of precursor proteins into mitochondria? It has been originally suggested that the precursor translocation may be promoted by Tom subunits via electrostatic interactions, a so-called “acid chain hypothesis”. The binding affinity of a precursor protein for the acidic patches of Tom20, Tom22, Tom5 and Tom40 increases sequentially along the import route. However, as hydrophobic interactions are also important, this model has been revised and renamed “binding chain hypothesis” that includes multiple interaction sites also containing different types of non-covalent bindings (Pfanner and Geissler, 2001; Truscott *et al.*, 2003).

Translocation of precursor proteins across the inner mebrane requires the membrane potential, □□, of about 40-60 mV and the hydrolysis of ATP. □□ is negative inside mitochondria and exerts an electrophoretic effect on positively charged presequences facilitating movement into the matrix. □□ also activates the Tim channel proteins. ATP is required for the ATP-dependent import motor, consisting of the mtHsp70 (and its co-chaperone Mge1) and Tim44 (Neupert and Brunner, 2002) that pull precursor proteins into mitochondria.

Two models have been proposed for the action of mtHsp70. In the trapping model, mtHsp70 is proposed to prevent backsliding of the precursor protein spontaneously oscillating in the import channel (Gaume *et al.*, 1998). In the pulling model, mtHsp70, in co-operation with Tim44 and Mge1, is believed to generate an active pulling force through ATP-dependent conformational changes (Matouschek *et al.*, 1997). Most likely, both mechanisms co-operate, and the choice of energy depends on the features of the precursor protein (Huang *et al.* 2002).

The regulation of protein import into mitochondria in plants represents a new aspect with a high potential for future studies and genetic engineering. Plants must be able to respond rapidly to changing circumstances, and therefore it is not surprising that they have additional levels of control not seen in other organisms. The control of protein import is particularly important for proteins with a dual location as the mechanism of control could determine the amount of the dual targeted protein in each organelle. Protein import into mitochondria has been reported to decline with plant development in concert with changes in the amount of mtHsp70 (Dudley *et al.* 1997) or Tom20 (Murcha *et al.*, 1999). Moreover, the overexpression of mtHsp70 in tobacco caused an increase in the growth rate, biomass accumulation and efficiency of protein import in engineered plants as compared to wild-type (Woods *et al.*, 1998). Protein import was also controlled by diurnal rhythms (Dessi *et al.*, 1997). Studies of the impact of various environmental stresses such as drought, chilling or herbicide treatment on protein import revealed differential effects on precursor proteins utilizing different import pathways (Taylor *et al.*, 2003). Whereas drought treatment stimulated the import of precursors via the general ATP-dependent import pathway, chilling and herbicide treatment caused inhibition. On the other hand, drought decreased the import of the F_Ad precursor using an ATP-independent import pathway.

The most thorough transcriptomic and proteomic analysis of the *Arabidopsis* mitochondrial protein import apparatus and its response to mitochondrial dysfunction was pursued in Whelan´s and Millar´s laboratories (Lister *et al.*, 2004). These studies showed that transcription of import component genes was induced when mitochondrial respiration was inhibited and that minor gene isoforms displayed a greater induction than the predominant isoforms. Microarray analysis indicated the up-regulation of genes involved in mitochondrial chaperone activity, protein degradation, respiratory chain assembly, and division. Under the same conditions the rate of protein import into isolated mitochondria was halved. These findings show that there exists an extra level of complexity in gene expression and import activities.

It will be of great interest to address in future studies further regulatory mechanisms of organellar protein import, organellar sorting mechanisms on molecular level and 3-D structures of translocase complexes and their interactions with precursor proteins.

5. ACKNOWLEDGMENTS

This work was supported by grants from The Swedish Research Council to EG and from the Deusche Forschungs-gemeinschaft to JS. We thank C. Rudhe for the graphical presentation of the chloroplast and plant mitochondrial import machineries.

6. REFERENCES

Abe, Y., Shodai, T., Muto, T., Mihara, K., Torii, H., Nishikawa, S. *et al.* (2000). Structural basis of presequence recognition by the mitochondrial protein import receptor Tom20. *Cell 100*, 551-560.

Andersson, S.G.E. & Kurland, C.G. (1999). Origins of mitochondria and hydrogenosomes. *Curr. Opin. Microbiol. 2*, 535-541.

Arabidopsis Genome Initiative, (2000). Analysis of the genome sequence of the flowering plant Arabidopsis thaliana. *Nature 408*, 796–815.

Bauer, M.F., Sirrenberg, C., Neupert, W. & Brunner, M. (1996). Role of Tim23 as voltage sensor and presequence receptor in protein import into mitochondria. *Cell 87*, 33-41.

Bauer, J., Chen, K., Hiltbrunner, A., Wehrli, E., Eugster, M., Schnell, D *et al.* (2000). The major protein import recoptor of plastids is essential for chloroplast biogenesis. *Nature 403*, 203-207.

Bauer, J., Hiltbrunner, A., & Kessler, F. (2001). Molecular biology of chloroplast biogenesis: gene expression, protein import and intraorganellar sorting. *Cell Mol. Life Sci. 58*, 420-433.

Becker, T., Jelic, M., Vojta, A., Radunz, A., Soll, J. & Schleiff, E. (2004). Preprotein recognition by the Toc complex. *EMBO J.* in press.

Beckmann, R., Mizzen, L., & Welch, W. (1990). Interaction of Hsp70 with newly synthesized proteins: implications for protein folding and assembly. *Science 248*, 850-854.

Bölter B., Soll J., Schulz A., Hinnah S. & Wagner R. (1998). Origin of a chloroplast protein importer. *Proc. Natl. Acad. Sci. USA 95*, 15831-15836.

Bhushan, S., Lefebvre B., Ståhl, A., Wright, S. J., Bruce, B.D., Boutry, M. *et al.* (2003). Signal peptide degrading Zinc-metalloprotease is dually targeted to both mitochondria and chloroplasts. *EMBO Rep. 4*, 1073-1078.

Boutry, M., Nagy, F., Poulsen, C., Aoyagi, K. & Chua, N.H. (1987). Targeting of bacterial chloramphenicol acetyltransferase to mitochondria in transgenic plants. *Nature* 328, 340-342.

Braun, H.-P. & Schmitz, U.K. (1999). The protein-import apparatus of plant mitochondria. *Planta 209*, 267-274.

Brink, S., Flugge, U.I., Chaumont, F., Boutry, M., Emmermann, M., Schmitz, U. *et al.* (1994). Preproteins of chloroplast envelope inner membrane contain targeting information for receptor-dependent import into fungal mitochondria. *J. Biol. Chem., 269*, 16478-16485.

Brink, S., Flugge, U.-I., Chaumont, F., Boutry, M., Emmermann, M. , Schmitz, U.K *et al.* (1994). Preproteins of chloroplast envelope inner membrane contain targeting information for receptor-dependent import into fungal mitochondria. *J. Biol. Chem. 269*, 16478-85.

Brix, J., Rudiger, S., Bukau, B., Schneider-Mergener, J. & Pfanner, N. (1999). Distribution of binding sequences for the mitochondrial import receptors Tom20, Tom22, and Tom70 in a presequence-carrying preprotein and a non-cleavable preprotein. *J. Biol. Chem. 274*, 16522-16530.

Bruce, B.D. (1998). The role of lipids in plastid protein transport. *Plant. Mol. Biol.* 38, 223-246.

Bruce, B.D. (2000). Chloroplast transit peptides: structure, function and evolution. *Trends Cell Biol., 10*, 440-447.

Bruce BD. (2001). The paradox of plastid transit peptides: conservation of function despite divergence in primary structure. *Biochim. Biophys. Acta. 1541*, 2-21.

Caliebe, A., Grimm, R., Kaiser, G., Lübeck, J., Soll, J. & Heins, L. (1997). The chloroplastic protein import machinery contains a Rieske-type iron-sulfur cluster and a mononuclear iron-binding protein. *EMBO 716*, 7342-7350.

Chabregas, S.M., Luche, D.D., Farias, L.P., Ribeiro, A.L., van Sluys, M.-A. & Menck, C.F.M. (2001). Dual targeting properties of the N-terminal signal sequence of Arabidopsis thaliana THI1 protein to mitochondria and chloroplasts. *Plant Mol. Biol. 46*, 639-650.

Chacinska, A., Pfanner, N. & Meisinger, C. (2002). How mitochondria import hydrophilic and hydrophobic proteins. *Trends Cell Biol. 12*, 299-303.

Chen, L.-J., & Li, H.A. (1998). A mutant deficient in the plastid lipid DGD is defective in protein import into chloroplasts. *Plant J: 16*, 33-39.

Chen, X., Smith, M.D., Fitzpatrick, L., & Schnell, D.J. (2002). *In vivo* analysis of the role of atTic20 in protein import into chloroplasts. *The Plant Cell 14*, 641-654.

Chew, O., Whelan, J. & Millar, A.H. (2003). Molecular definition of the ascorbate-glutathione cycle in Arabidopsis mitochondria reveals dual targeting of antioxidant defenses in plants. *J. Biol. Chem. 278*, 46869-46877.

Chew, O., Lister, R., Qbadou, S., Heazlewood, J.L., Soll, J., Schleiff, E. *et al.* (2004). A plant outer mitochondrial membrane protein with high amino acid sequence identity to a chloroplast protein import receptor. *FEBS Lett. 557*, 109-114.

Chew, O., Rudhe, C., Glaser, E. & Whelan, J. (2004). Characterisation of the targeting signal of dual targeted pea glutathione reductase. *Plant Mol. Biol.* in press.

Chou, M.-L., Fitzpatrick, L.M., Tu, S.L., Budziszewski, G., Potter-Lewis, S., Akita, M. *et al.* (2003). Tic40, a membrane-anchored co-chaperone homolog in the chloroplast protein translocon. *EMBO J. 22*, 2970-2980.

Chupin, V., van´t Hof, R., & de Kruijff, B. (1994). The transit sequence of a chloroplast precursor protein reorients the lipids in MGDG containing bilayers. *FEBS Lett. 350*, 104-108.

Chupin, V., Leenhouts, J. M., de Kroon, A. I. & de Kruijff, B. (1995). Cardiolipin modulates the secondary structure of the presequence peptide of cytochrome oxidase subunit IV: a 2D 1H-NMR study. *FEBS Lett. 373*, 239-244.

Cleary, S. P., Tan, F.-C., Nakrieko, K.-A., Thompson, S. J., Mullineaux, P. M., Creissen, G. P. *et al.* (2002). Isolated plant mitochondria import chloroplast precursor proteins in vitro with the same efficiency as chloroplasts. *J. Biol. Chem. 277*, 5562-5569.

Corral-Debrinski, M., Blugeon, C. & Jacq, C. (2000). In yeast, the 3' untranslated region or the presequence of ATM1 is required for the exclusive localization of its mRNA to the vicinity of mitochondria. *Mol Cell Biol, 20*, 7881-7892.

Creissen, G., Reynolds, H., Xue, Y. & Mullineaux, P. (1995). Simultaneous targeting of pea glutathione reductase and of a bacterial fusion protein to chloroplasts and mitochondria in transgenic tobacco. *Plant J. 8*, 167-175.

Curran, S.P., Leuenberger, D., Oppliger, W. & Koehler, C.M. (2002). The Tim9p-Tim10p complex binds to the transmembrane domains of the ADP/ATP carrier. *EMBO J. 21*, 942-953.

Daniell, H., Datta, R., Varma, S., Gray, S. & Lee, S.B. (1998). Containment of herbicide resistance through genetic engineering of the chloroplast genome. *Nat. Biotechnol. 1*, 345-348.

Dessi, P., Smith, M.K., Day, D.A. & Whelan, J. (1996). Characterization of the import pathway of the F_Ad subunit of mitochondrial ATP synthase into isolated plant mitochondria. *Arch. Biochem. Biophys 335*, 358-368.

Dessi, P. & Whelan, J. (1997). Temporal regulation of in vitro import of precursor proteins into tobacco mitochondria. *FEBS Lett. 415*, 173-178.

Dessi, P., Pavlov, P., Wållberg, F., Rudhe, R., Brack, S., Whelan, J. *et al.* (2003). Investigations on the in vitro import ability of mitochondrial precursor proteins synthesised in wheat germ transcription-translation extract. *Plant Mol. Biol. 52*, 259-271.

Dhingra, A., Portis, A.R. & Daniell, H. (2004). Enhanced translation of a chloroplast expressed *RbcS* gene restores SSU levels and photosynthesis in nuclear antisense *RbcS* plants. *Proc. Natl. Acad. Sci.*, U.S.A.101: 6315-6320.

Donzeau, M., Kaldi, K., Adam, A., Paschen, S., Wanner, G., Guiard, B. *et al.* (2000). Tim23 links the inner and outer mitochondrial membranes. *Cell 101*, 401-412.

Dudley, P., Wood, C.K., Pratt, J.R. & Moore, A.L. (1997). Developmental regulation of the plant mitochondrial matrix located HSP70 chaperone and its role in protein import. *FEBS Lett. 417*, 321-324.

Eckart, K., Eichacker, L., Sohrt, K., Schleiff, E., Heins, L. & Soll, J. (2002). A Toc75-like protein import channel is abundant in chloroplasts. *EMBO reports 3*, 557-562.

Emanuelsson, O., Nielsen, H. & von Heijne, G. (1999). ChloroP, a neural network-based method for predicting chloroplast transit peptides and their cleavage sites. *Protein Sci. 8*, 978-984.

Froehlich, J.E., Wilkerson, C.G., Ray, W.K., McAndrew, R.S., Osteryoung, K.W., Gage, D.A. *et al.* (2003). Proteomic study of the Arabidopsis thaliana chloroplastic envelope membrane utilizing alternatives to traditional two-dimensional electrophoresis. *J. Proteome Res. 2*, 413-425.

Frydman, J., Nimmesgern, E., Ohtsuka, K. & Hartl, F.U. (1994). Folding of nascent polypeptide chains in a high molecular mass assembly with molecular chaperonse. *Nature 370*, 111-117.

Funfschilling, U. & Rospert, S. (1999). Nascent polypeptide-associated complex stimulates protein import into yeast mitochondria. *Mol Biol Cell 10*, 3289-3299.

Gavel Y, von Heijne G. (1990). Cleavage-site motifs in mitochondrial targeting peptides. *Protein Eng. 4*, 33-37.

Gaume, B., Klaus, C., Ungermann, C., Guiard, B., Neupert, W. & Brunner, M. (1998). Unfolding of preproteins upon import into mitochondria. *EMBO J. 17*, 6497-6507.

Geissler, A., Chacinska, A., Truscott, K.N., Wiedemann, N., Brandner, K., Sickmann, A. *et al.* (2002). The mitochondrial presequence translocase: an essential role of Tim50 in directing preproteins to the import channel. *Cell 111*, 507-518.

Genevrois, S., Steeghs, L., Roholl, J., Letesson, J., & van der Ley, P. (2003). The Omp85 protein of *Neisseria meningitidis* is required for lipid export to the outer membrane. *EMBO J. 22*, 1780-1789.

Gentle, I., Gabriel, K., Beech, P., Waller, R. & Lithgow, T. (2004). The Omp85 family of proteins is essential for outer membrane biogenesis in mitochondria and bacteria. *J. Cell Biol. 164*, 19-24.

George, R., Beddoe, T., Landl, K. & Lithgow, T. (1998). The yeast nascent polypeptide-associated complex initiates protein targeting to mitochondria in vivo. *Proc. Natl. Acad. Sci. U S A, 95*, 2296-2301.

Ginsberg, M.D., Feliciello, A., Jones, J.K., Avvedimento, E.V. & Gottesman, M.E. (2003). PKA-dependent Binding of mRNA to the Mitochondrial AKAP121 Protein. *J Mol Biol. 327*, 885-897.

Glaser, E., & Dessi, P. (1999). Integration of the mitochondrial processing peptidase into the bc_1 complex of the respiratory chain in plants. *J. Bioenerg. Biomembr. 31*, 259-274.

Glaser, E., Sjoling, S., Tanudji, M. & Whelan, J. (1998). Mitochondrial protein import in plants. *Plant Mol. Biol. 38*, 311-338.

Gray, M.W., Burger, G. & Lang, B.F. (1999). Mitochondrial evolution. *Science 283*, 1476-1481.

Gutensohn, M., Schulz, B., Nicolay, P. & Flügge, U. (2000). Functional analysis of the two Arabidopsis homologues of Toc34, a component of the chloroplast protein import apparatus. *Plant J. 23*, 771-783.

Hachiya, N., Alam, R., Sakasegawa, Y., Sakaguchi, M., Mihara, K. & Omura, T. (1993). A mitochondrial import factor purified from rat liver cytosol is an ATP-dependant conformational modulator for precursor proteins. *EMBO J. 12*, 1579-1586.

Heazlewood, J.L., Tonti-Filippini, J.S., Gout, A.M., Day, D.A., Whelan, J. & Millar, A.H. (2004). Experimental analysis of the Arabidopsis mitochondrial proteome highlights signaling and regulatory components, provides assessment of targeting prediction programs, and indicates plant-specific mitochondrial proteins. *Plant Cell. 16*, 241-256.

Hedtke, B., Borner, T. & Weihe, A. (2000). One RNA polymerase serving two genomes. *EMBO Rep. 1*, 435-440.

Heins, L. & Soll, J. (1998). Chloroplast biogenesis: mixing the prokaryotic and the eukaryotic. *Curr. Biol. 8*, 215-217.

Heins L., Mehrle A., Hemmler R., Wagner R., Küchler M., Hörmann F.*et al.* (2002). The preprotein conducting channel at the inner envelope membrane of plastids. *EMBO J., 21*, 2616-2625.

Hell, K., Herrmann, J.M., Pratje, E., Neupert, W. & Stuart, R.A. (1998). Oxa1p, an essential component of the N-tail protein export machinery in mitochondria. *Proc Natl Acad Sci U S A 95*, 2250-2255.

Herrmann, J.M. & Neupert, W. (2000). What fuels polypeptide translocation? An energetical view on mitochondrial protein sorting. *Biochim. Biophys. Acta, 1459*, 331-338

Hill, K., Model, K., Ryan, M.T., Dietmeier, K., Martin, F., Wagner, R. *et al.* (1998). Tom40 forms the hydrophilic channel of the mitochondrial import pore for preproteins. *Nature 395*, 516-521.

Hiltbrunner, A., Bauer, J., Vidi, P.-A, Infanger, S., Weibel, P., Hohwy, M. *et al.* (2001a). Targeting of an abundant cytosolic form of the protein import receptor at Toc159 to the outer chloroplast membrane. *J. Cell Biol. 154*, 309-316.

Hiltbrunner, Al, Bauer, J., Alvarez-Huerta, M. & Kessler, F. (2001b). Protein translocon at the Arabidopsis outer chloroplast membrane. *Biochem. Cell Biol. 79*, 629-635.

Hinnah, S., Hill, K., Wagner, R., Schlicher, Th. & Soll, J. (1997). Reconstitution of a chloroplast protein import channel. *EMBO J. 16*, 7351-7360.

Hinnah, S., Wagner, R., Sveshnikova, N., Harrer, R. & Soll, J. (2002). The chloroplast protein import channel Toc75: pore properties and interaction with transit peptides. *Biophys. J. 83*, 899-911.

Hirohashi, T., Hase, T. & Nakai, M. (2001). Maize non-photosynthetic ferredoxin precursor is missorted to the intermembrane space of chloroplasts in the presence of light. *Plant Physiol. 125*, 2154-2163.

Hirsch, S., Muckel, E., Heermeyer, F., von Heijne, G. & Soll, J. (1994). A receptor component of the chloroplast protein translocation machinery. *Science 266*, 1989-1992.

Horniak L., Pilon M. & van´t Hof R: (1993). The secondary structure of the ferredoxin transit sequence is modulated by its interaction with negatively charged lipids. *FEBS Lett., 34*:241-246.

Huang, S., Ratliff, K.S. & Matouschek, A. (2002). Protein unfolding by the mitochondrial membrane potential. *Nat. Struct. Biol. 9*, 301-307.

Hugosson, M., Nurani, G., Glaser, E. & Franzén, L.G. (1995). Peculiar properties of the PsaF photosystem I protein from the green alga *Chlamydomonas reinhardtii*: presequence independent import of the PsaF protein into both chloroplasts and mitochondria. *Plant Mol. Biol. 28*, 525-535.

Inaba, T., Li, M., Avarez-Huerta, M., Kessler, F. & Schnell D. (2003). AtTic110 functions as a scaffold for coordinating the stromal events of protein import into chloroplasts. *J. Biol. Chem. 278*, 38617-38627.

Inoue, K., Demel, R., de Kruijff, B. & Keegstra, K. (2001). The N-terminal portion of the preToc75 transit peptide interacts with membrane lipids and inhibits binding and import of precursor proteins into isolated chloroplasts. *Eur. J. Biochem. 268*, 4036-4043.

Jackson, D., Froehlich, J. & Keegstra, K. (1998). The hydrophilic domain of Tic110, an inner envelope membrane component of the chloroplastic protein translocation apparatus, faces the stromal compartment. *J. Biol. Chem. 273*, 16583-16588.

Jackson-Constan, D. & Keegstra, K. (2001). Arabidopsis genes encoding components of the chloroplastic protein import apparatus. *Plant Physiol 125*, 1567-1576.

Jansch, L., Kruft, V., Schmitz, U.K.& Braun, H.P. (1998). Unique composition of the preprotein translocase of the outer mitochondrial membrane from plants. *J Biol Chem, 273*, 17251-17257.

Jensen, R.E. and Dunn, C.D. (2002). Protein import into and across the mitochondrial inner membrane: role of the TIM23 and TIM22 translocons. *Biochim. Biophys. Acta 1592*, 25-34.

Jarvis, P., Chen-L.-J., Li, H., Peto, C.A., Fankhauser, C. & Chory, J. (1998). An Arabidopsis mutant defective in the plastid general protein import apparatus. *Science 282*, 100-103.

Jelic, M., Sveshnikova, N., Motzkus, M., Hörth, P., Soll, J. & Schleiff, E. (2002). The chloroplast import receptor Toc34 functions as preprotein-regulated GTPase. *Biol. Chem. 383*, 1875-1883.

Jelic, M., Soll, J. & Schleiff, E. (2003). Two Toc34 homologues with different properties. *Biochemistry.42*, 5906-5916.

Karslake, C., Piotto, M. E., Pak, Y. K., Weiner, H. & Gorenstein, D. G. (1990). 2D NMR and structural model for a mitochondrial signal peptide bound to a micelle. *Biochemistry 29*, 9872-9878.

Keegstra, K. & Cline, K. (1999). Protein import and routing systems of chloroplasts. *Plant Cell 11*, 557-570.

Kerber, B., & Soll, J. (1992). Transfer of a chloroplast-bound recursor protein into the translocation apparaturs is impaired after phospholipase C treatment. *FEBS Lett. 306*, 71-74.

Kerscher, O., Holder, J., Srinivasan, M., Leung, R.S. & Jensen, R.E. (1997). The Tim54p-Tim22p complex mediates insertion of proteins into the mitochondrial inner membrane. *J. Cell Biol. 139*, 1663-1675.

Kerscher, O., Sepuri, N.B. & Jensen, R.E. (2000). Tim18p is a new component of the Tim54p-Tim22p translocon in the mitochondrial inner membrane. *Mol. Biol. Cell 11*, 103-116.

Kessler, F., Blobel, G., Patel, H.A. & Schnell, D.J. (1994). Identification of two GTP-binding proteins in the chloroplast protein import machinery. *Science 266*, 1035-1039.

Kessler, F. & Blobel, G. (1996) Interaction of the protein import and folding machineries of the chloroplast. *Proc. Natl. Acad. Scie. USA 93*, 7684-7689.

Koehler, C.M., Leuenberger, D., Merchant, S., Renold, A., Junne, T. & Schatz, G. (1999). Human deafness dystonia syndrome is a mitochondrial disease. *Proc. Natl. Acad. Sci. U S A 96*, 2141-2146.

Kouranov, A. & Schnell, D.J. (1997). Analysis of the interactions of preproteins with the import machinery over the course of protein import into chloroplasts. *J. Cell Biol. 139*, 1677-1685.

Kouranov, A., Chen, X., Fuks, B. & Schnell, D.J. (1998). Tic20 and Tic22 are new components of the protein import apparatus at the chloroplast inner envelope membrane. *J. Cell. Biol. 143*, 991-1002.

Kozjak, V., Wiedemann, N., Milenkovic, D., Lohaus, C., Meyer, H.E., Guiard, B. *et al.* (2003). An essential role of Sam50 in the protein sorting and assembly machinery of the mitochondrial outer membrane. *J. Biol. Chem. 278*, 48520-48523.

Krimm I, Gans P, Hernandez JF, Arlaud GJ, Lancelin JM. (1999). A coil-helix instead of a helix-coil motif can be induced in a chloroplast transit peptide from *Chlamydomonas reinhardtii. Eur. J. Biochem. 265*,171-180.

Küchler, M., Decker, S., Hörmann, F., Soll, J. & Heins, L. (2002.). Protein import into chloroplasts involves redox-regulated proteins. *Embo J. 21*, 6136-6145.

Kunkele, K.P., Heins, S., Dembowski, M., Nargang, F.E., Benz, R., Thieffry, M. *et al.* (1998). The preprotein translocation channel of the outer membrane of mitochondria. *Cell 93*, 1009-1019

Lancelin, J. M., Gans, P., Bouchayer, E., Bally, I., Arlaud, G. J. & Jacquot, J. P. (1996). NMR structures of a mitochondrial transit peptide from the green alga *Chlamydomonas reinhardtii. FEBS Lett. 391*, 203-208.

Lang, B.F., Gray, M.W. & Burger, G. (1999) Mitochondrial genome evolution and the origin of eukaryotes. *Annu. Rev. Genet. 33*, 351-397.

Lee, Y.J., Kim, D.H., Kim, Y-W. & Hwang I. (2001). Identification of a signal that distinguishes between the chloroplast outer envelope membrane and the endomembrane system *in vivo. The Plant Cell 13*, 2175-2190.

Lee, K. H., Kim, S. J., Lee, X. J. & Hwang, I. (2003). The M domain of atToc159 plays an essential role in the import of proteins into chloroplasts and chloroplast biogenesis. *J. Biol. Chem. 278*, 36794-36805.

Leister, D. (2003). *Trends Genet. 19*, 47-56.

Lister, R., Chew, O., Lee, M. & Whelan, J. (2001). Arabidopsis thaliana ferrochelatase-I and -II are not imported into Arabidopsis mitochondria. *FEBS Lett. 506*, 291-295.

Lister, R., Chew, O., Lee, M.N., Heazlewood, J.L., Clifton, R., Parker, K.L. *et al.* (2004). A Transcriptomic and Proteomic Characterization of the Arabidopsis Mitochondrial Protein Import Apparatus and Its Response to Mitochondrial Dysfunction. *Plant Physiol*., in press.

Lithgow, T. (2000). Targeting of proteins to mitochondria. *FEBS Lett, 476*, 22-26.

Lübeck, J., Soll, J., Akita, M., Nielsen, E. & Keegstra, K. (1996.). Topology of IEP110, a component of the chloroplastic protein import machinery present in the inner envelope membrane. *EMBO J. 15*, 4230-4238.

Lutz, T., Neupert, W. & Herrmann, J.M. (2003). Import of small Tim proteins into the mitochondrial intermembrane space. *EMBO J. 22*, 4400-4408.

Ma, Y., Kuranov, A., LaSala, S. & Schnell, D.J. (1996). Two components of the chloroplast protein import apparatus, IAP86 and IAP75, interact with the transit sequence during the recognition and translocation of precursor proteins at the outer envelope. *J. Cell. Biol. 134*, 315-327.

Macasev, D., Newbigin, E., Whelan, J. & Lithgow, T. (2000). How do plant mitochondria avoid importing chloroplast proteins? Components of the import apparatus Tom20 and Tom22 from Arabidopsis differ from their fungal counterparts. *Plant Physiol. 123*, 811-816.

Marc, P., Margeot, A., Devaux, F., Blugeon, C., Corral-Debrinski, M. & Jacq, C. (2002). Genome-wide analysis of mRNAs targeted to yeast mitochondria. *EMBO Rep. 3*, 159-164.

Mattingly, J.R., Youssef, J., Iriarte, A. & Martinez-Carrion, M. (1993). Protein folding in a cell-free translation system. The fate of the precursor to mitochondrial aspartate aminotransferase. *J. Biol. Chem. 268*, 3925-3937.

May, T. & Soll, J. (1998). Positive charges determine the topology and functionality of the transmembrane domain in the chloroplastic outer envelope protein Toc34. *J. Cell Biol. 141*, 895-904.

May, T. & Soll, J. (2000). 14-3-3 proteins form a guidance complex with chloroplast precursor proteins in plants. *The Plant Cell 12*, 53-63.

Miernyk, J., Duck, N., Shatters, R.J. & Folk, W. (1992). The 70-kilodalton heat shock cognate can act as a molecular chaperone during the membrane translocation of a plant secretory protein precursor. *Plant Cell 4*, 821-829.

Matouschek, A., Azem, A., Ratliff, K., Glick, B.S., Schmid, K. & Schatz, G. (1997). Active unfolding of precursor proteins during mitochondrial protein import. *EMBO J. 16*, 6727-6736.

Matouschek, A., Pfanner, N. & Voos, W. (2000). Protein unfolding by mitochondria. *EMBO Rep. 1*, 404-410.

May, T. & Soll, J. (2000). 14-3-3 proteins form a guidance complex with chloroplast precursor proteins in plants. *Plant Cell 12*, 53-63.

McFadden, G.I. (2001). Chloroplast origin and integration. *Plant Physiol. 125*, 50-53.

Moberg, P., Ståhl, A., Bhushan, S., Wright, S. J., Eriksson, A.C., Bruce, B.D. *et al.* (2003). Characterization of a novel zinc metalloprotease involved in degrading signal peptides in mitochondria and chloroplasts. *Plant J. 36*, 616-628.

Moberg, P., Nilsson, S., Ståhl, A., Eriksson, A.C., Glaser, E. & Mäler L. (2003). NMR solution structure of the mitochondrial F_1□ presequence from *Nicotiana plumbaginifolia. J. Mol. Biol.* in press.

Mokranjac, D., Paschen, S.A., Kozany, C., Prokisch, H., Hoppins, S.C., Nargang, F.E. *et al.* (2003). Tim50, a novel component of the TIM23 preprotein translocase of mitochondria. *EMBO J. 22*, 816-825.

Moore, M., Harrison, M.S., Peterson, E.C. & Henry, R. (2000). Chloroplast Oxa1p homolog albino3 is required for post-translational integration of the light harvesting chlorophyll-binding protein into thylakoid membranes. *J. Biol. Chem. 275*, 1529-1532.

Murakami, K., Tanase, S., Morino, Y. & Mori, M. (1992) Presequence binding factor-dependent and -independent import of proteins into mitochondria. *J. Biol. Chem. 267*, 13119-13122.

Murcha MW, Huang T, Whelan J. (1999). Import of precursor proteins into mitochondria from soybean tissues during development. *FEBS Lett. 464*, 53-59.

Murcha, M.W., Lister, R., Ho, A.Y. & Whelan, J. (2003). Identification, expression, and import of components 17 and 23 of the inner mitochondrial membrane translocase from Arabidopsis. *Plant Physiol. 131*, 1737-1747.

Muto, T., Obita, T., Abe, Y., Shodai, T., Endo, T. & Kohda, D. (2001). NMR identification of the Tom20 binding segment in mitochondrial presequences. *J. Mol. Biol. 306*, 137-43.

Okamoto, K., Brinker, A., Paschen, S.A., Moarefi, I., Hayer-Hartl, M., Neupert, W. *et al.* (2002). The protein import motor of mitochondria: a targeted molecular ratchet driving unfolding and translocation. *EMBO J. 21*, 3659-3671.

Ono, H. & Tuboi, S. (1990). Purification and identification of a cytosolic factor required for import of precursors of mitochondrial proteins into mitochondria. *Arch. Biochem. Biophys. 280*, 299-304.

Neupert, W. (1997). Protein import into mitochondria. *Annu Rev Biochem, 66*, 863-917.

Neupert, W. & Brunner, M. (2002). The protein import motor of mitochondria. *Nat. Rev. Mol. Cell. Biol. 3*, 555-565.

Nielsen, E., Akita, M., Davila-Aponte, J. & Keegstra, K. (1997). Stable association of chloroplastic precursors with protein translocation complexes that contain proteins from both envelope membranes and a stromal Hsp100 molecular chaperone. *EMBO J. 16*, 935-946.

Paschen, S.A., Waizenegger, T., Stan, T., Preuss, M., Cyrklaff, M., Hell, K. *et al.* (2003). Evolutionary conservation of biogenesis of □barrel membrane proteins. *Nature 426*, 862-867.

Peltier, J.B., Emanuelsson, O., Kalume, D.E., Ytterberg, J., Friso, G., Rudella, A- *et al.* (2002). Central functions of the lumenal and peripheral thylakoid proteome of Arabidopsis determined by experimentation and genome-wide prediction. *Plant Cell 14*, 211–236.

Perry, S.E. & Keegstra, K. (1994) Envelope membrane proteins that interact with chloroplastic precursor proteins. *Plant Cell 6*, 93-105.

Pfanner, N. & Chacinska, A. (2002). The mitochondrial import machinery: preprotein-conducting channels with binding sites for presequences. *Biochim. Biophys. Acta 1592*, 15-24.

Pilon, M., de Boer, D. A., Knols, S. L., Koppelman, M.H.G M., van der Graaf, R.M., de Kruijff, B.*et al.* (1990). Expression in *Escherichia coli* and purifaction of a translocation-competent precursor of the chloroplast protein ferredoxin. *J. Biol. Chem. 265*, 3358-3361.

Pilon, M, Rietveld, A.G., Weisbeck, P.J. & de Kruijff, B. (1992) Secondary structure and folding of a functional chloroplast precursor protein. *J. Biol. Chem. 267*, 19907-19913.

Pinnaduwage, P. & Bruce, B.D. (1996). *In vitro* interaction between a chloroplast transit peptide and chloroplast outer envelope lipids is sequence-specific and lipid-class dependent. *J. Biol. Chem. 271*, 32907-32915.

Qbadou, S., Tien, R., Soll. J. & Schleiff, E. (2003). Membrane insertion of the chloroplast outer envelope protein, Toc34: constrains for insertion and topology. *J. Cell Science 116*, 837-846.

Rassow, J., Dekker, P.J.T., van Wilpe, S., Meier, M. & Soll, J. (1999). The protein translocation of the mitochondrial inner membrane: function and evolution. *J. Mol. Biol. 286*, 105-120.

Rawlings ND, Barrett AJ. (1995). Evolutionary families of metallopeptidases. *Meth. Enzymol. 248*,183-228.

Rial, D.V., Ottado, J. & Ceccarelli, E.A. (2003). Precursors with altered affinity for Hsp70 in their transit peptides are efficiently imported into chloroplasts. *J. Biol. Chem. 278*, 46473-46481.

Rapaport, D. (2003). Finding the right organelle. Targeting signals in mitochondrial outer-membrane proteins. *EMBO Rep. 4*, 948-952.

Rehling, P., Model, K., Brandner, K., Kovermann, P., Sickmann, A., Meyer, H.E. *et al.* (2003). Protein insertion into the mitochondrial inner membrane by a twin-pore translocase. *Science 299*1747-1751.

Rehling, P., Wiedemann, N., Pfanner, N. & Truscott, K.N. (2001). The mitochondrial import machinery for preproteins. *Crit Rev Biochem Mol Biol 36*, 291-336

Reumann, S. & Keegstra, K. (1999). The endosymbiotic origin of the protein import machinery of chloroplastic envelope membranes. *Trends Plant Sci. 4*, 302-307.

Richter, S. & Lamppa, G.K. (1999). Stromal processing peptidase binds transit peptides and initiates their ATP-dependent turnover in chloroplasts. *J. Cell Biol. 147*, 33-44.

Rudhe, C., Chew, O., Whelan, J. & Glaser, E. (2002). A novel in vitro system for simultanous import of precursor proteins into chloroplast and mitochondria. *Plant J. 30*, 213-220.

Rudhe, C., Clifton, R., Whelan, J. & Glaser, E. (2002). N-terminal domain of the dual targeted pea glutathione reductase signal peptide controls organellar targeting efficiency. *J. Mol. Biol. 324*, 577-585.

Salomon, M., Fischer, K., Flügge, U.-I. & Soll, J. (1990). Sequence analysis and protein import studies of an outer chloroplast envelope polypeptide. *Proc. Natl. Acad. Sci. USA 87*, 5778-5782.

Schleiff, E. & Soll, J. (2000). Travelling of proteins through membranes: translocation into chloroplasts. *Planta 211*, 449-456.

Schleiff, E., Tien, R., Salomon, M. & Soll, J. (2001). Lipid composition of the outer leaflet of chloroplast outer envelope determines topology of OEP7. *Mol. Biol. Cell 12*, 4090-4102.

Schleiff, E., Motzkus, M. & Soll, J. (2002). Chloroplast protein import inhibition by soluble factor from wheat germ lysate. *Plant Mol. Biol. 50*: 177-185.

Schleiff, E., Soll, J., Küchler, M., Kühlbrandt, W. & Harrer, R. (2003a). Characterisation of the translocon of the outer envelope of chloroplasts. *J. Cell. Biol. 160*, 541-551.

Schleiff, E., Jelic, M. & Soll, J. (2003b). A GTP-driven motor moves proteins across the outer envelope of chloroplasts. *Proc. Natl. Acad. Sci. USA 100*, 4604-4609.

Schmitz, U.K. & Lonsdale, D.M. (1989). A yeast mitochondrial presequence functions as a signal for targeting to plant mitochondria in-vivo. *Plant Cell 1*, 783-791.

Schneider, G., Sjöling, S., Wallin, E., Wrede, P., Glaser, E. & von Heijne, G. (1998). Feature-extraction from endopeptidase cleavage sites in mitochondrial targeting sequences, Proteins: structure, function and genetics. *30*, 49-60.

Schnell, D.J., Kessler, F. & Blobel, G. (1994) Isolation of components of the chloroplast protein import machinery. *Science 266*, 1007-1012.

Schubert, M., Petersson, U.A., Haas, B.J., Funk, C., Schroder, W.P. & Kieselbach, T. (2002). Proteome map of the chloroplast lumen of Arabidopsis thaliana. *J. Biol. Chem. 277*, 8354–8365.

Schwartz, M.P. & Matouschek, A. (1999.) The dimensions of the protein import channels in the outer and inner mitochondrial membranes. *Proc Natl Acad Sci U S A 96*, 13086-13090.

Scott, S.V. & Theg, S.M. (1996) A new chloroplast protein import intermediate reveals distinct translocation machineries in the two envelope membranes: energetics and mechanistic implications. *J. Cell Biol. 132*, 63-75.

Seedorf, M. & Soll, J. (1995). Copper chloride, an inhibitor of protein import into chloroplasts. *FEBS Lett 367*, 19-22.

Seedorf, M., Waegemann, K. & Soll, J. (1995). A constituent of the chloroplast import complex represents a new type of GTP-binding protein. *Plant J. 7*, 401-411.

Sickmann, A, Reinders, J., Wagner, Y., Joppich, C., Zahedi, R., Meyer, H. E. *et al.* (2003). The proteome of *Saccharomyces cerevcisiae* mitochondria. *Proc. Natl. Acad. Sci. USA 100*, 13207-13212.

Silva-Filho, M.D.C., Wieers, M.-C., Flugge, U.-I., Chaumont, F. & Boutry, M. (1997). Different in vitro and in vivo targeting properties of the transit peptide of a chloroplast envelope inner membrane protein. *J. Biol. Chem. 272*, 15264-15269.

Silva-Filho, M.C. (2003). One ticket for multiple destinations: dual targeting of proteins to distinct subcellular locations. Curr. Opin. *Plant Biol. 6*, 589-595.

Sjöling, S. & Glaser, E. (1998). Mitochondrial targeting peptides in plants. *Trends Plant Sci., 3*, 136-140.

Small, I., Wintz, H., Akashi, K. & Mireau, H. (1998). Two birds with one stone: genes that encode products targeted to two or more compartments. *Plant Mol. Biol., 38*, 265-277.

Smith, M.K., Day, D.A. & Whelan, J. (1994). Isolation of a novel soybean gene encoding a mitochondrial ATP synthase subunit. *Arch. Biochem. Biophys. 313*, 235-240.

Sohrt, K. & Soll, J. (2000). Toc64, a new component of the protein translocon of chloroplasts *J. Cell Biol. 148*, 1213-1221.

Soll, J. (2002). Protein import into chloroplasts. *Curr. Opin. Plant Biol. 5*, 529-535.

Soll, J. & Schleiff, E. (2004). Protein import into chloroplasts. *Nature Rev. Mol. Cell Biol.*, in press. Soll, J. and Tien, R. (1998). Protein translocation into and across the chloroplastic envelope membranes. *Plant Mol. Biol. 38*, 191-207.

Stahl T, Glockmann C, Soll J & Heins L. (1999). Tic40, a new "old" subunit of the chloroplast protein import translocon. *J Biol Chem 274*, 37467-37472.

Stein, I., Peleg, Y., Even-Ram, S. & Pines, O. (1994). The single translation product of the FUM1 gene (fumarase) is processed in mitochondria before being distributed between the cytosol and mitochondria in Saccharomyces cerevisiae. *Mol Cell Biol, 14*, 4770-4778.

Ståhl, A. Moberg, P., Ytterberg, J., Panfilov, O., Brockenhuus von Löwenhielm, H., Nilsson, F. *et al.* (2002). Isolation and identification of a novel mitochondrial metalloprotease (PreP) that degrades targeting presequences. *J. Biol. Chem. 277*, 41931-41939.

Svesnikova, N., Grimm, R., Soll, J. & Schleiff, E. (2000a). Topology studies of the chloroplast protein import channel Toc75. *Biol. Chem. 381*, 687-693.

Svesnikova, N., Soll, J. & Schleiff, E. (2000b) Toc34 is a preprotein receptor regulated by GTP and phosphorylation. *Proc. Natl. Acad. Sci. USA 97*, 4973-4978.

Tanudji, M., Dessi, P., Murcha, M. & Whelan, J. (2001). Protein import into plant mitochondria: precursor proteins differ in ATP and membrane potential requirements. *Plant Mol. Biol., 45*, 317-325.

Taylor, N. L., Rudhe, C, Hulett J. M., Lithgow, T., Glaser, E., Day, D. A. *et al.* (2003). Environmental Stresses inhibit and stimulate Different Protein Import Pathways in Plant Mitochondria. *FEBS Lett. 547*, 125-130.

Tranel, P.J., Fröhlich, J., Goyal, A. & Keegstra, K.A. (1995) Titel *EMBO J. 14*, 2436-2446.

Truscott, K.N., Brandner, K. & Pfanner, N. (2003). Mechanisms of protein import into mitochondria. *Curr. Biol. 13*, 326-37.

Tsai, L-Y., Tu, S-L. & Li, H. (1999). Insertion of atToc34 into the chloroplastic outer membrane is assisted by at least two proteinaceous components in the import system. *J. Biol. Chem. 274*, 18735-18740.

Tu, S-L. & Li, H. (2000). Insertion of OEP14 into the outer envelope membrane is mediated by proteinaceous components of chloroplasts. *The Plant Cell 12*, 1951-1959.

von Heijne, G. (1986). Mitochondrial targeting sequences may form amphiphilic helices. *EMBO J. 5*, 1335-1342.

von Heijne, G., Steppuhn, J. & Herrmann, R.G. (1989). Domain structure of mitochondrial and chloroplast targeting peptides. *Eur. J. Biochem. 180*, 535-545.

von Heijne, G. (1992). Membrane protein structure prediction. Hydrophobicity analysis and the positive-inside rule. *J. Mol. Biol. 225*, 487-494.

Voulhoux, R., Bos, M.P., Geurtsen, J., Mols, M. & Tommassen, J. (2003) Role of a highly conserved bacterial protein in outer membrane protein assembly. *Science 299*, 262-265.

Waegemann, K., Paulsen, H. & Soll, J. (1990). Translocation of proteins into isolated chloroplasts requires cytosolic factors to obtain import competence. *FEBS Lett. 261*, 89-92.

Waegemann, K. & Soll, J. (1991). Characterization of the protein import apparatus in isolated outer envelopes of chloroplasts. *Plant J. 1*, 149-158.

Waegemann, K. & Soll, J. (1996). Phosphorylation of the transit sequence of chloroplast precursor proteins. *J.Biol. Chem. 271*, 6545-6554.

Watanabe, N., Che, F.S., Iwano, M., Takayama, S., Yoshida, S. & Isogai, A. (2001). Dual targeting of spinach protoporphyrinogen oxidase II to mitochondria and chloroplasts by alternative use of two in-frame initiation codons. *J. Biol. Chem. 276*, 20474-20481.

Werhahn, W., Niemeyer, A., Jansch, L., Kruft, V., Schmitz, U.K. & Braun, H. Purification and characterization of the preprotein translocase of the outer mitochondrial membrane from Arabidopsis. Identification of multiple forms of TOM20. *Plant Physiol . 125*, 943-954.

Wiedemann, N., Kozjak, V., Chacinska, A., Schonfisch, B., Rospert, S., Ryan, M.T. *et al.* (2003). Machinery for protein sorting and assembly in the mitochondrial outer membrane. *Nature, 424*, 565-571.

Wiedemann, N., Pfanner, M. & Ryan, M.T. (2001) The three modules of ADP/ATP carrier cooperate in receptor recruitment and translocation into mitochondria. *EMBO J. 20*, 951-960.

Wienk, H.L., Czisch, M. & de Kruijff, B. (1999). The structural flexibility of the preferredoxin transit peptide.*FEBS Lett. 453*,318-326.

Wienk, H. L. J., Wechselberger, R. W., Czisch, M. & de Kruijff, B. (2000). Structure, dynamics, and insertion of a chloroplast targeting peptide in mixed micelles. *Biochemistry 39*, 8219-8227.

Woods, C.K., Affourtit, C., Albury, M.S., Carre, J., Dudley, P., Gordon, J. *et al.* (1998). In Plant Mitochondria: from gene to function. (eds. Møller I.M., Gardeström, P., Glimelius, K. and Glaser, E.) Backhuys, Leiden.

Yano, M., Terada, K. & Mori, M. (2003). AIP is a mitochondrial import mediator that binds to both import receptor Tom20 and preproteins. *J. Cell Biol. 163*, 45-56.

Yen, M.R., Tsen, Y.H., Nguyen, E.H., Wu, L.F. & Saier, M.H.Jr. (2002). Sequence and phylogenetic analyses of the twin-arginine targeting (Tat) protein export system. *Arch. Microbiol. 177*, 441-450.

Young, J.C., Hoogenraad, N.J. & Hartl, F.U. (2003). Molecular chaperones Hsp90 and Hsp 70 deliver preproteins to the mitochondrial import receptor Tom70. *Cell 112*, 41-50.

Zhang, X.-P. (2001). Structure and function of mitochondrial and chloroplast signal peptides. PhD thesis. Department of Biochemistry. Stockholm University, Stockholm.

Zhang, X.P., Sjoling, S., Tanudji, M., Somogyi, L., Andreu, D., Eriksson, L.E. *et al.* (2001). Mutagenesis and computer modelling approach to study determinants for recognition of signal peptides by the mitochondrial processing peptidase. *Plant J. 27*, 427-438.

Zhang, X.P., Elofsson, A., Andreu, D. & Glaser, E. (1999). Interaction of mitochondrial presequences with DnaK and mitochondrial Hsp70. *J. Mol. Biol., 288*, 177-190.

Zhang, X.P. & Glaser, E. (2002). Interaction of plant mitochondrial and chloroplast signal peptides with Hsp70 molecular chaperone. *Trends Plant Sci. 7*, 14-21.

CHAPTER 15

PROTEOLYSIS IN PLANT ORGANELLES

ZACH ADAM

The Robert H. Smith Institute of Plant Sciences and Genetics in Agriculture, The Hebrew University of Jerusalem, Rehovot 76100, Israel

Abstract. Proteolytic processes are intimately involved in the biogenesis and maintenance of chloroplasts and mitochondria. As such, they influence photosynthesis, respiration and other functions of these organelles. The great majority of the proteins found in these organelles are imported post-translationally. The signal peptides directing the precursor proteins into the respective organelles are removed by specific metallo-processing peptidases to yield mature proteins. The released signal peptide is further degraded by another metalloprotease. Oxidatively damaged proteins are rapidly degraded to allow the incorporation of newly synthesized ones into the respective complexes. The size of the photosynthetic antenna is adjusted by proteolytic degradation due to changes in light intensity. Partially assembled complexes or proteins lacking their prosthetic groups are inactive, and thus are being rapidly turned over. Changes in plastid identity and senescence processes are accompanied by massive the degradation of proteins. Although it is not clear which proteases perform most of these activities, a number of organelle proteases have been identified in recent years. ClpCP is an ATP-dependent serine protease complex that is located in the stroma of chloroplasts. Its catalytic function is performed by the ClpP subunit; ClpC serves as the regulatory subunit, responsible for substrate recognition, unfolding and feeding into the catalytic chamber. Mutations in either ClpP or C affect chloroplast biogenesis and functions, and result in defective growth and development. Mitochondria also contain a Clp protease, but here the regulatory subunit is ClpX, an ATPase with a single ATP-binding domain. FtsH is a membrane-bound ATP-dependent metalloprotease found in both chloroplasts and mitochondria. Here the catalytic and the ATPase functions are found on the same polypeptide. This enzyme is involved in the repair cycle of photosystem II from photoinhibition by degrading the oxidatively damaged D1 protein of the reaction center. Mutations in FtsH isozymes lead to leaf variegation. DegP is a serine protease, peripherally attached to both sides of the thylakoid membrane, and expected to reside also in mitochondria. Another ATP-dependent serine protease, Lon protease, was identified in both mitochondria and chloroplasts. In addition, several amino- and endopeptidases are expected to reside in both organelles. Most proteases are encoded by multi-gene families. Specific products are targeted to either one of the organelles. However, the functional significance of this gene multiplication, i.e., whether they perform redundant or specific functions, is not clear yet.

H. Daniell and C.D. Chase (eds.), Molecular Biology and Biotechnology of Plant Organelles 419—440.

1. INTRODUCTION

Intracellular proteolytic processes are inherent to the function of any biological system. Chloroplasts and mitochondria are not an exception to this rule. Proteolysis is involved in the biogenesis and maintenance of these organelles, and hence is considered a vital factor that influences photosynthesis and respiration and other functions, under both optimal and adverse growth conditions. As the great majority of chloroplast and mitochondrial proteins are targeted to these organelles post-translationally, their signal peptides are removed after translocation by specific peptidases. This proteolytic cleavage of a single peptide bond is essential for the maturation of the precursor form of the protein. The free signal peptide, that now has no additional function, is completely degraded.

Some plastid-encoded proteins also need to be processed in order to become active. Once organelle proteins are in their final location, the level of different proteins is adjusted during development or in response to different environmental conditions. This adjustment involves either increased transcription and/or translation rates, resulting in an increase in the level of a specific protein, or proteolytic degradation that leads to a decrease in its level. For instance, under low-light conditions plants increase the size of their photosynthetic antenna in order to maximize the amount of light energy that is absorbed. When such plants are exposed to high light, they adjust the size of their main antenna complex, the light-harvesting complex (LHC) of photosystem II (PSII), by proteolytic degradation of specific chlorophyll *a/b*-binding proteins. A converse transition is accompanied by degradation of the early light-inducible protein (ELIP).

Functional proteins that are damaged during their activity and become non-functional are thus being degraded in a process that can be described as 'protein quality control'. Chloroplast and mitochondrial proteins, residing in highly oxidative environments, are very prone to such damage. In fact, oxidative damage is the primary reason for photoinhibition – inhibition of photosynthesis under increasing light intensities. Rapid degradation of oxidatively damaged proteins, such as the D1 protein of PSII reaction center, is a central component in the repair mechanism of photoinhibited PSII. Another manifestation of quality control is the removal of unassembled proteins that lack a protein partner or a prosthetic group. Whether the missing component is due to a mutation, inhibition of synthesis or nutrient deficiency, the consequence is proteolytic degradation of the existing partner. Leaf senescence and fruit ripening involve the breakdown of chloroplasts or the transition of chloroplasts to other types of plastids. In both cases, massive degradation of the existing protein repertoire, primarily of proteins involved in photosynthesis, occur. 'Timing proteins', usually involved in controlling gene expression, are expected to be short-lived, although their identity in plant organelles is not well established yet.

This chapter intends to review proteolytic processes in plant organelles, to describe organelle proteases and peptidases, to discuss the possible roles of proteases in developmental processes and maintenance, and to raise hypotheses regarding additional functions of these enzymes. Previous reviews, discussing some of these issues, primarily in chloroplasts, can be found in Adam, 1996; 2000; Andersson and Aro, 1997; Clarke, 1999; and Adam and Clarke, 2002.

2. PROCESSING OF PRECURSOR PROTEINS

2.1. Processing of chloroplast precursors

The chloroplast genome encodes only a small fraction of its proteome. In *Arabidopsis*, 87 proteins are encoded in the chloroplast genome (Sato et al., 1999), whereas more than 3500 of the nuclear gene products are predicted to target chloroplasts (The Arabidopsis Genome Initiative, 2000). All nuclear-encoded chloroplast-targeted proteins, with the exception of those targeted to the outer envelope membrane, are synthesized as precursors in the cytosol, and imported post-translationally into the organelle, where they are sorted to their sub-organellar final location. During this process, their N-terminal targeting sequence is removed by a single proteolytic cleavage. Most proteins are processed only once in the stroma, but those targeted to the thylakoid lumen are processed first in the stroma, and then once again in the lumen [for recent reviews, see Dalbey and Robinson, 1999; Jarvis and Soll, (2002); and Chapter 14 in this volume)]. A few chloroplast-encoded proteins, such as Cyt *f* and the D1 protein of PSII reaction center, are also synthesized with N- or C-terminal extensions, respectively. Although the function of these extensions is not known, they need to be removed before these proteins become functional (Trost et al., 1997).

The processing of chloroplast precursors is performed by the chloroplast stromal processing peptidase (SPP). It is a ~140 kDa metallopeptidase with a His-X-X-Glu-His zinc-binding motif at its catalytic site. It performs a single endoproteolytic cleavage step that releases the mature form of a wide range of chloroplast precursor proteins (Vandervere et al., 1995; Richter and Lamppa, 1998, 1999, and2003; Zhong et al., 2003). The bound transit peptide is then further cleaved once by SPP to release a subfragment that is degraded to completion by a protease with characteristics of a metallo-protease that requires ATP (Richter and Lamppa, 1999). A very recent report demonstrated that a recombinant zinc-metalloprotease, which could be targeted to both mitochondria and chloroplasts, was able to degrade both mitochondrial and chloroplast targeting sequences (Bhushan et al., 2003; Moberg et al., 2003). Although this reaction did not require ATP, this protease is a strong candidate for the protease that disposes of cleaved transit peptides after import of precursor proteins into the organelle. SPP is encoded by a single gene in *Arabidopsis*. Inhibiting its expression by anti-sense constructs resulted in a high proportion of lethal seedlings, suggesting that it functions as the general stromal peptidase (Zhong et al., 2003).

Proteins that are targeted to the thylakoid lumen are synthesized with a bipartite transit peptide. The N-terminal region of this peptide mediates translocation into the stroma, where it is cleaved by SPP to yield an intermediate form of the protein. This form is further translocated by the thylakoid translocation machinery, followed by removal of the thylakoid transfer domain by an enzyme designated the thylakoid processing peptidase (TPP) (Chaal et al., 1998). TPP is homologous to bacterial Type I leader peptidases that use a Ser-Lys catalytic dyad.

As mentioned above, some chloroplast-encoded proteins, such as the D1 protein of PSII reaction center or Cyt *f*, are synthesized with C- or N-terminal extensions, respectively. The function of these extensions is not known, but they need to be removed before the protein can assemble into the respective complex. The C-terminal extension of the D1 protein is processed by the CtpA peptidase that is located in the lumen (Shestakov et al., 1994; Inagaki et al., 1996; Oelmuller et al., 1996; Trost et al., 1997; Yamamoto et al., 2001). Similar to TPP, it uses a Ser-Lys catalytic dyad. It appears that the activity of CtpA is limited to this process only, and other lumenal proteins are insensitive to it. The enzyme responsible for the maturation of Cyt *f* is not known.

2.2. Processing of mitochondrial precursors

The products of almost 3000 genes in the *Arabidopsis* nuclear genome are expected to target mitochondria (The Arabidopsis Genome Initiative, 2000); only 33 proteins are encoded in the organelle genome itself (Marienfeld et al., 1999). Thus, the mitochondrion is heavily dependent on the post-translational import of most of its proteins. The N-terminal targeting sequences of the imported proteins are removed by a proteolytic cleavage of a single peptide bond to yield mature proteins. This cleavage event is carried out by the mitochondrial-processing peptidase (MPP) that is associated with the Cyt bc_1 complex (Glaser et al. 1996; Glaser and Dessi, 1999). MPP is a ~55 kDa metallopeptidase with a His-X-X-Glu-His zinc-binding motif. After the targeting sequence is removed, it is further degraded by the dually targeted metalloprotease that degrades also chloroplast transit peptides (Bhushan et al., 2003; Moberg et al., 2003) [for more details, see Chapter 14 in this volume].

3. SUBSTRATES FOR PROTEOLYTIC DEGRADATION

3.1. Degradation of oxidatively-damaged proteins

Not all the light energy absorbed by the photosynthetic antenna is dissipated by the photochemical process, especially under high light conditions. As a consequence, reactive oxygen species are generated. Although scavenging mechanisms exit in the organelle to neutralize their harmful effect, oxidative damage to chloroplast proteins is a common phenomenon. In fact, oxidative damage is considered the main cause for photoinhibition – the inhibition of photosynthesis under increasing light intensity (Barber and Andersson, 1992). The primary site of photoinhibition is PSII, but PSI is also damaged and inhibited (Tjus et al., 1998). Photoinhibition in PSII has been well characterized and reviewed [see Andersson and Aro (2001) and references therein]. The most sensitive component of this complex is the D1 protein of the reaction center, whose oxidation is believed to lead to a conformational change and loss of function. A repair cycle of PSII evolved that increased transcription and translation of the *psbA* gene encoding the D1 protein (Melis, 1999). However, a prerequisite for incorporation of the newly synthesized D1 copy into the inhibited complex is the proteolytic removal of the damaged copy. The operation of the repair cycle under photoinhibitory conditions is manifested by the rapid turnover rate of the D1 protein, a phenomenon that led to the discovery and characterization of photoinhibition (Kyle et al., 1984; Mattoo et al., 1984; Barber and Andersson, 1992; Prasil et al., 1992). Mitochondrial proteins are also expected to be prone to oxidative damage, and hence, are likely to be rapidly degraded. However, this was not documented yet in plant mitochondria.

3.2. Adjustment of antenna size

Plants have evolved mechanisms to minimize the potential damage of high-light intensity. Short-term mechanisms include leaf and chloroplast movement that reduce the amount of energy absorbed by the photosynthetic antenna (Wada et al., 2003), as well as state transition – the shuttle of LHCII antenna subunits between PSII and PSI, to balance the energy input between the two photosystems (Allen, 1992). A longer-term adaptation mechanism involves reducing the antenna size upon increase in light intensity by transfer from shade to full light or from a cloudy to a sunny period. In this case, the reduction in the antenna size involves proteolytic degradation of specific antenna subunits (Lindahl et al., 1995; Yang et al., 1998).

3.3. Degradation of partially assembled proteins

Inherently stable chloroplast proteins can be destabilized under certain conditions. When one subunit of a protein complex is missing, the other subunits will be degraded rapidly. This was shown for both soluble as well as membrane-bound complexes [e.g., Schmidt and Mishkind (1983); Leto et al. (1985)], and it could result from a mutation or inhibition of protein synthesis in either the chloroplast or the cytosol. Prosthetic groups have also a stabilizing effect on their cognate proteins. In the absence of chlorophyll for instance, chlorophyll-binding proteins are proteolytically degraded [e.g., Apel and Kloppstech (1980); Bennett (1981); Mullet et al. (1990); Kim et al. (1994)]. The lack of even a single copper ion also leads to the degradation of plastocyanin (Merchant and Bogorad, 1986; Li and Merchant, 1995). Thus, it appears that in the absence of major or minor components of a single protein or a protein complex, structural changes occur that lead to protein susceptibility to proteolysis. Degradation of unassembled mitochondrial proteins was documented also in yeast and mammal mitochondria, but not yet in plants.

3.4. Senescence and transition from chloroplasts to other types of plastids

Plant cell plastids, including chloroplasts, can change from one type of plastid to another. For instance, chloroplasts develop from either proplastids or etioplasts. Chromoplasts develop from either proplastids or chloroplasts. These changes, especially those involving transformation of chloroplasts, are accompanied by major changes in the respective proteome. The ripening of fruits or the senescence of leaves, is characterized by the transformation of chloroplasts into chromoplasts, a process that involves massive degradation of the protein components of the photosynthetic apparatus, although the details of these proteolytic processes are not known (Matile, 2001).

3.5. *Timing proteins*

A common theme in the control of gene expression is the appearance of certain regulators at a certain point in time. Their appearance is achieved by regulated transcription and translation of their genes. However, to limit their activity, these proteins need to be modified or to disappear. The half-life of many such regulators is inherently very short, ensuring that once their expression is ceased, their level drops down immediately. Such proteins are often designated 'timing proteins' (Gottesman, 1996). They are widely spread among different biological systems and are thus expected to function in chloroplasts and plant mitochondria as well (Adam, 2000). This is a class of potential chloroplast and mitochondrial substrates for proteolysis whose identity still needs to be revealed.

4. CHLOROPLAST AND MITOCHONDRIAL PROTEASES

Although examples for the degradation of specific chloroplast proteins have been accumulating for more than 20 years [for review, see Adam (1996)], the identity of the components of chloroplast proteases has been revealed only in the past 10 years or so. Given the prokaryotic origin of chloroplasts, it is not surprising that chloroplast proteases are all homologues of bacterial ones. However, whereas most bacterial proteases are encoded by single genes, many of the chloroplast ones are encoded by multi-gene families, a phenomenon whose functional significance is not clear yet. Interestingly, some specific members of these gene families are targeted not to chloroplasts, but to mitochondria. The following is a description of the biochemical properties of these enzymes and a discussion of their functions.

4.1. *Clp protease*

Clp protease is an ATP-dependent serine protease complex, composed of proteolytic subunits and associated ATPases. The *E.* coli form of this protease is best characterized; the core of the complex is composed of two heptameric rings of the 21 kDa proteolytic subunit ClpP. In this barrel-like structure, the active sites, composed of the catalytic triad Ser-His-Asp, are found within a central pore ~51 Å in diameter, whose entrance is only ~10 Å-wide (Wang et al., 1997). This implies that the catalytic chamber is inaccessible to most globular proteins unless they are unfolded prior to degradation. Indeed, *in vitro* assays demonstrated that the proteolytic sub-complex can degrade short peptides but is incapable of degrading proteins. These can be degraded only when the regulatory subunit is present. The pore of the catalytic sub-complex is capped on both sides by hexameric rings of ATPases, either ClpA or ClpX. These differ from each other by their size, 83- and 46 kDa, and the number of ATP-binding domains, two or one, respectively. These ATPases are responsible for substrate recognition and binding, and confer specificity to the complex. Once substrates are bound, they are unfolded and fed into the catalytic sub-complex in a process that requires the hydrolysis of ATP [for review, see Gottesman (1996); Porankiewicz et al., (1999)].

Subunits of the chloroplast Clp protease are found mostly solubilized in the stroma (Shanklin et al., 1995; Halperin and Adam, 1996; Ostersetzer et al., 1996), but a small fraction is also associated with thylakoid membranes (Peltier et al., 2001). ClpP is the only protease encoded in the chloroplast genome, but in addition to this copy, designated ClpP1, five more homologues are encoded in the nucleus (Clarke, 1999; Adam et al., 2001). Four of these, ClpP3, 4, 5 and 6, are targeted to the chloroplast, whereas ClpP2 is found in mitochondria (Halperin et al., 2001b; Millar et al., 2001; Zheng et al., 2002; Peltier et al., 2003). The *Arabidopsis* genome contains four additional homologues of ClpP, designated ClpR, but they do not contain a perfect catalytic triad (Adam et al., 2001). Thus, they are not expected to function as proteolytic subunits, and their function is still unknown. Nevertheless, they are found together with ClpP subunits in a 350 kDa complex (Peltier et al., 2001).

The regulatory subunit of the chloroplast Clp protease is encoded by two highly similar homologues of ClpA, designated ClpC1 and 2, each having two ATP-binding domains. Although they could not be found associated with complexes that were isolated from chloroplasts (Peltier et al., 2001; 2003), immunoprecipitation of ClpC from stromal fractions, supplemented with ATP-□S, could precipitate ClpP together with it (Halperin et al., 2001a). This observation suggests that the association between the proteolytic and the regulatory sub-complexes is dependent on the binding of ATP. Another plant homologue of ClpA is ClpD. Its transcript is up-regulated by different stress conditions (see Adam and Clarke, 2002), but similar to ClpB, the bacterial homologue of ClpA, there is no indication for its association with ClpP, and it might fulfil a chaperone function only. Plants have homologues of ClpX as well, but these are targeted to mitochondria where they might associate with ClpP2 (Halperin et al., 2001b). However, there is no experimental support for this suggestion yet.

The first circumstantial evidence for the importance of chloroplast proteases in plant viability came from the analysis of the plastid genome of the nonphotosynthetic parasitic plant *Epifagus virginiana*. This genome has lost all its photosynthetic genes, but one of the few remaining genes was *clpP* (Depamphilis and Palmer, 1990; Wolfe et al., 1992), suggesting that it plays an essential role in the function of plastids. This suggestion was corroborated when the *clpP* gene was found to be essential for the growth of *Chlamydomonas* (Huang et al., 1994). This is true also for chloroplasts of higher plants. Complete segregation was impossible in attempts to disrupt the chloroplast *clpP* gene (Shikanai et al., 2001). In this study, it was also demonstrated that etioplast development in the dark was inhibited, suggesting that ClpP was essential for non-photosynthetic plastid functions as well. Complete removal of the chloroplast-encoded ClpP could be recently achieved using the CRE-*lox* recombination system in tobacco (Kuroda and Maliga, 2003). This resulted in seedlings with white cotyledons that could not develop further even when grown on sucrose, suggesting a fundamental role for ClpP in shoot development.

Repression of nuclear-encoded ClpP expression also has negative effects on plant development. Expression of an antisense construct of *clpP4* in *Arabidopsis* resulted in severe growth inhibition and lack of chloroplast development, especially in the midrib region of the leaf (B. Zheng and A. K. Clarke, personal communication). The regulatory subunit of the Clp complex, ClpC, is also essential for proper development. Plants having T-DNA insertions in the *clpC1* gene, one of the two *clpC* genes in Arabidopsis, show a severe chlorotic phenotype (L. E. Sjogren and A. K. Clarke, personal communication). These lines contain ~ 30% of ClpC, compared with wild-type plants, which can be attributed to the expression of the *clpC2* gene.

Only little is known about ClpCP substrates at the biochemical level. Since the chloroplast-encoded ClpP could not be deleted from the chloroplast genome of *Chlamydomonas reinhardtii*, Majeran *et al.* (2000; 2001) attenuated the level of its expression by modifying its start codon. This manipulation resulted in a strain that was impaired in the degradation of subunits of the Cyt b_6-*f* complex, PSII subunits and Rubisco under nutrient starvation and light stress. Other possible substrates of ClpCP are mistargeted proteins in the stroma. Intentional targeting of the lumenal OE33 protein to the stroma led to its degradation, both *in vitro* and *in vivo*, by an enzyme with characteristics of ClpCP, or the Clp complex that was immunoprecipitated from the stroma (Halperin and Adam, 1996; Halperin et al., 2001a; Levy et al., 2003). There is currently no available information regarding the function of ClpP2 and ClpX in plant mitochondria.

4.2. FtsH protease

FtsH is a membrane-bound metalloprotease complex. Most available information on this protease comes from the *E. coli* enzyme and its three yeast mitochondrial orthologs, known as AAA proteases [reviewed by Langer (2000)]. In *E. coli,* FtsH is the only ATP-dependent protease that is essential for survival. Unlike Clp protease, FtsH proteolytic and ATPase domains are found on the same 71 kDa polypeptide. The N-terminus contains one or two transmembrane □-helices that anchor the protein to the respective membrane, the cytoplasmic one in *E. coli* and the inner membrane in yeast mitochondria. The hydrophobic domain is followed by the ATPase domain, and the zinc-binding motif, His-Glu-X-X-His, that serves as the active site of the enzyme is found toward the C-terminus of the protein. FtsHs form hexamers, but although the crystal structure of the ATPase domain was determined (Krzywda et al., 2002), the overall structure, and the relative arrangement of the proteolytic domain with respect to the ATPase one, is not known. Nevertheless, it is assumed that like other ATP-dependent proteases, the active sites of FtsH are self-compartmentalized, and access to them is regulated by the ATPase domain of the protein.

Plant FtsH was first found in the thylakoid membrane with its functional domains exposed to the stroma (Lindahl et al., 1996). Plant nuclear genomes contain multiple FtsH genes, with at least 12 in *Arabidopsis*. Four additional genes, encoding homologous proteins with impaired zinc-binding motifs, are also found in the *Arabidopsis* genome (Sokolenko et al., 2002). These could potentially function as chaperones, but not as proteases. Out of the 12 FtsH genes, the products of three, FtsH3, 4 and 10, are targeted to mitochondria, whereas the other nine are capable of entering the chloroplast, as revealed by transient expression assays of GFP-fusions (Sakamoto et al., 2003). However, separation of thylakoid membrane proteins by 2D-PAGE, followed by mass spectrometry analysis, revealed that only four isozymes were accumulated in *Arabidopsis* leaves grown under optimal conditions, FtsH1, 2, 5 and 8. Out of these, FtsH2 is by far the most abundant species (Adam et al., unpublished data). It is possible that the other isozymes are expressed under different environmental conditions, but this has not yet been demonstrated. Similar to bacterial and mitochondrial FtsHs, *Arabidopsis* FtsHs also form hexamers. These are composed of at least FtsH2 and 5 (Sakamoto et al., 2003; S. Rodermel et al., personal communication), and possibly other FtsHs as well. However, the stoichiometry of the different subunits in the complex is not known. The apparent differential abundance of the different FtsHs suggests that homomeric complexes, probably of FtsH2, may also exist, but it is not known whether these have different functions from the heteromeric complexes or not.

Analysis of variegated mutants in *Arabidopsis* revealed that mutations in the *ftsH2* and *ftsH5* genes were responsible for the *var2* and *var1* phenotypes, respectively (Chen et al., 2000; Takechi et al., 2000; Sakamoto et al., 2002; Sakamoto, 2003). The cotyledons of both mutants look like wild type, but the true leaves show different degrees of variegation. In *var2* they are yellow, and subsequent leaves demonstrate decreasing ratio of yellow to green sectors, with some yellow sectors even in mature plants. This phenotype suggests that FtsH is involved in early stages of chloroplast development, primarily in the development of thylakoids. However, the patchy nature of the phenotype suggests that the loss of FtsH2 can be compensated for, at least in the green sectors, probably by other FtsH proteins, and that the compensation mechanism becomes more and more efficient with time. The FtsH5 mutant phenotype is similar, but less pronounced (Sakamoto et al., 2002). Here, the first true leaves are already variegated, variegation decreases in subsequent younger leaves, and mature plants cannot be distinguished from wild type. Interestingly, mutations in other FtsH genes, including *ftsH1* and *8*, whose products are found in thylakoids, have no phenotypic effects (Sakamoto et al., 2003).

The first indication for chloroplast FtsH function came from an *in vitro* study. After import into isolated chloroplasts, unassembled Rieske Fe-S protein that accumulated on the stromal face of the thylakoid membrane, as well as a soluble mutant of this protein that accumulated in the stroma, were rapidly degraded (Ostersetzer and Adam, 1997). Characteristics of the degradation process were reminiscent of FtsH, and indeed, antibodies against the native protease could specifically inhibit degradation *in vitro*. Further support to the link between this proteolytic activity and FtsH came from column chromatography separation of solubilized thylakoid proteins. Activity assays and immunoblot analysis revealed that activity peaks always coincided with the presence of FtsH (Ostersetzer, 2001). Thus, FtsH is likely to play a central role in the degradation of unassembled thylakoid proteins.

Another critical role attributed to chloroplast FtsH protease is involvement in the degradation of oxidatively damaged thylakoid proteins. This has been demonstrated for the D1 protein of PSII reaction center, in the context of repair from photoinhibition. Oxidative damage to the D1 protein is considered the primary reason for photoinhibition of PSII. The repair cycle of PSII involves transcription and translation of new D1 protein copies, but these cannot be incorporated into the damaged complex unless the damaged copies of the D1 protein are degraded. Thus, D1 protein degradation represents a key step in the repair of PSII from photoinhibition. Degradation of PSII takes place in at least two steps. The first one is a single proteolytic cleavage in the stromal loop connecting the forth and the fifth transmembrane helices to yield 23- and 10-kDa fragments, followed by their complete degradation. An *in vitro* study suggested that FtsH was involved in the degradation of the 23-kDa fragment of the D1 protein. This degradation could be inhibited by removing FtsH from the membrane, and degradation could be restored by the addition of recombinant FtsH (Lindahl et al., 2000). The role of FtsH in the D1 protein degradation was supported by *in vivo* studies as well. An *Arabidopsis* FtsH2 mutant was found to be more susceptible to photoinhibition and exhibited slower repair as revealed by chlorophyll fluorescence measurements. In this plant, under conditions where synthesis of chloroplast-encoded proteins, including the D1 protein, was inhibited, damaged D1 protein accumulated in the mutant, whereas it rapidly disappeared in the wild-type plant (Bailey et al., 2002).

Plant mitochondrial FtsH protease was first demonstrated in pea (Kolodziejczak et al., 2002). This enzyme is found in the inner membrane with its functional domains facing the matrix. Expression of its cDNA in yeast cells could restore the growth of YTA10/12 mutants, suggesting that the pea enzyme was a functional homologue of the yeast protein. Using metalloprotease inhibitors and an ATP trap, both of which are known to inhibit the activity of FtsH proteases, it was shown that the accumulation of ATP9 in the inner membrane was inhibited. This observation suggests that mitochondrial FtsH is involved in the biogenesis and/or assembly of inner membrane proteins.

4.3. DegP protease

Bacterial DegP (or HtrA) is a serine protease complex peripherally attached to the periplasmic side of the plasma membrane. The *E.* coli form of this proteaseis characterized best, where it is essential for survival at elevated temperatures (Strauch et al., 1989; Lipinska et al., 1990; Skorko-Glonek et al., 1995; Pallen and Wren, 1997; Sassoon et al., 1999). Determination of its three-dimensional structure revealed that it forms a hexamer made of two staggered trimers (Clausen et al., 2002; Krojer et al., 2002). The 48 kDa monomer is composed of two domains. The N-terminal one is the proteolytic domain, where the typical catalytic triad of serine proteases, Ser-Asp-His, is found. The C-terminus of the protein contains two PDZ-domains, implicated in protein-protein interaction (Fanning and Anderson, 1996; Ponting, 1997), and substrate recognition and binding in the context of proteases (Levchenko et al., 1997). DegP has two homologues in *E. coli*, DegQ and DegS, but these are less characterized (Kolmar et al., 1996; Waller and Sauer, 1996). Another homologue of the DegP protease, designated HtrA2, is found in mitochondria, where it is involved in apoptosis. This enzyme forms a trimer whose three-dimensional structure was also determined (Gray et al., 2000; Li et al., 2002; Ramesh et al., 2002).

Although the proteolytic activity of bacterial DegP is independent of ATP, it has a chaperone activity as well. Interestingly, at normal growth temperature DegP is active as a chaperone, whereas the proteolytic activity dominates at elevated temperatures (Spiess et al., 1999). This temperature-dependent switch between the two different activities can now be explained by the structure of the protein. At normal growth temperature, the active site is blocked by segments of the protein itself, and only upon a thermal-induced conformational change it becomes accessible to substrates (Clausen et al., 2002; Krojer, et al., 2002).

The first plant homologue of DegP, designated DegP1, was found peripherally attached to the lumenal side of the thylakoid membrane (Itzhaki et al., 1998). Unlike the bacterial DegP that has two PDZ domains in tandem, DegP1 contains only one such domain. Similar to the bacterial enzyme, it forms hexamers and its activity is stimulated by high temperature (Chassin et al., 2002). Another two homologues, designated DegP5 and 8, were predicted to reside also in the lumen, and proteomic analyses confirmed this prediction (Adam et al., 2001; Peltier et al., 2002; Schubert et al., 2002). DegP5 is very similar to DegP1, but does not contain a PDZ domain, whereas DegP8 is somewhat less similar, but does have a PDZ domain, and hence, its mature size is almost identical to DegP1, ~35 kDa. Chloroplasts contain at least one more homologue, DegP2, peripherally attached to the stromal side of the thylakoid membrane (Haussuhl et al., 2001). Other plant homologues are predicted to reside in mitochondria, and maybe other cellular sites (Adam et al., 2001), but there is no experimental data to support these predictions yet.

Phenotypic consequences of disruption or inhibition of expression of chloroplast-targeted DegP gene products have not been documented yet. Thus, exploring the roles of DegP protease using mutant or transgenic plants is yet to come. The only reported activity of a chloroplast DegP to date is an *in vitro* study on DegP2 (Haussuhl et al., 2001). Recombinant DegP2 was shown to cleave the D1 protein of photoinhibited PSII and to yield the two typical fragments described above, whereas other thylakoid proteins were insensitive to the presence of protease. Thus, this enzyme is a strong candidate for performing the initial step in the D1 protein degradation.

Although recombinant DegP1 was capable of degrading lumenal proteins such as plastocyanin and OE33 *in vitro* (Chassin et al., 2002), its role *in vivo*, as well as the role of the other two lumenal DegP proteins, DegP5 and 8, is yet to be determined. An appealing role could be participation in the degradation of integral thylakoid membrane proteins. Degradation of these is poorly understood. Nevertheless, DegP proteases could theoretically facilitate degradation of these substrates by cleaving and degrading their lumenal exposed loops.

4.4. Lon protease

Lon is the first ATP-dependent protease that was discovered in *E. coli.* It is an 87 kDa serine protease, containing both its proteolytic and ATPase domains on the same polypeptide, that forms tetramers (for review, see Gottesman, 1996). It is responsible for the non-specific degradation of most abnormal proteins, as well as some specific substrates. Lon has homologues in yeast and human mitochondria (known also as PIM), where it is involved in biogenesis and control of gene expression (Wang et al., 1993; Suzuki et al., 1994; Rep et al., 1996; van Dyck et al., 1998; Fu and Markovitz, 1998).

Lon protease was discovered also in plant mitochondria. Expression of the maize Lon in a yeast *pim1* mutant could restore the maintenance of mitochondrial DNA integrity, but not the assembly of cytochrome *a-a3* complexes (Barakat et al., 1998), suggesting that at least some of the functions of the plant Lon are similar to those of the yeast protein. Lon protease was also implicated in the control of cytoplasmic male sterility in pea by degrading the mitochondrial ORF239 (Sarria et al., 1998). The *Arabidopsis* genome contains at least three genes encoding Lon proteases (Adam et al., 2001). Two of these are predicted to reside in chloroplasts, and indeed, proteins that cross-react with mitochondrial Lon antibodies were found associated with the stromal side of the thylakoid membrane (Ostersetzer, 2001). Interestingly, a Lon-encoding gene is absent from the *Synechocistis* genome, suggesting that today's plastid Lon was acquired from the mitochondrial progenitor, duplicated in plants and redirected to the chloroplast.

4.5. Other proteases and peptidases

In addition to the above proteases and peptidases, chloroplasts and mitochondria are likely to contain other less characterized enzymes. One such protease, whose existence in chloroplasts was already demonstrated, is SppA. This is a light-induced serine protease that is peripherally attached to the thylakoid membrane on its stromal side (Lensch et al., 2001). The existence of other proteases and peptidases in chloroplasts and mitochondria can be predicted from their N-terminal sequences by bioinformatic tools, but experimental evidence supporting these predictions is still lacking. Two copies of a processing metallopeptidase are expected to associate with chloroplast membranes, and one copy of a serine leader peptidase (Sokolenko et al., 2002). Homologues of the soluble metallopeptidases Prp1 and 3 are predicted to reside in the mitochondrial matrix. Several aminopeptidases, mostly methionine aminopeptidases, are predicted to reside in the chloroplast stroma and mitochondria as well (Sokolenko et al., 2002), but there is no further experimental support for these predictions.

Another novel peptidase that might be found in plant mitochondria and chloroplasts is the Rhomboid protease. It is an intramembrane serine protease with substrates that are usually cleaved within a membrane-spanning segment. It was first described in *Drosophila*, where it cleaves the membrane-bound growth factor Spitz, allowing it to activate the EGF receptor (Urban and Freeman, 2002, 2003; Urban et al., 2002a, 2002b). *Drosophila* Rhomboid is an integral endoplasmic reticulum membrane protein, with six transmembrane helices (Urban et al., 2001). Mutation analysis revealed that its catalytic triad is composed of Ser217, His281 and Asn169. In typical serine proteases, an Asp residue is found instead of an Asn. Interestingly, all these residues are located within transmembrane domains, suggesting that Rhomboid is an unusual serine protease with an active site within the membrane bilayer. This feature places Rhomboid within the framework of 'regulated intramembrane proteolysis', a new paradigm of signal transduction, together with Presenilin-1, Notch and S2P [see Koonin et al. (2003) and references therein].

Bioinformatic studies have revealed that Rhomboid proteases are ubiquitous, present in nearly all the sequenced genomes of Archaea, bacteria and eukaryotes (Koonin et al., 2003). They all share six trans-membrane helices and the catalytic triad Ser-His-Asn within the lipid bilayer. Whereas most prokaryotic species have a single gene coding for a Rhomboid protein, eukaryotes have evolved extended families, with seven members in *Drosophila* for instance, and at least eight in *Arabidopsis* (Koonin et al., 2003).

The function of Rhomboid proteases has been shown very recently to extend beyond signal transduction. Yeast mitochondria contain two Rhomboids, one of them (Rbd1) is involved in processing the bipartite signal peptides of Cytochrome *c* peroxidase (Ccp1) and a Dynamin-like GTPase (Mgm1), both located in the intermembrane space of the mitochondria (Esser et al., 2002; Herlan et al., 2003; McQuibban et al., 2003; van der Bliek and Koehler, 2003). Whereas typical residents of the intermembrane space such as Cyt b_2 are cleaved first by the matrix processing peptidase (MPP) and then by the intermembrane space processing peptidase, Ccp1 and Mgm1 are cleaved first by the mAAA protease and MPP, respectively, and then by Rbd1. Thus, in addition to mediating signal transduction and processing, Rhomboids may have other biological functions. A very plausible one would be to facilitate degradation of membrane proteins by cleaving their transmembrane □-helices.

The eight predicted Rhomboid proteins in *Arabidopsis* show homology in the regions of the three residues comprising the catalytic triad, and these residues are predicted to reside within the transmembrane □-helices. Out of the eight proteins, four appear to have chloroplast targeting sequences and two could be targeted to mitochondria. Thus, it is possible that both organelles contain intramembrane proteases that could function in precursor processing, like the aforementioned examples from mitochondria, but maybe to participate in the degradation of integral membrane proteins as well. This suggestion will have to be tested experimentally.

5. EVOLUTION OF ORGANELLE PROTEASES

The prokaryotic origin of chloroplasts and mitochondria is manifested also in their proteolytic machinery, as all components of this machinery are orthologs of bacterial proteases and peptidases. However, it is striking that whereas the great majority of bacterial proteases are encoded by single genes, chloroplast and mitochondrial proteases belong to gene families. It appears that this multiplication is associated, at least partially, with the evolution of photosynthesis. Comparison of the proteolytic machinery in *E. coli* to those found in the photosynthetic cyanobacterium *Synechocystis* sp. PCC 6803 and the higher plant *Arabidopsis* reveals an interesting trend. Whereas ClpP is encoded by a single gene in *E. coli*, three such genes are found in *Synechocystis* and six in *Arabidopsis*. Its non-proteolytic homologue ClpR, which is absent from *E. coli*, is found in *Synechocystis* and *Arabidopsis* in one and four copies, respectively. ClpC and X are found in single copies in *E. coli* and *Synechocystis*, whereas *Arabidopsis* has two copies of each. *E. coli* has a single copy of FtsH, *Synechocystis* has four and *Arabidopsis* 12. DegP has three related proteins in *E. coli*, and the same number is kept in *Synechocystis*. However, *Arabidopsis* has 14 homologues of this protease. It is not clear what were the driving forces that led to this multiplication of genes. A plausible explanation could be that the photosynthetic process is accompanied by an increasing risk of oxidative damage, generating a demand for more efficient and versatile proteolytic machinery to dispose off those proteins that are damaged beyond repair.

As already mentioned above, chloroplast and mitochondrial proteases belong to the same protein families. Phylogenetic analysis of *Arabidopsis* FtsH proteins, together with *Synechocystis* and yeast mitochondria ones, reveal an interesting phenomenon. Seven out of the nine FtsHs that can be targeted to chloroplasts, FtsH1, 2 and 5-9 (Sakamoto et al., 2003), are more related to the cyanobacterial proteins than to the yeast mitochondrial ones, suggesting that they are all descendents of the cyanobacterial progenitor of the chloroplast. Apparently, FtsHs already started to diverge in this progenitor and higher plants inherited three different forms of these. In almost all cases the higher plant ones were duplicated later on. The three FtsHs that were found in mitochondria, FtsH3, 4 and 10 (Sakamoto et al., 2003), are more related to those found in yeast mitochondria, suggesting that these were derived from □-proteobacteria, the progenitor of present-day mitochondria. However, re-targeting might also have occurred in plants as FtsH11 and 12, that are more related to the mitochondrial isozymes, are found in chloroplasts (Sakamoto et al., 2003).

Another interesting case is that of Lon protease. Whereas Lon is found in bacteria such as *E. coli*, it is absent from the *Synechocystis* genome. Nevertheless, it is found both in plant mitochondria and chloroplasts. Here again, the enzyme might have been inherited from □-proteobacteria, duplicated in plants, and than at least one of these gene products might have been redirected to chloroplasts instead of mitochondria. Such a redirection might have occurred for ClpP2 and the ClpXs as well.

6. BIOTECHNOLOGICAL IMPLICATIONS

During the past several years, plastids have become an attractive site for expression of recombinant proteins of economic value (see chapter 16 of this volume). This is primarily due to three attributes: 1) plastids are capable of accumulating large amounts of foreign protein, as much as 46% of total soluble protein (DeCosa et al., 2001; 2) the transgene is integrated into the genome by homologous recombination of chloroplast DNA flanking sequences; this eliminates position effect (Daniell et al., 2002; and 3) since plastids in most species are maternally inherited (see chapter 4 of this volume), the risk of transgene dissemination through pollen is eliminated (Daniell 2002). In order to turn plastids into efficient bioreactors, tight control over all aspects of their gene expression machinery -- from transcription through translation all the way to proper assembly and stability of the protein product -- is required.

In this respect, the 'protein quality control' system of plastids might be of utmost importance. Quality control of proteins is afforded by the combined action of molecular chaperones and proteases (Wickner et al., 1999). Chaperones assist in regulated protein folding by shielding hydrophobic sequences from aggregation; they can unfold misfolded proteins and dissolve aggregates. When all these options fail, damaged, misfolded or aggregated proteins will eventually be recognized and degraded by proteases. Thus, the quality control system in plastids is likely to be crucial in decreasing the ratio between improperly and properly-folded proteins, especially under conditions of overexpression, when the risk of aggregation is high. This might not be a problem when the desired proteins can be inactive, like in the case of immunogen production. However, when the overexpressed recombinant protein needs to retain its activity, proper folding is crucial. Here, well characterized, highly efficient and quantitatively balanced chaperones and proteases may have implications on the yield of the properly folded and active protein product.

Even though high levels of several foreign proteins have been achieved via plastid genetic engineering (see chapter 16 of this volume), it is quite evident that foreign proteins are highly susceptible to proteolytic degradation. Several approaches have been used to overcome proteolytic degradation of foreign proteins. One such approach is the use of chaperones to fold foreign proteins into structures that would protect them from proteases. This was demonstrated by using the CRY chaperone (encoded by *orf2* gene, part of the *cry* operon) to fold the insecticidal protein synthesized into cuboidal crystals; this increased CRY protein accumulation 128-fold (from 0.36% to 46.1% of total leaf protein, Decosa et al., 2001). Similarly, a 500-fold increase in translation of Human Serum Albumin was achieved when the *hsa* coding sequence was regulated by the chloroplast psbA 5' and 3' untranslated regions; rapid translation of *hsa* resulted in aggregates of Human Serum Albumin, which protected the newly synthesized protein from proteolytic degradation (Fernandez-San Millan et al., 2003). Similarly, when the pag gene coding for the anthrax protective antigen, regulated by the psbA 5' and 3' UTRs were expressed in transgenic chloroplasts, very high levels of expression was observed; one acre of chloroplast transgenic plants can produce up to 400 million doses of clean and fully functional anthrax vaccine (Chebolu and Daniell, 2004).

The chloroplast also contains machinery that allows for correct folding and disulfide bond formation, resulting in fully functional human blood proteins or vaccine antigens (Daniell et al., 2004a,b). For example, interferon □2b (IFN) functionality was demonstrated by the ability of IFN□2b to protect HeLa cells against the cytopathic effect of the encephalomyocarditis (EMC) virus and by RT PCR; these studies showed the chloroplast-derived interferon is properly folded with disulfide bonds (Daniell et al., 2004 a). Similarly, assembly of pentamers of cholera toxin β-subunit in plant extracts was confirmed by binding assays; chloroplast-synthesized and bacterial derived (commercial vaccine), both bound to the intestinal membrane GM_1-ganglioside receptor, confirming proper folding and formation of five disulfide bonds (Daniell et al., 2001).

7. CONCLUSIONS

In the past decade we have witnessed the identification of several chloroplast and mitochondrial proteases and peptidases. The plant sequences that have accumulated throughout these years, and the completion of the *Arabidopsis* and rice genome sequencing projects, suggest that the identity of most, if not all, chloroplast proteases and peptidases is now known. It appears that the upcoming challenge is to assign specific functions to each one of the identified ones. The function of two chloroplast peptidases, SPP and TPP, is apparently clear – they are responsible for the maturation of imported proteins in the stroma and the lumen, respectively. CtpA also functions in protein maturation. However, the function of a number of other peptidases, especially aminopeptidases, which were identified primarily by analysis of sequence data, is not known. One can assume that they are involved in degradation of short peptides, the products of ATP-dependent proteolysis, but such a function still needs to be demonstrated. Whether they have other more specific substrates and additional functions is another open question.

The degradation of the D1 protein of PSII reaction center, in the context of repair from photoinhibition, has been the focal point of numerous studies for more than 20 years. Only in recent years the identity of the proteases involved has started to unravel. Although the details of this process are still far from being fully understood, there is now experimental evidence for the involvement of two proteases, FtsH and DegP2, in this process. Another challenge in the field will be to link characterized proteolytic processes, such as degradation of LHCII or ELIP during acclimation to changes in light intensities, to the identified proteases. In this case, a reverse genetics approach might be useful. Specific protease mutants will have to be tested for their ability to degrade specific proteins under given conditions.

The multiplication of genes within the different protease families is now evident. However, the functional significance of this is not clear. Two alternative possibilities should be considered. Within each gene family, the products that are targeted to the same sub-compartment fulfil the same functions. The evolutionary driving force in this case might have been the essential functions fulfilled by a given protease that led to this redundancy. In this case, the differential importance of various gene products, as revealed by mutant analysis of FtsHs, is a result of differential expression levels, and not necessarily of different functions. Alternatively, different gene products within a family may have acquired specialized functions during evolution. It is possible that each one of these two alternatives describes better the relationship within a different family. A conclusive explanation will have to await a detailed examination of each gene family. Such studies are likely to shed additional light on the different functions of plant organelles and the role of proteases and peptidases in controlling and maintaining these functions.

8. REFERENCES

Adam, Z. (1996). Protein stability and degradation in chloroplasts. *Plant Mol. Biol., 32*, 773-783.

Adam, Z. (2000). Chloroplast proteases: Possible regulators of gene expression? *Biochimie, 82*, 647-654.

Adam, Z., Adamska, I., Nakabayashi, K., Ostersetzer, O., Haussuhl, K., Manuell, A., *et al.* (2001). Chloroplast and mitochondrial proteases in Arabidopsis. A proposed nomenclature. *Plant Physiol,. 125*, 1912-1918.

Adam, Z. & Clarke, A.K. (2002). Cutting edge of chloroplast proteolysis. *Trends Plant Sci,. 7*, 451-456.

Allen, J.F. (1992). How does protein phosphorylation regulate photosynthesis? *Trends Biochem. Sci., 17,* 12-17.

Andersson, B. & Aro, E.-M. (1997). Proteolytic activities and proteases of plant chloroplasts. *Physiol. Plant., 100*, 780-793.

Andersson, B. & Aro, E.-M. (2001). Photodamage and D1 protein turnover in photosystem II. In: *Regulation of Photosynthesis* (pp. 377-393). Dordrecht/Boston/London: Kluwer Academic Publishers.

Apel, K. & Kloppstech, K. (1980). The effect of light in the biosynthesis of the light-harvesting chlorophyll a/b protein. Evidence for the stabilization of the apoprotein. *Planta, 150,* 426-430.

Bailey, S., Thompson, E., Nixon, P.J., Horton, P., Mullineaux, C.W., Robinson, C., *et al.* (2002). A critical role for the Var2 FtsH homologue of *Arabidopsis thaliana* in the photosystem II repair cycle in vivo. *J. Biol. Chem., 277,* 2006-2011.

Barakat, S., Pearce, D.A., Sherman, F. & Rapp, W.D. (1998). Maize contains a Lon protease gene that can partially complement a yeast pim1-deletion mutant. *Plant Mol. Biol., 37,* 141-154.

Barber, J. & Andersson, B. (1992). Too much of a good thing: light can be bad for photosynthesis. *Trends Biochem. Sci., 17,* 61-66.

Bennett, J. (1981). Biosynthesis of the light-harvesting chlorophyll a/b protein. Polypeptide turnover in darkness. *Eur. J. Biochem., 118,* 61-70.

Bhushan, S., Lefebvre, B., Stahl, A., Wright, S.J., Bruce, B.D., Boutry, M., *et al.* (2003). Dual targeting and function of a protease in mitochondria and chloroplasts. *EMBO Rep., 4,* 1073-1078.

Chaal, B.K., Mould, R.M., Barbrook, A.C., Gray, J.C. & Howe, C.J. (1998). Characterization of a cDNA encoding the thylakoidal processing peptidase from *Arabidopsis thaliana*. Implications for the origin and catalytic mechanism of the enzyme. *J. Biol. Chem., 273,* 689-692.

Chassin, Y., Kapri-Pardes, E., Sinvany, G., Arad, T. & Adam, Z. (2002). Expression and characterization of the thylakoid lumen protease DegP1 from *Arabidopsis thaliana*. *Plant Physiol., 130,* 857-864.

Chebolu, S & Daniell, H. (2004). Chloroplast derived vaccine antigens and biopharmaceuticals: expression, folding, assembly and functionality. Current Trends in Microbiology and Immunology, in press.

Chen, M., Choi, Y., Voytas, D.F. & Rodermel, S. (2000). Mutations in the Arabidopsis VAR2 locus cause leaf variegation due to the loss of a chloroplast FtsH protease. *Plant J., 22,* 303-313.

Clarke, A.K. (1999). ATP-dependent Clp proteases in photosynthetic organisms - A cut above the rest! *Ann. Bot., 83,* 593-599.

Clausen, T., Southan, C. & Ehrmann, M. (2002). The HtrA family of proteases: implications for protein composition and cell fate. *Mol. Cell, 10,* 443-455.

Daniell, H. (2002). Molecular strategies for gene containment in transgenic crops. *Nat. Biotechnol., 20,* 581-586.

Daniell, D., Carmona-Sanchez, O., & Burns, B. B. (2004a). Chloroplast derived antibodies, biopharmaceuticals and edible vaccines. In R. Fischer & S. Schillberg (Eds.) *Molecular Farming* (pp.113-133). Weinheim: WILEY-VCH Verlag.

Daniell, H., Chebolu, S., Kumar, S., Singleton, M., Falconer, R. (2004b). Chloroplast-derived vaccine antigens and other therapeutic proteins. *Vaccine, in press.*

Daniell, H., Khan, M., & Allison, L. (2002). Milestones in chloroplast genetic engineering: an environmentally friendly era in biotechnology. *Trends Plant Sci., 7*(2), 84-91.

Daniell, H., Lee, S. B., Panchal, T., & Wiebe, P. O. (2001a). Expression of cholera toxin B subunit gene and assembly as functional oligomers in transgenic tobacco chloroplasts. *J. Mol. Biol., 311,* 1001-1009.

DeCosa, B., Moar, W., Lee, S. B., Miller, M., & Daniell, H. (2001). Overexpression of the *Bt* cry2Aa2 operon in chloroplasts leads to formation of insecticidal crystals. *Nat. Biotechnol., 19,* 71-74.

Dalbey, R.E. & Robinson, C. (1999). Protein translocation into and across the bacterial plasma membrane and the plant thylakoid membrane. *Trends Biochem. Sci., 24,* 17-22.

Depamphilis, C.W. & Palmer, J.D. (1990). Loss of photosynthetic and chlororespiratory genes from the plastid genome of a parasitic flowering plant. *Nature, 348,* 337-339.

Esser, K., Tursun, B., Ingenhoven, M., Michaelis, G. & Pratje, E. (2002). A novel two-step mechanism for removal of a mitochondrial signal sequence involves the mAAA complex and the putative rhomboid protease Pcp1. *J. Mol. Biol., 323,* 835-843.

Fanning, A.S. & Anderson, J.M. (1996). Protein-protein interactions: PDZ domain networks. *Curr. Biol., 6,* 1385-1388.

Fernandez-San Millan, A., Mingeo-Castel, A. M., Miller, M., & Daniell, H. (2003). A chloroplast transgenic approach to hyper-express and purify human serum albumin, a protein highly susceptible to proteolytic degradation. *Plant Biotechnology Journal 1,* 71-79.

Fu, G.K. & Markovitz, D.M. (1998). The human LON protease binds to mitochondrial promoters in a single- stranded, site-specific, strand-specific manner. *Biochemistry, 37,* 1905-1909.

Glaser, E. & Dessi, P. (1999). Integration of the mitochondrial-processing peptidase into the cytochrome bc1 complex in plants. *J. Bioenerg. Biomembr., 31,* 259-274.

Glaser, E., Sjoling, S., Szigyarto, C. & Eriksson, A.C. (1996). Plant mitochondrial protein import: precursor processing is catalysed by the integrated mitochondrial processing peptidase (MPP)/bc(1) complex and degradation by the ATP-dependent proteinase. *Biochim. Biophys. Acta, 1275,* 33-37.

Gottesman, S. (1996). Proteases and their targets in *Escherichia coli. Annu. Rev. Genet., 30,* 465-506.

Gray, C.W., Ward, R.V., Karran, E., Turconi, S., Rowles, A., Viglienghi, D., *et al.* (2000). Characterization of human HtrA2, a novel serine protease involved in the mammalian cellular stress response. *Eur. J. Biochem., 267,* 5699-5710.

Halperin, T. & Adam, Z. (1996). Degradation of mistargeted OEE33 in the chloroplast stroma. *Plant Mol. Biol., 30,* 925-933.

Halperin, T., Ostersetzer, O. & Adam, Z. (2001a). ATP-dependent association between subunits of Clp protease in pea chloroplasts. *Planta, 213,* 614-619.

Halperin, T., Zheng, B., Itzhaki, H., Clarke, A.K. & Adam, Z. (2001b). Plant mitochondria contain proteolytic and regulatory subunits of the ATP-dependent Clp protease. *Plant Mol. Biol., 45,* 461-468.

Haussuhl, K., Andersson, B. & Adamska, I. (2001). A chloroplast DegP2 protease performs the primary cleavage of the photodamaged D1 protein in plant photosystem II. *EMBO J., 20,* 713-722.

Herlan, M., Vogel, F., Bornhovd, C., Neupert, W. & Reichert, A.S. (2003). Processing of Mgm1 by the Rhomboid-type protease Pcp1 is required for maintenance of mitochondrial morphology and of mitochondrial DNA. *J. Biol. Chem., 278,* 27781-27788.

Huang, C., Wang, S., Lemieux, C., Otis, C., Turmel, M. & Liu, X.Q. (1994). The Chlamydomonas chloroplast *clpP* gene contains translated large insertion sequences and is essential for cell growth. *Mol. Gen. Genet., 244,* 151-159.

Inagaki, N., Yamamoto, Y., Mori, H. & Satoh, K. (1996). Carboxyl-terminal processing protease for the D1 precursor protein: cloning and sequencing of the spinach cDNA. *Plant Mol. Biol., 30,* 39-50.

Itzhaki, H., Naveh, L., Lindahl, M., Cook, M. & Adam, Z. (1998). Identification and characterization of DegP, a serine protease associated with the luminal side of the thylakoid membrane. *J. Biol. Chem., 273,* 7094-7098.

Jarvis, P. & Soll, J. (2002). Toc, tic, and chloroplast protein import. *Biochim. Biophys. Acta, 1590,* 177-189.

Kihara, A., Akiyama, Y. & Ito, K. (1999). Dislocation of membrane proteins in FtsH-mediated proteolysis. *EMBO J., 18,* 2970-2981.

Kim, J., Eichacker, L.A., Rudiger, W. & Mullet, J.E. (1994). Chlorophyll regulates accumulation of the plastid-encoded chlorophyll proteins P700 and D1 by increasing apoprotein stability. *Plant Physiol., 104,* 907-916.

Kolmar, H., Waller, P.R.H. & Sauer, R.T. (1996). The DegP and DegQ periplasmic endoproteases of *Escherichia coli*: specificity for cleavage sites and substrate conformation. *J. Bacteriol., 178,* 5925-5929.

Kolodziejczak, M., Kolaczkowska, A., Szczesny, B., Urantowka, A., Knorpp, C., Kieleczawa, J., *et al.* (2002). A higher plant mitochondrial homologue of the yeast m-AAA protease. Molecular cloning, localization, and putative function. *J. Biol. Chem., 277,* 43792-43798.

Koonin, E.V., Makarova, K.S., Rogozin, I.B., Davidovic, L., Letellier, M.C. & Pellegrini, L. (2003). The rhomboids: a nearly ubiquitous family of intramembrane serine proteases that probably evolved by multiple ancient horizontal gene transfers. *Genome Biol, 4,* R19.

Krojer, T., Garrido-Franco, M., Huber, R., Ehrmann, M. & Clausen, T. (2002). Crystal structure of DegP (HtrA) reveals a new protease-chaperone machine. *Nature, 416,* 455-459.

Krzywda, S., Brzozowski, A.M., Verma, C., Karata, K., Ogura, T. & Wilkinson, A.J. (2002). The crystal structure of the AAA domain of the ATP-dependent protease FtsH of *Escherichia coli* at 1.5 A resolution. *Structure, 10,* 1073-1083.

Kuroda, H. & Maliga, P. (2003). The plastid *clpP1* protease gene is essential for plant development. *Nature, 425,* 86-89.

Kyle, D.J., Ohad, I. & Arntzen, C.J. (1984). Membrane protein demage and repair: Selective loss of a quinone-protein function in chloroplast membranes. *Proc. Natl. Acad. Sci. USA, 81,* 4070-4074.

Langer, T. (2000). AAA proteases: cellular machines for degrading membrane proteins. *Trends Biochem. Sci., 25,* 247-251.

Lensch, M., Herrmann, R.G. & Sokolenko, A. (2001). Identification and characterization of SppA, a novel light-inducible chloroplast protease complex associated with thylakoid membranes. *J. Biol. Chem., 276,* 33645-33651.

Leto, K.J., Bell, E. & McIntosh, L. (1985). Nuclear mutation leads to an accelerated turnover of chloroplast-encoded 48 kd and 34.5 kd polypeptides in thylakoids lacking photosystem II. *EMBO J., 4,* 1645-1653.

Levchenko, I., Smith, C.K., Walsh, N.P., Sauer, R.T. & Baker, T.A. (1997). PDZ-like domains mediate binding specificity in the Clp/Hsp100 family of chaperones and protease regulatory subunits. *Cell, 91,* 939-947.

Levy, M., Bachmair, A. & Adam, Z. (2003). A single recessive mutation in the proteolytic machinery of Arabidopsis chloroplasts impairs photoprotection and photosynthesis upon cold stress. *Planta (in press).*

Li, H.H. & Merchant, S. (1995). Degradation of plastocyanin in copper-deficient *C. reinhardtii* - evidence for a protease-susceptible conformation of the apoprotein and regulated proteolysis. J. *Biol. Chem., 270,* 23504-23510.

Li, W., Srinivasula, S.M., Chai, J., Li, P., Wu, J.W., Zhang, Z., *et al.* (2002). Structural insights into the pro-apoptotic function of mitochondrial serine protease HtrA2/Omi. *Nat. Struct. Biol., 9,* 436-441.

Lindahl, M., Spetea, C., Hundal, T., Oppenheim, A.B., Adam, Z. & Andersson, B. (2000). The thylakoid FtsH protease plays a role in the light-induced turnover of the photosystem II D1 protein. *Plant Cell, 12,* 419-431.

Lindahl, M., Tabak, S., Cseke, L., Pichersky, E., Andersson, B. & Adam, Z. (1996). Identification, characterization, and molecular cloning of a homologue of the bacterial FtsH protease in chloroplasts of higher plants. *J. Biol. Chem., 271,* 29329-29334.

Lindahl, M., Yang, D.H. & Andersson, B. (1995). Regulatory proteolysis of the major light-harvesting chlorophyll a/b protein of photosystem II by a light-induced membrane-associated enzymic system. *Eur. J. Biochem., 231,* 503-509.

Lipinska, B., Zylicz, M. & Georgopoulos, C. (1990). The HtrA (DegP) protein, essential for *Escherichia coli* survival at high temperatures, is an essential endopeptidase. *J. Bact., 172,* 1791-1797.

Majeran, W., Olive, J., Drapier, D., Vallon, O. & Wollman, F.A. (2001). The light sensitivity of ATP synthase mutants of *Chlamydomonas reinhardtii*. *Plant Physiol, 126,* 421-433.

Majeran, W., Wollman, F.-A. & Vallon, O. (2000). Evidence for a role of ClpP in the degradation of the chloroplast cytochrome b6f complex. *Plant Cell, 12,* 137-149.

Marienfeld, J., Unseld, M. & Brennicke, A. (1999). The mitochondrial genome of Arabidopsis is composed of both native and immigrant information. *Trends Plant Sci, 4,* 495-502.

Matile, P. (2001). Senescence and cell death in plant development: chloroplast senescence and its regulation. In: *Regulation of Photosynthesis* (pp. 277-296). Dordrecht/Boston/London: Kluwer Academic Publishers.

Mattoo, A.K., Hoffman-Falk, H., Marder, J.B. & Edelman, M. (1984). Regulation of protein metabolism: coupling of photosynthetic electron transport to *in vivo* degradation of the rapidly metabolized 32-kilodalton protein of the chloroplast membranes. *Proc. Natl. Acad. Sci. USA, 81,* 1380-1384.

McQuibban, G.A., Saurya, S. & Freeman, M. (2003). Mitochondrial membrane remodelling regulated by a conserved rhomboid protease. *Nature, 423,* 537-541.

Melis, A. (1999). Photosystem-II damage and repair cycle in chloroplasts: what modulates the rate of photodamage ? *Trends Plant Sci., 4,* 130-135.

Merchant, S. & Bogorad, L. (1986). Rapid degradation of apoplastocyanin in Cu(II)-deficient cells of *Chlamydomonas reinhardtii*. *J. Biol. Chem., 261,* 15850-15853.

Millar, A.H., Sweetlove, L.J., Giege, P. & Leaver, C.J. (2001). Analysis of the Arabidopsis mitochondrial proteome. *Plant Physiol., 127,* 1711-1727.

Moberg, P., Stahl, A., Bhushan, S., Wright, S.J., Eriksson, A., Bruce, B.D., *et al.* (2003). Characterization of a novel zinc metalloprotease involved in degrading targeting peptides in mitochondria and chloroplasts. *Plant J., 36,* 616-628.

Mullet, J.E., Klein, P.G. & Klein, R.R. (1990). Chlorophyll regulates accumulation of the plastid-encoded chlorophyll apoprotein-CP43 and apoprotein-D1 by increasing apoprotein stability. *Proc. Natl. Acad. Sci. USA, 87,* 4038-4042.

Oelmuller, R., Herrmann, R.G. & Pakrasi, H.B. (1996). Molecular studies of CtpA, the carboxyl-terminal processing protease for the D1 protein of the photosystem II reaction center in higher plants. *J. Biol. Chem., 271,* 21848-21852.

Ostersetzer, O. (2001). Characterization of the proteolytic machinery in thylakoid membranes of higher plants. *The Hebrew University. Ph.D. Thesis.*

Ostersetzer, O. & Adam, Z. (1997). Light-stimulated degradation of an unassembled Rieske FeS protein by a thylakoid-bound protease: the possible role of the FtsH protease. *Plant Cell, 9,* 957-965.

Ostersetzer, O., Tabak, S., Yarden, O., Shapira, R. & Adam, Z. (1996). Immunological detection of proteins similar to bacterial proteases in higher plant chloroplasts. *Eur. J. Biochem., 236,* 932-936.

Pallen, M.J. & Wren, B.W. (1997). The HtrA family of serine proteases. *Mol. Microbiol., 26,* 209-221.

Peltier, J.-B., Ripoll, D.R., Friso, G., Rudella, A., Cai, Y., Ytterberg, J., *et al.* (2003). Clp protease complexes from photosynthetic and non-photosynthetic plastids and mitochondria of plants, their predicted 3-D structures and functional implications. *J. Biol. Chem. (in press).*

Peltier, J.-B., Emanuelsson, O., Kalume, D.E., Ytterberg, J., Friso, G., Rudella, A., *et al.* (2002). Central functions of the lumenal and peripheral thylakoid proteome of Arabidopsis determined by experimentation and genome-wide prediction. *Plant Cell, 14,* 211-236.

Peltier, J.-B., Ytterberg, J., Liberles, D.A., Roepstorff, P. & van Wijk, K.J. (2001). Identification of a 350 kDa ClpP protease complex with 10 different Clp isoforms in chloroplasts of *Arabidopsis thaliana*. *J. Biol. Chem., 276,* 16318-16327.

Ponting, C.P. (1997). Evidence for PDZ domains in bacteria, yeast, and plants. *Protein Sci., 6,* 464-468.

Porankiewicz, J., Wang, J. & Clarke, A.K. (1999). New insights into the ATP-dependent Clp protease: *Escherichia coli* and beyond. *Mol. Microbiol., 32,* 449-458.

Prasil, O., Adir, N. & Ohad, I. (1992). Dynamics of photosystem II: mechanism of photoinhibition and recovery processes. In: *Topics in Photosynthesis* (pp. 295-348). Amsterdam: Elsevier Science Publishers.

Ramesh, H., Srinivasula, S.M., Zhang, Z., Wassell, R., Mukattash, R., Cilenti, L., *et al.* (2002). Identification of Omi/HtrA2 as a mitochondrial apoptotic serine protease that disrupts inhibitor of apoptosis protein-caspase interaction. *J. Biol. Chem., 277,* 432-438.

Rep, M., Vandijl, J.M., Suda, K., Schatz, G., Grivell, L.A. & Suzuki, C.K. (1996). Promotion of mitochondrial membrane complex assembly by a proteolytically inactive yeast lon. *Science, 274,* 103-106.

Richter, S. & Lamppa, G.K. (1998). A chloroplast processing enzyme functions as the general stromal processing peptidase. *Proc. Natl. Acad. Sci. USA, 95,* 7463-7468.

Richter, S. & Lamppa, G.K. (1999). Stromal processing peptidase binds transit peptides and initiates their ATP-dependent turnover in chloroplasts. *J. Cell Biol., 147,* 33-43.

Richter, S. & Lamppa, G.K. (2003). Structural properties of the chloroplast stromal processing peptidase required for its function in transit peptide removal. *J. Biol. Chem., 278,* 39497-39502.

Sakamoto, W. (2003). Leaf-variegated mutations and their responsible genes in *Arabidopsis thaliana*. *Genes Genet. Syst., 78,* 1-9.

Sakamoto, W., Tamura, T., Hanba-Tomita, Y., Sodmergen & Murata, M. (2002). The VAR1 locus of Arabidopsis encodes a chloroplastic FtsH and is responsible for leaf variegation in the mutant alleles. *Genes to Cells, 7,* 769-780.

Sakamoto, W., Zaltsman, A., Adam, Z. & Takahashi, Y. (2003). Coordinated regulation and complex formation of VAR1 and VAR2, chloroplastic FtsH metalloproteases involved in the repair cycle of photosystem II in Arabidopsis thylakoid membranes. *Plant Cell, 15, (in press).*

Sarria, R., Lyznik, A., Vallejos, C.E. & Mackenzie, S.A. (1998). A cytoplasmic male sterility-associated mitochondrial peptide in common bean is post-translationally regulated. *Plant Cell, 10,* 1217-1228.

Sassoon, N., Arie, J.P. & Betton, J.M. (1999). PDZ domains determine the native oligomeric structure of the DegP (HtrA) protease. *Mol. Microbiol., 33,* 583-589.

Sato, S., Nakamura, Y., Kaneko, T., Asamizu, E. & Tabata, S. (1999). Complete structure of the chloroplast genome of Arabidopsis thaliana. *DNA Res., 6,* 283-290.

Schmidt, G.W. & Mishkind, M.L. (1983). Rapid degradation of unassembled ribulose 1,5-biphosphate carboxylase small subunit in chloroplasts. *Proc. Natl. Acad. Sci. USA, 80,* 2632-2636.

Schubert, M., Petersson, U.A., Haas, B.J., Funk, C., Schroder, W.P. & Kieselbach, T. (2002). Proteome map of the chloroplast lumen of Arabidopsis thaliana. *J. Biol. Chem., 277,* 8354-8365.

Shanklin, J., Dewitt, N.D. & Flanagan, J.M. (1995). The stroma of higher plant plastids contain ClpP and ClpC, functional homologs of *Escherichia coli* ClpP and ClpA: an archetypal two-component ATP-dependent protease. *Plant Cell, 7,* 1713-1722.

Shestakov, S.V., Anbudurai, P.R., Stanbekova, G.E., Gadzhiev, A., Lind, L.K. & Pakrasi, H.B. (1994). Molecular cloning and characterization of the ctpA gene encoding a carboxyl-terminal processing protease - analysis of a spontaneous photosystem II-deficient mutant strain of the cyanobacterium *Synechocystis* sp. PCC 6803. *J. Biol. Chem., 269,* 19354-19359.

Shikanai, T., Shimizu, K., Ueda, K., Nishimura, Y., Kuroiwa, T. & Hashimoto, T. (2001). The chloroplast clpP gene, encoding a proteolytic subunit of ATP- dependent protease, is indispensable for chloroplast development in tobacco. *Plant Cell Physiol., 42,* 264-273.

Skorko-Glonek, J., Wawrzynow, A., Krzewski, K., Kurpierz, K. & Lipinska, B. (1995). Site-directed mutagenesis of the HtrA(DegP) serine protease, whose proteolytic activity is indispensable for *Escherichia coli* survival at elevated temperatures. *Gene, 163,* 47-52.

Sokolenko, A., Pojidaeva, E., Zinchenko, V., Panichkin, V., Glaser, V.M., Herrmann, R.G., *et al.* (2002). The gene complement for proteolysis in the cyanobacterium *Synechocystis* sp. PCC 6803 and *Arabidopsis thaliana* chloroplasts. *Curr. Genet., 41,* 291-310.

Spiess, C., Beil, A. & Ehrmann, M. (1999). A temperature-dependent switch from chaperone to protease in a widely conserved heat shock protein. *Cell, 97,* 339-347.

Strauch, K.L., Johnson, K. & Beckwith, J. (1989). Characterization of *degP*, a gene required for proteolysis in the cell envelope and essential for growth of *Escherichia coli* at high temperatures. *J. Bacteriol., 171,* 2689-2696.

Suzuki, C.K., Suda, K., Wang, N. & Schatz, G. (1994). Requirement for the yeast gene *lon* in intramitochondrial proteolysis and maintenance of respiration. *Science, 264,* 273-276.

Takechi, K., Sodmergen, Murata, M., Motoyoshi, F. & Sakamoto, W. (2000). The YELLOW VARIEGATED (VAR2) locus encodes a homologue of FtsH, an ATP- dependent protease in Arabidopsis. *Plant Cell Physiol., 41,* 1334-1346.

The Arabidopsis Genome Initiative (2000). Analysis of the genome sequence of the flowering plant *Arabidopsis thaliana. Nature, 408,* 796-815.

Tjus, S.E., Moller, B.L. & Scheller, H.V. (1998). Photosystem I is an early target of photoinhibition in barley illuminated at chilling temperatures. *Plant Physiol., 116,* 755-764.

Trost, J.T., Chisholm, D.A., Jordan, D.B. & Diner, B.A. (1997). The D1 C-terminal processing protease of photosystem II from *Scenedesmus obliquus*. Protein purification and gene characterization in wild type and processing mutants. *J. Biol. Chem., 272,* 20348-20356.

Urban, S. & Freeman, M. (2002). Intramembrane proteolysis controls diverse signalling pathways throughout evolution. *Curr. Opin. Genet. Dev., 12,* 512-518.

Urban, S. & Freeman, M. (2003). Substrate specificity of rhomboid intramembrane proteases is governed by helix-breaking residues in the substrate transmembrane domain. *Mol. Cell, 11,* 1425-1434.

Urban, S., Lee, J.R. & Freeman, M. (2001). Drosophila rhomboid-1 defines a family of putative intramembrane serine proteases. *Cell, 107,* 173-182.

Urban, S., Lee, J.R. & Freeman, M. (2002a). A family of Rhomboid intramembrane proteases activates all Drosophila membrane-tethered EGF ligands. *EMBO J, 21,* 4277-4286.

Urban, S., Schlieper, D. & Freeman, M. (2002b). Conservation of intramembrane proteolytic activity and substrate specificity in prokaryotic and eukaryotic rhomboids. *Curr. Biol., 12,* 1507-1512.

Van der Bliek, A.M. & Koehler, C.M. (2003). A mitochondrial rhomboid protease. *Dev. Cell, 4,* 769-70.

Van Dyck, L., Neupert, W. & Langer, T. (1998). The ATP-dependent PIM1 protease is required for the expression of intron-containing genes in mitochondria. *Genes Dev., 12,* 1515-1524.

Vandervere, P.S., Bennett, T.M., Oblong, J.E. & Lamppa, G.K. (1995). A chloroplast processing enzyme involved in precursor maturation shares a zinc-binding motif with a recently recognized family of metalloendopeptidases. *Proc. Natl. Acad. Sci. USA, 92,* 7177-7181.

Wada, M., Kagawa, T. & Sato, Y. (2003). Chloroplast movement. *Annu. Rev. Plant Biol., 54,* 455-468.

Waller, P.R.H. & Sauer, R.T. (1996). Characterization of degQ and degS, *Escherichia coli* genes encoding homologs of the DegP protease. *J. Bacteriol., 178,* 1146-1153.

Wang, N., Gottesman, S., Willingham, M.C., Gottesman, M.M. & Maurizi, M.R. (1993). A human mitochondrial ATP-dependent protease that is highly homologous to bacterial Lon protease. *Proc. Natl. Acad. Sci. U S A, 90,* 11247-11251.

Wang, J., Hartling, J.A & Flanagan, J.M. (1997). The structure of ClpP at 2.3 A resolution suggests a model for ATP- dependent proteolysis. *Cell, 91,* 447-456.

Wickner, S., Maurizi, M.R & Gottesman, S. (1999). Posttranslational quality control: folding, refolding, and degrading proteins. *Science, 286,* 1888-1893.

Wolfe, K.H., Morden, C.W. & Palmer, J.D. (1992). Function and evolution of a minimal plastid genome from a nonphotosynthetic parasitic plant. *Proc. Natl. Acad. Sci. USA, 89,* 10648-10652.

Yamamoto, Y., Inagaki, N. & Satoh, K. (2001). Overexpression and characterization of carboxyl-terminal processing protease for precursor D1 protein: regulation of enzyme-substrate interaction by molecular environments. *J. Biol. Chem., 276,* 7518-7525.

Yang, D.H., Webster, J., Adam, Z., Lindahl, M. & Andersson, B. (1998). Induction of acclimative proteolysis of the light-harvesting chlorophyll a/b protein of photosystem II in response to elevated light intensities. *Plant Physiol., 118,* 827-834.

Zheng, B., Halperin, T., Hruskova-Heidingsfeldova, O., Adam, Z. & Clarke, A.K. (2002). Characterization of chloroplast Clp proteins in Arabidopsis: localization, tissue specificity and stress responses. *Physiol. Plant., 114,* 92-101.

Zhong, R., Wan, J., Jin, R. & Lamppa, G. (2003). A pea antisense gene for the chloroplast stromal processing peptidase yields seedling lethals in Arabidopsis: survivors show defective GFP import in vivo. *Plant J., 34,* 802-812.

Section 3

Organelle Biotechnology

CHAPTER 16

CHLOROPLAST GENETIC ENGINEERING

HENRY DANIELL[1*], PAUL R. COHILL[1], SHASHI KUMAR[1], AND NATHALIE DUFOURMANTEL[2]

*[1]Department of Molecular Biology and Microbiology, University of Central Florida, Biomolecular ScienceBldg #20, Room 336, 4000 Central Florida Blvd., Orlando, FL 32816-2364 *daniell@mail.ucf.edu.*
[2]CEA Cadarache, DSV, DEVM, Laboratoire d'Ecophysiologie de la Photosynthèse, UMR 6191 CNRS-CEA, Aix-Marseille II, F-13108 Saint-Paul-lez-Durance, France

Abstract. Chloroplasts are ideal hosts for the expression of transgenes. Once integrated via homologous recombination, these transgenes express large amounts of protein (up to 46% of total leaf protein) due to the high copy number of the chloroplast genome in each plant cell. Foreign proteins that are toxic when present in the cytosol, such as vaccine antigens, trehalose and xylanase, are non-toxic when sequestered within transgenic plastids. Because transgenes are maternally inherited in most crops, there is little danger of cross-polination with wild-type plants. By using chloroplast DNA sequences that flank transgenes, higher plants have efficiently and stably integrated transgenes imbuing important agronomic traits, including herbicide, insect and disease resistance, drought and salt tolerance, and phytoremediation. More recently, highly efficient, soybean, carrot and cotton plastid transformation have been accomplished via somatic embryogenesis using species-specific vectors. Chloroplast transgenic carrot plants withstand salt concentrations that only halophytes could tolerate. Previously an exclusively mitochondrial-encoded trait, cytoplasmic male sterility is now possible through □-ketothiolase expression via the chloroplast genome. This is a valuable tool towards transgene containment, in addition to the maternal inheritance of transgenes integrated into the chloroplast genome, in most crops. Crops such as tobacco have expressed transgenes for a variety of biopharmaceuticals, vaccines and biomaterials. Due to the high biomass of tobacco plants (~40 mtons/acre), large amounts of vaccines preventing anthrax, plague, tetanus and cholera, and pharmaceuticals like human somatotropin, serum albumin, interferons and insulin-like growth factor have been produced in transgenic chloroplasts. The chloroplast also contains machinery that allows for correct folding and disulfide bond formation, resulting in fully functional human blood proteins or vaccine antigens. Additionally, expression of the Rubisco small subunit gene (R*bc*S) via the chloroplast genome restored normal photosynthetic activity in a nuclear R*bc*S antisense line, a goal that has been elusive for decades. Multigene operons engineered into the chloroplast genome do not require processing of polycistrons to monocistrons for efficient translation. Secondary structures formed by intergenic spacer regions in bacterial operons are efficiently recognized by the chloroplast processing machinery; when such processing occurs, 3' UTRs are not required for transcript stability. Extension of chloroplast genetic engineering technology to other useful crops will depend on the availability of the plastid genome sequences and the ability to regenerate transgenic events and advance them towards homoplasmy. In addition to biotechnology applications, plastid transformation system has been extensively used to study chloroplast biochemistry and molecular biology.

H. Daniell and C. D. Chase (eds.), Molecular Biology and Biotechnology of Plant Organelles 443—490.

1. INTRODUCTION

The chloroplast genome is a circular molecule that self-replicates, varies in size from 120 to 220 kb in different plant species, and exists predominantly as a monomer along with other multimeric forms (Lilly *et al*., 2001). Because a typical plant cell contains approximately 100 chloroplasts, each with about 100 identical genomes, a single gene is represented perhaps 10,000 times within one cell. The copy number of the genes encoded in two inverted repeat regions present in higher plants reaches 20,000. One can therefore achieve high levels of transgene expression when transforming the plastid genome.

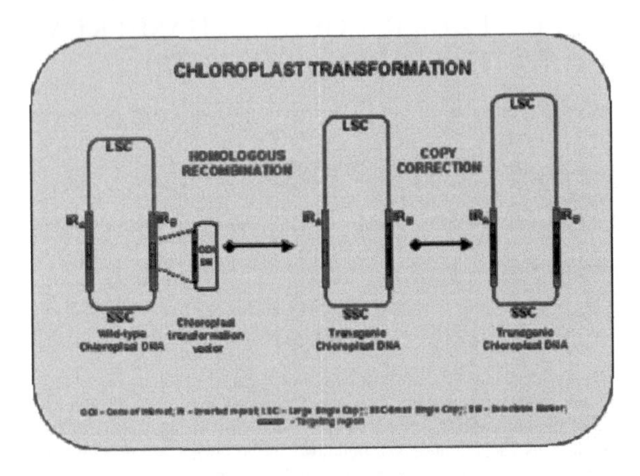

Figure 1. *Schematic representation of chloroplast transformation showing the phenomenon of homologous recombination and copy correction.*

Unlike the random integration that occurs during nuclear transformation, transgenes are integrated via homologous recombination within the plastid genome (Figure 1). Plastids have inherited from their cyanobacterial ancestors an efficient RecA-type system to facilitate homologous recombination (Cerutti *et al.*, 1992). Plastid transformation vectors are therefore designed to contain homologous flanking sequences on either side of transgenes and are introduced into plastids via particle bombardment (Sanford *et al.* 1993) or into protoplasts by PEG treatment (O'Neill et al., 1993; Golds *et al.,* 1993). After bombardment, the integration of transgenes occurs in a few plastid genomes, followed by approximately 15 to 20 cell divisions under selective pressure (in two or three rounds of selection), resulting in a homogenous population of plastid genomes. Integration into an inverted repeat region is followed by copy correction that inserts the introduced transgenes into the other inverted repeat region (Figure 1, Daniell et al., 2004a). Several sites of integration, including some in the inverted repeat region, have been used in the plastid genome (Tables 1 and 2). If a plastid transformation vector carries a plastid origin of replication, this increases the number of templates to be presented for homologous recombination and thus enhances the probability of transgene integration. Such chloroplast vectors have achieved homoplasmy, the replacement of all native plastid genomes with transformed genomes, even in the first round of selection (Daniell *et al.*, 1990, 1998; Guda *et al.*, 2000).

Table 1. *List of agronomic traits engineered via the chloroplast genome (only first reports are included).*

Agronomic trait	Gene	Site of integration	Promoter	5'/3' regulatory elements	Reference
Insect resistance	*Cry*1A(c)	*trn*V/*rps*12/7	Prrn	*rbc*L / T*rps*16	McBride *et al.*, 1995
Herbicide resistance	*aro*A (petunia)	*rbc*L/*acc*D	Prrn	ggagg / T*psb*A	Daniell *et al.*, 1998
Insect resistance	*Cry*2Aa2	*rbc*L/*acc*D	Prrn	ggagg (native) / T*psb*A	Kota *et al.*, 1999.
Herbicide resistance	*bar*	*rbc*L/*acc*D	Prrn	*rbc*L /T*psb*A	Iamtham and Day, 2000
Insect resistance	*Cry*2Aa2 operon	*trn*I/*trn*A	Prrn	native 5'UTRs / T*psb*A	De Cosa *et al.*, 2001.
Disease resistance	*MSI*-99	*trn*I/*trn*A	Prrn	ggagg / T*psb*A	DeGray *et al.*, 2001.
Drought tolerance	*tps*	*trn*I/*trn*A	Prrn	ggagg / T*psb*A	Lee *et al.*, 2003.
Phytoremediation	*mer*A[a]/*mer*B[b]	*trn*I/*trn*A	Prrn	ggagg[a, b] / T*psb*A	Ruiz *et al.*, 2003.
Salt tolerance	*badh*	*trn*I/*trn*A	Prrn-F	ggagg/rps16	Kumar *et al.*, 2004.

2. HISTORICAL ASPECTS

Chloroplast genetic engineering was first considered in the 1980s with the introduction of isolated intact chloroplasts into protoplasts (Daniell *et al.*, 2002; Devine and Daniell, 2004). Investigators then focused on the development of chloroplast systems capable of efficient, prolonged protein synthesis and the expression of foreign genes (Daniell and McFadden, 1987). Biolistic transformation made it soon possible to transform plastids without the need to isolate them (Klein *et al.*, 1987). In 1988, the first successful chloroplast genome complementation was reported for the unicellular green alga, *Chlamydomonas reinhardtii* (Boynton *et al.*, 1988). Mutants with a deletion in the *atp*B gene lacked chloroplast ATP synthase activity. Using tungsten microprojectiles, the wild type *atp*B gene was propelled into cells; the single large chloroplast provided an ideal target for DNA delivery. The gene introduced into the cells was able to correct the deletion mutant phenotype; the photoautotrophic growth was restored and the wild-type gene was successfully integrated into the *C. reinhardtii* chloroplast genome via homologous recombination. It was then demonstrated that foreign DNA, an *uid*A gene, flanked by chloroplast DNA sequences was incorporated and stably maintained in the *C. reinhardtii* chloroplast genome; although the introduced gene was transcribed, translated product could not be detected (Blowers *et al.*, 1989). At this time with higher plants, foreign genes were introduced and expressed only in isolated but intact plastids (Daniell and McFadden, 1987). The first expression of a foreign gene in plastids of cultured tobacco cells used autonomously replicating chloroplast vectors (Daniell *et al.*, 1990). This work was extended to wheat leaves, calli and somatic embryos (Daniell *et al.*, 1991). Simultaneously, the *C. reinhardtii* chloroplast genome was transformed with the *aad*A gene conferring spectinomycin or streptomycin resistance (Goldschmidt-Clermont, 1991); the majority of higher plants genetically transformed via the chloroplast genome now use this selectable

Table 2. *List of biopharmaceutical proteins expressed via the chloroplast genome.*

Therapeutic proteins	Gene	Site of integration	Promoter	5'/ 3' regulatory elements	% tsp expression	Reference
Elastin derived polymer	*EG*121	*trn*I/*trn*A	Prrn	T7gene10/T*psb*A	ND	Guda *et al.*, 2000.
Human somatotropin	*hST*	*trn*V/*rps*12/7	Prrn[a], P*psb*A[b]	T7gene10[a], *psb*A[b]/T*rps*16	7.0 %[a], 1.0%[b]	Staub *et al.*, 2000.
Cholera toxin	*Ctx*B	*trn*I/*trn*A	Prrn	Ggagg/T*psb*A	4.1%	Daniell *et al.*, 2001a.
Antimicrobial peptide	*MSI-99*	*trn*I/*trn*A	Prrn	Ggagg/T*psb*A	21.5%	De Gray *et al.*, 2001.
Insulin-like growth factor	*IGF*-1	*trn*I/*trn*A	Prrn	P*psb*A/T*psb*A	33%	Daniell et al., 2004b
Interferon alpha 5	*INFa*5	*trn*I/*trn*A	Prrn	P*psb*A/T*psb*A	ND	Torres, 2002.
Interferon alpha 2b	*INFa*2B	*trn*I/*trn*A	Prrn	P*psb*A/T*psb*A	19%	Daniell et al., 2004b
Human serum albumin	*hsa*	*trn*I/*trn*A	Prrn[a], P*psb*A[b]	ggagg[a], *psb*A[b]/T*psb*A	0.02%[a], 11.1%[b]	Fernandez *et al.*, 2003.
Interferon gamma	*IFN-g*	*rbc*L/*acc*D	P*psb*A	P*psb*A/T*psb*A	6%	Leelavathi, 2003.
Monoclonal antibodies	*Guy's 13*	*trn*I/*trn*A	Prrn	Ggagg/T*psb*A	ND	Daniell et al., 2004b.
Anthrax protective antigen	*Pag*	*trn*I/*trn*A	Prrn	P*psb*A/T*psb*A	18.1%	Chebolu & Daniell 2004.
Plague vaccine	*CaF1~L crV*	*trn*I/*trn*A	Prrn	P*psb*A/T*psb*A	14.8 %	Singleton, 2003.
CPV VP2	*CTB-2L21*[a] *GFP-2L21*[b]	*TrnI/trnA*	Prrn	*psb*A/T*psb*A	31.1%[a], 22.6%[b]	Molina *et al.*, 2003.
Tetanus toxin	*Tet C*	*Trnv/rps12/7*	Prrn	T7gene10[a], atpB[b]/Trbc L	25%[a], 10%[b]	Tregoning *et al.*, 2003.

marker. Stable integration of the *aad*A gene into the tobacco chloroplast genome was then demonstrated (Svab and Maliga, 1993). Since then, this field has advanced quite rapidly, and chloroplast genetic engineering technology is currently applied to other useful crops such as potato, tomato, soybean, cotton and carrot (Sidorov *et al.*, 1999; Ruf *et al.*, 2001; Kumar *et al.*, 2004 a, b; Dufourmantel *et al.*, 2004). Foreign genes expressed via the plastid genome now bestow useful agronomic traits (Table 1) as well as therapeutic proteins (Table 2). In one instance, the accumulation of

Bacillus thuringiensis (Bt) Cry2Aa2 crystals in transgenic tobacco chloroplasts was important for two reasons: highest accumulation (46.1% of total soluble protein, or tsp), and expression of a complete bacterial operon (DeCosa *et al.*, 2001).

3. UNIQUE FEATURES OF CHLOROPLAST GENETIC ENGINEERING

Chloroplast transformation provides not only higher gene expression capability than does nuclear genetic engineering, but also several other advantages. Even though transgenic chloroplasts may be present in pollen, a foreign gene will not escape to other crops because chloroplast DNA in not passed on to the egg cell. The chloroplast could also be a good place to accumulate proteins or their biosynthetic products that may be harmful if they were in the cytoplasm (Bogorad, 2000). Cholera toxin B subunit (CTB), a candidate oral subunit vaccine for cholera, was accumulated in large quantities within transgenic plastids and was nontoxic (Daniell *et al.*, 2001a; Molina *et al.*, 2004). Similarly, trehalose, a pharmaceutical industry preservative, was very toxic when it accumulated in the cytosol but was nontoxic when it was compartmentalized within plastids (Lee *et al.*, 2003). When expressed in the chloroplast, xylanase, an enzyme important in many industrial applications, did not cause cell wall degradation as seen in nuclear transformants; therefore, plant growth was not affected (Leelavathi *et al.*, 2003). Additionally, transgenes integrated site-specifically into spacer regions of the chloroplast genome eliminate concerns of position effects that are frequently observed in nuclear transgenic plants. All chloroplast transgenic lines express the same level of foreign protein, within the range of physiological variations (Daniell *et al.*, 2001a). Site-specific integration also eliminates the introduction of vector sequences and transgene silencing, which are serious concerns in nuclear transformation (Daniell et al., 2002). Transcripts do not silence genes in chloroplast transgenic lines despite their accumulation at a level 150-170 fold higher than nuclear transgenic plants (Lee *et al.*, 2003; Dhingra et al., 2004). Similarly, lack of post-transcriptional gene silencing was evident with the accumulation of over 46% tsp in chloroplast transgenic lines (DeCosa *et al.*, 2001).

4. MATERNAL INHERITANCE AND GENE CONTAINMENT

In most angiosperm plant species, plastid genes are inherited uniparentally in a strictly maternal fashion (Hagemann, 1992; Birky, 1995; Mogensen, 1996; Zhang *et al.*, 2003). For more information on inheritance of plastid genome, see Chapter 4. Although pollen from plants shown to exhibit maternal plastid inheritance contains metabolically active plastids, the plastid DNA itself is lost during the process of pollen maturation and hence is not transmitted to the next generation (Nagata *et al.*, 1999; Daniell, 1999, 2000). This minimizes the possibility of outcrossing transgenes to related weeds or crops (Daniell *et al.*, 1998; Scott and Wilkenson, 1999; Daniell, 2002) and reduces the potential toxicity of transgenic pollen to non-target insects (DeCosa *et al.*, 2001). Plastid DNA is eliminated from the male germ line at different points during sperm cell development, depending upon the plant species (chapter 4). Maternal inheritance thus offers containment of chloroplast transgenes due to lack of gene flow through pollen (Daniell, 2002b; Daniell and Parkinson, 2003). Pollination of wild-type plants with pollen from chloroplast transgenic lines results in the lack of growth of resultant seedlings on selective media; the pollinated plants remain sensitive to antibiotics (Figure 2). Conversely, the pollination of transgenic plants with pollen from wild-type plants results in progeny that will be antibiotic resistant. In the rare event where transgene escape occurs, engineering male sterility via the chloroplast genome should provide a failsafe mechanism; such a cytoplasmic male sterility system has been developed recently (see section 6.6 below). Chloroplast genetic engineering provides transgene containment, which ensures a much more ecologically safe transformation system when compared to nuclear transformation.

Figure 2. *Maternal inheritance of transgenes. (A) Wild-type or (B) chloroplast transgenic lines were germinated on MSO medium supplemented with 500 mg/l spectinomycin (see Color Plate 5).*

5. CROP SPECIES STABLY TRANSFORMED VIA THE PLASTID GENOME

A "universal vector" that utilizes the plastid DNA flanking sequences of one plant species to transform another species of unknown genome sequence (Daniell *et al.*, 1998) has been used to transform potato and tomato plastid genomes using flanking sequences from tobacco (Sidorov *et al.*, 1999; Ruf *et al.*, 2001). However, the transformation efficiency is lowered because the flanking sequences are not completely homologous. In general, tobacco plastid transformation is highly efficient when tobacco endogenous flanking sequences are used (Daniell et al., 2001; Fernandez San-Millan et al., 2003; Dhingra et al., 2004). When petunia flanking sequences were used to transform the tobacco plastid genome, however, transformation efficiency was lower than that observed using endogenous flanking sequences (DeGray *et al.*, 2001). Therefore, the concept of a universal vector is feasible but less efficient than species-specific chloroplast vectors.

5.1. Tobacco

Tobacco has by far been the most extensively used plastid transformation system because of its vast utility as a model organism. More transgenes have been expressed in tobacco through the nuclear and chloroplast genomes than all other plant species combined. A single tobacco plant is capable of producing a million seeds and one acre of tobacco produces more than 40 metric tons of leaves in multiple harvests per year (Cramer *et al.*, 1999). With such a large biomass, tobacco plants are ideal for scale up. The cost of producing recombinant proteins in tobacco leaves is estimated to be 50-fold lower than that of *E. coli* fermentation systems (Kusnadi, 1997). Most importantly, tobacco would not contaminate food sources as it is not a food or feed crop. It is thus an ideal bioreactor. Harvesting leaves before the appearance of reproductive structures also offers the complete containment of transgenes. For a recent list of foreign genes expressed via the tobacco plastid genome, please refer to Tables 1 and 2.

5.2. Potato

A successful potato plastid transformation protocol utilized a vector that was originally designed for tobacco (Sidorov *et al.,* 1999). The flanking sequences of tobacco and potato share ~98% homology. The vector contained genes for *aad*A, a selectable marker, and green fluorescent protein (GFP), a screenable marker. Quantification of GFP in regenerated shoots showed 5% tsp, but microtubers expressed only 0.05% GFP tsp. This was due to a lower plastid DNA copy number coupled with lower transcriptional/translational rates in microtuber plastids. No useful traits have since been introduced via the plastid genome in potato. This may be partly due to the very low efficiency of plastid transformation.

5.3. Tomato

Researchers have transformed plastids of the tomato (*Lycopersicon esculentum*) with the gene *aad*A, conferring antibiotic resistance (Ruf *et al.*, 2001). Transformed tomato plants produced fruits with viable seeds, which transmitted the transgene in a uniparentally maternal fashion as expected for a plastid-encoded trait. This indicates that it is possible to express a chloroplast transgene in a crop suitable for human consumption. No useful traits have since been introduced via the plastid genome perhaps due to extremely low plastid transformation efficiency.

5.4. Chloroplast genetic engineering through somatic embryogenesis

So far, transformation of the chloroplast in different plant species has been achieved preferably through organogenesis using green leaves as explants that contain large chloroplasts. However, most economically important crops regenerate *in vitro* through somatic embryogenesis instead of organogenesis. In order to extend the concept of chloroplast transformation in major crop species, which regenerate only via somatic embryogenesis, the understanding of tissue culture is essential. It has been pointed out that the foremost obstacles to extending this technology into major crop species include inadequate tissue culture and regeneration protocols and the inability to express the transgene in developing proplastids (Bogorad, 2000; Daniell *et al.*, 2002). Following are examples of recent breakthroughs in achieving transformation of three plant species (carrot, cotton and soybean) in which transformation of the plastid genome was achieved through somatic embryogenesis by bombarding non-green tissues.

5.4.1. Carrot

Carrot is one of the important vegetable crops used worldwide for human and animal consumption, as it is an excellent source of sugars, vitamins A and C and fiber in the diet. Carrot was the first crop among dicots and monocots in which somatic embryogenesis *in vitro* was demonstrated (Steward *et al.*, 1958; Reinert, 1958) and served as an excellent model for developing somatic embryos in other plant species. Using the carrot system, we have recently established the first stable plastid transformation through cell cultures with an adenylyl transferase (*aad*A) gene encoding resistance to spectinomycin (used as a selection agent) and betaine aldehyde dehydrogenase (*badh*) gene conferring salt tolerance. Homoplasmic transgenic plants exhibiting high levels of salt tolerance were regenerated from bombarded cell cultures via somatic embryogenesis, with a high frequency of transformation (Kumar et al., 2004a). *In vitro* transgenic carrot cells transformed with the *badh* transgene were visually green in color when compared to untransformed carrot cells and helped in visual selection for transgenic lines (Color Plate 6 A-D). Also, expression of the Green Fluorescent Protein (*gfp*) in non-green carrot cell cultures offered visual selection (Color Plate 6 H, I). Transgenic cell cultures grown in liquid medium in presence of 100 mM NaCl exhibited 8-fold enhanced BADH enzyme activity, 7-fold more embryogenic growth (Color Plate 6 E, F) and accumulated about 50-fold more betaine (93-101 μmol g^{-1} DW of β-alanine betaine and glycine betaine) than untransformed cells. Because of higher expression of *badh,* transgenic carrot plants grew well up to 400 mM NaCl (Figure 3). Such high levels of salt tolerance and betaine accumulation have been so far observed only among halophytes and this is the highest level of salt tolerance reported so far among genetically modified crop plants. Untransformed plants exhibited severe growth retardation in 200 mM or higher concentrations of NaCl. BADH expression was 74.8% in non-green edible parts (carrots) containing chromoplasts, 53% in proplastids of cultured cells when compared to chloroplasts (100%) in leaves (Color Plate 6 J).

The first successful plastid transformation via somatic embryogenesis and the first demonstration of a useful trait engineered via the plastid genome in a non-tobacco crop as well as plastid transformation from non-green tissues (Kumar *et al.*, 2004a) should pave the way to engineer other crops with useful agronomic traits. Being a biennial plant, carrot completes its life cycle in two years producing an edible fleshy taproot in the first year and flowers in the second year after passing through a cold season (Yan and Hunt, 1999). Strict maternal inheritance of the chloroplast genome in cultivated carrot crops has been reported (Vivek *et al.*, 1999). Carrot is environmentally safe and is doubly protected against transgene flow via pollen and seeds. This should help achieve zero-contamination of food crops by Pharm Crops, expected by various regulatory agencies. Carrot is also ideal to express therapeutic proteins because somatic embryos are single cell derived, they rapidly divide to produce large biomass in bioreactors, are viable for long term on culture medium, germinate from encapsulated embryos cryopreserved for many years (Tessereau *et al.*, 1994) and offer a single uniform source of cell culture for recombinant proteins.

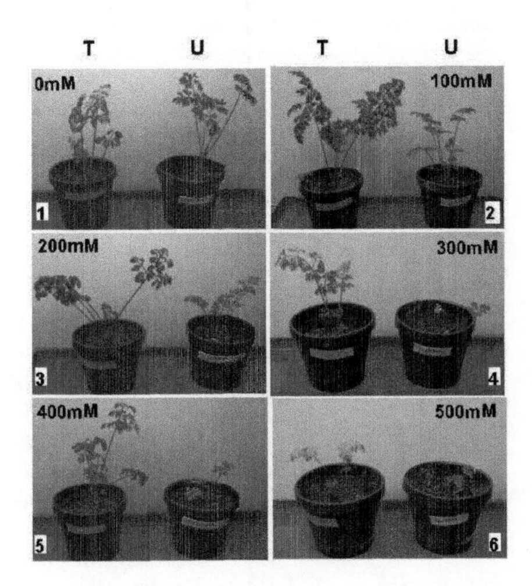

Figure 3. *Effect of salt (100–500 mM NaCl) on untransformed (**U**) and Transgenic (**T**) lines grown on different concentration of NaCl. Plants were irrigated with water containing different concentrations of NaCl on alternate days for up to four weeks (see Color Plate 6).*

5.4.2. Cotton

Cotton (*Gossypium hirsutum* L.) is an excellent natural source of textile fiber in the world, planted over 90 countries and more than 180 million people depend on it for their livelihood. The annual business revenue attributed to cotton in the U.S.

economy alone is about $120 billion. Such economic importance underscores the need to develop new environmentally friendly approaches for genetic modification of the cotton genome. Cotton plastid transformation was extensively attempted using the *aad*A gene containing species-specific cotton chloroplast vectors. However, no transgenic cultures or plants were recovered using spectinomycin as the selection agent due to its lethal effect on cotton cultures. Cotton plastid transformation was achieved for the first time using a "double barrel" plastid transformation vector that contains two selectable marker genes (*aph*A-6 and *npt*II) to detoxify the same antibiotic by two enzymes around the clock, irrespective of the type of tissues or plastids and an efficient regeneration system via somatic embryogenesis (Kumar *et al.*, 2004b). The double barrel transformation vector is at least 8-fold more efficient than single gene (*aph*A-6) based chloroplast vector. Transgenic cell lines were selected on a medium supplemented with kanamycin. Transformed callus cultures were multiplied on higher concentrations of kanamycin in order to increase the number of transgenic chloroplasts in cotton cultures. The maximum transformation efficiency (41.9%) was observed when cell cultures were bombarded using 650 psi rupture disc.

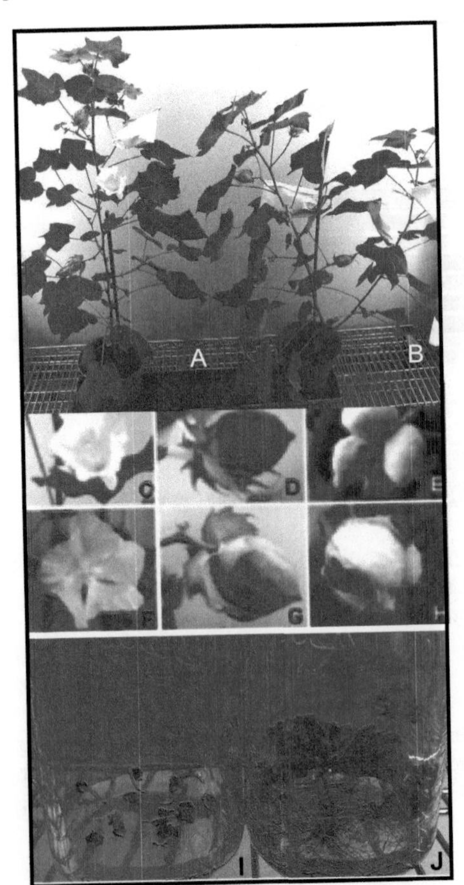

Figure 4. *(A) Transgenic and (B) non-transgenic control cotton plants at the stage of flowering and seed setting. (C-E) Different floral parts of transgenic cotton and (F-H) non-transgenic control cotton. (I-J) F1 Seedlings germinated on kanamycin (50 mg/l). (I) Cross between transgenic ♂ x ♀ non-transgenic cotton. (J) Transgenic cotton obtained after self pollination (see Color Plate 7).*

This suggests that higher cell death or production of phenolic compounds at sites of injury may be important determinants in transformation efficiency. Transgenic somatic embryos were matured and elongated into plantlets and transferred to growth chambers for flowering and setting seeds (Figure 4). Stable transgene integration into cotton chloroplast genomes was confirmed by PCR and Southern blot analyses. Several crosses were made between non-transgenic cotton and chloroplast transgenic lines. No germination of seeds from F1 crosses (non-transgenic female x male transgenic chloroplast) was observed on the kanamycin selection medium, whereas all self-pollinated transgenic chloroplast plants germinated on kanamycin medium, confirming maternal inheritance, lack of pollen transmission and Mendelian segregation of transgenes.

Transformation of cotton chloroplast not only addresses the concern of transgene escape to other plants via pollen but has also created a new way to engineer cotton with useful agronomic traits. For example, the *cry2Aa2* gene, when expressed in via the chloroplast genome, killed insects that developed resistance (up to 40,000-fold) to insecticidal proteins (Kota *et al.*, 1999). When expressed as cry2Aa2 operon via the chloroplast genome, insecticidal protein accumulated up to 46.1% of the total protein and killed insects that have been difficult to control, such as 10-day-old cotton bollworm or beet armyworm (DeCosa *et al.*, 2001). The transformation of the chloroplast genome should address concerns about transgene escape, insects developing resistance, inadequate insect control and promote public acceptance of genetically modified cotton (Kumar *et al.*, 2004b), which cannot be addressed through nuclear transgenic cotton. Upland cotton has the potential to hybridize with Hawaiian cotton, *G. tomentosum*, and feral populations of *G. hirsutum* in the Florida Keys, and *G. hirsutum* / *G. barbadense* on the U.S. Virgin Islands and Puerto Rico. For these reasons, restrictions on field plot experimental use permits and commercial planting of Bt-cotton have been instituted in these areas. Similarly, GM cotton is now planted only in regions of the world where there are no wild relatives, in order to avoid potential outcross with related weeds. The sub-optimal expression of Bt toxins has resulted in an increased risk of pests developing Bt resistance (Daniell, 2002).

5.5. Soybean

Soybean (*Glycine max*) is the most important crop provider of proteins and oil used in animal nutrition and for human consumption. It is also the most planted genetically modified crop, representing in 2003 more than half of the soybean cultivated area worldwide. This corresponds to the adoption by farmers of glyphosate tolerant cultivars, a trait that has been engineered via the nuclear genome. The engineering of such agronomic traits via the soybean plastid genome has so far not been described. Nuclear transformation in this species still remains difficult as in most grain legumes (Chandra and Pental, 2003), and this certainly contributed to the first description and generation of fertile chloroplast-transgenic plants (Dufourmantel *et al.*, 2004). The absence of the total plastome sequence in the databases until now may also have complicated this task.

The plastid transformation of soybean was first attempted by Zhang *et al.* (2001b) with the objective of modifying the photosynthetic potential. The transformation vector, targeting the insertion between the *atp*□ and *trn*L genes, contained the two *Chlamydomonas reinhardtii rbc*L and R*bc*S genes and an *aad*A selection cassette, allowing selection on spectinomycin. Two calli were selected from bombarded leaf pieces (93 bombardments) but were negative at the DNA and RNA levels for *aad*A, corresponding possibly to quite rare spontaneous plastid mutants. One heteroplasmic event that showed evidence by PCR for integration of the transgene cassettes at the expected location was recovered out of 984 bombardments performed on embryogenic suspension cultures. Unfortunately, this material could not be regenerated and characterized further. The reason for such a low transformation frequency is unclear.

The breakthrough in soybean plastid transformation was recently achieved by Dufourmantel *et al.* (2004), using embryogenic tissue as starting material (cv. Jack) and biolistic delivery. The transforming DNA carries the bacterial *aad*A antibiotic resistance gene under the control of tobacco plastid regulatory elements (16S rRNA promoter fused to the *rbc*L 5'UTR and the *psb*A 3' UTR). This selection cassette is flanked by two adjacent soybean plastome sequences allowing its insertion in the intergenic region between the *trn*V gene and the *rps*12/7 operon in the inverted repeated region. These regions were amplified from cv. Jack and substantial differences were noted between the sequences present in the public databases; mismatches and an additionnal sequence (Genbank AY575999) corresponded probably to allelic diversity in this species.

Figure 5. *In vitro germination of wild-type and transplastomic seeds on media containing 500 mg/l spectinomycin after 20 days (see Color Plate 8 B, C).*

In two independent experiments corresponding to a total of 8 bombardments, 16 spectinomycin-resistant calli were selected approximately 3 months after bombardment and analyzed by PCR. This analysis showed that all of them had integrated the selection cassette at the expected location in the soybean plastome. Southern analysis confirmed these results and did not detect any wild-type plastome molecules in 9 out of 10 calli analyzed, indicating that the selection process leads rapidly to homoplasmic material. No spontaneous mutant was selected, suggesting that the conformation of the *Glycine max* 16S rRNA, the target of spectinomycin, cannot be easily mutated to a resistant form, unlike those observed in other species (Skarjinskaia *et al.*, 2003; Sikdar *et al.*, 1998). Phenotypically normal transgenic soybean plants (T_0 generation, Color Plate 8A) were regenerated via somatic embryogenesis from each of those 16 calli, and were fully fertile. The T_1 progenies that were tested were uniformly resistant to spectinomycin as expected for plastid maternal inheritance (Figure 5, Plate 8C). The degree of homoplasmy was evaluated by PCR on transplastomic plants of T_0 and T_1 generation. While some wild-type plastome could still be detected in some calli under selection, it appears that all analyzed transplastomic plants, even those derived from heteroplasmic calli, were homoplasmic. The embryogenic step probably contributes to the rapid and general establishment of this homoplasmic state.

A high transformation frequency (per shot) similar to that obtained routinely with tobacco leaves was reported, despite the fact that undifferentiated tissue was used for starting material, which is known to contain fewer plastids and of smaller size (Zhang *et al.*, 2001b). The described protocol leads to homoplasmic fertile transplastomic soybean, and opens the door to the modification of the plastid genome of this crop and possibly that of other legumes.

5.6. Challenges Ahead in Achieving Plastid Transformation in Monocots

There are several factors that have impeded the extension of chloroplast transformation technology to other plant species. Chloroplast transformation vectors utilize homologous flanking regions for recombination and insertion of foreign genes. Transformation of *Arabidopsis*, potato and tomato chloroplast genomes were achieved via organogenesis by bombardment of green leaf tissues, but the efficiency was much lower than tobacco (Sikdar *et al.*, 1998, Sidorov *et al.*, 1999 and Ruf *et al.*, 2001). In *Arabidopsis*, 1 transgenic line resulted from either 40 or 151 bombarded plates. In potato, one chloroplast transgenic line per 35 bombarded plates was obtained, and in tomato, 1 transgenic line per 87 bombarded plates was obtained. In contrast, 15 tobacco chloroplast transgenic lines were obtained per bombarded plate (Fernandez-San Milan *et al.*, 2003). In case of *Lesquerella*, transgenic shoots had to be grafted onto *Brassica napus* rootstock to reconstruct transgenic plants (Skarjinskaia *et al.*, 2003). In oilseed rape, direct Southern blot analysis of the transgenic chloroplast genome was not presented as proof of stable integration (Hou *et al.*, 2003). The vectors employed for the chloroplast transformation of potato, tomato and *Lesquerella* contained the flanking sequences from tobacco or *Arabidopsis*. This may be one of the reasons for lower transformation efficiency. When petunia flanking sequences were used for chloroplast transformation of tobacco, the transformation efficiency decreased drastically (DeGray *et al.*, 2001). In contrast, highly efficient transformation of carrot, cotton and soybean chloroplast genomes was achieved using species-specific chloroplast vectors containing 100% homologous flanking sequences. Therefore, there is an urgent need to sequence plastid chloroplast genomes in order to facilitate the transformation of crop species. It is surprising that chloroplast genomes of only a few crops have been sequenced so far, despite the small size of the genome and availability of technology to sequence an entire genome within a single day.

It has been erroneously claimed that rice plastid transformation was achieved via somatic embryogenesis but no data was provided to support stable transgene integration into the plastid genome or homoplasmy by Southern analysis or maternal inheritance of transgenes; only transient integration of a transgene was observed (Khan and Maliga, 1999). Such transient integration of transgenes in the rice plastid genome has been reported as early as 1997 in the literature, when a tobacco chloroplast vector was used to transform the rice plastid genome (Figures 13, A and B, in patent WO 99/10513). Understanding of somatic embryogenesis and selection conditions to advance heteroplasmy to homoplasmy is essential to achieve stable plastid transformation. Although chloroplast genome sequences of several monocots, including rice, wheat and maize have been available for several years, none of their genomes have been transformed so far. Therefore, it is not adequate just to have the genome information to facilitate plastid transformation; a better understanding of DNA delivery, selection, regeneration, and progression towards homoplasmy is essential to achieve plastid transformation in monocots.

6. AGRONOMIC TRAITS EXPRESSED VIA THE PLASTID GENOME

Several useful genes have yielded valuable agronomic traits in tobacco via chloroplast transgene expression. For example, plants resistant to Bt toxin-sensitive insects were obtained by integrating the *cry*IAc gene into the chloroplast genome (McBride *et al.*, 1995). Plants resistant to toxin-resistant insects (Bt toxin concentrations up to 40,000 times higher) were obtained by hyperexpression of the *cry*2A gene within the tobacco chloroplast (Kota *et al.*, 1999). Plants have been transformed with a gene that confers herbicide resistance; this gene was maternally inherited, overcoming the problem of outcrossing with weeds (Daniell *et al.*, 1998, for additional information, see Chapter 17). More recently, plants exhibiting tolerance to bacterial and fungal diseases (DeGray *et al.*, 2001), drought (Lee *et al.*, 2003), salt (Kumar *et al.*, 2004a), and organomercurials (Ruiz *et al.*, 2003) have been reported. Some agronomic traits were engineered using multiple genes, novel pathways or bacterial operons (DeCosa *et al.*, 2001; Daniell and Dhingra, 2002; Ruiz *et al.*, 2003). Examples of these traits are detailed below.

6.1. Insect Resistance

Plants genetically modified to express Cry2Aa2, an insecticidal protein, have shown a significantly increased resistance against target insects. The Cry2Aa2 protein is encoded within an operon and has been expressed in the chloroplast genome both as a single gene (Kota *et al.*, 1999) and as an operon (DeCosa *et al.*, 2001; Daniell and Dhingra, 2002). The *Cry*Aa2 operon consists of the *cry*2Aa2 gene, *orf* 1 and *orf* 2. The function of *orf* 1 is unknown; *orf* 2 functions as a chaperone that has the ability to fold the Cry2Aa2 protein into cuboidal crystals. Expression of the *Cry*2Aa2 operon in tobacco chloroplasts resulted in the detection of cuboidal crystals using transmission electron microscopy (Figure 6, DeCosa *et al.*, 2001). Hyperexpression of the Cry2Aa2 protein resulted in levels up to 46.1% tsp, the highest expression levels on record.

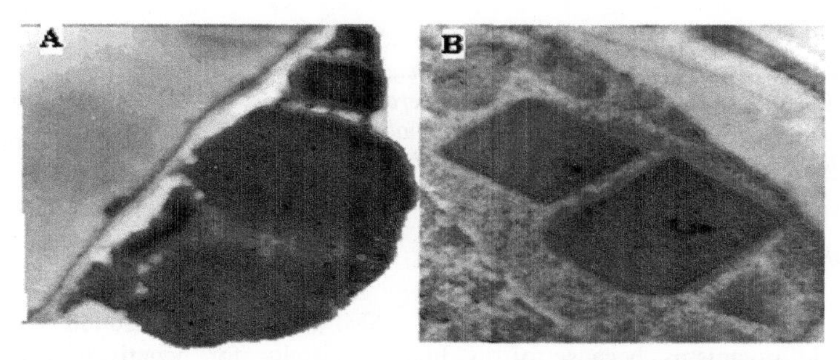

Figure 6. *Transmission electron micrographs. (A) Detection of Cry2A protein by immunogold-labeling using Cry2A antibody. (B) Accumulation of folded Cry2A protein as cuboidal crystals in transgenic chloroplasts.*

When wild-type control plant leaf material was fed to the tobacco budworm (*Heliothis virescens*), cotton bollworm (*Helicoverpa zea*) and beet armyworm (*Spodoptera exigua*), the leaves were totally devoured within 24 hours (Figure 7 A, D, G). When the tobacco budworm was fed leaves expressing the single gene, all the insects died after 5 days (Figure 7 B), whereas the insects fed leaves expressing the Bt operon died in 3 days (Figure 7 C). Similar results were obtained with the cotton bollworm (Figure 7 D-F) and the beet armyworm (Figure 7 G-I). Hyperexpression of the Cry2Aa2 protein in chloroplasts thus conferred 100% resistance to insects that feed on tobacco. Even old bleached senescent leaves contained high levels of the insecticidal protein, despite the high protease activity associated with senescence. Chaperone-assisted cuboidal crystal formation may have prevented proteolytic degradation of the protein. Even at lower levels of expression, chloroplast transgenic plants killed insects that could withstand insecticidal protein concentrations 40,000 times higher than normal (Kota *et al.*, 1999).

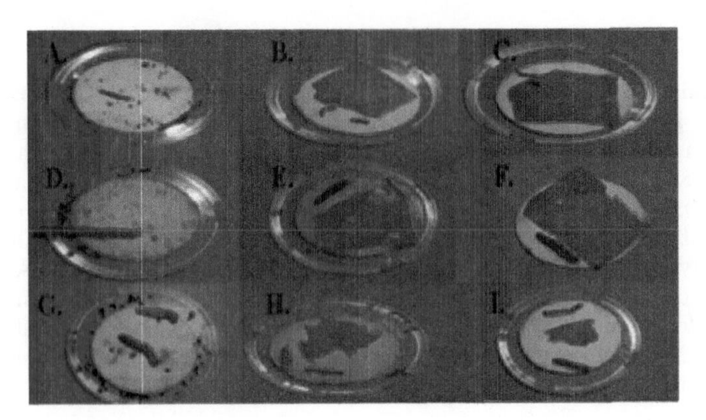

***Figure** 7. Insect bioassays. (A, D, G) Untransformed tobacco leaves; (B, E, H) single gene-derived Cry2Aa2 transformed leaves; (C,F,I) operon-derived Cry2Aa2 transformed leaves. (A-C) Bioassays with Heliothis virescens (tobacco budworm); (D-F) bioassays with Helicoverpa zea (cotton bollworm); (G-I) bioassays with Spodoptera exigua (beet armyworm). For each replicate samples from the same leaf were used (see Color Plate 9).*

6.2. Pathogen Resistance

MSI-99, a helical antimicrobial peptide (AMP), confers protection against prokaryotic organisms because of its high specificity for negatively charged phospholipids found mostly in bacteria. An *in planta* assay was performed on plants expressing MSI-99 in the chloroplast (DeGray *et al*., 2001). The leaves were inoculated with the phytopathogen *Pseudomonas syringae pv tabaci*, and then observed for necrosis around the site of the inoculation. No necrotic tissue could be observed in transgenic plants even when 8 x 10^5 cells were used; in wild-type plants inoculated with only 8 x 10^3 cells, a large necrotic area could be seen (Figure 8, C and D). These data suggest that the AMP expressed in the chloroplast of the transgenic plants is released during infection, coming in contact with the phytopathogen. When the same bioassays were performed with the plant fungal pathogen *Colletotrichum destructivum* in nontransgenic lines, the plant developed anthracnose lesions whereas the transgenic plants expressing MSI-99 did not (Figure 8, A and B). Additionally, leaf extracts from transgenic plants inhibited the growth of pregerminated spores from three fungal pathogens, *Aspergillis flavus*, *Fusarium moniliforme* and *Verticillium dahliae*, by 95% when compared with wild-type extracts. Because the lytic activity of AMPs is concentration dependent, the amount of AMP required to kill bacteria was used to estimate the level of expression in transgenic plants. Based on the minimum inhibitory concentration, it was estimated that transgenic tobacco plants expressed MSI-99 in the chloroplast at levels of at least 21.5% tsp (Daniell, 2004). These results show that transgenic chloroplasts expressing AMP can confer high levels of resistance to phytopathogenic organisms.

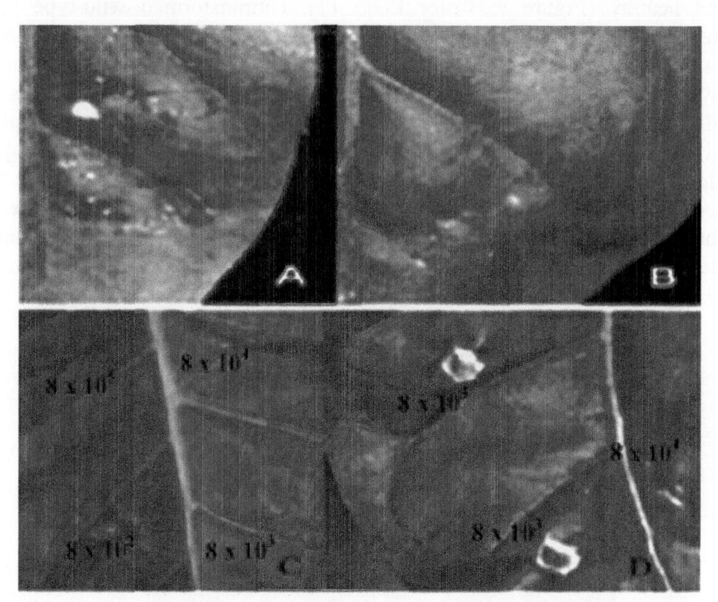

Figure 8. *In planta bioassays for disease resistance. (A, B) Fungal disease resistance. Leaves were inoculated on the adaxial surface with eight drops of 10 μl each of the culture containing 1 x 10^6 spores/ml of the fungal pathogen Colletotrichum destructivum. (A) Wild type leaf; (B) transgenic leaf. (C, D) Bacterial disease resistance: 8 · 10^5, 8 · 10^4, 8 · 10^3, and 8 · 10^2 cell cultures of bacterial pathogen Pseudomonas syringae pv tabaci were added to a 7-mm-scraped area in transgenic and nontransgenic tobacco lines. Photos were taken 5 days after inoculation (see Color Plate 10).*

6.3. *Drought Tolerance*

Because of their sessile way of life, plants face several environmental stress factors such as drought, salinity and freezing. Plants, yeast and other organisms produce osmoprotectants that confer resistance to several factors including drought. Trehalose phosphate synthase (encoded by the *TPS*1 gene) catalyzes the reaction to form the osmoprotectant trehalose. Hyperexpression of trehalose phosphate synthase in the chloroplast results in the accumulation of trehalose in transgenic plants (see Color Plate 14, Lee *et al.*, 2003). The phenotype of chloroplast transgenic plants was similar to untransformed control plants, and no pleiotropic effects were observed as seen in nuclear transgenic plants (see Color Plate 14; Lee *et al.*, 2003, Holmstrom *et al.*, 1996).

Drought tolerance bioassays in which transgenic and wild-type seeds were germinated in media containing concentrations of 3 to 6% PEG showed that the chloroplast transgenic plants expressing trehalose germinated, grew, and remained green and healthy (Figure 9; Color Plate 11). Untransformed wild-type seeds germinated under similar conditions showed severe dehydration, loss of chlorophyll (chlorosis), and retarded growth that resulted in the death of the seedlings. Loss of chlorophyll in the untransformed plants suggests that drought destabilizes the thylakoid membrane, but accumulation of trehalose in transgenic chloroplasts bestowed thylakoid membrane stability. Additionally, when potted chloroplast transgenics and untransformed plants were not watered for 24 days and were then rehydrated for 24 hours, the chloroplast transgenic plants recovered while the untransformed plants did not. These results indicate that expression of trehalose phosphate synthase in the chloroplast of transgenic plants confers drought tolerance.

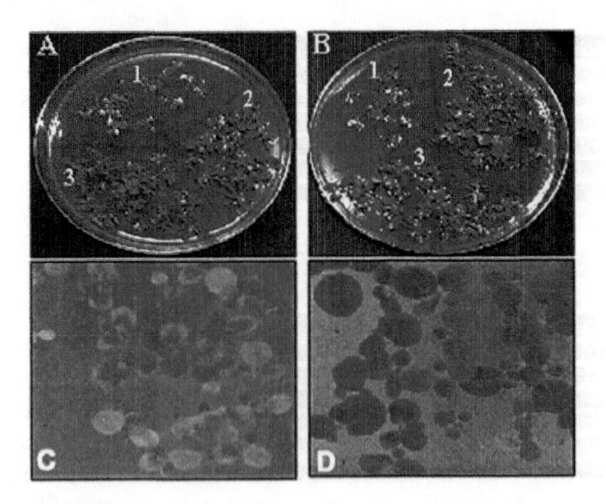

***Figure 9**. Drought tolerance assays. (A, B) Dehydration/rehydration assay. Three- week-old seedlings were dried for 7 hours and rehydrated in MS medium for 48 hours. (1) Untransformed; (2, 3) T_1 and T_2 chloroplast transgenic lines. (C, D) PEG growth assay. Four-week-old seedlings were grown on MS medium with 6% PEG. (C) Untransformed; (D) T_2 chloroplast transgenic line (see Color Plate 11).*

6.4. Phytoremediation

Organomercurials are the most toxic form of mercury, presenting a serious hazard to our environment. Current methods of chemical and physical remediation as well as bacterial bioremediation have thus far proven to be ineffective owing to the high cost and environmental concerns. An alternative method, phytoremediation, has been proposed as a safe and cost-effective system for the remediation of toxic chemicals in the environment. The chloroplast is the primary target for mercury and organomercurials in plants, making chloroplast genetic engineering advantageous for increasing resistance to and detoxifing organomercurials and metal mercury forms. Two bacterial enzymes that confer resistance to different forms of mercury, mercuric ion reductase (merA) and organomercurial lyase (merB) were expressed as an operon in chloroplasts of transgenic tobacco plants (Ruiz *et al.*, 2003). The transgenic lines were tested in a bioassay using the organomercurial phenyl mercuric acetate (PMA). The transgenic plants were shown to be substantially more resistant than wild-type tobacco plants grown under identical conditions. Transgenic 16-day-old tobacco seedlings were able to survive concentrations as high as 200 ∝M (Figure 10, Color Plate 12), even though shoots absorbed about one hundred times more PMA than untransformed plants that struggled to survive at a concentration of 50 ∝M (Figure 10). When nuclear transgenic seedlings containing *mer*A and *mer*B genes were germinated in a medium containing PMA, they were only resistant to concentrations of 5 ∝M (Bizily *et al.*, 2000). When transgenic plants were treated with concentrations of 100, 200, 300, and 400 ∝M PMA, they showed an increase in total dry weight at 100 ∝M; when compared with the untransformed wild type grown at similar concentrations, which progressively decreased with each increase in PMA concentration, the total dry weight of the transgenics remained much higher. The chlorophyll concentration of the leaf is an indicator of chloroplast structural and functional integrity. When leaf disks from the untransformed wild-type and chloroplast transgenic plants were grown for 10 days in 10 ∝M PMA, the chlorophyll concentration of the transgenic plants increased, while it decreased in untransformed plants (Ruiz et al., 2003). These bioassays show the efficiency and activity of the chloroplast-expressed *mer*A/*mer*B operon and establish that transgenic plants can be used successfully for phytoremediation.

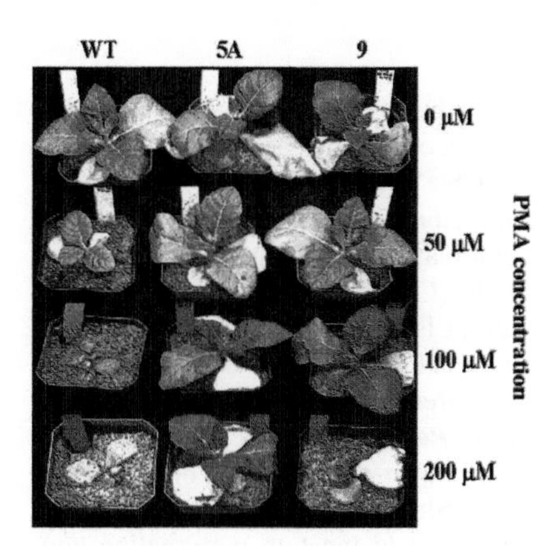

***Figure 10**. Effect of PMA concentration on the growth of wild-type or transgenic tobacco lines. Plants were treated with 200 mL Hoagland's nutrient solution supplemented with 0, 50, 100, or 200 ∝M PMA. Photographs were taken 14 days after treatment. WT: negative control Petit Havana, 5A: pLDR-MerAB transgenic line, 9: pLDR-MerAB-3'UTR transgenic line (see Color Plate 12).*

6.5. Salt Tolerance

Salinity is one of the major factors that limits the geographical distribution of plants, affecting crop productivity and quality. According to Gibberd *et al.* (2002), carrot has been classified as a salt sensitive plant and shows 7% growth reduction for every 10 mM increment in salinity above 20 mM salt. Salt stress results in reduced leaf gas exchange and a reduction in apparent photosynthetic capacity in cultivated carrot crops. In carrot, overexpression of the *badh* gene via engineering of the carrot chloroplast genome resulted in a significant enhancement of salt tolerance by accumulating the quaternary ammonium compounds glycine betaine and β-alanine betaine (Kumar *et al.*, 2004a). The quaternary ammonium compounds that accumulate in plants in response to salt stress protect the plant cell by maintaining an osmotic balance with the environment (Robinson and Jones, 1986; Rhodes and Hanson, 1993; Hanson *et al.*, 1991; Hanson and Gage, 1991; Rathinasabapathi *et al.*, 2001). Chloroplast transgenic carrot plants were able to grow well at 400 mM NaCl, a concentration at which only halophytes are able to thrive (Figure 3, Color Plate 6). This study demonstrated that overexpression of transgenes via the chloroplast genome could be a viable approach to considerably improve a crop with agronomic traits. Previously, several attempts were made using *badh* to manipulate the betaine biosynthesis pathway via nuclear genetic engineering in order to enhance salt tolerance in different plants (Flowers, 2004); however, no report has exhibited such high salt tolerance in plants as shown by chloroplast genetic engineering. In order to provide a broad range of salinity tolerance to plants, an alternate strategy may be to engineer genes that confer other mechanisms involved in signal transduction of salt stress in addition to osmoprotection. Therefore, engineering plants for salt tolerance either by large accumulation of betaine via the chloroplast genome (50-54 fold higher than untransformed control) or in combination with an antiport mechanism (Zhang and Blumwald, 2001) should be an attractive option for future strategies.

6.6. *Cytoplasmic Male Sterility*

Naturally occurring cytoplasmic male sterility has been reported for maize, oilseed rape, rice and *Beta beets* (Kriete *et al.*, 1996), but such systems are not available for most crops used in agriculture. In currently available cytoplasmic male sterile lines, the nuclear genome controls various restoration factors (often controlled by multiple loci) that are not fully understood. Also cytoplasmic male sterility has been associated with diseases like southern corn blight (CMS-T) and cold susceptibility (CMS Ogura); the infection site for sorghum ergot is the unfertilized stigma and it was feared that CMS hybrids could serve as an inoculum source for other sorghums. Male-sterility systems have been created by different mechanisms, most of these affecting tapetum and pollen development. In *Petunia hybrida,* an anti-sense approach led to the arrest of pollen maturation due to the depletion of flavonoid pigments in the anthers (Kriete *et al.*, 1996). Similarly, anti-sense expression inhibiting mitochondrial pyruvate dehydrogenase led to tapetum perturbation and male sterility (Yui *et al.*, 2003). The nuclear expression of the *barnase* gene, which was fused to a tapetum specific promoter, led to the degradation of the tapetum and lack of pollen formation (Kriete *et al.*, 1996). In tobacco, the expression of *rol*C via *Agrobacterium* transformation led to a male sterile phenotype (Kriete *et al.*, 1996); unfortunately, severe additional phenotypic alterations were also detected.

Additionally, Zheng *et al.* (2003) showed in *Arabidopsis* that disruption of the *AtGPAT*1 gene involved in the initial step of glycerolipid biosynthesis arrested pollen development. All these examples show that pollen development is a very complex process, sensitive to changes in cellular metabolism.

A major drawback of current male sterility systems is the possibility of the production of transgenic seeds that spread transgenic traits to non-transgenic plants. This is possible if a restorer line or wild relative is able to cross-pollinate the male sterile plant. Nuclear encoded male sterility systems are more vulnerable because through genetic segregation, the male sterility trait will be diluted out. Several sterility systems have been developed, interfering with the development (Goetz *et al.*, 2001) and metabolism of the tapetum as well as pollen maturation (Hernould *et al.*, 1998; Napoli *et al.*, 1999), but many of these present drawbacks, such as interference with general development and metabolism, are often times restricted to specific species. Therefore, an effective male sterility system able to inhibit pollen formation and viability, but not causing any detrimental effect to the plant, is highly desirable. Additionally, this system should inhibit any possible production of viable progeny (seeds) by inadvertent restoration or pollination by wild relatives.

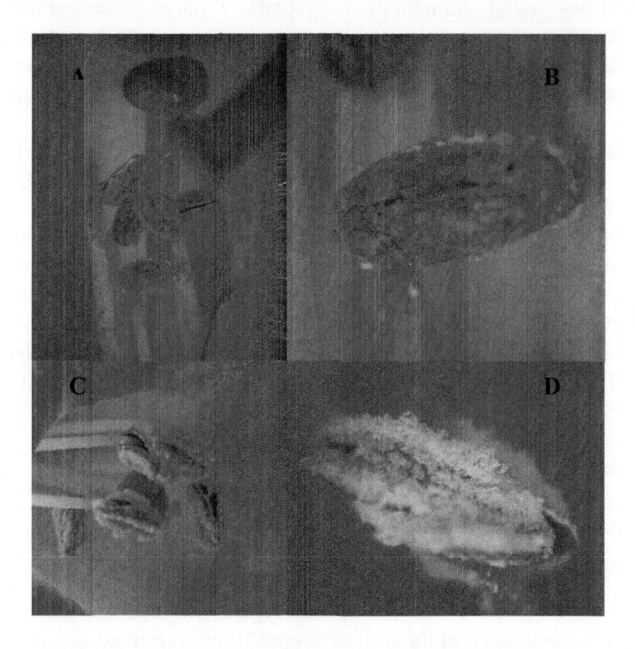

Figure 11. *(A, B) Flower from phaA transgenic plant showing shorter, undeveloped stamens without pollen. (C, D) Flower from wild-type plant showing normal development (see Color Plate 13).*

Ruiz and Daniell (2004) have reported the first engineered cytoplasmic male sterility via chloroplast genetic engineering to address some of these concerns related to male sterility systems. Stable integration into the chloroplast genome and transcription of the *pha*A gene coding for β-ketothiolase was confirmed by Southern and northern blots. Coomassie-stained gels and western blots confirmed hyperexpression of β-ketothiolase in leaves and anthers, with high enzyme activity. The transgenic lines were normal except for the male sterility phenotype, lacking pollen (Figure 11, Color Plate 13). SEM revealed a collapsed morphology of the pollen grains. Transgenic lines followed an accelerated anther developmental pattern, affecting their development and maturation, resulting in aberrant tissue patterns. Abnormal thickening of the outer wall, enlarged endothecium and vacuolation decreased the inner space of the locules, crushing pollen grains and resulting in the irregular shape and collapsed phenotype. This novel approach should overcome current challenges in transgene containment and is applicable to transgenes integrated into both nuclear and organelles genomes.

7. TRANSGENIC PLASTIDS AS PHARMACEUTICAL BIOREACTORS

Use of transgenic chloroplasts as bioreactors that produce foreign proteins offers several advantages over other systems. Plant systems are more economical than industrial facilities using fermentation systems. Large-scale harvesting and processing technology for plants is readily available. Also, extensive purification is not needed when the plant tissue containing the recombinant protein is used for oral delivery of therapeutic proteins. When proteins are expressed in chloroplasts, the amount of recombinant protein produced approaches industrial levels; however, health risks due to contamination with potential human pathogens or toxins are minimized. Stable expression of a pharmaceutical protein in chloroplasts was first reported for GVGVP, a protein-based polymer with medical uses such as wound coverings, artificial pericardia and programmed drug delivery (Guda *et al.*, 2000). Human somatotropin also expressed via the tobacco chloroplast genome was shown to be fully functional (Staub *et al.* 2000). Peptides as small as 20 amino acids (magainin; DeGray *et al.*, 2001) or as large as 83 kDa (anthrax protective antigen; Daniell *et al.*, 2004a-c) have been expressed in transgenic chloroplasts. All biopharmaceutical proteins thus far expressed in transgenic chloroplasts are listed in Table 2 and several examples are detailed below.

7.1. Human Somatotropin

Used in the treatment of hypopituitary dwarfism in children, Turner's syndrome, chronic renal failure, and HIV wasting syndrome, human somatotropin (hST) is produced in the pituitary gland and contains two disulfide bonds. Site-specific integration of the hST gene into the spacer region between the *trn*V gene and *rps*7/3'-*rps*12 operon located in the inverted repeat region resulted in a maximum of 7% hST tsp (Staub *et al.*, 2000). Upon purification, chloroplast-derived hST had proper disulfide bond formation and performed similar to native human hST. This work demonstrated that plastids possess the proper machinery to fold eukaryotic proteins and add disulfide bonds, possibly utilizing the chloroplast enzyme protein disulfide isomerase.

7.2. Human Serum Albumin

Human serum albumin (HSA) comprises approximately 60% of the protein in blood serum. Prescribed in multigram quantities to replace blood volume in trauma and in other clinical situations, HSA is a monomeric 66.5 kDa protein that contains 17 disulfide bonds. Although the current annual need for HSA worldwide exceeds 500 tons, with an approximate market value of $1.5 billion, current systems for mass production are not yet commercially feasible. Initial attempts at expressing HSA in nuclear transgenic plants have achieved disappointingly low levels of HSA (0.02% tsp). Current estimates suggest that the cost-effective yield for pharmaceutical production is 0.1 mg of HSA per gram of fresh weight (Farran *et al.*, 2002). HSA was expressed in transgenic chloroplasts using two different 5' regulatory sequences: the Shine Dalgarno (SD) ribosome binding site (ggagg) construct regulated by the *Prrn* promoter, and the light-regulated *psb*A 5' untranslated region (UTR), a translation enhancer, driven by its own promoter (Fernandez-San Millan *et al.,* 2003). Both transgenes were integrated between the *trn*I and *trn*A genes within the inverted repeat region of the plastid genome. Although the SD construct resulted in HSA levels of only 0.02% total protein (tp), the 5'UTR construct produced HSA up to 7.2% tp. Since the *psb*A 5'UTR is light regulated, chloroplast transgenic plants were then subjected to continuous illumination as opposed to light/dark cycles. The maximum HSA levels were observed when transgenic plants were exposed to 50 hours of continuous illumination; expression levels reached 11.2% tp in mature green leaves. The phenotypes of the chloroplast transgenic plants expressing HSA were identical to wild-type plants. High levels of HSA expression resulted in the formation of inclusion bodies (Figure 12) that not only protected HSA from proteolytic degradation but also facilitated single-step purification by centrifugation. Because HSA has a chemical and structural function rather than an enzymatic activity, complex studies are necessary to fully demonstrate its functionality (Watanabe *et al.*, 2001).

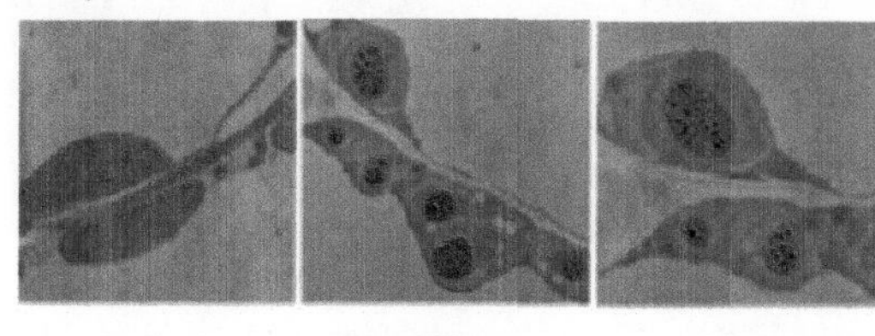

Figure 12. *HSA accumulation in transgenic chloroplasts. Electron micrographs of immunogold-labelled tissues from untransformed (A) or transformed (B, C) mature leaves using the chloroplast vector pLDApsbAHSA. Magnifications are (A) 10, 000, (B) 5000, and (C) · 6300.*

7.3. Antimicrobial Peptide

Many human pathogenic bacteria are resistant to known antibiotics or have acquired that trait over time. Magainin, and its analogues, have been examined for use as a systemic antibiotic and an anticancer agent. The magainin analog MSI-99, a synthetic lytic peptide, has been expressed via the tobacco chloroplast genome (DeGray *et al.*, 2001). This AMP contains an amphipathic alpha helix that possesses an affinity for negatively charged phospholipids present in the outer membrane of all bacteria. The probability that bacteria can adapt to the lytic activity of this synthetic peptide is very low. It was observed that this lytic peptide accumulated in transgenic chloroplasts at a high level of 21.5% tsp. *Pseudomonas aeruginosa*, a multidrug-resistant Gram-negative bacterium that acts as an opportunistic pathogen in plants, animals, and humans, was used for *in vitro* assays to determine the effectiveness of the lytic peptide expressed in tobacco chloroplasts (Figure 13). Cell extracts prepared from T_1 generation plants resulted in 96% inhibition of growth. The results observed are highly encouraging, especially for cystic fibrosis patients who are highly susceptible to *P. aeruginosa*. Previous studies have shown that analogs of magainin 2 were highly lethal to hematopoietic, melanoma, sarcoma, and ovarian teratoma cell lines, perhaps due to the presence of phosphatidylserine on the outer leaflets of the cells. The preference of this lytic peptide for negatively charged phospholipids makes MSI-99 a possible candidate for an anticancer agent.

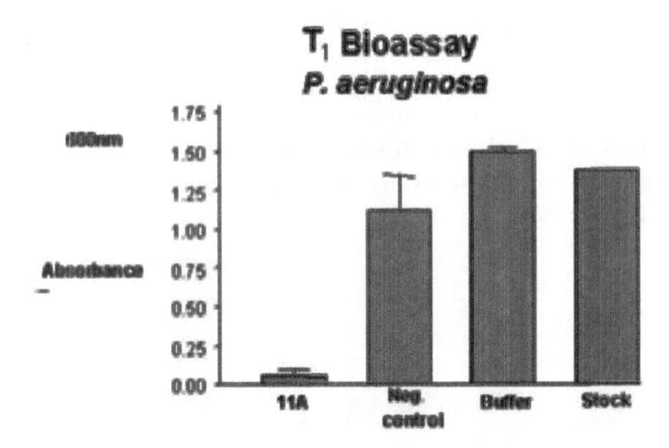

Figure 13. *In vitro bioassay for T1 generation (11A) against P. aeruginosa. Bacterial cells from an overnight culture were diluted to A_{600} 0.1–0.3 and incubated for 2 hours at 25°C with 100 µg of total protein extract. One ml of LB broth was added to each sample and incubated overnight at 26°C. Absorbance at 600 nm was recorded. Data were analyzed using GraphPad Prism.*

7.4. Human Interferon Alpha

Known to interfere with viral replication and cell proliferation, interferon alphas (IFN□s) also act as potent enhancers of the immune response and have many uses in clinical treatments. A specific subtype, IFN□2b, was first approved in 1986 by the Food and Drug Administration to treat patients with hairy cell leukemia; it has also shown efficacy in treatments for viral diseases and other malignancies. Because of the processing and purification involved, the average cost for treatment with *E. coli*-produced IFN□2b is $26,000 per year (Daniell et al., 2004b). IFN□2b is administered by injection and side effects may occur, such as when IFN□2b aggregates with human serum albumin in blood, which lessens the effectiveness of the treatment. Therefore, IFN□2b expressed via the chloroplast genome for oral delivery may eliminate some of the side effects. A recombinant IFN□2b containing a polyhistidine tag for single-step purification and a thrombin cleavage site was generated and introduced into the tobacco chloroplast genome of Petit Havana cultivar and into a low nicotine variety of tobacco, LAMD-609 (Daniell et al., 2004a; Falconer 2002). Along with correct processing, folding and disulfide bond formation, bioencapsulation by plant cells can protect recombinant proteins from degradation in the gastrointestinal tract, and plant-derived expression systems are free of human pathogens. Quantification of IFN□2b recombinant protein by ELISA showed 18.8% tsp in Petit Havana and up to 12.5% tsp in LAMD-609 T_0 transgenic plants. The next generation (T_1) of the Petit Havana chloroplast transgenic lines had accumulated even higher levels, clearly observable in a Commassie-stained agarose gel (Figure 14). In contrast, interferon gamma was barely detected (0.1% tsp) in transgenic chloroplasts unless it was fused with □-glucuronidase (GUS) (see below; Leelavathy and Reddy, 2003). IFN□2b functionality was demonstrated by the ability of IFN□2b to protect HeLa cells against the cytopathic effect of the encephalomyocarditis (EMC) virus. The mRNA levels of two genes induced by IFN□2b, 2'-5' oligoadenylate synthase and STAT-2, were measured with RT-PCR using primers specific for each gene. Chloroplast-derived IFN□2b behaved identically to commercially available IFN□2b in both assays (Daniell *et al.*, 2004b).

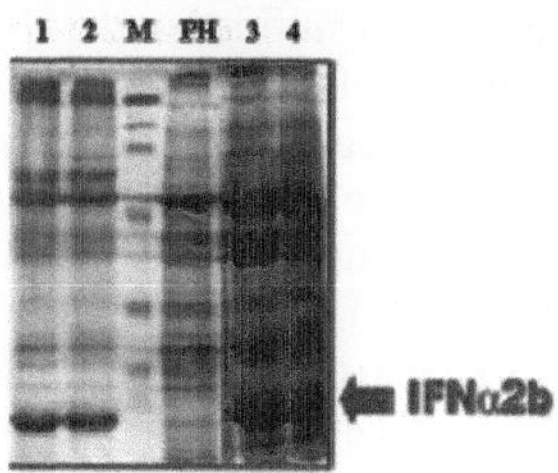

Figure 14*. Commassie-stained SDS-PAGE. Untransformed or transgenic lines expressing high levels of IFN α2b. Lanes 1 and 2: tsp (total soluble protein); Lane M: protein marker, Lane PH: untransformed Petit Havana; Lanes 3 and 4: tp (total protein).*

7.5. Human Interferon Gamma

Human interferon gamma (INF-□) is a major cytokine of the immune system that prevents viral proliferation and plays several roles in immunoregulatory actions regarding pathogenic bacteria. Researchers generated tobacco nuclear or chloroplast transgenic lines that expressed INF-□, either independently or as a GUS-fusion protein (Leelavathi and Reddy, 2003). Nuclear transgenic lines expressed INF-□ at the most 0.001% tsp. The estimated levels of GUS and INF-□ expressed in separate chloroplast transgenic plants were 3% and 0.1% (100-fold greater than nuclear transgenic lines), respectively. Pulse labeling experiments demonstrated that GUS had a ~48-hour half-life whereas INF-□ had a relatively short half-life of ~4-6 hours. A GUS/INF-□ fusion protein was then expressed in tobacco chloroplast transgenic lines. Here, the estimated level of the fusion protein was ~6% tsp. Pulse labeling experiments demonstrated that the GUS/INF-□ had a half-life similar to GUS alone, ~48 hours. The fusion protein was purified to homogeneity by a two-step His-tag based chromatography purification scheme. Upon cleavage of the fusion protein with factor Xa, GUS recovery was ~210 ∝g/g fresh weight tissue and INF-□ recovery was ~40 ∝g/g fresh weight tissue. As observed by silver stain, the INF-□ was highly pure, and the protein cross-reacted with anti-INF-□ antibody. Purified INF-□ was shown to offer complete protection to a human lung carcinoma cell line against infection with the EMC virus. The INF-□ produced in this study behaved identically to native human INF-□.

7.6. Insulin-like Growth Factor 1

Acting as a potent multifunctional anabolic hormone, human insulin-like growth factor 1 (IGF-1) is a 7.6 kDa protein composed of 70 amino acids with three disulfide bonds. IGF-1 regulates cell proliferation and differentiation and plays an important role in tissue renewal and repair. A cirrhotic patient requires 600 mg of IGF-1 per year at the cost of $30,000 per mg (Nilsson *et al.*, 1991). IGF-1 currently produced in *E. coli* lacks the requisite disulfide bonds. Using either the native human gene or one containing chloroplast-preferred codons to increase expression levels, the IGF-1 gene was expressed in the tobacco chloroplast genome. Quantification by ELISA of IGF-1 in tobacco chloroplast transformants showed that cells had accumulated up to 32% IGF-1 tsp, using either the human native gene or the synthetic gene (Figure 15; Ruiz, 2002). Unlike bacteria that can translate only the codon-optimized gene, the chloroplast translation machinery can translate both forms of the gene at similar levels (Daniell, 2004). However, observed expression levels should be confirmed with additional antibodies because of the interaction of IGF-1 antibody with a zz tag that was fused to the IGF-1 coding sequence.

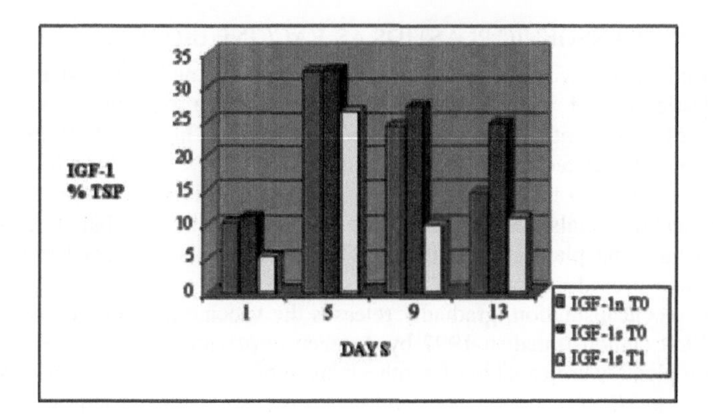

***Figure 15**. Expression of IGF-1 in transgenic chloroplasts under continuous light exposure (13 days). Plant tissue was collected at different times and IGF-1 expression is shown as a percentage of total soluble protein. IGF-1n is the native gene and IGF-1s is the chloroplast codon optimized gene. T_0 is the first generation and T_1 is the second generation.*

7.7. Guy's 13 Monoclonal Antibody

Monoclonal antibodies possess remarkable specificity for defined targets; therefore, they are emerging as therapeutic drugs at a fast rate. Guy's 13 monoclonal antibody targets the surface antigen of *Streptococcus mutans*, a bacterium that causes dental caries. In order to enhance translation, the Guy's 13 gene was codon optimized and placed under the control of a chloroplast 5' UTR. IgA-G, a humanized, chimeric form of Guy's 13 has been successfully synthesized in transgenic tobacco chloroplasts with the proper disulfide bonds (Daniell and Wycoff, 2001). Western blot analysis revealed the expression of heavy and light chains individually on a reducing gel, as well as the fully assembled antibody on a non-reducing gel (Figure 16). However, the levels of expression must be further enhanced to meet commercial feasibility. These results reiterate that within tobacco chloroplasts, chaperones allow for proper protein folding and enzymes catalyze the formation of disulfide bonds (Daniell et al, 2004b).

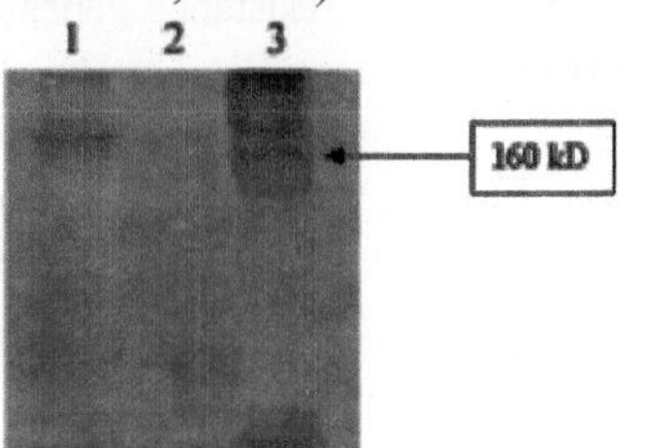

***Figure 16**. Western blot analysis of transgenic lines showing the assembled antibody in transgenic chloroplasts. Lane 1: extract from a transgenic line; Lane 2: negative control-extract from an untransformed plant; Lane 3: positive control-human IgA. The gel was run under non-reducing conditions. The blot was developed with AP-conjugated goat anti-human kappa antibody.*

8. TRANSGENIC PLASTIDS AS VACCINE BIOREACTORS

As opposed to injected subunit vaccines, oral delivery and low-cost purification make plastid-derived subunit vaccine production quite plausible. Developing countries often cannot afford vaccines because of the high cost of production, storage and refrigerated transportation. Subunit vaccines expressed in plants are capable of inducing a mucosal immune response when given parenterally or orally in experimental animals; these animals also withstand a pathogen challenge. Another advantage in using plant cells is that they have cell walls composed primarily of cellulose that provides protection for the vaccine in the stomach. This protection, known as bioencapsulation, gradually releases the vaccine into the gut (Mor *et al.*, 1998). First demonstrated in 1997 by Arntzen *et al.*, nuclear-transformed tomato and potato plants were capable of synthesizing subunit vaccines for Norwalk virus, enterotoxigenic *E. coli, V. cholerae*, and hepatitis B surface virus (for a review, see Langridge, 2000). The hepatitis B surface antigen and Norwalk virus capsid protein expressed via the nuclear genome in potato has been shown to elicit a serum immunoglobulin response when fed to mice and humans (Mason *et al.*, 1996; Kapusta *et al.*, 1999; Richter *et al.*, 2000; Tacket *et al.*, 2000; Kong *et al.*, 2001). The expression levels seen in nuclear transformants, however, are very low and an increased dosage is required for an effective immune response.

Plastids offer a logical alternative for the production of vaccines. Vaccine antigens that require oligomeric proteins with stable disulfide bridges (Daniell *et al.*, 2001a), large fusion proteins (for plague; Singleton, 2003), and viral epitopes fused with transmucosal carriers (for canine parvovirus; Molina *et al.*, 2004) have been expressed in the chloroplast. Both viral canine parvovirus 2L21 peptide (Molina *et al.*, 2004) and bacterial fragment C (TetC) of tetanus toxin (Tregoning *et al.*, 2003) vaccine antigens have been expressed at very high levels (up to 31.1% and 25% tsp, respectively) via chloroplast genetic engineering. Mice immunized intranasally with chloroplast-derived TetC were both immunogenic and immunoprotective against pathogen challenge, and the chloroplast-derived 2L21 peptide was shown to be immunogenic.

In order for the oral delivery of vaccines to be successful, two requirements must be fulfilled. First, large quantities of vaccine antigens should be expressed in edible parts of plants. As previously described, it is possible to transform chromoplasts of carrot, with high levels of transgene expression (Kumar et al., 2004a). Second, plastids must express the vaccines without the use of antibiotic resistance genes or selectable markers. Several examples are detailed below.

8.1. Antibiotic-Free Selection Using badh

Two major concerns of the public are that genetically engineered plants expressing antibiotic resistance genes may inactivate oral doses of that antibiotic (Daniell *et al.*, 2001c), and the transfer of antibiotic resistance genes to pathogenic microbes in the gastrointestinal tract or in soil may create antibiotic resistant pathogens (Daniell *et al.*, 2001b). The betaine aldehyde dehydrogenase (*badh*) gene from spinach encodes an enzyme that converts betaine aldehyde (BA), which is toxic to many plants, to nontoxic glycine betaine, which also serves as an osmoprotectant through drought or salt tolerance (Rathinasabapathy *et al.*, 1994). Chloroplast transformation efficiency was 25-fold higher using BA as a selective agent rather than spectinomycin; also, rapid regeneration of transgenic shoots was observed (Daniell *et al.*, 2001b). Because the *badh* gene is naturally present in plant species that are routinely consumed (*e.g.* spinach), the public concern regarding antibiotic resistance genes should be eased. In addition, BADH confers salt tolerance (Kumar *et al.*, 2004 a).

8.2. Selectable Marker Excision

Strategies have been developed to eliminate antibiotic resistance genes after transformation. Using endogenous chloroplast recombinases, marker genes are deleted via engineered direct repeats that flank them (Iamtham and Day, 2000). Early experiments with *C. reinhardtii* showed that it is possible to exploit these recombination events in this way. Homologous recombination between two direct repeats enabled marker removal under nonselective growth conditions (Fischer *et al.*, 1996). A similar approach applied to tobacco was effective in generating marker-free T_0 transplastomic lines while leaving a third unflanked transgene, *bar,* in the genome to confer herbicide resistance (Iamtham and Day, 2000). Using the P1 bacteriophage Cre/lox site-specific recombination system, a marker gene flanked by lox sites was removed after expression of the CRE protein was induced via the nuclear genome (Hajdukiewicz *et al.*, 2001; Corneille *et al.*, 2001). Alternatively, a transient co-integrative vector may be used to avoid the integration of selectable marker genes (Klaus *et al.*, 2004). Thus, efficient removal of selectable markers from chloroplast genomes is feasible.

8.3. CTB Vaccine

The CTB subunit encoded in the genome of *V. cholerae* forms a pentamer that binds to the GM_1-ganglioside receptor predominantly abundant in the intestinal epithelium. Acting as a very powerful adjuvant, the CTB subunit has potential as a candidate for a subunit vaccine. Expression of the native CTB gene lacking the leader sequence in the chloroplasts of transgenic tobacco plants resulted in CTB antigen accumulation up to 4.1% tsp (Daniell *et al.*, 2001a). Expression also resulted in the pentameric protein assuming the correct quaternary structure necessary for full activity. Binding aggregates of assembled pentamers in plant extracts were similar to purified bacterial antigen, and binding assays confirmed that chloroplast-synthesized and bacterial CTB both bound to the intestinal membrane GM_1-ganglioside receptor (Figure 17). Subsequent studies using translational enhancer elements resulted in hyperexpression of CTB (up to 31% tsp) with proper assembly of pentamers (Molina *et al.*, 2004). The tobacco chloroplast can thus correctly fold vaccine antigens and form native higher order structures.

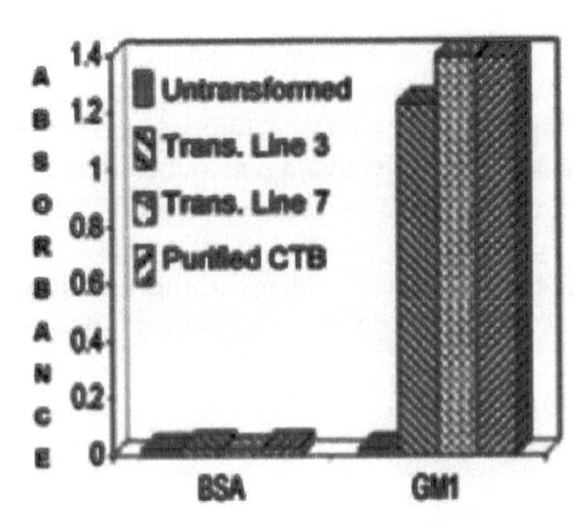

Figure 17. *CTB-GM_1-ganglioside binding ELISA assay. Plates, coated first with GM_1-gangliside and BSA, were plated with total soluble plant protein from chloroplast transgenic lines (3 and 7) and untransformed plant total soluble protein and 300 ng of purified bacterial CTB. The absorbance of the GM_1-ganglioside–CTB-antibody complex in each case was measured at 405 nm.*

8.4. Anthrax Vaccine

Bacillus anthracis, the causative agent of anthrax, is a spore-forming, aerobic bacterium. Spores produced by the organism can be spread quite readily using aerial bombs or even mail. Anthrax is almost always fatal unless treated immediately. In humans, anthrax protective antigen (PA) binds to a host cell receptor and forms a heptamer that can accommodate either the edema factor (EF) or the lethal factor (LF). When EF binds to PA, water homeostasis is disrupted, resulting in edema, or the accumulation of fluid. When LF binds to PA, macrophages release interleukin 1□, tumor necrosis factor □ and other cytokines, leading to shock and sudden death. The only vaccine currently licensed for human use in the U.S. is the Anthrax Vaccine Absorbed produced by BioPort Corporation. However, this vaccine is manufactured from a cell-free filtrate of a toxigenic, nonencapsulated strain of *B. anthracis* (Baillie, 2001). The immunogenic portion, PA, is bound to an adjuvant, but trace amounts of EF and LF are present in the filtrate.

In order to create a safer vaccine for anthrax, the PA gene was expressed in chloroplasts of tobacco plants (Chebolu and Daniell 2004). Transgenic plants contained up to 2.5 mg PA/g fresh weight. Macrophage lysis assays showed that chloroplast-derived PA could efficiently bind to anthrax toxin receptors and form heptamers, properly cleave, and bind LF, resulting in macrophage lysis (Figure 18). With observed expression levels, 7.2 billion doses of vaccine free of EF and LF could be produced per acre of transgenic tobacco using a commercial cultivar in the field.

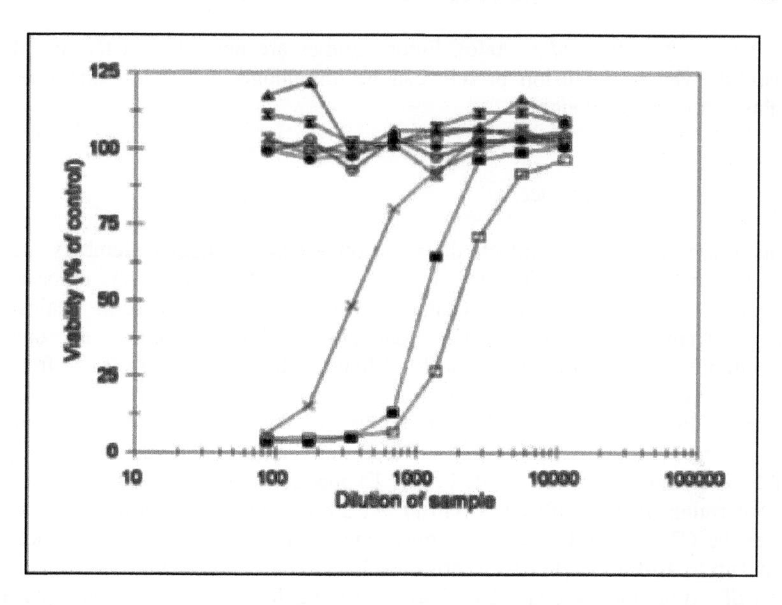

Figure 18. *Macrophage cytotoxic assays for extracts from untransformed and chloroplast transgenic plants. Supernatant samples from T_1 chloroplast transgenic lines expressing PA using the chloroplast vector pLD-JW1 (proteins extracted in buffer containing no detergent; MTT added after 5 h): (■) pLD-JW1 (extract stored 2 days); (□) pLD-JW1 (extract stored 7 days); (×) PA 5 ∝g/ml; (•) wild type (extract stored 2 days); (○) wild type (extract stored 7 days); (▲) wild type, no LF (extract stored 2 days); (□) wild type, no LF (extract stored 7 days);(⧗) pLD-JW1, no LF (extract stored 2 days); (⧖ pLD-JW1, no LF (extract stored 7 days.*

8.5. Plague Vaccine

The Gram-negative bacterium *Yersinia pestis* is the causative agent of three different forms of plague: bubonic, the most common, in which swollen lymph nodes form (buboes); septicemic, characterized by bacteremia without buboes; and pneumonic, in which the pathogen colonizes the alveolar spaces within the lung. The current vaccine for the plague is a killed whole-cell vaccine that is only moderately effective against the bubonic form and ineffective against the pneumonic and septicemic forms. Two *Y. pestis* subunits, the F1 capsular protein and the V antigen elicit a protective immune response individually, but the additive effect of both was more protective in immunized mice (Heath *et al.*, 1998). Against both subcutaneous and aerosolized challenges, an F1/V fusion protein potentially acts as a very effective subunit vaccine against pneumonic as well as bubonic plague. The F1/V fusion protein has been expressed in transgenic chloroplasts of tobacco plants; the maximum level of expression attained was 14.8% tsp (Singleton, 2003; Daniell *et al.*, 2004b; Daniell *et al.*, 2004c). Future studies are needed to determine if the plastid-derived F1/V fusion protein can be immunogenic and immunoprotective against pathogen challenge.

8.6. Canine Parvovirus Vaccine

Canine parvovirus (CPV) infects dogs and other Canidae family members such as wolves and coyotes. Infection leads to hemorrhagic gastroenteritis and myocarditis. Currently, young animals are vaccinated with an attenuated vaccine against the virus. Recent identification of CPV antigens capable of eliciting a protective immune response resulted in a gene for a linear antigenic peptide (2L21) from the VP2 capsid protein of CPV being used to transform the chloroplasts of tobacco plants (Molina *et al.*, 2004). The 2L21 peptide, coupled to a KLH carrier protein, has been demonstrated to effectively protect dogs and minks against parvovirus infection (Langeveld *et al.*, 1994, 1995). Expressed in nuclear transgenic plants as an N-terminal fusion with GUS (Gil *et al.*, 2001), the 2L21 peptide was fused to either the CTB subunit or GFP in transgenic chloroplasts and has accumulated at levels up to 10-fold greater than those levels in nuclear transgenics. The maximum level of expression of CTB-2L21 was 7.49 mg/g fresh weight, which equates to 31.1% tsp; the GFP-2L21 transgenic plants produced 5.96 mg/g fresh weight, or 22.6% tsp. The chloroplast-expressed 2L21 epitope was detected with a CPV-neutralizing monoclonal antibody, indicating that the epitope is correctly positioned at the C-terminus of the fusion protein. The tobacco chloroplast-derived CTB-2L21 fusion protein retained the ability to form pentamers and bind to GM_1-ganglioside receptors, just as native CTB does.

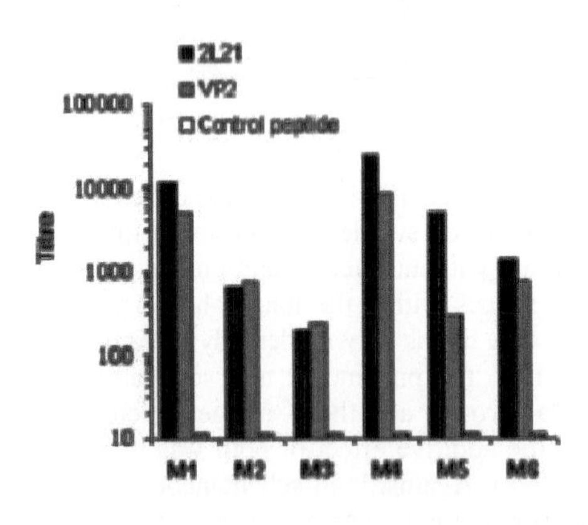

***Figure 19**. Titres of antibodies at day 50 induced by chloroplast-derived CTB-2L21 recombinant protein. Balb/c mice were intraperitoneally immunized with leaf extract from CTB-2L21 transgenic plants. Animals were boosted at days 21 and 35. Each mouse received 20 ∝g of CTB-2L21 recombinant protein. Individual mice sera were titrated against 2L21 synthetic peptide, VP2 protein and control peptide (amino acids 122-135 of hepatitis B virus surface antigen). Titres were expressed as the highest serum dilution to yield two times the absorbance mean of preimmune sera. M1-M6: mouse 1 to 6. 2L21: epitope from the VP2 protein of the canine parvovirus; CTB: cholera toxin B; VP2: protein of the canine parvovirus that includes the 2L21 epitope.*

Intraperitoneally immunized mice produced antibodies that recognized the VP2 protein from CPV (Figure 15). These data showed that chloroplast-derived CTB-2L21 fusion protein was immunogenic when administered intraperitoneally, confirmed by its ability to induce a humoral response that can cross react with native VP2 protein. Additional experiments are underway to determine if mucosal delivery of the fusion protein can produce immunogenicity as well.

9. BIOMATERIALS, ENZYMES AND AMINO ACIDS

9.1. Chorismate Pyruvate Lyase

Encoded by the *ubi*C gene in *E. coli,* chorismate pyruvate lyase (CPL) catalyzes the direct conversion of chorismate to pyruvate and p-hydroxybenzoic acid (pHBA). pHBA is the principle monomer found in all commercial thermotropic liquid crystal polymers. Chorismate, found in chloroplasts, is converted to pHBA in up to ten successive enzymatic reactions steps due to a lack of CPL. High levels of pHBA accumulated in tobacco plants transformed via the chloroplast genome to express the CPL enzyme. Despite the high accumulation of pHBA, transgenic plants were morphologically indistinguishable from wild-type and did not show any pleiotropic effects (Vitanen *et al.*, 2004). When grown under normal light/dark cycles, transgenic plants accumulated ~15 % dry weight pHBA conjugated to glucose in the vacuole after 100 days. When switched to continuous illumination, plants accumulated ~25% dry weight pHBA after 22 more days, demonstrating that the 5'UTR increases translation of the *ubi*C gene under continuous-light.

CPL activity, measured in picokatals (pkat), peaked at 50,783 pkat/mg of protein in the total leaf material of chloroplast transgenic T_1 lines; this equates to ~30% tsp. In contrast, nuclear transgenic plant CPL activity was only 208 pkat/mg protein. Subsequent pHBA accumulation in leaves was 50 times higher in transgenic chloroplasts compared to nuclear expression. The large difference between CPL enzyme activity and pHBA accumulation indicates that chorismate concentration may be a limiting factor; however, observed levels are well within commercial feasibility.

9.2. Polyhydroxy Butyrate Polymer

The enzymatic pathway for the synthesis of polyhydroxybutyrate (PHB), a polyester used in biodegradable plastics and elastomers, was expressed in the chloroplasts of transgenic tobacco plants (Lossl *et al.*, 2003). Three bacterial enzymes are encoded in the polycistronic *phb* operon that expresses PHB. First, condensation of two molecules of acetyl-CoA is catalyzed by □-ketothiolase to form acetoacetyl-CoA; acetyl-CoA reductase then reduces acetoacetyl-CoA to □-hydroxybutyryl-CoA, which is then polymerized by PHB synthase to form PHB. Chloroplast transgenic lines expressing this operon accumulated up to 1.7% dry weight PHB; however, even in mature plants expressing very low levels of PHB, male sterility and stunted growth were observed.

9.3. Xylanase

Industrial applications for xylanases are found in the paper, fiber, baking, brewing and animal feed industries (Biely, 1985). However, high production costs limit their routine use. Nuclear transgenic plants expressing xylanases encountered the degradation of the cell wall, limiting their growth . Xylanases were then targeted to the apoplast (Herbers *et al.*, 1995), seed oil body (Liu *et al.*, 1997) or were secreted into culture media through the roots (Borisjuk *et al.*, 1999). In nuclear transgenic approaches, expression levels were too low to make production economically feasible. However, the gene for alkali-resistant and thermostable xylanase (*xyn*A) from *Bacillus* sp. strain NG-6 has been expressed in the chloroplast of tobacco plants (Leelavathi *et al.*, 2003). Xylanase accumulated in the cells to ~6% tsp, and a zymography (fermentation) assay demonstrated that the estimated activity was 140,755 U/kg of fresh leaf tissue. Because the chloroplast-derived xylanase was thermostable, the enzyme was purified using heat in the first step in order to reduce proteolytic degradation by plant proteases. Enzyme activity correlated with enzyme quantities seen on an SDS-PAGE gel. Of the generations of transgenic lines tested, none appeared to be any different from untransformed controls in height, chlorophyll content, flowering time and biomass. Greater than 85% of the enzyme activity seen in fresh leaves was observed in leaves dried in the sun or at 42°C as well as leaves undergoing senescence. Characterization of the chloroplast-derived xylanase showed that the enzyme was as biologically as active as the bacterially derived enzyme in the pH range of 6 to11, with peak activity at pH 8.4 (Leelavathi *et al.*, 2003). The chloroplast-derived xylanase also retained substrate specificity. Chloroplasts can thus express industrially important cellulolytic enzymes that cannot be expressed at high levels via the nuclear genome because of adverse effects.

9.4. Anthranilate Synthase

Many enzymes involved in amino acid biosynthesis are encoded in the nucleus, synthesized in the plant cytosol, and then transported to plastids. Tryptophan (Trp) biosynthesis branches off the shikimate pathway at chorismate, which is the last common precursor of many aromatic compounds (Radwanski and Last, 1995). Anthranilate synthase (AS) catalyzes the first committed reaction for Trp biosynthesis, converting chorismate to anthranilate (Brotherton *et al.*, 1986; Haslam, 1993). Plant AS holoenzyme is comprised of a tetramer of two □ and two □ subunits. The □ subunit acts as an aminotransferase that cleaves glutamine to make ammonia, which the □ subunit utilizes to convert chorismate to anthranilate. The □ subunit also binds to Trp and acts as a feedback inhibitor (Bohlman *et al.*, 1995). The gene for AS □ subunit (*ASA2*) has been expressed within the chloroplast of transgenic tobacco plants (Zhang *et al.*, 2001a). Transgenic plants showed a high level of accumulation of *ASA2* mRNA, an increased expression of the AS □-subunit protein, and a 4-fold increase in AS enzyme activity that was less sensitive to feedback inhibition by Trp. A 10-fold increase in free Trp in leaves was also observed. Immunological data revealed a much higher level of the □ subunit compared to wild type, indicating that the abundance of the □ subunit encoded by *ASA2* may stabilize the □ subunit, resulting in an increase in holoenzyme levels.

10. RUBISCO ENGINEERING VIA THE PLASTID GENOME

Ribulose-1,5-bisphosphate carboxylase/oxygenase (Rubisco) is a key enzyme that converts atmospheric carbon to food and supports life on this planet. Its low catalytic activity and specificity for oxygen leads to photorespiration, severely limiting photosynthesis and crop productivity. Even a marginal increase in the efficiency of photosynthesis would result in a dramatic improvement in agricultural production. Rubisco is the most abundant protein on the planet, and it accounts for 30% to 50% tsp in the chloroplast (Ogren, 2003; Spreitzer & Saloucci, 2002). Large quantities of this enzyme are required because it works very slowly, catalyzing the reaction of only a few molecules per second. When Rubisco uses oxygen as a substrate instead of CO_2, ribulose bisphosphate is broken down and carbon dioxide is released. This is called photorespiration, a wasteful process that is of no use to the plant. Therefore, Rubisco has been a primary target for genetic engineering to improve its catalytic activity and substrate specificity. In higher plants, Rubisco consists of eight chloroplast encoded large subunits (LSU) and eight nuclear encoded small subunits (SSU). The SSU is imported from the cytosol and both subunits undergo several post-translational modifications before assembly into a functional holoenzyme (Houtz & Portis, 2003). Therefore, genetic manipulation of Rubisco to improve its function has involved both nuclear and chloroplast genetic engineering.

When the plastid *rbc*L gene encoding the large subunit was deleted and expressed via the nuclear genome, transgenic plants exhibited a severe Rubisco deficiency (Kanevski and Maliga, 1999). When the *rbc*L gene from *Chromatium vincosum* was expressed via the nuclear genome of Rubisco-deficient plants, it was poorly transcribed and no foreign large subunit was detected (Madgwick *et al.*, 2002). In yet another approach, a mutated *rbc*L gene, engineered into the chloroplast genome of tobacco, resulted in a Rubisco with a specificity factor and carboxylation rate of only 25% of the wild type (Whitney *et al.*, 1999). Also, when the tobacco *rbc*L gene was replaced with a heterologous gene from cyanobacteria or *Helianthus annus*, Rubisco-deficient plants were obtained (Kanevski *et al.*, 1999). Engineering the nuclear genome of *Arabidopsis* with a cDNA encoding a pea small subunit also resulted in a hybrid Rubisco that was compromised in catalysis (Getzoff *et al.*, 1998). Engineering the chloroplast genome with the *rbc*L-*rbc*S operon from a non-green alga resulted in a lack of Rubisco assembly (Whitney *et al.*, 2001). Engineering the L2 form of Rubisco from *Rhodospirillum rubrum* resulted in transgenic plants that were unable to survive at ambient CO_2 levels (Whitney and Andrews, 2001).

Table 3. *Previous attempts at Rubisco manipulation in tobacco (Nicotiana tabacum) via chloroplast genetic engineering*

Genetic change	**Plant phenotype/catalytic properties of Rubisco**	**References**
*Rbc*S insertion in *Rbc*S antisense plants	*Rbc*S protein accumulation and photosynthesis almost completely restored; complementation of the *Rbc*S antisense phenotype	Dhingra *et al.*, 2004
*Rbc*S insertion in *Rbc*S antisense plants	*Rbc*S protein accumulation not significantly increased; no complementation of the *Rbc*S antisense phenotype	Zhang *et al.*, 2002
insertion of *Rbc*S – his	low *Rbc*S protein accumulation from the plastid transgene (1% of total *Rbc*S)	Whitney & Andrews, 2001
Engineering of heterologous Rubisco subunits in tobacco via chloroplast genetic engineering		
replacement of *rbc*L with *rbc*M from *Rhodospirillum rubrum*	plant growth only in CO_2-enriched atmosphere; fully active form II Rubisco with catalytic properties identical to the *Rhodospirillum* enzyme	Whitney & Andrews, 2001, 2003
insertion of the *rbc*LS operon from non-green algae	high-level *rbc*L and *rbc*S protein accumulation, but no assembly of active Rubisco	Whitney *et al.*, 2001
*rbc*L replacement with *Helianthus rbc*L	accumulation of hybrid enzyme (tobacco SSU, sunflower LSU); CO2:O2 specificity similar to tobacco Rubisco; catalytic rate lowered to 25%	Kanevski *et al.*, 1999
*rbc*L replacement with Synechococcus *rbc*L	lack of photoautotrophic growth because of absence of *Rbc*L protein accumulation	Kanevski *et al.*, 1999

Yet another approach for Rubisco manipulation was to engineer the *Rbc*S gene into the chloroplast genome. Whitney and Andrews (2001) expressed two copies of a His-tagged *Rbc*S gene in transgenic chloroplasts and obtained less than ~ 1% accumulation of the small subunit. When a nuclear *Rbc*S antisense tobacco mutant (Rodermel *et al.*, 1988; Jiank and Rodermel, 1995; Tsai *et al.*, 1997) was used for engineering the *RbcS* gene via the chloroplast genome, it resulted in decreased photosynthetic activity and retarded growth (Zhang *et al.*, 2002). This mutant line is an ideal target for Rubisco engineering because any successful Rubisco assembly would be easily discernible. The *RbcS* gene was expressed under the regulation of a chloroplast RBS (GGAGG); the chloroplast transgenic plants accumulated abundant plastid *RbcS* transcript but it was not translated (Zhang *et al.*, 2002). The failure of several attempts to introduce into the chloroplast genome a *RbcS* gene that can be properly expressed to produce a fully functional Rubisco casts doubt on the usefulness of this approach for Rubisco engineering (see Table 3).

It is quite evident that a completely functional eukaryotic Rubisco has never been assembled in any foreign host. Therefore, we engineered *RbcS* cDNA into a transcriptionally active spacer region of the chloroplast genome of a nuclear *RbcS* antisense tobacco plant and demonstrated, for the first time, successful expression of this gene, proper assembly and restoration of photosynthesis (Dhingra et al., 2004). *RbcS* cDNA, when expressed under the regulation of heterologous (gene 10) or native (*psbA*) UTRs, resulted in the assembly of a functional holoenzyme and normal plant growth under ambient CO_2 conditions, fully short-circuiting nuclear control of gene regulation. There was about 150-fold more *RbcS* transcript in chloroplast transgenic lines when compared to the nuclear *RbcS* antisense line, while the wild type has 7-fold more transcript. The small subunit protein levels in the gene10/*RbcS* and *psbA*/*RbcS* plants were 60% and 106% respectively of the wild type. Photosynthesis of gene10/*RbcS* plants was about double that of the antisense plants while that of *psbA*/*RbcS* plants was almost completely restored to the wild-type rates. These results have opened an avenue for using chloroplast engineering for the evaluation of foreign Rubisco genes *in planta,* that can eventually result in achieving efficient photosynthesis and increased crop productivity.

11. MOLECULAR BIOLOGY AND BIOTECHNOLOGY

In addition to biotechnology applications, plastid transformation has become a powerful tool for the study of plastid biogenesis and function. This approach has been used to investigate plastid DNA replication origins (Chapter 3), RNA editing elements (Chapter 9), promoter elements (Chapter 8), RNA stability determinants (Chapter 10), intron maturases (Chapter 11), translation elements (Chapters 12 and 13), protein import machinery (chapter 14), proteolysis (Chapter 15), transgene movement and evolution (Chapters 4, 16) and transcription & translation of polycistrons (chapter 16).

11.1. Plastid genome sequence

As pointed out in chapter 5, among 30 chloroplast genomes sequenced so far, only five are crop chloroplast genomes. Chloroplast genome sequences are critical for identification of spacer regions (that are not often conserved) for integration of transgenes at desired sites via homologous recombination. Endogenous regulatory sequences are also essential to engineer transgenes because they are quite different among plant species. The concept of utilizing a "universal vector" employing plastid DNA flanking sequences from one plant species to transform another species (of unknown genome sequence) was proposed several years ago (Daniell et al.,1998). Using this concept, both potato and tomato plastid genomes have thus far been transformed using flanking sequences from tobacco (Ruf et al., 2001; Sidorov et al., 1999). However, since publication of these articles, no useful traits have been introduced via the plastid genome in potato or tomato. This is partly due to extremely low efficiency of plastid transformation. In potato, one chloroplast transgenic line per 35 bombarded plates was obtained, and in tomato, 1 transgenic line per 87 bombarded plates was obtained. In contrast, 15 tobacco chloroplast transgenic lines were obtained per bombarded plate (Fernandez-San Milan *et al.*, 2003). The flanking sequence between tobacco and potato or tomato was ~98% homologous. Similarly, when Petunia flanking sequences were used to transform tobacco plastid genome (which is highly efficient), the transformation efficiency was lower than that was observed with endogenous flanking sequences (DeGray et al., 2001). Furthermore, when Petunia chloroplast DNA was introduced into tobacco, with suitable mutations, the efficiency of homologous recombination was very low, in spite of 98% DNA sequence homology (Kavanagh et al., 1999). These examples highlight the requirement of species-specific vectors for efficient plastid transformation.

11.2. Transgene movement from plastid to nuclear genome

The chloroplast genome has undergone extreme reduction in size and gene content post endosymbiosis from a cynobacterial ancestor. Many of these losses occurred in parallel during diversification of photosynthetic eukaryotes, while a certain number of these losses were functionally transferred to the nucleus prior to land plant diversification 470 million years ago (chapter 5). However, after the evolution of land plants, there are only two documented cases of transfer of functional chloroplast genes to the nucleus. Intact gene transfer from the chloroplast to nuclear genome and expression requires acquiring regulatory elements for transcription, translation, and protein transport, the rate of which must be minimal (chapter 5). The incredibly high rate of chloroplast DNA transfer into the nucleus reported by Huang et al (2003), eighteen transposition events in a single generation from two transgenic lines, may be an overestimate caused by simultaneous integration of both chloroplast and nuclear transgene copies during bombardment.

An evolutionary perspective on the gene transfer rates adds further doubt to their conclusions based on the magnitude of selective cpDNA deletion required to accommodate such a transposition rate, if nuclear genome size exists at equilibrium. The 'conservative' estimates of Huang et al. (2003) for chloroplast to nuclear gene transfer rates were: 1/16000 offspring, multiple insertions with a mean of ~2 per, ~10kb cpDNA translocated (= 10/156 kb of the N. tabacum plastid genome), generation time = ~0.5. years. This requires selective removal of approximately 120 megabases of cpDNA (about the size of the Arabadopsis nuclear genome) over only 5 million years of evolution. Given such astronomical rates of cpDNA integration into the nucleus suggested, there is a conspicuous absence of any known, or heretofore suggested, mechanism capable of maintaining equilibrium in plant nuclear genomes, considering 470 million years of such an overwhelming influx of cpDNA. Without removal, this rate translates to the addition of ~40 bp/yr to the nuclear genome whereas evolution provides evidence for <0.045 bp/yr (~1000 fold higher rate, based on calculations from the Arabidopsis genome and Martin et al., 1998). Most importantly, from a biotech point of you, none of the transgenes that were functional within transgenic chloroplasts were functional in the nuclear genome. The challenges in transferred genes becoming fully functional are discussed in chapter 5 and it has been pointed out that it took 100 million years for rpl22 to be fully functional, after it was transferred to the nuclear genome. No study has yet determined how frequently plastid transgenes move to the nucleus, in a manner that supports their expression, and therefore, such transfer to the nucleus without expression is inconsequential in biotechnology applications (Daniell and Parkinson, 2003).

11.3. Role of transcription, translation, proteolysis, integration site and DNA replication in plastid genetic engineering

Newly synthesized proteins are highly susceptible to proteases and require protection from chloroplast proteases. One approach for such protection is the use of chaperones to fold foreign proteins into structures that would protect them from proteases. This was demonstrated by using the CRY chaperone (encoded by orf2 gene) to fold the insecticidal protein into cuboidal crystals; this increased CRY protein accumulation 128-fold (from 0.36% to 46.1% of total leaf protein, Decosa et al., 2001). Similarly, a 500-fold increase in translation of Human Serum Albumin was achieved when the *hsa* coding sequence was regulated by the chloroplast psbA 5' and 3' untranslated regions in the light; rapid translation of *hsa* resulted in aggregates of Human Serum albumin, which protected the newly synthesized protein from proteolytic degradation (Fernandez-San Millan et al., 2003). One acre of chloroplast transgenic plants can produce up to 400 million doses of clean and fully functional anthrax vaccine, when regulated by psbA UTRs in the light (Daniell et al., 2004c).

In previous studies, transcript abundance did not correlate with translation efficiency. For example, *tps1* gene transcripts were observed to be 169-fold higher than the best nuclear transgenic lines, even in the absence of a promoter for this transgene (Lee et al., 2003). Densitometry analysis of the chloroplast derived *Rbc*S transcripts confirmed 165-fold and 143-fold more transcript than the nuclear *Rbc*S antisense plants when the transgene translation was regulated by psbA 5' UTR/promoter or a promoterless gene 10 UTR; while the psbA UTR resulted in the first successful expression and assembly of a functional Rubisco in transgenic plants, gene 10 UTR transgenic lines performed poorly (Dhingra et al., 2004). Therefore, in all these studies, there was a lack of correlation between transcript abundance and translation efficiency suggesting that transcript abundance or stability is less of a concern than protein stability.

In aforementioned studies, the site of integration of transgene is the intergenic region between the *trn*I-*trn*A genes within the *rrn* operon present in the inverted repeat regions of the chloroplast genome. More transgenes have been successfully integrated and expressed at this site, in different crop species, than any other spacer region used so far in plastid transformation (Devine and Daniell, 2004; Kumar et al., 2004a,b). This site is transcriptionally active due to the read through transcription from the upstream 16S promoter (P*rrn*), which is responsible for transcribing six native chloroplast genes, in contrast to transcriptionally silent spacer regions that are located between divergent genes. Chloroplast DNA flanking region used in these vectors contain one of the two replication origins and promote replication of introduced chloroplast vectors and help achieve homoplasmy. The origin of replication has been mapped in tobacco chloroplast genome and it is located within the trnI gene that forms the left flank in the chloroplast transformation vector used for cotton chloroplast transformation (Kunnimalaiyaan *et al.* 1997; Lugo *et al.*, 2004). When chloroplast vectors with or without ori were bombarded into cultured tobacco cells, only the vector with ori located within the trnI gene showed prolonged and higher levels of CAT enzyme activity (Daniell *et al.*, 1990). When chloroplast vectors with or without ori were bombarded with the same transgenes, the vector with ori present within the trnI flanking region achieved homoplasmy even in the first round of selection (Guda et al., 2000). Therefore, the site of integration and flanking sequence used in chloroplast vectors could play a significant role in transgene integration and expression.

11.4. Heterologous multigene operons in transgenic chloroplasts

In higher plants, plastid genes are mainly organized as operons, which may group genes of related or unrelated functions (Sugita and Sugiura, 1996). Most of these genes are transcribed into polycistronic RNA precursors that may be later processed by mechanisms such as alternative splicing, editing or intercistronic processing, which make transcripts competent for translation (Barkan, 1988; Rochaix, 1992; Barkan & Goldschmidt-Clermont, 2000; Monde *et al.*, 2000). The general consensus is that most primary transcripts require processing in order to be functional (Barkan, 1988; Zerges, 2000; Meierhoff *et al.*, 2003), and that post-transcriptional RNA processing of primary transcripts represents an important control of chloroplast gene expression (chapters 10,11).

While the native operons exhibit different types of mechanisms for transcriptional and translational regulation, little is known about processing in the context of engineered foreign operons in the chloroplast. Ruiz *et al.* (2004) have recently analyzed the transcription and translation of chloroplast transgenic plants harboring following operons: *Bacillus thuringiensis* insecticidal protein (*cry*2Aa2) and associated chaperonin protein (*orf*2), trehalose phosphate synthase (*tps*1), human serum albumin (*hsa*), and cholera toxin β subunit (*ctx*B). Northern blot analyses confirmed that different transgenic lines harboring multigene operons generated polycistrons as the most abundant transcript form, which, in most lines, were not processed into monocistrons. Such unprocessed transcripts were translated very efficiently and resulted in the highest level of foreign gene expression reported in the transgenic literature (up to 47% of total leaf protein). Thus, multigene operons engineered into the chloroplast genome do not require processing of polycistrons to monocistrons for translation. Such high efficiency of translation of foreign transcripts may be the result of heterologous regulatory sequences that are not controlled by the plant cell. Furthermore, transcripts of bacterial operons, such as the native *Bacillus thuringiensis cry*2Aa2 operon could form stable secondary structures, which may play a critical role in their processing. When processed, transcripts were quite stable in the absence of 3' UTRs, which are necessary for stability of native transcripts. Addressing questions of the translation of polycistrons and sequences required for transcript processing and stability is essential for metabolic engineering. Knowledge of such factors would allow us to engineer pathways that will not be under the complex post-transcriptional regulatory machinery of the chloroplast.

12. CONCLUSIONS

Plastid transformation holds much potential in the introduction of important agronomic traits to plants, as well as the production of industrially valuable biomaterials and therapeutic proteins such as antibodies, biopharmaceuticals and vaccine antigens. High-level gene expression in transgenic chloroplasts can allow the generation of high quality, low-cost therapeutic proteins for people around the world. Purification methods such as chromatography are unnecessary, as oral delivery of therapeutic proteins should completely eliminate this need. In addition, plastid transformation has become a powerful tool for the study of plastid biogenesis and function. This approach has been used to investigate plastid DNA replication origins, RNA editing elements, promoters, RNA stability determinants, processing of polycistrons, intron maturases, translation elements and proteolysis, import of proteins and several other processes. Recent advancements augur well for the production of therapeutic proteins, vaccines, biomolecules in transgenic plastids, introduction of agronomic traits in crops using an environmentally friendly approach as well as increase our understanding of plastid biochemistry and molecular biology.

13. ACKNOWLEDGEMENTS

Results of investigations from the Daniell laboratory were supported in the past by USDA, NSF, DOE and DOD grants, and are currently supported by the United States Department of Agriculture (grant 3611-21000-017-00D) and the National Institutes of Health (grant R01 GM 63879). Henry Daniell is thankful to Todd Castoe (UCF) for his calculation on genome evolution discussed in section 11.2.

14. REFERENCES

Arntzen, C. J., Mason, H. S., Tariq, H. A., & Clements J. D. (1997). Oral immunization with transgenic plants. US patent WO96/12801.

Baillie, L. (2001). The development of new vaccines against *Bacillus anthracis*. *J. Applied Micro.*, *91*, 609-613.

Barkan, A. (1988). Proteins encoded by a complex chloroplast transcription unit are each translated from both monocistronic and polycistronic mRNAs. *EMBO J.* 7, 2637-2644.

Barkan, A., Goldschmidt-Clermont, M. (2000). Participation of nuclear genes in chloroplast gene expression. *Biochimie 82*, 559-572.

Biely, P. (1985). Microbial xylanolytic enzymes. *Trends Biotechnol.*, *3*, 286-290.

Birky, C. W. (1995). Uniparental inheritance of mitochondrial and chloroplast genes: mechanisms and evolution. *Proc. Natl. Acad. Sci. U.S.A.*, *92*, 11331-11338.

Bizily, S. P. Rugh, C. L. & Meagher, R. B. (2000). Phytodetoxification of hazardous organomercurials by genetically engineered plants. *Nat. Biotechnol.*, *18*, 213-217.

Blowers, A. D., Bogorad, L., Shark, K. B., & Sanford, J. C. (1989). Studies on *Chlamydomonas* chloroplast transformation: foreign DNA can be stably maintained in the chromosome. *Plant Cell, 1*, 123-132.

Bogorad, L. (2000). Engineering chloroplasts: an alternative site for foreign genes, proteins, reactions and products. *Trends in Biotechnology*, *18*, 257-263.

Bohlmann, J., De Luca, V., Eilert, U., & Martin, W. (1995). Purification and cDNA cloning of anthranilate synthase from *Ruta graveolens*: modes of expression and properties of native and recombinant enzymes. *Plant J.*, *7*, 491–501.

Borisjuk, N. V., Borisjuk, L. J., Logendra, S., Petersen, F., Gleba, Y., & Raskin, I. (1999). Production of recombinant proteins in plant root exudates. *Nat. Biotechnol.*, *17*, 466-469.

Boynton, J. E., Gillham, N. W., Harris, E. H., Hosler, J. P., Johnson, A. M., Jones A. M., *et al.* (1988). Chloroplast transformation in *Chlamydomonas* with high velocity microprojectiles. *Science, 240*, 1534-1538.

Brotherton, J. E., Hauptmann, R. M., & Widholm, J.M. (1986). Anthranilate synthase forms in plants and cultured cells of *Nicotiana tabacum* L. *Planta, 168*, 214–221.

Cerutti, H., Osman, M., Grandoni, P., & Jagendorf, A. T. (1992). A homolog of *Escherichia coli* RecA protein in plastids of higher plants. *Proc. Natl. Acad. Sci. U.S.A.*, *89*, 8068-72.

Chandra, A., & Pental, D. (2003). Regeneration and genetic transformation of grain legumes: An overview. *Current Science 84* (3), 381-387.

Chebolu, S & Daniell, H. (2004). Chloroplast derived vaccine antigens and biopharmaceuticals: expression, folding, assembly and functionality. *Current Trends in Microbiology and Immunology*, in press.

Corneille, S., Lutz, K., Svab, Z., & Maliga, P. (2001). Efficient elimination of selectable marker genes from the plastid genome by the CRE-lox site-specific recombination system. *Plant J.*, *27*, 171-178.

Cramer, C. L., Boothe, J. G., & Oishi, K. K. (1999). Transgenic plants for therapeutic proteins: linking upstream and downstream strategies. *Curr. Top. Microbiol. Immunol.*, *240*, 95-118.

Daniell, H. (1999). Environmentally friendly approaches to genetic engineering. *In Vitro Cell Dev. Biol. Plant, 35*, 361-368.

Daniell, H. (2000). Genetically modified crops: current concerns and solutions for next generation crops. *Biotechnol. and Gen. Engineer Rev.*, *17*, 327-352.

Daniell, H. (2002). Molecular strategies for gene containment in transgenic crops. *Nat. Biotechnol.*, *20*, 581-586.

Daniell, H. (2004a). Medical Molecular Pharming: Therapeutic recombinant antibodies, biopharmaceuticals, and edible vaccines in transgenic plants engineered via the chloroplast genome. in R. M. Goodman (Ed.), *Encyclopedia of Plant and Crop Science* (pp. 704-710). New York: Marcel Decker.

Daniell, H., Carmona-Sanchez, O., & Burns, B. B. (2004b). Chloroplast derived antibodies, biopharmaceuticals and edible vaccines. In R. Fischer & S. Schillberg (Eds.) *Molecular Farming* (pp.113-133). Weinheim: WILEY-VCH Verlag.

Daniell, H., Chebolu, S., Kumar, S., Singleton, M., & Falconer, R. (2004c). Chloroplast-derived vaccine antigens and other therapeutic proteins. *Vaccine*, in press.

Daniell, H., Datta, R., Varma, S., Gray, S., & Lee, S. B. (1998). Containment of herbicide resistance through genetic engineering of the chloroplast genome. *Nat. Biotechnol.*, *16*, 345-348.

Daniell, H., & Dhingra, A. (2002) Multigene engineering: dawn of an exciting new era in biotechnology. *Current Opinion in Biotechnol.*, *13*, 136-141.

Daniell, H., Khan, M., & Allison, L. (2002). Milestones in chloroplast genetic engineering: an environmentally friendly era in biotechnology. *Trends Plant Sci.*, *7*(2), 84-91.

Daniell, H., Krishnan, M., & McFadden, B. F. (1991). Expression of □-glucoronidase gene in different cellular compartments following biolistic delivery of foreign DNA into wheat leaves and calli. *Plant Cell Rep.*, *9*, 615-619.

Daniell, H., Lee, S. B., Panchal, T., & Wiebe, P. O. (2001a). Expression of cholera toxin B subunit gene and assembly as functional oligomers in transgenic tobacco chloroplasts. *J. Mol. Biol., 311*, 1001-1009.

Daniell, H., & McFadden, B. A. (1987). Uptake and expression of bacterial and cyanobacterial genes by isolated cucumber etioplasts. *Proc. Natl. Acad. Sci. U.S.A., 84*, 6349-6353.

Daniell, H., Muthukumar, B., & Lee, S. B. (2001b). Marker free transgenic plants: engineering the chloroplast genome without the use of antibiotic selection. *Curr. Genet., 39*, 109-116.

Daniell, H. & Parkinson, C. L. (2003). Jumping genes and containment. *Nature Biotechnol., 21*, 374-375.

Daniell, H., Ruiz, O. N., & Dhingra, A. (2004a) Chloroplast genetic engineering to improve agronomic traits. *Methods in Molecular Biology, 286*, 111-137.

Daniell, H., Vivekananda, J., Nielsen, B. L., Ye, G. N., Tewari, K. K., & Sanford, J.C. (1990). Transient foreign gene expression in chloroplasts of cultured tobacco cells after biolistic delivery of chloroplast vectors. *Proc. Natl. Acad. Sci. U. S. A., 87*, 88-92.

Daniell, H., Wiebe, P., & Fernandez-San Millan, A. (2001c). Antibiotic-free chloroplast genetic engineering – an environmentally friendly approach. *Trends in Plant Science, 6*, 237-239.

Daniell, H., & Wycoff, K. (2001). Production of antibodies in transgenic plastids. US Patent WO 01/64929.

DeCosa, B., Moar, W., Lee, S. B., Miller, M., & Daniell, H. (2001). Overexpression of the *Bt* cry2Aa2 operon in chloroplasts leads to formation of insecticidal crystals. *Nat. Biotechnol., 19*, 71-74.

DeGray, G., Rajasekaran, K., Smith, F., Sanford, J., & Daniell, H. (2001). Expression of an antimicrobial peptide via the chloroplast genome to control phytopathogenic bacteria and fungi. *Plant Physiology, 127*, 852-862.

Devine, A. L., & Daniell, H. (2004). Chloroplast genetic engineering. In S. Moller (Ed.), *Plastids* (pp. 283-320). United Kingdom: Blackwell Publishers.

Dhingra, A., Portis, A.R. & Daniell, H. (2004). Enhanced translation of a chloroplast expressed RbcS gene restores SSU levels and photosynthesis in nuclear antisense *RbcS* plants. *Proc. Natl. Acad. Sci., U.S.A., 101*, 6315-6320.

Dufourmantel, N., Pelissier, B., Garçon, F.,Peltier, J. M., & Tissot, G. (2004). Generation of fertile transplastomic soybean. *Plant Mol. Biol.*, in press.

Falconer, R. (2002). *Expression of interferon alpha 2b in transgenic chloroplasts of a low-nicotine tobacco.* M.S. thesis, University of Central Florida, Orlando, FL.

Farran, I., Sanchez-Serrano, J. J., Medina, J. F., Prieto, J., & Mingo-Castel, A. M. (2002). Targeted expression of human serum albumin to potato tubers. *Transgenic Res., 11*, 337-346.

Fernandez-San Millan, A., Mingeo-Castel, A. M., Miller, M., & Daniell, H. (2003). A chloroplast transgenic approach to hyper-express and purify human serum albumin, a protein highly susceptible to proteolytic degradation. *Plant Biotechnology Journal 1*, 71-79.

Fischer, N., Stampacchia, O., Redding, K., & Rochaix, J. D. (1996). Selectable marker recycling in the chloroplast. *Mol. Gen. Genet. 251*, 373-380.

Flowers, T. J. (2004). Improving crop salt tolerance. *J. Exp. Bot., 55*, 307-319.

Getzoff, T. P., Zhu, G. H., Bohnert, H. J. & Jensen, R. G. (1998). Chimeric *Arabidopsis thaliana* ribulose-1,5-bisphosphate carboxylase/oxygenase containing a pea small subunit protein is compromised in carbamylation. *Plant Physiol. 116*, 695-702.

Gibberd, M. R., Turner, N. C. & Storey, R. (2002). Influence of saline irrigation on growth, ion accumulation and partitioning, and leaf gas exchange of carrot (*Daucus carota* L.). *Ann. Bot., 90*, 715-724.

Gil, F., Brun, A., Wigdorovitz, A., Catala, R., Martinez-Torrecuadrada, J. L., Casal, I., Salinas, J., Borca, M. V., & Esscribano , J. M. (2001). High yield expression of a viral peptide vaccine in transgenic plants . *FEBS Lett. 488*, 13-17.

Goetz, M., Godt, D. E., Guivarch, A., Kahmann, U., Chriqui, D., Roitsch, T. (2001). Induction of male sterility in plants by metabolic engineering of the carbohydrate supply. *Proc. Natl. Acad. Sci. U S A. 98*, 6522-6527.

Golds, T., Maliga, P., & Koop, H. U. (1993) Stable plastid transformation in PEG-treated protoplasts of *Nicotiana tabacum. Bio/Technology, 11*, 95-97.

Goldschmidt-Clermont, M. (1991). Transgenic expression of aminoglycoside adenine transferase in the chloroplast: a selectable marker for site-directed transformation of *Chlamydomonas*. *Nucl. Acids Res., 19*, 4083-4089.

Guda, C., Lee, S. B., & Daniell, H. (2000). Stable expression of biodegradable protein based polymer in tobacco chloroplasts. *Plant Cell Rep.*, *19*, 257-262.
Hagemann, R. (1992). Plastid genetics in higher plants. In R. G. Herrmann (Ed.), *Cell Organelles* (pp. 65-69). Berlin: Springer-Verlag.
Hajdukiewicz, P. T., Gilbertson, L., & Staub, J. M. (2001). Multiple pathways for Cre/lox-mediated recombination in plastids. *Plant J.*, *27*, 161-170.
Hanson, A. D. & Gage, D. A. (1991). Identification and determination by fast atom bombardment mass spectrometry of the compatible solute choline-*O*-sulfate in *Limonium* species and other halophytes. *Austr. J. Plant Physiol.*, *18*, 317-327.
Hanson, A. D., Rathinasabapathi, B., Chamberlin, B. & Gage, D. A. (1991). Comparative physiological evidence that beta-alanine betaine and choline-*O*-sulfate act as compatible osmolytes in halophytic *Limonium* species. *Plant Physiol.*, *97*, 1199-1205.
Haslam, E. (1993). *Shikimic Acid: Metabolism and Metabolites.* Chichester, UK: John Wiley & Sons.
Heath, D. G., Anderson, J. W., Jr., Maurot, J. M., Welkos S. L., Andrews, J. P., Adamovicz, J. & Friedlander, A. M. (1998). Protection against experimental bubonic and pneumonic plague by a recombinant capsular F1~V antigen fusion protein vaccine. *Vaccine 16*, 1131-1137.
Herbers, K., Wilke, I., & Sonnewald, U. A. (1995). Thermostable xylanase from *Clostridium thermocellum* expressed at high levels in the apoplast of transgenic tobacco has no detrimental effects and is easily purified. *Bio/Technology*, *13*, 63-66.
Hernould, M., Suharsono, S., Zabaleta, E., Carde, J. P., Litvak, S., Araya, A., & Mouras, A. (1998). Impairment of tapetum and mitochondria in engineered male-sterile tobacco plants. *Plant Mol. Biol. 36*, 499-508.
Holmstrom, K.O., Mantyla, E., Welin, B., Mandal, A., Palva, E. T., Tunnela, O. E., *et al.* (1996) Drought tolerance in tobacco. *Nature, 379*, 683-684.
Hou, B. K., Zhou, Y. H., Wan, L. H., Zhang, Z. L., Shen, G. F., Chen, Z. H., *et al.* (2003). Chloroplast transformation in oilseed rape. *Transgenic Res. 12*, 111-114.
Houtz, R. L. & Portis, A. R. (2003). The life of ribulose 1,5-bisphosphate carboxylase/oxygenase—posttranslational facts and mysteries. *Arch. Biochem. Biophys. 414*, 150-158.
Huang, C. Y., Ayliffe, M. A. & Timmis, J. N. (2003). Direct measurement of the transfer rate of chloroplast DNA into the nucleus. *Nature 422*, 72-76.
Iamtham, S., & Day, A. (2000). Removal of antibiotic resistance genes from transgenic tobacco plastids. *Nature Biotechnol.*, *18*, 1172-1176.
Jiang, C. Z. & Rodermel, S. R. (1995). Regulation of photosynthesis during leaf development in *Rbc*S antisense DNA mutants of tobacco. *Plant Physiol. 107*, 215-224.
Kaneko, T., Matsubayashi, T., Sugita, M. & Sugiura, M. (1996). Physical and gene maps of the unicellular cyanobacterium *Synechococcus* sp strain PCC6301 genome. *Plant Mol. Biol. 31*, 193-201.
Kanevsky, I. & Maliga, P. (1994). Relocation of the plastid *rbc*L gene to the nucleus yields functional ribulose-1,5-bisphosphate carboxylase in tobacco chloroplasts. *Proc. Natl. Acad. Sci. USA, 91*, 1969-1973.
Kanevski, I., Maliga, P., Rhoades, D. F. & Gutteridge, S. (1999). Plastome engineering of ribulose-1,5-bisphosphate carboxylase/oxygenase in tobacco to form a sunflower large subunit and tobacco small subunit hybrid. *Plant Physiol. 116*, 133-141.
Kapusta, J., Modelska, A., Figlerowicz, M., Pniewski, T., Letellier, M., Lisowa, O., *et al.* (1999). A plant-derived edible vaccine against hepatitis B virus. *FASEB J.*, *13*, 1796-1799.
Kavanah,T.A., N.D. Thanh, N.T. Lao, N. McGrath, S.O. Peter, E.M. Horvath, P.J. Dix and P. Medgyesy. (1999). Homologous plastid DNA transformation is mediated by multiple recombination events. *Genetics* 152: 1111-1112
Khan, M. S. & Maliga, P. (1999). Fluorescent antibiotic resistance marker for tracking plastid transformation in higher plants. *Nature Biotenol. 17*, 910-915.
Klaus, S. M. J., Huang, F.-C., Golds, T. J. & Koop, H.-U. (2004). Generation of marker-free plastid transformants using a transiently cointegrated selection gene. *Nature Biotechnol. 22*, 225-229.
Klein, T. M., Wolf, E. D., & Sanford, J. C. (1987). High-velocity microprojectiles for delivering nucleic acids into living cells. *Nature, 327*, 70-73.
Kong, Q., Richter, L., Yang, Y. F., Arntzen, C., Mason, H., & Thanavala, Y. (2001). Oral immunization with hepatitis B surface antigen expressed in transgenic plants. *Proc. Natl. Acad. Sci. U.S.A., 98*, 11539-11544.

Kota, M., Daniell, H., Varma, S. Garczynski, S.F., Gould, F., & William, M. J. (1999). Overexpression of the *Bacillus thuringiensis* (*Bt*) Cry2Aa2 protein in chloroplasts confers resistance to plants against susceptible and *Bt*-resistant insects. *Proc. Natl. Acad. Sci. USA, 96*, 1840-1845.

Kriete, G., Niehaus, K., Perlick, A.M., Puhler, A., & Broer, I. (1996). Male sterility in transgenic tobacco plants induced by tapetum-specific deacetylation of the externally applied non-toxic compound N-acetyl-L-phosphinothricin. *Plant J. 9*, 809-818.

Kumar S, & Daniell H (2004) Engineering the chloroplast genome for hyper-expression of human therapeutic proteins and vaccine antigens. *Methods Mol Biol 267*: 365-383

Kumar, S., Dhingra, A. & Daniell, H. (2004a). Plastid expressed *betaine aldehyde dehydrogenase* gene in carrot cultured cells, roots and leaves confers enhanced salt tolerance. *Plant Physiol* in press.

Kumar, S., Dhingra, A., & Daniell, H. (2004b). Manipulation of gene expression facilitates cotton plastid transformation of cotton by somatic embryogenesis & maternal inheritance of transgenes. *Plant Mol. Biol.* in press.

Kunnimalaiyaan, M., Shi, F., & Nielsen, B. L. (1997). Analysis of the tobacco chloroplast DNA replication origin (*ori*B) downstream of the 23S rRNA gene. *J. Mol. Biol., 268*, 273-283.

Kusnadi, A., Nikolov, G., & Howard, J. (1997). Production of recombinant proteins in plants: practical considerations. *Biotechnology and Bioengineering, 56*, 473-484.

Langeveld, J. P., Casal, J. I., Osterhaus, A. D., Cortes, E., de Swart, R., Vela, C., *et al.* (1994). First peptide vaccine providing protection against viral infection in the target animal: studies of canine parvovirus in dogs. *J. Virology, 68*, 4506-4513.

Langeveld, J. P., Kamstrup, S., Uttenthal, A., Strandbygaard, B., Vela, C., Dalsgaard, K., *et al.* (1995). Full protection in mink against enteritis virus with new generation canine parvovirus vaccines based on synthetic peptide or recombinant protein. *Vaccine,* **13**, 1033-1037.

Langridge, W. (2000, Sept.). Edible vaccines. *Scientific American, 283*(3), 66-71.

Lee, S. B., Kwon, H. B., Kwon, S. J., Park, S. C., Jeong, M. J., Han, S. E., & Daniell, H. (2003). Accumulation of trehalose within transgenic chloroplasts confers drought tolerance. *Mol. Breeding, 11*, 1-13.

Leelavathi, S., & Reddy, V. S. (2003). Chloroplast expression of His-tagged GUS-fusions: a general strategy to overproduce and purify foreign proteins using transplastomic plants as bioreactors. *Mol. Breeding, 11*, 49-58.

Leelavathi, S., Gupta, N., Maiti, S., Ghosh, A., & Reddy, V. S. (2003). Overproduction of an alkali- and thermo-stable xylanase in tobacco chloroplasts and efficient recovery of the enzyme. *Mol.Breed., 11*, 59-67.

Lilly, J. W., Havey, M. J., Jackson, S. A., & Jiang, J. M. (2001). Cytogenomic analyses reveal the structural plasticity of the chloroplast genome in higher plants. *Plant Cell, 13*, 245-254.

Liu, J. H., Selinger, L. B., Cheng, K. J., Beauchemin, K. A., & Moloney, M. M. (1997). Plant seed oil-bodies as an immobilization matrix for a recombinant xylanase from the rumen fungus *Neocallimastix patriciarum*. *Mol. Breed., 3*, 463-470.

Lossl, A., Eibl, C., Harloff, H. J., Jung, C., & Koop, H. U. (2003). Polyester synthesis in transplastomic tobacco (*Nicotiana tabacum* L.): significant contents of polyhydroxybutyrate are associated with growth reduction. *Plant Cell Rep., 21*, 891-899.

Lugo, S.K., Kunnimalayaan, M., Singh, N.K., & Nielsen, B.L. (2004). Required sequence elements for chloroplast DNA replication activity in vitro and in electroporated chloroplasts. *Plant Science 166*: 151-161.

Martin, W., Stoebe, B., Goremykin, V., Hapsmann, S., Hasegawa, M., & Kowallik, K. V. (1998). Gene transfer to the nucleus and the evolution of chloroplasts. *Nature, 393*, 162-165.

Mason, H. S., Ball, J. M., Shi, J. J., Jiang, X., Estes, M. K., & Arntzen, C. J. (1996). Expression of Norwalk virus capsid protein in transgenic tobacco and potato and its oral immunogenecity in mice. *Proc. Natl. Acad. Sci. U.S.A., 93*, 5335-5340.

Mason, H. S., Haq, T. A., Clements, J. D. & Arntzen C. J. (1998). Edible vaccine protects mice against *Escherichia coli* heat-labile enterotoxin (LT): potatoes expressing a synthetic LT-B gene. *Vaccine, 16*, 1336-1343.

McBride, K. E., Svab, Z., Schaaf, D. J., Hogan, P. S., Stalker, D. M., & Maliga, P. (1995). Amplification of a chimeric *Bacillus* gene in chloroplasts leads to an extraordinary level of an insecticidal protein in tobacco. *Bio/Technology 13*, 362-365.

Meierhoff, K., Felder, S., Nakamura, T., Bechtold, N., Schuster, G. (2003). HCF152, an Arabidopsis RNA binding pentatricopeptide repeat protein involved in the processing of chloroplast psbB-psbT-psbH-petB-petD RNAs. *Plant Cell 15*, 1480-1495.

Mogensen, H. L. (1996). The hows and whys of cytoplasmic inheritance in seed plants. *Am. J. Bot., 83*, 383-404.

Molina, A., Herva-Stubbs, S., Daniell, H., Mingo-Castel, A. M., & Veramendi, J. (2004). High yield expression of a viral peptide animal vaccine in transgenic tobacco chloroplasts. *Plant Biotechnol. Journal, 2*, 141-153.

Monde, R., Schuster, G., & Stern, D.B. (2000). Processing and degradation of chloroplast mRNA. *Biochimie 82*, 573-582.

Mor, T. S., Gomez-Lim, M. A., & Palmer, K. E. (1998). Perspective: edible vaccines--a concept coming of age. *Trends Microbiol., 6*, 449-453.

Nagata, N., Saito, C., Sakai, A., Kuroiwa, H., & Kuroiwa, T. (1999). The selective increase or decrease of organellar DNA in generative cells just after pollen mitosis one controls cytoplasmic inheritance. *Planta, 209*, 53-65.

Napoli, C.A., Fahy, D., Wang, H.Y., & Taylor, L.P. (1999). White anther: A petunia mutant that abolishes pollen flavonol accumulation, induces male sterility, and is complemented by a chalcone synthase transgene. *Plant Physiol. 120*, 615-22.

Nilsson, B., Forsberg, G., & Hartmanis, M. (1991). Expression and purification of recombinant insulin-like growth factors from *E. coli. Methods in Enzymology, 198*, 3-16.

Ogren, W. L. (2003). Affixing the O to Rubisco: discovering the source of photorespiratory glycolate and its regulation. *Photosynth. Res. 76*, 53-63.

O'Neill, C., Horvath, G., Horvath, E., Dix, P., & Medgyesy, P. (1993). Chloroplast transformation in plants: polyethylene glycol (PEG) treatment of protoplasts is an alternative to biolistic delivery systems. *Plant J., 3*, 729-738.

Radwanski, E. R., & Last, R. L. (1995). Tryptophan biosynthesis and metabolism: biochemical and molecular genetics. *Plant Cell, 7*, 921–934.

Rathinasabapathi, B., Fouad, W. M. & Sigua, C. A. (2001). β-Alanine betaine synthesis in the Plumbaginaceae. Purification and characterization of a trifunctional, *S*-adenosyl-l-methionine-dependent *N*-methyltransferase from *Limonium latifolium* leaves. *Plant Physiol., 126,* 1241-1249.

Rathinasabapathy, B., McCue, K. F., Gage, D. A., & Hanson, A. D. (1994). Metabolic engineering of glycine betaine synthesis: plant betaine aldehyde dehydrogenasese lacking a typical transit peptides are targeted to tobacco chloroplasts where they confer aldehyde resistance. *Planta, 193*, 155-162.

Reinert J. (1958). Morphogenese und ihre kontrolle an gewebekulturen aus carotten. *Naturwissenchaften 45*, 344-345.

Rhodes, D. & Hanson, A. D. (1993). Quaternary ammonium and tertiary sulfonium compounds in higher plants. *Annu. Rev. Plant. Physiol., 44,* 357-384.

Richter, L. J., Thanavala, Y., Arntzen, C. J., & Mason, H. S. (2000). Production of hepatitis B surface antigen in transgenic plants for oral immunization. *Nat. Biotechnol., 18*, 1167-1171.

Robinson, S. P. & Jones, G. P. (1986). Accumulation of glycinebetaine in chloroplasts provides osmotic adjustment during salt stress. *Aust. J. Plant Physiol. 13*, 659-668.

Rochaix, J. D. (1996). Post-transcriptional regulation of chloroplast gene expression in *Chamydomonas reinhardtii*. *Plant Mol Biol 32*, 327-341.

Rodermel, S. R., Abbott, M. S. & Bogorad, L. (1988). Nuclear-organelle interactions – nuclear antisense gene inhibits ribulose bisphosphate carboxylase enzyme levels in transformed tobacco plants. *Cell 55*, 673-681.

Ruf, S., Hermann, M., Berger, I., Carrer, H., & Bock, R. (2001). Stable genetic transformation of tomato plastids and expression of a foreign protein in fruit. *Nat. Biotechnol., 19*, 870-875.

Ruiz, G. (2002) *Optimization of codon composition and regulatory elements for expression of the human IGF-1 in transgenic chloroplasts*. M.S. thesis, University of Central Florida, Orlando, FL.

Ruiz, O. & Daniell, H. (2004). Engineering cytoplasmic male sterility via the chloroplast genome. in review.

Ruiz, O. N., Hussein, H., Terry, N., & Daniell, H. (2003). Phytoremediation of organomercurial compounds via chloroplast genetic engineering. *Plant Physiol., 132*, 1344-1352.

Ruiz, O.N., Quesada-Vargas, T., & Daniell, H. (2004). Characterization of heterologous operons in transgenic chloroplasts: transcription, processing, translation. in review

Sanford, J. C., Smith, F. D., & Russell, J.A. (1993). Optimizing the biolistic process for different biological applications. *Methods Enzymol.*, *217*, 483-509.
Scott, S. E. & Wilkenson, M. J. (1999). Low probability of chloroplast movement from oilseed rape (*Brassica napus*) into wild *Brassica rapa*. *Nat. Biotechnol.*, *17*, 390-392.
Sikdar, S. R., Serino, G., Chaudhuri, S. & Maliga, P. (1998). Plastid transformation in *Arabidopsis thaliana*. *Plant Cell Rep.* *18*, 20-24.
Sidorov, V. A., Kasten, D., Pang, S. Z., Hajdukiewicz, P. T., Staub, J. M., & Nehra, N. S. (1999). Technical advance: stable chloroplast transformation in potato: use of green fluorescent protein as a plastid marker. *Plant J.*, *19*, 209-216.
Singleton, M. L. (2003). *Expression of CaF1 and LcrV as a fusion protein for a vaccine against* Yersinia pestis *via chloroplast genetic engineering*. M.S. thesis, University of Central Florida, Orlando, FL.
Skarjinskaia, M., Svab, Z., Maliga, P. (2003) Plastid transformation in *Lesquerella fendleri*, an oilseed Brassicacea. *Transgenic Res.* *12*, 115-122.
Spreitzer, R. J. & Salvucci, M. E. (2002). Rubisco: Structure, regulatory interactions, and possibities for a better enzyme. *Annu. Rev. Plant Biol.* *53*, 449-475.
Staub, J. M., Garcia, B., Graves, J., Hajdukiewicz, P. T. J., Hunter, P., Nehra, N., *et al.* (2000) High-yield production of a human therapeutic protein in tobacco chloroplasts. *Nat. Biotechnol.*, *18*, 333-338.
Steward, F. C., Mapes, M. O. & Mears, K. (1958). Growth and organized development of cultured cells: II. Organization in cultures grown from freely suspended cells. *Am. J. Bot.* *45*, 705-708.
Svab, Z. & Maliga, P. (1993). High-frequency plastid transformation in tobacco by selection for a chimeric *aad*A gene. *Proc. Natl. Acad. Sci. U. S. A.*, *90*, 913-917.
Tacket, C. O., Mason, H. S., Losonsky, G., Estes, M. K., Levine, M. M., & Arntzen, C. J. (2000). Human immune responses to a novel Norwalk virus vaccine delivered in transgenic potatoes. *J. Infect. Dis.*, *182*, 302-305.
Tessereau, H.., Florin, B., Meschine, M. C., Thierry, C. & Pétiard, V. (1994) Cryopreservation of somatic embryos: A tool for germplasm storage and commercial delivery of selected plants. *Ann. of Botany* *74*, 547-555.
Tregoning, J. S., Nixon, P., Kuroda, H., Svab, Z., Clare, S., Bowe, F., *et al.* (2003). Expression of tetanus toxin Fragment C in tobacco chloroplasts. *Nucleic Acids Res.*, *31*, 1174-1179.
Tsai, C. H., Miller, A., Spalding, M. & Rodermel, S. (1997). Source strength regulates an early phase transition of tobacco shoot morphogenesis. *Plant Physiol.* *115*, 907-914.
Vitanen, P.V., Devine, A. L., Kahn, S., Deuel, D. L., Van Dyk, D. E. & Daniell, H. (2004). Metabolic engineering of the chloroplast genome using the E. coli ubiC gene reveals that corismate is a readily abundant precursor for 4-hydroxybenzoic acid synthesis in plants. in review.
Vivek, B. S., Ngo, Q. A., & Simon P.W. (1999). Evidence for maternal inheritance of the chloroplast genome in cultivated carrot (*Daucus carota* L. ssp. *Sativus*). *Theor Appl Genet* **98**, 669-672.
Watanabe, H., Yamasaki, K., Kragh-Hansen, U., Tanase, S., Harada, K., Suenaga, A., *et al.* (2001) *In vitro* and *in vivo* properties of recombinant human serum albumin from *Pichia pastoris* purified by a method of short processing time. *Pharm. Res.*, *18*, 1775-1781.
Whitney, S. M. & Andrews T. J. (2001). Plastome-encoded bacterial ribulose-1,5-bisphosphate carboxylase/oxygenase (RubisCO) supports photosynthesis and growth in tobacco. *Proc. Natl. Acad. Sci. USA 98*, 14738-14743.
Whitney, S. M., von Caemmerer, S., Hudson, G. S., & Andrews, T. J. (1999). Directed mutation of the Rubisco large subunit of tobacco influences photorespiration and growth. *Plant Physiol.* *121*, 579-588.
Yan, W. K. & Hunt, L. A. (1999). Reanalysis of vernalization data of wheat and carrot. *Ann. Bot.* **84**, 615-619.
Yui, R., Iketani, S., Mikami, T., & Kubo, T. (2003). Antisense inhibition of mitochondrial pyruvate dehydrogenase E1alpha subunit in anther tapetum causes male sterility. *Plant J.* *34*, 57-66.
Zerges, W. (2000). Translation in chloroplasts. *Biochimie 82*, 583-601.
Zhang, H. X. & Blumwald, E. (2001). Transgenic salt-tolerant tomato plants accumulate salt in foliage but not in fruit. *Nat. Biotechnol.*, *19*, 765-768.
Zhang, Q., Liu, Y., & Sodmergen. (2003). Examination of the cytoplasmic DNA in male reproductive cells to determine the potential for cytoplasmic inheritance in 295 angiosperm species. *Plant Cell Physiol.* *44*(9), 941-951.

Zhang, X. H., Brotherton, J. E., Widholm, J. M., & Portis, A. R. (2001a). Targeting a nuclear anthranilate synthase subunit gene to the tobacco plastid genome results in enhanced tryptophan biosynthesis. Return of a gene to its pre-endosymbiotic origin. *Plant Physiol., 127*, 131-141.
Zhang, X. H., Ewy, R. G., Widholm, J. M. & Portis, A. R. (2002) Complementation of the nuclear antisense rbcS-induced photosynthesis deficiency by introducing an *rbc*S gene into the tobacco plastic genome. *Plant Cell Physiol. 43*, 1302-1313.
Zhang, X. H., Portis, A. R. & Wildholm, J. M. (2001b). Plastid transformation of soybean suspension cultures. *J. Plant Biotechnol. 3*, 39-44.
Zhang, X. H. Wildholm, J. M. & Portis, A. R. (2001c). Photosynthetic properties of two different soybean suspension cultures. *J. Plant Physiol. 158*, 357-365.
Zheng, Z. F., Xia, Q., Dauk, M., Shen, W. Y., Selvarej, G., Zou, V. T. (2003). *Arabidopsis* AtGPAT1, a member of the membrane-bound glycerol-3-phosphate acyltransferase gene family, is essential for tapetum differentiation and male fertility. *Plant Cell. 15*, 1872-87.

CHAPTER 17

ENGINEERING HERBICIDE RESISTANCE PATHWAYS IN PLASTIDS

AMIT DHINGRA AND HENRY DANIELL*

*Department of Molecular Biology and Microbiology, University of Central Florida, Biomolecular Science Bldg # 20, Room 336, 4000 Central Florida Blvd., Orlando, Florida 32816-2364. *daniell@mail.ucf.edu*

Abstract. Herbicides are an integral part of modern day agriculture as they facilitate efficient crop management. Most of the herbicides target specific enzymes involved in metabolic pathways that are vital for plant growth and survival. Since plastids are indispensable for plant survival most of the herbicide targets reside in the plastids and directly or indirectly affect plastid function. Several vital metabolic pathways including photosynthesis, biosynthesis of amino acids, pigments, purines, pyrimidines and fatty acids are carried out in the plastids. Most of the commercially available herbicides target the enzymes involved in the above-mentioned metabolic pathways. It is not a coincidence then that the first identified herbicide atrazine targeted the D-1 polypeptide of photosystem II.

In order to provide the crop species an advantage over the weeds, herbicides were traditionally selected on the basis of their selective lethality for the unwanted vegetation. Alternatively, cultivated crops have been genetically engineered with mutant version of the target enzymes rendering the plant insensitive to the herbicide or with genes that encode enzymes that metabolically detoxify the applied herbicide. The percentage of herbicide resistant crops planted in the world is on a rise; about 75% of the transgenic crops planted now are engineered for herbicide resistance.

Herbicide resistant transgenic crops grown in the field today harbor the transgene in their nuclear genome from there it finds its way into the pollen and the transgene might outcross with their wild relatives, thereby defeating the purpose of creating herbicide resistant transgenic lines. This phenomenon has fuelled public concerns and has made the acceptance of genetically modified food crops difficult. This requires alternate strategies for engineering plants for herbicide resistance, such as introducing male sterility in the transgenic lines or engineering plants in a way that transgene escape can be overcome. The latter approach has been achieved by integrating herbicide resistance transgenes into the chloroplast genome and demonstration of its maternal inheritance over the subsequent generations. Since target amplification is one of the major strategies for conferring herbicide resistance, integration of herbicide resistance gene into the chloroplast genome is ideal for this purpose. In addition chloroplast expressed transgenes do not encounter gene silencing and have no deleterious effect on the plant phenotype. Presence of herbicide targets within the chloroplast warrants integration of the herbicide resistance transgenes into the chloroplast genome. With the recent establishment of chloroplast transformation technology in major crops like cotton and soybean, herbicide resistant crops are poised to enter a new and an environmentally friendly era.

H. Daniell and C. D. Chase (eds.), Molecular Biology and Biotechnology of Plant Organelles, 491—511.

1. INTRODUCTION

Herbicides are chemicals that interact with specific molecular targets in a plant, resulting in disruption of normal metabolic processes. As a consequence the plant is unable to survive and eventually perishes. Herbicides are now a vital part of present day agriculture and are used primarily for three major reasons. Elimination of weeds reduces the competition for light, water & nutrients; they reduce the risk of cross infection from pests and pathogens from weeds to crops; and herbicide usage improves crop management especially during harvesting by raising the quality of the crop (Schulz et al., 1990). Since herbicides are not discriminative in their lethal action on plants, it is desirable to screen for compounds that selectively eliminate the unwanted weeds. Traditionally this has been done by screening for novel compounds that are active against a broad spectrum of weeds and ineffective on target crops (Mazur and Falco, 1989). More recently plants have been genetically engineered to counter the lethal effect of herbicides. There are several mechanisms that are responsible for natural herbicide tolerance in plants, prominent ones being insensitivity to the target site and breakdown of the toxic herbicide to non-toxic by-products. Both these mechanisms have been simulated in genetically engineered crops either by over expressing the target enzymes or by engineering foreign proteins that can rapidly detoxify the herbicides (Freyssinet, 2003). The latter approach allows for using a wider choice of herbicides that are also effective against a broad range of weeds, much to the advantage of farmers. Other mechanisms include mutations in the target site that render the plant insensitive to the herbicide or reduced herbicide uptake.

Resistance to herbicides was among the first traits to be genetically engineered into crops for practical and economic reasons (Freyssinet 2003). On a global level, about 75% of the transgenic crops are engineered for herbicide tolerance (Castle et al., 2004). In all these cases the trait has been engineered via the nuclear genome. A transgene integrated in the nuclear genome stays viable in the male gamete and is dispersed via the pollen. The incidence of transgene out-cross to wild relatives may be as high as 38% in sunflower and 50% in strawberries (Linder et al., 1998; Kling, 1996). Therefore out-cross of herbicide resistance genes from transgenic crops to wild type relatives has descended from the realm of prior predictions to present reality. Recently there has been a negative public perception against genetically modified crops due to the documented transfer of transgenes to wild relatives of a few plant species. In order to allay public fears and address other environmental concerns, there is an impending need for devising strategies that aid in the containment of an introduced transgene. Towards this goal there are several approaches available like use of male sterility, fertility barriers or maternal inheritance of the transgene.

Maternal inheritance of a transgene is possible if a transgene is integrated into the chloroplast genome. The chloroplast genome is maternally inherited in most crops (Zhang et al., 2003) therefore transgene escape via pollen may be effectively eliminated (Daniell, 2002). Detailed analyses of several inheritance parameters revealed that gene flow would be rare if a transgene is integrated into the chloroplast genome (Scott and Wilkinson, 1999). Integration of a herbicide resistant transgene into the chloroplast genome has other additional advantages. The expression level of the foreign protein is enhanced several 100 fold compared to the nuclear integrated transgene. As a result higher dosage of herbicides can be employed without any harm to the cultivated crop and effectively eliminate weeds in one or two applications. Since most of the herbicides target pathways are directly or indirectly related to the chloroplast function, in this review we discuss those target sites and the attempts to engineer those sites via nuclear genetic engineering to confer herbicide resistance. Further, we provide a brief introduction to the approach of chloroplast transformation, its advantages and the efforts made so far towards engineering herbicide resistance in plants utilizing this approach.

2. CHLOROPLAST SPECIFIC TARGETS AND ENGINEERING OF CROPS FOR HERBICIDE TOLERANCE BY ENGINEERING HERBICIDE INSENSITIVE ENZYMES

2.1. Introduction

Most of the herbicides target essential pathways that reside inside the chloroplast. An overview of the proteins that are targeted by herbicides, the pathways in which those enzymes participate, herbicides that target those proteins and the mechanism of tolerance or resistance are summarized in Table 1. One of the mechanisms to confer herbicide tolerance is target site amplification or expression of insensitive/mutant target enzymes/proteins. This section discusses the more commonly known chloroplast specific enzymes that are targeted by herbicides and the mechanism of conferring tolerance to susceptible plant species.

***Table 1**. List of chloroplast specific enzymes, biosynthetic pathways, herbicides and resistance mechanisms discussed in this chapter.*

Enzyme/gene	*Pathway*	*Herbicide/active ingredient*	*Resistance mechanism or gene*
5-enol-Pyruvylshikimate-3-phosphate synthase (EPSPS)	Shikimic acid	Glyphosate	Mutant EPSPS, CP4 EPSPS, Petunia EPSPS, *aro*A-M1 goxA, *gat*
Acetolactate synthase (ALS)	Branched chain amino acid	Sulfonylureas, Imidazolinones, Triazolopyrimidine sulfonamides	*csr1*-1 *imr*-1

Table 1 (cont.)

Table 1 (cont.)

Enzyme/gene	*Pathway*	*Herbicide/active ingredient*	*Resistance mechanism or gene*
Glutamine synthetase (GS)	Glutamine biosynthesis	Methionine sulfoximine, Phosphinothricin (glufosinate) and Tabtoxinine-□-lactam	*bar*, *pat*
Acetyl Co-A carboxylase (ACCase)	Lipid biosynthesis	Cyclohexanediones, Aryloxyphenoxy propionates	
D-1 polypeptide	Photosynthesis (PS II)	Substituted ureas, s-triazines and phenols Bromoxynil	Mutant *psb*A, *oxy*
Photosystem I	Photosynthesis	Paraquat and Diquat	Glutathione reductase, copper/zinc chloroplast superoxide dismutase
Protoporphyrinogen oxidase (Protox)	Tetrapyrrole biosynthesis	Diphenyl ethers	Protox (*A. thaliana*), Protox (*B. subtilis*)
Phytoene desaturase (PDS)	Carotenoid biosynthesis	Pyridazinones metflurazon, Norflurazon	crt*1*, *pds*
4-Hydroxyphenylpyruvate dioxygenase (HPPD)	Prenylquinone pathway	Isoxaflutole, Sulcotrione, NTBC	*HPPD*
Dihydroxypteroate Synthase (DHPS)	Folate biosynthesis	Asulam	*Sul*1

2.2. Enolpyruvylshikimatephosphate synthase (EPSPS)

Enolpyruvylshikimatephosphate synthase (EPSPS) is an enzyme of the shikimic acid pathway that links photosynthetic carbon reduction to synthesis of aromatic amino acids tyrosine, phenylalanine & tryptophan and several other secondary products in plants. Besides protein synthesis the aromatic amino acids are vital for hormone synthesis, production of compounds involved in energy transduction like plastoquinone, cell wall formation and also defense against pathogens and insects (Duke, 1988).

EPSPS is specifically targeted by the glyphosate (Figure 1) that binds the enzyme in a complex manner resulting in a conformational change that in turn prevents it from binding phosphoenolpyuruvate (PEP), one of the two substrates of EPSPS (Sikorski and Gruys, 1997). Inhibition of EPSPS by glyphosate results in depletion of aromatic amino acids and hyperaccumulation of shikimic acid (Hoagland et. al., 1978; Amrhein et. al., 1980). Glyphosate tolerance has been engineered successfully in major crops through engineering of modified EPSPS enzymes. Initial efforts focused on identifying glyphosate insensitive mutant versions of EPSPS from mutagenesis screens in bacteria. A mutant gene *aro*A, isolated from *Salmonella typhimurium* coding for an insensitive EPSPS (Comai et al 1983) was engineered into tobacco and tomato nuclear genome (Comai et al., 1985; Fillatti et al., 1987).

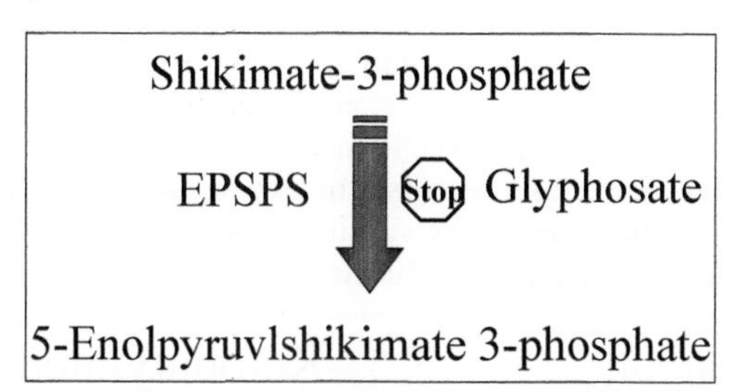

Figure 1: Shikimic acid pathway step catalyzed by enzyme EPSPS. Glyphosate is the specific inhibitor of EPSPS.

The transgenic plants did not have agronomically useful degree of herbicide tolerance mainly because the enzyme was not targeted to the chloroplast and it was compromised in its PEP binding capacity (Mousedale and Coggins, 1985; Padgette et al., 1996). The first successful glyphosate tolerant transgenic plant was obtained by engineering a mutant EPSPS obtained from petunia. Transgenic petunia cells and plants expressing the insensitive EPSPS were found to be tolerant to glyphosate (Shah et al., 1986). Later, a naturally occurring EPSPS from *Agrobacterium* sp. Strain CP4 was identified that had all the desirable properties for an ideal glyphosate resistance gene (Barry et al., 1992). This gene has been successfully used for engineering glyphosate resistance in soybean (Padgette et al., 1995) and cotton (Nida et al., 1996). This gene was not effective in case of maize where commercial level tolerance to glyphosate was conferred by a double mutant of maize EPSPS (Lebrun et al., 1997). With the help of novel strategies, mutated versions of *aro*A gene have been created that confer enhanced tolerance to glyphosate. Staggered extension process using the *aro*A genes from *Salmonella typhimurium* and *Escherichia coli*, resulted in the generation of four variants that were randomly mutated and recombined versions of the two genes. Three of these carried *de novo* mutations hitherto unknown in mediating glyphosate tolerance. Increased tolerance was due to a 2-10-fold increase in specific activity, 0.4-8-fold reduced affinity for glyphosate, and 2.5-19-fold decreased Km for phosphoenolpyuruvate (He et al., 2001). One of these mutants *aro*A-M1 was recently engineered into tobacco and it was demonstrated that the transgenic plants could survive up to 0.8 mM of glyphosate (Wang et al., 2003).

In the US, two lines of glyphosate resistant Rapeseed are commercially grown. Both lines express the EPSPS and *gox* gene; the former gene is derived from Agrobacterium in both cases but *gox* gene is derived from *Agrobacterium* and *Achromobacter*. Similarly three lines of glyphosate resistant corn are grown that express EPSPS alone or EPSPS in conjunction with *gox*. The EPSPS gene is either derived from *Agrobacterium* or corn itself. One line each of herbicide resistant cotton and soybean are also commercially grown that express *Agrobacterium* derived EPSPS (ISB 2004, http://www.isb.vt.edu/cfdocs/biopetitions3.cfm).

2.3. Acetolactate synthase (ALS)

Acetolactate synthase (ALS) catalyzes the first step in the biosynthesis of the branched chain amino acids valine, leucine and isoleucine that is carried out in the chloroplast (Schulze-Siebert et al., 1984). It condenses two molecules of pyruvate to form 2-acetolactate that serves as a precursor of leucine and valine. Similar reaction is catalyzed with 2-ketobutyrate as the substrate resulting in the formation of 2-acetohydroxybutyrate, a precursor of isoleucine.

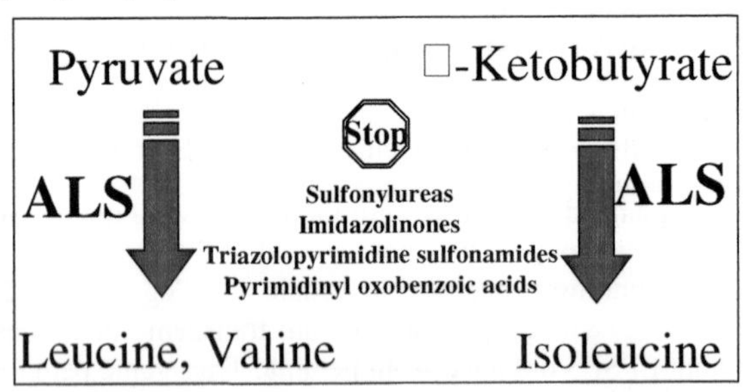

Figure 2: Branched chain amino acid biosynthesis pathway steps catalyzed by Acetolactate synthase. ALS inhibitors are listed.

ALS is specifically inhibited by four different classes of herbicides namely Sulfonylureas (nicosulfuron), Imidazolinones (imazaquin), Triazolopyrimidine sulfonamides and Pyrimidinyl oxobenzoic acids (Figure 2). Each class of herbicide has a unique mode of inhibiting ALS. They can generally be considered as growth inhibitors that act faster than glyphosate. Inhibition of ALS depletes the branched chain amino acids resulting in death of the susceptible species.

Resistance to ALS inhibitors has been conferred by engineering insensitive mutant genes like the *csr1-1* mutant ALS gene from *Arabidopsis thaliana* that has been successfully engineered in oilseed rape and flax (Miki et al., 1990; McHughen and Holm, 1995). Interestingly most ALS mutants are insensitive to only specific herbicides; therefore the strategy has been to use hybrid ALS mutants that can code for resistance to multiple herbicides. A chimeric ALS mutant derived from *csr1-1* and *imr1* mutant genes from *A. thaliana* resulted in conferring high-level tolerance to two different types of herbicides in transgenic tobacco (Hattori et al., 1992). There is one mutant gene that confers resistance to multiple classes of herbicides that specifically inhibit ALS. This mutant gene has been tested in transgenic tobacco and the plants were found to be tolerant to all the ALS specific herbicides tested (Hattori et al 1995). A similar mutant raised in cotton was used for conferring tolerance to multiple herbicides (Rajasekaran et al., 1996). Researchers are also investigating simulated crystal structures for ALS to identify mutations that would render the enzyme insensitive to herbicides. This has resulted in generation of one mutant that confers high-level tolerance to imidazolinone in transgenic tobacco (Ott et al., 1996). A mutant acetolactate synthase (ALS) gene has been recently engineered into carrot to control the parasitic *Orobanche* spp (Aviv et al., 2002). An ALS variant from chlorsulfuron tolerant *A. thaliana* (Haughn and Somerville 1986) was successfully used for engineering garlic that could withstand 3 mg/l of chlorsulfuron (Park et al., 2002).

There are three ALS expressing crops, resistant to sulfonylureas that are currently cultivated globally. Sulfonylurea herbicide tolerant flax, specifically triasulfuron and metsulfuron-methyl, is grown in Canada and the US. An ALS variant gene was isolated from chlorsulfuron tolerant *A. thaliana* and introduced into flax. Cotton expressing a variant form of ALS that is insensitive to sulfonylureas is grown in the US. Sulfonylurea resistant carnations, developed by Florigene Inc. are grown in Australia and the European Union (AGBIOS, 2004, http://www.agbios.com/dbase.php).

2.4. Glutamine synthetase (GS)

Glutamine synthetase (GS) is the first enzyme in the glutamine synthesis pathway that is instrumental in the assimilation of inorganic nitrogen into organic compounds. The enzyme helps in the utilization of ammonia produced by nitrite reductase and recycles the ammonia produced by photorespiration. Higher plants have a cytosolic and chloroplastic form of the enzyme (Lara et al., 1984). In leaves the chloroplastic form is predominant and this form is mainly responsible for most of the ammonia assimilation as evidenced by studies on developing wheat seedlings (Tobin et al., 1985). Using immunogold labeling the enzyme has been localized in the stromal region of the chloroplast (Botella et al., 1988).

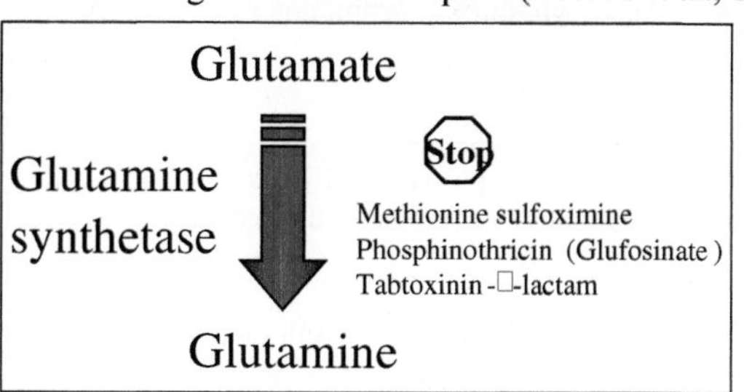

Figure 3: Glutamine biosynthesis pathway step catalyzed by glutamine synthetase. Specific inhibitors of GS are listed.

Glutamine synthetase is inhibited by naturally occurring peptides that are variations of substituted glutamate. There are three major compounds that inhibit glutamine synthetase namely, methionine sulfoximine (MSO), phosphinothricin (glufosinate) and tabtoxinine-□-lactam (Figure 3). Inhibition of glutamine synthetase activity leads to rapid accumulation of ammonia, cessation of photosynthesis, disruption of chloroplast structure and vesiculation of the stroma (Devine et al., 1993).

Resistance to glutamine synthetase has been selected in culture conditions from which lines with gene amplification or insensitive to glutamine synthetase inhibitors have been isolated. The other approach has been to genetically engineer the plant with genes that metabolically detoxify the herbicides. This approach has been discussed later in section 3.3.

2.5. Acetyl Co-A carboxylase (ACCase)

The very first step of *de novo* fatty acid biosynthesis in plants is catalyzed by acetyl Co-A carboxylase. The reaction takes place inside the chloroplast and results in the formation of malonyl-Co-A from acetyl Co-A and bicarbonate.

ACCase is inhibited by two groups of herbicides, the cyclohexanediones (CHD) and aryloxyphenoxy propionates (AOPP) (Figure 4). Two forms of ACCase are known to exist in dicots. One is referred to as a prokaryotic form that is heteromeric and the other is referred to as a eukaryotic form, which is homomeric. The prokaryotic heteromeric form is resistant to herbicides while the eukaryotic homomeric form is sensitive. Since members of the graminae family lack the plastid *acc*D gene that codes for one of the subunits of ACCase, the prokaryotic form is absent from grasses making them susceptible to herbicides. This ensures selectivity for the control of grass weeds in a variety of broad leaf and cereal crops (Konishi and Sasaki, 1994).

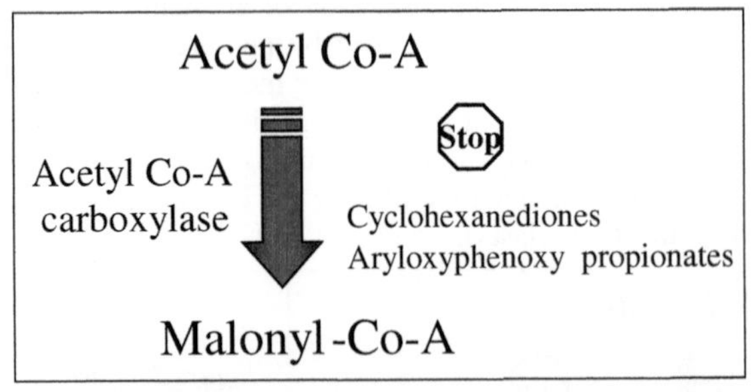

Figure 4: Lipid biosynthesis pathway step catalyzed by Acetyl CO-A carboxylase. Specific inhibitors of ACCase are listed.

Resistance to CHD and AOPP herbicides is conferred by mutations in the 400 amino acid fragment within the carboxyltransferase □-subdomain (Nikolskaya et al., 1999). Several such mutations can confer unique patterns of cross-resistance to these herbicides. Based on observations in susceptible grasses it was proposed that

AOPP herbicides act by rapidly depolarizing the plasma membrane and in resistant species the toxic effect is countered by altering the plasma membrane electromagnetic potential (Hausler et al 1991). Currently 6 field trials, 4 for corn and 2 for Rapeseed have been reported in the US where the crops express the ACCase gene (ISB, 2004).

2.6. *Photosynthesis*

Photosynthesis is a vital metabolic process carried out only by plants. All the light reactions of this process are carried out in the chloroplast and two major participants in the process are photosystem II and photosystem I. Herbicides affecting photosynthesis have been found since the 1950s. Some herbicides act on the electron flow of photosystem II and others act by diverting the electron flow at Photosystem I.

2.6.1. *D-1 (Q_B) protein*

The D-1 protein is a 32-kDa polypeptide of the Photosystem II encoded by the *psb*A gene in the chloroplast genome and has a high rate of turnover that is light dependent. The D-1 protein acts as an apoprotein for Q_B that is a specialized form of plastoquinone, which mediates electron flow towards plastoquinone pool within the thylakoid membrane (Dodge, 1991).

Substituted ureas, s-triazines and phenols specifically target the D-1 polypeptide and block the transport of the mobile electron carrier, plastoquinone (Devine et al., 1993). Disruption of electron flow between the two photosystems results in blocking of the photosynthetic output of reducing byproducts that are utilized in carbon fixation. A side effect of this is the apparent reduction in beta-carotene that exposes chloroplasts to oxidative damage.

Resistance mechanism against PSII inhibitors is already present in some crops species that is mediated by metabolic detoxification or decreased uptake and translocation. The most common resistance mechanism is mediated by a mutation in the *psb*A gene (Golden and Haselkorn, 1985). In most cases a Ser_{264} to Gly mutation confers resistance to the herbicides (Shukla and Devine, 2000). This is a mutation that affects the stromal side of the protein and decreases the binding of *s*-triazine herbicides. Also, this mutation results in impaired photosynthesis and plant growth and as a result plant vigor is compromised (McCloskey and Holt, 1990). Alternatively, resistance mechanism can be engineered with genes for metabolic detoxification of herbicides. This approach is discussed in section 3.2.

2.6.2. *Photosystem I (PS I)*

Photosystem I functions as a light-driven plastocyanin-ferredoxin oxidoreductase that is comprised of several intermediate redox components, which appear to be associated to a PS I core complex constituted by seven polypeptides.

Bipyridinium herbicides paraquat and diquat divert the normal electron flow between iron sulfur centers A, B and $NADP^+$ (Zweig et al., 1965). The bipyridyl radical reacts with molecular oxygen present in the grana to form superoxide that in turn leads to the formation of hydroxyl radical, which is highly toxic and is ultimately responsible for the cellular damage.

Natural resistance to PS I specific herbicides is reported but elucidation of the precise molecular mechanism for the same remains elusive. It is proposed that resistance could be either due to amplification of the enzyme involved in detoxification of oxygen radicals or by sequestration of the toxic chemical compound away from the chloroplast (Preston, 1994). Resistance to PSI specific herbicides has been engineered in model system tobacco by expressing glutathione reductase from *E. coli* and a copper/zinc chloroplast superoxide dismutase from pea (Aono et al., 1993; Sen Gupta et al., 1993).

2.7. Protoporphyrinogen oxidase (Protox)

Protoporphyrinogen oxidase (Protox) is the final enzyme in the tetrapyrrole biosynthesis pathway before it bifurcates to chlorophyll or heme synthesis (Figure 5). Applied as foliar spray, Protox inhibitors cause desiccation and rapid cellular disintegration. Protox inhibition results in production of large amounts of protoporphyrin IX in the cytoplasm, which causes photodynamic damage in presence of light and oxygen (Devine and Preston, 2000).

Resistance to a Protox targeting herbicide has been engineered by overexpressing the plastidic *Arabidopsis* Protox gene in tobacco (Lermontova and Grimm, 2000). Rice and soybean are naturally resistant to most Protox inhibiting herbicides. Most of the eukaryotic Protox genes identified so far are sensitive to diphenyl ether herbicides but a Protox gene from *Bacillus subtilis* is resistant to this class of herbicides. It was recently demonstrated that a plastid targeted *Bacillus subtilis* derived Protox gene, expressed in rice conferred high level tolerance to diphenyl ether herbicide oxyfluorofen (Lee et al., 2000). Expression of a plastidic form of Protox gene isolated from *Arabidopsis* with or without a chloroplast transit peptide in rice also resulted in tolerance to oxyfluorofen. Surprisingly the protein without the transit peptide was also localized in the chloroplast (Ha et al., 2004).

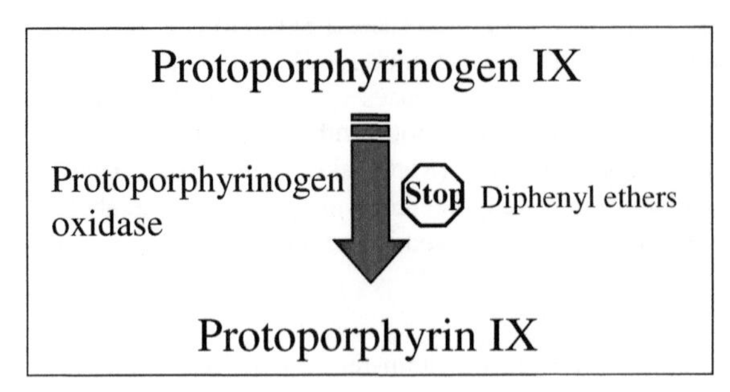

Figure 5: Tetrapyrrole biosynthesis pathway step catalyzed by Protox. Diphenyl ethers specifically inhibit Protox.

2.8. Phytoene desaturase (PDS)

Phytoene desaturase is an enzyme involved in the biosynthesis of carotenoids that provide protection to plants from photo-oxidative damage. Inhibition of PDS results in phytoene accumulation that has short chromophores and it is unable to protect from photooxidation resulting in the death of plant by light and oxygen.

Herbicide pyridazinones metflurazon and norflurazon inhibit carotenoid biosynthesis by specifically blocking desaturation that results in the accumulation of phytoene (Bartels and Watson, 1978). There is a reasonable level of natural resistance to PDS inhibitors in some crops like cotton, red beet and soybeans and the resistance is conferred by reduced absorption and translocation. PDS derived from bacteria are insensitive to the conventional bleaching herbicides. One such PDS encoded by *crtI* gene was isolated from *Erwinina uredovora* and engineered into tobacco nuclear genome. The protein was targeted to the chloroplast where carotenoid biosynthesis takes place and it could confer a good level of tolerance to the herbicides (Misawa et al., 1994). Tobacco was recently transformed with a mutated Cyanobacterial phytoene desaturase gene that was isolated from the Synechococcus PCC 7942 mutant NFZ4. Transgenic plants possessed 58-fold higher norflurozan resistance in comparison to wild type and the transgenic plants maintained a higher level of D1 protein of photosystem II suggesting lower susceptibility to photooxidative damage in the presence of norflurozan (Wagner et al., 2002).

2.9. 4-Hydroxyphenylpyruvate dioxygenase (HPPD)

4-Hydroxyphenylpyruvate dioxygenase (HPPD) is involved in the prenylquinone pathway in plants as it provides homogentisic acid that is an aromatic precursor of all prenylquinones namely tocopherols and plastoquinones. HPPD activity has been observed both in the chloroplasts and the peroxisomes where it performs anabolic and catabolic functions respectively (Feidler et al., 1982).

HPPD is inhibited by Isoxaflutole and triketone family of bleaching herbicides namely Sulcotrione and NTBC. Tolerance to HPPD specific herbicides can be engineered in plants as demonstrated by the expression of the *Pseudomonas* HPPD in tobacco. Again, much higher tolerance was achieved when the HPPD was targeted to the plastids (Sailland et al., 1996). Arabidopsis 4-hydroxyphenylpyruvate dioxygenase was expressed in tobacco along with the yeast (*Saccharomyces cerevisiae*) prephenate dehydrogenase gene. Transgenic plants accumulated high levels of tocotrienols in leaves. In these plants increased resistance against the herbicidal p-hydroxyphenylpyruvate dioxygenase inhibitor diketonitril was observed (Rippert et al., 2004). Therefore overexpression of HPPD should help in raising transgenic crops resistant to HPPD specific herbicides.

2.10. Dihydropteroate synthase (DHPS)

Dihydropteroate synthase (DHPS) is involved in the folate synthesis pathway (Veerasekaran et al., 1981). It is specifically targeted by the herbicide asulam. Asulam is a substrate analog of p-aminobenzoate for 7,8-dihydropteroate synthase. An asulam tolerance gene *sul1* that encodes for a mutant DHPS was used for transformation of tobacco. The enzyme was targeted to the chloroplast. Expression of *sul*1 enabled selection of transgenic plants on sulphadiazine (Guerineau et al., 1990) and the resulting plants could withstand excess foliar application rates of asulam (Cole, 1994). The *sul1* gene has further been engineered into potato for the control of parasitic weeds (Surov et al., 1998).

3. HERBICIDE RESISTANT CROPS USING METABOLIC DETOXIFICATION

Another mechanism for engineering herbicide resistance is to express genes that code for enzymes, which detoxify the toxic herbicides.

3.1. Glyphosate tolerance

In addition to expressing insensitive forms of EPSPS enzyme, tolerance to glyphosate can be conferred by expressing enzymes that actively break down the herbicide. A glyphosate detoxification gene *gox*, identified from a glyphosate waste stream facility, encodes for glyphosate oxidoreductase and helps the bacterium *Achromobacter* spp strain LBAA utilize glyphosate as a carbon or phosphorous source (Hallas et al., 1988; Barry et al., 1992). The *gox* gene is not employed alone but rather in combination with C4 EPSPS and when both are targeted to the plastids the effectiveness of conferred tolerance is higher (Zhou et al., 1995; Mannerlof et al., 1997). Transgenic canola and maize expressing both these genes are now grown commercially (Saroha et al., 1998).

An alternative method to achieve detoxification is by N-acetylation of glyphosate that converts it into its non-toxic form N-acetylglyphosate. Recently enzymes exhibiting glyphosate N-acetyltransferase activity have been discovered from Bacillus sps. One such gene *gat* from *B. licheniformis* was subjected to repeated DNA shuffling to obtain a gene that provides increased glyphosate tolerance when engineered into E. coli, *Arabidopsis*, tobacco and maize (Castle et al., 2004). Although, DNA shuffling resulted in a 10,000 fold higher K_{cat}/K_m improvement over the parental enzyme, the gene conferred only moderate levels of resistance to glyphosate when engineered in transgenic plants. This may be due to the cytosolic location of the enzyme instead of plastids.

3.2. Bromoxynil tolerance

The herbicide bromoxynil acts as a photosynthetic electron transport inhibitor and an uncoupler (Sanders and Pallett, 1985). It is not well tolerated by dicots but in wheat fields it efficiently eliminates broad leaf weeds. A nitrilase-encoding gene *oxy* that uses bromoxynil as a substrate was isolated from *Klebsiella ozaenae* (Stalker and McBride, 1987; Stalker et al., 1988). The enzyme detoxifies bromoxynil to nontoxic benzoic acid. The *oxy* gene has been successfully expressed in tobacco cotton, potato and oilseed rape where a high level of tolerance was observed (Freyssinet et al., 1989). The nitrilase detoxifies bromoxynil in the cytosol even before it reaches its site of action in the chloroplast. Currently two lines of cotton resistant to bromoxynil are commercially grown in the US (ISB 2004).

3.3. Glufosinate tolerance

Glufosinate inhibits glutamine synthetase and nitrogen metabolism in plants (Leason et al., 1982; Tachibana et al., 1986). Glufosinate tolerance has been engineered in more than 36 plant species by the expression of *pat* and *bar* genes isolated from two streptomyces sps (Murakami et al., 1986; Strauch et al., 1988). The first plant to be engineered for glufosinate tolerance was tobacco where expression of the *bar* gene conferred efficient tolerance to glufosinate without any deleterious effect on flowering or seed set (De Block et al., 1987). Since then glufosinate tolerance has been engineered in several dicot and monocot crops.

Glufosinate resistant crops are grown commercially in over 8 countries worldwide. The crops include 9 lines of *Zea mays*, 8 lines of *Brassica napus*, 4 lines of *Glycine max* and 1 line each of Sugarbeet, *Brassica rapa*, Chicory and Cotton (AGBIOS, 2004).

4. NOVEL APPROACHES FOR ENGINEERING OF HERBICIDE TOLERANCE IN HIGHER CROPS

4.1. Chloroplast transformation

As is evident from the preceding discussions targeting a herbicide insensitive gene or a herbicide-detoxifying gene to the plastids proves to be more effective against the herbicide. This should not be surprising as the site of action of these herbicides resides within plastids. Therefore expressing these candidate genes in the chloroplast by engineering the gene directly in the chloroplast genome should prove to be more efficient. Chloroplast transcription and translation machinery is partly prokaryotic in nature; therefore expressing prokaryotic genes coding for herbicide resistance in the chloroplast seems to be an attractive option. Chloroplast transformation has been used for the hyper expression of transgenes (Daniell et al., 2002a). It also enables single step engineering of multiple genes (DeCosa et al., 2001; Daniell and Dhingra, 2002; Ruiz et al., 2003). Most importantly there is transgene containment due to maternal inheritance of the transgenes (Daniell, 2002). Also, position effect due to site-specific integration (Daniell et al., 2002b) and pleiotropic effects due to sub-cellular compartmentalization of transgenes (Daniell et al., 2001; Lee et al., 2003) are avoided. Thus far 6 agronomic traits including herbicide resistance and 11 biopharmaceutical proteins have been successfully engineered via chloroplast transformation mostly in tobacco (Kumar and Daniell, 2004, Daniell et al., 2004a; Dhingra and Daniell, 2004; Devine and Daniell, 2004; Daniell et al., 2004b). Also, engineering of RUBISCO has been successfully achieved for the first time using the chloroplast transformation approach (Dhingra et al., 2004). For a detailed discussion on these aspects please refer to chapter 17 in this book. Recently reproducible protocols for chloroplast transformation have been established for cotton, soybean and carrot that should enable expression of desired agronomic traits in these species as well (Kumar et al., 2004a, b; Dhingra and Daniell, 2004, Dufourmantel et al., 2004).

4.2. Herbicide resistance via chloroplast genetic engineering

Herbicide resistance has been engineered in tobacco via chloroplast genetic engineering where considerable level of herbicide tolerance has been achieved. Chloroplast transgenic glyphosate tolerant tobacco plants were obtained for the first time by expressing an EPSPS gene from Petunia where maternal inheritance of the transgene was also demonstrated (Daniell et al., 1998). Amplification of the target site i.e. EPSPS, facilitated high-level tolerance to the herbicide. In the same study it was proposed that mutant forms of EPSPS that are insensitive to glyphosate would be ideal candidates for engineering herbicides. A petunia derived EPSPS gene was engineered into the transcriptionally active spacer region between the *trn*I-*trn*A genes in the *rrn* operon of the tobacco chloroplast genome or in the transcriptionally silent spacer region between *rbc*L-*acc*D genes (Daniell et al., 1998). The copy number of the transgene varied depending on the site of integration; ~ 20000 when integrated in between the *trn*I-*trn*A genes and ~10000 when integrated between the *rbc*L-*acc*D genes as the former is present in the inverted repeat region and the latter in the large single copy region of the chloroplast genome. Successful integration of the EPSPS gene into the chloroplast genome was confirmed by PCR followed by Southern analysis. The chloroplast transgenic plants expressing EPSPS could survive up to 5 mM of sprayed glyphosate compared to wild type plant that perished at even 0.5 mM of glyphosate (Figure 6). Even though the petunia EPSPS gene and the wild type EPSPS have similar susceptibility to glyphosate, overexpression of EPSPS in the chloroplast probably out titrated the substrate and resulted in glyphosate tolerance in the chloroplast transgenic plants.

***Figure 6**: Herbicide resistance assay. 18-week-old chloroplast transgenic and wild type plants were sprayed with 5 mM glyphosate solution. A. Chloroplast transgenic line; B. Wild type control. (see Color Plate 15).*

This report also unequivocally demonstrated maternal inheritance of the introduced transgene (Figure 7). This is especially relevant in the context of transgene escape via pollen. There is a high rate of gene flow from crops to wild relative. In wild sunflower the frequency of marker gene outcross is 28-38%, in wild strawberries, 50% of the wild type plants harbored the marker genes from cultivated strawberries and the herbicide resistance trait from transgenic oilseed rape crossed over to field brassica under field conditions (Kling, 1996; Mikkelsen et al., 1996; Linder et al., 1998). It seems that under such high rates of gene flow, engineering herbicide resistance transgenes via chloroplast engineering should be an obvious alternative.

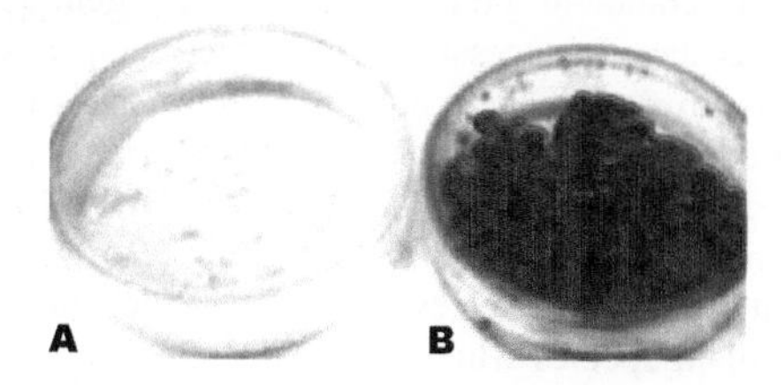

***Figure** 7*: Maternal Inheritance of transgenes. A. Wild type and B. Chloroplast transgenic seeds expressing EPSPS gene were germinated on MSO medium supplemented with 500 mg/l spectinomycin.

In a subsequent study three mutant EPSPS genes were engineered into the chloroplast genome of tobacco (Ye et al., 2000). Even though the EPSPS enzyme constituted 10% of tsp, glyphosate tolerance levels were similar to the tolerance in the nuclear-engineered plant used as a control (Ye et al., 2001). Recently, EPSPS encoding *aro*A gene from *S. typhimurium* was expressed in tobacco with the N-terminus encoding region integrated in the chloroplast genome and the C-terminus encoding region integrated in the nuclear genome. With the help of intein proteins this split gene was reconstituted into a functional gene in the chloroplast. The chloroplast transgenic plants were tolerant to higher levels of glyphosate compared to the wild type (Chin et al., 2003). The authors proposed this approach for preventing outcrossing of functional genes from transgenic plants to their wild relatives.

Chloroplast integrated *bar* gene was demonstrated to confer high-level tolerance to glufosinate. Seeds obtained from T1 chloroplast transgenic plants germinated on 25 ∝g/ml glufosinate while the wild type seeds could not survive even at 5 ∝g/ml glufosinate. Further, soil-grown chloroplast transgenic plants could survive 2.5% herbicide solution where as WT plants were killed with a 0.1% solution of the herbicide (Iamtham and Day, 2000). Tolerance to phosphinothricin was demonstrated in chloroplast transgenic tobacco plants expressing bacterial *bar* gene or a synthetic codon-optimized *bar* gene. These transgenic plants exhibited field level tolerance to phosphinothricin (Lutz et al., 2001).

5. CONCLUSIONS

Chloroplast is the site where various metabolic processes, including photosynthesis, biosynthesis of chlorophyll, carotenoids, purines, pyrimidines, fatty acids and reduction of nitrite and sulfate are carried out. Several of the enzymes involved in these biosynthetic pathways are sensitive to herbicide action and thus most commercially available herbicides, directly or indirectly, affect one or the other chloroplast functions. So far all the commercial transgenic herbicide resistant crops harbor the transgene in the nuclear genome. The foreign protein is usually targeted to the chloroplast and in several examples, this has improved the degree of herbicide tolerance. This observation suggests that it would be desirable to introduce the herbicide resistance transgenes directly into the chloroplast genome where overexpression of the insensitive enzyme or a herbicide detoxifying enzyme would be able to confer an improved degree of tolerance to the herbicide. Nuclear genetic engineering for herbicide resistance has been successfully employed and has markedly improved the modern day agriculture but there are impending issues of transgene containment and development of herbicide resistance in wild relatives.

Chloroplast transformation technology has the unique advantage of alleviating the concern of herbicide resistance transgenes outcrossing to wild relatives. It has been demonstrated that transgenes expressed via the chloroplast genome are maternally inherited and in plant species where there is a rare event of paternal inheritance and seed or fruit is not the commercial product, methods are now available to engineer male sterility (Ruiz and Daniell, 2004). It has been proposed that for improved tolerance to herbicides, engineering of multiple genes would be desirable. This is exemplified by the glyphosate tolerant maize and canola that have been engineered with CP4 EPSPS gene and *gox* gene, which codes for a glyphosate-detoxifying enzyme. It is well known that engineering of the beta-carotene pathway in rice that involved four genes was accomplished in seven years (Ye et al., 2000, Daniell and Dhingra, 2002) but improved methods are now available. Multigene engineering is routinely done in case of chloroplast transformation (DeCosa et al., 2001, Ruiz et al., 2003, Lossl et al., 2003). Therefore new strategies involving engineering of multiple transgenes could be incorporated in the ongoing endeavor of generating herbicide resistant crops.

There is also a need for identifying new targets and novel herbicides that provide enhanced performance and greater environmental compatibility, as the toxic effect of agrochemicals is yet another concern. Recently, a functional genomics approach in tobacco has resulted in the identification of 46 potential herbicide targets whose partial inhibition results in chlorosis and necrosis leading to growth defects (Lein et al., 2004). Some of the cDNAs identified in this study encode hitherto unidentified proteins. This necessitates functional characterization of these proteins and utilization of those in transformation strategies. Generation of novel mutant versions of target enzymes by staggered PCR approach or repeated DNA shuffling could also help in the ongoing endeavor of engineering herbicide tolerance. It seems that the next phase of genetic engineering for herbicide resistance would greatly benefit from the plethora of information that would be generated from the functional genomics and the information already available from nuclear-engineered plants. Combination of the knowledge gained from these efforts with chloroplast transformation approach should help in creating environmentally friendly herbicide resistant crops.

6. REFERENCES

AGBIOS. (2004) Biotech crop database. http://www.agbios.com/dbase.php (accessed May 2004).

Amrhein, N., Deus, B., Gehrke, P., & Steinrucken, H. C. (1980). The site of the inhibition of the Shikimate pathway by Glyphosate .2. Interference of glyphosate with Chorismate formation *in vivo* and in vitro. *Plant Physiology*, 66, 830-834.

Aono, M., Kubo, A., Saji, H., Tanaka, K., & Kondo, N. (1993). Enhanced tolerance to photooxidative stress of transgenic *Nicotiana tabacum* with high chloroplastic glutathione-reductase activity. *Plant and Cell Physiology*, 34, 129-135.

Aviv, D., Amsellem, Z., & Gressel, J. (2002). Transformation of carrots with mutant acetolactate synthase for *Orobanche* (broomrape) control. *Pest Management Science,* 58, 1187-1193.

Barry, G., Kishore, G., Padgette, S., Taylor, M., Kolacz, K., Weldon, M., et al. (1992). Inhibition of amino acid biosynthesis: strategies for imparting glyphosate tolerance to crop plants. In B. K. Singh, H. E. Flores & J. C. Shannon (Eds.), *Biosynthesis and Molecular Recognition of Amino Acids in Plants* (pp. 139-145). Rockville, MD: American Society of Plant Physiologists.

Bartels, P. G., & Watson, C. W. (1978). Inhibition of carotenoid synthesis by fluridone and norflurazon. *Weed Science,* 26, 198-203.

Botella, J. R., Verbelen, J. P., & Valpuesta, V. (1988). Immunocytolocalization of glutamine synthetase in green leaves and cotyledons of *Lycopersicon esculentum*. *Plant Physiology,* 88, 943-946

Castle, L. A., Siehl, D. L., Gorton, R., Patten, P. A., Chen, Y. H., Bertain, S., Cho, H. J., Duck, N., Wong, J., Liu, D., Lassner, M. W. (2004) Discovery and directed evolution of a glyphosate tolerance gene. *Science* 304,1151-1154.

Chin, H. G., Kim, G. D., Marin, I., Mersha, F., Evans, T. C., Chen, L. X., et al. (2003). Protein trans-splicing in transgenic plant chloroplast: Reconstruction of herbicide resistance from split genes. *Proceedings of the National Academy of Sciences of the United States of America,* 100, 4510-4515.

Cole, D. J. (1994). Detoxification and Activation of Agrochemicals in Plants. *Pesticide Science,* 42, 209-222.

Comai, L., Sen, L. C. and Stalker, D. M. (1983) An altered a*ro*A gene-product confers resistance to the herbicide glyphosate. *Science* 221, 370-371.

Comai, L., Facciotti, D., Hiatt, W. R., Thompson, G., Rose, R. E., & Stalker, D. M. (1985). Expression in plants of a mutant aroA gene from *Salmonella typhimurium* confers tolerance to glyphosate. *Nature,* 317, 741-744.

Daniell, H. (2002). Molecular strategies for gene containment in transgenic crops. *Nature Biotechnology,* 20, 581-586.

Daniell, H., Datta, R., Varma, S., Gray, S., & Lee, S. B. (1998). Containment of herbicide resistance through genetic engineering of the chloroplast genome. *Nature Biotechnology,* 16, 345-348.

Daniell, H., & Dhingra, A. (2002). Multigene engineering: dawn of an exciting new era in biotechnology. *Current Opinion in Biotechnology,* 13, 136-141.

Daniell, H., Dhingra, A., & Allison, L. (2002). Chloroplast transformation: from basic molecular biology to biotechnology. *Reviews in Plant Physiology & Biochemistry,* 1, 1-20.

Daniell, H., Khan, M. S., & Allison, L. (2002). Milestones in chloroplast genetic engineering: an environmentally friendly era in biotechnology. *Trends in Plant Science,* 7, 84-91.

Daniell, H., Lee, S. B., Panchal, T., & Wiebe, P. O. (2001). Expression of the native cholera toxin B subunit gene and assembly as functional oligomers in transgenic tobacco chloroplasts. *Journal of Molecular Biology,* 311, 1001-1009.

Daniell, H., Ruiz, O. N., & Dhingra, A. (2004a). Chloroplast genetic engineering to improve agronomic traits. *Methods in Molecular Biology,* 286, 111-137.

Daniell, H., Carmona-Sanchez, O., & Burns, B. (2004b). Chloroplast derived antibodies, biopharmaceuticals and edible vaccines. In S. Schillberg (Ed.), *Molecular Farming* (pp. 113-133). Germany: Wiley -VCH Verlag publishers.

De Cosa, B., Moar, W., Lee, S. B., Miller, M., & Daniell, H. (2001). Overexpression of the Bt *cry*2Aa2 operon in chloroplasts leads to formation of insecticidal crystals. *Nature Biotechnology,* 19, 71-74.

DeBlock, M., Botterman, J., Vandewiele, M., Dockx, J., Thoen, C., Gossele, V., et al. (1987). Engineering herbicide resistance in plants by expression of a detoxifying enzyme. *EMBO Journal,* 6, 2513-2518.

Devine, A. L., & Daniell, H. (2004). Chloroplast genetic engineering. In S. Moller (Ed.), *Plastids* (283-320). United Kingdom: Blackwell Publishers.

Devine, M. D., Duke, S. O., & Fedtke, C. (1993). *Physiology of herbicide action*. Englewood Cliffs, NJ: Prentice Hall.

Devine, M. D., & Preston, C. (2000). The molecular basis of herbicide resistance. In A. H. Cobb & R. C. Kirkwood (Eds.), *Herbicides and their mechanisms of action* (pp. 72-104). Boca Raton, FL: CRC Press.

Dhingra, A., & Daniell, H. (2004). Chloroplast genetic engineering via organogenesis or somatic embryogenesis. *Arabidopsis protocols, 2*, in press.

Dhingra, A., Portis Jr, A. R. and Daniell, H. (2004) Enhanced translation of a chloroplast expressed *Rbc*S gene restores SSU levels and photosynthesis in nuclear antisense *Rbc*S plants. *Proceedings of the National Academy of Sciences of the United States of America,* 101, 6315-6320.

Dodge, A. D. (1991). Photosynthesis. In R. C. Kirkwood (Ed.), *Target sites for herbicide action* (pp. 1-28). New York: Plenum Press.

Dufourmantel, N., Pelissier, B., Garcon, F., Peltier, G., Ferullo, J.-M., & Tissot, G. (2004). Generation of fertile transplastomic soybean. *Plant Molecular Biology, In press.*

Duke, S. O. (1988). Glyphosate. In P. C. Kearney & D. D. Kaufman (Eds.), *Herbicides - Chemistry, Degradation and Mode of Action* (Vol. III, pp. 1-70). New York: Marcel Dekker.

Fiedler, E., Soll, J., & Schultz, G. (1982). The formation of homogentisate in the biosynthesis of tocopherol and plastoquinone in spinach chloroplasts. *Planta,* 155, 511-515.

Fillatti, J. J., Kiser, J., Rose, R., & Comai, L. (1987). Efficient transfer of a glyphosate tolerance gene into tomato using a binary *Agrobacterium tumefaciens* vector. *Bio-Technology,* 5, 726-730.

Freyssinet, G. (2003). Herbicide-resistant transgenic crops - A benefit for agriculture. *Phytoparasitica,* 31, 105-108.

Freyssinet, G., Leroux, B., Lebrun, M., Pelissier, B., Sailland, A., & Pallett, K. E. (1989). *Transfer of bromoxynil resistance into crops.* Paper presented at the Brighton Crop Protection Conference, Farnham, UK.

Golden, S. S., & Haselkorn, R. (1985). Mutation to herbicide resistance maps within the *psb*A gene of *Anacystis nidulans* R2. *Science,* 229, 1104-1107.

Guerineau, F., Brooks, L., Meadows, J., Lucy, A., Robinson, C., & Mullineaux, P. (1990). Sulfonamide resistance gene for plant transformation. *Plant Molecular Biology,* 15, 127-136.

Ha, S. B., Lee, S. B., Lee, Y., Yang, K., Lee, N., Jang, S. M., et al. (2004). The plastidic Arabidopsis protoporphyrinogen IX oxidase gene, with or without the transit sequence, confers resistance to the diphenyl ether herbicide in rice. *Plant Cell and Environment,* 27, 79-88.

Hallas, L. E., Hahn, E. M., & Korndorfer, C. (1988). Characterization of microbial traits associated with glyphosate biodegradation in industrial activated sludge. *Journal of Industrial Microbiology,* 3, 377-385.

Hattori, J., Brown, D., Mourad, G., Labbe, H., Ouellet, T., Sunohara, G., et al. (1995). An acetohydroxy acid synthase mutant reveals a single-site involved in multiple herbicide resistance. *Molecular & General Genetics,* 246, 419-425.

Hattori, J., Rutledge, R., Labbe, H., Brown, D., Sunohara, G., & Miki, B. (1992). Multiple resistance to sulfonylureas and imidazolinones conferred by an acetohydroxy acid synthase gene with separate mutations for selective resistance. *Molecular & General Genetics,* 232, 167-173.

Haughn, G. W., & Somerville, C. (1986). Sulfonylurea resistant mutants of *Arabidopsis thaliana. Molecular & General Genetics,* 204, 430-434.

Hausler, R. E., Holtum, J. A. M., & Powles, S. B. (1991). Cross resistance to herbicides in annual ryegrass (*Lolium rigidum*) .4. correlation between membrane effects and resistance to graminicides. *Plant Physiology,* 97, 1035-1043.

He, M., Yang, Z. Y., Nie, Y. F., Wang, J., & Xu, P. L. (2001). A new type of class I bacterial 5-enopyruvylshikimate-3-phosphate synthase mutants with enhanced tolerance to glyphosate. *Biochimica Et Biophysica Acta-General Subjects,* 1568, 1-6.

Hoagland, R. E., Duke, S. O., & Elmore, D. (1978). Effects of glyphosate on metabolism of phenolic compounds .2. influence on soluble hydroxyphenolic compound, free amino-acid and soluble protein levels in dark-grown maize roots. *Plant Science Letters,* 13, 291-299.

Iamtham, S., & Day, A. (2000). Removal of antibiotic resistance genes from transgenic tobacco plastids. *Nature Biotechnology,* 18, 1172-1176.

ISB. (2004) Information Systems for Biotechnology. http://www.isb.vt.edu (accessed May 2004).

Kling, J. (1996). Agricultural ecology - Could transgenic supercrops one day breed superweeds? *Science,* 274, 180-181.

Konishi, T., & Sasaki, Y. (1994). Compartmentalization of 2 forms of acetyl-coa carboxylase in plants and the origin of their tolerance toward herbicides. *Proceedings of the National Academy of Sciences of the United States of America,* 91, 3598-3601.

Kumar, S., & Daniell, H. (2004). Engineering the chloroplast genome for hyperexpression of human therapeutic proteins and vaccine antigens. In P. Balbas & A. Lorence (Eds.), *Recombinant Protein Protocols* (Vol. 267, pp. 365-383). New Jersey: The Humana Press Inc.

Kumar, S., Dhingra, A., & Daniell, H. (2004a). Plastid expressed betaine aldehyde dehydrogenase gene in carrot cultured cells, roots and leaves confers enhanced salt tolerance. *Plant Physiology*, in press.

Kumar, S., Dhingra, A., & Daniell, H. (2004b). Stable cotton plastid transformation using one or two genes driven by green and non-green regulatory signals, conferring resistance to the same selection agent. *Plant Molecular Biology*, in press.

Lara, M., Porta, H., Padilla, J., Folch, J., & Sanchez, F. (1984). Heterogeneity of glutamine synthetase polypeptides in *Phaseolus vulgaris* L. *Plant Physiology,* 76, 1019-1023.

Leason, M., Cunliffe, D., Parkin, D., Lea, P. J., & Miflin, B. J. (1982). Inhibition of pea leaf glutamine-synthetase by methionine sulfoximine, phosphinothricin and other glutamate analogs. *Phytochemistry,* 21, 855-857.

Lebrun, M., Sailland, A., Freyssinet, G., & Degryse, E. (1997). Mutated EPSPS, gene coding for said protein and transformed plants containing said gene (pp. WO 97/04103).

Lee, H. J., Lee, S. B., Chung, J. S., Han, S. U., Han, O., Guh, J. O., et al. (2000). Transgenic rice plants expressing a Bacillus subtilis protoporphyrinogen oxidase gene are resistant to diphenyl ether herbicide oxyfluorfen. *Plant and Cell Physiology,* 41, 743-749.

Lee, S. B., Kwon, H. B., Kwon, S. J., Park, S. C., Jeong, M. J., Han, S. E., Daniell, H. (2003). Accumulation of trehalose within transgenic chloroplasts confers drought tolerance. *Molecular Breeding,* 11, 1-13.

Lein, W., Bornke, F., Reindl, A., Ehrhardt, T., Stitt, M., & Sonnewald, U. (2004). Target-based discovery of novel herbicides. *Current Opinion in Plant Biology,* 7, 219-225.

Lermontova, I., & Grimm, B. (2000). Overexpression of plastidic protoporphyrinogen IX oxidase leads to resistance to the diphenyl-ether herbicide acifluorfen. *Plant Physiology*, 122, 75-83.

Linder, C. R., Taha, I., Seiler, G. J., Snow, A. A. & Rieseberg, L. H. (1998). Long-term introgression of crop genes into wild sunflower populations. *Theoretical and Applied Genetics,* 96, 339-347.

Lossl, A., Eibl, C., Harloff, H. J., Jung, C. and Koop, H. U. (2003) Polyester synthesis in transplastomic tobacco (Nicotiana tabacum L.): significant contents of polyhydroxybutyrate are associated with growth reduction. *Plant Cell Reports* 21, 891-899.

Lutz, K. A., Knapp, J. E., & Maliga, P. (2001). Expression of *bar* in the plastid genome confers herbicide resistance. *Plant Physiology,* 125, 1585-1590.

Mannerlof, M., Tuvesson, S., Steen, P., & Tenning, P. (1997). Transgenic sugar beet tolerant to glyphosate. *Euphytica,* 94, 83-91.

Mazur, B. J., & Falco, S. C. (1989). The development of herbicide resistant crops. *Annual Review of Plant Physiology and Plant Molecular Biology,* 40, 441-470.

McCloskey, W. B., & Holt, J. S. (1990). Triazine resistance in S*enecio vulgaris* parental and nearly isonuclear backcrossed biotypes is correlated with reduced productivity. *Plant Physiology,* 92, 954-962.

McHughen, A., & Holm, F. A. (1995). Transgenic flax with environmentally and agronomically sustainable attributes. *Transgenic Research,* 4, 3-11.

Miki, B. L., Labbe, H., Hattori, J., Ouellet, T., Gabard, J., Sunohara, G., et al. (1990). Transformation of *Brassica napus* canola cultivars with *Arabidopsis thaliana* acetohydroxyacid synthase genes and analysis of herbicide resistance. *Theoretical and Applied Genetics,* 80, 449-458.

Mikkelsen, T. R., Andersen, B., & Jorgensen, R. B. (1996). The risk of crop transgene spread. *Nature,* 380, 31-31.

Misawa, N., Masamoto, K., Hori, T., Ohtani, T., Boger, P., & Sandmann, G. (1994). Expression of an *Erwinia phytoene desaturase* gene not only confers multiple resistance to herbicides interfering with carotenoid biosynthesis but also alters xanthophyll metabolism in transgenic plants. *Plant Journal,* 6, 481-489.

Mousdale, D. M., & Coggins, J. R. (1985). Subcellular localization of the common shikimate pathway enzymes in *Pisum sativum* L. *Planta,* 163, 241-249.

Murakami, T., Anzai, H., Imai, S., Satoh, A., Nagaoka, K., & Thompson, C. J. (1986). The bialaphos biosynthetic genes of *Streptomyces hygroscopicus* – molecular cloning and characterization of the gene cluster. *Molecular & General Genetics,* 205, 42-50.

Nida, D. L., Kolacz, K. H., Buehler, R. E., Deaton, W. R., Schuler, W. R., Armstrong, T. A., et al. (1996). Glyphosate-tolerant cotton: Genetic characterization and protein expression. *Journal of Agricultural and Food Chemistry,* 44, 1960-1966.

Nikolskaya, T., Zagnitko, O., Tevzadze, G., Haselkorn, R., & Gornicki, P. (1999). Herbicide sensitivity determinant of wheat plastid acetyl-CoA carboxylase is located in a 400-amino acid fragment of the carboxyltransferase domain. *Proceedings of the National Academy of Sciences of the United States of America,* 96, 14647-14651.

Ott, K. H., Kwagh, J. G., Stockton, G. W., Sidorov, V., & Kakefuda, G. (1996). Rational molecular design and genetic engineering of herbicide resistant crops by structure modeling and site-directed mutagenesis of acetohydroxyacid synthase. *Journal of Molecular Biology,* 263, 359-368.

Padgette, S. R., Re, D. B., Barry, G. F., Eichholtz, D. E., Delanny, X., Fuchs, R. L., et al. (1996). New weed control opportunities: development of soybeans with a Roundup ready gene. In S. O. Duke (Ed.), *Herbicide-Resistant Crops: Agricultural, Environmental, Economic, Regulatory and Technical Aspects* (pp. 53-84). Boca Raton, FL: CRC Press.

Park, M. Y., Yi, N. R., Lee, H. Y., Kim, S. T., Kim, M., Park, J. H., et al. (2002). Generation of chlorsulfuron-resistant transgenic garlic plants (*Allium sativum* L.) by particle bombardment. *Molecular Breeding,* 9, 171-181.

Preston, C. (1994). Resistance to photosystem I disrupting herbicides. In S. B. Powles & J. A. M. Holtum (Eds.), *Herbicide Resistance in Plants: Biology and Biochemistry* (pp. 61-82). Boca Raton, FL: Lewis Publishers.

Rajasekaran, K., Grula, J. W., Hudspeth, R. L., Pofelis, S., & Anderson, D. M. (1996). Herbicide-resistant Acala and Coker cottons transformed with a native gene encoding mutant forms of acetohydroxyacid synthase. *Molecular Breeding,* 2, 307-319.

Rippert, P., Scimemi, C., Dubald, M., & Matringe, M. (2004). Engineering plant shikimate pathway for production of tocotrienol and improving herbicide resistance. *Plant Physiology,* 134, 92-100.

Ruiz, O. N., & Daniell, H. (2004). Engineering male sterility via chloroplast genome. *In Review.*

Ruiz, O. N., Hussein, H. S., Terry, N., & Daniell, H. (2003). Phytoremediation of organomercurial compounds via chloroplast genetic engineering. *Plant Physiology,* 132, 1344-1352.

Sailland, A., Rolland, A., Matringe, M., & Pallett, K. E. (1996). DNA sequence of a gene of hydroxy-phenylpyruvate dioxygenase and production of plants containing a gene of hydroxy-phenylpyruvate dioxygenase and which are tolerant to certain herbicides (pp. WO patent 96/38567).

Sanders, G. E., & Pallett, K. E. (1985). *In vitro* activity and binding characteristics of the hydroxybenzonitriles in chloroplasts isolated from *Matricaria inodora* and *Viola arvensis. Pesticide Biochemistry and Physiology,* 24, 317-325.

Saroha, M. K., Sridhar, P., & Malik, V. S. (1998). Glyphosate-tolerant crops: Genes and enzymes. *Journal of Plant Biochemistry and Biotechnology,* 7, 65-72.

Schulz, A., Wengenmayer, F., & Goodman, H. M. (1990). Genetic engineering of herbicide resistance in higher plants. *Critical Reviews in Plant Sciences,* 9, 1-15.

Schulze-Siebert, D., Heineke, D., Scharf, H., & Schultz, G. (1984). Pyruvate derived amino acids in spinach chloroplasts - synthesis and regulation during photosynthetic carbon metabolism. *Plant Physiology,* 76, 465-471.

Scott, S. E., & Wilkinson, M. J. (1999). Low probability of chloroplast movement from oilseed rape (*Brassica napus*) into wild *Brassica rapa. Nature Biotechnology,* 17, 390-392.

Sen Gupta, A., Webb, R. P., Holaday, A. S., & Allen, R. D. (1993). Overexpression of superoxide-dismutase protects plants from oxidative stress - induction of ascorbate peroxidase in superoxide dismutase overexpressing plants. *Plant Physiology,* 103, 1067-1073.

Shah, D. M., Horsch, R. B., Klee, H. J., Kishore, G. M., Winter, J. A., Tumer, N. E., et al. (1986). Engineering herbicide tolerance in transgenic plants. *Science,* 233, 478-481.

Shukla, A., & Devine, M. D. (2000). Mechanisms of selectivity and resistance to triazine herbicides. In H. M. LeBaron, L. P. Gianessi & J. McFarland (Eds.), *The Triazine Herbicides.* Boca Raton FL: Lewis Publishers.

Sikorski, J. A., & Gruys, K. J. (1997). Understanding glyphosate's molecular mode of action with EPSP synthase: Evidence favoring an allosteric inhibitor model. *Accounts of Chemical Research,* 30, 2-8.

Stalker, D. M., Malyj, L. D., & McBride, K. E. (1988). Purification and properties of a nitrilase specific for the herbicide bromoxynil and corresponding nucleotide-sequence analysis of the *bxn* gene. *Journal of Biological Chemistry,* 263, 6310-6314.

Stalker, D. M., & McBride, K. E. (1987). Cloning and expression in *Escherichia coli* of a *Klebsiella ozaenae* plasmid-borne gene encoding a nitrilase specific for the herbicide bromoxynil. *Journal of Bacteriology,* 169, 955-960.

Strauch, E., Wohlleben, W., & Puhler, A. (1988). Cloning of a Phosphinothricin N-Acetyltransferase gene from *Streptomyces viridochromogenes* Tu494 and its expression in *Streptomyces lividans* and *Escherichia coli. Gene,* 63, 65-74.

Surov, T., Aviv, D., Aly, R., Joel, D. M., Goldman-Guez, T., & Gressel, J. (1998). Generation of transgenic asulam-resistant potatoes to facilitate eradication of parasitic broomrapes (*Orobanche* spp.), with the sul gene as the selectable marker. *Theoretical and Applied Genetics,* 96, 132-137.

Tachibana, K., Watanabe, T., Sekizawa, Y., & Takematsu, T. (1986). Action mechanism of bialaphos .2. Accumulation of ammonia in plants treated with bialaphos. *Journal of Pesticide Science,* 11, 33-37.

Tobin, A. K., Ridley, S. M., & Stewart, G. R. (1985). Changes in the activities of chloroplast and cytosolic isoenzymes of glutamine synthetase during normal leaf growth and plastid development in wheat. *Planta,* 163, 544-548.

Veerasekaran, P., Kirkwood, R. C., & Parnell, E. W. (1981). Studies of the mechanism of action of asulam in plants .2. Effect of asulam on the biosynthesis of folic-acid. *Pesticide Science,* 12, 330-338.

Wagner, T., Windhovel, U., & Romer, S. (2002). Transformation of tobacco with a mutated cyanobacterial phytoene desaturase gene confers resistance to bleaching herbicides. *Journal of Biosciences,* 57, 671-679.

Wang, H. Y., Li, Y. F., Xie, L. X., & Xu, P. L. (2003). Expression of a bacterial *aro*A mutant, *aro*A-M1, encoding 5-enolpyruvylshikimate-3-phosphate synthase for the production of glyphosate resistant tobacco plants. *Journal of Plant Research,* 116, 455-460.

Ye, G. N., Hajdukiewicz, P. T. J., Broyles, D., Rodriguez, D., Xu, C. W., Nehra, N., et al. (2001). Plastid-expressed 5-enolpyruvylshikimate-3-phosphate synthase genes provide high level glyphosate tolerance in tobacco. *Plant Journal,* 25, 261-270.

Ye, X. D., Al-Babili, S., Kloti, A., Zhang, J., Lucca, P., Beyer, P., et al. (2000). Engineering the provitamin A (beta-carotene) biosynthetic pathway into (carotenoid-free) rice endosperm. *Science,* 287, 303-305.

Zhang, Q., Liu, Y., & Sodmergen. (2003). Examination of the cytoplasmic DNA in male reproductive cells to determine the potential for cytoplasmic inheritance in 295 angiosperm species. *Plant and Cell Physiology,* 44, 941-951.

Zhou, H., Arrowsmith, J. W., Fromm, M. E., Hironaka, C. M., Taylor, M. L., Rodriguez, D., et al. (1995). Glyphosate-tolerant CP4 and GOX genes as a selectable marker in wheat transformation. *Plant Cell Reports,* 15, 159-163.

Zweig, G., Shavit, N., & Avron, M. (1965). Diquat (1,1'-Ethylene-2,2'-Dipyridylium Dibromide) in Photoreactions of Isolated Chloroplasts. *Biochimica Et Biophysica Acta,* 109, 332.

CHAPTER 18

METABOLIC ENGINEERING OF CHLOROPLASTS FOR ABIOTIC STRESS TOLERANCE

JOE H. CHERRY[1] AND BRENT L. NIELSEN[2]

[1]Department of Biological Sciences, Auburn University, Auburn, AL. 36849-5407. jcherry@acesag.auburn.edu; [2]Department of Microbiology and Molecular Biology, Brigham Young University, 775 WIDB, Provo, UT, 84602. brent_nielsen@byu.edu

Abstract. Plants must be capable of coping with various environmental stresses. These stresses range from high and low temperatures to salinity, drought, oxidative, and radiation damage. Much of the biochemical machinery localized in the chloroplast makes it possible to regulate specific reactions associated with stress tolerance in this organelle. In the case of heat stress, it has been noted that there are chloroplast-localized heat shock proteins that may be involved in the acquisition of thermotolerance in plants. Following an initial heat episode and recovery period, HSP21 was localized in the soluble fraction of the chloroplast. Furthermore, heat stress resulted in a temperature-dependant redistribution of HSP21 from a soluble form to an insoluble chloroplast protein fraction. Other proteins, such as components of the PSII complex, are thought to be most sensitive to both heat and light. Thermotolerance and acclimation of PSII varies widely among species in general. Even in the case of PSII thermotolerance, small heat shock proteins appear to offer thermo-protection.

Glycine betaine accumulates in plants that are drought tolerant and so it appears to act as an osmoprotectant. Protection may occur as a result of the stabilization of the quaternary structure of enzymes and enzyme complexes. Accumulation of trehalose in chloroplasts also has a positive effect on protein against drought. Thus, the insertion of genes into the chloroplast genome that resulted in an over-production of glycine betaine or trehalose, conferred high levels of salt and drought tolerance. Exposure of plants to light intensities higher than that required to saturate photosynthesis may cause light damage or photoinhibition of the thylakoid proteins of the chloroplast. To maintain normal functions under light stress, plants have developed several repair and protective mechanisms. The induction of a family of proteins, called early light-induced proteins (elips) consist of low molecular proteins localized in the thylakoid membranes of chloroplasts. These proteins may be genetically engineered to provide greater stability against photoinhibition. Therefore, there is a great potential for the genetic manipulation of key enzymes involved in stress metabolism in plants within plastids.

One set of metabolic activities involves the over-production of glutathione in transgenic plants. There are several enzymes that play an important role in the oxidation/reduction of compounds that are involved in oxidative stress. When transgenic maize plants were created by targeting iron superoxide dismutase (FeSOD) to chloroplasts, enhanced oxidative stress tolerance was observed. Similarly, MnSOD, localized in chloroplasts of bundle sheath cells conferred stress tolerance.

Transgenes that confer tolerance to abiotic stress should not outcross with weeds and permanently transfer these valuable traits to the weed nuclear genome. Therefore, maternal inheritance of transgenes that confer abiotic stress tolerance via the chloroplast genome may be ideal for this purpose. Because chloroplast genomes of major crops including cotton and soybean have been successfully transformed, this offers an exciting new approach to create transgenic plants with abiotic stress tolerance. While there

H. Daniell and C. D. Chase (eds.), Molecular Biology and Biotechnology of Plant Organelles 513—525.

may be some limitations (including substrates or intermediates in complex pathways) to chloroplast genetic engineering for stress tolerance, there appears to be tremendous potential for increasing tolerance in plants to a number of stresses by expression of appropriate genes within chloroplasts.

1. INTRODUCTION

Chloroplasts are a unique class of intracellular organelles found in green plants. They contain their own DNA molecules organized into discrete protein-associated nucleoids, and chloroplast genome structure and overall gene order in higher plants are highly conserved. The chloroplast genome encodes genes for ribosomal and transfer RNA and some of the proteins involved in chloroplast functions, while nuclear DNA encodes the rest. The chloroplast genome generally varies from 120-200 KB and contains a large and a small single copy region separated by two large (~25 KB) inverted repeat regions. The genome has a capacity to code for 4 subunits of RNA polymerase, 4 ribosomal RNAs, 23 ribosomal proteins, 30 tRNAs, 30 thylakoid proteins for light reactions and the large subunit of Rubisco (Tyangi, 1999). In addition, there are over a dozen genes capable of coding for NADH dehydrogenase subunits and other products. There are also a few open reading frames that have the capacity to code for proteins of other as yet unidentified functions.

2. HEAT STRESS

In response to hyperthermia stress, most cells activate a small set of genes and preferentially synthesize heat shock proteins (HSP) encoded by those genes. The heat shock response was first described in *Drosophila* (Tissieres et al., 1974), and has since been reported in essentially all living eukaryotic and prokaryotic cells. In the past several years, an understanding of the mechanism of heat shock gene activation and expression has increased considerably, but the biological response to heat shock is still not well understood. While there have been numerous molecular and biochemical studies concerning the response of plant cells to induced heat stress, ultrastructural studies have indicated (a) the formation and distribution of heat shock granules, (b) morphological alterations of nucleoli with accumulation of pre-ribosomal ribonucleoprotein, and (c) the intracellular localization of plant HSP (Dylewski et al., 1991). Other experiments (Neumann et al., 1998) indicate that at the onset of heat shock there is a loss of ribosomes, dictyosomes, and endoplasmic reticulum (Belanger et al., 1986) accompanied by morphological alterations of chloroplasts and mitochondria.

Accumulation of small heat shock proteins (sHSPs) in response to high temperature stress is thought to contribute to the development of thermotolerance in eukaryotic organisms, but the mechanism of action is unknown. Osteryoung and Vierling (1994) have investigated the chloroplast-localized HSP21 to define how it contributes to the acquisition of thermotolerance in plants. Following an initial heat stress and recovery period, HSP21 was found to be localized in the soluble fraction of the chloroplast. Further heat stress resulted in a temperature-dependent redistribution of HSP21 from a soluble form to an insoluble fraction of the chloroplast. A nonionic detergent under conditions that release all pigments and proteins from the thylakoid membranes did not solubilize this insoluble HSP21. This indicates that the insoluble form of HSP21 is not attached to membranes. The temperature-dependent redistribution of HSP21 is affected by light intensity but occurs in both leaf and root plastids, which indicates that this process is not dependent on the presence of a photosynthetic apparatus. It appears likely that the HSP21 plays a functional role in thermotolerance of chloroplasts, and that expression of this protein in transgenic plants may enhance heat tolerance.

Other HSPs have been related to biochemical functions besides heat stress. Kropat et al., (1997) have shown that the expression of the nuclear heat-shock genes HSP70A and HSP70B, encoding cytosolic and plastid-localized heat-shock proteins, respectively, can be induced by light as well as heat. The pathway for light induction of typical HSPs is different than that for heat induction. It was observed that the growth of *Chlamydomonas* cells in the dark followed by a shift to a light regime caused photoinhibition that resulted in less damage to photosystem II (the H_2O-oxidizing, quinine-reducing complex) when the cells were first exposed to dim light for 60 minutes. When the HSP70 genes were induced during pre-induction, it was learned that the plastid localized HSP70B levels were detected only after light induction. Based on the use of mutant cells, it was suggested that HSP70B is involved in the recovery of photosystem II after photoinhibition damage by increasing the level of chaperones (Kropat et al., 1997). It is thought that PSII is the most sensitive of the chloroplast thylakoid-membrane protein complexes in photosynthetic electron transfer and ATP synthesis and is one of the most thermolabile photosynthetic processes in general (Heckathorn et al., 1998).

Within PSII, the O_2-evolving-complex proteins are frequently the most susceptible to heat stress, although both the reaction center and the light harvesting complexes can be disrupted by elevated temperatures. PSII's thermotolerance and its acclimation to heat vary widely among species. Other than the possible protective effect of various carotenoids, there is accumulating evidence that chloroplast HSPs are involved in photosynthetic and PSII thermotolerance. Heckathorn et al. (1998) have shown that certain low molecular weight (lmw) HSPs are an important determinant of photosynthetic thermotolerance. The results indicate that chloroplast lmw HSPs (a) increase PSII activity by as much as sevenfold, (b) completely account for all of the observed heat accumulation of PSII and whole-chain electron transport in pre-heat-stressed plants, and (c) impart heat protection to PSII very rapidly. This is important considering PSII is a highly thermolabile component of photosynthetic electron transport.

Recently, Gustavsson et al., (2002) reported that methionine residues in HSP21 are frequently oxidized to methionine sulfoxides, and that reduction of the methionine sulfoxides by peptide methionine sulfoxide reductase (PMSR) can mediate protein repair. Five PMSR-like genes were identified in *Arabidopsis thaliana,* including one plastidic isoform (pPMSR) that is localized in the chloroplast. In this study, a set of highly conserved methionine residues found in HSP21 are chloroplast-localized small heat shock proteins, which can be sulfoxidized and therefore reduced back to methionine pPMSR; when the oxidized form of HSP21 is reduced by pPMSR, the protein regains chaperone-like activity.

Other studies on the turnover of chloroplast ribulose-1,5-bisphosphate carboxylate (Rubisco) show that enzyme activity may be controlled by the redox state of certain cysteine residues. In this case, we know neither 1) whether HSPs are involved in the oxidation/reduction state of Rubisco, or 2) the role of various stressful conditions on CO_2 fixation.

3. COLD STRESS

Many plants from temperate regions are able to increase their freezing tolerance in response to low, non-freezing temperatures. This is an adaptive process known as acclimation (Levitt, 1980); Sakai and Larcher, 1987) that involves a large number of biochemical and physiological changes, including ultrastructural modifications in cellular organelles (Fujikawa and Takabe, 1996) and the plasma membrane (Steponkus, 1994; Griffith and Antikainen, 1996). Many cold-induced alterations are regulated through changes in gene expression, in which a number of these genes from several plants have been isolated and characterized (Thomashow, 1999). While the expression of some cold-inducible genes seems to be specifically regulated by low temperature, many of them respond to abscisic acid (ABA) and water stress. Both ABA and water stress treatments simulate the effects of low temperature by increasing freezing tolerance. Additionally, several cold-inducible genes have been reported to be regulated by salt stress (Kurkela and Borg-Franck, 1992) and light stress (Leyva et al.,1995; Capel et al.,1998).

Plant tissues' developmental stage affects their freezing tolerance. Since low temperatures result in a large multitude of expressed gene products, it is interesting to note that information regarding the regulation of cold-inducible genes that protect the photosynthetic apparatus during growth and development is very limited. Apparently there is no information on cold-inducible genes located in the chloroplasts. Kreps et al. (2002), using Gene-Chip microarray analysis, have identified 2,409 unique stress-regulated genes that display a greater than twofold change in expression with environmental stresses. 42 genes are exclusively regulated by cold stress in both root and leaf tissues that represent only 2% of all the cold-induced changes. After 27 hours of cold treatment, 78 genes were identified in roots while 115 were found in leaves. It was not determined whether any of the gene products are chloroplast localized.

The interaction between plant development and environmental conditions implies that some genes must be regulated by both environmental factors and developmental cues. To understand this situation Medina et al., (2001), isolated two genomic clones for RC12A and RC12B. These two genes from *Arabidopsis* whose expression is induced in response to low temperature, dehydration, salt stress, and ABA transgenic plants with either the RC12A or RC12B promoter fused to the GUS reporter gene, showed strong expression during the first stage of seed development and germination, in vascular bundles, pollen, and most interestingly in guard cells. When transgenic plants were exposed to low temperature, dehydration, salt stress, or

ABA, reporter gene expression was induced in most tissues. These results indicate that RC12A and RC12B are regulated at the transcriptional level during plant development and induced by environmental stimulus. The signaling pathway implicated in the perception and in the transduction of cold signals in the plant cells is poorly understood. Vigh and his collaborators (Vigh et al., 1993) proposed that rigidification of the plasma membrane might be the event that initiates the downstream signaling cascade. In this regard, Ruelland et al., (2002) began to analyze the possible activation of enzymes of the phosphoinositide signaling pathway. While using *Arabidopsis*, they learned that phospholipase C and phospholipase D activation occurred immediately after plants were transferred to low temperature. Also, unsaturation of the membrane lipids of chloroplasts stabilizes the photosynthetic machinery against low-temperature photoinhibition in transgenic tobacco plants (Yokoi et al., 1998).

In this regard, tobacco plants transformed with cDNA for glycerol-3-phosphate acyltransferase demonstrated that chilling tolerance is affected by levels of unsaturated membrane lipids. Of four major lipid classes isolated from thylakoid membranes, phosphatidylglycerol decreased the most in unsaturation during transformation. The isolated thylakoid membranes from wild-type and transgenic plants did not significantly differ from each other in terms of the sensitivity of photosystem II to high and low temperatures or photoinhibition. However, leaves of the transformed plants were more sensitive to photoinhibition than those of wild-type plants. Moreover, the recovery of photosynthesis from photoinhibition in leaves of wild-type plants was faster than leaves of transgenic plants. It appears from these results that unsaturation of fatty acids of phosphatidylglycerol in thylakoid membranes stabilizes the photosynthetic machinery against low-temperature photoinhibition.

Cold induces the expression of a number of genes that encode proteins that enhance tolerance to low temperatures as well as freezing temperatures in plants. Some of the proteins expressed are antifreeze proteins (Atici and Nalbantoglu, 2003), plasma membranes proteins associated with increased cryostability (Kawamura and Uemura, 2003), and transcriptional activators and transcription factors that play a central role in the synthesis of proteins involved in cold tolerance (Wang et el., 2003). However, at this time we have not found information that indicates specific genes localized in the chloroplast that regulate cold tolerance.

4. DROUGHT STRESS

Water substantially alters plant metabolism and photosynthesis, and decreases plant growth and productivity. Water stress decreases photosynthetic assimilation of CO_2 mainly as a result of stomatal closure. In this regard, stress reduces the amounts of ATP and ribulose bisphosphate produced in leaves (Tezara et al., 1999).

Glycine betaine appears to be a critical determinant of stress tolerance in plants, acting as a protectant under drought and/or salinity environments (Rhodes and Hansen, 1993). Also glycine betaine accumulation is induced under stressful conditions (Gordon,1995), and the degree of drought tolerance correlates to the amount of glycine betaine accumulation (Saneoka et al., 1995). Exogenous applications of glycine betaine improves growth and survival of a large number of plants under various stressful conditions (Allard et al., 1998). Of major interest, glycine betaine is very effective in the *in vitro* stabilization of the quaternary structure of enzymes and complex proteins, as well as the conformational state of membranes (Gordon, 1995).

Hayashi and Murata (1998) have demonstrated that one of the enzymes (Choline oxidase, COD) in a soil bacteria *Arthobacter globiformis* is involved in the conversion of choline to glycine betaine. This and other enzymes involved in the COD pathway have been genetically engineered in cyanobacteria (Sakamoto and Mutata, 2001). The enzymes required for glycine betaine synthesis have been stably expressed in various photosynthetic organisms that normally do not accumulate glycine betaine, and the results indicate that the production of glycine betaine enhances tolerance to osmotic stress including drought stress. Sakamoto and Murata (2001) recently demonstrated that glycine betaine protected the photosynthetic system against *in vitro* salt and heat stress. Expression of the *Atriplex hortensis* choline monooxygenase gene in transgenic tobacco was shown to lead to increased protection of plants from drought and salt stress (Shen et al., 2002). This introduced gene was found to be expressed throughout the plant, including in organs lacking functional chloroplasts, and is regulated by drought, abscisic acid and circadian rhythm.

Trehalose is another osmoprotectant produced by many organisms. The introduction of the yeast trehalose phosphate synthase gene into tobacco chloroplasts and nuclear genomes confers drought tolerance (Lee et al., 2003). PCR and Southern blots confirm stable integration of the synthase gene in transgenic plants. Northern blot analysis of transgenic plants showed that the chloroplast transformant expressed several thousand-fold increase over the best surviving nuclear transgenic plant. Nuclear transgenic plants that expressed significant amounts of trehalose accumulation showed stunted phenotype, sterility and other pleiotropic effects while the chloroplast transgenic plants showed normal growth and increased resistance to drought (Lee et al., 2003).

5. SALINITY STRESS

Salinity and drought stresses in both dry land and irrigated agricultural settings are the most important factors limiting modern agricultural systems (Cushman and Bohnert, 2000). Genetic and molecular studies have shown that many gene products contribute to salinity as well as drought tolerance determinants including: osmoprotectants; ion carriers and channels; transporters and symporters; water channels; reactive oxygen scavengers; heat shock proteins and other stress proteins; transcription activators; and signaling molecules (Hasegawa et al., 2000). The common ice plant (*Mesembryanthemum crystallium*, a halophyte) has adapted to extreme environments of high salt. These plants are thought to have evolved novel structural or regulatory adaptations that allow them to survive in hostile environments. These plants are useful as a source of salt stress determinants in genetic engineering studies. Agarie et al., (2003) used the ice plant to make a cDNA microarray analysis of the temporal gene expression profile in response to high salt. They noted pronounced changes in gene expression under salt stress over time for an estimated 15% of all genes in the ice plant genome.

Previously, we discussed the importance of glycine betaine as an osmoprotectant for stress tolerance. Certain plants synthesize glycine betaine in chloroplasts via a two-step oxidation of choline (Cho). Previously, a chloroplastic glycine betaine synthesis pathway from spinach was inserted into tobacco lacking glycine betaine that did not express choline monooxygenase (CMO). Transformants had low CMO enzyme activity and produced little glycine betaine. In contrast, tobacco expressing a cytosolic glycine betaine pathway accumulated much more (up to a hundred-fold increase) glycine betaine. Nuccio et al., (1998) and Nuccio et al., (2000) have suggested that the import of choline into chloroplasts may be a limiting factor for the flux to glycine betaine in the chloroplastic pathway. Kumar et al (2004a) have reported that stable expression of the *badh* gene via the carrot plastid genome resulted in very high levels of betaine accumulation within plastids (as high as in halophytes) and resulted in the highest levels of salt tolerance (up to 400 mM NaCl) reported in the literature.

The additional engineering of a high-activity choline transporter in the chloroplast envelope may be required for obtaining high levels of glycine betaine in chloroplasts engineered to express choline monooxygenase (McNeil et al., 2000). Sakamoto et al., (1998) have genetically engineered rice plants to synthesize glycine betaine by introducing the codA gene for choline oxidase from the soil bacterium *Arthobacter globiformis.* Treatment of wild-type and transformed plants with 0.15 M NaCl showed that both types of plants grew slowly in the beginning, but only the transgenic plants began to grow normally within a short time. Using photosynthetic inactivation as a measure of cellular damage, it was determined that CHLCOD plants were more tolerant to salt stress than were the CytCOD plants. These results indicate that the subcellular compartmentalization of the biosynthesis of glycine betaine is a critical element in the control of salt tolerance.

The glyoxalase pathway involving glyoxalase I (gly I) and glyoxalase II (gly II) enzymes are required for glutathione-based detoxification of methylglyoxal. This pathway, particularly gly I, has been indicated as a potential candidate gene for the regulation of salinity tolerance. Abiotic stresses have been shown to have a quantitative character, and thus they are controlled by multiple genes (Cushman and Bohnert, 2000; Hasegawa et. al., 2000). The glyoxalase system has long been known in animal systems (Thornalley et al., 1983) to be involved in cell division and proliferation, microtubule assembly, and protection against oxoaldehyde toxicity. Singla-Pareek et al. (2003) have investigated the effect of genetic engineering of glyoxalase with respect to salinity tolerance by inserting gly I and gly II in tobacco plants. Their results show that the transgenic plants maintain sustained growth and produce seeds under salinity stress, indicating that the glyoxalase pathway is potentially involved in salinity tolerance. These genes are thus potential candidates for expression in chloroplasts to increase tolerance to salinity stress.

6. LIGHT STRESS

Exposure of plants to light intensities higher than those required to saturate photosynthesis leads to a reduction in photosynthetic capacity. This effect is known as photoinhibition or light induced stress syndrome. Photoinhibition is generated by the production of reactive oxygen species (Kasahara et. al., 2002) followed by destruction of proteins, lipids and pigments by generated free radicals. The main target of light stress in plants is the chloroplast; however, other cell compartments are also severely affected. To maintain normal functions under light stress conditions, plants have developed multiple repair and protection mechanisms. The induction of specific light stress proteins or the degradation of damaged cell components and their replacement by *de novo* synthesized molecules is part of the protective mechanism. In one case, a family of early light-induced proteins (elips) consisted of low molecular mass proteins localized in the thylakoid membranes of chloroplasts. Five of these elips have been cloned in *Arabidopsis thaliana* and at this time are predicted to have protective roles within the thylakoids under light stress conditions by scavenging free chlorophyll molecules (Adamksa et. al., 1999).

A daily occurrence in the life of a green plant is the function of photosystem II (PSII) damage and repair cycle in chloroplasts. In *Dunaliella salina* (a green alga), growth under low light results in damage, degradation, and replacement of D1 (the 32-kDa reaction-center protein of PSII) every 7 hours while high light intensity (irradiance stress) entails damage to D1 every 20 minutes. These changes contribute to a shift in the PSII/PSI ratio from 1.4:1 under low-light conditions to 15:1 under irradiance stress. Even under irradiance stress, organization of the altered thylakoid membrane ensures that a small fraction of PSII centers remain functional and sustain electron flow from H_2O to ferredoxin with rates sufficient for chloroplast photosynthesis and cell growth (Vasilikiotis and Melis, 1994). Monod et al., (1994) have identified a novel protein that appears to be a photosystem II subunit that is impaired under irradiation stress. Maintaining electron flow through the photosynthetic apparatus, even in the absence of a sufficient amount of $NADP^+$ as an electron acceptor, is essential for chloroplast protection from photooxidative stress (Rizhsky et al., 2003). Expression of one or more elips in chloroplasts may enhance light stress tolerance.

7. OXIDATIVE STRESS

The formation of oxygen radicals by partial reduction of molecular oxygen is an unfortunate consequence of aerobic life. Active oxygen species (AOS), such as the superoxide anion and hydrogen peroxide (H_2O_2) are natural byproducts of metabolism but they can increase to toxic levels (oxidative stress) during a wide range of environmental stresses (such as chilling, ozone, drought, and salt stress). Defense mechanisms against environmental changes that induce oxidative stress are critical for plant growth and survival. However, the molecular, biochemical and cellular mechanisms that regulate these environmental responses are poorly characterized, and the signaling networks involved are not understood. A central role for reactive oxygen species (ROS) during both biotic and abiotic stress responses is well recognized, although under these conditions ROS can either exacerbate damage or act as signal molecules that activate multiple defense responses. In recent years, a new role for ROS has been identified: the control and regulation of biological processes such as programmed cell death, hormonal signaling, stress responses and development. These studies extend our understanding of ROS and suggest a dual role for ROS in plant biology: 1) to remove of toxic byproducts of aerobic metabolism, and 2) to function as key regulators of metabolic and defense pathways. Aerobic organisms have a battery of enzymatic and non-enzymatic antioxidants that scavenge ROS. Superoxide dismutases (SOD), ascorbate peroxidases (APX) and catalases are the main components within the enzymatic defense system.

Evidence obtained from a number of research laboratories (Van Breusegem et al.,1999) has demonstrated the importance of these enzymes in the protection of plants against environmental stresses. In other studies, transgenic maize plants have been generated by particle gun bombardment that overproduce an *Arabidopsis thaliana* iron superoxide dismutase (FeSOD). To target this enzyme into chloroplasts, the mature FeSOD coding sequence was fused to a chloroplast transit peptide from a pea ribulose-1-5-bisphosphate carboxylase gene. Growth characteristics and *in vitro* oxidative stress tolerance of the transgenic lines indicated that plants with the highest transgenic FESOD activities had enhanced stress tolerance and increased growth rates (Van Breusegem et al., 1999). Also from the same laboratory (Van Breusegem et al., 1998), transgenic maize and tobacco plants have been generated which overproduce a mitochondrial manganese superoxide dismutase (MnSOD) in chloroplasts. The MnSOD gene product was predominately localized in the chloroplasts of the bundle sheath cells of maize.

There is a great potential for the genetic manipulation of key enzymes linked with stress metabolism in transgenic plants. One set of metabolic activities involves the over-production of glutathione in transgenic plants. Futhermore, dehydroascorbate reductase (DHAR) reduces dehydroascorbate (DHA) to ascorbate with glutathione as the electron donor (Shimaoka et al., 2003). Other evidence is emerging that glutathione provides protection against oxidative stress (Creissen et al.,1996). In another situation (Konig et al., 2002), the 2-cysteine peroxiredoxins (2-Cys Prx) constitute an ancient family of peroxide detoxifying enzymes and have acquired a plant-specific function in the oxygenic environment of the chloroplast. The 2-Cys Prx has a broad substrate specificity with activity toward hydrogen peroxides and complex alkyl hydroperoxides. During the peroxide reduction reaction, 2-Cys Prx is alternatively oxidized and reduced as it catalyzes an electron flow from an electron donor to peroxide. The activity of 2-Cys Prx also is linked to chloroplastic NAD(P)H metabolism as indicated by the presence of the reduced form of the enzyme after feeding dihydroxyacetone phosphate to intact chloroplasts. From these data, it is suggested that 2-Cys Prx mediates peroxide detoxification in the plastids during the dark phase of photosynthesis and may be a suitable gene for overexpression in chloroplasts to protect plants from oxidative stress.

8. CONCLUSIONS

The objective of this chapter is to survey possible genes for expression in chloroplasts to genetically engineer stress tolerance via the nuclear or chloroplast genome. In the case of heat stress, it has been noted that there are chloroplast-localized heat shock proteins that may be involved in the acquisition of thermotolerance in plants. Following an initial heat episode and a recovery period, HSP21 was localized in the soluble fraction of the chloroplast. Furthermore, heat stress resulted in a temperature-dependant redistribution of HSP21 from a soluble form to an insoluble chloroplast protein fraction. Other proteins, such as components of the PSII complex, are thought to be most sensitive to heat as well as to light. Thermotolerance and acclimation of PSII varies widely among species in general. Even in the case of PSII thermotolerance, small heat shock proteins appear to offer thermo-protection. These data indicate that a logical avenue for the enhancement of thermotolerance in plants would be to express and characterize the protective mechanism provided by small heat shock proteins.

Cold stress induces the expression of a number of genes that results in the synthesis of new proteins localized in the cytosol. However, it appears that none of these proteins are produced in the chloroplast nor is the induction phase controlled by any factor localized in the chloroplast. Thus, at this point there are no clear candidate genes for expression in chloroplasts to increase cold stress tolerance. Since the function of chloroplast genes are for photosynthetic activity and the expression of proteins that make up the synthetic apparatus, it is not surprising that cold stress genes are not found in the chloroplast genome.

Glycine betaine is a compound that is critically involved in various osmotic stresses, including drought and salinity stresses. The application of glycine betaine improves growth and survival of a large number of plants under various stressful conditions. Glycine betaine accumulates in drought-tolerant plants and as such appears to act as an osmoprotectant. Protection may occur as a result of the stabilization of the quaternary structure of enzymes and enzyme complexes. Expression of trehalose in chloroplasts also has a positive effect in protein against drought. Thus, the insertion of genes into the chloroplast genome that cause an over-production of glycine betaine or trehalose resulted in plants that were highly tolerant to salt or drought stress. In addition, transgenes that confer tolerance to abiotic stress should not outcross with weeds and permanently transfer these valuable traits to the weed nuclear genome. Therefore, maternal inheritance of transgenes (Daniell 2002, see also chapter 4 in this book) that confer abiotic stress tolerance via the chloroplast genome may be ideal for this purpose. Because chloroplast genomes of major crops including cotton (Kumar et al., 2004b) and soybean (Dufourmantel et al., 2004) have been successfully transformed, this offers an exciting new approach to create transgenic plants with abiotic stress tolerance (Devine and Daniell, 2004). For more information on chloroplast genetic engineering, the reader is directed to chapter 16 of this book.

Exposure of plants to light intensities higher than that required to saturate photosynthesis may cause light damage or photoinhibition of the chloroplast's thylakoid proteins. To maintain normal functions under light stress, plants have developed several repair and protective mechanisms. The induction of a family of proteins, called early light-induced proteins (elips), consists of low molecular proteins localized in the thylakoids membranes of chloroplasts. These proteins may be genetically engineered to provide greater stability against photoinhibition.

There is a great potential for the genetic manipulation of key enzymes involved in stress metabolism in plants. One set of metabolic activities involves the over-production of glutathione in transgenetic plants. There are several enzymes that play an important role in the oxidation/reduction of compounds that are involved in oxidative stress. Genetic engineering of these enzymes within the chloroplast matrix may provide protection against oxidative stress. While there may be some limitations (including substrates or intermediates in complex pathways) to chloroplast genetic engineering for stress tolerance, there appears to be tremendous potential for increasing tolerance in plants to a number of stresses by expression of appropriate genes in chloroplasts.

9. REFERENCES

Adamska, I., Roobol-Boza, M., Lindahl, M., & Andersson, B. (1999). Isolation of pigment-binding early-inducible proteins from pea. *Eur. J. Biochem.*, 259, 453-460.

Agarie, S., Cushman, M.A., Kore-eda, S., Deyholos, M., Galbraith, D., & Cushman, C. (2003). Using expressed sequence tag (EST) based microarrays in the common ice plant, *Mesembryanthemum crystallium. PAG Poster.*

Allard, F., Houde, M., Krol, M., Ivanov, A., Hunter, N.P.A., & Sarham, F. (1998). Betaine improves freezing tolerance in wheat. *Plant Cell Physiol.*, 30, 1194-1202.

Atici, O. & Nalbantoglu, B. (2003). Antifreeze proteins in higher plants. Phytochem. 64, 1187-1196.

Belanger, F.C., Brodl, M.R., & Ho, T.H.D. (1986). Heat shock causes destabilization of specific mRNAs and destruction of endoplasmic reticulum in barley aleurone cells. *Proc. Natl. Acad. Sci. USA*, 83, 1354-1358.

Capel, J., Jarillo, J.A., Salinas, J., & Martinex-Zapater, J.M. (1997). Two homologous low-temperature-inducible genes from Arabidopsis encode highly hydrophobic proteins. *Plant Physiol.*, 115, 569-576.

Creissen, G., Broadbent, P., Stevens, R., Wellburn, A.R., & Mullineaux, P. (1996). Manipulation of glutathione metabolism in transgenic plants. *Biochem. Soc. Trans.*, 24, 465-469.

Cushman, J.C. and Bohnert, N.J. (2000). Genomic approaches to plant stress tolerance. *Opin Plant Biol.*, 3, 117-124.

Daniell, H. (2002). Molecular strategies for gene containment in transgenic crops. *Nat. Biotechnol.*, *20*, 581-586.

Devine, A. L., & Daniell, H. (2004). Chloroplast genetic engineering. In S. Moller (Ed.), *Plastids* (pp. 283-320). United Kingdom: Blackwell Publishers.

Dufourmantel, N., Pelissier, B., Garçon, F.,Peltier, J. M., & Tissot, G. (2004). Generation of fertile transplastomic soybean. *Plant Mol. Biol.*, in press.

Dylewski, D.P., Singh, N.K., & Cherry, J.H. (1991). Effects of heat shock and thermoadaptation on the ultrastructure of cowpea (Vigna unguicultata) cells. *Protoplasma*, 163, 125-135.

Fujikawa, S. & Takabe, K. (1996). Formation of multiplex lamella by equilibrium slow freezing of cortical parenchyma cells of mulberry and its possible relationship to freezing tolerance. *Protoplasma*, 190, 189-203.

Gorham, J. (1995). Betaines in higher plants. In R.M. Wallsgrove (Ed.) *Amino acids and their derivatives in higher plants* (pp 171-203). Cambridge, England: Cambridge University Press.

Griffith, N.J. & Antikainen, M. (1996). Extracellular ice formation in freezing-tolerant plants. Steponkus, P.L. (Ed.) *Advances in Low-Temperature Biology. Vol 3* (pp 107-139). London: JAI.

Gustavsson, N., Kokke, B.P., Harndahl, U., Silow, M., Bechtold, U., Poghosuan, Z., Murphy, D., Boelens, W.C., & Sundby, C. (2002). A peptide methionine sulfoxide reductase highly expressed in

photosythentic tissue in Arabidopsis thaliana can protect the chaperone-like activity of a chloroplast-localized small heat shock protein. *Plant J.*, 29, 545-553.
Hasegawa, P.M., Bressan, R.A., Zhu, J-K., & Bohnert, H.J. (2000). Plant cellular and molecular responses to high salinity. *Annu. Rev. Plant Physiol Mol. Biol.*, 51, 463-499.
Hayashi H. & Murata, N. (1998). Genetically engineered enhancement of salt tolerance in higher plants. In Satoh. K. and Murata, N. (Eds.) *Stress Responses of Photosynthetic Organisms.* (pp.133-148). Amsterdam: Elsevier Press.
Heckathorn, S.A., Downs, C.A., Sharkey, T.D., & Coleman, J. S. (1998). The small, methionine-rich chloroplast heat-shock protein protects photosystem II electron transport during heat stress. *Plant Physiol.*, 116, 439-444.
Kasahara, M., Kagawa, T., Oikawa, K., Suetsugu, N., Miyao, M., & Wade, M. (2002). Chloroplast avoidance movement reduces photodamage in plants. *Nature*, 420, 829-832.
Kawamura, Y. & Uemura, M. (2003). Mass spectrometric approach for identifying putative membrane proteins of Arabidopsis leaves associated with cold acclimation. *Plant J.*, 36, 141-145.
Konig, J., Baier, M., Horling, F., Kahmann, U., Harris, G., Schurmann, P., & Dietz, K-J. (2002). The plant- specific function of 2-Cys peroxiredoxin-mediated detoxification of peroxides in the redox-hierarchy of photosynthetic electron flux. *Proc. Natl. Acad. Sci. USA.*, 99, 5738-5743.
Kreps, J.A., Wu, Y., Chang, H-S., Zhu, T., Wang, X., & Harper, J. F. (2002). Transcriptome Changes for Arabidopsis in Response to Salt, Osmotic, and Cold Stress. *Plant Physiol.*, 130, 2129-2141.
Kropat, J., Oster, U., Rudiger, W., & Beck, C.F. (1997). Chlorophyll precursors are signals of chloroplast origin involved in light induction of nuclear heat-shock genes. *Proc. Natl. Acad. Sci.*, 94, 14168-14172.
Kurkela, S. & Borg-Franck, M. (1999). Structure and expression of *Kin* 2, one of two cold- and ABA-induced genes of *Arabidopsis thaliana. Plant Mol. Biol.*, 19, 689-692.
Kumar, S., Dhingra, A. & Daniell, H. (2004a). Plastid expressed *betaine aldehyde dehydrogenase* gene in carrot cultured cells, roots and leaves confers enhanced salt tolerance. *Plant Physiol* in press.
Kumar, S., Dhingra, A., & Daniell, H. (2004b). Manipulation of gene expression facilitates cotton plastid transformation of cotton by somatic embryogenesis & maternal inheritance of transgenes. *Plant Mol. Biol.* in press.
Lee, S.B., Kwon, H.B., Kwon, S.J., Park, S.C., Jeong, M.J., Han, S.E., Byun, M.O. & Daniell, H. (2003). Accumulation of trehalose within transgenic chloroplasts confers drought tolerance. *Mol. Breeding*, 11, 1-13.
Levitt, J. (1980). *Responses of Plants to Environmental Stresses: Chilling, Freezing and High Temperature Stresses.* New York: Academic Press.
McNeil, S.D., Rhodes, D., Russell, B.L., Nuccio, M.L., Shachar-Hill, Y. & Hanson, A.D. (2000) Metabolic modeling identifies key constraints on an engineered glycine betaine synthesis pathway in tobacco. *Plant Physiol.*, 124, 153-162.
Medina, J., Catala, R., and Salinas, J. (2001). Developmental and Stress Regulation of RC12A & RC12B, Two cold-inducible genes of Arabidopsis encoding highly conserved hydrophobic proteins. *Plant Physiol.*, 125, 1655-1666.
Monod, C., Takahashi, Y., Goldschmidt-Clermont, M., & Rochaix, J.D. (1994). The chloroplast ycf8 open reading frame encodes a photosystem II polypeptide which maintains photosynthetic activity under adverse growth conditions. *EMBO J.*, 13, 2747-2754.
Neumann, D., Scharf K.D., & Nover, L. (1984). Heat shock induced changes of plant cell ultrastructure and autoradiographic localization of heat shock proteins. *Eur. J. Cell Biol.*, 34, 254-264.
Nuccio, M., Russell, B.L., Nolte, K.D., Rathinasabapathi, B., Gage,D.A., & Hanson, A. (1998).The endogenous choline supply limits betaine synthesis in transgenic tobacco expressing choline monooxygenase. *Plant J.*, 16, 487-496.
Nuccio, M.L., McNiel, S.D., Ziemak, M.J., Hanson, A.D., Jain. R.K., & Selvaraji, G. (2000). Choline import into chloroplasts limits glycine betaine synthesis in tobacco: Analysis of plants engineered with a chloroplastic or a cytosolic pathway. *Metab. Eng.*, 4, 300-311.
Osteryoung, K.W. & Vierling, E. (1994). Dynamics of small heat shock protein distribution within the chloroplasts of higher plants. *J. Biol. Chem.*, 269, 28676-28682.
Rhodes, D. & Hanson, A.D. (1993). Quaternary ammonium and tertiary sulfonium compounds in higher plants. *Annu. Rev. Plant Physiol. Plant Mol. Biol.*, 44, 357-384.
Rizhsky, L., Liang, H., & Mittler, R. (2003). The water-water cycle is essential for chloroplast protection in the absence of stress. *J Biol Chem.*, 278, 38921-38925.

Ruelland, E., Cantrel, C., Gawer, M., Kader, J-C., & Zachowski, A. (2002). Activation of phospholidases C and D is an early response to cold exposure in Arabidopsis suspension cells. *Plant Physiol.*, 130, 999-1007.

Sakai, A. & Larcher, W. (1987). Frost survival of plants. In *Responses and Adaptation to Freezing Stress.* Berlin: Springer-Verlag.

Sakamoto, A. & Murata, N. (2001). The use of bacterial choline oxidase, a glycine betaine-synthesizing enzyme, to create stress-resistant transgenic plants. *Plant Physiol.*, 125. 180-188.

Sakamoto, A., Murata, N., & Murata, A. (1998). Metabolic engineering of rice leading to biosynthesis of glycine betaine and tolerance to salt and cold. *Plant Mol. Biol.* 38.1011-1019.

Saneoka, H., Nagasaka, C., Hahn, D.T., Yang, W-J, Premachandra, G.S., Joly, R.J., & Rhodes, D. (1995). Salt tolerance of glycinebetaine-deficient and -containing maize lines. *Plant Physiol.*, 107. 631-638.

Shen, Y.G., Du, B.X., Zhang, W.K., Zhang, J.S. & Chen, S.Y. (2002) *AhCMO*, regulated by stresses in *Atriplex hortensis*, can improve drought tolerance in transgenic tobacco. *Theor. Appl. Genet.*, 105, 815-821.

Shimaoka, T., Miyake, C., and Yokota, A. (2003). Mechanism of the reaction catalyzed by dehydroascrobate reductase from spinach chloroplasts. *Eur. J. Biochem.*, 270, 921-928.

Singla-Pareek, S.L., Reddy, M.K., & Sopory, S.K. (2003). Genetic engineering of the glyoxalase pathway in tobacco leads to enhanced salinity tolerance. *Proc. Natl. Acad. Sci. USA*, 100, 14672-14677.

Steponkus, P.L. (1984). Role of the Plasma Membrane in Freezing Injury and Cold Acclimation. Annu. *Rev. Plant Physiol. Plant Mol. Biol.*, 35,543-584.

Tezara, W., Mitchell, V.J., Driscoll S.D., & Lawlor, D.W. (1999). Water stress inhibits photosynthesis by decreasing coupling factor and ATP. *Nature*, 401. 914-917.

Thomashow, M.L. (1999). Plant cold acclimation: freezing tolerance genes and regulatory mechanisms. *Annu. Rev. Plant Physiol. Plant Mol. Biol.*, 59, 571-599.

Thornally, P.J., Trotta, R.J., & Stern, A. (1983). Free radical involvement in the oxidative phenomena induced by tert-butyl hydroperoxide in erythrocytes. *Biochim. Biophys. Acta.*, 759, 16-22.

Tissiers, A., Mitchell, H.K. & Tracy, U.M. (1974). Protein synthesis in salivary glands of *D. melanogaster*. Relation to chromosomal puffs. *J. Mol. Biol.*, 84, 389-398.

Tyangi, A. (1999). Regulation of Plastid Gene Expression. In Singhal, G.S., Renger, G., Sopory, S.K., Irrgang, K-D., & Govindjee (Eds.) *Concepts in Photobiology: Photosynthesis and Photomorphogenesis.* Dordrecht, Netherlands: Kluwer Academic Publishers and Narosa Publishing House.

Van Breusegem, F., Kushnir, S., Slooten, L., Brauw, G., Botterman, J., Van Montagu, M., & Inze, B. (1998). Processing of a chimeric protein in chloroplasts is different in transgenic maize and tobacco plants. *PLANT Mol. Biol.*, 38, 491-496.

Van Breusegem, F., Slooten, L., Stassart, J.-M., Moens, T., Van Montagu, M., & Inze, D. (1999). Overproduction of *Arabidopsis thaliana* FeSOD confers oxidative stress tolerance to transgenic maize. *Plant Cell Physiol.*, 40, 98-129.

Vasilikiotis, C. & Melis, A. (1994). Photosystem II reaction center damage and repair cycle: Chloroplast Acclimation Strategy to irradiance stress. *Proc. Natl. Acad. Sci. USA*, 19, 7222-7226.

Vigh, L., Da, L., & Murata, N. (1993). The primary signal in the biological perception of temperature: Pd-catalyzed hydrogenation of membrane lipids stimulated by the expression of the des A gene in Synechocystis PCC6803. *Proc. Natl. Acad. Sci. USA*, 90, 369-374.

Wang, Y. J., Zhang, Z.G., He, X.J., Zhou, H.L., Wen, Y.X., Dai, J.X., Zhang, J.S., & Chen, S.Y. (2003). A rice transcription factor OsbHLH1 is involved in cold stress response. *Theor. Appl. Genet.*, 107, 1409-1409.

Yokoi, S., Higashi, S-I., Kishitani, S., Murata, N., & Toriyama, K. (1998). Introduction of the cDNA for shape Arabidopsis glycerol-3-phosphate acyltransferase (GPAT) confers unsaturation of fatty acids chilling tolerance of photosynthesis on rice. *Molec. Breeding*, 4, 269-275.

CHAPTER 19

METABOLIC PATHWAY ENGINEERING FOR NUTRITION ENRICHMENT

NIRANJAN BAISAKH AND SWAPAN DATTA

Plant Breeding, Genetics, and Biochemistry Division
International Rice Research Institute, DAPO Box 7777, Metro Manila, Philippines,
Email: *N.Baisakh@cgiar.org, S.Datta@cgiar.org*

Abstract. Plants form the basis of life and are the major sources of nutrition in the human diet. They have constantly been improved to meet the food demand of the ever-burgeoning world population. Although food sufficiency has been attained in many major food crops, concern about worldwide malnutrition has led scientists to aim at improving the food crops to alleviate the hidden hunger. Recently, research began to understand the biosynthesis of metabolites that have significant value for human health, to elucidate the regulation, storage, sink, and degradation of the metabolites and the interaction of the different metabolic biosynthesis pathways. This research focuses on manipulating the biosynthetic pathway to develop value-added food crops (popularly known as functional food) with enhanced health benefits. So far, through metabolic engineering, transgenic plants have been developed to contain enhanced levels of vitamins (provitamin A, vitamins E and C), lysine, polyhydroxybutyrate, inulin, and flavonoids etc. Most of these metabolic pathways occur in the plastid compartment of the plant cell, though enzymes involved in the pathway are encoded by nuclear genes. Here we describe the efforts that are underway to extend nuclear transgenic technologies from model crops such as *Arabidopsis* to staple and non-staple food crops including rice, maize, rapeseed, soybean, potato and sugarbeet.

1. INTRODUCTION

International agricultural research has made a significant contribution to increasing food supplies in developing nations by improving productivity, following the 1960s - during the post-Green Revolution era. This resulted in self-sufficiency and food security in many developing countries among the major cereal crops considered the principal staple foods. Plants, besides forming the basis of the human food chain, are also an important means of improving human health and well-being. For centuries, rice

H. Daniell and C.D. Chase (eds.), Molecular Biology and Biotechnology of Plant Organelles.
527—542.

has been the main livelihood for millions of people in Asia. Until recently, much research has focused on attaining food security and sustainability by stabilizing the yield or elevating the yield ceiling of rice. In Asia, rice is the principal staple food crop and provides 40%-70% of the caloric value and is the source of mineral and micronutrients. About 1.2 billion people worldwide suffer from malnutrition and deficiency of minerals and micronutrients. Interventions like diet diversification, supplementation and fortification have been used with limited success. Biotechnology, especially plant genetic engineering, has the potential to augment the development of nutritionally enriched crops. They are known as second-generation transgenic crops, and offer a more sustainable strategy to combat human micronutrient deficiencies (Zimmerman and Hurrel, 2002). Metabolic engineering is generally defined as the redirection of one or more enzymatic reactions to produce new compounds in an organism, to improve the production of existing compounds, or to mediate the degradation of compounds. In this chapter we highlight the recent progress and achievements in transgenic crops, with an emphasis on rice, engineered with genes involved in the metabolic pathway for modification/improvement of crop quality (e.g., vitamins and protein quality) with potential impacts on human nutrition.

2. VITAMIN A

Vitamin A deficiency (VAD) manifestations range from early night-blindness symptoms to corneal xerosis, keratomalacia and total blindness in severe cases. It can also cause increased morbidity and mortality especially in children (Gerster, 1997). In Southeast Asia, an estimated one quarter of a million people become blind each year because of VAD. Diet is the only source of vitamin A since mammals cannot manufacture it on their own. Most of the dietary vitamin A is of plant food origin in the form of provitamin A that is converted to vitamin A in the body (Sivakumar, 1998). Unfortunately, rice germplasm screened so far do not possess □-carotene in the polished seeds (milled rice) that are usually consumed, although a trace amount is present in the brown rice (Tan et al., 2004). Carotenoids, besides being the accessory pigment in photosynthesis act as photoreceptors in protecting against photooxidation and as precursors to phytohormone, abscisic acid, and are important components for human health. Carotenoids are synthesized by all photosynthetic organisms as well as by some bacteria and fungi. In plant cells, the carotenoid biosynthetic pathway takes place in the plastid compartment of the cell and the enzymes in the pathway are encoded by the nuclear genes (Cunnigham and Gantt, 1998). Biosynthesis of provitamin A carotenoid beta-carotene involves the coordinated activity of four enzymes - phytoene synthase (PSY), phytoene desaturase (PDS), zeta-carotene desaturase (ZDS) and lycopene cyclase (LCY-B). However, the enzyme CRTI from the bacterium *Erwinia uredovora* has a dual function for PDS and ZDS (Figure 1). So, manipulation of the beta-carotene accumulation in any organism requires engineering of the three genes (*psy*, *crtI*, and *lcy*) encoding the above enzymes. From the screening of a large number of germplasm accessions, it was found that provitamin A is not accumulated in the plastid tissues of the endosperm in rice and other cereals. However, the precursor molecule geranylgeranyl pyrophosphate (GGPP, the carotenoid building block) was present in the endosperm, making the endosperm biochemically competent to biosynthesize carotenoids.

The hypothesis that rice could accumulate □-carotene in the endosperm tissue was proved by the accumulation of phytoene through the heterologous expression of phytoene synthase from a daffodil (Burkhardt et al., 1997). Then came the success story of "golden rice" in a japonica cultivar, T-309, through *Agrobacterium*-mediated transformation accumulating beta-carotene with a characteristic yellow color in the endosperm (Ye et al., 2000). This was a landmark proof-of-concept concerning the installation of a complete functional metabolic pathway for beta-carotene biosynthesis by the coordinated expression of three transgenes (*psy*, *crtI*, and *lcy*). The *psy* and *lcy* genes were under the control of an endosperm-specific glutelin promoter and the bacterial *crtI* gene was under the control of the constitutive CaMV35S promoter which is fused to the plastid-specific pea rubisco small subunit transit peptide. The endosperm could accumulate up to 1.6 ∝g of □-carotene in 1 g of the endosperm dry matter. This technology was immediately extended to the indica rice cultivars, including IR64, consumed by 90% of the Asian population adapted to different agroecological zones of several tropical Asian countries (Datta et al., 2003). The transgenic indica rice, developed through biolistic transformation accumulated up to 1.05 ∝g of carotenoids in the endosperm tissue. Datta et al (2003) used both antibiotic and Positech™ selection systems using *hph* (hygromycin phosphotransferase) and *pmi* (phosphomannose isomerase) as selectable marker genes, respectively, whereas Ye et al (2000) used the antibiotic system. The accumulation of □-carotene was further evidenced from the characteristic yellow color of the endosperm, HPLC profiling (Fig. 2), and the blue color of the endosperm tissue microtome section with Carr-Price reaction (Krishnan et al., 2003).

Given the increasing public and environmental concern about the use of antibiotic resistant marker genes, Hoa et al (2003) developed transgenic indica and japonica rice with mannose as a selection agent and reported accumulation of total carotenoid to the level of 0.8 ∝g per g endosperm dry mass in indica rice cultivar IR64. Very recently, in our laboratory, transgenic indica rice lines have been developed in the background of IR64 and BR29 (from Bangladesh) through *Agrobacterium*-mediated transformation using mannose-based selection that accumulated up to 1.6 ∝g of □-carotene alone per g of polished seed (K. Datta, unpublished data). The loss of the total carotenoids due to cooking under laboratory conditions was also found to be minimal though beta-carotene level decreased by nearly 10% (Datta et al., 2003). These published results show that provitamin A can be engineered in the endosperm by only two genes, *psy* and *crtI*, in the absence of heterologous *lcy*. It is presumed that either the lycopene cyclase is constitutively present in the endosperm or the activation of this enzyme is through a feedback-signaling loop by all-trans-configured lycopene that is delivered by the CRTI in the transgenic process - instead of the poly-*cis*-lycopene that is produced by the two plant desaturases occurring naturally (Bartley et al., 1999).

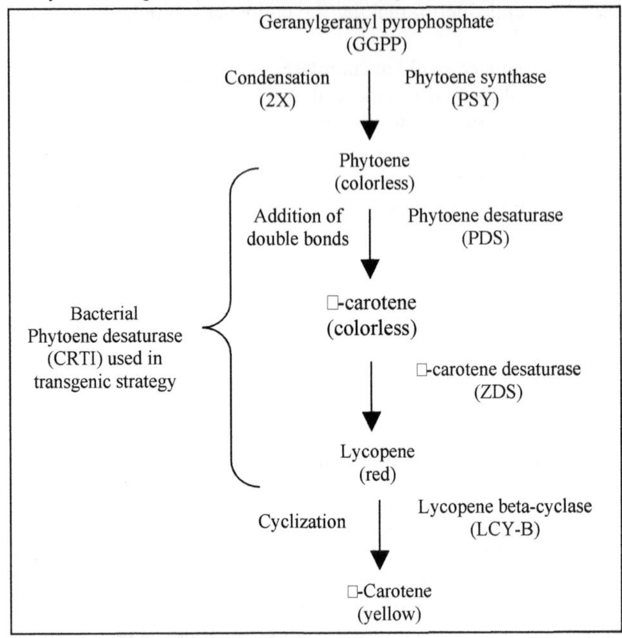

***Figure 1**. Carotenoid biosynthetic pathway in normal plant and transgenic rice plant*

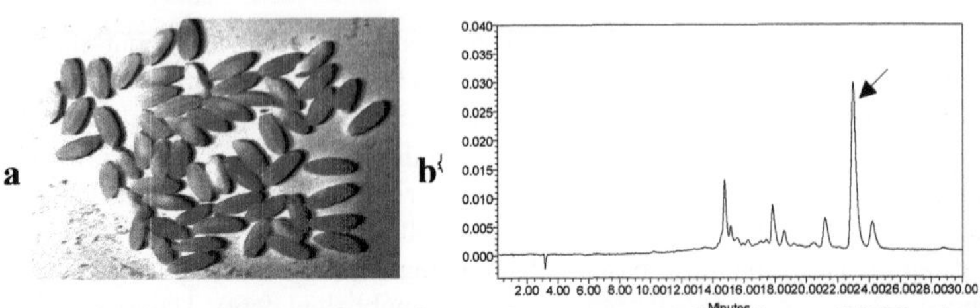

***Figure 2**. Polished transgenic golden indica rice seeds of cv. BR29 showing characteristic yellow endosperm (a) (see Color Plate 16) due to accumulation of □-carotene (arrow) as revealed from the HPLC chromatogram (b).*

Efforts are underway using different promoters and modified or codon-optimized genes for enhancing the amount of beta-carotene in the rice endosperm tissue. Besides rice there is considerable research on enhancing carotenoid level, especially beta-carotene in some non-staple food crops (Table 1) that are necessary for good health and a productive life.

***Table 1**. Metabolic engineering approaches for enhanced □-carotene/carotenoid levels in transgenic plants*

Genes	*Source*	*Promoter*	*Transgenic plant*	*Tissue*	*Enhanced products*	*References*
Phytoene synthase (*psy*)	Daffodil	Rice glutelin (Gt1)	Rice	Endosperm	Lutein, Zeaxanthin, □-carotene, □-carotene	Ye at al (2000) Datta et al (2003) Hoa et al (2003)
Phytoene desaturase (*crtI*)	*Erwinia uredovora*	CaMV 35S				
Lycopene cyclase (*lcy*)	Daffodil	Rice glutelin (Gt1) CaMV 35S				
Phytoene synthase (*crtB*)	*E. uredovora*	*Brassica* napin	Rapeseed	Embryo	□-carotene, □-carotene	Shewmaker et al (1999)
Phytoene desaturase (*crtI*)	*E. uredovora*	CaMV 35S	Tobacco	Leaves	Violaxanthin, □-carotene, reduced lutein	Misawa et al (1993)
Phytoene desaturase (*crtI*)	*E. uredovora*	CaMV 35S	Tomato	Fruit	□-carotene	Römer et al (2000)
□-cyclase	*Arabidopsis*	Tomato *Pds*	Tomato	Fruit	□-carotene	Rosati et al (2000)
Phytoene synthase (*psy*)	Tomato	CaMV 35S	Tomato	Fruit	Lycopene, phytoene	Fray et al (1995)

3. VITAMIN E

Natural vitamin E comprises of a group of eight hydrophobic compounds with different antioxidant activity. Of these, □-tocopherol is physiologically the most important for humans, with maximum vitamin E activity. The tocopherols and tocotrienols differ in their saturation at the side carbon chain. Collectively, these eight compounds are called tocols. Tocopherols possess antioxidant activity, with the ability to quench free oxygen radicals in cell membranes and protect polyunsaturated fatty acids from damage. Plants are the only organisms that produce tocopherols with both photosynthetic and non-photosynthetic functions.

The tocopherol biosynthetic pathway in plants has not yet been fully characterized, but a putative pathway is shown in Figure 3. Studies show that both the cyclization and methylation of dimethylphytylquinol to □-tocopherol and finally □-tocopherol are localized in the plastid membranes (chloroplast-envelope and thylakoids, chromoplast-envelope and *a*-chl lamellae) (Camara et al., 1982).

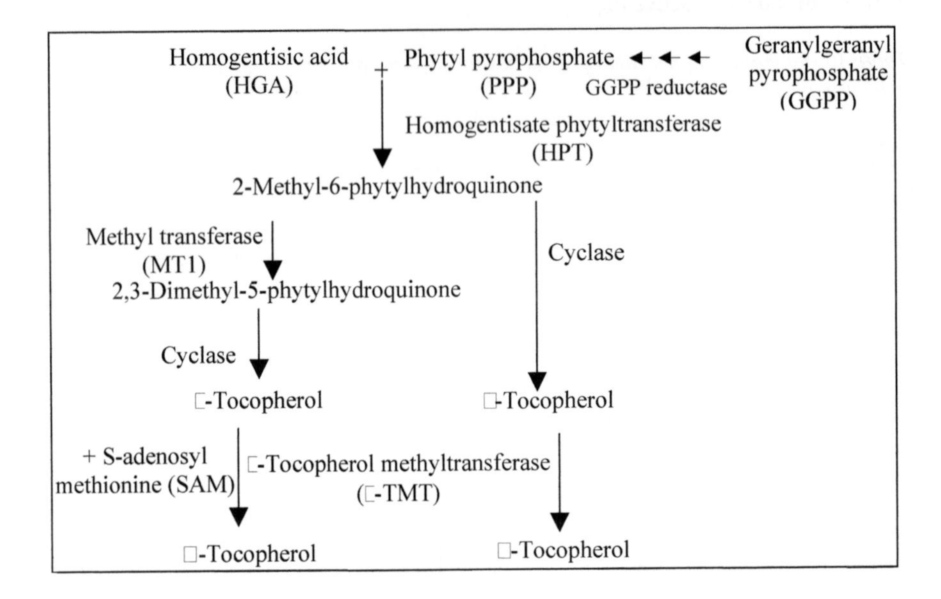

Figure 3. *Schematic diagram of a putative pathway of tocopherol biosynthesis*

Four out of at least six proposed genes involved in the biosynthetic pathway have been cloned in *Arabidopsis* (DellaPenna et al., 1999; Shintani and DellaPenna, 1998). Different strategies including marker-assisted breeding and transgenesis have been used to manipulate the tocopherol levels in different crops either qualitatively or quantitatively. Some publications even mention successful alteration of the overall levels of vitamin E.

The gene □-TMT - involved in the final conversion of □-tocopherol into □-tocopherol - was cloned from *Arabidopsis* using a map-based cloning strategy. The *Arabidopsis* □-TMT under the control of carrot DC3 was overexpressed in transgenic *Arabidopsis*. Although there was no alteration in the total vitamin E content, there was a near complete shift from □-tocopherol to □-tocopherol (from 10% to 90% □-tocopherol) and from □-tocopherol to □-tocopherol.

Synthetic zinc finger transcription factors (ZFP-TFs) were designed to upregulate the expression of the endogenous □-TMT in transgenic *Arabidopsis*, thereby increasing the □-tocopherol levels (Van Enennaam et al., 2004).

The *Arabidopsis* vitamin E pathway gene (*At-vte3*) was coexpressed with *At-vte4* (gamma-tocopherol methyltransferase) in transgenic soybean in which the seeds accumulated to >95% alpha-tocopherol, a dramatic change from the normal 10%, resulting in a greater than eight-fold increase in alpha-tocopherol and up to five-fold increase in seed vitamin E activity (Van Enennaam et al., 2003). The overexpression of *Arabidopsis* homogentisate phytyltransferase (HPT) under a seed-specific napin promoter resulted in an increase of up to 100% vitamin E in homozygous (T_3) transgenic seeds of *Arabidopsis* (Savidge et al., 2002).

Transgenic technology is now being applied for enhancing vitamin E levels in cereal crops such as maize (Rocheford et al., 2002) and rice (unpublished data).

4. VITAMIN C

The role of vitamin C (L-ascorbic acid: AsA) in human nutrition is well understood in cardiovascular functions, immune cell development, tissue connectivity and iron utilization. Its most vital role in the human body is its water-soluble antioxidant activity. In plants, AsA acts as an antioxidant in addition to its other major physiological roles (Smirnoff, 1996). AsA is also crucially involved in the regeneration of α-tocopherol and zeaxanthin as well as pH-mediated modulation of PS II activity. The critical importance of AsA in photosynthetic metabolism is emphasized by its abundance in chloroplasts. Plants and animals are capable of synthesizing AsA but via different pathways. Humans, though lack the L-gluono-1,4-lactone oxidoreductase, the key enzyme in the final step of AsA synthesis. In plants, AsA is synthesized from D-glucose through L-galactose and L-galactono-1,4-lactone (Fig. 4).

Because AsA cannot be stored in the body, a regular supply through plant dietary sources is required. However, not many foods are rich in AsA, metabolic engineering seeks to increase the level of vitamin C, and improve the nutritive value of the crops.

Transgenic tobacco and lettuce plants expressing L-glucono-1,4-lactone oxidoreductase (GLO) from rats contained increased levels (4-7-fold) of AsA in the leaves. This increased level of AsA, in addition to improving the nutritional quality of lettuce, might also help in reducing the commercial need for bisulfides required to prevent the browning of lettuce leaves (Jain and Nessler, 2000). However, the mechanism of such an increase and the ascorbate redox state were not mentioned.

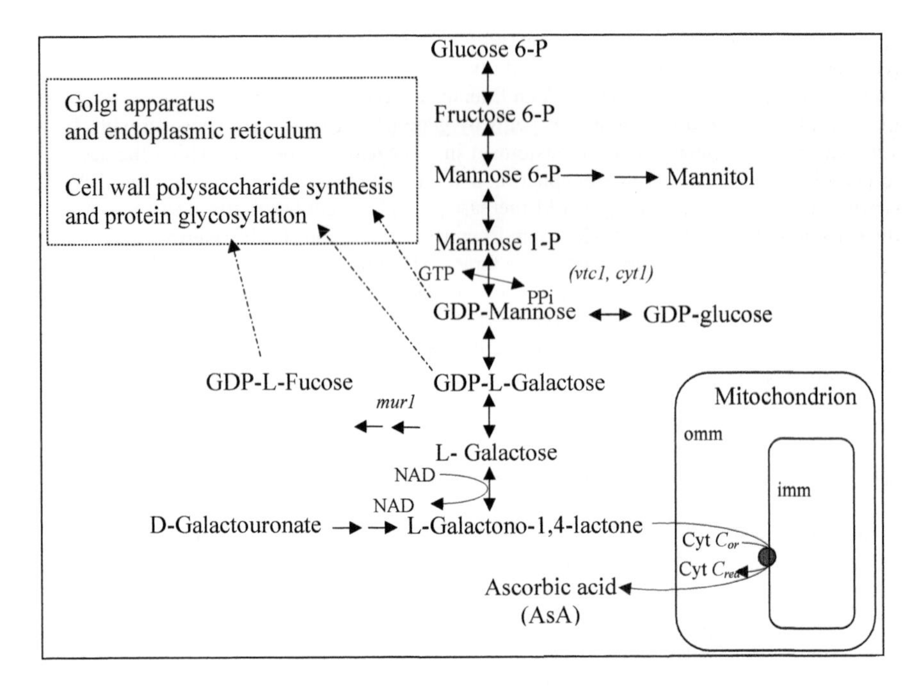

Figure 4. *Biosynthetic pathway of ascorbic acid biosynthesis in plants*

To prove another proposed pathway of D-galacturonic acid as the precursor of L-galactono-1,4-lactone in ascorbic acid biosynthesis, a gene, aldo-keto reductase (*AKR2*) encoding D-galcturonic acid reductase (GalUR), was cloned from strawberry and overexpressed in transgenic *Arabidopsis*. The transgenic *Arabidopsis* contained 2 to 3 times higher vitamin C content than the control (Agius et al., 2003).

The overexpression of His-tagged wheat dehydroascorbate reductase cDNA (DAHR) under the control of the CaMV 35S promoter in transgenic tobacco leaves resulted in an increase in AsA levels up to 2.4-fold in expanding leaves, 3.9-fold in mature leaves and 2.2-fold in prenascent leaves and a higher ratio (from 1.5 in control leaves to 4 in transgenics) of reduced to oxidized form of AsA. The increase in the AsA level was accompanied by an increase in the glutathione (GSH) level due to the oxidative stress response by the AsA level (Chen et al., 2003). This increase in AsA level resulted from the increased activity of L-galactono-1,4-lactone dehydrogenase (GLDH), which catalyzes the terminal step in the ascorbate biosynthesis.

Similarly, the overexpression of wheat DAHR cDNA (without His-tag) under the control of the maize ubiquitin promoter or *Shrunken* 2 (*Sh* 2) promoter resulted in a 1.8-fold and 1.9-fold increases in the AsA level in transgenic maize leaves and kernels, respectively. This was accompanied by a 30%-40% increase in the ascorbate redox state and 1.9-fold increase in GSH. All these results suggested that DHAR was responsible for the ascorbate recycling, thereby recapturing AsA before it is lost (Chen et al., 2003).

5. PROTEIN QUALITY AND QUANTITY

Plant proteins are an important source of essential amino acids in the human diet. Vegetarians depend on plants and plant-derived products to obtain needed protein. Conventional breeding methods have had limited success in improving both the quantity and quality of proteins because of tedious and time-consuming selection procedures. Genetic engineering techniques provide a way to make directional changes in both the quantity and quality of proteins.

In this section, we discuss improving lysine, an aminoacid, which is inadequate in the cereals that are the basis of the diet of majority of the world's population. A plant-based diet is the only source of this aminoacid since humans and other monogastric animals cannot synthesize it.

Lysine is synthesized from aspartate through a branch in the aspartate-family pathway (Fig. 5). Another branch of this pathway leads to the synthesis of two other essential aminoacids, threonine and methionine (Fig. 5). The entire aspartate-family pathway, except the last step of methionine production by methionine synthase, occurs in the plastid and is regulated by several feedback inhibition loops (Galili, 2002). Several isozymes of aspartate kinase (AK) are feedback-inhibited by either lysine or threonine. Similarly, the first enzyme, dihydrodipicolinate synthase (DHPS), committed to lysine biosynthesis is also feedback-inhibited by lysine. *In vitro* studies revealed DHPS to be more sensitive to lysine than AK, and hence expression of feedback-insensitive DHPS results in overproduction of lysine, whereas expression of AK results in threonine overproduction. Several forms of bacterial genes *lysC* (from *Escherichia coli*) and *dapA* (from *Corynebacterium*) encoding AK and DHPS, respectively, are highly insensitive to lysine concentration.

Transgenic rice was produced with a lysine-rich protein gene (*lys*) cloned from a winged bean (*Psophocarpus tetragonoloba*) placed under the control of the maize ubiquitin promoter. The lysine content in the transgenic seeds was higher than in the wild-type seeds. One of the transgenic lines was reported to have a 16.04% increase in the lysine content in the seeds (Gao et al., 2001).

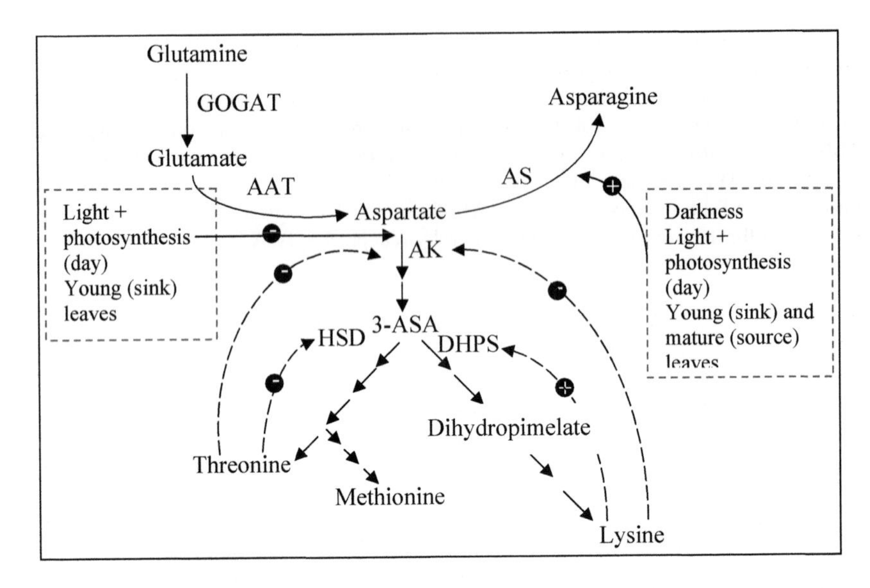

Figure 5. *Biosynthetic pathway for aspartate-family aminoacids: lysine, threonine, methionine*

GOGAT = glutamate synthase, AAT = aspartate aminotransferase, AS = asparagine synthase, AK = aspartate kinase, DHPS = dihydrodipicolinate synthase,
HSD = homoserine dehydrogenase, 3-ASA = 3-aspartic semialdehyde

Transgenic plants have been developed with the *dapA* gene that showed increased lysine accumulation. The lysine-feedback-insensitive *dapA* gene and a mutant *lysC* gene were linked to a chloroplast transit peptide and expressed from a seed-specific promoter in transgenic canola and soybean seeds (Falco et al., 1995). Expression of DHDPS resulted in more than a hundred-fold increase in the accumulation of free lysine in the seeds of canola, doubling the total seed lysine content. Expression of DHPS plus lysine-insensitive AK in soybean caused several hundred fold increases in free lysine and a five-fold increase in total seed lysine content (Falco et al., 1995). The lysine level in transgenic tobacco plants constitutively expressing the bacterial DHPS was regulated by growth, development and by light (Shaul and Galili, 1992; Zhu-Shimoni and Galili, 1998). In the case of maize, the expression of DHPS driven by the endosperm-specific promoter did not lead to lysine accumulation, whereas expression with an embryo-specific promoter resulted in higher concentrations (up to 50% to 100% of the seed), with a minor accumulation of catabolic products (Mazur et al., 1999).

In transgenic potato, expression of bacterial DHPS resulted in a six-fold increase in lysine content and expression of AK resulting in an eight- and two-fold increase in methionine and threonine content, respectively. Further, the overexpression of a potato feedback insensitive DHPS gene (modified with one aminoacid change) caused a dramatic increase in the lysine content from 1% to 15% of the total aminoacid level (Sévenier et al., 2002).

Overexpression of the glutelin gene increased the glutelin protein content in wheat, leading to an overall increase in the total protein. Likewise, engineering protein quality by mobilizing heterologous genes such as the 2S storage protein from *Amaranthus* into potato has been reported (Chakraborty et al., 2000). It has been demonstrated that this storage protein from *Amaranthus* has higher levels of some essential aminoacids than those recommended by World Health Organization (WHO).

6. INULIN

Inulin (also known as fructan) is a polymer of fructose synthesized by 15% of the angiosperms and the type of inulin depends on the nature of the link between the fructose moieties (□ 2-1 or □ 2-6). The chemical properties of inulin molecules vary with the length of the polymer. The short molecules with a degree of polymerization (DP) of 3 to 6 are used as low-calorie sweeteners, with the sweetening power decreasing with the extension of the molecule. These short molecules are prepared mainly by the fermentation of sucrose by *Aspergillus*-derived fructosyl transferase enzyme. On the other hand, the long molecules (DP 6 to 60) are used to mimic fat to replace it in food such as ice cream. Inulin is extracted mainly from chicory (*Chicorium intybus*).

Awareness of the importance of inulin in human health is increasing and this importance has been documented (Boeckner et al., 2001). The health-promoting effect of inulin is due to the production of short-chain fatty acids by the fermentation of endogenous flora in the colon, where 85% of the non-digested inulin is not accumulated from the upper part of the digestive tract. In addition, consumption of inulin improves blood lipid composition and mineral uptake, and reduces the chance of colon cancer.

In the inulin biosynthetic pathway, long-chain molecules are catalyzed by two enzymes: (1) sucrose sucrose:fructosyl transferase (SST) that catalyzes the transfer of the fructose moiety from one sucrose molecule onto another sucrose molecule, resulting in the synthesis of 1-kestose (DP3) and in the release of free glucose; and (2) fructan fructan:fructosyl transferase (FFT) that catalyzes the fructosyl moiety from one fructan molecule onto another fructan molecule, resulting in the synthesis of long-chain inulin molecules.

The two cDNAs (*1-sst* and *1-fft*) were cloned from the Jerusalem artichoke tuber crop, and their functionality was assessed in both petunia and potato (Van Deer Meer et al., 1998; Sévenier et al., 2002).

Sugar beet; well-known for its agronomy, with production as high as 10 tons of sucrose per hectare in the western world and with an established processing industry; is considered to be an ideal and efficient candidate for the concept of "plant as factory" by genetic engineering to produce a large quantity of inulin.

To test the feasibility of inulin production, transgenic sugar beet was developed with the *1-sst* gene from *Helianthus* using guard cell protoplast as the target explant. The transgenic sugar beet accumulated 37.2, 22.5, and 4.8 mg/g fresh weight (FW) of DP3, 4, and 5, respectively. However, the sucrose concentration in the tap roots of the transgenic plant was only 7.9 mg/g FW vis-à-vis 84.2 mg/g FW in the control. This demonstrated that around 90% of the sucrose was channeled into inulin synthesis. At the same time the amount of released glucose re-metabolized increased from 4.5 to 29.6 mg/g (Sévenier et al., 1998). Similarly, the introduction of both genes from *Helianthus* resulted in the accumulation of long-chain inulin.

7. OTHERS

Edible vegetable oils, except coconut and palm oil, possess relatively less saturated fatty acid. The composition of the fatty acids in these oils vary from crop to crop. Genetic engineering provides an easy way for selectively altering the fatty acid composition such that the modified oil is better suited to human health. With the help of antisense technology, Calgene, Inc. (USA) has succeeded in developing rape seed with high stearate levels by blocking the stearoyl ACP-desaturase (SAD) activity in seeds. The seed stearate content increased from a mere 2% to as high as 40%.

The high-stearate rape offers a novel and safer alternative to industrially hydrogenated oils for human consumption. Oils stored for a relatively longer period undergo spontaneous changes, making them unacceptable for human consumption. An increase in oleic acid content in oil reduces these changes and improves the stability and shelf life of edible oils. Antisense repression of 1-12 oleoyl desaturase in transgenic rapeseed resulted in an increase in oleic acid content of up to 83%.

Genetic engineering techniques have been effectively used to modify starch content in several crop plants. Overexpression of the enzyme ADP-glucose pyrophosphorylase (ADPGPP) resulted in an increased amount of starch in potato (Stark et al., 1992). The results of careful consideration of enzyme kinetics in metabolic engineering were elegantly demonstrated by research directed at modifying starch synthesis by manipulating ADPGPP. Plant ADPGPP is sensitive to allosteric effectors and has been proposed as a key regulator limiting starch synthesis. *Escherichia coli* ADPGPP is involved in glycogen synthesis and is also sensitive to allosteric effectors. Mutations affecting allosteric regulation cause an increase in glycogen levels in *E. coli.*

Stark et al (1992) engineered wild-type and mutant *E. coli* ADPGPP for expression in plants and assayed the effect on starch accumulation. Tubers from potato plants transformed with the wild-type *E. coli* enzyme had starch levels similar to those of wild-type plants, whereas those transformed with the allosterically insensitive *E. coli* ADPGPP enzyme had starch levels up to 60% higher than the wild type. The effect was observed only when the mutant protein was targeted to the chloroplast and driven by a tuber-specific promoter; constitutive expression was lethal. Such results demonstrate the importance of considering the target tissues, subcellular localization and kinetics of enzymes when engineering plant metabolism. Chloroplast engineering with targeting genes may markedly enhance transgene expression (Daniell et al., 2002; Daniell and Dhingra, 2002; Devine and Daniell, 2004).

Homozygous rice lines have also been developed using the *glgc* gene, and they showed increased grain filling due to higher starch synthesis (unpublished results).

8. CONCLUSIONS

Biotechnology offers the unique advantage of substantially improving the nutritional quality of staple food. Once the details of the biosynthetic pathways leading to the formation of desired end-products are known, biotechnological tools could be used to improve the quality of human food. The plant system has greater potential to be used as a factory for metabolites of food and non-food value than other production systems. Rice, being the staple food of more than two-thirds of the world population and the most accessible source of human nutrition/calories, is an ideal candidate for developing second-generation and/or third-generation transgenics. However, definitive proof of the benefit of nutritionally enriched transgenic foods (functional foods) depends on further development, release and eventual impact evaluation. Plant scientists and human nutritionists need to work together to realize the full potential of the genomic studies leading to the elucidation of different biosynthetic pathways and emerging biotechnological tools to develop functional food. In this context, hyper-expression capabilities of chloroplasts (Devine and Daniell, 2004; Daniell et al., 2004) and transgene containment offered by maternal inheritance (Daniell, 2002) make chloroplast genetic engineering an attractive new option (for further details see chapter 16 of this book). Metabolic engineering of rice holds great promise for its use as a human food and also for industrial applications.

9. ACKNOWLEDGEMENTS

We thank the HarvestPlus research program “Breeding for nutrition improvement” for partial support of the research. Thanks are due to Dr. K. Datta for the Golden Rice picture and unpublished data, and to Dr. Bill Hardy for editorial assistance.

10. REFERENCES

Agius, F., González-Lamothe, R., Caballero, J.L., Muñoz-Blanco J., Botella, M.A., & Valpuesta, V. (2003) Engineering increased vitamin C levels in plants by overexpression of a D-galacturonic acid reductase. *Nature Biotechnol. 21,* 177-181.

Bartley, G.E., Vitanen, P.V., Pecker, I., Chamovitz, D., Hirschberg, J., & Scolnik, P.A. (1991) Molecular cloning and expression in photosynthetic bacteria of a soybean cDNA coding for phytoene desaturase, an enzyme of the carotenoid biosynthesis pathway. *Proc. Natl. Acad. Sci. USA 88*, 6532-6536.

Boeckner, L.S., Schnepf, M.I., & Tungland, B.C. (2001) Inulin: a review of nutritional and health implications. *Adv. Food Nutr. Res.* 43, 1-63.

Burkhardt, P., Beyer, P., Wünn, J., Klöti, A., Armstrong, G.A., Schledz, M., von Lintig, J., & Potrykus, I. (1997) Transgenic rice (*Oryza sativa*) endosperm expressing daffodil (*Narcissus pseudonarcissus*) phytoene synthase accumulates phytoene, a key intermediate of provitamin A biosynthesis. *Plant J. 11,* 1071-1078.

Camara, B., Bardat, F., Seye, A., D'Halrlingue, A., & Monéger, R. (1982) Terpenoid metabolism in plastids: localization of □-tocopherol synthesis in *Capsicum* chromoplasts. *Plant Physiol. 70*, 1562-1563.

Chakraborty, S., Chakraborty, N., & Datta, A. (2000) Increased nutritive value of transgenic potato by expressing a non-allergenic seed albumin from *Amaranthus hypochondriacus*. *Proc. Natl. Acad. Sci. USA 97*, 3724-3729.

Chen, Z., Young, T.E., Ling, J., Chang, S.C., & Gallie, D.R. (2003) Increasing vitamin C content of plants through enhanced ascorbate recycling. *Proc. Natl. Acad. Sci. USA 100*, 3525-3530.

Cunningham, F.X., & Gantt, E. (1998) Genes and enzymes in carotenoid biosynthsis. *Annu. Rev. Plant Physiol. Plant Mol. Biol. 49*, 557-583.

Daniell, H. (2002). Molecular strategies for gene containment in transgenic crops. *Nat. Biotechnol., 20*, 581-586.

Daniell, D., Carmona-Sanchez, O., & Burns, B. B. (2004). Chloroplast derived antibodies, biopharmaceuticals and edible vaccines. In R. Fischer & S. Schillberg (Eds.) *Molecular Farming* (pp.113-133). Weinheim: WILEY-VCH Verlag.

Daniell, H., & Dhingra, A. (2002) Multiple gene engineering. *Curr. Opin. Biotechnol. 13,* 136-141.

Daniell, H., Khan, M., & Allison, L. (2002). Milestones in chloroplast genetic engineering: an environmentally friendly era in biotechnology. *Trends Plant Sci.,* 7(2), 84-91.

Datta, K., Baisakh, N., Oliva, N., Torrizo, L., Abrigo, E., Tan, J., Rai, M., Rehana, S., Al-Babili, S., Beyer, P., Potrykus, I., & Datta, S.K. (2003) Bioengineered 'golden' indica rice cultivars with beta-carotene metabolism in the endosperm with hygromycin and mannose selection systems. Plant Biotechnol. J. **1,** 81-90.

DellaPenna, D., Collakova, E., Coughlan, S., & Helentjaris, T. (1999) Phytyl/prenyltransferase nucleic acids, polypeptides and uses thereof. WO 00/68393 (Pioneer Hi-Bred, University of Nevada).

Devine, A. L., & Daniell, H. (2004). Chloroplast genetic engineering. In S. Moller (Ed.), *Plastids* (pp. 283-320). United Kingdom: Blackwell Publishers.

Falco, S.C., Guida, T., Locke, M., Mauvais, J., Sandres, C., Ward, R.T., & Webber, P. (1995) Transgenic canola and soybean seeds with increased lysine. *Bio/Technology 13*, 577-582.

Fray, R.G., Wallace, A., Fraser, P.D., Valero, D., Hedden, P., Bramley, P.M., & Grierson, D. (1995) Constitutive expression of a fruit phytoene synthase gene in transgenic tomatoes causes dwarfism by redirecting metabolites from the *gibberellin* pathway. *Plant J. 8,* 693-701.

Galili, G. (2002). New insights into the regulation and functional significance of lysine metabolism in plants. *Annu. Rev. Plant Physiol. Plant Mol. Biol. 53,* 27-43.

Gao, F.Y., Jing, Y.X., Shen, S.H., Tian, S.P., Kuang, T.Y., & Sun, S.S.M. (2001) Transfer of lysine-rice protein gene into rice and production of fertile transgenic plants. *Acta Bot. Sin. 43,* 506-511.

Gerster, H. (1997) Vitamin A functions, dietary requirements and safety in humans. *Int. Vit. Nutr. Res. 67,* 71-90.

Hoa, T.T.C., Al-Babili, S., Schaub, P., Potrykus, I., & Beyer, P. (2003) Golden indica and japonica rice lines amenable to deregulation. *Plant Physiol. 133,* 161-169.

Jain, A.K., & Nessler, C.L. (2000) Metabolic engineering of an alternative pathway for ascorbic acid biosynthesis in plants. *Mol. Breed. 6,* 73-78.

Krishnan, S., Datta, K., Baisakh, N., de Vasconcelos, M., & Datta, S.K. (2003) Tissue-specific localization of □-carotene and iron in transgenic indica rice (*Oryza sativa* L.). *Curr. Sci. 84,* 1232-1234.

Mazur, B., Krebbers, E., & Tingey, S. (1999) Gene discovery and product development for grain quality traits. *Science 285,* 372-375.

Misawa, N., Yamona, H., Linden, H., de Felipe, M.R., Lucas, M., Ikenga, H., & Sandmann G. (1993) Functional expression of the *Erwinia uredovora* biosynthesis gene *crtI* in transgenic plants showing an increase of □-carotene biosynthesis activity and resistance to the bleaching herbicide Norflurazon. *Plant J. 4,* 833-838.

Rocheford, T.R, Wong, J.C., Egesel, C.O., & Lambert, R.J. (2002) Enhancement of vitamin E levels in corn. *J Am. Col. Nutr. 21,* 191S-198S.

Römer, S., Fraser, P.D., Kiano, J.W., Shipton, C.A., Misawa, N., Schuch, W., & Bramley, P.M. (2000) Elevation of provitamin A content of transgenic tomato plants. *Nature Biotechnol. 18,* 666-669.

Rosati, C., Aquilani, R., Dharmapuri, S., Pallara, P., Marusic, C., Tavazza, R., Bouvier, F., Camara, B., & Giuliano, G. (2000) Metabolic engineering of beta-carotene and lycopene content in tomato fruit. *Plant J. 24,* 413-419.

Savidge, B., lassner, M.W., Weis, J.D., Mitsky, T.A., Shewmaker, C.K., Post-Beittenmiller, D., & Valentin, H.E. (2002) Isolation and characterization of homogentisate phytyltransferase genes from *Synechocystis* sp. PCC 6803 and *Arabidopsis. Plant Physiol. 129,* 321-322.

Sévenier, R., van der Meer, I., Bino, R., & Koops, A.J. (2002) Increased production of nutriments by genetically engineered crops. *J. Am. Col. Nutr. 21,* 199S-204S.

Sévenier, R., Hall, R.D., van der Meer, I., Hakkert, J.C., van Tunen, A.J., & Koops, A.J. (1998) High level of fructan accumulation in a transgenic sugar beet. *Nature Biotechnol. 16,* 843-846.

Shaul, O., & Galili, G. (1992) Increased lysine synthesis in transgenic tobacco plants expressing a bacterial dihydropicolinate synthase in their chloroplasts. *Plant J. 2,* 203-209.

Shewmaker, C.K., Sheehy, J.A., Daley, M., Colburn, S., & Ke, D.J. (1999) Seed-specific over-expression of phytoene synthase: increase in carotenoids and other metabolic effects. *Plant J. 20,* 401-412.

Shintani, D., & DellaPenna, D. (1998) Elevating the vitamin E content of plants through metabolic engineering. *Science 282,* 2098-2100.

Sivakumar, B. (1998) Current controversies in carotene nutrition. *Ind. J. Med. Res. 108,* 157-166.

Smirnoff, N. (1996) The function and metabolism of ascorbic acid in plants. *Ann. Bot. 78,* 661-669.

Stark, D.M., Timmerman, K.P., Barry, G.F., Preiss, J., & Kishore, G.M. (1992) Regulation of the amount of starch in plant tissues by ADP-glucose pyrophosphorylase. *Science 258,* 287-292.

Tan, J., Baisakh, N., Oliva, N., Torrizo, L., Abrigo, E., Datta, K., & Datta, S.K. (2004) Screening rice germplasm for carotenoid profile, Including □-carotene, by HPLC analysis. *Int. J. Food Sci. Technol. (in press).*

Van Der Meer, I.M., Koops, A.J., Hakkert, J.C., & van Yunen, A.J. (1998) Cloning of the fructan biosynthesis pathway of Jerusalem artichoke. *Plant J. 15,* 489-500.

Van Enennaam, A.L., Li, G., Venkatramesh, M., Levering, C., Gong, X., Jamieson, A.C., Rebar, E.J., Shewmaker, C.K., & Case, C.C. (2004) Elevation of seed □-tocopherol levels using plant-based transcription factors targeted to an endogenous locus. *Metabol. Eng. 6,* 101-108.

Van Enennaam, A.L., Lincoln, K., Durrett, T.P., Valentin, H.E., Shewmaker, C.K., Thorne, G.M., Jiang, J., Baszis, S.R., Levering, C.K., Aasen, E.D., Hao, M., Stein, J.C., Norris, S.R., & Last, R.L. (2003) Engineering vitamin E content: from *Arabidopsis* mutant to soy oil. *Plant Cell 15,* 3007-3019.

Ye, X., Al-Babili, S., Kloti, A., Zhang, J., Lucca, P., Beyer, P., & Potrykus, I. (2000) Engineering the provitamin A (□-carotene) biosynthetic pathway into (carotenoid-free) rice endosperm. *Science 287,* 303-330.

Zhu-Shimoni, X.J., & Galili, G. (1998) Expression of an *Arabidopsis* aspartate kinase/homoserine dehydrogenase gene is metabolically regulated by photosynthesis-related signals, but not by nitrogenous compounds. *Plant Physiol. 116,* 1023-1028.

Zimmerman, M., & Hurrel, R. (2002) Improving iron, zinc, and vitamin A nutrition through plant biotechnology. *Curr. Opin. Biotechnol. 13,* 142-145.

CHAPTER 20

PLASTID METABOLIC PATHWAYS FOR FATTY ACID METABOLISM

IKUO NISHIDA

Department of Biological Sciences, Graduate School of Science, The University of Tokyo, Tokyo, 113-0033, Japan.

Abstract. Fatty acid metabolism is important for storage oil synthesis as well as membrane biogenesis in plant cells. In the past few decades, attention has been paid to the pathways, mechanisms, regulation and genetic manipulation of fatty acid metabolism in oil-accumulating tissues, because plant oil provides nutritious fatty acids and even renewable feedstock for industry. The fatty acid synthesis in plant cells takes place almost exclusively within plastids. The plastidial fatty acid synthesis requires acetyl-CoA, ATP and NAD(P)H as substrates, and multiple routes exist to provide these substrates, although the relative importance of these routes depends on the sources of plastids. Pyruvate and phosphoenolpyruvate derived from cytosolic glycolysis are suggested to be converted to acetyl-CoA in plastids, and imported malate also provides acetyl-CoA via pyruvate in some plant species. Chain extension of fatty acids takes place by C2 unit derived from malonyl-CoA, which is synthesized from acetyl-CoA by acetyl-CoA carboxylase (ACCase) in an ATP-dependent manner. The fact that ACCase is a regulatory enzyme in fatty acid synthesis has stimulated a number of molecular and transgenic studies to elucidate regulatory mechanisms and also to improve oil production. Plastids contain "so-called" type II fatty acid synthase (FAS) that comprises a set of discrete enzymes catalysing one of essential reactions for chain extension and requiring acyl-carrier protein (ACP) for all the catalytic activities. Three major classes of condensing enzymes, termed KAS I, KAS II and KAS III, are involved in chain-length determination. KAS III, a cerulenin-insensitive enzyme, initiates the chain extension by condensing acetyl-CoA with malonyl-ACP and is known to be feedback-regulated by the product acyl-ACP. KAS I, a cerulenin-sensitive enzyme, is responsible for the production of palmitoyl-ACP, whereas KAS II is less sensitive to this drug and involved in elongation of myristoyl- and palmitoyl-ACPs to stearoyl-ACP. In addition, the fourth KAS belonging to a subclass of KAS II is involved in medium-chain acyl-ACP synthesis in some plant tissues. Although the plastid FAS produces saturated acyl-ACPs as long as C18 carbon atoms, stearoyl-ACP desaturase in the plastid stroma is essential for unsaturated fatty acid synthesis and regulates total cellular amounts of unsaturated fatty acids. Variants of this desaturase are known to produce monounsaturated fatty acids having unusual double-bond positions. Acyl-ACP thioesterase (TE) in the plastid stroma, which terminates the FAS reaction by hydrolysing acyl-ACPs into free fatty acids, plays a central role in chain-length determination, especially in medium-chain fatty acid synthesis in some plant tissues. Although current knowledge of plant fatty acid metabolism enables us to modify the quality of plant oils in common oil crops, further studies are required for increasing the plant oil yield. Furthermore, for a dramatic enhancement of plant oil production, the understanding of the basic mechanism(s) for converting the carbon allocation from starch to fatty acids in starch-accumulating crops is challenging but essential.

H. Daniell and C. D. Chase (eds.), Molecular Biology and Biotechnology of Plant Organelles 543—564.

1. INTRODUCTION

Fatty acids are ubiquitous components of plant oil as well as plant membrane lipids. Plant oils (or triacylglycerols; TGs) contain nutritious fatty acids, such as lenoleate ($18{:}2^{\square 9,12}$)* and □-linolenate ($18{:}3^{\square 9,12,15}$), although composition of these fatty acids in common crop oils are not fully optimised by traditional breeding and selection. Some plant tissues uniquely produce unusual fatty acids, such as medium-chain fatty acids (Graham et al., 1981) or unusual monounsaturated fatty acids (Kleiman and Spencer, 1982), which often accumulate over 80% of total fatty acids in these plant tissues. Some of these fatty acids are potentially important as renewable feedstock for industry due to their useful chemical or physical properties (Ohlrogge 1994; Töpfer et al., 1995).

In the past few decades, the pathways, mechanisms and regulation of plant fatty acid metabolism have been elucidated, which clearly shows the importance of plastid metabolism in producing these important fatty acids. Based on such knowledge, it is possible to genetically manipulate the composition of long-chain and medium-chain fatty acids in some of common oil crops, such as *Brassica napus* (oilseed rape), *Glycine max* (soybean), *Helianthus annuus* (sunflower) and *Zea mays* (maize) (Drexler et al., 2003). However, difficulty still exists in producing a large quantity of plant oils with particular fatty acid compositions, or increasing the composition of unusual fatty acids, such as unusual monounsaturated fatty acids (Suh et al., 2002), to the levels that are comparable to those found in the native species producing these fatty acids.

In this chapter, I summarize the basic knowledge of fatty acid metabolism in plastids, where *de novo* fatty acid synthesis takes place almost exclusively. I also introduce some past attempts to increase nutritional and industrial values of plant oil and a prospect in lipid biotechnology. There are a number of excellent previous reviews dealing with similar topics (Ohlrogge and Browse, 1995; Harwood, 1996; Ohlrogge and Jaworski, 1997; Voelker and Kinney, 2001).

2. PLANT FATTY ACIDS

In the plant kingdom, there occur various kinds of fatty acids of different chain lengths, double-bond positions and oxygenated functional groups (van de Loo et al., 1993). Like other cellular organisms, long-chain fatty acids having 16 and 18 carbon atoms occur in plants in both membrane and storage lipids. These are plamitate (16:0), stearate (18:0), oleate ($18{:}1^{\square 9}$), $18{:}2^{\square 9,12}$ and $18{:}3^{\square 9,12,15}$. 16:0 is a typical fatty acid in the oil from *Elaeis guineensis* (oil palm) mesocarps. Although 16:0 occurs in small quantities in membrane lipids, it is enriched in the *sn*-2 position of phosphatidylglycerol (PG) in plastids. 18:0 also occurs in high amounts in some seed oils (e.g., peanut oil), but negligibly in most membrane lipids. Very long-chain fatty acids having >18 carbon atoms serve as precursors to cuticular waxes and other

*All fatty acids are defined by $M{:}N^{\square m,n}$, where M and N represent carbon and double-bond numbers, respectively, and the *cis* and *trans* double-bond positions from the carboxyl terminus are indicated by the superscripts □m,n and □mt,nt, respectively.

surface lipids such as cutin and suberin and are rarely found in membrane lipids, except phosphatidylserine (Murata et al., 1984). Medium-chain saturated fatty acids ranging from 8 to 14 carbon atoms occur in seed oils from palms (Araceae) and the genus *Cuphea* (Lythraceae) (Graham et al., 1981; Wolf et al., 1983; Hirsinger and Knowles, 1984). The members of Ulmaceae accumulate caprylate (8:0) and caprate (10:0), many Lauraceae species predominantly accumulate 10:0 and laurate (12:0), and nutmeg (*Myristica fragrans*) accumulates predominantly myristate (14:0) in their seed oils (Davies, 1993; Voelker et al., 1997).

$18{:}1^{\Delta 9}$ occurs in large amounts in some plant oils (e.g., olive oil), but usually in small quantities in plant membrane lipids. *cis*-Vaccenate ($18{:}1^{\Delta 11}$), a typical bacterial unsaturated fatty acid, also occurs in plants in very small quantities (Shibahara et al., 1990). Some tree seed oils contain unusual monounsaturated fatty acids, such as petroselinic acid ($18{:}1^{\Delta 6}$) in seed oils from the members of Umbelliferae [e.g., coriander (*Coriandrum sativum* L.) or dill (*Anethum graveolens* L.)], Araliaceae, and Ganyaceae (Kleiman and Spencer, 1982), and $16{:}1^{\Delta 6}$ as a major fatty acid (>80%) in the seed oil of *Thunbergia alata*. Oil from glandular trichomes of *Pelargonium x hortorum* (garden geranium) contains $16{:}1^{\Delta 11}$, $18{:}1^{\Delta 13}$ and their elongation products (Schultz et al., 1996). Seed oil of *Macadamia integrifolia* contains 30% $16{:}1^{\Delta 9}$ (Gummeson et al., 2000).

Plant lipids contain large amounts of polyunsaturated fatty acids, such as $18{:}2^{\Delta 9,12}$ and $18{:}3^{\Delta 9,12,15}$, which are abundant in monogalactosyldiacylglycerol (MGDG) in plastids. In some dicots, such as in Solanaceae, Brassicaceae and Chenopodiaceae, $16{:}3^{\Delta 7,10,13}$ occurs exclusively in the *sn*-2 position of MGDG and, hence, these plants are called "16:3 plants". Other plants called "18:3 plants" include most of dicots and all monocots (Mongrand et al., 1998). *Borago officinalis* contains γ-linolenate ($18{:}3^{\Delta 6,9,12}$) and $18{:}4^{\Delta 6,9,12,15}$ [see Drexler et al. (2003) and references therein]. Seeds of *Momordica charantia* and *Impatiens balsamina* contain fatty acids of conjugated double bonds, α-eleostearic ($18{:}3^{\Delta 9,11t,13t}$) and α-parinaric acids ($18{:}4^{\Delta 9,11t,13t,15}$), respectively (Cahoon et al., 1999).

3. FATTY ACID SYNTHESIS IN PLASTIDS

Fatty acids are synthesized almost exclusively within plastids in plant cells [for a review, see Harwood (1996)]. Plastids contain type II FAS that comprises discrete enzymes catalysing essential reactions for fatty acid synthesis. These include malonyl-CoA:ACP transacylase (MCAT), the condensing enzymes (or 3-ketoacyl-ACP synthases I, II and III; KAS I, II and III), NADPH-dependent 3-ketoacyl-ACP reductase, 3-hydroxyacyl-ACP dehydratase, and NADH-dependent enoyl-ACP reductase (ENR). Acyl carrier protein (ACP) is required for all the catalytic activities, and fatty acids are extended while esterified on ACP [for details of each reaction, see Harwood (1996)]. All the above components are thought to form a complex *in vivo*, facilitating the efficient utilization of low concentration of substrates, process called substrate channelling (Roughan, 1997).

Mitochondria appear to provide another place for the synthesis of short- or medium-chain fatty acids (Shintani and Ohlrogge, 1994), which is required for lipoic acid biosynthesis (Wada et al, 1997; Gueguen et al., 2000). Mitochondria appear to contain all the FAS components required for incorporation of [2-^{14}C]malonate into fatty acids (Wada et al, 1997). Mitochondrial ACP appears to direct premature synthesis of short- or medium-chain fatty acids in mitochondria (Shintani and Ohlrogge, 1994), although mitochondrial FAS system has potential to produce a broad chain-length of fatty acids, such as 8:0-ACP, 16:0-ACP and 18:0-ACP, when coupled with other ACPs (Gueguen et al., 2000; Yasuno et al., 2004).

4. REQUIREMENTS FOR FATTY ACID SYNTHESIS

Plastids require acetyl-CoA and malonyl-CoA as the carbon substrates for fatty acid synthesis. Malonyl-CoA is synthesized from acetyl-CoA by acetyl-CoA carboxylase (ACCase), which requires ATP. Malonyl-CoA condenses with the primer acetyl-CoA or nascent acyl-ACPs, providing 2 carbon unit, and the resultant oxygenated acyl-ACPs are reduced to saturated acyl-ACPs, consuming one each of NADPH and NAPH per C2 chain extension (Slabas and Fawcett, 1992). Thus, plastids require acetyl-CoA, ATP and the reducing power NAD(P)H for fatty acid synthesis. Multiple routes have been elucidated for providing these substrates, which are briefly summarised in this section. The relative importance of different routes depends on the sources as well as the physiological and developmental status of plastids [for the recent review, see Ohlrogge et al. (2000); Rawsthorne (2002)].

4.1. Acetyl-CoA

Four possible routes have been proposed for the provision of acetyl-CoA in plastids: these are: 1) the conversion of passively imported acetate to acetyl-CoA by acetyl-CoA synthetase (ACS); 2) the conversion of pyruvate to acetyl-CoA by plastidial pyruvate dehydrogenase complex (plPDC); 3) the transport of the acetyl group from acetylcarnitine to acetyl-CoA in plastids; and 4) the synthesis of acetyl-CoA via an ATP-dependent citrate lyase (ACL), although the third and fourth routes to acetyl-CoA are not widely accepted [see Rawsthorne (2002)]. Acetate produced by mitochondria was thought to be the best substrate for fatty acid synthesis *in vivo* (Roughan et al., 1979). However, Ohlrogge and his colleagues argued that endogenous acetate is not the source of acetyl-CoA for fatty acid biosynthesis (Bao et al., 2000; Ohlrogge et al., 2000). Under the conditions that ensure the maximum rate of *in vivo* fatty acid synthesis in Arabidopsis and other plants, $^{14}CO_2$ was rapidly incorporated into fatty acids (within 2-3 min) but scarcely into acetate for 40 min (Bao et al., 2000; Ohlrogge et al., 2000).

Although the pathways from CO_2 to acetyl-CoA have not been elucidated, it seems unlikely that free acetate serves as a major intermediate in fatty acid synthesis from CO_2 in photosynthetic tissues. Results of *in situ* hybridisation analysis of ACS mRNA accumulation in developing Arabidopsis seeds (Ke et al., 2000), as well as the normal phenotypes of ACS-antisensed plants (Nikolau et al. 2000), also do not support the role of ACS in the acetyl-CoA synthesis for fatty acid synthesis in photosynthetic and non-photosynthetic tissues.

4.2. Pyruvate

Several lines of evidence suggest that pyruvate in plastids is the most important source of acetyl-CoA for fatty acid biosynthesis [see Rawsthorne (2002) and references therein]. First, pyruvate is efficiently incorporated into fatty acids by isolated plastids from various sources. Second, plastids contain the pyruvate dehydrogenase complex (plPDC) (Tovar-Méndez et al., 2003) which converts pyruvate to acetyl-CoA. Third, the temporal and spatial accumulation patterns of mRNAs for the E1□ subunit of plPDC and plastidial ACCase are mutually correlated in developing Arabidopsis seeds (Ke et al., 2000). Fourth, disruption of the E2 subunit of plPDC causes an early embryo lethal phenotype (Lin et al., 2003). The abundance of expressed sequence tags (the method is termed “digital northern”) (White et al., 2000) as well as microarray profiles (Ruuska et al., 2002) also support the role of plPDC during oil accumulation in Arabidopsis seeds.

Pyruvate is available from the cytosolic glycolysis, and pyruvate transport activity has been characterized in plastids isolated from *Brassica napus* (oilseed rape) embryos and *Ricinus communis* (castor bean) endosperm (Eastmond and Rawsthorne, 2000). In developing Arabidopsis seeds, however, pyruvate synthesis in the cytoplasm might be limited by a low expression level of cytosolic pyruvate kinase relative to that of other glycolytic enzymes, as predicted by “digital northern” (White et al., 2000) and microarray (Ruuska et al., 2002) analysis.

Plastids from developing *B. napus* embryos can synthesize fatty acids from exogenously supplied glucose, glucose 6-phosphate (G6P), dihydroxyacetone-phosphate (DHAP) and phosphoenolpyruvate (PEP), all of which are converted to pyruvate via plastidial glycolysis (Kang and Rawsthorne, 1994). These plastids contain a complete set of the glycolytic enzymes (Kang and Rawsthorne, 1994; Eastmond and Rawsthorne, 2000). However, in developing seeds of Arabidopsis and other plants, the plastidial glycolytic pathway appears to be limited by the low abundance or absence of particular enzymes, like phosphoglyceromutase and enolase [see also relevant discussions in White et al. (2000); Rawsthorne (2002)].

Identification of a cDNA for plastid-targeted PEP/phosphate translocator (plPPT, Flügge, 1998) has drawn attention to cytosolic PEP as the source of plastidial pyruvate. The activity of plPPT is comparable to those of G6P and pyruvate translocators in developing *B. napus* embryos (Kubis and Rawsthorne, 2000). High transcript scores of plastidial pyruvate kinase in developing Arabidopsis seeds, as revealed by “digital northern” (White et al., 2000) and microarray (Ruuska et al., 2002) techniques, also support the conversion of PEP to pyruvate in plastids. However, the plPPT pathway is not essential for fatty acid synthesis by Arabidopsis plastids, because an Arabidopsis *cue1* mutant defective in a plPPT gene (At5g33320) shows normal overall lipid levels in both leaf and seeds (Streatfield et al., 1999). The mutant shows a reticulate leaf phenotype (the pale yellow leaf phenotype, leaving green parts only along veins) due to the shortage of aromatic amino acids, whose synthesis totally depends on the shikimate pathway from PEP in plastids. This result suggests alternative routes for pyruvate synthesis in Arabidopsis plastids, most probably via direct import of pyruvate.

Malate is converted to stoichiometric amounts of pyruvate and NADPH by NADP-malic enzyme in plastids, and the subsequent synthesis of acetyl-CoA from pyruvate provides stoichiometric amounts of CO_2 and NADH. Therefore, the import of malate supplies stoichiometric amounts of all the substrates but ATP that are required for C2 chain extension. This pathway, though, appears to be confined to some plant tissues, such as castor bean endosperm and some species of *Brassica* (Singal et al., 1987; Smith et al., 1992; Eastmond et al., 1997; Möhlmann and Neuhaus, 1997; Schwender and Ohlrogge, 2002).

4.3. ATP and NAD(P)H

ATP is provided by photophosphorylation in chloroplasts. In non-green plastids, however, ATP is imported by ATP/ADP translocators (Kampfenkel et al., 1995), by DHAP/phosphoglycerate and oxaloacetate/malate metabolite shuttles, or produced by plastidial metabolism such as glycolysis. Conversion of 1,3-bisphosphoglycerate to 3-phosphoglycerate by phosphoglycerokinase, and that of PEP to pyruvate by pyruvate kinase, produce ATP in plastids [see Rawsthorne (2002) and references therein].

Multiple routes provide reductant for plastids, including steps in plastidial glycolysis, photochemical reactions, the import of reducing equivalents from the mitochondria or cytosol, or the cytosolic and plastidial oxidative pentose phosphate (OPP) pathway (Rawsthorne, 2002; Schwender et al., 2003). In plastids from a late-developing stage embryos of *B. napus*, an interaction between fatty acid synthesis and the OPP pathway has been suggested, exemplified by pyruvate stimulation in utilization of glucose 6-phosphate in the OPP pathway (Eastmond and Rawsthorne, 2000). However, in *in vitro*-cultured developing *B. napus* embryos, the reductant produced by the OPP pathway accounts for at most 44% of the NADPH and only 22% of total reductant needed for fatty acid synthesis, suggesting a significant contribution from other reductant sources *in vivo* (Schwender et al., 2003).

4.4. Coordinate regulation of multiple substrate pathways in plastids

As described above, multiple routes support the provision of carbon and energy substrates for fatty acid synthesis in plastids. Moreover, even a single metabolite, like G6P, can be processed by multiple routes, such as starch synthesis, glycolysis and the OPP pathway. How are these different fluxes regulated, and how are the metabolic fates determined in order to support fatty acid synthesis by plastids? For the first question, Eastmond and Rawsthorne (2000) suggested the importance of selective substrate import processes by plastids, using plastids from developing *B. napus* embryos, although experiments using isolated plastids should be confirmed by *in vivo* experiments. For the second question with respect to the metabolic fates of G6P *in vivo*, Schwender and Ohlrogge (2002) cultured detached *B. napus* embryos in a specially optimized medium supplemented with [1,2-^{13}C]- or [U-$^{13}C_6$]-labeled glucose in the presence and absence of cold amino acids. In these experiments, relative contribution of glycolysis and the OPP pathway in the conversion of hexose to pyruvate can be estimated from the relative incorporation ratios into [^{13}C]pyruvate isotopomers. They concluded that direct glycolysis (probably cytosolic) accounts for at least 90% of precursors of plastidial acetyl-CoA, and amino acids do not provide carbons for plastidial fatty acid synthesis.

5. KEY GENES INVOLVED IN SUBSTRATE SYNTHESIS

5.1. Metabolite translocators

Plastids import a number of metabolites using specific translocators situated in the inner envelope membrane (Flügge, 1998). Three major translocators that are relevant to fatty acid metabolism have been cloned to date. These are glucose-6-phosphate/phosphate translocators (GPT), chloroplast triose-phosphate/phosphate translocators (cTPT) and plPPT. The deduced amino-acid sequences of GPT proteins are similar to each other, but share ~38% homology with members of cTPT and plPPT families. GPT activity of the plastids form rapeseed embryos is in excess of the combined rates of G6P metabolism via fatty acid and starch synthesis and through the OPP pathway (Kang and Rawsthorne, 1994, 1996). The *B. napus* embryo GPT is sensitive to inhibition by long-chain acyl-CoA at 50% inhibition concentrations (IC_{50}) of 200–300 nM (Johnson et al., 2000; Fox et al., 2000).

However, since *in vivo* concentrations of acyl-CoAs are estimated to range from 10 to 15 μM, the involvement of acyl-CoA binding protein is suggested in regulation of free acyl-CoA concentrations and in feedback regulation of GPT activity (Fox et al., 2000). Because GPT accepts G6P, triose phosphate and phosphate as substrates, but neither glucose 1-phosphate nor PEP, it is proposed that plastid-targeted GPT is responsible for triose phosphate transport in non-green plastids, where expression of cTPT is negligible (Kammerer et al., 1998). Recent analysis of expressed sequence tags identified a TPT tag in developing Arabidopsis seeds, which may reflect the fact that developing Arabidopsis seeds are green and still retain some CO_2 assimilation activity which is suggested to play a role in refixation of CO_2 released by plPDC (White et al., 2000).

5.2. plPDC

PDC converts pyruvate to acetyl-CoA, producing stoichiometric amounts of NADH and CO_2 [for the recent review, see Tovar-Méndez et al. (2003)]. Plant cells contain the PDC activity in both mitochondria and plastids. The mitochondrial PDC (mtPDC), a regulatory protein in mitochondrial respiration, consists of three catalytic subunits: pyruvate dehydrogenase (PDH or E1), dihydrolipoyl acetyltransferase (E2), and dihydrolipoyl dehydrogenase (E3), plus tow regulatory subunits: PDH kinase (PDHK) and phosphoPDH phosphatase. PDH has an □2□2 structure and the E1□ component is inactivated by phosphorylation by PDHK. plPDC comprises the three catalytic subunits similar to those of mtPDC, but none of the regulatory subunits for phosphorylation/dephosphorylation. Since biosynthesis of isoprenoids and some amino acids in plastids start from pyruvate (Lichtenthaler, 1999), plPDC plays a role in shifting the carbon flux towards acetyl-CoA synthesis.

To date, all of the subunits of plPDC have been cloned, and their plastidial localization has been verified. The activity of plPDC is sensitive to product inhibition by NADH and acetyl-CoA. Interestingly, down-regulation of PDHK by antisense technology caused increased seed oil content and seed weight at maturity in transgenic Arabidopsis seeds (Marillia et al., 2003). Increased mtPDC activity may cause increased synthesis of acetyl-CoA from pyruvate, which may lead to increased free acetate that could be utilized for fatty acid synthesis by plastids. Alternatively, increased respiration in PDHK-antisensed plants may cause increased levels of cellular ATP, which may stimulate fatty acid synthesis by plastids.

6. KEY GENES FOR FATTY ACID SYNTHESIS IN PLASTIDS

Except for the plastome gene *accD* encoding a subunit of ACCase, all genes required for fatty acid synthesis by plastids are encoded by the nuclear genome [for the latest information, visit http://www.canr.msu.edu/lgc/]. Genes for ACPs from Chromista and KAS III from the red alga *Porphyra umbilicalis* are also encoded by the plastome (Wang and Liu, 1991; Hwang and Tabita, 1991; Reith, 1993).

6.1. ACCase

Although plastidial acetyl-CoA serves as the substrate for fatty acid and cysteine syntheses, ACCase directs the carbon flux into fatty acid synthesis (Ohlrogge and Jaworski, 1997). Plants contain two types of ACCase: heteromeric ACCase in plastids and homomeric ACCase in the cytoplasm [see Sasaki and Nagano (2004) and references therein]. Heteromeric ACCases, like those found in bacteria, comprise three dissociable components: a biotin carboxyl carrier protein (BCCP) subunit, a biotin carboxylase (BCase) subunit, and a carboxyltransferase (CT) complex that comprises □-CT and □-CT subunits. BCase and CT catalyse two sequential reactions: an ATP-dependent carbonylation of biotin from bicarbonate and a transfer of carboxyl groups to acetyl-CoA, respectively. BCCP is required for both of these reactions. Homomeric ACCases, like those found in yeast and mammals, are a homodimmer of a $> \sim 200$ kD subunit that contains a BCCP domain, a BCase domain and a CT domain. Gramineae plants exceptionally contain homomeric ACCase in both the cytoplasm and the plastid but no heteromeric ACCase (Sasaki and Nagano, 2004). *B. napus* also exceptionally contains one of homomeric ACCases in the plastid (Schulte et al., 1997), and *Hordeum vulgare* (barley) is reported to contain ACCase in mitochondria (Focke et al., 2003). Homomeric ACCase is a target of aryloxyphenoxypropionate herbicides such as diclofop, but heteromeric ACCase is insensitive to these herbicides.

Although it was previously thought that homomeric ACCase was ubiquitous in plastids from all plants, identification of *E. coli accD* (encoding □-CT) homologs in several plastome sequences led to the first discovery of heteromeric ACCase from pea in plants (Sasaki et al., 1993). Pea □-CT has a long C-terminal extension with unknown function when compared with *E. coli* accD. Heteromeric ACCase is suggested to be associated with inner envelope membranes from pea

(Thelen and Ohlrogge, 2002a), although conflicting results are reported (Sasaki et al., 1993; Shorrosh et al., 1996; Ke et al., 2000). To date, all genes for the subunits of heteromeric ACCase and for homomeric ACCase have been cloned [see Sasaki and Nagano (2004) and references therein]. Arabidopsis contains the nuclear-encoded, single-copy genes *CAC1*, *CAC2* and *CAC3* for BCCP1, BCase, and □-CT, respectively, and the plastid-encoded gene *accD* for □-CT (Bao et al., 1997; Choi et al., 1995; Ke at al., 1997; Sun et al., 1997).

In addition, a second gene for BCCP (*BCCP2*) has been identified in Arabidopsis and other Brassicaceae plants (Thelen et al., 2001). *BCCP1* expresses ubiquitously in all tissues and is particularly abundant in developing leaves and seeds, whereas *BCCP2* expresses in flowers and developing seeds (Thelen et al., 2001). Arabidopsis has two copies of homomeric ACCase genes, which are encoded by the tandem-repeated genes *ACC1* and *ACC2* in chromosome 1, and *ACC1* is essential for embryo development (Baud et al., 2003). Although ACCase activity in tissue homogenates is too low for reliable assays, detection of BCCP by immunoblotting, using ^{125}I-streptavidin or anti-biotin antibodies, provides a reliable estimate for the relative protein content. In Arabidopsis leaf extracts, BCCP1 and BCCP2 are detected as a 35-38 kD protein and a 25 kD protein, respectively, together with a 240 kD protein for homomeric ACCase and a 78 kD protein for a subunit of methylcrotonyl-CoA carboxylase (Nikolau et al., 2003).

In monocots, the three copies of *Acc-1* gene encoding plastid-targeted homomeric ACCase have been cloned from wheat [see Zuther et al. (2004) and references therein]. Interestingly, in each of these genes, nested alternative promoters produce two transcripts of different length, such that the second (shorter) transcripts is initiated within the first intron located in the 5' non-coding region of the first (longer) transcripts, and these transcripts accumulate differentially in sectors along young leaf blades.

Heteromeric ACCase has been suggested to play a regulatory role in the light-driven fatty acid synthesis by isolated spinach chloroplasts (Ohlrogge and Jaworski, 1997). The activity of heteromeric ACCase in chloroplasts may be regulated by light/dark-induced changes in stromal pH, ATP and Mg^{2+}, which largely reflect the enzyme characteristics of BCase (Sun et al., 1997). However, it is not known if such mechanism also operates in non-green plastids. In addition to such fine regulation, heteromeric ACCase activity appears to be coarsely regulated post-translationally by redox regulation [see Sasaki and Nagano (2004) and references therein], reversible phosphorylation (Savage and Ohlrogge, 1999), and feedback inhibition by extra-plastidial long-chain fatty acids (Shintani and Ohlrogge, 1995). Homomeric ACCase activity purified from developing pea embryos is strongly inhibited competitively by ADP and malonyl-CoA (Bettey et al, 1992).

In developing Arabidopsis siliques (Ke et al., 2000), mRNAs for *CAC1*, *CAC2*, *CAC3* and *accD* genes peaked at both 2 days after flowering (DAF) (most rapidly expanding siliques) and 7 DAF (siliques with near-maximal oil accumulation), and each mRNA level subsequently declined between 9 and 12 DAF. *In situ* hybridisation analysis also revealed coordinate changes in the spatial and temporal patterns of mRNAs for all the ACCase subunits in the developing siliques

(Ke et al., 2000). Coordination between the nuclear-encoded *CAC* genes could be regulated by a common set of sequence motifs in the promoter regions of these genes (Bao et al., 1997; Sun et al., 1997; Ke et al., 1997, 2000), although their significance remains to be evaluated experimentally.

However, ACCase content in plastids appears to be regulated at the step of subunit assembly directed by accD protein (Madoka et al., 2002). In transgenic tobacco overexpressing endogenous *accD* by promoter replacement, the protein levels of each of heteromeric ACCase subunits increased with no significant increase in the mRNA level of *CAC2* (or *accC*) (Madoka et al., 2002). By contrast, neither sense nor antisense expression of BCase in transgenic tobacco plants affected BCCP levels (Shintani et al., 1997). Overexpression of BCCP2 in Arabidopsis also did not affect the expression levels of other ACCase subunits but produced unbiotinylated BCCP2, which appeared to be assembled into inactive ACCase, causing a 23% decrease of the average seed fatty acid content when compared with the wild-type seeds (Thelen and Ohlrogge, 2002b).

Overexpression of homomeric ACCase in the plastids of *B. napus* seeds successfully enhanced the yield of seed oils by 5% (Roesler et al., 1997), whereas transgenic enforcement of plastidial ACCase in tobacco plastids caused increased seed yield and extended leaf longevity (Madoka et al., 2002), although the reason for the latter phenotypes remain to be elucidated in future research.

6.2. ACP

ACP binds fatty acids through the 4'-phosphopantetheine prosthetic group (Ohlrogge, 1987). ACP appears to form a complex with FAS and other enzymes in plastids (Post-Beittenmiller et al., 1989). The role of ACP structure in the activity of desaturase, fatty acid synthase, acyl-ACP thioesterase, and acyltransferases has been suggested by *in vitro* experiments (Guera et al., 1986; Suh et al., 1999; Schütt et al., 1998; Salas and Ohlrogge, 2002), although the results should be confirmed by *in vivo* experiments. Coriander ACP is preferred by Δ4 palmitoyl-ACP desaturase (Δ4 16:0-ACP desaturase) from the same plants when compared with spinach and *E. coli* ACP (Suh et al., 1999), and spinach ACP is preferable to *E. coli* ACP as a substrate for Arabidopsis *FATB* thioesterase (Salas and Ohlrogge, 2002). However, co-expression of coriander Δ4 16:0-ACP desaturase with coriander ACP did not improve the production of $16{:}1^{\Delta 4}$ in transgenic Arabidopsis (Suh et al., 2002).

6.3. KASs

Plants usually require KAS I, KAS II and KAS III for long-chain fatty acid synthesis. KAS III is a dimer with subunit sizes of 37–41 kD, completely insensitive to cerulenin, and commits the extension from acetyl-CoA to 4:0-ACP [see Harwood (1996) and references therein; Schuch et al., 1994]. KAS I is a dimer with subunit sizes of 48–50 kD, sensitive to cerulenin (an IC_{50} at ~2 μM) and is responsible for the extension from 4:0-ACP to 16:0-ACP. KAS II is a 46 kD protein, relatively insensitive to cerulenin (an IC_{50} at ~50 μM) and elongates 14:0-ACP and 16:0-ACP to 18:0-ACP. Genes for KAS III (*KAS3*) and KAS I (*KAS1*) have been cloned from several plants [see Slabaugh et al. (1998) and references therein]. KAS II genes designated *KASA* were first identified as castor and barley clones that belong to neither *KAS3* nor *KAS1* (Wissenbach, 1994; Slabaugh et al., 1998). KAS II identity has been verified by complementation of the *fab1* mutant defective in 16:0 elongation (Carlsson et al., 2002).

KAS II-type condensing enzymes are involved in medium-chain fatty acid synthesis in *Cuphea* plants. These include Cw KASA1 and Cw KASA2 from immature embryos of *C. wrightii* (Slabaugh et al., 1998) and KAS IV from *Cuphea* sp. (Dehesh et al., 1998). Seed-specific expression of cDNAs for these condensing enzymes in transgenic Arabidopsis or *B. napus* did not change the fatty acid composition of the seed oil from that in the wild type. However, expression of *Cw KASA1* cDNA in Arabidopsis seeds, together with a cDNA for medium-chain specific *Cuphea FATB* thioesterase (see the section 6.6 of this chapter) enhanced the level of medium-chain fatty acids over that achieved by *Cuphea FATB* expression alone (Leonard et al., 1998). Similar co-expression effects are obtained in transgenic *Brassica* plants overexpressing a *Cuphea KAS IV* cDNA together with either 12:0-ACP thioesterase (*Uc FATB1*) or 8:0/10:0-ACP thioesterase (*Ch FATB2* or *Cp FATB1*) (Dehesh et al., 1998)

KAS III activities from the 10:0-acccumulating species *Cuphea lanceolata* and the 10:0/12:0-accumulating species *C. wrightii* are sensitive to 10:0-ACP and 12:0-ACP, respectively. The inhibition is uncompetitive to acetyl-CoA and non-competitive to malonyl-ACP (Brück et al., 1996; Abbadi et al., 2000). A putative regulatory acyl-ACP binding motif has been identified, and transgenic plants overexpressing a modified *Cuphea* KAS III without this motif are deserved for testing effects on medium-chain fatty acid synthesis (Abbadi et al., 2000), although KAS III overexpression in transgenic plants caused rather complicated effects on fatty acid composition and lipid content (Verwoert et al., 1995; Dehesh et al., 2001).

KASs also play important roles in the biosynthesis of other unusual fatty acids. Accumulation of 14:0 in *Myristica fragrans* (nutmeg) seed oil is most likely governed by the elongation capacity of the fatty acid synthase (Voelker et al., 1997). $18:1^{\square 11}$ biosynthesis occurs via elongation of $16:0^{\square 9}$ in higher plants or double bond shift from $18:1^{\square 9}$ in *Diospyros kaki* (kaki) pulp (Shibahara et al., 1990). $18:1^{\square 6}$ biosynthesis also occurs via elongation of $16:0^{\square 4}$ in coriander (Cahoon and Ohlrogge, 1994).

6.4. ENR

ENR catalyses the NADH-specific reduction of a *trans* double bond to produce saturated fatty acids. This enzyme plays a determinant role in completing cycles of fatty acid biosynthesis in *E. coli* (Heath and Rock, 1995). ENRs from plants and bacteria are a target for the anti-microbial agent isoniazid which inhibits the growth of *Mycobacterium tuberculosis*, as well as for diazaborines which inhibit the growth of *E. coli* [see Fawcett et al. (2000) and references therein]. The three dimensional structure of *B. napus* ENR has been resolved (Baldock et al., 1998).

6.5. Stearoyl-ACP desaturase and its variants

Stearoyl-ACP desaturase (□9 18:0-ACP desaturase) and other variant acyl-ACP desaturases in the stroma of plastids are a class of soluble enzymes that introduce a double bond into saturated acyl chains esterified on ACP. □9 18:0-ACP desaturase catalyses the first step of unsaturated fatty acid biosynthesis in plants, and the Arabidopsis *fab2/ssi2* mutant that is defective in □9 18:0-ACP desaturase shows dwarfing, cell death, and enhanced disease resistant phenotypes (Nandi et al., 2003). This enzyme was first purified from avocado mesocarp, and the first cDNA clones were obtained from cucumber, castor bean and safflower [see Harwood (1996) and references therein]. Antisense expression of a □9 18:0-ACP desaturase gene in *Brassica rapa* caused up to 40% increase of 18:0 content in seed oil (Knutzon et al., 1992).

Variant acyl-ACP desaturases form some species showed unusual specificities to the double-bond position and the chain length. These are *C. sativum* □4 16:0-ACP desaturase, *T. alata* □6 16:0-ACP desaturase, *P. x hortorum* □9 14:0-ACP desaturase, and □9 16:0-ACP desaturases from *Asclepias syriaca* (milkweed) and *Doxantha unguis-cati* (cat's claw) (see Schultz et al., 1996; Cahoon et al. (1998) and references therein). Developing seeds of *Brassica scoparia* are suggested to have □5 16:0-ACP desaturase (Whitney et al., 2004). Acyl-ACP desaturases require ferredoxin as the electron-donating cofactor, preferring strongly the heterotrophic rather than the photosynthetic isoforms (Schultz et al., 2000).

Specificities of the *T. alata* □6 16:0-ACP desaturase toward the □6 position regardless of C16 or C18 chain length was enhanced by the substitution A188G/Y189F and also by domain swapping with castor bean □9 18:0-ACP desaturase (Cahoon et al., 1997). In another example, the growth of an *E. coli* auxotroph for unsaturated fatty acids is complemented by a genetically screened G188L variant for the castor bean □9 18:0-ACP desaturase that is originally unable to use *in vivo* acyl-ACP substrates and to complement the growth of the *E. coli* auxotroph (Cahoon and Shanklin, 2000). The plastid-targeted overexpression of this variant cDNA in developing seeds of Arabidopsis *fab1* increased $16{:}1^{\square 9}$+ $18{:}1^{\square 11}$+ $20{:}1^{\square 13}$ up to the amounts of >25% of the seed oil over the amounts of <5% in the wild-type seed oil. Similarly, the substitutions L118W and T117R/G188L of the castor bean □9 18:0-ACP desaturase are also suggested to increase the chain-length specificity toward 16:0-ACP (Cahoon et al., 1998; Whittle and Shanklin, 2001). The three-dimensional positions of these functional amino acid residues are also available (Lindqvist et al., 1996).

6.6. Acyl-ACP thioesterases

Fatty acid biosynthesis in plastids is terminated by either acyltransferases that incorporate acyl chains of acyl-ACPs into glycerolipids in plastids (the prokaryotic or the plastidial pathway) or acyl-ACP thioesterases (TEs) that release free fatty acids from acyl-ACPs. Free fatty acids are then exported in the form of acyl-CoA to the endoplasmic reticulum (ER) for lipid synthesis (the eukaryotic or the cytoplasmic pathway). Thus, TEs determine the amount of fatty acids directed toward the eukaryotic pathway [for a review, see Ohlrogge and Browse, 1995].

Higher-plant TE genes are classified into two homologous families of nuclear genes termed *FATA* (*fatty acyl-ACP thioesterase A*) and *FATB*, (Jones et al., 1995). FATA homologs ubiquitously exhibit the highest activity *in vitro* toward 18:1-ACP, but much less activity towards saturated acyl-ACPs [see Salas and Ohlrogge (2002) and references therein]. However, a Garm FATA from *Garcinia mangostana* (mangosteen) is responsible for a high 18:0 proportion (~50 %) in seed oil (Hawkins and Kridl, 1998), and the specific activity toward 18:0-ACP was successfully increased by genetic engineering (Facciotti et al., 1999). It was initially thought that FATB homologs were responsible for the production of medium-chain fatty acids in some plant oils, because the first *FATB* cDNA clone (*Uc FATB1*) isolated from the 12:0-accumulating *Umbellularia californica* endosperm encoded a TE specific to 12:0-ACP (Pollard et al., 1991; Davies et al., 1991) and TE activities preferring medium-chain acyl-ACPs were also identified from developing seeds of other plants (Davies, 1993; Dörmann et al., 1993).

However, further studies have confirmed that FATB orthologs hydrolysing acyl-ACP substrates ranging from 14:0-ACP to 18:1-ACP with strong preference for 16:0-ACP are ubiquitous in plants [see Salas and Ohlrogge (2002) and references therein]. Unique FATBs preferring medium-chain acyl-ACPs appear to have evolved from ubiquitous FATBs several times during evolution of angiosperms (Jones et al., 1995). Arabidopsis FATB1 contributes to 16:0 production especially in flowers (Dörmann et al., 2000), and disruption of this gene causes reduction in 16:0 and 18:0 levels and the wax content (Bonaventure et al., 2003).

FATBs exhibit a variety of chain-length specificities and, hence, are extensively explored by genetic manipulation. Expression of *Ch FATB1* and *Ch FATB2* cDNAs from *C. hookeriana* in *B. napus* produced oils enriched in 16:0, and 8:0 and 10:0, respectively (Dehesh et al., 1996). *Ulmus americana* (elm) FATB shows a binary specificity towards 10:0-ACP and 14:0-ACP (Voelker et al., 1997). FATB orthologs from *U. californica* (Uc FATB1) and *Cinnamomum camphorum* specifically hydrolyse 12:0-ACP and 14:0-ACP, respectively, although their amino acid sequences share 92% identity with each other (Yuan et al., 1995). Uc FATB1 can be converted to the one that prefers 14:0-ACP by site-directed mutagenesis of only three amino acids (Yuan et al., 1995).

An active-site cysteine (C320) and a histidine (H285) for Uc FATB1 are located within conserved active-site motifs of N-Q(K)-H-V-N(S)-N and Y-R-R(K)-E-C-G/Q(T), respectively (the amino acids in parenthesis occur less frequently), which contrasts with a lipase-like catalytic triad made up of S, H and D in *Vivrio harveyi* 14:0-ACP thioesterase (Yuan et al., 1996). Plant TEs have a catalytic machinery that is basically similar to that of the cystein proteinase papain, whereas the active sites of bacterial and animal thioesterases resemble that of serine proteinases such as trypsin (Yuan et al., 1996).

6.7. Acyl-CoA synthetase

Long-chain acyl-CoA synthetase (LACS, EC 6.2.1.3) in the outer plastid envelope converts fatty acids released by TE, to acyl-CoAs in an ATP-dependent manner [see Schnurr et al. (2002) and references therein]. LACSs belong to the AMP-binding protein superfamily and are characterized by the presence of a linker domain of 30-40 amino acid residues located between two conserved domains LS1 and LS2 (Shockey et al., 2002). Based on these criteria, nine *LACS* genes have been identified in Arabidopsis (Shockey et al., 2002), and one of the genes designated *LACS9* has been identified as the major LACS in the chloroplast envelope (Schnurr et al., 2002). The T-DNA knockout mutant *lacs9-1* is indistinguishable from the wild type in growth and appearance and retains 10% levels of the wild-type LACS activity in the mutant chloroplasts, suggesting another LACS isoform in the chloroplasts.

7. GENES REQUIRED FOR BIOSYNTHESIS OF POLYUNSATURATED FATTY ACIDS IN PLASTIDS

Polyunsaturated fatty acids in plastids are produced by acyl-lipid desaturases located in the ER and the plastid inner envelope [for reviews, see Somerville and Browse 1991; Ohlrogge and Browse (1995)]. In the eukaryotic pathway, $18{:}1^{\Delta 9}$ assembled into phosphatidylcholine in the ER is desaturated to $18{:}2^{\Delta 9,12}$ and then to $18{:}3^{\Delta 9,12,15}$ by FAD3 and FAD4 acyl-lipid desaturases, respectively. These fatty acids are then imported back to plastids, conserving a diacylglycerol configuration, such that C18 or C16 fatty acids bind to the *sn*-1 position and C18 fatty acids bind to the *sn*-2 position. In plastids, $18{:}2^{\Delta 9,12}$ is further desaturated to $18{:}3^{\Delta 9,12,15}$ by FAD7 and FAD8 desaturases. Thus, highly unsaturated molecular species in the plastids, such as 18:3/18:3 MGDG, are synthesized via the eukaryotic pathway.

By contrast, $18{:}1^{\Delta 9}$ is specifically assembled into the *sn*-1 position of phosphatidic acid (PA) in the prokaryotic pathway and then converted to $18{:}1^{\Delta 9}$-PG in all plants and to $18{:}1^{\Delta 9}$-MGDG in 16:3 plants. $18{:}1^{\Delta 9}$ on MGDG (and also that on PG) is desaturated to $18{:}2^{\Delta 9,12}$ by FAD6 acyl-lipid desaturase and then to $18{:}2^{\Delta 9,12,15}$ by FAD7 and FAD8 desaturases. In the prokaryotic pathway, however, only 16:0 esterified to the *sn-2* position of MGDG is desaturated to $16{:}3^{\Delta 7,10,13}$ in 16:3 plants. Thus, highly unsaturated molecular species in plastids, such as 18:3/16:3 MGDG, are produced via the prokaryotic pathway. Recently, a putative Δ7 16:0-MGDG desaturase was annotated in the Arabidopsis genome, which shows a significant homology to 18:0-CoA desaturase (Mekhedov et al., 2000). Although PG contains $16{:}1^{\Delta 3t}$ in the *sn-2* position, the gene for Δ3t 16:0-PG desaturase has not been identified to date. Plants also contain unusual acyl-lipid desaturases in the ER that introduce the second double bond in positions other than Δ12 [see Drexler et al. (2003) and references therein].

8. GENETIC ANALYSIS OF THE OIL ACCUMULATION

Isolation of Arabidopsis mutants with altered seed oil content should provide useful information about the regulation of fatty acid metabolism in plastids. The *wri1* mutant of Arabidopsis is a low-seed-oil mutant that exhibits abnormal accumulation of carbohydrates due to unidentified downstream components in the regulation of carbohydrate metabolism (Focks and Benning, 1998; Ruuska et al., 2002). On the other hand, the *pkl* mutant shows abnormal oil accumulation in primary roots (Ogas et al., 1997). Fatty acid composition of the *pkl* root oil is similar to that in the embryo.

9. PERSPECTIVE OF PLASTID FATTY ACID METABOLISM IN PLANT LIPID BIOTECHNOLOGY

Compilation of the basic knowledge of plastid fatty acid metabolism has been stimulating activities to improve the composition and the yield of plant oils by genetic manipulation [for the recent reviews, see Millar et al. (2000); Voelker and Kinney, 2001; Thelen and Ohlrogge (2002c); Suh et al. (2002); Drexler et al., 2003]. For these activities, *B. napus* has been extensively used among common oil crops, because of its superior oil productivity and ease of genetic transformation. Studies with *A. thaliana* are complementary to that with *B. napus*, due to their similarity in genetic organization and regulation (Girke et al., 2000). To date, *B. napus* seeds have been genetically engineered to accumulate one of the following levels of fatty acids: 38% of 8:0, 58% of 12:0, 34% of 16:0, 22%-40% of 18:0, 89% of 18:1 and 47% of □-18:3 [for compiled examples of plant oil genetic engineering, see Thelen and Ohlrogge (2002c); Drexler et al. (2003)].

Different accumulation levels between different fatty acids appear to be ascribed to the substrate specificities of other enzymes related to the biosynthesis and catabolism of membrane and storage lipids in the ER rather than to insufficient enforcement of the plastid fatty metabolism [see Thelen and Ohlrogge (2002c) and references therein; Larson et al., 2002].

B. napus might have been explored by classical breeding and selection; therefore, genetic engineering for biochemical enhancement would not further increase the seed oil content. Therefore, to increase the plant oil productivity, Thelen and Ohlrogge (2002c) have pointed out the use of other oil-accumulating crops that have relatively low oil content. On the other hand, the transformation of starch-accumulating crops into oil-accumulating crops is ambitious but challenging (Somerville and Bonetta, 2001). For these attempts to be feasible, understanding of the basic mechanism(s) for conversion of starch to fatty acids or for selective allocation of translocated carbons toward fatty acid synthesis is essential (da Silva et al., 1997).

The recent attempt by Klaus et al. (2004), using Arabidopsis *ACC1* as a transgene, resulted in a 5-fold increase of oil content in transgenic potato tubers. However, the triacylglycerol content in the transgenic potato tuber is so small (<0.006% of the fresh weight) that the choice of plant materials might be important for initial attempts. In addition, it might be necessary to isolate a master key transcription factor that would coordinately regulate all the necessary genes for oil accumulation. Alternatively, transformation of the plastome with polycistronic transgenes is an interesting approach for expressing multiple genes (DeCosa et al., 2001; Ruiz et al., 2003; Daniell and Dhingra, 2002; Nakashita et al., 2001). As described above, heteromeric ACCase overexpression in tobacco by plastome transformation doubled the seed yield and improved leaf longevity in tobacco (Madoka et al., 2002). Although it should be tested in other plants such as *B. napus*, the plastome transformation has advantage over the nuclear transformation, alleviating the risk of gene dispersal (Daniell 2002). These attempts are challenging but essential to dramatic enhancement of plant oil productivity. For further information on chloroplast genetic engineering, the reader is referred to chapter 16 of this book.

10. REFERENCES

Abbadi, A., Brummel, M., & Spener, F. (2000) Knock out of the regulatory site of 3-ketoacyl-ACP synthase III enhances short- and medium-chain acyl-ACP synthesis. *Plant J. 24*, 1-9.

Baldock, C., Rafferty, J.B., Stuitje, A.R. Rice, & D.W. (1998) Molecular structure of a reductase component of fatty acid synthase. In: Harwood, J.L. (ed) Plant Lipid Biosynthesis. Fundamentals and Agricultural Applications. Cambridge University Press, Cambridge, pp73-92.

Bao, X., Shorrosh, B.S., & Ohlrogge, J.B. (1997) Isolation and characterization of an *Arabidopsis* biotin carboxylase gene and its promoter. *Plant Mol. Biol. 35*, 539-550.

Bao, X.M., Focke, M., Pollard, M., & Ohlrogge, J. (2000) Understanding in vivo carbon precursor supply for fatty acid synthesis in leaf tissues. *Plant J. 22*, 39-50.

Baud, S., Guyon, V., Kronenberger, J., Wuillème, S., Miquel, M., Caboche, M., Lepiniec, L., & Rochat, C. (2003) Multifunctional acetyl-CoA carboxylase 1 is essential for very long chain fatty acid elongation and embryo development in Arabidopsis. *Plant J. 33*, 75-86.

Bettey, M., Ireland, R.J., & Smith, A.M. (1992). Purification and characterization of acetyl-CoA carboxylase from developing pea embryos. *J. Plant Physiol.* 140, 513-520.

Bonaventure, G., Salas, J.J., Pollard, M.R., & Ohlrogge, J.B. (2003). Disruption of the *FATB* gene in Arabidopsis demonstrates an essential role of saturated fatty acids in plant growth. *Plant Cell 15*, 1020-1033.

Brück, F.M., Brummel, M., Schuch, R. & Spener, F. (1996) In-vitro evidence for feed-back regulation of □-ketoacyl-acyl carrier protein synthase III in medium-chain fatty acid biosynthesis. *Planta 198*, 271-278.

Cahoon, E.D., & Ohlrogge, J.B. (1994) Metabolic evidence for the involvement of a □^{4}-palmitoyl-acyl carrier protein desaturase in petroselinic acid synthesis in coriander endosperm and transgenic tobacco cells. *Plant Physiol. 104,* 827-837.

Cahoon, E.B., & Shanklin, J. (2000) Substrate-dependent mutant complementation to select fatty acid desaturase variants for metabolic engineering of plant seed oils. *Proc. Natl. Acad. Sci. 97,* 12350-12355.

Cahoon, E.B., Lindqvist, Y., Schneider, G., & Shanklin, J. (1997) Redesign of soluble fatty acid desaturases from plants for altered substrate specificity and double bond position. *Proc. Natl. Acad. Sci. USA 94*, 4872-4877.

Cahoon, E.B., Shah, S., Shanklin, J., & Browse, J. (1998) A determinant of substrate specificity predicted from the acyl-acyl carrier protein desaturase of developing cat's claw seed. *Plant Physiol. 117,* 593-598.

Cahoon, E.B., Carlson, T.J., Ripp, K.G., Schweiger, B.J., Cook, G.A., Hall, S.E., & Kinney, A. (1999) Biosynthetic origin of conjugated double bonds: production of fatty acid components of high-value drying oils in transgenic soybean embryos. *Proc. Natl. Acad. Sci. USA 96*, 12935-12940.

Carlsson, A.S., LaBrie, S.T., Kinney, A.T., von Wettstein-Knowles, P., & Browse, J. (2002) A *KAS2* cDNA complements the phenotypes of the *Arabidopsis fab1* mutant that differs in a single residue bordering the substrate binding pocket. *Plant J., 29*, 761-770.

Choi, J.-K., Yu, F., Wurtele, E.S., & Nikolau, B.J. (1995) Molecular cloning and characterization of the cDNA coding for the biotin-containing subunit of the chloroplastic acetyl-coenzyme A carboxylase. *Plant Physiol. 109*, 619-625.

Daniell, H. (2002). Molecular strategies for gene containment in transgenic crops. *Nat. Biotechnol., 20*, 581-586.

Daniell, H., & Dhingra, A. (2002) Multigene engineering: dawn of an exciting new era in biotechnology. *Current Opinion in Biotechnol., 13*, 136-141.

DeCosa, B., Moar, W., Lee, S. B., Miller, M., & Daniell, H. (2001). Overexpression of the *Bt* cry2Aa2 operon in chloroplasts leads to formation of insecticidal crystals. *Nat. Biotechnol., 19*, 71-74.

da Silva, P.M.F.R., Eastmond, P.J., Hill, L.M., Smith, A.M., & Rawsthorne, S. (1997) Starch metabolism in developing embryos of oilseed rape. *Planta 203*, 480-487.

Davies, H.M., Anderson, L., Fan, C. & Hawkins, D.J. (1991) Developmental induction, purification, and further characterization of 12:0-ACP thioesterase from immature cotyledons of *Umbellularia californica. Arch. Biochem. Biophys. 290*, 37-45.

Davies, H.M. (1993) Medium chain acyl-ACP hydrolysis activities of developing oilseeds. *Phytochemistry 33*, 1353-1356.

Dehesh, K., Edwards, P., Fillatti, J., Slabaugh, M., & Byrns, J. (1998) KAS IV: a 3-ketoacyl-ACP synthase from *Cuphea* sp. is a medium chain specific condensing enzyme. *Plant J. 15*, 383-390.

Dehesh, K., Jones, A., Knutzon, D.S. & Voelker, T.A. (1996) Production of high levels of 8:0 and 10:0 fatty acids in transgenic canola by overexpression of Ch FatB2, a thioesterase cDNA from *Cuphea hookeriana. Plant J. 9*, 167-172.

Dehesh, K., Tai, H., Edwards, P., Byrne, J. & Jaworski, J.G. (2001) Overexpression of 3-ketoacyl-acyl-carrier protein synthase IIIs in plants reduces the rate of lipid synthesis. *Plant Physiol. 125*, 1103-1114.

Drexler, H., Spiekermann, P., Meyer, A., Dormergue, F., Zank, T., Sperling, P., Abbadi, A., & Heinz, E. (2003) Metabolic engineering of fatty acids for breeding of new oilseed crops: strategies, problems and first results. *J. Plant Physiol. 160*, 779-802.

Dörmann, P., Spener, F., & Ohlrogge, J.B. (1993) Characterization of two acyl-acyl carrier protein thioesterases from developing *Cuphea* seeds specific for medium-chain and oleoyl-acyl carrier protein. *Planta 189*, 425-432.

Dörmann, P., Voelker, T.A., & Ohlrogge, J.B. (2000) Accumulation of palmitate in Arabidopsis mediated by acyl-acyl carrier protein thioesterase FATB1. *Plant Physiol. 123*, 637-643.

Eastmond, P.J., & Rawsthorne, S. (2000) Coordinate changes in carbon partitioning and plastidial metabolism during the development of oilseed rate embryos. *Plant Physiol. 122*, 767-774.

Eastmond, P.J., Dennis, D.T., & Rawsthorne, S. (1997) Evidence that a malate/inorganic phosphate exchange translocator imports carbon across the leucoplast envelope for fatty acid synthesis in developing castor seed endosperm. *Plant Physiol. 114*, 851-856.

Facciotti, M.T., Bertin, P.B., & Yuan, L. (1999) Improved stearate phenotype in transgenic canola expressing a modified acyl-acyl carrier protein thioesterase. *Nat. Biotechnol. 7*, 593-597.

Fawcett, T., Copse, C.L., Simon, W., & Slabas, A.R. (2000) Kinetic mechanism of NADH-enoyl-ACP reductase from *Brassica napus. FEBS Lett. 484*, 65-68.

Flügge, U.-I. (1998) Metabolite transporters in plastids. Curr. Opin. Plant Biol. 1, 201-206.

Focke, M., Gieringer, E., Schwan, S., Jänsch, L., Binder, S. and Braun, H.-P. (2003) Fatty acid biosynthesis in mitochondria of grasses: malonyl-coenzyme A is generated by a mitochondrial-localized acetyl-Coenzyme A carboxylase. *Plant Physiol. 133*, 875-884.

Focks, N., & Benning, C. (1998). *Wrinkled*1: A novel, low-seed-oil mutant of Arabidopsis with a deficiency in the seed-specific regulation of carbohydrate. *Plant Physiol. 118*, 91-101.

Fox, S.R., Hill, L.M., Rawsthorne, S., & Hills, M.J. (2000) Inhibition of the glucose-6-phosphate transporter in oilseed rape (*Brassica napus* L.) plastids by acyl-CoA thioesters reduces fatty acid synthesis. *Biochem. J. 352*, 525-532.

Girke, T., Todd, J., Ruuska, S., White, J., Benning, C., & Ohlrogge, J. (2000) Microarray analysis of developing Arabidopsis seeds. *Plant Physiol. 124*, 1570-1581.

Graham, S., Hirsinger, F., & Robbelen, G. (1981) Fatty acids of *Cuphea* (Lythraceae) seed lipids and their systematic significance *Am. J. Bot. 68*, 908-917.

Gueguen, V., Macheral, D., Jaquinod, M., Douce, R., & Bourguignon, J. (2000) Fatty acid and lipoic acid biosynthesis in higher plant mitochondria. *J.Biol. Chem. 257*, 5016-5025.

Guerra, D.J., Ohlrogge, J.B., & Frentzen, M. (1986) Activity of acyl carrier protein isoforms in reactions of plant fatty acid metabolism. *Plant Physiol. 82,* 448-453.

Gummeson, P.O., Lenman, M., Lee, M., Singh, S.,& Stymne, S. (2000) Characterization of acyl-ACP desaturases from *Macadamia integrifolia* Maiden & Betche and *Nerium oleander* L. *Plant Sci. 154,* 53-60.

Harwood, J.L. (1996) Recent advances in the biosynthesis of plant fatty acids. *Biochim. Biophys. Acta 1301*, 7-56.

Hawkins, D.J., & Kridl, J.C. (1998) Characterization of acyl-ACP thioesterases of mangosteen (*Garcinia mangostana*) seed and high levels of stearate production in transgenic canola. *Plant J. 13*, 743-752.

Heath, R.J., & Rock, C.O. (1995) Enoyl-acyl carrier protein reductase (*fabI*) plays a determinant role in completing cycles of fatty acid elongation in *Escherichia coli*. **J. Biol. Chem. 270**, 26538-26542.

Hirsinger, F., & Knowles, P.F. (1984) Morphological agronomic description of selected *Cuphea* germplasm. *Econ. Bot. 38*, 439-451.

Hwang, S.-R., & Tabita, F.R. (1991) Acyl carrier protein-derived sequence encoded by the chloroplast genome in the marine diatom *Cylindrotheca* sp. Strain N1. *J. Biol. Chem. 266*, 13492-13494.

Johnson, P.E., Fox, S.R., Hills, M.J., & Rawsthorne, S. (2000) Inhibition by long-chain acyl-CoAs of glucose 6-phosphate metabolism in plastids isolated from developing embryos of oilseed rape (*Brassica napus* L.). *Biochem. J. 348*, 145-150.

Jones, A., Davies, H.M., & Voelker, T.A. (1995) Palmitoyl-acyl carrier protein (ACP) thioesterase and the evolutionary origin of plant acyl-ACP thioesterases. *Plant Cell 7*, 359-371.

Kammerer, B., Fischer, K., Hilpert, B., Schubert, S., Gutensohn, M., Weber, A., & Flügge, U.I. (1998) Molecular characterization of a carbon transporter in plastids from heterotrophic tissues: the glucose 6-phosphate/phosphate antiporter. *Plant Cell 10,* 105-117.

Kampfenkel, K., Möhlmann, T., Batz, O., van Montagu, M., Inzé, D., & Neuhaus, H.E. (1995) Molecular characterization of an Arabidopsis thaliana cDNA encoding a novel putative adenylate translocator of higher plants. *FEBS Lett. 374*, 351-355.

Kang, F., & Rawsthorne, S. (1994) Starch and fatty acid synthesis in plastids from developing embryos of oilseed rape (*Brassica napus* L.). *Plant J 6*, 795-805.

Kang, F.,& Rawsthorne, S. (1996) Metabolism of glucose-6-phosphate and utilization of multiple metabolites for fatty acid synthesis by plastids from developing oilseed rape embryos. *Planta 199,* 321-327.

Ke, J., Behal, R.H., Back, S.L., Nikolau, B.J., Wurtele, E.S., & Oliver, D.J. (2000). The role of pyruvate dehydrogenase and acetyl-coenzyme A synthetase in fatty acid synthesis in developing Arabidopsis seeds. *Plant Physiol. 123*, 497-508.

Ke, J., Choi, J.-K., Smith, M., Horner, H.T., Nikolau, B.J., & Wurtele, E.S. (1997) Structure of the *CAC1* gene and in situ characterization of its expression. The *Arabidopsis thaliana* gene coding for the biotin-containing subunit of the plastidic acetyl-coenzyme A carboxylase. *Plant Physiol. 13*, 357-365.

Klaus, D., Ohlrogge, J.B., Neuhaus, H.E., & Dörmann, P. (2004) Increased fatty acid production in potato by engineering of acetyl-CoA carboxylase. *Planta,* in press.

Kleiman, R., & Spencer, G.F. (1982) Search for new industrial oils: XVI. Umbelliflorae–Seed oils rich in petroselinic acid. *J. Am. Oil Chem. Soc. 59*, 29-38.

Knutzon, D.S., Thompson, G.A., Radke, S.E., Johnson, W.B., Knauf, V.C., & Kridl, J.C. (1992) Modification of Brassica seed oil by antisense expression of a stearoyl-acyl carrier protein desaturase gene. *Proc. Natl. Acad. Sci. USA 89*, 2624-2628.

Kubis, S.E., & Rawsthorne, S. (2000). The role of plastidial transporters in developing embryos of oilseed rape (*Brassica napus* L.) for fatty acid synthesis. *Biochem. Soc. Trans. 28*, 665-666.

Larson, T.R., Edgell, T., Byrne, J., Dehesh, K., & Graham, I.A. (2002) Acyl CoA profiles of transgenic plants that accumulate medium-chain fatty acids indicate inefficient storage lipid synthesis in developing oilseeds. *Plant J. 32*, 519-527.

Leonard, J.M., Knapp, S.J., & Slabaugh, M.B. (1998) A *Cuphea* □-ketoacyl-ACP synthase shifts the synthesis of fatty acids towards shorter chains in *Arabidopsis* seeds expressing *Cuphea* FATB thioesterases. *Plant J. 13,* 621-628.

Lichtenthaler, H.K. (1999) The 1-deoxy-D-xylulose-5-phosphate pathway of isoprenoid biosynthesis in plants. *Annu. Rev. Plant Physiol. Plant Mol. Biol. 50*, 47-65.

Lin, M., Behal, R., & Oliver, D.J. (2003) Disruption of *plE2*, the gene for the E2 subunit of the plastid pyruvate dehydrogenase complex, in *Arabidopsis* causes an early embryo lethal phenotype. *Plant Mol. Biol. 52*, 865-872.

Lindqvist, Y., Huang, W., Schneider, G., & Shanklin, J. (1996) Crystal structure of □[9] stearoyl-acyl carrier protein desaturase from castor seed and its relationship to other di-iron proteins. *EMBO J 15*, 4081-4092.

Madoka, Y., Tomizawa, K.-I., Mizoi, J., Nishida, I., Nagano, Y., & Sasaki, Y. (2002) Chloroplast transformation with modified *accD* operon increases acetyl-CoA carboxylase and causes extension of leaf longevity and increase in seed yield in tobacco. *Plant Cell Physiol. 43*, 1518-1525.

Marillia, E.-R., Micallef, B.J., Micallef, M., Weninger, A., Pedersen, K.K., Zou, J., & Taylor, D.C. (2003) Biochemical and physiological studies of *Arabidopsis thaliana* transgenic lines with repressed expression of the mitochondrial pyruvate dehydrogenase kinase. *J. Exp. Bot. 54*, 259-270.

Mekhedov, S., Martínez de Ilárduya, O., & Ohlrogge, J. (2000) Toward a functional catalog of the plant genome: a survey of genes for lipid biosynthesis. *Plant Physiol. 122*: 389-402.

Millar, A.A., Smith, M.A., & Kunst, L. (2000) All fatty acids are not equal: discrimination in plant memebrane lipids. *Trends Plant Sci. 5*, 95-101.

Möhlmann, T., & Neuhaus, H.E. (1997). Precursor and effector dependency of lipid synthesis in amyloplasts isolated from developing maize and wheat endosperm. *J. Cereal Sci. 26*, 161-167.

Mongrand, S., Bessoule, J.J., Cabantous, F., & Cassagne, C. (1998). The $C_{16:3}/C_{18:3}$ fatty acid balance in photosynthetic tissues from 468 plant species. *Phytochemistry 49*, 1049-1064.

Murata, N., Sato, N., & Takahashi, N. (1984) Very-long-chain saturated fatty acids in phosphatidylserine from higher plant tissues. *Biochem. Biophys. Acta 795*, 147-150.

Nakashita, H., Arai, Y., Shikanai, T., Doi, Y., & Yamaguchi, I. (2001) Introduction of bacterial metabolism into higher plants by polycistronic transgene expression. *Biosci. Biotechnol. Biochem. 65*, 1688-1691.

Nandi, A., Krothapalli, K., Buseman, C.M., Li, M., Welti, R. Enyedi, A., & Shah, J. (2003) Arabidopsis *sfd* mutants affect plastidic lipid composition and suppress dwarfing, cell death, and the enhanced disease resistance phenotypes resulting from the deficiency of a fatty acid desaturase. *Plant Cell 15*, 2383-2398.

Nikolau, B.J., Oliver D.J., Schnable, P.S., & Wurtele, E.S. (2000) Molecular biology of acetyl-CoA metabolism. *Biochem. Soc. Trans. 28*, 591-593.

Nikolau, B.J., Ohlrogge, J.B., & Wurtele, E.S. (2003) Plant biotin-containing carboxylases. *Arch. Biochem Biophys. 414*, 211-222.

Ogas, J., Cheng, J-C., Sung, A.R., & Somerville, C.R. (1997) Cellular differentiation regulated by gibberellin in the Arabidopsis thaliana pckle mutant. *Science 277*, 91-94.

Ohlroge, J.B. (1987) Biochemistry of plant acyl carrier protein. In: Stumpf, P.K., Conn, E.E. (eds.) The Biochemistry of Plants. Vol. 9. Academic Press, Orlando, pp 137-157.

Ohlrogge, J. B., & Jaworski, J.G. (1997) Regulation of fatty acid synthesis. *Annu. Rev. Plant Physiol. Plant Mol. Biol. 48*, 109-136.

Ohlrogge, J., Pollard, M., Bao, X., Focke, M., Girke, T., Ruuska, S., Mekhedov, S., & Benning, C. (2000) *Biochem. Soc. Trans. 28*, 567-573?.

Ohlrogge, J.B. (1994) Design of new plant products: engineering of fatty acid metabolism. *Plant Physiol. 104*, 821-826.

Ohlrogge, J., & Browse, J. (1995) Lipid Biosynthesis. *Plant Cell 7*, 957-970.

Pollard, N.R., Anderson, L., Fan, C., Hawkins, D.H. and Davies, H.M. (1991) A specific acyl-ACP thioesterase implicated in medium-chain fatty acid production in immature cotyledons of *Umbellularia californica*. *Arch. Biochem. Biophys. 284*, 306-312.

Post-Beittenmiller, D., Schmid, K.M., & Ohlrogge, .B. (1989) Expression of holo and apo forms of spinach acyl carier protein-I in leaves of transgenic tobacco plants. *Plant Cell 1*, 889-899.

Rawsthorne, S. (2002). Carbon flux and fatty acid synthesis in plants. *Prog. Lipid Res. 41*, 182-196.

Reith, M. (1993) A □-ketoacyl-acyl carrier protein III gene (*fabH*) is encoded on the chloroplast genome of the red alga *Porphyra umbilicalis*. *Plant Mol. Biol. 21*, 185-189.

Roesler, K.R., Shintani, D., Savage, L., Boddupalli, S., & Ohlrogge, J.B. (1997) Targeting of the Arabidopsis homomeric acetyl-coenzyme A carboxylase to plastids of rapeseeds. *Plant Physiol. 113*, 75-81.

Roughan, P.G. (1997) Stromal concentrations of coenzyme A and its esters are insufficient to account for rates of chloroplast fatty acid synthesis: evidence for substrate channelling within the chloroplast fatty acid synthase. *Biochem J. 327*, 267-273.

Roughan, P.G., Holland, R., Slack, C.R., & Mudd, J.B. (1979) Acetate is the preferred substrate for long-chain faty acid synthesis in isolated spinach chloroplasts. *Biochem. J. 184*, 565-569.

Ruiz, O. N., Hussein, H., Terry, N., & Daniell, H. (2003). Phytoremediation of organomercurial compounds via chloroplast genetic engineering. *Plant Physiol.*, *132*, 1-9.

Ruuska, S.A., Girke, T., Benning, C., & Ohlrogge, J.B. (2002) Contrapuntal networks of gene expression during Arabidopsis seed filling. *Plant Cell 14,* 1191-1206.

Salas, J.J., & Ohlrogge, J.B. (2002) Characterization of substrate specificity of plant FatA and FatB acyl-ACP thioesterases. *Arch. Biochem. Biophys. 403*, 25-34.

Sasaki, Y., Hakamada, K., Suama, Y., Nagano, Y., & Furusawa, I. and Matsuno, R. (1993) Chloroplast-encoded protein as a subunit of acetyl-CoA carboxylase in pea plant. *J. Biol. Chem. 268*, 25118-25123.

Sasaki, Y., & Nagano, Y. (2004) Plant acetyl-CoA carboxylase: structure, biosynthesis, regulation, and gene manipulation for plant breeding. *Biosci. Biotechnol. Biochem.*, in press.

Savage, L., & Ohlrogge, J.B. (1999) Phosphorylation of pea chloroplast acetyl-CoA carboxylase. *Plant J. 18,* 521-527.

Schnurr, J.A., Shockey, J.M., de Boer, G.-J., & Browse, J.A. (2002) Fatty acid export from the chloroplast. Molecular characterization of a major plastidial acyl-coenzyme A synthetase from Arabidopsis. *Plant Physiol. 129*, 1700-1709.

Schuch, R., Brummel, M., & Spener, F. (1994) □-Ketoacyl-acyl carrier protein (ACP) synthase III in *Cuphea lanceolata* seeds: Identification and analysis of reaction products. *J. Plant Physiol. 143*, 556-560.

Schulte, W., Töpfer, R., Stracke, R., Schell, J., & Martini, N. (1997) Multi-functional acetyl-CoA carboxylase from Brassica napus is encoded by a multi-gene family: identification for plastidic localization of at least one isoform. *Proc. Natl. Acad. Sci. USA 94*, 3465-3470.

Schultz, D.J., Cahoon, E.B., Shanklin, J., Craig, R., Cox-Foster, D.L., Mumma, R.O., & Medford, J.I. (1996) Expression of a $\square^{9}$ 14:0-acyl carrier protein fatty acid desaturase gene is necessary for the production of $\square^{5}$ anacardic acids found in pest-resistant geranium (*Pelargonium xhortorum*). *Proc. Natl. Acad. Sci. USA 93*, 8771-8775.

Schultz, D.J., Suh, M.C., & Ohlrogge, J.B. (2000) Stearoyl-acyl carrier protein and unusual acyl-acyl carrier protein desaturase activities are differentially influenced by ferredoxin. *Plant Physiol. 124*, 681-692.

Schwender, J., & Ohlrogge, J.B. (2002). Probing in vivo metabolism by stable isotope labeling of storage lipids and proteins in developing *Brassica napus* embryos. *Plant Physiol. 130*, 347-361.

Schwender, J., Ohlrogge, J.B., & Shachar, H.Y. (2003). A flux model of glycolysis and the oxidative pentose phosphate pathway in developing *Brassica napus* embryos. *J. Biol. Chem. 278*, 29442-29453.

Schütt, B.S., Brummel, M., Schuch, R., & Spener, F. (1998) The role of acyl carrier protein isoforms from *Cuphea lanceolata* seeds in the de-novo biosynthesis of medium-chain fatty acids. *Planta 205*, 263–268.

Shibahara, A., Yamamoto, K., Takeoka, M., Kinoshita, A., Kajimoto, G., Nakayama, T., & Noda, M. (1990) Novel pathways of oleic and *cis*-vaccenic acid biosynthesis by an enzymatic double-bond shifting reaction in higher plants. *FEBS Lett. 264*, 228-230.

Shintani, D., Roesler, K., Shorrosh, B., Savage, L., & Ohlrogge, J. (1997) Antisense expression and overexpression of biotin carboxylase in tobacco leaves. *Plant Physiol. 114*, 881-886.

Shintani, D.K., & Ohlrogge, J.B. (1994) The characterization of a mitochondrial acyl carrier protein isoform isolated from *Arabidopsis thaliana. Plant Physiol. 104*, 1221-1229.

Shintani, D.K., & Ohlrogge, J.B. (1995) Feedback inhibition of fatty acid synthesis in tobacco suspension cells. *Plant J. 7*, 577-587.

Shockey, J.M., Fulda, M.S., & Browse, J.A. (2002) Arabidopsis contains nine long-chain acyl-coenzyme A synthetase genes that participate in fatty acid and glycerolipid metabolism. *Plant Physiol. 129*, 1710-1722.

Shorrosh, B.S., Savage, L.J., Soll, J., & Ohlrogge, J.B. (1996) The pea chloroplast membrane-associated protein, IEP96, is a subunit of acetyl-CoA carboxylase. *Plant J 10*, 261-268.

Singal, H.R., Sheoran, I.S., & Singh, R. (1987) Photosynthetic carbon fixation characteristics of fruit structures of *Brassica campestris* L. *Plant Physiol. 83*, 1043-1047.

Slabas, A.R., & Fawcett, T. (1992) The biochemistry and molecular biology of plant lipid biosynthesis. *Plant Mol. Biol. 19*, 161-191.

Slabaugh, M.B., Leonard, J.M., & Knapp, S.J. (1998) Condensing enzymes from *Cuphea wrightii* associated with medium chain fatty acid biosynthesis. *Plant J. 13*, 611-620.

Smith, R.G., Gauthier, D.A., Dennis, D.T., & Turpin, D.H. (1992) Malate- and pyruvate-dependent fatty acid synthesis in leucoplasts from developing castor endosperm. *Plant Physiol. 98*, 1233-1238.

Somerville, C.R., & Bonetta, D. (2001) Plants as factories for technical materials. *Plant Physiol. 125*, 168-171.

Somerville, C., & Browse, J. (1991) Plant Lipids: metabolism, mutants, and membranes. *Science 252*: 80-87.

Streatfield, S.J., Weber, A., Kinsman, E.A., Häusler, R.E., Li, J., Post-Beittenmiller, D., Kaiser, W.M., Pyke, K.A., Flügge, U.-I., & Chory, J. (1999) The phosphoenolpyruvate/translocator is required for phenolic metabolism, palisade cell development, and plastid-dependent nuclear gene expression. *Plant Cell 11*, 1609-1621.

Suh, M. C., Schultz, D.J., & Ohlrogge, J.B. (1999) Isoforms of acyl carrier protein involved in seed-specific fatty acid synthesis. *Plant J. 17*, 679-688.

Suh, M. C., Schultz, D.J., & Ohlrogge, J.B. (2002) What limits production of unusual monoenoic fatty acids in transgenic plants? *Planta 215*, 584-595.

Sun, J., Ke, J., Johnson, J.L., Nikolau, B.J., & Wurtele, E.S. (1997) Biochemical and molecular biological characterization of *CAC2*: the *Arabidopsis thaliana* gene coding for the biotin carboxylase subunit of plastidic acetyl-coenzyme A carboxylase. *Plant Physiol. 115*, 1371-1383.

Thelen, J.J., & Ohlrogge, J.B. (2002a). The multisubunit acetyl-CoA carboxylase is strongly associated with the chloroplast envelope through non-ionic interactions to the carboxyltransferase subunits. *Arch. Biochem. Biophys. 400*, 245-257.

Thelen, J.J., & Ohlrogge, J.B. (2002b) Both antisense and sense expression of biotin carboxyl carrier protein isoform 2 inactivates the plastid acetyl-coenzyme A carboxylase in *Arabidopsis thaliana*. *Plant J. 32*, 419-431.

Thelen, J.J., & Ohlrogge, J.B. (2002c) Metabolic engineering of fatty acid biosynthesis in plants. *Metab. Eng. 4*, 12-21.

Thelen, J.J., Mekhedov, S., & Ohlrogge, J.B. (2001) Brassicaceae express multiple isoforms of biotin carboxyl carrier protein in a tissue-specific manner. *Plant Physiol. 125*, 2016-2028.

Töpfer, R., Martini, N., & Schell, J. (1995) Modification of plant lipid synthesis. Science 268, 681-686.

van de Loo, F.J., Fox, B.G. and Somerville, C. (1993) Unusual fatty acids. In: TS Moore, Jr (ed) Lipid metabolism in plants. CRC Press, Boca Raton, pp 91-126.

Tovar-Méndez, A., Miernyk, J.A., & Randall, D.D. (2003) Regulation of pyruvate dehydrogenase complex activity in plant cells. *Eur J. Biochem. 270*, 1043-1049.

Verwoert, I.I.G.S., van der Linden, K.H., Walsh, M.C., Nijkamp, H.J.J., & Stuitje, A.R. (1995). Modification of *Brassica napus* seed oil by expression of the *Escherichia coli fabH* gene, encoding 3-ketoacyl-acyl carrier protein synthase III. *Plant Mol. Biol. 27*, 875-886.

Voelker, T., & Kinney, A.J. (2001) Variations in the biosynthesis of seed-storage lipoids. *Annu. Rev. Plant Physiol. Plant Mol. Biol. 52*, 335-361.

Voelker, T.A., Jones, A., Cranmer, A.M., Davies, H.M., & Knutzon, D.S. (1997) Broad-range and binary-range acyl-acyl-carrier-protein thioesterase suggest an alternative mechanism for medium-chain production in seeds. *Plant Physiol. 114*. 669-677.

Wada, H., Shintani, D., & Ohlrogge, J. (1997) Why do mitochondria synthesize fatty acids? Evidence for involvement in lipoic acid production. Proc. Natl. Acad. Sci. USA 94, 1591-1596.

Wang, S., & Liu, X.-Q. (1991) The plastid genome of *Cryptomonas* □ encodes an hsp70-like protein, a histone-like protein, and an acyl carrier protein. *Proc. Natl. Acad. Sci. USA 88*, 10783-10787.

White, J.A., Todd, J., Newman, T., Focks, N., Girke, T., de Ilarduya, O.M., Jaworski, J.G., Ohlrogge, J.B., & Benning, C. (2000) A new set of Arabidopsis expressed sequence tags from developing seeds. The metabolic pathway from carbohydrates to seed oil. *Plant Physiol. 124*, 1582-1594.

Whitney, H.M., Sayanova, O., Pickett, J.A., & Napier, J.A. (2004) Isolation and expression pattern of two putative acyl-ACP desaturase cDNA from *Brassica scoparia*. *J. Exp. Bot. 55*, 787-789.

Whittle, E., & Shanklin, J. (2001) Engineering □^{9}-16:0-acyl carrier protein (ACP) desaturase specificity based on combinatorial saturation mutagenesis and logical redesign of the castor □^{9}-18:0-ACP desaturase. *J. Biol. Chem. 276*, 21500-21506.

Wissenbach, M. (1994) New members of the barley *Kas* gene family encoding □-ketoacyl-acyl carrier protein synthases. *Plant Physiol. 106*, 1711-1712.

Wolf, R.B., Graham, S.A., & Kleiman, R. (1983) Fatty acid composition of *Cuphea* seed oils. J Am. Oil Chem Soc. 60, 103-104.

Yasuno, R, von Wettstein-Knowles, P., & Wada, H. (2004) Identification and molecular characterization of the □-ketoacyl-[acyl carrier protein] synthase component of the *Arabidopsis* mitochondrial fatty acid synthase. *J. Biol. Chem. 279*, 8242-8251.

Yuan, L., Nelson, B.A., & Caryl, G. (1996) The catalytic cysteine and histidine in the plant acyl-acyl carrier protein thioesterases. *J Biol. Chem. 271*, 3417.

Yuan, L., Voelker, T.A., & Hawkins, D.J. (1995) Modification of the substrate specificity of an acyl-acyl carrier protein thioesterase by protein engineering. *Proc. Natl. Acad. Sci. USA 92*, 10639-10643.

Zuther, E., Huang, S., Jelenska, J., Eilenberg, H., Arnold, E.M., Su, X., Sirikhachornkit, A., Podkowinski, J., Zilberstein, A., Haselkorn, R., & Gornicki, P. (2004) Complex nested promoters control tissue-specific expression of acetyl-CoA carboxylase genes in wheat. *Proc. Natl. Acad. Sci. USA 101*, 1403-1408.

CHAPTER 21

METABOLIC ENGINEERING: PLASTIDS AS BIOREACTORS

KAREN BOHMERT, OLIVER P. PEOPLES AND KRISTI D. SNELL

Metabolix, Inc.,
21 Erie St., Cambridge, MA 02139

Abstract. Plastids are the site of biosynthesis of fatty acids, starch, pigments and amino acids in plants and contain a variety of natural precursors that can be used as building blocks to construct novel biopolymers. This chapter will focus on plastids as the site of production for polyhydroxyalkanoates (PHAs), bacterially derived biopolymers composed of repeating units of (R)-hydroxyacids. Efforts to engineer PHAs in plastids, as well as proposed plastidial routes for formation of precursor acetyl-CoA, will be discussed. Efforts to produce other biopolymers in plastids will be briefly mentioned to illustrate the utility of the plastid as a bioreactor to produce a range of materials.

1. INTRODUCTION

A variety of novel biomaterials have been produced within plastids of transgenic plants, including starches with modified properties, fructans, cyclodextrins, the protein based polymer GVGVP and polyhydroxyalkanoates (Figure 1). Starch is composed of amylose, a linear chain of glucose units and amylopectin, a highly branched chain of glucose units, and naturally forms in chloroplasts as a transitory material or in amyloplasts of tubers, roots, and seeds as a storage material (Kok-Jacon et al., 2003; Slattery et al., 2000). Starches altered in branching or phosphorylation have been produced through genetic engineering efforts (Table I) yielding novel properties that may further the material's industrial uses (Jobling, 2004). Fructans consist of one or more □-linked fructose chains attached to sucrose (Figure 1b) and are naturally synthesized within vacuoles of plant cells (Ritsema & Smeekens, 2003). Metabolic engineering efforts have enabled high level fructan production in plastids (Table I) altering the structure of starch granules (Gerrits et al., 2001). GVGVP (Figure 1c) is a protein based polymer useful as a biomedical or biodegradable plastic material, designed to contain similarities to elastin (Urry,

H. Daniell and C.D. Chase (eds.), Molecular Biology and Biotechnology of Plant Organelles, 565—591. © 2004 *Springer. Printed in the Netherlands.*

1995). GVGVP has been successfully produced in chloroplasts of tobacco leaves (Table 1) by expression of a synthetic gene coding for the biomaterial. The formation of □- and □-cyclodextrins, cyclic chains of six or seven □ – 1,4 glucopyranose units (Figure 1d), has been achieved from starch in plastids upon expression of a microbial gene encoding cyclodextrin glycosyltransferase (Table I). The formation of cyclodextrins with starch may yield a product with novel properties (Oakes et al., 1991).

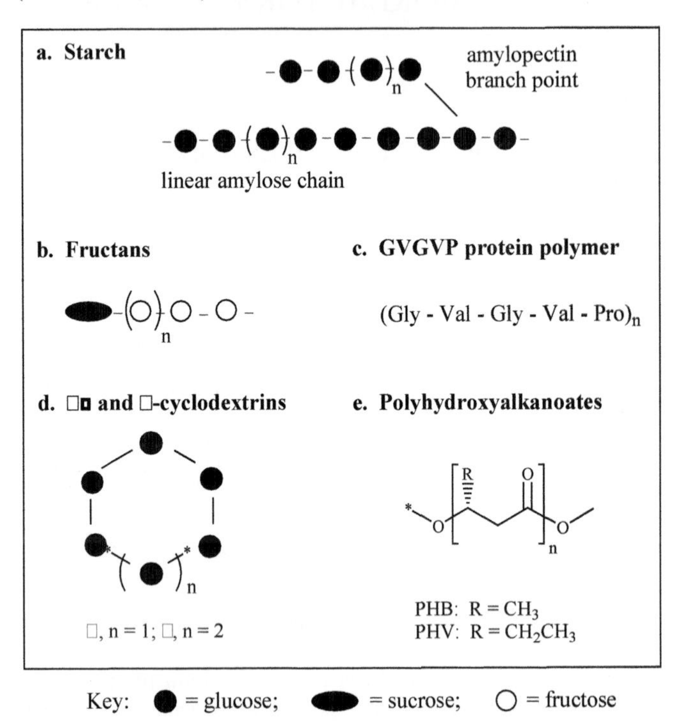

Figure 1. *Structures of Biomaterials Produced in Plastids of Transgenic Plants*

Polyhydroxyalkanoates (PHAs) are a family of high molecular weight polyesters of (R)-hydroxyacids (Figure 1e). They are useful as renewable replacements for at least half of the petroleum based plastics used today and accumulate as a granular storage materials in many bacteria (Madison & Huisman, 1999). The production of these materials has been engineered into plants by expression of microbial genes encoding pathways for polymer synthesis (Table 2). Direct production of PHA polymers in plants provides a unique opportunity to link low cost large-scale agricultural production with the 300 billion lb/year industrial polymer market (growing at 4%-5% per year) currently served through the depletion of fossil resources. This chapter will focus on production of PHAs within plastids as well as plastidial routes for formation of acetyl-CoA, the substrate for PHA biosynthesis.

Table 1. *Examples of Biomaterials Produced in Plastids*

Material *Tissue/engineered protein(s)[method]**	*Result[reference]ɸ*
GVGVP	
tobacco leaves/synthetic GVGVP[a, b]	GVGVP production[1]
Starch	
potato tuber/glycogen synthase[a]	↑ branching; ↓ phosphorylation[2]
regulatory variant of AGPase[a]	↑ starch yield[3]
ATP-ADP transporter[c]	↑ starch & amylose content[4]
SS II & III[d]	altered amylopectin chain length[5, 6]
GBSS[d]	↓ amylose[7,8] & amylose-free starch[7]
GBSS, SS II & III[d]	amylose-free starch/↓ length amylopectin[9]
SBE[d]	↑ phosphorylation[10]
SBE A[d] †	↑ phosphorylation/amylopectin length[11]
SBE I & II[d]	↓ amylopectin, ↑ phosphorylation[12]
antibody fragment to SBE II[a]	↑ amylose[13]
starch R1 protein[d] §	↓ phosphorylation[14]
starch R1 protein[d,e] §	↓ phosphorylation; ↑ amylose[15]
ATP-ADP transporter[d]	↓ starch & amylose content[4]
adenylate kinase[d]	↑ starch and tuber yield[16]
E. coli branching enzyme[a] ‡	↑ starch branching[17]
sweet potato/GBSS I[e]	amylose-free starch[18]
maize kernel/AGPase[f]	↑ seed weight[19]
wheat/regulatory variant of AGPase[c]	↑ seed yield[20]
barley/SS IIa[g]	↑ amylose/ ↓ amylopectin length[21]
rice/SBE I[g]	altered amylopectin structure[22]
Fructans	
potato tuber/levansucrase[a]	fructan production/altered starch[23]
tobacco leaves/levansucrase[a]	fructan production[23]
Cyclodextrins	
potato tuber/cyclodextrin glycosyltransferase[a]	□- and □-cyclodextrin production[24]

Table abbreviations: AGPase, ADP glucose pyrophosphorylase; GBSS, granule bound starch synthase; SS, starch synthase; SBE, starch branching enzyme. *Methods used to engineer proteins or enzyme activities: (a) nuclear transformation of gene with engineered plastid targeting sequence; (b) direct plastid transformation of gene; (c) nuclear transformation of gene; (d) nuclear transformation of anti-sense construct; (e) nuclear transformation of sense construct, gene silencing; (f) site specific mutagenesis; (g) chemical mutagenesis. ɸTable references: [1](Guda et al., 2000); [2](Shewmaker et al., 1994); [3](Stark et al., 1992); [4](Tjaden et al., 1998); [5](Edwards et al., 1999); [6](Lloyd et al., 1999); [7](Visser et al., 1991); [8](Kuipers et al., 1994); [10](Safford et al., 1998); [11](Jobling et al., 1999); [12](Schwall et al., 2000); [14](Lorberth et al., 1998); [15](Vikso-Nielsen et al., 2001); [16](Regierer et al., 2002); [17](Kortstee et al., 1996); [19](Giroux et al., 1996); [23](Gerrits et al., 2001); [24](Oakes et al., 1991); [18](Kimura et al., 2001); [9](Jobling et al., 2002); [13](Jobling et al., 2003); [21](Morell et al., 2003); [20](Smidansky et al., 2002); [22](Satoh et al., 2003); [23](Gerrits et al., 2001); [24](Oakes et al., 1991). †SBE A is also known as SBE II (Ball & Morell, 2003). §Starch R1 protein was later found to be an □-glucan water dikinase (Ritte et al., 2002). ‡ Experiment performed in amylose-free potato mutant.

Table 2. *PHAs Produced in Plastids of Transgenic Plants*

Material; *Tissue* *engineered protein(s)*	*Result*[(reference)] ф
PHB; *Arabidopsis* leaves	
PhaA$_{(R.e)}$, PhaB$_{(R.e)}$, PhaC$_{(R.e)}$	14% dwt[1];13% dwt[2]; 4% fwt[3]
tobacco leaves	
PhaA$_{(R.e)}$, PhaB$_{(R.e)}$, PhaC$_{(A.c.)}$	0.09% dwt[4]
PhaA$_{(R.e)}$, PhaB$_{(R.e)}$, PhaC$_{(R.e)}$	0.3% dwt [5]
PhaA$_{(R.e)}$, PhaB$_{(R.e)}$, PhaC$_{(R.e)}$*	0.04% dwt [6], 0.002% dwt[7], 1.7% dwt[8]
potato leaves	
PhaA$_{(R.e)}$, PhaB$_{(R.e)}$, PhaC$_{(R.e)}$	0.09% dwt[5]
alfalfa leaves	
PhaA$_{(R.e)}$, PhaB$_{(R.e)}$, PhaC$_{(R.e)}$	0.18% dwt[9]
BktB$_{(R.e)}$, PhaB$_{(R.e)}$, PhaC$_{(R.e)}$	0.034% dwt[9]
corn stover	
PhaA$_{(R.e)}$, PhaB$_{(R.e)}$, PhaC$_{(R.e)}$	0.1-5.66% dwt[10]; 0.1-5.73% dwt[11]
hairy roots of sugar beet	
PhaA$_{(R.e)}$, PhaB$_{(R.e)}$, PhaC$_{(R.e)}$	5.5% dwt[12]
sugarcane leaves	
PhaA$_{(R.e)}$, PhaB$_{(R.e)}$, PhaC$_{(R.e)}$	Polymer observed by TEM[13]
Brassica seeds	
PhaA$_{(R.e)}$, PhaB$_{(R.e)}$, PhaC$_{(R.e)}$	7.7% dwt[14]
oil palm leaves	
BktB$_{(R.e)}$, PhaB$_{(R.e)}$, PhaC$_{(R.e)}$	Polymer observed by TEM[15]
flax	
PhaA$_{(R.e)}$, PhaB$_{(R.e)}$, PhaC$_{(R.e)}$	0.005% fwt [16]
PHBV; *Arabidopsis* leaves	
IlvA$_{(E.c.)}$, BktB$_{(R.e)}$, PhaB$_{(R.e)}$, PhaC$_{(R.e)}$	1.6% dwt, 2% HV; 0.84%dwt, 4%HV[17]
IlvA$_{(E.c.)}$, BktB$_{(R.e)}$, PhaB$_{(R.e)}$, PhaC$_{(N.c.)}$	0.07-0.38% dwt, 5-17% HV[17]
IlvA466$_{(E.c.)}$, BktB$_{(R.e)}$, PhaB$_{(R.e)}$,PhaC$_{(N.c.)}$	0.08-0.09% dwt, 8-9% HV[17]
IlvA466$_{(E.c.)}$, BktB$_{(R.e)}$, PhaB$_{(R.e)}$, PhaC$_{(R.e)}$	0.2-0.8% dwt, 4-17 mol% HV[18]
Brassica seeds	
IlvA466$_{(E.c.)}$, BktB$_{(R.e)}$, PhaB$_{(R.e)}$, PhaC$_{(R.e)}$	0.7-2.3% dwt, 2.3-6.4% HV[17] 0.7-2.3% dwt, 2.3-6.4 mol% HV[18]
MCL-PHA; potato leaves	
PhaG$_{(P.p.)}$, PhaC$_{(P.o.)}$	polymer production reported[19]

Table abbreviations: dwt, dry weight; fwt, fresh weight; TEM, transmission electron microscopy; %HV, % of hydroxyvalerate monomer fraction in copolymer; mol% HV, mol% of hydroxyvalerate in copolymer; (R.e.), *Ralstonia eutropha;* (A.c.), *Aeromonas caviae*; (E.c.), *Escherichia coli;* (N.c.), *Norcardia carollina*; (P.p.), *Pseudomonas putida*; (P.o.), *Pseudomonas oleovorans*. Genes inserted by direct plastid transformation where indicated (*). All other experiments performed via nuclear transformation, plastid targeting. ФTable references: [1](Nawrath et al., 1994); [2](Valentin et al., 1999); [3](Bohmert et al., 2000); [4](Arai et al., 2001); [5](Bohmert et al., 2002); [6](Lössl et al., 2000); [7](Nakashita et al., 2001); [8](Lössl et al., 2003); [9](Saruul et al., 2002); [10](Mitsky et al., 2000); [11](Poirier & Gruys, 2002); [12](Menzel et al., 2003); [13](Brumbley et al., 2003); [14](Houmiel et al., 1999); [15](Abdullah et al., 2003); [16](Wrobel et al., 2004); [17](Slater et al., 1999); [18](Valentin et al., 1999); [19](Romano et al., 2002).

Metabolic engineering efforts have enabled the production of PHAs in various organelles and tissues of a wide range of plants (Poirier & Gruys, 2002; Snell & Peoples, 2002). Production of the homopolymer polyhydroxybutyrate (PHB) and the copolymer poly-3-hydroxybutyrate-co-3-hydroxyvalerate (PHBV) has been most successful in the plastid, perhaps due to the natural presence of precursor acetyl-CoA for fatty acid biosynthesis. The synthesis of PHA has been engineered into other plant organelles but has yielded only low levels of polymer accumulation (Poirier & Gruys, 2002; Snell & Peoples, 2002)

2. PHB PRODUCTION IN PLASTIDS

Production of PHB requires the presence of three enzyme activities - thiolase, reductase, and synthase - to convert acetyl-CoA to polymer (Figure 2). PHB production via nuclear transformation and plastid targeting of transgene encoded enzymes has been described in dicots (*Arabidopsis*, tobacco, potato, *Brassica*, alfalfa, sugar beet and flax) as well as monocots (corn, sugarcane, and oil palm) (Table 2). PHB production via direct plastid transformation has only been described in tobacco (Lössl et al., 2000; Lössl et al., 2003; Nakashita et al., 2001).

2.1. Nuclear Transformation of Genes

2.1.1. Arabidopsis (Arabidopsis thaliana)

The PHB biosynthetic pathway of *R. eutropha* was targeted to chloroplasts of *Arabidopsis* (Nawrath et al., 1994) after previous attempts to produce PHB in the cytosol yielded low levels of polymer (Poirier et al., 1992a). A genetic fragment encoding a transit peptide was fused to the N-terminus of each PHB gene and separate vectors for thiolase, reductase and synthase were transformed into *Arabidopsis*. Lines containing all three enzymes were obtained by cross pollination yielding plants producing up to 14% dry weight (dwt) PHB (Nawrath et al., 1994), a level much higher than the 20 to 100 ∝g of PHB per gram of fresh weight (fwt) observed in cytosolic experiments (Poirier et al., 1992a). High levels of plastidial PHB were observed to lead to chlorosis in leaves (Nawrath et al., 1994).

Similar experiments with separate transformation plasmids for each PHB gene yielded PHB levels up to 1.7% dwt (Valentin et al., 1999), a level much lower than the 14% dwt previously observed (Nawrath et al., 1994). While reasons for these lower yields are not known, it has been speculated (Poirier & Gruys, 2002) that multiple insertions within the lines of Nawrath et al. (Nawrath et al., 1994) may have yielded higher polymer levels.

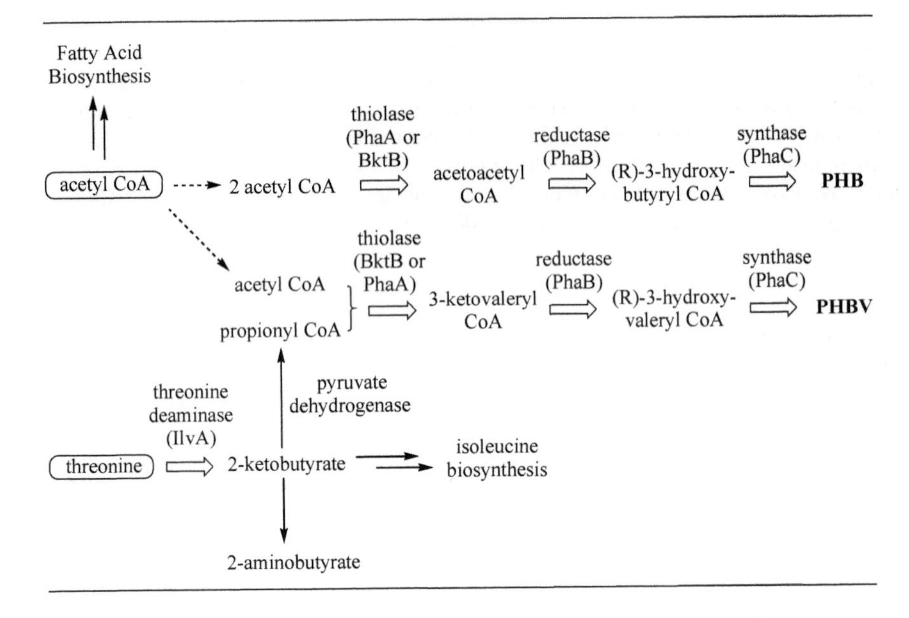

Figure 2. *Pathways for PHA Synthesis in Plastids of Transgenic Plants. Open block arrows indicate enzyme activities engineered for PHA synthesis.*

Subsequent efforts to improve PHB production involved transformation of multi-gene constructs. This strategy facilitated the generation of stable lines since only one transformation was required to generate a PHB producing plant and multiple integration sites and/or time consuming sexual crossing procedures were eliminated. Transformation of *Arabidopsis* with a multi-gene construct containing the expression cassettes of Nawrath et al. (Nawrath et al., 1994) yielded plants producing PHB at up to 4% fwt, a number roughly equivalent to 40% dwt (Bohmert et al., 2000). Whereas only slight chlorosis was observed in lines producing more than 3% dwt in previous studies (Nawrath et al., 1994), a correlation between high PHB accumulation and stunted growth and/or decreased fertility was observed in the lines of Bohmert et al. producing greater than 3% dwt PHB (Bohmert et al., 2000). Chloroplasts are naturally able to cope with accumulation of starch granules and it was expected (Poirier & Gruys, 2002) that some accumulation of osmotically inert PHB granules should be tolerated. However limited storage capacity, depletion of plastidial carbon sources, or other effects could result in the observed phenotypes. Metabolite profiling of high PHB producing lines (Bohmert et al., 2000) indicated reduced levels of fumarate and isocitrate, possibly indicating reduced activity of the tricarboxylic acid cycle and reduced acetyl-CoA availability. Interestingly, transformation of a gene encoding plastid-targeted thiolase unexpectedly led to a significantly decreased transformation efficiency (Bohmert et al., 2000), suggesting over expression of plastid targeted thiolase is harmful.

In similar experiments, transformation of a multi-gene vector containing expression cassettes for plastid-targeted thiolase, reductase and synthase produced up to 7.6% and 13% PHB dwt in heterozygous and homozygous plants, respectively (Valentin et al., 1999). It should be noted that the creation of homozygous lines has not always led to increased PHB production. Homozygous plants have been isolated containing lower levels of polymer (Bohmert, unpublished data; Poirier & Gruys, 2002), possibly due to gene silencing.

Attempts to reduce altered plant phenotypes associated with PHB formation have included co-expression of phasins with the PHB biosynthetic pathway or use of regulated promoters (Bohmert et al., 2002). Co-expression of phasins, proteins that naturally associate with PHB granules in bacteria (Jurasek & Marchessault, 2002; Wieczorek et al., 1995), was attempted in order to prevent interference of granules with plastid structures such as the thylakoid membranes. Co-expression of the *R. eutropha* phasin gene with the PHB genes yielded 5.8% PHB dwt but did not prevent negative growth and developmental effects (Bohmert et al., 2002). Use of regulated promoters for PHB gene expression did however reduce production of altered phenotypes (Bohmert et al., 2002). Inducible expression of thiolase with the prp-1 promoter (Martini et al., 1993), which can be induced with salicylic acid, was leaky but improved PHB production in *Arabidopsis* such that 6% PHB dwt could be produced before changes in plant growth were observed, a significant increase from the 3% dwt previously reported with constitutive gene expression (Bohmert et al., 2000). In experiments with a 35S-Ac promoter, in which gene expression is expected only after random excision of the maize transposable element activator (*Ac*), PHB yields of up to 1.7% dwt were obtained (Bohmert et al., 2002).

2.1.2. Tobacco (Nicotiana tabacum)

It was initially expected that the high yields of PHB, achieved in *Arabidopsis* with nuclear transformation and plastid targeting of the PHB pathway, might easily be transferred to other plants. However, similar strategies in tobacco have only yielded low levels of polymer. Expression of plastid targeted *A. caviae* synthase, and *R. eutropha* reductase and thiolase yielded healthy tobacco plants producing 0.09% PHB dwt (Arai et al., 2001), an amount several orders of magnitude lower than the levels previously achieved in plastids of *Arabidopsis* (Bohmert et al., 2000; Nawrath et al., 1994; Valentin et al., 1999). Similar experiments with a construct previously shown to produce 40% PHB dwt in *Arabidopsis* (Bohmert et al., 2000) (Section 2.1.1) yielded healthy tobacco plants containing only 0.08% PHB dwt (Bohmert et al., 2002). As with *Arabidopsis* (Section 2.1.1), transformation of a gene encoding plastid targeted thiolase significantly decreased transformation efficiency (Bohmert et al., 2002).

Experiments to increase acetyl-CoA availability for PHB formation in tobacco were performed by treating PHB lines with quizalofop, an inhibitor of cytosolic but not plastidial acetyl-CoA carboxylase (ACCase), and mevastatin, an inhibitor of 3-hydroxy-3-methylglutaryl-CoA reductase (HMG-CoA reductase) (Suzuki et al., 2002). PHB production in transgenic tobacco lines increased 150% upon treatment of plants with quizalofop, however the best PHB producing lines still possessed extremely low amounts of PHB (less than 0.01% PHB dwt). Inhibition of HMG-CoA reductase resulted in little increase in plastidial PHB formation (Suzuki et al., 2002).

To avoid possible harmful effects associated with expression of the PHB pathway during tobacco transformation, the chemically and somatically activated constructs previously described in Section 2.1.1, in which thiolase expression was under the control of the prp-1 promoter or the 35S promoter fused to the transposable element Ac, were transformed into tobacco (Bohmert et al., 2002). Both the chemically and somatically activated constructs led to normal transformation efficiencies in tobacco. The somatically activated construct yielded up to 0.3% PHB dwt, a level 3.7 fold greater than the 0.08% PHB dwt observed with constitutive plastid targeted expression (Bohmert et al., 2000). Several of the isolated tobacco lines showed phenotypic alterations (i.e. chlorosis or morphological changes) but no correlation between phenotype and PHB accumulation was observed. PHB production using the inducible prp-1 promoter yielded up to 0.04% PHB dwt.

2.1.3. Potato (Solanum tuberosum)

Transformation of potato with the constitutive expression construct that previously yielded up to 40% PHB dwt in *Arabidopsis* (Section 2.1.1) (Bohmert et al., 2000) failed to produce any transgenic lines (Bohmert et al., 2002). Regulated expression of thiolase, using the chemically or somatically activated constructs previously described for *Arabidopsis* (Section 2.1.1) or tobacco (Section 2.1.2), enabled isolation of transformants but yielded only low levels of PHB in potato leaves (Bohmert et al., 2002). The salicylic acid inducible prp-1 promoter and somatically activated constructs yielded 0.009% and 0.002% PHB dwt, respectively. The inability to transform the constitutive expression construct or obtain higher levels of PHB in potato with regulated expression of thiolase is currently not understood.

2.1.4. Brassica (Brassica napus)

Plastids of oilseeds have been suggested as a particularly suitable site for PHB production since fatty acid biosynthesis from acetyl-CoA occurs efficiently in this tissue (Poirier et al., 1992b). Plastids of *Brassica napus* were engineered for PHB production by transforming individual plasmids containing expression cassettes for plastid targeted *R. eutropha* thiolase, reductase or synthase under the control of the *Brassica* seed specific 7s promoter (Mitsky et al., 2000; Valentin et al., 1999). Crossing of individual lines yielded plants containing up to 2% PHB fwt in seeds (seeds were reported to contain approximately 15% water) (Mitsky et al., 2000; Valentin et al., 1999). Transformation of *Brassica* using a multi-gene vector containing expression cassettes for plastid targeted *R. eutropha* thiolase *(phaA)*, reductase (*phaB*) and synthase (*phaC*) under the control of the seed specific promoter from the *Lesquerella fendleri* oleate 12-hydroxylase gene yielded up to 7.7% PHB dwt (Houmiel et al., 1999; Mitsky et al., 2000; Valentin et al., 1999). The *Lesquerella* promoter was chosen for these experiments since its expression coincides with accumulation of storage lipids (Broun & Somerville, 1997), a time in seed development when large amounts of acetyl-CoA are converted to fatty acids. All of the PHB producing lines were reported to produce viable seed (Mitsky et al., 2000). In transmission electron micrographs of high PHB producing lines, leucoplasts were distorted but intact, and they appeared to expand in response to polymer production (Houmiel et al., 1999; Valentin et al., 1999). This is similar to starch storing amyloplasts whose size increases with starch accumulation (Houmiel et al., 1999; Mitsky et al., 2000).

2.1.5. Alfalfa (Medicago sativa)

Transformation of alfalfa with multi-gene constructs previously used for PHB production in *Arabidopsis* (Mitsky et al., 2000; Valentin et al., 1999) (Section 2.1.1 yielded PHB granules in plastids, albeit at low levels (Saruul et al., 2002). Constructs contained plastid targeted *R. eutropha* thiolase (*phaA*), synthase (*phaC*), and reductase (*phaB*) each in individual expression cassettes with the 35S promoter. Transformation of the multi-gene construct containing *phaA*, *phaB*, and *phaC* into alfalfa yielded plants accumulating up to 0.18% PHB dwt, a number significantly less than the levels of up to 13 % PHB dwt observed with the same construct in *Arabidopsis* (Mitsky et al., 2000; Valentin et al., 1999). PHB granules were shown to accumulate exclusively in the chloroplasts by transmission electron microscopy, and phenotypical changes were not observed in the plants upon polymer production. The reasons for the low levels of PHB accumulation in alfalfa are currently not understood (Saruul et al., 2002).

2.1.6. Sugar beet (Beta vulgaris).

PHB production in hairy root cultures of sugar beet has been reported (Menzel et al., 2003) upon transformation of the multi-gene construct previously used to produce up to 40% PHB dwt in *Arabidopsis* (Bohmert et al., 2000) (Section 2.1.1). Levels of up to 3.4% PHB dwt were observed in growing hairy root cultures with levels increasing to 5.5% PHB dwt in the final growth stages of the cultures. Transmission electron microscopy demonstrated that granules were localized in leucoplasts. Growth retardation was observed upon PHB production in cultures. In contrast to wild-type cultures, PHB producing hairy root cultures did not contain plastoglobules for lipid storage in their leucoplasts, possibly indicating insufficient acetyl-CoA availability (Menzel et al., 2003) despite the supplementation of culture medium with carbon.

2.1.7. Corn (Zea mays)

Strategies for PHB formation in corn were designed such that polymer formation would occur primarily in the corn stover (leaves and stalk) providing additional value from the residual biomass that is usually discarded after harvest of corn seed. Multi-gene constructs containing individual expression cassettes for plastid targeted PHB genes under the control of the enhanced 35S (e35S), the enhanced figwort mosaic virus (eFMV), the maize chlorophyll A/B binding protein (P-ChlA/B) (Mitsky et al., 2000; Poirier & Gruys, 2002), or the rice actin (rACT) promoter

(Poirier & Gruys, 2002) were prepared for corn transformation. To further enhance expression of the PHB pathway genes, the HSP70 intron was inserted downstream of the promoters. The highest levels of PHB production were achieved with the ChlA/B promoter yielding up to 5.73 % PHB dwt. Constructs with the e35S, FMV, or rACT promoters yielded up to 4.81%, 4.84%, and 0.8 % PHB dwt, respectively. The highest levels of PHB accumulation were detected in leaves that tended to be chlorotic. Only low amounts of PHB were observed in seeds. It was observed by transmission electron microscopy that plants transformed with constructs containing the e35S, or the ChlA/B promoter contained more granules in plastids of vascular tissue than plastids of mesophyll cells.

2.1.8. Sugarcane (Saccharum spp.)

The production of PHB granules in plastids of sugarcane leaves has been observed by transmission electron microscopy upon insertion of expression cassettes for plastid targeted *R. eutropha* thiolase, reductase and synthase (Brumbley et al., 2003). Treatment of leaf tissue samples with acid and subsequent HPLC detection of PHB derived crotonoic acid confirmed the identity of the polymer.

2.1.9. Oil Palm (Elaeis guineensis)

Oil palm has been targeted as a crop for PHB production due to its high level of oil productivity (Moffat, 1999; Sambanthamurthi et al., 2000) suggesting large amounts of acetyl-CoA flowing through the plastids for fatty acid biosynthesis. Transformation of expression cassettes containing a modified 35S promoter and genes encoding plastid targeted *R. eutropha* thiolase (*bktB*), reductase and synthase yielded polymer granules in chloroplasts of oil palm leaves that were detectable by transmission electron microscopy (Abdullah et al., 2003). No negative effects were observed upon PHB formation. Yields of polymer have not yet been reported for these plants.

2.1.10. Flax (Linum usitatissimum)

PHB production has been reported in flax transformed with a constitutive multi-gene expression construct, previously shown to produce up to 40% PHB dwt in transgenic *Arabidopsis* (Bohmert et al., 2000) (Section 2.1.1), and with a construct containing constitutive expression of reductase and synthase and stem specific expression of thiolase (Wrobel et al., 2004). Transformation of flax with the constitutive construct yielded transgenic plants with up to 50 ∝g PHB per gram fwt. These plants had significantly reduced growth and were senescent. Transgenic plants that were screened for normal phenotypes contained much lower levels of PHB with up to 0.5 ∝g PHB per gram fwt observed.

In contrast, constructs containing a cassette for stem specific expression of thiolase only, or cassettes for stem specific thiolase expression and constitutive reductase and synthase expression, yielded healthy plants. For stem specific expression, the potato 14-3-3 gene promoter (Szopa et al., 2003), a promoter yielding predominantly stem specific expression in flax (Wrobel et al., 2004), was used. PHB yields of up to 4.62 ∝g PHB per gram fwt were obtained, levels 10 fold less than the 50 ∝g PHB per gram fwt observed in the top producers isolated upon transformation of the constitutive expression construct. This reduced yield may be due to the approximately 10 fold lower expression levels observed with the 14-3-3 promoter compared to the constitutive CaMV 35S promoter in flax stems (Wrobel et al., 2004). Interestingly, some lines with increased seed production, seeds with altered lipid composition or decreased amounts of starch in stems and leaves were observed. Decreased amounts of glucose were also reported in the highest PHB producing lines, a result that may be indicative of increased glucose consumption for acetyl-CoA production.

2.1.11. Discussion

There are apparent benefits to using multi-gene constructs for PHB production. In studies with identical gene expression cassettes, more PHB was produced in plants using a multi-gene construct [up to 40% PHB dwt, (Bohmert et al., 2000)] compared to plants in which single gene constructs were transformed and crossed [up to 14% PHB dwt (Nawrath et al., 1994)]. Although it is not known why higher levels of PHB were obtained with the multi-gene construct, it has been speculated that concerted expression of genes, locally high levels of transcription factors, or reduced gene silencing due to tight linkage of genes might promote increased product formation (Mitsky et al., 2000; Poirier & Gruys, 2002).

It is unclear why levels of PHB produced by Bohmert *et al.* in *Arabidopsis* (Bohmert et al., 2000) were significantly higher than levels observed by other researchers using similar multi-gene transformation constructs (Mitsky et al., 2000; Nawrath et al., 1994; Poirier & Gruys, 2002; Valentin et al., 1999). Possible explanations include the presence of more than one transgene integration site in the highest PHB producers (Bohmert et al., 2000) or differences in tissue-specific PHB production. Plants producing 40% PHB dwt contained granules in chloroplasts of mesophyll cells (Bohmert et al., 2000), whereas lower producing lines from independent experiments in other labs contained granules primarily in epidermal cells and cells associated with vascular tissue (Gruys, 2000; Poirier & Gruys, 2002). While differences in host ecotype used in the experiments (Poirier & Gruys, 2002) would not be expected to alter the cellular location of PHB production, different metabolite compositions have been detected in different ecotypes of *Arabidopsis* (Taylor et al., 2002). Two of the metabolites that can be used to distinguish ecotypes are malate and citrate, components of the tricarboxylic acid cycle that oxidizes pyruvate, via acetyl-CoA, to produce NADH and $FADH_2$ within the mitochondria.

Transformation of the identical constitutive expression construct into *Arabidopsis* (Bohmert et al., 2000), tobacco (Bohmert et al., 2002), potato (Bohmert et al., 2002), and flax (Wrobel et al., 2004) suggests the level of PHB production and the effect of PHB production on the host differs amongst plant species. In flax, comparably low PHB production (50 ∝g PHB per gram of flax fwt) is accompanied with severe growth reduction (Wrobel et al., 2004) whereas the same construct is well-tolerated in transgenic *Arabidopsis* and tobacco plants at levels up to 3 mg PHB per gram of fwt (equivalent to approximately 3% PHB dwt) or 90 ∝g PHB per gram of fwt (equivalent to 0.08% PHB dwt), respectively. In potato, regeneration of transgenic lines was not possible with a constitutive multi-gene construct and only chemically inducible or somatically activated PHB gene expression constructs yielded transgenic PHB producing potato plants (Bohmert et al., 2002). It is currently unknown whether gene expression, acetyl-CoA availability, or some other species-specific metabolic factor is responsible for the differences observed in PHB formation and phenotype in different plant species. There are some indications that carbon flow may be an issue in plants. Tobacco lines engineered for PHB production were found to yield higher levels of polymer when grown heterotrophically in tissue culture supplemented with carbon than when grown autotrophically in soil (Bohmert et al., 2002; Lössl et al., 2003). However differences in expression of the PHB genes can not be eliminated. A detailed analysis of carbon flow and gene expression would greatly aid the interpretation of differences in PHB production in different host plants.

2.2. Direct Plastid Transformation

Poor gene expression has been suggested as a possible cause for the low levels of PHB observed in tobacco expressing a nuclear-encoded, plastid-targeted PHB pathway (Nawrath, 1995). Since direct plastid transformation typically yields higher levels of transgene expression than nuclear-encoded strategies (Daniell et al., 2004; Devine and Daniell, 2004), plastid transformation is particularly attractive for attempts to increase PHB formation in tobacco (for more information on plastid genetic engineering see chapter 16 of this volume). Insertion of the *R. eutropha* operon, with its native promoter, untranslated regions, and termination signals, into the plastome of tobacco yielded plants containing up to 0.04% PHB dwt (Lössl et al., 2000). Interestingly, these plants were stunted (Lössl et al., 2000) whereas plants obtained via nuclear transformation and plastid targeting techniques exhibited normal phenotypes with slightly higher levels of polymer (Arai et al., 2001; Bohmert et al., 2002). Subsequent attempts at plastid-encoded PHB formation focused on modification of expression cassettes with regulatory elements more amenable to plastidial expression (Lössl et al., 2003; Nakashita et al., 2001). The strong, constitutive chloroplast 16S ribosomal RNA gene promoter (*Prrn*) and the 3' UTR from the gene encoding the D1 protein of photosystem II (*psbA*) were used to create an expression cassette for the *R. eutropha* thiolase, reductase and synthase (Nakashita et al., 2001). Plants containing up to 0.002% PHB dwt (Nakashita et al., 2001), a level 20 fold less than values reported with the unoptimized transformation construct (Lössl et al., 2000) were obtained. Transformation of tobacco with a construct containing a DNA fragment with synthase, thiolase and reductase genes as well as the native bacterial 3' regulatory element, behind the plastidial *psbA* promoter and 5' UTR yielded plants containing up to 1.7% PHB dwt in early regeneration stages of plantlets in tissue culture (Lössl et al., 2003). Deficiencies in growth were observed in these plants and even low amounts of PHB were associated with male sterility. Interestingly, PHB production in these lines decreased upon transfer of the plants to autotrophic growth conditions in soil, yet Southern analysis confirmed the plants were still homozygous.

Several factors could account for the observed differences in product formation in the plastid-encoded experiments (Lössl et al., 2000; Lössl et al., 2003; Nakashita et al., 2001). The expression of many plastid genes is primarily regulated post-transcriptionally (Eibl et al., 1999) such that translation efficiency may be lower for genes regulated by a bacterial 5' UTR (Lössl et al., 2000; Nakashita et al., 2001) than for the *psbA* 5'UTR (Lössl et al., 2003) that is associated with strong transgene expression from the plastid genome (Dhingra et al., 2004; Eibl et al., 1999; Fernandez-San Millan et al., 2003; Staub and Maliga, 1994). However, the highest level of foreign protein expression in chloroplasts has been reported with a transgene regulated by a bacterial UTR (up to 46% of total protein; DeCosa et al., 2001). The use of the strong constitutive *Prrn* promoter in the experiments of Nakashita et al. (2001) could have led to high expression of PHB genes throughout the transformation procedure possibly hindering the isolation of efficient PHB producers. Constitutive nuclear-encoded expression of plastid targeted PHB genes, or the thiolase gene alone, has been shown to significantly reduce efficiencies of nuclear transformation in tobacco (Section 2.1.2) and potato (Section 2.1.3) (Bohmert et al., 2002). The expression of the PHB pathway from the light inducible *psbA* promoter by Lössl et al. (2003) might have led to more moderate gene expression during dark incubations of freshly transformed tissue allowing the isolation of plants capable of producing higher levels of PHB.

Differences in PHB production have been observed with heterotrophic and autotrophic growth of tobacco engineered with plastid (Lössl et al., 2003) or nuclear genes encoding plastid-targeted enzymes (Bohmert et al., 2002). These differences may indicate that PHB producing tobacco plants are limited in total carbon supply when grown under autotrophic conditions, possibly decreasing the availability of the acetyl-CoA precursor for PHB formation.

3. COPOLYMER PRODUCTION

The production of copolymers of PHB is of commercial interest to expand the range of polymer properties that can be obtained (Feng et al., 2002). Most copolymer studies in plastids have been performed with poly-3-hydroxybutyrate-co-3-hydroxyvalerate (PHBV) (Mitsky et al., 2000; Slater et al., 1999; Valentin et al., 1999). The pathway for PHBV synthesis is similar to PHB synthesis, requiring thiolase, reductase, and synthase activities to convert propionyl CoA and acetyl-CoA to PHBV (Figure 2). The *bktB* thiolase gene is often used instead of *phaA* since it encodes a thiolase with a substrate specificity more amenable to condensation of acetyl-CoA and propionyl-CoA (Slater et al., 1998) (Figure 2). One of the key challenges for PHBV production in plastids is generation of sufficient propionyl CoA for 3-hydroxyvalerate (3-HV) formation. Propionyl-CoA is not readily available in plant plastids since expression of *bktB* with reductase and synthase yields only PHB in *Arabidopsis* (Mitsky et al., 2000), *Brassica* (Houmiel et al., 1999; Mitsky et al., 2000) and alfalfa (Saruul et al., 2002). While the plastidial pyruvate dehydrogenase complex from *Brassica* leucoplasts is able to convert 2-

ketobutyrate to propionyl-CoA (Figure 2), the reaction is ten times less efficient than the conversion of pyruvate to acetyl-CoA (Mitsky et al., 2000; Slater et al., 1999). 2-ketobutyrate is also an intermediate of isoleucine biosynthesis (Valentin et al., 1999), such that pyruvate dehydrogenase must compete for available substrate to form propionyl CoA. There have also been attempts to produce medium chain length PHAs in plastids using a 3-hydroxyacyl-acyl carrier protein-coenzyme A transferase (*phaG*) and a medium chain length synthase. These efforts have been reported to yield some production of polymer in potato leaves (Table II) (Romano et al., 2002), however similar efforts have not been successful in *Arabidopsis* (Mittendorf et al., 2000; Snell & Peoples, 2002). The following sections outline strategies that have resulted in PHBV production in plastids.

3.1. *Arabidopsis*

In efforts to increase 2-ketobutyrate for PHBV formation, *E. coli* threonine deaminase (IlvA), an enzyme that converts threonine to 2-ketobutyrate, was used (Figure 2) (Slater et al., 1999). Transgenic *Arabidopsis* lines containing plastid targeted *ilvA* and *phaA* (*R. eutropha*), each under the control of the 35S promoter, produced elevated levels of 2-ketobutyrate, isoleucine and 2-aminobutyrate, a compound that is believed to be formed by transamination of 2-ketobutyrate (Figure 2). Transgenic lines expressing *phaB* (*R. eutropha*), *phaC (R. eutropha* or *Nocardia corallina), bktB (R. eutropha*), and wild-type *ilvA*, each under control of the 35S promoter, yielded polymer that ranged from 0.07%-1.6% PHBV dwt with an HV component between 2% and 17 % (Slater et al., 1999). In further attempts to increase 2-ketobutyrate, two mutant *ilvA* alleles that partially [*ilvA466,* (Taillon et al., 1988)] or completely [*ilvA219,* (Eisenstein et al., 1995)] removed the natural feedback inhibition of IlvA with isoleucine were used (Slater et al., 1999). While transformation of *ilvA219* was not successful, transgenic *Arabidopsis* lines with *ilvA466* were recovered, albeit with reduced frequency (Slater et al., 1999). Co-expression of *ilvA466* with *phaB, phaC (N. corallina),* and *bktB* resulted in lower polymer yields (0.08%-0.09% dwt PHBV) and lower HV levels (8%-9% 3-HV) than similar lines expressing wild-type *ilvA* (0.07%-0.38% dwt PHBV; 5%-17% 3-HV), providing further evidence that deregulated threonine deaminase activity is harmful, possibly due to increased amounts of ammonia produced as a by product of the threonine deaminase reaction (Slater et al., 1999). Interestingly, no phenotypic effects were observed in these plants. Pyruvate dehydrogenase was assumed to be the bottleneck to 3-HV synthesis since levels of pyruvate were higher than 2-ketobutyrate, and previous studies have demonstrated the *bktB, phaB and phaC* pathway to be efficient (Slater et al., 1999). The low levels of total polymer formed compared to previous studies with plastid-based PHB production (Section 2.1.1) were attributed to an increased metabolic burden upon *ilvA* expression.

Similar results have been reported in *Arabidopsis* lines expressing plastid targeted *bktB (R. eutropha)*, *phaB* (*R. eutropha*), *phaC* (*R. eutropha*) and *ilvA466* (Valentin et al., 1999). A 20 fold increase in 2-ketobutyrate was detected in these lines whereas levels of pyruvate and threonine remained unchanged. Since levels of

2-aminobutyrate and isoleucine (Figure 2) also increased, it was assumed that the majority of 2-ketobutyrate was utilized to form these compounds rather than propionyl CoA for PHBV synthesis. Levels of 0.2%-0.8% PHBV dwt were produced with 3-HV levels ranging from 4% to 17 mol%.

3.2. *Brassica*

Expression of plastid targeted *R. eutropha bktB*, *phaB*, *phaC* and *E. coli ilvA466* (Section 3.1) under control of the *Lesquerella* hydroxylase promoter (Broun et al., 1998) yielded total polymer levels in seeds ranging from 0.7% to 2.3% PHBV dwt, with a 3-HV content of 2.3% to 6.4% (Slater et al., 1999). A negative correlation between 3-HV composition and total polymer yield was observed in these plants and one fourth of the transgenic plants possessed phenotypic alterations like low seed recovery, sterility, and plant mortality before seed set. Half of the progeny of these lines exhibited chlorosis and stunted growth shortly after germination, but recovered during later growth stages.

In similar experiments (Valentin et al., 1999), expression of plastid targeted *ilvA466* with the *R. eutroha bktB, phaB* and *phaC* genes under the control of the *Lesquerella* hydroxylase promoter produced total polymer yields and 3-HV contents almost identical to those reported previously (Slater et al., 1999) (Table II).

3.3. Discussion

The low germination rates of PHBV producing *Brassica* seeds (Slater et al., 1999) have also been observed in high PHB producing lines of *Arabidopsis* (Bohmert et al., 2000) (Section 2.1.1). Male sterility has been observed in both PHB producing *Arabidopsis* (Bohmert et al., 2000) and transplastomic tobacco lines (Lössl et al., 2003) (Section 2.2). These phenotypes suggest the presence of PHA accumulation may interfere with pollen and/or seed production.

Efforts to increase total PHBV polymer levels and 3-HV content in transgenic plants will likely require further metabolic engineering. As mentioned in Section 3.1, studies suggest pyruvate dehydrogenase as a likely bottleneck for formation of the HV monomer unit (Slater et al., 1999). Increasing levels of this activity by transgene insertion is not however, trivial, since pyruvate dehydrogenases are composed of three subunits that require either three [i.e. *E. coli*, (Guest & Stephens, 1980)] or four [i.e. *Arabidopsis*, (Johnston et al., 1997; Lutziger & Oliver, 2000; Mooney et al., 1999)] separate gene products. Improving PHBV formation may also require optimization of thiolase, reductase and synthase activities. Mathematical simulations of PHBV production in plastids have suggested that increases in thiolase activity may have the greatest impact on increasing carbon flow to total polymer synthesis (Daae et al., 1999). This increase is, however, at the expense of the 3HV to 3HB ratio. Increases in acetoacetyl-CoA reductase and PHB synthase activities are predicted to increase the 3HV/3HB monomeric ratio (Daae et al., 1999).

4. PLASTID BIOCHEMISTRY RELEVANT TO PHA FORMATION

Studies with PHB producing *Arabidopsis* leaves (Bohmert et al., 2000) (Section 2.1.1), hairy root cultures of sugar beet (Menzel et al., 2003) (Section 2.1.6) and tobacco leaves (Lössl et al., 2003) (Section 2.2) have suggested total carbon supply may be limiting in PHB producing plants. PHB synthesis competes with plastidial fatty acid biosynthesis for the common precursor acetyl-CoA (Figure 2), and diversion of too much acetyl-CoA to PHB formation may affect plant health. A detailed understanding of acetyl-CoA biosynthesis would allow the rational design of strategies to produce healthy plants containing high levels of PHB.

Acetyl-CoA is not transported through plastid membranes (Fischer & Weber, 2002; Masterson & Wood, 2000; Ratledge et al., 1997) and must be generated within plastids. The pathway for its formation within plastids has been the subject of much study and debate (Bao et al., 2000; Nikolau et al., 2000; Ohlrogge et al., 2000; Rawsthorne, 2002), with suggested precursors including pyruvate, acetate, citrate and acetylcarnitine (Figures 3 and 4) (Rawsthorne, 2002). *In vitro* import studies with isolated plastids have demonstrated the uptake and conversion of most of these metabolites into fatty acids suggesting necessary enzyme activities are present. The role of some of these precursors has been studied in more detail with *in vivo* labelling experiments. In this section, we will briefly update the current understanding of plastidial acetyl-CoA formation, giving preference to *in vivo* studies where available.

4.1. Pyruvate

The plastidial pyruvate dehydrogenase complex (Tovar-Mendez et al., 2003) possesses activities capable of converting pyruvate, CoA, and NAD+ to acetyl-CoA, CO_2, NADH, and H+ (Figures 3 & 4, reaction a). Experimental results have provided some evidence that pyruvate may significantly contribute to acetyl-CoA formation for fatty acid biosynthesis in both leaves and seeds. In *Arabidopsis* leaves, the rapid incorporation of carbon from labelled CO_2 into fatty acids and isoprenoids is consistent with the formation of fatty acids from newly fixed carbon (Bao et al., 2000). The existence of a complete glycolytic pathway within the chloroplasts has been debated and there is some evidence that a complete pathway may only be functional in immature chloroplasts (Heintze et al., 1990) (Figure 3, reaction b). An alternative pathway in which glycolytic reactions are located in both the cytosol and chloroplast (Fischer et al., 1997) has also been proposed (i.e. Figure 3, reactions d, f, h). In developing embryos of *Brassica napus*, ^{13}C labelling experiments suggest at least 90% of carbon for fatty acid synthesis is formed from sugars by glycolysis providing support for fatty acid formation, and thus acetyl-CoA formation, from pyruvate (Schwender & Olhlrogge, 2002). The expression of pyruvate dehydrogenase in developing seeds of *Arabidopsis* has also been found to correlate to both acetyl-CoA carboxylase expression (Nikolau et al., 2000) and lipid formation (Ke et al., 2000).

4.2. Acetate

Acetyl-CoA synthetase converts acetate, CoA and ATP to acetyl-CoA and AMP (Figure 3, reaction i; Figure 4, reaction c). Acetate has previously been suggested as a principal precursor to plastidial acetyl-CoA formation (Liedvogel & Stumpf, 1982; Murphy & Leech, 1981; Murphy & Stumpf, 1981; Roughan, 1995; Springer & Heise, 1989) and has been proposed to originate in the mitochondria and travel to the plastid via diffusion (Figure 3, reactions j, k) (Liedvogel & Stumpf, 1982; Murphy & Stumpf, 1981). Other potential routes to acetate formation (Ke et al., 2000) include a pathway converting pyruvate to acetate using pyruvate decarboxylase and acetaldehyde dehydrogenase (op den Camp & Kuhlemeier, 1997; Tadege & Kuhlemeier, 1997) or a pathway converting serine and acetyl-CoA to acetate and cysteine using serine acetyltransferase and *O*-acetyl-Ser thiol-lyase, enzymes involved in cysteine biosynthesis (Leustek & Saito, 1999). Recent *in vivo* studies however have questioned the contribution of acetate as a precursor for acetyl-CoA formation for fatty acid biosynthesis (Bao et al., 2000; Behal et al., 2002). Treatment of *Arabidopsis,* barley and pea leaves with $^{14}CO_2$ yielded little labelling of free acetate, despite the rapid labelling of fatty acids, suggesting a pool of free acetate does not contribute to acetyl-CoA synthesis (Bao et al., 2000). Similarly, whole leaf labelling studies of *Arabidopsis* possessing different acetyl-CoA synthetase activities suggest very little carbon enters plastidial fatty acid biosynthesis through acetate *in vivo* (Behal et al., 2002). In developing seeds of *Arabidopsis,* little correlation has been observed with the expression of acetyl-CoA synthetase and acetyl-CoA carboxylase (Nikolau et al., 2000). Other studies have shown that the production of acetyl-CoA synthetase does not correlate with lipid accumulation (Ke et al., 2000). It is currently assumed that acetate does not play a major role as a precursor for plastidial acetyl-CoA synthesis.

4.3. Citrate

Citrate, originating from the mitochondria (Figure 3, reactions l, m), has been suggested as a precursor to plastidial acetyl-CoA formation in both leaves (Rangasamy & Ratledge, 2000a) and seeds (Ratledge et al., 1997). Citrate has been shown to be incorporated into fatty acids, albeit at low rates, in experiments with isolated chloroplasts from pea leaves (Masterson et al., 1990a). However, to our knowledge, similar incorporation experiments have not been performed in non-green plastids. ATP citrate lyase, an enzyme that catalyzes the conversion of citrate, CoA and ATP to acetyl-CoA, oxaloacetate and ADP (Figure 3, reaction n; Figure 4 reaction h) has been suggested as an activity that could form plastidial acetyl-CoA. Experiments with ATP citrate lyase have often yielded conflicting results such that the enzyme's role in metabolism has been difficult to elucidate.

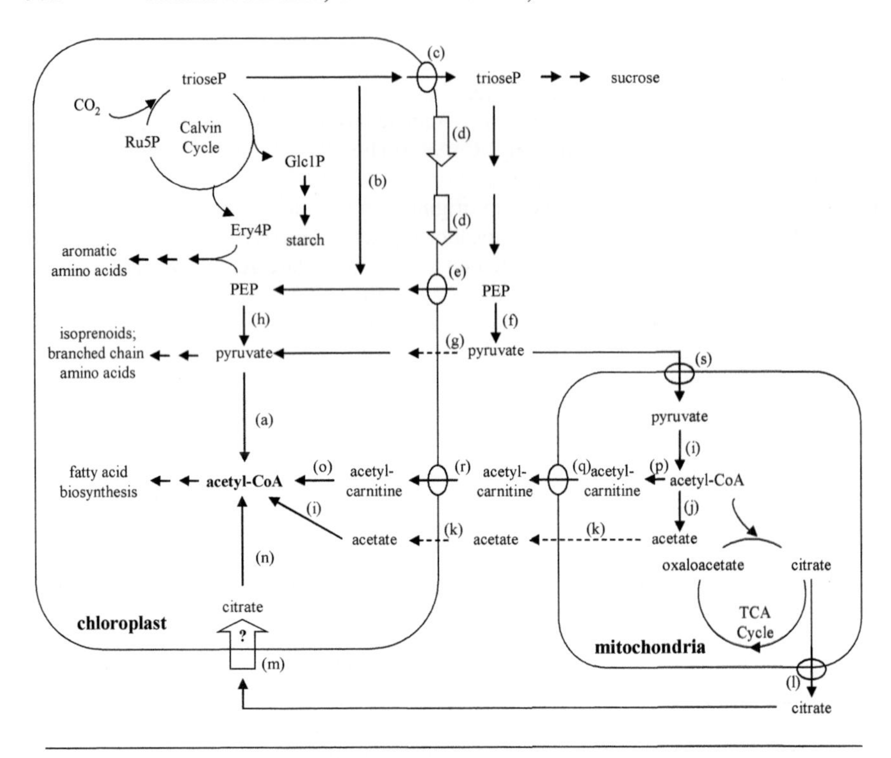

Figure 3. Enzymatic reactions relevant to Acetyl-CoA in chloroplasts.*

*Illustration of the multiple reactions proposed to contribute to acetyl-CoA supply in chloroplasts of C3 type plants. See text (Section 4) for discussion of most probable pathways for carbon flow. Figure abbreviations: PEP, phosphoenolpyruvate; Glc1P, glucose 1-phosphate; Ery4P, erythrose 4-phosphate; Ru5P, ribulose 5-phosphate; TCA cycle, tricarboxylic acid cycle. Pathways and enzymatic conversions: (a) plastidial pyruvate dehydrogenase complex; (b) propopsed plastidial glycolytic pathway in immature chloroplasts (Heintze et al., 1990); (c) triose phosphate/phosphate translocator, (Bowsher & Tobin, 2001; Flugge et al., 1989); (d) glycolytic reactions split between cytosol and plastid (Fischer et al., 1997) (e) PEP/phosphate translocator (Bowsher & Tobin, 2001; Fischer et al., 1997); (f) cytosolic pyruvate kinase; (g) diffusion, (Fischer & Weber, 2002; Proudlove & Thurman, 1981); (h) plastid pyruvate kinase; (i) acetyl-CoA synthetase, (Roughan & Ohlrogge, 1994); (j) acetyl-CoA hydrolase activity (Liedvogel & Stumpf, 1982; Murphy & Stumpf, 1981; Zeiher & Randall, 1990); (k) (Fischer & Weber, 2002); (l) tricarboxylate transporter, (Laloi, 1999; McIntosh & Oliver, 1992); (m) citrate uptake observed in isolated chloroplasts, (Masterson et al., 1990a); (n) ATP citrate lyase; (o) chloroplast carnitine acetyl-transferase (Masterson & Wood, 2000; Wood et al., 1992); (p) mitochondrial carnitine acetyltransferase (Burgess & Thomas, 1986; Schwadbedissen-Gerbling & Gerhardt, 1995; Thomas & Wood, 1982); (q) mitochondrial carnitine/acyl carnitine translocase (Burgess & Thomas, 1986; Wood et al., 1992); (r) chloroplast carnitine/acyl carnitine translocase (Wood et al., 1992); (s) pyruvate transport protein (Laloi, 1999; Proudlove & Moore, 1982).

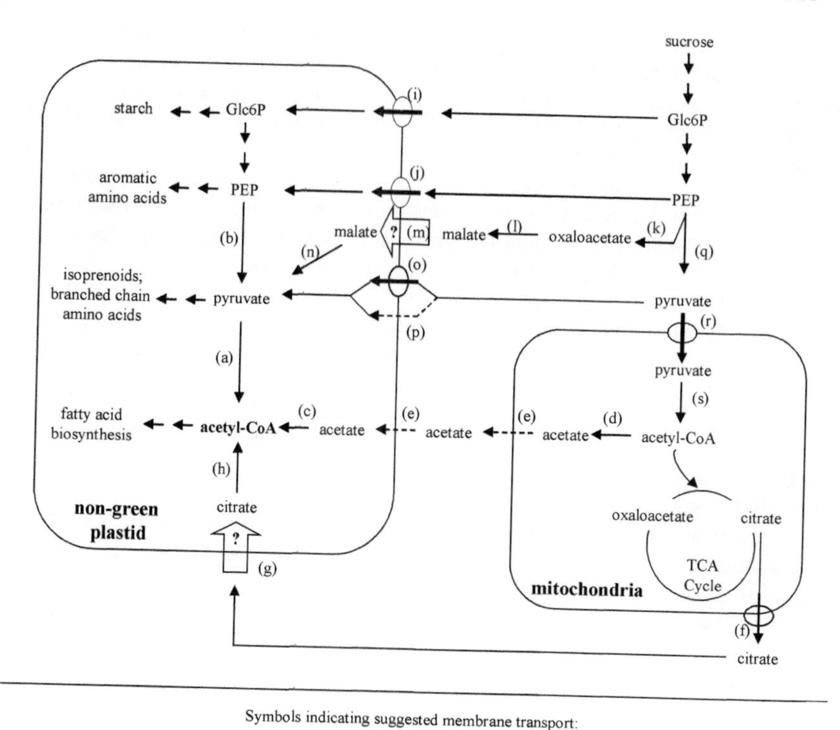

***Figure 4**. Enzymatic reactions relevant to Acetyl-CoA in non-green plastids.**

*Illustration of the multiple reactions proposed to contribute to acetyl-CoA supply in non-green plastids. See text (Section 4) for discussion of most probable pathways for carbon flow. Figure abbreviations: PEP, phosphoenolpyruvate; Glc6P, glucose-6-phosphate. Pathways and enzymatic conversions: (a) plastidial pyruvate dehydrogenase complex; (b) plastidial pyruvate kinase; (c) acetyl-CoA synthetase; (d) acetyl-CoA hydrolase activity?; (e) acetate transport, (Fischer & Weber, 2002); (f) tricarboxylate transporter, (Laloi, 1999; McIntosh & Oliver, 1992); (g) citrate uptake unknown (Fischer & Weber, 2002); (h) ATP citrate lyase; (i) glucose-6-phosphate-phosphate translocator (Bowsher & Tobin, 2001; Fischer & Weber, 2002; Kammerer et al., 1998); (j) PEP/phosphate translocator; (Bowsher & Tobin, 2001; Fischer et al., 1997; Fischer & Weber, 2002); (k) phosphoenolpyruvate carboxylase; (l) malate dehydrogenase; (m) proposed transport in castor bean endosperm, (Eastmond et al., 1997; Smith et al., 1992); (n) malic enzyme; (o) transport proposed for oil seed rape embryos, (Eastmond & Rawsthorne, 2000); (p) transport proposed for leucoplasts of castor bean, (Eastmond et al., 1997); (q) cytosolic pyruvate kinase; (r) pyruvate transport protein (Laloi, 1999; Proudlove & Moore, 1982); (s) mitochondrial pyruvate dehydrogenase

While a correlation between ATP citrate lyase activity has been observed with both lipid formation and acetyl-CoA carboxylase activity in developing seeds of *Brassica napus* (Ratledge et al., 1997), similar studies in developing seeds of *Arabidopsis* have reported little correlation with acetyl-CoA carboxylase and ATP citrate lyase expression (Nikolau et al., 2000). The subcellular location of ATP citrate lyase activity has also been debated. ATP citrate lyase has been reported to be predominantly located in the chloroplast in rape and spinach leaves (Rangasamy & Ratledge, 2000a), predominantly cytosolic in pea (Kaethner & ap Rees, 1985; Rangasamy & Ratledge, 2000a) and tobacco leaves (Rangasamy & Ratledge, 2000a), and completely cytosolic in pea seedlings (Fatland et al., 2000; Fatland et al., 2002). Other studies with developing seeds of *Brassica napus* (Ratledge et al., 1997) and germinating castor bean endosperm (Fritsch & Beevers, 1979) have detected activity in gradient fractions corresponding to both plastid and cytosolic proteins. The recently isolated DNA sequences of the ATP citrate lyase genes from *Arabidopsis* do not encode plastid targeting signals (Fatland et al., 2002). A role for the enzyme in cytosolic acetyl-CoA formation has recently been proposed (Fatland et al., 2002).

Whether or not endogenous ATP citrate lyase activity is present within plastids and contributes to acetyl-CoA formation, this activity may still be a good target for engineering increased plastidial acetyl-CoA levels. Expression of a transgene encoding the cytosolic rat ATP citrate lyase, engineered to encode an N-terminal chloroplast targeting signal, increased fatty acid levels in tobacco leaves by 16%, suggesting the presence of increased acetyl-CoA levels upon expression of the transgene (Rangasamy & Ratledge, 2000b).

4.4. Acetylcarnitine

Chloroplast carntine acetyltransferase (Figure 3, reaction o) has been proposed to convert acetylcarnintine, originating in the mitochondria and transported to the chloroplast (Figure 3, reactions p, q, r), to plastidial acetyl-CoA (Masterson et al., 1990a, 1990b; Wood et al., 1992). The presence of carnitine acetyltransferase in chloroplasts has been debated, with some studies reporting measurable enzyme activity in the chloroplasts of young peas (Masterson & Wood, 2000; McLaren et al., 1985) and others reporting no activity in chloroplasts isolated from pea shoots or spinach leaves (Roughan et al., 1993). Experiments comparing rates of L-acetylcarnitine incorporation into fatty acids in isolated chloroplasts have also yielded conflicting results. L-acetylcarnitine was observed to have the highest rate of incorporation into fatty acids in an *in vitro* study in which citrate, pyruvate, acetate and L-acetylcarnitine were fed to isolated pea leaf chloroplasts (Masterson et al., 1990a). Yet, acetyl-carnitine possessed significantly lower rates of incorporation than acetate in a separate study using isolated chloroplasts from spinach leaves, pea shoots, amaranthus leaves and maize leaves (Roughan et al., 1993). Clearly more work is needed, preferably with *in vivo* studies, to clarify the role of acetylcarnitine in plastidial acetyl-CoA formation.

4.5. Discussion

Conversion of pyruvate to acetyl-CoA is currently the favored pathway for plastidial acetyl-CoA formation in the literature (Bao et al., 2000; Fischer & Weber, 2002; Ke et al., 2000; Nikolau et al., 2000; Ohlrogge et al., 2000; Rawsthorne, 2002). However, additional *in vivo* studies are required to verify the roles of all the potential acetyl-CoA precursors. If pyruvate is a predominant precursor to acetyl-CoA biosynthesis in plastids, pathways that consume pyruvate and its precursor, phosphoenolpyruvate (PEP) (Figures 3 & 4), must be considered as potential pathways competing with PHB synthesis for carbon. Pyruvate is used within the plastids for branched chain amino acid formation and isoprenoid synthesis whereas PEP is used by the shikimate pathway for the production of aromatic amino acids and related aromatic compounds.

5. CONCLUSIONS

The production of PHAs in plastids illustrates the utility of this organelle for novel biopolymer production. The highest levels of PHA production in transgenic plants have been achieved in plastids, perhaps due to the natural presence of acetyl-CoA for fatty acid biosynthesis. Attempts to produce PHAs in other cellular compartments, including the cytosol and the peroxisome, have not yielded significant PHA production (Snell & Peoples, 2002). Future efforts to optimize PHA production in plastids for commercial use will require the formation of economical levels of polymer in crops of agronomic interest. High levels of PHB have been produced within plastids of transgenic *Arabidopsis*, however it has been more challenging to produce comparable levels in crop plants. Whether this is due to differences in achieving efficient gene expression or to inherent metabolic differences between plant types is currently not understood. Recent advances have been made with methodology for metabolic profiling (Sumner et al., 2003; Trethewey, 2004) and detection of metabolites related to PHA biosynthesis (Larson & Graham, 2001). Metabolic profiling of subcellular fractions has also been demonstrated (Farré et al., 2001) by combining previously developed methods for non-aqueous fractionation (Heineke et al., 1997; Stitt et al., 1989) with gas chromatography-mass spectrometry (GC/MS) metabolite analysis (Fiehn et al., 2000; Roessner, 2000). These developments should allow a more detailed study of PHA production in plants providing a greater understanding of metabolic issues that need to be addressed in future engineering efforts.

6. REFERENCES

Abdullah, R., Joseph, J., Rashdan, Y., & Azma, S. (2003). Metabolic Engineering of Oil Palm for Bioplastics Production. Paper presented at the 7th International Congress on Plant Molecular Biology (ISPMB 2003), Barcelona, Spain.

Arai, Y., Nakashita, H., Doi, Y., & Yamaguchi, I. (2001). Plastid Targeting of Polyhydroxybutyrate Biosynthetic Pathway in Tobacco. *Plant Biotechnology 18,* 289-293.

Ball, S. G., & Morell, M. K. (2003). From Bacterial Glycogen to Starch: Understanding the Biogenesis of the Plant Starch Granule. *Annu. Rev. Plant Biol. 54,* 207-233.

Bao, X., Focke, M., Pollard, M., & Ohlrogge, J. (2000). Understanding *in vivo* carbon precursor supply for fatty acid synthesis in leaf tissue. *Plant J. 22,* 39-50.

Behal, R. H., Lin, M., Back, S., & Oliver, D. J. (2002). Role of acetyl-coenzyme A synthetase in leaves of *Arabidopsis thaliana. Arch. Biochem. Biophys. 402,* 259-267.

Bohmert, K., unpublished data.

Bohmert, K., Balbo, I., Kopka, J., Mittendorf, V., Nawrath, C., Poirier, Y., Tischendorf, G., Trethewey, R. N., & Willmitzer, L. (2000). Transgenic *Arabidopsis* plants can accumulate polyhydroxybutyrate to up to 4% of their fresh weight. *Planta 211,* 841-845.

Bohmert, K., Balbo, I., Steinbuchel, A., Tischendorf, G., & Willmitzer, L. (2002). Constitutive Expression of the b-Ketothiolase Gene in Transgenic Plants. A Major Obstacle for Obtaining Polyhydroxybutyrate-Producing Plants. *Plant Physiol. 128,* 1282-1290.

Bowsher, C. G., & Tobin, A. K. (2001). Compartmentation of metabolism within mitochondria and plastids. *J. Exp. Bot. 52,* 513-527.

Broun, P., & Somerville, C. (1997). Accumulation of ricinoleic, lesquerolic, and densipolic acids in seeds of transgenic *Arabidopsis* plants that express a fatty acid acyl hydroxylase cDNA from castor bean. *Plant Physiol. 113,* 933-942.

Brumbley, S. M., Petrasovits, L. A., Bonaventura, P. A., O' Shea, M. J., Purnell, M. P., & Nielsen, L. K. (2003, April 7 -14, 2003). Production of polyhydroxyalkanoates in sugarcane. Paper presented at the ISSCT Molecular Biology Workshop, Montpelier, France.

Burgess, N., & Thomas, D. R. (1986). Carnitine acetyltransferase in pea cotyledon mitochondria. *Planta 167,* 58-65.

Daae, E. B., Dunnill, P., Mitsky, T. A., Padgette, S. R., Taylor, N. B., Valentin, H. E., & Gruys, K. J. (1999). Metabolic Modeling as a Tool for Evaluating Polyhydroxyalkanoate Copolymer Production in Plants. *Metab. Eng. 1,* 243-254.

Daniell, D., Carmona-Sanchez, O., & Burns, B. B. (2004). Chloroplast derived antibodies, biopharmaceuticals and edible vaccines. In R. Fischer & S. Schillberg (Eds.) *Molecular Farming* (pp.113-133). Weinheim: WILEY-VCH Verlag.

DeCosa, B., Moar, W., Lee, S. B., Miller, M., & Daniell, H. (2001). Overexpression of the *Bt* cry2Aa2 operon in chloroplasts leads to formation of insecticidal crystals. *Nat. Biotechnol., 19,* 71-74.

Devine, A. L., & Daniell, H. (2004). Chloroplast genetic engineering. In S. Moller (Ed.), *Plastids* (pp. 283-320). United Kingdom: Blackwell Publishers.

Dhingra, A., Portis, A.R. & Daniell, H. (2004). Enhanced translation of a chloroplast expressed *RbcS* gene restores SSU levels and photosynthesis in nuclear antisense *RbcS* plants. *Proc. Natl. Acad. Sci.,* U.S.A.101: 6315-6320.

Eastmond, P. J., Dennis, D. T., & Rawsthorne, S. (1997). Evidence that a malate/inorganic phosphate exchange translocator imports carbon across the leucoplast envelope for fatty acid synthesis in developing castor seed endosperm. *Plant Physiol. 114,* 851-856.

Eastmond, P. J., & Rawsthorne, S. (2000). Coordinate changes in carbon partitioning and plastidial metabolism during the development of oilseed rape embryos. *Plant Physiol. 122,* 767-774.

Edwards, A., Fulton, D. C., Hylton, C. M., Jobling, S. A., Gidley, M., Rossner, U., Martin, C., & Smith, A. M. (1999). A combined reduction in activity of starch synthases II and III of potato has novel effects on the starch of tubers. *Plant J. 17,* 251-261.

Eibl, C., Zou, Z., Beck, A., Kim, M., Mullet, J., & Koop, H.-U. (1999). In vivo analysis of plastid psbA, rbcL and rpl32 UTR elements by chloroplast transformation: tobacco plastid gene expression is controlled by modulation of transcript levels and translation efficiency. *Plant J. 19*(3), 333-345.

Eisenstein, E., Yu, H. D., Fisher, K. E., Iacuzio, D. A., Ducote, K. R., & Schwarz, F. P. (1995). An expanded two-state model accounts for homotropic cooperativity in biosynthetic threonine deaminase from *Escherichia coli. Biochemistry 34,* 9403-9412.

Farré, E. M., Tiessen, A., Roessner, U., Geigenberger, P., Trethewey, R. N., & Willmitzer, L. (2001). Analysis of the compartmentation of glycolytic intermediates, nucleotides, sugars, organic acids, amino acids and sugar alcohols in potato tubers using a non-aqueous fractionation method. *Plant Physiol. 127,* 685-700.

Fatland, B., Anderson, M., Nikolau, B. J., & Wurtele, E. S. (2000). Molecular biology of cytosolic acetyl-CoA generation. *Biochem. Soc. Trans. 28,* 593-595.

Fatland, B. L., Ke, J., Anderson, M. D., Mentzen, W. I., Cui, L. W., Allred, C. C., Johnston, J. L., Nikolau, B. J., & Wurtele, E. S. (2002). Molecular Characterization of a Heteromeric ATP-Citrate Lyase That Generates Cytosolic Acetyl-Coenzyme A in *Arabidopsis. Plant Physiol. 130,* 740-756.

Feng, L., Watanabe, T., Wang, Y., Kichise, T., Fukuchi, T., Chen, G.-Q., Doi, Y., & Inoue, Y. (2002). Studies on Comonomer Compositional Distribution of Bacterial Poly(3-hydroxybutyrate-*co*-3-hydroxyhexanoate)s and Thermal Characteristics of Their Factions. *Biomacromolecules 3,* 1071-1077.

Fernandez-San Millan, A., Mingeo-Castel, A. M., Miller, M., & Daniell, H. (2003). A chloroplast transgenic approach to hyper-express and purify human serum albumin, a protein highly susceptible to proteolytic degradation. *Plant Biotechnology Journal 1,* 71-79.

Fiehn, O., Kopka, J., Doermann, P., Altmann, T., Trethewey, R. N., & Willmitzer, L. (2000). Metabolic profiling for plant functional genomics. *Nat. Biotechnol. 18,* 1157-1161.

Fischer, K., Kammerer, B., Gutensohn, M., Arbinger, B., Weber, A., Hausler, R. E., & Flugge, U.-I. (1997). A new class of plastidic phosphate translocators: a putative link between primary and secondary metabolism by phosphoenolpyurvate/phosphate antiporter. *Plant Cell 9,* 453-462.

Fischer, K., & Weber, A. (2002). Transport of carbon in non-green plastids. *Trends Plant Sci. 7,* 345-351.

Flugge, U.-I., Fischer, K., Gross, A., Sebald, W., Lottspeich, F., & Eckerskorn, C. (1989). The triose phosphate-3-phosphoglycerate-phosphate translocator from spinach chloroplasts: nucleotide sequence of a full-length cDNA clone and import of the *in vitro* synthesized precursor protein into chloroplasts. *EMBO J 8,* 39-46.

Fritsch, H., & Beevers, H. (1979). ATP: Citrate lyase from germinating castor bean endosperm. *Plant Physiol.* 63, 687-691.

Gerrits, N., Turk, S. C. H. J., van Dun, K., Hulleman, S. H. D., Visser, R. G. F., Weisbeek, P. J., & Smeekens, S. C. M. (2001). Sucrose metabolism in plastids. *Plant Physiol. 125,* 926-934.

Giroux, M. J., Shaw, J., Barry, G., Cobb, B. G., Greene, T., Okita, T., & Hannah, L. C. (1996). A single gene mutation that increases maize seed weight. *Proc. Natl Acad. Sci. USA 93,* 5824-5829.

Gruys, K. J. (2000, July 17-19). Metabolic Engineering and Multigene Transformation of Plants for Bioplastic Production. Paper presented at the Plant Protein Club, Added Value Products from Plants, 1: Starches, Proteins and Plastics, York, UK.

Guda, C., Lee, S.-B., & Daniell, H. (2000). Stable expression of a biodegradable protein-based polymer in tobacco chloroplasts. *Plant Cell Rep. 19,* 257-262.

Guest, J. R., & Stephens, P. E. (1980). Molecular Cloning of the Pyruvate Dehydrogenase Complex Genes of *Escherichia coli. J. Gen. Microbiol. 121,* 277-292.

Heineke, D., Lohaus, G., & Winter, H. (1997). Compartmentation of C/N Metabolism. In C. H. Foyer & W. P. Quick (Eds.), *A Molecular Approach to Primary Metabolism in Higher Plants* (pp. 205-217). London: Tailor & Francis Ltd.

Heintze, A., Gorlach, J., Leuschner, C., Hoppe, P., Hagelstein, P., Schulze-Siebert, D., & Schultz, G. (1990). Plastidic isoprenoid synthesis during chloroplast development. Change from metabolic autonomy to division-of-labor stage. *Plant Physiol 93,* 1121-1127.

Houmiel, K. L., Slater, S., Broyles, D., Casagrande, L., Colburn, S., Gonzalez, K., Mitsky, T. A., Reiser, S. E., Shah, D., Taylor, N. B., Tran, M., Valentin, H. E., & Gruys, K. J. (1999). Poly(beta-hydroxybutyrate) production in oilseed leukoplasts of Brassica napus. *Planta 209,* 547-550.

Jobling, S. (2004). Improving starch for food and industrial applications. *Curr. Opin. Plant Biol. 7,* 210-218.

Jobling, S. A., Jarman, C., Teh, M. M., Holmberg, N., Blake, C., & Verhoeyer, M. E. (2003). Immunomodulation of enzyme function in plants by single-domain antibody fragments. *Nat Biotechnol. 21,* 77-80.

Jobling, S. A., Schwall, G. P., Westcott, R. J., Sidebottom, C. M., Debet, M., Gidley, M. J., Jeffcoat, R., & Safford, R. (1999). A minor form of starch branching enzyme in potato (*Solanum tuberosum* L.) tubers has a major effect on starch structure: cloning and characterisation of multiple forms of SBE A. *Plant J. 18,* 163-171.

Jobling, S. A., Westcott, R. J., Tayal, A., Jeffcoat, R., & Schwall, G. P. (2002). Production of a freeze-thaw-stable potato starch by antisense inhibition of three starch synthase genes. *Nat. Biotechnol. 20,* 295-299.

Johnston, M. L., Luethy, M. H., Miernyk, J. A., & Randall, D. D. (1997). Cloning and molecular analyses of the *Arabidopsis thaliana* plastid pyruvate dehydrogenase subunits. Biochim. Biophys. *Acta 1321,* 200-206.

Jurasek, L., & Marchessault, R. H. (2002). The Role of Phasins in the Morphogenesis of Poly(3-hyrdroxybutyrate) Granules. *Biomacromolecules 3,* 256-261.

Kaethner, T. M., & ap Rees, T. (1985). Intracellular location of ATP citrate lyase in leaves of *Pisum sativum* L. *Planta 163*, 290-294.

Kammerer, B., Fischer, K., Hilpert, B., Schubert, S., Gutensohn, M., Weber, A., & Flugge, U.-I. (1998). Molecular characterization of a carbon transporter in plastids from heterotrophic tissues: the glucose-6-phosphate/phosphate antiporter. Plant *Cell 10,* 105-117.

Ke, J., Behal, R. H., Back, S. L., Nikolau, B. J., Wurtele, E. S., & Oliver, D. J. (2000). The role of pyruvate dehydrogenase and acetyl-CoA synthetase in fatty acid synthesis in developing *Arabidopsis* seeds. *Plant Physiol. 123*, 497-508.

Kimura, T., Otani, M., Noda, T., Ideta, O., Shimada, T., & Saito, A. (2001). Absense of amylose in sweet potato (*Ipomea batatas* (L.) Lam.) following the introduction of granule-bound starch synthase I cDNA. *Plant Cell Rep. 20,* 663-666.

Kok-Jacon, G. A., Ji, Q., Vincken, J.-P., & Visser, R. G. F. (2003). Towards a more versatile a-glucan biosynthesis in plants. *J. Plant Physiol. 160* (765-777).

Kortstee, A. J., Vermeesch, A. M. B., de Vries, B. J., Jacobsen, E., & Visser, R. G. F. (1996). Expression of the *Escherichia coli* branching enzyme in tubers of amylose-free transgenic potato leads to an increased branching degree of amylopectin. *Plant J. 10*, 83-90.

Kuipers, A. G. J., Jacobsen, E., & Visser, R. G. F. (1994). Formation and deposition of amylose in the potato tuber starch granules are affected by the reduction of granule-bound starch synthase gene expression. *Plant Cell 6,* 43-52.

Laloi, M. (1999). Plant mitochondrial carriers: an overview. *Cell. Mol. Life Sci. 56*, 918-944.

Larson, T. R., & Graham, I. A. (2001). Technical Advance: A novel technique for the sensitive quantification of acyl CoA esters from plant tissues. *Plant J. 25*, 115-125.

Leustek, T., & Saito, K. (1999). Sulfate Transport and assimilation in plants. *Plant Physiol. 120*, 637-644.

Liedvogel, B., & Stumpf, P. K. (1982). Origin of acetate in spinach leaf cell. *Plant Physiol 69*, 897-903.

Lloyd, J. R., Landschutze, V., & Kossmann, J. (1999). Simultaneous antisense inhibition of two starch-synthase isoforms in potato tuber leads to accumulation of grossly modified amylopectin. *Biochem J. 338*, 515-521.

Lorberth, R., Ritte, G., Willmitzer, L., & Kossmann, J. (1998). Inhibition of a starch-granule-bound protein leads to modified starch and repression of cold sweetening. *Nat. Biotechnol. 16*, 473-477.

Lössl, A., Eibl, C., Dovzhenko, A., Winterholler, P., & Koop, H.-U. (2000, September 11-15). Production of Polyhydroxybutyric Acid (PHB) Using Chloroplast Transformation. Paper presented at the The 8th International Symposium of Biological Polyesters, Cambridge, MA, USA.

Lössl, A., Eibl, C., Harloff, H. J., Jung, C., & Koop, H.-U. (2003). Polyester Synthesis in Transplastomic Tobacco (*Nicotiana tabacum* L.): Significant Contents of Polyhydroxybutyrate are Associated with Growth Reduction. *Plant Cell Rep. 21,* 891-899.

Lutziger, I., & Oliver, D. J. (2000). Molecular evidnece of a unique lipoamide dehydrogenase in plastids: analysis of plastidic lipoamide dehydrogenase from *Arabidopsis thaliana. FEBS Lett. 484*, 12-16.

Madison, L. L., & Huisman, G. W. (1999). Metabolic Engineering of Poly(3-Hydroxyalkanoates): From DNA to Plastic. Microbiol. *Mol. Biol. Rev. 63*, 21-53.

Martini, N., Egen, M., Runtz, I., & Strittmatter, G. (1993). Promoter sequences of a potato pathogenesis-related gene mediate transcriptional activation selectively upon fungal infection. *Mol. Gen. Genet. 236*, 179-186.

Masterson, C., & Wood, C. (2000). Pea chloroplast carnitine acetyltransferase. *Proc. R. Soc. Lond. B 267*, 1-6.

Masterson, C., Wood, C., & Thomas, D. R. (1990a). L-acetylcarnitine, a substrate for chloroplast fatty acid synthesis. *Plant Cell Environ. 13*, 755-765.

Masterson, C., Wood, C., & Thomas, D. R. (1990b). Inhibition studies on acetyl group incorporation into chloroplast fatty acids. *Plant Cell Environ. 13*, 767-771.

McIntosh, C. A., & Oliver, D. J. (1992). Isolation and characterization of the tricarboxylate transporter from pea mitochondria. *Plant Physiol. 100*, 2030-2034.

McLaren, I., Wood, C., Jalil, M. N. H., Yong, B. C. S., & Thomas, D. R. (1985). Carnitine acyltransferases in pea chloroplasts. *Planta 163*, 197-200.

Menzel, G., Harloff, H.-J., & Jung, C. (2003). Expression of bacterial poly(3-hydroxybutyrate) synthesis genes in hairy roots of sugar beet (*Beta vulgaris* L.). *Appl. Microbiol. Biotechnol.* 60, 571-576.

Mitsky, T. A., Slater, S., Reiser, S. E., Hao, M., & Houmiel, K. L. (2000). Multigene expression vectors for the biosynthesis of products via multienzyme biological pathways. PCT application WO 00/52183.

Mittendorf, V., Leisse, A., Bohmert, K., & Willmitzer, L. (2000). Diversion of plastidial fatty acid biosynthesis intermediates for MCL PHA biosynthesis in *Arabidopsis*. Presentation at The 8th Annual International Symposium on Biological Polyesters, Sept. 11-15, 2000, Cambridge, MA.

Moffat, A. S. (1999). Plant Biotechnology: Food and Feed, Crop Engineering Goes South. *Science 285*, 370-371.

Mooney, B. P., Miernyk, J. A., & Randall, D. D. (1999). Cloning and characterization of the dihydrolipoamide S-acetyltransferase subunit of the plastid pyruvate dehydrogenase complex (E2) from *Arabidopsis*. *Plant Physiol. 120*, 443-452.

Morell, M. K., Kosar-Hashemi, B., Cmiel, M., Samuel, M. S., Chandler, P., Rahman, S., Buleon, A., Batey, I. L., & Li, Z. (2003). Barley *sex6* mutants lack starch synthase IIa activity and contain a starch with novel properties. *Plant J. 34*, 173-185.

Murphy, D. J., & Leech, R. M. (1981). Photosynthesis of Lipids from $^{14}CO_2$ in *Spinacia oleracea*. *Plant Physiol 68*, 762-765.

Murphy, D. J., & Stumpf, P. K. (1981). The origin of chloroplastic acetyl-CoA. *Arch Biochem Biophys* 212, 730-739.

Nakashita, H., Arai, Y., Shikanai, T., Doi, Y., & Yamaguchi, I. (2001). Introduction of Bacterial Metabolism into Higher Plants by Polycistronic Transgene Expression. Biosci. *Biotechnol. Biochem.* 65, 1688-1691.

Nawrath, C., Poirier, Y., & Somerville, C. (1994). Targeting of the polyhydroxybutyrate biosynthetic pathway to the plastids of Arabidopsis thaliana results in high levels of polymer accumulation. *Proc. Natl. Acad. Sci. USA 91*, 12760-12764.

Nawrath, C., Poirier, Y., and Somerville, C. (1995). Plant polymers for biodegradable plastics: cellulose, starch and polyhydroxyalkanoates. *Mol. Breeding 1*, 105-122.

Nikolau, B. J., Oliver, D. J., Schnable, P. S., & Wurtele, E. S. (2000). Molecular Biology of Acetyl-CoA Metabolism. *Biochem. Soc. Trans. 28*, 591-593.

Oakes, J. V., Shewmaker, C. K., & Stalker, D. M. (1991). Production of cyclodextrins, a novel carbohydrate, in the tubers of transgenic potato plants. *Biotechnol. 9*, 982-986.

Ohlrogge, J., Pollard, M., Bao, X., Focke, M., Girke, T., Ruuska, S., Mekhedov, S., & Benning, C. (2000). Fatty acid synthesis: from CO_2 to functional genomics. Biochem. Soc. Trans. 28, 567-573.

op den Camp, R. G., & Kuhlemeier, C. (1997). Aldehyde dehydrogenase in tobacco pollen. *Plant. Mol. Biol. 35*, 355-365.

Poirier, Y., Dennis, D., Klomparens, K., Nawrath, C., & Somerville, C. (1992b). Perspectives on the production of polyhydroxyalkanoates in plants. *FEMS Microbiology Reviews 103*, 237-246.

Poirier, Y., Dennis, D. E., Klomparens, K., & Somerville, C. (1992a). Polyhydroxybutyrate, a Biodegradable Thermoplastic, Produced in Transgenic Plants. *Science 256*, 520-523.

Poirier, Y., & Gruys, K. J. (2002). Production of polyhydroxyalkanoates in transgenic plants. In A. Steinbüchel (Ed.), *Biopolymers* (Vol. 3a, pp. 401-435). Weinheim: Wiley-VHC Verlag GmbH.

Proudlove, M. O., & Moore, A. L. (1982). Movement of amino acids into isolated plant mitochondria. *FEBS Lett. 147*, 26-30.

Proudlove, M. O., & Thurman, D. A. (1981). The uptake of 2-oxoglutarate and pyruvate by isolated pea chloroplasts. *New Phytol. 88*, 255-264.

Rangasamy, D., & Ratledge, C. (2000a). Compartmentation of ATP:Citrate Lyase in Plants. *Plant Physiol. 122*, 1225-1230.

Rangasamy, D., & Ratledge, C. (2000b). Genetic Enhancement of Fatty Acid Synthesis by Targeting Rat Liver ATP:Citrate Lyase into Plastids of Tobacco. *Plant Physiol. 122*, 1231-1238.

Ratledge, C., Bowater, M. D. V., & Taylor, P. N. (1997). Correlation of ATP: citrate lyase activity with lipid accumulation in developing seeds of *Brassica napus L. Lipids 32*, 7-12.

Rawsthorne, S. (2002). Carbon flux and fatty acid synthesis in plants. *Prog. Lipid Res. 41*, 182-196.

Regierer, B., Fernie, A. R., Springer, F., Perez-Melis, A., Leisse, A., Koehl, K., Willmitzer, L., Geigenberger, P., & Kossmann, J. (2002). Starch content and yield increase as a result of altering adenylate pools in transgenic plants. *Nat. Biotechnol. 20*, 1256-1260.

Ritsema, T., & Smeekens, S. (2003). Fructans: beneficial for plants and humans. *Curr. Opin. Plant Biol. 6*, 223-230.

Ritte, G., Lloyd, J. R., Eckermann, N., Rottman, A., Kossmann, J., & Steup, M. (2002). The starch-related R1 protein is an a-glucan, water dikinase. *Proc. Natl. Acad. Sci. USA 99*, 7166-7171.

Roessner, U. (2000). Simultaneous analysis of metabolites in potato tuber by gas chromatography-mass spectrometry. *Plant J 23*, 131-142.

Romano, A., de Roo, G., Vreugdenhil, D., van der Plas, L. H. W., Witholt, B., Eggink, G., & Mooibroek, H. (2002). Accumulation of medium-chain length polyhydroxyalkanoate in intact transgenic potato plants. Presentation at The 9th Annual International Symposium on Biological Polyesters, Sept. 22-26, 2002, Munster, Germany.

Roughan, G., Post-Beittenmiller, D., Ohlrogge, J., & Browse, J. (1993). Is Acetylcarnitine a Substrate for Fatty Acid Synthesis in Plants? *Plant Physiol. 101*, 1157-1162.

Roughan, P. G. (1995). Acetate concentrations in leaves are sufficient to drive in vivo fatty acid synthesis at maximum rates. *Plant Sci 107*, 49-55.

Roughan, P. G., & Ohlrogge, J. B. (1994). On the assay of acetyl-CoA synthetase activity in chloroplasts and leaf extracts. *Anal. Biochem. 216*, 77-82.

Safford, R., Jobling, S. A., Sidebottom, C. M., Westcott, R. J., Cooke, D., Tober, K. J., Strongitharm, B. H., Russell, A. L., & Gidley, M. J. (1998). Consequences of antisense RNA inhibtion of starch branching enzyme activity on properties of potato starch. *Carbohyd. Polym. 35,* 155-168.

Sambanthamurthi, R., Sundram, K., & Tan, Y.-A. (2000). Chemistry and biochemistry of palm oil. Prog. Lipid Res. 39, 507-558.

Saruul, P., Srienc, F., Somers, D. A., & Samac, D. A. (2002). Production of a Biodegradable Plastic Polymer, Poly-b-Hydroxybutyrate, in Transgenic Alfalfa. *Crop Sci. 42,* 919-927.

Satoh, H., Nishi, A., Yamashita, K., Takemoto, Y., Tanaka, Y., Hosaka, Y., Sakurai, A., Fujita, N., & Nakamura, Y. (2003). Starch-branching enzyme-I-deficient mutation specifically affects the structure and properties of starch in rice endosperm. *Plant Physiol. 133,* 1111-1121.

Schwadbedissen-Gerbling, H., & Gerhardt, B. (1995). Purification and Characterization of Carnitine Acyltransferase from Higher Plant Mitochondria. *Phytochem. 39,* 39-44.

Schwall, G. P., Safford, R., Westcott, R. J., Jeffcoat, R., Tayal, A., Shi, Y.-C., Gidley, M. J., & Jobling, S. A. (2000). Production of very-high-amylose potato starch by inhibition of SBE A and B. *Nat. Biotechnol. 18*, 551-554.

Schwender, J., & Olhlrogge, J. B. (2002). Probing in Vivo Metabolism by Stable Isotope Labeling of Storage Lipids and Proteins in Developing *Brassica napus* Embryos. *Plant Physiol. 130*, 347-361.

Shewmaker, C. K., Boyer, C. D., Wiesenborn, D. P., Thompson, D. B., Boersig, M. R., Oakes, J. V., & Stalker, D. M. (1994). Expression of *Escherichia coli* glycogen synthase in the tubers of transgenic potatoes (*Solanum tuberosum*) results in highly branched starch. *Plant Physiol. 104*, 1159-1166.

Slater, S., Houmiel, K., Tran, M., Mitsky, T. A., Taylor, N. B., Padgette, S. R., & Gruys, K. J. (1998). Multiple b-ketothiolases mediate poly(b-hydroxyalkanoate) copolymer synthesis in *Ralstonia eutropha. J. Bacteriol. 180*, 1979-1987.

Slater, S., Mitsky, T. A., Houmiel, K. L., Hao, M., Reiser, S. E., Taylor, N. B., Tran, M., Valentin, H. E., Rodriguez, D. J., Stone, D. A., Padgette, S. R., Kishore, G., & Gruys, K. J. (1999). Metabolic engineering of Arabidopsis and Brassica for poly(3-hydroxybutyrate-co-3-hydroxyvalerate) copolymer production. *Nat. Biotech. 17*, 1011-1016.

Slattery, C. J., Kavakli, I. H., & Okita, T. W. (2000). Engineering starch for increased quantity and quality. *Trends Plant Sci. 5*, 291-298.

Smidansky, E. D., Clancy, M., Meyer, F. D., Lanning, S. P., Blake, N. K., Talbert, L. E., & Giroux, M. J. (2002). Enhanced ADP-glucose pyrophosphorylase activity in wheat endosperm increases seed yield. *Proc. Natl. Acad. Sci. USA 99,* 1724-1729.

Smith, R. G., Gauthier, D. A., Dennis, D., & Turpin, D. H. (1992). Malate- and pyruvate-dependent fatty acid synthesis in leucoplasts from developing castor endosperm. *Plant Physiol. 98*, 1233-1238.

Snell, K. D., & Peoples, O. P. (2002). Polyhydroxyalkanoate Polymers and Their Production in Transgenic Plants. *Metab. Eng. 4*, 29-40.

Springer, J., & Heise, K.-P. (1989). Comparison of acetate- and pyruvate-dependent fatty-acid synthesis by spinach chloroplasts. *Planta 177*, 417-421.

Stark, D. M., Timmerman, K. P., Barry, G. F., Preiss, J., & Kishore, G. M. (1992). Regulation of the amount of starch in plant tissues by ADP glucose pyrophosphorylase. *Science 258*, 287-292.

Staub, J. & Maliga, P. (1994). Translation of the psbA mRNA is regulated by light via the 5'-untranslated region in tobacco plastids. *Plant J. 6,* 547-553.

Stitt, M., Lilley, R. M., Gerhardt, R., & Heldt, H. W. (1989). Metabolite levels in specific cells and subcellular compartments of plant leaves. *Methods Enzymol. 174*, 518-550.

Sumner, L. W., Mendes, P., & Dixon, R. A. (2003). Plant metabolomics: large-scale phytochemistry in the functional genomics era. *Phytochem. 62,* 817-836.

Suzuki, Y., Kurano, M., Arai, Y., Nakashita, H., Doi, Y., Usami, R., Horikoshi, K., & Yamaguchi, I. (2002). Enzyme Inhibitors to Increase Poly-3-hydroxybutyrate Production by Transgenic Tobacco. Biosci. Biotechnol. *Biochem. 66,* 2537-2542.

Szopa, J., Lukaszewicz, M., Aksamit, A., Korobczak, A., & Kwiatkowska, D. (2003). Structural organisation, expression, and promoter analysis of a 16R isoform of 14-3-3 protein gene from potato. Plant Physiol. *Biochem. 41*, 417-423.

Tadege, M., & Kuhlemeier, C. (1997). Aerobic fermentation during tobacco pollen development. *Plant Mol. Biol. 35*, 343-354.

Taillon, B. E., Little, R., & Lawther, R. P. (1988). Analysis of the functional domains of biosynthetic threonine deaminase by comparison of the amino acid sequences of three wild-type alleles to the amino acid sequence of biodegradative threonine deaminase. *Gene 63*, 245-252.

Taylor, J., King, R. D., Altmann, T., & Fiehn, O. (2002). Application of metabolomics to plant genotype discrimination using statistics and machine learning. *Bioinformatics 18*, 241-248.

Thomas, D. R., & Wood, C. (1982). Oxidation of acetate, acetyl CoA and acetylcarnitine by pea mitochondria. *Planta 154*, 145-149.

Tjaden, J., Möhlmann, T., Kampfenkel, K., Henrichs, G., & Neuhaus, H. E. (1998). Altered plastidic ATP/ADP transporter activity influences potato (*Solanum tuberosum* L.) tuber morphology, yield and composition of tuber starch. *Plant J. 16*, 531-540.

Tovar-Mendez, A., Miernyk, J. A., & Randall, D. D. (2003). Regulation of pyruvate dehydrogenase complex activity in plant cells. *Eur. J. Biochem. 270*, 1043-1049.

Trethewey, R. N. (2004). Metabolite profiling as an aid to metabolic engineering in plants. *Curr. Opin. Plant Biol. 7*, 196-201.

Urry, D. W. (1995). Elastic biomolecular machines. *Sc. Am. 272*, 64-69.

Valentin, H. E., Broyles, D. L., Casagrande, L. A., Colburn, S. M., Creely, W. L., DeLaquil, P. A., Felton, H. M., Gonzalez, K. A., Houmiel, K. L., Lutke, K., Mahadeo, D. A., Mitsky, T. A., Padgette, S. R., Reiser, S. E., Slater, S., Stark, D. M., Stock, R. T., Stone, D. A., Taylor, N. B., Thorne, G. M., Tran, M., & Gruys, K. (1999). PHA production, from bacteria to plants. *Int. J. Biol. Macromol. 25,* 303-306.

Vikso-Nielsen, A., Blennow, A., Jorgensen, K., Kristensen, K. H., Jensen, A., & Moller, B. L. (2001). Structural, physicochemical, and pasting properties of starches from potato plants with repressed *r1-gene. Biomacromolecules 2*, 836-843.

Visser, R. G. F., Somhorst, I., Kuipers, G. J., Ruys, N. J., Feenstra, W. J., & Jacobsen, E. (1991). Inhibition of expression of the gene for granule-bound starch synthase in potato by anti-sense constructs. *Mol. Gen. Genet.s 225*, 289-296.

Wieczorek, R., Pries, A., Steinbuchel, A., & Mayer, F. (1995). Analysis of a 24-Kilodalton Protein Associated with the Polyhydroxyalkanoic Acid Granules in *Alcaligenes eutrophus. J. Bacteriol. 177,* 2425-2435.

Wood, C., Masterson, C., & Thomas, D. R. (1992). The role of carnitine in plant cell metabolism. In A. K. Tobin (Ed.), *Society of Experimental Biology Seminar Series 50: Plant organelles: compartmentation of metabolism in photosynthetic cells*. Cambridge: Cambridge University Press.

Wrobel, M., Zebrowski, J., & Szopa, J. (2004). Polyhydroxybutyrate synthesis in transgenic flax. *J. Biotechnol. 107*, 41-54.

Zeiher, C. A., & Randall, D. D. (1990). Identification and characterization of mitochondrial acetyl coenzyme A hydrolase from *Pisum sativum* L. seedlings. *Plant Physiol. 94*, 20-27.

CHAPTER 22

CYTOPLASMIC MALE STERILITY AND FERTILITY RESTORATION BY NUCLEAR GENES

CHRISTINE D. CHASE[1] AND S. GABAY-LAUGHNAN[2]

[1]*Horticultural Sciences Department, University of Florida Institute of Food and Agricultural Sciences, Gainesville, FL USA*
[2]*Department of Plant Biology, University of Illinois at Urbana-Champaign, Urbana, IL USA*

Abstract. Cytoplasmically inherited male sterility (CMS) the maternally inherited failure to produce or shed functional pollen, has been reported in a wide range of plant species. In many cases, the genetic basis for this trait has been traced to the mitochondrial genome. Although male fertility is affected in all cases, the developmental features are varied, as are the molecular mechanisms that produce them. While mitochondrial genomes encode CMS, nuclear *restorer-of-fertility* genes can suppress or compensate for the expression of mitochondrial CMS genes, thereby conditioning male fertility. Molecular mechanisms of fertility restoration are also varied. CMS and fertility restoration provide a unique set of molecular - genetic tools for the investigation of interactions between nuclear and mitochondrial genetic systems.

1. INTRODUCTION

Early observations of cytoplasmically inherited male sterility (CMS) (reviewed by Edwardson, 1956; Duvick, 1959) were made in *Satureja hortensis* (savory) (Correns, 1904; vonWettstein, 1924), *Linum usitatis-simum* (flax) (Bateson and Gairdner, 1921) and *Zea mays* (maize) (Rhoades, 1931; 1933). Rhoades (1950) also observed maternally inherited male sterility while studying plastid mutations arising in the presence of the nuclear *iojap* mutation of maize. Because the maternally inherited male sterility and the maternally inherited mutant plastids were transmitted independently, Rhoades postulated a mitochondrial basis for CMS. The CMS trait was exploited to produce uniform populations of male-sterile plants for the production of commercial F1 hybrid seed (reviewed by Duvick, 1959; Havey, Chapter 23 of this volume). Through the use of CMS-T maize, the link between CMS and mitochondria was further established. CMS-T maize was widely used in the production of U.S. hybrid seed corn until it became apparent that CMS-T plants

H. Daniell and C.D. Chase (eds.), Molecular Biology and Biotechnology of Plant Organelles, 593—621.

were susceptible to fungal pathogens *Bipolaris (Helminothsporium) maydis* race T and *Phyllosticta maydis* (reviewed by Ullstrup, 1972; Pring and Lonsdale, 1989; Ward, 1995; Wise et al., 1999). Hooker et al. (1970) established the susceptibility of CMS-T maize plants to a toxin (HmT) produced by the *B. maydis* race T fungus. HmT was subsequently shown to uncouple oxidative phosphorylation in mitochondria isolated from CMS-T plants. Mitochondria from normal-cytoplasm maize were, however, relatively insensitive to the toxin (Miller and Koeppe, 1971; Gengenbach et al., 1973). Microscopic observations of pollen development in CMS-T and normal-cytoplasm maize revealed mitochondrial abnormalities to be early indicators of male sterility in CMS-T anthers (Warmke and Lee, 1977).

The application of emerging molecular techniques linked CMS to specific mitochondrial genes in two different plant species, confirming Rhoades' original hypothesis. Plant mitochondrial genomes are large and complex with a high degree of intraspecific variation in genome organization (reviewed by Pring and Lonsdale, 1985; Hanson and Folkerts, 1992; Bendich, 1993; Fauron et al., 1995; Backert et al., 1997). Differences between CMS and normal mitochondrial genomes are not always relevant to CMS. Nevertheless, comparisons of mitochondrial genome organization and expression in selected genetic variants revealed mitochondrial loci associated with the CMS trait. A mitochondrial gene (*pcf*) co-segregates with the CMS phenotype in cybrid plants recovered after fusion of protoplasts prepared from male-fertile and CMS genotypes of Petunia (*Petunia parodii* or *P. hybrida*) (Boeshore et al., 1985; Young and Hanson, 1987). A different mitochondria open reading frame (*urf13*) (Dewey et al., 1986) was associated with the CMS-T trait in maize. This association was made by differences in the expression of *urf13* in the presence and absence of nuclear *restorer-of-fertility* alleles (Dewey et al. 1987) and by a cytoplasmic mutation that disrupted *urf13* and conditioned a male-fertile phenotype (Wise et al., 1987a).

The CMS trait has been observed in a wide range of plant species. Different examples of CMS reveal a surprising diversity in the timing, location and nature of developmental events leading to male sterility (reviewed by Laser and Lersten, 1972). Failed pollen development can result from abnormalities in diploid (sporophytic) or haploid (gametophytic) tissues. This phenotypic variation is reflected in a diversity of associated molecular events. The wide-spread occurrence of this trait among seed plants suggests a selective advantage of male-sterility inducing cytoplasms, at least in some circumstances (reviewed by Bailey et al., 2003; Budar et al., 2003; Saur-Jacobs and Wade, 2003).

CMS systems provide a unique opportunity to investigate interactions between mitochondrial and nuclear genetic systems. Although the mitochondrial genome encodes CMS, nuclear *restorer-of-fertility* genes can suppress or compensate for the mitochondrial genotype to condition male fertility (reviewed by Hanson and Conde, 1985; Schnable and Wise, 1998; Wise and Pring, 2002; Hanson, 2004). In some cases, the mitochondrial genotype has been shown to cause male sterility and other floral abnormalities through an influence on nuclear gene expression (reviewed by Zubko, 2004). Genetic mutations that alter mitochondrial function are often lethal in obligate aerobes (reviewed by Newton and Gabay-Laughnan, 1998; Wallace, 2001; Newton et al., 2004). CMS and nuclear fertility restoration systems therefore provide an unusual opportunity to investigate molecular-genetic aspects of mitochondrial function in higher eukaryotes.

2. MITOCHONDRIAL CMS LOCI

2.1. Chimeric CMS genes

Mitochondrial loci associated with CMS are summarized in Table 1. The first CMS genes to be identified, the *pcf* gene of Petunia (Young and Hanson, 1987) and the *urf13* gene of CMS-T maize (Dewey et al., 1986; Wise et al., 1987a) are completely different in sequence. Both are, however, chimeric genes, containing coding and noncoding segments of mitochondrial or unknown origin, spliced together to form functional open reading frames (orfs). In contrast to *pcf*, which includes segments of known protein coding genes ATP synthase subunit 9 (*atp9*) and cytochrome oxidase subunit II (*coxII*) fused to a segment of unknown origin (*urfS*), *urf13* is comprised entirely of sequences that do not ordinarily encode a protein product – 26S rRNA (*rrn26*) gene flanking and coding sequences (Dewey et al., 1986; Wise et al., 1987a). The Petunia and maize CMS genes are both transcribed and translated into protein products (Dewey et al., 1987; Wise et al., 1987b; Nivison and Hanson, 1989) although only the *urfS* portion of the *pcf* translation product is stably accumulated (Nivison and Hanson, 1994). Since the discovery of *pcf* and *urf13*, chimeric mitochondrial genes have been associated with CMS in many different plant species (reviewed by Hanson, 1991; Schnable and Wise, 1998; Hanson, 2004).

In most cases, chimeric orfs associated with CMS are unrelated in sequence. There are, however, some interesting exceptions to this generalization. *Orf224* and *orf222*, associated with the nap and pol male-sterility-inducing cytoplasms of *Brassica napus* (Brassica) predict proteins sharing 79% identity in amino acid sequence (L'Homme et al., 1997). *Orf138* and *orf125*, associated with the ogura (ogu) and kosena (kos) male-sterility-inducing cytoplasms of *Raphanus sativus* (radish), predict proteins identical except for two amino acid substitutions and a thirteen amino acid deletion (Iwabuchi et al., 1999). While all four loci condition male sterility in Brassica, the radish-derived reading frames are unrelated to the nap and pol reading frames. Another interesting case of similarity between CMS-associated orfs involves *orf107* from the A3 cytoplasm of *Sorghum bicolor* (Sorghum) and *orf79* from the BT (also called Bo) cytoplasm of *Oryza sativa* (rice). The 49 C-terminal amino acids predicted by *orf107* share 57% identity with *orf79* (Akagi et al., 1994; Tang et al., 1996).

A number of chimeric CMS genes predict proteins carrying segments derived from one or more mitochondrial genes encoding respiratory chain components. Segments derived from genes encoding subunits of the mitochondrial F1/F0 ATP synthase are common, and the mitochondrial locus associated with petaloid CMS in *Daucus carota* (carrot) predicts a full-length ORFB (ATP8) protein with a C-terminal extension of 57 amino acids (Nakajima et al., 2001). Collectively, chimeric CMS genes support models of CMS resulting from gain-of-function mutations that interfere with normal mitochondrial respiratory machinery or metabolism.

Table 1. Mitochondrial loci associated with CMS phentoypes

Species / cytoplasm	*CMS phenotype*	*Gene or locus*	*Predicted protein features*	*Genome context (5' – 3')*	*Observed protein*	*References*
Allium Schoeno-prasum / CMS_1	Tapetal hypertrophy; microspore degeneration	*orf260*	Amino acids (aa) 13-61 ~ ATP9; C-terminal (ter) aa unknown	*atp9 5' flank / orf260*		(Singh and Kobabe, 1969; Engelke and Tatlioglu, 2002; 2004)
Beta vulgaris ssp. *maritima* / G		*coxII*	COXII minus 8 C-ter aa	*coxII*	31 kDa COXII in leaves	(Ducos et al., 2001)
Brassica napus / nap	No anther locule or meiocytes	*orf222*	58 N-ter aa ~ ATP8	*orf222 / nad5C / orf139 / atp6*		(L'Homme et al., 1997; reviewed by Brown, 1999)
Brassica napus / pol	No anther locule or meiocytes	*orf224*	~ ORF222	*nad5C / orf139 / orf224 / ψtrnfM / atp6*		(Singh and Brown, 1993; Handa et al., 1995; reviewed by Brown, 1999)
Daucus carota	Petaloid or carpeloid anthers	*atp8*	ATP8 with 57 aa C-ter extension	*atp8*	25 kDa in flowers, not in leaves	(Nakajima et al., 2001)
Helianthus annuus / PET1	Tapetal collapse; microspore degeneration	*orf522*	18 N-ter aa ~ ATP8	*atpA / orf522*	15 kDa in seedlings and male florets	(Köhler et al., 1991; Laver et al., 1991; Monéger et al., 1994)
Nicotiana sylvestris / CMS I	Normal tapetum; microspore degeneration	*nad7*	Truncated NAD7	*nad7* exons 3 and 4	No NAD7 or NAD9 in leaves or pollen	(Pla et al., 1995; Gutierres et al., 1997)

Table 1. (cont.)

Table 1. (cont.)

Species / cytoplasm	CMS phenotype	Gene or locus	Predicted protein features	Genome context (5' – 3')	Observed protein	References
Nicotiana sylvestris / CMS II	Normal tapetum; microspore degeneration	*nad7*	Deleted NAD7	*nad7* / *nad1B* / *rrn26* / *orf87* / *nad3*	No NAD7 or NAD9 in leaves or pollen	(Gutierres et al., 1997; Lelandais et al., 1998)
Nicotiana tabacum / *N. repanda*	Stigmatoid anthers; anthers fused to carpels	*orf274*	62 N-ter aa unknown; C-ter ~ ORF209 / 214 bean / soybean	*orf274* / *atpA*		(Bergman et al., 2000; Farbos et al., 2001)
Oryza sativa / BT (Bo)	Gameto-phytic; pollen abortion after mitosis I	*orf79*	28 N-ter aa ~ COXI	*atp6* / *orf79*		(Akagi et al., 1994)
Petunia parodii or *P. hybrida* / CMS	Tapetal degeneration; failure of meiosis	*pcf*	35 N-ter aa ~ ATP9; 158 aa ~ COXII; 157 aa unknown	*pcf* / *nad3* / *rps12*	25 kDa in anthers and cell cultures	(Nivison et al., 1989; 1994; Conley and Hanson, 1994; reviewed by Hanson et al., 1999)
Phaseolus vulgaris / Ci	Failure of callose dissolution; pollen abortion after mitosis I	*orf239*	Unknown	*atpA* / *orf209* / *orf98* / *orf239*	27 kDa in flowers, not in roots, leaves or seedlings	(Chase and Ortega, 1992; Johns et al., 1992; Abad et al., 1995)
Raphanus sativus or *B. napus* / kosena		*orf125*	Unknown	*trnfM* / *orf125* / unknown	17 kDa in flower buds	(Iwabuchi et al., 1999; Koizuka et al., 2000)

Table 1. (cont.)

Table 1. (cont.)

Species / cytoplasm	CMS phenotype	Gene or locus	Predicted protein features	Genome context (5' – 3')	Observed protein	References
Raphanus sativus or *B. napus* / ogura	Tapetal collapse; microspore degeneration	*orf138*	~ ORF125	*trnfM* / *orf138* / *atp8*	20 kDa in flowers, leaves and roots	(Krishnasamy and Makaroff, 1993; Grelon et al., 1994; reviewed by Makaroff, 1995)
Sorghum bicolor / A3	Gametophytic; pollen arrest just before or after mitosis I	*orf107*	31N-ter aa ~ ATP9; 49 aa ~ CMS BT ORF79	*atp9* 5' region / *orf107*		(Tang et al., 1996; reviewed by Pring et al., 1999)
Triticum aestivum / *T. timopheevi*		*orf256*	11 N-ter aa ~ COXI; 245 aa unknown	*coxI* 5' region / *orf256* / *coxI*	7 kDa	(Song and Hedgcoth, 1994a; 1994b)
Zea mays / T	Tapetal degeneration; microspore degeneration	*urf13*	88 aa *rrn26* 3' flank; 9 aa unknown; 18 aa *rrn26*	*atp6* 5' flank / *urf13-T* / *orf25* (*atp4*)	13 kDa in roots, shoots, leaves and anthers	(Dewey et al., 1986; Wise et al., 1987a; Hack et al. , 1991; reviewed by Wise et al., 1999)
Zea mays / S	Gametophytic; pollen collapse after mitosis I	*orf355* / *orf77* (edited to *orf17*)	243 aa R1 episome; 112 aa unknown / 17 N-ter aa ~ ATP9	*coxI* / *orf355* / *orf77* (*orf17*)		(Zabala et al., 1997; Gallagher et al., 2002)

2.2. *The origin and maintenance of chimeric mitochondrial genes*

Chimeric mitochondrial genes are likely the consequence of plant mitochondrial recombination systems (reviewed by Hanson 1991; Hanson and Folkerts, 1992). Normal plant mitochondrial genes are relatively conserved in DNA sequence, but plant mitochondrial genomes evolve rapidly in terms of gene order and organization (Palmer and Herbon, 1988; reviewed by Fauron et al. in Chatper 6 of this volume).

These rearrangements implicate recombination as a key element in plant mitochondrial genome evolution. Physical maps of plant mitochondrial genomes (reviewed by Hanson and Folkerts, 1992; Bendich, 1993; Fauron et al., 1995; Backert et al., 1997) revealed repeated regions, in either direct or inverted orientation. Many of these repeats are substrates for recombination, demonstrated by the presence of subgenomic molecules and by the presence of molecules carrying segments in inverted orientation. Such recombination events result in complex, multipartite genomes.

Of particular relevance to chimeric mitochondrial genes are recombination events between relatively short similar sequences of 10-200 nucleotides. These events are known to generate novel, low-abundance genome configurations termed sublimons (Small et al., 1989). Recombination events of this nature are also likely to be involved in the creation of chimeric mitochondrial genes (reviewed by Hanson and Folkerts, 1992). For example, recombination through a short region of similar sequence within the normal Petunia *atp9* and *coxII* genes would generate the *atp9-coxII* junction observed in the *S-pcf* gene (Pruitt and Hanson, 1991). Furthermore, short regions of similar sequence have been identified in the segments recombined to form many of the chimeric CMS genes listed in Table 1.

Recombination events are also relevant to the maintenance of CMS genes. Recombination is implicated in the loss of *urf13-T* from the maize mitochondrial genome during tissue culture (Rottman et al., 1987; Fauron et al. 1995). Recombination through direct repeats flanking *urf13* excised this gene from the mitochondrial genome. The excised subgenomic fragment was either lost from the genome entirely or retained as a sublimon that is not expressed at levels sufficient to condition CMS.

The factors determining the relative copy number of plant mitochondrial subgenomes are also of relevance to the maintenance and expression of CMS genes. Small et al. (1989) suggest that nuclear genotype determines the relative abundance of the various mitochondrial subgenomes. In *Phaseolus vulgaris* L. (common bean), the nuclear allele *Fr* conditions loss of the mitochondrial subgenome carrying the *orf239* CMS gene (Mackenzie and Chase, 1990; Janska and Mackenzie, 1993; He et al., 1995). This subgenome is expendable because all essential mitochondrial genes it carries are duplicated elsewhere in the genome (Janska and Mackenzie, 1993). While the nature of the *Fr* gene is currently unknown, a nuclear homolog of the *Escherichia coli MutS* gene was recently shown to influence mitochondrial subgenome copy number in *Arabidopsis thaliana* (Arabidopsis) (Abdelnoor et al., 2003).

2.3. *Loss-of-function CMS mutations*

CMS can also result from loss-of-function mutations in essential mitochondrial genes. CMS in *Beta vulgaris* ssp. *maritima* (wild beet) is associated with a *coxII* gene missing eight highly conserved, C-terminal amino acids. Furthermore, the activity of respiratory complex IV is decreased by 50% in vegetative tissues of CMS plants. This loss may be compensated to some extent by an increased expression of the alternative oxidase (AOX) branch of the plant mitochondrial respiratory chain

(Ducos et al., 2001). In the CMS I and CMS II mutants of *Nicotiana sylvestris* L., male sterility results from truncation or deletion of the mitochondrial NADH dehydrogenase subunit 7 (*nad7*) gene (Pla et al., 1995; Lelandais et al., 1998). These mutations are likely viable because *nad7* encodes a subunit of respiratory complex I, and there are other points of electron entry into the respiratory chain that by-pass this complex.

3. EXPRESSION OF MITOCHONDRIAL CMS GENES

3.1. Transcripts of CMS loci

The chimeric CMS loci (summarized in Table 1) are often expressed by means of normal mitochondrial gene transcription and translation signals (reviewed in Chapters 8 and 12 of this volume, respectively). The 5' termini of the Petunia *pcf* transcripts match those observed for the Petunia *atp9* gene (Young and Hanson, 1987), due to duplicate sequences that flank these loci. A similar situation is observed at the *orf256* locus associated with male sterility-inducing *Triticum timopheevi* cytoplasm in cultivated wheat (*T. aestivum*). The 5' flanking and 5' coding region of *orf256* is similar to that of the *T. aestivum coxI* gene. *T. timopheevi orf256* and *T. aestivum coxI* transcripts utilize an identical upstream promoter sequence (Song and Hedgecoth, 1994b). CMS genes are almost always part of dicistronic or polycistronic transcripts, with the CMS gene located either 5' or 3' to normal mitochondrial genes. Normal mitochondrial gene promoters are therefore instrumental in transcribing CMS genes that lie 3' to a normal gene. The *Helianthus annuus* (sunflower) CMS gene, *orf522*, for example, lies 3' to *atpa* sequences on a di-cistronic transcript (Monéger et al., 1994).

Frequently, multiple promoters 5' to CMS (and other) loci result in multiple transcript initiation points. The *S-pcf* locus of Petunia (Young and Hanson, 1987), the *orf107* locus of Sorghum (Yan and Pring, 1997), and the *orf224* locus of Brassica (Singh and Brown, 1993) are each transcribed from three different 5' promoters. Transcript accumulation patterns can be further modified by plant mitochondrial RNA processing activities (reviewed by Hoffmann et al., 2001; Binder and Brennicke, 2003). For example, a monocistronic *atp6* transcript is likely the result of processing a dicistronic *orf224-atp6* transcript in the Brassica pol cytoplasm (Singh and Brown, 1991). Further processing of CMS locus transcripts is frequently associated with fertility restoration, described below.

3.2. Editing of CMS-associated transcripts

The chimeric nature of CMS genes sometimes places plant mitochondrial RNA editing sites within these coding sequences. Plant mitochondrial RNA editing (reviewed by Mulligan and Maliga, 1998; Brennicke et al., 1999; Hoffmann et al., 2001; Binder and Brennicke, 2003; Mulligan, Chapter 9 of this volume) results in selective, post-transcriptional C-to-U or U-to-C conversions. In most cases, these conversions are required for normal mitochondrial transcripts to encode the proper amino acid sequence. Because segments of plant mitochondrial genes can be edited even when taken out of context (Gualberto et al., 1991; Kumar and Levings, 1993; Nivison et al., 1994; Kubo and Kadowaki, 1997; Williams et al., 1998), CMS genes are potentially changed by plant mitochondrial RNA editing.

A number of CMS genes are altered at one or more coding sites as a result of RNA editing. Sunflower *orf522* is edited in two positions; one of these corresponds to an editing site shared with the normal mitochondrial *atp8* gene (Monéger et al., 1994). The *atp8*-derived CMS gene of carrot undergoes the same RNA editing events observed in the normal *atp8* gene (Nakajima et al., 2001). The Petunia *pcf* transcripts are edited in both the *atp9*- and *coxII*-derived segments, but only the unedited *urf-S* portion of the transcript is represented in the mature protein product of this locus (Nivison et al., 1994). The Sorghum *orf107* transcript exhibits editing at four sites. A site shared with the normal mitochondrial *atp9* gene is edited with high efficiency, whereas a downstream site is edited at low efficiency. Two additional downstream sites are differentially edited, and these edits may influence subsequent transcript processing conditioned by nuclear *restorer-of-fertility* alleles (Pring et al. 1998; 1999).

The most striking change conferred to a CMS-associated transcript by RNA editing is the truncation of *orf77* to *orf17* in CMS-S maize (Gallagher et al., 1999). Here, RNA editing creates a termination codon truncating *orf77* immediately 3' to a segment derived from the mitochondrial *atp9* gene. *Orf17* therefore predicts a protein sharing significant similarity with the C-terminal transmembrane domain of the normal ATP9 protein, although the relevance of this RNA edit to the expression of the CMS trait is currently unknown.

Transcripts of two other CMS genes are not edited. Although the maize *urf13* sequence is not edited, the *orf25* (*atp4*) gene co-transcribed with *urf13* is edited efficiently (Ward and Levings, 1991). The absence of editing in *urf13* is not surprising, as this gene does not contain sites edited in normal context. Transcripts of *T. timopheevi orf256* are also unedited (Song and Hedgecoth, 1994a). Only the 33 5' coding nucleotides of *orf256* are derived from a known mitochondrial gene (*coxI*), and the editing status of these nucleotides within the normal *coxI* transcripts has not been reported.

3.3. Protein products of CMS loci

Through the use of specific antibodies, many CMS loci have been shown to encode proteins that exhibit stable accumulation in one or more plant organs (Table 1). In some cases the size of the observed protein is in agreement with the protein that is predicted by the CMS gene. Examples include URF13, an ATP8 protein with a C-terminal extension in CMS carrot (Nakajima et al. 2001), a truncated COXII protein in CMS wild beet (Ducos et al., 2001), a 20-kDa protein in the CMS ogu radish (Krishnasamy and Makaroff, 1994; Grelon et al., 1994), a 17-kDa protein in CMS kos radish (Koizuka et al., 2000), a 15-kDa protein associated with CMS in sunflower (Monéger et al., 1994), and a 27-kDa polypeptide associated with CMS in common bean (Abad et al., 1995). In other cases, the accumulated protein represents only a segment of the CMS gene. The *pcf* locus of Petunia predicts a 49-kDa polypeptide (Young and Hanson, 1997), but a protein of 25-kDa, derived from the *urfS* portion of *pcf* accumulates (Nivison et al., 1994). *Orf254* of *T. timopheevi* predicts a protein of 28 kDa, but a 7-kDa protein derived from the central coding region of *orf254* is stably accumulated (Song and Hedgecoth, 1994b).

Many CMS-associated proteins have been observed to accumulate in association with mitochondrial membranes (Dewey et al., 1987; Hack et al., 1991; Krishnasamay and Makaroff, 1994; Song and Hedgecoth, 1994). The 25-kDa *pcf* gene product is found in both soluble and membrane fractions of Petunia mitochondrial protein preparations (Nivison et al., 1994). The 27-kDa protein associated with CMS in common bean exhibits an unusual spatial distribution. This protein is found not only in mitochondria of floral buds, but also in extra-mitochondrial locations. These include the callose and cell walls that surround the developing micrspores (Abad et al., 1995).

Numerous CMS-associated proteins accumulate in vegetative as well as reproductive tissues (Table 1) (See section 4.1 below for specific examples). This observation is not surprising considering that many of these gene products result from transcription and translation signals that drive expression of normal mitochondrial genes throughout the plant. This expression pattern does, however, raise the question of why the most obvious phenotypic effects of CMS genes are restricted to male reproductive tissues and organs. At least two CMS gene products do exhibit tissue or organ-specific patterns of accumulation. The 27-kDa protein encoded by *orf239* of common bean is synthesized in mitochondria from vegetative and anther tissues. Stable accumulation of this protein is, however, restricted to anthers because the protein is degraded in vegetative tissues (Sarria et al., 1996). The ATP8-related, CMS-associated protein of carrot also accumulates preferentially in anther tissues. This pattern of accumulation results from a post-transcriptional mechanism that is not yet identified (Nakajima et al., 2001).

4. CMS PHENOTYPES

4.1. Phenotypes conditioned by CMS loci

CMS can result from a wide variety of reproductive abnormalities (reviewed by Laser and Lersten, 1972; Kaul, 1988; Conley and Hanson, 1995; Zubko 2004). In a few cases, anthers undergo homeotic transformation to petaloid or carpeloid structures (reviewed by Zubko, 2004), similar to the phenotypes that result from mutations in nuclear genes specifying floral organ identity (reviewed by Coen and Meyerowitz, 1991). Many CMS loci result in failed pollen development due to abnormalities in either the anther (sporophytic) or the developing pollen (gametophytic) tissues. Laser and Lersten (1972) note that the CMS effects have been seen at virtually every stage of pollen development, and that sporophytic CMS phenotypes can be associated with abnormalities in anther vascular tissue, the tapetal cell layer that lines the anther locule, or the callose layer that surrounds the developing microspores.

While the primary phenotypic effect of mitochondrial CMS genes is the failure to produce or shed functional pollen, the protein products of many CMS loci accumulate in vegetative and reproductive tissues. The URF13 protein of CMS-T maize accumulates throughout the maize plant (Dewey et al., 1987). This CMS gene product is unusual in that it conditions an obvious vegetative phenotype – susceptibility to fungal pathogens (reviewed by Ullstrup, 1972; Pring and Lonsdale, 1989; Ward, 1995; Wise et al. 1999). Expression of the *pcf* locus in suspension culture cells of CMS Petunia conditions changes in mitochondrial respiratory electron transfer partitioning through cytochrome and alternative pathways (Connett and Hanson, 1990). Other CMS gene products known to accumulate in vegetative tissues or organs (listed in Table 1) potentially condition similar changes in vegetative physiology. The variant COXII subunit associated with the male sterility-inducing G cytoplasm of wild beet conditions a 50% decrease in the activity of leaf mitochondrial respiratory complex I, along with increased accumulation of AOX (Ducos et al. 2001). The CMS I and CMS II mutations of *N. sylvestris* exhibit vegetative, slow-growth phenotypes in addition to male sterility (Gutierres et al., 1997).

4.2. The molecular basis of degenerative CMS phenotypes

The mechanisms by which CMS gene products condition reproductive failure are poorly understood, even in examples such as CMS-T maize, where many features of the URF13 protein are known. This protein functions as a pore-forming receptor for HmT produced by the fungal pathogen *B. maydis* race T (reviewed by Rhoades et al., 1995). The accumulation of URF13 in leaf tissues of CMS-T maize plants therefore explains the susceptibility of this genotype to the fungal pathogen. Interaction of CMS-T mitochondria with the toxin results in mitochondrial swelling and an uncoupling of oxidative phosphorylation (Miller and Koeppe, 1971; Gengenbach et al., 1973). More puzzling is how accumulation of URF13 in anther tissues conditions failure of pollen development (Levings, 1993; Wise et al., 1999). Analysis of CMS-T and normal anther development by electron microscopy has led to one hypothesis. The earliest developmental abnormality observed in CMS-T anthers is a swelling of mitochondria in the tapetal cell layer of the anther (Warmke and Lee, 1977). These workers also noted a rapid, 20- to 40-fold increase in the number of mitochondria in tapetal cells and microspores of both normal and T-cytoplasm anthers. This increase in mitochondrial number occurs just prior to the mitochondrial abnormalities associated with the CMS-T genotype. Perhaps URF13 compromises mitochondrial function such that energy demands of the reproductive process cannot be met (Warmke and Lee, 1978). The response of CMS-T mitochondria to HmT led to the alternative hypothesis that a toxin-like compound accumulates specifically in anther tissues and causes swelling and dysfunction of CMS-T mitochondria (Flavell, 1974). The hypothesis of Warmke and Lee (1978) could be generalized to explain CMS in a number of systems where tapetal or microspore degeneration is observed.

The role of mitochondria in regulating plant and animal cell death (reviewed by Green and Reed, 1998; Jones, 2000; Lam et al., 2001; Granville and Gottlieb, 2002; Hoeberichts and Woltering, 2003; Kim et al., 2003; Mayer and Oberbauer, 2003) raises additional considerations regarding the mechanisms of CMS. Release of cytochrome *c* from the mitochondrial inter-membrane space is a feature of programmed cell death (PCD) in plants and animals. In animals, cytosolic cytochrome *c* can activate a caspase protease cascade, resulting in an orderly program of cell disassembly know as apoptosis (reviewed by Reed, 1997; Raff, 1998). Although a role for cytosolic cytochrome *c* in plant PCD is far from clear, a mitochondrial nuclease has been implicated in plant (Balk et al., 2003) and animal PCD (Li et al., 2001). PCD does, however, require ATP (reviewed by Kim et al., 2003). In contrast, cell death by necrosis results when cells cannot maintain mitochondrial function or ATP levels sufficient to support PCD. Necrosis is manifested by the swelling and lysis of cells and organelles (Kerr et al., 1972).

Perhaps CMS gene products are capable of disrupting mitochondrial function to the extent that PCD or necrosis follows. While this response could potentially occur in any tissue expressing a CMS gene, many factors, including the level of ATP, the balance of positive and negative cell death regulators, the presence or absence of downstream cell death machinery, and the level of intracellular calcium regulate the life/death decision. Reduced concentrations of ATP predispose cells to undergo PCD in response to calcium signals (reviewed by Jones, 2000; Rizzuto et al., 2003; Orrenius et al., 2003). PCD has been implicated in the degeneration of CMS sunflower anther tissues (Balk and Leaver, 2001). Sunflower ORF225 is related to ATP8, and the accumulation of the ATP synthase complex is reduced in CMS sunflower compared to male-fertile sunflower (Sabar et al., 2003). It will be interesting to determine whether other CMS systems exhibit hallmarks of PCD. The cytology of degenerating CMS-T maize anthers (Warmke and Lee, 1977) appears to be necrotic, while the morphology of collapsed CMS-S maize pollen (Lee et al., 1980) is much like that of apoptotic animal cells (Kerr et al., 1972). Mitochondria purified from collapsed CMS-S maize pollen are depleted of cytochrome *c* and exhibit reduced levels of ATP relative to mitochondria purified from normal pollen (our unpublished observations). Hence there may be parallels between CMS sunflower and CMS-S maize.

4.3 The molecular basis of homeotic CMS phenotypes

Several CMS systems involve the transformation of anthers to petals (petaloid) or carpels (carpeloid) (reviewed by Zubko, 2004). Homeotic transformation of floral organ identity has been described in CMS carrot (Kitagawa et al., 1994), wheat (Murai et al., 2002) and tobacco (Kofer et al., 1991; Zubko et al., 2001). In the wheat and tobacco examples, the CMS is alloplasmic, resulting from novel, interspecific or intergeneric combinations of mitochondrial and nuclear genomes. Homeotic CMS phenotypes are similar to those conditioned by mutations in nuclear genes that encode MADS-box transcription factors specifying floral organ identity (reviewed by Coen and Meyerowitz, 1991). Recent investigations in CMS carrot (Linke et al., 2003), wheat (Murai et al., 2002) and tobacco (Zubko et al., 2001) (reviewed by Zubko, 2004) revealed decreased expression of nuclear MADS box genes similar to those specifying identity of second and third whorl floral organs (petals and anthers).

In CMS plants with the *N. tabacum* nucleus and *N. repanda* cytoplasm, stigmatoid anthers and anthers fused with carpels do not result from altered expression of floral organ identity genes (Farbos et al., 2001). Instead, this phenotype appears to result from failure to express the *N. tabacum* homolog of the Arabidopsis *SUPERMAN* gene (*AtSUP*) (Berterbide et al., 2002). *AtSUP* is necessary for the specification of floral organ boundaries (Sakai et al., 1995), and male fertility is partially restored to alloplasmic *N. tabacum* plants by ectopic expression of *AtSUP* (Berterbide et al., 2002).

The retrograde regulation of nuclear genes by mitochondrial signals has been described for some plant nuclear genes (Vanlerberghe and McIntosh, 1994; Maxwell et al., 2002). A well-studied example is the plant nuclear gene encoding mitochondrial AOX (reviewed by Vanlerberghe and McIntosh, 1997; McIntosh et al., 1998). The *aox* gene is upregulated in response to inhibition of mitochondrial electron transfer (Vanlerberghe and McIntosh, 1994), and is therefore part of plant response to mitochondrial stress. In the homeotic CMS systems, changes in mitochondrial physiology could result from the expression of a specific CMS gene, such as the altered *atp8* gene in carrot (Nakajima et al., 2001), or from subtle incompatibilities among nuclear and mitochondrial-encoded respiratory subunits in alloplasmic combinations. Regardless of the genetic basis, physiological changes within the mitochondria could result in a retrograde signal affecting expression of floral organ identity genes or other nuclear genes required for normal flower development.

5. EVOLUTION OF CMS

The diversity of CMS genotypes and phenotypes indicates that this trait arose independently in many plant lineages. The recombinogenic nature of plant mitochondrial genomes (reviewed by Hanson and Folkerts, 1992; Bendich, 1993; Fauron et al., 1995; Backert et al., 1997) may be the reason that actively expressed CMS genes arose repeatedly. How then, does natural selection shape and maintain this trait? Land plant mitochondrial genomes are frequently, although not always, maternally inherited (reviewed by Hagemann, this volume). Any mitochondrial dysfunction that causes necrosis or apoptosis in vegetative or female reproductive organs would be quickly eliminated by natural selection. In contrast, mitochondrial mutations affecting only male fertility could potentially be maintained in natural populations. CMS is the basis for gyondioecy observed in many natural populations (reviewed by Budar et al., 2001; Bailey et al., 2003; Saur-Jacobs and Wade, 2003). In some cases, CMS is associated with increased seed production and could potentially encounter positive selection (reviewed by Budar et al. 2001). Budar et al. (2001) further propose the interesting hypothesis that expression of CMS genes in vegetative tissues may alter physiology such that plants have improved stress tolerance, thereby providing an additional basis for positive selection and the maintenance of CMS in natural populations.

6. FERTILITY RESTORATION BY NUCLEAR GENES

6.1. Genetics of fertility restoration

Nuclear *restorer-of-fertility* alleles suppress or compensate for expression of CMS mitochondrial genes, resulting in male fertility. With only one exception (see below), restorer genes produce no heritable changes in the mitochondrial DNA. There are two modes of fertility restoration, sporophytic and gametophytic. In addition, one-gene and two-gene systems have been characterized in both cases. In sporophytic restoration, the genotype of the diploid sporophyte determines whether or not functional pollen is produced. Therefore, in a one-gene sporophytic system, a CMS plant heterozygous for a restorer produces all functional pollen even though only 50% of the pollen grains carry a restoring allele. An example of such a system is Petunia CMS and its restoring allele *Rf,* recently renamed *Rf-PPR59* (reviewed by Hanson et al., 1995; Bentolila et al., 2002). In a two-gene sporophytic system, a CMS plant heterozygous at both restorer loci also produces all normal pollen even though only 25% of the pollen grains carry both restoring alleles. The CMS-T system of maize, in which *Rf1* and *Rf2* are both required for restoration, is an example of such a system (reviewed by Ward, 1995; Wise et al., 1999).

When restoration is gametophytic, the genotype of the haploid gametophyte, i.e. the pollen grain, determines whether or not the pollen functions. In a gametophytic one-gene system, a CMS plant that is heterozygous for a restorer produces 50% normal and 50% aborted pollen grains, the former carrying the restoring allele and the later carrying a nonrestoring allele. Restoration of CMS-S maize by *Rf3* represents this kind of system (Buchert, 1961; reviewed by Gabay-Laughnan et al., 1995). In a two-gene gametophytic system, a CMS plant heterozygous at both restorer loci produces only 25% viable pollen if the restorers are unlinked; only those grains carrying both restoring alleles function. Restoration of Sorghum A3 cytoplasm by *Rf3* and *Rf4* is an example of such a system (Tang et al., 1998; reviewed by Kempken and Pring, 1999; Pring et al., 1999). Dominance of gametophytic restoring alleles is often assumed but must be tested through the use of tetraploid stocks (Kamps et al., 1996).

6.2. Molecular mechanisms of fertility restoration

Through study of the different CMS systems, numerous mechanisms of fertility restoration have been identified. The copy number of the mitochondrial CMS locus is decreased by the *Fr restorer-of-fertility* allele in common bean (Mackenzie and Chase, 1990; Janska et al., 1993). Restoration by *Fr* therefore resembles cytoplasmic reversion to fertility (reviewed by Newton et al., 2004). Fertility restoration frequently results from changes to CMS gene transcripts, as described for a number of systems below. In other cases, no transcript changes are observed, but the encoded protein fails to accumulate. In contrast, the *Rf2* restoring allele of CMS-T maize does not alter *urf13* gene expression, but instead alters mitochondrial metabolism to compensate for the effects of the URF13 protein (reviewed by Schnable and Wise, 1998).

The most common mechanism of fertility restoration involves the processing of CMS gene transcripts, often internal to the CMS gene itself. In the presence of *Rfp1,* the restorer gene for pol CMS in Brassica, the relative abundance of the 2.2- and 1.9-kb *orf224-atp6* co-transcripts is decreased and two shorter transcripts of 1.4 and 1.3 kb are present (Singh and Brown, 1991). The restorer gene *Rfn* for nap CMS of Brassica alters the accumulation of *orf222/nad5c/orf139* co-transcripts (Singh et al., 1996; Li et al., 1998). In the presence of *Rfn*, nap-cytoplasm plants exhibit reduced levels of *orf222* transcripts and express an additional transcript detected by an *orf139* probe (L'Homme et al. 1997). The two restorers *Rfn* and *Rfp* are allelic or very tightly linked (Li et al., 1998; reviewed by Brown,1999).

In BT CMS rice and A3 CMS Sorghum, fertility restoration is also associated with RNA processing, and links between RNA editing and processing are revealed. The *Rf-1* restoring allele of BT CMS rice is associated with processing of di-cistronic mitochondrial CMS transcripts. In addition, the *atp6* gene co-transcribed with the CMS gene is more efficiently edited in processed transcripts than in unprocessed transcripts (Kadowaki et al., 1990; Iwabuchi et al., 1993). Restoration of male fertility to A3 cytoplasm Sorghum requires the complementary action of two restorer loci, *Rf3* and *Rf4*. *Rf3* confers enhanced RNA processing internal to the CMS-associated *orf107* (Tang et al., 1998). Residual, unprocessed transcripts in the restored line show a reduced frequency of RNA editing compared to unprocessed transcripts in the *rf3rf3* genotype. RNA editing is predicted to alter RNA secondary structure to favor the processing conditioned by *Rf3* (Pring et al., 1998). The function of *Rf4* in this system is unknown (Pring et al., 1999; Wen et al., 2002).

Restoration of fertility to CMS-T maize requires the action of two unlinked restorers, *Rf1* on chromosome 3 and *Rf2* on chromosome 9. An altered pattern of *urf13* transcripts is exhibited by CMS-T plants carrying *Rf1*. Novel 1.6- and 0.6-kb transcripts are detected, with only a slight decrease in the accumulation of the other major transcripts (Dewey et al., 1986; 1987; Kennell and Pring, 1989). The amount of URF13 protein is reduced by 80% in the presence of *Rf1* (Dewey et al. 1987). Possibly RNA processing associated with *Rf1* results as a secondary consequence of the failure to translate the unprocessed message. *Rf8* and *Rf**, tightly linked restorers on the long arm of chromosome 2, can partially substitute for *Rf1* in fertility restoration. These alleles also condition the appearance of novel *urf13* transcripts (Wise et al., 1996; Wise and Pring, 2002). In plants carrying *Rf8*, additional 1.42- and 0.42-kb transcripts are observed, while novel 1.4- and 0.4-kb transcripts are detected in the presence of *Rf** (Dill et al., 1997). A reduction in the abundance of URF13 protein is also observed in *Rf8* plants, however, this reduction is not as pronounced as that brought about by *Rf1* (Dill et al., 1997). The expression of URF13 is unaffected by *Rf2* (Dewey et al., 1987). This restorer was the first to be cloned and appears to act by a "metabolic compensation" mechanism, discussed below.

In CMS-S maize, 2.8- and 1.6-kb transcripts are associated with the expression of the chimeric mitochondrial *orf355-orf77* region (Zabala et al., 1997). The restoring allele *Rf3* on the long arm of chromosome 2 co-segregates with internal processing and decreased accumulation of these transcripts (Zabala et al., 1997; Wen and Chase, 1999). In both sporophytic (leaf or immature ear) and gametophytic (microspore or pollen) tissues carrying an *Rf3* allele the 2.8-kb *orf355-orf77* transcript is processed to a 2.1-kb RNA.

Other *restorer-of-fertility* alleles alter the accumulation of CMS gene transcripts without obvious evidence of endonucleolytic processing. The Petunia *Rf* allele (now *Rf-PPR592*) is associated with failure to accumulate the *pcf* transcripts that initiate from the promoter proximal to the *pcf* translation initiation codon (Pruitt and Hanson, 1991). Although other *pcf* transcripts accumulate in the presence of *Rf-PPR592*, the PCF protein fails to accumulate (Nivison and Hanson, 1989; Nivison et al., 1994). In the case of sunflower, the restoring allele is associated with a flower-specific decrease in the accumulation of *orf522* transcripts, and a concomitant decrease in the accumulation of the ORF522 protein product. Mitochondrial run-on transcription studies indicate that post-transcriptional events account for the failure to accumulate *orf522* transcripts (Monéger et al., 1994).

In several cases, fertility restoration does not alter the accumulation of CMS transcripts, but does reduce the accumulation of the corresponding protein products. This situation occurs in the presence of the *Rfo* and *Rfk* restoring alleles for ogu and kos radish (Krishnasamy and Makaroff, 1994; Iwabuchi et al., 1999; Koizuka et al., 2000), and in the presence of the *Fr2* restoring allele of common bean. In the bean system, pulse labeling experiments demonstrate turn-over of the ORF239 protein product in the presence of *Fr2* (Sarria et al., 1998).

6.3. Modifiers of mitochondrial transcripts

The processing of normal mitochondrial gene transcripts has been associated with *restorer-of-fertility* alleles or loci in several CMS systems. Makaroff and Palmer (1988) were the first to comment on a possible linkage of restorer genes and genes affecting the accumulation of essential mitochondrial gene transcripts when they observed that many lines that restored ogu CMS in radish had altered *atpA* transcript patterns. While the *Rfp1* allele for pol CMS of Brassica is associated with processing of *orf224/atp6* transcripts (Singh and Brown, 1991), a recessive nonrestoring allele at this locus acts as a dominant **m**odifier of **m**itochondrial **t**ranscripts (*Mmt)* in that it conditions processing of the *nad4* gene transcripts (Singh et al., 1996). The restorer gene for nap CMS of Brassica, *Rfn,* has been determined to be identical to *Mmt,* and is therefore redesignated *Rfn-Mmt* (Li et al., 1998). Thus *Rfp-mmt*, *Rfn-Mmt,* and *rf-mmt* appear to represent different alleles at a single nuclear locus that functions in mitochondrial RNA processing. Alternatively, these three activities result from the action of three different, but tightly linked genes (Li et al., 1998). The processing of Sorghum mitochondrial *urf209* transcripts is controlled by a single dominant *Mmt1* gene (Tang et al., 1998). This *Mmt1* gene is linked to the nuclear restorer *Rf3* that regulates processing of the CMS-associated *orf107* transcripts (discussed above). In CMS-S maize, the *Rf3* restoring allele cosegregates with altered patterns of *cob* and *atp6* gene transcripts in addition to processed forms of the male sterility-associated *orf355-orf77* transcripts (Wen and Chase, 1999). The restorer locus may encode or regulate *Mmt* activity or, alternatively, *Mmt* and the restorer may represent closely linked loci (Gabay-

Laughnan et al., 2004). In these examples, fertility restoration may result because nuclear *Mmt* gene products by chance process mitochondrial CMS locus transcripts in a manner that disrupts expression of CMS.

6.4. Fertility restoration in CMS - S maize

An exceptional feature of the CMS-S system is the identification and recovery of *restorers-of-fertility* that arise by spontaneous mutation at many different nuclear loci (Laughnan and Gabay, 1973; Laughnan and Gabay, 1978; Gabay-Laughnan et al., 1995). These are observed as sectors of male fertility or fully fertile tassels that occur in populations of S male-sterile plants. Restorer alleles are then recovered by performing crosses with pollen from the male-fertile sectors or tassels. Although these restoring alleles rescue CMS-S pollen from collapse, many are homozygous-lethal for seed development (Laughnan and Gabay, 1975; reviewed by Gabay Laughnan et al., 1995; Chase and Gabay-Laughnan, 2003). These observations led to the hypothesis that *restorer-of-fertility* alleles in CMS-S maize arise as loss-of-function mutations in nuclear genes essential for mitochondrial gene expression. Fertility restoration therefore results from failure to express mitochondrial genes, including the CMS-associated *orf355-orf77* locus, in haploid CMS-S pollen (Wen et al., 2003). These alleles have been designated as *restorer-of-fertility lethal* (*rfl*). Molecular-genetic analysis of one such allele (*rfl*1) confirms that fertility restoration results from a recessive mutation conditioning not only decreased accumulation of *orf355-orf77* transcripts but also failure to accumulate wild-type levels of mitochondrial-encoded ATPA protein in developing pollen (Wen et al., 2003).

Because CMS-S is a gametophytic system, the molecular-genetic events that determine male fertility or sterility take place in developing pollen (Buchert, 1961; Wen and Chase, 1999b). While pollen development is an energetically demanding process associated with an increase in mitochondrial numbers and mitochondrial gene expression (Warmke and Lee, 1977; 1978; Monéger et al., 1994), maturing pollen has active glycolytic fermentation pathways (reviewed by Tadege et al., 1999) that potentially enable pollen to survive and function despite reduced mitochondrial function. CMS-S maize therefore allows the recovery and analysis of fertility restorers having essential roles in normal mitochondrial gene expression (Wen et al., 2003; reviewed by Chase and Gabay-Laughnan, 2003).

7. *RESTORER-OF-FERTILITY* LOCI

7.1. The rf2 locus of CMS-T maize

The first *restorer-of-fertility* gene to be cloned was *Rf2*, required to restore male fertility in CMS-T maize (Cui et al., 1996). The *Rf2* restoring allele encodes a functional mitochondrial aldehyde dehydrogenase that compensates in some way for the effects of the mitochondrial URF13 protein causal to male sterility (Cui et al., 1996; Liu et al., 2001) but does not alter URF13 protein accumulation (Dewey et al., 1987). This is not typical of most restorers characterized and supports the hypothesis (Schnable and Wise, 1994) that the *rf2* locus has a function other than restoration of fertility. In fact, normal cytoplasm maize plants lacking *Rf2* function exhibit reduced fertility in lower florets, demonstrating that anther development in normal cytoplasm maize requires *Rf2* function (Liu et al., 2001).

7.2. Pentatricopeptide repeat (ppr) genes and proteins

Several restorer genes that do alter the expression of mitochondrial CMS genes have recently been cloned (Table 2), and all encode proteins belonging to the pentatricopeptide repeat (PPR) family. PPR protein genes are widespread in eukaryotes and comprise a large gene family of over 200 members in Arabidopsis (Small and Peeters, 2000). Two-thirds of the *ppr* genes in Arabidopsis encode proteins that are likely targeted to organelles, either mitochondria or chloroplasts (Small and Peeters, 2000). In general, *ppr* genes encode an organelle targeting sequence, followed by a short, variable sequence that is potentially responsible for functional specificity. The remaining portion of these genes encodes tandem arrays of a degenerate 35-amino-acid repeat, ranging in number from 2-26 with a mean just over 9. PPR proteins have been implicated in the expression of organelle genes in a diversity of organisms (reviewed by Barkan and Goldschmidt-Clermont, 2000).

Table 2. *Restorer loci encoding PPR proteins*

Genus	Cytoplasm	Locus	Predicted number of amino acids	Predicted number of repeat units	Number of tandem *ppr* genes	References
Petunia	CMS	*Rf-PPR592*	592	14	≥ 2	(Bentolila et al., 2002)
Oryza	CMS BT	*Rf-1*	791	16 or 18	4	(Kazama and Toriyama, 2003; Komori et al., 2004)

Table 2 (cont.)

Table 2 cont.

Genus	Cytoplasm	Locus	Predicted number of amino acids	Predicted number of repeat units	Number of tandem *ppr* genes	References
Raphanus	CMS kosena	*Rfk1*	687	16		(Koizuku et al., 2003)
Raphanus	CMS ogura	*Rfo*	687	16	3	(Brown et al., 2003; Desloire et al., 2003)

Small and Peeters (2000) propose that the PPR motifs each consist of two alpha helices and that the tandem PPR repeats form a superhelix with a central groove or tunnel. The central groove is lined with side chains that are predominantly hydrophilic and the residues at the bottom of the groove are positively charged. Based upon this model, these researchers suggest that *ppr* genes could encode sequence-specific RNA-binding proteins.

An RNA binding function for PPR proteins would be consistent with our present knowledge of restorer gene functions. Petunia *Rf-PPR592*, rice *Rf1*, radish *Rfo* and radish *Rfk* are all associated with failure to accumulate CMS locus transcripts and/or gene products. In radish, the restorer affects protein but not RNA accumulation (Krishasamy and Makaroff, 1994; Iwabuchi et al., 1999; Koizuka et al., 2000), whereas rice *Rf1* is associated with CMS transcript processing and an increased efficiency of RNA editing in processed, monocistronic *atp6* transcripts (Kadowaki et al., 1990; Iwabuchi et al., 1993). Petunia *Rf-PPR592* conditions failure to accumulate one of three *pcf* transcripts (Pruitt and Hanson, 1991) and the PCF protein (Nivison and Hanson, 1989; Nivison et al., 1994). These observations raise the question of whether PPR proteins have diverse functions in plant organelle RNA metabolism or if the effects of the different restoring alleles all stem from a common molecular mechanism. Characterization of the nuclear-encoded PPR proteins and their functions will help to elucidate mechanisms of plant organelle expression (Small and Peeters, 2000; Bentolila et al., 2002). Since most of the cloned *restorers-of-fertility* for CMS are *ppr* genes, it is possible that restorers in other plant species are *ppr* genes as well (Bentolila et al., 2002; Wise and Pring, 2002; Desloire et al., 2003).

7.3. ppr gene clusters

The cloned *restorer-of-fertility* genes for Petunia, radish and rice are located in tandem arrays of *ppr* genes (Table 2). The restorer locus of Petunia is composed of at least two *ppr* genes, designated *Rf-PPR591* and *Rf-PPR592* with a possible third *ppr* gene nearby (Bentolila et al., 2002). The *Rfo* restorer gene of CMS ogu radish is flanked by two similar genes that do not appear to function as restorers (Brown et al., 2003; Desloire et al., 2003). The predicted proteins encoded by the three genes are similar but differ in the number of PPR repeats (Brown et al., 2003). In rice, three additional *ppr* genes that do not function in fertility restoration are located upstream of *Rf-1* (Komori et al. 2004).

The clustering of *ppr* genes raises a hypothesis to explain the *Mmt* activity associated with *restorer-of-fertility* loci (Gabay-Laughnan et al., 2004). While the product of a restorer gene could perform multiple RNA processing functions within the mitochondria, *Mmt* genes may actually be *ppr* genes that are tightly linked to *restorer-of-fertility* genes but that have no direct role in the restoration of fertility to CMS plants.

8. EVOLUTION OF FERTILITY RESTORATION

8.1. Evolutionary implications of ppr gene clusters

The clustering of *ppr* genes has implications for the evolution of restoring and nonrestoring alleles. Perhaps restoring alleles evolve by modification of pre-existing *ppr* genes having roles in mitochondrial gene expression (Bentolila et al., 2002). Modifications in expression and function could result from duplication and divergence. Brown et al. (2003) propose that the three similar *ppr* genes in the *rfo* region of the radish genome arose via gene duplication. Gain or loss of PPR domains following duplication could explain their differences. Komori et al. (2004) propose a similar hypothesis, citing recombination as a means to confer or remove restorer function. In Petunia, a nonrestoring *rf-PPR529* allele has a 530-nucleotide deletion in the promoter region compared to the restoring *Rf-PPR592* allele. Recombination between tandem *Rf-PPR591* and *Rf-PPR592* genes is proposed to have given rise to *rf-PPR592.* While *rf*-PPR592 predicts a protein similar to *Rf-PPR592, rf-PPR592* is not transcribed in floral buds (Bentolila et al. 2002).

In contrast with the nonrestoring allele that was likely generated by recombination between two *ppr* genes in Petunia, (Bentolila et al., 2002), nonrestoring alleles of radish (Koizuka et al., 2003) and rice (Kazama and Toriyama, 2002; Komori et al., 2004) apparently result from coding sequence changes. In the case of rice, the nonrestoring *rf-1* allele lacks one nucleotide relative to the restoring allele. This creates a frame shift mutation such that the *rf-1* allele predicts a truncated PPR protein (Kazama and Toriyama, 2002; Komori et al., 2004). The *rfk1* nonrestoring allele of radish differs by 11 nucleotide substitutions that create coding differences between the restoring and nonrestoring allele (Koizuka et al., 2003).

The resemblance of nuclear *restorer-of-fertility* loci and disease resistance loci has not gone unnoticed (Li et al., 1998; Brown et al., 2003). Multiple, related tightly linked loci, and variation arising as a result of recombination between these loci, are also features of disease resistance gene clusters (reviewed by Richter and Ronald, 2000).

8.2. Fertility restorers in wild populations, races, varieties and inbred lines

The question was raised long ago whether restoring alleles for CMS are present in plants with normal cytoplasm and, if so, why (Duvick, 1965). Might they be nuclear alleles that are essential to normal mitochondrial gene expression and function (Schnable and Wise, 1994; Wise et al., 1999; Hanson et al., 1999; Kempken and Pring, 1999; Wise et al., 1999; Gabay-Laughnan, 2000; Gabay-Laughnan et al., 2004)? Alternatively, in naturally occurring gynodioecious populations, females result from the presence of CMS genes and hermaphrodites result from the introduction of nuclear restoring alleles. To maintain these breeding systems, both restoring and nonrestoring alleles must be maintained in the population. Negative fitness of restoring alleles is one factor that has been invoked to explain the stability of these breeding systems (reviewed by Bailey et al., 2003; Budar et al., 2003; Saur-Jacobs and Wade, 2003).

Some wild populations, races, varieties, and inbred lines have been surveyed for the presence of restoring alleles. Fifty crucifer varieties were tested for the presence of restorers for the nap CMS of Brassica (Li et al., 1998). Only in plants carrying nap cytoplasm was the restoring allele present. This is an indication of the co-evolution of the restoring allele and the male-sterility-inducing cytoplasm.

Similar surveys conducted in maize had different outcomes. Surveys of over 90 normal-cytoplasm inbred lines and nine varieties of maize for the presence of the CMS-T restoring allele *Rf2* revealed that this allele is present in all but inbred Wf9 and Wf9-related lines (Wise et al., 1999; J. B. Beckett, personal communication). The restoring allele *Rf1,* also required for fertility restoration in T cytoplasm, is relatively rare, occurring in about 20% of cultivars from Mexico, Central America, the Caribbean and the southeast U.S., and in only 5% of cultivars from the central and northern U.S. (Shaver, 1956; reviewed by Ward, 1995), but it is also is present in strains with normal cytoplasm. A survey of 30 Mexican races of maize and 13 accessions of its wild relative teosinte revealed that 27 maize races and seven teosinte accessions carried restoring alleles for CMS-S (Gabay-Laughnan, 2000). Most of these alleles are likely independent isolates of the *Rf3* restoring allele (Gabay-Laughnan et al., 2004). Although prevalent in Mexican maize and teosinte, CMS-S restorers are present in only about 42% of U. S. inbred lines (Beckett, 1971; Gracen, 1982; Gabay-Laughnan and Laughnan, 1994; our unpublished data). CMS-EP results from the incompatibility of certain maize nuclear backgrounds and the cytoplasm of *Zea perennis* (Gracen and Grogan, 1974; Laughnan and Gabay-Laughnan, 1983). Gabay-Laughnan (2001) tested over 50 inbred lines of maize for the presence of a CMS-EP restoring allele. A restorer was carried by all but 10 related inbred lines. Additionally, a restoring allele for CMS-EP was also present in six strains belonging to four varieties of maize and the teosinte *Z. mays* ssp. *mexicana.*

Thus, in at least three of the cases from maize, it appears that the nonrestoring alleles may have arisen as mutations and that the restorers are the naturally occurring alleles. Therefore, restoring alleles probably have functions in mitochondrial gene expression other than restoration of fertility. The hypothesis that restorers represent the wild-type alleles is supported by studies of the recently cloned *restorer-of-fertility ppr* genes. The nonrestoring alleles in Petunia (Bentolila et al. 2002), radish (Koizuka et al. 2003) and rice (Kazama and Toriyama 2003) represent loss-of-function genes allelic to the restorers. In contrast, the restoring alleles recovered as spontaneous mutants in CMS-S maize result from loss-of-function alleles in nuclear genes that normally function in mitochondrial gene expression (Wen et al., 2003). Furthermore, these alleles are frequently associated with deleterious effects on seed and plant development (Laughnan and Gabay, 1975; reviewed by Gabay-Laughnan et al., 1995). Both types of restoring alleles provide insights into the functions of nuclear genes with respect to mitochondrial biogenesis. Furthermore, the molecular-genetic diversity of CMS and fertility restoration systems potentially explains the diverse array of sex ratio patterns observed in natural gynodioecious populations (Bailey et al., 2003; Budar et al., 2003; Saur-Jacobs and Wade, 2003).

9. ACKNOWLEDGEMENTS

We thank Daryl Pring for his critical review of this manuscript. Our research on CMS and fertility restoration is funded by the United States Department of Agriculture National Research Initiative (USDA NRI) Award 00-35300-9409 to S. G. –L., USDA NRI Award 2001-0534-10888 to C. D. C., Illinois Foundation Seeds, and the Florida Agricultural Experiment Station. This work was approved for publication as Journal Series No. N-02500.

10. REFERENCES

Abad, A. R., Mehrtens, B. J., & Mackenzie, S. A. (1995). Specific expression in reproductive tissues and fate of a mitochondrial sterility-associated protein in cytoplasmic male-sterile bean. *Plant Cell, 7,* 271-285.

Abdelnoor, R. V., Yule, R., Elo, A., Christensen, A., Meyer-Gauen, G., & Mackenzie, S. A. (2003). Substoichiometric shifting in the plant mitochndrial genome is influenced by a gene homologous to *MutS. Proc. Natl. Acad. Sci. USA, 100,* 5968-5973.

Akagi, H., Sakamoto, M., Shinjyo, C., Shimada, H., & Fujimura, T. (1994). A unique sequence located downstream from the rice mitochondrial *atp6* may cause male sterility. *Curr. Genet., 25,* 52-58.

Backert, S., Nielsen, B. L., & Borner, T. (1997). The mystery of the rings: Structure and replication of mitochondrial genomes from higher plants. *Trends Plant Sci., 2,* 477-483.

Bailey, M. F., Delph, L. F., & Lively, M. (2003). Modeling gynodioecy: Novel scenarios for maintaining polymorphism. *Am. Nat., 161,* 762-776.

Balk, J., & Leaver, C. J. (2001). The PET1-CMS mitochondrial mutation in sunflower is associated with premature programmed cell death and cytochrome *c* release. *Plant Cell, 13,* 1803-1818.

Balk, J., Chew, S. K., Leaver, C. J., & McCabe, P. F. (2003). The intermembrane space of plant mitochondria contains a DNase activity that may be involved in programmed cell death. *Plant J., 34,* 573-583.

Barkan, A., & Goldschmidt-Clermont, M. (2000). Participation of nuclear genes in chloroplast gene expression. *Biochimie, 82,* 559-572.

Bateson, W., & Gairdner, A. E. (1921). Male sterility in flax subject to two types of segregation. *Jour. Genet., 11,* 269-275.

Beckett, J. B. (1971). Classification of male sterile cytoplasms in maize (*Zea mays* L.). *Crop Sci., 11,* 724-726.

Bendich, A. (1993). Reaching for the ring: The study of mitochondrial genome structure. *Curr. Genet.,* 24, 279-290.

Bentolila, S., Alfonso, A. A., & Hanson, M. R. (2002). A pentatricopeptide repeat-containing gene restores fertility to cytoplasmic male-sterile plants. *Proc. Natl. Acad. Sci. USA, 99*, 10887-10892.

Bergman, P., Edqvist, J., Farbos, I., & Glimelius, K. (2000). Male-sterile tobacco displays abnormal mitochondrial *atp1* transcript accumulation and reduced floral ATP/ADP ratio. *Plant Mol. Biol., 42*, 531-544.

Berterbide, A., Hernould, M., Faros, I., Glimelius, A., & Mouras, A. (2002). Restoration of stamen development and production of functional pollen in an alloplsmic CMS tobacco line by ectopic expresssion of the *Arabidopsis thaliana* SUPERMAN gene. *Plant J., 29*, 607-615.

Binder, S., & Brennicke, A. (2003). Gene expression in plant mitochondria: Transcriptional and post-transcriptional control. *Philos. Trans. R. Soc. Lond. B. Biol. Sci., 358*, 181-188.

Boeshore, M. L., Hanson, M. R., & Izhar, S. (1985). A variant mitochondrial DNA rearrangement specific to Petunia stable sterile somatic hybrids. *Plant Mol. Biol., 4*, 125-132.

Brennicke A, Marchfelder, A., & Binder, S. (1999). RNA editing. *FEMS Microbiol. Rev., 23*, 297-316.

Brown, G. G. (1999). Unique aspects of cytoplasmic male sterility and fertility restoration in Brassica napus. *J. Hered., 90*, 351-356.

Brown, G. G., Formanova, N., Jin, H., Warzachuk, R., Dandy, C., Patil, P., et al. (2003). The radish *Rfo* restorer gene of Ogura cytompasmic male sterility encodes a protein with multiple pentatricopeptide repeats. *Plant J., 35*, 262-272.

Buchert, J. G. (1961). The stage of the genome-plasmon interaction in the restoration of fertility to cytoplasmically pollen-sterile maize. *Proc. Natl. Acad. Sci. USA, 47*, 1436-1440.

Budar, F., Touzet, P., & De Paepe, R. (2003). The nucleo-mitochondrial conflict in cytoplasmic male sterilities revisited. *Genetica, 117*, 3-13.

Chase, C. D., & Ortega, V. M. (1992). Organization of ATPA coding and 3' flanking sequences associated with cytoplasmic male sterility in *Phaseolus vulgaris* L. *Curr. Genet., 22*, 147-153.

Chase, C. D., & Gabay-Laughnan, S. (2003). Exploring mitochondrial-nuclear genome interactions with S male-sterile maize. In S. G. Pangali (Ed.), *Recent Research Developments in Genetics* (vol. 3) (pp. 31-41). Kerala: Research Signpost.

Coen, E. S., & Meyerowitz, E. (1991). The war of the whorls: Genetic interactions controlling flower development. *Nature, 353*, 31-37.

Conley, C. A., & Hanson, M. R. (1994). Tissue-specific protein expression in plant mitochondria. *Plant Cell, 6*, 85-91.

Conley, C. A. & Hanson, M. R. (1995). How do alterations in plant mitochondrial genomes disrupt pollen development? *J. Bioenerg. Biomembr., 27*, 447-457.

Connett, M. B., & Hanson, M. R. (1990). Differential mitochondrial electron transport through the cyanide-sensitive and cyanide insensitive pathways in isonuclear lines of cytoplasmic male sterile, male fertile, and restored Petunia. *Plant Physiol., 93*, 1634-1640.

Correns, C. (1904). Experimentelle Untersuchungen über die Bynodioecie. *Ber. Deut. Bot. Ges., 22*, 506-517.

Cui, X., Wise, R. P., & Schnable, P. S. (1996). The *rf2* nuclear restorer gene of male-sterile T-cytoplasm maize. *Science, 272*, 1334-1336.

Desloire, S., Gerbhi, H., Wassila, L., Marhadour, S., Cloute, V., Cattolico, L., et al. (2003). Identification of the fertility restoration locus, *Rfo*, in radish, as a member of the pentatricopeptide-repeat protein family. *EMBO Rep., 4*, 588-594.

Dewey, R. E., Levings, C. S. III, & Timothy, D. H. (1986). Novel recombinations in the maize mitochondrial genome produce a unique transcriptional unit in the texas male-sterile cytoplasm. *Cell, 44*, 439-449.

Dewey, R. E., Timothy, D. H., & Levings, C. S. III. (1987). A mitochondrial protein associated with cytoplasmic male sterility in the T cytoplasm of maize. *Proc. Natl. Acad. Sci. USA, 84*, 5374-5378.

Dill C. L., Wise, R. P., & Schnable, P. S. (1997) *Rf8* and *Rf** mediate unique *T-urf13* transcript accumulation, revealing a conserved motif associated with RNA processing and restoration of pollen fertility in T-cytoplasm maize. *Genetics, 147*, 1367-1379.

Ducos, E., Touzet, P., & Boutry, M. (2001). The male sterile G cytoplasm of wild beet displays modified mitochondrial respiratory complexes. *Plant J., 26*, 171-180.

Duvick, D. N. (1959). The use of cytoplasmic male-sterility in hybrid seed production. *Econ. Bot.,* 13, 167-195.

Duvick, D. N. (1965). Cytoplasmic pollen sterility in corn. *Adv. Genet., 13*, 1-56.

Edwardson, J. N. (1956). Cytoplasmic male sterility. *Bot. Rev., 22*, 696-738.

Engelke, T., & Tatlioglu, T. (2002). A PCR-marker for the CMS_1 inducing cytoplasm in chives derived from recombination events affecting the mitochondrial gene *atp9*. *Theor. Appl. Genet., 104*, 698-702.

Engelke, T., & Tatlioglu, T. (2004). The fertility restorer genes *X* and *T* alter the transcripts of a novel mitochondrial gene implicated in CMS_1 in chives (*Allium scheonoprasum* L.). *Mol. Gen. Genomics, 271*, 150-160.

Farbos, I., Mouras, A., Berterbide, A., & Glimelius, K. (2001). Defective cell proliferation in the floral meristem of alloplasmic plants of *Nicotiana tabacum* leads to abnormal floral organ development and male sterility. *Plant J., 26*, 131-142.

Fauron, C. M.-R., Havlik, M., & Brettell, R. I. S. (1990). The mitochondrial genome organization of a maize fertile cms-T revertant line is generated through recombination between two sets of repeats. *Genetics, 124*, 423-428.

Fauron, C. M., Moore, B., & Casper, M. (1995). Maize as a model of higher plant mitochondrial genome plasticity. *Plant Sci., 112*, 11-32.

Flavell, R. B. (1974). A model for the mechanism of cytoplasmic male sterility with special reference to maize. *Plant Sci. Lett., 3*, 259-263.

Gabay-Laughnan, S. (2000). *Restorers_of_fertility* for CMS-S are present in maize and teosinte from Mexico. *Maydica, 45*, 117-124.

Gabay-Laughnan, S. (2001) High frequency of *restorers-of-fertility* for CMS-EP in *Zea mays* L. *Maydica, 46*, 125-132.

Gabay-Laughnan, S., & Laughnan, J. R. (1994). Male sterility and restorer genes in maize. In M. Freeling and V. Walbot (Eds.), *The maize handbook* (pp. 418-423). New York: Springer-Verlag.

Gabay-Laughnan, S., Zabala, G., & Laughnan, J. R. (1995). S-type cytoplasmic male sterility in maize. In C. S. Levings III, & I .K. Vasil (Eds.), *Advances in cellular and molecular biology of plants: Molecular biology of the mitochondria* (pp. 395-432). Dordrecht: Kluwer.

Gabay-Laughnan, S., Chase, C. D., Ortega, V. M., & Zhao, L. (2004). Molecular-genetic characterization of CMS-S *restorer-of-fertility* alleles identified in Mexican races of maize and teosinte. *Genetics, 166*, 959-970.

Gallagher, L., Betz, S. K., & Chase, C. D. (2002). Mitochondrial RNA editing truncates a chimeric open reading frame associated with S male-sterility in maize. *Curr. Genet., 42*, 179-184.

Gengenbach, B., Koeppe, D., & Miller, R. (1973). A comparison of mitochondria isolated form male-sterile and nonsterile cytoplasm etiolated corn seedlings. *Physiol. Plant., 29,* 103-107.

Gracen, V. E. (1982). Types and availability of male sterile cytoplasms. In W. F. Sheridan (Ed.), *Maize for biological research* (pp. 221-224). Charlottesville: Plant Molecular Biology Association.

Gracen, V. E., & Grogan, C. O. (1974). Diversity and suitability for hybrid production of different sources of cytoplasmic male sterility in maize. *Agron. J., 66*, 654-657.

Granville, D. J., & Gottlieb, R. A. (2002). Mitochondria: Regulators of cell death and survival. *Sci. World J., 2,* 1569-1578.

Green, D.R., & Reed, J. C. (1998). Mitochondria and apoptosis. *Science, 281*, 1309-1312.

Grelon, M., Budar, F., Bonhomme, S., & Pelletier, G. (1994). Ogura cytoplasmic male-sterility (CMS)-associated *orf138* is translated into a mitochondrial membrane polypeptide in male-sterile Brassica cybrids. *Mol. Gen. Genet., 243*, 540-547.

Gualberto, J. M., Bonnard, G., Lamattina, L., & Grienenberger, J. M. (1991). Expression of the wheat mitochondrial *nad3-rps12* transcription unit: correlation between editing and mRNA maturation. *Plant Cell, 3,* 1109-1120.

Gutierres, S., Savar, M., Lelandais, C., Chetrit, P. Diolez, P., Degand, H., et al. (1997). Lack of mitochondrial and nuclear-encoded subunits of complex I and alteration of the respiratory chain in *Nicotiana sylvestris* mitochondrial deletion mutants. *Proc. Natl. Acad. Sci. USA, 94*, 3436-3441.

Hack, E., Lin, C., Yang, H., & Horner, H. (1991). T-URF13 protein from maitochondria of Texas male-sterile maize (*Zea mays* L.). *Plant Physiol., 95*, 861-870.

Handa, H., Gualberto, J. M., & Grienenberger, J.-M. (1995). Characterization of the mitochondrial *orfB* gene and its derivative, *orf224*, a chimeric open reading frame specific to one mitochondrial genome fo the "Polima" male-sterile cytoplasm in rapeseed (*Brasica napus* L.). *Curr. Genet., 28*, 546-552.

Hanson, M. R. (1991). Plant mitochondrial mutations and male sterility. *Annu. Rev. Genet., 25*, 461-486.

Hanson, M. R., & Bentolila, S. (2004). Interactions of mitochondrial and nuclear genes that affect male gametophyte development. *Plant Cell, 16*, S154-S169.

Hanson, M. R., & Conde, M. F. (1985). Functioning and variation of cytoplasmic genomes: Lessons from cytoplasmic-nuclear interactions affecting male fertility in plants. *Intl. Rev. Cytol., 94*, 214-267.

Hanson, M. R. and O. Folkerts. (1992). Structure and function of the higher plant mitochondrial genome. *Intl. Rev. Cytol., 141*, 129-172.

Hanson, M. R., Nivison, H. T., & Conley, C. A. (1995). Cytoplasmic male sterility in Petunia. In C. S. Levings III, & I .K. Vasil (Eds.), *Advances in cellular and molecular biology of plants: Molecular biology of the mitochondria* (pp. 497-514). Dordrecht: Kluwer.

Hanson, M. R., Wilson, R. K., Bentolila, S. Köhler, R. H., & Chen, H.-C. (1999). Mitochondrial gene organization and expression in Petunia male- fertile and sterile plants. *J. Hered., 90*, 362-368.

He, S., Yu, Z. H., Vallejos, C. E., & Mackenzie, S. A. (1995). Pollen fertility restoration by nuclear gene *Fr* in CMS common bean: an *Fr* linkage map and the mode of *Fr* action. *Theor. Appl. Genet., 90*, 1056-1062.

Hoeberichts, F. A., & Woltering, E. J. (2003). Multiple mediators of plant programmed cell death: Interplay of conserved cell death mechanisms and plant-specific regulators. *Bioessays, 25*, 47-57.

Hoffmann, M., Kuhn, J., Daschner, K., & Binder, S. (2001). The RNA world of plant mitochondria. *Prog. Nucleic Acid Res. Mol. Biol., 70*, 119-154.

Hooker, A. L., Smith, D. R., Lim, S. M., & Beckett, L. B. (1970). Reaction of corn seedlings with male-sterile cytoplasm to *Helminthosporium maydis*. *Plant Dis. Reptr., 54*, 708-712.

Iwabuchi, M., Kyozuka, J., & Shimamoto, K. (1993). Processing followed by complete editing of an altered mitochondrial *atp6* RNA restores fertility of cytoplasmic male sterile rice. *EMBO J., 12*, 1437-1446.

Iwabuchi, M., Koizuka, N., Fujimoto, H., Sakai, T., & Imamura, J. (1999). Identification and expression of Kosena rasish (*Raphanus sativus* L. cv. Kosena) homologue of the *ogura* radish CMS-associated gene, *ORF138*. *Plant Mol. Biol., 39*, 183-188.

Janska, H., & Mackenzie, S. A. (1993). Unusual mitochondrial genome organization in cytoplasmic male sterile common bean and the nature of cytoplasmic reversion to fertility. *Genetics, 135*, 869-879.

Johns, C., Lu, M., Lyznik, A., & Mackenzie, S. (1992). A mitochondrial DNA sequence is associated with abnormal pollen development in cytoplasmic male sterile bean plants. *Plant Cell, 4*, 435-449.

Jones, A. (2000). Does the plant mitochondrion integrate cellular stress and regulate programmed cell death? *Trends Plant Sci., 5*, 225-230.

Kadowaki, K., Suzuki, T., & Kazama, S. (1990). A chimeric gene containing the 5' portion of *atp6* is associated with cytoplasmic male sterility of rice. *Mol. Gen. Genet., 224*, 10-16.

Kamps, T. L., McCarty, D. R., & Chase, C. D. (1996). Gametophyte genetics in *Zea mays* L.: Dominance of a restoration of fertility allele *(Rf3)* in diploid pollen. *Genetics, 142*, 1001-1007.

Kaul, M. L. H. (1988) *Male Sterility in higher plants*. Berlin: Springer-Verlag

Kazama, T., & Toriyama, K. (2003). A pentatricopeptide repeat-containing gene that promotes the processing of aberrant *atp6* RNA of cytoplasmic male-sterile rice. *FEBS Lett., 544*, 99-102.

Kempken, F., & Pring, D. R. (1999) Male sterility in higher plants – fundamentals and applications. *Prog. Bot., 60*, 140-166.

Kerr, J. F., Wyllie, A. H., & Currie, A. R. (1972). Apoptosis: A basic biological phenomenon with wide-ranging implications in tissue kinetics. *Br. J. Cancer, 26*, 239-257.

Kennell, J. C., & Pring, D. R. (1989). Initiation and processing of *atp6*, *Turf-13* and ORF 25 transcripts from mitochondria of T-cytoplasm maize. *Mol. Gen. Genet., 216*, 16-24.

Kim, J. S., He, L., & Lemasters, J. J. (2003). Mitochondrial permeability transition: A common pathway to necrosis and apoptosis. *Biochem. Biophys. Res. Commun., 304*, 463-470.

Kitagawa, J., Posluszny, W., Gerrath, J. M., & Wolyn, D. J. (1994). Developmental and morphological analyses of homeotic cytoplasmic male sterile and fertile carrot flowers. *Sex. Plant Reprod.*, 7, 41-50.

Kofer, W., Glimelius, K., & Bonnett, H. T. (1991). Modifications of mitochondrial DNA cause changes in floral development in homeotic-like mutants of tobacco. *Plant Cell, 3*, 759-769.

Köhler, R., Horn, R., Lossl, A., & Zetsche, K. (1991). Cytoplasmic male sterility in sunflower is correlated with the co-transcription of a new open reading frame with the *atpA* gene. *Mol. Gen. Genet., 227*, 369-376.

Koizuka, N., Imai, R., Iwabuchi, M., Sakai T., & Imamura, J. (2000). Genetic analysis of fertility restoration and accumulation of ORF125 mitochondrial protein in the Kosena radish (*Raphanus sativus* L. cv. Kosena) and a *Brassica napus* restorer line. *Theor. Appl. Genet., 100*, 949-955.

Koizuka, N., Imai, R., Fujimoto, H., Hayakawa, T., Kimura, Y., Kohno-Murase, J., et al. (2003). Genetic characterization of a pentatricopeptide repeat protein gene, *orf687*, that restores fertility in the cytoplasmic male-sterile Kosena radish. *Plant J., 34*, 407-415.

Komori, T., Ohta, S., Murai, N., Takakura, Y., Kuraya, Y., Suzuki, S., et al. (2004). Map-based cloning of a fertility restorer gene, *Rf-1* in rice (*Oryza sativa* L.). *Plant J., 37*, 315-325.

Krishnasamy S., & Makaroff, C. A. (1994). Organ-specific reduction in the abundance of a mitochondrial protein accompanies fertility restoration in cytoplasmic male-sterile radish. *Plant Mol. Biol., 26*, 935-946.

Kubo, N., & Kadowaki, K. (1997). Involvement of 5' flanking sequence for specifying RNA editing sites in plant mitochondria. *FEBS Lett., 413*, 40-44.

Kumar, R., & Levings, C. S. III. (1993). RNA editing of a chimeric maize mitochondrial gene transcript is sequence specific. *Curr. Genet., 23*, 154-159.

Lam, E., Katon, N., & Lawton, M. (2001) Programmed cell death, mitochondria and the plant hypersensitive response. *Nature, 411*, 848-853.

Laser, K. D., & Lersten, N. R. (1972). Anatomy and cytology of microsporogenesis in cytoplasmic male sterile angiosperms. *Bot. Rev., 38*, 425-454.

Laughnan, J. R., & Gabay, S. J. (1973). Mutations leading to nuclear restoration of fertility in S male-sterile cytoplasm in maize. *Theor. Appl. Genet., 43*, 109-116.

Laughnan, J. R., & Gabay, S. J. (1975). An episomal basis for instability of S male sterility in maize and some implications for plant breeding. In C. W. Birky, Jr., P. S. Perlman, & T. J. Byers (Eds.), *Genetics and the biogenesis of mitochondria and chloroplasts* (pp. 330-349). Columbus: Ohio State University Press.

Laughnan, J. R., & Gabay, S. J. (1978). Nuclear and cytoplasmic mutations to fertility in S male-sterile maize. In B. D. Walden (Ed.), *Maize breeding and genetics* (pp. 427-446). New York: John Wiley.

Laver, H., Reynolds, S. J., Monéger, F., & Leaver, C. J. (1991). Mitochondrial genome organization and expression associated with cytoplasmic male sterility in sunflower (*Helianthus annuus*). *Plant J., 1*, 185-193.

Lee, S. J., Earle, E. D., & Gracen, V. E. (1980). The cytology of pollen abortion in S cytoplasmic male-sterile corn anthers. *Amer. J. Bot., 67*, 237-245.

Lelandais, C., Albert, B., Gutierres, S., De Paepe, R., Godelle, B., Vedel, F., & Chétrit, P. (1998). Organization and expression of the mitochondrial genome in *Nicotiana sylvestris* CMSII mutant. *Genetics, 150*, 873-882.

Levings, C. S. III (1993).Thoughts on cytoplasmic male sterility in cms-T maize. *Plant Cell, 5*, 1285-1290.

L'Homme, Y., Stahl, R. J., Li, X. Q., Hameed, A., & Brown, G.G. (1997). Brassica nap cytoplasmic male sterility is associated with expression of a mtDNA region containing a chimeric gene similar to the pol CMS-associated *orf224* gene. *Curr. Genet., 31*, 325-335.

Li, L. Y., Luo, X., & Wang, X. (2001). Endonuclease G is an apoptotic DNase when released from mitochondria. *Nature, 412*, 95-99.

Li, X. Q., Jean, M., Landry, B. S., & Brown, G. G. (1998). Restorer genes for different forms of Brassica cytoplasmic male sterility map to a single nuclear locus that modifies transcripts of several mitochondrial genes. *Proc. Natl. Acad. Sci. USA, 95*, 10023-10037.

Linke, B., Nothnagel, T., & Börner, T. (2003). Flower development in carrot CMS plants: mitochondria affect the expression of MADS box genes homologous to GLOBOSA and DEFICIENS. *Plant J., 34*, 27-37.

Liu, F., Cui, X., Horner, H. T., Weiner, H., & Schnable, P. S. (2001). Mitochondrial aldehyde dehydrogenase activity is required for male fertility in maize. *Plant Cell, 13*, 1063-1078.

Mackenzie, S. A., & Chase, C. D. (1990). Fertility restoration is associated with loss of a portion of the mitochondrial genome in cytoplasmic male-sterile common bean. *Plant Cell, 2*, 905-912.

Makaroff, C. A. (1995) Cytoplasmic male sterility in Brassica species. In C. S. Levings III, & I. K. Vasil (Eds.), *Advances in cellular and molecular biology of plants: Molecular biology of the mitochondria* (pp. 395-432). Dordrecht: Kluwer.

Makaroff. C. A., and Palmer, J. D. (1988) Mitochondrial DNA rearrangements and transcriptional alterations in the male-sterile cytoplasm of Ogura radish. *Mol. Cell. Biol., 8*, 1474-1480.

Maxwell, D. P., Nickels, R., & McIntosh, L. (2002). Evidence of mitochondrial involvement in the transduction of signals required for the induction of genes associated with pathogen attack and senescence. *Plant J., 29*, 269-279.

Mayer, B., & Oberbauer, R. (2003) Mitochondrial regulation of apoptosis. *News Physiol. Sci., 18*, 89-94.

McIntosh, L., Eichler, T., G. Gray, Maxwell, D., Nickels, R., & Wang, Y. (1998). Biochemical and genetic controls exerted by plant mitochondria. *Biochem. Biophys. Acta, 1365*, 278-284.

Miller, R. J., & Koeppe, D. E. (1971). Southern corn leaf blight: Susceptible and resistant mitochondria. *Science, 173*, 67-69.

Mulligan, R. M., & Maliga, P. (1998). RNA editing in mitochondria and plastids. In J. Bailey-Serres, & D. R. Gallie (Eds.), *A look beyond transcription: Mechanisms determining mRNA stability and translation in plants* (pp. 153-161). Rockville: American Society of Plant Physiologists.

Monéger, F., Mandaron, P., Niogret, M. F., Freyssinet, G., & Mache, R. (1992). Expression of mitochondrial genes during microsporogenesis in maize. *Plant Physiol., 99*, 396-400.

Monéger, F., Smart, S. J., & Leaver, C. J. (1994). Nuclear restoration of cytoplasmic male sterility in sunflower is associated with the tissue-specific regulation of a novel mitochondrial gene. *EMBO J., 13,* 8-17.

Murai , K., Takumi, S., Koga, H., & Ogihara, Y. (2002). Pistilloidy, homeotic transformation of stamens into pistil-like structures, caused by nuclear-cytoplasm interaction in wheat. *Plant J., 29*, 169-181.

Nakajima, Y., Yamamoto, T., Muranaka, T., & Oeda, K. (2001). A novel *orfB*-related gene of carrot mitochondrial genomes that is associated with homeotic cytoplasmic male sterility (CMS). *Plant Mol. Biol., 46*, 99-107.

Newton, K. J., & Gabay-Laughnan, S. J. (1998). Abnormal growth and male sterility associated with mitochondrial DNA rearrangements in plants. In K. K. Singh (Ed.), *Mitochondrial DNA mutations in aging, disease and cancer*, (pp. 365-381). Georgetown: R. G. Landes Company.

Newton, K. J., Gabay-Laughnan, S., & De Paepe, R. (2004). Mitochondrial mutations in plants. In D. Day, H. Millar, & J. Whelan, (Eds.), *Advances in photosynthesis and respiration* (vol. 17) *Plant mitochondria from genome to function* (in press). Dordrecht: Kluwer.

Nivison, H. T., & Hanson, M. R. (1989). Identification of a mitochondrial protein associated with cytoplasmic male sterility in Petunia. *Plant Cell, 1*, 1121-1130.

Nivison H. T., Sutton, C. A., Wilson, R. K., & Hanson, M. R. (1994). Sequencing, processing, and localization of the Petunia CMS-associated mitochondrial protein. *Plant J., 5*, 613-623.

Orrenius, S., Zhivotovsky, B., & Nicotera, P. (2003). Regulation of cell death: the calcium-apoptosis link. *Nat. Rev. Mol. Cell. Biol., 4*, 552-565.

Palmer, J. D. & Herbon, L. A. (1988). Plant mitochondrial DNA evolves rapidly in structure, but slowly in sequence. *J. Mol. Evol., 28*, 87-97.

Pla, M., Mathieu, C., De Paepe, R., Chetrit, P., & Vedel, F. (1995). Deletion of the last two exons of the mitochondrial *nad7* gene results in lack of the NAD7 polypeptide in a *Nicotiana sylvestris* CMS mutant. *Mol. Gen. Genet., 248*, 79-88.

Pring, D. R., & Lonsdale, D. M. (1985). Molecular biology of higher plant mitochondrial DNA. *Intl. Rev. Cytol., 97*, 1-46.

Pring, D.R., & Lonsdale, D. M. (1989). Cytoplasmic male sterility and maternal inheritance of disease susceptibility in maize. *Annu. Rev. Phytopathol., 27*, 483-502.

Pring, D. R., Chen, W., Tang H. V., Howad, W., & Kempken, F. (1998). Interaction of mitochondrial RNA editing and nucleolytic processing in the restoration of male-fertility in sorghum. *Curr. Genet. 33*, 429-436.

Pring, D. R., Tang, H. V., Howad, S., & Kempken, F. (1999). A unique two-gene gametophytic male sterility system in Sorghum involving a possible role of RNA editing in fertility restoration. *J. Hered., 90*, 386-393.

Pruitt, K. D., & Hanson, M. R. (1991). Transcription of the Petunia mitochondrial CMS-associated *Pcf* locus in male sterile and fertility-restored lines. *Mol. Gen. Genet., 227*, 348-355.

Raff, M. (1998). Cell suicide for beginners. *Nature, 396*, 119-122.

Reed, J. C. (1997). Cytochrome *c*: Can't live with it – can't live without it. *Cell, 91*, 559-562.

Rhoades, D. M., Levings, C. S. III, & Siedow, J. N. (1995). URF13, a ligand-gated, pore-forming receptor for T-toxin in the inner membrane of cms-T mitochondria. *J. Bioenerg. Biomembr., 27*, 437-445.

Rhoades, M. M. (1931). Cytoplasmic inheritance of male sterility in *Zea mays. Science, 73*, 340-341.

Rhoades, M. M. (1933). The cytoplasmic inheritance of male sterility in *Zea mays. J. Genet., 27*, 71-93.

Rhoades, M. M. (1950). Gene induced mutation of a heritable cytoplasmic factor producing male sterility in maize. *Proc. Natl. Acad. Sci. USA, 36,* 634-635.

Richter, T. E., & Ronald, P. C. (2000). The evolution of disease resistance genes. *Plant Mol. Biol., 42*, 195-204.

Rottman, W. H., Brears, T., Hodge, T. P., & Lonsdale, D. M. (1987). A mitochondrial gene is lost via homologous recombination during reversion of CMS-T maize to fertility. *EMBO J., 6,* 1541-1546.

Rizzuto, R., Pinton, P., Ferrari, D., Chami, M., Szabadkai, B., Magalhaes, P. J., et al. (2003). Calcium and apoptosis: facts and hypotheses. *Oncogene, 22*, 8619-8627.

Sabar, M., Gagliardi, D., Balk, J., & Leaver, C. J. (2003). ORFB is a subunit of F1F0-ATP synthase: insight into the basis of cytoplasmic male sterility in sunflower. *EMBO Rep., 4*, 381-386.

Sakai, H., Medrano, L. J., & Meyerowitz, W. M. (1995). Role of SUPERMAN in maintaining Arabidopsis floral whorl boundaries. *Nature, 378*, 199-203.

Sarria, R., Lyznik, A., Vallejos, C. E., and Mackenzie, S.A. (1998). A cytoplasmic male sterility-associated mitochondrial peptide in common bean is post-translationally regulated. *Plant Cell, 10*, 1217-1228.

Saur-Jacobs, M., & Wade, M. J. (2003). A synthetic review of the theory of gynodioecy. *Am. Nat., 161*, 837-851.

Schnable, P. S., & Wise, R. P. (1994). Recovery of heritable, transposon-induced, mutant alleles of the *rf2* nuclear restorer of T-cytoplasm maize. *Genetics, 136*, 1171-1185.

Schnable P. S., & Wise, R. P. (1998). The molecular basis of cytoplasmic male sterility and fertility restoration. *Trends Plant. Sci., 3*, 175-180.

Shaver, D. L. (1956). A summary of an extensive screening project for "T" and "S" sterile cytoplasm restorers. *Maize Genet. Coop. Newslett., 30*, 155-157. (cited by permission).

Sing, M., & Brown, G. G. (1991). Suppression of cytoplasmic male sterility by nuclear genes alters expression of a novel mitochondrial gene region. *Plant Cell, 12*, 1349-1362.

Singh, M., & Brown, G. G. (1993). Characterization of expression of a mitochondrial gene region associated with the Brassica "Polima" CMS: developmental influences. *Curr. Genet., 24*, 316-322.

Singh, M., Hamel, N., Menassa R., Li, X-Q , Young, B., Jean, M. et al., (1996). Nuclear genes associated with a single Brassica CMS restorer locus influence transcripts of three different mitochondrial gene regions. *Genetics, 143*, 505-516.

Singh, V. P., & Kobabe, G. (1969). Cyto-morphological investigation on male-sterility in *Allium schoenoprasum* L. *Indian J. Genet. Plant Breed., 29*, 241-247.

Small, I., & Peeters, N. (2000). The PPR motif - a TPR-related motif prevalent in plant organellar proteins. *Trends Biochem. Sci., 25*, 46-47.

Small I., Suffolk, R., & Leaver, C. J. (1989). Evolution of plant mitochondrial genomes via substoichiometric intermediates. *Cell, 58*, 69-76.

Song, J., & Hedgcoth, C. (1994a). Influence of nuclear background on transcription of a chimeric gene (*orf256*) and *coxI* in fertile and cytoplasmic male sterile wheats. *Genome, 37*, 203-209.

Song, J., & Hedgcoth, C. (1994b). A chimeric gene (*orf256*) is expressed as protein only in cytoplasmic male-sterile lines of wheat. *Plant Mol. Biol., 26*, 535-539.

Tadege, M., Dupuis, I. I., & Kuhlemeier, C. (1999). Ethanolic fermentation: new functions for an old pathway. *Trends Plant Sci., 4*, 320-325.

Tang H. V, Chang, R., & Pring, D. R. (1998). Cosegregation of single genes associated with fertility restoration and transcript processing of Sorghum mitochondrial *orf107* and *urf209*. *Genetics, 150*, 383-391.

Tang, H. V., Pring, D. R., Shaw, L. C., Salazar, R. A., Muza, F. R., Yan, B., & Schertz, K. F. (1996). Transcript processing internal to a mitochondrial open reading frame is correlated with fertility restoration in male-sterile Sorghum. *Plant J., 10*, 123-133.

Ullstrup, A. J. (1972). The impacts of the Southern corn leaf blight epidemics of 1970-1971. *Annu. Rev. Phytopathol. 10*, 37-50.

Vanlerberghe, G. C., & McIntosh, L. (1994). Mitochondrial electron transport regulation of nuclear gene expression: studies with the alternative oxidase gene of tobacco. *Plant Physiol., 105*, 867-874.

Vanlerberghe G. C., & McIntosh, L. (1997). Alternative oxidase: From gene to function. *Annu. Rev. Plant Physiol. Plant Mol. Biol., 48*, 703-734.

von Wettstein, F. (1924). Morphologie und physiologie des formwechsels der moose auf genetischer grundlage. I. *Zeits. Ind. Abst. Vererb., 33*, 1-236.

Wallace, D. C. (2001). A mitochondrial paradigm for degenerative diseases and ageing. *Novartis Found. Symp., 235*, 247-263.

Ward, G. C. (1995). The Texas male-sterile cytoplasm of maize. In C. S. Levings III, & I. K. Vasil (Eds.), *Advances in cellular and molecular biology of plant: Molecular biology of the mitochondria* (pp. 433-460). Dordrecht: Kluwer.

Ward, G. C., & Levings, C. S. III. (1991). The protein-encoding gene *T-urf13* is not edited in maize mitochondria. *Plant Mol. Biol., 17*, 1083-1088.

Warmke, H. E., & Lee, S. -L. J. (1977). Mitochondrial degeneration in Texas cytoplasmic male-sterile corn anthers. *J. Hered., 68*, 213-222.

Warmke, H. E., & Lee, S. -L. J. (1978). Pollen abortion in T cytoplasmic male-sterile corn (*Zea mays*): a suggested mechanism. *Science, 200*, 561-562.

Wen, L., & Chase, C. D. (1999a). Pleiotropic effects of a nuclear *restorer-of-fertility* locus on mitochondrial transcripts in male-fertile and S male-sterile maize. *Curr. Genet., 35*, 521-526.

Wen, L., & Chase, C. D. (1999b). Mitochondrial gene expression in developing male gametophytes of male-fertile and S male-sterile maize. *Sex. Plant. Reprod., 11*, 323-330.

Wen, L., Tang, H. V., Chen, W., Chang, R., Pring, D. R., Klein, P. E., et al. (2002). Development and mapping of AFLP markers linked to the sorghum fertility restorer gene *rf4*. *Theor. Appl. Genet., 104*, 577-585.

Wen, L., Ruesch, K. L. Ortega V. M., Kamps T. L., Gabay-Laughnan, S., & Chase, C. D. (2003). A nuclear *restorer-of-fertility* mutation disrupts accumulation of mitochondrial ATP synthase subunit □ in developing pollen of S male-sterile maize. *Genetics, 165*, 771-779.

Williams, M. A., Kutcher, B. M., & Mulligan, R. M. (1998). Editing site recognition in plant mitochondria: the importance of 5' flanking sequences. *Plant mol. Biol., 36*, 229-237.

Wise, R. P., & Pring, D. R. (2002). Nuclear-mediated mitochondrial gene regulation and male fertility in higher plants: Light at the end of the tunnel? *Proc. Natl. Acad. Sci. USA., 99*, 10240-10242.

Wise, R. P., Pring, D. R., & Gengenbach, B.G. (1987a). Mutation to male fertility and toxin insensitivity in Texas (T)-cytoplasm maize is associated with a frame shift in a mitochondrial open reading frame. *Proc. Natl. Acad. Sci. USA, 84*, 2858-2862.

Wise, R. P., Fliss, A. E,, Pring, D. R., & Gengenbach, B.G. (1987b). *Urf13-T* of T cytoplasm maize mitochondria encodes a 13-kD polypeptide. *Plant Mol. Biol., 9*, 121-126.

Wise, R. P., Dill, C. L., & Schnable, P. S. (1996). Mutator-induced mutations of the *rf1* nuclear fertility restorer of T-cytoplasm maize alter the accumulation of the T-*urf13* mitochondrial transcripts. *Genetics, 143*, 1383-1394.

Wise, R. P., Bronson, C. R., Schnable, P. S., & Horner, H. T. (1999). The genetics, pathology, and molecular biology of T-cytoplasm male sterility in maize. *Adv. Agron., 65*, 79-130.

Yan, B. & Pring, D. R. (1997). Transcriptional initiation sites in sorghum mitochondrial DNA indicate conserved and variable features. *Curr. Genet., 32*, 287-295.

Young, E. G. & Hanson, M. R. (1987). A fused mitochondrial gene associated with cytoplasmic male sterility is developmentally regulated. *Cell, 50*, 41-49.

Zabala, G., Gabay-Laughnan, S., & Laughnan, J. R. (1997). The nuclear gene *Rf3* affects the expression of the mitochondrial chimeric sequence R implicated in S-type male sterility in maize. *Genetics, 147*, 847-850.

Zubko, M. K. (2004). Mitochondrial tuning fork in nuclear homeotic functions. *Trends Plant Sci., 9*, 61-64.

Zubko, M. K., Zubko, E. I., Ruban, A. B., Adler, K., Mock, H. -P., Misera, S., et al. (2001). Extensive developmental and metabolic alterations in cybrids *Nicotiana tabacum* (+ *Hyoscyamus niger*) are caused by complex nucleo-cytoplasmic incompatibility. *Plant J., 25*, 627-639.

CHAPTER 23

THE USE OF CYTOPLASMIC MALE STERILITY FOR HYBRID SEED PRODUCTION

MICHAEL J. HAVEY

USDA-ARS, Department of Horticulture, University of Wisconsin, 1575 Linden Drive, Madison, WI USA

Abstract Cytoplasmically inherited male sterility (CMS) results from an interaction between the organellar and nuclear genomes that conditions the failure to produce functional pollen . CMS provides an expedient mechanism to produce large populations of male-sterile plants for commercial F1 hybrid seed production. In cases where the F1 hybrid crop must produce pollen and set seed, male sterility can be reversed by nuclear-encoded restorer-of-fertility alleles. Although the unfortunate epidemic of Southern Corn Leaf Blight on T-cytoplasmic maize revealed the dangers of hybrid-seed production using a single source of CMS, no other genetic vulnerability to disease or stress has been attributed directly to a CMS gene, in spite of the worldwide use of CMS. In some cases, mitochondrial-encoded CMS genes have been linked to undesirable, plastid-encoded traits, and nuclear restorer-of-fertility alleles have been linked to undesirable nuclear-encoded traits, but these linkages were successfully broken by somatic cell genetics and conventional plant breeding, respectively. The use of CMS to produce hybrid seed is very cost effective and has been widely exploited in a plethora of agronomic and horticultural crops. The purpose of this chapter is to list and review the major sources of CMS used commercially to produce hybrid seed.

1. INTRODUCTION

Male-sterility-inducing cytoplasms were recognized early in the 1900s. Bateson and Gairdner (1921) reported that male sterility in flax was inherited from the female parent, although genes passed from both the male and female parents affected its expression. Chittenden and Pellow (1927) recognized that male sterility in flax was due to an interaction between the cytoplasm and nucleus, with the male sterility conditioned by the homozygous recessive nuclear genotype in combination with the sterility-inducing cytoplasm. Rhoades (1931, 1933) observed that male sterility in maize was inherited only through the female parent, although occasional male fertile progeny were observed. In a seminal paper, Jones and Clarke (1943) established in

H. Daniell and C.D. Chase (eds.), Molecular Biology and Biotechnology of Plant Organelles 623—634. © 2004 *Springer. Printed in the Netherlands.*

onion that male sterility is conditioned by the interaction of the male-sterile (S) cytoplasm with the homozygous recessive genotype at a single male-fertility restoration locus (*Ms*) in the nucleus, written in onion as S *msms*. A dominant allele at *Ms* conditions male fertility in plants possessing S cytoplasm (S *Ms*--). Plants possessing the normal (N) male-fertile cytoplasm are male fertile regardless of the genotype at *Ms*. In the same manuscript, Jones and Clarke (1943) described the technique used today to exploit cytoplasmic-genic male sterility (CMS) for the production of hybrid seed. A male-sterile (S *msms*) inbred line [termed the A line by Jones and Clarke (1943)] is seed propagated by growing it in isolation with a male-fertile maintainer (B) line that is N cytoplasmic and homozygous recessive at the nuclear male-fertility restoration locus (N *msms*). All seed harvested off of the A line is male sterile (S *msms*); all seed harvested off of the maintainer line is male fertile (N *msms*). The A line then becomes the female parent for hybrid-seed production. This system to seed propagate CMS inbred lines has been widely used for hybrid production of many crops.

The use of CMS for hybrid seed production received a "black eye" after the epidemic of *Bipolaris maydis* on T-cytoplasmic maize (Pring and Lonsdale, 1989). This epidemic is often cited as a classic example of genetic vulnerability of our major crop plants. However the *B. maydis* epidemic on T cytoplasmic maize appears to be the exception rather than the rule, and such a serious epidemic has never been observed in any other CMS system, even though single sources of CMS are widely used around the world to produce hybrid seed. This does not mean that breeders should ignore cytoplasmic uniformity, only that the epidemic of Southern Corn Leaf Blight on T-cytoplasmic maize appears to date to be an anomaly associated with this specific source of CMS. Onion is an example of continued cytoplasmic uniformity in a major crop plant. The majority of hybrid-onion seed is produced using S cytoplasm (Havey, 1995; 2000), which traces back to a single onion plant identified in Davis, CA, in 1925 (Jones and Emsweller, 1936). RFLPs in the organellar genomes have demonstrated that, in addition to almost all hybrids, many significant open-pollinated populations of onion also possess S cytoplasm (Courcel et al., 1989; Havey, 1993; Satoh et al., 1993; Havey and Bark, 1994). Bulb and seed production can occur simultaneously, such as in the Treasure Valley of Idaho or the Central Valley of California, resulting in year round cultivation of cytoplasmically uniform onion.

Numerous hybrid crops representing significant (>50%) components of world production are generated without CMS, such as tomato and maize by hand emasculation, spinach and the cucurbits by monoecy, and the Brassicas by self incompatibility (Janick, 1998). Hybrids of other crop plants may be produced using systems of genic (nuclear) male sterility. An example is leek (*Allium ampeloprasum* L.), a vegetable crop for which growers demand high uniformity. Because leek is an autotetraploid, it shows extreme inbreeding depression and relatively poor uniformity. A source of genic male sterility was identified in leek (Smith and Crowther, 1995). Leek hybrids are produced by propagating *in vitro* male-sterile plants conditioned by this single source of genic male sterility. The asexually propagated leek plants are used as the female to produce hybrids that show significantly greater uniformity than open pollinated (OP) cultivars (Smith and

Crowther, 1995). Engineered sources of nuclear male sterility have been developed in model systems (Mariani et al., 1990; Hernould et al., 1993; Perez-Prat and van Lookeren Campagne, 2002). A problem with these nuclear transformants is that they segregate for male fertility or sterility and must be over planted and rogued by hand or sprayed with herbicides to remove male-fertile plants. Nevertheless *Brassica* hybrid-seed is produced using the Bayer SeedLink[tm] system, in which a transgene conditioning male sterility is linked to herbicide resistance. Hybrid Brassicas produced using the SeedLink[tm] system have reached a significant market share by combining herbicide tolerance with consistent high performance.

An important consideration for hybrid development using CMS systems is the requirement, or not, for male-fertility restoration. For vegetable, fruit, or forage crops, nuclear restoration of male fertility in the hybrid is not necessary. This simplifies the production of hybrids because effort can concentrate on maintainer line development, without concern whether the pollinator restores male fertility in the hybrid. For crops with seeds as the economically important product, such as canola, sunflower, or maize, one or both of the hybrid's parents must bring in male-fertility restoration factors or the male-sterile hybrid seed must be blended with male-fertile hybrid seed (see maize section below).

In this chapter, my intention is to document and review the use of CMS to produce hybrids for a wide range of economically important plants. Plants were chosen to illustrate the various uses of CMS to produce hybrid seed, and the omission of any specific crop is unintentional. I contacted leading plant researchers to estimate the predominance of hybrids in these crops and types of CMS used to produce the hybrids. Some researchers in the private sector were willing to share their opinions, but asked that I not report their names. Molecular-genetic aspects of CMS and fertility restoration, including many of the systems discussed below, are reviewed in chapter 22 of this volume.

2. CMS IN HYBRID CROP PRODUCTION

2.1. Alfalfa

Alfalfa or lucerne (*Medicago sativa* L.) is an outcrossing, autotetraploid legume widely grown as forage for ruminant animals. Historically, alfalfa cultivars or populations were developed by selecting and interpollinating among superior plants. Sources of CMS in alfalfa have been described (Pedersen and Stucker, 1970; Brown and Bingham, 1984) and significant specific combining abilities among plants demonstrated (Pedersen and Hill, 1972). Although I could find no example of true single-cross hybrids in alfalfa, there exist on the North American market "hybrid swarms". These populations are created by planting together in seed fields mixtures of CMS plants with male-fertile pollinators. My colleague, Dr. Edwin Bingham at the University of Wisconsin, estimates that these seed-production fields possess 10 to 20% CMS plants. The CMS plants increase outcrossing, contributing to superior forage yields by reduced inbreeding.

2.2. Alliums

Hybrid-onion cultivars dominate the North American, European, and Japanese markets, although specific regions still grow OP populations for specific markets (such as the Walla Walla Sweets in Washington, USA). Numerous areas of the world with significant onion production also grow OP cultivars (e.g., Argentina, Turkey, Australia, India, and New Zealand). Reasons for the prevalence of OP cultivars in these regions include lower seed prices, or that adapted indigenous populations possess exclusively or predominantly the male-sterile cytoplasm, making the extraction of maintainer lines difficult to impossible (Havey, 1993). Two sources of CMS are commercially used to produce hybrid onion. The most widely used source of CMS is S cytoplasm, as described by Jones and Clarke (1943). T cytoplasm is a second source of CMS (Berninger, 1965) used to produce hybrids in Europe and Japan (Havey, 2000). The vast majority of onion hybrids are produced using S cytoplasm, likely because it was the first source of onion CMS released in 1951 and was made available both in long- and short-day germplasms (Goldman et al., 2000). CMS has also been described in chive (*A. schoenoprasum* L.) and Japanese bunching onion (*A. fistulosum* L.) and exploited for hybrid production in both crops. CMS in chive shows unique sensitivity to tetracycline, which restores male fertility (Tatlioglu, 1986) and tetracycline susceptibility is conditioned by recessive alleles at a single locus (*aa*) (Tatlioglu and Wricke, 1988). CMS in Japanese bunching onion is conditioned by the male-sterile cytoplasm and recessive alleles at two nuclear restorer loci (Moue and Uehara, 1985). Although these sources of CMS are used to produce hybrid and Japanese bunching onion cultivars, I was not able to acquire information on the prevalence of hybrid versus OP cultivars for these Alliums. Alloplasmic sources of CMS conditioned by the cytoplasm of *A. galanthum* have been transferred to onion, shallot, and the Japanese bunching onion (*A. fistulosum* L.) (Havey, 1999; Yamashita and Tashiro, 1999; Yamashita et al., 1999). The main advantage of the galanthum CMS system is that nuclear male-fertility restoration loci appear to be rare or non-existent (Havey, 1999), indicating that many populations may be used to maintain this source of CMS.

2.3. Brassicas

Almost 100% of broccoli and cabbage cultivars are hybrid, with fewer (approximately 50%) cauliflower and collard cultivars being hybrid. For the Brassicas, hybrids were historically produced using self-incompatibility (SI). Although the SI system is very effective for hybrid production, some self pollination can occur, allowing competitors to screen for and isolate one, or in some cases both, of the hybrid's parental inbred lines. This has been an impetus for the development of an effective CMS system to produce hybrid Brassicas. Ogura (1968) identified CMS in radish, and this cytoplasm was transferred by backcrossing to *B. oleracea*. Unfortunately the Ogura CMS was associated with cold susceptibility, conditioned by the chloroplast genome. This defect was overcome by several different laboratories after protoplast fusion and organellar sorting to combine the CMS conditioned by the radish mitochondrial genome with cold tolerance conditioned by the *Brassica* chloroplast genome. A cold-tolerant form of Ogura CMS was patented by Syngenta and is used to produce hybrid Brassicas. Other companies and public labs (Walters and Earle, 1993) have developed similar cold-tolerant CMS lines independent of the Syngenta source. One problem with the Brassica CMS is that the flower structure of some male-sterile lines is not as attractive to insects, which can negatively affect seed production (Mark Farnham, USDA, personal communication).

The proportion of hybrid canola or rapeseed is approximately 80% in China, 50% in Germany and the USA, 35% in Canada and Austria, 20% in the Czech Republic, and France, 10% to 15% in Australia, Denmark, Poland, and the UK. In Canada and Europe, approximately 60% of the hybrid seed is produced using the nuclear SeedLink^{tm} system. The other 40% is produced using CMS, largely based on the INRA Ogura system the was discovered in *Raphanus sativus*, transferred to *B. napus* after protoplast fusion, and later improved upon to make ogu-INRA CMS. A small part of the hybrid market is based on the Polima cytoplasm and MSL (male-sterile Lembke) systems. The Polima CMS was identified by Dr. Fu in China in the Polima cultivar of Polish *B. napus*. In the USA, approximately 50% of the hybrid market is produced using the Polima cytoplasm, while the SeedLink^{tm} and Ogura systems are increasing in market share registered (Dale Burns, personal communication). In China, approximately 95% of hybrids are based on Polima type CMS with the remaining 5% produced using genic male sterility (Dr. Fu, personal communication). In Australia, all hybrid seed production is based on Ogura-INRA CMS (Dr. Radisa Gjuric, DSV Canada Inc., Winnipeg, Manitoba).

Canola and rapeseed hybrids produced using the SeedLink^{tm} system continue to increase in market share. Both the Ogura and Polima sources of CMS have been successfully commercialised. Polima hybrids are in decline due to problems with sterility breakdown under both high and low temperatures. Flowers are also not attractive to pollinators due to flower morphology and absence of nectaries. Ogura CMS is more stable than Polima, but possesses similar problems with flower morphology. Under lower temperatures (<12 C), Ogura CMS plants develop severe chlorosis. A major improvement made by INRA in France produced stable CMS and monogenic restoration. Initially, the restoration factor introduced from radish (*Raphanus sativus*) was linked with high glucosinolate contents in the seed. Because canola is low in erucic acid and glucosinolates, it was not possible to use Ogura CMS to develop commercial hybrids for registration in Canada, but possible for the US and Australia. This explains the relative higher percentage of CMS hybrids in these two markets. The first low glucosinolates restorers were developed first by Advanta Seeds in 1996 (first hybrid in 2001), followed by other breeders in Canada and EU. At the moment the majority of breeding organizations are focused on ogu-INRA and this system is expected to become the dominant hybrid system worldwide (Dr. Radisa Gjuric and Dr. Dale Burns, personal communications).

2.4. Carrot

For carrot, two sources of CMS are used to produce hybrid seed. The predominant CMS is the petaloid male-sterile cytoplasm, in which the anthers are replaced by a whorl of petals (Eisa and Wallace, 1969). The second source of CMS is brown anther, in which complete flowers produce shrivelled anthers with no pollen (Welch and Grimball 1947). Male-fertility restoration for these sources of CMS is complexly inherited with up to five loci affecting this trait (Peterson and Simon, 1986). Hybrid carrot represents approximately 50% of the world market, with petaloidy as the predominant source of CMS used for hybrid-seed production in the US (over 90%) and world (70%) markets. The rest of hybrid-carrot seed is produced using the brown anther source of CMS (Philipp Simon, USDA, personal communication). The pentaloid CMS is generally preferred because of less frequent reversion to male fertility; however seed yields on the brown-anther CMS are generally higher (Peterson and Simon, 1986).

2.5. Lolium

Two sources of CMS have been developed in ryegrass (*Lolium perenne* L.), one from an interspecific cross between *L. multiflorum* and *L. perenne* (Wit, 1974) and one from an intergeneric cross between *Festuca pratensis* and *L. perenne* (Connolly and Wright-Turner, 1984). However no ryegrass hybrids are commercially available at this time (Michael Casler, USDA, personal communication).

2.6. Maize

Maize was the first hybrid crop to be produced on a large scale. In Europe, Japan, Australia, New Zealand, and the US, the field and sweet-corn production is based exclusively on hybrid cultivars. Open-pollinated cultivars are still widely cultivated in areas where the cost of hybrid seed is prohibitive. The prevalence of CMS in the production of hybrids is not clear. In the case of sweet corns, the relatively short life (three to five years) of a specific hybrid means that there may not be enough time to convert new inbreds to CMS by backcrossing and the hybrid will be produced by emasculation. When CMS is used to produce hybrid seed, C and S cytoplasms are common; T cytoplasm is not used due to its susceptibility to *B. maydis*. The preference of C versus S cytoplasm is dependent on the specific population and stability of the male sterility over environments. Many maize hybrids do not carry nuclear male-fertility restoration alleles and, as a result, are male sterile. The hybrid seed is then blended with N-cytoplasmic, hand emasculated hybrid seed that is obviously male fertile and provides enough pollen in the production field. Most maize breeders reported that the blended hybrids are 50% CMS with 50% N-cytoplasmic plants. The recent incorporation of transgenes in maize hybrids has decreased the use of CMS to produce hybrid seed, because of the time requirement to backcross sources of CMS into transgenic lines.

2.7. Millets

The prevalence of pearl-millet hybrids ranges from nearly 100% in the US, to approximately 50% in India, to very low in some regions of Africa (Dr. Wayne Hanna, University of Georgia, personal communication). CMS in pearl millet (*Pennisectum glaucum* L. R. Br.) was discovered in the 1950s (Burton, 1958; Menon, 1959) and first released as the male-sterile inbred Tifton 23A, termed the A1 or milo cytoplasm. Most, if not all, of the world's hybrid pearl millet is produced using the A1 cytoplasm (Smith and Chowdhury, 1989). Other sources of CMS were identified (Burton and Athwal, 1967), but were not stable enough for commercial use (Smith and Chowdhury, 1989) or were difficult to find nuclear restorers (A4). The A4 cytoplasm is useful for production of forage hybrids that do not require male-fertility restoration.

2.8. Pepper

Hybrid cultivars dominate the production of some market classes of peppers. For example, approximately 100% of bell peppers are hybrid, as compared to 50-75% of Jalapenos. The other major types of pepper are almost exclusively open-pollinated (Paul Bosland, New Mexico State University, personal communication). The only source of CMS used to produce hybrid-pepper cultivars was described by Peterson, (1958).

2.9. Rye

Rye is an important grain crop in Germany, Poland, and parts of Russia. In Germany, hybrid cultivars represent approximately 60% of rye production; hybrid rye is starting to be cultivated in Poland at about 5 to 10% of production (Dr. Thomas Miedaner, University of Hohenheim, Germany, personal communication). Rye is naturally outcrossing due to gametophytic self-incompatibility; however self-fertile plants have been identified allowing for inbred line development. Hybrid rye is produced using the Pampa (P) cytoplasmic source of CMS, which was introgressed at the University of Hohenheim, Germany, from a primitive Argentina population (Geiger and Schnell, 1970). The P cytoplasm is widely used because its male sterility is stable across environments and research has focused on identification effective restorer genes (Geiger and Miedaner, 1996; Miedaner et al., 2004). A second CMS (G cytoplasm) was developed in the former German Democratic Republic and is used to produce one registered hybrid in Germany (Novus). Other sources of CMS have been described and their relationships to the P and G cytoplasms are unclear.

2.10. Sorghums

Hybrid sorghum predominates the US and world markets. Hybrid sudangrass cultivars also exist, however OP populations still represent a significant component of production. CMS is routinely used to produce hybrid seed of sorghum and sundangrass. Although several sources of CMS have been described (A1 through A4), the predominant male-sterile cytoplasm is the A1 (milo) cytoplasm (Holland and Stephans, 1954; Pring et al., 1995; Schertz et al., 1997). Fertility restoration of the A1 cytoplasm requires the action of two complementary nuclear male-fertility restorers (*Rf1* and *Rf2*), similar to T-cytoplasmic maize. There was some effort in producing male sterile sorghum x sudangrass hybrids in systems with fewer fertility restorers. However the emergence of sorghum ergot (*Claviceps africana*) in the Western Hemisphere in the 1990s caused abandonment of this effort. The infection site for sorghum ergot is the unfertilized stigma and it was feared that large populations of CMS sudangrass hybrids could serve as an inoculum source for other sorghums (Dr. Jeff Peterson, University of Nebraska, personal communciation).

2.11. Sugar and table beets

CMS is used to produce hybrids of both table and sugar beets. Sugar-beet cultivars are almost exclusively hybrids in the US and Europe, with some open-pollinated cultivars grown in regions of the world with lower inputs such as Morocco and Egypt (Mitch McGrath, USDA, personal communication). Both diploid and triploid sugar-beet hybrids are widely grown. Approximately 50% of table-beet cultivars are hybrid; OP cultivars are still produced with the advantage of cheaper seed (I. Goldman, personal communication). The sole source of CMS used to produce hybrid beets was described by Owen in 1945. In beets, male sterility is conditioned by the interaction of the S cytoplasm with recessive alleles at two nuclear loci (*xx zz*). Owen (1945) recognized that the approach of Jones and Clarke (1943) could be used to produce beet hybrids. A second alloplasmic source of CMS from *Beta maritima* has been described (Boutin et al., 1987), but has not been used to date to produce commercial beet hybrids.

2.12. Sunflower

Hybrid sunflower dominates many of the word's production areas, including 100% in Argentina, Australia, Europe, India, South Africa, Turkey, and USA. Other countries, such as China, have significant (>35%) hybrid production. Hybrid sunflower is produced using the PET1 cytoplasm, which was developed by transferring the cytoplasm of *Helianthus petiolaris* to *H. annuus* (LeClerq, 1969). Although other sources of CMS in sunflower can be distinguished by organellar polymorphisms (Crouzillat et al., 1991), PET1 is the only cytoplasm used to produce sunflower hybrids. The predominance of the PET1 source of CMS is due primarily to its stable expression and complete male-fertility restoration by one dominant locus (*Rf1*) (Jerry Miller, USDA, personal communication).

3. FUTURE ASPECTS

This chapter documents the worldwide use of CMS to produce competitive hybrid cultivars. Major investments of time and resources are required to backcross a male-sterility-inducing cytoplasm into elite lines. These generations of backcrossing could be avoided by transformation of an organellar genome of the elite male-fertile inbred to produce female inbred lines for hybrid seed production. Because the male-fertile parental and male-sterile transformed lines would be developed from the same inbred, they should be highly uniform and possess the same nuclear genotype (excluding mutations and residual heterozygosity). Therefore, the male-fertile parental line becomes the maintainer line to seed-propagate the newly transformed male-sterile line (Havey, in press). A few generations of seed increases would produce a CMS-maintainer pair for hybrid seed production. An additional advantage of organellar transformation would be the diversification of CMS sources used in commercial hybrid-seed production. Transformation of the mitochondrial genome would allow breeders to introduce different male-sterility-inducing factors into superior inbred lines. These male-sterility-inducing factors could be from the same species, such as T-cytoplasmic onion (Berninger, 1965), or possibly from another species, such as the mitochondrial factors conditioning CMS from Petunia (Nivison et al., 1994), sorghum (Tang et al., 1996), or sunflower (Moneger et al., 1994). Although the nuclear genome has been successfully transformed with male-sterility-inducing factors (Mariani et al., 1990; Hernould et al., 1993), these nuclear transformants segregate for male fertility or sterility and must be rogued by hand or with herbicides to remove male-fertile plants (Perez-Prat and van Lookeren Campagne, 2002).

Transformation of the mitochondrial genome to produce CMS lines is more practical for hybrid-seed production because removal of male-fertile plants is eliminated or greatly reduced. The chloroplast genome of *Chlamydomonas* (Boynton et al., 1988; Kindle et al., 1991) was the first organellar genome to be successfully transformed, followed by tobacco (Svab and Maliga, 1993; O'Neill et al., 1993; Koop et al., 1996; Daniell et al., 2004; Devine and Daniell, 2004, see also chapter 16 of this book), *Arabidopsis* (Sikdar et al., 1998), carrot (Kumar et al., 2004a), Cotton (Kumar et al., 2004b), soybean (Dufourmantel et al., 2004) and tomato (Ruf et al., 2001). Although transformation of the mitochondrial genome has been reported for *Chlamydomonas* (Randolph-Anderson et al., 1993) and yeast (Butow et al., 1996), there is no report of mitochondrial transformation of a higher plant. Introduction of a male-sterility inducing transgene into one of the organellar genomes of a higher plant would be a major breakthrough in the production of male-sterile inbred lines. This has been indeed accomplished recently by expressing the phaA gene coding for β-ketothiolase via the tobacco chloroplast genome (see chapter 16). The transgenic lines were normal except for the male sterility phenotype, lacking pollen. Scanning electron microscopy revealed a collapsed morphology of the pollen grains. Transgenic lines followed an accelerated anther developmental pattern, affecting their development and maturation, resulting in aberrant tissue patterns. Abnormal thickening of the outer wall, enlarged endothecium and vacuolation, which decreased the inner space of the locules, affecting pollen grain and resulting in the irregular shape and collapsed phenotype. This is the first report of engineered cytoplasmic male sterility and offers a new tool for transgene containment for both nuclear and organelle genomes. This technique would be of great potential importance in the production of hybrid crops by avoiding generations of backcrossing, an approach especially advantageous for crop plants with longer generation times (Havey, 2004).

4. REFERENCES

Bateson, W., & Gairdner, A. E. (1921). Male sterility in flax subject to two types of segregation. *Jour. Genet., 11*, 269-275.

Berninger, E. (1965). Contribution a l'etude de la sterilite male de l'oignon (*Allium cepa* L.). *Ann. Amelior. Plantes, 15*, 183-199.

Boutin, V., Pannenbecker, G., Ecke, W., Schewe, G., Saumitou-Laprade, P., Jean, R., et al. (1987). Cytoplasmic male sterility and nuclear restorer genes in a natural population of *Beta maritima*: genetical and molecular aspects. *Theor. Appl. Genet., 73*, 625-629.

Boynton, J. E., Gillham, N. W., Harris, E. H., Hosler, J. P., Johnson, A. M., Jones, A. R. et al. (1988). Chloroplast transformation in *Chlamydomonas* with high velocity microprojectiles. *Science, 240*, 1534-1537.

Brown, D. E., & Bingham, E. T. (1984). Hybrid alfalfa seed production using a female-sterile pollenizer. *Crop Sci., 24*, 1207-1208.

Burton, G. W. (1958). Cytoplasmic male sterility in pearl millet (*Pennisectum glaucum* L. R. Br.). *Agron. J., 50*, 230.

Burton, G. W., & Athwal, D. S. (1967). Two additional sources of cytoplasmic male-sterility in pearl millet and their relationship to Tift 23A. *Crop Sci., 7*, 209-211.

Butow, R. A., Henke, R. M., Moran, J. V., Selcher, S. M., & Perlman, P. S. (1996). Transformation of *Saccharomyces cerevisiae* mitochondria using the biolistic gun. *Methods Enzymol., 264*, 265-278.

Chittenden, R. J., & Pellow, C. A. (1927). A suggested interpretation of certain cases of anisogeny. *Nature, 119*, 10-12.

Connolly, V., & Wright-Turner, R. (1984). Induction of cytoplasmic male sterility into rye grass (*Lolium perenne* L.). *Theor. Appl. Genet., 68*, 449-453.

Courcel, A. de, Veder, F., & Boussac, J. (1989). DNA polymorphism in *Allium cepa* cytoplasms and its implications concerning the origin of onions. *Theor. Appl. Genet., 77*, 793-798.

Crouzillat, D., De La Canal, L., Perrault, A., Ledoigt, G., Vear, F., & Serieys, H. (1991). Cytoplasmic male sterility in sunflower: comparison of molecular biology and genetic studies. *Plant Mol. Biol., 16*, 415–426.

Daniell, D., Carmona-Sanchez, O., & Burns, B. B. (2004a). Chloroplast derived antibodies, biopharmaceuticals and edible vaccines. In R. Fischer & S. Schillberg (Eds.) *Molecular Farming* (pp.113-133). Weinheim: WILEY-VCH Verlag.

Devine, A. L., & Daniell, H. (2004). Chloroplast genetic engineering. In S. Moller (Ed.), *Plastids* (pp. 283-320). United Kingdom: Blackwell Publishers.Dufourmantel, N., Pelissier, B., Garçon, F.,Peltier, J. M., & Tissot, G. (2004). Generation of fertile transplastomic soybean. *Plant Mol. Biol.*, in press.

Eisa, H. M., & Wallace, D. H. (1969). Morphological and anatomical aspects of petaloidy in the carrot, *Daucus carota* L. *J. Amer. Soc. Hort. Sci., 94*, 545-548.

Geiger, H. H., & Schnell, F. W. (1970). Cytoplasmic male sterility in rye (*Secale cereale* L.). *Crop Sci., 10*, 590 –593.

Geiger, H. H., & Miedaner, T. (1996). Genetic basis and phenotypic stability of male fertility restoration in rye. *Vortr. Pflanzenzüchtg, 35*, 27-38.

Goldman, I. L., Schroeck, G., & Havey, M. J. (2000). History of public onion breeding programs and pedigree of public onion germplasm releases in the United States. *Plant Breed. Rev., 20*, 67-103.

Havey, M. J. (1993). A putative donor of S-cytoplasm and its distribution among open-pollinated populations of onion. *Theor. Appl. Genet., 86*, 128-134.

Havey, M. J. (1999). Seed yield, floral morphology, and lack of male-fertility restoration of male-sterile onion (*Allium cepa*) populations possessing the cytoplasm of *Allium galanthum*. *J. Amer. Soc. Hort. Sci., 124*, 626-629.

Havey, M. J. (2000). Diversity among male-sterility-inducing and male-fertile cytoplasms of onion. *Theor. Appl. Genet., 101*, 778-782.

Havey, M. J. (2004). A new paradigm for the breeding of longer-generation hybrid crops. *Acta Hort.* in press.

Havey, M. J., & Bark, O. H. (1994). Molecular confirmation that sterile cytoplasm has been introduced into open-pollinated populations of Grano-type onion. *J. Amer. Soc. Hort. Sci., 119*, 90-93.

Hernould, M, Suharsono, S. Litvak, S. Araya, A. & Mouras, A. (1993). Male-sterility induction in transgenic tobacco plants with an unedited atp9 mitochondrial gene from wheat. *Proc. Natl. Acad. Sci. USA, 90*, 2370-2374.

Janick, J. (1998). Hybrids in horticultural crops. In K. R. Lamkey and J. E. Staub (Eds.), Concepts and *Breeding of Heterosis in Crop Plants, Crop Science Society of America Special Publication 25* (pp. 45-57). Madison: American Society of Agronomy.

Jones, H., & Clarke, A. (1943). Inheritance of male sterility in the onion and the production of hybrid seed. *Proc. Amer. Soc. Hort. Sci., 43*, 189-194.

Jones, H., & Emsweller, S. (1936). A male sterile onion. *Proc. Amer. Soc. Hort. Sci., 34*, 582-585.

Kindle, K. L., Richards, K. L., & Stern, D. B. (1991). Engineering the chloroplast genome: techniques and capabilities for chloroplast transformation of *Chlamydomonas reinhardtii*. *Proc. Natl. Acad. Sci. USA, 88*, 1721-1725.

Koop, H., Steinmuller, K., Wagner, H., Robler, C., Eibl, C., & Sacher, L. (1996). Integration of foreign sequences into the tobacco plastome via polyethylene glycol-mediated protoplast transformation. *Planta, 199*, 193-201.

Kumar, S., Dhingra, A. & Daniell, H. (2004a). Plastid expressed *betaine aldehyde dehydrogenase* gene in carrot cultured cells, roots and leaves confers enhanced salt tolerance. *Plant Physiol.*, in press.

Kumar, S., Dhingra, A., & Daniell, H. (2004b). Manipulation of gene expression facilitates cotton plastid transformation of cotton by somatic embryogenesis & maternal inheritance of transgenes. *Plant Mol. Biol.*, in press.

LeClerq, P. (1969). Une stérilité mâle cytoplasmique chez le tournesol. *Ann. Amelior. Plant. (Paris), 19*, 99-106.

Mariani, C., de Beuckeleer, M., Truettner, J., Leemans, J., & Goldberg, R. B. (1990). Induction of male sterility in plants by a chimaeric ribonuclease gene. *Nature, 347*, 737-741.

Menon, P. M. (1959). Occurrence of cytoplasmic male sterility in pearl millet (*Pennisectum typhoides* S. and H.). *Curr. Sci., 28*, 165-167.

Miedaner, T., Wilde, P., & Wortmann, H. (2004). Combining ability of non-adapted sources for male-fertility restoration in Pampa and alternative CMS cytoplasms of hybrid rye. *Plant Breed.*, in press.

Moneger, F., Smart, C. J., & Leaver, C. J. (1994). Nuclear restoration of cytoplasmic male sterility in sunflower is associated with the tissue-specific regulation of a novel mitochondrial gene. *EMBO J., 13*, 8-17.

Moue, T., & Uehara, T. (1985). Inheritance of cytoplasmic male sterility in *Allium fistulosum* L. (Welsh onion). *J. Japan. Soc. Hort. Sci., 53*, 432-437.

Nivison, H. T., Sutton, C. A., Wilson, R. K., & Hanson, M. R. (1994). Sequencing, processing, and localization of the petunia CMS-associated mitochondrial protein. *Plant J., 5*, 613-623.

O'Neill, C., Horvath, G., Horvath, E., Dix, P., & Medgyesy, P. (1993). Chloroplast transformation in plants: polyethylene glycol (PEG) treatment of protoplasts is an alternative to biolistic delivery systems. *Plant J., 3*, 729-738.

Ogura, H. (1968). Studies on the new male sterility in Japanese radish, with special references to the utilization of this sterility towards the practical raising of hybrid seeds. *Mem. Fac. Agric. Kagoshima Univ., 6*, 39–78.

Owen, F.V. (1945). Cytoplasmically inherited male sterility in sugar beets. *J. Agric. Res., 71*, 423-440.

Pedersen, M. W., & Stucker, R. E. (1970). Evidence of cytoplasmic male sterility in alfalfa. *Crop Sci., 9*, 767-770.

Pedersen, M. W., & Hill, R. R. (1972). Combining ability in alfalfa hybrids made with cytoplasmic male sterility. *Crop Sci., 12*, 500-502.

Perez-Prat, E., & van Lookeren Campagne, M. M. (2002). Hybrid seed production and the challenge of propagating male-sterile plants. *Trends Plant Sci., 7*, 199-203.

Peterson, C. E., & Simon, P. W. (1986). Carrot breeding. In M. J. Bassett (Ed.), *Breeding Vegetable Crops*) (pp. 321-356). Westport: AVI Publishing.

Peterson, P. A. (1958). Cytoplasmically inherited male-sterility in *Capsicum*. *Amer. Natural., 92*, 111-119.

Pring, D. R., Tang, H. V., & Schertz, K. F. (1995). Cytoplasmic male sterility and organelle DNAs of sorghum. In C. S. Levings, III, and I. K. Vasil (Eds.), *The Molecular Biology of Plant Mitochondria* (pp. 461-495). Dordrecht: Kluwer Academic Publishers.

Pring, D., & Lonsdale, D. M. (1989). Cytoplasmic male sterility and maternal inheritance of disease susceptibility in maize. *Ann. Rev. Phytopathol., 27*, 483–502.

Randolph-Anderson, B. L., Boynton, J. E., Gillham, N. W., Harris, E. H., Johnson, A. M., Dorthu, M. P. & Matagne, R. F. (1993). Further characterization of the respiratory deficient *dum-1* mutation of *Chlamydomonas reinhardtii* and its use as a recipient for mitochondrial transformation. *Mol. Gen. Genet., 238*, 235-244.

Rhoades, M. M. (1931). Cytoplasmic inheritance of male sterility in *Zea mays*. *Science, 73*, 340-341.

Rhoades, M. M. (1933). The cytoplasmic inheritance of male sterility in *Zea mays*. *Jour. Genet., 27*, 71-95.

Ruf, S., Hermann, M., Berger, I. J., Carrer, H., & Bock, R. (2001). Stable genetic transformation of tomato plastids and expression of a foreign protein in fruit. *Nat. Biotech., 19*, 870-875.

Satoh, Y., Nagai, M., Mikami, T., & Kinoshita, T. (1993). The use of mitochondrial DNA polymorphism in the classification of individual plants by cytoplasmic genotypes. *Theor. Appl. Genet., 86*, 345-348.

Schertz, K.F., Sivaramakrishnan, S., Hanna W., Mullet, J., Sun, Y., Murty, U., et al. (1997). Alternate cytoplasms and apomixis of sorghum and pearl millet. In *Proceedings of the International Conference on Genetic Improvement of Sorghum and Pearl Millet. USAISD Title XII Collaborative Research Support Program on Sorghum and Pearl Millet, and International Crops Research Institute for the Semi-Arid Tropics* (pp. 213-223).

Sikdar, S. R., Serino, G., Chaudhuri, S., & Maliga, P. (1998). Plastid transformation in *Arabidopsis thaliana*. *Plant Cell Rep., 18*, 20-24.

Smith, B., & Crowther, T. (1995). Inbreeding depression and single cross hybrids in leeks (*Allium ampeloprasum* ssp. *porrum*). *Euphytica, 86*, 87-94.

Smith, R. L., & Chowdhury, M. K. U. (1989). Mitochondrial DNA polymorphism in male-sterile and fertile cytoplasms of pearl millet. *Crop Sci., 29*, 809-814.

Svab, Z., & Maliga, P. (1993). High frequency plastid transformation by selection for a chimeric *aadA* gene. *Proc. Natl. Acad. Sci. USA, 90*, 913-917.

Tang, H. V., Pring, D. R., Muza, F. R., & Yan, B. (1996). Sorghum mitochondrial *orf25* and a related chimeric configuration. *Curr. Genet., 29*, 265-274.

Tatlioglu, T. (1986). Influence of tetracycline on the expression of cytoplasmic male sterility (cms) in chives (*Allium schoenoprasum* L.). *Plant Breed., 97*, 46-55.

Tatlioglu, T., & Wricke, G. (1988). Genetic control of tetracycline-sensitivity of cytoplasmic male sterility (cms) in chives (*Allium schoenoprasum* L.). *Plant Breed., 100*, 34-40.

Walters, T. W., & Earle, E. D. (1993). Organellar segregation, rearrangement and recombination in protoplast fusion-derived *Brassica oleracea* calli. *Theor. Appl. Genet., 85*, 761-769.

Welch, J. E., & Grimball, E. L. (1947). Male sterility in carrot. *Science, 106*, 594.

Wit, F. (1974). Cytoplasmic male sterility in rye grasses (*Lolium* spp.) detected after intergeneric hybridization. *Euphytica, 23*, 31-38.

Yamashita, K., & Tashiro, T. (1999). Possibility of developing male sterile line of shallot (*Allium cepa* L. *Aggregatum* group) with cytoplasm from *A. galanthum* Kar. et Kir. *J. Japan. Soc. Hort. Sci., 68*, 256-262.

Yamashita, K., Arita, H., & Tashiro,Y. (1999). Cytoplasm of a wild species *Allium galanthum* Kar. et Kir. is useful for developing male sterile line of *A. fistulosum* L. *J. Japan. Soc. Hort. Sci., 68*, 788-797.

CHAPTER 24

SOMATIC CELL CYBRIDS AND HYBRIDS IN PLANT IMPROVEMENT

W. W. GUO[1], X. D. CAI[1], AND J. W. GROSSER[2]

[1]*National Key Laboratory of Crop Genetic Improvement, Huazhong Agricultural University, Wuhan 430070, PR China*
[2]*Citrus Research and Education Center, IFAS, University of Florida, Lake Alfred, FL 33850, USA*

Abstract. Somatic hybridization and cybridization have great potential in plant improvement. In this chapter, types of somatic hybrids and cybrids, somatic fusion methods, selection schemes and characterization methods are reviewed. The inheritance patterns of cytoplasmic genomes in plant somatic hybrids and cybrids, as revealed by molecular markers from known examples, are summarized. The exploitation of somatic hybridization and cybridization for the recovery of organelle-encoded traits for plant improvement is discussed, with focus on examples from rice, rapeseed, tomato and potato, citrus and other higher plants.

1. INTRODUCTION

Somatic hybridization of plants by protoplast fusion is a technique that combines somatic cells from two different cultivars, species or genera in an effort to regenerate novel germplasm. This technique can circumvent such problems as sexual incompatibility, polyembryony, and male or female sterility encountered in conventional sexual crossing. Since the first successful report on somatic hybridization with tobacco (Charlson et al., 1972) hundreds of reports have been published during the past three decades; these extend the procedures to additional plant genera and evaluate the utilization potential of somatic hybrids in many crops including rice, rapeseed, tomato, potato and citrus. Excellent reviews of the subject are available. Kumar and Cocking (1987) reviewed organelle genetics in higher plants through protoplast fusion. Waara and Glimelius (1995) reviewed somatic hybridization through the early 1990s. Somatic hybrids have been reviewed three times previously in Plant Breeding Reviews, twice generally (Bravo and Evans,

H. Daniell and C.D. Chase (eds.), Molecular Biology and Biotechnology of Plant Organelles, 635—659.

1985; Johnson and Veilleux, 2001b) and once specifically for citrus (Grosser and Gmitter 1990). A more recent review of *Citrus* somatic hybridization was provided by Grosser et al. (2000). Potato somatic hybridization was reviewed by Waclaw et al. (2003). The present review focuses primarily on literature published in the 1990s and 2000s, with particular emphasis on the patterns of cytoplasmic inheritance in somatic hybridization as revealed by novel molecular markers, and the successful exploitation of this technology for the recovery of organelle-encoded traits for plant improvement.

1.1. Types of somatic hybrids

Somatic hybrids can be classified into three types: symmetric somatic hybrids, asymmetric somatic hybrids, and cytoplasmic hybrids (cybrids). Symmetric somatic hybridisation is defined as the combination of nuclear and cytoplasmic genetic information of both parents. Asymmetric somatic hybridization is incomplete, with the loss of some cytoplasmic or nuclear DNA, and this type of hybridization has been used to introduce fragments of the nuclear genome from one parent (donor) into the intact genome of another one (recipient). Cybrids harbor only one parental nuclear genome and either the cytoplasmic genome of the other (non-nuclear) parent or a combination of both parents. Both symmetric and asymmetric fusion experiments can generate these three types of somatic hybrids. With the development of somatic hybridization technology, many new avenues have been adopted to create somatic hybrids. The evolution of such techniques is continuing, as Binsfeld et al. (2000) recently obtained asymmetric hybrids in sunflower via microprotoplast fusion with partial chromosome transfer from the micronuclear parent.

1.2. Methods to produce cybrids

Symmetric hybrids often have no economic value because of the associated increase in ploidy level, and the combining of all nuclear encoded traits of both parents. Cybridization is a more attractive alternative for crop improvement because one or more traits can be added while maintaining cultivar integrity (just as with genetic transformation). Three methods are routinely used to create cybrids.

1.2.1. Asymmetric fusion treatment

Cybrids can be obtained by asymmetric fusion between irradiated donor protoplasts whose nuclei have been destroyed, and recipient protoplasts whose organelle genomes usually have been metabolically inhibited by iodoacetate (IOA). As a result, the heterokaryons combine vital cytoplasm from the donor parent with the intact nucleus from the recipient parent, resulting in the creation of asymmetric hybrids or cybrids (Vardi et al., 1987; Varotto et al., 2001). In addition to donor-recipient asymmetric hybridization, IOA treatment of one parent (or irradiation of one parent) and keeping the other parent intact can also be applied to create cybrids. Kochevenko et al. (1999) once obtained cybrids via protoplast fusion between mesophyll protoplasts of a chlorophyll deficiency mutant *Lycopersicon peruvianum* var. dentatum and gamma-irradiated mesophyll protoplasts of *L. esculentum.*

1.2.2. Cytoplast isolation and fusion

Cytoplast-protoplast fusion was introduced by Maliga et al. (1982). Presently, two procedures for eliminating the nuclear DNA are used, one is by cytochalasin B treatment (Wallin et al., 1978), and the other is by a discontinuous percoll/mannitol gradient ultracentrifugation (Lorz et al., 1981). This method can also realize transfer of organelle-encoded traits to obtain cybrids (Spangenberg et al. 1990; Sakai and Imamura, 1990). For example, Sigareva and Earle (1997) used this method to isolate cytoplasts. Because many nucleated protoplasts were present, the cytoplast/protoplast fraction was then subjected to gamma-irradiation, and finally they successfully transferred a desirable male-sterile cytoplasm into cabbage.

1.2.3. Cybrids produced by symmetric fusion

Besides asymmetric fusion and cytoplast-protoplast fusion, intraspecific, interspecific or intergeneric symmetric hybridization can spontaneously produce cybrids in higher plants. This is a common phenomenon in some species, especially tobacco and citrus. In interspecific symmetric somatic hybridization in tobacco, half of all regenerated plants were confirmed to be cybrids (Gleba and Sytnik, 1984). Citrus cybrids can sometimes be produced as a byproduct from the application of standard symmetric somatic hybridization procedures. To date, more than 40 of 250 parental combinations produced cybrids via symmetric fusion (Grosser et al., 2000; Guo et al., 2004a).

1.3. Somatic fusion methods

Among hundreds of published documents with regard to plant protoplast fusion, the two primary methods are fusion induced by polyethylene glycol (PEG-induced fusion) and electrofusion (Grosser and Gmitter, 1990; Guo et al., 1998). PEG-induced fusion has advantages of not requiring special equipment, low cost, and high frequency of heterokaryon formation. Electrofusion relies on two different electrical pulses. Protoplasts are brought into intimate contact during the first pulse called dielectrophoresis; and the second pulse is a very short burst of intense direct current, which results in membrane fusion. Electrofusion has the advantages of convenience, no cell toxicity, and high frequency heterokaryon formation.

1.4. Selection schemes

For successful somatic hybrid regeneration, it is necessary to select the hybrid products from among the unfused and homo-fused protoplasts. An efficient selection system avoids the tedious identification of somatic hybrids among large numbers of regenerated calli or plants. Several schemes have been developed for somatic hybrid selection. These schemes include selective media (Hossain et al., 1994); metabolic inhibitors (Shimonaka et al., 2002; Xiang et al., 2003); complementation systems such as chlorophyll deficiency complementation (Dragoeva et al., 1999), auxotroph complementation (Wolters et al., 1995), resistance markers (Ilcheva et al., 2001) and double mutants (Conner et al., 1995); individual selection and culture (Saha et al., 2001); and application of the green fluorescent protein (GFP) marker gene (Olivares-Fuster et al., 2002).

The GFP gene has been a newly exploited marker to select somatic hybrids. It derives from the aquatic jellyfish *Aequorea victora,* and emits stable and distinctive green fluorescence when expressed by living cells, without any cofactors or subtrates but oxygen. For this reason, transgenic plants expressing the GFP gene have been recently used as a parent in somatic hybridization. The potential of GFP as a somatic hybridization marker was first documented by Olivares-Fuster et al. (2002), by using a transgenic citrange plant expressing GFP as a parent in a somatic fusion experiment. GFP was shown to be useful for the continuous monitoring of the fusion process, identification of hybrid colonies, and selection of somatic hybrid embryos or plants. Guo and Grosser (2004) further used the GFP marker in citrus somatic fusion and provided direct evidence of somatic hybrid vigor.

1.5. Methods to characterize somatic hybrids and their organellar genomes

Methods for characterizing regenerated somatic hybrid lines include morphological evaluation, cytological evaluation by chromosome counting and flow cytometry analysis, isozyme analysis, and molecular characterization by DNA markers. Of course, none of these methods alone could sufficiently characterize the somatic hybrid or cybrid nature, and a combination of these methods is generally required.

In recent years, molecular markers have been exploited, especially to characterize the organellar genomes of somatic hybrids and cybrids. The widely used markers with the ability to characterize the nature of somatic hybrids include RAPD (random amplified polymorphic DNA), RFLP (restriction fragment length polymorphism), AFLP (amplified fragment length polymorphism), SSR (simple sequence repeat), GISH (genomic *in situ* hybridization), PCR-RFLP or CAPS (cleaved amplified polymorphic sequence) and more recently cpSSR (chloroplast microsatellites or chloroplast simple sequence repeats). Often multiple techniques are used because results from one method are not sufficient to draw unequivocal conclusions about hybridity/cybridity.

1.5.1. RAPD analysis

RAPD was the first molecular marker to identify somatic hybrids because it requires only a small amount of DNA and can characterize somatic hybridity at an early stage. It is still widely used (Szczerbakowa et al., 2003; Xiang et al., 2003; Ishikawa et al., 2003). Total genomic DNA was usually used in RAPD analysis.

1.5.2. RFLP analysis

RFLP has been successfully applied to identify nuclear and cytoplasmic genomes of somatic hybrids. RFLP with mitochondrial DNA (mtDNA)-specific probes and chloroplast DNA (cpDNA)-specific probes is routinely used to study inheritance patterns of cytoplasmic genomes in plant somatic hybrids and cybrids. However, RFLP requires large amounts of DNA and is relatively expensive to assay.

1.5.3. GISH analysis

GISH has proven to be efficient for identification of somatic hybrids, especially it has the ability to display evidence for interspecific chromosome recombination and was used to determine chromosome origin (Xiang et al., 2003; Szarka et al., 2002; Escalante et al., 1998; Collonnier et al., 2003). GISH is also very powerful for studying somatic hybrid meiosis (Garriga-Calderé et al., 1999; Gavrilenko et al., 2001).

1.5.4. SSR analysis

SSR or microsatellites are tandem repeats interspersed throughout the genome and can be amplified using primers that flank these regions. SSR loci enabled distinction of somatic hybrids from parental somaclones, and rapid DNA extraction with SSR analysis enabled screening of calluses (Provan et al., 1996) to identify somatic hybrid tissue prior to plant regeneration (Johnson et al., 2001a). In somatic hybrids between *Solanum tuberosum* and *Solanum sanctae-rosae*, SSR analysis revealed that somatic hybrids contained the genetic background of *S. tuberosum* with some specific markers from *S. sanctae-rasae* (Harding and Millam, 2000). In *Citrus*, SSR analysis of seven randomly selected tetraploids and three triploids showed that they had specific fragments from both fusion parents, thereby confirming their hybridity (Fu et al., 2003).

1.5.5. PCR-RFLP or CAPS markers

Isolation of sufficient DNA for RFLP analysis is time-consuming and labor intensive. However, PCR can be used to amplify very small amounts of DNA to the level required for RFLP analysis, usually in 2-3 h. Therefore, more samples can be analyzed in a shorter time. CAPS analysis using mitochondrial or chloroplast specific primers (universal primer pairs), which is simpler, less expensive and more efficient compared with RFLPs, has been extensively applied to study the cytoplasmic constitution of somatic hybrids arising from intergeneric and interspecific combinations (Bastia et al., 2001; Cheng et al., 2003b; Guo et al., 2002; Lotfy et al., 2003; Collonnier et al., 2003).

1.5.6. Other molecular markers

AFLP analysis was applied by several researchers to characterize somatic hybrids in higher plants (Brewer et al., 1999; Guo et al., 2002; Ilcheva et al., 2001; Wang et al., 2003). Inter-simple sequence repeat (ISSR) -PCR is a rapid and sensitive method for detecting the introgression of alien DNA into cultivars via somatic hybridization. Matthews et al. (1999) applied it to characterize potato somatic hybrids. Scarano et al. (2002) concluded that ISSR-PCR technique was a useful method to characterize allotetraploid somatic hybrids of mandarins.

Chloroplast microsatellites or simple sequence repeats (cpSSR) analysis, is an even more efficient and simpler technique than CAPS. The technique was applied by Cheng et al. (2003a) to verify the chloroplast genome origin of citrus somatic hybrids, which was the first report on cytoplasmic inheritance analysis of somatic hybrids in higher plants by cpSSR. By this method, even intraspecific citrus somatic hybrids from closely related parents ('Page' tangelo + 'Murcott' tangor) could be distinguished (Guo et al., 2004b).

2. INHERITANCE PATTERNS OF CYTOPLASMIC GENOMES IN PLANT SOMATIC HYBRIDS

Unlike sexual hybridization of most plant species, somatic hybridization can not only realize the recombination of nuclear genomes, but also achieve the recombination of cytoplasmic genomes. This makes it possible to obtain asymmetric hybrids or cybrids transferring useful cytoplasm-controlled agronomic traits from donor to recipient parent. Chloroplast and mitochondria are two main organelles in the cytoplasm. In the past decade, various reports on somatic hybridization have put particular emphasis on the segregation pattern analysis of chloroplast and mitochondria genomes.

2.1. Inheritance of chloroplast genomes

Chloroplast genomes of both fusion partners coexist in the preliminary product after protoplast fusion, but with the division of hybrid cells, callus formation and plant regeneration, somatic hybrids/cybrids subsequently have been found in most cases to possess only one parental chloroplast type. Random segregation of chloroplast genomes generally occurs following protoplast fusion (Escalante et al., 1998; Wolters et al., 1995; Bonnema et al., 1992; Mohapatra et al., 1998), especially in the somatic hybrids derived from symmetric fusion. In the somatic hybrid plants of *Diplotaxis catholica* + *B. juncea*, Mohapatra et al. (1998) found that one of the five hybrid plants analysed derived its chloroplast from *D. catholica*, two hybrids had chloroplasts of *B. juncea* origin, and two hybrid plants maintained mixed population of chloroplasts. An appropriately 1:1 cpDNA segregation ratio was revealed by RFLP in the somatic hybrids between tomato and *Solanum lycopersicon* or tobacco (Wolters et al., 1993a; Escalante et al., 1998). Lossl et al. (1994) also found chloroplast segregation was with a 1:1 ratio, in all but one fusion combination.

In addition to the random segregation of the chloroplast genome in most fusion experiments, non-random or biased segregation of cpDNA was also detected in somatic hybrids/cybrids from some other fusions. The inheritance of the chloroplast genome in some asymmetric hybrids or cybrids was biased towards the recipient parent (Shimonaka et al., 2002; Samoylov et al., 1996; Spangenberg et al., 1995; Xu et al. 2003), whereas others only possess the chloroplast genome of donor parent (Ratushnyak et al., 1995). For example, in asymmetric somatic hybrid plants between an interspecific tomato hybrid (EP) as donor and eggplant (E) as recipient, all plants have chloroplast DNA fragments specific for the recipient (Samoylov et al., 1996). Besides asymmetric fusion, the chloroplast genomes of many symmetric somatic hybrids also showed to be biased to one or the other parent (Donaldson et al., 1993; Buiteveld et al., 1998). Fock et al. (2000) have analyzed the chloroplast genome of hybrids obtained from protoplast fusion between *Solanum tuberosum* and a wild species. Most of the hybrids possessed the chloroplast genome of the wild species, and only two contained the *Solanum tuberosum* chloroplast type. In symmetric and asymmetric fusions between *Brassica napus* and *Lesquerella fendleri* (Skarzhinskaya et al., 1996), chloroplast and mitochondrial DNA analysis revealed a biased segregation that favored *B. napus* organelles in the hybrids from symmetric fusion. The bias was even stronger in the hybrids from the asymmetric fusion, where no hybrids with *L. fendleri* organelles were found. Defined numbers (1-5) of (donor) chloroplasts were transferred into (recipient) protoplasts of plastid albino mutants by subprotoplast/protoplast microfusion, and Eigel et al. (1992) observed a high frequency of cell lines and regenerated shoots recovered containing only the donor plastome, even when only a single chloroplast was transferred. In *Citrus*, Grosser et al. (1996) revealed that the cybrid plants contained the chloroplast genome of either one or the other parent. Similar results were obtained in somatic hybrids arised from protoplast fusions of *Allium cepa* + *A. sativum* (Yamashita et al., 2002), *Diospyros glandulosa* + *D. kaki* (Tamura et al., 1998), *Sinapis alba* + *Brassica juncea* (Gaikwad et al., 1996) and 'Seminole' tangelo + Indian Atalantia (or Chinese box-orange) (Motomura et al., 1995).

Some recent reports have cited some preliminary hypothesis that may explain the phenomenon of non-random chloroplast genome segregation (i.e. unidirectional to one parent type). For example, Bonnema et al. (1991) believed that cpDNA would be damaged when protoplasts were exposed to a high dosage irradiation (i.e., X-ray and UV), which can explain the reason that some cybrids only possess the chloroplast genome of the recipient parent. Kisaka et al. (1994) believed that the chloroplast genome of somatic hybrids segregated for streptomycin resistance because *N. tabacum* chloroplasts had been eliminated when a kanamycin- and streptomycin-resistant *Daucus carota* was fused with a chlorophyll-deficient *Nicotiana tabacum* cell suspension culture. Malone et al. (1992) studied the impact of the stringency of cell selection on plastid segregation in protoplast fusion-derived *Nicotiana* regenerates, and found that the different culture conditions could lead to differential segregation of plastids. Smith et al. (1989) considered that the chloroplasts of the mesophyll parent gained an advantage over the proplastids of the suspension parent regarding vigor after protoplast fusion, which affected the segregation of chloroplasts. For example, mesophyll (m)- and suspension culture (s)-derived protoplasts of both tomato and its wild relative were fused as s + m, m + m and s + s combinations, respectively. Results showed that the mesophyll cpDNA was preferentially transmitted to 96% of the plants in m + s and s + m fusion combinations, but for the m + m combination there was an equal distribution of either tomato cpDNA or that of its wild relative among the 34 hybrid plants. The cpDNA type was correlated with the nuclear DNA composition among regenerated hybrids from protoplast fusion of *L. esculentum* or *L. Peruvianum* + *N. tabacum* or *N. Plumbaginifolia*. Wolters et al. (1993b) found that hybrids with more than 2c *Nicotiana* nuclear DNA possessed *Nicotiana* chloroplasts, whereas hybrids with 2c or less *Nicotiana* nuclear DNA contained *Lycopersicon* chloroplasts. Contrarily, Hung et al. (1993) found that chloroplast segregation in the somatic hybrids from protoplast fusion of *N. plumbaginifolia* + *N. Sylvestris* is independent of chloroplast input of the fusion partners and of nuclear background of fusion products. These hypotheses might explain the phenomenon of non-random chloroplast genome segregation in some sense, but all these hypotheses need further confirmation because some of them are even in contradiction to each other.

Although clear evidence for the recombination of cpDNA in higher plants is very limited, in recent years, a few reports support that the recombination of chloroplast genomes exists in some fusion combinations. For example, chloroplast DNA recombinations have occurred in products of protoplast fusion between *Trachystoma ballii* and *B. juncea*. When the chloroplast genome was digested by restriction endonuclease, Baldev et al. (1998) found that the recombinant plastome gave rise to novel fragments in addition to the parent-specific fragments, and these recombinant events did not cause any imbalance in terms of chloroplast-related functions, which have remained stable over generations. Sequencing of the junction fragment of *Nicotiana* somatic hybrids showed extensive homologous recombination between the two parental chloroplast genomes (Fejes et al., 1990). The same result was also observed in fusions between *Hordeum vulgare* + *Daucus carota* (Kisaka et al., 1997), wheat + *Haynaldia villosa* (Zhou et al., 2002) and *Allium ampeloprasum* + *A. cepa* (Buiteveld et al., 1998) respectively.

As mentioned, chloroplast random segregation is a common phenomenon in somatic hybridization, whereas chloroplast genome recombination rarely occurs. It is possible that the frequency of cpDNA recombination is too low or recombination only occurs in regions too small to be easily detected. Kanno et al. (1997) found the restriction patterns in one region near the *rps16* region gene of the chloroplast genome from the hybrid arising from protoplast fusion between cabbage and radish were not similar to that of radish, in contrast to a previous report indicating that such hybrids possess the chloroplast genome of radish.

Kumar and Cocking (1987) reviewed reports of fusion products with a mixture of chloroplast types that have been detected among the regenerated intraspecific and interspecific somatic hybrids, and also pointed out two reasons why parental chloroplasts in cells of somatic hybrids could coexist: either parental chloroplast genome similarity or no selection pressure favoring one parental type. In the past ten years, several experiments have also found chloroplasts from both partners coexisting for a long time (Derks et al., 1991; Saha et al., 2001; Moreira et al. 2000a, b). For example, all somatic hybrids of a cytoplasmic albino tomato and monoploid potato contained potato cpDNA, but one out of eighteen hybrids between a nitrate reductase-deficient tomato and monoploid potato contained both tomato and potato cpDNA (Wolters et al., 1995). In interspecific somatic fusion of jute, evidence for the presence of both types of cpDNA in the hybrid cell lines was obtained when the total genomic DNA of 4- to 7-month-old hybrid cell lines was challenged with the cpDNA marker through Southern analysis, which showed that the early segregation of the parental chloroplasts did not occur in jute (Saha et al., 2001).

2.2. Inheritance of mitochondrial genomes

Mitochondria of the fusion partners also initially coexist in the cytoplasm of heterokaryons following somatic hybridization. Nevertheless, compared with the segregation of the chloroplast genome, the mitochondrial genome of somatic hybrids or cybrids undergoes a much more complex segregation, recombination and rearrangement. Cardi et al. (1999) pointed out that the difference of plastid genomes between two partners is very tiny, but many polymorphisms exist among their mitochondrial genomes. In most cases, parental mitochondrial genomes undergo recombination and rearrangement during the development of heterokaryons. There are only a few reports of random sorting to homogeneity in resulting somatic hybrids.

In contrast to cpDNA, mtDNA has been found to be highly variable and have a high frequency of recombination and rearrangement after cell fusion. Wolters et al. (1995) using RFLP analysis showed that mtDNA of almost 75% regenerated cybrids possessed different RFLP fragments. Landgren et al. (1993) studied mitochondrial segregation and rearrangements of regenerated somatic hybrids from seven different species combinations including intrageneric, intergeneric and intertribal combinations, and found among the various species combinations, 43-95% of the hybrids exhibited mtDNA rearrangements. Similar results were recently reported for *Solanum* (Cardi et al., 1999; Rasmussen et al., 2000), *Nicotiana* (Donaldson et al., 1995; Raineri et al., 1992), *Lycopersicon* (Kochevenko et al., 1999, 2000), *Citrus* (Motomura et al., 1995; Moriguchi et al., 1997), *Brassica* (Liu et al., 1995; Walters et al., 1993), for intergeneric fusions such as *N. tabacum* + *Petunia hybrida* (Dragoeva et al., 1999), *N. tabacum* + *Hyoscyamus niger* (Zubko et al., 2003), *Sinapis alba* + *Brassica juncea* (Gaikwad et al., 1996), *B. napus* + *A. thaliana* (Yamagishi et al., 2002), *Diplotaxis catholica* + *B. juncea* (Mohapatra et al., 1998), *Oryza sativa* + *Hordeum vulgare* (Kisaka et al. 1998), *Festuca arundinacea* + *Lolium multiflorum* (Takamizo et al., 1991), *Cichorium intybus* + *Helianthus annuus* (Varotto et al., 2001), and for interfamily somatic hybrids between barley and carrot (Kisaka et al., 1997). Moreover, Xu et al., (1993) suggested that in spite of symmetric or asymmetric protoplast fusion, and regardless of fusion methods, the probability of recombination or rearrangement of mitochondrial genomes in somatic hybrids is almost the same.

With the development of new molecular markers, more detailed information of cytoplasmic inheritance can be revealed resulting in more in-depth studies. RFLP analysis of mitochondrial genes in somatic hybrids and their parental lines, namely *Medicago sativa* and three *Medicago* species, showed various degrees of rearrangement. Pupilli et al. (2001) presumed that the different outcomes were attributed mainly to differences in the genetic distance between the parents and each hybrid. Evidence for inter-molecular recombination between parental mitochondrial genomes was revealed by Akagi et al. (1995b). They analyzed more than 100 independent rice cybrids using PCR analysis around the *atp6* gene, and results indicated that inter-parental recombination occurred in practically all cybrid calli within 2 weeks after protoplast fusion. At this point, parental and recombinant mitochondrial genomes coexisted within the callus. Over the course of further cultivation, however, mitochondrial genome diversity decreased as parental and/or recombinant genomes segregated out. Besides these results, recombination hotspots were also found by Mohapatra et al. (1998). Using detailed molecular analysis of somatic hybrid plants, they found similar mitochondrial genome organization in two hybrids, which suggested that intergenomic recombination might be preferred at specific sites. In *Brassica* hybrids or cybrids, some sites on the mitochondrial genome seem to be preferentially involved in fusion-induced rearrangements of mtDNA (Stiewe and Röbbelen, 1994; Liu et al., 1995). The *cox2* coding region may serve as an active site for inter- or intra-mtDNA homologous recombination. The distribution of information obtained from each fusion partner was not random. Temple et al. (1992) found several regions, including the *cox1* gene and a major recombination repeat sequence, were always derived from the *Brassica campestris* fusion partner, and some regions were always derived from the *B. napus* mitochondrial genome.

It is worth noting that segregation of mitochondrial genomes in some hybrids or cybrids, like that of the chloroplast genome, reveals a bias toward one or the other parent. As mentioned in citrus, cybrids occur frequently from symmetric fusions of protoplasts derived from embryogenic callus or suspension cultures of one parent with leaf protoplasts of a second parent (Grosser et al., 2000). Such cybrids usually exhibit the phenotype of the leaf parent. Grosser et al. (1996) analyzed three citrus combinations by RFLP analysis using mtDNA-specific probes, and showed that these plants contained the mitochondria of the embryogenic callus parent. Similar results have been reported with other citrus combinations (Saito et al., 1993, 1994). Subsequent reports on this phenomenon in citrus indicated that the mitochondrial genome of the embryogenic callus parent has dominance in somatic hybrids and cybrids (Kobayashi et al., 1991; Motomura et al., 1995; Yamamoto and Kobayashi, 1995; Moriguchi et al., 1997; Moreira et al., 2000a, b; Cabasson et al., 2001). It was theorized that since embryogenic culture cells have higher quantities of mtDNA than leaf cells (approximately 4 times more), that their mtDNA is preferentially inherited in resulting somatic hybrids or cybrids as necessary to provide the high level of enery required for regeneration from protoplasts (Moreira et al., 2000b). In terms of cpDNA, there was no variation in the relative abundance of cpDNA present in leaves versus suspension cells, and cybrids inherited cpDNA from either fusion partner (Moreira et al., 200b). In cybrids derived from somatic hybridization with one parent being a CMS (cytoplasmic male sterility) line, the segregation of mitochondria in the regenerated cybrids was slightly biased towards the CMS parent (Mukhopadyay et al., 1994; Liu et al., 1995). Cardi and Earle (1997) infered this could be due to an inherent better ability of the "Anand" (CMS line) mitochondria to replicate in the hybrid cytoplasm and/or to a reduced contribution of the *B. oleracea* mitochondria as a consequence of the iodoacetate treatment of the recipient protoplasts.

Interestingly, although recombination in mtDNA during somatic cybridization and hybridization has been reported in many higher plants, including the citrus cybrids/hybrids mentioned above, a lack of recombination within the mitochondria genome has also been reported in other higher plants. For example, following symmetric fusion between leek and onion, Buiteveld et al. (1998) found 5 out of 7 hybrids possessed a rearranged mitochondrial genome, but with a predominance of mtDNA fragments from leek, and 1 out of the 7 hybrids contained the unaltered mitochondrial genome of the leek parent. The same results were also reported in fusions as follows: *Nicotiana tabacum* + *N. megalosiphon* (Donaldson et al., 1995), *L. pennellii* + *L. esculentum* (Bonnema et al., 1991), *B. nigra* + *B.*

oleracea (Narasimhulu et al., 1992), *S. alba* L + *B. napus* (Lelivelt et al., 1993), *Moricandia arvensis* + *B. oleracea* (Ishikawa et al., 2003), *Nicotiana tabacum* + *Daucus carota* (Kisaka et al., 1994) and *Brassica napus* + *Lesquerella fendleri* (Skarzhinskaya et al., 1996).

Several factors could influence mitochondrial segregation, such as differences in replication rates, protoplast origin, pre-treatment and nuclear-organelle incompatibility. In protoplast fusion-derived *Brassica oleracea* calli, mitochondrial and chloroplast segregation were independent but biased. Most calli had *B. oleracea* chloroplasts, but more calli had Ogura mitochondria. Protoplast source and pre-treatment could affect organelle segregation (Walters et al. 1993). Nuclear-organelle incompatibility may interpret why only those which carry a majority of chicory mitochondria genes will be selected to the chicory nuclear context in the somatic hybridization between chicory and sunflower (Dubreucq et al. 1999). In addition, differences in mitochondria replication rates could have influenced the progressive elimination of the sunflower's mitochondria genome, while recombination events maintained the sunflower sequences that were introduced into the chicory mitochondria genome. In contrast, the chloroplast type and mitochondrial composition of 17 somatic hybrids between a cytoplasmic albino tomato and monoploid potato (a7-hybrids) and 18 somatic hybrids between a nitrate reductase-deficient tomato and monoploid potato (c7-hybrids) were independent of nuclear DNA composition (Wolters et al., 1995).

Kumar and Cocking (1987) presented a theoretical model depicting the enormous cytoplasmic genomic diversity that can be created via somatic hybridization. In their model, 46 different results could occur, but in fact, because of the interaction between nuclear and cytoplasmic genomes and other factors mentioned above, the resulting types were much less. For example, many protoplast fusion protocols have been developed to create somatic cybrids that possess a new nuclear-cytoplasm combination. In addition to the fusion protocols, alloplasmic compatibility, namely the functional interaction between the nuclear genome of a given species with chloroplast genome and mitochondrial genome of another species, also has a big impact on the production of cybrids (Perl et al., 1991a). Zubko et al. (2001) studied the cybrids of *Nicotiana tabacum* + *Hyoscyamus niger* combining the nuclear genome of *N. tabacum*, plastome of *H. niger* and recombinant mitochondria, and results showed that the plants possess a complex, maternally inheritable syndrome of nucleocytoplasmic incompatibility, severely affecting growth, metabolism and development. Wolters et al. (1993a) described the analysis of cpDNA and mtDNA in 21 somatic hybrid calli of *Solanum tuberosum* and *Nicotiana plumbaginifolia* by Southern blot, and the results suggested that a strong nuclear-cytoplasmic incongruity affects the genomic composition of somatic hybrids between distantly related species.

In addition to the interaction between nuclear and cytoplasmic genomes, several reports demonstrated the existence of a connection between chloroplast and mitochondrial genomes. With mtDNA probes, the potato cultivars and di-haploid clones used for protoplast fusion could be grouped according to their mitochondrial types α, β, γ, δ and ε. Potato plastome-type W was mostly found in combinations with mitochondrial types α, γ and δ. Mitochondrial type β was the only chondriome in linkage with plastid type T, reflecting a co-evolution of both organelles and a common uniparental inheritance (Lössl et al., 1999). Wang et al. (1993b) probed Southern blots of the digested total DNA from maize-*Triticum* hybrids with a mitochondrial gene, i.e. *atp*A from pea, or a chloroplast gene, i.e. *trn*K from rice, and revealed that all hybrids carried the organellar DNAs of only one parent. The same phenomenon also reported in fusions between *Sinapis alba* and *Brassica napus* (Lelivelt et al., 1993).

3. EXPLOITATION OF SOMATIC HYBRIDIZATION FOR RECOVERY OF ORGANELLE-ENCODED TRAITS FOR PLANT IMPROVEMENT

Due to the maternal inheritance of cytoplasmic genomes in sexual crossing (reviewed by Hagemann in Chapter 4 of this volume) in most cases it is impossible to extensively study the cytoplasmic control of inherited traits. In contrast to sexual crossing, somatic hybridization can create considerable variability in cytoplasmic genomes and generate novel nuclear-cytoplasmic interactions. The fact that new combinations of nuclei and organelles can be obtained after protoplast fusion may have interesting applications to the creation of novel plant materials of potential agronomic value (Grosser et al., 1996), which provides a new strategy for transfer of organelle-encoded traits to cultivars. Of course, compared with symmetric hybrids that possess higher ploidy levels, cybrids are particularly attractive when one is interested in the transfer of organelle-encoded traits into a particular genetic background. Chloroplast and mitochondria genomes, or their interaction with nuclear backgrounds, are involved in the control of several agronomic traits for plant improvement.

3.1. Cytoplasmic male sterility (CMS)

CMS (reviewed in Chapters 22 and 23 of this volume) is a maternally inherited phenotype characterized by the inability of a plant to produce functional pollen. CMS has been considered as a mitochondria controlled trait (Kumar and Cocking, 1987; Nivision et al., 1989), and CMS is probably the most efficient type of heritable male sterility for breeders to utilize. For example, the introduced CMS was stably transmitted to progenies through at least eight backcross generations in *Brassica* (Akagi et al., 1995c). Transfer of CMS to new genetic backgrounds is a routine requirement for hybrid seed production, and transfer of CMS to seedy citrus cultivars provides a novel method to create potential seedless diploid fruits (Yamamoto and Kobayashi, 1995; Guo et al., 2004a). Through conventional breeding, it can require a series of backcrosses (potentially time-consuming) for the recovery of the original parental character. In addition, transfer of CMS through genetic transformation is limited in some crops like citrus because of poor molecular biological information about CMS and because there is currently no technology for the stable genetic transformation of plant mitochondrial genomes. In such cases, protoplast fusion offers a more rapid method for the transfer of CMS to new lines. As early in 1978, Zelcer et al. (1978) successfully transferred CMS to *Nicotiana sylvestris* by fusion between protoplasts of normal *N. sylvestris* and X-ray irradiated protoplasts of male sterile *N. tabacum*. Transfer of CMS has been successfully realized in *Nicotiana*, *Brassica*, carrot, *Petunia* and rice mostly by asymmetric fusion, cytoplast-protoplast fusion or symmetric fusion.

3.1.1. Brassicaceae

Among *Brassicaceae*, sexual backcrosses may be the simplest approach to transfer CMS in annuals such as cauliflower and broccoli. However, for biennials, like cabbage and rapeseed, it will take many years of backcrosses for the recovery of the original parental character. Rapeseed, *Brassica napus* L., is an important oil crop in China, Northern Europe and Canada. Rapeseed has a small, hermaphrodite flower, which makes hand pollination impossible for hybrid seed production (Stiewe et al., 1994). In such cases, the production of cybrids via somatic hybridization may offer a more rapid method for the transfer of CMS to new *Brassica lines*.

Several types of CMS have been known in *Brassica* spp. and related genera (Cardi and Earle, 1997). Nevertheless, the preservation and the enlargement of genetic variability at the cytoplasmic level are highly desirable objectives. To provide breeders with new and improved CMS, somatic hybridization has proven to be a powerful and efficient method to create new CMS lines such as 'Polima CMS' line (Yarrow et al., 1990), Oxy CMS line (Prakash and Chopra, 1990), Spines CMS line (Kirti et al., 1991), Tour CMS line (Stiewe and Röbbelen, 1994; Arumugam et al. 1996, 2000), Annard CMS line (Cardi and Earle. 1997) and Ogura CMS line (Kirti et al., 1995a). Kao et al. (1992) created two male-sterile somatic hybrids from the fusion of broccoli protoplasts with those of an Ogura CMS/triazine-resistant cybrid *B. napus*, while four male-sterile plants recovered from the fusion of broccoli with Ogura CMS *B. napus*. Cold-tolerant CMS cabbage (*Brassica oleracea* var. *capitata*) was produced by the fusion of leaf protoplasts from male-fertile cabbage and cold-tolerant Ogura CMS broccoli lines (Sigareva et al. 1997). Kirti et al. (1995b) created a stable CMS line of *Brassica juncea* carrying restructured organellar genomes from the somatic hybrid *Trachystoma ballii* + *B. juncea.* Intergeneric somatic hybrids produced via asymmetric fusion between *Brassica napus* and *Arabidopsis thaliana* by Forsberg et al. (1998), were backcrossed to *B. napus* cv. Hanna, and the BC_1 generation was screened for male sterility and aberrant flower phenotypes, among which nine hybrids were selected and backcrossed recurrently to *B. napus.* Finally Leino et al. (2003) created a new *Brassica napus* line with rearranged *Arabidopsis* mitochondria displaying CMS and a range of developmental aberrations. Of course, not all recombinations of mtDNA lead to CMS. In protoplast fusion between radish and rapeseed, Sakai and Imamura (1990) found 4 out of 10 cybrids showed mtDNA recombination, but only one was male sterile and the remaining 3 were fertile. In order to produce an alloplasmic, CMS *Brassica napus* line, somatic cybrids combining the nucleus of *B. napus* and the mitochondria of *Brassica tournefortii* were produced (Liu et al., 1996).

With the continuing evolution of molecular biology techniques, researchers are getting an increasingly deeper insight into CMS via somatic hybridization. *Brassicaceae* is the model plant and many reports about CMS are related to it. To identify regions of the mitochondrial genome that are genetically correlated with the Ogura CMS in *Brassica napus*, Grelon et al. (1994) revealed that mitochondria isolated from male-sterile rapeseed cybrids synthesized a polypeptide of 19 kDa (ORF138), which was absent in fertile revertants. Wang et al. (1995) analyzed mtDNAs of three male-sterile somatic hybrids derived from the fusion of *B. oleracea* and Polima CMS *B. napus* protoplasts. Results indicated that the *orf224/atp6* gene region is genetically correlated with male sterility and provided significant additional support for the view that this gene region may be involved in specifying the CMS trait.

3.1.2. Rice

Wild abortive (WA) CMS has been extensively used in hybrid seed production in rice in the tropical region. By asymmetric protoplast fusion between CMS and fertile maintainer lines, Bhattacharjee et al. (1999) transferred WA CMS to the nuclear background of RCPL1-2C, an advance rice breeding line that also served as maintainer of this cytoplasm. The donor-recipient protoplast fusion method was used to produce cybrid plants and to transfer CMS from two CMS lines, MTC-5A and MTC-9A, into a fertile japonica cultivar. The CMS was expressed in the cybrid plants and was stably transmitted to their progenies (Akagi et al., 1995a). Another study by Akagi et al. (1995c) indicated that the unique region downstream from *atp6* is tightly linked with the CMS phenotype.

3.1.3. Potato and tomato

In contrast to tuber-seeds, F_1 hybrid true-seeds of potato have potential advantages. Transfer of CMS to potato cultivars via somatic hybridization offers an alternative for the production of F1 seeds, which can eliminate laborious hand emasculation. The donor-recipient protoplast-fusion is a routine method to convert fertile potato cultivars into useful seed-parents for F1 hybrid-seed production. Perl et al. (1990) successfully used this method to transfer CMS from an alloplasmic male sterile donor into two male fertile potato cultivars that were previously used as seed-parents in F_1 hybrid seed-production.

Though one-step production of CMS tomato was established by the fusion of tomato protoplasts with protoplasts of *S. acaule* or potato as the mtDNA donor, the introduction of CMS into tomato proved not easily achieved (Melchers et al. 1992). For example, the fusion of IOA-treated tomato protoplasts with irradiated protoplasts from *S. lycopersicoides* or *N. tabacum* did not produce CMS tomato, and Melchers et al. (1992) suggested the fusion partners for CMS production in tomato should not be too closely or too remotely related.

3.1.4. Tobacco

An effective selection system preceded by double inactivation of parental protoplasts was used by Matibiri et al. (1994) to transfer *Nicotiana suaveolens* Leh. CMS into a commercial tobacco (*N. tabacum* L.), and out of forty-four regenerated plants, four putative cytoplasmic male-sterile (CMS) plants transferred to glasshouse, were obtained. Dragoeva et al. (1999) created cytoplasmic hybrids (cybrids) between two sexually incompatible species *Nicotiana tabacum* and *Petunia hybrida*, and *in vitro* culture of ovules from one cybrid plant pollinated by

N. tabacum resulted in the regeneration of CMS progeny plants. Twenty-nine green CMS regenerants were obtained after fusion of protoplasts from a tobacco cytoplasmic chlorophyll-deficient mutant and gamma irradiation-inactivated leaf protoplasts of *N. alata*, and analysis of the progenies from three successive backcrosses of the studied cybrids with *N. tabacum* demonstrated a strict cytoplasmic inheritance of the male-sterile phenotype (Atanassov et al. 1998). Novel CMS lines could also be created between two fertile species via somatic hybridization. For example, cybrids that contain the nuclear genome from *Nicotiana tabacum* and cytoplasm from *Hyoscyamus niger* or *Scopolia carniolica* were constructed by protoplast fusion, and both types of hybrids exhibited CMS (Zubko et al., 1996).

3.1.5. Citrus

Seedlessness is a prerequisite for new fresh-market citrus cultivars. Seedlessness in diploid citrus generally relates to male and/or female sterility. The seedless Satsuma mandarin (*Citrus unshiu*) is typically male sterile and its male sterility has been identified to be a CMS type (Yamamoto et al., 1997). Due to complex citrus biology, it is possible, but not easy, to transfer the sterility character from Satsuma mandarin to other seedy citrus cultivars by conventional breeding. However, it may be possible to transfer the CMS trait of Satsuma into other commercially important diploid seedy cultivars via cybridization.

Yamamoto and Kobayashi (1995) produced a cybrid plant between CMS "Juman" Satsuma mandarin and "Washington" navel orange. Though sterile pollen and seedless fruit were harvested from the cybrid plant, since navel orange is also a male sterile type, it is difficult to determine whether the seedlessness was due to the sterile cytoplasm of Satsuma. More cybrid plants were then produced from the fusion of "Juman" Satsuma with *Citrus junos* and *C. limon* cv. Eureka (Tokunaga et al., 1999). More recently, by using an embryogenic suspension culture of 'Guoqing No.1' Satsuma mandarin, diploid cybrid plants of 'Hirado Buntan Pink' pummelo and putative cybrid plants of 'Sunburst' mandarin hybrid and "Lee" mandarin hybrid ('Clementine' · 'Murcott') were produced by standard symmetric fusion (Guo et al., 2004a). These plants will be flowered and fruited as soon as possible to determine if the substitution of Satsuma mtDNA can result in a new mitochondria-nuclear interaction that could result in making these cultivars seedless without otherwise altering their cultivar integrity. If successful, this strategy could be applied to remove seed from many superior diploid citrus cultivars.

With the exception of potential seedlessness, the agronomic value of citrus cybrids is currently unknown. The evaluation of citrus cybrids in the field will allow characterization of agronomic traits encoded by the cytoplasmic genomes. Tusa et al. (2000) suggested from cybrid evaluation that specific mechanisms of resistance against Mal Secco could be activated in these genotypes. Mandarin and sweet orange cybrids in the field at the University of Florida / Institute of Food and Agricultural Sciences Citrus Research and Education Center are showing significant variation in agronomically important traits including fruit maturity date and seed content, indicating that cybridization is a potential source of genetic variation for citrus cultivar improvement.

3.1.6. Other crops

Sunflower. Rambaud et al. (1993) performed protoplast fusions between fertile industrial chicory protoplasts and CMS PET1 sunflower protoplasts in efforts to induce or transfer CMS into chicory. Later, Dubreucq et al. (1999) backcrossed three of these CMS chicory cybrids with different witlof chicories in order to transfer the three male-sterile cytoplasms from an industrial chicory nuclear environment to a witlof chicory nuclear context. With Southern blot analysis using different mitochondrial genes as probes, it was revealed that mitochondrial genomes of the three cybrids were different and they were stable throughout backcross generations regardless of the pollinator. Asymmetric fusion between *Cichorium intybus* and *Helianthus annuus* also produced cybrids, which presented mitochondrial patterns of both the fertile chicory and the male-sterile sunflower fragments (Varotto et al. 2001).

Ryegrass. Creemers-Molenaar et al. (1992) fused a CMS breeding line of perennial ryegrass (B200) with iodoacetamide-treated protoplasts of a fertile breeding line (Jon 401), and obtained 15 calli possessing mtDNA of both parents.

3.1.7. Modification of CMS lines and restoration of fertility

CMS lines in several crop species have been created via somatic hybridization as mentioned. However, in addition to male sterility, these resulting alloplasmics often exhibited developmental or floral abnormalities, which include leaf chlorosis of varying degrees, transformation of anthers into petaloid structures, and low female fertility (Kirti et al., 1998a; Prakash et al., 1998). Therefore, backcrosses to either parent are needed to eliminate the nuclear genome of the other parent while retaining the CMS and novel cytoplasmic composition established in the original hybrid plants (Bohman et al., 1999). The male-sterile *B. juncea* arising from somatic fusion between male-sterile *B. napus* and *B. juncea* cv. 'RLM 198', after repetitive backcrossing, however, is highly chlorotic and late, and it has low female (seed) fertility and small, contorted pods. To rectify these defects, Kirti et al. (1995a) fused it with normal 'RLM 198'; four dark green, completely male-sterile plants were obtained and identified as putative cybrids. In order to modify floral and vegetative phenotypes, CMS cybrids of *Nicotiana tabacum* L. + *Hyoscyamus niger* L. were backcrossed to tobacco by Zubko et al. (2003). All the above-mentioned studies confirmed that transfer of cytoplasmic encoded traits via somatic hybridization, followed by conventional backcrossing, is a useful tool for generating alternative CMS sources with novel nuclear-cytoplasmic compositions.

In addition, it is necessary to establish a CMS restored fertility (Rf) system for F1 seed production (reviewed in Chapters 22 and 23). The restoration mechanism, which is under the influence of nuclear genes, indicates the influence of the nuclear genome on the function of mitochondria. In *Brassica*, protoplast fusion not only can produce CMS lines but also can establish Rf ones. For example, Sakai et al. (1996) transferred a gene from fertility-restored radish to *B. napus* by protoplast fusion, and obtained six male-fertile regenerated plants. The gene for fertility restoration has also been introgressed by Prakash et al. (1998) following the development of a *M. arvensis* monosomic addition line on CMS *B. juncea* (Dubreucq et al., 1999). In rice, the CMS of cybrid plants produced by Akagi et al. (1995a) could be restored completely by crossing with MTC-10R, which had the single dominant gene *Rf-1* for restoring fertility. In tobacco, the 'donor-recipient' fusion method was proven a good way to restore fertility of *Nicotiana* CMS lines (Raineri et al. 1992). Fusion of two CMS cultivars of *Nicotiana tabacum*, one with *N. bigelovii* cytoplasm and the other with *N. undulata* cytoplasm, resulted in the restoration of male fertility in cybrid plants (Kofer et al., 1991).

3.2. Other mitochondrial-encoded traits

Through analysis of mitochondrial composition with different probes homologous to coding regions, Lossl et al. (1994) found a relationship between the homogeneity of the mitochondrial genome and the yield level. In addition, Perl et al. (1991b) successfully bridged the intergeneric barriers between *Nicotiana*, with putativly oligomycin-resistant mitochondria, and *Solanum* with respect to chondriome transfer, and at last verified that chondriome components could be transferred from a *N. sylvestris* donor into the recipient fusion partner (*S. tuberosum*).

3.3. Chloroplast-encoded traits

Crop plants have severe sensitivity to herbicides inhibiting photosynthesis. Triazine herbicides, such as atrazine and terbutryn, have been widely used for weed control because of their low cost and their effectiveness against a broad spectrum of weeds. However, some agronomically important crops are sensitive to triazine (Farooqui et al. 1997). Herbicide resistance is often controlled by chloroplast-encoded proteins (reviewed in Chapter 17 of this volume). (See Malone and Dix, 1990 for an example.) Somatic hybridization can provide a new avenue to create resistant plants. For example, Xia et al. (1992) transferred atrazine resistance from black nightshade into tomato via somatic hybridization.

3.4. Somatic hybridization between C_3 and C_4 crops

The greater efficiency of CO_2 recapture has been observed in C_4 species, and therefore there is an interest in transferring chloroplasts of C_4 or C_3-C_4 species into C_3 crop species via somatic hybridization. If this is realized, it will improve water-use efficiency compared with C_3 forms of the same crop, especially under conditions of water stress, which should bring enormous economical and social benefits. O'Neil et al. (1996) revealed the expression of the C_3-C_4 intermediate character in somatic hybrids between *Brassica napus* and the C_3-C_4 species *Moricandia arvensis*. Somatic hybrids between *Brassica oleracea* and *Moricandia nitens*, a C_3-C_4 intermediate wild species, were also produced by Yan et al. (1999), and measurement of the CO_2 compensation point revealed that six out of the eight hybrid plants expressed a gas-exchange character that was intermediate between the C_3-C_4 *M. nitens* and C_3 *B. oleracea* parents.

4. CONCLUSIONS AND PROSPECTS

The application of emerging molecular biology techniques for the characterization of new somatic hybrid and cybrid plant combinations is greatly increasing our understanding of organelle inheritance patterns and interaction with nuclear genomes. The production of novel nuclear/cytoplasmic combinations is increasing as somatic hybridization technology continues to evolve. Chloroplast genomes (reviewed by Maier and Schmitz-Linneweber in Chapter 5 of this volume) are highly conserved and are sometimes difficult to follow in fusion studies. Chloroplast genome segregation in somatic hybrids and cybrids is generally random, but there are some examples of biased inheritance. Recombination in chloroplasts is rare and difficult to detect. In contrast, the inheritance of mitochondrial genomes in somatic hybrids and cybrids is frequently biased, and there are numerous documented cases of mitochondrial genome rearrangement. Reports on manipulation of organelle transfer for plant improvement are increasing. CMS has been transferred to important genotypes in several species to enhance hybrid seed production or to develop seedless fruit cultivars. Progress with the transfer of cytoplasmically controlled herbicide or disease resistance is also evident. Opportunities to improve yield and water use efficiency via organelle manipulation are also being explored. The future of organelle manipulation as a tool for plant improvement looks quite promising, as our ability to create and understand novel nuclear/cytoplasmic combinations continues to evolve.

5. ACKNOWLEDGEMENTS

This work was approved for publication as Florida Agricultural Experiment Station Journal Series No. R-10191. W. W. Guo thanks the National Natural Science Foundation of China (NSFC) and the International Foundation for Science (IFS) for financial support to conduct the citrus somatic hybrid/cybrid production and molecular characterization research.

6. REFERENCES

Akagi, H., Nakamura, A., Sawada, R., Oka, M., & Fujimura, T. (1995a). Genetic diagnosis of cytoplasmic male sterile cybrid plants of rice. *Theor. Appl. Genet., 90*, 948-951.

Akagi, H., Shimada, H., & Fujimura, T. (1995b). High-frequency inter-parental recombination between mitochondrial genomes of rice cybrids. *Curr. Genet., 29*, 58-65.

Akagi, H., Taguchi, T., & Fujimura, T. (1995c). Stable inheritance and expression of the CMS traits introduced by asymmetric protoplast fusion. *Theor. Appl. Genet., 91*, 563-567.

Arumugam, N., Mukhopadhyay, A., Gupta, V., Pental, D., & Pradhan, A. K. (1996). Synthesis of hexaploid AABBCC somatic hybrids: a bridging material for transfer of 'tour' cytoplasmic male sterility to different *Brassica* species. *Theor. Appl. Genet., 92*, 762-768.

Arumugam, N., Mukhopadhyay, A., Gupta, V., Sodhi, Y. S., Verma, J. K., Pental, D., & Pradhan, A. K. (2000). Somatic cell hybridization of 'oxy' CMS *Brassica juncea* (AABB) with *B. oleracea* (CC) for correction of chlorosis and transfer of novel organelle combinations to allotetraploid brassicas. *Theor. Appl. Genet., 100*, 1043-1049.

Atanassov, I. I., Atanassova, S. A., Dragoeva, A. I., & Atanassov, A. I. (1998). A new CMS source in *Nicotiana* developed via somatic cybridization between *N. tabacum* and *N. alata*. *Theor. Appl. Genet., 97*, 982-985.

Baldev, A., Gaikwad, K., Kirti, P. B., Mohapatra, T., Prakash, S., & Chopra, V.L. (1998). Recombination between chloroplast genomes of *Trachystoma ballii* and *Brassica juncea* following protoplast fusion. *Mol. Gen. Genet., 260*, 357-361.

Bastia, T., Scotti, N., Cardi, T. (2001). Organelle DNA analysis of *Solanum* and *Brassica* somatic hybrids by PCR with 'universal primers'. *Theor. Appl. Genet., 102*, 1265-1272.

Bhattacharjee, B., Sane, A. P., & Gupta, H. S. (1999). Transfer of wild abortive cytoplasmic male sterility through protoplast fusion in rice. *Mol. Breed., 5*, 319-327.

Binsfeld, P. C., Wingender, R., & Schnabl, H. (2000). Characterization and molecular analysis of transgenic plants obtained by microprotoplast fusion in sunflower. *Theor. Appl. Genet., 101*, 1250-1258.

Bohman, S., Forsberg, J., Glimelius, K., & Dixelius, C. (1999). Inheritance of Arabidopsis DNA in offspring from *Brassica napus* and *A. thaliana* somatic hybrids. *Theor. Appl. Genet., 98*, 99-106.

Bonnema, A. B., Melzer, J. M., Murray, L. W., & O'connell, M. A. (1992). Non-random inheritance of organellar genomes in symmetric and asymmetric somatic hybrids between *Lycopersicon esculentum* and *L. Pennellii*. *Theor. Appl. Genet., 84*, 435-442.

Bonnema, A. B., Melzer, J. M., & O'Connell, M. A. (1991). Tomato cybrids with mitochondrial DNA from *Lycopersicon pennelli*. *Theor. Appl. Genet., 81*, 339-348.

Bravo, J. E., & Evans, D. A. (1985). Protoplast fusion for crop improvement. *Plant Breed. Rev., 3*, 193-218.

Brewer, E. P., Saunders, J. A., Angle, J. S., Chaney, R. L., & McIntosh, M. S. (1999). Somatic hybridization between the zinc accumulator *Thlaspi caerulescens* and *Brassica napus*. *Theor. Appl. Genet., 99*, 761-771.

Buiteveld, J., Kassies, W., Geels, R., Campagne, M. M. V., Jacobsen, E., & Creemers-Molenaar, J. (1998) Biased chloroplast and mitochondrial transmission in somatic hybrids of *Allium ampeloprasum* L. and *Allium cepa* L. *Plant Science, 131*, 219-228.

Cabasson, C. M., Luro, F., Ollitrault, P., & Grosser, J. W. (2001). Non-random inheritance of mitochondrial genomes in *Citrus* hybrids produced by protoplast fusion. *Plant Cell Rep., 20*, 604-609.

Cardi, T, Bastia, T., Monti, L., & Earle, E. D. (1999). Organelle DNA and male fertility variation in *Solanum* spp. and interspecific somatic hybrids. *Theor. Appl. Genet., 99*, 819-828.

Cardi, T., Earle, E. D. (1997). Production of new CMS *Brassica oleracea* by transfer of 'Anand' cytoplasm from *B. rapa* through protoplast fusion. *Theor. Appl. Genet., 94*, 204-212.

Carlson, P.S., Smith, H. H., & Dearing, R. D. (1972). Parasexual interspecific plant hybridisation. *Proc. Natl. Acad. Sci., USA, 69*, 2292-2294 .

Cheng, Y. J., Guo, W. W., Deng, X.X. (2003a). cpSSR: a new tool to analyze chloroplast genome of Citrus somatic hybrids. *Acta Bot. Sin., 45*, 906-909.

Cheng, Y. J., Guo, W. W., & Deng, X. X. (2003b) Molecular characterization of cytoplasmic and nuclear genomes in phenotypically abnormal Valencia orange (*Citrus sinensis*) + Meiwa kumquat (*Fortunella crassifolia*) intergeneric somatic hybrids. *Plant Cell Rep., 21*, 445-451.

Clark, E., Gafni, Y., & Izhar, S. (1988). Loss of CMS-specific mitochondrial DNA arrangement in fertile segregants of *Petunia* hybrids. *Plant Mol. Biol., 11*, 249-253.

Collonnier, C., Fock, I., Daunay, M. –C., Servaes, A., Vedel, F., Siljak-Yakovlev, S. et al. (2003). Somatic hybrids between *Solanum melongena* and *S. sisymbrifolium*, as a useful source of resistance against bacterial and fungal wilts. *Plant Sci., 164*, 849-861.

Conner, A. J., & Tynan, J. L. (1995). A maternally inherited, chlorophyll-deficient mutant of a transgenic kanamycin-resistant *Nicotiana-plumbaginifolia* plant. *Intl. J. Plant Sci., 156*, 320-325.

Creemers-Molenaar. J., Hall, R. D., & Krens, F. A. (1992). Asymmetric protoplast fusion aimed at intraspecific transfer of cytoplasmic male sterility (CMS) in *Lolium perenne* L. *Theor. Appl. Genet., 84*, 763-770.

Derks, F. H. M., Wijbrandi, J., Koornneef, M., & Colijn-Hooymans, C. M. (1991). Organelle analysis of symmetrical and asymmetrical hybrids between *Lycopersion peruvianum* and *Lycopersion esculentum*. *Theor. Appl. Genet., 81*,199-204.

Donaldson, P, Sproule, A., Bevis, E., Pandeya, R., Keller, W.A., & Gleddie, S. (1993). Non-random chloroplast segregation in *Nicotiana tabacum* (+) *N.rustica* somatic hybrids selected by dual nuclear-encoded resistance. *Theor. Appl. Genet., 86*, 465-473.

Donaldson, P. A., Bevis, E., Pandeya, R., & Gleddie, S. (1995). Rare symmetric and asymmetric *Nicotiana tabacum* (+) *N. megalosiphon* somatic hybrids recovered by selection for nuclear-encoded resistance genes and in the absence of genome inactivation. *Theor. Appl. Genet., 91*, 747-755.

Dragoeva, A., Atanassoy. I., & Atanassov, A. (1999). CMS due to tapetal failure in cybrids between *Nicotiana tabacum* and *Petunia hybrida*. *Plant Cell Tiss. Orgn. Cult., 55*, 67-70.

Dubreucq, A., Berthe, B., Asset, J. F., Boulidard, L., Budar, F., Vasseur, J., & Rambaud, C. (1999). Analyses of mitochondrial DNA structure and expression in three cytoplasmic male sterile chicories originating from somatic hybridisation between fertile chicory and CMS sunflower protoplasts. *Theor. Appl. Genet., 99*,1094–1105.

Eigel, L., & Koop, H. U. (1992). Transfer of defined numbers of chloroplasts into albino protoplasts by subprotoplast/protoplast microfusion: chloroplasts can be "cloned", by using suitable plastome combinations or selective pressure. *Mol. Gen. Genet., 233*, 479-482.

Escalante, A., Imanishi, S., Hossain, M., Ohmido, N., & Fukui, K. (1998). RFLP analysis and genomic in situ hybridization (GISH) in somatic hybrids and their progeny between *Lycopersicon esculentum* and *Solanum lycopersicoides*. *Theor. Appl. Genet., 96*, 719-726.

Farooqui, M. A., Rao, A. V., Jayasree, T., & Sadanandam, A. (1997). Induction of atrazine resistance and somatic embryogenesis in *Solanum melongena*. *Theor. Appl. Genet., 95*, 702-705.

Fejes, E., Engler, D., & Maliga, P. (1990). Extensive homologous chloroplast DNA recombination in the pt14 *Nicotiana* somatic hybrid. *Theor. Appl. Genet., 79*, 28–32.

Fock, I., Collonnier, C., Purwito, A., Luisetti, J., Souvannavong, V., Vedel, F., et al. (2000). Resistance to bacterial wilt in somatic hybrids between *Solanum tuberosum* and *Solanum phureja*. *Plant Sci., 160*, 165-176.

Forsberg, J., Dixelius, C., Lagercrantz, U., & Glimelius, K. (1998). UV dose-dependent DNA elimination in asymmetric hybrids between *Brassica napus* and *Arabidopsis thaliana*. *Plant Sci., 131*, 65–76.

Fu, C. H., Guo, W. W., Liu, J. H., & Deng, X. X. (2003). Regeneration of *Citrus sinensis* (+) *Clausena lansium* intergeneric triploid and tetraploid somatic hybrids and their identification by molecular markers. *In Vitro Cell. Devel. Biol. Plant, 39*, 360-364.

Gaikwad, K., Kirti, P. B., Sharma, A., Prakash, S., & Chopra, V. L. (1996). Cytogenetic and molecular investigations on somatic hybrids of *Sinapis alba* and *Brassica juncea* and their backcross progeny. *Plant Breed., 115*, 480-483.

Garriga-Calderé, F., Huigen, D. J., Jacobsen, E., & Ramanna, M. S. (1999). Prospect for introgressing tomato chromosomes into potato genome: an assessment through GISH analysis. *Genome, 42*, 282-288.

Gavrilenko, T., Thieme, R., & Rokka, V. M. (2001). Cytogenetic analysis of *Lycopersicon esculentum* (+) *Solanum etuberosum* somatic hybrids. *Theor. Appl. Genet., 103*, 231-239.

Gleba, Y. Y., & Sytnik, K. M. (1984). Protoplast fusion. In R. Shoeman (ed.), *Genetic engineering in higher plants*. Berline, Heidelberg: Springer-Verlag.

Grelon, M., Budar, F., Bonhomme, S., & Pelletier, G. Ogura cytoplasmic male-sterility (CMS)-associated *orf138* is translated into a mitochondrial membrane polypeptide in male-sterile *Brassica* cybrids. *Mol. Gen. Genet., 243*, 540-547.

Grosser, J. W., Gmitter, F. G., Tusa, N., Recupero, G. R., & Cucinotta, P. (1996). Further evidence of a cybridization requirement for plant regeneration from citrus leaf protoplasts following somatic fusion. *Plant Cell Rep., 15*, 672-676.

Grosser, J. W., Gmitter, F. G. (1990). Protoplast fusion and citrus improvement. *Plant Breed. Rev., 8*, 339-374.

Grosser, J. W., Ollitrault, P., & Olivares-Fuster, O. (2000). Somatic hybridization in *Citrus*: An effective tool to facilitate variety improvement. *In Vitro Cell. Dev. Biol. Plant, 36*, 434-449.

Guo, W. W., Cheng, Y. J., & Deng, X. X. (1998). Regeneration and molecular characterization of intergeneric somatic hybrids between *Citrus reticulata* and *Poncirus trifoliata*. *Plant Cell Rep., 20*, 829-834.

Guo, W. W., Deng, X. X., & Shi, Y. Z. (1998). Optimization of electrofusion parameters and interspecific somatic hybrid regeneration in *Citrus*. *Acta Bot. Sin., 40*, 417-424.

Guo, W. W., & Grosser, J. W. (2004). Somatic hybrid vigor in *Citrus*: direct evidence from protoplast fusion of an embryogenic callus line with a transgenic mesophyll parent expressing the GFP gene. In *Abstracts for 10th International Citrus Congress*, Riverside: International Society for Citriculture.

Guo, W. W., Prasad, D., Cheng, Y. J., Serrano, P., Deng, X. X., & Grosser, J. W. (2004a). Targeted cybridization in *Citrus*: transfer of Satsuma cytoplasm to seedy cultivars for potential seedlessness. *Plant Cell Rep., 22*, 752-758.

Guo, W. W., Prasad, D., Serrano, P., Gmitter, F. G. Jr., & Grosser, J. W. (2004b) Citrus somatic hybridization with potential for direct tetraploid scion cultivar development. *J. Hort. Sci. Biotechnol., 79*, 400-405.

Harding, K., & Millam, S. (2000). Analysis of chromatin, nuclear DNA and organelle composition in somatic hybrids between *Solanum tuberosum* and *Solanum sanctae-rosae*. *Theor. Appl. Genet., 101*, 939-947.

Hossain, M., Imanishi, S., & Matsumoto, A. (1994). Production of somatic hybrids between tomato (*Lycopersicon esculentum*) and night shade (*Solanum lycopersicoides*) by electrofusion. *Breed. Sci., 44*, 405-412.

Hung, C. Y., Lai, Y. K., Feng, T. Y., & Chen, C. C. (1993). Chloroplast segregation in somatic hybrids of *Nicotiana plumbaginifolia* and *N. Sylvestris* having different ratios of parental nuclear genomes. *Plant Cell Rep., 13*, 83-86.

Ilcheva, V., San, L. H., Zagorska, N., & Dimitrov, B. (2001). Production of male sterile interspecific somatic hybrids between transgenic *N. Tabacum* (bar) and *N. Rotundifolia* (*nptII*) and their identification by AFLP analysis. *In Vitro Cell. Dev. Biol. Plant, 37*, 496-502.

Ishikawa, S., Bang, S. W., Kaneko, Y., & Matsuzawa, Y. (2003). Production and characterization of intergeneric somatic hybrids between *Moricandia arvensis* and *Brassica oleracea*. *Plant Breed., 122*, 233-238.

Johnson, A. A. T., Piovano, S. M., Ravichandran, V., & Veilleux, R. E. (2001a). Selection of monoploids for protoplast fusion and generation of intermonoploid somatic hybrids of potato. *Amer. J. Potato Res., 78*, 19-29.

Johnson, A. A. T., & Veilleux, R. E. (2001b). Somatic hybridization and applications in plant breeding. *Plant Breed. Rev. , 20*, 167-225.

Kanno, A., Kanzaki, H. & Kameya, T. (1997). Detailed analyses of chloroplast and mitochondrial DNAs from the hybrid plant generated by asymmetric protoplast fusion between radish and cabbage. *Plant Cell Rep., 16*, 479-484.

Kao, H. M., Keller, W. A., Gleddie, S., & Brown, G. G. (1992). Synthesis of *Brassica oleracea + B. napus* somatic hybrid plants with novel organelle DNA compositions. *Theor. Appl. Genet., 83*, 313-320.

Kirti, P. B., Banga, S. S., Prakash, S., & Chopra, V. L. (1995a). Transfer of Ogu cytoplasmic male sterility to Brassica and improvement of the male sterile line through somatic cell fusion. *Theor. Appl. Genet., 91*, 517-521.

Kirti, P. B., Kumar, V. D., Prakash, S., Chopra, V. L., & Mohapatra, T. (1998). Random chloroplast segregation and mitochondrial genome recombination in somatic hybrid plants of *Diplotaxis catholica+Brassica juncea, Plant Cell Rep., 17*, 814-818.

Kirti, P. B., Mohapatra, T., Baldev. A., Prakash, S., & Chopra, V. L. (1995b). A stable cytoplasmic male-sterile line of *Brassica juncea* carrying restructured organelle genomes from the somatic hybrid *Trachystoma ballii + B. juncea*. *Plant Breed., 114*, 434-438.

Kirti, P. B., Prakash, S., & Chopra, V. L. (1991). Interspecific hybridization between *Brassica juncea* and *B.spinescens* through protoplast fusion. *Plant Cell Rep., 9*, 639-642.

Kisaka, H., & Kameya, T. (1994). Production of somatic hybrids between *Daucus carota* L. and *Nicotiana tabacum*. *Theor. Appl. Genet., 88*, 75-80.

Kisaka, H., Kisaka, M., Kanno, A., & Kameya, T. (1998). Intergeneric somatic hybridization of rice (*Oryza sativa* L.) and barley (*Hordeum vulgare* L.) by protoplast fusion. *Plant Cell Rep., 17*, 362-367.

Kisaka, H., Kisaka, M., Kanno, A., & Kameya, T. (1997). Production and analysis of plants that are somatic hybrids of barley (*Hordeum vulgare* L.) and carrot (*Daucus carota* L.). *Theor. Appl. Genet., 94*, 221-226 .

Kobayashi, S., Ohgawara, T., Fujiwara, K., & Oiyama, I. (1991). Analysis of cytoplasmic genomes in somatic hybrids between navel orange (*Citrus sinensis* Osb.) and "Murcott tangor". *Theor. Appl. Genet., 82*, 6-10.

Kochevenko, A., Ratushnyak, Y., Kornyeyev, D., Stasik, O., Porublyova, L., Kochubey, S., Suprunova, T., & Gleba, Y. (2000). Functional cybrid plants of *Lycopersicon peruvianum* var 'Dentatum' with chloroplasts of *Lycopersicon esculentum*. *Plant Cell Rep., 19*, 588-597.

Kochevenko, A. S., Rudas, V. A., & Ratushnyak, Y. I. (1999). Use of cytoplasmic hybrids for analysis of a chlorophyll deficiency mutation in *Lycopersicon peruvianum* var. Dentatum. *Russ. J. Genet., 35*, 52-57.

Kofer, W., Glimelius, K., & Bonnett, H. T. (1991). Restoration of normal stamen development and pollen formation by fusion of different cytoplasmic male-sterile cultivars of *Nicotiana tabacum*. *Theor. Appl. Genet., 81*, 390-396.

Kumar, A., & Cocking, E. C. (1987). Protoplast fusion: a novel approach to organelle genetics in higher plants. *Amer. J. Bot., 74*, 1289-1303.

Landgren, M., & Glimelius, K. A. (1993). High frequency of intergenomic mitochondrial recombination and an overall biased segregation of *B. Campestris* or recombined *B. Campestris* mitochondria were found in somatic hybrids made within Brassicaceae. *Theor. Appl. Genet., 87*, 854-862.

Leino, M., Teixeira, R., Landgren, M., & Glimelius, K. (2003). *Brassica napus* lines with rearranged Arabidopsis mitochondria display CMS and a range of developmental aberrations. *Theor. Appl. Genet., 106*, 1156–1163.

Lelivelt, C. C., Leunissen, E. H. M., Frederiks, H. J, Helsper, J. P. F. G., & Krens, F. A. (1993). Transfer of resistance to the beet cyst from *Sinapis alba* L. (white mustard) to the *Brassica napus* L gene pool by means of sexual and somatic hybridisation. *Theor. Appl. Genet., 85*, 688-696.

Liu, J. H., Dixelius, C., Erikson, I., & Glimelius, K. (1995). *Brassica napus* (+) *B. tournefortii*, a somatic hybrid containing traits of agronomic importance for rapeseed breeding. *Plant Sci., 109*, 75–86.

Liu, J. H., Landgren, M., Glimelius, K. (1996). Transfer of the *Brassica tournefortii* cytoplasm to *B. napus* for the production of cytoplasmic male sterile *B. napus*. *Physiolog. Plant., 96*, 123-129.

Lorz, H., Paszkowshi, J., Dierks-Ventling, C., & Potrykus, I. (1981). Isolation and characterization of cytoplasts and miniprotoplasts derived from protoplasts of cultured cells. *Physiolog. Plant., 53*, 385-391.

Lössl, A., Adler, N., Horn, R., Frei, U., & Wenzel, G. (1999). Chondriome type characterization of potato: mt α, β, γ, δ, ε and novel plastid-mitochondrial configurations in somatic hybrids. *Theor. Appl. Genet., 99*, 1–10

Lossl. A., Frei, U., & Wenzel, G. (1994). Interaction between cytoplasmic composition and yield parameters in somatic hybrids of *S. tuberosum* L. *Theor. Appl. Genet., 89*, 873-878.

Lotfy, S., Luro, F., Carreel, F., Froelicher, Y., Rist, D., & Ollitrault, P. (2003). Application of cleaved amplified polymorphic sequence method for analysis of cytoplasmic genome among Aurantioideae intergeneric somatic hybrids. *J. Amer. Soc. Hort. Sci., 128*, 225-230.

Maliga, P., Lorz, H., Larzar, G., & Nagy, F. (1982). Cytoplast protoplast fusion for interspecific chloroplast transfer in *Nicotiana*. *Mol. Gen. Genet., 185*, 211-215.

Malone, R., Horvath, G. V., Cseplo, A., Buzas, B., Dix, P. J., & Medgyesy, P. (1992). Impact of the stringency of cell selection on plastid segregation in protoplast fusion-derived *Nicotiana* regenerates. *Theor. Appl. Genet., 84*, 866-873.

Malone, R. P., & Dix, P. J. (1990). Mutagenesis and triazine herbicide effects of strawberry shoot cultures. *J. Exp. Bot., 41*, 463–469.

Matibiri, E. A., & Mantell, S. H. (1994). Cybridization in *Nicotiana tabacum* L. using double inactivation of parental protoplasts and post-fusion selection based on nuclear-encoded and chloroplast-encoded marker genes. *Theor. Appl. Genet., 88*, 1017-1022.

Matthews, D., McNicoll, J., Harding, K., & Millam, S. (1999). 5'-Anchored simple-sequence repeats primers are useful for analyzing potato somatic hybrids. *Plant Cell Rep., 19*, 210–212.

Melchers, G., Mohri, Y., Watanabe, K., Wakabayashi, S., & Harada, K. (1992). One-step generation of cytoplasmic male sterility by fusion of mitochondrial-inactivated tomato protoplasts with nuclear-inactivated Solanum protoplasts. *Proc. Natl. Acad. Sci. USA, 89*, 6832-6836.

Mohapatra, T., Kirti, P. B., Kumar V. D., Prakash, S., & Chopra, V. L. Random chloroplast segregation and mitochondrial genome recombination in somatic hybrid plants of *Diplotaxis catholica* + *Brassica juncea*. *Plant Cell Rep., 17*, 814–818.

Moreira, C. D., Chase, C. D., Gmitter, F. G. Jr., & Grosser, J. W. (2000a). Inheritance of organelle genomes in citrus somatic cybrids. *Mol. Breed., 6*, 401-405.

Moreira, C. D., Chase, C. D., Gmitter, F. G. Jr, & Grosser, J. W. (2000b). Transmission of organelle genomes in citrus somatic hybrids. *Plant Cell, Tiss. Orgn. Cult., 61*, 165-168.

Moriguchi, T., Motomura, T., Hidaka, T., Akihama, T., & Omura, M. (1997). Analysis of mitochondrial genomes among *Citrus* plants produced by the interspecific somatic fusion of "Seminole" tangelo with rough lemon. *Plant Cell Rep., 16*, 397-400.

Motomura, T., Hidaka, T., Moriguchi, T., Akihama, T., & Omura, M. (1995). Intergeneric somatic hybrids between *Citrus* and *Atalantia* or *Severinia* by electrofusion, and recombination of mitochondrial genomes. *Breed. Sci., 45,* 309-314.

Narasimhulu, S. B., Kirti, P. B., Prakash, S., Chopra, V. L. (1992). Resynthesis of *Brassica carinata* by protoplasts fusion and recovery of a novel cytoplasmic hybrid. *Plant Cell Rep., 11*, 428-432.

Olivares-Fuster, O., Pena, L., Duran-Vila, N., Navarro, L. (2002). Green fluorescent protein as a visual marker in somatic hybridisation. *Ann. Bot., 89*, 491-497.

Perl, A., Aviv, D., & Galun, E. (1991a). Nuclear-organelle interaction in *Solanum*: Interspecific cybridizations and their correlation with a plastome dendrogram. *Mol. Gen. Genet., 228,* 193-200.

Perl, A., Aviv, D., & Galun, E. (1991b). Protoplast fusion mediated transfer of oligomycin resistance from *Nicotiana sylvestris* to *Solanum tuberosum* by intergeneric cybridization. *Mol. Gen., Genet., 225*, 11-16.

Perl, A., Aviv, D., & Galun, E. (1990). Protoplast-fusion-derived CMS potato cybrids: potential seed-parents for hybrid, true-potato-seeds. *J. Hered., 81*, 438-442.

Prakash, S. & Chopra, V. L. (1990). Male sterility caused by cytoplasm of *Brassica oxyrrhina* in *B.acmpestris* and *B.juncea*. *Theor. Appl. Genet., 79*, 285-287.

Prakash, S., Kirti, P. B., Bhat, S. R., Gaikwad, K., Kumar, V. D., & Chopra, V. L. (1998). A Moricandia arvensis–based cytoplasmic male sterility and fertility restoration system in *Brassica juncea*. *Theor. Appl. Genet., 97*, 488-492.

Provan, J., Kumar, A., Shepard, L., Powell, W., & Waugh, R. (1996). Analysis of intra-specific somatic hybrids of potato (*Solanum tuberosum*) using simple sequence repeats. *Plant Cell Rep., 16*, 196-199.

Pupilli, F., Labombarda, P., & Arcioni, S. (2001). New mitochondrial genome organization hybrids of *Medicago sativa* including the substoichiometric mitochondrial DNA in three interspecific somatic parent-specific amplification units. *Theor. Appl. Genet., 103*, 972-978.

Raineri, D., Jordan, P., & Kumar, A. (1992). Restoration of fertility in cytoplasmic male-sterile *Nicotiana tabacum* (Cytoplasm *N. bigelovii*) by protoplast fusion with X-irradiated protoplasts of *N. tabacum*, SR-1. *J. Exp. Bot., 43*, 195-203.

Rambaud, C., Dubois, J., & Vasseur, J. (1993). Male-sterile chicory cybrids obtained by intergeneric protoplast fusion. *Theor. Appl. Genet., 87*, 347–352.

Rasmussen, J. O., Lossl, A., & Rasmussen, O. S. (2000). Analysis of the plastome and chondriome origin in plants regenerated after asymmetric Solanum ssp protoplast fusions. *Theor. Appl. Genet., 101*, 336-343.

Ratushnyak, Y. I., Kochevenko, A. S., Cherep, N. N., Zavgorodnyaya, A. V., Latypov, S. A., & Gleba, Y. Y. (1995). Alloplasmatic incompatibility in cybrid plants possessing a *Lycopersicon esculentum* Mill genome and *Lycopersicon peruvianum* var Dentatum dun plasmagenes. *Genetika, 31*, 660-667.

Saha, T., Majumdar, S., Banerjee, N. S., & Sen, S. K. (2001). Development of interspecific somatic hybrid cell lines in cultivated jute and their early characterization using jute chloroplast RFLP marker. *Plant Breed., 120*, 439-444.

Saito, W., Ohgawara, T., Shimizu, J., Ishii, S., & Kobayashi, S. (1993). Citrus cybrid regeneration following cell-fusion between nucellar cells and mesophyll-cells. *Plant Science, 88*, 195-201.

Saito, W., Ohgawara, T., Shimizu, J., & Kobayashi, S.(1994). Somatic hybridization in citrus using embryogenic cybrid callus. *Plant Sci., 99*, 89-95.

Sakai, T., & Imamura, J. (1990). Intergeneric transfer of cytoplasmic male sterility between *Raphanus sativus* CMS line and *Brassica napus* through cytoplast-protoplast fusion. *Theor. Appl. Genet., 80*, 421-427.

Sakai, T., Liu, H. J., Iwabuchi, M., Kohno-Murase, J., & Imamura, J. (1996). Introduction of a gene from fertility restored radish (*Raphanus sativus*) into *Brassica napus* by fusion of X-irradiated protoplasts from a radish restorer line and iodacetoamide-treated protoplasts from a cytoplasmic male-sterile cybrid of *B. napus*. *Theor. Appl. Genet., 93*, 373-379.

Samoylov, V. M., Izhar, S., & Sink, K. C. (1996). Donor chromosome elimination and organelle composition of asymmetric somatic hybrid plants between an interspecific tomato hybrid and eggplant. *Theor. Appl. Genet., 93*, 268-274.

Scarano, M. T., Abbate, L., Ferrante, S., Lucretti, S., & Tusa, N. (2002). ISSR-PCR technique: a useful method for characterizing new allotetraploid somatic hybrids of mandarin. *Plant Cell Rep., 20*, 1162–1166.

Shimonaka, M., Hosoki, T., Tomita, M., & Yasumuro, Y. (2002). Production of somatic hybrid plants between Japanese bunching onion (*Allium fistulosum L.*) and bulb onion (*A. cepa L.*) via electrofusion. *J. Japan Soc. Hort. Sci., 71*, 623-631.

Sigareva, M. A., & Earle E. D. (1997). Direct transfer of a cold-tolerant Ogura male-sterile cytoplasm into cabbage (*Brassica oleracea* ssp. capitata) via protoplast fusion. *Theor. Appl. Genet., 94*, 213-220.

Skarzhinskaya, M., Landgren, M., & Glimelius, K. (1996). Production of intertribal somatic hybrids between *Brassica napus* L. and *Lesquerella fendleri* (Gray) wats. *Theor. Appl. Genet., 93*, 1242-1250.

Smith, M. V., Pay, A., & Dudits, D. (1989). Analysis of chloroplast and mitochondrial DNAs in asymmetrical hybrids between tobacco and carrot. *Theor. Appl. Genet., 77*, 641-644.

Spangenberg, G., Osusky, M., Oliveira, M. M., Freydl, E., Nagel, J., Pais, M. S., & Potrykus, I. (1990). Somatic hybridization by microfusion of defined protoplast pairs in *Nicotiana*: morphological, genetic and molecular characterization. *Theor. Appl. Genet., 80*, 577-587.

Spangenberg, G., Wang, Z. Y., Legris, G., Montavon, P., Takamizo, T., Perezvicente, R., et al. (1995). Intergeneric symmetric and asymmetric somatic hybridization in festuca and Lolium. *Euphytica, 85*, 235-245.

Stiewe, G., & Röbbelen, G. (1994). Establishing cytoplasmic male sterility in *B. napus* by mitochondrial recombination with *B. tournefortii*. *Plant Breed., 113*, 294-304.

Szarka, B., Gonter, I., Molnar-Lang, M., Morocz, S., & Dudits, D. (2002). Mixing of maize and wheat genomic DNA by somatic hybridization in regenerated sterile maize plants. *Theor. Appl. Genet., 105*, 1-7.

Szczerbakowa, A., Maciejewska, U., Zimnoch-Guzowska, E., Wielgat, B. (2001). Somatic hybrids *Solanum nigrum* (+) *S. tuberosum*: Morphological assessment and verification of hybridity. *Plant Cell Rep., 21*, 577-584.

Takamizo, T., Spangenberg, G., Suginobu, K. I., & Potrykus, I. (1991). Intergeneric somatic hybridization in Gramineae: somatic hybrid plants between tall fescue (*Festuca arundinacea* Schreb.) and Italian ryegrass (*Lolium multiflorum* Lam.). *Mol. Gen. Genet., 231*, 1-6.

Tamura, M., Tao, R., & Sugiura, A. (1998). Production of somatic hybrids between *Diospyros glandulosa* and *D. kaki* by protoplast fusion. *Plant Cell Tiss. Orgn. Cult., 54*, 85-91.

Temple, M., Makaroff, C. A., Mutschler, M. A., & Earle, E. D. (1992). Novel mitochondrial genomes in *Brassica napus* somatic hybrids. *Curr. Genet., 22*, 243–249.

Tokunaga, T., Yamao, M., Takenaka, M., Akai, T., Hasebe, H., & Kobayashi, S. (1999). Cybrid plants produced by electrofusion between satsuma mandarin (*Citrus unshiu*) and yuzu (*C. junos*) or lemon (*C. limon*), and recombination of mitochondrial genomes. *Plant Biotechnol., 16*, 297-301.

Vardi, A., Breiman, A., & Galun, E. (1987). Citrus cybrids: Production by donor-recipient protoplast-fusion and verification by mitochondrial-DNA restriction profiles. *Theor. Appl. Genet., 75*, 51-58.

Varotto S., Nenz, E., Lucchin, M., & Parrini, P. (2001). Production of asymmetric somatic hybrid plants between *Cichorium intybus* L. and *Helianthus annuus* L. *Theor. Appl. Genet., 102*, 950–956.

Waara, S., & Glimelius, K. (1995). The potential of somatic hybridisation in crop breeding. *Euphytica, 85*, 217-233.

Wallin, A., Glimelius, K., & Eriksson, T. (1978). Enucleation of plant protoplast by cytochalasin B. *Z. Pflanzenphysiol., 78*, 333-340.

Walters, T. W., & Earle, E. D. (1993). Organellar segregation, rearrangement and recombination in protoplast fusion-derived *Brassica oleracea* calli. *Theor. Appl. Genet., 85*, 761-769.

Wang, H. M., Ketela, T., Keller, W. A., Gleddie, S. C., & Brown, G. G. (1995). Genetic correlation of the *orf224/atp6* gene region with Polima CMS in *Brassica* somatic hybrids. *Plant Mol. Biol., 27*, 801-807.

Wang, T. B., Niizeki, M., Harada, T., Ishikawa, R., Oian, Y. O., & Saito, K. (1993). Establishment of somatic hybrid cell-lines between *Zea-mays* L. (maize) and *Triticum sect. trititrigia* Mackey (trititrigia). *Theor. Appl. Genet., 86*, 371-376.

Wang, Y. P., Sonntag, K., & Rudloff, E. (2003). Development of rapeseed with high erucic acid content by asymmetric somatic hybridisation between *Brassica napus* and *Crambe abyssinic*. *Theor. Appl. Genet., 106*, 1147-1155.

Wolters, A. M. A., Koornneef, M., & Gilissen, L. J. W. (1993a). The chloroplast and mitochondrial-DNA type are correlated with the nuclear composition of somatic hybrid calli of *Solanum-tuberosum* and *Nicotiana-plumbaginifolia*. *Curr. Genet., 24*, 260-267.

Wolters, A. M. A., Schoenmakers, H. C. H., & Koornneef, M. (1995). Chloroplast and mitochondrial DNA composition of triploid and tetraploid somatic hybrids between *Lycopersicon esculentum* and *Solanum tuberosum*. *Theor. Appl. Genet., 90*, 285-293.

Wolters, A. M. A., Vergunst, A. C., Vanaerwerff, F., & Koornneef, M. (1993b). Analysis of nuclear and organellar DNA of somatic hybrid calli and plants between *Lycopersicon* spp. and *Nicotiana* spp. *Mol. Gen. Genet., 241*, 707-718.

Xia, Z. A., An, X. S., Wang, F. D., Ye, X. F. (1992). Somatic hybridisation between tomato (*Nicotiana tabacum* L.) and black nightshade (*Solanum nigrum* L.) and selection of a new strain 694-L. In *Agriculture Biotechnology* (pp. 505-508). Beijing: China Science and Technology Press.

Xiang, F. N., Xia, G. M. (2003). Effect of UV dosage on somatic hybridization between common wheat (*Triticum aestivum L.*) and *Avena sativa* L. *Plant Sci., 164*, 697-707.

Xu, C. H., Xia, G. M., Zhi, D. Y., Xiang, F. N., & Chen, H. M. (2003). Integration of maize nuclear and mitochondrial DNA into the wheat genome through somatic hybridization. *Plant Sci., 165*, 1001-1008.

Xu, Y. S., Jones, M. G. K., Karp, A., & Pehu, E. (1993). Analysis of the mitochondrial DNA of the somatic hybrids of *Solanum brevidens* and *S. tuberosum* using non-radioactive digoxigenin-labelled DNA probes. *Theor. Appl. Genet., 85*, 1017-1022.

Yamagishi. H., Landgren, M., Forsberg, J., & Glimelius, K. (2002). Production of asymmetrical hybrids between *Arabidopsis thaliana* and *Brassica napus* utilizing an efficient protoplast culture system. *Theor. Appl. Genet., 104*, 959-964.

Yamamoto, M., & Kobayashi, S. (1995). A cybrid plant produced by electrofusion between *Citrus unshiu* and *C. sinensis*. *Plant Tiss. Cult. Lett. 12*, 131-137.

Yamashita, K., Hisatsune, Y., Sakamoto, T., Ishizuka, K., & Tashiro, Y. (2002). Chromosome and cytoplasm analyses of somatic hybrids between onion (*Allium cepa* L.) and garlic (*A. sativum* L.). *Euphytica, 125*, 163-167.

Yan, Z., Tian, Z., Huang, B., Huang, R., & Meng, J. (1999). Production of somatic hybrids between *Brassica oleracea* and the C3–C4 intermediate species *Moricandia nitens*. *Theor. Appl. Genet., 99*, 1281-1286.

Yarrow, S. A., Burnett, L. A., Wildeman, R. P, & Kemble, R. J. (1990). The transfer of 'Polima' cytoplasmic male sterility from oilseed rape (*Brassica napus*) to broccoli (*B. oleracea*) by protoplast fusion. *Plant Cell Rep., 9*, 185-188.

Zelcer, A., Avid, D., & Galun, E.. (9178). Interspecific transfer cytoplasmic male sterility by fusion between protoplasts of normal *Nicotiana sylvestris* and X-ray irradiated protoplasts of male sterile *N. tabacum*. *Z. Pflanzenphysiol., 90*, 397-407.

Zhou, A. F., Xia, G. M., Chen, X. L., &Chen, H. M. (2002). Production of somatic hybrid plants between two types of wheat protoplasts and the protoplasts of *Haynaldia villosa*. *Acta Bot. Sin., 44*, 1004-1008.

Zubko, M. K., Zubko, E. I., Patskovsky, Y. V., Khvedynich, O. A, Fisahn, J., Gleba, Y. Y., & Schieder, O. (1996). Novel 'homeotic' CMS patterns generated in *Nicotiana* via cybridization with *Hyoscyamus* and *Scopolia*. *J. Exp. Bot., 47*, 1101-1110.

Zubko, M. K., Zubko, E. I., Ruban, A. V., Adler, K., Mock, H. P, Misera, S., Gleba, Y. Y, & Grimm, B. (2001). Extensive developmental and metabolic alterations in cybrids (*Nicotiana tabacum* + *Hyoscyamus niger*) are caused by complex nuclear cytoplasmic incompatibility. *Plant J., 25*, 627-639.

Zubko, M. K., Zubko, E. I., Adler, K., Grimm, B., & Gleba, Y. Y. (2003). New CMS-associated phenotypes in cybrids *Nicotiana tabacum* L. + *Hyoscyamus niger* L. *Ann. Bot.*, 92, 281-288.